KIRK-OTHMER ENCYCLOPEDIA OF

CHEMICAL TECHNOLOGY

Fifth Edition

VOLUME 4

KIRK-OTHMER ENCYCLOPEDIA OF CHEMICAL TECHNOLOGY, FIFTH EDITION
EDITORIAL STAFF

Vice President, STM Books: **Janet Bailey**

Executive Editor: **Jacqueline I. Kroschwitz**

Editor: **Arza Seidel**

Managing Editor: **Michalina Bickford**

Director, Book Production and Manufacturing: **Camille P. Carter**

Production Manager: **Shirley Thomas**

Senior Production Editor: **Kellsee Chu**

Illustration Manager: **Dean Gonzalez**

Editorial Assistant: **Liam Kuhn**

KIRK-OTHMER ENCYCLOPEDIA OF

CHEMICAL TECHNOLOGY

Fifth Edition

VOLUME 4

Kirk-Othmer Encyclopedia of Chemical Technology
is available Online in full color and with additional content at
http://www3.interscience.wiley.com/cgi-bin/mrwhome/104554789/HOME.

⟨⟩WILEY-INTERSCIENCE

A John Wiley & Sons, Inc., Publication

Library of Congress Cataloging-in-Publication Data:

Kirk-Othmer encyclopedia of chemical technology. – 5th ed.
　　　p.　cm.
Editor-in-chief, Arza Seidel.
"A Wiley-Interscience publication."
Includes index.
　　ISBN 0-471-48494-6 (set) – ISBN 0-471-48519-5 (v. 4)
　．1. Chemistry, Technical–Encyclopedias. I. Title: Encyclopedia of
chemical technology. II. Kroschwitz, Jacqueline I.
　　TP9.K54　2004
　　660′.03–dc22　　　　　　　　　　　　　　2003021960

CONTENTS

CONTRIBUTORS

Fazlul Alam, *U.S. Borax Research Corporation, Anaheim, CA,* Boron Halides

Peter Bacher, *Loyola University Medical Center, Maywood, IL,* Blood Coagulation and Anticoagulant Drugs

Frederick Baker, *Westvaco Corporation, Charleston, SC,* Carbon, Activated

James N. BeMiller, *Whistler Center for Carbohydrate Research, Purdue University, West Lafayette, IN,* Carbohydrates

Ernst Billig, *Union Carbide Corporation, South Charleston, WV,* Butyl Alcohols; Butyraldehydes

Hans Joachim Breunig, *University of Bremen, Bremen, Germany,* Bismuth Compounds

Michael Briggs, *U.S. Borax Research Corporation, Anaheim, CA,* Boron Oxides, Boric Acid, and Borates

N. Calamur, *Amoco Chemical Company, Naperville, IL,* Butylenes

William Cameron, *William Cameron Consulting, Niagara Falls, ON, Canada,* Calcium Carbide

F. Patrick Carr, *OMYA, Inc., Proctor, VT,* Calcium Carbonate

Martin E. Carrera, *Amoco Chemical Company, Naperville, IL,* Butylenes

Artur Cavaco-Paulo, *University of Minito, Guimaraes, Portugal,* Bleaching Agents

Mark Chagnon, *Atlantic Metals and Alloy, Stratford, CT,* Bismuth and Bismuth Alloys

J. M. Criscione, *UCAR Carbon Company Inc., Columbia, TN,* Carbon

Purnendu K. Dasgupta, *Texas Tech University, Lubbock, TX,* Capillary Separations

Francis Evans, *Honeywell International, Morristown, NJ,* Boron Halides

James P. Farr, *The Clorox Company, Pleasanton, CA,* Bleaching Agents

Jawed Fareed, *Loyola University Medical Center, Maywood, IL,* Blood Coagulation and Anticoagulant Drugs

David K. Frederick, *OMYA, Inc., Proctor, VT,* Calcium Carbonate

Mira Freiberg, *Dead Sea Bromine Group, Beer Sheva, Israel,* Bromine; Bromine, Inorganic Compounds

Charles A. Gray, *Cabot Corporation, Billerica, MA,* Carbon Black

Baruch Grinbaum, *IMI (TAMI) Institute for Research and Development, Haifa Bay, Israel,* Bromine

Norman Herron, *E. I. Du Pont de Nemours & Co., Inc., Wilmington, DE,* Cadmium Compounds

Linda H. Jansen, *Callery Chemical Co., Pittsburgh, PA,* Boron, Elemental

Walter P. Jeske, *Loyola University Medical Center, Maywood, IL,* Blood Coagulation and Anticoagulant Drugs

Arie Kampf, *Dead Sea Bromine Group, Beer Sheva, Israel,* Bromine, Organic Compounds

Richard Kieffer, *Technical University of Vienna, Vienna, Austria,* Carbides, Survey

Petr Kuban, *Texas Tech University, Lubbock, TX,* Capillary Separations

Yakov Kutsovsky, *Cabot Corporation, Billerica, MA,* Carbon Black

Amedeo Lancia, *University of Napoli, Naples, Italy,* Calcium Sulfate

J. C. Long, *UCAR Carbon Company Inc., Columbia, TN,* Carbon

Khaled Mahmud, *Cabot Corporation, Billerica, MA,* Carbon Black

Ganpat Mani, *Honeywell International, Morristown, NJ,* Boron Halides

Charles E. Miller, *Westvaco Corporation, Charleston, SC,* Carbon, Activated

Hugh Morrow, *International Cadmium Council, Great Falls, VA,* Cadmium and Cadmium Alloys

Dino Musmarra, *University of Napoli, Naples, Italy,* Calcium Sulfate

S. Ted Oyama, *Virginia Polytechnic Institute & State University, Blacksburg, VA,* Carbides, Survey

John Papcun, *Atotech, Rock Hill, SC,* Boron Halides

Ronald Pierantozzi, *Air Products and Chemicals, Inc., Allentown, PA,* Carbon Dioxide

Marina Prisciandaro, *University of Napoli, Naples, Italy,* Calcium Sulfate

Albert J. Repik, *Westvaco Corporation, Charleston, SC,* Carbon, Activated

Steve R. Reznek, *Cabot Corporation, Billerica, MA,* Carbon Black

A. T. Santhanam, *Kennametal, Inc., Latrobe, PA,* Carbides, Cemented

David M. Schubert, *U.S. Borax Research Corporation, Valencia, CA,* Boron Hydrides, Heteroboranes, and their Metalla Derivatives

Timothy D. Shaffer, *ExxonMobil, Baytown, TX,* Butyl Rubber

David E. Smith, *FMC Corporation, Philadelphia, PA,* Carbon Disulfide

William L. Smith, *The Clorox Company, Pleasanton, CA,* Bleaching Agents

Dale Steichen, *The Clorox Company, Pleasanton, CA,* Bleaching Agents

William M. Stoll, *Consultant, Ligonier, PA,* Carbides, Industrial Hard

H. N. Sun, *ExxonMobil, Baytown, TX,* Butadiene

Robert W. Timmerman, *FMC Corporation, Philadelphia, PA,* Carbon Disulfide

E. Donald Tolles, *Westvaco Corporation, Charleston, SC,* Carbon, Activated

Andy H. Tsou, *ExxonMobil, Baytown, TX,* Butyl Rubber

Tzanko Tzanov, *University of Minito, Guimaraes, Portugal,* Bleaching Agents

Schmuel Ukeles, *IMI (TAMI) Institute for Research and Development, Haifa Bay, Israel,* Bromine, Inorganic Compounds

Lisa Vrana, *Consultant, Columbus, OH,* Calcium and Calcium Alloys; Calcium Chloride; Calcium Fluoride

Jeanine Walenga, *Loyola University Medical Center, Maywood, IL,* Blood Coagulation and Anticoagulant Drugs

Meng-Jiao Wang, *Cabot Corporation, Billerica, MA,* Carbon Black

Robert N. Webb, *ExxonMobil, Baytown, TX,* Butyl Rubber

Richard A. Wilsak, *Amoco Chemical Company, Naperville, IL,* Butylenes

Robert M. Winslow, *Sangart Inc., San Diego, CA,* Blood Substitutes

J. P. Wristers, *Exxon Chemical Company, Baytown, TX,* Butadiene

David Yoffe, *IMI (TAMI) Institute for Research and Development, Haifa Bay, Israel,* Bromine, Organic Compounds

CONVERSION FACTORS, ABBREVIATIONS, AND UNIT SYMBOLS

SI Units (Adopted 1960)

The International System of Units (abbreviated SI), is implemented throughout the world. This measurement system is a modernized version of the MKSA (meter, kilogram, second, ampere) system, and its details are published and controlled by an international treaty organization (The International Bureau of Weights and Measures) (1).

SI units are divided into three classes:

BASE UNITS

length	meter[†] (m)
mass	kilogram (kg)
time	second (s)
electric current	ampere (A)
thermodynamic temperature[‡]	kelvin (K)
amount of substance	mole (mol)
luminous intensity	candela (cd)

SUPPLEMENTARY UNITS

plane angle	radian (rad)
solid angle	steradian (sr)

DERIVED UNITS AND OTHER ACCEPTABLE UNITS

These units are formed by combining base units, suplementary units, and other derived units (2–4). Those derived units having special names and symbols are marked with an asterisk in the list below.

[†] The spellings "metre" and "litre" are preferred by ASTM; however, "-er" is used in the *Encyclopedia*.

[‡] Wide use is made of Celsius temperature (t) defined by

$$t = T - T_0$$

where T is the thermodynamic temperature, expressed in kelvin, and $T_0 = 273.15$ K by definition. A temperature interval may be expressed in degrees Celsius as well as in kelvin.

Quantity	Unit	Symbol	Acceptable equivalent
*absorbed dose	gray	Gy	J/Kg
acceleration	meter per second squared	m/s^2	
*activity (of a radionuclide)	becquerel	Bq	1/s
area	square kilometer	km^2	
	square hectometer	hm^2	ha (hectare)
	square meter	m^2	
concentration (of amount of substance)	mole per cubic meter	mol/m^3	
current density	ampere per square meter	A/m^2	
density, mass density	kilogram per cubic meter	kg/m^3	g/L; mg/cm^3
dipole moment (quantity)	coulomb meter	$C \cdot m$	
*dose equivalent	sievert	Sv	J/kg
*electric capacitance	farad	F	C/V
*electric charge, quantity of electricity	coulomb	C	$A \cdot s$
electric charge density	coulomb per cubic meter	C/m^3	
*electric conductance	siemens	S	A/V
electric field strength	volt per meter	V/m	
electric flux density	coulomb per square meter	C/m^2	
*electric potential, potential difference, electromotive force	volt	V	W/A
*electric resistance	ohm	Ω	V/A
*energy, work, quantity of heat	megajoule	MJ	
	kilojoule	kJ	
	joule	J	$N \cdot m$
	electronvolt[†]	eV[†]	
	kilowatt-hour[†]	$kW \cdot h$[†]	
energy density	joule per cubic meter	J/m^3	
*force	kilonewton	kN	
	newton	N	$kg \cdot m/s^2$

[†]This non-SI unit is recognized by the CIPM as having to be retained because of practical importance or use in specialized fields (1).

Quantity	Unit	Symbol	Acceptable equivalent
*frequency	megahertz	MHz	
	hertz	Hz	1/s
heat capacity, entropy	joule per kelvin	J/K	
heat capacity (specific), specific entropy	joule per kilogram kelvin	$J/(kg \cdot K)$	
heat-transfer coefficient	watt per square meter kelvin	$W/(m^2 \cdot K)$	
*illuminance	lux	lx	lm/m^2
*inductance	henry	H	Wb/A
linear density	kilogram per meter	kg/m	
luminance	candela per square meter	cd/m^2	
*luminous flux	lumen	lm	$cd \cdot sr$
magnetic field strength	ampere per meter	A/m	
*magnetic flux	weber	Wb	$V \cdot s$
*magnetic flux density	tesla	T	Wb/m^2
molar energy	joule per mole	J/mol	
molar entropy, molar heat capacity	joule per mole kelvin	$J/(mol \cdot K)$	
moment of force, torque	newton meter	$N \cdot m$	
momentum	kilogram meter per second	$kg \cdot m/s$	
permeability	henry per meter	H/m	
permittivity	farad per meter	F/m	
*power, heat flow rate, radiant flux	kilowatt	kW	
	watt	W	J/s
power density, heat flux density, irradiance	watt per square meter	W/m^2	
*pressure, stress	megapascal	MPa	
	kilopascal	kPa	
	pascal	Pa	N/m^2
sound level	decibel	dB	
specific energy	joule per kilogram	J/kg	
specific volume	cubic meter per kilogram	m^3/kg	
surface tension	newton per meter	N/m	
thermal conductivity	watt per meter kelvin	$W/(m \cdot K)$	
velocity	meter per second	m/s	
	kilometer per hour	km/h	
viscosity, dynamic	pascal second	$Pa \cdot s$	
	millipascal second	$mPa \cdot s$	
viscosity, kinematic	square meter per second	m^2/s	
	square millimeter per second	mm^2/s	

Quantity	Unit	Symbol	Acceptable equivalent
volume	cubic meter	m^3	
	cubic diameter	dm^3	L (liter) (5)
	cubic centimeter	cm^3	mL
wave number	1 per meter	m^{-1}	
	1 per centimeter	cm^{-1}	

In addition, there are 16 prefixes used to indicate order of magnitude, as follows

Multiplication factor	Prefix	symbol	Note
10^{18}	exa	E	
10^{15}	peta	P	
10^{12}	tera	T	
10^9	giga	G	
10^6	mega	M	
10^3	kilo	k	
10^2	hecto	h^a	[a] Although hecto, deka, deci, and
10	deka	da^a	centi are SI prefixes, their use
10^{-1}	deci	d^a	should be avoided except for SI
10^{-2}	centi	c^a	unit-multiples for area and
10^{-3}	milli	m	volume and nontechnical use of
10^{-6}	micro	μ	centimeter, as for body and
10^{-9}	nano	n	clothing measurement.
10^{-12}	pico	p	
10^{-15}	femto	f	
10^{-18}	atto	a	

For a complete description of SI and its use the reader is referred to ASTM E380 (4) and the article UNITS AND CONVERSION FACTORS which appears in Vol. 24.

A representative list of conversion factors from non-SI to SI units is presented herewith. Factors are given to four significant figures. Exact relationships are followed by a dagger. A more complete list is given in the latest editions of ASTM E380 (4) and ANSI Z210.1 (6).

Conversion Factors to SI Units

To convert from	To	Multiply by
acre	square meter (m^2)	4.047×10^3
angstrom	meter (m)	1.0×10^{-10}[†]
are	square meter (m^2)	1.0×10^{2}[†]
astronomical unit	meter (m)	1.496×10^{11}

[†]Exact.

To convert from	To	Multiply by
atmosphere, standard	pascal (Pa)	1.013×10^5
bar	pascal (Pa)	$1.0 \times 10^{5\dagger}$
barn	square meter (m²)	$1.0 \times 10^{-28\dagger}$
barrel (42 U.S. liquid gallons)	cubic meter (m³)	0.1590
Bohr magneton (μ_B)	J/T	9.274×10^{-24}
Btu (International Table)	joule (J)	1.055×10^3
Btu (mean)	joule (J)	1.056×10^3
Btu (thermochemical)	joule (J)	1.054×10^3
bushel	cubic meter(m³)	3.524×10^{-2}
calorie (International Table)	joule (J)	4.187
calorie (mean)	joule (J)	4.190
calorie (thermochemical)	joule (J)	$4.184^\dagger$
centipoise	pascal second (Pa·s)	$1.0 \times 10^{-3\dagger}$
centistokes	square millimeter per second (mm²/s)	$1.0^\dagger$
cfm (cubic foot per minute)	cubic meter per second (m³s)	4.72×10^{-4}
cubic inch	cubic meter (m³)	1.639×10^{-5}
cubic foot	cubic meter (m³)	2.832×10^{-2}
cubic yard	cubic meter (m³)	0.7646
curie	becquerel (Bq)	$3.70 \times 10^{10\dagger}$
debye	coulomb meter (C·m)	3.336×10^{-30}
degree (angle)	radian (rad)	1.745×10^{-2}
denier (international)	kilogram per meter (kg/m)	1.111×10^{-7}
	tex‡	0.1111
dram (apothecaries')	kilogram (kg)	3.888×10^{-3}
dram (avoirdupois)	kilogram (kg)	1.772×10^{-3}
dram (U.S. fluid)	cubic meter (m³)	3.697×10^{-6}
dyne	newton (N)	$1.0 \times 10^{-5\dagger}$
dyne/cm	newton per meter (N/m)	$1.0 \times 10^{-3\dagger}$
electronvolt	joule (J)	1.602×10^{-19}
erg	joule (J)	$1.0 \times 10^{-7\dagger}$
fathom	meter (m)	1.829
fluid ounce (U.S.)	cubic meter (m³)	2.957×10^{-5}
foot	meter (m)	$0.3048^\dagger$
footcandle	lux (lx)	10.76
furlong	meter (m)	2.012×10^{-2}
gal	meter per second squared (m/s²)	$1.0 \times 10^{-2\dagger}$
gallon (U.S. dry)	cubic meter (m³)	4.405×10^{-3}
gallon (U.S. liquid)	cubic meter (m³)	3.785×10^{-3}
gallon per minute (gpm)	cubic meter per second (m³/s)	6.309×10^{-5}
	cubic meter per hour (m³/h)	0.2271

†Exact.
‡See footnote on p. ix.

To convert from	To	Multiply by
gauss	tesla (T)	1.0×10^{-4}
gilbert	ampere (A)	0.7958
gill (U.S.)	cubic meter (m^3)	1.183×10^{-4}
grade	radian	1.571×10^{-2}
grain	kilogram (kg)	6.480×10^{-5}
gram force per denier	newton per tex (N/tex)	8.826×10^{-2}
hectare	square meter (m^2)	$1.0 \times 10^{4\dagger}$
horsepower (550 ft · lbf/s)	watt (W)	7.457×10^{2}
horsepower (boiler)	watt (W)	9.810×10^{3}
horsepower (electric)	watt (W)	$7.46 \times 10^{2\dagger}$
hundredweight (long)	kilogram (kg)	50.80
hundredweight (short)	kilogram (kg)	45.36
inch	meter (m)	$2.54 \times 10^{-2\dagger}$
inch of mercury (32°F)	pascal (Pa)	3.386×10^{3}
inch of water (39.2°F)	pascal (Pa)	2.491×10^{2}
kilogram-force	newton (N)	9.807
kilowatt hour	megajoule (MJ)	$3.6^{\dagger}$
kip	newton (N)	4.448×10^{3}
knot (international)	meter per second (m/S)	0.5144
lambert	candela per square meter (cd/m^3)	3.183×10^{3}
league (British nautical)	meter (m)	5.559×10^{3}
league (statute)	meter (m)	4.828×10^{3}
light year	meter (m)	9.461×10^{15}
liter (for fluids only)	cubic meter (m^3)	$1.0 \times 10^{-3\dagger}$
maxwell	weber (Wb)	$1.0 \times 10^{-8\dagger}$
micron	meter (m)	$1.0 \times 10^{-6\dagger}$
mil	meter (m)	$2.54 \times 10^{-5\dagger}$
mile (statue)	meter (m)	1.609×10^{3}
mile (U.S. nautical)	meter (m)	$1.852 \times 10^{3\dagger}$
mile per hour	meter per second (m/s)	0.4470
millibar	pascal (Pa)	1.0×10^{2}
millimeter of mercury (0°C)	pascal (Pa)	$1.333 \times 10^{2\dagger}$
minute (angular)	radian	2.909×10^{-4}
myriagram	kilogram (Kg)	10
myriameter	kilometer (Km)	10
oersted	ampere per meter (A/m)	79.58
ounce (avoirdupois)	kilogram (kg)	2.835×10^{-2}
ounce (troy)	kilogram (kg)	3.110×10^{-2}
ounce (U.S. fluid)	cubic meter (m^3)	2.957×10^{-5}
ounce-force	newton (N)	0.2780
peck (U.S.)	cubic meter (m^3)	8.810×10^{-3}
pennyweight	kilogram (kg)	1.555×10^{-3}
pint (U.S. dry)	cubic meter (m^3)	5.506×10^{-4}

†Exact.

To convert from	To	Multiply by
pint (U.S. liquid)	cubic meter (m^3)	4.732×10^{-4}
poise (absolute viscosity)	pascal second (Pa · s)	0.10†
pound (avoirdupois)	kilogram (kg)	0.4536
pound (troy)	kilogram (kg)	0.3732
poundal	newton (N)	0.1383
pound-force	newton (N)	4.448
pound force per square inch (psi)	pascal (Pa)	6.895×10^3
quart (U.S. dry)	cubic meter (m^3)	1.101×10^{-3}
quart (U.S. liquid)	cubic meter (m^3)	9.464×10^{-4}
quintal	kilogram (kg)	$1.0 \times 10^{-2†}$
rad	gray (Gy)	$1.0 \times 10^{-2†}$
rod	meter (m)	5.029
roentgen	coulomb per kilogram (C/kg)	2.58×10^{-4}
second (angle)	radian (rad)	$4.848 \times 10^{-6†}$
section	square meter (m^2)	2.590×10^6
slug	kilogram (kg)	14.59
spherical candle power	lumen (lm)	12.57
square inch	square meter (m^2)	6.452×10^{-4}
square foot	square meter (m^2)	9.290×10^{-2}
square mile	square meter (m^2)	2.590×10^6
square yard	square meter (m^2)	0.8361
stere	cubic meter (m^3)	1.0†
stokes (kinematic viscosity)	square meter per second (m^2/s)	$1.0 \times 10^{-4†}$
tex	kilogram per meter (kg/m)	$1.0 \times 10^{-6†}$
ton (long, 2240 pounds)	kilogram (kg)	1.016×10^3
ton (metric) (tonne)	kilogram (kg)	$1.0 \times 10^{3†}$
ton (short, 2000 pounds)	kilogram (kg)	9.072×10^2
torr	pascal (Pa)	1.333×10^2
unit pole	weber (Wb)	1.257×10^{-7}
yard	meter (m)	0.9144†

†Exact.

Abbreviations and Unit Symbols

Following is a list of common abbreviations and unit symbnols used in the Encyclopedia. In general they agree with those listed in *American National Standard Abbreviations for Use on Drawings and in Text (ANSI Y1.1)* (6) and *American National Standard Letter Symbols for Units in Science and Technology (ANSI Y10)* (6). Also included is a list of acronyms for a number of private and government organizations as well as common industrial solvents, polymers, and other chemicals.

Rules for Writing Unit Symbols (4):

1. Unit symbols are printed in upright letters (roman) regardless of the type style used in the surrounding text.
2. Unit symbols are unaltered in the plural.
3. Unit symbols are not followed by a period except when used at the end of a sentence.
4. Letter unit symbols are generally printed lower-case (for example, cd for candela) unless the unit name has been derived from a proper name, in which case the first letter of the symbol is capitalized (W, Pa). Prefixes and unit symbols retain their prescribed form regardless of the surrounding typography.
5. In the complete expression for a quantity, a space should be left between the numerical value and the unit symbol. For example, write 2.37 lm, *not* 2.37 lm, and 35 mm, *not* 35 mm. When the quantity is used in an adjectival sense, a hyphen is often used, for example, 35-mm film. *Exception:* No space is left between the numerical value and the symbols of degree, minute, and second of plane angle, degree Celsius, and the percent sign.
6. No space is used between the prefix and unit symbol (for example, kg).
7. Symbols, not abbreviations, should be used for units. For example, use "A," not "amp," for ampere.
8. When multiplying unit symbols, use a raised dot:

$$N \cdot m \text{ for newton meter}$$

In the case of W·h, the dot may be omitted, thus:

$$Wh$$

An exception to this practice is made for computer printouts, automatic typewriter work, etc, where the raised dot is not possible, and a dot on the line may be used.

9. When dividing unit symbols, use one of the following forms:

$$m/s \quad or \quad m \cdot s^{-1} \quad or \quad \frac{m}{s}$$

In no case should more than one slash be used in the same expression unless parentheses are inserted to avoid ambiguity. For example, write:

$$J/(mol \cdot K) \quad or \quad J \cdot mol^{-1} \cdot K^{-1} \quad or \quad (J/mol)/K$$

but *not*

$$J/mol/K$$

10. Do not mix symbols and unit names in the same expression. Write:

$$\text{joules per kilogram} \quad or \quad \text{J/kg} \quad or \quad \text{J} \cdot \text{kg}^{-1}$$

but *not*

$$\text{joules/kilogram} \quad nor \quad \text{Joules/kg} \quad nor \quad \text{Joules} \cdot \text{kg}^{-1}$$

ABBREVIATIONS AND UNITS

A	ampere	AOAC	Association of Official Analytical Chemists
A	anion (eg, HA)		
A	mass number	AOCS	American Oil Chemists' Society
a	atto (prefix for 10^{-18})		
AATCC	American Association of Textile Chemists and Colorists	APHA	American Public Health Association
		API	American Petroleum Institute
ABS	acrylonitrile–butadiene–styrene	aq	aqueous
abs	absolute	Ar	aryl
ac	alternating current, *n*.	*ar-*	aromatic
a-c	alternating current, *adj*.	*as-*	Asymmetric(al)
ac-	alicyclic	ASHRAE	American Society of Heating, Refrigerating, and Air Conditioning Engineers
acac	acetylacetonate		
ACGIH	American Conference of Governmental Industrial Hygienists		
		ASM	American Society for Metals
ACS	American Chemical Society	ASME	American Society of Mechanical Engineers
AGA	American Gas Association		
Ah	ampere hour	ASTM	American Society for Testing and Materials
AIChE	American Institute of Chemical Engineers	at no.	atomic number
AIME	American Institute of Mining, metallurgical, and Petroleum Engineers	at wt	atomic weight
		av(g)	average
		AWS	American Welding Society
		b	bonding orbital
AIP	American Institute of Physics	bbl	barrel
		bcc	body-centered cubic
AISI	American Iron and Steel Institute	BCT	body-centered tetragonal
		Bé	Baumé
alc	alcohol(ic)	BET	Brunauer-Emmett-Teller (adsorption equation)
Alk	alkyl		
alk	alkaline (not alkali)	bid	twice daily
amt	amount	Boc	*t*-butyloxycarbonyl
amu	atomic mass unit	BOD	biochemical (biological) oxygen demand
ANSI	American National Standards Institute		
		bp	boiling point
AO	atomic orbital	Bq	becquerel

C	coulomb	dil	dilute
°C	degree Celsius	DIN	Deutsche Industrie Normen
C-	denoting attachment to carbon	*dl*-; DL-	racemic
c	centi (prefix for 10^{-2})	DMA	dimethylacetamide
c	critical	DMF	dimethylformamide
ca	circa (Approximately)	DMG	dimethyl glyoxime
cd	candela; current density; circular dichroism	DMSO	dimethyl sulfoxide
		DOD	Department of Defense
CFR	Code of Federal Regulations	DOE	Department of Energy
		DOT	Department of Transportation
cgs	centimeter-gram-second		
CI	Color Index	DP	degree of polymerization
cis-	isomer in which substituted groups are on some side of double bond between C atoms	dp	dew point
		DPH	diamond pyramid hardness
		dstl(d)	distill(ed)
cl	carload	dta	differential thermal analysis
cm	centimeter		
cmil	circular mil	(*E*)-	entgegen; opposed
cmpd	compound	ε	dielectric constant (unitless number)
CNS	central nervous system		
CoA	coenzyme A	*e*	electron
COD	chemical oxygen demand	ECU	electrochemical unit
coml	commerical(ly)	ed.	edited, edition, editor
cp	chemically pure	ED	effective dose
cph	close-packed hexagonal	EDTA	ethylenediaminetetra-acetic acid
CPSC	Consumer Product Safety Commission		
		emf	electromotive force
cryst	crystalline	emu	electromagnetic unit
cub	cubic	en	ethylene diamine
D	debye	eng	engineering
D-	denoting configurational relationship	EPA	Environmental Protection Agency
d	differential operator	epr	electron paramagnetic resonance
d	day; deci (prefix for 10^{-1})		
d	density	eq.	equation
d-	*dextro*-, dextrorotatory	esca	electron spectroscopy for chemical analysis
da	deka (prefix for 10^{-1})		
dB	decibel	esp	especially
dc	direct current, *n*.	esr	electron-spin resonance
d-c	direct current, *adj*.	est(d)	estimate(d)
dec	decompose	estn	estimation
detd	determined	esu	electrostatic unit
detn	determination	exp	experiment, experimental
Di	didymium, a mixture of all lanthanons	ext(d)	extract(ed)
		F	farad (capacitance)
dia	diameter	*F*	fraday (96,487 C)

f	femto (prefix for 10^{-15})		hyd	hydrated, hydrous
FAO	Food and Agriculture Organization (United Nations)		hyg	hygroscopic
			Hz	hertz
			^{i}(eg, Pr^i)	iso (eg, isopropyl)
fcc	face-centered cubic		i-	inactive (eg, i-methionine)
FDA	Food and Drug Administration		IACS	international Annealed Copper Standard
FEA	Federal Energy Administration		ibp	initial boiling point
			IC	integrated circuit
FHSA	Federal Hazardous Substances Act		ICC	Interstate Commerce Commission
fob	free on board		ICT	International Critical Table
fp	freezing point			
FPC	Federal Power Commission		ID	inside diameter; infective dose
FRB	Federal Reserve Board			
frz	freezing		ip	intraperitoneal
G	giga (prefix for 10^9)		IPS	iron pipe size
G	gravitational constant $= 6.67 \times 10^{11} \text{N} \cdot \text{m}^2/\text{kg}^2$		ir	infrared
			IRLG	Interagency Regulatory Liaison Group
g	gram		ISO	International Organization Standardization
(g)	gas, only as in $H_2O(g)$			
g	gravitatonal acceleration			
gc	gas chromatography			
gem-	geminal		ITS-90	International Temperature Scale (NIST)
glc	gas–liquid chromatography			
g-mol wt; gmw	gram-molecular weight		IU	International Unit
			IUPAC	International Union of Pure and Applied Chemistry
GNP	gross national product			
gpc	gel-permeation chromatography			
			IV	iodine value
GRAS	Generally Recognized as Safe		iv	intravenous
			J	joule
grd	ground		K	kelvin
Gy	gray		k	kilo (prefix for 10^3)
H	henry		kg	kilogram
h	hour; hecto (prefix for 10^2)		L	denoting configurational relationship
ha	hectare			
HB	Brinell hardness number		L	liter (for fluids only) (5)
Hb	hemoglobin		l-	levo-, levorotatory
hcp	hexagonal close-packed		(l)	liquid, only as in NH_3(l)
hex	hexagonal		LC$_{50}$	conc lethal to 50% of the animals tested
HK	Knoop hardness number			
hplc	high performance liquid chromatography		LCAO	linear combnination of atomic orbitals
HRC	Rockwell hardness (C scale)		lc	liquid chromatography
			LCD	liquid crystal display
HV	Vickers hardness number		lcl	less than carload lots

LD_{50}	dose lethal to 50% of the animals tested	N	newton (force)
LED	light-emitting diode	N	normal (concentration); neutron number
liq	liquid	N-	denoting attachment to nitrogen
lm	lumen		
ln	logarithm (natural)	n (as n_D^{20})	index of refraction (for 20°C and sodium light)
LNG	liquefied natural gas		
log	logarithm (common)		
LOI	limiting oxygen index	n (as Bu^n), n-	normal (straight-chain structure)
LPG	liquefied petroleum gas		
ltl	less than truckload lots	n	neutron
lx	lux	n	nano (prefix for 10^9)
M	mega (prefix for 10^6); metal (as in MA)	na	not available
		NAS	National Academy of Sciences
M	molar; actual mass		
$\overline{M}_w$	weight-average mol wt	NASA	National Aeronautics and Space Administration
$\overline{M}_n$	number-average mol wt		
m	meter; milli (prefix for 10^{-3})	nat	natural
		ndt	nondestructive testing
m	molal	neg	negative
m-	meta	NF	*National Formulary*
max	maximum	NIH	National Institutes of Health
MCA	Chemical Manufacturers' Association (was Manufacturing Chemists Association)		
		NIOSH	National Institute of Occupational Safety and Health
MEK	methyl ethyl ketone	NIST	National Institute of Standards and Technology (formerly National Bureau of Standards)
meq	milliequivalent		
mfd	manufactured		
mfg	manufacturing		
mfr	manufacturer		
MIBC	methyl isobutyl carbinol	nmr	nuclear magnetic resonance
MIBK	methyl isobutyl ketone		
MIC	minimum inhibiting concentration	NND	New and Nonofficial Drugs (AMA)
min	minute; minimum	no.	number
mL	milliliter	NOI-(BN)	not otherwise indexed (by name)
MLD	minimum lethal dose		
MO	molecular orbital	NOS	not otherwise specified
mo	month	nqr	nuclear quadruple resonance
mol	mole		
mol wt	molecular weight	NRC	Nuclear Regulatory Commission; National Research Council
mp	melting point		
MR	molar refraction		
ms	mass spectrometry	NRI	New Ring Index
MSDS	material safety data sheet	NSF	National Science Foundation
mxt	mixture		
μ	micro (prefix for 10^{-6})	NTA	nitrilotriacetic acid

NTP	normal temperature and pressure (25°C and 101.3 kPa or 1 atm)	pwd	powder
		py	pyridine
		qv	quod vide (which see)
NTSB	National Transportation Safety Board	R	univalent hydrocarbon radical
O-	denoting attachment to oxygen	(R)-	rectus (clockwise configuration)
o-	ortho	r	precision of data
OD	outside diameter	rad	radian; radius
OPEC	Organization of Petroleum Exporting Countries	RCRA	Resource Conservation and Recovery Act
o-phen	o-phenanthridine	rds	rate-determining step
OSHA	Occupational Safety and Health Administration	ref.	reference
		rf	radio frequency, n.
owf	on weight of fiber	r-f	radio frequency, adj.
Ω	ohm	rh	relative humidity
P	peta (prefix for 10^{15})	RI	Ring Index
p	pico (prefix for 10^{-12}	rms	root-mean square
p-	para	rpm	rotations per minute
p	proton	rps	revolutions per second
p.	page	RT	room temperature
Pa	Pascal (pressure)	RTECS	Registry of Toxic Effects of Chemical Substances
PEL	personal exposure limit based on an 8-h exposure	^{s}(eg, Bu^{s}); sec-	secondary (eg, secondary butyl)
pd	potential difference	S	siemens
pH	negative logarithm of the effective hydrogen ion concentration	(S)-	sinister (counterclockwise configuration)
		S-	denoting attachment to sulfur
phr	parts per hundred of resin (rubber)	s-	symmetric(al)
p-i-n	positive-intrinsic-negative	S	second
pmr	proton magnetic resonance	(s)	solid, only as in H_2O(s)
p-n	positive-negative	SAE	Society of Automotive Engineers
po	per os (oral)		
POP	polyoxypropylene	SAN	styrene-acrylonitrile
pos	positive	sat(d)	saturate(d)
pp.	pages	satn	saturation
ppb	parts per billion (10^9)	SBS	styrene–butadiene–styrene
ppm	parts per milion (10^6)	sc	subcutaneous
ppmv	parts per million by volume	SCF	self-consistent field; standard cubic feet
ppmwt	parts per million by weight		
PPO	poly(phenyl oxide)	Sch	Schultz number
ppt(d)	precipitate(d)	sem	scanning electron microscope(y)
pptn	precipitation		
Pr (no.)	foreign prototype (number)	SFs	Saybolt Furol seconds
pt	point; part	sl sol	slightly soluble
PVC	poly(vinyl chloride)	sol	soluble

soln	solution	*trans-*	isomer in which
soly	solubility		substituted groups are
sp	specific; species		on opposite sides of
sp gr	specific gravity		double bond between
sr	steradian		C atoms
std	standard	TSCA	Toxic Substances Control
STP	standard temperature and		Act
	pressure (0°C and	TWA	time-weighted average
	101.3 kPa)	Twad	Twaddell
sub	sublime(s)	UL	Underwriters' Laboratory
SUs	Saybolt Universal seconds	USDA	United States Department
syn	synthetic		of Agriculture
^{t} (eg, Bu^t),	tertiary (eg, tertiary	USP	*United States*
t-, tert-	butyl)		*Pharmacopeia*
T	tera (prefix for 10^{12}); tesla	uv	ultraviolet
	(magnetic flux density)	V	volt (emf)
t	metric to (tonne)	var	variable
t	temperature	*vic-*	vicinal
TAPPI	Technical Association of	vol	volume (not volatile)
	the Pulp and Paper	vs	versus
	Industry	v sol	very soluble
TCC	Tagliabue closed cup	W	watt
tex	tex (linear density)	Wb	weber
T_g	glass-transition	Wh	watt hour
	temperature	WHO	World Health Organization
tga	thermogravimetric		(United Nations)
	analysis	wk	week
THF	tetrahydrofuran	yr	year
tlc	thin layer chromatography	(*Z*)-	zusammen; together;
TLV	threshold limit value		atomic number

Non-SI (Unacceptable and Obsolete) Units		Use
Å	angstrom	nm
at	atmosphere, technical	Pa
atm	atmosphere, standard	Pa
b	barn	cm^2
bar†	bar	Pa
bbl	barrel	m^3
bhp	brake horsepower	W
Btu	British thermal unit	J
bu	bushel	m^3; L
cal	calorie	J
cfm	cubic foot per minute	m^3/s
Ci	curie	Bq
cSt	centistokes	mm^2/s
c/s	cycle per second	Hz
cu	cubic	exponential form

†Do not use bar (10^5 Pa) or millibar (10^2 Pa) because they are not SI units, and are accepted internationally only in special fields because of existing usage.

Non-SI (Unacceptable and Obsolete) Units   Use

D	debye	$C \cdot m$
den	denier	tex
dr	dram	kg
dyn	dyne	N
dyn/cm	dyne per centimeter	mN/m
erg	erg	J
eu	entropy unit	J/K
°F	degree Fahrenheit	°C; K
fc	footcandle	lx
fl	footlambert	lx
fl oz	fluid ounce	m^3; L
ft	foot	m
ft · lbf	foot pound-force	J
gf den	gram-force per denier	N/tex
G	gauss	T
Gal	gal	m/s^2
gal	gallon	m^3; L
Gb	gilbert	A
gpm	gallon per minute	(m^3/s); (m^3/h)
gr	grain	kg
hp	horsepower	W
ihp	indicated horsepower	W
in.	inch	m
in. Hg	inch of mercury	Pa
in. H_2O	inch of water	Pa
in.-lbf	inch pound-force	J
kcal	kilo-calorie	J
kgf	kilogram-force	N
kilo	for kilogram	kg
L	lambert	lx
lb	pound	kg
lbf	pound-force	N
mho	mho	S
mi	mile	m
MM	million	M
mm Hg	millimeter of mercury	Pa
$m\mu$	millimicron	nm
mph	miles per hour	km/h
μ	micron	μm
Oe	oersted	A/m
oz	ounce	kg
ozf	ounce-force	N
η	poise	$Pa \cdot s$
P	poise	$Pa \cdot s$
ph	phot	lx
psi	pounds-force per square inch	Pa
psia	pounds-force per square inch absolute	Pa
psig	pounds-force per square inch gage	Pa
qt	quart	m^3; L
°R	degree Rankine	K
rd	rad	Gy
sb	stilb	lx
SCF	standard cubic foot	m^3
sq	square	exponential form
thm	therm	J
yd	yard	m

BIBLIOGRAPHY

1. The International Bureau of Weights and Measures, BIPM (Parc Saint-Cloud, France) is described in Ref. 4. This bureau operates under the exclusive supervision of the International Committee for Weights and Measures (CIPM).

2. *Metric Editorial Guide (ANMC-78-1)*, latest ed., American National Metric Council, 900 Mix Avenue, Suite 1 Hamden CT 06514-5106, 1981.

3. *SI Units and Recommendations for the Use of Their Multiples and of Certain Other Units (ISO 1000-1992)*, American National Standards Institute, 25 W 43rd St., New York, 10036, 1992.

4. Based on IEEE/ASTM-SI-10 *Standard for use of the International System of Units (SI): The Modern Metric System* (Replaces ASTM380 and ANSI/IEEE Std 268-1992), ASTM International, West Conshohocken, PA., 2002. See also www.astm.org

5. *Fed. Reg.*, Dec. 10, 1976 (41 FR 36414).

6. For ANSI address, see Ref. 3. See also www.ansi.org

B

Continued

BISMUTH AND BISMUTH ALLOYS

1. Introduction and History

Bismuth [7440-69-9] (Bi) is a very brittle, silvery metal having a high metallic luster. The element is found in the periodic table of the elements under Group 15 (Va). The atomic number is 83 and the atomic weight is 208.98. Bismuth is next to lead on the periodic table and exhibits many characteristics of lead except that bismuth is considered nontoxic despite its heavy-metal status.

Bismuth was first mentioned in the literature by a German monk named Basil Valentine in the fifteenth century. During the sixteenth century, another German, a scientist, Georgius Agricola, detailed the smelting of bismuth from ore (1). Compounds of bismuth were first used in that century for the relief of stomach disorders. Bismuth compounds are still widely used today for the same purpose, in particular bismuth subsalicylate.

Before the 1800s bismuth was usually referred to as one of the elements with which it was associated, such as antimony, lead, tin, or silver. During the 1800s the metal was refined and proven to be an element. Since that time, many varied uses have been found for the element bismuth from metallurgical applications to fusible alloys to chemical applications.

The Bismuth Institute, located in Belgium, is a nonprofit organization whose sole purpose is to provide information on bismuth and its uses. The Institute publishes papers and abstracts that report advances in the uses of bismuth in *The Bulletin of the Bismuth Institute*. The Bismuth Institute has a Web site: http://www.bismuth.be.

2. Occurrence

Bismuth is a minor metal, that is, it is a mining by-product and is therefore not mined for its own intrinsic value. Usually, bismuth is a by-product of lead or copper ores. However, tungsten ore in China contains bismuth. Until the 1980s some bismuth was mined for its own value from the Tasna mine, which is owned by Comibol, the state-owned mining company of Bolivia. Unfortunately, the fluctuation of world bismuth prices over the years has made it economically impossible to keep that mine open.

Since the late 1990s, however, Corriente Resources of Canada has leased the Tasna mine and has initiated studies into the feasibility of reopening it. This will probably not happen before 1999 and only if bismuth prices rise and remain over the $4.00 level during 1998.

The concentration of bismuth in the earth's crust has not been clearly determined. Estimates range from 0.008 to 0.1 ppm. Oceanic manganese nodules contain bismuth in a range of 0.5–24 ppm. (see OCEAN RAW MATERIALS). Silicic rock contains the next highest concentration of bismuth, at 0.02–0.9 ppm (2).

3. Properties

Bismuth has unique properties that make it a valuable metal for certain industrial applications. These properties are characterized by a low melting point, a high density, and expansion on solidification. Bismuth is one of only two metals that expand on solidification. The solid metal floats on molten metal just as ice floats on water. Gallium is the other metal that does the same. Bismuth is a poor electrical conductor and is the most diamagnetic of the metals. The thermal conductivity of bismuth is lower than that of most other metals. Tables 1 and 2 list many of the properties of bismuth.

4. Production

Bismuth is a mining by-product, primarily from the mining of ores such as copper and lead. Tungsten, tin, and molybdenum ores also may contain bismuth. Significant quantities of bismuth are mined in Australia, Bolivia, Canada, China, Japan, Mexico, Peru, and the United States. Mining & Chemical Products, Ltd., in the United Kingdom and Sidech in Belgium are major refiners of bismuth in the West operating with concentrates. World production is summarized in Table 3 (3).

In Australia, bismuth is mined as a by-product of copper ores and exported for refining. As mentioned earlier, the Tasna mine in Bolivia is the only location in the world where the concentration of bismuth is high enough to make mining for bismuth even remotely worthwhile. That mine was closed during most of the 1980s because the price of bismuth was so low that it was not economical to keep the mine open. Only during the late 1990s was there been renewed interest in opening the Tasna mine.

Table 1. **Physical Properties of Bismuth**

Property	Value
boiling point °C	1,560
Bi–Bi bond length at 25°C, nm	0.309
crystal ionic radius, nm	
$\quad$ Bi$^+$	0.098
$\quad$ Bi^{3+}	0.096
$\quad$ Bi^{5+}	0.074
crystal structure	Rhombohedral
density, kg/m^3	
$\quad$ 20°C	9800
$\quad$ 271°C^a	9740
$\quad$ 271°C^b	10,070
$\quad$ 600°C	9660
electrical resistivity, Ω·cm	
$\quad$ 0°C	106×10^{-6}
$\quad$ 20°C	120×10^{-6}
expansion on freezing, % by volume	3.3
hardness, Mohs scale	2.5
magnetic susceptibility	
$\quad$ solid	-280×10^{-13}
$\quad$ liquid	-10.5×10^{-13}
melting point, °C	271.3
vapor pressure, kPa^c	
$\quad$ 400°C	1.013×10^{-4}
$\quad$ 600°C	1.013×10^{-1}
$\quad$ 880°C	1.013×10^{2}
$\quad$ 1420°C	1.013×10^{5}
viscosity, mPa·s	
$\quad$ 285°C	1.610
$\quad$ 304°C	1.662
$\quad$ 365°C	1.460
$\quad$ 451°C	1.280
$\quad$ 600°C	0.998

^{a} Solid.
^{b} Liquid.
^{c} To convert kPa to psi, divide by 6.895×10^3.

Mined bismuth in China is a by-product of tungsten mining operations. Most of the bismuth produced in China is refined from concentrates.

5. Manufacture and Processing

Four basic forms of bismuth are readily available commercially: ingot, needle, pellet, and powder. Bismuth ingots range from 4.5 to 20 kg each depending on the producer. Ingots are used mostly in metallurgical applications and in making fusible alloys. Bismuth needle is typically 0.16 cm (0.0625 in.) in diameter by nominally 2.54 cm (1 in.) in length. Needle is primarily used in the production of bismuth compounds for pharmaceutical and catalyst applications. The high surface area of the needle makes it easy to dissolve in various acids. Bismuth powder is produced in varying mesh sizes for the electronics industry.

Table 2. Thermochemical and Thermodynamic Properties of Bismuth

Parameter	Value
entropy at 298 K, $\Delta S°$, J/(mol·K)[a]	56.9
entropy of transition, ΔS, J/(mol·K)[a] solid	20.2
liquid	90.4
heat capacity, ΔC_p, J/(kg·K)[a] −173.15°C	108
25°C	122
25–271°C	$4.49–5.40 \times 10^{-3}\ T$[b]
271–1027°C[c]	235[d]
heat of fusion, ΔH_{fus}, kJ/kg[a]	51.816
heat of transition, ΔH, J/mol[a] solid	11.0
liquid	172
heat of vaporization, ΔH_{vap}, kJ/kg[a]	858.29

	Temperature		
	0°C	25°C	100°C
thermal conductivity, W/(m·K)			
parallel to triagonal axis	5.54	5.30	4.81
perpendicular to triagonal axis	9.53	9.19	8.44
polycrystalline	8.22	7.92	7.22
thermal expansion coefficient, 20–100°C	(1.34×10^{-6})/K		

[a] To convert from J to cal, divide by 4.184.
[b] Temperature in degrees kelvin.
[c] Liquid.
[d] Units are J/(mol·K). To convert from J to cal, divide by 4.184.

Bismuth pellets range in size from 4.5 to 50 g and are used for metallurgical additives. Their convenient size and specific weights make them particularly useful as feedstock when a given quantity of bismuth must be added regularly to a melt. They are used primarily in cast iron additions, but this use is declining.

5.1. Fabrication. The principal portion of the *bismuth* in *copper ores* follows the copper into the matte. During the conversion of the matte to blister copper, most of the bismuth fumes off. The fumes are caught in the baghouse or Cottrell system along with other elements such as lead, arsenic, and antimony. The dusts are transferred to the lead-smelting operation. The portion of the bismuth remaining with the blister copper is separated during the electrolytic refining in the slimes. The procedure for handling the slimes results in collection of the bismuth in the lead bullion (4).

Table 3. Annual Bismuth Production in tons[a]

	Year				
World Production[b]	1992	1993	1994	1995	1996
mines	2870	3220	3060	3490	3440[c]
refineries	3710	4360	4080	4260	4230[c]

[a] Reference 3.
[b] Excluding the United States.
[c] Estimated.

The bismuth that is found in the lead ore accompanies the lead through the smelting operation right up to the last refining steps. The removal of bismuth then requires special techniques; the most common are the Betterton–Kroll and the Betts processes (5).

5.2. Betterton–Kroll Process. Metallic calcium and magnesium are added to the lead bullion in a melt and form ternary compounds that melt higher than lead and are lower in density. On cooling of the lead bath to a temperature close to the melting point of lead, the intermetallic compounds high in bismuth content solidify and float to the top, where they are removed by skimming.

The bismuth–calcium–magnesium dross also contains lead that must be removed. The dross is heated in a kettle to free any entrapped lead that melts and forms a pool under the dross. This lead is cast and returned to the bismuth separation cycle. The dross is then melted and treated with chlorine and/or lead chloride to remove the calcium and magnesium. The resulting molten metal is an alloy of bismuth and lead, high in bismuth, which is then treated to produce refined bismuth metal.

5.3. Betts Electrolytic Process. The Betts process starts with lead bullion, which may carry tin, silver, gold, bismuth, copper, antimony, arsenic, selenium, tellurium, and other impurities, but should contain at least 90% lead (6,7). If more than 0.01% tin is present, it is usually removed from the bullion first by means of a tin-drossing operation (see TIN AND TIN ALLOYS DETERMINING). The lead bullion is cast as plates or anodes, and numerous anodes are set in parallel in each electrolytic cell. Between the anodes, thin sheets of pure lead are hung from conductor bars to form the cathodes. Several cells are connected in series.

The electrolyte is a solution of lead fluosilicate [25808-74-6] ($PbSiF_6$), and fluosilicic acid [16961-83-4] (H_2SiF_6), containing a small amount of glue or other suitable agent. Direct current is passed through the cells to dissolve the lead from the anodes and deposit it on the cathodes. The impurities in the lead anodes are insoluble under the conditions of normal cell operation and remain on the face of the anodes as a porous slime blanket. The finished cathodes are withdrawn from the cells, washed, and melted to refined lead. The scrap anodes are withdrawn, and the slime is washed free of soluble matter, either while still on the anode or after it has been removed by cleaning. The cleaned anode scrap is returned to the anode-casting kettle for recasting.

The washed slime is dried and melted to produce slag and metal. The slag is usually purified by selective reduction and smelted to produce antimonial lead. The metal is treated in the molten state by selective oxidation for the removal of arsenic, antimony, and some of the lead. It is then transferred to a cupel furnace, where the oxidation is continued until only the silver–gold alloy (doré) remains. The bismuth-rich cupel slags are crushed, mixed with a small amount of sulfur, and reduced with carbon to a copper matte and impure bismuth metal; the latter is transferred to the bismuth refining plant.

The gases from the several furnaces treating the slimes carry bismuth, silver, gold, and other values as particulates, which are recovered via Cottrell precipitation, baghouses, or scrubbers.

5.4. Recovery of Bismuth from Tin Concentrates. Bismuth is leached from roasted tin concentrates and other bismuth-bearing materials by means of hydrochloric acid. The acid leach liquor is clarified by settling or filtration, and the bismuth is precipitated as bismuth oxychloride [7787-59-9] (BiOCl), when the liquors are diluted using large volumes of water. The impure bismuth oxychloride is usually redissolved in hydrochloric acid and re-precipitated by diluting several times. It is then dried, mixed with soda ash and carbon, and reduced to metal. The wet bismuth oxychloride may also be reduced to metal by means of iron or zinc in the presence of hydrochloric acid. The metallic bismuth produced by the oxychloride method requires additional refining.

The Sperry process for making white lead in an electrolytic cell recovers bismuth as a by-product in the anode slimes.

The crystallization process for concentrating bismuth in lead by squeezing the eutectic (high in bismuth) liquid out of the solidified high-lead portion at a temperature within the melting range of the alloy is seldom used.

5.5. Refining.   The alloy of bismuth and lead from the separation procedures is treated with molten caustic soda to remove traces of such acidic elements as arsenic and tellurium (4). It is then subjected to the Parkes desilverization process to remove the silver and gold present. This process is also used to remove these elements from lead.

The desilverized alloy now contains bismuth as well as lead and zinc. Removed of the lead and zinc is facilitated by the fact that the formation of zinc and lead chlorides precedes the formation of bismuth chloride [7787-60-2] ($BiCl_3$), when the alloy is treated at 500°C with chlorine gas. Zinc chloride [7646-85-7] ($ZnCl_2$), forms first, and after its removal lead chloride [7758-95-4] ($PbCl_2$), forms preferentially. This process is continued until the desired level of lead removal has been reached. The bismuth is given a final oxidation with air and caustic soda; the refined product typically has a purity of 99.997%.

6. Economic Aspects

In order to adequately discuss the economics of bismuth, it must first be reiterated that bismuth is a minor metal. The supply of bismuth is not easily changed because it is dependent on the supply of the associated metals with which it is mined. Operation of the normal principles of supply and demand with minor metals differs from that with major metals. When the need for copper or zinc increases, it is possible to mine more copper or zinc. This is not the case with a minor metal. To increase the supply of bismuth, it would be necessary to mine more copper or lead to get more bismuth from the ore. It does not make economic sense for copper and lead mines to increase production in order to produce more of what is in effect a contaminant in the ore. It would also cause the prices of those metals to fall as excess material would become available on the market.

The amount of bismuth in the world is finite, of course, as are all nonrenewable resources. Bismuth reserves can be divided into categories described as "currently economic," "marginally economic," and "subeconomic" resources on the basis of bismuth concentration, current mining techniques, and current refining capabilities. With the crustal abundance of bismuth ranging from 0.008 to 24 ppm, there are clearly reserves where it is not possible to recover the bismuth present. Table 4 lists the world currently economic bismuth reserves as determined by a joint U.S. Bureau of Mines Geological Survey in 1992.

Table 4. **Bismuth Reserves, Currently Economic**[a]

Continent	Country	Tons	Subtotal[b]
Asia	China	20,000	—
	Japan	9,000	—
	Republic of Korea	4,000	33,000
Australia	Australia	18,000	18,000
Europe		16,000	16,000
North America	Canada	5,000	—
	Mexico	10,000	—
	USA	9,000	24,000
South America	Bolivia	5,000	—
	Peru	11,000	16,000

[a] Source: Reference 8
[b] World Total Reserves 107,000 tonnes.

With the growing interest in the uses of bismuth, especially as a replacement option for lead in many applications, a new assessment of the world bismuth reserves would be warranted. However, the figures from the 1992 survey (Table 4) are the most recent available In addition to the currently economic reserves of 107,000 tonnes, it is estimated that there is a reserve base including economic, marginally economic, and subeconomic reserves totaling in excess of 200,000 tons. The Tasna mine, if reopened, could in conjunction with Adex Mining/Mount Pleasant, New Brunswick, Canada, add in excess of 50,000 tons to this figure (9).

The world production of refined bismuth for 1996 was approximately 5500 tons. Belgium, China, Mexico and Peru were the largest producers, with Japan not far behind (see Fig. 1).

China's production has remained relatively stable over the last several years. During that time, however, world demand for bismuth has increased, so China's market share has been reduced, in effect from 30 to 20%. The Chinese influence on bismuth market prices is disproportionate to their market share,

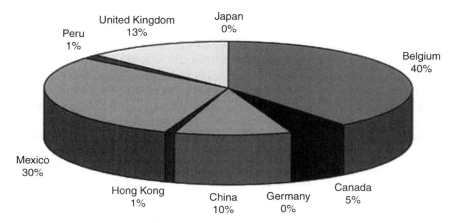

Fig. 1. 1996 bismuth imports by country (in tons).

Table 5. **World Bismuth Mine and Refinery Production**[a]

Country	Mine 1992	1993	1994	1995	1996	1992	1993	1994	1995	1996
Belgium[b]	—	—	—	—	—	800	950	900	800	800
Bolivia	—	—	—	121	125	30	7	36	19	20
Bulgaria[b]	40	40	40	40	40	40	40	40	40	40
Canada	224	144	129	187	185	—	—	—	—	—
China[b]	820	740	610	740	700	1060	1050	850	800	800
Italy	—	—	—	—	—	20	—	5[b]	5[b]	5
Japan	159[b]	149[b]	152[b]	177[b]	169	530	497	505	591	563
Kazakhstan[b]	160	160	160	155	155	170	170	170	166	160
Korea, Republic of	9[b]	5[b]	—	—	—	9	5	—	—	—
Mexico	807	908	900[b]	995	1000	550[b]	650[b]	650[b]	924	925
Peru	550[b]	1000[b]	1000[b]	1000[b]	1000	419	937	871	870[b]	870
Romania[b]	50	40	40	40	40	50	35	35	35	35
Russia[b]	5	4	4	4	4	10	9	9	10	10
Serbia and Montenegro	10	5	5[b]	5[b]	5	20	10	<0.5	<0.5	<0.5
Tajikistan[b]	20	16	12	11	5	—	—	—	—	—
Uzbekistan[b]	15	10	10	10	11	—	—	—	—	—
Total	*3870*	*3220*	*3060*	*3490*	*3440*	*3710*	*4360*	*4070*	*4260*	*4230*

[a] *Source*: Reference 10. *Note*: World totals and estimated data are rounded and may not add to totals shown. United States values have been withheld to avoid disclosing company proprietary information.
[b] Estimated.

however. By multichannelling material through Hong Kong traders, the availability of material is exaggerated. Some Chinese suppliers renege on delivery when they see bismuth prices rising, thus causing short-term shortage of metal (9).

In Japan, production was down in 1996 from 1995. In Peru, Centromin produced their highest output ever at 939 tons. This was a substantial increase over 1995 production. At Penoles of Mexico, higher bismuth concentrations caused increased yields in production over the previous year. Table 5 shows world production for the late 1990s.

The United States is and has always been highly dependent on bismuth imports because domestic usage has greatly exceeded domestic production. That situation worsened during 1997. Asarco, the only U.S. producer of bismuth, closed its Omaha, Nebraska refinery and smelter. In 1996, the United States imported 1490 tonnes of bismuth (3). Of this amount, approximately 40% came from Belgium; 30%, from Mexico; and 13%, from the United Kingdom. Through November 1997, the United States had imported 2040 tonnes of bismuth, representing a 36% increase over the previous year. Part of this was due to the need to replace the Asarco production.

Also in the United States, the federal government has been considered a producer of sorts since the late 1990s. The U.S. Defense Logistics Agency (DLA) built up a stockpile of about 994 tonnes of bismuth over the years. Beginning in the early 1990s the government began to sell off its stockpile, selling over

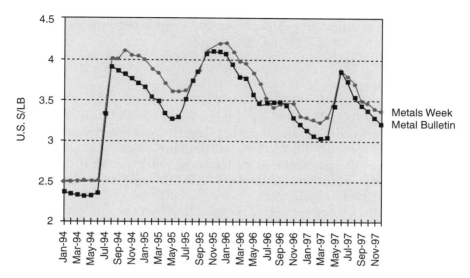

Fig. 2. Bismuth prices 1994–1997.

100 tonnes each year. Although this amount does not include the DLA in the list of top producers, it is a significant quantity of material to be sold each year. At the end of 1997, the DLA sold the last of its stockpile.

6.1. Pricing. There are two 6.1. published prices for bismuth. In Europe, the Metal Bulletin publishes prices twice each week listing both a high and a low for bismuth. In the United States, Platt's Metals Week magazine publishes a weekly price that also lists both highs and lows. The American price is usually about $0.10–$0.20/lb higher than the London price. Since the 1970s bismuth prices have ranged from an all time low of less than $1.50/lb ($3.30/kg) in summer 1982 to an all-time high of over $20.00/lb ($44.09/kg) in May 1974.

In 1994–1997, bismuth prices were more stable than during the 1970s and 1980s. During that 4-year period, the price averaged $3.56/lb ($7.848/kg) for Metals Week and $3.40/lb ($7.496/kg) for the London Metal Bulletin price. The lowest price for this period was the January 1994 average for Metals Week at $2.48/lb ($5.467/kg) and the April 1994 price for LMB at $2.325/lb ($5.126/kg). The highest prices were in January 1996 at $4.20/lb ($9.259/kg) for Metals Week and November 1995 for the London Metal Bulletin at $4.097/lb ($9.032/kg). Figure 2 shows the bismuth price averages of those four years.

7. Grades and Specifications

The purity of bismuth ranges from 99.99 to 99.999% depending on the use. Ingot for metallurgical and fusible alloy applications is commercially available at 99.99% pure. Needle for catalyst and other chemical applications is usually supplied as 99.99% pure. Pharmaceuticals require bismuth to be minimum 99.997% pure. This is typically supplied as needle. Needle of 99.999% purity is useful for the manufacture of high-purity bismuth compounds for medical, electronic and ceramic applications.

8. Analytical Methods

Many of the methods available for determination of bismuth are not very selective; thus it is necessary to separate the bismuth from interfering substances. Titration of bismuth with EDTA (ethylenediamine tetraacetic acid) has been found to be one of the best general methods for determining both macroscale and semimicroscale quantities of bismuth (11). The method is fast, accurate, and convenient. Few foreign ions in moderate amounts interfere. The titration is best carried out at a pH between 1.5 and 2.0. When the endpoint is detected by photometric methods, the titration of bismuth as dilute as 10^{-6} molar is feasible.

For the determination of trace amounts of bismuth, atomic absorption spectrometry is probably the most sensitive method. A procedure involving the generation of bismuthine by the use of sodium borohydride followed by flameless atomic absorption spectrometry has been described (12). The sensitivity of this method is given as 10 ppb/0.0044 A, where A is an absorbance unit; the precision is 6.7% for 25 pg of bismuth. The low neutron cross section of bismuth virtually rules out any determination of bismuth based on neutron absorption or neutron activation.

9. Environmental Concerns

Bismuth is considered both nontoxic and noncarcinogenic despite the fact that it is a heavy metal. It exhibits many of the characteristics of lead, which is next to it on the periodic table of the elements, yet it is not poisonous as lead. The Bismuth Institute published a report written by Yves Palmieri entitled *Bismuth: An Amazingly Green Environmentally-Minded Element* (13). In this report, Mr. Palmieri lists many of the characteristics as well as the many uses of bismuth, describes instances where bismuth has replaced lead because of the toxicity of lead, and explains why bismuth is more than just nonhazardous. It actually has a beneficial influence on human health (13).

9.1. Recycling and Disposal.   Bismuth has been involved in recycling since long before recycling became mandated by the Environmental Protection Agency. Because bismuth alloys melt at relatively low temperatures, it is relatively easy to reclaim and reprocess them. In a simplified application, the bismuth alloy to be reclaimed is simply melted and slowly cooled to allow some of the contaminants to float out and the dross skimmed. The resulting alloy can then be analyzed for metal ratios and used to make other bismuth alloys. Bismuth-containing materials should not be disposed of because of the ease in refining and the inherent value of the bismuth contained in the material.

9.2. Health and Safety Factors.   No industrial poisoning from bismuth has been reported (14). The use of bismuth compounds in the medical field for several hundred years indicates the safety of the material. However, precautions

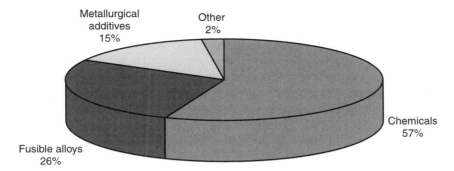

Fig. 3. Bismuth usage in 1996.

should be taken against the careless handling of bismuth and its compounds; ingestion and inhalation of dusts and fumes should be avoided.

10. Uses and Market Demand

The uses of bismuth can be broken down into three primary categories: chemical, metallurgical additive, and fusible alloy (see Fig. 3). The individual categories can be subdivided into numerous lesser categories as listed in Table 6 and described in the following paragraphs.

The chemical category is broken down into pharmaceuticals, cosmetics, catalysts, industrial pigments, and electronics. As stated earlier, bismuth salts have been used for medicinal purposes for hundreds of years. About 90% of bismuth usage was for pharmaceuticals prior to the 1930s (10,15–17). From that point until the 1970s, research produced new applications that greatly expanded the uses of the metal. At that point the pharmaceutical use accounted for about 50% of total bismuth consumption. By 1991, the fusible alloy category along with the other subdivisions each accounted for about 10% of the bismuth usage. Pharmaceutical usage was down to 20% by the 1990s.

Table 6. **U.S. Consumption by Category in tons**

Use	Year				
	1993[a]	1994[a]	1995[b]	1996[b]	1997[c]
chemicals[d]	750	841	1320	855	841
fusible alloys	256	276	544	401	490
metallurgical additives	232	325	257	231	258
other	59	26	27	35	25
Total	*1300*	*1470*	*2150*	*1520*	*1610*

[a] Reference 16.
[b] Reference 10.
[c] Reference 17.
[d] Includes pharmaceuticals, cosmatics, catalysts, and electronics.

The use of bismuth in metallurgical applications increased greatly in the 1990s. With the many environmental concerns about the use of lead, bismuth has been found to be a suitable replacement for lead in many areas. Bismuth is used to replace the lead in some brass water fittings. In conjunction with tin, bismuth is used in replacing lead in bronze bearings (18). It is also used in place of lead in heat-treating applications and in galvanizing operations. Because of recent developments, bismuth, alloyed with a small amount of tin, is being used to make shotgun ammunition.

As a metallurgical additive, bismuth is used in the manufacturing of free-machining steel and free-machining aluminum. The amount of bismuth added can range from 0.003 to 0.10% depending on the type of steel or aluminum. The tiny droplets of bismuth dispersed within the alloy melt when they come in contact with the cutting edge lubricating it. The bismuth addition enhances the action of lead in the steel by producing less frictional resistance at the cutting edge of the tool. This produces thinner and smaller chips, higher productivity, and better surface finish.

11. Bismuth Alloys

Because bismuth expands on solidification and because it alloys with certain other metals to give low-melting alloys, bismuth is particularly well suited for a number of uses. Alloys of bismuth can be made that expand, shrink, or remain dimensionally stable on solidification. So, bismuth alloys have lent themselves to a wide range of industrial applications. Composition and uses are summarized in Table 7.

11.1. Uses of Alloys.   *Anchoring.*   Bismuth alloys that expand on solidification are particularly useful for aligning and setting punches in a die plate. It is much easier to melt and pour an alloy around a punch than to machine the entire die plate and punch at the same time. This method also makes it easier to relocate parts or change dies. The low temperatures involved do not cause distortion.

Radiation Shielding.   Bismuth, like lead, absorbs radiation. Therefore, bismuth alloys are widely used in the medical industry during radiation therapy.

The alloy is cast into a block in the shape of various organs that are to be shielded. Then the alloy block is mounted on a glass plate located between the patient and the radiation source. The bismuth alloy absorbs the radiation

Table 7. **Alloy Compositions and Uses**

Melting Point, °C	Bismuth	Lead	Tin	Cadmium	Indium	Uses[a]
47	44.7	22.6	8.3	5.3	19.1	FSD, LB
70	50	26.7	13.3	10	0	FSD, LB, RS, T, W
95	52.5	32	15.5	0	0	FSD, RS
138	58	0	42	0	0	FC, FSD, SMF, W
138/170	40	0	60	0	0	FSD, IC

[a]*Abbreviations*: FC = fusible core; FSD = fusible safety devices; IC = investment casting; LB = lens blocking; RS = radiation shielding; SMF = sheetmetal forming; T = tube bending; W = work holding.

aimed at the organs it is intended to protect, thus shielding vital organs from unnecessary radiation exposure. The same block is stored and reused for the same patient repeatedly until treatment is complete. Since these alloys melt at low temperatures, usually 158 or 203°F, they are easily melted and recast for use on other patients.

Electromagnetic and Radiofrequency Shielding. Because bismuth is highly diamagnetic, its alloys are quite useful in applications where electronic equipment must be protected from outside interference or where equipment can cause outside interference. This is an important application in the design and construction of medical testing equipment that uses high-powered electromagnetics. These alloys can be easily cast into or sprayed onto the surface of the area needed for maximum shielding.

Tube Bending. The search for high-strength, low-weight structural materials produced the use of hollow tubes of many metals and alloys as structural components. These materials must often be bent and shaped to fit. Bending an empty tube causes distortion of the shape of the tube by flattening or wrinkling. These tubes can be filled with a low-melting bismuth alloy that allows the tube to be bent as if it were solid, thus eliminating distortion. Then the alloy can be easily melted out of the tube and reused.

Fusible Safety Devices. Low-melting bismuth alloys, especially those that are eutectic, have found numerous uses in safety devices. These alloys can be cast into any shape necessary in order to be used in a plug or switch that must function at a given temperature. Sprinkler heads, high-pressure valve fittings, and meat timers are among the devices that contain low-melting bismuth alloys.

Lens Blocking. Bismuth alloys have become particularly useful in the optical industry for securing lenses for grinding processes. The low melting point provides that the lenses may be secured without preheating. The alloys have high strength so that good control is maintained during the grinding process. The density of the alloy provides for a dampening of the vibrations caused by the grinding. After processing, the lenses are immersed in hot-water bath to remove the alloy, which is reused.

Fusible Core Technology. This is currently the fastest growing area of bismuth alloy use today. Low melting bismuth alloys make it possible to produce items having complex internal cavities that cannot be produced using conventional core molds. These alloys are dimensionally stable so that when they are cast they result in a core piece that has the most exact detail and surface finish required. These alloys are being used in the electroforming industry as well as the plastics industry, where cost and weight reduction have become critical. Once molding or electroforming is complete, the part is immersed in a heat bath that melts out the alloy for reuse. This process is especially useful in the automotive industry for the manufacture of lightweight plastic air-intake manifolds.

Steel Quenching. Bismuth and bismuth–lead alloys are used in the processing of some steel products. The thermal conductivity of bismuth makes it ideal for use in tempering steel. The use of bismuth or bismuth–lead alloy in place of pure lead has the advantage of lowering the operating temperature of the bath as well as reducing adherence of alloy to the steel.

Proof Casting of Dies and Molds.   Low-melting alloys make the process of diemaking faster and easier. The low-temperature alloys can be cast into a mold pattern at virtually any point in the manufacturing process without long delays in production and without the possibility of heat distortion. These alloys produce a casting that is exact in detail, requiring no curing time. The casting can then be inspected and measured to gauge the accuracy of the dies. After the alloy has been measured and tested, it can be melted and is completely reusable.

Work Holding and Work Supporting.   Low-melting bismuth alloys have been used to solve two problems in machining operations: (1) these alloys are used to hold parts that need machining but have no regular side that can be clamped, and (2) bismuth alloys provide a method of support for parts such as turbine blade vanes that are not stiff enough to stand alone for machining.

Sheetmetal Forming Dies.   Engineers have found that the low melting alloys, because they are tough and durable, are suitable for making castings that produce hundreds of pressings in sheetmetal of normal materials from aluminum to titanium. Once a given short run of pressings is complete, the tooling can be melted easily and used for another tooling job.

Lead-Free Waterfowl Shot.   Waterfowl poisoning from lead shot has been a conservationists concern for years. Research has indicated that a large percentage of waterfowl have been poisoned by ingestion of lead shot. This prompted the search for an alternative to lead shot.

The first alternative was steel shot. This posed two problems to hunters. The first is that since lead shot is much denser than steel shot, the velocity

Table 8. Properties of Low Melting Bismuth Alloys

	Alloy			
Property	47°C	70°C	138°C	138/170°C
melting point or range, °C (°F)	47.5 (117)	70 (158)	138.5 (281)	138.5–170 (281–338)
density, lb/in.^{3} (g/cm^3)	0.34 (9.36)	0.35 (9.67)	0.31 (8.58)	0.30 (8.21)
specific heat, cal/g·C[a]				
solid	0.039	0.035	0.040	0.043
liquid	0.047	0.044	0.048	0.051
heat of fusion, cal/g	8.8	9.5	10.7	10.6
coefficient of thermal expansion, 1×10^{-6} °C	25	22	15	15
thermal conductivity	0.035	0.043	0.044	0.071
electric conductivity (% of pure copper)	3.09	3.54	2.88	5.00
resistivity, μΩ·cm	55.0	48.0	59.0	34.0
Brinell hardness (2-mm ball, 4-kg load)	14.5/16.5	13/14.5	23/23	23.5/24
tensile strength, lb/in.^{2} (Pa)[b]	4868–5337	2668–3775	8701–9013	8459–9041
maximum sustained load, lb/in.^{2} (Pa)[b]				
30 s	NA[c]	10,000	15,000	15,000
300 s	NA[c]	4,000	9,000	9,500

[a] To convert calories to joules, multiply by 4.184.
[b] To convert pounds force per square inch to pascals, multiply by 6.895×10^3.
[c] Not available.

Table 9. **Growth and Shrinkage of Low-Melting Alloys**[a]

Time after Casting	Alloy			
	47°C	70°C	138°C	138/170°C
6 min	−0.00025	0.00490	−0.00010	0.00030
20 min	−0.00030	0.00565	.00000	0.00035
1 h	−0.00025	0.00570	0.00015	0.00060
8 h	−0.00020	0.00600	0.00045	0.00095
1 day	−0.00015	0.00615	0.00060	0.00105
1 month	0.00025	0.00635	0.00090	0.00120

[a] Cumulative changes, inches per inch relative to cold mold dimensions. Test bar $\frac{1}{2} \times \frac{1}{2} \times 10$ in. $(1.27 \times 1.27 \times 25.4$ cm$)$.

and accuracy of shot changed drastically from that of lead. The second problem is that since steel shot is much harder than lead shot, hunters were finding that the shot was damaging the gun barrels. The development of a bismuth–tin alloy shot has provided the means to solve all of these problems: (1) the bismuth–tin alloy used in the new shot is not poisonous to waterfowl, so the conservationists' problem is solved; (2) bismuth–tin shot is close to lead in density, so accuracy and velocity are not drastically changed from those of lead shot; and (3) the new alloy is not as hard as steel, so there is no damage to gun barrels.

Table 8 lists some of the properties of low-melting bismuth alloys. It is the various combinations of these properties that make the appropriate for the uses described above.

Bismuth's almost unique property of expansion on solidification makes the alloys of bismuth particularly useful for various applications. For example, in the work-holding applications, it is useful for the alloy to expand on solidification in order to adequately hold the piece in question. In the fusible core application, it would be a problem for the alloy to expand significantly on solidification as this would distort the dimensions of the final product. Therefore an alloy with minimal final expansion would be best suited for this use. Table 9 below lists the growth and shrinkage of some of the alloys.

BIBLIOGRAPHY

"Bismuth and Bismuth Alloys" in *ECT* 1st ed., Vol. 2, pp. 526–533, by W. C. Smith, Cerro de Pasco Copper Co.; in *ECT* 2nd ed., Vol. 3, pp. 527–535, by H. E. Howe, American Smelting and Refining Co.; in *ECT* 3rd ed., Vol. 3, pp. 912–921, by S. C. Carapella, Jr. and H. E. Howe, American Smelting and Refining Co. "Bismuth and Bismuth Alloys" in *ECT* 4th ed., Vol. 4, pp. 237–245, by Mark J. Chagnon, Metalspecialties, Inc.; "Bismuth and Bismuth Alloys" in *ECT* (online), posting date: December 4, 2000, by Mark J. Chagnon, Metalspecialties, Inc.

CITED PUBLICATIONS

1. B. Dibner, *Agricola on Metals*, Burndy Library, Norwalk, Conn., 1958, p. 41.
2. J. W. Hasler, M. H. Miller, and R. M. Chapman, *United States Mineral Resources*, Geological Survey Professional Paper 820, U.S. Department of the Interior, Washington, D.C., 1973, p. 96.

3. *Mineral Industry Surveys, Bismuth Annual Review*, U.S. Bureau of Mines, Washington, D.C., 1990.

4. A. R. Powell, in A. R. Powell, *papers of the Institute of Mining and Metallurgy Symposium*, Institute of Mining and Metallurgy, London, UK, 1950, 245–257.

5. J. O. Betterton and Y. E. Levebeff, *Metallurgy of Lead and Zinc*, American Institute of Mining and Metallurgy Engineers, New York, 1936.

6. P. M. Gruzensky and W. J. Crawford, *U.S. Bur. Mines Inf. Circ.*, 7681 (1954).

7. F. Vogel, *Metall Berlin* **21**(2), 122 (1967).

8. M. D. Fickling, Review of Bismuth in 1993, *Mining Journal* Annual Review, 1994.

9. M. D. Fickling, Review of Bismuth in 1996, *Mining Journal* Annual Review, 1997 .

10. *Mineral Industry Surveys, Bismuth Annual Review*, U.S. Department of the Interior. Bureau of Mines, Washington, D.C., 1996.

11. J. S. Fritz, in I. M. Kolthoff and P. J. Elving, eds., *Treatise on Analytical Chemistry*, Part II **Vol. 8**, J Wiley & Sons, Inc., New York, 1963, pp. 147–175.

12. D. S. Lee, *Anal. Chem.* **54**, 1682 (1982).

13. Y. Palmieri, *An Amazingly "Green" Environmentally-Minded Element*, The Bismuth Institute, 1992.

14. N. I. Sax, *Dangerous Properties of Industrial Materials*, 4th ed., Van Nostrand Reinhold Co., New York, 1975, p. 459.

15. J. F. Carlin, Jr., *Mineral Facts and Problems*, U.S. Bureau of Mines, Washington, D.C., 1985, p. 1.

16. *Mineral Industry Surveys, Bismuth Annual Review*, U.S. Department of the Interior, Bureau of Mines, Washington, D.C. 1994.

17. *Mineral Industry Surveys, Bismuth In The Fourth Quarter*, U.S. Department of the Interior, U.S. Geological Survey, Reston, Va., 1997.

18. P. Koslosky, Jr., in *The Bulletin of the Bismuth Institute* **Vol. 69**, Bismuth Institute, Grimbergen, Belgium, 1996, pp. 1–2.

General References

R. C. Weast, ed., *CRC Handbook of Chemistry and Physics*, 71st ed., CRC Press, Boca Raton, Fla., 1991.

L. G. Stevens and C. E. T. White, *The Metals Handbook*, Vol. **2**, 10th ed., ASM International, 1990, 750–757.

Mark J. Chagnon
Atlantic Metals & Alloys

BISMUTH COMPOUNDS

1. Introduction

Bismuth is the fifth member of the nitrogen family of elements and, like its congeners, possesses five electrons in its outermost shell, $6s^2 6p^3$. In many compounds, the bismuth atom utilizes only the three $6p$ electrons in bond formation and retains the two $6s$ electrons as an inert pair. Compounds are also known where bismuth is bonded to four, five, or six other atoms. Many bismuth compounds do not have simple molecular structures and exist in the solid state as polymeric chains or sheets.

The +3 oxidation state is exhibited by bismuth in the vast majority of its compounds. A few inorganic and a variety of organic compounds, however, contain the element in the +5 state. Other rarer oxidation states reported for bismuth include +2, +1, and −3. Bismuth also forms polynuclear ionic species with oxidation states that are usually fractional and range from −1 to +1.

Technical information concerning bismuth and its compounds is distributed periodically by the Bismuth Institute, a nonprofit organization incorporated in La Paz, Bolivia, that has an information center in Brussels. Information on the production and consumption of bismuth is available from the U.S. Geological Survey Minerals Yearbook on the World Wide Web. World production of bismuth in 2000 stands at ~4000 tons/year. Major applications of bismuth compounds include pharmaceuticals (eradicating, eg, *Helicobacter pyloris*, a bacteria inflicting ulcers), additives to ceramics, plastics, catalysts for use in industrial organic chemistry, pigments in cosmetics (BiOCl), and paints (Bi-vanadate colors). Metallic bismuth is widely used in metallurgy (steel additives, fusible alloys etc). Ecological considerations favor applications of bismuth because it is considered nontoxic and noncarcinogenic notwithstanding its heavy metal status.

2. Analysis

Dissolution of Bi in various environments for analytical purposes is effected by digestion with HNO_3 or HNO_3/H_2SO_4. Separations from interfering substances are achieved by extraction with dithizone and cupferron (1) or ion exchange separations. Gravimetric methods for the determination of bismuth involve weighing as $BiPO_4$, BiOCl, or Bi_2O_3 (2). The best general method for determining macro and semimicro quantities of bismuth is the direct titration of bismuth, using the disodium salt of ethylenediaminetetraacetic acid (EDTA) (3). The method is fast, convenient, and accurate. Few foreign ions in moderate amounts interfere. The titration is best carried out at a pH between 1.5 and 2.0. When the end point is detected by photometric methods, the titration of bismuth as dilute as 10^{-6} M is feasible.

Trace amounts of bismuth in ores or in nonferrous alloys can be determined with photometric methods using reagents such as dithizone, thiourea, potassium iodide, or dimercaptothiopyrone (4). Atomic absorption spectrometry (AAS) is most frequently used for the analyses of low concentrations of Bi in a variety of samples including ores and concentrates, superconductors, geological and biological materials (5). Bismuth can be determined by AAS by using flame, hydride generation (HG), electrothermal (ET) atomization, and other methods. Detection limits of 50 ng/mL, 0.2 ng/mL, and 4 pg of Bi have been reported for flame AAS, HG AAS, and GF AAS, respectively. Extremely low concentrations of Bi (detection limit 0.4 pg/mL) can be determined by the technique of inductively coupled plasma mass spectrometry (ICP MS) (5).

3. Inorganic Compounds of Bismuth

3.1. Bismuthine. Bismuthine, BiH_3, [18288-22-7], is a colorless gas, unstable at room temperature, but isolatable as a colorless liquid at lower temperatures. Owing to its instability and difficulty of preparation, no more than a few hundred milligrams of the pure compound have been available for any single study. Vapor pressure data from -116 to $-43°C$ have been determined, and by extrapolation, a normal boiling point of $+16.8°C$ has been indicated; ΔH_v, calculated from the same data, is 25.15 kJ/mol (6.01 kcal/mol) (6). Experimental quantities are best prepared by disproportionation of either methylbismuthine, CH_3BiH_2, [66172-95-0] or dimethylbismuthine, $(CH_3)_2BiH$ [14381-45-4] (6):

$$(CH_3)_{3-n}BiH_n \longrightarrow n/3BiH_3 + (3-n)/3(CH_3)_3Bi \tag{1}$$

In the case of methylbismuthine, the disproportionation occurs upon keeping the compound at $-45°C$ for several hours; 389.1 mg of methylbismuthine yields 241.1 mg of BiH_3.

At room temperature, bismuthine rapidly decomposes into its elements. The rate of decomposition increases markedly at higher temperatures (7). Bismuthine decomposes when bubbled through silver nitrate or alkali solutions but is unaffected by light, hydrogen sulfide, or 4 N sulfuric acid solution. There is no evidence for the formation of BiH_4^+, though the methyl derivative, $(CH_3)_4Bi^+$, is known. The existence of BiH_4^+ would not be anticipated on the basis of the trend found with other Group 15 "onium" ions.

3.2. Bismuthides. Many intermetallic compounds of bismuth with alkali metals and alkaline earth metals have the expected formulas M_3Bi and M_3Bi_2, respectively. These compounds are not salt-like but have high coordination numbers, interatomic distances similar to those found in metals, and metallic electrical conductivities. They dissolve to some extent in molten salts (eg, $NaCl-NaI$) to form solutions that have been interpreted from cryoscopic data as containing some Bi^{3-}. Both the alkali and alkaline earth metals form another series of alloy-like bismuth compounds that become superconducting at low temperatures (Table 1). The MBi_2 compounds are particularly noteworthy as having extremely short bond distances between the alkali metal atoms.

3.3. Bismuth Halides. Bismuth forms subhalides that have or approximate the composition BiX, trihalides, BiX_3 and a single pentahalide, BiF_5.

Table 1. **Alloy-like Superconducting Bismuth Compounds**

Compound	CAS Registry Number	Formula	Temperature, K^a
lithium bismuthide	[12048-27-0]	LiBi	2.47
sodium bismuthide	[12258-63-8]	NaBi	2.22
potassium dibismuthide	[12431-17-3]	KBi_2	3.58
rubidium dibismuthide	[55127-10-1]	$RbBi_2$	4.25
cesium dibismuthide	[12233-24-8]	$CsBi_2$	4.75
calcium tribismuthide	[66271-89-4]	$CaBi_3$	1.7
strontium tribismuthide	[12589-81-0]	$SrBi_3$	5.62
barium tribismuthide	[12047-02-8]	$BaBi_3$	5.69

^{a}Temperature below which the compound is superconducting.

Bismuth Subhalides. Vapors above solutions of a bismuth trihalide in molten bismuth contain the species BiX and/or $(BiX)_n$, where X = Cl, Br, or I (8). At temperatures <323°C, a black, diamagnetic, orthorhombic solid of the overall composition $BiCl_{1.167}$ may be isolated from solutions of bismuth trichloride in molten bismuth. The crystal structure consisting of bismuth cluster cations and two different bismuth chloroanions corresponds to the formula $[(Bi_9{}^{5+})_2(BiCl_5{}^{2-})_4(Bi_2Cl_8{}^{2-})]$ (9). The bromo analogue $BiBr_{1.167}$ (10) has a similar structure. Bismuth also forms true crystalline mono halides, BiBr (10) and BiI (11), which have polymeric mixed-valent structures, $[Bi(0)Bi(II)X_2]_n$.

Bismuth Trihalides. Reaction of bismuth trioxide with aqueous HF, HCl, or HBr or direct halogenation of metallic bismuth in the case of Cl_2, Br_2, I_2 yields the corresponding bismuth trihalides. Physical and thermochemical properties of the more important bismuth halides appear in Table 2.

Bismuth Trifluoride. Bismuth(III) fluoride is a white to gray-white powder, density 8.3 g/mL, that is essentially isomorphous with orthorhombic YF_3, requiring nine-coordination about the bismuth (19). It has been suggested that BiF_3 is best considered an eight-coordinate structure with the deviation from the YF_3 structure resulting from stereochemical activity of the bismuth lone-pair of electrons. In accord with its structure, the compound is more ionic than the other bismuth halides. It is almost insoluble in water ($5.03 \pm 0.05 \times 10^{-3}$ M at pH 1.15) and dissolves only to the extent of 0.010 g/100 g of anhydrous HF at 12.4°C.

Bismuth trifluoride is usually prepared by dissolving either Bi_2O_3 or BiOF in hydrofluoric acid to yield the addition compound bismuth trifluoride trihydrofluoride $BiF_3 \cdot HF$ or $H_3(BiF_6)$, [66184-11-0]. Careful evaporation of the solution permits isolation of a gray solid, which upon heating loses HF to yield BiF_3. It may be purified by sublimation in a stream of HF at 500°C. Bismuth trifluoride may also be prepared by (*1*) reaction of Bi_2O_3 with sulfur tetrafluoride, SF_4; (*2*) treatment of metallic bismuth with HF at 350°C; and (*3*) reduction of BiF_5 in a dilute stream of hydrogen. Bismuth trifluoride is not readily hydrolyzed even by boiling water. However, addition of HF causes the formation of $BiF_3 \cdot 3HF$, which is readily hydrolyzed to bismuth oxyfluoride, BiOF, [13520-72-4]. Heating BiF_3 at 200–300°C in air results in the formation of some oxide or oxyfluoride. Between 600 and 800°C fluorine is gradually replaced by oxygen yielding phases such as $BiO_{0.1}F_{2.8}$, BiOF, and, on prolonged heating (60 h at 670°C), Bi_2O_3. The so-called cubic phase of BiF_3 probably contains some oxygen.

Complexes of BiF_3 are almost unknown, but crystallization from a hot solution of ammonium fluoride that has been saturated with freshly precipitated bismuth trioxide yields crystals of ammonium tetrafluorobismuthate(III), NH_4BiF_4, [13600-76-5]. This complex is readily decomposed by water.

Bismuth Trichloride. Bismuth(III) chloride is a colorless, crystalline, deliquescent solid made up of pyramidal molecules (20). The nearest intermolecular Bi–Cl distances are 0.3216 and 0.3450 nm. The density of the solid is 4.75 g/mL and that of the liquid at 254°C is 3.851 g/mL. The vapor density corresponds to that of the monomeric species. The compound is monomeric in dilute ether solutions, but association occurs at concentrations > 0.1 M. The electrical conductivity of molten $BiCl_3$ is of the same order of magnitude as that found for ionic substances.

Table 2. **Physical Properties of Bismuth Compounds**

Bismuth compound	CAS Registry Number	Formula	mp, °C	bp, °C	$\Delta H^{\circ}_{f,298}$, kJ/mol[a]	S°_{298}, J/mol·K[a]	$\Delta H^{\circ}_{fusion}$, kJ/mol[a]	$\Delta S^{\circ}_{fusion}$, J/mol·K[a]	$\Delta H^{\circ}_{subl,298}$, kJ/mol[a]	$\Delta S^{\circ}_{subl,298}$, J/mol·K[a]	Bi–X bond energy, kJ/mol[a]	Reference
bismuth trifluoride	[7787-61-3]	BiF$_3$	649[b]	900 ± 10	−900 ± 13	123 ± 4	21.6 ± 0.6	23.4 ± 0.8	201 ± 3	195 ± 3	381	12
bismuth pentafluoride	[7787-62-4]	BiF$_5$	151	230								
bismuth trichloride	[7787-60-2]	BiCl$_3$	233.5	447	−379	174 ± 6	23.9		114 ± 1	183 ± 2	279	13,14
bismuth monochloride	[14899-70-8]	BiCl			−131	94.5					300 ± 4	13,15
bismuth oxychloride	[7787-59-9]	BiOCl			−367	120						13
bismuth tribromide	[7787-58-8]	BiBr$_3$	219	441	−276 ± 2	190 ± 1	21.8		115 ± 1	182 ± 1	233.0 ± 1.4	16
bismuth triiodide	[7787-64-6]	BiI$_3$	408.6	542[c]	−151 ± 4	224.8 ± 0.8	39.1 ± 0.3	57.3 ± 0.4	134.3 ± 0.5	183.4 ± 0.8	181 ± 5	17,18
bismuth trioxide	[1304-76-3]	Bi$_2$O$_3$[d]	824		−574	151.5						13
bismuth trisulfide	[1345-07-9]	Bi$_2$S$_3$	850		−143	200						13
bismuth tritelluride	[1304-82-1]	Bi$_2$Te$_3$			−77.4	260.9						13

[a]To convert J to cal, divide by 4.184.
[b]The mp frequently cited is 120°C higher than this and is, apparently, for material contaminated with oxyfluoride.
[c]The normal bp has been extrapolated from vapor-pressure data.
[d]Monoclinic.

20

Bismuth trichloride is usually prepared by chlorination of the molten metal or by dissolving bismuth metal in aqua regia; evaporation of the solution yields the bismuth trichloride dihydrate, $BiCl_3 \cdot 2H_2O$ [66172-88-1], and upon distillation, it decomposes to give anhydrous bismuth trichloride. The commercial product frequently is not anhydrous. Bismuth trichloride shows considerable tendency to form addition compounds. Reaction with ammonia yields the colorless, easily volatilized bismuth trichloride triammine, $BiCl_3 \cdot 3NH_3$, [66172-89-2] as well as the red, thermally unstable bismuth trichloride hemiammine, $2BiCl_3 \cdot NH_3$, [66172-90-5]. Compounds of formula $BiCl_3 \cdot NO$, $BiCl_3 \cdot 2NO_2$, and $BiCl_3 \cdot NOCl$ may be isolated; these are stable in dry air but are decomposed by moisture. Bismuth trichloride is soluble in organic solvents like diethyl ether, acetone, or trimethyl amine; solution is presumably accompanied by the formation of complexes between the organic donor molecules and $BiCl_3$ as acceptor. Recently, complexes of $BiCl_3$ with dimethylsulfoxide (dmso) (21), tetrahydrofuran (thf) (22), bipyridine (bpy) (23), phenantroline (24), bis(diphenylphosphino)methane (24), or crown ethers (25) have been characterized by X-ray crystallography. Also complexes between $BiCl_3$ (or $BiBr_3$) and aromatic hydrocarbons, like benzene, mesitylene, or paracyclophanes have been prepared. Usually the aromatic ring adopts a centroid η^6-coordination on the bismuth center (26). A number of complex bismuth halides are also well known, eg, disodium bismuth pentachloride, Na_2BiCl_5, [66184-10-9]; sodium dibismuth heptachloride, $NaBi_2Cl_7$, [66184-09-6]; and trisodium bismuth hexachloride, Na_3BiCl_6, [66114-82-7]. The acid, hydrogen dibismuth heptachloride trihydrate, $HBi_2Cl_7 \cdot 3H_2O$, [66124-39-9] is a crystalline substance, stable at room temperature, that may be isolated by cooling a solution of $BiCl_3$ in concentrated hydrochloric acid to $0°C$. Also other polynuclear anions, eg, $Bi_2Cl_{11}^{5-}$, $Bi_4Cl_{18}^{6-}$, $Bi_8Cl_{13}^{6-}$ are known (27).

Bismuth Tribromide. Bismuth(III) bromide is a hygroscopic, golden-yellow, crystalline solid that exists in two modifications, α-$BiBr_3$ stable below and β-$BiBr_3$ above $158°C$. The structure of the former consists of trigonal pyramidal $BiBr_3$ molecules with Bi–Br 2.66 Å and three intermolecular contacts Bi$\cdots$Br 3.32 Å per bismuth atom (28). The high temperature modification, β-$BiBr_3$ is polymeric with Bi in octahedral holes of a kind of hexagonal dense packing of the Br atoms (28).

Bismuth tribromide may be prepared by dissolving Bi_2O_3 in excess concentrated hydrobromic acid. The slurry formed is allowed to dry in air, then gently heated in a stream of nitrogen to remove water, and finally distilled in a stream of dry nitrogen. Bismuth tribromide is soluble in aqueous solutions of KCl, HCl, KBr, and KI but is decomposed by water to form bismuth oxybromide, BiOBr, [7787-57-7]. It is soluble in acetone and ether, and practically insoluble in alcohol. It forms complexes with NH_3 and dissolves in hydrobromic acid from which dihydrogen bismuth pentabromide tetrahydrate, $H_2BiBr_5 \cdot 4H_2O$, [66214-38-8], may be crystallized at $-10°C$. Polynuclear bromobismuthates, eg, $Bi_2Br_{10}^{4-}$ or $Bi_4Br_{16}^{4-}$ (27) and complexes of $BiBr_3$ with N- (23), P- (29), O- (30), or S- (31) donor ligands are also known.

Bismuth Triiodide. Bismuth(III) iodide is a greenish black crystalline powder. The iodines are in a hexagonal close-packed array with each bismuth having six nearest-neighbor iodines at 0.32 nm (32). This suggests that the lone pair on bismuth is stereochemically inactive and that the compound is largely ionic in character.

Bismuth triiodide may be prepared by heating stoichiometric quantities of the elements in a sealed tube. It undergoes considerable decomposition at 500°C and is almost completely decomposed at 700°C. However, it may be sublimed without decomposition at 3.3 kPa (25 mmHg). Bismuth triiodide is essentially insoluble in cold water and is decomposed by hot water. It is soluble in absolute alcohol (3.5 g/100 g), benzene, toluene, xylene and in liquid ammonia, in which it forms a red triammine complex. It dissolves in hydroiodic acid solutions from which hydrogen tetraiodobismuthate(III), $HBiI_4 \cdot 4H_2O$, [66214-37-7] may be crystallized, and it dissolves in potassium iodide solutions to yield the red compound, potassium tetraiodobismuthate(III), $KBiI_4$, [39775-75-2]. Compounds of the type tripotassium bismuth hexaiodide, K_3BiI_6, [66214-36-6] salts with polynuclear or polymeric anions, eg, $Bi_3I_{12}{}^{3-}$ or $(BiI_4)_n{}^{n-}$ (27) and complexes of BiI_3 with $OP[N(CH_3)_2]_3$ (33) are also known.

Bismuth Pentafluoride. Bismuth(V) fluoride consists of long white needles that have been shown to have the same structure as the body-centered, tetragonal α-polymorph of uranium hexafluoride. The density of the solid is 5.4 g/mL at 25°C. The solid consists of infinite chains of trans-bridged BiF_6 polyhedra; dimers and trimers are present in the vapor phase (34). Bismuth pentafluoride may be prepared by the fluorination of BiF_3 or metallic bismuth at 600°C. For purification, it may be separated from BiF_3 by repeated sublimation at 120°C. At higher temperatures it decomposes to yield BiF_3 and fluorine.

Bismuth pentafluoride is an active fluorinating agent. It reacts explosively with water to form ozone, oxygen difluoride, and a voluminous chocolate-brown precipitate, possibly a hydrated bismuth(V) oxyfluoride. A similar brown precipitate is observed when the white solid compound bismuth oxytrifluoride, $BiOF_3$, [66172-91-6], is hydrolyzed. Upon standing, the chocolate-brown precipitate slowly undergoes reduction to yield a white bismuth(III) compound. At room temperature BiF_5 reacts vigorously with iodine or sulfur; > 50°C it converts paraffin oil to fluorocarbons; at 150°C it fluorinates uranium tetrafluoride to uranium pentafluoride; and at 180°C it converts Br_2 to bromine trifluoride, BrF_3, and bromine pentafluoride, BrF_5, and chlorine to chlorine fluoride, ClF. It apparently does not react with dry oxygen.

Treatment of BiF_5 with BrF_3 results in the formation of fluorobromonium hexafluorobismuthate(V), $[BrF_2][BiF_6]$, [36608-81-8], which may be isolated as white hygroscopic crystals upon removal of excess BrF_3 under vacuum. This compound is relatively stable but at reduced pressures undergoes decomposition to BrF_3 and BiF_5. Adducts are formed between BiF_5 and the fluorides of lithium, sodium, potassium, or silver by heating equimolar quantities of the respective compounds to 85–150°C. The resulting hexafluorobismuthate(V) compounds are more stable to reduction than BiF_5. The X-ray diffraction pattern of potassium hexafluorobismuthate(V), $KBiF_6$, [26914-71-6], is similar to that of potassium hexafluoroantimonate(V), $KSbF_6$, [16893-92-8]. Silver hexafluorobismuthate(V), $AgBiF_6$, [66184-08-5], may also be formed by warming elemental silver and $[BrF_2][BiF_6]$ dissolved in BrF_3. This compound is hygroscopic, soluble in HF, and reacts with water to form a chocolate-brown precipitate and probably some ozone.

3.4. Bismuth Oxide Halides. Hydrolysis of a bismuth trihalide yields the corresponding bismuth(III) oxide halide (oxyhalide). Bismuth oxyfluoride, BiOF, [13520-72-4], and bismuth oxyiodide, BiOI, [7787-63-5], may also be formed by heating the corresponding bismuth trihalide in air. When either bismuth oxychloride [7787-59-9] or bismuth oxybromide, BiOBr, [7787-57-7] is heated >700°C, complex bismuth oxyhalides of composition $Bi_{24}O_{31}X_{10}$ are formed. Bismuth oxychloride, BiOCl, is a white, lustrous, crystalline powder (density, 7.72 g/mL) that is practically insoluble in water, alcohol, acids, and bases. Hot concentrated alkali solutions convert it to bismuth trioxide. It is used in fingernail polishes, lipsticks, and face powders to give a nacreous effect. A study (35) in which rats were fed 1, 2, or 5% BiOCl for 2 years showed no carcinogenic or toxic effects. Bismuth oxybromide, a white, amorphous powder, and bismuth oxyiodide, a brick-red, amorphous powder having a density of 7.92 g/mL, are essentially insoluble in water, alcohol, acids, and bases and have been used in the manufacture of dry-cell cathodes.

3.5. Bismuth Oxides and Bismuthates. The only oxide of bismuth that has been definitely isolated in a pure state is bismuth trioxide. An acidic oxide that approximates the composition Bi_2O_5 certainly exists. However, there is considerable question as to the exact nature of this material and the species involved. A number of other oxides have been reported, eg, bismuth oxide (1:1) [1332-64-5], bismuth oxide (1:2), bismuth oxide (2:4) [12048-50-9], bismuth oxide (3:5), and bismuth oxide (4:9), but the evidence for their existence as chemical entities is meager at best.

Bismuth Trioxide. Bismuth(III) oxide [1304-76-3] has a complicated polymorphism. At times some of the reported phases deviate from Bi_2O_3 by having too little or too much oxygen; at least in one instance, because of the ready contamination of Bi_2O_3 melts with silicon, the reported phase probably has the composition of bismuth oxide silicate, $Si_2Bi_{24}O_{40}$, [66256-73-3]. The common oxide, α-Bi_2O_3, is a pale-yellow, monoclinic solid, density = 9.32 ± 0.02 g/mL, which is stable up to 710°C. Half of the bismuth atoms are five-coordinate, and half are six coordinate. The lone-pair electrons on bismuth presumably occupy the sixth position for the five-coordinate bismuth and are responsible for the distortion of the oxygen about the six-coordinate bismuth (36).

Bismuth trioxide may be prepared by the following methods: (*1*) the oxidation of bismuth metal by oxygen at temperatures between 750 and 800°C; (*2*) the thermal decomposition of compounds such as the basic carbonate, the carbonate, or the nitrate (700–800°C); (*3*) precipitation of hydrated bismuth trioxide upon addition of an alkali metal hydroxide to a solution of a bismuth salt and removal of the water by ignition. The gelatinous precipitate initially formed becomes crystalline on standing; it has been represented by the formula $Bi(OH)_3$ and called bismuth hydroxide [10361-43-0]. However, no definite compound has been isolated. Bismuth trioxide is practically insoluble in water; it is definitely a basic oxide, and hence dissolves in acids to form salts. Acidic properties are just barely detectable, eg, its solubility slightly increases with increasing base concentration, presumably because of the formation of bismuthate(III) ions, such as $Bi(OH)_6^{3-}$ and related species.

Bismuth trioxide forms numerous, complex, mixed oxides of varying composition when fused with CaO, SrO, BaO, and PbO. If high purity bismuth, lead, and copper oxides and strontium and calcium carbonates are mixed together with metal ratios Bi:Pb:Sr:Ca:Cu = 1.9:0.4:2:2:3 or 1.95:0.6:2:2:3 and calcined at 800–835°C, the resulting materials have the nominal composition $Bi_{1.9}Pb_{0.4}Sr_2Ca_2Cu_3O_x$ and $Bi_{1.95}Pb_{0.6}Sr_2Ca_2Cu_3O_x$ and become superconducting at ~110 K (37). Also lead free high Tc superconducting Bi-Sr-Ca-Cu-O oxides (38,39) are known.

3.6. Higher Oxides of Bismuth and Related Compounds.

Oxidation of either a fused mixture of sodium oxide and bismuth trioxide or a suspension of bismuth trioxide in 40–50% sodium hydroxide solution yields a product in which much of the bismuth is apparently in the +5 oxidation state. Air or oxygen are suitable oxidizing agents for the molten mixture; sodium hypochlorite, chlorine, bromine, or sodium persulfate may be used for the aqueous suspension. The reactions are favored by excess alkali, and though >90% of the bismuth is oxidized to the pentacovalent state, the product is contaminated with considerable excess alkali. Extraction with methanol at 0°C removes the excess alkali and yields pure sodium metabismuthate(V), $NaBiO_3$, [12232-99-4]. Addition of nitric or perchloric acid produces a material ranging in composition from Bi_2O_4 to Bi_2O_5, which slowly loses oxygen. This material and sodium metabismuthate(V) are very good oxidizing agents. The sodium bismuthate of commerce varies in color from yellow to brown to black. It has about two molecules of water per bismuth atom and is insoluble in water. It is capable of oxidizing manganese(II) compounds in nitric acid solution to permanganate, a reaction commonly used as a qualitative test for manganese. $NaBiO_3$ oxidizes Fe^{3+} in basic medium to $FeO_4{}^{2-}$.

3.7. Sulfides and Related Compounds.

Bismuth(III) sulfide, [1345-07-9], bismuth sesquisulfide, Bi_2S_3, is a dark-brown to grayish black crystalline solid, mp 850°C and density of 6.78 g/mL. It occurs naturally as the mineral bismuth glance and is isostructural with stibnite, Sb_2S_3. It may be prepared by heating sulfur and bismuth or by the addition of a soluble sulfide to an aqueous solution of bismuth(III). It is almost insoluble in water or alkaline solutions but dissolves in concentrated nitric acid or hot concentrated hydrochloric acid. Concentrated alkali metal sulfide solutions or melts dissolve Bi_2S_3 to yield crystalline compounds such as potassium thiobismuthate(III), $KBiS_2$, [12506-13-7]. These compounds are rapidly oxidized in air; similar compounds of silver, copper, and lead occur naturally. Fusion of Bi_2S_3 with a bismuth halide yields air-stable compounds, bismuth chlorosulfide, BiClS, [19264-19-8], bismuth bromosulfide, BiBrS, [14794-86-6], and bismuth iodosulfide, BiIS [15060-32-9]. Other mixed halide sulfide compounds have been reported: bismuth bromide sulfide $Bi_{19}S_{27}Br_3$, (19:3:27) [51185-13-8], has been shown to consist of $(Bi_4S_6)_\infty$ chains in which bismuth atoms in neighboring chains are linked by bromide ions (40). Bismuth disulfide, BiS_2, [12323-18-1], can be prepared at 5000 MPa at 1250°C. The compound is a soft, gray, needlelike, crystalline solid. Bismuth trisulfide has been used as a high temperature lubricant and has been of interest for its photo- and thermoelectric properties.

Related Compounds. Bismuth triselenide, Bi_2Se_3 [12068-69-8] and bismuth tritelluride, Bi_2Te_3, [1304-82-1], are known, and in addition to the stoichiometric compounds, preparations can be made containing excess Bi, Se, or Te. Compounds are also known in which some of the Te in Bi_2Te_3 is replaced by S or Se and some of the Se in Bi_2Se_3 by S, eg, dibismuth ditellurium selenide, Bi_2Te_2Se [12010-72-9] and dibismuth ditellurium sulfide, Bi_2Te_2S [1304-78-5]. All of these materials are of interest for their semiconducting properties.

3.8. Bismuth Salts. Bismuth trioxide dissolves in concentrated solutions of strong oxyacids to yield bismuth salts. In more dilute solutions of strong acids or in solutions of weak acids, the oxide reacts to form bismuthyl or basic salts. The normal salts are very susceptible to hydrolysis.

Bismuth Triperchlorate Pentahydrate. Bismuth(III) perchlorate pentahydrate, $Bi(ClO_4)_3 \cdot 5H_2O$ [66172-92-7], is prepared by dissolving Bi_2O_3 in 70% $HClO_4$. Anhydrous bismuth triperchlorate, $Bi(ClO_4)_3$, [14059-45-1] may be prepared by heating bismuthyl perchlorate monohydrate, $BiOClO_4 \cdot H_2O$ [66172-93-8] between 80 and 100°C. Attempts to dissolve bismuth metal in concentrated perchloric acid have resulted in explosions. Treatment of bismuth or Bi_2O_3 with dilute solutions of perchloric acid yields hydrates of bismuthyl perchlorate.

Bismuth Trinitrate Pentahydrate. Bismuth(III) nitrate pentahydrate, $Bi(NO_3)_3 \cdot 5H_2O$ [10035-06-0] is obtained by dissolving bismuth metal, Bi_2O_3, or $(BiO)_2CO_3$ in nitric acid. Attempts to remove the water of hydration yield monoclinic crystals of bismuthyl nitrate hemihydrate, $BiONO_3 \cdot 1/2H_2O$ [10361-64-3] (41). Addition of bismuth trinitrate pentahydrate to alkali yields a product termed bismuth subnitrate that is widely used in pharmaceuticals. This material approximates the composition $6Bi_2O_3 \cdot 5N_2O_5 \cdot 9H_2O$.

Bismuth Trisulfate. Bismuth(III) sulfate, $Bi_2(SO_4)_3$ [7787-68-0], is a colorless, very hygroscopic compound that decomposes >405°C to yield bismuthyl salts and Bi_2O_3. The compound hydrolyzes slowly in cold water and rapidly in hot water to the yellow bismuthyl sulfate, $(BiO)_2SO_4$ [12010-64-9]. The normal sulfate is isomorphous with the sulfates of yttrium, lanthanum, and praseodymium.

Numerous other bismuth and bismuthyl salts are known, eg, bismuth triacetate, $Bi(C_2H_3O_2)_3$ [22306-37-2]; bismuth phosphate, $BiPO_4$ [10049-01-1]; bismuth trithiocyanate, $Bi(SCN)_3$ [43384-63-0]; bismuthyl nitrite hemihydrate, $BiO(NO_2) \cdot 1/2H_2O$ [66172-94-9]; bismuthyl carbonate hemihydrate, $(BiO)_2 CO_3 \cdot 1/2H_2O$ [5798-45-8]; etc. Bismuth is present in the anion in the oxalato complex sodium dioxalatobismuthate(III), $NaBi(C_2O_4)_2$ [19033-91-1], the nitrito complex trisodium hexanitrobismuthate(III), $Na_3Bi(NO_2)_6$ [18515-86-1], and the yellow, water-soluble thiosulfato complex trisodium trithiosulfatobismuthate(III), $Na_3Bi(S_2O_3)_3$ [66256-75-5].

3.9. Alkoxides, Thiolates, Carboxylates, and Related Compounds. Alkoxides of bismuth, $Bi(OR)_3$ (R = alkyl, aryl) are potential precursors for bismuth-containing superconductors. They are obtained by reaction of $BiCl_3$ with NaOR or by other methods (42). Depending on the size of the R group the alkoxides are monomeric or polymeric (43,44). A thiolate with a known crystal structure is $Bi(SR)_3$ (R = 2,4,6-$[(CH_3)_3C]_3C_6H_2$) (45). Complexes of Bi(III) with the SCH_2CH_2OH ligand have been studied with respect to structural aspects and to the antimicrobial potential of these compounds (46,47). Recently studied carboxylates of Bi(III) include mono or bidentate complexes with the ligands CH_3COO (48), $CH_3CH(OH)COO$ (49), $(CH_3)_3COO$ (50), CF_3COO (51). Also complexes with bidentate thio ligands, eg, $S_2P(C_6H_5)_2$ (52), or $S_2COC_2H_5$ (53) are known.

3.10. Compounds with Bismuth Metal Bonds. Bonds between Bi and transition metals including Cr, Mo, W, Mn, Fe, Co, Ru, Ir, Os, Ni exist in clusters. The structures and chemistry of these compounds have been reviewed recently (54). Also compounds with bonds between Bi main group metals or semimetals, eg, Ge (55) or Si (56) are known.

4. Organobismuth Compounds

In a manner similar to phosphorus, arsenic, and antimony, the bismuth atom can be either tri- or pentacovalent. However, organobismuth compounds are less stable thermally than the corresponding phosphorus, arsenic, or antimony compounds, and there are fewer types of organobismuth compounds. For example, with R_4MX, R_3MX_2, R_2MX_3, and RMX_4, where M is a Group 15 (VA) element and X is a halogen, only the first two types have been prepared where M = Bi, but all four types are known where M = P, As, or Sb. The chemistry of organobismuth compounds has been described in several publications (26,57-63). The use of organobismuth compounds, as well as organoantimony ones, in organic synthesis has been exhaustively reviewed (64).

4.1. Primary and Secondary Bismuthines. Unstable methylbismuthine, CH_3BiH_2, and dimethylbismuthine, $(CH_3)_2BiH$ (6) are prepared by the lithium aluminum hydride reduction of methyldichlorobismuthine, CH_3BiCl_2 [105309-90-8] and dimethylchlorobismuthine, respectively, in a nitrogen atmosphere at $-150°C$. On being warmed to $-45°C$, methylbismuthine disproportionates to trimethylbismuthine, $(CH_3)_3Bi$ [593-91-9], and bismuthine. Dimethylbismuthine undergoes a similar disproportionation at $-15°C$. Both methyl- and dimethylbismuthine are stable, colorless liquids at $-60°C$, but at room temperature they decompose to trimethylbismuthine, bismuth, and hydrogen. An attempt to prepare phenylbismuthine and diphenylbismuthine, $(C_6H_5)_2BiH$ [14381-43-2] by reduction of phenyldibromobismuthine, $(C_6H_5)_2$ $BiBr$ [39110-02-6] and diphenylbromobismuthine, $(C_6H_5)_2BiBr$, [39248-62-9], respectively, with $LiAlH_4$ or sodium borohydride at low temperatures yielded only black polymeric substances of empirical formula C_6H_5Bi (65). A stable organobismuth hydride, $(2,6-Mes_2C_6H_3)_2BiH$ (Mes = $2,4,6-(CH_3)_3C_6H_2$) (66) is obtained by reduction of the corresponding chloride with $LiAlH_4$. The signal for ν Bi$-$H appears at 1759 cm^{-1} in the ir spectrum. $(2,6-Mes_2C_6H_3)_2BiH$ decomposes at 135 °C with formation of the dibismuthene, $(2,6-Mes_2C_6H_3Bi)_2$ (66).

4.2. Tertiary Bismuthines. A large number of trialkyl- and triarylbismuthines are known (58). They are usually prepared by the interaction of a reactive organometallic compound and a bismuth trihalide, a halobismuthine, or a dihalobismuthine. The Grignard reagent (see GRIGNARD REAGENTS) is the type of organometallic compound most widely employed in these syntheses (67–69). Organolithium (70–72), and other organometallic compounds (73–79) have also been used. Triphenylbismuthine, $C_{18}H_{15}Bi$ [603-33-8], has been obtained in a 50% yield by the addition of phenyltrifluorosilane and ammonium fluoride to a solution of bismuth hydroxide in hydrofluoric acid (80). The interaction of organomercury compounds and metallic bismuth has also been employed for the preparation of tertiary bismuthines (81).

A number of unsymmetrical tertiary bismuthines have been prepared by the interaction of a sodium diaryl- or dialkylbismuthide and an alkyl or aryl halide (82). There have been several reports of the formation of tertiary bismuthines by the action of free radicals on metallic bismuth. One method of generating the radicals involves cleavage of ethane or hexafluoroethane in a radiofrequency glow discharge apparatus; the radicals thus formed are allowed to oxidize the metal at $-196°C$ (83). Trimethylbismuthine and tris(trifluoromethyl)bismuthine, $(CF_3)_3Bi$ [5863-80-9] have been obtained by this procedure.

Trialkylbismuthines are colorless or pale yellow liquids or solids, and most of them are spontaneously flammable in air. Except for trimethylbismuthine, these compounds cannot be distilled at ordinary pressures without decomposition (84). In general, trialkylbismuthines are not affected by water or aqueous bases but are readily hydrolyzed by many inorganic and organic acids (85,86). Trialkylbismuthines generally react with chlorine and bromine at low temperatures to form dialkylhalobismuthines (77,78). The reaction of tris(trifluoromethyl)bismuthine with chlorine, bromine, or iodine, however, has been found to yield the corresponding bismuth trihalide and trifluoromethyl halide (87). Triarylbismuthines are solids, which usually have sharp melting points. Most of these compounds are unaffected by oxygen or water at ordinary temperature and are, in general, much less reactive than their trialkyl counterparts. Triphenylbismuthine can be readily distilled, bp 242°C at 1.9 kPa without decomposition, and it has been obtained so pure that it has been used in measurements of the atomic weight of bismuth (88).

Most triarylbismuthines readily undergo oxidative addition with bromine or chlorine to yield the corresponding triarylbismuth dihalides:

$$Ar_3Bi + X_2 \longrightarrow Ar_3BiX_2 \tag{2}$$

Triphenylbismuth difluoride, $(C_6H_5)_3BiF_2$ [2023-48-5] has been obtained in a similar manner (89). All three carbon–bismuth bonds of tribenzylbismuthine, $C_6H_5CH_2)_3Bi$ [99715-52-3] (90) and triphenylbismuthine (91) can be cleaved by alkali metals. Under some conditions, however, tertiary bismuthines react with sodium or potassium to yield secondary bismuthides. Thus a number of sodium dialkylbismuthides have been obtained by the interaction of a trialkylbismuthine and sodium in liquid ammonia (92–95):

$$R_3Bi + 2\,Na \longrightarrow R_2BiNa + RNa \tag{3}$$

where $R = CH_3, C_2H_5, n\text{-}C_3H_7, iso\text{-}C_3H_7,$ or $n\text{-}C_4H_9$. Triphenylbismuthine reacts with potassium in tetrahydrofuran (thf) in a similar manner (96):

$$(C_6H_5)_3Bi + 2\,K \longrightarrow (C_6H_5)_2BiK + C_6H_5K \tag{4}$$

Treatment of 1-phenyl-2,5-dimethylbismole, $C_{12}H_{13}Bi$ [88635-81-8], with sodium in liquid ammonia results in cleavage of the phenyl–bismuth bond and formation of 1-sodio-2,5-dimethylbismole, C_6H_8BiNa [88644-52-4] (97):

$$\tag{5}$$

The secondary bismuthides discussed herein are useful for preparing several types of organobismuth compounds, eg, tertiary bismuthines and dibismuthines. Reaction of diphenylmethyl bismuthine with gaseous HCl or HBr results in the substitution of the phenyl groups by halogen atoms (98).

Although trialkyl- and triarylbismuthines are much weaker donors than the corresponding phosphorus, arsenic, and antimony compounds, they have nevertheless been employed to a considerable extent as ligands in transition metal complexes. The metals coordinated to the bismuth in these complexes include chromium (99–104), cobalt (105,106), iridium (107), iron (104,108,109), manganese (110,111), molybdenum (99,102–108,112–116), nickel (102,106,117, 118), niobium (119), rhodium (120,121), silver (122–124), tungsten (99,102–104,114,115), uranium (125), and vanadium (126). The coordination compounds formed from tertiary bismuthines are less stable than those formed from tertiary phosphines, arsines, or stibines. Several types of compounds containing an organometallic R_nBi ($n = 1$ or 2) moiety attached to a transition metal atom are also known (26). Examples are $Ph_2BiMn(CO)_5$, $MeBi(Fe(CO)_4)_2$ (127), or the cluster $[W_2(CO)_8(\mu\text{-}\eta^2\text{-}Bi_2)\text{-}\mu\text{-}MeBiW(CO)_5]\cdot C_6H_6$ (128).

Tertiary bismuthines appear to have a number of uses in synthetic organic chemistry (64), eg, they promote the formation of 1,1,2-trisubstituted cyclopropanes by the interaction of electron-deficient olefins and dialkyl dibromomalonates (129). They have also been employed for the preparation of thin films (qv) of superconducting bismuth strontium calcium copper oxide (130), as cocatalysts for the polymerization of alkynes (131), as inhibitors of the flammability of epoxy resins (132), and for a number of other industrial purposes.

4.3. Halobismuthines, Dihalobismuthines, and Related Compounds. Chloro-, dichloro-, bromo-, and dibromobismuthines are prepared by the reaction of a tertiary bismuthine and bismuth trichloride or tribromide (6,75,77,78,133–137):

$$R_3Bi + 2\,BiX_3 \longrightarrow 3\,RBiX_2 \tag{6}$$

$$2\,R_3Bi + BiX_3 \longrightarrow 3\,R_2BiX \tag{7}$$

Pure samples of CH_3BiCl_2 or CH_3BiBr_2 were obtained by reaction of CH_3BiPh_2 with HCl or HBr (98). Iodo- and diiodobismuthines are easily obtained by the reaction of the corresponding chloro-, dichloro-, bromo-, or dibromobismuthine with sodium or potassium iodide (137–140). Several halo- and dihalobismuthines have been prepared by the partial alkylation or arylation of bismuth trichloride or tribromide with an organocadmium (141), organolead (142,143), organotin (144,145), or organozinc reagent (77). The reaction of a Grignard reagent with a bismuth trihalide usually leads to the formation of a tertiary bismuthine. In a few cases, however, it has been possible to isolate a halo or dihalobismuthine from this type of reaction (146). The reaction of RLi [R = (Me_3Si)_2CH] with $BiCl_3$ in the 2:1 molar ratio gives R_2BiCl (147). $RBiCl_2$ [R = (Me_3Si)_2CH] is obtained from $RBiPh_2$ and HCl (147). Apparently, no fluoro- or difluorobismuthine has ever been isolated. The formation of bis(trifluoromethyl)fluorobismuthine [124252-79-5] by the interaction of tris(trifluoromethyl)bismuthine and iodine pentafluoride has, however, been suggested by ^{19}F nmr spectroscopy (148).

Halo- and dihalobismuthines are crystalline solids, many of which have polymeric structures (26) with melting points >100°C. They are, in general, decomposed by moisture, alcohols, and ammonia (149). Dialkylhalobismuthines are especially sensitive substances. The diaryl compounds are more stable, but they should also be handled with caution. Some of them are powerful sternutators (150).

Diphenylhalobismuthines and phenyldihalobismuthines react with tetraorgano ammonium, phosphonium, or arsonium salts to give compounds of the type $[R_4E][Ph_2BiX_4]$ or $[R_4E][PhBiX_3]$ where the bismuth moiety is anionic (151–153).

A number of compounds of the types $RBiY_2$ or R_2BiY, where Y is an anionic group other than halogen, have been prepared by the reaction of a dihalo- or halobismuthine with a lithium, sodium, potassium, ammonium, silver, or lead alkoxide (154,155), amide (156,157), azide (158,159), carboxylate (155,160), cyanide (159,161), dithiocarbamate (162,163), dithiophosphinate (133), mercaptide (164,165), nitrate (138), phenoxide (154), selenocyanate (159), silanolate (166), thiocyanate (159), or xanthate (167). Dialkyl- and diarylhalobismuthines can also be readily converted to secondary bismuthides by treatment with an alkali metal (82,135,168):

$$R_2BiX + 2\,M \longrightarrow R_2BiM + MX \tag{8}$$

4.4. Ionic and Neutral Organobismuth Complexes with π-Bonding or Hypervalency.

Cyclopentadienyl bismuth dihalides and related compounds (26,169,170,171) belong to the group of compounds with π-bonding between the metal and the organic ligands. With two substituted cyclopentadienyl ligands coordinated on bismuth the formation of a bismocenium salt, $Cp^x_2Bi^+AlCl_4^-$ ($Cp^x = 1,2,4\text{-}t\text{-}Bu_3C_5H_2$) is achieved (171). Other known cations derived from organobismuth halides are $R_2BiL_2^+$ and $RBiL_4^+$ [R = Ph, L = $(Me_2N)_3PO$] (172). A cationic hypervalent complex stabilized through intermolecular coordination is $[2\text{-}(Me_2NCH_2)C_6H_4]_2Bi^+$, where both nitrogen atoms are strongly coordinated to the bismuth center (173). Also, neutral hypervalent complexes with intramolecular coordination are known. Examples are 2-$(Me_2NCH_2)C_6H_4BiC_6H_4\{C(CF_3)_2O\}$-2 and related compounds (174).

4.5. Dibismuthines, Dibismuthenes, and Cyclo-Bismuthines

About a dozen dibismuthines are known (175). These compounds can be obtained by the reaction of a sodium dialkyl- or diarylbismuthide and a 1,2-dihaloethane (92–95):

$$2\,R_2BiNa + XCH_2CH_2X \longrightarrow R_2BiBiR_2 + H_2C{=}CH_2 + 2\,NaX \tag{9}$$

where X = Cl or Br. Several dibismuthines have also been obtained by the addition of the stoichiometric amount of sodium to a solution of a halobismuthine in liquid ammonia (135,176–179):

$$2\,R_2BiX + 2\,Na \longrightarrow R_2BiBiR_2 + 2\,NaX \tag{10}$$

where X = Cl or I. The best method for the synthesis of tetraphenyldibismuthine, $(C_6H_5)_4Bi_2$ [7065-21-6] involves the reduction of diphenyliodobismuthine,

$(C_6H_5)_2BiI$ [95825-92-6] using bis(cyclopentadienyl)cobalt(II), $(C_5H_5)_2Co$ [1277-43-6], (180):

$$2\,(C_6H_5)_2BiI + 2\,Co(C_5H_5)_2 \longrightarrow (C_6H_5)_2BiBi(C_6H_5)_2 + 2\,(C_5H_5)_2CoI \qquad (11)$$

Dibismuthines tend to be thermally unstable. Thus tetramethyldibismuthine, $C_4H_{12}Bi_2$ [82783-70-8] decomposes at 25°C to yield trimethylbismuthine and metallic bismuth (92):

$$3\,(CH_3)_2BiBi(CH_3)_2 \longrightarrow 4\,(CH_3)_3Bi + 2\,Bi \qquad (12)$$

Tetraethyldibismuthine, $C_8H_{20}Bi_2$, [81956-27-6], undergoes a similar reaction at 40°C (93). At 0°C this dibismuthine as well as tetra-4-tolyldibismuthine, $C_{28}H_{28}Bi_2$ [114245-28-2] (135) decompose to form dark polymeric solids:

$$R_2BiBiR_2 \longrightarrow R_3Bi + 1/x(RBi)_x \qquad (13)$$

where $R = C_2H_5$ or $4\text{-}CH_3C_6H_4$. Tetraphenyldibismuthine and 2,2′,5,5′-tetramethylbibismole, $C_{12}H_{16}Bi_2$ [88635-82-9] (97), however, are stable to 100°C.

Dibismuthines are very sensitive to oxidation. Thus tetramethyldibismuthine fumes in air, and tetraphenyldibismuthine in toluene solution rapidly absorbs oxygen. Under controlled conditions, dibismuthines react with chalcogens resulting in cleavage of the bismuth–bismuth bond and insertion of a chalcogen atom (135,178,180–182):

$$R_2BiBiR_2 + Y \longrightarrow R_2BiYBiR_2 \qquad (14)$$

where $Y = O$, S, Se, or Te. Dibismuthines undergo a variety of other interesting reactions and have attracted considerable attention because a number of these substances are thermochromic (175,183).

Only two cyclobismuthines, a trimer $(RBi)_3$ and a tetramer $(RBi)_4$ have been reported in the literature. In benzene solution, the two rings exist in an equilibrium. On crystallization, the equilibrium is shifted to the tetramer (184,185).

Dibismuthenes, RBi=BiR, involving double bonds between the bismuth centers, exist as stable compounds with bulky aryl substituents (R = 2,4,6-$[(Me_3Si)_2CH]_3C_6H_2$, (Tbt); 2,6-$[(Me_3Si)_2CH]_2$-4-$(Me_3Si)_3C\text{-}C_6H_2$, (Bbt); R = 2,6-$Mes_2C_6H_3$, 2,6-$Trip_2C_6H_3$ (Trip = 2,4,6-i-Pr_3-C_6H_2); Me = CH_3, i-Pr = $(CH_3)_2CH$. RBi=BiR (R = Tbt) was formed by deselenation of $(RBiSe)_3$ (R = Tbt) with $(Me_2N)_3P$ (186–188). Dehalogenation of $RBiBr_2$ with Mg in thf gave RBi=BiR (R = Bbt) (188). The terphenyl substituted dibismuthenes, RBi=BiR (R = 2,6-$Mes_2C_6H_3$, 2,6-$Trip_2C_6H_3$); Mes = 2,4,6-$(CH_3)_3C_6H_2$ are synthesized by reduction of the corresponding aryl bismuth dichlorides with potassium in toluene/hexane (189,190) or by decomposition of R_2BiH (R = 2,6-$Mes_2C_6H_3$) (66). The dibismuthenes are purple solids soluble without dissociation in organic solvents. Attempts to trap RBi momomers with dienes failed. The length of the Bi–Bi double bond in RBi=BiR [R = Tbt, 2.8206 (8) Å] is 6 % shorter than the Bi–Bi single bond length of $Ph_2Bi–BiPh_2$ [2.990(2)Å] (178). The condensation reaction of $RSbH_2$ with $RBiBr_2$ (R = Bbt) in the presence of 1,8-diazabicyclo[5.4.0]undec-7-ene

(DBU) in thf afforded a stibabismuthene RSb=BiR (R = Bbt) (191), which in benzene at 70°C reacts to RSb=SbR and RBi=BiR (R = Bbt). This reaction does not involve the formation of RBi or RSb units, instead the formation of a bismuth antimony ring compound as intermediate is assumed.

Thermolysis of bismuth trifluoracetate in presence of Me_6C_6 gives $[Bi_2(O_2CCF_3)_4] \cdot C_6Me_6$. Crystals of this compound consist of stacks o trifluoro-acetato bridged dibismuth units with Bi—Bi bond lengths of 2.947(1) Å and η^6-coordinated Me_6C_6 units in bridging positions (192).

4.6. Bismin and Its Derivatives.

Bismin (bismabenzene), C_5H_5Bi [289-52-1], the bismuth analogue of pyridine, has never been isolated, but it can be formed in solution by the dehydrohalogenation of 1-chloro-1,4-dihydrobismin, C_5H_6BiCl [39553-69-0], using 1,8-diazabicyclo[5.4.0]undec-7-ene (DBU) at low temperatures (144,193,194):

$$\text{(structure)} + \text{DBU} \longrightarrow \text{(structure)} + \text{DBU·HCl} \qquad (15)$$

4-Methylbismin, $CH_3C_5H_4Bi$ [82995-62-8] and 4-tert-butylbismin, $(CH_3)_3CC_5H_4Bi$ [82995-64-0], have been obtained in solution by similar reactions. The presence of bismin and its 4-alkyl derivatives in these solutions has been demonstrated both by spectroscopy and via chemical trapping with 1,1,1,4,4,4-hexafluoro-2-butyne to yield Diels–Alder adducts. The potential aromaticity of the bismin ring system has aroused considerable interest and has been investigated with a variety of spectral methods. There seems to be little doubt that bismin does possess aromatic character.

4.7. Diarylbismuthinic Acids and Their Esters.

A number of methyl diarylbismuthinates (diarylmethoxybismuth oxides) $Ar_2Bi(O)OCH_3$ (Ar = C_6H_5, 4-$CH_3C_6H_4$, 3,4-$(CH_3)_2C_6H_3$, 1-naphthyl, or 4-CH_3-1-naphthyl) were reported in 1988 (195). The methyl esters underwent ester exchange when recrystallized from ethyl or isopropyl alcohols. Methyl diphenylbismuthinate, $C_{13}H_{13}BiO_2$ [124066-62-2], was readily hydrolyzed in water to diphenylbismuthinic acid [124066-70-2]. Because organobismuth(V) compounds have found considerable use as oxidizing agents, the oxidizing ability of methyl di-1-naphthylbismuthinate, $C_{21}H_{17}BiO_2$ [124066-66-6] was investigated. Benzoin yields benzil, naphthalene, and metallic bismuth; hydrazobenzene yields azobenzene, and 1,1,2,2-tetraphenylethanediol yields benzophenone. 1,2-Diphenyl-1,2-ethane-dione dihydrazone gives diphenylacetylene in 50% yield. Cyclohexane-1,2-diol and 1-phenylethane-1,2-diol, however, were unaffected (62,195,196).

4.8. Trialkyl- and Triarylbismuth Dihalides and Related Compounds.

The triarylbismuth dihalides constitute an important class of organo-bismuth compounds. Only very few trialkylbismuth dihalides are kown. Trimethylbismuth dichloride, prepared by oxidative chlorination of Me_3Bi with SO_2Cl_2 at −78°C is unstable at room temperature. The geometry of the C_3BiCl_2 core in crystals of $Me_3BiCl_2 \cdot Me_2CO$ is distorted trigonal bipyramidal (197). The formation of the cis and trans-isomers of tripropenylbismuth dibromide, $C_9H_{15}BiBr_2$, [66173-00-0, 66212-22-4] by oxidative bromination of the corresponding bismuthines at $-55°C$ was reported in 1963 (198). With other trialkylbismuthines studied, cleavage of one carbon–bismuth bond occurs on halogenation:

$$R_3Bi + X_2 \longrightarrow R_2BiX + RX \qquad (16)$$

By contrast, triarylbismuthines are readily chlorinated or brominated to the corresponding dichlorides or dibromides using chlorine or bromine. Other chlorinating agents include sulfur dichloride, sulfur monochloride, thionyl chloride, and iodine trichloride. Triarylbismuth difluorides have been prepared from the dichlorides by metathesis with potassium fluoride or by direct fluorination of triarylbismuthines with fluorine diluted with argon. No triarylbismuth diiodides are known. However, the two compounds triphenylbismuth iodide azide, $(C_6H_5)_3Bi(I)N_3$, [106112-77-0] and triphenylbismuth iodide isocyanate, $(C_6H_5)_3Bi(I)NCO$ [106112-78-1], have been prepared from triphenylbismuthine and iodine azide or iodine isocyanate, respectively (199). The triarylbismuth dihalides are stable crystalline solids with high melting points. X-ray studies, conductivities, and other physical measurements suggest that the bismuth atom in these compounds has trigonal–bipyramidal geometry (26).

In addition to the halides Ar_3BiX_2, a large number of compounds of the type Ar_3BiY_2, where Y is NO_3, N_3, CN, OCN, CH_3CO_2, CF_3CO_2, ½ CO_3, ½ SO_4, O_3SR, etc, can be prepared by metathesis from the dihalides and a silver, sodium, or potassium salt of the desired anion. The disulfonates $(C_6H_5)_3Bi(O_3SR)_2$ have been prepared from the carbonate and the appropriate sulfonic acid (200). A number of mixed carbonates $Ar_2Ar'BiCO_3$, as well as mixed dichlorides $Ar_2Ar'BiCl_2$, have been prepared (201). Triphenylbismuth dicarboxylates have been obtained by several different methods including the following (202):

$$(C_6H_5)_3Bi + (CH_3)_3COOH + (RCO)_2O \longrightarrow$$
$$(C_6H_5)_3Bi(O_2CR)_2 + (CH_3)_3COH \qquad (17)$$

The majority of the compounds of the type Ar_3BiY_2 are stable crystalline solids, soluble in organic solvents. They give normal molecular weights in solvents such as benzene and have more or less distorted trigonal bipyramidal structures (26).

Triphenylbismuth oxide, $C_{18}H_{15}BiO$ [7173-99-1] has been prepared from triphenylbismuth dicyanide, $C_{20}H_{15}BiN_2$ [41083-16-3] and mercuric oxide (203), and from triphenylbismuth dichloride and moist silver oxide (204). The ir and Raman spectra of this compound suggest that it is polymeric and has Bi–O–Bi bonds (205). Triphenylbismuth dihydroxide and triarylbismuth hydroxide halides, eg, triphenylbismuth hydroxide chloride have been reported in the earlier chemical literature (206). There is, however, no modern research on these types of compounds, and they may or may not exist.

In recent years organobismuth(V) compounds have found increasing use as reagents in organic synthesis. Thus they have been used for the oxidation of primary and secondary alcohols to the corresponding aldehydes (qv) and ketones (qv), for the oxidative cleavage of vicinal glycols (qv), and for the O-, C-, and N-arylation of a wide variety of organic compounds. Because most of these reactions occur under relatively mild conditions, organobismuth(V) reagents have proved to be of particular value when the substrates are sensitive natural products. Several review articles on this subject have been published (64,207–209), and a patent on the oxidation of steroids, terpene, and sugar alcohols to the corresponding aldehydes (qv) and ketones (qv) has been issued (210) (see SUGAR ALCOHOLS). Although other types of organobismuth(V) compounds, Ar_4BiY and Ar_5Bi, have been used, the triaryl compounds Ar_3BiY_2 are the reagents of choice because of their ease of preparation and chemical stability.

In addition to use in organic synthesis, triarylbismuth dihalides and related compounds have found limited industrial use. A patent has been issued for the use of such compounds as antifungal agents on plastics or fibrous material (211). Compounds of the type $(C_6H_5)_3Bi(O_2CR)_2$, eg, triphenylbismuth dimethacrylate, $C_{26}H_{25}BiO_4$ [3371-98-0] and triphenylbismuth bis(4-vinylbenzoate), $C_{36}H_{29}BiO_4$ [2181-48-8] are claimed to be effective agents against *Staphylococcus aureus* infections (212). Triphenylbismuth dichloride, $(C_6H_5)_3BiCl_2$ [594-30-9] is active against bean rust (213). Triarylbismuth dihalides have been used as catalysts for the carbonation of epoxides to form cyclic carbonates (214).

4.9. Quaternary Bismuth Compounds. Although earlier attempts had been made to prepare quaternary bismuth compounds, it was not until 1952 that tetraphenylbismuth bromide, $(C_6H_5)_4BiBr$ [66173-02-2] was obtained from pentaphenylbismuth, $(C_6H_5)_5Bi$ [3049-07-8] and 1 M equivalent of bromine at $-70°C$ (215):

$$(C_6H_5)_5Bi + Br_2 \longrightarrow (C_6H_5)_4BiBr + C_6H_5Br \qquad (18)$$

In a similar manner, tetraphenylbismuth chloride, $(C_6H_5)_4BiCl$ [42967-53-3] (215) and tetraphenylbismuthonium tetrafluoroborate, $(C_6H_5)_4Bi[BF_4]$ [36682-02-7] (216) are obtained from pentaphenylbismuth and hydrogen chloride or hydrogen tetrafluoroborate, respectively. When triphenylboron is used, the tetraphenylborate is obtained (215):

$$(C_6H_5)_5Bi + (C_6H_5)_3B \longrightarrow [(C_6H_5)_4Bi]\,[(C_6H_5)_4B] \qquad (19)$$

A number of other tetraarylbismuth compounds Ar_4BiY, where Y is a group, such as NO_3^-, ClO_4^-, OCN^-, N_3^-, etc, have been prepared from the chloride by metathesis. Tetramethylbismuthonium trifluoromethylsulfonate, a crystalline compound stable at $150°C$ is formed from $(CH_3)_3Bi$ and $CH_3OSO_2CF_3$ (198). Also the formation of $[CH_3(C_2H_5)_3Bi][BF_4]$ (217) and $[(C_6H_5)_3BiCH_2COCH_3][ClO_4]$ (218) was reported. When triphenylbismuth dichloride, in acetone, is treated with silver perchlorate, in absolute ethanol, tetraphenylbismuthonium perchlorate, $(C_6H_5)_4BiClO_4$ [43047-28-5] is formed (219). Many quaternary bismuth compounds are unstable. When the anionic ligand is chloride or bromide, the compounds decompose spontaneously on standing; azides and selenocyanates decompose more rapidly. The perchlorates, tetrafluoroborates, and hexafluorophophates, however, are considerably more stable but eventually decompose. Quaternary bismuth compounds have not found extensive use in industry or in organic synthesis. In manifold studies of organobismuth(V) compounds as oxidizing and arylating agents, such quaternary bismuth compounds as $(C_6H_5)_4BiO_2$CCH_3$, $(C_6H_5)_4BiO_3SC_6H_4CH_3$-4, and $(C_6H_5)_4BiO_2CCF_3$ have been employed (64).

There seems to be no marked advantage of the quaternary compounds over the more stable and more easily prepared compounds of the type Ar_3BiY_2.

4.10. Bismuthonium Ylides. Prior to 1988 the only bismuthonium ylides known were (1) and (2).

Compound (1) is an unstable blue solid that cannot be obtained in a pure state (220); structure (2) [105071-90-7], however, is stable (221). Compound (2) was obtained from triphenylbismuth carbonate and dimedone. More recently a number of bismuthonium ylides, eg, (3) [119016-81-8] (222), have been prepared and their reactions studied.

Reactions of these compounds with phenylisothiocyanate and a number of aromatic and aliphatic aldehydes have been investigated (222,223). Only where R is $4\text{-}CH_3OC_6H_4$ or $C_6H_5CH=CH$ are the normal Wittig products obtained:

$$(20)$$

By using a number of other aldehydes, more complicated products result. Structure (2) was also found to react with alkynes in the presence of copper(I) chloride to give furans:

$$(21)$$

where R is C_6H_5, $4\text{-}CH_3C_6H_4$, $4\text{-}ClC_6H_5$, hexyl, and $C_6H_5COCH_2$. An excellent general method for the preparation of bismuthonium ylides from diazo compounds has been devised in which bis(hexafluoroacetylacetonato)copper(II) is employed as a catalyst (224). Two relatively stable ylides prepared by this procedure are (4) [117968-30-6] and (5) [117968-31-7]. The latter compound did not react with 2,4-dinitrobenzaldehyde.

$$(C_6H_5SO_2)_2C{=}Bi(C_6H_5)_3 \qquad\qquad (4)$$

$(C_6H_5SO_2)_2C{=}Bi(C_6H_5)_3$

(4)   **(5)**

4.11. Qinquenary Bismuth Compounds.

A number of pentaarylbismuth compounds and pentamethylbismuth are known. Pentaphenylbismuth, $C_{30}H_{25}Bi$ [3049-07-8] was first prepared by means of the following reaction (215):

$$(C_6H_5)_3BiCl_2 + 2\ C_6H_5Li \longrightarrow (C_6H_5)_5Bi + 2\ LiCl \qquad (22)$$

It can also be prepared by the reaction of phenyllithium with tetraphenylbismuth chloride or the *N*-triphenylbismuth derivative of 4-toluenesulfonamide (225):

$$4 - CH_3C_6H_4SO_2N{=}Bi(C_6H_5)_3 + 2C_6H_5Li \longrightarrow$$
$$(C_6H_5)_5Bi + 4 - CH_3C_6H_4SO_2NLi_2 \qquad (23)$$

Pentaphenylbismuth is a violet-colored, crystalline compound that decomposes spontaneously after standing for several days in a dry nitrogen atmosphere. With a variety of agents, eg, hydrohalic acids, halogens, and triphenylboron, one phenyl group is cleaved to form quaternary bismuth compounds (62). The deep violet color of pentaphenylbismuth and certain other pentaarylbismuth compounds has been the subject of considerable speculation. It has been shown by X-ray diffraction (226) that the bismuth atom in pentaphenylbismuth is square–pyramidal. Well-formed crystals are dichromic, appearing violet when viewed in one plane but colorless in another plane. The nature of the chromophore has been suggested to be a charge-transfer transition by excitation of the four long equatorial bonds:

$$(C_6H_5)_4Bi - C_6H_5 \longrightarrow (C_6H_5)_4Bi^-C_6H_5^+ \qquad (24)$$

In support of this suggestion, it has been shown that strong electron-withdrawing substituents on the aryl groups, which would make the charge-transfer transition more difficult, result in less highly colored compounds. Thus bis(pentafluorophenyl)triphenylbismuth, $C_{30}H_{15}BiF_{10}$ [111210-36-7] is an orange-colored solid. A number of pentaarylbismuth compounds that vary in color from violet, bis(4-fluorophenyl)tri-4-tolylbismuth, $C_{33}H_{29}BiF_2$ [124652-38-6] to yellow, bis(pentafluorophenyl)tri-4-tolylbismuth, $C_{33}H_{21}BiF_{10}$ [118798-77-9] have been prepared (227). The majority of structures that could be determined by x-ray diffraction exhibit square pyramidal geometry. Pentamethylbismuth, a violet crystalline compound, decomposing with explosion at room temperature is obtained by reaction of Me_3BiCl_2 with MeLi at $-95°C$ (198). The crystal structure of Me_5Bi is trigonal bipyramidal (198).

Pentaphenylbismuth has been studied as a reagent in organic synthesis where it can act either as an oxidizing or an arylating agent. Thus it can be used for the oxidation of primary or secondary alcohols to aldehydes or ketones, respectively. Unlike compounds of the type Ar_3BiY_2 or Ar_4BiY that require the presence of a strong base for the oxidation of alcohols, pentaphenylbismuth oxidizes alcohols under neutral conditions. Thus benzyl alcohol is oxidized to benzaldehyde in 45% yield by pentaphenylbismuth; 3β-cholestanol gives the corresponding ketone in 70% yield. Pentaphenylbismuth can also act as an arylating agent. 2-Naphthol reacts to give 1-phenyl-2-naphthol in 61% yield

(C-arylation), but phenol gives diphenyl ether in 42% yield (O-arylation). Often both oxidation and phenylation occur. Thus estradiol gives a mixture of 2,4-diphenylestrone, 4-phenylestrone, and 2,4-diphenylestradiol. When 2-phenyl-ethanol is treated with pentaphenylbismuth, triphenylacetaldehyde is produced in 69% yield (62,209).

5. Bismuth Compounds Used in Medicine

Therapeutic properties were first attributed to bismuth during the seventeenth century, and bismuth compounds were tried for the treatment of both syphilis and gonorrhea before the end of the eighteenth century (149). During the 1920s, it was shown that bismuth compounds were comparable in efficacy to the best antisyphilic drugs then available (228). During the next quarter of a century, bismuth compounds became widely used as adjuncts to the arsenical therapy of syphilis (229). However, antibiotics (qv), especially penicillin, have made both arsenic and bismuth compounds completely obsolete for the treatment of this disease (230,231). Bismuth compounds were employed for the treatment of amoebic dysentery, certain skin diseases, and several spirochetal diseases besides syphilis, but these substances are now seldom considered the drugs of choice. Various insoluble preparations of bismuth, especially the subcarbonate, subnitrate, subgallate, subcitrate, and subsalicylate, are employed for the treatment of gastrointestinal disorders. By the time *H. pyloris* was recognized as being central for the formation of ulcers in the stomach the use of bismuth preparations was appreciated. Eradication of *H. pyloris* is achieved by means of a combination therapy that uses bismuth compounds and other substances. Reviews of the biological activity of organobismuth compounds have been published (232–236). Bismuth subsalicylate, [14882-18-9], Pepto-Bismol, is a basic salt of varying composition, corresponding approximately to o-$HOC_6H_4CO_2(BiO)$. It does appear to be effective for the relief of mild diarrhea and for the prevention of travelers' diarrhea (233) in the symptomatic treatment of isosporiasis, a disease caused by the intracellular parasite *Isospora belli* (237). Bismuth subcarbonate (basic bismuth carbonate) [5892-10-4] is a white or pale yellow powder that is prepared by interaction of bismuth nitrate and a water-soluble carbonate. The exact composition of this drug depends on the conditions of precipitation; it corresponds approximately to the formula $(BiO)_2CO_3$. It has been widely used as an antacid (238). Tripotassium dicitratobismuthate, (bismuth subcitrate), [57644-54-9], De-Nol is a buffered aqueous suspension of a poorly defined, water-insoluble bismuth compound. It is said to be very effective for the treatment of gastric and duodenal ulcers (233,238). There have not yet been any reports of bismuth encephalopathy following the use of this drug. Bismuth subnitrate (basic bismuth nitrate), [1304-85-4] can be prepared by the partial hydrolysis of the normal nitrate with boiling water. It has been used as an antacid and in combination with iodoform as a wound dressing (238). Taken internally, the subnitrate may cause fatal nitrite poisoning because of the reduction of the nitrate ion by intestinal bacteria. Bismuth subgallate, (basic bismuth gallate) [12552-60-2], dermatol, is a bright yellow powder that can be prepared by the interaction of bismuth nitrate and gallic acid in an acetic acid medium. It has been employed as a dusting powder in some skin disorders and as an ingredient of suppositories for the treatment of hemorrhoids (237,238). It has been taken orally for many years by colostomy patients in order to control fecal odors, but the drug may cause serious neurological problems (239).

BIBLIOGRAPHY

"Bismuth Compounds" in *ECT* 1st ed., Vol. 2, pp. 533–538, by R. Pasternack, Charles Pfizer & Co., Inc.; "Bismuth Preparations" in *ECT* 1st ed., Vol. 2, pp. 538–540, by G. O. Doak, U.S. Public Health Service Laboratory of Experimental Therapeutics, Johns Hopkins School of Hygiene; "Bismuth Compounds" in *ECT* 2nd ed., Vol. 3, pp. 535–549, by G. O. Doak, L. D. Freedman, and G. G. Long, North Carolina State University; "Bismuth Compounds" in *ECT* 3rd ed., Vol. 3, pp. 921–937, by G. G. Long, L. D. Freedman, and G. O. Doak, North Carolina State University; "Bismuth Compounds" in *ECT* 4th ed., Vol. 4, pp. 246–270, by G. Gilbert, Leon D. Freedman, G. O. Doak, North Carolina State University; "Bismuth Compounds" in *ECT* (online), posting date: December 4, 2000, by G. Gilbert, Leon D. Freedman, G. O. Doak, North Carolina State University.

CITED PUBLICATIONS

1. H. Onishi, *Photometric determination of traces of Metals, Part IIA: Individual Metals, Aluminium to Lithium*, 4th ed. John Wiley, and Sons, Inc., New York.
2. W. F. Hillebrand, G. E. F. Lundell, H. A. Bright, and J. I. Hoffman, *Applied Inorganic Analysis*, 2nd ed. John Wiley, and Sons, Inc., New York, 1953.
3. G. H. Jeffery, J. Bassett, J. Mendham, and R. C. Denney, *Vogel's Textbook of Quantitative Chemical Analysis*, 5th ed., Longman Scientific and Technical, Harlow, Essex, 1989.
4. E. Upor, M. Mohai, and G. Y. Novák in G. Svehla, ed., *Wilson and Wilson's Comprehensive Analytical Chemistry*, Vol. XX, Elsevier, Amsterdam, The Netherlands, 1985, pp. 164–167.
5. H. Onishi in *Chemistry of Arsenic, Antimony and Bismuth*, N. C. Norman ed. Blackie Academic and Professional, Thomson Science, London 1998.
6. E. Amberger, *Chem. Ber.* **94**, 1447 (1961).
7. F. Paneth and co-workers, *Ber. Deut. Chem. Ges.* **51**, 1704, 1728 (1918); **55**, 769 (1922); **58**, 1138 (1925).
8. D. Cubicciotti, *J. Phys. Chem.* **64**, 791, 1506 (1966); **65**, 521 (1961).
9. J. D. Corbett and co-workers, *J. Phys. Chem.* **62**, 1149 (1958); *J. Am. Chem. Soc.* **80**, 4757 (1958); *J. Chem. Phys.* **36**, 551 (1962); *Inorg. Chem.* **2**, 979 (1963).
10. H. von Benda, A. Simon, and W. Bauhofer, *Z. Anorg. Allg. Chem.* **438**, 53 (1978).
11. H. G. von Schnering, H. von Benda, and C. Kalveram, *Z. Anorg. Allg. Chem.* **438**, 37 (1978).
12. D. Cubicciotti, *J. Electrochem. Soc.* **115**, 1138 (1968).
13. D. D. Wagman and co-workers, *Natl. Bur. Stand. Tech. Note 270-3*, U.S. Department of Commerce, Government Printing Office, Washington, D.C., 1968.
14. D. Cubicciotti, *J. Phys. Chem.* **70**, 2410 (1966).
15. D. Cubicciotti, *J. Phys. Chem.* **71**, 3066 (1967).
16. D. Cubicciotti, *Inorg. Chem.* **7**, 208 (1968).
17. D. Cubicciotti and H. Eding, *J. Phys. Chem.* **69**, 3621 (1965).
18. D. Cubicciotti, *Inorg. Chem.* **7**, 211 (1968).

19. K. Cheetham and N. Norman, *Acta Chem. Scand. Ser. A* **28A**, 55 (1974).
20. S. C. Nyburg, G. A. Ozin, and J. T. Szymanski, *Acta Crystallogr.* **B27**, 2298 (1971); **B28**, 2885 (1972).
21. P. G. Jones, D. Henschel, and A. Blaschette, *Z. Anorg. Allg. Chem.* **620**, 1037 (1994).
22. J. R. Eveland and K. H. Whitmire, *Inorg. Chim. Acta*, **249**, 41 (1996).
23. G. A. Bowmaker, F. M. M. Hannaway, P. C. Junk, A. M. Lee, B. W. Skelton, and A. H. White, *Aust. J. Chem.* **51**, 325 (1998); *Aust. J. Chem.* **51**, 331 (1998).
24. G. R. Wiley, L. T. Daly, and M. G. B. Drew. *J. Chem. Soc. Dalton Trans.* 1063 (1996).
25. R. D. Rogers, A. H. Bond, S. Aguinaga, and A. Reyes, *J. Am. Chem. Soc.* **114**, 2967 (1992).
26. C. Silvestru, H. J. Breunig, and H. Althaus, *Chem. Rev.* **99**, 3277 (1999).
27. G. A. Fisher and N. C. Norman, *Adv. Inorg. Chem.* **41**, 233 (1994).
28. H. von Benda, *Z. Kristallogr.* **151**, 271 (1980).
29. W. Clegg, M. R. J. Elsegood, V. Graham, N. C. Norman, N. L. Pickett, and K. Tavakkoli, *J. Chem. Soc. Dalton Trans.* 1743 (1994).
30. C. J. Carmalt, W. Clegg, M. R. J. Elsegood, R. J. Errington, J. Avelock, P. Lightfoot, N. C. Norman, and A. J. Scott, *Inorg. Chem.* **35**, 3709 (1996).
31. J. Beech, P. J. Cragg, and M. G. B. Drew, *J. Chem. Soc. Dalton Trans.* 719 (1994).
32. M. Ruck, *Z. Kristallogr.* **210**, 650 (1995).
33. W. Clegg, L. J. Farrugia, A. Mc Camley, N. C. Norman, A. G. Orpen, N. L. Pickett, and S. E. Stratford, *J. Chem. Soc. Dalton Trans.* 2579 (1993).
34. M. J. Vasile, G. R. Jones, and W. E. Falconer, *Adv. Mass Spectrom.* **6**, 557 (1974).
35. A. Preussman and S. Ivankovic, *Food Cosmet. Toxicol.* **13**, 543 (1975).
36. G. Malmros, *Acta Chem. Scand.* **24**, 384 (1970).
37. N. Ichinose and H. Maiwa, *Bull. Bismuth Inst.* **58**, 1 (1989).
38. I. J. Polmear, in *Chemistry of Arsenic, Antimony, and Bismuth*, N. C. Norman, ed. Blackie Academic and Professional, London 1998, p. 60.
39. H. Maeda, Y. Tanaka, M. Fukutorni, and T. Asano, *Jpn. J. Appl. Phys.* **27**, L 209 (1988).
40. K. Mariolacos, *Acta Crystallogr.* **B32**, 1947 (1976).
41. G. Gattow and G. Kiel, *Naturwissenschaften* **54**, 18 (1967); **55**, 389 (1968).
42. W. J. Ewans, J. H. Hain, Jr., and J. W. Ziller, *J. Chem. Soc. Chem. Commun.* 1628 (1989).
43. M. A. Matchett, M. Y. Chiang, and W. E. Buhro, *Inorg. Chem.* **29**, 358 (1990).
44. C. M. Jones, M. D. Burkart, and K. H. Whitmire, *Angew. Chem. Int. Ed. Engl.* **31**, 451 (1992).
45. D. A. Atwood, A. H. Cowley, R. D. Hernandez, R. A. Jones, L. L. Rand, S. G. Bott, and J. L. Atwood, *Inorg. Chem.* **32**, 2972 (1993).
46. Y. Akamine, T. Fukami, R. Nukada, M. Mikuriya, S. Deguchi, and Y. Yokota, *Bull. Chem. Soc. Jpn.* **70**, 639 (1997).
47. L. Agocs, G. G. Briand, N. Burford, T. S. Cameron, W. Kwiatkowski, and K. N. Robertson, *Inorg. Chem.* **36**, 2855 (1997).
48. W. Bensch, E. Blazso, and H. R. Oswald, *Acta Crystallogr* **C43**, 1699 (1987).
49. P. Kiprof, W. Scherer, L. Pajdla, E. Herdtweck, and W. A. Herrmann, *Chem. Ber.* **125**, 43 (1992).
50. S. I. Troyanov and A. P. Pisarevsky, *J. Chem. Soc. Chem. Commun.* 335 (1993).
51. L. K. Thompson, *Inorg. Chim. Acta*, **250**, 163 (1996).
52. M. J. Begley, D. B. Sowerby, and I. Haiduc, *J. Chem. Soc. Dalton Trans.* 145 (1987).
53. A. M. Hounslow, S. F. Lincoln, and E. R. T. Tiekink, *J. Chem. Soc. Dalton Trans.* 233, (1989).
54. K. H. Whitmire, *Adv. Organometal. Chem.* **42**, 1 (1998).

55. M. N. Bochkarev, G. A. Razuvaev, L. N. Zhakarov, and Y. T. Struchkov, *J. Organometal. Chem.* **199**, 205 (1980).
56. G. M. Kolleger, H. Siegl, K. Hassler, and K. Gruber, *Organometallics*, **15**, 4337 (1996).
57. S. Samaan, *Methoden der Organischen Chemie. Metallorganische Verbindungen As, Sb, Bi, Band XIII, Teil 8*, Georg Thieme Verlag, Stuttgart, Germany, 1978.
58. M. Wieber, *Gmelin Handbuch der Anorganischen Chemie, Band 47, Bismut-Organische Verbindungen*, Springer-Verlag, Berlin, 1977.
59. L. D. Freedman and G. O. Doak, *Chem. Rev.* **82**, 15 (1982).
60. G. O. Doak and L. D. Freedman, *Organometallic Compounds of Arsenic, Antimony, and Bismuth*, Wiley-Interscience, New York, 1970.
61. H. Suzuki and Y. Matano eds., *Organobismuth Chemistry*, Elsevier, Amsterdam, 2001.
62. H. Suzuki and Y. Matano in *Chemistry of Arsenic, Antimony, and Bismuth*, N. C. Norman, ed. Blackie Academic and Professional, London, 1998, p. 283.
63. S. Patai, ed., *The Chemistry of organic arsenic, antimony and bismuth compounds*, John Wiley & Sons, Inc., Chichester, UK, 1994.
64. L. D. Freedman and G. O. Doak, in F. R. Hartley, ed., *The Chemistry of the Metal–Carbon Bond*, Vol. 5, John Wiley & Sons, Inc., Chichester, UK, 1989, Chapt. 9.
65. E. Wiberg and K. Mödritzer, *Z. Naturforsch., Teil B* **12B**, 132 (1957).
66. N. J. Hardman, B. Twamley, and P. P. Power, *Angew. Chem. Int. Ed. Engl.* **39**, 2771 (2000).
67. A. Banfi, M. Bartoletti, E. Bellora, and co-workers, *Synthesis* 775, (1994).
68. W. Levason, B. Sheikh, and F. P. McCullough, *J. Coord. Chem.* **12**, 53 (1982).
69. P. Bras, A. Van der Gen, and J. Wolters, *J. Organomet. Chem.* **256**, C1 (1983).
70. W. Levason, C. A. McAuliffe, and R. D. Sedgwick, *J. Organomet. Chem.* **122**, 351 (1976).
71. V. V. Sharutin and A. E. Ermoshkin, *Izv. Akad. Nauk SSSR, Ser. Khim.*, 187 (1987).
72. P. J. Fagan and W. A. Nugent, *J. Am. Chem. Soc.* **110**, 2310 (1988).
73. R. S. Dickson and B. O. West, *Aust. J. Chem.* **15**, 710 (1962).
74. D. Naumann and W. Tyrra, *J. Organomet. Chem.* **334**, 323 (1987).
75. F. Challenger and C. F. Allpress, *J. Chem. Soc.* **119**, 913 (1921).
76. E. O. Fischer and S. Schreiner, *Chem. Ber.* **93**, 1417 (1960).
77. A. Marquardt, *Ber. Deut. Chem. Ges.* **20**, 1516 (1887).
78. A. Marquardt, *Ber. Deut. Chem. Ges.* **21**, 2035 (1888).
79. A. N. Nesmeyanov, V. A. Sazonova, and N. N. Sedova, *Dokl. Akad. Nauk SSSR* **198**, 590 (1971).
80. R. Müller and C. Datke, *Chem. Ber.* **99**, 1609 (1966).
81. R. E. Humphries, N. A. A. Al-Jabar, D. Bowen, A. G. Massey, and G. B. Deacon, *J. Organometal. Chem.* **319**, 59 (1987).
82. M. Wieber and K. Rudolph, *Z. Naturforsch., Teil B* **43B**, 739 (1988).
83. T. J. Juhlke, R. W. Braun, T. R. Bierschenk, and R. J. Lagow, *J. Am. Chem. Soc.* **101**, 3229 (1979).
84. T. N. Bell, B. J. Pullman, and B. O. West, *Aust. J. Chem.* **16**, 636 (1963).
85. H. Gilman and J. F. Nelson, *J. Am. Chem. Soc.* **59**, 935 (1937).
86. J. F. Nelson, *Iowa State Coll. J. Sci.* **12**, 145 (1937).
87. W. Tyrra and D. Naumann, *Can. J. Chem.* **67**, 1949 (1989).
88. A. Classen and G. Strauch, *Z. Anorg. Allg. Chem.* **141**, 82 (1924).
89. I. Ruppert and V. Bastian, *Angew. Chem. Int. Ed. Engl.* **17**, 214 (1978).
90. G. Bähr and G. Zoche, *Chem. Ber.* **90**, 1176 (1957).
91. R. A. Rossi and J. F. Bunnett, *J. Am. Chem. Soc.* **96**, 112 (1974).
92. A. J. Ashe, III, and E. G. Ludwig, Jr., *Organometallics* **1**, 1408 (1982).

93. H. J. Breunig and D. Müller, *Angew. Chem. Int. Ed. Engl.* **21**, 439 (1982).

94. A. J. Ashe, III, E. G. Ludwig, Jr., and J. Oleksyszyn, *Organometallics* **2**, 1859 (1983).

95. H. J. Breunig and D. Müller, *Z. Naturforsch. Teil B* **38B**, 125 (1983).

96. H. J. Breunig and D. Müller, *J. Organomet. Chem.* **253**, C21 (1983).

97. A. J. Ashe, III and F. J. Drone, *Organometallics* **3**, 495 (1984).

98. H. Althaus, H. J. Breunig, and E. Lork, *Organometallics* **20**, 586 (2001).

99. R. A. Brown and G. R. Dobson, *Inorg. Chim. Acta* **6**, 65 (1972).

100. A. J. Carty, N. J. Taylor, A. W. Coleman, and M. F. Lappert, *J. Chem. Soc. Chem. Commun.*, 639 (1979).

101. E. O. Fischer and K. Richter, *Chem. Ber.* **109**, 1140 (1976).

102. H. Schumann and H. J. Breunig, *J. Organomet. Chem.* **87**, 83 (1975).

103. R. A. Brown and G. R. Dobson, *J. Inorg. Nucl. Chem.* **33**, 892 (1971).

104. H. J. Breunig and U. Gräfe, *Z. Anorg. Allg. Chem.* **510**, 104 (1984).

105. J. F. White and M. F. Farona, *Inorg. Chem.* **10**, 1080 (1971).

106. W. Levason, C. A. McAuliffe, and S. G. Murray, *J. Chem. Soc. Dalton Trans.*, 711, (1977).

107. L. Vaska and J. Peone, Jr., *Suom. Kemistil., B* **44B**, 317 (1971).

108. F. Hein and H. Pobloth, *Z. Anorg. Allg. Chem.* **248**, 84 (1941).

109. H. Schumann, *Chem.-Ztg.* **110**, 121 (1986).

110. A. E. Ermoshkin, N. P. Makarenko, and K. I. Sakodynskii, *J. Chromatogr.* **290**, 377 (1984).

111. C. Barbeau, *Can. J. Chem.* **45**, 161 (1967).

112. L. W. Houk and G. R. Dobson, *Inorg. Chem.* **5**, 2119 (1966).

113. D. Benlian and M. Bigorgne, *Bull. Soc. Chim. Fr.*, 1583 (1963).

114. P. K. Baker and S. G. Fraser, *J. Coord. Chem.* **16**, 97 (1987).

115. H. Schumann, *J. Organomet. Chem.* **323**, 193 (1987).

116. P. K. Baker, S. G. Fraser, and T. M. Matthews, *Inorg. Chim. Acta* **150**, 217 (1988).

117. W. Levason, C. A. McAuliffe, and S. G. Murray, *J. Chem. Soc. Chem. Commun.*, 164 (1975).

118. G. M. Bodner, C. Gagnon, and D. N. Whittern, *J. Organomet. Chem.* **243**, 305 (1983).

119. J. Desnoyers and R. Rivest, *Can. J. Chem.* **43**, 1879 (1965).

120. E. B. Boyar and S. D. Robinson, *Inorg. Chim. Acta* **76**, L137 (1983).

121. E. B. Boyar and S. D. Robinson, *J. Chem. Soc. Dalton Trans.* 629 (1985).

122. R. H. Nutall, E. R. Roberts, and D. W. A. Sharp, *J. Chem. Soc.* 2854 (1962).

123. S. Ahrland, T. Berg, and P. Trinderup, *Acta Chem. Scand. Ser. A* **A31**, 775 (1977).

124. F. Hultén and I. Persson, *Inorg. Chim. Acta* **128**, 43 (1987).

125. J. Selbin, N. Ahmad, and M. J. Pribble, *J. Inorg. Nucl. Chem.* **32**, 3249 (1970).

126. R. Talay and D. Rehder, *Chem. Ber.* **111**, 1978 (1978).

127. J. Cassidy and K. H. Whitmire, *Inorg. Chem.* **28**, 3164 (1989); **30**, 2788 (1991).

128. A. M. Arif, A. H. Cowley, N. C. Norman, and M. Pakulski, *Inorg. Chem.* **25**, 4836 (1986).

129. C. Chen, Y. Liao, and Y.-Z. Huang, *Tetrahedron* **45**, 3011 (1989).

130. J. Zhang and co-workers, *Appl. Phys. Lett.* **54**, 1166 (1989).

131. T. Masuda and co-workers, *Macromolecules* **21**, 281 (1988).

132. F. J. Martin and K. R. Price, *J. Appl. Polym. Sci.* **12**, 143 (1968).

133. K. H. Ebert, R. E. Schulz, H. J. Breunig, C. Silvestru, and I. Haiduc, *J. Organometal. Chem.* **470**, 93 (1994).

134. R. Okawara, K. Yasuda, and M. Inoue, *Bull. Chem. Soc. Jpn.* **39**, 1823 (1966).

135. M. Wieber and I. Sauer, *Z. Naturforsch. Teil B* **42B**, 695 (1987).

136. M. Ali, S. P. Bond, S. A. Mbogo, W. R. McWhinnie, and P. M. Watts, *J. Organomet. Chem.* **371**, 11 (1989).

137. H. Gilman and H. L. Yablunky, *J. Am. Chem. Soc.* **63**, 207 (1941).

138. F. Dünhaupt, *Liebigs Ann. Chem.* **92**, 371 (1854).
139. F. F. Blicke, U. O. Oakdale, and F. D. Smith, *J. Am. Chem. Soc.* **53**, 1025 (1931).
140. F. Dünhaupt, *J. Prakt. Chem.* **61**, 399 (1854).
141. D. Hellwinkel and M. Bach, *J. Organometal. Chem.* **17**, 389 (1969).
142. A. E. Goddard, J. N. Ashley, and R. B. Evans, *J. Chem. Soc.* **121**, 978 (1922).
143. L. Maier, *Tetrahedron Lett.* **6**, 1 (1959).
144. A. J. Ashe, III, *Acc. Chem. Res.* **11**, 153 (1978).
145. N. I. Anishchenko, E. M. Panov, O. P. Syutkina, and K. A. Kocheshkov, *Zh. Obshch. Khim.* **49**, 1185 (1979).
146. G. Grüttner and M. Wiernik, *Ber. Deut. Chem. Ges.* **48**, 1473 (1915).
147. H. Althaus, H. J. Breunig, R. Rösler, and E. Lork, *Organometallics*, **18**, 328 (1999).
148. W. Tyrra and D. Naumann, *Can. J. Chem.* **67**, 1949 (1989).
149. H. Gilman and H. L. Yale, *Chem. Rev.* **30**, 281 (1942).
150. H. McCombie and B. C. Saunders, *Nature (London)* **159**, 491 (1947).
151. W. Clegg, R. J. Errington, G. A. Fisher, D. C. R. Hockless, N. C. Norman, A. G. Orpen, and S. E. Stratfort, *J. Chem. Soc. Dalton Trans.*, 1967 (1992).
152. W. Clegg, R. J. Errington, G. A. Fisher, G. A. Flynn, and N. C. Norman, *J. Chem. Soc. Dalton Trans.*, 637 (1993).
153. G. Faraglia, *J. Organomet. Chem.* **20**, 99 (1969).
154. M. Wieber and U. Baudis, *Z. Anorg. Allg. Chem.* **439**, 134 (1978).
155. F. Huber and S. Bock, *Z. Naturforsch., Teil B* **37B**, 815 (1982).
156. O. J. Scherer, P. Hornig, and M. Schmidt, *J. Organomet. Chem.* **6**, 259 (1966).
157. P. Krommes and J. Lorberth, *J. Organomet. Chem.* **93**, 339 (1975).
158. W. T. Reichle, *J. Organomet. Chem.* **13**, 529 (1968).
159. R. G. Goel and H. S. Prasad, *Spectrochim. Acta, Part A* **35A**, 339 (1979).
160. M. Wieber, D. Wirth, and K. Hess, *Z. Anorg. Allg. Chem.* **505**, 138 (1983).
161. F. Challenger and J. F. Wilkinson, *J. Chem. Soc.* **121**, 91 (1922).
162. E. J. Kupchik and C. T. Theisen, *J. Organomet. Chem.* **11**, 627 (1968).
163. M. Wieber and A. Basel, *Z. Anorg. Allg. Chem.* **448**, 89 (1979).
164. M. Wieber, I. Fetzer-Kremling, D. Wirth, and H. G. Rüdling, *Z. Anorg. Allg. Chem.* **520**, 59 (1985).
165. T. Klapötke, *J. Organomet. Chem.* **331**, 299 (1987).
166. H. Schmidbaur and M. Bergfeld, *Z. Anorg. Allg. Chem.* **363**, 84 (1968).
167. M. Wieber, H. G. Rüdling, and Ch. Burschka, *Z. Anorg. Allg. Chem.* **470**, 171 (1980).
168. D. D. Davis and C. E. Gray, *Organomet. Chem. Rev. Sect. A* **6**, 283 (1970).
169. W. Frank, *J. Organomet. Chem.* **177**, 386 (1990).
170. H. Sitzmann and G. Wolmershäuser, *Chem. Ber.* **127**, 1335 (1994).
171. H. Sitzmann and G. Wolmershäuser, *Z. Naturforsch.* **52b**, 398 (1997).
172. C. J. Carmalt, L. I. Farrugia, and N. C. Norman, *J. Chem. Soc. Dalton Trans.* 443 (1996).
173. C. J. Carmalt, D. Walsh, A. H. Cowley, and N. C. Norman, *Organometallics* **16**, 3597 (1997).
174. K. J. Akiba and co-workers, *J. Am. Chem. Soc.*, **117**, 3922 (1995); *Heteroatom. Chem.* **4**, 293 (1995).
175. A. J. Ashe, III, *Adv. Organomet. Chem.* **30**, 77 (1990).
176. F. Calderazzo, A. Morvillo, G. Pelizzi, and R. Poli, *J. Chem. Soc. Chem. Commun.*, 507 (1983).
177. A. J. Ashe, III, C. M. Kausch, and O. Eisenstein, *Organometallics* **6**, 1185 (1987).
178. F. Calderazzo, R. Poli, and G. Pelizzi, *J. Chem. Soc. Dalton Trans.*, 2365 (1984).
179. M. Wieber and I. Sauer, *Z. Naturforsch., Teil B* **39B**, 887 (1984).
180. F. Calderazzo, A. Morvill, G. Pelizzi, R. Poli, and F. Ungari, *Inorg. Chem.* **27**, 3730 (1988).

181. H. J. Breunig and D. Müller, *Z. Naturforsch., Teil B* **41B**, 1129 (1986).
182. M. Wieber and I. Sauer, *Z. Naturforsch., Teil B* **39B**, 1668 (1984).
183. O. Mundt, G. Becker, M. Rössler, and C. Witthauer, *Z. Anorg. Allg. Chem.* **506**, 42 (1983).
184. H. J. Breunig, R. Rösler, and E. Lork, *Angew. Chem. Int. Ed. Engl.* **37**, 3175 (1998).
185. H. J. Breunig and R. Rösler, *Chem. Soc. Rev.* **29**, 403 (2000).
186. N. Tokitoh, Y. Arai, S. Okazaki, and S. Nagase, *Science* **277**, 78 (1997).
187. N. Tokitoh, Y. Arai, and S. Okazaki, *Phosphorus Sulfur Silicon* **124**, 371 (1997).
188. N. Tokitoh, *J. Organomet. Chem.* **611**, 217 (2000).
189. B. Twamley, C. D. Sofield, M. M. Olmstead, and P. P. Power, *J. Am. Chem. Soc.* **121**, 3357 (1999).
190. P. P. Power, *Chem. Rev.* **99**, 3463 (1999).
191. T. Sasamori, N. Takeda, and N. Tokitoh, *Chem. Commun.* 1353 (2000).
192. W. Frank, V. Reiland, and G. J. Reiß, *Angew. Chem. Int. Ed. Engl.* **37**, 2984 (1998).
193. A. J. Ashe, III, *Top. Curr. Chem.* **105**, 125 (1982).
194. A. J. Ashe, III, T. R. Diephouse, and M. Y. El-Sheikh, *J. Am. Chem. Soc.* **104**, 5693 (1982).
195. T. Ogawa, T. Murafuji, K. Iwata, and H. Suzuki, *Chem. Lett.*, 2021 (1988).
196. T. Ogawa, T. Miyazaki, and H. Suzuki, *Chem. Lett.*, 1651 (1990).
197. S. Wallenhauer and K. Seppelt, *Angew. Chem. Int. Ed. Engl.* **33**, 976 (1994).
198. A. E. Borisov, M. A. Osinova, and A. N. Nesmeyanov, *Izv. Akad. Nauk SSSR Ser. Khim.*, 1507 (1963).
199. P. Raj, K. Singhal, and R. Rastogi, *Polyhedron* **5**, 677 (1986).
200. R. Rüther, F. Huber, and H. Preut, *Z. Anorg. Allg. Chem.* **539**, 110 (1986).
201. D. H. R. Barton, N. Y. Bhatnagar, J.-P. Finet, and W. B. Motherwell, *Tetrahedron* **42**, 311 (1986).
202. V. A. Dodonov, A. V. Gushchin, and M. B. Ezhova, *Zh. Obshch. Khim.* **58**, 2170 (1988).
203. R. G. Goel and H. S. Prasad, *J. Organomet. Chem.* **50**, 129 (1973).
204. R. G. Goel and H. S. Prasad, *J. Organomet. Chem.* **36**, 323 (1972).
205. R. G. Goel and H. S. Prasad, *Spectrochim. Acta, Part A* **32A**, 569 (1976).
206. F. Challenger and A. E. Goddard, *J. Chem. Soc.* **117**, 762 (1930).
207. D. H. R. Barton and J.-P. Finet, *Pure Appl. Chem.* **59**, 937 (1987).
208. D. H. R. Barton, W. B. Motherwell, B. da Silva, and M. Teresa, *Rev. Port. Quim.* **26**, 177 (1984).
209. J.-P. Finet, *Chem. Rev.* **89**, 1487 (1989).
210. Fr. Demande 2,441,602 (June 13, 1980), D. Barton and W. B. Motherwell (to Agence Nationale de Valorisation de la Recherche).
211. U.S. Pat. 3,239,411 (Mar. 8, 1966), J. R. Leebrick (to M&T Chemicals, Inc.).
212. Neth. Appl. 6,405,309 (Nov. 16, 1964), (to M&T Chemicals, Inc.).
213. P. Evrard and P. Lepoivre, *Parasitica* **39**, 117 (1983).
214. Jpn. Kokai Tokkyo Koho 82 49614 (Mar. 23, 1982), (to Mitsubishi Petrochemical Co., Ltd.).
215. G. Wittig and K. Clauss, *Liebigs Ann. Chem.* **578**, 136 (1952).
216. O. A. Ptitsyna, M. E. Gurskii, T. D. Maiorova, and O. Reutov, *Izv. Akad. Nauk SSSR Ser. Khim.*, 2618 (1971).
217. R. Dötzer, *Abstracts of Papers of the 3rd International Symposium on Organometallic Chemistry*, Munich, Germany, Aug. 28–Sept. 1, 1967, p. 196.
218. R. G. Goel and H. S. Prasad, *J. Chem. Soc. A*, 562 (1971).
219. G. O. Doak, G. G. Long, S. K. Kakar, and L. D. Freedman, *J. Am. Chem. Soc.* **88**, 2342 (1966).
220. D. Lloyd and M. I. C. Singer, *J. Chem. Soc. Chem. Commun.*, 1042 (1967).

221. D. H. R. Barton and co-workers, *J. Chem. Soc. Perkin Trans.* **1**, 2667 (1985).
222. H. Suzuki, T. Murafuji, and T. Ogawa, *Chem. Lett.*, 847 (1988).
223. T. Ogawa, T. Murafuji, and H. Suzuki, *Chem. Lett.*, 849 (1988).
224. C. Glidewell, D. Lloyd, and S. Metcalfe, *Synthesis*, 319 (1988).
225. G. Wittig and D. Hellwinkel, *Chem. Ber.* **97**, 789 (1964).
226. A. Schmuck, J. Buschmann, J. Fuchs, and K. Seppelt, *Angew. Chem. Int. Ed. Engl.* **26**, 1180 (1987).
227. A. Schmuck and K. Seppelt, *Chem. Ber.* **122**, 803 (1989).
228. C. Levaditi, *Lancet* **204**, 639 (1923).
229. J. E. Moore, *The Modern Treatment of Syphilis*, 2nd ed., Charles C. Thomas, Springfield, Ill., 1941, pp. 134–153.
230. F. N. Judson, in Y. M. Felman, ed., *Sexually Transmitted Diseases*, Churchill Livingstone, New York, 1986, pp. 23–37.
231. N. J. Fiumara, in Ref. 230, pp. 39–50.
232. Th. Klapötke, *Biol. Met.* **1**, 69 (1988).
233. L. L. Brunton, in A. G. Gilman, T. W. Rall, A. S. Niews, and P. Taylor, eds., *Goodman and Gilman's The Pharmacological Basis of Therapeutics*, 8th ed., Pergamon Press, New York, 1990, pp. 910, 911.
234. J. Reglinsky in *Chemistry of Arsenic, Antimony and Bismuth*, N. C. Norman, ed. Blackie Academic and Professional, Thomson Science; London, 1998.
235. U. Wormser and I. Nir in *The Chemistry of Organic Arsenic, Antimony, and Bismuth Compounds*, S. Patai ed., J. Wiley & Sons, Inc., Chichester, 1994.
236. G. G. Briand and N. Burford, *Chem. Rev.* **99**, 2601 (1999).
237. *Drug Evaluations*, 6th ed., American Medical Association Department of Drugs, W. B. Saunders Company, Philadelphia, Pa., 1986, p. 1568.
238. J. E. F. Reynolds, ed., Martindale. *The Extra Pharmacopoeia*, 29th ed., The Pharmaceutical Press, London, 1989, p. 1548.
239. F. P. Morgan and J. J. Billings, *Med. J. Aust.* (2), 662 (1974).

HANS JOACHIM BREUNIG
University of Bremen

BLEACHING AGENTS

1. Introduction

A bleaching agent is a material that lightens or whitens a substrate through chemical reaction. The bleaching reactions usually involve oxidative or reductive processes that degrade color systems. These processes may involve the destruction or modification of chromophoric groups in the substrate as well as the degradation of color bodies into smaller, more soluble units that are more easily removed in the bleaching process. The most common bleaching agents generally fall into two categories: chlorine and its related compounds (such as sodium hypochlorite) and the peroxygen bleaching agents such as hydrogen peroxide and sodium perborate. Reducing bleaches represent another category. Bleaching agents are used for textile, paper, and pulp bleaching as well as for home laundering. The textile industry is continuously under pressure to improve the environmental acceptability of its processes, and the related discharged effluents. The introduction of new enzyme-based technology to replace existing processes is thus particularly attractive. In processing natural fibers for textiles, the bleaching process is not only aimed at brightening the fibers, removing noncellulosic natural matter, eg, fats, waxes, pectines, proteins, and pigments (1,2), but is also directly related to subsequent wet processing operations such as dyeing, printing, and finishing (3). Whitening textiles is achieved traditionally by using various oxidizing or reducing agents at acidic or alkaline conditions, and at a wide range of temperatures. The desired whiteness level depends on the ultimate use of the fabrics. When higher level of whiteness is required it is necessary to perform a repeated oxidizing treatment.

Currently, hydrogen peroxide replaces almost entirely the traditional chlorine oxidizing agents (4,5). Accordingly, hydrogen peroxide precursors, such as perborates and percarbonates, are major constituents of commercial detergents. Normally the bleaching agents are dosed in excess to the fibers, resulting in repeated washing to remove the residual oxidants, which are destructive to the proceeding processes. This method renders the bleaching process high in chemicals, water, energy, and time, as well as posing environmental hazard because of the discharged waste liquor. The introduction of a new, more environmentally friendly alternative enzyme-based technology to replace existing processes is therefore especially appealing.

2. The History of Bleaching

2.1. Textile Bleaching.
There is evidence of chemical bleaching of cloth prior to 300 BC (6). Soda ash prepared from the burning of seaweed was used to clean the cloth followed by souring, ie, treatment with soured milk to neutralize the alkalinity remaining on the cloth. The cloth was then exposed to the sun to complete the bleaching process. Sun bleaching, which became known as crofting, occurred over a matter of weeks during which time the cloth was kept moist to enhance the bleaching process (7). During the eighteenth century, improvements were developed including the use of sulfuric acid in the souring process and the use of lime in the cleaning process, though crofting still required large tracts of primarily coastal land. With the onset of mechanized weaving, the production of cloth was outstripping the availability of land, which set the stage for the introduction of chemical bleaching.

Scheele, a Swedish chemist, discovered chlorine gas in 1784 and demonstrated its use in decolorizing vegetable dyes. Berthollet first produced solutions of hypochlorite by combining chlorine gas with alkalies and suggested using the gas for bleaching. A Scottish bleacher followed the suggestion and introduced chlorine into a bleach works in Glasgow. The efficiency of the process lead to its widespread use, though the low pH resulted in fabric damage and worker health problems. Two chemists, Valette and Tennant, developed chlorinated lime solutions that minimized these difficulties.

Tennant received a patent in 1799 for bleaching powder formed by the absorption of chlorine gas by dry hydrate of lime. Although this eliminated the need for on-site manufacture of chlorine, evidence suggests its use by bleachers caught on slowly. The bleaching powder was the chief source of textile bleaches over the next century and was the impetus for much of the early chemical and chemical engineering developments. Tropical bleach was developed by the addition of quicklime to bleaching powder to make a material suitable for use under tropical conditions. After World War I, technology for shipping liquid chlorine and caustic economically was developed allowing for the on-site manufacture of sodium hypochlorite solutions at the textile mills. As a result, use of bleaching powder diminished.

After World War I, other chlorine-based bleaches were developed. In 1921, the use of chlorine dioxide for bleaching fibers was reported followed by the development of the commercial process for large-scale production of sodium chlorite. In 1928, the first dry calcium hypochlorite containing 70% available chlorine was produced in the United States. This material largely replaced bleaching powder as a commercial bleaching agent.

Although hydrogen peroxide was prepared as early as 1818 by Thenard, the peroxides received little use as textile bleaches. Hydrogen peroxide was first prepared by the action of dilute sulfuric acid on barium peroxide, but later sodium peroxide and dilute acids were used. The prices of peroxides were high initially, and they found use only as a specialty chemical. Electrolytic methods in the 1920s allowed for the synthesis of less costly, strong (~30%) solutions of hydrogen peroxide. By 1930, hydrogen peroxide was being used to bleach cotton goods, wool, and silk on a limited scale. Shortly thereafter, the J-Box was developed by the FMC Corp. allowing for continuous bleaching of textiles with hydrogen peroxide (8). By 1940, 65% of all cotton bleaching was done with hydrogen peroxide.

2.2. Pulp Bleaching. The development of pulp bleaching parallels textile bleaching in many respects partially because early paper was generally made from rags. In the 1700s, sunlight was used to bleach paper. After the turn of the century, bleaching powder was used to whiten the rags used to make paper. During the nineteenth century, wood began to be used as a source of paper and sulfite pulping was developed. Although the Kraft process was discovered not long after, the sulfite process dominated for many years, since it yielded a whiter more easily bleached pulp. Calcium hypochlorite continued to be the bleaching agent used but multistage bleaching processes began to be employed. After World War I, compressed chlorine gas became available and its well-established properties as a delignifying agent ultimately resulted in its use in a chlorine-caustic extraction-hypochlorite (CEH) bleaching sequence. By the 1950s, chlorine dioxide generators were developed leading to the extensive use of this chemical as a bleaching agent particularly for the hard to bleach Kraft pulp. More recently peroxygens, particularly hydrogen peroxide, have been utilized (see PULP BLEACHING).

2.3. Household and Commercial Laundering. Prior to the turn of the twentieth century home bleaching in the United States was accomplished by the same method used by the ancient Romans and Gauls. Clothes were laundered in a mildly alkaline bath and then subjected to sunlight bleaching. In the period from 1910 to 1920, 5.25% sodium hypochlorite solutions were developed and distributed regionally in the United States. By the mid-1930s these solutions were available nationwide. This formula has remained essentially unchanged since its initial introduction. In the 1950s laundry products containing dry sources of hypochlorite were introduced into the United States. However, by the late 1960s the dry chlorine products had disappeared probably because of lower efficacy compared to liquid hypochlorite and fabric damage resulting from placement of the product on wet fabric. In Europe, laundry detergents containing sodium perborate as a bleaching agent were introduced in the early 1900s (9). The perborate dissolves during the laundering process and releases hydrogen peroxide. Sodium perborate continues to be heavily used in European laundry detergents because of the high (up to 95°C) wash temperatures. In the 1950s, laundry products containing sodium perborate were introduced in the United States. In the late 1970s, tetraacetylethylenediamine (TAED), a perborate activator, was introduced into European detergents. TAED with perborate generates peracetic acid in the wash, which is more effective than hydrogen peroxide. TAED is currently contained in >50% of European detergents (10). In the United States in 1982 a dry bleach containing diperoxydodecanedioic acid was test marketed but not expanded. In the late 1980s, a detergent product containing the perborate activator nonanoyloxybenzene sulfonate was introduced. This activator generates pernonanoic acid when combined with hydrogen peroxide generated from sodium perborate monohydrate.

Commercial laundries have used and continue to use sodium hypochlorite as the primary bleaching agent because of its whitening and disinfectant properties.

3. The Mechanism of Bleaching

Bleaching is a decolorization or whitening process that can occur in solution or on a surface. The color-producing materials in solution or on fibers are typically organic compounds that possess extended conjugated chains of alternating single and double bonds and often include heteroatoms, carbonyl, and phenyl rings in the conjugated system. The portion of molecule that absorbs a photon of light is referred to as the chromophore (Greek: *color bearer*). For a molecule to produce color, the conjugated system must result in sufficiently delocalized electrons such that the energy gap between the ground and excited states is small enough so that photons in the visible portion of the light spectrum are absorbed (see COLOR).

Bleaching and decolorization can occur by destroying one or more of the double bonds in the conjugated chain, by cleaving the conjugated chain, or by oxidation of one of the other moieties in the conjugated chain. The result of any one of the three reactions is an increase in the energy gap between the ground and excited states, so that the molecule then absorbs light in the ultraviolet region, and no color is produced. Bleaching may also increase the water solubility of organic compounds after reaction. Conversion of an olefin to a vicinal diol, eg, dramatically increases the polarity and consequently water solubility of the compound. A variety of bleaching agents can affect this transformation. The increased solubility allows actual removal of the bleached substance from a surface.

Chlorine bleaches react with more chromophores than oxygen bleaches. They react irreversibly with aldehydes, alcohols, ketones, carbon–carbon double bonds, acidic carbon–hydrogen bonds, nitrogen compounds, sulfur compounds, and aromatic compounds. Mixtures of products are usually formed because of the variety of active forms in equilibrium with each other (11). Also, many reactions occur in a series of steps, of which the first is usually ratelimiting. With hypochlorous acid, the first step often involves electrophilic addition to carbon–carbon double bonds to form chlorohydrins and epoxides. Or, it may involve electrophilic substitution of aromatic or acidic hydrogens by chlorine. Free-radical reactions are also possible. Nucleophilic addition also occurs with hypochlorite anion, which forms oxygenated products via the elimination of HCl. Aliphatic compounds usually react further to form acids, carbon dioxide, and ketones. Also, most carbon–carbon double bonds and carbon–carbon bonds with adjacent carbonyl or hydroxyl groups can be cleaved. Similar results are often obtained with chlorine dioxide. However, chlorine dioxide reacts with slightly fewer functional groups and often at slower rates than chlorine or hypochlorous acid. Chlorine dioxide also reacts by different mechanisms, which are generally less well understood. With some substrates the main products are chlorinated, but the percentage is usually less than with chlorine or hypochlorous acid (12–14).

The mechanism of bleaching of hydrogen peroxide is not well understood. It is generally believed that the perhydroxyl anion (HOO^-) is the active bleaching species since both the concentration of this anion and the rate of the bleaching process increase with increasing pH (15). Whereas the role of free-radical reactions in the bleaching process remains speculative, mechanisms involving heavy-metal catalyzed reactions are generally undesirable, since they often reduce the effective bleaching because of the rapid loss of peroxide and may also result in fabric damage if the metal is entrapped in the fabric (16). Hydrogen peroxide and other peroxygen compounds can destroy double bonds by epoxidation. This involves addition of an oxygen atom across the double bond usually followed by hydrolysis of the epoxide formed to 1,2-diols under bleaching conditions.

Peracids undergo a variety of reactions which result in bleaching. Peracids can add an oxygen across a double bond to give an epoxide, which can undergo further reactions including hydrolysis to give a vicinal diol. Peracids can oxidize aldehydes to acids, sulfur compounds to sulfoxides and sulfones, and nitrogen compounds to amine oxides, hydroxylamines, and nitro compounds (17). Peracids can also oxidize alpha-diketone compounds to anhydrides and ketones to esters. The protonated and deprotonated forms of the peracids are both effective bleaches. The protonated form acts as an electrophile whereas the deprotonated form is a nucleophilic oxidant.

Reducing agents are thought to work by reduction of the chromophoric carbonyl groups in textiles or pulp.

4. Chlorine-Containing Bleaching Agents

Chlorine-containing bleaching agents are the most cost-effective bleaching agents known. They are also effective disinfectants, and water disinfection is often the largest use of many chlorine-containing bleaching agents. They may be divided into four classes: chlorine, hypochlorites, N-chloro compounds, and chlorine dioxide.

The first three classes are called available chlorine compounds and are related to chlorine by the equilibria in equations 1–4. These equilibria are rapidly established in aqueous solution (18), but the dissolution of some hypochlorite salts and N-chloro compounds can be quite slow.

$$Cl_2 \text{ (gas)} \rightleftharpoons Cl_2 \text{ (aq)} \tag{1}$$

$$Cl_2 \text{ (aqueous)} + H_2O \rightleftharpoons HOCl + H^+ + Cl^- \tag{2}$$

$$HOCl \rightleftharpoons H^+ + OCl^- \tag{3}$$

$$RR'NCl + H_2O \rightleftharpoons HOCl + RR'NH \tag{4}$$

The total concentration or amount of chlorine-based oxidants is often expressed as available chorine or less frequently as active chlorine. Available chlorine is the equivalent concentration or amount of Cl_2 needed to make the oxidant according to equations 1–4. Active chlorine is the equivalent concentration or amount of Cl atoms that can accept two electrons. This is a convention, not a description of the reaction mechanism of the oxidant. Because Cl_2 only accepts two electrons as does HOCl and monochloramines, it only has one active Cl atom according to the definition. Thus the active chlorine is always one-half of the available chlorine. The available chlorine is usually measured by iodometric titration (19,20). The weight of available chlorine can also be calculated by equation 5.

$$\text{weight available chlorine} = 70.9 \times \text{moles of oxidant} \times \frac{\text{number active Cl atoms}}{\text{molecule}} \tag{5}$$

where 70.9 represents the mol wt of Cl_2 and moles of oxidant can be represented wt oxidant/mol wt of oxidant.

In solutions, the concentration of available chlorine in the form of hypochlorite or hypochlorous acid is called free-available chlorine. The available chlorine in the form of undissociated N-chloro compounds is called combined-available chlorine. Several analytical methods can be used to distinguish between free- and combined-available chlorine (20). Bleaches that do not form hypochlorite in solution like chlorine dioxide and nonchlorine bleaches can be characterized by their equivalent available chlorine content. This can be calculated from equation 5 by substituting the number of electrons accepted divided by two for the number of active chlorine atoms. It can also be measured by iodometric titration.

The actual form of an available chlorine bleach in solution must be determined from equations 1–4. The equilibrium constants for equations 2 and 3 are $3.94 \times 10^{-4}\ M^2$ (21) and $2.88 \times 10^{-8}\ M$ (22) at 25°C, respectively. Thus, above pH 9.5 > 99% of the available chlorine is present as hypochlorite ions. The ratio of hypochlorous acid to hypochlorite ion increases with decreasing pH until pH 5.5, below which <1% of the available chlorine will be hypochlorite ions. Below pH 6, Cl_2 may be present. Its amount increases with decreasing pH

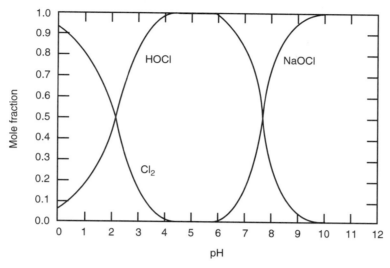

Fig. 1. Distribution of 0.5 wt% available chlorine with equimolar chloride at 25°C in a closed system.

and increasing total available chlorine. With an available chlorine concentration of 0.1%, Cl_2 begins to appear ~pH 4 and becomes dominant ~pH 2.5. At an available chlorine concentration of 10%, Cl_2 appears ~pH 6, and becomes dominant ~pH 4.5. Figure 1 shows the distribution of available chlorine species as a function of pH in an aqueous solution of 0.5% available chlorine. Other species that may be present in insignificant amounts are Cl_3^-; from Cl_2 (aq) and H_2OCl^+ and Cl_2O from HOCl (11).

With N-chloro compounds, the species in solution also depends on their individual hydrolysis constants as shown in equation 4. Those with larger constants allow a higher hypochlorite concentration, which usually gives stronger bleaching because N-chloro compounds are weaker oxidants than hypochlorite (see CHLORAMINES AND BROMAMINES). Although N-chloro compounds react quickly with other N-compounds and may remove some stains faster than hypochlorite (23), the undissociated N-chloro compound usually serves as a reservoir to replenish hypochlorite as it is consumed. This reduces reactivity compared to the higher hypochlorite concentration required to deliver the same amount of available chlorine (24–26). Stability of the available chlorine may also be improved (27,28). Even when the N-chloro compound completely dissociates at the use concentration, it can reduce the hypochlorite concentration in a concentrated solution or product. This may reduce the damage done by spills, improve stability, or allow the use of a wider variety of ingredients. N-Chloro compounds may also be formed *in situ* as when N-compounds are used in dry compositions with active chlorine compounds to scavenge free hypochlorite to protect other materials (29).

Solutions of available chlorine bleaches decompose on standing at a rate that depends on the conditions described below. Hypochlorous acid [7790-92-3] and hypochlorite anions decompose according to equations 6 and 7 (30,31).

$$3\ HOCl \longrightarrow HClO_3 + 2\ HCl \tag{6}$$

$$3\ OCl^- \longrightarrow ClO_3^- + 2\ Cl^- \tag{7}$$

The solutions are most stable >pH 11 where the decomposition rate is nearly independent of pH. In this region, the decomposition rate has a second-order dependence on the concentration of hypochlorite. It also increases with increasing ionic strength. Thus concentrated solutions decompose much faster than dilute solutions. Because of an unusually high activation energy, the decomposition rate increases greatly with temperature. Nevertheless, solutions with less than ~6% available chlorine and a pH > 11 have acceptable long-term stability below ~30°C.

Below pH 11, the decomposition rate becomes dependent on pH and the mechanism becomes more complicated. The rate increases greatly as the pH decreases from 11 to 7 where the rate reaches a maximum. As the pH decreases from 7 to 3, the rate decreases. Below ~pH 3 the rate becomes reasonably slow but still considerably faster than at pH 11. The mechanisms are not well known, and they may be different in each of these pH ranges. In all cases, the decomposition rate increases greatly with temperature and with a second- or higher-order dependence on the concentration of hypochlorous acid and/or hypochlorite.

Decomposition also occurs by equation 8, which can usually be ignored unless it is catalyzed by transition-metal ions (32,33):

$$2\ OCl^- \longrightarrow O_2 + 2\ Cl^- \tag{8}$$

Even very small amounts of transition-metal ions like cobalt, nickel, and copper cause rapid decomposition. They form reactive intermediates that can decrease the stability of oxidizable compounds in the bleach solution and increase the damage to substrates. Hypochlorite is also decomposed by ultraviolet (uv) light (34,35). Acidic solutions also lose available chlorine by the reverse of equations 1 and 2.

Commercially important solid available chlorine bleaches are usually more stable than concentrated hypochlorite solutions. They decompose very slowly in sealed containers. But most of them decompose quickly as they absorb moisture from air or from other ingredients in a formulation. This may release hypochlorite that destroys other ingredients as well.

4.1. Chlorine.   Except to bleach wood pulp and flour, chlorine [7782-50-5] itself is rarely used as a bleaching agent. Chlorine is almost always first converted into one of the bleaching agents described, and they are almost always used at a pH where Cl_2 is not present. However, it has been the practice to use acid chlorination where Cl_2 is the active species in the first step of pulp bleaching. Since chlorine reacts primarily by chlorination, large amounts of chlorinated organic by-products are formed (36). Environmental concerns about discharging these by-products in waste effluents are rapidly changing this process (37–39).

4.2. Hypochlorites.   *Sodium Hypochlorite.*   The principal form of hypochlorite produced is sodium hypochlorite [7681-52-9], NaOCl. It is invariably made and used as an aqueous solution and is usually prepared by the chlorination of sodium hydroxide solutions as shown in equation 9, though other bases such as sodium carbonate can be used (40).

$$Cl_2 + 2\,NaOH \rightleftharpoons NaOCl + NaCl + H_2O \tag{9}$$

Chlorine gas is usually used, but electrolysis of alkaline salt solutions in which chlorine is generated *in situ* is also possible and may become more important in the future. The final pH of solutions to be sold or stored is always adjusted >11 to maximize stability. The salt is usually not removed. However, when the starting solution contains >20.5% sodium hydroxide some salt precipitates as it is formed. This precipitate is removed by filtration to make 12–15% NaOCl solutions with about one-half of the normal amount of salt. Small amounts of such solutions are sold for special purposes. Solutions with practically no salt can be made by reaction of high purity hypochlorous acid with metal hydroxides.

A 5–6% sodium hypochlorite solution is sold for household purposes, of which the largest use is in laundry. Solutions of 10–15% NaOCl are sold for swimming pool disinfection, institutional laundries, and industrial purposes. Solutions of various strengths are used in household and industrial and institutional (I & I) cleaners, disinfectants, and mildewcides. A small amount is used in textile mills. Sodium hypochlorite is also made on site with 30–40 g/L available chlorine for pulp bleaching, but its use is decreasing in order to reduce chloroform emissions (see CHLORINE OXYGEN ACIDS AND SALTS).

Calcium Hypochlorite. The principal form of solid hypochlorite produced commercially is calcium hypochlorite [7778-54-3], Ca(OCl)$_2$. It decomposes rapidly and exothermically gives off oxygen and chlorine monoxide gases when heated >175°C. It also reacts vigorously or explosively with oxidizable materials. The most common form contains 6–12% water and 65% available chlorine. The water reduces the risk of self-sustained decomposition because of organic contaminants or ignition. The older variety contains about 1% water and 70–74% available chlorine. Both forms also contain sodium chloride [7647-14-5] and small amounts of calcium hydroxide [1305-62-0], calcium chloride [10043-52-4], calcium chlorate [10137-74-3], and calcium carbonate [471-34-1]. They are made by chlorination of hydrated lime (calcium hydroxide) in a way that minimizes the amounts of unwanted salts. The resulting product contains much fewer insoluble materials and is more stable than bleaching powder.

The largest use of calcium hypochlorite is for water treatment. It is also used for I & I and household disinfectants, cleaners, and mildewcides. Most of the household uses have been limited to in-tank toilet bowl cleaners. In areas where chlorine cannot be shipped or is otherwise unavailable, calcium hypochlorite is used to bleach textiles in commercial laundries and textile mills. It is usually first converted to sodium hypochlorite by mixing it with an aqueous solution of sodium carbonate and removing the precipitated calcium carbonate. Or, it can be dissolved in the presence of sufficient sodium tripolyphosphate to prevent the precipitation of calcium salts. However, calcium hypochlorite is not usually used to bleach laundry and textiles because of problems with insoluble inorganic calcium salts and precipitation of soaps and anionic detergents as their calcium salts.

Bleach Liquor.   Bleach liquor or lime bleach liquor is an aqueous solution of calcium hypochlorite and calcium chloride. It typically contains 30–35 g/L of available chlorine, though it may be as high as 85 g/L. It has been used in pulp bleaching, when it can be made more cheaply than sodium hypochlorite. It is prepared on site by chlorinating lime solutions.

Bleaching Powder and Tropical Bleach.   Bleaching powder [64175-94-6], also known as chlorinated lime and chloride of lime, is an indefinite, complex mixture of calcium hypochlorite, calcium hydroxide, calcium chloride, and their hydrates. The proportions of these species vary with the manufacturer as does the available chlorine, which usually ranges between 24 and 37%. It is usually made by chlorinating slightly moist hydrated lime (calcium hydroxide). It has also been made by chlorinating a slurry or solution of calcium chloride (41). Bleaching powder readily decomposes in moist air through the absorption of water and carbon dioxide. Its stability can be improved by adding calcium oxide. Since this is needed especially in hot, humid climates, such mixtures are known as tropical bleach, super tropical bleach, or stabilized tropical bleach. They typically contain 15–30% available chlorine.

Historically, bleaching powder and tropical bleach were significant sources of available chlorine but very little are used today. This is because of the greater availability of sodium hypochlorite solutions and the development of calcium hypochlorite. They are still used to sanitize fields, drainage ditches, and reservoirs where its insoluble portion is not important. And, they are important sources of available chlorine within some less developed tropical countries.

Dibasic Magnesium Hypochlorite.   This salt $Mg(OCl)_2 \cdot 2Mg(OH)_2$, [11073-21-5], is safer than calcium hypochlorite because of its higher thermal stability and its endothermic rather than exothermic decomposition. Its preparation as a solid with 50–58% available chlorine is patented (42,43) but not sold commercially.

Lithium Hypochlorite.   Commercial lithium hypochlorite [13840-33-0], LiOCl, is a solid with ~35% available chlorine. It is made from concentrated solutions of sodium hypochlorite and lithium chloride. It consists of 30% lithium hypochlorite and various other salts (44).

Lithium hypochlorite is used in I & I laundry detergents and I & I dry laundry bleaches. Like sodium hypochlorite, it does not precipitate soaps and other anionic detergents. However, lithium hypochlorite is an expensive source of available chlorine and not much is used for bleaching. Its principal use is as a shocking agent for swimming pool disinfection.

Chlorinated Trisodium Phosphate.   Chlorinated trisodium phosphate [11084-85-8] is a crystalline complex of hydrated trisodium orthophosphate and sodium hypochlorite that releases hypochlorite when mixed with water. Its formula is $(Na_3PO_4 \cdot 11H_2O)_4 \cdot NaOCl$. Commercial products have 3.5–3.7% available chlorine. They are probably a mixture of phosphate salts, and they contain some sodium chloride.

The use of chlorinated trisodium phosphate is declining. It has been largely replaced by chlorinated isocyanurates in powdered abrasive cleansers and automatic dishwash detergents to reduce cost, improve performance, or comply with restrictions on the use of phosphates. Some chlorinated trisodium phosphate is still used in commercial laundries and in disinfectant cleaners.

Hypochlorous Acid. Hypochlorous acid [7790-92-3] solutions are made for immediate use as chemical intermediates from chlorine monoxide or in bleaching and water disinfection by adjusting the pH of hypochlorite solutions. Salt-free hypochlorous acid solutions have been economically made from steam and chlorine (45). These solutions may have sufficient stability at 0°C to be sold for industrial use.

Oxidation of Chlorides. Hypochlorite can also be formed by the *in situ* oxidation of chloride ions by potassium peroxymonosulfate [25482-78-4] (46). Ketones like acetone catalyze the reaction (47). The triple salt of potassium peroxymonosulfate is a stable powder that has been combined with chloride salts and sold as toilet bowl cleaners. Bromides can be used in place of chlorides to form hypobromites, and such combinations are used to disinfect spas and hot tubs.

Hypobromites. The chemistry of hypobromite is similar to hypochlorite. It reacts faster than hypochlorite and gives better bleaching at higher pH and lower temperatures. It also decomposes according to equations 6 and 7 much faster than hypochlorite. The most stable solutions decompose quickly and must be freshly prepared. This can be done by adding a bromide salt to a hypochlorite solution, since hypochlorite oxidizes bromide to hypobromite. Usually a catalytic amount of bromide is used since much of it will be regenerated as the hypobromite is reduced during bleaching (48). Dry compositions containing a bromide salt and a solid available chlorine compound can also be used (49). A few *N*-bromo compounds are also available.

4.3. N-Chloro Compounds. *Chlorinated Isocyanurates.* The principal solid chlorine bleaching agents are the chlorinated isocyanurates. The one used most often for bleaching applications is sodium dichloroisocyanurate dihydrate [51580-86-0] with 56% available chlorine. It is the most water-soluble, the fastest to dissolve, and the least hazardous. It has good stability and compatibility with other dry ingredients. Anhydrous sodium [2893-78-9] (1) and potassium [2244-21-5] salts with 63 and 59% available chlorine are also available. The potassium salt is less soluble in water and is claimed to have the best stability in dry products containing other ingredients (50). Trichloroisocyanuric acid [87-90-1] (2) with 90% available chlorine is the most economical and the one used most often for water disinfection. Not much is used for bleaching because it dissolves slowly and has poor stability in dry products containing other ingredients. It also decomposes with moisture to give off explosive nitrogen trichloride gas.

        (1)   (2)   (3)

Halogenated Hydantoins. These are stable solids with limited use as bleaches. They dissolve too slowly to use in household laundry and automatic dishwashing. 1,3-Dichloro-5,5-dimethylhydantoin [118-52-5] (3) is sold with 65–75% available chlorine. It is used as a bleach in hospital and other industrial laundries and in disinfectant cleaners. Some 1-bromo-3-chloro-5,5-dimethylhydantoin [6079-88-2] is also used. It is a more effective bleach and disinfectant at lower temperatures and higher alkalinities than 1,3-dichloro-5,5-dimethylhydantoin because it hydrolyzes to hypobromite.

Other N-Chloro Compounds. Sodium N-chlorobenzenesulfonamide (chloramine B) [127-52-6] (4), sodium N-chloro-p-toluenesulfonamide (chloramine T) [127-65-1] (5), N-chlorosuccinimide [128-09-6] (6), and trichloromelamine [12379-38-3] (7) have also had minor roles as bleaching agents.

(4)   (5)   (6)   (7)

They are mainly sold as disinfectants, because they display poor bleaching as a result of low hydrolysis constants or poor solubility. N-Chloro compounds with low hydrolysis constants like chloramine T can be used to boost the bleaching of peroxide laundry bleaches (51–54). The bleaching remains inferior to using sodium hypochlorite laundry bleach, however. Tetrachloroglycoluril [776-19-2] (55) sodium trichlorometaphosphimate [67651-15-14] (56), sodium N-chloroimidodisulfonate [67700-32-7] (57), 1,3-dichlorotetrahydroquinazoline-2,4-dione [23767-45-5] (58,59), and N-chlorophenylbiguanidino compounds (60,61) have been unsuccessfully marketed as bleaching agents. Many other N-chloro compounds have been patented as bleaching agents, the most notable are mono-[17172-27-9] and dichlorosulfamic acid [17085-87-9] (62–64).

4.4. Chlorine Dioxide. Chlorine dioxide [10049-04-4], ClO_2, is a gas that is more toxic than chlorine. It can explode at concentrations >10% in air. The liquid boils at 11°C but explodes above −40°C. It can be stored and transported as its octahydrate if kept frozen, but almost all chlorine dioxide is made on site for immediate use. Large amounts for pulp bleaching are made by several processes (65,66) in which sodium chlorate [7775-09-9] is reduced with chloride, methanol, or sulfur dioxide in highly acidic solutions by complex reactions. For most other purposes, chlorine dioxide is made from sodium chlorite [7758-19-2]. Acidic solutions of sodium chlorite are oxidized by chlorine as in equation 10:

$$2\,NaClO_2 + Cl_2 \longrightarrow 2\,NaCl + 2\,ClO_2 \qquad (10)$$

Hypochlorous acid can also be used, but the reaction is slower. Chlorine dioxide is also made by adding acid to sodium chlorite solutions by the overall reaction in equation 11:

$$5\,NaClO_2 + 4\,HCl \longrightarrow 4\,ClO_2 + 5\,NaCl + 2\,H_2O \qquad (11)$$

Some chlorine and chlorate also form through competing reactions. Chlorine dioxide is also evolved from mixtures of powdered sodium chlorite and acidic clays or alumina.

The reactivity of acidified chlorite solutions is reduced for bleaching some textiles by adding compounds like polyamines, pyrophosphates, and hydrogen peroxide that suppress the formation of chlorine dioxide (12). Another method is to buffer the solution at pH 5–6 to reduce the rate of chlorine dioxide formation. Hydrolysis of anhydrides and esters or oxidation of alcohols can be used to slowly generate acids to promote chlorine dioxide formation (67). Aldehydes also promote chlorine dioxide generation from neutral chlorite solutions, but the effect is greater than simply lowering the pH as they oxidize to acids (68).

Chlorine dioxide is usually used in aqueous solution. It is a weaker oxidant than hypochlorite. Unlike chlorine it does not react with water to form hypochlorite or with amines to form *N*-chloro compounds. Thus chlorine dioxide is easily removed from solutions by passing air through the solution or its headspace. Chlorine dioxide solutions decompose by equation 12:

$$2\,ClO_2 + 2\,NaOH \longrightarrow NaClO_2 + NaClO_3 + H_2O \qquad (12)$$

This reaction is very slow in acid but rapid above pH 10. Chlorine dioxide solutions are also decomposed by light.

The biggest use of chlorine dioxide is in bleaching wood pulp. In some mills, much of the chlorine and hypochlorite has been replaced by chlorine dioxide to reduce the amount of chlorinated by-products. Chlorine dioxide is also used to bleach textiles, flour, and edible fats and oils.

5. Peroxygen Compounds

Peroxygen compounds contain the peroxide linkage $(-O-O-)$ in which one of the oxygen atoms is active. This activity, referred to as active oxygen (AO), is measured by the oxidation of iodide to iodine under acidic conditions or by a ceric sulfate titration (69). Active oxygen content, usually expressed as a percent, is the atomic weight of active oxygen divided by the molecular weight of the compound (eq. 13):

active oxygen, $\% = 100 \times$ number of active oxygens $\times$ 16 mol wt of compound

5.1. Hydrogen Peroxide. Hydrogen peroxide [7722-84-1] is one of the most common bleaching agents (see HYDROGEN PEROXIDE). It is the primary bleaching agent in the textile industry, and is also used in pulp, paper, and home laundry applications. In textile bleaching, hydrogen peroxide is the most common bleaching agent for protein fibers, and is also used extensively for cellulosic fibers.

Pure hydrogen peroxide has an active oxygen content of 47%. It is the least expensive source of active oxygen commercially available. Moreover, it is a liquid, making it convenient for many bleaching applications. Hydrogen peroxide is usually sold in solutions containing 30–35%, 50% or 65–70 wt % of the active material. More concentrated solutions (80–85%, 90%) are available in limited quantities. Concentrated solutions of hydrogen peroxide are hazardous and must be handled with extreme care (70).

Hydrogen peroxide is a very weak acid and in aqueous solutions only dissociates slightly (eq. 14); $K_a = 1.78 \times 10^{-12}$. Undissociated hydrogen peroxide is relatively stable, and for this reason all commercial products are adjusted to an acid pH (71).

$$H_2O_2 \rightleftharpoons H^+ + HO_2^- \tag{14}$$

A considerable amount of energy is liberated when hydrogen peroxide undergoes decomposition to oxygen and water (eq. 15): $\Delta H_{25°C} = -94.64$ kJ/mol (-22.62 kcal/mol); activation energy $= 209$ kJ/mol (50 kcal/mol).

$$H_2O_2 \longrightarrow H_2O + \frac{1}{2}O_2 \tag{15}$$

This decomposition may be considered a self-oxidation and occurs most rapidly in basic solutions. Decomposition of hydrogen peroxide is also greatly accelerated in the presence of heavy metals and easily oxidizable substances (72). The presence of low concentrations of heavy metals (such as Fe and Cu) in hydrogen peroxide can increase the rate of decomposition by many orders of magnitude over the entire pH range for typical uses. Therefore, commercial hydrogen peroxide solutions are stabilized with additives that provide protection against decomposition (73–77). Typically these additives are metal chelating agents that bind free metal ions and significantly reduce the catalytic rate of decomposition.

Hydrogen peroxide bleaching is performed in alkaline solution where part of the hydrogen peroxide is converted to the perhydroxyl anion (eq. 14). The perhydroxyl anion is generally believed to be the active bleaching species and its concentration in solution increases with hydrogen peroxide concentration, alkalinity, and temperature (78). The alkaline agents most commonly used to generate HO_2^- are caustic soda, carbonates, silicates, pyrophosphates, and polyphosphates. Better bleaching is obtained at these alkaline conditions by increasing the temperature and by adding stabilizers to prevent the uncontrolled decomposition reactions of hydrogen peroxide. Common stabilizers include silicates, pyrophosphates, and polyaminocarboxylates. These stabilizers may be different from those used to stabilize commercial acidic hydrogen peroxide (15).

As a bleaching agent, hydrogen peroxide is much less effective than chlorine or hypochlorite; however, it does have several advantages over these bleaching agents. Hydrogen peroxide causes less textile fiber damage, is much gentler on fabric dyes, and does not have a strong odor. Attempts have been made to increase the bleaching power of hydrogen peroxide-based laundry bleaches by the addition of heavy-metal catalysts (79–83). The effectiveness of these systems remains controversial and these catalysts have not been incorporated into commercial products.

5.2. Solid Peroxygen Compounds.   Hydrogen peroxide reacts with many compounds, such as borates, carbonates, pyrophosphates, sulfates, silicates, and a variety of organic carboxylic acids, esters, and anhydrides to give peroxy compounds or peroxyhydrates. A number of these compounds are stable solids that hydrolyze readily to give hydrogen peroxide in solution.

Perborates. Sodium perborate [7632-04-4] is the most widely used solid peroxygen compound. Commercially, it is available as a tetrahydrate [10486-00-7] and a monohydrate [10322-33-9]. The tetrahydrate is produced by treating a borax solution with hydrogen peroxide and sodium hydroxide:

$$Na_2B_4O_7 + 2\,NaOH + 4\,H_2O_2 + 11\,H_2O \longrightarrow 4\,NaBO_3 \cdot 4\,H_2O \qquad (16)$$

The tetrahydrate has the structure (8) (84):

$$2Na^+ \left[\begin{array}{c} HO \\ HO \end{array} \!\!> B <\!\! \begin{array}{c} O-O \\ O-O \end{array} \!\!> B <\!\! \begin{array}{c} OH \\ OH \end{array} \right]^{2-} \cdot 6H_2O$$

(8)

It has good stability and can be used in formulations with many compounds without serious loss of active oxygen. The monohydrate is made by dehydration of the tetrahydrate. The active oxygen contents of the tetrahydrate and monohydrate are 10.5 and 16.0%, respectively. The tetrahydrate is the perborate salt most commonly used in bleaching applications (85). However, as consumer trends move toward more concentrated products, monohydrate is growing in demand because of its higher AO content (86). Because sodium perborate has much greater stability than sodium hypochlorite, it can be formulated into a wide variety of products, including detergents. In the United States, perborates are used in all-fabric bleach formulations, detergents, denture cleaners, tooth powders, and other special cleaners. Sodium perborate is used extensively in Europe in detergent formulations.

Sodium Carbonate Peroxyhydrate. Sodium carbonate peroxyhydrate [15630-89-4], which contains about 14 wt% of active oxygen, has the composition $2Na_2CO_3 \cdot 3H_2O_2$. A white, free-flowing solid, it generally can be used in all applications where perborate is used. Despite the fact that sodium carbonate peroxyhydrate has a greater rate of dissolution than sodium perborate tetrahydrate, the latter is usually favored for its good storage stability and better compatibility with the various materials used in formulations (87).

Peroxymonosulfate. Peroxymonosulfuric acid (Caro's acid) [7722-86-3], the peroxygen product of hydrogen peroxide and sulfuric acid, is a powerful oxidizing agent; however, because of its instability, it is hazardous (88). It is commercially available in Europe but not in the United States. The salt, potassium permonosulfate [25482-78-4] is commercially available under the trade name Oxone. This monopersulfate compound is a white solid having a satisfactory shelf life and an active oxygen content of $\sim$4.5%. It is a triple salt with the composition $2KHSO_5 \cdot K_2SO_4 \cdot KHSO_4$. Oxone is used as a bleaching agent and in several other applications where a solid peroxygen source is required; however, the extent of its use is limited.

5.3. Peracids. Peracids are compounds containing the functional group $-OOH$ derived from an organic or inorganic acid functionality. Typical structures include $CH_3C(O)OOH$ derived from acetic acid and $HOS(O)_2OOH$ (peroxymonosulfuric acid) derived from sulfuric acid. Peracids have superior cold water bleaching capability versus hydrogen peroxide because of the greater electrophilicity of the peracid peroxygen moiety (89–91). Lower wash temperatures and phosphate reductions or bans in detergent systems account for the recent utilization and vast literature of peracids in textile bleaching (17,92).

Peracids can be introduced into the bleaching system by two methods. They can be manufactured separately and delivered to the bleaching bath with the other components or as a separate product. Peracids can also be formed *in situ* utilizing the perhydrolysis reaction shown in equation 17.

$$
\underset{\substack{\text{O}\\||\\ \text{R}-\text{C}-\text{Z}}}{} + \; ^-\text{OOH} \;\longrightarrow\; \underset{\substack{\text{O}\\||\\ \text{RCOOH}}}{} + \; \text{Z}^- \tag{17}
$$

R can be a variety of structures. The leaving group is Z^- and is typically the conjugate base of a weak acid whose pK_a can range from 5 to 20 (93). The hydrogen peroxide is typically incorporated into the bath by adding a solid source of peroxide such as sodium percarbonate or the mono- or tetrahydrate of sodium perborate (93).

Peracid Analysis.   Peracid concentrations can be measured in a product or in the bath by use of a standard iodide/thiosulfate titration (69). With preformed peracids or peracids formed via perhydrolysis care must be exercised to minimize the interference of hydrogen peroxide, present intentionally as a component of the perhydrolysis reaction or as a result of the hydrolysis of the peracid (94,95) as shown in equation 18.

$$
\underset{\substack{\text{O}\\||\\ \text{RCOOH}}}{} + \; \text{H}_2\text{O} \;\rightleftharpoons\; \underset{\substack{\text{O}\\||\\ \text{RCOH}}}{} + \; \text{H}_2\text{O}_2 \tag{18}
$$

This is typically accomplished by cooling the titration solution with ice, determining the blank, and titrating rapidly. Another method utilizes determination of the total peroxide and peracid content by use of a ceric sulfate titration to measure hydrogen peroxide followed by a iodide/thiosulfate titration to measure total active oxygen (69).

Peracid Classification.   Peracids can be broadly classified into organic and inorganic peracids, based on standard nomenclature. The limited number of inorganic peracids has required no subclassification scheme (9). However, the tremendous number of new organic peracids developed (17) has resulted in proposals for classification. For example, a classification scheme based on liquid chromatography retention times and critical micellization constants (CMC) of the parent acids has been proposed (96). The parent acids are used because of the instability of the peracids under chromatographic and micellization measurement conditions. This classification scheme is shown in Table 1.

The technique used to classify the peracids is artificial with respect to bleaching, but the classification of a peracid does relate to its location in the bleaching bath microenvironment. Hydrophilic peracids are quite water soluble and as such they are located in the bulk phase and their bleaching performance is the result of random collisions with the fabric surface (9). Since collisional frequency is increased at higher temperature the hydrophilic peracids are useful in high temperature washing conditions. The inorganic peracids are exclusively of the hydrophilic type. Hydrophobic peracids are similar in structure and behavior

Table 1. **Classification of Peracids**

Peracid type	Retention time[a] of parent acid	CMC	Typical structures
hydrophilic	<5 min	>0.5 M	peracetic, perpropionic, perbenzoic acids[b]
hydrophobic (surface active)	na	<0.5 M	c
hydrotropic	>5 min	none, or >0.5 M	c

[a] Chromatographic conditions: elution with 50:50 methanol/water solvent at the rate of 1.5 mL/min through a DuPont Zorbax ODS column using a Waters R-401 Refractive Index Detector.
[b] That is, in equations 17 and 18 R = CH_3, CH_2CH_3, or C_6H_5, respectively.
[c] See Figure 2.

to common detergent surfactants because they possess a hydrophilic head group [$-C(O)OOH$] and a hydrophobic tail [$CH_3(CH_2)_n-$]. These peracids have a defined CMC and as such are likely to be located in micelles (9,97). The ability of these peracids to partition to an interface make them more suitable for cold water washing than hydrophilic peracids (98,99). At cooler temperatures improved stain removal by the hydrophobic bleaches versus the hydrophilic bleaches has been demonstrated. No reports of the hydrotropic peracid bleach microenvironment locale have been published, but because of their oily character and based on Table 1 they are likely dissolved into the detergent micelle.

5.4. Peracid Precursor Systems. Compounds that can form peracids by perhydrolysis are almost exclusively amide, imides, esters, or anhydrides (17). Two compounds were commercially used for laundry bleaching as of 1990. Tetraacetylethylenediamine [10543-57-4] is utilized in >50% of western European detergents (10). The perhydrolysis reaction of this compound is shown in equation 19. TAED generates two moles of peracid and one mole of diacetylethylenediamine per mole of imide (100).

$$(CH_3\overset{O}{\overset{\|}{C}})_2NCH_2CH_2N(\overset{O}{\overset{\|}{C}}CH_3)_2 + 2\,H_2O_2 \longrightarrow 2\,CH_3\overset{O}{\overset{\|}{C}}OOH + CH_3\overset{O}{\overset{\|}{C}}NHCH_2CH_2NH\overset{O}{\overset{\|}{C}}CH_3 \qquad (19)$$

$$CH_3(CH_2)_n\overset{O}{\overset{\|}{C}}OOH$$

$$n = 5\text{–}14$$
$$(9)$$

$$CH_3(CH_2)_n\text{—}\bigcirc\text{—}\overset{O}{\overset{\|}{C}}OOH$$

$$n = 8\text{–}16$$
$$(10)$$

$$HOO\overset{O}{\overset{\|}{C}}(CH_2)_n\overset{O}{\overset{\|}{C}}OOH$$

$$n = 7\text{–}12$$
$$(11)$$

$$CH_3(CH_2)_n\overset{O}{\overset{\|}{C}}OOH$$

$$n = 14\text{–}20$$
$$(12)$$

Fig. 2. Typical peracids; see Table 1. Compounds (9) and (10) are hydrophobic; (11) and (12) are hydrotropic.

The perhydrolysis reaction could theoretically continue to give four moles of peracid per mole of TAED but stops at this stoichiometry because of the substantial increase in the conjugate acid pK_a of the leaving group going from an amide ($pK_a = 17$) to an amine ($pK_a = 35$) (101,102). Nonanoyloxybenzene sulfonate (NOBS) [101482-85-3] is used in detergent products in the United States and Japan. The NOBS perhydrolysis reaction is shown in equation 20 (103).

$$CH_3(CH_2)_7\overset{\overset{O}{\|}}{C}O-\!\!\!\langle\bigcirc\rangle\!\!\!-SO_3Na + H_2O_2 \longrightarrow CH_3(CH_2)_7\overset{\overset{O}{\|}}{C}OOH + HO-\!\!\!\langle\bigcirc\rangle\!\!\!-SO_3Na \quad (20)$$

The NOBS system undergoes an additional reaction that forms a diacyl peroxide as a result of the nucleophilic attack of the peracid anion on the NOBS precursor as shown in equation 21. This undesirable side reaction can be minimized by the use of an excess molar quantity of hydrogen peroxide (98,103) or by the use of shorter dialkyl chain acid derivatives. However, the use of these acid derivatives also appears to result in less efficient bleaching. The dependence of the acid group on the side product formation is apparently the result of the proximity of the newly formed peracid to unreacted NOBS in the micellar environment (91). A variety of other peracid precursor structures can be found (104–125).

$$R\overset{\overset{O}{\|}}{C}OO^- + R\overset{\overset{O}{\|}}{C}O-\!\!\!\langle\bigcirc\rangle\!\!\!-SO_3Na \longrightarrow R\overset{\overset{O}{\|}}{C}OO\overset{\overset{O}{\|}}{C}R + {}^-O-\!\!\!\langle\bigcirc\rangle\!\!\!-SO_3Na \quad (21)$$

The pK_a of the leaving group and the hydrophobe chain length can dramatically affect the efficiency of the perhydrolysis reaction. Additionally, the structure of the acid portion of the precursor can affect the yield and sensitivity of the reaction to pH. The mono-4-hydroxybenzenesulfonic acid ester of α-decylsuccinic acid (13) undergoes extremely efficient perhydrolysis at much lower pH values than other peracid precursors, eg, decanoyloxybenzene sulfonate (14). This may be because of the neighboring group participation of the adjacent carboxylate as shown in Table 2 (122).

$$CH_3(CH_2)_9\overset{\overset{O}{\|}}{\underset{\underset{\overset{\|}{O}}{CH_2COH}}{C}H}CO-\!\!\!\langle\bigcirc\rangle\!\!\!-SO_3Na \qquad CH_3(CH_2)_8\overset{\overset{O}{\|}}{C}O-\!\!\!\langle\bigcirc\rangle\!\!\!-SO_3Na$$

(13)   (14)

Electronic effects within the acid portion of the precursor have also been utilized for enhanced reactivity. The 4-hydroxybenzenesulfonate ester of

Table 2. Effect of Alpha-Carbonyl Group on Peracid Yield

Compound	Molecular formula[a]	pH	Peracid yields %
13	$C_{20}H_{30}O_7S$	9.5	97
14	$C_{16}H_{24}O_5S$	9.5	26

[a] Of the sulfonic acid.

octanoyloxyacetic acid, 15, undergoes efficient perhydrolysis at lower pH values because of the activation of the susceptible carbonyl by the beta-oxygen of the hydrophobic tail (107).

$$CH_3(CH_2)_6\overset{\overset{O}{\|}}{C}OCH_2\overset{\overset{O}{\|}}{C}O \longrightarrow \bigcirc \longrightarrow SO_3Na$$

(15)

Attempts have also been made to reduce the odor associated with the per-acid in the home laundry. Use of a precursor that generates the peracid of a fatty acid can result in an objectionable odor in the wash bath (113). This odor is exacerbated by the higher pK_a of the peracid versus its parent acid resulting in a greater proportion of the peracid in the unionized and therefore less water-soluble form. To mitigate this circumstance, functionalization of the fatty tail typically alpha to the carbonyl has been utilized (119). The modifications include alpha-chloro and alpha-methoxy substituents on the parent acid portion of the precursor ester.

The peracid precursors can be susceptible to hydrolysis or perhydrolysis in the solid state particularly when incorporated into a detergent product that is exposed to high humidity conditions (125). To minimize the loss of precursor over time the material can be incorporated into granules to minimize the surface area to volume ratio, which minimizes storage instability. A variety of granulation techniques have been described in the literature of which extrusion and agglomeration have been commercialized (126–138). A limited number of references also discuss the incorporation of a precursor into a liquid matrix that contains either hydrogen peroxide or an insoluble source of hydrogen peroxide such as perborate (139,140).

5.5. Preformed Peracids. Peracids can be generated at a manufacturing site and directly incorporated into formulations without the need for *in situ* generation. Two primary methods are utilized for peracid manufacture. The first method uses the equilibrium shown in equation 22 to generate the peracid from the parent acid.

$$\overset{\overset{O}{\|}}{R}COH + H_2O_2 \rightleftharpoons \overset{\overset{O}{\|}}{R}COOH + H_2O \tag{22}$$

The equilibrium is shifted by removal of the water (141) or removal of the peracid by precipitation (142,143). Peracids can also be generated by treatment of an anhydride with hydrogen peroxide to generate the peracid and a carboxylic acid.

$$\overset{\overset{O}{\|}}{R}CO\overset{\overset{O}{\|}}{C}R + H_2O_2 \longrightarrow \overset{\overset{O}{\|}}{R}COOH + \overset{\overset{O}{\|}}{R}COH \tag{23}$$

The latter method typically requires less severe conditions than the former because of the labile nature of the organic anhydride (94,144). Both of these reactions can result in explosions and significant precautions should be taken prior to any attempted synthesis of a peracid (94). For solid peracids the reaction mixture can be neutralized with sodium hydroxide and the resulting filtercake washed with water. In the case of the sulfuric acid mediated reaction, the peracid has sodium sulfate incorporated in the cake (142). The water of hydration present in the sodium sulfate is desirable to prevent detonation or deflagration of the solid peracid when isolated in a dry state (94,145,146).

The water of hydration, particularly that incorporated in sodium sulfate decahydrate, however, can cause instability when the peracid–sulfate blend is subjected to elevated temperatures. To mitigate this problem materials known as exotherm control agents have been incorporated into peracid formulas. These materials release their water only when conditions occur that could lead to detonation or deflagration. Magnesium sulfate heptahydrate, magnesium sulfate–sodium sulfate tetrahydrate, and boric acid each release water near 100°C (145,147). The release of the water prevents the propagation of the decomposition of the peracid by removing heat via evaporation (147). These materials are incorporated into the peracid composition at concentrations up to equal weight by weight to the peracid.

The peracid–exotherm control agent mixtures can be granulated using a variety of techniques common in the industry, including agglomeration. As with peracid precursors, the surface area to volume ratio can impact the stability of the peracid. Particles are thus made as large as possible to maintain stability (148).

Two solid organic peracids have been utilized in textile bleaching products. Diperoxydodecanedioic acid, 16, [66280-55-5], a hydrotropic peracid, and the magnesium salt [78948-87-5] of monoperoxyphthalic acid, 17, [2311-91-3], a hydrophilic peracid, were contained in bleaching products for a short period of time (149).

$$HOOC(CH_2)_{10}COOH$$

(16)   (17)

Peracids are also available as aqueous solutions that contain the peracid in equilibrium with hydrogen peroxide and the parent acid. Peracetic acid [79-21-0] is commercially available as a 40% solution in dilute acetic acid. The water and dilution of the peracid make these solutions easier to handle than their solid counterparts, but they still require careful handling and protection from heat.

Peracid Decomposition. Peracids, whether preformed or formed *in situ* via the perhydrolysis reaction, are susceptible to decomposition in an aqueous bleaching bath. The decomposition is caused by the occurrence of one of four reactions. The peracid can decompose as a result of oxidation of the bleachable material. Transition metals present even at extremely low concentration in the bath from the incoming water can decompose the peracid catalytically (150,151). To minimize this effect, metal-sequestering agents have been proposed to prevent the degradation of the peracid in solution (150,151). Peracids can also hydrolyze to the parent acid and hydrogen peroxide because of the large excess of water present in the aqueous bleaching bath. This is generally a kinetically slow process (94). A final decomposition mechanism involves the reaction of 2 mol of peracid generating 2 mol of parent acid and 1 mol of oxygen.

$$RCOOH + RCOO^- \longrightarrow RC{-}OO{-}H \longrightarrow RCO^- + RCOH + O_2 \tag{24}$$

The reaction involves the nucleophilic attack of a peracid anion on the unionized peracid giving a tetrahedral diperoxy intermediate that then eliminates oxygen giving the parent acids. The observed rate of the reaction depends on the initial concentration of the peracid as expected in a second-order process. The reaction also depends on the structure of the peracid (specifically whether the peracid can micellize) (9). Micellization increases the effective second-order concentration of the peracid because of the proximity of one peracid to another. This effect can be mitigated by the addition of an appropriate surfactant, which when incorporated into the peracid micelle, effectively dilutes the peracid, reducing the rate of decomposition (9,97).

6. Reducing Bleaches

The reducing agents generally used in bleaching include sulfur dioxide, sulfurous acid, bisulfites, sulfites, hydrosulfites (dithionites), sodium sulfoxylate formaldehyde, and sodium borohydride. These materials are used mainly in pulp and textile bleaching (see SULFUR COMPOUNDS; BORON COMPOUNDS).

6.1. Sulfur Dioxide, Sulfites, and Bisulfites. Sulfur dioxide [7446-09-5] and its derivatives have been used to bleach textiles since earliest times. Sulfur dioxide is a gas formed by burning sulfur in air. Besides being an important bleaching agent in the pulp and paper industry, sulfur dioxide is also integral to some processes for chlorine dioxide, sodium hydrosulfite, and sodium sulfite. Sulfur dioxide is used in both Kraft and mechanical pulp processes, and it is unique in that its bleaching effect is independent of pH over the range 3–10 (152). When SO_2 is dissolved in water, it yields a complex mixture given the trivial name sulfurous acid [7782-99-2] (H_2SO_3), which contains SO_2, H_3O^+, $S_2O_5^{2-}$, and HSO_3^-. The composition of the mixture depends on the concentration of the sulfur dioxide in the water, the pH, and the temperature (153). Although sulfurous acid does not exist in the free state, it forms stable salts (the neutral sulfite, SO_3^{2-}, and the hydrogen sulfite or bisulfite HSO_3^-) which are good reducing agents (eqs. 25 and 26).

$$2\,SO_3^{2-} + O_2 \longrightarrow 2\,SO_4^{2-} \tag{25}$$

$$HSO_3^- + Cl_2 + 2\,OH^- \longrightarrow HSO_4^- + 2\,Cl^- + H_2O \tag{26}$$

Sodium sulfite [7757-83-7], which is used in pulp and paper bleaching, is usually produced by the reaction of sulfur dioxide with either caustic soda or soda ash.

$$SO_2 + 2\,NaOH \longrightarrow Na_2SO_3 + H_2O \qquad (27)$$

$$SO_2 + Na_2CO_3 \longrightarrow Na_2SO_3 + CO_2 \qquad (28)$$

Dithionites.   Although the free-dithionous acid, $H_2S_2O_4$, has never been isolated, the salts of the acid, in particular zinc [7779-86-4] and sodium dithionite [7775-14-6] have been prepared and are widely used as industrial reducing agents. The dithionite salts can be prepared by the reaction of sodium formate with sodium hydroxide and sulfur dioxide or by the reduction of sulfites, bisulfites, and sulfur dioxide with metallic substances such as zinc, iron, or zinc or sodium amalgams, or by electrolytic reduction (154).

$$2\,HSO_3^- + SO_2 + Zn \longrightarrow ZnSO_3\downarrow + S_2O^{2-}{}_4 + H_2O \qquad (29)$$

Aqueous solutions of dithionite are not stable in the presence of oxygen, low pH, or elevated temperatures. The decomposition of dithionite occurs by the following equation:

$$2\,S_2O^{2-}{}_4 + H_2O \longrightarrow 2\,HSO_3^- + S_2O_3^{2-} \qquad (30)$$

Both sodium dithionite and zinc dithionite are produced commercially, though the uses of this latter salt have declined because of the regulatory constraints on pollution of water by zinc. The zinc salt is used under those conditions of pH and temperature where the sodium dithionite would be unstable. The principal applications of these compounds are in bleaching of mechanical pulp and in dyeing, printing, and stripping in the textile industry. A derivative of sodium dithionite is sodium sulfoxylate formaldehyde, which is prepared by the reaction of formaldehyde with the dithionite. Its applications are like those of dithionite except that it is less reactive and more stable thermally. When the sulfoxylate is used, a pH range of 3.2–3.5 produces the best results. For both the dithionite salts and sulfoxylate, the higher the temperature, the greater the reducing strength. The sulfoxylates can be used at temperatures as high as 100°C.

The principal bleaching applications of sodium dithionite are in the bleaching of mechanical or CTM pulps and the bleaching of kaolin clays for use as filler for fine paper. Other applications include the bleaching of glues, gelatin, soap, and food products. A significant new application for dithionite is in the bleaching of recycled paper.

7. Enzymes for Bleaching in the Textile Industry

7.1. Bleaching of Textile Substrates.   Enzymes are used for bleaching purposes in the textile industry for decolorization of dyehouse effluents, bleaching of released dyestuff, and inhibiting dye transfer (155–158). However, only a few enzymes, ie, xylanses, cellulases, and proteases (159–169), exert some

bleaching effect on the textile substrate itself. Enzymatic systems suitable for bleaching of textile materials are further classified as: enzymes for production of the bleaching agent and enzymes acting directly on the textile substrate.

Biogeneration of Hydrogen Peroxide for Bleaching. Hydrogen Peroxide Generation with Free Glucose Oxidase. The most common bleaching agent, hydrogen peroxide, may be produced enzymatically by glucose oxidase (EC 1.1.34), which catalyzes the conversion of β-D-glucose in aqueous solutions in the presence of oxygen as electron acceptor (170). Similar enzymatic systems are used as constituents in detergent formulations to generate controlled rates of hydrogen peroxide (171,172).

The gluconic acid formed in the enzymatic reaction acts as a chelator for metal ions, thus, addition of peroxide stabilizing agents may be avoided. Activation of the enzymatically produced peroxide is performed at elevated temperature and in the alkaline medium. Aeration is crucial for the production of hydrogen peroxide as it increases considerably the activity of glucose oxidase. However, enzyme concentration has to be optimized, as higher concentrations speed up reaction rate; but at additional costs.

One of the drawbacks of biobleaching is inactivation of free enzymes, which is caused by foam formation in the aerated solution, by shear forces acting on the enzyme molecule during stirring (173) or by high concentration of hydrogen peroxide formed during the reaction. Another problem is that the amount of enzymatically produced peroxide that is required is twofold that of the standard bleaching amount (174). This might be due to the added glucose, which is assumed to be acting as a stabilizer or as a substrate for oxidation, thus competing with the cotton fiber and lowering the bleaching rate. Another possible reason for the increased required amount of peroxide is the elevated process temperature this temperature cause protein denaturation and facilitates hydrophobic fabric–enzyme interaction (175), thereby preventing effective bleacing rates. The latter problems are avoided by using immobilized instead of free enzyme.

An example for a process with free glucose oxidase is a closed-loop enzymatic desizing–scouring–bleaching process. Using residual starch desizing baths containing glucose as an additional source for substrate in the enzymatic peroxide generation is an economically attractive option. Advantageously, the entire process is carried out in the residual desizing bath, so no fresh water is added, this results in considerable water savings (174). The combined desizing and bleaching process is suitable when there is no need for very high level of whiteness, since the high contamination of the reused desizing–scouring–bleaching liquor decreases the bleaching effect of the peroxide.

Hydrogen Peroxide Generation with Immobilized Glucose Oxidase. One reason that a bleaching process using enzymatically produced peroxide is not yet commercial is that the enzymes are still quite expensive. Immobilization of the enzymes provides long-term application of the enzyme at lower process cost. Immobilization accounts for easy recovery and recycling, reudced enzyme dosage, and continuous operations.

Enzymes have been immobilized on various insoluble carriers for a large number of applications inlcuding research, diagnostics, food, pharmaceutical, medical, and industrial uses. The methods of immobilization include adsorption, ionic, and covalent binding, which is most appropriate when the enzymes are used under extreme conditions when high stability is required. When choosing a support for enzyme immobilization for industrial application, the most important criteria are the stability of the carrier and the cost.

Glucose oxidases have been immobilized on polyethylene-*g*-acrylic acids graft copolymer membranes (176), silicone supports (177) silk fibroin membranes (178) activated carbon, glass, collagen, polycarbonate, polyurethane, polypyrrole films, and cellulose (179). The main application is as electrochemical biosensors (180). For textile bleaching purposes (181) glucose oxidase is covalently immobilized on an inexpensive, commercially available porous alumina carrier using glutaraldehyde as cross-linking agent. Alumina is recommended as carrier in the pH range of pH 5–11 (182). The pH range provides the slightly acidic environment required for the enzymatic reaction of glucose oxidase and glucose.

Immobilized glucose oxidase shows good operational stability in the textile bleaching process, and is reusable for at least three cycles without significant loss of activity. Sufficient amount of peroxide for bleaching can be produced with relatively low enzyme concentration. As a whole, the rate of peroxide generation from immobilized glucose oxidase is nearly twofold lower compared with the free enzyme. Immobilized enzyme increases slightly the level of whiteness of the fabrics bleached compared with the free enzyme, by eliminating the effect of the protein in the bleaching liquor. The recycling capacity of the enzyme might be improved by appropriate stabilization techniques. A prolonged storage, however, has a clear adverse influence on the enzyme activity.

When using immobilized glucose oxidase higher enzyme loadings on the support do not produce higher levels of hydrogen peroxide (183,184), as the amount of bound enzyme does not necessarily correlate with the recovered activity (185). Binding the glucose oxidase onto the support results normally in significant alterations of enzyme conformation and microenvironment (186). Moreover, fluoresence microscopy studies show (187) that the immobilized enzyme is in the outer shell of the alumina carrier pellets, and that not all the pore volume is efficiently accessible to immobilization. Thus, diffusion constraint, which is known as one of the most dominant effects, influences the performance of the immobilized enzyme (188).

Bleaching of Textiles Chloroperoxidases. The haloperoxidases (EC 1.11.1.10) form a class of heme-containing enzymes capable of oxidizing halides in the presence of hydrogen peroxide to the corresponding hypohalous acid (189–193). If an appropriate nucleophile is present, a reaction occurs with formation of hypohalous acid, whereby bleaching takes place at lower than the conventional bleaching temperature and pH. This type of bleaching reduces damage to fiber and avoids the use of bleaching auxiliaries. A recent patent reported the use of haloperoxidases for textile treatment (194).

The mechanism of chloroperoxidase catalyzed chlorination reactions is considerably controversial. Some reports favor the direct transfer of the chlorine atom from the enzyme to the substrate (191), others insist on the involvement of free oxidized chlorine intermediate Cl_2 in a reaction catalyzed by chloroperoxidases (195). The differenece between these two reaction pathways is based on the rate of oxidation of the chloride to its respective molecular species and the rate of enzymatic chloination of the substrates (191). In a haloperoxidase-catalyzed halogenation reactions both hypohalous acid intermediate and hydrogen peroxide are reactive toward the enzyme, and can cause significant inactivation at high concentrations (194). High halide concentration might also inhibit the enzyme.

Chloroperoxidases application alone cannot replace chemical bleaching (196). The enzymatic pretreatment might be an alternative to the repeated conventional bleaching process or a way to reduce significantly the peroxide dosage in subsequent chemical bleaching. However, this enzymatic system might face problems with environmental restrictions regarding the absorbable organic halogens by-products in industrial effluents.

Bleaching of Textiles with Laccases. Laccases (EC 1.10.3.2) are multi-copper enzymes that catalyze the oxidation of a wide range of inorganic substances using oxygen as an electron acceptor (197). The oxidation of a reducing substrate involves typically formation of a free (cation) radical after the transfer of a single electron to laccase (198). Laccases have found various biotechnological and environmental applications eg, removal of toxic compounds from polluted effluents through oxidative enzymatic coupling of the contaminants leading to insoluble complex structures, or as biosensors for phenols (199–204). Laccases have been extensively used in bleaching of craft pulp in delignification and demethylation (197,204–208). Capability of laccases to act on chromophore compounds suggested their application in industrial decolorization processes (155–157,208,209). However, these enzymes have not yet been used for bleaching of textiles despite promising experimental results. Apparently, laccase pretreatment alone does not improve the whiteness of the textile material; moreover, it generates color. However, after hydrogen peroxide bleaching, the whiteness of the enzymatically pretreated fabrics reaches whiteness index enhancement comparable with the whiteness achieved in two consecutive peroxide bleaching runs (210). It is well documented that laccase produces colored substances when suitable substrate is present (211). Although the nature of the coloring matter, in the case of cotton, is not fully characterized, it is believed to be related to nitrogen-free flavone pigments. These compounds are based on the flavonoid skeleton, which is a three-ring molecule, two of them aromatic, connected by a heterocyclic central ring. These pigments are normally removed only after oxidizing bleaching; however, they might be subjected to laccase-mediated oxidation. Pectin substances, remaining in the scoured cotton, might be a substrate for laccase as well. The mechanism of laccase bleaching action is not fully understood; however, experiments support the hypothesis that the enzyme transforms the cellulose coloring matter into other colored compounds, which are more easily oxidized by peroxide. The advantage of the enzymatic process over conventional bleaching is in terms of reduced time, energy, and chemicals consumption.

In a different process, the bleaching is carried out at elevated temperature (1–5 min at 100°C), after impregnation of the fabrics with laccase on foulard. These short-time pretreatments render the enzymatic approach suitable for continuous operations.

7.2. Enzymes for Treatment of Textile Bleaching Effluents.

Removal of Residual Peroxide in Bleaching Effluents with Free Catalases. The washing process after bleaching consumes large amounts of water, since any residual hydrogen peroxide has to be removed to avoid problems in subsequent dyeing processes. Minor modifications of the dye molecule can result in color loss (212,213). The demands to reduce water consumption by reducing or eliminating the washing cycle after bleaching while ensuring good reproducibility of dyeing can be met by application of catalases. Catalase is widely distributed enzyme in nature and well known for its ability to catalyze the conversion of hydrogen peroxide to water and gaseous oxygen. It has found numerous applications in food science, industrial food production, and medical any analytical fields (214,215). Commercial products containing catalase for textile applications are also available. These have been used to decompose residual hydrogen peroxide in fabric prior to dyeing, and are normally applied after draining the bleaching bath and refilling it with fresh water (216). A new, unconventional dyeing technique, ie, dyeing within the bleaching bath, is now implemented (217). In this technique, the bleaching bath, containing the fabric, is treated directly with catalase to destroy the residual hydrogen peroxide, and then being reused for reactive dyeing.

There are two particular limitations to this approach: the low of stability of the catalase at high temperature and pH, and the influence of the bleaching bath components on enzyme efficiency and on dye uptake. Though hydrogen peroxide is completely destroyed by the enzyme, the bleaching bath formulation caused unacceptable color changes of the dyed fabrics. Temperature increase causes significant changes in the structure of catalase, which is related to thermal unfolding and denaturing of the enzyme (175). Hence, color changes of the dyed textiles are attributed to the temperature-dependent dye–enzyme precipitation as well as to the bleaching bath composition. Bath liquors contain various substances that are extracted from the cotton, eg, oils and waxes, pectins, proteins, organic acids, mineral, and natural coloring matter. Optimizing the dyeing process, ie, dye, salt, alkali, and enzyme concentrations are the key to successful dyeing in the bleaching bath, and thereby to a considerable water and energy conservation gained by avoiding an extensive washing cycle after bleaching.

Removal of Residual Peroxide in Bleaching Effluents with Immobilized Catalases.   Major problems in use of catalases arise from the high temperature and alkalinity of the bleaching and washing liquors. The sensitivity of catalytically active protein structures to high temperatures, extreme pH, and other denaturing causes is one of the most important factors in the commercialization and industrial application of enzymes (218,219). In general, stabilization of the enzymes can be achieved in several ways: screening for stable ones such as thermophiles and extremophiles, chemical modification, protein engineering, immobilization, or stabilizing additives use (220–223). Interactions between dye and protein renders the use of soluble catalase inappropriate (175,224) alternatively, immobilized catalase can be used (219,225). Covalent binding of catalase to the support improves the resistance of the enzyme to inactivation presumably by restricting the protein unfolding process as a result of both intra- and intermolecular cross-linking.

Immobilized catalase is used for hydrogen peroxide degradation in column and tank reactor. Column reactor with substrate recirculation is more appropriate due to faster decompostion of the peroxide. For both types, reactor inhibition of the enzyme occurs when substrate concentrations exceeds 1 M. Dyeing in the recycled water provides a greater consistency of color than that obtained by using the free enzyme.

8. Economic Aspects

The chemicals used for bleaching have a variety of uses outside of bleaching technology. As a consequence, detailed information regarding production of these materials for bleach use is limited.

Sodium hypochlorite accounts for 92% of global use of the hypochlorite bleaches. Calcium hypochlorite accounts for the remaining 8%.

Residential use of sodium hypochlorite breaks down into the following categories: laundry bleaching (80%); sanitizers (18%); and residential pool and spa treatment (2%). Industrial uses are as follows: industrial and municipal water treatment (45%); commercial and municipal swimming pool treatment (33%); commercial laundry bleach (5%); liquid dishwashing detergent (5%); textile bleaching (4%), chemical (4%), and miscellaneous (4%).

Household demand in 2002 was 540×10^6 gal, projection for 2006 is 569×10^6 gal. Industrial demand in 2002 was 278×10^6 gal. The projected demand for 2002 is 292×10^6 gal. Growth is expected to continue at the rate of 1.3% through 2006 (226).

Calcium hypochlorite is used as a shock treatment in swimming pools. Shock treatments boost the chlorine levels in pools. Swimming pool treatment accounts for 75% of calcium hypochlorite use, the remaining is used for municipal and industrial water treatment.

Demand projected for 2003 is ~79,000 tons (227). Demand is expected to grow in the United States at the rate of 2–4%. Demand in Japan is expected to rise ~ 1% (228). Displacement of calcium hypochlorite by chlorinated isocyanurates has leveled off. Consumption of the chlorinated isocyanurates in 2000 was 180,000 t (229).

9. Health and Safety Factors

Much new information regarding the toxicities of sodium chlorite and sodium hypoclorite has become available, primarily as a result of safety concerns about chlorinated drinking water. In general, human population studies and animal bioassays have not found an association between exposure to these compounds and an increased risk of cancer, reproductive, or teratogenic effects.

The International Agency for Research on Cancer has concluded that there is inadequate evidence for the carcinogenicity of sodium hypochlorite in animals, and that sodium hypochlorite is not classifiable as to its carcinogenicity in humans (group 3).

The current OSHA PEL for perchlorates as nuisance dust is 15 mg/m^3. The California groundwater standard is 18 µg/L (230).

Hydrogen peroxide is a confirmed carcinogen and is moderately toxic by inhalation, ingestion, and skin contact. OSHA PEL TWA is ppm no as is the ACG IH TLV (231).

Sulfur dioxide OSHA PEL TWA and ACGIH TLV TWA are both 2ppm, STEL. ACGIH notes not classifiable as a human carcinogen (231).

10. Uses

10.1. Laundering and Cleaning. *Home and Institutional Laundering.* The most widely used bleach in the United States is liquid chlorine bleach, an alkaline aqueous solution of sodium hypochlorite. This bleach is highly effective at whitening fabrics and also provides germicidal activity at usage concentrations. Liquid chlorine bleach is sold as a 5.25% solution and 1 cup provides 200 ppm of available chlorine in the wash. Liquid chlorine bleaches are not suitable for use on all fabrics. Dry and liquid bleaches that deliver hydrogen peroxide to the wash are used to enhance cleaning on fabrics. They are less efficacious than chlorine bleaches but are safe to use on more fabrics. The dry bleaches typically contain sodium perborate in an alkaline base whereas the liquid peroxide bleaches contain hydrogen peroxide in an acidic solution. Detergents containing sodium perborate tetrahydrate are also available.

The worldwide decreasing wash temperatures, which decrease the effectiveness of hydrogen peroxide based bleaches, have stimulated research to identify activators to improve bleaching effectiveness. Tetraacetylethylenediamine is widely used in European detergents to compensate for the trend to use lower wash temperatures. TAED generates peracetic acid in the wash in combination with hydrogen peroxide. TAED has not been utilized in the United States where one activator nonanoyloxybenzene sulfonate (NOBS) has been commercialized and incorporated into several detergent products. NOBS produces pernonanoic acid when combined with hydrogen peroxide in the washwater and is claimed to provide superior cleaning to perborate bleaches.

In industrial and institutional bleaching either liquid or dry chlorine bleaches are used because of their effectiveness, low cost, and germicidal properties. Dry chlorine bleaches, particularly formulated chloroisocyanurates, are used in institutional laundries.

Hard Surface Cleaners and Cleansers. Bleaching agents are used in hard surface cleaners to remove stains caused by mildew, foods, etc, and to disinfect surfaces. Disinfection is especially important for many industrial uses. Alkaline solutions of 1–5% sodium hypochlorite that may contain surfactants and other auxiliaries are most often used for these purposes. These are sometimes thickened to increase contact times with vertical surfaces. A thick, alkaline cleaner with 5% hydrogen peroxide is also sold in Europe. Liquid abrasive cleansers with suspended solid abrasives are also available and contain ∼1% sodium hypochlorite. Powdered cleansers often contain 0.1–1% available chlorine and they may contain abrasives. Sodium dichloroisocyanurate is the most common bleach used in powdered cleansers, having largely replaced chlorinated trisodium phosphate. Calcium hypochlorite is also used. Dichloroisocyanurates are also used in effervescent tablets that dissolve quickly to make cleaning solutions. In-tank toilet cleaners use calcium hypochlorite, dichloroisocyanurates, or *N*-chlorosuccinimide to release hypochlorite with each flush to prevent stains from forming. One powdered toilet bowl cleaner uses potassium peroxymonosulfate and sodium chloride to generate hypochlorite in *in situ*.

Automatic Dishwashing and Warewashing. The primary role of bleach in automatic dishwashing and warewashing is to reduce spotting and filming by breaking down and removing the last traces of adsorbed soils. They also remove various food stains such as tea. All automatic dishwashing and warewashing detergents contain alkaline metal salts or hydroxides. Liquids, gels, and slurries contain 1–3% sodium hypochlorite. Powders and tablets almost always use 1–4% sodium dichloroisocyanurate dihydrate. Trichloroisocyanurate and the once popular chlorinated trisodium phosphate are also used. A few powders use sodium percarbonate or sodium perborate. They are less effective than chlorine bleaches, but this is largely overcome by increased amounts of alkaline metal salts or hydroxides. Enzymes (qv) also work with peroxygen bleaches but are deactivated by chlorine bleaches. Sodium hypochlorite or chloramine T are also used as sanitizers in the last rinse of low temperature warewashing.

10.2. Textile Bleaching. Many textiles are bleached to remove any remaining soil and colored compounds before dyeing and finishing (see TEXTILES). Bleaching is usually preceded by washing in hot alkali to remove most of the impurities in a process called scouring. Bleaching is usually done as part of a continuous process, but batch processes are still used. Not all fabrics are bleached, but natural fibers and their blends usually are. To minimize fiber damage, a minimum of bleaching agent is used. Making white and lightly colored fabrics requires the most bleaching. Bleaching conditions vary widely, depending on the equipment, the bleaching agent, the type of fiber, and the amount of whiteness required for the end use (67,68,232–235).

Cotton and Cotton–Polyester. Cotton is the principal fiber bleached today, and almost all cotton is bleached. About 80–90% of all cotton and cotton–polyester fabric is bleached with hydrogen peroxide. With hydrogen peroxide the fabric does not need to be scoured before bleaching and there is little risk of overbleaching. Typically, bleaching with 0.3–0.6% hydrogen peroxide solutions at pH 10.5–11.5 is done for 1–3 h at 90–95°C. The time can be reduced to 2–15 min by increasing the severity of the scouring step, impregnating the fabric with larger amounts of hydrogen peroxide, and using steam to attain temperatures of 95–100°C. In pressurized vessels at 130–145°C the time is reduced to 0.75–2 min. In order to save energy, some plants combine scouring and bleaching with hydrogen peroxide into a single step. Other plants use cold bleaching in which the textiles are scoured, bleached with sodium hypochlorite, and then bleached with hydrogen peroxide at room temperature for 4–5 h with a catalyst (236), or for 12–16 h without a catalyst. With all processes, bleaching with hypochlorite or chlorine dioxide may precede peroxide bleaching when white or lightly colored textiles are desired.

In the past, sodium hypochlorite solution (called chemic by textile workers) was the most commonly used bleach. Some is still used today because of its lower cost and better whitening ability than hydrogen peroxide. However, bleaching with hypochlorite needs to be carefully controlled to prevent fiber damage. Also, when only hypochlorite bleach is used, the fabrics need to be well scoured first to remove soils that consume hypochlorite. Otherwise higher hypochlorite concentrations and longer bleaching times are needed, which increase the risk of fabric damage. Protein soils may also form chloramines that cause color reversion. Solutions of 0.1–0.5% sodium hypochlorite are typically used at pH 10–11.5 for 0.5–4 h at 20°C or for 15–30 min at 40–50°C. The fabric is then bleached with hydrogen peroxide, or it is washed in a solution of a reducing agent (antichlor), such as bisulfite or sulfur dioxide, to remove residual hypochlorite and chloramines.

Sodium chlorite is also used to bleach some cotton and cotton–polyester fabrics. Unlike peroxide and hypochlorite, which only whiten the cotton portion of cotton–polyester blends, sodium chlorite also whitens polyester. However, the polyester portion usually does not need to be whitened. Sodium chlorite whitens better than hydrogen peroxide, does not damage fibers, and can be used with unscoured fabrics. However, it is more expensive, more corrosive to metals, and more hazardous than peroxide or hypochlorite. Typically, solutions of 0.1–3% sodium chlorite with sufficient dihydrogen phosphate and formic acid to give pH 3.8–4.2 are used at 80–95°C for 1–6 h. In a room temperature process, fabric is treated overnight with a neutral solution of sodium chlorite that is activated by formaldehyde. Minor bleaching agents that are used in a manner similar to hydrogen peroxide are sodium peroxide [1313-60-6], sodium perborate, and sodium percarbonate [20745-24-8]. Perborate or percarbonate are most frequently used as additives to scouring solutions in place of a separate bleaching step when a fully whitened fabric is not needed.

Other Cellulosics. Rayon is bleached similarly to cotton but under milder conditions since the fibers are more easily damaged and since there is less colored material to bleach. Cellulose acetate and triacetate are not usually bleached. They can be bleached like rayon, except a slightly lower pH is used to prevent hydrolysis. The above fibers are most commonly bleached with hydrogen peroxide. Linen, flax, and jute require more bleaching and milder conditions than cotton, so multiple steps are usually used. Commonly, an acidic or neutral hypochlorite solution is followed by alkaline hypochlorite, peroxide, chlorite, or permanganate, or a chlorite step is done between two peroxide steps. A one-step process with sodium chlorite and hydrogen peroxide is also used.

Synthetic Fibers. Most synthetic fibers are sufficiently white and do not require bleaching. For white fabrics, unbleached synthetic fibers with fluorescent whitening agents are usually used. When needed, synthetic fibers and many of their blends are bleached with sodium chlorite solutions at pH 2.5–4.5 for 30–90 min at concentrations and temperatures that depend on the type of fiber. Solutions of 0.1% peracetic acid are also used at pH 6–7 for 1 h at 80–85°C to bleach nylon.

Wool and Silk. Wool must be carefully bleached to avoid fiber damage. It is usually bleached with 1–5% hydrogen peroxide solutions at pH 8–9 for several hours at 40–55°C or at pH 5.5–8 for 20–60 min at 70–80°C. Silk is bleached similarly, but at slightly higher temperatures.

Wool with dark pigmented fibers is treated with ferrous sulfate, sodium dithionite, and formaldehyde before it is bleached with hydrogen peroxide. The ferrous ions are absorbed by the dark pigments where they increase the bleaching done by the peroxide.

Wool may also be bleached with reducing agents, usually after bleaching with hydrogen peroxide. This is the normal practice with wool blends. In the reducing step, 0.2–0.5% sodium dithionite solutions are often used at pH 5.5–7 for 1–2 h at 45–65°C. Faster bleaching is obtained with zinc hydroxymethane-sulfinate[24887-06-7] below pH 3 and >80°C.

The ancient process of stoving is still occasionally used to bleach wool and silk with sulfur dioxide. In this process, wet fabrics are hung in chambers of burning sulfur or sulfur dioxide gas for at least 8 h. The fabrics are then washed with sodium sulfite to remove excess sulfur dioxide. Fabric so treated may have unpleasant odors, and the original color eventually returns, but the process is simple and inexpensive.

10.3. Bleaching of Other Materials. *Hair.* Hydrogen peroxide is the most satisfactory bleaching agent for human hair. Solutions containing 3–4% hydrogen peroxide, available from drug stores and supermarkets, are commonly used. In beauty shops, more rapid bleaching is desired and a 5–6% solution is used. Ammonium hydroxide is usually the source of alkalinity in both systems (see COSMETICS; HAIR PREPARATIONS).

Fur. Fur is bleached to permit dyeing to the desired shade. The coloring matter in fur is usually bleached using hydrogen peroxide stabilized with sodium silicate. For difficult to bleach dark hairs it is necessary to add a step using a reducing agent with a catalyst such as ferrous sulfate. The formula and procedures are the same as those used for wool.

Foodstuffs, Oils. Sulfur dioxide is used to preserve grapes, wine (qv), and apples; the process also results in a lighter color. During the refining of sugar (qv), sulfur dioxide is added to remove the last traces of color. Flour can be bleached with a variety of chemicals including chlorine, chlorine dioxide, oxides of nitrogen, and benzoyl peroxide [94-36-0]. Bleaching agents such as chlorine dioxide or sodium dichromate [10588-01-9] are used in the processing of non-edible fats and fatty oils for the oxidation of pigments to colorless forms (see FOOD PROCESSING).

11. General Consideration for Application of Enzymes

There are two applications of the biocatalysts in textile bleaching operations, ie, bleaching of the textile material itself and treating the bleaching effluents. For substrate in solution, the enzymes may be applied either in free or immobilized form.

Production of high quality textile materials should be carried out through processes that are low in bath volume, reagents dosage, as well as short in processing time. The feasibility of an enzymatic laboratory-scale process for industrial application depends on the specific technology cycle adopted in the textile plant and on the available equipment. The guidelines to set the operating conditions for enzymes application can be obtaines from pH and temperature profiles of enzymes activity. For practical application of biocatalysts it is important to know the rate of acivity decay during process. The activity of enzmes is determined normally for soluble substrates in a homogenous catalysis reaction media, which differs significantly from textile-industry conditions (in the textile industry). In the textile industry, the substrate is insoluble and the wet processes are caried out at high turbulence and mechanical agitation in the textile baths.

Industrial implementation of new biotechnological processes and products in the bleaching stage of textiles not only will replace hazardous chemicals by biodegradable, naturally based products, but will also respond to the contemporary social needs by contributing to the quality of life, health, and safety of citizens of involved communities.

BIBLIOGRAPHY

"Bleaching Agents" in *ECT* 2nd ed., Vol. 3, pp. 550–567, by H. L. Robson, Olin Mathieson Chemical Corp.; in *ECT* 3rd ed., Vol. 3, pp. 938–958, by B. M. Baum, and co-workers, FMC Corp.; in *ECT* 4th ed., Vol. 4, pp. 271–300, by James P. Farr, William L. Smith, and Dale S. Steichen, The Clorox Company, 1–5 on p. 52; "Bleaching Agents, Survey" in *ECT* (online), posting date: December 4, 2000, by James P. Farr, William L. Smith, and Dale S. Steichen, The Clorox Company.

CITED PUBLICATIONS

1. S. H. Batra, in M. Lewin and E. M. Pearce, eds., *Handbook of Fiber Service and technology*, Vol. 4, Marcel Dekker, New York, 1985.
2. J. N. Etters, P. A. Husain, and N.K. Lange, *Textile Asia* **5**, 83 (1999).
3. Y. Hsieh, J. Thompson, and A. Miller, *Textile Res. J.* **66**, 456 (1996).
4. M. Weck, *Text. Praxis Int.* **2**, 144 (1991).
5. M. C. Spirro, and P. W. Criffith, *Text. Chem. Color.* **29**, 11, 12 (1997).
6. B. K. Easton, *CIBA Geigy Rev.* **3**, 3 (1971).
7. S. H. Higgins, *A History of Bleaching*, Longmans, Green and Co.,1924, p. 10.
8. U.S. Pats. 2,353,615 (July 11, 1944), H. O. Kauffmann and E. S. Shanley (to Buffalo Electro Chemical Co., Inc.); 2,391,905 (June 1, 1946), H. O. Kauffmann, E. S. Shanley, and R. L. McEwen (to Buffalo Electro Chemical Co., Inc.).
9. D. S. Steichen, R. A. Fong, R. J. Wiersema, and S. B. Kong, *Laundry Bleaches: History and New Frontiers*, paper presented at American Oil Chemists' Society Meeting, New Orleans, La., 1987.
10. *Soap Cosmet. Chem. Spec.* **63**, 73 (Nov. 1987).
11. D. H. Rosenblatt, in J. D. Johnson, ed., *Disinfection Water and Wastewater*, Ann Arbor Science Publishers Inc., Ann Arbor, Mich.,1975, pp. 249–276.
12. G. Gordon, R. G. Kieffer, and D. H. Rosenblatt, in S. J. Lippard, ed., *Progress in Inorganic Chemistry*, Vol. 15, John Wiley & Sons, Inc., New York, 1972, p. 202.
13. W. J. Masschelein and R. G. Rice, *Chlorine Dioxide Chemistry and Environmental Impact of Oxychlorine Compounds*, Ann Arbor Science Publishers Inc., Ann Arbor, Mich., 1979, pp. 59–87.
14. H. A. Ghanbari, W. B. Wheeler, and J. R. Kirk, in R. L. Jolley and co-eds., *Water Chlorination: Environmental Impact and Health Effects*, Vol. 4, Book 1, Ann Arbor Science Publishers Inc., Ann Arbor, Mich., 1981, pp. 167–177.
15. E. R. Trotman, *Dyeing and Chemical Technology of Textile Fibres*,Wiley-Interscience, New York, 1984, pp. 193–200.
16. J. Cegarra and J. Galen, *Wool Sci. Rev.* **59**, 6–7 (1983).
17. R. Fong, D. Steichen, and R. Wiersema, *Recent Developments in Perborate Activation Chemistry*, paper presented at American Oil Chemists' Society Meeting, New Orleans, La., 1987.

18. A. J. Downs and C. J. Adams, *The Chemistry of Chlorine, Bromine, Iodine, and Astatine*, Pergamon Press, Oxford, U.K., 1973, pp. 1400–1410.
19. I. M. Kolthoff, E. B. Sandell, E. J. Meehan, and S. Bruckenstein, *Quantitative Chemical Analysis*, 4th ed., The Macmillan Co., New York, 1969, pp. 851, 852.
20. A. E. Greenberg, R. R. Trussell, L. S. Clesceri, and M. A. H. Franson, eds., *Standard Methods for the Examination of Water and Wastewater*, 16th ed., American Public Health Association, Washington, D.C., 1985, 294–315.
21. R. E. Connick and Y.-T. Chia, *J. Am. Chem. Soc.* **81**, 1280 (1959).
22. J. C. Morris, *J. Phys. Chem.* **70**, 3798 (1966).
23. U.S. Pat. 3,583,922 (June 8, 1971), H. K. McClain and L. E. Meyer (to The Procter & Gamble Co.).
24. U.S. Pat. 3,177,111 (Apr. 6, 1965), L. E. Larson (to Weyerhaeuser Co.).
25. U.S. Pat. 4,759,852 (July 26, 1988), M. G. Trulear (to Nalco Chemical Co.).
26. U.S. Pat. 4,148,742 (Apr. 10, 1979), M. M. Crutchfield, R. P. Langguth, and J. M. Mayer (to Monsanto Co.).
27. U.S. Pat. 4,187,293 (Feb. 5, 1980), G. D. Nelson (to Monsanto Co.).
28. U.S. Pat. 2,988,471 (June 13, 1961), R. J. Fuchs and I. A. Lichtman (to Food Machinery and Chemical Corp.).
29. U.S. Pat. 4,279,764 (July 21, 1981), G. R. Brubaker (to FMC Corp.).
30. U.S. Pat. 4,909,956 (Mar. 20, 1990), R. S. Webber (to Olin Corp.).
31. M. W. Lister and R. C. Petterson, *Can. J. Chem.* **40**, 729 (1962).
32. M. W. Lister, *Can. J. Chem.* **34**, 479 (1956).
33. E. T. Gray, R. W. Taylor, and D. W. Margerum, *Inorg. Chem.* **16**, 3047 (1977).
34. K. W. Young and A. J. Allmand, *Can. J. Res.* **27B**, 318 (1949).
35. G. V. Buxton and R. J. Williams, *Proc. Chem. Soc.*, 141 (1962).
36. K. P. Kringstad and K. Lindström, *Environ. Sci. Technol.* **18**, 236A (1984).
37. S. A. Heimburger, D. S. Blevins, J. H. Bostwick, and J. P. Bonni, *Tappi J.* **71**, 51 (Oct. 1988); **71**, 69 (Nov. 1988).
38. R. M. Berry and co-workers, *Pulp Pap. Can.* **92**, T155 (1991).
39. I. J. Bowen and J. C. L. Hsu, *Tappi J.* **73**, 205 (Sept. 1990).
40. B. Mildwidsky, *HAPPI*, 92 (Nov. 1988).
41. U.S. Pat. 4,849,201 (July 18, 1989), R. K. Smith, E. R. Zamejc, and J. F. Miller (to the U.S. Army).
42. U.S. Pat. 3,582,265 (June 1, 1971), J. J. Bishop and S. I. Trotz (to Olin Corp.).
43. U.S. Pat. 4,071,605 (Jan. 31, 1978), J. A. Wojtowicz (to Olin Corp.).
44. R. B. Ellestad, *Soap Chem. Spectrosc.* **37**, 77 (Sept. 1961).
45. U.S. Pat. 4,584,178 (Apr. 22, 1986), R. E. Yant and R. J. Galluch(to Quantum Technologies Inc.).
46. *OXONE Monopersulfate Compound*, E. I. du Pont de Nemours & Co., Inc., Wilmington, Del., 1984.
47. W. Adam, R. Curci, and J. O. Edwards, *Acc. Chem. Res.* **22**, 205 (1989).
48. U.S. Pat. 2,989,519 (June 20, 1961), J. H. E. Herbst and H. A. Kraessig (to Canadian International Paper Co.).
49. U.S. Pat. 2,815,311 (Dec. 3, 1957), J. G. Ellis and V. Dvorkovitz(to The Diversey Corp.).
50. *Industrial Uses of ACL Chlorinated s-Trazine Triones*, Monsanto Co., St. Louis, Mo., 1979.
51. U.S. Pat. 4,820,437 (Apr. 11, 1989),Y. Akabane,T. Tamura, and M. Fujiwara (to Lion Corp.).
52. Eur. Pat. Appl. 315,204 (May 10, 1989), S. Nishida, T. Tamura, and T. Toda (to Lion Corp.).
53. Jpn. Pat. 63,161,088 (May 7, 1988) (to Lion Corp.).

54. Ger. Pat. 159,180 (Feb. 23, 1983), K. R. Hepke, B. Liedloff, and D. Gisbier (to VEB Stickstoffwerk Piesteritz).
55. Brit. Pat. 831,853 (Apr. 6, 1960), F. B. Slezak and I. Rosen (to Diamond Alkali Co.).
56. U.S. Pats. 2,796,321 and 2,796,322 (June 18, 1957), M. C. Taylor.
57. U.S. Pat. 4,201,687 (May 6, 1980), M. M. Crutchfield, R. P. Langguth, and J. M. Mayer (to Monsanto Co.).
58. Brit. Pat. 847,566 (Sept. 7, 1960), (to Thomas Hedley & Co., Ltd.).
59. U.S. Pat. 3,007,876 (Nov. 7, 1961), J. R. Schaeffer (to The Procter & Gamble Co.).
60. U.S. Pat. 2,684,924 (July 27, 1954), F. L. Rose and G. Swain (to Imperial Chemical Industries Ltd.).
61. U.S. Pat. 2,841,474 (July 1, 1958), R. R. Dorset (to Mangels Herold Co., Inc.).
62. U.S. Pat. 3,749,672 (July 31, 1973), W. C. Golton and A. F. Rutkiewic (to E. I. du Pont de Nemours & Co., Inc.).
63. Eur. Pat. Appl. 119,560 (Sept. 26, 1984), J. Schroeder and H. Stabl (to Intermedicat GmbH).
64. Jpn. Pat. 63,108,099 (May 12, 1988), Y. Sugahara, Y. Toma, and K. Yokoi (to Lion Corp.).
65. R. J. Gall, in R. G. Rice and J. A. Cotruvo, eds., *Ozone/Chlorine Dioxide Oxidation Products of Organic Materials*, Ozone Press International, Cleveland , Ohio, 1978, pp. 356–382.
66. W. J. Masschelein and R. G. Rice, *Chlorine Dioxide Chemistry and Environmental Impact of Oxychlorine Compounds*, Ann Arbor Science Publishers Inc., Ann Arbor, Mich., 1979, pp. 111–145.
67. M. Lewin, in M. Lewin and S. B. Sello, eds., *Hand book of Fiber Science and Technology: Vol. 1; Chemical Processing Fibers and Fabrics. Fundamentals and Preparation Part B*, Marcel Dekker Inc., New York, 1984, pp. 92–256.
68. E. R. Trotman, *Dyeing and Chemical Technology of Textile Fibres*, 6th ed., John Wiley & Sons, Inc., New York, 1984, pp. 187–217.
69. S. N. Lewis, in R. Augustine, ed., *Oxidation*, Marcel Dekker Inc., New York, 1969, 213–258.
70. *Hydrogen Peroxide Solutions, Storage and Handling*, E. I. du Pont de Nemours & Co., Inc., Wilmington, Del., 1983.
71. W. C. Schumb, L. N. Satterfield, and R. C. Wentworth, *Hydrogen Peroxide*, Reinhold Publishing Corp., New York, 1955.
72. F. A. Cotton and G. Wilkinson, *Advanced Inorganic Chemistry*, 5th ed.,Wiley-Interscience, New York, 1988, pp. 458, 459.
73. U.S. Pat. 3,591,341 (July 6, 1971), V. Reilly (to E. I. du Pont de Nemours & Co., Inc.).
74. U.S. Pat. 3,607,053 (July 6, 1971), V. Reilly (to E. I. du Pont de Nemours & Co., Inc.).
75. U.S. Pat. 3,681,022 (Aug. 1, 1972), W. Kibbel and E. O'Neil (to FMC Corp.).
76. U.S. Pat. 4,362,706 (Dec. 7, 1982), P. E. Willard (to FMC Corp.).
77. U.S. Pat. 4,534,945 (Aug. 13, 1985), Q. G. Hopkins and J. N. Browning (to FMC Corp.).
78. J. Falbe, ed., *Surfactants in Consumer Products*, Springer-Verlag, New York, 1987, pp. 265–273.
79. U.S. Pat. 4,536,183 (Aug. 20, 1985), J. Namnath (to Lever Brothers Inc.).
80. U.S. Pat. 4,539,132 (Sept. 3, 1985), J. Oakes (to Lever Brothers Inc.).
81. U.S. Pat. 4,626,373 (Dec. 2, 1986), T. D. Finch and R. J. Wilde (to Lever Brothers Inc.).
82. U.S. Pat. 4,430,243 (Feb. 7, 1984), C. D. Bragg (to The Procter & Gamble Co.).
83. U.S. Pat. 4,810,410 (Mar. 7, 1989), E. Diakun and C. T. Wright (to Interox Chemicals Ltd.).
84. A. Hansson, *Acta Chem. Scand.* **15**, 934 (1961).

85. J. Gunter and A. Lohr, *Detergents and Textile Washing*,VCH Publishers, New York, 1987, pp. 78, 79.

86. M. Walker, *Recent Advances in Detergent Industry*, Book of Abstracts, SCI, University of Cambridge, 1990, pp. 48–51.

87. A. R. Baldwin, ed., *Proceedings of Second World Conference on Detergents*, American Oil Chemist's Society, Champaign, Ill., 1987, 177–180.

88. J. O. Edwards, *Chem. Eng. News* **133**, 3336 (1955); *Chem. Eng. News* **138**, 59 (1960).

89. Brit. Pat. 847,702 (Sept. 14, 1960), J. C. Van Embden and J. Boldingh (to Unilever).

90. U.S. Pat. 3,075,921 (Jan. 29, 1963), P. Brocklehurst and P. Pengilly (to The Procter & Gamble Co.).

91. A. P. James and I. S. Mackirdy, in Ref. 86, pp. 52–55.

92. P. Jurges, J. Gethofer, and G. Reinhardtk, in Ref. 86, pp. 44–47.

93. U.S. Pat. 4,367,156 (Jan. 4, 1983), F. Diehl (to The Procter & Gamble Co.).

94. *Organic Peracids, Their Preparation, Properties, Reactions and Uses*,Technical bulletin, FMC Corporation Inorganic Chemicals Division, New York, 1964.

95. G. Greenspan and G. Mackellar, *Anal. Chem.* **20**, 1061 (1948).

96. U.S. Pat. 4,391,723 (July 5, 1983), D. Bacon and F. Bossu (to The Procter & Gamble Co.).

97. U.S. Pat. 4,655,781 (Apr. 7, 1987), J. Hsieh, S. Kong, D. Steichen, and H. Wheeler (to The Clorox Co.).

98. U.S. Pat. 4,412,934 (Nov. 1, 1983), G. Spandini and S. Chung (to The Procter & Gamble Co.).

99. J. Parker, *Chem. Times Trends* **8**, 23 (Oct. 1985).

100. P. Kuzel and T. Lieser, *Tenside Surfactants Deterg.* **27**, 23 (1990).

101. J. March, *Advanced Organic Chemistry*, 2nd ed., McGraw-Hill Book Co., Inc., New York, 1977.

102. J. Dean, ed., *Lange's Hand book of Chemistry*,12th ed., McGraw-Hill Book Co., Inc., New York, 1979.

103. J. K. Grime, A. Clauss, and K. Leslie, in Ref. 86, pp. 12–14.

104. U.S. Pat. 4,681,592 (July 21, 1987), F. Hardy and B. Ingram (to The Procter & Gamble Co.).

105. U.S. Pat. 4,686,061 (Aug. 11, 1987), A. Nollet, J. Meijer, and J. Overkamp (to AKZO Co.).

106. Eur. Pat. Appl. 267,046 (Nov. 5, 1988), A. Zielske (to The Clorox Co.).

107. U.S. Pat. 4,778,618 (Oct. 18, 1988), R. Fong, S. Lewis, R. Wiersema, and A. Zielske (to The Clorox Co.).

108. U.S. Pat. 4,751,015 (June 14, 1988), R. Humphreys and S. Madison (to Lever Brothers Inc.).

109. U.S. Pat. 4,686,063 (Aug. 11, 1987), M. Burns (to The Procter & Gamble Co.).

110. U.S. Pat. 4,859,800 (Aug. 22, 1989), A. Zielske and R. Fong (to The Clorox Co.).

111. U.S. Pat. 4,378,300 (Mar. 29, 1983), F. Gray (to Colgate Palmolive Co.).

112. U.S. Pat. 4,293,301 (Aug. 1, 1981), F. Diehl (to The Procter & Gamble Co.).

113. U.S. Pat. 4,486,327 (Dec. 4, 1984), A. Murphy, A. Kassamali, and J. Curry (to The Procter & Gamble Co.).

114. U.S. Pat. 3,061,550 (Oct. 30, 1962), M. Baevsky (to E. I. du Pont de Nemours & Co., Inc.).

115. U.S. Pat. 4,483,778 (Nov. 20, 1984), J. E. Thompson and C. D. Broaddus (to The Procter & Gamble Co.).

116. Eur. Pat. Appl. 300,462 (Jan. 25, 1989), C. Venturello and C. Cavaliotti (to Ausimont SpA).

117. U.S. Pat. 4,487,723 (Dec. 11, 1984), J. Mayer (to Monsanto Co.).

118. U.S. Pat. 4,536,314 (Aug. 20, 1985), F. E. Hardy, D. J. Kitko, and C. M. Cambre (to The Procter & Gamble Co.).

119. U.S. Pat. 4,606,838 (Aug. 19, 1986), M. E. Burns (to The Procter & Gamble Co.).

120. Eur. Pat. Appl. 284,292 (Sept. 28, 1988), M. Aoyagi and co-workers (to Kao Corp.).

121. U.S. Pat. 4,814,110 (Mar. 21, 1989), R. Fong and R. Rowland (to The Clorox Co.).

122. U.S. Pat. 4,790,952 (Dec. 13, 1988), D. Steichen, S. Lewis, and H. Ku (to The Clorox Co.).

123. U.S. Pat. 4,735,740 (Apr. 5, 1988), A. Zielske (to The Clorox Co.).

124. Eur. Pat. Appl. 185,522 (June 25, 1986), R. Fong and S. Kong (to The Clorox Co.).

125. U.S. Pat. 4,111,651 (Sept. 5, 1978), J. Blumbergs, J. Finley, and B. Baum (to FMC Corp.).

126. U.S. Pat. 4,444,674 (Apr. 24, 1984), I. Gray (to The Procter & Gamble Co.).

127. Eur. Pat. Appl. 106,634 (Apr. 25, 1984), S. Chung and G. Spandini (to The Procter & Gamble Co.).

128. Brit. Pat. 2,053,998 (Feb. 11, 1981), R. Foret and P. Van Deer Hoeven (to Unilever).

129. U.S. Pat. 4,290,903 (Sept. 22, 1981), N. Macgilp and D. Mann (to The Procter & Gamble Co.).

130. U.S. Pat. 4,372,868 (Feb. 8, 1983), H. Saran, M. Witthaus, E. Smulders, and H. Schwadtke (to Henkel KGaA).

131. U.S. Pat. 4,009,113 (Feb. 22, 1977), R. Green and R. Johnson (to Lever Brothers Inc.).

132. U.S. Pat. 3,925,234 (Dec. 9, 1975), K. Hachmann, H. Saran, and G. Sperling (to Henkel & Cie GmbH).

133. Brit. Pat. 2,178,075 (Feb. 4, 1987), E. Parformak and W. Uchiyama (to Colgate Palmolive Co.).

134. U.S. Pat. 4,003,841 (Jan. 18, 1977), K. Hachmann, R. Puchta, and G. Sperling (to Henkel & Cie GmbH).

135. U.S. Pat. 3,789,002 (Jan. 29, 1974), R. Weber and A. Opgenoorth (to Henkel & CIE GmbH).

136. U.S. Pat. 4,591,450 (May 27, 1986), U. Nistri, R. Baroffio, and P. Colombo (to Mira Lanza SpA).

137. U.S. Pat. 4,457,858 (July 3, 1984), H. Saran and M. Witthaus (to Henkel KGaA).

138. U.S. Pat. 4,422,950 (Dec. 27, 1983), H. Kemper and P. Versluis (to Lever Brothers Inc.).

139. U.S. Pat. 4,900,469 (Feb. 13, 1990), D. Carty and J. Farr (to The Clorox Co.).

140. U.S. Pat. 4,772,290 (Sept. 20, 1988), D. Carty, A. Zielske, and J. Mitchell (to The Clorox Co.).

141. U.S. Pat. 4,244,884 (Jan. 13, 1981), J. Hutchins and D. Winn (to The Procter & Gamble Co.).

142. U.S. Pat. 4,337,213 (June 29, 1982), C. Marynowski and M. Geigel (to The Clorox Co.).

143. U.S. Pat. 4,233,235 (Nov. 11, 1980), J. Camden and M. McCarty (to The Procter & Gamble Co.).

144. U.S. Pat. 4,659,519 (Apr. 21, 1987), H. Ku (to The Clorox Co.).

145. U.S. Pat. 4,100,095 (July 11, 1978), J. Hutchins, D. Julian, and M. Burns (to The Procter & Gamble Co.).

146. U.S. Pat. 3,494,787 (Feb. 10, 1970), J. Lund and D. Nielsen (to PPG Industries Inc.).

147. U.S. Pat. 4,865,759 (Sept. 12, 1989), T. Coyne and co-workers (to The Clorox Co.).

148. U.S. Pat. 4,225,451 (Sept. 30, 1980), J. McCrudden and A. Smith (to Interox Chemicals Ltd.).

149. J. Parker and O. Raney, "The Use of Peroxygen Compounds as a Means of Improving Detergent Performance," in C. Bapa, ed., *New Horizons 1989 An AOCS/CSMA Detergent Industry Conference*, Hershey, Pa., 1989.

150. U.S. Pat. 4,091,544 (May 30, 1978), J. Hutchins (to The Procter & Gamble Co.).
151. Eur. Pat. Appl. 373,743 (June 20, 1990), S. Bolkan and co-workers (to The Clorox Co.).
152. J. T. Burton, *Pulp Pap.*, 75–77 (July 1986).
153. K. A. Kolbe and K. C. Hellwig, *Ind. Eng. Chem.* **47**, 1116 (1955).
154. J. C. Bailar,H. J. Emeleus, R. Nyholm, and A. F. Trotman-Dickenson, eds., *Comprehensive Inorganic Chemistry*, Vol. 2, Pergamon Press Ltd., Oxford, UK, 1973.
155. E. Abadulla, T. Tzanov, S. Costa, K.-H. Robra, A. Cavaco-Paulo, and G. M. Gübitz, *Appl. Environ. Microbiology*, **66**, 3357 (2000).
156. M. Morita, R. Ito, T. Kamidate, and H. Watanabe, *Textile Res. J.* **66**, 470 (1996).
157. WO Pat. 9105839, (1991), T. Damhus, O. Kirk, G. Pedersen, and M. G. Venegas (Novo Nordisk a/s The Pocter & Gamble Company).
158. WO Pat. 9218687, (1992), G. Pedersen, and M. Schmidt (to Novo Nordisk).
159. A. B. Kundu, B. S. Ghosh,S. K. Chakrabarti, and B. L. Ghosh, *Textile Res. J.* **61**, 720 (1991).
160. A. B. Kundu, B. S. Ghosh, and S. K. Chakrabarti, *Textile Res. J.* **63**, 451 (1993).
161. G. Buschle-Dileer, Y. E. Mogahzy, M. K. Inglesby, and S. H. Zeronian, *Textile Res. J.* **68**, 920 (1998).
162. M. Traore, and G. Buschle-Diller, *Book of Paper, Int. AATCC Conference*, Charlotte, N.C., 183 (1999).
163. J. Buchert, J. Pere, A. Miettinen-Oinonen, A. Puolakka, P. Nousiainen, and L. Johansson, *Book of Papers, Int. AATCC Conference,*Charlotte, 479 (1999).
164. J. Buchert, J. Pere, A. Puolakka, and P. Nousiainen, *Text. Chem. Colorist* **32**, 48 (2000).
165. P. Husain, N. Lange, L. Henderson, J. Liu, and B. Condox, *Book of Papers, Int. AATCC Conference*, Charlotte, N.C., 170 (1999).
166. N. Lange, J. Liu, P. Husain, and B. Condox, *Book of Papers, Int. AATCC Conference*, Philadelphia, 463 (1998).
167. Y. Hsieh, and L. Cram, *Textile Res., J.* **69**, 590 (1999).
168. L. Yonghua, and I. Hardin, *Text. Chem. Colorist* **29**, 71 (1997).
169. U. Röbner, *Melliand Textilberichte* **74**, 144 (1993).
170. H. Schacht, W. Kesting, and E. Schollmeyer, *Textilveredlung* **30**, 237 (1995).
171. U.S. Pat. 5288746 (1994), K. Pramod.
172. WO Patent 98/28400, (1998), Van der Helm.
173. C. R. Thomas and P. Dunnill, *Biotechnol. Biochem.* **25**, 14 (1979).
174. T. Tzanov, M. Calafell, G. M. Guebitz, and A. Cavaco-Paulo, *Enzyme Microb. Technol.* **29**, 357 (2001).
175. T. Tzanov, S. Costa, G. M. Guebitz, and A. Cavaco-Paulo, *Color. Technol.* **117**, 28 (2001).
176. G. Hsiue, and C. Wang, *Biotechnol. Bioeng.* **36**, 811 (1990).
177. A. Subramanian, S. Kennel, P. Oden, B. K. Jacobson, J. Woodward, and M. J. Doktycz, *Enzyme Microb. Technol.* **24**, 26 (1999).
178. T. Asakura, H. Yoshimizu, A. Kuzuhara, and T. Matsunaga, *J. Seric. Sci. Jpn.* **57**, 203 (1988).
179. R. F. Taylor, *Anal. Chim. Acta* **172**, 241 (1985).
180. R. Wilson, and A. P. F. Turner, *Biosens. Bioelectr.* **7**, 165 (1992).
181. T. Tzanov, S. Costa, G. M. Gübitz, and A. Cavaco-Paulo *J. of Biotechnology*, **93** (1), 87 (2001).
182. R. A. Messing, *Methods Enzymol.* **44**, 148 (1976).
183. P. C. Oliveira, G. M. Alves, and H. F. De Castro, *Biochem. Eng. J.* **5**, 63 (2000).
184. J. A. Bosley, and A. D. Peilow, *J. Am. Oil Chem. Soc.* **74**, 107 (1997).
185. R. F. Taylor, *Protein Immobilization. Fundamentals and Applications*, Marcel Dekker, Inc., New York, 1991.

186. C. M. F. Soares, H. F. De Castro, F. F. De Moraes, and G. M. Zanin, *Appl. Biochem. Biotechnol.* **77–79**, 745 (1999).
187. P. T. Vasudevan and D. S. Thakur, *Appl. Biochem. Biotechnol.* **49**, 173 (1994).
188. W. Hartmeier, *Immobilized Biocatalysts*, Springer-Verlag Berlin Heidelberg, l986.
189. R. Libby, J. Thomas, L. Kaiser, and L. Hager, *J. Biol. Chem.* **257**, 5030 (1982).
190. R. Libby, A. Shedd, A. Phipps, T. Beachy, and S. Gerstberg, *J. Biol. Chem.* **267**, 1769 (1992).
191. R. Libby, T. Beachy, and A. Phipps, *J. Biol. Chem.* **271**, 21820 (1996).
192. J. Kanofsky, *J. Biol. Chem.* **259** (1984).
193. U.S. Pat. 4,707,446, (1987), J. Geigert, E. T. N. Liu, and T. N'timkulu (to Cetus Corporation, Emeryville, Calif)
194. U.S. Pat. 5,928,380, (1999), J. Winlder, and L. S. Conrad, (Novo Nordisk).
195. J. Geigert, S. L. Neidleman, and D. J. Dalietos, *J. Biol. Chem.* **258**, 2273 (1983).
196. T. Tzanov, H. Shin, and A. Cavaco-Paulo, *Colourage*, 25 Annual (2001).
197. R. Bourbonnais, M. G. Paice, B. Freiermuth, E. Bodie, and S. Borneman, *Appl. Environ. Microbiol.* **12**, 4627 (1997).
198. A. Robles, R. Lucas, A. G. De Cienfuegos, and A. Galvez, *Enz. Microb. Technol.* **26**, 484 (2000).
199. A. Gardiol, R. Hernand ez, Reinhammar, and B. Harte, *Enz. Microb. Technol.* **18**, 347 (1996).
200. A. L. Ghindilis, A. Makower, C. G. Bauer, F. F. Bier, and F. W. Scheller, *Anal. Chim. Acta*, **304**, 25 (1995).
201. G. Hublik and F. Schinner, *Enz. Microb. Technol.* **27**, 330 (2000).
202. A. Lante, A. Crapisi, P. Krastanov, and P. Spettale, *Process Biochem*, **36**, 51, (2000).
203. F. Xu, *Biochemistry*, **35**, 7608 (1996).
204. U.S. Pat. 5,795,855, (1998), P. Schneider, and A. H. Pedersen (to Novozymes A/S).
205. U.S. Pat. 4,485,016, (1984), Hopkins and R. Thomas.
206. R. Bourbonnais, and M. G. Paice, *Appl. Environ. Microbiol.* **36**, 823 (1992).
207. R. Bourbonnais, M. G. Paice, I. D. Reid, P. Lanthier, and M. Yaguchi, *Environ. Microbiol.* **6l**, 1876 (1995).
208. K. Li,F. Xu, and K. E. L. Eriksson, *Appl. Environ. Microbiol.* **65**, 2654 (1999).
209. WO Patent 9,6l2,845, (1996), A. H. Pedersen, and J. V. Kierulff (Novo Nordisk A/S).
210. Portuguese Pat. Appl. 102,779 (2002), T. Tzanov and A. Cavaco-Paulo.
211. U.S. Pat. 2,001,037,532, (2001), M. Barfoed, O. Kirk, and S. Salmon (Novozymes A/S).
212. U. Sewekow, *Melliand Textilberichte* **74**, 153 (1993).
213. A. Uygur, *J.S.D.C.* **113**, 111 (1997).
214. L. Goldstein and G. Manecke, in L. B. Wingard, Jr., E. Katchalski-Katzir and L. Goldstein eds., *Applied Biochemistry and Bioengineering*, Vol. 1, New York: Academic (1976) p 23.
215. E. Akertek, and L. Tarhan, *Appl. Biochem. Biotechnol.* **50**, 291 (1995).
216. U.S. Pat. 5,071,439, (1991), K. H. Weible.
217. T. Tzanov, S. Costa, G. M. Guebitz, and A. Cavaco-Paulo, *Color. Technol.* **117**, 1 (2001).
218. M. Daumantas, W. Charles, V. P. Tong, G. Chandra, and L. Ref, *J. Mol. Cat. B: Enz.* **7**, 21 (1999).
219. H. U. Renate and A. J. M. Ulrich, *J. Mol. Cat B: Enz.* **7**, 125 (1999).
220. S. Costa, T. Tzanov, A. Paar, M. Gudelj, G. M. Gübitz, and A. Cavaco-Paulo, *Enz. Microb. Technol.* **28**, 815 (2001).
221. J. M. S. Rocha, M. H. Gil, and F. A. P. Garcia, *J. Biotechnol.* **66**, 61 (1998).
222. A. Paar, S. Costa,T. Tzanov, M. Gudelj, K. H. Robra, A. Cavaco-Paulo, and G. M. Gübitz, *J. Biotechnol.* **89**, 142 (2001).

223. S. Costa, T. Tzanov, A. F. Carneiro, A. Paar, G. M. Gübitz, and A. Cavaco-Paulo, *Enz. Microb. Technol.* **24**, 173 (2002).

224. T. Tzanov, S. Costa, M. Calafell, G. M. Gübitz, and A. Cavaco-Paulo *Colourage*, **65**, Annual (2000).

225. E. Amar, K. Tadasa, H. Fujita, and H. Kayahara, *Biotechnol. Lett.* **22**, 295 (2000).

226. "Sodium Hypochlorite," *Chemical Profiles, Chemical Market Reporter*, 13 (March 24, 2003).

227. "Calcium Hypochlorite," *Chemical Profiles, Chemical Market Reporter*,Oct. 31, 2000.

228. E. Linak,A. Leder, and N. Takei, *Chemical Economics Hand book*,Stanford Research Institute, Menlo Park, Calif., May 2000.

229. E. Linak, F. Dubas, and A. Kishi, *Chemical Economic Hand book*, Stanford Research Institute, Menlo Park, Calif, Feb. 2001.

230. T. Teitelbaum In E. Bingham, B. Cohrssen, and C. H. Powell, eds., *Patty's Toxicology*, 5th ed.,Vol. 3, John Wiley & Sons, Inc., New York, 2001, pp. 797–799.

231. R. J. Lewis, Sr, Sax's *Dangerous Properties of Industrial Materials*, 10th ed., Vol. 3, John Wiley & Sons, Inc., New York, 2000.

232. R. Levene, in Ref. 67, pp. 305–337.

233. K. Dickinson, in C. Preston, ed., *The Dyeing of Cellulosic Fibers*, Dyers' Company Publications Trust, Bradford, UK, 1986, pp. 55–105.

234. K. V. Datye and A. A. Vaidya, *Chemical Processing of Synthetic Fibers and Blends*, John Wiley & Sons, Inc., New York, 1984, pp. 139–152.

235. J. E. Nettles, *Handbook of Chemical Specialties; Textile Fiber Processing, Preparation, and Bleaching*, John Wiley & Sons, Inc., New York, 1983, pp. 391–457.

236. H. U. Mehta and M. N. Mashruwala, *Colourage* 9 (Mar. 1982).

JAMES P. FARR
WILLIAM L. SMITH
DALE S. STEICHEN
The Chlorox Company
TZANKO TZANOV
ARTHUR CAVACO-PAULO
University of Minito

BLOOD COAGULATION AND ANTICOAGULANT DRUGS

1. Introduction

Cardiovascular disease, including intravascular clot formation, represents the primary cause of death in the Western world. Blood coagulation is essential to our health; however, when it proceeds abnormally, myocardial infarction (heart attack), stroke, or pulmonary embolism can result. Pharmacologic interventions to control and correct these thromboembolic disorders have recently made much progress as the mechanisms of blood clotting have become better understood. This chapter will review the components of the hemostatic system

Table 1. **Components of the Hemostatic System**

The function of the hemostatic system is to maintain blood flow
throughout the body and to react immediately to repair
vascular damage to avoid blood loss. This is accomplished by
an integrated balance between several cellular elements
and plasma-based components.

cellular elements
blood vessel
endothelial cells
platelets
leukocytes
erythrocytes
plasma-based components
coagulation system
activators
cofactors
inhibitors
fibrinolytic system
activators
inhibitors
blood flow—viscosity

and the process of blood coagulation. The anticoagulant and antiplatelet
drugs used to treat thrombotic conditions as well as research trends will be
reviewed.

To maintain blood in a fluid state is vital in order to deliver oxygen, nutrients and physiological messengers throughout the body. When vascular damage
occurs the body reacts with an immediate response to preserve normal physiology. The hemostatic system achieves this balance between the fluid and solid
states of blood. The components of the hemostatic system include blood flow,
blood vessels, platelets, the coagulation system and the fibrinolytic system
(Table 1) (1,2).

When the integrity of the vascular system has been compromised, the blood
clots to preserve the continuity of the vasculature and the blood supply (Fig. 1).
The initial response is the formation of the platelet aggregate. Platelets in the
flowing blood rapidly adhere to the exposed subendothelial vessel wall matrix
and become activated at the site where the endothelial cells have been damaged.
During this activation process, products from the platelets are released causing
further platelet activation and platelet aggregation. The platelet plug initially
arrests the loss of blood. This, however, is not a permanent block. Negatively
charged phospholipids on the outer membrane of activated platelets create a
procoagulant surface on which coagulation activation takes place. The formation
of a fibrin clot stabilizes the platelet plug.

The coagulation system is a network of proteins that work together to ultimately form fibrin, the physical structure of the blood clot (Fig. 2). Traditionally,
coagulation has been viewed as having two distinct branches, the intrinsic and
the extrinsic pathways depending on the initiating source of activation. The two
pathways are linked at the level of factor Xa.

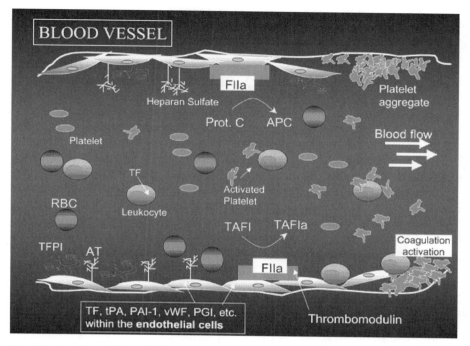

Fig. 1. Illustration of the cellular components of the hemostatic system: endothelial cells on the blood vessel wall, platelets (quiescent and activated), leukocytes, and erythrocytes. These cells normally express surface mediators that regulate coagulation, fibrinolysis, and platelet activation. Upon activation the cells express and/or release substances that modulate the physiological responses of cells and proteins in their environment and cause cell-cell interactions. These dynamic reactions take place under the physical conditions of flowing blood with vasoconstriction and relaxation of the blood vessel wall.

The extent to which each component of the hemostatic system contributes to the final clot is dependent on where in the circulation the clot is formed. In the venous circulation, where blood flow is relatively sluggish, clots contain a higher proportion of fibrin and fewer trapped blood cells. In the arterial circulation, where flow rates are higher and the presence of a stenosis leading to areas of high shear stress is more likely, clots tend to be richer in platelets. When a blood clot is no longer needed, it is broken down (lysed) by activated components of the fibrinolytic system.

Both the coagulation system and the fibrinolytic system are composed of several activators and inhibitors that provide for efficient physiological checks and balances. If any one component is over- or underactivated due to congenital or acquired abnormalities, pathologic blood clotting (thrombosis) occurs. As the components of hemostasis are many, there are multiple targets for therapeutic intervention. Targeting the mechanism that initiated the cardiovascular disorder will enhance the efficacy of the antithrombotic treatment. Bleeding complications can arise if the balance is pushed to the other extreme with drug treatment or due to physiologic abnormalities.

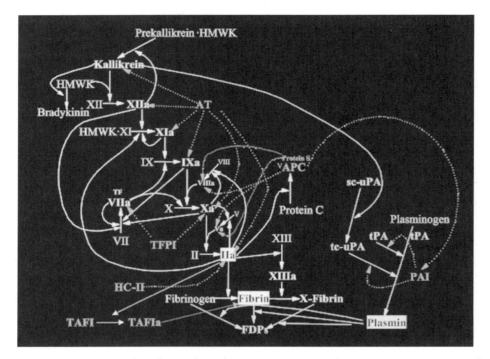

Fig. 2. Illustration of the plasma-based components of the hemostatic system: coagulation system, fibrinolytic system, and major inhibitors. Each system has its own series of activators, enzymes, inhibitors and feedback cycles. These systems interact intimately with each other. Platelets provide the phospholipid needed to activate factor X and factor II. This figure may be viewed in color online: http://www.interscience.wiley.com/cgi-bin/mrwhome/104554789/HOME

2. The Hemostatic System

2.1. Vascular Endothelium. The vascular endothelium plays an important role in hemostasis in that quiescent endothelial cells act as a barrier separating the flowing blood from subendothelial components such as tissue factor (activation of coagulation) and collagen (activation of platelets) (Fig. 1). More than just a passive barrier, the endothelium also produces a variety of substances that modulate platelet, coagulation, fibrinolytic, and vascular contraction processes (Figs. 1 and 2). Endothelial cells play a regulatory role to balance cellular and plasmatic reactions. The functional interactions of endothelial cells can be either procoagulant or anticoagulant in nature, as summarized in the following. These actions can lead to either maintenance of normal hemostasis or to pathologic occlusive disorders (stenosis, thrombosis).

Antithrombotic actions of endothelium:

1. Release of prostaglandin derivatives to control platelet activation.
2. Synthesis and release of TFPI to control coagulation activation.
3. Regulation of thrombin function through thrombomodulin.

4. Release of fibrinolytic mediators to regulate the fibrinolytic system.
5. Release of nitric oxide to promote vascular dilatation.
6. Presence of antithrombotic glycosaminoglycans (heparin-like molecules).

Prothrombotic actions of endothelium:

1. Release of tissue factor to initiate the clotting process.
2. Release of PAI-1 to inhibit the fibrinolytic response.
3. Generation of procoagulant proteins.
4. Expression of von Willebrand factor to promote platelet adhesion.

2.2. Platelets.   Platelets are disk-shaped, anuclear cells that contain a contractile system, storage granules, and cell surface receptors. Platelets normally circulate in a nonactivated state in the blood but are extremely reactive to changes in their environment. Platelet membranes contain receptors for a variety of agonists including adenosine diphosphate (ADP), thromboxane A_2, platelet activating factor, immune complexes, and thrombin. Serotonin and epinephrine synergistically promote aggregation induced by other agents.

Upon activation the expression of cell receptors and procoagulant phospholipids on the platelet surface is upregulated (Figs. 1 and 2). A number of glycoproteins (GP) present on the membrane serve as receptors for collagen (GPIa/IIa), fibrinogen (GPIIb/IIIa), and von Willebrand factor (GPIb). These receptors belong to the superfamily of adhesive protein receptors known as integrins as they integrate cell–cell and cell–matrix interactions. Stimulation of these processes allows for bidirectional signaling between the intracellular and extracellular compartments of the platelet. Of the platelet-associated integrins, GPIIb/IIIa is the most abundant. Lack of GPIIb/IIIa receptors leads to the congenital bleeding disorder known as Glanzmann's thrombasthenia. Platelet GPIb binding to von Willebrand factor, which acts as a bridge to collagen binding in the blood vessel, is important as it serves to anchor the platelets to the blood vessel. Lack of the GPIb receptor leads to the congenital bleeding disorder known as Bernard-Soulier syndrome.

Platelet aggregation is another of the fundamental platelet functions. Fibrin(ogen) binding to platelet GPIIb/IIIa receptors is important as it serves as a bridge that links individual platelets together to form a large platelet aggregate. During the activation process, there is a morphologic shape change in the overall platelet structure as pseudopods are formed. This change facilitates the platelet aggregation process. Platelet aggregates serve to plug the damage to the vascular wall. Platelet granule release products promote vasoconstriction. In normal pathology, this decreases blood loss; in abnormal pathology, this causes stenosis of a blood vessel that can result in downstream tissue ischemia.

An increase in cytosolic calcium levels leads to activation of internal platelet enzymes with the subsequent release of platelet granule contents. The α-storage granules contain platelet factor 4 (PF4), β-thromboglobulin, platelet-derived growth factor, fibrinogen, factor V, von Willebrand factor, and plasminogen activator inhibitor-1 (PAI-1). The dense or β-granules contain adenosine triphosphate (ATP), ADP, and serotonin. The release of platelet granule contents leads to further platelet activation and aggregation as well as coagulation activation.

Platelet activation also leads to the formation of platelet-derived microparticles. These are small pieces of the platelet membrane cleaved off from the platelet surface. Platelet microparticles promote activation of the coagulation system and further platelet activation (Figs. 1 and 2). The role of platelets bound to leukocytes in coagulation activation is under study.

Of particular interest in the study of thrombosis and antithrombotic drugs is the acute coronary syndrome (ACS), which encompasses unstable angina, non-ST segment myocardial infarction (MI), and acute ST elevation (transmural) myocardial infarction (AMI). The ACS stems from rupture of atherosclerotic plaque leading to intravascular thrombosis. Disruption of the protective cap exposes procoagulant materials (tissue factor, thrombin) that activate platelets and coagulation (factor Xa and thrombin generation). The role of platelets in ACS, in addition to the role of thrombin and the coagulation system, has been the focus of extensive drug development.

2.3. The Coagulation System.   The plasma proteins that comprise the coagulation system are referred to as coagulation factors. Most coagulation proteins are zymogens (nonactivated enzymes) that upon activation are converted into active serine proteases. A schematic of the coagulation cascade is depicted in Figure 2. Several of the coagulation factors are dependent on vitamin K for structural formation required for activity.

In the intrinsic pathway of the coagulation system, activation occurs when the complex of factor XII, factor XI, prekallikrein, and high molecular weight kininogen come together on a negatively charged surface. This is referred to as contact activation. Factor XII is converted to its active form, factor XIIa, which in turn converts prekallikrein to kallikrein. Kallikrein can convert factor XII to its active form thereby setting up a positive feedback loop that amplifies the activation of the coagulation system. Kallikrein also activates urokinase, an activator of the fibrinolytic system.

Factor XIIa converts factor XI to factor XIa, which, in turn, activates factor IX. Factor IXa bound to the negatively charged phospholipid (on activated platelet membranes) along with its cofactor factor VIIIa and calcium ions form the "tenase" complex. Through this complex, factor X is converted to factor Xa initiating activation of the common pathway of the coagulation system.

The extrinsic pathway of coagulation is activated when circulating factor VII comes into contact with tissue factor. Tissue factor is a transmembrane glycoprotein that is expressed by subendothelial cells that surround the blood vessel. Tissue factor expression can also be induced on activated monocytes and activated endothelial cells.

Factor VII exhibits a weak procoagulant activity on its own, typically accounting for $\sim 1-2\%$ of the total factor VII/VIIa activity. Upon binding to tissue factor, a 10,000,000-fold increase in factor VIIa enzymatic activity occurs. Both factor VII and factor VIIa bind to tissue factor with equal affinity. The factor VIIa-tissue factor complex can then activate factor X. The tissue factor-factor VIIa complex also activates factor IX to factor IXa.

The small amounts of factor Xa initially generated are sufficient to cleave prothrombin and generate a small amount of thrombin. In a feedback loop, thrombin activates factors V, VIII, and possibly XI, thereby sustaining continued activation of the coagulation cascade. Factors V and VIII are activated through direct proteolytic cleavage by factor Xa or thrombin; they are not active proteases as are the other coagulation factors.

The majority of factor Xa joins with its cofactor factor Va, calcium ions and phospholipid (on surface membranes of activated platelets) to form the "prothrombinase" complex. The prothrombinase complex acts to convert prothrombin (factor II) into the active enzyme thrombin. Thrombin (factor IIa) serves many functions in coagulation as well as in various physiological processes. In the coagulation cascade, thrombin holds the key position in that it cleaves soluble fibrinogen to generate an insoluble fibrin clot (thrombus).

Fibrinogen circulates as a disulfide-linked dimer containing two A-α chains, two B-β chains, and two γ chains. Cleavage of fibrinogen by thrombin results in the release of fibrinopeptides A and B and the exposure of charged domains at opposite ends of the molecule. Exposure of these charged domains leads to polymerization of the fibrin monomers. These monomers are cross-linked by the transaminase factor XIIIa and calcium to form the physical meshwork of the fibrin clot.

Thrombin augments its own generation through several feedback loops in the coagulation cascade activating factors XII, XI, VIII, and V. Thrombin also activates platelets, it activates the coagulation inhibitor protein C through binding with thrombomodulin, and it stimulates activated endothelial cells to release the profibrinolytic enzyme tissue plasminogen activator (see Figs. 1 and 2).

The outcome of activation of the various factors that comprise the coagulation system is to generate thrombin. Excessive thrombin generation (a hypercoagulable state) results in unwanted blood clots that cause tissue ischemia. Depending on the location of the thrombus, skeletal muscle, heart, lung, brain, or other organs are affected. There are several anticoagulant drugs that target one or another of the coagulation factors to reduce thrombin generation. Inhibition of one or more of the coagulation factors that excessively reduces thrombin generation, such as by a congenital factor deficiency or overdose of anticoagulant treatment, may result in bleeding.

2.4. Natural Inhibitors of Coagulation. Antithrombin (AT) is a single chain glycoprotein with a molecular weight of ~58,000 Da. Normal plasma levels of AT are ~2–3 μM. AT is the primary inhibitor of coagulation and targets most coagulation factors as well as trypsin, plasmin and kallikrein (Fig. 2). Inhibition takes place when a stoichiometric complex between the active site serine of the enzyme and the Arg393-Ser394 bond of AT forms.

The efficient inhibition of proteases by AT requires heparin as a cofactor. In the presence of heparin, the inhibition rate constants for thrombin and factor Xa have been estimated to be accelerated 1000-fold to 3×10^7 and 4×10^6 L mol^{-1} s^{-1}, respectively. Deficiency of AT, due to low protein levels or to functionally abnormal molecules predisposes an individual to thrombotic complications.

Heparin cofactor II (HCII) is another plasma inhibitor that resembles AT in that it is activatable by glycosaminoglycan binding. HCII has a molecular weight of ~68,000 Da. The normal plasma level of HCII is ~1.0–1.4 μM. Two patients to date have been described as having HCII deficiency related to thrombosis.

HCII has a higher protease specificity than AT. Of the coagulation enzymes, it only inhibits thrombin (Fig. 2). However, it has also been shown to inhibit chymotrypsin and leukocyte cathepsin G. Like AT, HCII inhibits proteases by forming a 1:1 stoichiometric complex with the enzyme. Whereas AT contains an Arg-Ser bond as its active site, HCII is unique in containing a Leu-Ser bond suggesting that another portion of the HCII molecule may be required for protease binding.

Although the inhibition of protease activity by HCII is promoted by glycosaminoglycan binding, it can be activated by a wide variety of agents unlike AT, which is dependent on the presence of a specific heparin chain sequence. Heparins, heparans, and dermatan sulfate all bind to HCII and promote thrombin inhibition. Agents with relatively little sulfation such as chondroitin 4-O- or 6-O-sulfate, keratan sulfate, or hyaluronic acid do not activate HCII.

Tissue factor pathway inhibitor (TFPI) is a 42-kDa inhibitor that contains three Kunitz domains tandemly linked between a negatively charged amino terminus and a positively charged carboxy terminus. It serves an important function to control coagulation activation. The active site of the first Kunitz domain binds to the active site of the VIIa-tissue factor complex; the active site of the second Kunitz domain binds to the active site of factor Xa. The second domain appears to facilitate the inhibitory action of the first domain, and the carboxy-terminus appears to facilitate the action of the second domain. The third Kunitz domain has been shown to contain a heparin-binding site. Mutation of the active site of the third Kunitz domain has no effect on the inhibition of either factor VIIa or factor Xa.

TFPI is produced by megakaryocytes and the endothelium (Fig. 1). Small amounts of TFPI are stored in platelets (<2.5%) and can be released upon platelet activation. Plasma TFPI accounts for 10–50% of the total pool. Most plasma TFPI is bound to lipoproteins, only ~5% of the plasma pool of TFPI circulates in the free form. Lipoprotein bound TFPI is of relatively low inhibitory activity. The largest pool of TFPI is bound to the endothelial surface. The TFPI bound to the endothelium can be released into the plasma by heparin and low molecular weight heparin treatment.

Protein C is another important natural anticoagulant. Circulating thrombin can bind to a high affinity receptor on the endothelium known as thrombomodulin (Fig. 1). The complex of thrombin bound to thrombomodulin is a 20,000-fold better activator of protein C than is free thrombin. Thrombomodulin-bound thrombin no longer cleaves fibrinogen, is not able to activate other coagulation proteases such as factors V and VIII and does not activate platelets.

Protein C is a vitamin K-dependent zymogen. It is made up of disulfide linked heavy and light chains and has a molecular weight of approximately 62,000 Da. Protein C derives its anticoagulant properties from its ability to cleave and inactivate membrane bound forms of factors Va and VIIIa. Protein C requires two cofactors to express its anticoagulant activity, protein S and factor V.

2.5. The Fibrinolytic System.   The fibrinolytic system keeps the formation of blood clots in check. Like the coagulation cascade, this system consists of a number of serine protease activators and inhibitors (Fig. 2). The zymogen plasminogen normally circulates in the blood in micromolar concentrations.

Two endogenous activators of plasminogen, tissue-type plasminogen activator (tPA) and urokinase-type plasminogen activator (uPA), are produced primarily by the endothelium and circulate in sub-picomolar amounts. Both tPA and uPA convert plasminogen to the active fibrinolytic enzyme plasmin. Plasmin ultimately cleaves fibrin into smaller fibrin degradation products.

Regulation of the fibrinolytic pathway occurs at the level of several inhibitors. Plasminogen activator inhibitor-1 (PAI-1) inhibits the enzymatic activity of the activators tPA and uPA. PAI-1 covalently binds to the active site of these plasminogen activators thereby preventing the generation of plasmin. Activated platelets are an important source of PAI-1. Plasmin also can be directly inhibited by the serine protease inhibitor α_2-antiplasmin.

Thrombin activatable fibrinolytic inhibitor (TAFI) is a third recently identified inhibitor that has a different type of inhibitory function. TAFI is a procarboxypeptidase that is activated by the thrombin/thrombomodulin complex. Activated TAFI (TAFIa) catalyzes the cleavage of carboxy-terminal basic amino acids (such as arginine and lysine) from fibrin, plasmin, and other proteins. Without these end structures plasmin loses it ability to digest fibrin. Thus, fibrinolytic activity is suppressed leaving procoagulant activity to proceed unopposed. New studies have revealed that certain antithrombotic drugs in addition have a pro-fibrinolytic effect mediated by the drug's interaction with and blockade of TAFIa.

2.6. Leukocytes.   Recent studies suggest more and more that the line between coagulation and inflammation is less distinct (3). Studies have indicated that leukocytes, alone or bound to platelets, play a role in coagulation activation (Fig. 1). Cytokines elicit the expression of tissue factor (extrinsic coagulation system activator) on mononuclear cells, and procoagulant activity associated with leukocytes is not limited to the expression of tissue factor (Fig. 2). Several monocyte/macrophage derived procoagulant activities have been characterized including factor VII, factor XIII, factor V/Va, and binding sites for factor X and for the factor IXa/VIII complex. Prothrombin can be activated on the cell surface of monocytes and lymphocytes. Monocyte procoagulant activity is also induced by endotoxin, complement and prostaglandins.

Coagulation that takes place on the surface of endothelial cells is affected by the inflammatory process. Cytokines released from activated leukocytes, such as interleukin-1 (IL-1), IL-6, and tumor necrosis factor (TNF), upregulate the procoagulant and down regulate the fibrinolytic nature of endothelial cells.

In addition, products of the coagulation process such as thrombin, fibrinopeptides, and fibrin degradation products have chemotactic and mitogenic properties.

2.7. Autonomic Nervous System.   Although limited research has been undertaken in this area, there is supportive evidence that the autonomic nervous system may impart control on the regulation of hemostasis and activation mechanisms leading to thrombogenesis. Circadian variations with peak incidences of coronary events in the morning hours has been known. This has been shown to be associated with an increase in blood pressure, heart rate, platelet aggregability, and a decrease in fibrinolytic activity. These physiological responses reflect sympathetic activity largely induced by increased levels of plasma noradrenaline (4,5). In combination with an increase in sympathetic mediated vasoconstriction, these factors can lead to atherosclerotic plaque rupture. During hemorrhage the hemostatic mechanisms controlling hemostasis are also partly controlled by the autonomic nervous system (6,7).

3. Therapeutic Intervention of Thromboembolic Disorders

Thrombosis is associated with a high degree of morbidity and mortality. There are numerous risk factors for thrombosis (Table 2). Anticoagulant drugs (heparin, and warfarin-derivatives) have been used clinically since the late 1930s. These drugs are not specific in mechanism; they target the inhibition of thrombin, thrombin generation and the initiation of coagulation, among other factors. Historically, heparin and warfarin have been rather easily monitored as they prolong the time to clot of global clotting assays [activated partial thromboplastin time (aPTT) and prothrombin time (PT), respectively] in a dose-dependent manner. Thus, these have been called anticoagulant drugs.

With an increased understanding of the mechanisms involved in the pathogenesis of thrombosis, specific plasma, and cellular sites within the hemostatic network are now targeted by a host of newly developed anticoagulant, antithrombotic and antiplatelet drugs (Table 3). These drugs are collectively referred to as antithrombotic drugs since their mechanisms and their effect on coagulation lab assays differ from the anticoagulant drugs heparin and warfarin. There is now a division between *in vitro / ex vivo* clot inhibition per se (as determined by traditional coagulation assays) and control of thrombogenesis *in vivo*. The pharmacology of heparin has also advanced. In this section, the growing area of antithrombotic agents will be reviewed.

The category of fibrinolytic agents, which differ from antithrombotic drugs in their targets and mechanisms of action, will not be covered.

3.1. Heparin.   Heparin, discovered in 1916 by Jay McLean, is "... a family of polysaccharide species whose chains are made up of alternating 1–4 linked and variously sulfated residues of uronic acid and D-glucosamine" (8). It is a

Table 2. **Risk Factors of Thrombosis (Partial List)**[a]

congenital deficiencies/abnormalities of the hemostatic components
   (eg, factor V Leiden, prothrombin 20210, mutations of the AT molecule)
antiphospholipid antibodies/lupus anticoagulant
hyper-homocysteinemia
heparin-induced thrombocytopenia
heart failure
malignancy
burn
previous thrombosis
smoking
oral contraceptives
obesity
age
surgery
physical inactivity/stasis/immobilization

[a] Usually two or more risk factors need to be present for thrombosis to occur.

Table 3. **Antithrombotic Drugs**

heparin
low molecular weight heparin
synthetic heparin pentasaccharides (fondaparinux, idraparinux)
oral heparins
non-heparin glycosaminoglycans (eg, dermatan sulfates, intimatan)
vitamin K antagonists (oral)
direct thrombin inhibitors
oral thrombin inhibitors
factor Xa inhibitors (direct and indirect FXa inhibitors)
anti-tissue factor agents
other protease inhibitors (other than thrombin and Xa inhibitors, eg,
    FVIIa and FIXa inhibitors, protein Ca)
antiplatelet drugs

strongly anionic glycosaminoglycan that contains three functional side groups: $-OSO_3^-$, $-NHSO_3^-$, and $-COO^-$. The chemical structure of heparin is depicted in Figure 3. Heparin is largely derived from porcine intestinal mucosa. The average molecular weight of heparin is 15,000, but the individual molecules range from 3,000 to 30,000 Da. Thus, heparin is not one molecule but a heterogeneous

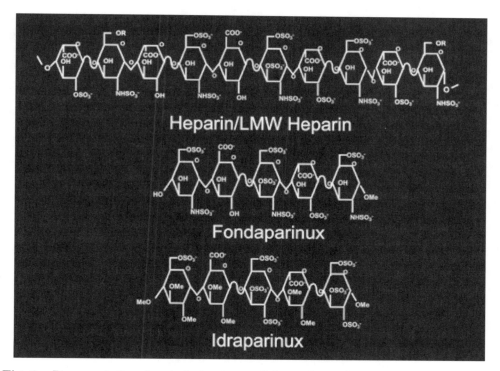

Fig. 3. Representative chemical structures of the anticoagulant heparin: the heterogeneous unfractionated heparin/low molecular heparin (differ by molecular weight and end groups), the synthetic heparin pentasaccharide (fondaparinux) with high affinity binding to AT that only produces inhibition of factor Xa, and a sulfated/methylated modification of pentasaccharide (idraparinux) designed for an extended half-life.

mixture of different molecules. Owing to its structural heterogeneity, heparin exhibits a number of pharmacologic properties. Among these are antilipemic and antiviral properties; it can also inhibit tumor growth (9,10).

Foremost among the actions of heparin is its ability to inhibit blood clotting. Heparin produces little anticoagulant or antithrombotic effect directly. Rather, its effects are mediated through specific saccharide sequences that bind to one of several endogenous plasma proteins that include AT, HCII, and TFPI. The AT–heparin complex inhibits several of the coagulation factors. The major antithrombotic activity of heparin, that which is used for pharmacologic evaluation, is the ability of heparin to inhibit thrombin (anti-thrombin or anti-factor IIa activity) and factor Xa (anti-factor Xa activity). Administration of heparin causes an increase in the plasma levels of TFPI that adds to its antithrombotic action. In addition, heparin has numerous antithrombotic properties derived from its components that have low or no affinity to AT.

Heparin is administered either by intravenous infusion or subcutaneous injection. Heparin binds to a variety of plasma proteins in the blood, thereby lowering its bioavailability and producing a variable anticoagulant response. These proteins include histidine rich glycoprotein, platelet factor 4 (PF4), vitronectin and von Willebrand factor. Heparin is eliminated by receptor-mediated internalization into endothelial cells and macrophages and by a nonsaturable renal mechanism. The anticoagulant effect of heparin is, therefore, not linearly related to dose when in the therapeutic range. The biologic half-life of heparin increases from 30 min following an IV bolus dose of 25 U kg^{-1} to 150 min following a dose of 400 U kg^{-1} (9). Subcutaneous bioavailability of heparin is limited to only 20–30%. Heparin administered by inhalation exhibits a prolonged elimination half-life, but does not exhibit significant bioavailability following oral administration.

Clinical Uses of Heparin. Heparin is the drug of choice for effective treatment of venous thrombosis and pulmonary embolism (PE) (Table 4) (11,12). Mortality is reduced in patients receiving heparin for the treatment of PE. Heparin is also used for prophylaxis in patients at risk of developing deep venous

Table 4. Clinical Uses of Anticoagulant/Antithrombotic Drugs

prophylaxis of venous thrombosis
treatment of established venous thromboembolic events
treatment of acute coronary syndromes
adjunct to atrial fibrillation treatment
treatment of thrombotic stroke
alternative anticoagulant for heparin compromised patients
disseminated intravascular coagulation associated with sepsis
adjunct to chemotherapy (cancer associated thrombosis)
posttransplant veno-occlusive disease
interventional cardiology procedures
surgical anticoagulation
anticoagulation for extracorporeal devices (eg, heart–lung and
    dialysis machines)
surface coating of biomedical devices
adjunct to anti-inflammatory agents
modulatory agent for growth factors

thrombosis (DVT) and PE (11). The risk of developing DVT and PE is reduced by 60–70% compared to patients not receiving prophylaxis. Both congenital and acquired risk factors associated with the development of thrombosis are many (Table 2). Postsurgical patients constitute the largest single group that routinely receives thrombosis prophylaxis. General medicine patients have risk factors for thrombosis that include cancer, bed rest, heart failure, and severe lung disease.

Heparin is also used to anticoagulate patients with ACS. This includes unstable angina, non-Q wave MI and AMI with or without thrombolytic therapy (11). Heparin can prevent AMI and recurrent refractory angina in patients with unstable angina. In patients with a previous MI, heparin significantly reduces reinfarction and death. Heparin as an adjunct to thrombolytic therapy increases patency during the initial stages of recanalization by preventing rethrombosis. Heparin is also used to treat thrombotic stroke.

ACS is now and treated more frequently by percutaneous coronary intervention (PCI). Since these procedures can cause intraarterial thrombus formation at the site of vessel wall damage, intravenous administration of unfractionated heparin is most frequently used to inhibit such a complication (13).

Extracorporeal devices are used in multiple clinical situations. Blood in contact with a foreign surface will clot within minutes if left without anticoagulant. Heparin has been used as a flush solution for most catheters inserted in hospitalized patients. Heparin-coated devices are now being produced that eliminate the need for heparin administration directly to the patient. Heparin is used with renal dialysis. The most extreme case where the highest level of anticoagulation is needed is with the heart–lung machine used for cardiopulmonary bypass in cardiac surgery. Heparin is used to prevent blood clotting here.

Heparin can be used safely in pregnancy because it does not cross the placental barrier and does not cause unwanted effects on the fetus (14). Heparin is effectively used in the pediatric population for the same indications as in the adult, but dosing regimens are different (15).

Part of heparin's attractiveness for use as an anticoagulant in surgical situations relates to the relative ease in which it can be neutralized upon completion of the procedure or in the event of an overdose (10). The anticoagulant actions of heparin are neutralized with a protamine salt that binds heparin in a charge-dependent manner. Protamine reduces both the anti-thrombin and anti-factor Xa activities of heparin. Heparinase, a bacterial enzyme that can cleave heparin chains into components as small as disaccharides, is being developed as a heparin-neutralizing agent. Heparinase effectively neutralizes the anti-thrombin activity of heparin but is less effective against its anti-factor Xa activity.

Monitoring heparin levels is necessary in order that drug concentrations remain in the safe and effective therapeutic range. This minimizes bleeding from overdosing or clotting from underdosing. Therapeutic intravenous heparin is monitored by the aPTT assay. This assay is performed in a laboratory using patient's blood plasma. In situations where high concentrations of heparin are required such as in surgery, the activated clotting time (ACT) is used. This assay can be done at the point of patient care on whole blood. For both the aPTT and the ACT, several reagents and instruments are commercially available. Results from each system can vary.

Heparin has been considered by some to be an old, inefficient, and clinically suboptimized drug. This statement is not justified. Heparin has provided reliable thromboprophylaxis for many years and it remains a useful and very effective drug that is easily dosed, monitored and neutralized. Had it not been for heparin, it would not have been feasible to have such surgical procedures as open-heart surgery, organ transplantation, and medical treatments of heart attack, deep vein thrombosis, and pulmonary embolism. As a polypharmaceutical, heparin is a unique drug with multiple beneficial effects. Whether the new antithrombotic agents or the newer versions of heparin (represented today by the low molecular weight heparins, and the synthetic pentasaccharide), as discussed below, will prove to be better clinical options than the original (unfractionated) heparin remains to be determined for different clinical settings.

Side Effects of Heparin Therapy. The most common side effect of heparin therapy is hemorrhage (16). The hemorrhagic effect associated with heparin therapy can range from minor to life threatening and is related to the total administered dose and the degree of prolongation of the aPTT.

Heparin-induced thrombocytopenia (HIT), which occurs in ~3% of patients exposed to heparin, is perhaps the worst of all drug-induced allergic reactions (17). HIT Type I occurs early in heparin treatment, causes a transient reduction in platelet count and patients remain asymptomatic. This is due to a direct effect of heparin on platelets. HIT Type II is a more severe thrombocytopenia and typically occurs with a delayed onset. This form of HIT often results in thrombosis and is associated with an increased mortality. While the mechanism of HIT has not been completely identified, it is known that antibodies are generated against the heparin–PF4 complex. Antibodies bind to the FcγRIIA receptor on the platelet surface resulting in platelet activation. Due to the severity of the clinical events in patients with HIT Type II, all exposure to heparin must be stopped including heparin for catheter flushes, etc. Alternative antithrombotic agents have recently become available for treatment of patients with HIT (see thrombin inhibitors).

Heparin therapy is associated with transient elevations in serum transaminase levels. Whether this is of clinical importance in terms of liver dysfunction is unknown. Long-term heparin therapy has been shown to produce osteoporosis. Heparin-induced skin necrosis is a rare complication of subcutaneously administered heparin (18).

3.2. Low Molecular Weight Heparin. The depolymerization of heparin (the original unfractionated heparin) either chemically (nitrous acid degradation, benzylation-alkaline hydrolysis, peroxidative cleavage), enzymatically (heparinase), or by physicochemical means (γ irradiation) results in the production of clinically useful drugs known as low molecular weight (LMW) heparins. This depolymerization process produces a material whose molecular weight is approximately one-third that of the parent heparin (average 5000 Da; range 2000–6000 Da), and also modifies some structural elements (19). Chemical depolymerization results in partial desulfation, reduction in charge density, a reduction in the number of AT binding sites and other changes in the consensus sequences. End-residues of fragments are typical of the specific depolymerization method.

The bioavailability of LMW heparin is nearly 80% as determined by anti-factor Xa activity. The LMW heparin is dosed subcutaneously once or twice

Table 5. **Low Molecular Weight Heparins**

Generic name	Trade name	Currently approved indications in the United States
enoxaparin	Lovenox	DVT/PE prophylaxis, extended outpatient use, DVT/PE treatment, ACS
dalteparin	Fragmin	DVT/PE prophylaxis, ACS
ardeparin	Normiflo	DVT/PE prophylaxis (inpatient use only)
tinzaparin	Innohep	DVT/PE treatment
Synthetic ultra-low molecular weight heparin		
fondaparinux (pentasaccharide)	Arixtra	DVT/PE prophylaxis
idraparinux (methylated derivative of pentasaccharide)		DVT prophylaxis

daily. The specific activity of LMW heparins ranges from 35 to 45 anti-factor IIa units/mg; anti-factor Xa activity ranges from 80 to 120 units mg^{-1}. Thus, as heparin exhibits a 1:1 ratio of anti-thrombin (160 units mg^{-1}): anti-factor Xa activity (160 units mg^{-1}), the ratio for LMW heparins range from 1:2 to 1:4 depending upon the molecular composition of the given LMW heparin (19). The LMW heparins have a lower anticoagulant potency (aPTT activity) than unfractionated heparin as a reflection of the lower anti-thrombin activity. Reduced protein binding of LMW heparin results in a more predictable dose-response. The LMW heparins cause the release of TFPI as does heparin.

Clinical Uses of LMW Heparins. It is important to know that each LMW heparin is a different chemical entity, as well each has different pharmacological behaviors (20). It is for these reasons that their dosing regimens differ. Therefore, each individual LMW heparin should only be used as described in its corresponding package insert.

Four LMW heparins have been approved for use in the United States (Table 5). These include enoxaparin (Lovenox, Aventis), ardeparin (Normiflo, Wyeth-Ayerst), dalteparin (Fragmin, Pharmacia & Upjohn) and tinzaparin (Innohep, Pharmion). Clinical trials have established their safety and efficacy in a number of indications: prevention of venous thrombosis in patients undergoing abdominal surgery, hip/knee repair/replacement (21) or medically ill patients with restricted mobility; treatment of existing venous thrombosis with or without PE (12); and prevention of ischemic complications in patients with unstable angina/non-Q-wave MI (11).

LMW heparins can also be used as anticoagulants in patients with end-stage renal disease requiring extracorporeal hemodialysis treatment. They can be used in children (15) and are the drug of choice in pregnant women requiring anticoagulation (14).

Because LMW heparins are safe and effective as thromboprophylactic agents, routine monitoring is not required for this clinical use. If monitoring is requested, a special chromogenic anti-factor Xa assay has to be used since the aPTT does not detect these drugs. The chromogenic assay is not routine, but can be found in clinical–research laboratories associated with medical centers.

There is a debate regarding the duration of postsurgical thrombotic risk and the appropriate duration of prophylaxis. Several studies have suggested that prolonged prophylaxis results in improved clinical outcome (22). At-home dosing with LMW heparin is as safe and effective as in-hospital treatment by heparin infusion. Another debate focuses on the timing for the initiation of therapy in surgical patients (23). There are arguments both for beginning before and after surgery.

Newer indications for possible uses of LMW heparins are in the management of thrombotic stroke and in cancer patients. LMW heparins may not only decrease the incidence of cancer associated thrombosis but they may also positively impact all-cause mortality.

In cardiology, LMW heparins are effective for the reduction of restenosis after interventional cardiologic procedures, maintenance of peripheral arterial and coronary graft patency, and as adjunct anticoagulants in stenting and other interventional cardiologic procedures (24). In this setting where drug levels are higher, the ACT as used for heparin, has also been used for LMW heparins. However, a definitive monitoring system with optimal performance characteristics for all LMW heparins is not available.

LMW heparins will not become the drug of choice in surgical settings where a short half-life anticoagulant is required. Additional disadvantages of LMW heparins in surgery are that reversal agents such as protamine do not completely block the antithrombotic activity of LMW heparin, and there are no commonly available devices/assays to effectively monitor the high drug levels required.

Side Effects of LMW Heparin Therapy.   Data from clinical trials has shown that LMW heparins are less likely to cause hemorrhagic complications than unfractionated heparin during treatment of venous thrombosis (16).

Retrospective data suggests that LMW heparins are less likely to cause clinical symptoms of HIT Type II (17). However, the generation of the antibody to heparin-PF4 that causes HIT occurs and LMW heparins can cross-react with a pre-formed antibody (25,26). Thus, LMW heparin should not be given to a patient suspected of having HIT.

Other side effects of heparin, such as osteoporosis are reduced with LMW heparins.

3.3. Synthetic Heparins.

Heparin exerts its antithrombotic activity mainly via binding to AT thereby inhibiting thrombin and factor Xa. Investigations into the structure–activity relationships of heparin revealed a molecular weight dependence of heparin–AT on the inhibition of the coagulation proteins. Of particular interest was the finding that a saccharide sequence of 18 units or longer was necessary to produce thrombin inhibition. Inhibition of factor Xa could be produced with heparin chains of smaller length. Eventually studies focused on a decasaccharide and an octasaccharide possessing high anti-factor Xa activity with no detectable inhibitory action against thrombin. Careful study of the structures by ^{13}C NMR revealed that a pentasaccharide was the minimal heparin sequence that would bind AT and elicit a high anti-factor Xa activity.

The original pentasaccharide sequence was identified from natural heparin by fractionation procedures (27). A specific pentasaccharide of a predetermined sequence was subsequently synthesized (28). This synthetic pentasaccharide was composed of a regular region (units G and H) and an irregular region of heparin (units D, E and F) (Fig. 3). The relative positioning of the sulfated monosaccharides was of critical importance. Four specific sulfate groups within the pentasaccharide were also shown to be critical for optimal binding to AT, ie, the 6-O sulfate on the D unit, the 3-O sulfate on the F unit, and 2-N sulfates on the F and H units (Fig. 3). Particularly important for binding to AT, was a unique 3-O sulfate group within the glucosamine residues in the irregular region.

This pentasaccharide is the first synthetic heparin (29). However, unlike heparin and LMW heparins derived from natural material, the chemically synthesized pentasaccharide is free of viral or other animal contaminants. It represents a homogeneous, single targeting entity. It does not bind to plasma proteins. Because of its minimal size, it possesses only the ability to inhibit factor Xa, via binding to AT. It has a high specific activity of ~650 anti-factor Xa units mg^{-1}. It is devoid of other therapeutic effects of heparins such as the release of TFPI, anti-thrombin activity, pro-fibrinolytic actions and antiinflammatory actions. It has 100% subcutaneous bioavailability and a half-life of ~18 h. These characteristics may make this drug useful for long-term prophylaxis such as for home therapy.

In clinical evaluation, pentasaccharide (fondaparinux, Arixtra; Sanofi-Organon) was well tolerated in healthy individuals (30). The PT and aPTT were not significantly prolonged even at excessive doses. Clinical studies have been performed with once daily subcutaneous dosing of fondaparinux in patients undergoing hip fracture repair or hip/knee replacement. These studies revealed that fondaparinux is more effetive but comparable to the safety of enoxaparin or unfractionated heparin for the treatment of DVT or PE, respectively, in these surgical populations (31).

Whether fondaparinux produces enhanced clinical efficacy compared to LMW heparin treatment remains a point of debate as the study endpoints may have been influenced by the timing of test drug administration. A potential limiting factor in the use of fondaparinux is the somewhat higher rate of hemorrhage compared to LMW heparin observed in some studies. The long half-life, lack of an effective antidote and lack of an easy to perform monitoring assay may be additional limitations. Because of the high affinity of pentasaccharide for AT and the limited amount of AT in plasma, decreased levels of AT with some disease states and congenital deficiencies, studies will be needed to relate the effects of pentasaccharide to endogenous AT plasma concentrations.

The synthesis of the pentasaccharide has opened the door for the possibility of synthesizing other heparin-like agents that exhibit specific pharmacologic profiles. Modified pentasaccharides with varying degrees of sulfation and/or methylation have been described. Idraparinux is one such agent that is in clinical trial (32). Such agents exhibit higher affinity to AT, more potent anti-factor Xa activity and extended half-life. In addition, larger molecules have been synthesized that incorporate the high AT affinity pentasaccharide with a thrombin-binding domain.

3.4. Oral Heparin. Recent attempts to produce heparin formulations that exhibit oral bioavailability have met with varying degrees of success. The use of diamine salts as counterions, bile acids, and surfactants promotes heparin absorption in various animal species. More recently, oral administration of heparin-loaded biodegradable nanoparticles has been shown to produce increases in plasma anti-factor Xa levels and to prolong the aPTT. An absolute bioavailability of 23% was observed. The most studied means of delivering heparin orally is through the coadministration of N-[8-(2-hydroxybenzoyl)amino] caprylate (SNAC) (Emisphere) (33). It was recently shown in patients undergoing elective hip surgery that antithrombotic protection could be achieved with the administration of SNAC–heparin. The incidence of DVT/PE in patients receiving SNAC–heparin was comparable to the incidence in patients treated with subcutaneous LMW heparin; however, further studies need to be conducted to better assess the clinical usefulness of this compound. Several issues remain at this time that limit the development of SNAC–heparin including safety and patient compliance (taste).

3.5. Non-Heparin Glycosaminoglycans. Dermatans, heparans and chondroitin sulfates represent non-heparin glycosaminoglycans (GAGs) that are used mainly in the intravenous management of DVT prophylaxis. These drugs can be given to patients who are heparin compromised.

Dermatan Sulfate. Dermatan sulfate is a glycosaminoglycan polymer of iduronic acid and N-acetylated galactosamine. Due to a difference in the molecular backbone, dermatan sulfate is unable to interact with AT, but rather complexes with HCII to mediate thrombin inhibition. Dermatan sulfate inhibits thrombin as it is formed rather than preventing its generation. It has been shown that thrombin generation inhibition by dermatan sulfate is much less than for an equigravimetric amount of heparin.

Dermatan sulfate is active *in vivo* as an antithrombotic agent in the rabbit stasis thrombosis model, but to a lesser extent than heparin. The advantage dermatan sulfate has over heparin as an antithrombotic agent is a lower risk of bleeding complications.

Intimatan. Intimatan, iduronate → N-acetyl-D-galactosamine 4,6-O-disulfate, is a newly developed semisynthetic dermatan sulfate prepared by site-specific sulfation of a highly purified dermatan sulfate derived from porcine mucosa. Due to its unique structure, Intimatan has higher anti-thrombin potency than the naturally occurring, parent dermatan sulfate. It has been suggested that a unique benefit of Intimatan as an antithrombotic agent is its ability to inhibit surface-bound thrombin (eg, on extracorporeal devices) more effectively than heparin–AT (34). Since surface-bound thrombin contributes to coagulant activity and thrombus formation as fluid-phase thrombin does, and clot-bound thrombin remains active and catalyzes the generation of systemic thrombin promoting further clot growth, this agent may prove to be better than other antithrombotics. However, Intimatan has not been studied in humans yet.

3.6. Vitamin K Antagonists. Long-term prophylaxis against thrombosis is typically achieved using vitamin K antagonists (VKA). In the United States, warfarin (Coumadin) is most commonly used. Acenocoumarol and phenprocoumon with shorter and longer half-lives than warfarin, respectively, are used in other countries. Warfarin was first isolated as the substance from moldy sweet clover that induced hemorrhage in cattle. The VKAs interfere with blood coagulation by inhibiting vitamin K reductase and vitamin K epoxide reductase (Fig. 4). These enzymes are involved in the recycling of vitamin K

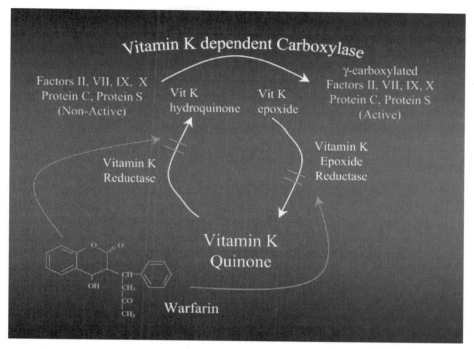

Fig. 4. The oral vitamin K antagonists: chemical structure of warfarin, the conversion of the coagulation factors from an inactive to an active enzymatic state, and the mechanism of action of warfarin to inhibit the γ-carboxylation of the coagulation factors, thus producing a physiologic anticoagulant effect. The γ-carboxylation of vitamin K-dependent coagulation factors makes possible the binding of calcium that promotes the attachment of the molecules to a lipid surface where coagulation activation takes place.

(quinone) to vitamin KH_2 (hydroquinone), which is a required cofactor for the specific carboxylation of glutamic acid residues on vitamin K-dependent coagulation factors (factors II, V, VII, IX, X, protein C, and protein S). Inhibition of this γ-carboxylation of glutamic acid residues results in a loss of calcium-binding ability with resultant decreases in binding to phospholipid surfaces (9–11).

The primary benefit of VKAs over the heparins is their ability to be administered orally. As such, VKAs are commonly used for the prophylaxis and treatment of DVT (in particular when long-term treatment is needed) (35), for anticoagulation of patients with atrial fibrillation (36), and for anticoagulation of patients with mechanical heart valves (37,38). Patients experiencing AMI may also benefit from anticoagulant treatment with VKAs.

The use of oral anticoagulants is associated with several limitations and management of patients requires a knowledgeable care provider (39). Unlike with heparin, the onset of anticoagulation with VKAs takes several days as the half-lives of affected coagulation factors range from 6 to 72 h. Plasma VKA levels are altered by a number of factors including diet, gastrointestinal and metabolic factors, vitamin K levels, coadministration of a wide variety of different drugs, and patient compliance. Hemorrhage is by far the most frequent complication of VKA therapy (16). As this class of drugs has a relatively narrow

Table 6. **Direct Thrombin Inhibitor Drugs**

Generic name	Trade name	Molecular weight	Clinical uses
lepirudin (hirudin)	Refludan	6980	anticoagulation for HIT
argatroban	Argatroban	527	anticoagulation for HIT; in PCI
bivalirudin (hirulog)	Angiomax	2180	PCI
ximelagatran/melagatran	Exanta	474/430	DVT/PE prophylaxis

safety–efficacy margin, frequent monitoring of drug levels using the prothrombin time test and the calculated international normalized ratio (PT/INR) is necessary.

The anticoagulant actions of VKAs are not easily reversed, complicating the anticoagulant management of treated patients who require surgical intervention. Vitamin K given with fresh frozen plasma or prothrombin complex concentrate if the bleeding is severe is used to reverse over-anticoagulation. Skin necrosis can occur during treatment in individuals with protein C deficiency, activated protein C resistance (factor V Leiden), and protein S deficiency. Teratogenic effects of VKAs on fetuses preclude its use in the anticoagulant management of pregnant patients.

3.7. Direct Thrombin Inhibitors. Direct thrombin inhibitors are a new class of antithrombotic agents that includes the following agents (Table 6, Fig. 5). The first thrombin inhibitor to be developed was hirudin, a leech-derived protein of 65 amino acids (6980 Da). The synthetic peptide inhibitor hirulog (bivalirudin) is a 20 amino acid polypeptide bioengineered as an analogue of hirudin. It consists specifically of the two thrombin binding sites within hirudin (2180 Da). There are two small molecule inhibitors. Argatroban (527 Da) is a derivative of

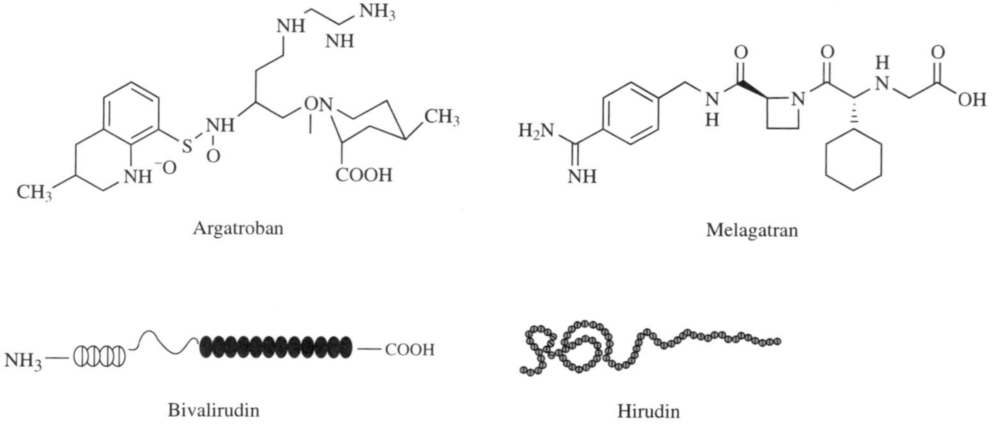

Argatroban

Melagatran

Bivalirudin

Hirudin

Fig. 5. The chemical structures of the new direct thrombin inhibitors: the arginine derivative argatroban that hinders the access of substrates to the catalytic pocket of thrombin (527 Da); melagatran a modified dipeptide that mimics the peptide sequence preceding the thrombin cleavage site in the αA-chain of fibrinogen (430 Da); bivalirudin a 20 amino acid polypeptide engineered as an analogue of hirudin consisting of its two binding sites to thrombin (2180 Da); and the leech-derived hirudin (65 amino acids; 6980 Da).

arginine. It acts by hindering the access of substrates to the catalytic pocket of thrombin. Melagatran (430 Da) is a modified dipeptide that mimics the peptide sequence preceding the thrombin cleavage site in the αA-chain of fibrinogen. Melagatran bound thrombin is blocked from producing its biological effects.

Thrombin inhibitors offer advantages over heparin in that they selectively inhibit thrombin. They directly inhibit thrombin without the need for plasmatic cofactors. They can inhibit both fluid-phase and clot-bound thrombin. Because they are not neutralized by plasma proteins, they have predictable and consistent pharmacodynamics, which can translate into fast therapeutic control and fewer treatment failures.

While all direct thrombin inhibitors have high affinity for thrombin they differ in their mechanisms of binding, with some binding only the active site of the enzyme, whereas others also bind to various exosites on thrombin. In addition, the binding to thrombin is so strong that it is nearly irreversible in some cases (hirudin) but quickly reversible in others (bivalirudin, argatroban, melagatran). As a class, direct thrombin inhibitors exhibit a short half-life; however, the specific half-lives differ among the drugs. There are reports of antibodies developing to hirudin in up to 40% of treated patients. These affect the pharmacokinetics of the drug, increasing the anticoagulant effect in the majority of cases. In several patients, anaphylactic reactions upon reexposure have resulted in death.

Thrombin inhibitors have a relatively narrow therapeutic window, and there is no known effective means of neutralizing their anticoagulant activity. Direct thrombin inhibitors not only inhibit the procoagulant actions of thrombin, but also thrombin's many physiological regulatory actions. The clinical relevance of this, however, has not been shown yet.

Three direct thrombin inhibitors have been approved by the FDA for use in United States, Canada, and Europe (clinical indications vary by country). These include argatroban (Argatroban), hirudin (lepirudin, Refludan) and hirulog (bivalirudin, Angiomax). Both argatroban (for prophylaxis and treatment) and lepirudin (for treatment) are approved for managing thrombotic complications in patients with HIT Type II (40–42). The approval of thrombin inhibitors was an extremely important achievement for providing treatment to these patients in whom heparin is contraindicated yet require anticoagulation.

Argatroban and bivalirudin are also approved for use as an anticoagulant during percutaneous coronary interventions in patients with ACS.

For prophylaxis and treatment of thrombosis in patients with HIT, the aPTT can be effectively used to monitor the thrombin inhibitor treatment. In the setting of PCI, the ACT can be used to monitor the higher doses of these drugs. In some settings and with some thrombin inhibitors, the aPTT and ACT do not provide an optimal dose-response (particularly with higher doses of the thrombin inhibitors). The ecarin clotting time assay (ECT) has been proposed as a replacement. This test is not standardized and is not widely available for purchase.

These inhibitors are currently limited to use by parenteral administration. Patients treated with thrombin inhibitors who require long-term anticoagulation are switched to a VKA. Bridging between a thrombin inhibitor and an oral anticoagulant is complicated by the fact that thrombin inhibitors prolong all clot-based assays including an artificial increase in the PT/INR level used to monitor dosing of the oral anticoagulant. Specific dosing protocols are provided by the manufacturers to guide one through this process.

Ongoing clinical studies are evaluating additional indications for the use of direct thrombin inhibitors. Some of these indications include prevention of DVT following orthopedic surgery, as an anticoagulant for coronary interventions, for the treatment of ACS, as an adjunct anticoagulant to thrombolysis for the treatment of AMI, in patients undergoing peripheral and cerebral vascular procedures, for patients with intermittent claudication and during renal dialysis. Several large clinical trials using thrombin inhibitors for treatment of ACS have been completed with mixed results (43). Although cardiac events have been reduced, significant bleeding occurred. Some have suggested that thrombin inhibitors may be used for anticoagulation in cardiopulmonary bypass surgery. This is not an approved indication and must be used with extreme caution as dosing regimens and monitoring systems have not been established and bleeding has been reported.

Oral Thrombin Inhibitors.   Orally administered drugs facilitate long-term prophylaxis of venous thrombosis, PE, stroke in patients with atrial fibrillation, as well as treatment of acute venous thrombosis, PE, AMI, and ACS. Today VKAs are the only oral antithrombotics. The high risk of bleeding with VKAs and the need for frequent monitoring, however, have prompted the development of new oral antithrombotic agents. Agents developed to date have not had sufficient activity. Thrombin inhibitors may prove otherwise.

The agent that has progressed the farthest in development is the prodrug ximelagatran (Exanta, AstraZeneca). The active form of the drug, melagatran, is a competitive, reversible, selective inhibitor of thrombin. This agent is a modified dipeptide mimicking the peptide sequence preceding the thrombin cleavage site on the αA-chain of fibrinogen. When first pass metabolism is taken into consideration, ximelagatran exhibits an ∼20% bioavailability. As this drug exhibits low protein binding, no active transport during absorption or excretion and is not metabolized by the cytochrome P450 system, ximelagatran may produce a predictable anticoagulant response to fixed doses of the drug. If this is true it may not require routine monitoring.

In patients undergoing total knee arthroplasty, a fixed dose of 24-mg ximelagatran twice daily started the morning of surgery was at least as effective as warfarin in preventing venous thromboembolism without the need for monitoring or dose adjustment (44). Additional studies on ximelagatran in comparison to either VKAs or subcutaneous enoxaparin are in progress in which increased doses are being studied, as well as a combination of subcutaneous melagatran and oral ximelagatran.

3.8. Factor Xa Inhibitors.   Antithrombotic drugs that target factor Xa may have certain advantages over drugs that target thrombin. Factor Xa is a key enzyme involved in the generation of thrombin. It is common to both the extrinsic and intrinsic coagulation pathways. Factor Xa is formed at an earlier stage than thrombin in the coagulation pathway, and the procoagulant effect of factor Xa is strongly amplified by the prothrombinase complex. Factor Xa has no known activity other than as a procoagulant, in contrast to thrombin, which has multiple physiological roles. Factor Xa has relatively slow activation kinetics, in contrast to thrombin. Opposing its function could result in easier management of the balance between the therapeutic and bleeding effects of a drug.

Factor Xa inhibitors are a diverse class of agents, each with distinct characteristics. It was learned from early experience that a drug that targets factor Xa requires high potency to produce an effective antithrombotic response. Several agents have been identified that are able to either directly bind to and inhibit factor Xa or indirectly inhibit factor Xa (45). Thus, these agents derive their antithrombotic activity from their ability to inhibit thrombin generation. Proteins and small peptides derived from natural sources including ticks, leeches, snakes, and hookworms, the original factor Xa inhibitors identified, have been discontinued from clinical development. The synthetic heparin pentasaccharide, fondaparinux, can be considered a factor Xa inhibitor. It differs from all other factor Xa inhibitors in that it has an indirect inhibitory effect requiring binding to AT to produce its anti-factor Xa activity.

A number of synthetic or recombinant, direct factor Xa inhibitors are in development. Unlike heparin, LMW heparin and fondaparinux, the direct factor Xa inhibitors can inhibit clot-bound factor Xa and factor Xa that has been incorporated within the prothrombinase complex. This is an advantage for drug efficacy. It may also explain why direct factor Xa inhibitors can be monitored by the aPTT, whereas fondaparinux cannot. The agents in Phase II clinical development planned for thrombosis prophylaxis or in cardiology include recombinant TFPI (tifacogin; Pharmacia/Chiron) and the synthetic agents DX-9065a (Daiichi), DPC-423 (DuPont), ZK-807834 (Berlex/Pfizer), and BAY 59-7939 (Bayer). The therapeutic potential and safety issues of all factor Xa inhibitors remain to be defined.

3.9. Other Inhibitors.   Several other proteases of the coagulation system are also targeted for drug development. In addition to the thrombin and factor Xa inhibitors already discussed, agents that specifically target either factor VIIa or tissue factor are in development. The inhibitor to the factor VIIa/tissue factor complex, rNAPc2 (Corvas), is in phase II development planned for DVT prophylaxis in post hip and knee surgery and in arterial thrombosis treatment. Inhibitors to factor XIIa/XIa, factor IXa, and factor XIIIa are also under development as potential antithrombotic drugs.

The natural inhibitors of the coagulation system are also targets for potential drug development. In addition to TFPI already discussed, AT, HCII, C_1-esterase inhibitor, and PAI-1 are under development. Protein Ca concentrate, which targets the inhibition of factor Va and factor VIIIa, has shown successful outcomes in patients with sepsis and disseminated intravascular coagulation (DIC).

3.10. Antiplatelet Drugs.   The ability to inhibit platelet activation or the aggregation of previously activated platelets is of significant importance in the treatment of coronary artery disease (43), cerebrovascular, peripheral vascular disease (46), following coronary interventions (13), and following cardiac surgery (47). A wide variety of antiplatelet agents that inhibit different aspects of the platelet activation response have proven clinical efficacy or are currently under development (Table 7) (48).

Aspirin, the most widely used antiplatelet agent, blocks platelet activation by inhibiting cyclooxygenase (COX) thereby limiting thromboxane (a potent

Table 7. **Antiplatelet Drugs**

Generic name	Trade name	Mechanism	Clinical uses
aspirin		COX inhibitor	ACS, PCI, stroke
clopidogrel	Plavix	ADP receptor blockers	aspirin substitute
abciximab	ReoPro	fibrinogen receptor blocker	ACS, PCI
tirofiban	Aggrastat	"	
eptifibatide	Integrilin	"	
dipyridamole	Persantine	phosphodiesterase inhibitor	stroke, aspirin adjunct
cilostazol	Pletaal	type III phosphodiesterase inhibitor	intermittent claudication

platelet aggregation activator) generation. Aspirin offers clinical benefit in both the primary and secondary prevention of cardiovascular events, prevention of DVT and PE, and is used in the treatment of AMI, stable and unstable angina, carotid artery stenosis, ischemic stroke, and placental insufficiency. Aspirin is also used in combination with other antiplatelet drugs during PCI and in the prophylaxis of thrombotic complications following PCI. After coronary bypass grafting (CABG surgery) patients are put on life-long aspirin therapy.

The minimum effective dose of aspirin ranges between 50 and 100 mg day^{-1}. For high risk patients, doses from 50 to 1500 mg day^{-1} have been shown to be effective. Aspirin has been shown in vascular patients to reduce the risk of stroke, MI and death by ~25%. However, it has been estimated that up to 40% of patients may experience thromboembolic events despite aspirin therapy. This has been referred to as aspirin resistance (49). The true significance of the problem remains unknown because of differences in the definition of resistance, variations in detection methods and lack of controlled trials. Multiple mechanisms have been proposed, including increased reactivity to platelet aggregating factors, genetic polymorphisms and alternate pathways for thromboxane synthesis. Strategies are needed to identify patients at risk for aspirin resistance who might benefit from alternative or combined antiplatelet therapy.

Another specific inhibitor of platelet activation is dipyridamole (Persantine). This drug is believed to work by inhibition of a phosphodiesterase enzyme that degrades cyclic adenosine monophosphate (AMP). An accumulation of cyclic AMP inhibits platelet activation. The clinical efficacy of dipyridamole has been questioned; however, it may be that doses have been too low.

The thienopyridines, ticlopidine and clopidogrel (Plavix), are ADP receptor antagonists. Platelets have two distinct ADP receptors, $P2X_1$ and $P2Y_1$. It is not clear, but suggested that these drugs block a third ADP receptor that mediates the inhibition of stimulated adenylyl cyclase activity ($P2T_{AC}$). These are prodrugs that require hepatic transformation into the active state. Ticlopidine has been associated with neutropenia, thrombocytopenia, aplastic anemia, and thrombotic thrombocytopenic purpura. Clopidogrel, a drug chemically similar to ticlopidine is associated with a lower incidence of these side effects and, therefore, is more commonly used. While initially approved for use in patients with symptomatic atherosclerosis, clopidogrel also has proven benefit in the treatment of unstable angina, AMI, and is used extensively in combination with aspirin for the prevention of stent thrombosis. Clopidogrel is often used in combination with aspirin.

There are multiple activation pathways in platelets. Glycoprotein (GP) IIb/IIIa receptor blockers are potent antiplatelet agents that inhibit the final common pathway of platelet aggregation. The GPIIb/IIIa inhibitors are effective regardless of the platelet activation stimulus. Several such drugs are currently available. The Fab fragment of a monoclonal antibody against the GPIIb/IIIa receptor, abciximab (ReoPro), was the first GPIIb/IIIa inhibitor to be clinically developed. The development of synthetic small molecules and peptidomimetic inhibitors that compete with fibrinogen and other ligands for occupancy of the platelet receptor followed. Tirofiban (Aggrastat) is a non-peptide derivative of tyrosine. Eptifibatide (Integrilin) is a synthetic heptapeptide based on the amino acid sequence Lys-Gly-Asp found in the venom of the *Sistrurus m barbouri* snake. GPIIb/IIIa inhibitors have been shown to be effective in reducing late restenosis and preventing mortality in patients undergoing PCI and in preventing mortality in patients with ACS.

One of the main weaknesses of the current GPIIb/IIIa inhibitors is the need to administer them intravenously, thus precluding their use for long-term platelet inhibition. The optimal level of GPIIb/IIIa inhibition to maximize efficacy without inducing an enhanced risk of bleeding is unknown. Unlike aspirin and clopidogrel where complete inhibition of cyclooxygenase or ADP receptors, respectively, is desirable, complete inhibition of GPIIb/IIIa leads to an unacceptably high incidence of bleeding. To some degree, the lack of knowledge concerning the optimal degree of GPIIb/IIIa inhibition is related to a lack of efficient assays for monitoring the antiplatelet effects of these agents.

Antibodies that cause thrombocytopenia to <50,000/μL platelets in up to 2% of treated patients is an important side effect of abciximab. In 1% of these cases, the thrombocytopenia is rapid beginning within 2 h of initiation of drug administration.

For chronic arterial insufficiency in the extremeties aspirin, dipyridamole and clopidogrel have been found useful. Cilostazol (Pletaal) is a type III phosphodiesterase inhibitor with both antiplatelet and vasodilatory properties. It has been recently approved for the treatment of intermittent claudication.

Trials with orally available GPIIb/IIIa inhibitors have not been successful to date. An oral P2T ADP receptor inhibitor, CS-747 (Sankyo/Lilly), is in phase I development planned for arterial thrombosis treatment. Other antiplatelet drugs under development include thromboxane and serotonin receptor antagonists.

3.11. Pharmacologic Considerations. The future holds promise for effective new antithrombotic drugs in individual and specific indications. It is likely that each drug will have a role in specific clinical indications and that one drug will not be optimal for all thrombotic situations as these drugs do not exhibit a polytherapeutic spectrum. As more is learned of the mechanisms of thrombosis, there will be drugs that target a patient's individual needs to combat the various type of clinical thrombosis that he/she is experiencing. Combination approaches may be more beneficial in the overall management of thrombotic disorders.

The development of LMW heparin began in the 1980s. Today these are the drugs of choice for several thrombotic indications. From identified structure–

Table 8. **Pharmacologic Considerations for New Antithrombotic Drugs**

mechanism of action (dependence on plasma factors, pro-drug)
subcutaneous bioavailability
oral bioavailability
endogenous modulations
endothelial/vascular interactions
metabolic transformations
patient-to-patient variability in dose-response
drug interactions
bleeding risk
ease of monitoring drug levels
ability to generate antibodies (heparin-induced thrombocytopenia,
    anti-platelet antibodies, antibodies that cause leukopenia, antibodies
    that alter the pharmacokinetics) other unwanted side effects

activity relationships, such heparinomimetics as fondaparinux have been developed. Additionally, studies to develop drugs with antithrombotic activities but without anticoagulant aspects are in progress at this time. Direct acting factor Xa inhibitors are currently in clinical development. However, many of these drugs do not have anticoagulant actions and may be limited in scope for different situations. The development of direct thrombin inhibitors has moved quickly. This was spurred by the obvious need for alternative antithrombotic treatment in patients with HIT. These patients have the highest risk of thrombosis but cannot receive heparin products and no other fast-acting, strong anticoagulant was available. There are potential advantages for factor Xa inhibitors over thrombin inhibitors, particularly a higher safety margin in prophylactic regimens and less frequent dosing requirements. On the other hand, factor Xa inhibitors may be less potent than thrombin inhibitors and could thus have limited clinical application.

As with any new drug there are certain issues to be considered during development (Table 8). Synthetic agents have certain advantages over naturally derived products, not the least of which is their specific chemical design to target desired biological effects. Drug interactions have to be considered as patients are often on multiple antithrombotic as well as other types of drugs. For example, heparin/antiplatelet drugs, antithrombin/antiplatelet drugs, and antiplatelet drugs of different mechanisms are often combined. In the future, antithrombin/anti-factor Xa/antiplatelet drugs may be combined. Cost is an obvious issue that can prevent widespread use of any new drug.

How and where each drug is used clinically, and how each will compete with the standard heparin, warfarin, and aspirin treatments remains to be determined. The newly developed drugs are mostly monotherapeutic and do not mimic the polytherapeutic actions of heparins. It is therefore, important to stress that heparin and its derived/modified forms will continue to play an important role in the management of thrombosis (50).

BIBLIOGRAPHY

"Coagulants and Anticoagulants" in *ECT* 2nd ed., Vol. 5, pp. 586–605, by D. M. Stuart, Neisler Laboratories, Inc.; "Blood, Coagulants and Anticoagulants" in *ECT* 3rd ed., Vol. 4, pp. 1–24, by D. M. Stuart and J. K. Hruschka, Ohio Northern University; in *ECT* 4th ed., Vol. 4, pp. 333–360, by William R. Bell, Jr., The Johns Hopkins University School of Medicine; "Blood Coagulation and Anticoagulant Drugs" in *ECT* (online), posting date: December 4, 2000, by William R. Bell, Jr., The Johns Hopkins University School of Medicine.

CITED PUBLICATIONS

1. R. W. Colman, J. Hirsh, V. J. Marder, and E. W. Salzman, *Hemostasis and Thrombosis. Basic Principles and Clinical Practice*, 3rd ed., J.B. Lippincott, Philadelphia, Pa., 1994.
2. J. Loscalzo and A. I. Schafer, *Thrombosis and Hemorrhage*, 2nd ed., Williams and Wilkins, Baltimore, 1998.
3. A. Celi, R. Lorenzet, B. Furie, and B. C. Furie, *Seminars Hematol.* **34**(4), 327 (1997).
4. W. Kiowski and S. Osswald, *J. Cardiovascular Pharmacol.* **21** (Suppl 2), S45 (1993).
5. J. Kawahara, H. Sano, H. Fukuzaki, K. Saito, and H. Hirouchi, *Am. J. Hypertension* **2**(9), 724 (1989).
6. G. DiPasquale, A. Andreoli. A. M. Lusa, S. Urbinati, S. Biancoli, E. Cere, M. L. Borgatti, and G. Pinelli, *J. Neurosurg. Sci.* **42** (Suppl 1), 33 (1998).
7. V. Svigelj, A. Grad, and T. Kiauta, *Acta Neurolog. Scand.* **94**(2), 120 (1996).
8. B. Casu, *Heparin: Chemical and Biological Properties, Clinical Applications*, Edward Arnold, London, 1989, pp. 25–50.
9. P. W. Majerus, G. J. Broze, J. P. Miletich, and D. M. Tollefsen, *Anticoagulant, Thrombolytic, and Antiplatelet Drugs, in Goodman and Gilman's, The Pharmacological Basis of Therapeutics*, 8th ed., Pergamon Press, New York, Chapt. 55, 1990, pp. 1311–1331.
10. W. Jeske, H. L. Messmore, and J. Fareed, Pharmacology of Heparin and Oral Anticoagulants, in *Thrombosis and Hemorrhage*, 2nd ed., Williams and Wilkins, Baltimore, Chapt. 55, 1998, pp. 1193–1213.
11. J. Hirsh, T. E. Warkentin, S. G. Shaughnessy, S. S. Anand, J. L. Halperin, R. Raschke, C. Granger, E. M. Ohman, and J. E. Dalen, *Chest* **119**(1), 64S (2001).
12. T. M. Hyers, G. Agnelli, R. D. Hull, T. A. Morris, M. Samama, V. Tapson, and J. G. Weg, *Chest* **119**(1), 176S (2001).
13. J. J. Popma, E. M. Ohman, J. Weitz, A. M. Lincoff, R. A. Harrington, and P. Berger, *Chest* **119**(1), 321S (2001).
14. J. S. Ginsberg, I. Greer, and J. Hirsh, *Chest* **119**(1), 122S (2001).
15. P. Monagle, A. D. Michelson, E. Bovill, and M. Andrew, *Chest* **119**(1), 344S (2001).
16. M. N. Levine, G. Raskob, S. Landefeld, and C. Kearon, *Chest* **119**(1), 108S (2001).
17. T. E. Warkentin and A. Greinacher, *Heparin-Induced Thrombocytopenia*, 2nd ed., Marcel Dekker, New York, 2001.
18. J. M. Walenga and R. L. Bick, *Med. Clinics North America* **82**(3), 635 (1998).
19. J. Fareed, J. M. Walenga, D. Hoppensteadt, X. Huan, and R. Nonn, *Ann. N. Y. Acad. Sci.* **556**, 333 (1989).
20. J. Fareed, W. Jeske, D. Hoppensteadt, R. Clarizio, and J. M. Walenga, *Am. J. Cardiol.* **82**(5B), 3L (1998).
21. W. H. Geerts, J. A. Heit, G. P. Clagett, G. F. Pineo, C. W. Colwell, F. A. Anderson, Jr., and H. B. Wheeler, *Chest* **119**(1), 132S (2001).
22. R. D. Hull, G. F. Pineo, P. D. Stein, A. F. Mah, S. M. MacIsaac, O. E. Dahl, M. Butcher, R. F. Brant, W. A. Ghali, and D. Bergqvist, *Ann. Intern. Med.* **135**, 858 (2001).
23. R. D. Hull, G. F. Pineo, P. D. Stein, A. F. Mah, S. M. MacIsaac, O. E. Dahl, W. A. Ghali, M. S. Butcher, R. F. Brant, D. Bergqvist, K. Hamulyak, C. W. Francis, V. J. Marder, and G. E. Raskob, *Arch. Intern. Med.* **161**(16), 1952 (2001).

24. J. P. Zidar, *J. Invasive Cardiol.* **12** (Suppl B), 16B (2000).
25. J. M. Walenga, M. J. Koza, B. E. Lewis, and R. Pifarré, *Clin. Appl. Thromb. Hemost.* **2** (Suppl 1), S21 (1996).
26. J. M. Walenga, W. P. Jeske, A. R. Fasanells, J. J. Wood, S. Ahmad, and M. Bakhos, *Clin. Appl. Thromb. Hemost.* **5** (Suppl 1), S21 (1999).
27. J. Choay, M. Petitou, J. C. Lormeau, P. Sinay, B. Casu, and G. Gatti, *Biochem. Biophys. Acta.* **116**(2), 492 (1983).
28. M. Petitou, P. Duchaussoy, I. Lederman, J. Choay, P. Sinay, J. C. Jacquinet, and G. Torri, *Carbohydr. Res.* **147**, 221 (1986).
29. J. M. Walenga, W. P. Jeske, L. Bara, M. M. Samama, and J. Fareed, *Thromb. Res.* **86**(1), 1 (1997).
30. B. Boneu, J. Necciari, R. Cariou, P. Sie, A. M. Gabaig, G. Kieffer, J. Dickinson, G. Lamnd, H. Moelker, T. Mant, and H. Magnani, *Thromb. Haemost.* **74**(6), 1468 (1995).
31. J. M. Walenga, W. P. Jeske, M. M. Samama, F. X. Frapaise, R. L. Bick, and J. Fareed, *Expert Opin. Investig. Drugs* **11**(3), 397 (2002).
32. J. M. Herbert, J. P. Herault, A. Bernat, R. G. M. van Amsterdam, J. C. Lormeau, M. Petitou, C. van Boeckel, P. Hoffmann, and D. G. Meuleman, *Blood* **91**(11), 4197 (1998).
33. R. A. Baughman, S. C. Kapoor, R. K. Agarwal, J. Kisicki, F. Catella-Lawson, and G. A. FitzGerald, *Circulation* **98**(16), 1610 (1998).
34. M. R. Buchanan, G. A. Maclean, and S. J. Brister, *Thromb. Haemost.* **86**, 909 (2001).
35. J. Hirsh, J. E. Dalen, D. R. Anderson, L. Poller, H. Bussey, J. Ansell, and D. Deykin, *Chest* **119**, 8S (2001).
36. G. W. Albers, J. E. Dalen, A. Laupacis, W. J. Mannin, P. Petersen, and D. E. Singer, *Chest* **119**(1), 194S (2001).
37. D. N. Salem, D. H. Daudelin, H. J. Levine, S. G. Pauker, M. H. Eckman, and J. Riff, *Chest* **119**(1), 207S (2001).
38. P. D. Stein, J. S. Alpert, H. I. Bussey, J. E. Dalen, and A. G. G. Turpie, *Chest* **119**(1), 220S (2001).
39. J. Ansell, J. Hirsh, J. Dalen, H. Bussey, D. Anderson, L. Poller, A. Jacobson, D. Deykin, and D. Matchar, *Chest* **119**(1), 22S (2001).
40. B. E. Lewis, D. E. Wallis, S. D. Berkowitz, W. H. Matthai, J. Fareed, J. M. Walenga, J. Bartholomew, R. Sham, R. G. Lerner, Z. R. Zeigler, P. K. Rustagi, I. K. Jang, S. D. Rifkin, J. Moran, M. J. Hursting, and J. G. Kelton, *Circulation* **103**, 1838 (2001).
41. A. Greinacher, U. Janssens, G. Berg, M. Bock, H. Kwasny, B. Kemkes-Matthes, P. Eichler, H. Volpel, B. Pötzsch, and M. Luz, *Circulation* **100**, 587 (1999).
42. A. Greinacher, H. Völpel, U. Janssens, V. Hach-Wunderle, B. Kemkes-Matthes, P. Eichler, H. G. Mueller-Velten, and B. Pötzsch, *Circulation* **99**, 73 (1999).
43. J. A. Cairns, P. Théroux, H. D. Lewis, Jr., M. Ezekowitz, and T. W. Meade, *Chest* **119**(1), 253S (2001).
44. C. W. Francis, B. L. Davidson, S. D. Berkowitz, P. A. Lotke, J. S. Ginsberg, J. R. Lieberman, A. K. Webster, J. P. Whipple, G. R. Peters, and C. W. Colwell, *Ann. Int. Med.* **137**(8), 648 (2002).
45. J. M. Walenga, W. P. Jeske, D. Hoppensteadt, and J. Fareed, *Curr. Opin. in Invest. Drugs* **4**(3) (2003).
46. M. R. Jackson and G. P. Clagett, *Chest* **119**(1), 283S (2001).
47. P. D. Stein, J. E. Dalen, S. Goldman, and P. Théroux, *Chest* **119**(1), 278S (2001).
48. C. Patrono, B. Coller, J. E. Dalen, G. A. FitzGerald, V. Fuster, M. Gent, J. Hirsh, and G. Roth, *Chest* **119**(1), 39S (2001).

49. D. L. Bhatt and E. J. Topol, *Nature Reviews / Drug Discovery* **2**, 15 (2002).

50. J. Fareed, D. A. Hoppensteadt, R. L. Bick, *Seminars in Thrombosis &* Hemostasis **26** (Suppl. 1), 5 (2000).

JEANINE M. WALENGA
WALTER P. JESKE
PETER BACHER
JAWED FAREED
Loyola University

BLOOD SUBSTITUTES

1. Introduction

Artificial blood is herein defined as consisting of red cell substitutes. Red cell substitutes are solutions intended for use in patients whose red cells are either not available or their use is to be avoided for other reasons. Efforts to produce red cell substitutes began when the circulation of the blood was discovered (1), and the current group of potential products have been under development for the last one-half of the twentieth century.

In 1983, the move to develop red cell substitutes intensified when recognition that the acquired immune deficiency syndrome (AIDS) could be transmitted by the blood-borne human immunodeficiency virus (HIV) produced grave concern for the nation's blood supply. Since that time modernized blood bank methods have dramatically reduced the risk of transfused blood. Furthermore, indications for transfusion have been reevaluated, and the use of blood products has become much more efficient. More careful screening of donors, testing of all donated units, and a general awareness in the donor population have all contributed to a decreased risk from transfusion-contracted AIDS.

The idea of red cell substitutes is not new. In Ovid's Metamorphosis the witch Medea restored Jason's aged father, Acson by slitting his throat to let out old blood, replacing it with a magic brew she had concocted (2). Sir Christopher Wren was one of the first to apply the new knowledge about circulation to blood substitutes. In 1656, he infused ale, wine, scammony, and opium into dogs and from these efforts conceived the idea of transfusing blood from one animal to another. Lower actually carried out the first transfusion experiments (3). The early history of blood substitute research has been summarized (4).

2. Historical Blood Substitutes

2.1. Milk.   Milk, one of the first materials to be used as a red cell substitute (5,6), was used in cases of Asiatic cholera in 1854. It was suggested that milk could regenerate white blood cells (5). Two patients were given 340 g or more of cow's milk and did well, but two others died (7). In all, 12 cases of injection of milk into the circulatory system were reported, and it was concluded that using milk in place of blood was a feasible, safe, and legitimate procedure. These results were met with excitement, and it was thought that milk injections would supplant the dangerous transfusion of blood (8). Milk was also shown to support function in isolated, perfused hearts from a variety of mammals (9). However, the transfusion of milk never gained widespread favor and soon disappeared from the literature.

2.2. Normal Saline.   In the laboratory, the search for a red cell substitute was directed at understanding the physiologic role of blood and its many components. Some of the early work involved frogs. Salzfrosche-frogs, where the blood was completely washed out and replaced with a pure sodium chloride solution, survived for some hours (10). Urea-frogs and sugar-frogs lived longer; if a small amount of red cells remained, they could survive indefinitely (11). But frogs are simple animals, and a frog's nervous system can be kept alive for some time without any circulation at all. Normal saline is, however, used widely as a plasma volume expander.

2.3. Ringer's Solution.   In 1883, the excised ventricle of the frog was found to beat for some hours if supplied with an aqueous solution of sodium, potassium, and calcium salts (12). The concentration of potassium and calcium was critical, whereas the amounts of the anions had little effect on the frog heart. The composition of this saline, coined Ringer's solution, after the English physiologist Sydney Ringer, is given in Table 1. Many years later it was shown to be very close to that of frog plasma.

Ringer's lactate, in which salts of lactic acid (in Europe, salts of acetic acid) are added to Ringer's solution, is probably the most popular crystalloid (ie, noncolloid) solution for intravenous use in humans. The lactate is gradually converted to sodium bicarbonate within the body so that an uncompensated alkalosis is prevented (13). These crystalloid solutions cannot support life without red cells; saline passes rather quickly into the tissue spaces of various organs (14), especially the liver (15).

2.4. Gum-Saline.   Gum is a galactoso-gluconic acid having a molecular weight of ~1500. First used in kidney perfusion experiments (16), gum-saline enjoyed great popularity as a plasma expander starting from the end of World War I. The aggregation state of gum depends on concentration, pH, salts, and temperature, and its colloid osmotic pressure (often just "oncotic" pressure, or COP) and viscosity are quite variable. Conditions were identified under which the viscosity would be the same as that of whole blood (17).

Table 1. **The Composition of Ringer's Solution**[a]

	Ringer's solution		Frog plasma mEq/L
	g/100 mL	mEq/L	
NaCl	0.6	102	104
KCl	0.0075	1	2.5
$CaCl_2$	0.01	1.8	1.0
$NaHCO_3$	0.01	1.2	25.4

[a] See Ref. (4).

In early animal studies, gum was found to coat the surface of all blood cells and to promote coagulation. The use of gum-saline became popular in World War 1, but it was soon discovered that if the hematocrit was <25%, gum-saline was not efficacious in hemorrhagic shock. In fact, gum-saline was not as efficacious in treating hemorrhagic shock as was saline alone (14), but it was useful in temporarily stabilizing blood volume (18). Through the 1920s, reports of anaphylaxis and other untoward reactions appeared. Many of these were attributed to impurities and when properly purified, gum-saline was safe for human use. Pharmacologic studies in the 1930s (19) showed that gum was deposited in the liver and spleen and could remain there for many years. Its half-life in the circulation was ~30 h and anaphylaxis occasionally occurred. Success with gum-saline became common in the 1930s, but the need for it decreased with the increased availability of plasma.

2.5. Blood Plasma and Serum. The terms "plasma" and "serum" are frequently confused. Plasma is the liquid component of blood in which cellular elements are suspended. After removal from the body, and after a coagulum formed and was removed, the liquid that remained was serum. Thus, serum contains no coagulation factors or cellular elements.

As early as 1871, it was noted that frog hearts could be maintained by perfusion using sheep and rabbit serum (20). This solution was superior to 0.6% aqueous NaCl (21). Over ensuing years it was recognized that serum exerts a colloid osmotic pressure, contains bicarbonate, and may ensure capillary integrity. After dismissing a physiological role for plasma lipids, it was eventually agreed that albumin added to a balanced salt solution was superior to salt solution alone in maintaining the frog heart (22).

In the first one-half of the twentieth century, much work was devoted to the study of plasma and serum as blood substitutes. One problem in this field was the recognition of toxic substances (23). Reports were published of intravascular coagulation and vasotonins that appeared mysteriously after the infusion of serum or plasma. It was suggested that this activity could be reduced by heating the serum or by filtering it before use. Platelets, insulin, and also adenosine triphosphate (ATP) were implicated. Once the red cell surface antigens were elucidated, these effects were attributed to immune reactions, and the use of serum from donors of blood group AB markedly reduced the vasoconstrictor activity.

World War II ushered in the modern era of blood fractionation. It was shown that plasma could be administered directly to humans (24,25). Although cases of serum sickness frequently occurred 5–7 days after the infusion, the procedure could be life-saving in eases of hemorrhagic shock (26) (see, FRACTIONATION, BLOOD).

2.6. Albumin. Investigation into the safety of bovine plasma for clinical use was undertaken in the early 1940s in anticipation of wartime need (27). Modern methods of protein chemistry, including electrophoresis and ultracentrifugation, have shown that many of the human adverse reactions to blood plasma were caused by the globulin fraction, and that albumin was generally safe for parenteral use. Human albumin is now used extensively as a plasma expander in many clinical settings.

2.7. Starches. The use of nonanimal sources of colloidal plasma expanders arose from the earlier use of gum, but with more controlled and predictable chemistry. In the United States, the commonly used starch derivative is Heta-starch, a plant-derived waxy starch composed mainly of amylopectin. Hydroxyethyl ether groups are introduced into the glucose units of the starch, and the resultant material is hydrolyzed to yield a product with a molecular weight suitable for use as a plasma volume expander. For Hetastarch, the molecular substitution is ~0.75, which means there are ~75 hydroxyethyl groups for every 100 glucose units. The molecular weight ranges from 450 to 800 kDa. Recently, the U.S. FDA approved a novel formulation of Hetastarch. Its commercial name is Hextend, which combines Hetastarch with salts, including calcium, glucose, and lactate. Clinical trials with Hextend showed a significant improvement of the major concern with starches in general, prolonged, or increased bleeding (28,29).

2.8. Perfluorocarbons. In 1966, Clark and Gollan demonstrated that a laboratory mouse could survive total immersion in a perfluorochemical (PFC) solution (30). This material, similar to commercial Teflon, is almost completely inert and is insoluble in water. A water-soluble emulsion was prepared that could be mixed with blood (31), and in 1968 (32) the blood volume in rats was completely replaced with an emulsion of perfluorotributylamine [311-89-71], $C_{12}F_{27}N$. The animals survived in an atmosphere of 90–100% O_2 and went on to long-term recovery. However, the O_2 content of the perfluorochemicals has a linear dependence on the partial pressure of oxygen (pO_2), as shown in Figure 1. The very high pO_2 required to transport physiologic amounts of O_2 and the propensity of the perfluorocarbon to be taken up by the reticuloendothelial cells were considered to be severe limitations to the development of clinically useful perfluorocarbon blood substitutes (33).

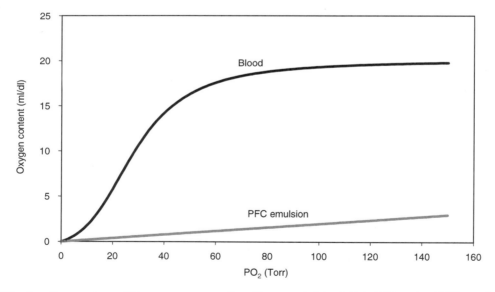

Fig. 1. Comparison of the oxygen capacity of isooncotic blood (14 g/dL) and a PFC emulsion. Note that the tetermeric structure of hemoglobin and its cooperativity lead to nearly complete saturation at the arterial oxygen partial pressure of ~100 mmHg.

One perfluorocarbon emulsion, Fluosol-DA, a 20% by weight emulsion, was licensed for use in coronary angioplasty. However, due to poor market acceptance and limited efficacy, it was subsequently withdrawn from the market by its manufacturer. An emulsion (oxygent) with a greater amount of perfluorocarbon (and, hence, O_2 capacity) was tested extensively, but also discontinued because of poor results in clinical trials (34).

2.9. Cell-Free Hemoglobin. Hemoglobin seems to be the logical choice for a red cell substitute because of its high capacity to carry oxygen (Fig. 1) and its oncotic properties. Probably the first treatment of anemic patients with hemoglobin solution occurred in 1898 (35). Although the results were encouraging, stable solutions could not be prepared and the studies were not pursued further. Better preparations were reported in 1916 (36) when very small amounts of hemoglobin were administered in an effort to discover its renal threshold. No untoward reactions were reported in 33 subjects.

After these reports, there were many attempts to administer hemoglobin solutions to humans. Some patients did well, but others demonstrated hypertension, bradycardia, oliguria, and even anaphylaxis. These untoward effects were not correlated with specific biochemical properties of the solutions themselves.

2.10. Modified Hemoglobin. Whereas interest in hemoglobin-based red cell substitutes remained extremely high, particularly in wartime, difficult problems impaired progress. First, hemoglobin in dilute solution is rapidly cleared by the kidney, as the tetrameric protein dissociates into smaller dimers. Second, dilute hemoglobin has a very high affinity for oxygen, and it was feared that little of the bound oxygen would be released in tissue capillary beds. Third, even exceedingly small amounts of stromal, ie, cell membrane, contaminants, or endotoxin in hemoglobin solutions, appeared to be toxic.

The rapid disappearance of hemoglobin from the circulation was solved when it was discovered that cross-linking with bis(N-maleimidomethyl) ether (BME) prolonged plasma retention (37,38). It was concluded that cross-linking reduced the tendency to form dimers, and therefore the hemoglobin was not filtered by the kidney. Accordingly, most of the hemoglobin was found in various tissues, but not in the urine.

The problem of the high oxygen affinity of cell-free human hemoglobin was solved when reagents were discovered that could bind to the 2,3-diphosphoglycerate (2,3-DPG) binding site and thus reduce the affinity for oxygen (39). This discovery led to a variety of hemoglobin modifications that not only reduce oxygen affinity but also stabilize the hemoglobin tetrameric structure so that vascular retention was prolonged. The most widely used of these agents is pyridoxal 5′-phosphate [54-47-7] (PLP), $C_8H_{10}NO_6P$ (39).

In the 1960s, it was believed that contradictory toxicity reports could be explained by contamination of the solutions with foreign materials. Novel ways to remove stroma from red cell hemolysates were studied, and the phrase stroma-free hemoglobin (SFH) was coined (40). These included filtration methods that could be applied to large volumes of hemolysate, making possible physiologic studies in large animals. As a result, it was generally accepted that the toxic effects of hemoglobin could be prevented by rigorous purification.

Several pure hemoglobin solutions were later produced on a large scale for experimental use. A procedure was described for crystallization of hemoglobin and the product was evaluated in a series of animal trials (41–44). A 6-g/dL hemoglobin solution that had a p50 of (18–20 Torr) was used in studies of tissue distribution (45) as was a similar solution of stroma-free hemoglobin. This hemoglobin was used for many basic studies of O_2 transport (46) as well as for a clinical trial in humans (47). The term p50 corresponds to the partial pressure of oxygen at which 50% of the oxygen binding sites are filled.

The polymerization of proteins using the tissue fixative glutaraldehyde [111-30-8]. The compound $C_5H_8O_2$ was described in 1973 (48). Soon, a process for polymerizing hemoglobin with the agent was patented (49). This material demonstrated a markedly prolonged intravascular retention time. Although the reaction is extremely difficult to control, products for infusion have been developed (45,50,51). For example, PLP-polyhemoglobin, obtained by reaction with PLP followed by polymerization with glutaraldehyde, was the first modified hemoglobin to be used in published human trials (52). Now, many preparations of modified hemoglobin have been tested in animals and humans. It appears that most are efficacious in transporting oxygen, and the nature of the specific modification affects biological properties such as plasma retention time, oxygen affinity, and colloid osmotic pressure, more than the toxicity.

Many other variations of modified hemoglobin have been studied, including those stabilized using various cross-linkers. Some products are derived from hemoglobin conjugated to synthetic materials such as dextran [9004-54-0] or poly(ethylene glycol) [25322-63-3]. Sources other than human outdated blood also have been investigated, including cow and recombinant hemoglobins produced in bacteria, yeast, and even transgenic mammals.

2.11. Encapsulated Hemoglobin. Because hemoglobin is normally packaged inside a membrane, encapsulated hemoglobin is thought to be the ultimate red cell substitute. The use of microencapsulated hemoglobin as artificial red blood cells was reported in 1957 (53). Since that time, dramatic results have been reported in the complete exchange transfusion of laboratory animals (54), but progress toward development of an artificial red cell for human use has been slow. The reason is because of reticuloendothelial and macrophage stimulation problems (55). Other problems include maintaining sterility and endotoxin contamination.

2.12. Synthetic Heme. Synthetic compounds that bind or chelate O_2 have been produced. These are commercially attractive because manufacture and licensure might be developed as a drug, rather than as a biological product. Synthetic heme can be used to transfuse animals (56). Although synthetic O_2 carriers would avoid any potential limitations of hemoglobin as raw material, the synthetic procedures are very tedious, and the possibility of scale up seems remote.

3. Hemoglobin Modifications

3.1. Reactivity. Hemoglobin can exist in either of two structural conformations, corresponding to the oxy (R, relaxed) or deoxy (T, tense) states. The key differences between these two structures are that the constrained T state has a much lower oxygen affinity than the R state and the T state has a lower tendency

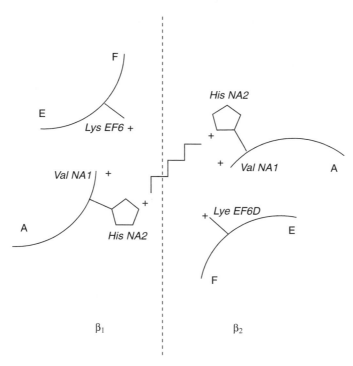

Fig. 2. Reaction of 2,3-DPG and deoxyhemoglobin. The molecule fits into the central cavity of hemoglobin and forms salt bridges with valine NA1(1)β, histidines NA2(2)β, H121(143)β, and lysine EF6(82)β. A, E, and F refer to specific hemoglobin helices and NA is the sequence from the amino-terminals to the A helix.

to dissociate into subunits that can be filtered in the kidneys. Therefore, stabilization of the T conformation would be expected to both reduce renal filtration and maintain oxygen affinity similar to that of red cells. The transition between the T and R states of hemoglobin is also deeply involved in the critical for the Bohr (pH) effect and cooperativity. Therefore stabilization of either of the two structures should diminish these effects, which might have important physiologic consequences.

Stabilization of the T conformation under normal conditions is illustrated by the reaction of 2,3-diphosphoglycerate, (2,3-DPG) (Fig. 2). The negative charges on this polyphosphate form electrostatic, reversible interactions with eight positive charges on hemoglobin: two α-amino groups of valine NA1(l)β, two ε-amino groups of lysine EF6(82)β and four histidines, NA2(2)β and HC3(l43)β. In the R, state the dimensions of the pocket change enough so that 2,3-DPG does not fit as well, and it drops out. Thus 2,3-DPG preferentially stabilizes the T conformation and has an overall effect of reducing oxygen affinity and increasing cooperativity. Analogues of 2,3-DPG, used to modify hemoglobin by forming permanent covalent bonds, are variously effective, depending on molecular dimensions and charge. Some of the compounds bind to only one of the reactive amino groups in the 2,3-DPG pocket; others react with all four.

Table 2. **Classes of Hemoglobin Modification**[a]

Class	Examples
amino-terminal modification	carbamylation
	carboxymethylation
	pyridoxylation
	acetaldehyde
Lys EF6(82)β modification	mono(3,5-dibromosalicyl)fumarate
Val NA1(1)β-Lys EF6(82)β cross-link	2-nor-2-formylpyridoxal 5′-phosphate
	(NFPLP)
	bis(pyridoxal) tetraphosphate (bis-PL)P$_4$
Lys G6(99)α$_1$-Lys G6(99)α$_2$ cross-link	bis(3,5-dibromosalicyl)fumarate
2,3-DPG analogue	pyridoxal 5′-phosphate
surface, multisite	glutaraldehyde
	polyaldehydes
	ring-open dials
	diimidate esters
conjugated hemoglobin	dextran, starch aldehydes
	poly(ethylene glycol)

[a] See Ref. (4).

In addition to the 2,3-DPG pocket, human hemoglobin contains 40 reactive lysines, ie, ε-amino groups, two α-chain N-terminal α-amino groups, and two sulfhydryl groups, ie, cysteine F9(93)β. Most of the lysines are on the surface of the molecule, but some are internal, such as lysine G6(99)α. Thus the groups can be accessed by various cross-linkers and polymerizing agents, especially aldehydes. Although the lysine groups provide many potential sites for modification, their large number also means that such reactions are difficult to control.

All of the reactions considered to be useful in the production of hemoglobin-based blood substitutes use chemical modification at one or more of the sites discussed above. Table 2 lists the different types of modifications with examples of the most common reactions for each. Differences in the reactions are determined by the dimensions and reactivity of the cross-linking reagents. Because the function of hemoglobin in binding and releasing oxygen is intricately connected to the transition between T and R conformations, it is not surprising that the p50 yields are highly variable. Even small differences among structures of the reagents can yield products having very different properties. In addition, the conditions of the reaction are very important, not only in regard to the state of ligation, ie, oxygen saturation, but also in regard to the presence of agents or molecules that block or compete for certain reactive sites.

A further complication of these reactions is that many nonhemoglobin proteins also contain reactive groups that may also be co-modified. These molecules, if present at the time of reaction, could affect the properties of the final solution. For this reason, derivatives prepared for studies of the hemoglobin molecule per se, must start with highly purified stroma-free hemoglobin.

3.2. Amino-Terminal Modification. *Carbamylation.* Modification of the amino-terminal groups of hemoglobin (Fig. 3) by the carbamylation reaction using isocyanic acid [75-13-8] was used to show that valine NA1(1)a is one of the residues involved in the alkaline Bohr effect, and that the sickling of cells containing hemoglobin S could be inhibited specifically (57). It was also used to

$$Hb-NH_2 \ + \ HN=C=O \ \longrightarrow \ Hb-NH-\underset{\underset{O}{\|}}{C}-NH_2$$

Fig. 3. Carbamylation of the α-amino groups of hemoglobin.

show the site of binding of carbon dioxide. The products of these reactions have increased oxygen affinity if the α-chains are carbamylated and decreased affinity if the β-chains are carbamylated. The carbamylation reaction held great promise in the control of sickling, but orally administered sodium cyanate was toxic.

Carboxymethylation. Other modifications of the amino-terminal groups of hemoglobin have also been studied. It was reasoned that a covalent adduct at the amino-terminal amino group might reduce oxygen affinity by lowering or neutralizing the net positive charge in the 2,3-DPG pocket (58). Carboxymethylation using glyoxylic acid [298-12-4] followed by reduction with sodium borohydride [16940-66-2], $NaBH_4$, resulted in a product that demonstrated lowered oxygen affinity and nearly intact Bohr and carbon dioxide effects (59). X-Ray and nuclear magnetic resonance (nmr) studies confirmed that the introduced group occupies nearly the same position as the naturally occurring carbamino group, ie, a carbon dioxide adduct (60).

Acetaldehyde. Acteldehyde [75-07-0], C_2H_4O, has a slightly different reaction mechanism with hemoglobin (61). Although this reagent reacts with surface lysines under some conditions, the principal products are derivatives of the amino-terminal groups, both of the α and β chains. These products are not reduced with sodium borohydride, and therefore do not involve an intermediate Schiff's base. Instead, a stable cyclic imidazolidinone derivative is formed.

Pyridoxal Derivatives. Various aldehydes of pyridoxal react with hemoglobin at sites that can be somewhat controlled by the state of oxygenation (62). It is thereby possible to prepare derivatives having a wide range of functional properties. The reaction of PLP with hemoglobin first involves the formation of a Schiff's base between the amino groups of hemoglobin and the aldehyde(s) of the pyridoxal compound, followed by reduction of the Schiff's base with sodium borohydride, to yield a covalently linked pyridoxyl derivative in the form of a secondary amine. This reaction has been used widely to reduce the oxygen affinity of the final product (62).

Pure diPLP-hemoglobin, in which both β-chain amino termini are modified, was isolated by column chromatography. The structure was confirmed by X-ray diffraction (63) and peptide analysis (64). An electrostatic interaction of the 5′-phosphate with the 2,3-DPG binding site, lysine EF6(82)β, was shown, so that this modification closely mimics the action of 2,3-DPG in stabilizing the deoxy conformation. The oxygen affinity of the derivative was found to be about one-half of that of unmodified hemoglobin under similar conditions, but a degree of cooperativity was preserved. Equilibrium and kinetic ligand-binding studies on this derivative (65) showed a perturbed R state.

The reaction of hemoglobin with PLP was scaled up (66,67) to batches of 20 L yielding 70–80% modified hemoglobin. Methemoglobin was <10%, and the material was apparently unchanged after infusion into baboons. This solution was effective in resuscitation from hemorrhagic shock (68,69), but the plasma retention was thought to be too short and colloid osmotic pressure (COP) too high to be a definitive red cell substitute (70–72). A major problem with the large-scale preparation of pyridoxylated hemoglobin was the heterogeneity of reaction products, probably representing modifications at either or both of the α- and β-amino-terminal residues as well as surface lysines.

3.3. Lysine EF6(82)β Modification. In this reaction, sometimes called a "pseudolink" (73) hemoglobin reacts with the monofunctional reagent, mono-(3,5-dibromosalicyl)fumarate, in oxygenated conditions. The product is specifically acylated at lysine EF6(82)β, in about 70% yield. Although cooperativity is reduced somewhat. ie, to a Hill coefficient of 2.0, the p50 under physiologic conditions is ~25 Torr, and carbon dioxide binding is intact, because the sites for carbon dioxide binding are unaffected. It is of particular interest that the tetramer–dimer dissociation is retarded, possibly by stabilization at the β–β interface (74). The resulting plasma retention half-time in the rat is also prolonged by about fourfold for this acylated material, as compared to unmodified hemoglobin.

3.4. Valine NA1(1)β-Lysine EF6(82)β Cross-link. *2-Nor-2-Formylpyridoxal 5′-Phosphate.* 2-Nor-2-formylpyridoxal 5′-phosphate, CHNOP, (NFPLP), is of special interest because it contains two reactive aldehyde groups and reacts as shown in Figure 4 at two sites: at the amino-terminal group of one β-chain and at lysine 82 of the other (63,75). Thus in one modification reaction, this reagent both reduces the oxygen affinity of native hemoglobin and prevents its dissociation into αβ-dimers.

NFPLP has been studied extensively. Because hemoglobin dimerization is prevented, NFPLP is not eliminated in the urine (76–78) and the plasma retention of the modified material is at least three times that of either unmodified hemoglobin or pyridoxylated hemoglobin (79). Tissue distribution and elimination have been documented in detail (80–82). Accumulation of this modified

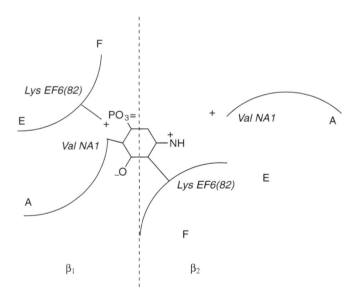

Fig. 4. Reaction of NFPLP and hemoglobin. A, E, and F refer to helices; NA is the sequence from the amino-terminals to segment A.

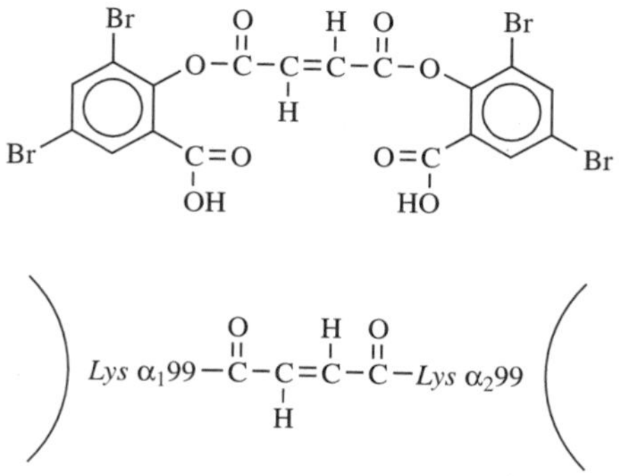

Fig. 5. Structures of (**a**) bis DBBF and (**b**) the αα-hemoglobin cross-link (4).

hemoglobin derivative in the kidney is much reduced as compared to unmodified hemoglobin (83). The oxygen affinity of the derivative under physiologic conditions is ~47 Torr, with cooperativity retained. When used to perfuse isolated organs, the derivative supports a higher tissue oxygen tension in both the rabbit heart (84) and rat liver (85).

Although the 60–80% yield of the NFPLP product may have been satisfactory for commercialization (86), the main drawback was difficulty in preparation of the reagent itself.

Bis(pyridoxal) tetraphosphate. A second class of bifunctional reagents, described in 1988, involves two pyridoxal groups linked by phosphates of different lengths (87) (Fig. 5). As shown in Table 3, the yield of intramolecularly cross-linked hemoglobin increases dramatically with increasing length of the phosphate backbone. It is believed that the site of reaction of (bis-PL)P_4 is between the amino-terminal amino group of one β-chain and the lysine 82 of the other β-chain, as for NFLP (87). However, the distance between these two residues is only 1.1 nm, and the reagent is much longer. Therefore it is concluded that the cross-linker must fold back upon itself to form a stacked pyridine ring conformation.

Further study of (bis-PL)P_4 modified hemoglobin (88) showed its P50 to be 31 Torr (pH 7.4, pCO$_2$ 40 Torr, 37°C), with a Bohr effect about one-half of that of unmodified hemoglobin. Its plasma retention is prolonged threefold in the rat,

Table 3. **The Effect of Polyphosphate Backbone Size on the Resulting Hemoglobin Polymer Size**[a]

Compound	R	% Tetramer
(bis-PL)P_2	0	16
(bis-PL)P_3	orthophosphate	18
(bis-PL)P_4	pyrophosphate	68
CH$_2$-(bis-PL)P_4	methylene diphosphate	53
fructose (bis-PL)P_4	fructose 1,6-diphosphate	41
PLP-DPG-PLP	2,3-diphosphoglycerate	70

[a] Refer to Figure 5.

and there was no apparent toxicity in screening studies. An attractive feature of (bis-PL)P$_4$ is that its synthesis is much simpler than that of NFPLP–hemoglobin (89).

Other 2,3-Diphosphoglycerate Pocket Cross-Linkers. The reactivity of the valine NA1(l)α and lysine EF6(82)β residues in the 2,3-DPG pocket shown by NFPLP and (bis-PL)P$_4$ stimulated the search for other reagents that react similarly but have potentially greater efficiency and ease of scaleup. The systematic study of four different dicarboxylic acid derivatives, cross-linked in both oxygenated and deoxygenated conditions, has been reported (90). Each of these derivatives presents problems in purification, and proof of the sites of reaction is tedious.

3.5. Lysine G6(99)α$_1$-Lysine G6(99)α$_2$ Cross-Link. A class of bifunctional reagents that cross-link human hemoglobin internally to preserve the native dimensions of the molecule has been very useful in the production of a very well characterized product for research purposes (91–94). The derivatives increased the oxygen affinity of native hemoglobin and were thought to have potential in preventing sickling in patients having sickle-cell anemia. When oxyhemoglobin was cross-linked using his(3,5-dihromosalicyl)fumarate [71337-53-6], (DBBF) C$_{18}$H$_8$Br$_4$O$_6$, the reaction site was shown to be between lysine EF6(82)β$_1$ and lysine EF6(82)β$_2$(93). However, when cross-linking was carried out in deoxyhemoglobin, the α-chains were modified (95,96) (Fig. 5).

This hemoglobin derivative was proposed to be developed as a blood substitute (96,97) because a single modification could achieve the dual goals of reduced oxygen affinity and restricted tetramer–dimer dissociation. The product, called αα-Hb by the U.S. Army and DCLHb by Baxter, was formulated in Ringer's lactate. The p50 under physiologic conditions was approximately that of human blood (98), Hill's parameter was 2.2, and the Bohr effect was reduced (99). Plasma retention was increased, and the product appeared to be less heterogeneous than some of the other derivatives under study. An interesting property of αα-Hb is its thermal stability, which was exploited to achieve both a partial purification of the crude reaction mixture after cross-linking and inactivation of viruses in the final product (100,101). Its production was scaled up by Baxter Healthcare, under contract to the U.S. Army, but later abandoned because of its propensity to cause vasoconstriction in animals and humans (102).

3.6. Surface, Multisite Reagents. Surface modification of hemoglobin with multifunctional aldehydes has been one of the most popular modifications because it results in large aggregates of molecules with potentially prolonged intravascular retention time. An inherent problem is that the extent of polymerization may be both nonspecific and difficult to control. Glutaraldehyde is a prime example of such a reagent (Fig. 6).

Glutaraldehyde. Polymerization of pyridoxylated human hemoglobin using glutaraldehyde was first reported in 1980 (103). In the years that followed,

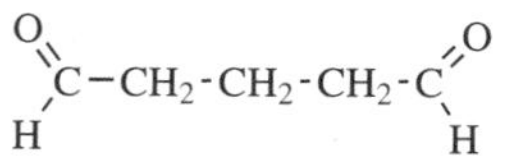

Fig. 6. Glutaraldehyde.

Table 4. **Properties of a Typical Pyridoxylated Glutaraldehyde-Polymerized Hyman Hemoglobin**[a]

hemoglobin concentration, g/dL	12–14
methemoglobin (%)	<5
molecular weight (kDa)	64–400
mean molecular mass (kDa)	150
p50 (Torr)	~18
n (Hill parameter)	1.5–2.2
Bohr factor (Δp50/ΔpH)	−0.12 to −0.25
colloid osmotic pressure, mmHg	~25
viscosity (cPs)	~2
endotoxin (EU/mL)	<0.6
rabbit pyrogen test	Pass

[a] see Ref. 4.

this polyhemoglobin was studied intensively. A research team at Michael Reese Hospital in Chicago commercialized a glutaraldehyde-polymerized human hemoglobin product that has been tested extensively in humans (104).

The production process begins with pyridoxylated hemoglobin (14–16 g/dL) that is then polymerized using a 12.5% solution of glutaraldehyde. When the COP of the reaction mixture reaches normal values of 20–25 mmHg, the reaction is quenched by the addition of an amino acid such as 1.3 M lysine. The resulting product has a distribution of molecular masses from 120 to 600 kDa, a p50 of 2.1 ~16 Torr, a Bohr effect reduced by one-half, and a Hill coefficient of 1.7. The viscosity is ~2 cP, compared to 4 cPs for human blood, and the solution has no effect on coagulation, as measured by the prothrombin and partial thromboplastin times. Some of the properties of glutaraldehyde hemoglobin are shown in Table 4. Physiologic studies with a glutaraldehyde derivative with a molecular weight of 124 kDa (51) showed that the product transports oxygen as expected and that the reduced p50 did not diminish its usefulness (105). The plasma half-life in baboons was up to 46 h, compared with ~6 h for PLP–hemoglobin (70).

Reexamination of the products of the glutaraldehyde reaction of pyridoxylated hemoglobin revealed extreme heterogeneity (106–108) and showed that the products are unstable on storage at 4°C. Rearrangements of polymeric species occur so that it is difficult to prepare a predictably modified species. This heterogeneity and instability are regarded as serious drawbacks to the product because reactions with plasma proteins in vivo would he impossible to predict and toxicity would be difficult to understand. Concern has been raised (108) that the low molecular weight material might be preferentially lost through the kidneys, leaving the inherently less stable polymers with the less favorable oxygen-transport properties.

Glutaraldehyde treatment of hemoglobin has the effect of making the tetrametric structure of the molecule more rigid. Indeed, it seems that the more highly modified the polymerized hemoglobin molecules are, the more rigid they become, as reflected by increasing oxygen affinity and decreasing cooperativity. Studies using the very sensitive Mossbaur technique (109) have shown that glutaraldehyde-treated hemoglobin has an increased rate of autooxidation and increased thermal stability. These properties could be explained by a weakening of the hemoglobin linkage.

Toxicology studies with glutaraldehyde products are of great concern because glutaraldehyde can leach out of prosthetic devices (110,111). Glutaraldehyde is also used as a tissue fixative, and even small amounts have been found to have cytotoxic activity (112).

Other Polyaldehydes.   Other dialdehyde reagents can be prepared by oxidizing the ring structures of sugars or nucleotides (113). These reagents can react with hemoglobin at any of its amino groups, and therefore form a variety of modifications including intramolecular and intermolecular links. One example of this type of modification involves opening the ring of inositol tetraphosphate. Another example, involves the opening of the pyridine ring of ATP (114) to form modified ATP−hemoglobin. This latter product was reported to have an elevated p50 and normal cooperativity.

Optimization of the ATP−hemoglobin reaction conditions produced a derivative with a reduced oxygen affinity. Five fractions from a reaction mixture, when isolated, were found to have P50 values ranging from 8 to 38 Torr. Most fractions have little cooperativity (115). These results are consistent with those found with other polyfunctional reagents that react on the surface of hemoglobin.

Diimidate Esters.   Diimidate esters bifunctional reagents have been used in cross-linking a variety of proteins including hemoglobin. In a typical reaction, a lysyl ε-amino group reacts with the ester. The reagent is specific for surface ε-amino groups and forms polymers of varying size. One of the advantages of the reaction is that it replaces the $-NH_3^+$ group with an $=NH_2^+$ group, so the overall charge is unchanged. One reported product (116) had 30 of the 44 surface lysyl residues modified and had a molecular mass ranging from 68 to 600 kDa. Intravascular retention time was increased about fourfold in rabbits.

Zero-Link Polymerization.   Zero-link polymerization is a variety of polymerization developed by researchers at the University of Maryland (117). This product is called "zero-linked" hemoglobin (ZL-HbBv) because the chemistry involves direct coupling hemoglobin molecules together without using polymerizing agents such as glutaraldehyde or other bifunctional agents (Fig. 7). The key property of ZL-HbBv appears to be its very large molecular size, which results in reduced extravasation compared to native hemoglobin. The developers of ZL-HbBv believe that this reduced extravasation avoids NO binding that would cause vasoconstriction. This was based on the theory that hemoglobin in interstitial space more effectively binds NO than intravascular hemoglobin.

3.7. Conjugated Hemoglobin.   An alternative approach to prolongation of the plasma retention is to conjugate hemoglobin to a larger molecule, which was first done by coupling hemoglobin to dextran (118−120). The coupling reaction is carried out using a lysate of human red cells and bromodextran, molecular weight 20 kDa. The product was shown to support life in the absence of red cells in dogs and cats (121), and it did not appear to be immunogenic (122). Because the oxygen affinity of dextran-hemoglobin was essentially that of hemoglobin, it was modified further by covalently linking an analogue of inositol hexaphosphate (123,124). This new derivative had a p50 of 55 Torr, compared to 23 Torr) for dextran-hemoglobin, and the oxygenation curve showed cooperativity. These modifications were demonstrated to reduce renal toxicity of unpurified hemoglobin.

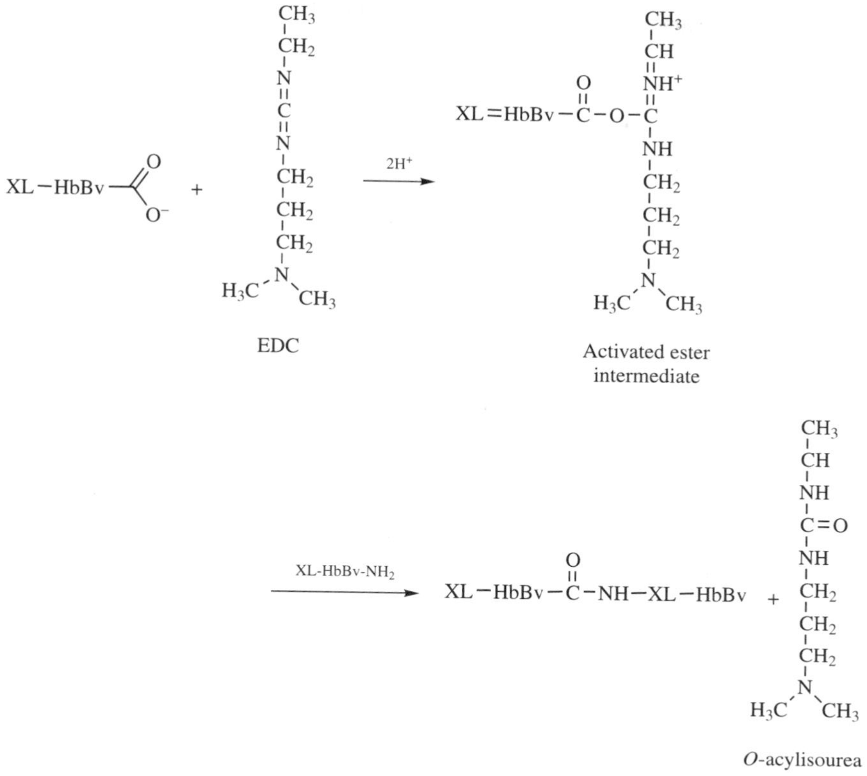

Fig. 7. Production of ZL-HbBv. Deoxygenated bovine hemoglobin is first internally cross-linked with bis(3,5-dibormosalicyl)-adipoate to form XL-HbBv. These molecules are then polymerized at free carboxyl groups after activation with 1-ethyl-3-(3-dimethyl-aminopropyl)carbodiimide (EDC)(117).

Hemoglobin also can be conjugated to other synthetic polymers such as inulin [9005-80-5] (125), poly(vinylpyrrolidinone) [9003-39-81], $(C_6H_9NO)_x$ (126), and poly(ethylene glycol) [25322-68-3], $(C_2H_4O_n)H_2O)$ (127,128). One of the most studied of these conjugates is pyridoxylated hemoglobin- polyoxyethylene (PHP) (127). PHP has a molecular mass of ∼90 kDa and a p50 of ∼22 Torr, compared to 15 Torr for hemoglobin under the same conditions, but is has reduced cooperativity (129). The plasma retention half-life in dogs is ∼36 h, and apparently causes no renal, hepatic or coagulation toxicity. Histologic examination of the lungs and spleen of transfused dogs at 2 weeks and later showed no abnormalities (130).

A new conjugated hemoglobin product is based on a novel view of the physiology governing microvascular blood flow. According to this theory, microvascular vasoconstriction in the case of cell-free hemoglobin does not result entirely from NO scavenging, but from oversupply of O_2 to regulatory arterioles because of facilitated diffusion of hemoglobin in the plasma space (131). The product MP4, controls facilitated diffusion by its large molecular size and high O_2 affinity. It is produced by novel attachment of six strands of maleimide-poly- (ethylene glycol) (5 kDa) to human hemoglobin in a site-specific

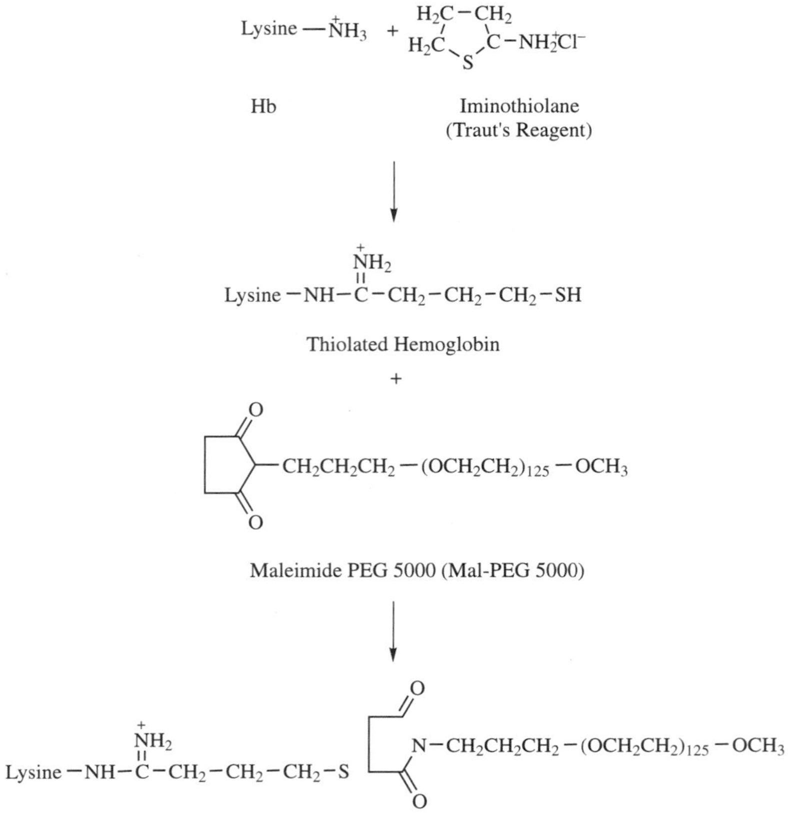

Fig. 8. Schematic of the maleimide-PEG modification of human hemoglobin (132).

reaction (132) (Fig. 8). The MP4 has been shown to be free of vasoconstriction in the hamster microcirculation and does not elevate systemic or pulmonary vascular resistance in pigs. It also does not cause myocardial lesions in rhesus monkeys. The chemistry, as shown in Figure 8, involves the surface modification of ε-amino groups of lysine followed with site-specific coupling of maleimide-activated PEG (133). The product is extremely homogeneous.

3.8. Recombinant Hemoglobin. Human hemoglobin has been expressed in a variety of recombinant systems, but in large scale in Escherichia coli (134,135). A novel molecule that was modified to genetically fuse the two α chains preventing subunit dissociation was called rHb1.1. The properties were almost identical to DCLHb and ααHb. This product was abandoned as a blood substitute candidate because of pronounced gastrointestinal side effects and hypertension. These effects were attributed to NO scavenging. Consequently, a series of mutants with altered NO affinity was developed (136). A product, rHb 2.2, shows reduced NO affinity, but has intact O_2 binding. It is free of hypertension in rats and did not produce microscopic necrotic lesions in monkey hearts (137). Some information about this product is available (138), but data have not yet been published to prove that reduced vasoactivity is a result of altered NO binding (139).

4. Hemoglobin Sources

4.1. Purification.　Hemoglobin is provided by the red blood cell in highly purified form. However, the red cell contains many enzymes and other proteins, and red cell membranes contain components, such as phospholipids, that could be toxic. Furthermore, plasma proteins and other components could trigger immune reactions in recipients. The chemical modification reactions discussed here are not specific for hemoglobin they may modify other proteins as well. Indeed, multifunctional reagents could actually couple hemoglobin to nonhemoglohin proteins.

Rabiner's method (40) for the filtration purification of hemoglobin was thought to be a significant advance over older centrifugation methods (140,141). However, hemoglobin prepared in this way still caused unwanted reactions in human recipients (47). The crystallization method (142) showed fewer toxic effects in animals (142), but batch-to-batch reproducibility was uneven. Ultrapurification of hemoglobin using ion-exchange chromatographic technique is possible but is tedious and expensive (143).

4.2. Outdated Human Blood.　If clinical efficacy and safety of hemoglobin solutions can be shown, the demand for product would soon outstrip the supply of outdated human blood. About 12 million units of blood (1 unit is ~480 mL) are used in the United States each year, and only about 1% outdate. The primary use of blood is in intraoperative and emergency settings. The quantity of blood available for use in production of blood substitutes depends on the willingness of donors who are qualified to donate, and the efficient matching of donor blood to the recipients.

4.3. Bovine Hemoglobin.　One solution to the hemoglobin supply problem is to use nonhuman sources of blood as a starting material such as cows (bovine hemoglobin). The ultimate success of bovine, or any other hemoglobin, depends on demonstration of safety, not supply. One problem in using bovine hemoglobin is the fear of bovine spongiform encephalitis (BSE) virus. This virus, related to the Scrapie organism, has been detected in cows in Europe as well as other mammals in North America. The variant Creuzfield-Jacob Disease (vCJD) has been associated with BSE in Europe, and it is known that BSE can be transmitted by blood in animals. Although there are no known cases of human transmission by blood transfusion at this time, the FDA has placed restrictions on the importation of blood from Europe into the United States. At this time, there is no adequate test for BSE in donated blood that could be implemented on a large scale.

Under conditions found in the red cell, bovine hemoglobin has a lower oxygen affinity than human hemoglobin because of its greater sensitivity to anions (144–147). Thus instead of regulation by 2,3-DPG, as is the ease with human hemoglobin, bovine hemoglobin oxygen affinity is regulated by chloride ion.

Bovine hemoglobin has been cross-linked using the bifunctional reagent DBBF to obtain a product with a p50 in excess of 40 Torr under physiologic conditions (148). Although the reaction mixture was somewhat heterogeneous, no uncross-linked material was detected by sedimentation velocity analysis. The plasma retention in rats was prolonged 10-fold as compared to unmodified hemoglobin. It also has been found that the pyridoxylation reaction raises the p50 from 28 to 38 Torr, and glutaraldehyde polymerization drops the p50 to 18 Torr (149). The polymerized material had essentially the same plasma retention time as human hemoglobin modified in the same way, and rats could also be supported at zero hematocrit.

4.4. Recombinant Hemoglobin.

An alternative and novel source of hemoglobin, which is used for modification, is from microorganisms, the genome of which has been modified to contain globin genes for recombinant hemoglobin (rHb) production (see GENETIC ENGINEERING, MICROBES). Significant strides have been made in this approach, and it is possible to express both human α- and β-globin chains in E. coli (134,135).

Much of the development work with recombinant hemoglobin for commercial purposes has been done in commercial laboratories, so not all details of the process are available. However, it is likely that production on the scale needed for a viable red cell substitute product could be a problem. One unit of blood (500 mL) contains ~15 g/dL of hemoglobin, so a total of 75 g of hemoglobin would be needed to produce a unit. If the yield is 0.1 g/L of culture medium, 750 L of cell culture would be needed. In the future, it might be possible to express hemoglobin genes in higher organisms; synthesis of functional human hemoglobin has already been reported in yeast, transgenic mice (150), and pigs (151), however, these approaches present additional purification, logistic, and economic problems. Purification of rHb could also be a significant challenge, since it would need to be separated from media components and other microorganism products. Endotoxin contamination could be a serious problem for bacteria.

5. Current Status of Artificial Blood

The magnitude of the undertaking to produce "Artificial Blood" cannot be overestimated. In the past 20 years, >$2 billion have been expended on attempts to do so. To date, only one product has been presented to the FDA for licensure approval. At the same time, several prominent attempts, fueled by significant investments from large pharmaceutical companies have failed.

Some of the major problems that have plagued the development of hemoglobin solutions appear to be solved. For example, renal failure, a consequence of renal filtration of dissociated hemoglobin molecules has been eliminated by cross-linking or polymerization chemistry and strict elimination of unreacted hemoglobin. Modern perfluorocarbon emulsions can be stored at refrigerator temperature and need not be reconstituted prior to use. Optimal sizing of the emulsion particles has decreased some of the side effects that have hampered development of earlier emulsions. However, some significant scientific problems remain to be solved or clarified in the field.

For hemoglobin-based products, there is still no agreement on the mechanism of vasoconstriction. This problem does not affect all products. Most studies in the published literature are based on αα-Hb or DCL-Hb, mainly because these products have been readily available to researchers. Understandably, this has been a sensitive subject with commercial developers, and fewer basic studies are available with polymerized products that are in advanced clinical trials.

Both theories of vasoconstriction, NO binding and O_2 autoregulation, are being studied. To date it is not certain if they are mutually exclusive or what their physiological and clinical consequences are.

The mechanism of the frequently reported gastrointestinal (GI) side effects of hemoglobin solutions is not clear. Studies have been done in animals (152) showing that some hemoglobin solutions interfere with esophageal motility and gastric emptying. However, how plasma hemoglobin can interfere with smooth muscle function is not established. Other possible concerns center around the effect of hemoglobin on cellular function. Some preparations stimulate monocytes and macrophages (153) and others degranulate basophils and reduce eosinophil mobility (154). Still other reports suggest that neutrophil function and platelet-endothelial adhesion can be affected. Work is still in the early stages on these phenomena, and it is not yet established whether or not they have clinical implications.

The relatively short intravascular persistence of any of the products remains a problem for certain clinical applications such as chronic anemia, but should not impact use in many elective surgical procedures or in trauma, where the patient's own marrow should replenish red cells in a few days. This issue is complicated by the fact that detailed metabolic pathways for many of the products are still not completely defined, so the impact of the rate of breakdown is not known.

On the positive side, clinical trials with all of the products studied so far have shown the ability to reduce the number of patients that receive allogeneic transfusions and the number of units transfused in elective surgical procedures. The U.S. FDA has stated publicly that such reduction or elimination of allogeneic blood could be a basis for approval of these products. Establishing reduced mortality as an end point is a more difficult task, and the FDA has stated that while it will not require such demonstration as a prerequisite to approval, it will require phase II clinical trials in trauma to establish safety in this application (155).

Finally, the cost of new products will be an important issue once regulatory approval has been won. Some commercial developers and financial analysts believe that a safe and effective product that can reduce the number of patients who receive transfusions or the number of units of allogeneic blood transfused, will command a price significantly higher than that of banked blood. However, as no products are as yet approved, this remains to be seen.

BIBLIOGRAPHY

"Blood, Artificial", in *ECT* 4th ed., Vol. 4, pp. 312–333, by Robert M. Winslow, University of California, San Diego; "Blood, Artificial" in *ECT* (online), posting date: December 4, 2000, by Robert M. Winslow, University of California, San Diego.

CITED PUBLICATIONS

1. W. Harvey, *Exercitatio Anatomica de Motu Cordiset Sanguinis in Animalibus*, Leach, London, 1653.

2. L. Diamond, A history of blood tranfsusion, in *Blood, Pure and Eloquent*, M. Wintrobe, ed., McGraw-Hill, New York, 1980, pp. 659–690.

3. M. Hollingsworth, *Ann. Med. Hist.* **X**, 213 (1928).

4. R. Winslow, *Hemoglobin-based Red Cell Substitutes*, Johns Hopkins University Press, Baltimore, M.D, 1992.

5. T. Thomas, *N.Y. State J. Med.* **27**, 449 (1878).

6. C. Jennings, *Brit. Med. J.* **1**, 1147 (1885).

7. J. Bovell, *Can. J.* **3**, 188 (1955).

8. J. Brinton, *Med. Rec.* **14**, 344 (1878).

9. C. Guthrie and F. Pike, *Am. J. Physiol.* **18**, 14 (1907).

10. W. Kühne, *Arch. Anat. Physiol. Med.* 769, 1859.

11. W. von Moraczewski, *Pflug. Arch. Physiol.* **77**, (1899).

12. S. Ringer, *J. Physiol. Cambridge* 380 (1880).

13. A. Hartmann and M. Senn, *Trans. Am. Pediat. Soc.* **44**, 56 (1932).

14. J. Miller and C. Poindexter, The effects observed following the intravenous and subcutaneous administration of fluid. An experimental study on dogs, *J. Lab. Clin. Med.* **18**, 287 (1932).

15. P. Lamson, B. Robbins, and M. Greig, *J. Pharm. Expe. Therapeutics* **83**, 225 (1945).

16. C. Ludwig, *S. B. Akad. Wiss. Wien* **48**, 731 (1863).

17. W. Bayliss, *Brit. J. Exper. Path.* **1**, 1 (1920).

18. Y. Henderson and H. Haggard, *J. Am. Med. Ass.* **78**, 697 (1922).

19. W. Amberson, *Biological Reviews* **12**, 48–86 (1937).

20. H. Bowditch, *Arbeit. Physiolog. Anstalt Leipzig* **6**, 139 (1871).

21. H. Kronecker and W. Stirling, *Beitrage zur Anatomie und Physiologie al Festgabe (av)*, Ludwig Gewidmet, Leipzig, 1875.

22. S. Ringer, *J. Physiol.* **6**, 361 (1885).

23. J. Moldovan, *Dtsch. Med. Wschr.* **2**, 2422 (1910).

24. O. Wangensteen, H. Hall, A. Kremen, and B. Stevens, *Proc. Soc. Exp. Biol. Med.* **43**, 616 (1940).

25. A. Kremen, H. Hall, H. Koschnitzke, B. Stevens, and O. Wangensteen. *Surgery* **11**, 333 (1942).

26. J. Dunphy and J. Gibson, *Surgery* **14**, 509 (1943).

27. E. Cohn, *Ann. Intern. Med.* **26**, 341 (1947).

28. J. Boldt, G. Haisch, S. Suttner, B. Kumle, and A. Schellhaass, *Brit. J. Anaesth.* **89**, 722 (2002).

29. T. Gan, E. Bennett-Guerrero, B Phillips-Bute, H. Wakeling, D. Moskowitz, Y. Olufolabi, S. Konstadt, C. Bradford, P. Glass, S. Machin, and M. Mythen, *Anesth. Analg.* **88**, 992 (1999).

30. L. Clark and F. Gollan, *Science* **152**, 1755 (1966).

31. H. Sloviter and T. Kamimoto, *Nature (London)* **216**, 458 (1967).

32. R. Geyer, R. Monroe, and K. Taylor, *Fead. Proc.* **27**, 384 (1968).

33. S. Gould, A. Rosen, L. Sehgal, H. Sehgal, L. Langdale, and L. Krause, *N. Engl. J. Med.* **314**, 1653 (1986).

34. P. Keipert, *Japan II(2)*, 47 (2003).

35. G. VonStark, *Deuts. Med. Wochenschr.* **24**, 805 (1898).

36. A. Sellards and G. Minot, *J. Med. Res.* **34**, 469 (1916).

37. H. Bunn and J. Jandl, *J. Exp. Med.* **129**, 925 (1969).

38. H. Bunn and J. Jandl, *Trans. Assoc. Am. Phys.* **81**, 147 (1968).

39. R. Benesch, R. Benesch, R. Renthal, and N. Maeda, *Biochemistry* **11**(19), 3576 (1972).

40. S. Rabiner, J. Helbert, H. Lopas, and L. Friedman, *J. Exp. Med.* **126**, 1127 (1967).

41. F. DeVenuto, A. Zegna, and K. Busse, *Surgery, Gynecol., and Obstet.* **148**, 69 (1979).

42. F. DeVenuto, H. Friedman, R. Neville, and C. Peck, *Surgery, Gynecol., Obstet.* **149**, 417 (1979).
43. F. DeVenuto, K. Busse, and A. Zegna, *Surg. Gynecol. Obstet.* **153**(3), 332 (1981).
44. F. DeVenuto, *Crit. Care Med.* **10**, 238 (1982).
45. K. Bonhard, *Experimentelle Erfahrungen mit Ha(umlaut)moglobin—Infusion-slo(umlaut)sungen.* S. Rappaport and F. Jung. Akademie-Verlag, Berlin (1975), pp. 221–225.
46. L. Djordjevich, J. Mayoral, and A. Ivankovich, *Anesthesiology* **63**(3A), A109 (1985).
47. J. Savitsky, J. Doczi, J. Black, and J. Arnold, *Clin. Pharmacol. Therap.* **23**, 73 (1978).
48. J. Payne, *Biochem. J.* **135**, 867 (1973).
49. U.S. Pat. 554-051, M. Laver, P. Boysen, and K. Morris, 1975.
50. L. Sehgal, A. Rosen, S. Gould, H. Sehgal, L. Dalton, J. Mayoral, and G. Moss, *Fed. Proc.* **39**, 718 (1979).
51. F. DeVenuto, and A. Zegna, *Transfusion* **21**, 599 (1981).
52. S. A. Gould, E. E. Moore, D. B. Hoyt, J. M. Burch, J. B. Haenel, J. Garcia, R. DeWoskin, and G. S. Moss, *J. Am. Coll. Surg.* **187**, 113 (1998).
53. T. Chang, *Biomat., Art Cells., Art. Org.* **16**(1–3), 1 (1988).
54. C. Hunt, R. Burnette, R. MacGregor, A. Strubbe, D. Lau, and N. Taylor, *Science* **230**, 1165 (1985).
55. R. Rabinovici, W. T. Phillips, G. Z. Feuerstein, and A. S. Rudolph, Encapsulation of hemoglobin in liposomes, In: *Red Blood Substitutes. Basic Principles and Clinical Application*, A. S. Rudolph, R. Rabinovici, and G. Z. Feuerstein, eds., Marcel Dekker, New York, 1998, pp. 263–286.
56. E. Tsuchida, D. Eng, and H. Nishide, *Biomat., Art Cells, Art Org.* **16**(1–3), 313 (1988).
57. J. Manning, *Preparation of hemoglobin carbamylated at specific NH2-terminal residues* E. Antonini, L. Rossi-Bernardi, and E. Chiancone, eds., Academic Press, New York, 1981, pp. 159–167.
58. W. Fantl, L. Manning, H. Ueno, A. DiDonato, and M. Maning, *Biochemistry* **26**, 5755 (1987).
59. J. Manning, W. Fantl, L. Manning, and H. Ueno, *Biomat., Art Cells, Art Org.* **17**(5), 639 (1989).
60. W. Fantl, A. Donato, J. Manning, P. Rogers, and A. Arnone, *J. Biol. Chem.* **26**, 12700 (1987).
61. R. San George and H. Hoberman, *J. Biol. Chem.* **261**(15), 6811 (1986).
62. R. Benesch and R. Benesch, *Preparation and properties of hemoglobin midified with derivatives of Pyridoxal*, E. Antonini, L. Rossi-Bernardi, and E. Chiancone, eds., Academic Press, New York, 1981, pp. 147–158.
63. A. Arnone, R. Benesch, and R. Benesch, *JMB* **115**, 627 (1977).
64. R. Benesch, R. Benesch, and S. Kwong, *JBC* **257**, 1320 (1982).
65. A. Bellelli and M. Brunori, *Ann. Ital. Med. Int.* **1**, 275 (1986).
66. L. Sehgal, A. Rosen, G. Noud, H. Sehgal, S. Gould, R. DeWoskin, C. L. Rice, and G. S. Moss, *J. Surg. Res.* **30**, 14 (1981).
67. F. DeVenuto and A. Zegna, *J. Surg. Res.* **34**, 205 (1983).
68. D. Hoyt, A. Greenburg, G. Peskin, S. Forbes, and H. Reese, *Surg. Forum* **31**, 15 (1980).
69. F. Jesch, W. Peters, J. Hobbhahn, M. Schoenberg, and K. Mesmer, *Crit. Care. Med.* **10**, 270 (1982).
70. L. Sehgal, S. Gould, A. Rosen, H. Sehgal, and G. Moss, *Surgery* **95**(4), 433 (1984).
71. S. Gould, L. Sehgal, A. Rosen, H. Sehgal, and G. Moss, *Ann. Emer. Med.* **15**, 1416 (1986).

72. G. Moss, S. Gould, L. Sehgal, H. Sehgal, and A. Rosen, *Biomat., Art Cells, Art Org.* **16**(1–3), 57 (1988).

73. E. Bucci, A. Razynska, B. Urbaitis, and C. Fronticelli, *J. Biol. Chem.* **264**, 6191 (1989).

74. E. Bucci, C. Fronticelli, A. Razynska, and B. Urbaitis, *Biomat., Art Cells, Art Org.* **17**(5), 637 (1989).

75. R. Benesch, R. Benesch, S. Yung, and R. Edalji, *Biochem. Biophys. Res. Commun.* **63**, 1123 (1975).

76. H. Sloviter, B. Mukherji, R. Benesch, R. Benesch, and S. Kwong, *Fed. Proc.* **42**, 652 (1981).

77. P. Keipert, M. Verosky, and L. Triner, *Biomat., Art Cells, Art Org.* **16**, 643 (1988).

78. L. Triner, R. Benesch, R. Benesch, S. Kwong, and M. Verosky, *Physiologist* **26**, A69, 1983.

79. J. VanderPlas, V-V. A. Rossen, W. Bleeker, and J. Bakker, *J. Lab. Cun. Med.* **108**, 253 (1986).

80. P. Keipert and L. Triner, *Catabolism and excretion of crosslinked hemoglobin*, G. Brewer, ed., Alan R. Liss, New York, 1989, pp. 383–405.

81. G. Ayer and A. Gauld, *Arch. Pathol.* **33**, 513 (1942).

82. W. Bleeker, J. VanderPlas, R. Feitsma, J. Agterberg, G. Rigter, A. de Vries-van Rossen, E. Pauwels, and J. Bakker, *J. Lab. Clin. Med.* **113**, 151 (1989).

83. W. Bleeker, J. VanderPlas, R. Feitsma, J. Agterberg, G. Rigter, E. Pauwels, and J. Bakker, *Nucl. Med. Biol.* **16**(4), 365 (1989).

84. R. Benesch, L. Triner, R. Benesch, S. Kwong, and M. Verosky, *Proc. Natl. Acad. Sci. U.S.A.* **81**, 2941 (1984).

85. J. Bakker, W. Bleeker, and J. VanderPlas, *Prog. Clin. Biol. Res.* **211**, 49 (1986).

86. J. VanderPlas, J. Damm, and J. Bakker, *Transfusion* **27**, 425 (1987).

87. R. Benesch and S. Kwong, *Biochem. Biophys. Res. Commun.* **156**, 9 (1988).

88. P. Keipert, A. Aderiran, S. Kwong, and R. Benesch, *Transfusion* **29**, 768 (1989).

89. R. Winslow, *Transfusion* **29**(9), 753 (1989).

90. R. Jones, C. Head, T. Fujita, A. Grant, and R. Kluger, *Biomat., Art Cells, Art Org.* **17**(5), 643 (1989).

91. R. Zaugg, L. King, and I. Klotz, *BBRC* **64**(4), 1192 (1975).

92. J. Walder, R. Zaugg, R. Walder, J. Steele, and I. Klotz, *Biochemistry* **18**, 4265 (1979).

93. J. Walder, R. Walder, and A. Arnone, *J. Mol. Biol.* **141**, 195 (1980).

94. L. Wood, D. Haney, J. Patel, S. Clare, G.-Y. Shi, L. King, and I. Klotz, *JBC* **256**, 7046 (1981).

95. J. Walder, R. Chatterjee, and A. Arnone, *Fed. Proc.* **41**, 651(abstr 2228) (1982).

96. S. Snyder, E. Welty, R. Walder, L. Williams, and J. Walder, *PNAS* **84**, 7280 (1987).

97. R. Tye, F. Medina, R. Bolin, G. Knopp, G. Irion, and S. McLaughlin, *Modification of hemoglobin—Tetrameric stabilization*, R. Bolin, R. Geyer, and G. Nemo, eds., Alan R. Liss, New York, 1983, pp. 41–49.

98. V. Macdonald and R. Winslow, *J. Appl. Physiol.* **72**, 476 (1992).

99. K. Vandegriff, F. Medina, M. Marini, and R. Winslow, *J. Biol. Chem.* **264**, 17824 (1989).

100. T. Estep, M. Bechtel, T. Miller, and A. Bagdasarian, *Artif. Cells Blood Substit Immobil Biotechnol.* **16**(1–3), 129 (1988).

101. T. Estep, M. Bechtel, S. Bush, T. Miller, S. Szeto, and L. Webb, *The purification of hemoglobin solutions by heating*, G. Brewer, ed., Alan R. Liss, New York, 1989, pp. 325–338.

102. R. M. Winslow, *Vox sang* **79**, 1 (2000).

103. L. Sehgal, A. Rosen, S. Gould, H. Sehgal, L. Dalton, J. Mayoval, and G. Moss, *Fed. Proc.* **39**, 718 (1980).

104. J. Carson, H. Noveck, J. Berlin, and S. Gould, *Transfusion* **42**, 812 (2002).
105. A. Rosen, L. Sehgal, S. Gould, H. Sehgal, and G. Moss, *Crit. Care Med.* **14**(2), 147 (1986).
106. M. Marini, G. Moore, R. Fishman, R. Jesse, F. Medina, S. Snell, and A. Zegna, *Biopolymers* **29**, 871 (1990).
107. M. Marini, G. Moore, R. Fishman, R. Jesse, F. Medina, S. Snell, and A. Zegna, *Biopolymers* **28**, 2071 (1989).
108. A. Agostoni, R. Stabilini, M. Perrella, P. Pietta, M. Pace, V. Russo, and L. Sabbioneda, *Trans. Am. Soc. Artif. Intern. Org.* **33**, 372 (1987).
109. D. Guillochon, M. Vijayalakshmi, A. Sow-Thiam, and D. Thomas, *Biochemistry* **64**(1), 29 (1986).
110. E. Gendler, S. Gendler, and M. Nimni, *J. Biomed. Mater. Res.* **18**, 727 (1984).
111. J. McPherson, S. Sawamura, and R. Armstrong, *J. Biomed. Mater. Res.* **20**, 93 (1986).
112. D. Speer, M. Chvapil, C. Eslekson, and J. Ulreich, *J. Biomed. Mater. Res.* **14**, 753 (1980).
113. P. Scannon, *Crit. Care Med.* **10**, 261 (1982).
114. A. Greenburg and P. Maffuid, *Modification of hemoglobin—ring opened dials*, R. Bolin, R. Geyer, and G. Nemo, eds., Alan R. Liss, New York, 1983, pp. 9–17.
115. M. McGarrity, K. Nightingale, and J. Hsia, *J. Chromatogr, Biomed App.* **415**, 136 (1987).
116. W. Mok, D. Chen, and A. Mazur, *Fed. Proc.* **34**, 1458 (1975).
117. B. Matheson, A. Razynaka, H. Kwansa, and E. Bucci, *J. Appl. Physiol.* **93**, 1479 (2002).
118. S. Tam, J. Blumenstein, and J. Wong, *PNAS* **73**(6), 2128 (1976).
119. J. Chang and J. Wong, *Can. J. Biochem.* **55**, 398 (1977).
120. J. Blumenstein, S. Tam, J. Chang, and J. Wong, *Experimental transfusion of dextran-hemoglobin*, G. Jamieson and T. Greenwalt, eds., Alan R. Liss, New York, 1978, pp. 205–212.
121. R. Humphries, P. Killingback, J. Mann, J. Sempik, and J. Wilson, *Brit. J. Pharm.* **74**, 266P (1981).
122. P. G. Cunnington, S. N. Jenkins, S. C. Tam, and J. T. Wong, *Biochem. J.* **193**, 261 (1981).
123. J. Wong, *Artif Cells Blood Substit Immobil Biotechnol* **16**(1–3), 237 (1988).
124. X. Wu, L. Porter, and J. Wong, *Biomat., Art Cells, Art Org.* **17**(5), 645 (1989).
125. K. Iwasaki, K. Ajisaka, and Y. Iwashita, *BBRC* **113**, 513 (1983).
126. K. Schmidt, *Klin. Wochenschr.* **57**, 1169 (1979).
127. K. Ajisaka and Y. Iwashita, *BBRC* **97**, 1076 (1980).
128. M. Leonard and E. Dellacherie, *Biochim. Biophys. Acta* **791**(2), 219 (1984).
129. M. Matsushita, A. Yabuki, P. Malchesky, H. Harasaki, and Y. Nose, *Biomat., Art Cells, Art Org.* **16**(1–3), 247 (1988).
130. M. Matsushita, A. Yabuki, J. Chen, T. Takahashi, H. Harasaki, and P. Malchesky, Y. Iwashita, and Y. Nose, *ASAIO Trans* **34**(3), 280 (1988).
131. M. R. McCarthy, K. D. Vandegriff, and R. M. Winslow, *Biophys. Chem.* **92**, 103 (2001).
132. K. D. Vandegriff, A. Malavalli, J. Wooldridge, J. Lohman, and R. M. Winslow, *Transfusion* **43**, 509 (2003).
133. A. S. Acharya, B. N. Manjula, and P. Smith, *Hemoglobin crosslinkers* Albert Einstein College of Medicine of Yeshiva University, New York (5,585,484), 1996, pp. 1–16.
134. S. Hoffman, D. Looker, J. Roehrich, P. Cozart, S. Durfee, J. Tedesco, and G. Stetler, *Proc. Natl. Acad. Sci., U.S.A.* **87**, 8521 (1990).
135. D. Looker, D. Abbott-Brown, P. Cozart, S. Durfee, S. Hoffman, and A. Mathews, *Nature (London)* **356**, 258 (1992).

136. D. H. Doherty, M. P. Doyle, S. R. Curry, R. J. Vali, T. J. Fattor, J. S. Olson, and D. D. Lemon, *Nature Biotechnol.* **16**, 672 (1998).

137. K. E. Burhop and T. E. Estep, *Art Cells, Blood Subs., Immob. Biotech.* **29**, 101 (2001).

138. K. Burhop, M. Doyle, M. Schick, and M. Matthews, *The Development and Preclinical Testing of a Novel Second Generation Recombinant Hemoglobin Solution.* The 9th International Symposium on Blood Substitutes Program II(1), Vol. 50, 2003.

139. T. Resta, B. Walker, M. Eichinger, and M. Doyle, *J. Appl. Physiol.* **93**, 1327 (2002).

140. G. Brown and H. Dale, *J. Physiol.* **86**, 42P (1936).

141. P. Hamilton, A. Hiller, and D. Van Slyke, *J. Exp. Med.* **86**, 477 (1947).

142. F. DeVenuto, T. Zuck, A. Zegna, and W. Moores, *J. Lab. Clin. Med.* **89**, 509 (1977).

143. S. Christensen, F. Medina, R. Winslow, S. Snell, A. Zegna, and M. Marini, *J. Biochem. Biophys. Methods* **17**, 143 (1988).

144. H. Bunn, *Science* **172**, 1049 (1971).

145. M. Feola, H. Gonzalez, P. Canizaro, D. Bingham, and P. Periman, *Surgery, Gynecolosy, Obstetrics* **157**, 399 (1983).

146. C. Fronticelli, E. Bucci, and C. Orth, *J. Biol. Chem.* **259**, 10841 (1984).

147. C. Fronticelli, T. Sato, C. Orth, and E. Bucci, *Biochim. Biophys. Acta* **874**, 76 (1986).

148. H. Friedman, F. DeVenuto, B. Schwartz, and T. Nemeth, *Surgery* **159**, 429 (1984).

149. F. DeVenuto, *Biomat., Art Cells, Art Org.* **16**(1–3), 77 (1988).

150. R. Behringer, T. Ryan, M. Reilly, T. Asakura, R. Palmiter, R. Brinster, and T. Townes, *Science* **245**, 971 (1989).

151. M. Swanson, M. Martin, K. O'Donnell, K. Hoover, W. Lago, and V. Huntress, *Biotechnology* **10**, 557 (1992).

152. C. D. Conover, L. Lejeune, K. Shum, and R. G. Shorr, *Life Sci.* **59**, 1861 (1996).

153. S. McFaul, P. Bowman, and V. Villa, *Blood* **84**, 3175 (1994).

154. A. L. Baldwin, *Am. J. Physiol.* **277** (2 Pt 2), H650 (1999).

155. T. Silverman, *Regulation of Blood Substitutes*, The 9th International Symposium on Blood Substitutes Program II(1), Tokyo, Vol. 34, 2003.

ROBERT M. WINSLOW
Sangart, Inc., and University of California

BORON, ELEMENTAL

1. Introduction

Boron [7440-42-8], B, is unique in that it is the only nonmetal in Group 13 (IIIA) of the Periodic Table. Boron, at wt 10.81, at no. 5, has more similarity to carbon and silicon than to the other elements in Group 13 (III A). There are two stable boron isotopes, ^{10}B and ^{11}B, which are naturally present at 19.10–20.31% and 79.69–80.90%, respectively. The range of the isotopic abundances reflects a variability in naturally occurring deposits such as high ^{10}B ore from Turkey and low ^{10}B ore from California. Other boron isotopes, ^{8}B, ^{12}B, and ^{13}B, have half-lives of <1 s. The ^{10}B isotope has a very high cross-section for absorption of thermal neutrons, 3.835×10^{-25} m^2 (3835 barns). This neutron absorption produces alpha particles.

Table 1. **Boron Minerals of Commercial Importance**

Mineral[a]	Chemical composition	CAS Registry no.	Boron oxide, wt%
Boracite (stassfurite)	$Mg_6B_{14}O_{26}C_{12}$	[1318-33-8]	62.2
Colemanite	$Ca_2B_6O_{11} \cdot 5H_2O$	[12291-65-5]	50.8
Datolite	$CaBSiO_4 \cdot OH$	[1318-40-7]	24.9
Hydroboracite	$CaMgB_6O_{11} \cdot 6H_2O$	[12046-12-7]	50.5
Kernite (rasortie)	$Na_2B_4O_7 \cdot 4H_2O$	[12045-87-3]	51.0
Priceite (pandermite)	$CaB_{10}O_{19} \cdot 7H_2O$	[61583-61-7]	49.8
Probertite (kramerite)	$NaCaB_3O_9 \cdot 5H_2O$	[12229-14-0]	49.6
Sassolite (natural boric acid)	H_3BO_3	[10043-35-3]	56.3
Szaibelyite (ascharite)	$MgBO_2 \cdot 20H_2O$	[12447-04-0] [36564-04-2]	41.4
Tincal (natural borax)	$Na_2B_4O_7 \cdot 10H_2O$	[1303-96-4]	36.5
Tincalconite (mohavite)	$Na_2B_4O_7 \cdot 5H_2O$	[12045-88-4]	47.8
Ulexite (boronatrocalcite)	$NaCaB_5O_9 \cdot 8H_2O$	[1319-33-1]	43.0

[a]Parentheses include common names.

There is a very low cosmic abundance of boron, but its occurrence at all is surprising for two reasons. First, boron's isotopes are not involved in a star's normal chain of thermonuclear reactions, and second, boron should not survive a star's extreme thermal condition. The formation of boron has been proposed to arise predominantly from cosmic ray bombardment of interstellar gas in a process called spallation (1).

Boron is the 51st most common element present in the earth's crust at a concentration of three grams per metric ton. A widespread boron mineral is tourmaline [1317-93-7], a complex borosilicate of aluminum containing ~10% boron. However, the most common ores are alkali and alkaline-earth borates. Table 1 is a list of boron minerals of commercial importance (2). Borax is the most important ore of boron. Commercial deposits are rare; the two principal ones are in the Mojave desert in California and in Turkey (see BORON COMPOUNDS, BORON OXIDES).

2. Properties

Elemental boron has a diverse and complex chemistry, primarily influenced by three circumstances. First, boron has a high ionization energy, 8.296, 23.98, and 37.75 eV for first, second, and third ionization potentials, respectively. Second, boron has a small size. Third, the electronegativities of boron (2.0), carbon (2.5), and hydrogen (2.1) are all very similar resulting in extensive and unusual covalent chemistry.

Boron has electronic structure $1s^2 2s^2 2p$ and an expected valence of three. Because of the high ionization energies there is no formation of univalent compounds as for the other Group 13 elements. Boron forms planar tricovalent compounds, BX_3, X = halides, alkyls, etc, having the expected 120° bonding angles. The empty p orbital makes these compounds electron-pair acceptors or Lewis acids. Alkyls and halides of aluminum dimerize to make up for the deficiency of electrons, but the boron atom is too small to coordinate strongly.

Boron also has a high affinity for oxygen-forming borates, polyborates, boro-silicates, peroxoborates, etc. Boron reacts with water at temperatures above 100°C to form boric acid and other boron compounds (qv).

Boron is electron deficient relative to carbon. Therefore, small amounts of boron, replacing carbon in a diamond lattice, causes electron holes. As electrons move to fill these lattice vacancies, infrared light is absorbed causing the blue color of the Hope diamond and other blue diamonds.

Boron forms B−N compounds that are isoelectronic with graphite (see BORON COMPOUNDS, REFRACTORY BORON COMPOUNDS). The small size also has a significant role in the interstitial alloy-type metal borides boron forms. Boron forms borides with metals that are less electronegative than itself including titanium, zirconium, and hafnium.

Boron's electron deficiency does not permit conventional two-electron bonds. Boron can form multicenter bonds. Thus the boron hydrides have structures quite unlike hydrocarbons. The ^{11}B nucleus, which has a spin of 3/2, which has been employed in boron nuclear magnetic resonance (nmr) spectroscopy.

Crystalline boron is very inert. Low purity, higher temperatures, and changes in or lack of crystallinity all increase the chemical reactivity. Hot concentrated $H_2SO_4–HNO_3$ at 2:1 ratio can be used to dissolve boron for chemical analysis but boron is not soluble in boiling HF or HCl. Boron is also unreactive toward concentrated NaOH up to 500°C. At room temperature, boron reacts with F_2, but only superficially with O_2.

The physical properties of elemental boron are significantly affected by purity and crystal form. In addition to being an amorphous powder, boron has four crystalline forms: α-rhombohedral, β-rhombohedral, α-tetragonal, and β-tetragonal. The α-rhombohedral form has mp 2180°C, sublimes at ∼3650°C, and has a density of 2.45 g/mL. Amorphous boron, by comparison, has mp 2300°C, sublimes at ∼2550°C, and has a density of 2.35 g/mL.

Boron is an extremely hard refractory solid having a hardness of 9.3 on Mohs' scale and a very low ($1.5 \times 10^{-6}\,\Omega^{-1}\,cm^{-1}$) room temperature electrical conductivity so that boron is classified as a metalloid or semiconductor. These values are for the α-rhombohedral form.

The electron-deficient character of boron also affects its allotropic forms. The high ionization energies and small size prevent boron from adopting metallic bonding to compensate for its electron deficiency and that of other hypoelectronic elements. The structural unit dominating boron's covalent bonding is the B_{12} icosahedron.

The α-rhombohedral form of boron has the simplest crystal structure with slightly deformed cubic close packing. At 1200°C α-rhombohedral boron degrades, and at 1500°C converts to β-rhombohedral boron, which is the most thermodynamically stable form. The unit cell has 104 boron atoms, a central B_{12} icosahedron, and 12 pentagonal pyramids of boron atom directed outward. Twenty additional boron atoms complete a complex coordination (1).

The α-tetragonal form of boron has a unit cell $B_{50}C_2$ or $B_{50}N_2$; it always has a carbon or nitrogen in the crystal. The cell is centered: a single boron atom is coordinated to four icosahedrons ($4B_{12} + 2B$). The β-tetragonal form is believed to have a unit cell of 192 boron atoms.

3. Preparation

Amorphous boron, discovered and named by Sir Humphry Davy in 1807, was first made by electrolyzing boric acid. Then in 1808, boron was produced by using potassium to reduce boric acid. The initial reactions resulted in boron that was <50% pure. A process to produce boron of >90% purity was developed in 1892 by reducing boric oxide with magnesium, and by 1909, >99% boron was obtained by the decomposition of boron trichloride in hydrogen using an alternating current arc. These three methods, electrolytic reduction, chemical reduction, and thermal decomposition, are still used on a laboratory scale. A high purity (>99%) boron comes from the direct thermal decomposition of boron hydrides such as diborane [19287-45-7], B_2H_6. The kinetics of boron formation is discussed in an excellent review (3). Less pure boron from other methods can be purified by zone-refining (qv) or progressive recrystallization.

4. Production

The Moissan process, the reduction of boric oxide with magnesium, is the most widely used commercial process for producing boron. Although boric oxide can be reduced by many other agents, including calcium and potassium, the most efficient is magnesium. This process yields material from 90–92% pure. The boron is then leached with acid to separate it from the magnesium oxide formed in the process followed by multiple washes and final drying. Chemical processing can increase this purity to 95–97% pure. Boron is ground and made available in a particle size of about one micrometer. Multiple steps require an increase in handling of chemical waste, which must be recycled or disposed.

Another commercial process yields high purity boron of >99%. In this process boron hydrides, such as diborane, are thermally decomposed (4). Because only boron and hydrogen are present in the starting material, contamination is minimal, and a very uniform, submicrometer powder is formed by the gas nucleation process.

5. Health and Safety Factors

Boron is a trace element that is essential to human health and behavior. Evidence points to the fact that boron may reduce either the symptoms or incidence of arthritis. Researchers have discovered a relationship between the amount of boron in the soil and drinking water and the incidence of arthritis in an area. Postmenopausal women who were magnesium deficient, benefited from 3 mg/day of boron added to their diet, which resulted in calcium and magnesium retention and elevations in circulating serums of testosterone and a form of estrogen. Similar improvements were seen in Vitamin D deficient postmenopausal women. No recommended daily allowance has been set for humans, but females are recommended to take 2.0 mg/day (5).

There have been no reports of toxicity in humans. Since the element is very insoluble, the dust should be treated as such, with a respirable dust limit applied (6).

6. Uses

Elemental boron is used in very diverse industries from metallurgy (qv) to electronics. Other areas of application include ceramics (qv), propulsion, pyrotechnics, and nuclear chemistry.

Dispersed mixtures of boron and another metal are used as deoxidizing and degassing agents to harden steel (qv) (7,8), to increase the conductivity of copper (qv) in turbojet engines, and in the making of brass and bronze (see COPPER ALLOYS). Two examples are alloys of ferroboron and manganese boron.

Another metallurgical application is in amorphous magnetic alloys that are based on boron and iron, nickel, or cobalt (see MAGNETIC MATERIALS, BULK). The boron is used in power transformers as a soft magnet to convert from high to lower voltage; this material is commercially available from Allied-Signal, Inc. under the trademark METGLAS (see GLASSY METALS).

Another material that has permanent magnetic properties is neodymium–iron–boron, $Nd_2Fe_{14}B$. For an in-depth discussion see Ref. 9. Aimants Ugima, a member of the Pechiney Group, is a leading producer of rare-earth magnets in both the United States and Europe. Sumitomo Special Metals of Japan also produces these rare-earth magnets under the trade name of Neomax, and General Motors has commercial-scale production of a material called MAGNEQUENCH. The first World Solar Challenge Race, in 1987, was won using an electric motor fabricated using Nd–Fe–B. More mundane applications for cars include fuel pump motors, headlight door motors, starter motors, and heater motors. A patent has also been awarded for Nd–Fe–B bonded in polymer material that can be stamped and shaped easily for making electronic components for applications including stereo speakers and computer chip switches (10) (see MAGNETIC MATERIALS, THIN-FILMS AND PARTICLES).

The Federal Railroad Administration of the U.S. Department of Transportation intends to construct a magnetic levitation train (maglev). Maglev is an advanced transportation technology in which magnetic forces lift, propel, and guide a vehicle over a specially designed guide way. Boron is used in the superconducting and other high intensity magnets of this system. Maglev could reduce the need for many mechanical parts and thereby minimizing resistance permitting excellent acceleration and cruising speeds of 386 km/h (240 mi/h) (11).

The ceramic, polycrystalline silicon carbide [409-21-2], SiC, is processed using β-silicon carbide and boron (12). The boron is a sintering aid used at 0.3–3% by weight to densify the sintered body to at least 85% of theoretical. The increased density improves strength (see ADVANCED CERAMICS).

Boron filaments are formed by the chemical vapor deposition of boron trichloride on tungsten wire. High performance reinforcing boron fibers are available from 10 to 20 mm in diameter. These are used mainly in epoxy resins and aluminum and titanium. Commercial uses include golf club shafts, tennis and squash racquets, and fishing rods.The primary use is in the aerospace industry.

Boron has been studied as a possible fuel for solid fuel ramjets (13,14). Fine particle sized boron, where the average particle size is 0.3 μm, has been studied for use as a gas-generating agent for solid fuel ram rockets (15).

DaimlerChrysler AG's Natrium Town and Country minivan is powered by fuel technology using water, borax, and hydrogen. Hydrogen derived from sodium borohydride is the most promising of all energy carriers. It is entirely free of carbon monoxide, which can damage cells. It is efficient, clean, abundant, and renewable energy. The fuel cells were to be used in buses in 2002 and are expected to be used in passenger cars in 2004 (16).

Boron mixed with an oxidizer is used as a pyrotechnic. This ordnance application for missiles and rockets is predominantly military. However, boron is also used in air bags, placed in automobiles as safety devices, for initiating the sodium azide [26628-22-8] that fills the bag with nitrogen (17). Other boron compounds are also used in the air-bag pyrotechnic application.

Boron creates an electron deficiency in the silicon lattice resulting in a p-type semiconductor for $p-n$ junctions. Boron compounds are more commonly used as the dopant, however (see BORON HYDRIDES).

The high cross-section for thermal neutrons results in the use of boron and boron compounds for radiation shielding (18). The ease of detecting the α-particle produced when boron absorbs thermal neutrons results in the use of boron for neutron counters as well. The U.S. Department of Energy plans to store spent nuclear fuel encased in boron-containing glass inside stainless steel containers underground at Yucca Mountain in Nevada in 2009 (2).

The U.S. Department of Energy has studied boron–aluminum–magnesium alloys and found its hardness nearly equal to that of diamond. Diamond wears quickly when cutting steel. The new alloy cuts without getting hot because of its fine grain size and complex crystal structure (19).

Millenium Cell has patented a design for boron-based longer life batteries. The batteries are potentially several times better than zinc batteries and can last as long as traditional batteries (20).

Boron is one of 16 nutrients essential to all plants. It is essential to plant growth and can be applied as a spray and incorporated in fertilizers, herbicides, and irrigation water.

BIBLIOGRAPHY

"Elemental Boron" under "Boron and Boron Alloys" in *ECT* 1st ed., Vol. 2, pp. 584–588, by W. Crafts, Union Carbide and Carbon Research Laboratories, Inc.; in *ECT* 2nd ed., Vol. 3, pp. 602–605, by J. G. Bower, U.S. Borax Research Corporation; *ECT* 3rd ed., Vol. 4, pp. 62–66, by J. G. Bower, U.S. Borax and Chemical Corporation; in *ECT* 4th ed., Vol. 4, pp. 360–365, by Linda H. Jansen, Callery Chemical Company; "Boron, Elemental" in *ECT* (online), posting date: December 4, 2000 by Linda H. Jansen, Callery Chemical Company.

CITED PUBLICATIONS

1. N. N. Greenwood and A. Earnshaw, *Chemistry of the Elements*, Pergamon Press, Oxford, U.K., 1984.

2. P. A. Lyday, "Boron," *Minerals Yearbook*, U.S. Geological Survey, Reston, Va., 2001.

3. S. H. Bauer, in J. F. Liebman, ed., *Advances in Boron and the Boranes*, Vol. 19, VCH Publishers, Inc., New York, 1988, p. 391.

4. *Elemental Boron*, Technical Report AFAPL-TR-65-88, U.S. Air Force Contract No. AF33(615)2258 Callery Chemical, Callery, Pa., 1988.

5. A. G. Schauus, "Boron," http://www.traceminerals.com/products/boron.html, undated, accessed May 13, 2002.

6. B. W. Culver, P. L. Strong, and J. F. Murray, in E. Bingham, B. Cohrssen, and C. H. Powell, eds., *Patty's* Toxicology, 5th ed., Vol. 3, John Wiley & Sons, Inc., New York, 2001, p. 546.

7. *Met. Fabric. News* **24**(2), 1 (Mar.–Apr. 1985).

8. H. E. Boyer and T. L. Gall, eds., *Metal Handbook, Desk Edition*, American Society for Metals, Metals Park, Ohio, 1985, pp. 4–11.

9. E. P. Wohlfarth and K. H. J. Buschow, eds., *Ferromagnetic Materials*, North-Holland, Amsterdam, The Netherlands, 1988, p. 1.

10. U.S. Pat. 4,873,504 (Oct. 10, 1989), W. S. Blume (to Electrodyne Co.).

11. Federal Railroad Administration, Jan. 18, 2001, http://www.fta.dot.gov/research/equip/raileq/unmag/unmag.html, accessed Jan. 14, 2002.

12. U.S. Pat. 3,993,602 (Nov. 23, 1976), S. Prochazka (to General Electric Co.).

13. A. Gany and D. W. Netzer, *J. Propulsion* **2**(5), 423 (1986).

14. S. C. Li, F. A. Williams, and F. Takahshi, *Proceedings of the 22nd International Symposium on Combustion, 1951–1960*, Seattle, Wash., 1988.

15. Jpn. Pat. 192,787 (Aug. 2, 1989), N. Kubota and M. Mitsuno (to Nissan Motor Co., Ltd.).

16. J. Grenville–Robinson, "Borax to Fuel the Future?" *Review no. 61*, 7–9 (March 2002).

17. F. P. Watkins, *Borax Rev.* (7), 10 (1990).

18. Rom. Pat. 82,282 (Jan. 22, 1983), M. M. Fanica (to Polycolor Dye and Paint Corp.).

19. Advanced Material and Process, *Advanced Materials* **160**(2), 13 (Feb. 2002).

20. Millenium Cell plc, undated, http://www.milleniumcell.com/solutions/white.html.

Linda H. Jansen
Callery Chemical Company

BORON HALIDES

1. Introduction

The boron trihalides boron trifluoride [7637-07-2], BF_3, boron trichloride [10294-34-5], BCl_3, and boron tribromide [10294-33-4], BBr_3, are important industrial chemicals having increased usage as Lewis acid catalysts and in chemical vapor deposition (CVD) processes (see Electronic materials). Boron halides are widely used in the laboratory as catalysts and reagents in numerous types of organic reactions and as starting material for many organoboron and inorganic boron compounds. An exhaustive review of the literature on boron halides up to 1984 is available (1–5). Of particular interest are review articles on BCl_3 (1), BBr_3 (2), and boron triiodide [13517-10-7], BI_3 (3). An excellent review on diboron tetrahalides and polyhedral boron halides is available (6).

2. Physical Properties

Boron trihalides, BX_3, are trigonal planar molecules which are sp^2 hybridized. The X–B–X angles are 120°. Important physical and thermochemical data are presented in Table 1 (7–13). Additional thermodynamic and spectroscopic data may be found in the literature (1–5).

Table 1. **Physical Properties of the Boron Trihalides**

Property	BF$_3$	BCl$_3$	BBr$_3$	BI$_3$	References BCl$_3$	BBr$_3$	BI$_3$	BF$_3$
mp, °C	−128.37	−107	−46	−49.9	7	7	7	14
bp, °C	−99.9	12.5	91.3	210	7	7	7	14
density^a, g/mL (liq)		1.434$_4^0$ 1.349$_4^{11}$	2.643$_4^{18}$	3.35	8	9		
critical temperature, °C	−12.25 ± 0.03	178.8	300		7	7		14
critical pressure, kPa^b	4984	3901.0			10			14
vapor pressure, kPa^b					10	9	11	
−80°C		0.53	^{c}					
−40°C		8.9						
0°C		63.5						
40°C		243						
80°C		689						
at 145 K^d	8.43							
at 170 K^d	80.19							
at 220 K^d	1156							
at 260 K^d	4842							14
density								
critical, d_c, g/cm^3	ca 0.591							16
gas at STP, g/L	3.07666							17
gas limiting, L_N, g/L	3.02662							17
liquid, for 148.9 to 170.8 K, g/cm^3	1.699− 0.00445 $(t + 125.0)$							18
viscosity, mPa·s (= cP)		^{e}	^{e}		8	9		
ΔH_f^0, kJ/mol, gas^f		−403	−206	+18	12	12	12	
ΔH_{vap}, kJ/mol^f		23.8	34.3		12	12		
C_p, J/(mol·°C), for gas at 25°C^f		62.8	67.78		10	10		
C_p, J/(mol·°C), for liquid at 25°C^f		121	128		10	10		
ΔH_{hydrol}, kJ/mol, liquid at 25°C^f		−289	−351		10	10		
ΔH_{fus}, J/g at mp^f		18			10			
ethalpy of fusion, $\Delta H_{144.45}$, kJ/mol	4.2417							19
ethalpy of vaporization, $\Delta H_{154.5}$, kJ/mol^f	18.46							20
entropy, $S_{298.15}$, J/(mol·K)^f	254.3							

Table 1 (*Cotinued*)

Property	BF$_3$	BCl$_3$	BBr$_3$	BI$_3$	BCl$_3$	BBr$_3$	BI$_3$	BF$_3$
					\multicolumn{4}{c}{References}			
Gibbs free energy of formation, ΔG_f298.15' kJ/mol^f	−1119.0							
enthalpy of formation, ΔH_f298.15' kJ/mol^f	−1135.6							21
infrared absorption frequencies, cm^{-1}								
ν_1	888							
ν_2	696.7							
ν_3	1463.3							
ν_4	480.7							
B–X bond energy, kJ/mol^f		−443.9	368.2	266.5	13	13	13	
B–X distance, nm	0.1307±0.002	0.173	0.187	0.210	13	13	13	21

^{a} For BCl$_3$: $\rho = 1.3730 − 2.159 \times 10^{-3}°C − 8.377 \times 10^{-7}°C$; from −44 to 5°C. Superscript indicates temperature of measurement. For BBr$_3$: $\rho = 2.698 − 2.996 \times 10^{-3}°C$; from −20 to 90°C.
^{b} To convert kPa to mm Hg, multiply by 7.50.
^{c} For BBr$_3$: log(pressure) = $[6.9792 − 1311/(°C + 230)] − 0.8752$; from 0–90°C.
^{d} Liquid.
^{e} For BCl$_3$: $\eta = 0.34417/(1 − 6.9662 \times 10^{-3}°C − 5.9013 \times 10^{-6}°C)$; from −40 to 10°C. For BBr$_3$: $\log \eta = (333/K) − 1.257$; from 0–90°C.
^{f} To convert J to cal, divide by 4.184.

Nuclear magnetic resonance ^{11}B spectral studies of BF$_3$ have given a value of 9.4 ± 1.0 ppm for the chemical shift relative to BF$_3$·O(C$_2$H$_5$)$_2$ as the zero reference (22). Using methylcyclohexane as a solvent at 33.5°C and BF$_3$·O(CH$_2$CH$_3$)$_2$ as the internal standard, a value of 10.0 ± 0.1 ppm was obtained for the chemical shift (23). A value for the ^{19}F chemical shift of BF$_3$ in CCl$_3$F relative to CCl$_3$F is reported to be 127 ppm (24). The coupling constant $J_{11_B}−19_F$ is reported to be 15 ± 2 Hz for BF$_3$ (25). Additional constants are available (26,27). See Table 2 for solubilities.

Table 2. Solubilities of Boron Trifluoride

BF$_3$, g	Solvent, g	Temperature, °C	Product	CAS Registry Number	Reference
369.4	water,^{a} 100^b	6	BF$_3$·H$_2$O	[15799-89-0]	
			HBF$_3$(OH)	[16903-52-9]	28
2.06	sulfuric acid, conc, 100%	25			29
	nitric acid^a	20	HNO$_3$·2BF$_3$	[20660-63-3]	30
	orthophosphoric acid^a	25	H$_3$PO$_4$·BF$_3$	[13699-76-6]	31
2.18	hydrofluoric acid,^{c}	4.4			33
	hydrochloric acid, anhydrous (l)	24	miscible		34

^{a} Dissolves with reaction to form complexes and other species.
^{b} A higher dilution results in a mixture of H[BF$_2$(OH)$_2$], HBF$_4$, and H$_3$BO$_3$.
^{c} Equations for the solubility of BF$_3$ in liquid HF at 24, 49, and 90°C and up to 6.8 kPa (51 mm Hg) may be found in Reference 32.

3. Chemical Properties

The boron trihalides are strong Lewis acids, however, the order of relative acid strengths, $BI_3 \geq BBr_3 > BCl_3 > BF_3$, is contrary to that expected based on the electronegativities and atomic sizes of the halogen atoms. This anomaly has been explained in terms of boron–halogen π-bonding, which increases from BI_3 to BF_3 (35,36). The Lewis acidity of the boron trihalides strongly influences their chemistry (37–40). The trihalides react with Lewis bases containing O, S, N, P, or As atoms to form donor–acceptor complexes. For donor compounds containing active hydrogen, such as NH_3, PH_3, AsH_3, primary and secondary amines, and lower alcohols, BCl_3, BBr_3, and BI_3 react to liberate the corresponding hydrogen halide. Tertiary alcohols and the boron trihalides yield the alkyl halide and boric acid. The boron trihalides hydrolyze readily in water or moist air to produce boric acid and hydrogen halides.

BCl_3, BBr_3, and BI_3 undergo exchange reactions to yield mixed boron halides. Exchange reactions also occur with trialkyl, triaryl, trialkoxy, or triaryloxy boranes and with diborane. Anhydrous metal bromides and iodides can be prepared by the exchange reaction of the metal chloride or oxide and BBr_3 or BI_3 (41).

Boron trihalides can be reduced to elemental boron by heating and presence of alkali metals, alkaline-earth metals, or H_2 (42–46); such reductions can also yield boron subhalides, eg, chloroborane [20583-55-5], BCl, B_xCl_x, dichloroborane [10325-29-0], $HBCl_2$, (47–50), and/or diborane [19287-45-7], B_2H_6 (47–50). Metal hydrides also react with BX_3 to yield diborane (51–53).

Some of the general reactions for the boron trihalides where X represents Cl, Br, or I and X′ a different halogen, are

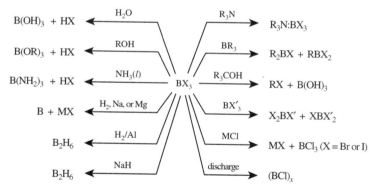

Reactions of boron trihalides that are of commercial importance are those of BCl_3, and to a lesser extent BBr_3, with gases in chemical vapor deposition (CVD). CVD of boron by reduction, of boron nitride using NH_3, and of boron carbide using CH_4 on transition metals and alloys are all technically important processes (54–58). The CVD process is normally supported by heating or by plasma formed by an arc or discharge (59,60).

Aqueous mineral acids react with BF_3 to yield the hydrates of BF_3 or the hydroxyfluoroboric acids, fluoroboric acid, or boric acid. Solution in aqueous alkali gives the soluble salts of the hydroxyfluoroboric acids, fluoroboric acids,

Table 3. **Reactions of Boron Trifluoride**

Reactant	Temperature,°C	Products	Formula	Reference
sodium[a]		boron, amorphous, sodium fluoride	NaF	61
magnesium, molten alloys	no reaction			
calcium	1600	calcium hexaboride	CaB_6	
aluminum	1200	aluminum boride (1:12), tetragonal boron	AlB_{12}	
	1650[b]	β-rhombohedral boron		
titanium	1600	titanium boride	TiB_2	62
copper, mercury, chromium, iron	RT or below	no reaction[c]		
sodium nitrate, sodium nitrite	180	sodium fluoroborate, boric oxide	$NaBF_4$	63

[a] With incandescence.
[b] Further reaction.
[c] Even when subjected to pressure for a considerable length of time; also no reaction with red-hot iron.

or boric acid. Boron trifluoride, slightly soluble in many organic solvents including saturated hydrocarbons (qv), halogenated hydrocarbons, and aromatic compounds, easily polymerizes unsaturated compounds such as butylenes (qv), styrene (qv), or vinyl esters, as well as easily cleaved cyclic molecules such as tetrahydrofuran. Other molecules containing electron-donating atoms such as O, S, N, P, etc, eg, alcohols, acids, amines, phosphines, and ethers, may dissolve BF_3 to produce soluble adducts.

In addition to the reactions listed in Table 3, boron trifluoride reacts with alkali or alkaline-earth metal oxides, as well as other inorganic alkaline materials, at 450°C to yield the trimer trifluoroboroxine [13703-95-2], $(BOF)_3$, MBF_4, and MF (64) where M is a univalent metal ion. The trimer is stable below $-135°C$ but disproportionates to B_2O_3 and BF_3 at higher temperatures (65).

$$
\begin{array}{c}
\text{F} \\
| \\
\text{B} \\
\text{O}^{\diagup \quad \diagdown}\text{O} \\
| \qquad\qquad | \\
\text{F}-\text{B} \diagdown_{\text{O}}\diagup \text{B}-\text{F}
\end{array}
$$

The reaction of metal hydrides and BF_3 depends on the stoichiometry as well as the nature of the metal hydride. For example, LiH and $BF_3 \cdot O(C_2H_5)_2$ may form diborane (14) or lithium borohydride (66,67):

$$6\,\text{LiH} + 8\,\text{BF}_3 \cdot \text{O}(\text{C}_2\text{H}_5)_2 \longrightarrow \text{B}_2\text{H}_6 + 6\,\text{LiBF}_4 + 8\,(\text{C}_2\text{H}_5)_2\text{O}$$

$$4\,\text{LiH} + 4\,\text{BF}_3 \cdot \text{O}(\text{C}_2\text{H}_5)_2 \longrightarrow \text{LiBH}_4 + 3\,\text{LiBF}_4 + 4\,(\text{C}_2\text{H}_5)_2\text{O}$$

The first method is commonly used for preparing diborane.

Metal halides react with BF_3 (68) when heated to form BX_3 and the metal fluoride. For example,

$$\text{AlBr}_3 + \text{BF}_3 \longrightarrow \text{BBr}_3 + \text{AlF}_3$$

The reaction of BF$_3$ with alkali halides yields the respective alkali fluoroborates (69):

$$3\,KCl + 4\,BF_3 \longrightarrow 3\,KBF_4 + BCl_3$$

Alkyl and arylboranes are obtained (70) from BF$_3$ using the appropriate Grignard reagent, alkylaluminum halide, or zinc alkyl, using diethyl ether as the solvent:

$$BF_3 + 3\,RMgX \longrightarrow BR_3 + 3\,MgXF$$

Tetraorganylborate complexes may be produced when tetrahydrofuran is the solvent (71).

Alkylfluoroboranes result from the reaction of the appropriate alkylborane and BF$_3$ under suitable conditions (72):

$$BR_3 + 2(C_2H_5)_2O \cdot BF_3 \longrightarrow 3\,RBF_2 + 2\,(C_2H_5)_2O$$

Adducts of BF$_3$ and some organic compounds having labile hydrogen atoms in the vicinity of the atom bonding to the boron atom of BF$_3$ may form a derivative of BF$_3$ by splitting out HF. For example, β-diketones such as acetylacetone or benzoylacetone react with BF$_3$ in benzene (73):

$$BF_3 + CH_3COCH_2COCH_3 \longrightarrow CH_3COCH{=}C(CH_3)OBF_2 + HF$$

In Group 14 (IV), carbon serves as a Lewis base in a few of its compounds. In general, saturated aliphatic and aromatic hydrocarbons are stable in the presence of BF$_3$, whereas unsaturated aliphatic hydrocarbons, such as propylene or acetylene, are polymerized. However, some hydrocarbons and their derivatives have been reported to form adducts with BF$_3$. Typical examples of adducts with unsaturated hydrocarbons are 1:1 adducts with tetracene and 3,4-benzopyrene (74), and 1:2 BF$_3$ adducts with α-carotene and lycopene (75).

In Group 15 (V), nitrogen compounds readily form molecular compounds with BF$_3$. Phosphorus compounds also form adducts with BF$_3$. Inorganic or organic compounds containing oxygen form many adducts with boron trifluoride, whereas sulfur and selenium have been reported to form only a few (76–78).

Boron trifluoride forms two hydrates, BF$_3$·H$_2$O and boron trifluoride dihydrate [13319-75-0], BF$_3$·2H$_2$O, (also BF$_3$·D$_2$O [33598-66-2] and BF$_3$·2D$_2$O [33598-66-2]). According to reported nmr data (78,79), the dihydrate is ionic, H$_3$O$^+$F$_3$BOH$^-$. The trihydrate has also been reported (80). Acidities of BF$_3$–water systems have been determined (81). Equilibrium and hydrolysis of BF$_3$ in water have been studied (82–84).

Most of the coordination compounds formed by trifluoroborane are with oxygen-containing organic compounds (Table 4). Although the other boron halides frequently react to split out hydrogen halide, boron trifluoride usually forms stable molecular compounds. The reason is attributed to the back coordination of electrons from fluorine to boron forming a strong B–F bond which is 28% ionic (85).

Table 4. Boron Trifluoride Adducts with Oxygen-Containing Compounds

Donor	Adduct name	CAS Registry Number	Molecular formula
alcohols	ethanol trifluoroborane	[353-41-3]	$C_2H_5OH \cdot BF_3$
	bis(ethanol)trifluoroborane	[373-59-1]	$2C_2H_5OH \cdot BF_3$
	bis(2-chloroethanol)trifluoroborane	[72985-81-0]	$2ClCH_2CH_2OH \cdot BF_3$
	benzyl alcohol trifluoroborane	[456-31-5]	$C_6H_5CH_2OH \cdot BF_3$
acids	acetic acid trifluoroborane	[753-53-7]	$CH_3COOH \cdot BF_3$
	bis(acetic acid)trifluoroborane	[373-61-5]	$2CH_3COOH \cdot BF_3$
	stearic acid trifluoroborane	[60274-92-2]	$CH_3(CH_2)_{16}COOH \cdot BF_3$
	bis(phenol)trifluoroborane	[462-05-5]	$2C_6H_5OH \cdot BF_3$
ethers	diethyl ether trifluoroborane	[109-63-7]	$(C_2H_5)_2O \cdot BF_3$
	tetrahydrofuran trifluoroborane	[462-34-0]	$(CH_2)_4O \cdot BF_3$
	anisole trifluoroborane	[456-31-5]	$CH_3OC_6H_5 \cdot BF_3$
acid anhydride	acetic anhydride trifluoroborane	[591-00-4]	$(CH_3CO)_2O \cdot BF_3$
esters	ethyl formate trifluoroborane	[462-33-9]	$HCOOC_2H_5 \cdot BF_3$
	phenyl acetate trifluoroborane	[30884-81-6]	$CH_3COOC_6H_5 \cdot BF_3$
ketones	acetone trifluoroborane	[661-27-8]	$(CH_3)_2CO \cdot BF_3$
	benzophenone trifluoroborane	[322-21-4]	$(C_6H_5)_2CO \cdot BF_3$
	acetphenone trifluoroborane	[329-25-9]	$C_6H_5COCH_3 \cdot BF_3$
aldehydes	acetaldehyde trifluoroborane	[306-73-0]	$CH_3CHO \cdot BF_3$
	neopentanal trifluoroborane	[306-78-5]	$(CH_3)_3CCHO \cdot BF_3$
	benzaldehyde trifluoroborane	[456-30-4]	$C_6H_5CHO \cdot BF_3$

It has been reported (86) that some adducts of alkyl ethers and/or alcohols are unstable and decompose at $-80°C$ to yield BF_3, H_2O, and the polyalkene. Adducts of BF_3 have been reported with hydrogen sulfide, sulfur dioxide, thionyl fluoride, and the sulfur analogues of many of the kind of oxygen-containing organic molecules cited in Table 4. The carbonyl oxygen or the carbonyl sulfur is the donor to BF_3 in 1:1 adducts such as $CH_3COOCH_3 \cdot BF_3$ [7611-14-5], $CH_3COSCH_3 \cdot BF_3$ [52913-04-9], and $CH_3CSOCH_3 \cdot BF_3$ [52912-98-8] (87).

Compounds containing fluorine and chlorine are also donors to BF_3. Aqueous fluoroboric acid and the tetrafluoroborates of metals, nonmetals, and organic radicals represent a large class of compounds in which the fluoride ion is coordinating with trifluoroborane. Representative examples of these compounds are given in Table 5. Coordination compounds of boron trifluoride with the chlorides of sodium, aluminum, iron, copper, zinc, tin, and lead have been indicated (88); they are probably chlorotrifluoroborates.

Table 5. Boron Trifluoride Adducts with Compounds Containing Chlorine and Fluorine

Name	CAS Registry Number	Molecular formula
potassium tetrafluoroborate	[14075-53-7]	KBF_4
hexaamminenickel(II) tetrafluoroborate	[13877-20-8]	$[Ni(NH_3)_6](BF_4)_2$
nitrosyl tetrafluoroborate	[14635-75-7]	$NOBF_4$
acetylium tetrafluoroborate	[2261-02-1]	CH_3COBF_4
tetramethylammonium tetrafluoroborate	[661-36-9]	$(CH_3)_4NBF_4$
difluorobromine tetrafluoroborate	[14282-83-8]	BrF_2BF_4
anilinium tetrafluoroborate	[15603-97-1]	$C_6H_5NH_2HBF_4$

Trifluoroborane may form adducts with some of the transition elements. See Reference 89 for a detailed discussion of complexes of trifluoroborane with various Group 6–10 (VI, VII, and VIII) species.

4. Manufacture

4.1. Boron Trifluoride.   Boron trifluoride is prepared by the reaction of a boron-containing material and a fluorine-containing substance in the presence of an acid. The traditional method used borax, fluorspar, and sulfuric acid.

In another process fluorosulfonic acid is treated with boric acid:

$$3\,HSO_3F + H_3BO_3 \longrightarrow BF_3 + 3\,H_2SO_4$$

Numerous other reactions are available for the preparation of small quantities of boron trifluoride, some of which are of high purity (90,91). A process for recovering boron trifluoride from oligomerization mix have been reported (92).

4.2. Boron Trichloride.   Boron trichloride is prepared on a large scale by the reaction of Cl_2 and a heated mixture of borax [1303-96-4], $Na_2B_4O_7{\cdot}10H_2O$, and crude oil residue (93) in a rotary kiln heated to 1038°C. Borax is fed at the rate of 1360 kg/h and sprayed with 635 kg/h of 17% residue crude oil. The heated mixture then reacts with Cl_2 at 760°C in a separate reactor to yield BCl_3. On a smaller scale, BCl_3 can be prepared by the reaction of Cl_2 and a mixture of boron oxide [1303-86-2], B_2O_3, petroleum coke, and lampblack in a fluidized bed (94). Other methods for the preparation of BCl_3 from oxygen-containing boron compounds are also known (1,95–98).

A convenient laboratory method for the preparation of BCl_3 is by the reaction of $AlCl_3$ and BF_3 or BF_4^- (99–101). A patent describes the preparation of BCl_3 by halogenating $B(OH)_3$ or esters of $B(OH)_3$ using an excess of the oxychloride of S or P in the presence of a dessicant and catalytic amounts of Fe, Co, or Ni, at temperatures below 100°C was issued (102). This process eliminates formation of carbonic dichloride [75-44-5], $COCl_2$, a common impurity in large-scale production of BCl_3. Other common impurities associated with the preparation of BCl_3 are CO, CO_2, Cl_2, HCl, $FeCl_3$, $SiCl_4$, $AsCl_3$, and SO_2. Methods for purification include distillation (103–105), sometimes in the presence of KCl; activated charcoal or polyphenylene dioxide (106); adsorption desorption on silica gel (107,108); countercurrent crystallization (105); and passage of impure gas through Ti sponge (109), molten Zn (110), Cu or charcoal at elevated temperatures (111). $COCl_2$ can also be destroyed by pyrolysis (112), by a discharge, or irradiation using a radio frequency or an electron beam (113,114), or by uv photolysis using a laser (112,115,116). A system for purification of boron trichloride so that there is less than 10 ppm of phosgene, chlorine, and HCl has been reported (117).

4.3. Boron Tribromide.   Boron tribromide is produced on a large scale by the reaction of Br_2 and granulated B_4C at 850–1000°C (118) or by the reaction of HBr with CaB_6 at high temperatures (119). Reaction of Br_2 and a mixture of B_4C and CaB_6 at 900–1200°C is used to prepare high purity BBr_3 (120). Another method for preparing high purity BBr_3 is the reaction of the two elements, B and Br_2, at 750°C in N_2 atmosphere, followed by fractional distillation (121).

Most of the methods for preparing BBr_3 are similar to those for preparation of BCl_3. A convenient laboratory preparation involves reaction of $AlBr_3$ and BF_3 or BF_4^- (2). A procedure for the preparation of labeled $^{10}BBr_3$ from the reaction of $^{10}BF_4^-$ and $AlBr_3$ has also been described (122).

4.4. Boron Triiodide. Boron triiodide is not manufactured on a large scale. Small-scale production of BI_3 from boron and iodine is possible in the temperature range 700–900°C (123–125). Excess I_2 can be removed as SnI_4 by reaction with Sn, followed by distillation (124). The reaction of metal tetrahydroborates and I_2 is convenient for laboratory preparation of BI_3 (126,127). BI_3 can also be synthesized from B_2H_6 and HI in a furnace at 250°C (128), or by the reaction of B with excess AgI or CuI between 450–700°C, under vacuum (129). High purity BI_3 has been prepared by the reaction of I_2 with mixtures of boron carbide and calcium carbide at elevated temperatures.

In addition to distillation (126), BI_3 can be purified by sublimation under reduced pressure (130).

5. Shipment and Handling

Boron trifluoride gas is nonflammable and is shipped in DOT 3A and 3AA steel cylinders at a pressure of approximately 12,410 kPa (1800 psi). Boron trifluoride is classified as a poison gas, both domestically and internationally. Cylinders must have a poison gas diamond and an inhalation hazard warning label. Tube trailers carry both a poison gas placard and an inhalation hazard warning. Cylinders containing 27.2 kg and tube trailers containing 4.5–10 metric tons are available. If boron trifluoride is compressed using oil as a compressor lubricant, it must not be used with oxygen under pressure nor with gauges, valves, or lines that are to be used with oxygen.

In as much as the gas hydrolyzes readily, all equipment should be purged repeatedly using inert dry gas before admitting boron trifluoride. Under anhydrous conditions, carbon steel equipment is satisfactory. Stainless steel and aluminum silicon bronze may also be used. Stainless steel tubing is recommended for both temporary and permanent connections.

In the presence of moisture, boron trifluoride may be handled in polytetrafluoroethylene (PTFE), polyethylene, Pyrex glass (limit to atmospheric pressure), or Hastelloy C containers. At 600°C, stainless steel (304 L) and Hastelloy N are attacked by BF_3; Hastelloy C is more resistant (131). Kel F and PTFE serve as satisfactory gasket and packing materials, whereas rubber, fiber, polymerizable materials, or organic oxygen- and nitrogen-containing compounds must be avoided. Because boron trifluoride is soluble in, and reacts with, many liquids, the gas must not be introduced into any liquid unless a vacuum break or similar safety device is employed.

BCl_3 is shipped in steel cylinders (0.9-, 45-, 590-, and 817-kg, net); BBr_3 is shipped in glass bottles (0.45- and 2.3-kg, net) and 91-kg (net) monel drums (10). Both BCl_3 and BBr_3 are classed as corrosive liquids, and must be shipped by private carriers.

Table 6. **Specifications for BCl$_3$ and BBr$_3$**[a]

	BCl$_3$, wt %		BBr$_3$, wt %	
Assay	Specification	Typical	Specification	Typical
BX$_3$	99.9 min	99.95	99.9 min	99.98
Br$_2$			0.05 max	0.01
Cl$_2$	0.01 max			
COCl$_2$	0.09 max			
Si	0.001 max			0.0002[b]

[a] Ref. 134.
[b] Other impurities in wt % are typically P, 0.002; Fe, 0.003; and Mg, 0.0002.

6. Specifications and Analytical Methods

Commercial boron trifluoride is usually approximately 99.5% pure. The common impurities are air, silicon tetrafluoride, and sulfur dioxide. An excellent procedure for sampling and making a complete analysis of gaseous boron trifluoride has been developed (132).

Analysis for boron, halide, free halogen, and silicon is carried out by standard methods following hydrolysis of BX$_3$ (10,133). Specifications for BCl$_3$ and BBr$_3$ supplied by Kerr-McGee Corp. are given in Table 6.

7. Health and Safety Factors

Boron trifluoride is primarily a pulmonary irritant. The toxicity of the gas to humans has not been reported (135), but laboratory tests on animals gave results ranging from an increased pneumonitis to death. The TLV is 1 ppm (136,137). Inhalation toxicity studies in rats have shown that exposure to BF$_3$ at 17 mg/m^3 resulted in renal toxicity, whereas exposure at 6 mg/m^3 did not result in a toxic response (138). Prolonged inhalation produced dental fluorosis (139). High concentrations burn the skin similarly to acids such as HBF$_4$ and, if the skin is subject to prolonged exposure, the treatment should be the same as for fluoride exposure and hypocalcemia. No chronic effects have been observed in workers exposed to small quantities of the gas at frequent intervals over a period of years.

Boron trichlorides are highly reactive, toxic, and corrosive; these trihalides (BCl$_3$, BBr$_3$, BI$_3$) react vigorously, even explosively, with water. High temperature decomposition of BX$_3$ can yield toxic halogen-containing fumes. Boron trichloride is a poison by inhalation and a severe irritant to skin, eyes and mucous membranes. BCl$_3$, BBr$_3$, and BI$_3$ emit toxic fumes when heated to decomposition (141). Safe handling, especially of BCl$_3$, has been reviewed (10,140).

8. Uses

8.1. Boron Trifluoride.
Boron trifluoride is an excellent Lewis acid catalyst for numerous types of organic reactions. Its advantages are ease of

handling as a gas and the absence of undesirable tarry by-products. As an electrophilic molecule, it is an excellent catalyst for Friedel-Crafts and many other types of reactions (142–144) (see FRIEDEL-CRAFTS REACTIONS).

$BF_3 \cdot HF$ compositions have been reported to act as super acids in catalyzing condensation reactions (145). BF_3-catalyzed preparation of 1- or 2-naphthol is reported to be regioselective (146). Dehydration reactions may also be regioselective (147). Selected fluorinations may be catalyzed by BF_3 using HF as the fluoride source (148). BF_3 is widely used for the preparation of hydrocarbon resins (149), tall oil (qv) resins (150), and tackifier resins (151). Alpha olefin-based synthetic lubricants are commonly made using BF_3-based catalysts (152–154). BF_3 is widely used as a polymerization catalyst (155–157). A developing use for BF_3 is as an ion implant medium for semiconductor materials (158). BF_3 may be used as a chemical reagent for the manufacture of fluoroboro complexes (159), boron nitride [10043-11-5] (160), and boron trichloride [10294-34-5] (161). Carboxylic acids and esters may be prepared by reacting CO with olefins in the presence of BF_3-containing catalysts (162).

In addition, boron trifluoride and some of its adducts have widespread application as curing agents for epoxy resins (qv), and in preparing alcohol-soluble phenolic resins (qv) (76).

Boron trifluoride catalyst is used under a great variety of conditions either alone in the gas phase or in the presence of many types of promoters. Many boron trifluoride coordination compounds are also used.

Boron trifluoride catalyst may be recovered by distillation, chemical reactions, or a combination of these methods. Ammonia or amines are frequently added to the spent catalyst to form stable coordination compounds that can be separated from the reaction products. Subsequent treatment with sulfuric acid releases boron trifluoride. An organic compound may be added that forms an adduct more stable than that formed by the desired product and boron trifluoride. In another procedure, a fluoride is added to the reaction products to precipitate the boron trifluoride which is then released by heating. Selective solvents may also be employed in recovery procedures (see CATALYSTS, REGENERATION).

Boron trifluoride is also employed in nuclear technology by utilizing several nuclear characteristics of the boron atom. Of the two isotopes, ^{10}B and ^{11}B, only ^{10}B has a significant absorption cross section for thermal neutrons. It is used in $^{10}BF_3$ as a neutron-absorbing medium in proportional neutron counters and for controlling nuclear reactors (qv). Some of the complexes of trifluoroborane have been used for the separation of the boron isotopes and the enrichment of ^{10}B as $^{10}BF_3$ (163).

Boron trifluoride is used for the preparation of boranes (see BORON COMPOUNDS). Diborane is obtained from reaction with alkali metal hydrides; organoboranes are obtained with a suitable Grignard reagent.

Boron trifluoride has been used in mixtures to prepare boride surfaces on steel (qv) and other metals, and as a lubricant for casting steel (see LUBRICATION AND LUBRICANTS).

8.2. Boron Trichloride. Much of the BCl_3 consumed in the United States is used to prepare boron filaments by CVD. These high performance fibers are used to reinforce composite materials (qv) made from epoxy resins and metals

(Al, Ti). The principal markets for such composites are aerospace industries and sports equipment manufacturers.

Another important use of BCl_3 is as a Friedel-Crafts catalyst in various polymerization, alkylation, and acylation reactions, and in other organic syntheses (see FRIEDEL-CRAFTS REACTION). Examples include conversion of cyclo-phosphazenes to polymers (164,165); polymerization of olefins such as ethylene (128,166–171); graft polymerization of vinyl chloride and isobutylene (172); stereospecific polymerization of propylene (173); copolymerization of isobutylene and styrene (174,175), and other unsaturated aromatics with maleic anhydride (176); polymerization of norbornene (177), butadiene (178); preparation of electrically conducting epoxy resins (179), and polymers containing B and N (180); and selective demethylation of methoxy groups ortho to OH groups (181).

BCl_3 is also used for the production of halosilanes, in the preparation of many boron compounds (1,4,5), in the production of optical wave guides (182), and for the prevention of solid polymer formation in liquid SO_3 (10); for the removal of SiO_2 from SiC powders (183), carbochlorination of kaolinitic ores (184), and removal of impurities from molten Mg (182). It is also used as a critical solvent in metal recovery from chlorination processes (185), for the removal of potential catalyst poisons from hydrocarbon oils (186), and in the production of lithium–thionyl chloride batteries (187). Other than production of boron fibers, important CVD processes involving BCl_3 include: production of boron carbide-coated carbon fiber (188,189); doping Si or Ge with B and for doping electric or photoconducting polymers, in the preparation of scratch-resistant silicon-based coatings, and in glass-fiber technology (5); production of boron nitride (5,190–192), and metal borides (5). BCl_3 is also used in reactive ion etching and plasma etching in the production of silicon-integrated circuits and devices, in the dry etching of boron nitride, gallium arsenide, and SnO_2 (5), and Al–Si (193,194).

8.3. Boron Bromide. A large portion of BBr_3 produced in the United States is consumed in the manufacture of proprietory pharmaceuticals (qv). BBr_3 is used in the manufacture of isotopically enriched crystalline boron, as a Friedel-Crafts catalyst in various polymerization, alkylation, and acylation reactions, and in semiconductor doping and etching. Examples of use of BBr_3 as a catalyst include copolymerization of butadiene with olefins (195); polymerization of ethylene and propylene (196), and N-vinylcarbazole (197); in hydroboration reactions and in tritium labeling of steroids and aryl rings (5).

BBr_3 is a very useful reagent for cleaving ethers, esters, lactones, acetals, and glycosidic bonds; it is used to deoxygenate sulfoxides and in the preparation of image-providing materials for photography (5).

8.4. Boron Triiodide. There are no large-scale commercial uses of boron triiodide. It can cleave ethers without affecting aldehyde groups and thus finds use in the synthesis of the antibiotic frustulosin (198,199). BI_3 is used to prepare SnI_4, SbI_3, and TiI_4 (200) in 99–100% yield. It is used to clean equipment for handling UF_6 (201) and in the manufacture of lithium batteries (202).

9. Derivatives

9.1. Fluoroboric Acid and the Fluoroborate Ion.
Fluoroboric acid [16872-11-0], generally formulated as HBF_4, does not exist as a free, pure

substance. The acid is stable only as a solvated ion pair, such as $H_3O^+BF^-_4$; the commercially available 48% HBF_4 solution approximates $H_3O^+BF_4^- \cdot 4H_2O$. Other names used infrequently are hydrofluoroboric acid, hydroborofluoric acid, and tetrafluoroboric acid. Salts of the acid are named as fluoroborates or occasionally borofluorides. Fluoroboric acid and its salts were investigated as early as 1809 (203,204). The acid and many transition-metal salts are used in the electroplating (qv) and metal finishing industries. Some of the alkali metal fluoroborates are used in fluxes.

Properties. Fluoroboric acid is stable in concentrated solutions, and hydrolyzes slowly in aqueous solution to hydroxyfluoroborates. For the stability of the fluoroborate species, see Reference 205. The equilibrium quotients Q (206,207) in 1 molal NaCl at 25°C show the strong affinity of boron for fluoride:

$$B(OH)_3 + F^- \rightleftharpoons BF(OH)_3^- \qquad \log Q = -0.36 \pm 0.19$$

$$B(OH)_3 + 2\,F^- + H^+ \rightleftharpoons BF_2(OH)_2^- + H_2O \qquad \log Q = 7.06 \pm 0.02$$

$$B(OH)_3 + 3\,F^- + 2\,H^+ \rightleftharpoons BF_3OH^- + 2\,H_2O \qquad \log Q = 13.689 \pm 0.003$$

$$B(OH)_3 + 4\,F^- + 3\,H^+ \rightleftharpoons BF_4^- + 3\,H_2O \qquad \log Q = 19.0 \pm 0.1$$

The hydrolysis of BF_4^- occurs stepwise to BF_3OH^-, $BF_2(OH)_2^-$, and $BF(OH)_3^-$. By conductivity measurements the reaction of boric acid and HF was found to form $H[BF_3(OH)]$ [15433-40-6] rapidly; subsequently HBF_4 formed much more slowly from HBF_3OH. These studies demonstrate that BF_4^- is quite stable to hydrolysis yet is slow to form from BF_3OH^- and HF:

$$BF_4^- + H_2O \rightleftharpoons BF_3OH^- + HF$$

Kinetic results (207) and ^{19}F nmr experiments (208) illustrate clearly that the hydroxyfluoroborates are in rapid equilibrium and easily exchange fluoride.

Table 7 lists some of the physical properties of fluoroboric acid. It is a strong acid in water, equal to most mineral acids in strength and has a pK_{H_2O} of -4.9 as

Table 7. Physical Properties of Fluoroboric Acid

Property	Value	Reference
heat of formation, kJ/mol[a]		
aqueous, 1 molal, at 25°C	-1527	
from boric oxide and HF (aq)	-123.34	204
BF_4^-, gas	-1765 ± 42	209
entropy of the BF_4^- ion, J/(mol·K)[a]	167	
specific gravity		
48% soln	1.37	
42% soln	1.32	
30% soln	1.20	
surface tension, 48% soln at 25°C, mN/m(= dyn/cm)	65.3	
ir absorptions,[b] cm^{-1}	ca 1100	210
	ca 530	

[a] To convert J to cal, divide by 4.184.
[b] Generally observed as strong absorptions.

compared to -4.3 for nitric acid (211). The fluoroborate ion contains a nearly tetrahedral boron atom with almost equidistant B–F bonds in the solid state. Although lattice effects and hydrogen bonding distort the ion, the average B–F distance is 0.138 nm; the F–B–F angles are nearly the theoretical $109°$ (212,213). Raman spectra on molten, ie, liquid $NaBF_4$ agree with the symmetrical tetrahedral structure (214).

The fluoroborate ion has traditionally been referred to as a noncoordinating anion. It has shown little tendency to form a coordinate–covalent bond with transition metals as do nitrates and sulfates. A few exceptional cases have been reported (215) in which a coordinated BF_4^- was detected by infrared or visible spectroscopy.

Hydroxyfluoroborates are products of the reaction of BF_3 with water; $BF_3 \cdot 2H_2O$ [13319-75-0] is actually $H_3O^+BF_3OH^-$. Salts such as sodium hydroxyfluoroborate [13876-97-6], $NaBF_3OH$, are made by neutralizing the acid. The BF_3OH^- anions are distorted tetrahedra (216). In the HBO_2–HF system, $HBO_2 \cdot 2HF$ was found to be $HBF_2(OH)_2$, dihydroxyfluoroboric acid [17068-89-2] (217).

Manufacture, Shipping, and Waste Treatment. Fluoroboric acid (48%) is made commercially by direct reaction of 70% hydrofluoric acid and boric acid, H_3BO_3. The reaction is exothermic and must be controlled by cooling.

The commercial product is usually a 48–50% solution which contains up to a few percent excess boric acid to eliminate any HF fumes and to avoid HF burns. Reagent-grade solutions are usually 40%. A 61% solution can be made from metaboric acid, HBO_2, and 70% HF, and a lower grade by direct combination of fluorospar, CaF_2, sulfuric acid, and boric acid (218). The product contains a small amount of dissolved calcium sulfate. A silica-containing (0.11% SiO_2) fluoroboric acid is produced from inexpensive fluorosilicic acid (219). Boric acid is added to a 10% H_2SiF_6 solution and then concentrated in several steps to 45% HBF_4. Granular silicon dioxide must be filtered from the product.

Vessels and equipment must withstand the corrosive action of hydrofluoric acid. For a high quality product the preferred materials for handling HBF_4 solutions are polyethylene, polypropylene, or a resistant rubber such as neoprene (see ELASTOMERS, SYNTHETIC). Where metal must be used, ferrous alloys having high nickel and chromium content show good resistance to corrosion. Impregnated carbon (Carbate) or Teflon can be used in heat exchangers. Teflon-lined pumps and auxilliary equipment are also good choices. Working in glass equipment is not recommended for fluoroboric acid or any fluoroborate.

Fluoroboric acid and some fluoroborate solutions are shipped as corrosive material, generally in polyethylene-lined steel pails and drums or in rigid nonreturnable polyethylene containers. Acid spills should be neutralized with lime or soda ash.

Waste treatment of fluoroborate solutions includes a pretreatment with aluminum sulfate to facilitate hydrolysis, and final precipitation of fluoride with lime (220). The aluminum sulfate treatment can be avoided by hydrolyzing the fluoroborates at pH 2 in the presence of calcium chloride; at this pH, hydrolysis is most rapid at elevated temperature (221).

Analysis. Fluoroboric acid solutions and fluoroborates are analyzed gravimetrically using nitron or tetraphenylarsonsium chloride. A fluoroborate ion-selective electrode has been developed (222).

Table 8. General Properties of Metal Fluoroborates

Compound	CAS Registry Number	Molecular weight	Color	Physical form	Mp, °C	Density,[a] g/cm³	Solubility H₂O g/100 mL[b]	Other	References
$LiBF_4$	[14283-07-9]	93.74	white				very soluble		204,233
$NaBF_4$	[13755-29-8]	109.79	white	orthorhombic <240°C $a = 0.68358$, $b = 0.62619$, $c = 0.67916$ nm noncubic >240°C	406 dec	2.47 210[c]	108 (26°C)	sl alcohol	215,233,234
KBF_4	[14075-53-7]	125.92	colorless	rhombic <283°C $a = 0.7032$, $b = 0.8674$, $c = 0.5496$ nm cubic >283°C	530 dec	2.498	0.45 (20°C) 6.27 (100°C)	sl ethanol insol alkali	233–235
$RbBF_4$	[18909-68-7]	172.27		orthorhombic <245°C $a = 0.7296$, $b = 0.9108$, $c = 0.5636$ nm cubic >245°C	612 dec	2.820 10[c]	0.6 (17°C)		233–235
$CsBF_4$	[18909-69-8]	219.71	white	orthorhombic <140°C $a = 0.7647$, $b = 0.9675$, $c = 0.5885$ nm cubic >245°C	555 dec	3.20 30[c]	1.6 (17°C)		233–235
NH_4BF_4	[13826-83-0]	104.84	white	orthorhombic <205°C $a = 0.7278$, $b = 0.9072$, $c = 0.5678$ nm	487 dec	1.871[d]	3.09 (−1.0°C) 5.26 (−1.5°C) 10.85 (−2.7°C) 12.20 (0°C) 25 (16°C) 25.83 (25°C) 44.09 (50°C) 67.50 (75°C) 98.93 (100°C) 113.7 (108.5°C)	HF[e]	233,235–237
$NaBF_3OH$	[13876-97-6]			hexagonal $a = 0.8084$, $c = 0.7958$ nm		2.46			212

[a] Unless otherwise stated, at 20°C. [b] Temperature given in parentheses. [c] At 100°C. [d] At 15°C. [e] Value at 0°C is 19.89%.

Health and Safety Factors. Fluoroborates are excreted mostly in the urine (223). Sodium fluoroborate is absorbed almost completely into the human bloodstream and over a 14-d experiment all of the $NaBF_4$ ingested was found in the urine. Although the fluoride ion is covalently bound to boron, the rate of absorption of the physiologically inert BF_4^- from the gastrointestinal tract of rats exceeds that of the physiologically active simple fluorides (224).

Uses. Printed circuit tin–lead plating is the main use of fluoroboric acid (225). However, the Alcoa Alzak process for electropolishing aluminum requires substantial quantities of fluoroboric acid. A 2.5% HBF_4 solution is used to produce a highly reflective surface (226). The high solubility of many metal oxides in HBF_4 is a decided advantage in metal finishing operations (see METAL SURFACE TREATMENTS). Before plating or other surface treatment, many metals are cleaned and pickled in fluoroboric acid solution; eg, continuous strip pickling of hot-rolled low carbon steel is feasible in HBF_4 solutions (227). Nontempered rolled steel requires 80°C for 60 s in HBF_4 130 g/L, whereas tempered rolled steel requires only 65°C for 60 s in 65 g/L. The spent pickling solution is recovered by electrodialysis.

Fluoroboric acid is used as a stripping solution for the removal of solder and plated metals from less active substrates. A number of fluoroborate plating baths (228) require pH adjustment with fluoroboric acid (see ELECTROPLATING).

A low grade fluoroboric acid (218) is used in the manufacture of cryolite for the electrolytic production of aluminum:

$$4\,Na_2SO_4 \cdot NaF + 5\,HBF_4 + 2\,Al_2O_3 + 9\,H_2O \longrightarrow 4\,Na_3AlF_6 + 5\,H_3BO_3 + 4\,H_2SO_4$$

The boric and sulfuric acids are recycled to a HBF_4 solution by reaction with CaF_2. As a strong acid, fluoroboric acid is frequently used as an acid catalyst, eg, in synthesizing mixed polyol esters (229). This process provides an inexpensive route to confectioner's hard-butter compositions which are substitutes for cocoa butter in chocolate candies (see CHOCOLATE AND COCOA). Epichlorohydrin is polymerized in the presence of HBF_4 for eventual conversion to polyglycidyl ethers (230) (see CHLOROHYDRINS). A more concentrated solution, 61–71% HBF_4, catalyzes the addition of CO and water to olefins under pressure to form neo acids (231) (see CARBOXYLIC ACIDS). Deprotection of polymers prepared with selyoxyl protected functional initiators by reaction with agreous fluoroboric aid has been reported (232).

9.2. Main Group. *Properties.* A summary of the chemical and physical properties of alkali-metal and ammonium fluoroborates is given in Tables 8 and 9. Chemically these compounds differ from the transition-metal fluoroborates usually separating in anhydrous form. This group is very soluble in water, except for the K, Rb, and Cs salts which are only slightly soluble. Many of the soluble salts crystallize as hydrates.

Lithium fluoroborate crystallizes from aqueous solutions as $LiBF_4 \cdot 3H_2O$ [39963-05-8] and $LiBF_4 \cdot H_2O$ [39963-03-6]. The heat of dehydration of the monohydrate at 91°C is 70.9 kJ/mol (16.95 kcal/mol); the melting point is 117°C (246). Magnesium, calcium, strontium, and barium fluoroborates crystallize as hydrates: $Mg(BF_4)_2 \cdot 6H_2O$ [19585-07-0], $Ca(BF_4)_2 \cdot 2H_2O$ [27860-81-7], $Sr(BF_4)_2 \cdot 4H_2O$ [27902-05-2], and $Ba(BF_4)_2 \cdot 2H_2O$ [72259-09-7], respectively.

Table 9. Thermodynamic Data[a] for Metal Fluoroborates, kJ/mol[b]

Compound	ΔH_{diss}	Lattice energy, $-U$	ΔH_{fus}	ΔH_f	Other	$\log P_{\text{Pa}} = -aT^{-1} + b$			References
						a	b^c	T, °C	
LiBF$_4$	15.9	699		−1838.4	$\Delta H^d = -89.54$	833	6.40	210–320	238–240
NaBF$_4$	69.83	657.3	13.6	−1843.5	$\Delta H^e = -134.1$	3650	8.75	400–700	234,239–241
KBF$_4$	121	598	18.0	−1881.5	$\Delta H^f = -180.5$ $\Delta H_{\text{sub}} = 330$ $S = 130^g$ $C_p = 112.1$	6317	8.15	510–830	234,239,240,242,243
RbBF$_4$	112.8	577	19.6			5960	9.57	600–1000	234,239,242
CsBF$_4$	112.5	556	19.2			5880	9.47	610–1040	234,239,242
NaBF$_3$OH	77.0			−1754		4024	9.11	400–700	241
NH$_4$BF$_4$		607^h			$\Delta H_{\text{sub}} = 47.3$	2469	8.94		244,245

[a] ΔH_{diss} = heat of dissociation, ΔH_{fus} = heat of fusion, ΔH_f = heat of formation, ΔH_{sub} = sublimation. All thermodynamic data at 25°C, unless otherwise stated.
[b] To convert J to cal, divide by 4.184.
[c] To convert $\log P_{\text{Pa}}$ to $\log P_{\text{mm Hg}}$, subtract 2.12 from b.
[d] LiF(s) + BF$_3$(g) ⟶ LiBF$_4$(s).
[e] NaF(s) + BF$_3$(g) ⟶ NaBF$_4$(s).
[f] KF(s) + BF$_3$(g) ⟶ KBF$_4$(s).
[g] Units are in J/(mol·K).
[h] At 260°C.

These hydrated fluoroborates can be dehydrated completely to the anhydrous salts, which show decreasing stabilities: Ba > Sr > Ca > Mg.

The anhydrous magnesium salt is least stable thermally. It forms MgF_2, which has the highest lattice energy. This has been confirmed by differential thermal analysis (dta) of the crystalline hydrates (247). Aluminum fluoroborate [14403-54-4], $Al(BF_4)_3 \cdot (H_2O)_n$, is soluble in strongly acid solutions and displays a tendency for fluoride exchange with BF_4^- to form aluminum fluorides. The aluminyl compound, $AlO^+BF_4^-$, is extremely hygroscopic and is prepared by the reaction of AlOCl, BF_3, and HF (248). Differential thermal analysis experiments show thermal decomposition beginning at 85°C, corresponding to removal of BF_3 and formation of AlOF.

Differential thermal analysis studies of ammonium fluoroborate showed the orthorhombic to cubic transition at 189 ± 5°C and BF_3 generation from 389 to 420°C (249). Sodium hydroxide reacts with NH_4BF_4 liberating ammonia and forming $NaBF_4$. When sodium fluoroborate was studied by infrared spectroscopy, sodium hydroxyfluoroborate, $NaBF_3OH$, was found to be present (250). Although pure sodium hydroxyfluoroborate is thermally unstable, decomposing to $Na_2B_2F_6O$ [18953-03-2] and H_2O, in a melt of $NaBF_4$-NaF no instability of the small amount of $NaBF_3OH$ present was detected. Fusion of $NaBF_4$ or KBF_4 with boric oxide generates BF_3 and complex borates such as KFB_4O_6 (251). Most fluoroborates decompose readily to give BF_3 when treated with sulfuric acid or when calcined (see Table 9 for dissociation pressure). Under strongly basic conditions the chemical equilibrium is shifted away from BF_4^- to borates and fluorides.

Manufacture. Fluoroborate salts are prepared commercially by several different combinations of boric acid and 70% hydrofluoric acid with oxides, hydroxides, carbonates, bicarbonates, fluorides, and bifluorides. Fluoroborate salts are substantially less corrosive than fluoroboric acid but the possible presence of HF or free fluorides cannot be overlooked. Glass vessels and equipment should not be used.

Sodium Fluoroborate. Sodium fluoroborate is prepared by the reaction of NaOH or Na_2CO_3 with fluoroboric acid (252), or by treatment of disodium hexafluorosilicate with boric acid.

Potassium Fluoroborate. Potassium fluoroborate is produced as a gelatinous precipitate by mixing fluoroboric acid and KOH or K_2CO_3. Alternatively, fluorosilicic acid is treated with H_3BO_3 in a 2:1 molar ratio to give HBF_3OH, which reacts with HF and KCl to yield 98% of KBF_4 in 98.5% purity (253). Commercial KBF_4 normally contains less than 1% KBF_3OH.

Ammonium and Lithium Fluoroborates. Ammonia reacts with fluoroboric acid to produce ammonium fluoroborate (254). An alternative method is the fusion of ammonium bifluoride and boric acid (255):

$$2\,NH_4HF_2 + H_3BO_3 \longrightarrow NH_4BF_4 + 3\,H_2O + NH_3$$

The water and ammonia must be removed from the melt. Lithium hydroxide or carbonate react with HBF_4 to form $LiBF_4$.

Magnesium Fluoroborate. Treatment of magnesium metal, magnesium oxide, or magnesium carbonate with HBF_4 gives magnesium fluoroborate [14708-13-5]. The MgF_2 is filtered and the product is sold as a 30% solution.

Uses. Alkali metal and ammonium fluoroborates are used mainly for the high temperature fluxing action required by the metals processing industries (see METAL SURFACE TREATMENTS; WELDING). The tendency toward BF_3 dissociation at elevated temperatures inhibits oxidation in magnesium casting and aluminum alloy heat treatment.

The molten salts quickly dissolve the metal oxides at high temperatures to form a clean metal surface. Other uses are as catalysts and in fire-retardant formulations (see FLAME RETARDANTS).

Potassium Fluoroborate. The addition of potassium fluoroborate to grinding wheel and disk formulations permits lower operating temperatures (256). Cooler action is desirable to reduce the burning of refractory materials such as titanium and stainless steels. Excellent results in grinding wheels are also obtained with $NaBF_4$ (257). A process for boriding steel surfaces using B_4C and KBF_4 as an activator improves the hardness of the base steel (258). Fluxes for aluminum bronze and silver soldering and brazing contain KBF_4 (259) (see SOLDERS AND BRAZING FILLER METALS). Fire retardance is imparted to acrylonitrile polymers by precipitating KBF_4 within the filaments during coagulation (260). In polyurethanes, KBF_4 and NH_4BF_4 reduce smoke and increase flame resistance (261). Both the potassium and ammonium salts improve insulating efficiency of intumescent coatings (262). The endothermic characteristics of these fillers (qv) (release of BF_3) counteract the exothermic nature of the intumescent agents (nitroaromatic amines) in the coating. The sodium and potassium salts are claimed to have a synergistic effect with polyhalogenated aromatics that improve flame-retardant properties of polyesters (263). Elemental boron is prepared by the Cooper electrolysis of a KBF_4 melt with B_2O_3 and KCl (264). The boron may be up to 99.5% purity and, if KBF_4 containing the ^{10}B isotope is used, the product is ^{10}B which is used in the nuclear energy field as a neutron absorber (see NUCLEAR REACTORS).

Sodium Fluoroborate. Sodium fluoroborate can be used in the transfer of boron to aluminum alloys but the efficiency is lower than for KBF_4 (265). Sodium fluoroborate in an etching solution with sulfamic acid and H_2O_2 aids in removing exposed lead in printed circuit manufacture (266). During the annealing of galvanized iron (galvannealing), the surface becomes oxidized. The resulting oxide coating, which causes difficulty in soldering, can be removed by aqueous $NaBF_4$ or NH_4BF_4 (267). Work at Oak Ridge National Lab (Tennessee) has shown that a $NaBF_4$, with 8 mol % NaF, salt mixture could be used as the coolant in the molten breeder reactor (268); in this molten salt at nearly 600°C the corrosion rate of Hastelloy N is about 8 μm/yr. Sodium fluoroborate acts as a catalyst for cross-linking cotton cellulose with formaldehyde (269); transesterification in the preparation of polycarbonates (270); and preparation of cyclic oligoethers from ethylene oxide (271). Sodium and lithium fluoroborates are effective flame retardants for cotton and rayon (272).

Ammonium Fluoroborate. Ammonium fluoroborate blends with antimony oxide give good results in flame-retarding polypropylene (273). The complete thermal vaporization makes ammonium fluoroborate an excellent gaseous flux for inert-atmosphere soldering (274). A soldering flux of zinc chloride and ammonium fluoroborate is used in joining dissimilar metals such as Al and Cu (275). Ammonium fluoroborate acts as a solid lubricant in cutting-oil emulsions for aluminum rolling and forming.

Lithium Fluoroborate. Lithium fluoroborate is used in a number of batteries (qv) as an electrolyte, for example in the lithium–sulfur battery (276).

Miscellaneous. Flame-resistant cross-linked polyethylene can be made with a number of fluoroborates and antimony oxide. This self-extinguishing material may contain the fluoroborates of NH_4^+, Na^+, K^+, Ca^{2+}, Mg^{2+}, Sr^{2+}, or Ba^{2+} in amounts of 4–20% (277). Magnesium fluoroborate catalyzes the epoxy treatment of cotton fabrics for permanent-press finishes (278) (see TEXTILES).

9.3. Transition-Metal and Other Heavy-Metal Fluoroborates.

The physical and chemical properties are less well known for transition metals than for the alkali metal fluoroborates (Table 10). Most transition-metal fluoroborates are strongly hydrated coordination compounds and are difficult to dry without decomposition. Decomposition frequently occurs during the concentration of solutions for crystallization. The stability of the metal fluorides accentuates this problem. Loss of HF because of hydrolysis makes the reaction proceed even more rapidly. Even with low temperature vacuum drying to partially solve the decomposition, the dry salt readily absorbs water. The crystalline solids are generally soluble in water, alcohols, and ketones but only poorly soluble in hydrocarbons and halocarbons.

Differential thermal analysis in air on the crystalline hexahydrates of Zn, Cd, Fe, Co, and Ni fluoroborates show the loss of BF_3 and H_2O simultaneously at

Table 10. **Properties of Metal Fluoroborates**^{a}

Compound	CAS Registry Number	Color	Specific gravity	Solubility	Miscellaneous
$Mn(BF_4)_2 \cdot 6H_2O$	[26044-57-5]	pale pink	1.982	water, ethanol	
$Fe(BF_4)_2 \cdot 6H_2O$	[13877-16-2]	pale green	2.038	water, ethanol	
$Co(BF_4)_2 \cdot 6H_2O$	[15684-35-2]	red	2.081	water, alcohol	
$Ni(BF_4)_2 \cdot 6H_2O$	[14708-14-6]	green	2.136	water, alcohol	
$Cu(BF_4)_2 \cdot 6H_2O$	[72259-10-0]	blue	2.175	water, alcohol	
$AgBF_4 \cdot H_2O$	[72259-11-1]	colorless		water, less sol in alcohol, sol benzene, sol ether	dec 200°C, light sensitive
$Zn(BF_4)_2 \cdot 6H_2O$	[27860-83-9]	white	2.120	water, alcohol	dehydrates at 60°C
$Cd(BF_4)_2 \cdot 6H_2O$	[27860-84-0]	white	2.292	water, alcohol	
$In(BF_4)_3 \cdot xH_2O$	[27765-48-6]	colorless		water	
$TlBF_4 \cdot H_2O$	[72259-12-2]	colorless		water	orthorhombic, $a = 0.947$, $b = 0.581$, $c = 0.740$ nm, light sensitive
$Sn(BF_4)_2 \cdot xH_2O$	[72259-13-3]	white		water	$Sn(BF_4)_2 \cdot SnF_2 \cdot 5H_2O$ crystallizes from soln
$Pb(BF_4)_2 \cdot H_2O$	[26916-34-7]	colorless			

^{a} Crystalline solids (204,279).

195, 215, 180, 185, and 205°C, respectively (247,280). The dta curves also indicate initial melting at 107, 117, and 150°C for Zn, Cd, and Fe fluoroborates, respectively. The anhydrous metal fluoride and/or oxide is usually isolated. The copper salt also decomposes with liberation of BF_3 and H_2O (281).

The water of hydration of these complexes can be replaced with other coordinating solvents. For example, the ethanol and methanol solvates were made by dissolving the hydrates in triethyl and trimethyl orthoformate, respectively (282,283). The acetic acid solvates are made by treating the hydrates with acetic anhydride (284). Conductivity and visible spectra, where applicable, of the Co, Ni, Zn, and Cu fluoroborates in N,N-dimethylacetamide (L) showed that all metal ions were present as the ML_6^{2+} cations (285). Solvated fluoroborate complexes of Cr^{3+}, Fe^{2+}, Co^{2+}, Ni^{2+}, Cu^{2+}, Cu^+, and Zn^{2+} in diethyl ether, nitromethane, and benzene solutions have been prepared. Solutions of $Ti(BF_4)_3$, $V(BF_4)_3$, and $Fe(BF_4)_3$ could not be prepared probably because of formation of BF_3 and the metal fluoride (286). Ammonia easily replaces the coordinated water; the products are usually tetrammine or hexammine complexes (204) (see COORDINATION COMPOUNDS). The hexahydrate of $Ni(BF_4)_2$ was found to be stable from 25 to 100°C; solubility also was determined to 95°C (287). At 120°C the solid decomposed slowly to NiF_2 with loss of HF, H_3BO_3, and H_2O.

Manufacture.   The transition- and heavy-metal fluoroborates can be made from the metal, metal oxide, hydroxide, or carbonate with fluoroboric acid. Because of the difficulty in isolating pure crystalline solids, these fluoroborates are usually available as 40–50% solutions, $M(BF_4)_x$. Most of the solutions contain about 1–2% excess fluoroboric acid to prevent precipitation of basic metal complexes. The solutions are usually sold in 19 and 57 L polyethylene containers.

In some cases, particularly with inactive metals, electrolytic cells are the primary method of manufacture of the fluoroborate solution. The manufacture of Sn, Pb, Cu, and Ni fluoroborates by electrolytic dissolution (288,289) is patented. A typical cell for continous production consists of a polyethylene-lined tank with tin anodes at the bottom and a mercury pool (in a porous basket) cathode near the top (289). Fluoroboric acid is added to the cell and electrolysis is begun. As tin fluoroborate is generated, differences in specific gravity cause the product to layer at the bottom of the cell. When the desired concentration is reached in this layer, the heavy solution is drawn from the bottom and fresh HBF_4 is added to the top of the cell continuously. The direct reaction of tin with HBF_4 is slow but can be accelerated by passing air or oxygen through the solution (290). The stannic fluoroborate is reduced by reaction with mossy tin under an inert atmosphere. In earlier procedures, HBF_4 reacted with hydrated stannous oxide.

Anhydrous silver fluoroborate [1404-20-2] is made by the addition of BF_3 gas to a suspension of AgF in ethylbenzene (291). An $AgBF_4 \cdot C_8H_{10}$ complex is precipitated with pentane and the complex is washed with pentane to give anhydrous $AgBF_4$.

Uses.   Metal fluoroborate solutions are used primarily as plating solutions and as catalysts. The Sn, Cu, Zn, Ni, Pb, and Ag fluoroborates cure a wide range of epoxy resins at elevated or ambient room temperature (292,293). In the textile industry zinc fluoroborate is used extensively as the curing agent in applying resins for crease-resistant finishes (294). Emulsions of epoxy resins (295), polyoxymethylene compounds (296), or aziridinyl compounds (297) with $Zn(BF_4)_2$ and other additives are applied to the cloth. After the excess is removed, the cloth is dried and later cured at a higher temperature. Similarly treated acrylic textiles using epoxy resins take on an antistatic finish (298), or the acrylic textiles can be coated with 20% $Zn(BF_4)_2$ which results in up to 5.5% added solids for a fire-resistant finish (299).

The use of silver fluoroborate as a catalyst or reagent often depends on the precipitation of a silver halide. Thus the silver ion abstracts a Cl^- from a rhodium chloride complex, $((C_6H_5)_3As)_2(CO)RhCl$, yielding the cationic rhodium fluoroborate [30935-54-7] hydrogenation catalyst (300). The complexing tendency of olefins for $AgBF_4$ has led to the development of chemisorption methods for ethylene separation (301,302). Copper(I) fluoroborate [14708-11-3] also forms complexes with olefins; hydrocarbon separations are effected by similar means (303).

The manufacture of linear polyester is catalyzed by Cd, Sn (304), Pb, Zn, or Mn (305) fluoroborates. The Beckmann rearrangement of cyclohexanone oxime to caprolactam is catalyzed by $Ba(BF_4)_2$ [13862-62-9] or $Zn(BF_4)_2$ [13826-88-5] (306). The caprolactam is polymerized to polyamide fibers using $Mn(BF_4)_2$ [30744-82-2] catalyst (307). Nickel and cobalt fluoroborates appear to be good catalysts for the polymerization of conjugated dienes to *cis*-1,4-polydienes; the cis configuration is formed in up to 96% yields (308–310).

Metal fluoroborate electroplating (qv) baths (228,311,312) are employed where speed and quality of deposition are important. High current densities can be used for fast deposition and near 100% anode and cathode efficiencies can be expected. Because the salts are very soluble, highly concentrated solutions can be used without any crystallization. The high conductivity of these solutions reduces the power costs. The metal content of the bath is also easily maintained and the pH is adjusted with HBF_4 or aqueous ammonia. The disadvantages of using fluoroborate baths are treeing, lack of throwing power, and high initial cost. Treeing and throwing power can be controlled by additives; grain size of the deposits can also be changed. As of this writing, metals being plated from fluoroborate baths are Cd, Co, Cu, Fe, In, Ni, Pb, Sb, and Zn. Studies on Fe (313,314), Ni (314), and Co (314) fluoroborate baths describe the compositions and conditions of operation as well as the properties of the coatings. Iron foils electrodeposited from fluoroborate baths and properly annealed have exceptionally high tensile strength (314).

The Fe, Co, and Ni deposits are extremely fine grained at high current density and pH. Electroless nickel, cobalt, and nickel–cobalt alloy plating from fluoroborate-containing baths yields a deposit of superior corrosion resistance, low stress, and excellent hardenability (315). Lead is plated alone or in combination with tin, indium, and antimony (316). Sound insulators are made as lead–plastic laminates by electrolytically coating Pb from a fluoroborate bath to 0.5 mm on a copper-coated nylon or polypropylene film (317) (see INSULATION, ACOUSTIC). Steel plates can be simultaneously electrocoated with lead and poly(tetrafluoroethylene) (318). Solder is plated in solutions containing $Pb(BF_4)_2$ and $Sn(BF_4)_2$; thus the lustrous solder-plated object is coated with a Pb–Sn alloy (319).

BIBLIOGRAPHY

"Boron Halides" under "Boron Compounds" in *ECT* 1st ed., Vol. 4, pp. 592–593, by M. H. Pickard; in *ECT* 2nd ed., Vol. 3, pp. 680–683, by M. L. Iverson, Atomics International, and S. M. Dragonor, U.S. Borax Research Corp.; in *ECT* 3rd ed., Vol. 4, pp. 129–135, by L. D. Lower, U.S. Borax Research Corp.; in *ECT* 4th ed., Vol. 4, pp. 430–439, by Fazlul Alam, U.S. Borax Research Corp.; "Boron Trifluoride" under "Fluorine Compounds, Inorganic" in *ECT* 1st ed., Vol. 6, pp. 678–684, by D. R. Martin, University of Illinois; "Boron Trifluoride" under "Boron" under "Fluorine Compounds, Inorganic" in *ECT* 2nd ed., Vol. 9, pp. 554–562, by D. R. Martin, The Harshaw Chemical Co.; in *ECT* 3rd ed., Vol. 10, pp. 685–693 by D. R. Martin, University of Texas at Arlington; in *ECT* 4th ed., Vol. 11, pp. 300–308, by Francis Evans and Ganpat Mani, Allied Signal, Inc.; "Fluoroboric Acid" under "Fluorine Compounds, Inorganic," in *ECT* 1st ed., Vol. 6, pp. 684–688, by F. D. Loomis, Pennsylvania Salt Manufacturing Co.; "Fluoroboric Acid and Fluoroborates" under "Fluorine Compounds, Inorganic" in *ECT* 2nd ed., Vol. 9, pp. 562–572, by H. S. Halbedel, The Harshaw Chemical Co.; in *ECT* 3rd ed., Vol. 10, pp. 693–706, by H. S. Halbedel and T. E. Nappier, The Harshaw Chemical Co.; in *ECT* 4th ed., Vol. 11, pp. 309–323, by John R. Papcun, Atotech; "Boron Halides" in *ECT* (online), posting date: December 4, 2000, by Fazlul Alam, U.S. Borax Research Corporation.

CITED PUBLICATIONS

1. B. R. Gragg, in K. Niedenzu, K. C. Buschbeck, and P. Merlet, eds., *Gmelin Handbook of Inorganic Chemistry, Borverbindungen*, Vol. 19, Springer-Verlag, Berlin, Germany, 1978, pp. 109, 168.
2. M. K. Das, in Ref. 1, pp. 169, 206.
3. B. R. Gragg, in Ref. 1, pp. 207, 223.
4. A. Meller, "Boron Compounds," in *Gmelin Handbook of Inorganic Chemistry*, 2nd Suppl., Vol. 2, Springer-Verlag, Berlin, Germany, 1982, pp. 77, 154.
5. Ref. 4, 3rd Suppl., Vol. 4, 1988, pp. 1, 102.
6. J. A. Morrison, *Chem. Rev.* **91**, 35 (1991).
7. G. Urry, in E. Muetterties, ed., *The Chemistry of Boron and Its Compounds*, John Wiley & Sons, Inc., New York, 1967, p. 325.
8. T. J. Ward, *J. Chem. Eng. Data* **14**, 167 (1969).
9. W. F. Barber, C. F. Boynton, and P. E. Gallagher, *J. Chem. Eng. Data* **9**, 137 (1964).
10. *Technical Bulletin*, Kerr-McGee Chemical Corp., Oklahoma City, Okla., 1973, No. 0211.
11. P. D. Ownby and R. D. Gretz, *Surface Sci.* **9**, 37 (1968).
12. A. Finch and P. J. Gardner, in R. J. Brotherton and H. Steinberg, eds., *Progress in Boron Chemistry*, Pergamon Press, New York, 1970, p. 177.
13. N. N. Greenwood and B. S. Thomas, in A. F. Trotman-Dickerson, ed., *Comprehensive Inorganic Chemistry*, Pergamon Press, Oxford, UK, 1973, p. 956.
14. E. Pohland and W. Harlos, *Z. Anorg. Allgem. Chem.* **207**, 242 (1932).
15. H. S. Booth and J. M. Carter, *J. Phys. Chem.* **36**, 1359 (1932).
16. R. F. Smith, U.S. Atomic Energy Commission, *NAA-SR-5286*, 1960.
17. C. F. Rumold, PhD. dissertation, Case Western Reserve University, Cleveland, Ohio, 1931.
18. E. Wiberg and W. Mäthing, *Ber. Dtsch. Chem. Ges. B.* **70B**, 690 (1937).
19. A. Eucken and E. Schröder, *Z. Physik. Chem.* **341**, 307 (1938).
20. H. M. Spencer, *J. Chem. Phys.* **14**, 729 (1946).

21. D. R. Stull and H. Prophet, *Natl. Stand. Ref. Data Ser. Natl. Bur. Stand.* **37** (1971).
22. T. P. Onak and co-workers, *J. Phys. Chem.*, 63 (1959).
23. M. F. Lappert and co-workers, *J. Chem. Soc. A.*, 2426 (1971).
24. T. D. Coyle, S. L. Stafford, and F. G. Stone, *J. Chem. Soc.*, 3103 (1961).
25. T. D. Coyle and F. G. A. Stone, *J. Chem. Phys.* **32**, 1892 (1960); I. S. Jaworiwsky and co-workers, *Inorg. Chem.* **18**, 56 (1979).
26. *Gmelins Handbuch der Anorganischen Chemie*, Vol. 13, 8th ed., Verlag Chemie, GmbH, Weinheim/Bergstrasse, Germany, 1954, 167–196.
27. H. C. Brown and R. R. Holmes, *J. Am. Chem. Soc.* **78**, 2173 (1956).
28. S. Pawlenko, *Z. Anorg. Allegem. Chem.* **300**, 152 (1959).
29. N. N. Greenwood and A. Thompson, *J. Chem. Soc.*, 3643 (1959).
30. H. Gerding and co-workers, *Rec. Trav. Chim.* **71**, 501 (1952).
31. N. N. Greenwood and A. Thompson, *J. Chem. Soc.*, 3493 (1959).
32. R. J. Mikovsky, S. D. Levy, and A. L. Hensley, Jr., *J. Chem. Eng. Data* **6**, 603 (1961).
33. E. C. Hughes and S. M. Darling, *Ind. Eng. Chem.* **43**, 746 (1951).
34. H. S. Booth and D. R. Martin, *J. Am. Chem. Soc.* **64**, 2198 (1942).
35. D. B. Beach and W. L. Jolly, *J. Phys. Chem.* **88**, 4647 (1984).
36. J. F. Liebman, *Struct. Chem.* **1**, 395 (1990).
37. A. G. Massey, in H. J. Emeleus and A. G. Sharpe, eds., *Inorganic and Radiochemistry*, Vol. 10, Academic Press, New York, 1967, p. 1.
38. W. Gerrard and M. F. Lappert, *Chem. Rev.* **58**, 1081 (1958).
39. D. R. Martin, *Chem. Rev.* **42**, 581 (1948).
40. D. R. Martin and J. M. Canon, in G. Olah, ed., *Friedel-Crafts and Related Reactions*, Vol. 1, Interscience Publishers, a division of John Wiley & Sons, Inc., New York, 1963, p. 399.
41. P. M. Druce and M. F. Lappert, *J. Chem. Soc. A*, 3595 (1971).
42. D. Z. Hobbs, T. T. Campbell, and F. E. Block, *U. S. Bur. Mines Rept. Invest.* (6456), 16 1964.
43. B. Kamenar, *Z. Anorg. Allgem. Chem.* **342**, 108 (1966).
44. P. Pichat and D. Forest, *Bull. Soc. Chim. France*, 3825 (1967).
45. I. Ahmad and W. J. Hefferman, *J. Electrochem. Soc.* **118**, 1670 (1971).
46. D. J. Cameron and R. A. J. Shelton, *J. Inorg. Nucl. Chem.* **28**, 77 (1966).
47. P. L. Timms, *Chem. Commun.*, 258 (1968).
48. J. L. Down, J. Lewis, B. Moore, and G. Wilkinson, *Proc. Chem. Soc.*, 209 (1957).
49. P. L. Timms, *J. Chem. Soc. Dalton Trans.*, 830 (1972).
50. J. H. Murib, D. Horvitz, and C. A. Vonecutter, *Ind. Eng. Chem. Prod. Res. Develop.* **4**, 273 (1965).
51. T. Hanslik, J. Plesek, and S. Hermanek, *Collection Check. Chem. Commun.* **31**, 4215 (1966).
52. H. D. Batha, C. D. Good, and J. P. Faust, *J. Appl. Chem. (London)* **14**, 257 (1964).
53. J. Cueilleron and J. Bouix, *Bull. Soc. Chim. France*, 1941 (1966).
54. B. A. Jacob, F. C. Douglas, and F. S. Galasso, *Am. Ceram. Soc. Bull.* **52**, 896 (1973).
55. H. E. Hintermann, R. Bonatti, and H. Breiter, *Chem. Vap. Deposition, Int. Conf.*, 536 (1973).
56. J. Guilly, M. Pennaneach, and G. Lassau, in M. I. Buolos and R. J. Munz, eds., *Proceedings of the 6th International Symposium on Plasma Chemistry*, McGill University, Montreal, Quebec, Canada, Vol. 1, 1983, p. 91.
57. L. Vandenbulcke, *Proc. Electrochem. Soc.* **79**(3), 315 (1979).
58. H. Hannache, R. Naslain, and C. Bernard, *J. Less-Common Met.* **95**, 221 (1983).
59. T. L. Wen, W. L. Li, and S. C. Li, *Guisuanyan Xuebao* **8**, 351 (1980).
60. U.S. Pat. 4,436,762 (1984), W. P. Lapatovich, J. M. Proud, and L. A. Riseberg (to GTE Laboratories).

61. K. L. Khachishvile and co-workers, *Zh. Neorg. Khim.* **6**, 1493 (1961).
62. P. Pichat, *C. R. Acad. Sci. Paris Ser. C* **265**, 385 (1967).
63. R. N. Scott and D. F. Shriver, *Inorg. Chem.* **5**, 158 (1966).
64. P. Baumgarten and W. Bruns, *Ber. Dtsch. Chem. Ges. B.* **B72**, 1753 (1939); *Ibid.* **B74**, 1232 (1941).
65. H. D. Fishcher, W. J. Lehmann, and I. Shapiro, *J. Phys. Chem.* **65**, 1166 (1961).
66. H. I. Schlesinger and co-workers, *J. Am. Chem. Soc.* **75**, 195 (1953).
67. *Ibid.*, p. 199.
68. E. L. Gamble, *Inorg. Synth.* **3**, 27 (1950).
69. Brit. Pat. 226,490 (Dec. 20, 1923), A. F. Meyerhofer.
70. E. Krause and R. Nitsche, *Chem. Ber.* **54B**, 2784 (1921).
71. H. C. Brown and U. S. Racherla, *Organometallics 1986* **5**(2), 391–393 (1986).
72. B. M. Mikhailov and T. A. Schhegoleva, *J. Gen. Chem. U.S.S.R.* **29**, 3404 (1959).
73. G. T. Morgan and R. B. Tunstall, *J. Chem. Soc.* **125**, 1963 (1924).
74. W. I. Aalbersberg and co-workers, *J. Chem. Soc.*, 3055 (1959).
75. W. V. Bush and L. Zechmeister, *J. Am. Chem. Soc.* **80**, 2991 (1958).
76. H. S. Booth and D. R. Martin, *Boron Trifluoride and Its Derivatives*, John Wiley & Sons, Inc., New York, 1949.
77. P. Baumgarten and H. Henning, *Chem. Ber.* **72B**, 1743 (1939).
78. C. Gascard and G. Mascherpa, *J. Chim. Phys. Phys. Chim. Biol.* **70**, 1040 (1973).
79. R. J. Gillespie and J. L. Hartman, *Can. J. Chem.* **45**, 859 (1967).
80. H. S. Booth and D. R. Martin, *Boron Trifluoride and its Derivatives*, John Wiley & Sons, Inc., New York, 1948.
81. D. Farcasiu and A. Ghenciu, *J. Catal.* **134**(1), 126–133 (1992).
82. J. S. McGrath and co-workers, *J. A. C. S.*, **66**, 126 (1944).
83. C. A. Wamser, *J. Am. Chem. Soc.* **73**, 409 (1951).
84. C. A. Wamser, *J. Am. Chem. Soc.* **70**, 1209 (1948).
85. V. I. Durkov and S. S. Batsanov, *Zh. Strukt. Khim.* **2**, 456 (1961).
86. E. F. Mooney and M. A. Qaseem, *Chem. Commun.*, 230 (1967).
87. M. J. Bula, J. S. Hartman, and C. V. Raman, *J. Chem. Soc. Dalton Trans.*, 725 (1974).
88. Brit. Pat 486,887 (June 13, 1938), (to E. I. du Pont de Nemours & Co., Inc.).
89. D. R. Martin and J. M. Canon, in G. A. Olah, ed., *Friedel-Crafts and Related Reactions*, Vol. 1, Wiley-Interscience, New York, 1963, 399–567.
90. H. S. Booth and K. S. Wilson, *Inorg. Synth.* **1**, 21 (1939).
91. U.S. Pat. 4,830,842 (May 16, 1989), B. Leutner and H. H. Reich.
92. U.S. Pat. 5,371,152 (Dec. 6, 1994), T. Kawamura, N. Akatsu, and A. Ishimoto.
93. U.S. Pat. 2,876,076 (1959), C. W. Montgomery and W. A. Pardee (to Gulf Research and Development Co.).
94. Ger. Pat. 1,079,609 (1960), R. Kumagai, W. F. Robinson, and J. C. Slaughter (to Olin Mathieson Chemical Corp.).
95. Eur. Pat. Appl. 34,897 (1981), T. Iwai, H. Mizuno, and M. Miura (to Ube Industries, Ltd.).
96. B. C. Pai, B. E. Ramachandran, and V. Velpari, *Indian J. Technol.* **22**, 233 (1984).
97. U.S. Pat. 4,239,738 (1980), K. W. Richardson (to PPG. Industries, Inc.).
98. Ger. Offen. 3,011,246 (1980), N. R. Delue and J. C. Crano (to PPG. Industries, Inc.).
99. K. Higashi, *Kyushu Kozan Gakkai-Shi* **29**, 312 (1961).
100. H. Ito, T. Yanagase, and K. Higashi, *Trans. Japan Inst. Met.* **4**, 205 (1963).
101. E. L. Gamble, *Inorg. Synth.* **3**, 27 (1950).
102. Ger. Pat. DD285,766 A5 (Jan. 1, 1991), K. Stengel, D. Klemm, and K. Eckadt (to Academie der Wissenschaften der DDR).
103. N. Sevryugora and N. Zhavoronko, *Kim. Prom.* **46**, 433 (1970).

104. N. Sevryugora, *Metody, Poluch. Anql. Veshehestv. Osobi Chist. Tr. Vscs. Knof.*, 64 (1970).
105. G. Devyatykh and co-workers, *Zh. Prikl. Khim.* **49**, 1280 (1976).
106. Czech. Pat. 223,494 (1984), J. Plesek, S. Hermanek, and J. Mostecky.
107. S. Mohan and R. Manickkavachagam, *Indian J. Pure Appl. Chem.* **16**, 55 (1978).
108. A. A. Efremov and Y. Zelvenskii, *Tr. Mosk. Khim. Tekhnol. Inst.* **96**, 4 (1977).
109. U.S. Pat. 3,207,581 (Sept. 21, 1965), D. R. Stern and W. W. Walker (to American Potash and Chemical Corp.).
110. U.S. Pat. 3,043,665 (July 10, 1962), J. R. Gould and D. M. Gardner (to Thiokol Chemical Corp.).
111. U.S. Pat. 3,126,256 (Mar. 24, 1964), J. N. Haimsohn, L. A. Smalheiser, and B. J. Luberoff (to Stauffer Chemical Co.).
112. R. C. Hyer, S. M. Freund, A. Hartford, Jr., and J. H. Atencio, *J. Appl. Phys.* **52**, 6944 (1981).
113. U.S. Pat. 4,204,926 (1980), H. C. Meyer, G. A. Tanton, R. I. Greenberg, and J. E. Williams (to United States Dept. of the Army).
114. U.S. Appl. 21,037 (1980), H. C. Meyer (to United States Dept. of the Army).
115. R. C. Hyer, A. Hartford, Jr., and J. H. Atencio, *Proc. Intl. Conf. Lasers*, 293 (1980–1981).
116. U.S. Pat. 4,405,423 (1983), S. M. Freund.
117. U.S. Pat. 6,361,750 (March 26, 2002), D. Zhou and co-workers (to Air Liquide America Corp.)
118. U.S. Pat. 2,989,375 (1961), F. H. May and J. L. Bradford (to American Potash and Chemical Corp.).
119. Ger. Offen. 2,113,591 (1972), G. Wiebke, G. Stohr, G. Vogt, and G. Kratel (to Elektroschmelzwerke Kempten GmbH).
120. Ger. Offen. 1,957,949 (1971), G. Kratel and G. Vogt (to Elektroschmelzwerke Kempten GmbH).
121. Q. Yuan, K. Zhang, and J. Wang, *Huaxue Shijie* **23**, 194 (1982).
122. H. Noth and R. Staudigl, *Chem. Ber.* **111**, 3280 (1978).
123. A. F. Farmington, J. T. Buford, and R. J. Starks, Papers of the 2nd *Intern. Symp. Boron Prep. Properties Appl.*, Paris, 1964, p. 21.
124. B. N. Ivanov-Emin, L. A. Niselson, and I. V. Petrusevich, *Zh. Prikl. Khim.* **34**, 2378 (1961).
125. L. V. McCarty and D. R. Carpenter, *J. Electrochem. Soc.* **107**, 3 (1960).
126. T. Renner, *Agnew. Chem.* **69**, 478 (1957).
127. J. Cueilleron and H. Mongeot, *Bull. Soc. Chim. France*, 76 (1966).
128. M. Berkenbilt and A. Reisman, *J. Electrochem. Soc.* **117**, 1100 (1970).
129. Jpn. Kokai 78 05093 (1978), N. Kano, S. Tanaka, and S. Okano (to Mitsubishi Chemical Industries Co., Ltd.).
130. M. F. Lappert, M. R. Kitzow, J. B. Pedley, T. R. Spalding, and H. Noth, *J. Chem. Soc. A*, 383 (1971).
131. J. W. Koger, Oak Ridge National Laboratory, TM-4172, 1972; *Nucl. Sci. Abstr.* **28**, 11,211 (1973).
132. C. F. Swinehart, A. R. Bumblish, and H. F. Flisik, *Anal. Chem.* **19**, 28 (1947); *Ann. Proc.* 35-0049, internal document, Harshaw Chemical Co., Mar. 23, 1964.
133. F. D. Snell and C. L. Hilton, eds., *Encyclopedia of Industrial Chemical Analysis*, Interscience Publishers, New York, 1968, p. 332.
134. A. Ferguson and D. Treskon, in *Chemical Economics Handbook*, Stanford Research Institute, Menlo Park, Calif., 1977, p. 717.1000a.
135. K. H. Jacobson, R. A. Rhoden, and R. L. Roudabush, *HEW Pub. (NIOSH) Publ. 77*, (1976).

136. Code of Fed. Reg. 29, part. 1901, U.S. Govt. Printing Office, Washington, D.C., 1988.
137. A.C.G.I.H., *Threshold Limit Values for Chemical Substances*, 1989–1990.
138. G. M. Rusch and co-workers, *Toxicology and Applied Pharmacology* **83**, 69–78 (1986).
139. C. J. Spiegl, *Natl. Nucl. Energy Ser. Div. VI 1 (Book 4)*, 2291 (1953).
140. T. Shirai, *Oyo Butsuri* **52**, 597 (1983).
141. R. J. Lewis, Sr., *Sax's Properties of Dangerous Materials*, Vol. 2, 10th ed., John Wiley & Sons, Inc., New York, 2000.
142. G. A. Olah, ed., in Ref. 89, 228–235.
143. Ref. 76, Chapt. 6.
144. A. V. Topchiev, S. V. Zavgorodnii, and Y. M. Paushkin, *Boron Fluoride and Its Compounds as Catalysts in Organic Chemistry*, Pergamon Press, New York, 1959.
145. Fr. Pat. 2,647,108 (Nov. 23, 1990), L. Gilbert and co-workers (to Rhône-Poulenc).
146. U.S. Pat. 4,419,528 (Dec. 6, 1983), G. A. Olah (to PCUK Ugine Kuhlman).
147. G. H. Posner and co-workers, *Tetrahedran Lett.* **32**(45) 6489–6492 (1991).
148. U.S. Pat. 4,962,244 (Oct. 9, 1990), M. Y. Elsheikh (to Atochem, N. Amer. Inc.).
149. U.S. Pat. 4,657,773 (Apr. 14, 1987), S. C. Durkee (to Hercules Inc.).
150. U.S. Pat. 4,657,706 (Apr. 14, 1987), S. C. Durkee (to Hercules Inc.).
151. U.S. Pat. 5,051,485 (Sept. 24, 1991), J. J. Schmid and J. W. Booth (to Arizona Chem.).
152. U.S. Pat. 4,434,309 (Feb. 28, 1984), J. M. Larkin and W. H. Brader (to Texaco Inc.).
153. U.S. Pat. 4,484,014 (Nov. 20, 1984), W. I. Nelson and co-workers (to Phillips Pet. Co.).
154. U.S. Pat. 4,935,570 (June 19, 1990), M. B. Nelson and co-workers (to Ethyl Corp.).
155. U.S. Pat. 5,068,490 (Feb. 29, 1988), B. E. Eaton (to Amoco Corp.).
156. U.S. Pat. 5,071,812 (Mar. 31, 1989), D. R. Kelsey (to Shell Oil Co.).
157. M. C. Throckmorton, *J. Appl. Polym. Sci.* **42**(11), 3019–3024 (1991).
158. M. H. Juang and H. C. Cheng, *J. Appl. Phys.* **71**(3), 1265–1270 (1992).
159. B. K. Mohapatra and co-workers, *Indian. J. Chem., Sect. A* **30A**(11), 944–947 (1991).
160. W. Ahmed and co-workers, *J. Phys. IV*, **1**(C2) 119–126 (1991).
161. Jpn. Pat. 03,218,917[91,218,917] (Sept. 26, 1991), (to Hashimoto Chem. Ind. Co. Ltd.).
162. U.S. Pat. 5,034,368 (July 23, 1991), E. Drent (to Shell Int. Res. MIJ BV).
163. A. A. Palko and J. S. Drury, *J. Chem. Phys.* **47**, 2561 (1967).
164. Eur. Appl. 4877 (1979), L. V. Snyder, J. W. Kang, and J. W. Fieldhouse (to Firestone Tire & Rubber Co.).
165. U.S. Pat. 4,867,957 (Sept. 19, 1989), M. S. Sennett (to U.S. Dept. of the Army).
166. Jpn. Kokai Tokkyo Koho 78 108184 (1978), T. Tanaka and O. Kishiro (to Mitsubishi Chemical Industries Co., Ltd.).
167. Jpn. Kokai 77 127990 (1977), K. Tsubaki, H. Morinaga, and S. Yamamoto (to Nissan Chemical Industries, Ltd.).
168. Ger. Offen. 2,752,577 (1979), R. Franke and B. Diedrich (to Hoechst A.-G.).
169. Brit. Pat. 1,539,375 (1979), M. J. Todd (to BP Chemicals, Ltd.).
170. Jpn. Kokai Tokkyo Koho 87 149710 (July 3, 1987), T. Kashiwada, S. Ito, K. Sugihara, and A. Irizumi (to Toyo Soda Mfg. Co., Ltd.).
171. N. Saxena and M. M. Husain, *Indian J. Chem., Sect. B*, **30B**, 616 (1991).
172. S. N. Gupta and J. P. Kennedy, *Polym. Bull. (Berlin)* **1**, 253 (1979).
173. Jpn. Kokai Tokkyo Koho 79125194 (1979), M. Miya, H. Sakurayi, H. Morita, K. Takaya, and H. Yoneda (to Asahi Chemical Industry Co., Ltd.).
174. J. P. Kennedy, S. C. Feinberg, and S. Y. Huang, *Polym. Prepr. Am. Chem. Soc. Div. Polym. Chem.* **17**, 194 (1976).
175. J. P. Kennedy, S. C. Feinberg, and S. Y. Huang, *J. Polym. Sci. Polym. Chem. Ed.* **16**, 243 (1978).
176. Austrian Pat. 347,679 (1979), I. Zurbov and co-workers (to "Kristall" Resins Plant).

177. Fr. Demande 2,323,709 (1977), (to Showa Denko K. K.).

178. Jpn. Kokai Tokkyo Koho 91-91506 (Apr. 17, 1991), S. Kamata, H. Ishiwatari, M. Suzuki, and T. Arao (to Japan Synthetic Rubber Co., Ltd.).

179. Jpn. Kokai Tokkyo Koho 79-16696 (1979), (to Elektrofein-mechanische Werke Jakob Pleh).

180. Eur. Appl. 364,323 (Apr. 18, 1990), P. Ardaud, G. Mignani, and M. Charpenel (to Rhone-Poulenc Chemie).

181. J. P. O'Brien and S. Teitel, *Heterocycles* **11**, 347 (1978).

182. Fr. Demande 2,516,940 (1983), A. Mena, J. M. Charriere, and J. Desbrest (to Societe Francaise d'Electrometallurgie).

183. J. Brynestad, C. E. Bamberger, D. E. Heatherly, and L. F. Land, *J. Am. Ceram. Soc.* **67**, 184 (1984).

184. Austrian Pat. 523,570 (1982), R. Wyndham (to Toth Aluminum Corp.).

185. U.S. Pat. 4,457,812 (1984), T. A. Rado (to Kerr-McGee Chemical Corp.).

186. U.S. Pat. 2,941,940 (1960), C. D. Shiah.

187. J. Bressan, G. Feuillade, and R. Wiart, *J. Electrochem. Soc.* **129**, 2649 (1982).

188. Eur. Appl. 68,752 (1983), R. V. Sara (to Union Carbide Corp.).

189. Ger. Offen. 2,303,407 (1973), F. S. Galasso and R. D. Veltri (to United Aircraft Corp.).

190. Eur. Appl. 392,941 (Oct. 17, 1990), P. Ardaud and G. Mignani (to Rhone-Poulenc Chimie).

191. Fr. Demande 2,620,454 (Mar. 17, 1989), G. Mignani and J. J. Leburn (to Rhone-Poulenc Chimie).

192. Eur. Appl. 432,007 (June 12, 1991), A. Iltis (to Rhone-Poulenc Chimie).

193. H. B. Bell, H. M. Anderson, and R. W. Light, *J. Electrochem. Soc.* **135**, 1184 (1988).

194. Jpn. Kokai Tokkyo Koho 91 34318 (Feb. 14, 1991), J. Sato (to Sony Corp.).

195. Jpn. Pat. 7,251,834 (1972), J. Furukawa and R. Hirai (to Research Institute for Production Development).

196. Ger. Offen. 2,506,270 (1975), A. Shiga, Y. Fukui, and K. Matsumura (to Sumitomo Chemical Co., Ltd.).

197. T. Matsuda, T. Higahimura, and S. Okamura, *Kobunshi Kagaku* **24**, 165 (1967).

198. J. M. Lansinger and R. C. Ronald, *Syn. Commun.* **9**, 341 (1979).

199. R. C. Ronald and J. M. Lansinger, *J. Chem. Soc., Chem. Commun.*, 124 (1979).

200. P. M. Druce and M. F. Lappert, *J. Chem. Soc. A*, 3595 (1971).

201. Ger. Offen. 3,009,933 (1981), W. Bacher and E. Jakob (to Maschinefabrik Augsburg-Nuernberg A.-G.).

202. U.S. Pat. 4,298,664 (1981), A. V. Joshi and A. D. Jatkar (to Ray-O-Vac Corp.).

203. J. W. Mellor, *Comprehensive Treatise on Inorganic and Theoretical Chemistry*, Vol. 5, Longman, Green and Co., New York, 1929, 123–129.

204. H. S. Booth and D. R. Martin, *Boron Trifluoride and Its Derivatives*, John Wiley & Sons, Inc., New York, 1949, 87–165.

205. R. E. Mesmer, K. M. Palen, and C. F. Baes, *Inorg. Chem.* **12**(1), 89 (1973).

206. I. G. Ryss, *The Chemistry of Fluorine and Its Inorganic Compounds*, State Publishing House for Scientific, Technical, and Chemical Literature, Moscow, USSR, 1956; F. Haimson, English trans., *AEC-tr-3927*, U.S. Atomic Energy Commission, Washington, D. C., 1960, 505–579.

207. C. H. Wamser, *J. Am. Chem. Soc.* **70**, 1209 (1948); **73**, 409 (1951).

208. R. E. Mesmer and A. C. Rutenberg, *Inorg. Chem.* **12**(3), 699 (1973).

209. R. D. Srinastava, M. O. Uy, and M. Faber, *J. Chem. Soc. Farad. Trans. 1* **70**, 1033 (1970).

210. H. Bonadeo and E. Silberman, *J. Mol. Spect.* **32**, 214 (1969).

211. J. Bessiere, *Anal. Chim. Acta* **52**(1), 55 (1970).

212. M. J. R. Clark, *Can. J. Chem.* **47**, 2579 (1969).

213. G. Brunton, *Acta Crystallogr. Sect. B* **24**, 1703 (1968).

214. A. S. Quist and co-workers, *J. Chem. Phys.* **54**, 4896 (1971); **55**, 2836 (1971).

215. M. R. Rosenthal, *J. Chem. Ed.* **50**(5), 331 (1973).

216. M. J. R. Clark and H. Linton, *Can. J. Chem.* **48**, 405 (1970).

217. I. Pawlenko, *Z. Anorg. Allgem. Chem.* **340**(3–4), 201 (1965).

218. H. W. Heiser, *Chem. Eng. Prog.* **45**(3), 169 (1949); U. S. Pats. 2,182,509–11 (Dec. 5, 1939), (to Alcoa).

219. U.S. Pat. 2,799,559 (July 16, 1957), T. J. Sullivan, C. H. Milligan, and J. A. Grady.

220. U.S. Pat. 3,959,132 (May 25, 1976), J. Singh (to Gilson Technical Services, Inc.).

221. U.S. Pat. 4,045,339 (Aug. 30, 1977), T. F. Korenowski, J. L. Penland, and C. J. Ritzert (to Dart Industries Inc.).

222. D. C. Cornish and R. J. Simpson, *Meas. Contr.* **4**(11), 308 (1971).

223. E. J. Largent, "Metabolism of Inorganic Fluoride" in *Fluoridation as a Public Health Measure*, American Association for the Advancement of Science, Washington, D.C., 1954, 49–78.

224. I. Zipkin and R. C. Likens, *Am. J. Physiol.* **191**, 549 (1957).

225. U.S. Pat. 3,888,778 (Mar. 13, 1973), M. Beckwith and G. F. Hau.

226. J. F. Jumer, *Met. Finish.* **56**(8), 44 (1958); **56**(9), 60 (1958).

227. R. M. Hudson, T. J. Butler, and C. J. Warning, *Met. Finish.* **74**(10), 37 (1976); U.S. Pat. 3,933,605 (Jan. 20, 1976), T. J. Butler, R. M. Hudson, and C. J. Warning (to U.S. Steel Corp.).

228. R. D. Mawiya and K. P. Joshi, *Indian Chem. J.* **6**(2), 19 (1971).

229. U.S. Pat. 3,808,245 (Apr. 30, 1974), D. E. O'Connor and G. R. Wyness (to Procter & Gamble Co.).

230. U.S. Pat. 3,305,565 (Feb. 21, 1967), A. C. Mueller (to Shell Oil Co.).

231. U.S. Pat. 3,349,107 (Oct. 24, 1967), S. Pawlenko (to Schering Akliengessellshaft).

232. U.S. Pat. 6,228,947 (May 8, 2001), D. K. Schesta (to Steel Oil Company).

233. R. C. Weast, ed., *Handbook of Chemistry and Physics*, Vol. 59, The Chemical Rubber Co., Cleveland, Ohio, 1978.

234. A. S. Dworkin and M. A. Bredig, *J. Chem. Eng. Data* **15**, 505 (1970).

235. M. J. R. Clark and H. Lynton, *Can. J. Chem.* **47**, 2579 (1969).

236. V. S. Yatlov and E. N. Pinaevskays, *Zh. Obshch. Khim.* **15**, 269 (1945).

237. H. Boch, *Z. Naturforsch.* **17b**, 426 (1962).

238. L. J. Klinkenberg, doctoral thesis, Leiden, Germany, 1937.

239. T. C. Waddington, *Adv. Inorg. Chem. Radiochem.* **1**, 158 (1959).

240. P. Gross, C. Hayman, and H. A. Joel, *Trans. Faraday Soc.* **64**, 317 (1968).

241. L. J. Klinkenberg, *Rec. Trav. Chim.* **56**, 36 (1937).

242. J. H. de Boer and J. A. H. Van Liempt, *Rec. Trav. Chim.* **46**, 24 (1927).

243. *JANAF Thermochemical Tables*, Clearinghouse for Federal Scientific and Technical Information, U.S. Dept. of Commerce, Springfield, Va., Dec. 1963.

244. A. W. Laubengayer and G. F. Condike, *J. Am. Chem. Soc.* **70**, 2274 (1948).

245. A. P. Altschuller, *J. Am. Chem. Soc.* **77**, 6515 (1955).

246. V. N. Plakhotnik, V. B. Tul'chinski, and V. K. Steba, *Russ. J. Inorg. Chem.* **22**, 1398 (1977).

247. T. V. Ostrovskaya and S. A. Amirova, *Russ. J. Inorg. Chem.* **15**, 338 (1970).

248. A. V. Pankratov and co-workers, *Russ. J. Inorg. Chem.* **17**, 47 (1972).

249. R. T. Marano and J. L. McAtee, *Thermochimica Acta* **4**, 421 (1972).

250. J. B. Bates and co-workers, *J. Inorg. Nucl. Chem.* **34**, 2721 (1972).

251. L. Maya, *J. Am. Ceram. Soc.* **60**(7–8), 323 (1977).

252. V. Pecak, *Chem. Prum.* **23**(2), 71 (1973).

253. Ger. Pat. 2,320,360 (Nov. 7, 1974), H. K. Hellberg, J. Massonne, and O. Gaertner (to Kali-Chemie Fluor GmbH).

254. U.S. Pat. 2,799,556 (Feb. 1, 1954), T. J. Sullivan and C. G. Milligan (to American Agriculture Chemical Co.).

255. H. S. Booth and S. Rhemar, *Inorganic Synthesis*, Vol. 2, McGraw-Hill Book Co., New York, 1946, p. 23.

256. U.S. Pat. 3,541,739 (Nov. 24, 1970), J. P. Bryon and A. G. Rolfe (to English Abrasives Limited).

257. U.S. Pat. 3,963,458 (June 15, 1976), M. T. Gladstone and S. J. Supkis (to Norton Co.).

258. G. von Matuschka, *Kunstofftechnik* **11**(11), 304 (1972).

259. USSR Pat. 495,178 (Dec. 15, 1975), V. Boiko.

260. U.S. Pat. 3,376,253 (Apr. 2, 1968), E. V. Burnthall and J. J. Hirshfeld (to Monsanto Co.).

261. Ger. Pat. 2,121,821 (Dec. 2, 1971), K. C. Frisch (to Owens Corning Fiberglass Co.).

262. P. M. Sawko and S. R. Riccitiello, *Tech. Brief ARC-11043*, NASA-Ames Research Center, Moffett Field, Calif., July 1977.

263. U.S. Pat. 3,909,489 (Sept. 30, 1975), D. D. Callander (to Goodyear Tire and Rubber Co.).

264. U.S. Pats. 2,572,248-9 (Oct. 23, 1951), H. S. Cooper (to Walter M. Weil).

265. J. D. Donaldson, C. P. Squire, and F. E. Stokes, *J. Mater. Sci.* **13**, 421 (1978).

266. U.S. Pat. 3,305,416 (Feb. 21, 1967), G. J. Kahan and J. L. Mees (to International Business Machines Corp.).

267. U.S. Pat. 3,540,943 (Nov. 17, 1970), E. M. Grogan (to U. S. Steel Corp.).

268. W. R. Huntley and P. A. Gnadt, *Report ORNL-TM-3863*, Oak Ridge National Laboratory, Oak Ridge, Tenn., 1973.

269. L. Kravetz and G. R. Ferrante, *Text. Res. J.* **40**, 362 (1970).

270. Fr. Pat. 1,578,918 (Aug. 22, 1968), J. Borkowski.

271. J. Dale and K. Daasvet'n, *J. Chem. Soc. Chem. Commun.*, (8), 295 (1976).

272. M. A. Kasem and H. R. Richard, *Ind. Eng. Chem. Prod. Res. Dev.* **11**(2), 114 (1972).

273. *Technical Bulletin FR175*, Harshaw Chemical Co., Cleveland, Ohio, 1975.

274. U.S. Pat. 2,561,565 (July 24, 1951), A. P. Edson and I. L. Newell (to United Aircraft Corp.).

275. Br. Pat. 1,181,753 (Feb. 18, 1970), (to Aluminum Co. of America).

276. Ger. Pat. 2,334,660 (Jan. 23, 1975), H. Lauck.

277. U.S. Pat. 3,287,312 (Nov. 22, 1966), T. H. Ling (to Anaconda Wire and Cable Co.).

278. T. Hongu, *S. Gakkaishi* **26**(1), 38 (1970).

279. D. W. A. Sharp, in M. Stacy, J. C. Tatlow, and A. G. Sharpe, eds., *Advances in Fluorine Chemistry*, Vol. 1, Academic Press, Inc., New York, 1960, 68–128.

280. T. V. Ostrovskaya, S. A. Amirova, and N. V. Startieva, *Russ. J. Inorg. Chem.* **12**, 1228 (1967).

281. R. T. Marano and J. L. McAtee, *Therm. Anal. Proc. Int. Conf. 3rd, 1971* **2**, 335 (1972).

282. A. D. Van Ingen Schenau, W. L. Groenveld, and J. Reedijk, *Recl. Trav. Chim. Pays-Bas* **9**, 88 (1972).

283. P. W. N. M. Van Leeuwen, *Recl. Trav. Chim. Pays-Bas* **86**, 247 (1967).

284. U.S. Pat. 3,672,759 (July 4, 1972), T. Yamawaki and co-workers.

285. E. Kamienska and I. Uruska, *Bull. Akad. Pol. Sci. Ser. Sci. Chim.* **21**, 587 (1973).

286. D. W. A. Sharp and co-workers, *Proc. Int. Conf. Coord. Chem. 8th, Vienna*, 322 (1964).

287. V. N. Plakhotnik and V. V. Varekh, *Izv. Vyssh. Uchebn. Zaved. Khim. Khim. Tekhnol.* **16**, 1619 (1973).

288. U.S. Pat. 3,795,595 (Mar. 5, 1974), H. P. Wilson (to Vulcan Materials Co.).

289. U.S. Pat. 3,300,397 (Jan. 24, 1967), G. Baltakmens and J. P. Tourish (to Allied Chemical Corp.).

290. U.S. Pat. 3,432,256 (Mar. 11, 1969), H. P. Wilson (to Vulcan Materials Co.).

291. S. Buffagni and I. M. Vezzosi, *Gazz. Chim. Ital.* **97**, 1258 (1967).

292. U.S. Pat. 4,092,296 (May 30, 1978), R. A. Skiff.

293. U.S. Pat. 3,432,440 (Mar. 11, 1969), D. A. Shimp, W. F. McWhorter, and N. G. Wolfe (to Celanese Coatings Co.).

294. *Technical Bulletin ZBF873*, Harshaw Chemical Co., Solon, Ohio.

295. A. Zemaitaitis and J. Zdanavicius, *Cellul. Chem. Technol.* **4**, 621 (1970).

296. U.S. Pat. 3,854,869 (Dec. 17, 1974), Y. Yanai (to Nisshin Spinning Co., Ltd.).

297. C. E. Morris and G. L. Drake, Jr., *Am. Dyestuff Rep.* **58**(4), 31 (1969).

298. Jpn. Pat. 71 11,080 (Mar. 20, 1971), S. Hiroaka and K. Mitsumura (to Mitsubishi Rayon Co., Ltd.).

299. U.S. Pat. 3,577,342 (May 4, 1971), L. I. Fidell (to American Cyanamid Co.).

300. U.S. Pat. 3,697,615 (Oct. 10, 1972), W. B. Hughes (to Phillips Petroleum Co.).

301. E. Rausz and S. Hulisz, *Chemik* **28**(7), 256 (1975).

302. H. W. Quinn and R. L. Van Gilder, *Can. J. Chem.* **48**, 2435 (1970).

303. U.S. Pat. 3,514,488 (May 26, 1970), C. E. Uebele, R. K. Grasselli, and W. C. Nixon (to Standard Oil Co. of Ohio).

304. Jpn. Pat. 714,030 (Dec. 3, 1971), Y. Fujita and T. Morimoto (to Mitsui Petrochemical Industries, Ltd.).

305. Jpn. Pat. 70 19,514 (July 3, 1970), I. Hiroi (to Toho Rayon Co., Ltd.).

306. Jpn. Pat. 76 04,163 (Jan. 14, 1976), J. Takeuchi, F. Iwata, and K. Kubo (to Ube Industries, Ltd.).

307. Jpn. Pat. 72 18,227 (May 26, 1972), S. Sugiura and co-workers (to Ube Industries, Ltd.).

308. Jpn. Pat. 73 06,185 (Feb. 23, 1973), T. Yamawaki, T. Suzuki, and S. Hino (to Mitsubishi Chemical Industries Co. Ltd.).

309. Fr. Pat. 2,039,808 (Jan. 15, 1971), (to Mitsubishi Chemical Industries, Co., Ltd.).

310. Jpn. Pat. 72 06,411 (May 4, 1972), T. Yamawaki and co-workers (to Mitsubishi Chemical Industries Co., Ltd.).

311. *Plating Processes*, Harshaw Chemical Co., Solon, Ohio, Mar. 1977.

312. Y. M. Faruq Marikan and K. I. Vasu, *Met. Finish.* **67**(8), 59 (1969).

313. F. Wild, *Electroplat. Met. Finish.* **13**, 331 (Sept. 1960).

314. E. M. Levy and G. J. Hutton, *Plating* **55**(2), 138 (1968).

315. U.S. Pat. 3,432,338 (Mar. 11, 1969), R. E. Sickles (to Diamond Shamrock Corp.).

316. N. J. Spiliotis, *Galvanotech. Oberflaechenschutz* **7**(8), 192 (1966).

317. Jpn. Pat. 76 02,633 (Jan. 10, 1976), J. Hara, R. Miyashata, and Y. Fukuoka (to Nippon Kayaku Co., Ltd.).

318. Ger. Pat. 2,146,908 (Mar. 23, 1972), K. Ishiguro and H. Shinohara (to Toyota Motor Co., Ltd.).

319. *Plating Processes, Tin-Lead Solder Alloy Fluoborate Plating Process for Printed Circuit Applications, HTPB5N 0272*, Harshaw Chemical Co., Solon, Ohio.

FAZLUL ALAM
U.S. Borax Research Corporation

FRANCIS EVANS
GANPAT MANI
Allied Signal, Inc.

JOHN R. PAPCUN
Atotech

BORON HYDRIDES, HETEROBORANES, AND THEIR METALLA DERIVATIVES

1. Introduction

The boron hydrides, including the polyhedral boranes, heteroboranes, and their metalla derivatives, encompass an amazingly diverse area of chemistry. This class contains the most extensive array of structurally characterized cluster compounds known. Included here are many unique clusters possessing idealized molecular geometries ranging over every point group symmetry from identity (C_1) to icosahedral (I_h). Because boron hydride clusters may be considered in some respects to be progenitorial models of metal clusters, their development has provided a framework for the development of cluster chemistry in general as well as for chemical bonding theory.

The first definitive studies of boron hydrides were carried out by Alfred Stock in Germany starting ~1912 (1). Through extensive and now classic synthetic studies, the field of boron hydride chemistry was founded with the isolation of a series of highly reactive, air-sensitive, and volatile compounds of general composition B_nH_{n+4} and B_nH_{n+6}. This accomplishment required the development of basic vacuum line techniques for the manipulation of air-sensitive compounds. An American effort in boron hydride research was subsequently initiated by Hermann I. Schlesinger. His students, including Anton Burg and Herbert C. Brown, went on to make great contributions to boron chemistry. Brown pursued the practical development of hydroboration procedures that are now so important to organic and pharmaceutical chemistry and for which, in 1979, he received a Nobel Prize (2).

Following World War II, activity in boron hydride research increased tremendously in the United States A and Russia as a result of classified government research programs on high energy fuels. In the United States, these included projects HERMES, HEF, and ZIP (3). Research was initially directed toward the development of new fuels for strategic aircraft, such as the B-70A Valkyrie long-range bomber, and eventually led to research on solid rocket propellants. The United States and Russia nearly simultaneously declassified these programs in the early 1960s revealing many important findings, including the discovery of carboranes. Although much information from these enormous governmental efforts has never been published, boron hydride materials have since resulted from government stockpiles produced during this period.

The structural and theoretical aspects of boron hydrides were delineated through X-ray diffraction, theoretical analyses of bonding, and structure and reactivity studies. William N. Lipscomb received the 1976 Nobel Prize for his definitive work on the structure and bonding of boron hydride chemistry, leading to the development of valence bond theory (4). The emergence of a theoretical understanding of boron hydrides and the residual momentum of the high energy fuels program led to a rapid proliferation of significant new boron hydride compounds and the elaboration of their chemistry. The polyhedral borane anions, carboranes, and metallacarboranes were discovered a half-century after the first report of highly reactive boron hydrides, the most stable classes of boron hydrides. The isomeric icosahedral carborane $C_2B_{10}H_{12}$ was reported almost simultaneously in 1963 by American industrial chemists and by workers in Russia (5). The highly fruitful marriage of transition metal and carborane cluster chemistry leading to the metallacarboranes soon followed (6).

2. Nomenclature

The nomenclature of boron hydride derivatives has been somewhat confusing and many inconsistencies exist in the literature. The structures of some reported boron hydride clusters are so complicated that only a structural drawing or graph, often accompanied by explanatory text, is used to describe them. Nomenclature systems often can be used to describe compounds unambiguously, but the resulting descriptions may be so unwieldy that they are of little use. The International Union of Pure and Applid Chemistry (IUPAC) (7) and the Chemical Abstract Service (CAS) (8) have made recommendations, and nomenclature methods have now been developed that can adequately handle nearly all cluster compounds; however, these methods have yet to be widely adopted. For the most part, the nomenclature used in the original literature is retained herein.

The neutral boron hydrides are termed boranes. The molecule BH_3 is called borane or borane(3) [13283-31-3]. For more complex polyboranes, the number of boron atoms is indicated by the common prefixes di-, tri-, tetra-, etc, and the number of hydrogens (substituents) is given by an arabic numeral in parentheses following the name. For example, B_5H_{11} is named pentaborane(11) [18433-84-6], $B_{20}H_{16}$ is named eicosaborane(16) [12008-84-3], and $B_{10}H_{12}I_2$ is diiododecaborane(14) [23835-60-1]. The position of the substituents can be designated precisely from framework numbering conventions. The numbering conventions for selected polyhedra are given in Figure 1. Because other numbering systems are often also used, it is advisable to refer to structural diagrams.

Borane polyhedra have both closed and open skeletons and it has become common practice to include the appropriate structural classification in the compound's name. Closed polyhedra having only triangular faces are termed *closo*, and open structures are designated *nido*, *arachno*, or *hypho*. For example, more complete names for the previous examples are *arachno*-pentaborane(11), *closo*-icosaborane(16), and 2,4-diiodo-*nido*-decaborane(14). Boron hydride anions are generally termed hydroborates using prefixes to designate the number of hydrogens and borons; the charge follows the name in parentheses. For example, $NaBH_4$ is sodium tetrahydroborate(1−) [16940-66-2], $K_2B_{10}H_{10}$ is potassium decahydro-*closo*-decaborate(2−) [12447-89-1], and the 2,4-dichlorododecahydro*nido*-decaborate(2−) anion [51668-03-2] is $(2,4-Cl_2B_{10}H_{12})^{2-}$.

When a boron atom of a borane is replaced by a heteroelement, the compounds are called carbaboranes, phosphaboranes, thiaboranes, azaboranes, etc, by an adaptation of organic replacement nomenclature. The original term carborane has been widely adopted in preference to carbaborane, the more systematic name. The numbering of the skeleton in heteroboranes is such that the

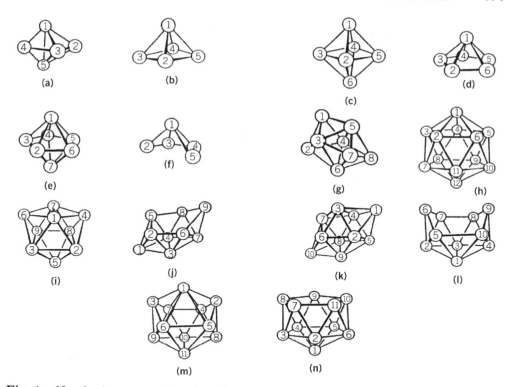

Fig. 1. Numbering conventions for selected borane polyhedra (7) discussed in the text.

heteroelement is given the lowest possible number consistent with the conventions of the parent borane. Thus $C_2B_3H_5$ is dicarba-*closo*-pentaborane(5) and could occur as the 1,2-, 2,3-, or 1,5-isomeric forms (1,2-dicarba-*closo*-pentaborane(5) [23777-70-0], 2,3-dicarba-*closo*-pentaborane(5) [30396-61-3], and 1,5-dicarba-*closo*-pentaborane(5) [20693-66-7]) (see Fig. 1a). When different heteroelements occur in combination in a polyhedron, *Chemical Abstracts* gives priority by descending group number and increasing atomic number within a group, eg, 1,2-$PCB_{10}H_{11}$ is 1-phospha-2-carba-*closo*-decaborane(11) [30112-97-1]; however, the hierarchy in the original literature often gives the lowest number to the element of lowest atomic number, eg, 1,2-$CPB_{10}H_{11}$ or 1-carba-2-phospha-*closo*-dodecaborane(11) [30112-97-1]. This convention carries over to the metallaboranes and metallaheteroboranes when a metal occupies a polyhedral vertex. Examples are 9,9-bis(triphenylphosphine)-6-thia-9-platina-*nido*-decaborane(10) [52628-81-6], 9,9-$[P(C_6H_5)_3]_2$-6,9-$SPtB_8H_{10}$, 3-η^5-cyclopentadienyl-2-dimethyl-1,2-dicarba-3-ferra-*closo*-dodecaborane(11) [66750-82-1], 3-$(\eta^5$-$C_5H_5)$-1,2-$(CH_3)_2$ -3,1,2-$FeC_2B_9H_9$. The arabic numeral in parentheses following the name does not include exopolyhedral ligands bonded to the metal, only the total of the hydrogen atoms plus other substituents bonded to boron and main group heteroelements in the skeletal framework. Examples of metallaborane anions are the 1-η^5-cyclopentadienyl-1-nickela-undecahydro-*closo*-undecaborate(1−) ion, $[(\eta^5$-$C_5H_5)Ni(B_{11}H_{11})]^-$, and the 2-$\eta^5$-cyclopentadienyl-2-cobalta-heptahydro-*nido*-tetraborate (1−) ion, $[2$-$(\eta^5$-$C_5H_5)$-2-$CoB_4H_7]^-$.

A variety of heteroboranes, metallaboranes, and metallaheteroboranes exist that contain more than one interconnected polyhedral cluster. These complex clusters are referred to as conjuncto-boranes. Conjuncto-boranes may be interconnected by sharing a single common boron atom, having a direct $B-B$ bond between two clusters, sharing two boron atoms at a polyhedral edge or three boron atoms at a face, or more extensive polyhedral fusion by the sharing of four or more boron atoms between clusters. Examples include the *commo*-7,7′-bis(dodecahydro-7-nickela-*nido*-undecaborate)(2−) dianion [31388-28-0], [7,7′(-Ni(B$_{10}$H$_{12}$)$_2$]$^{2-}$, and the decahydro-2,11-bis(η^5-cyclopentadienyl)-2,11-dicobalt-1-carba-*closo*-dodecaborate(1−) ion [59422-34-3], [2,11-(η^5-C$_5$H$_5$)$_2$-2,11,1-Co$_2$-CB$_9$H$_{10}$]$^-$. The *commo* prefix is often used to indicate that the metal vertex is shared by two polyhedra. The *commo* nomenclature of metallaboranes and metallaheteroboranes is a widely used special case of the IUPAC recommended conjuncto nomenclature.

3. Structural Systematics

Because the polyhedral boron hydrides are cage molecules, which usually possess triangular faces, their idealized geometries can be described accurately as deltahedra or deltahedral fragments. The left-hand column of Figure 2 illustrates the deltahedra containing $n = 6-12$ vertices: the octahedron, pentagonal bipyramid, bisdisphenoid, symmetrically tricapped trigonal prism, bicapped square antiprism, octadecahedron, and icosahedron. These idealized structures are convex deltahedra except for the octahedron, which is not a regular polyhedron. The left-hand column of Figure 2 also represents the class of deltahedral *closo* molecules from which the other idealized structures (deltahedral fragments) can be generated systematically. Any *nido* or *arachno* cluster can be generated from the appropriate deltahedron by ascending a diagonal from left to right. This progression generates the *nido* structure (center column) by removing the most highly connected vertex of the deltahedron, and the *arachno* structure (right column) by removal of the most highly connected atom of the open (nontriangular) face of the *nido* cluster. The structural correlations shown in Figure 2 were formulated in 1971 (9), and subsequently elaborated (10,11). The terms *closo, nido, arachno*, and *hypho* are derived from Greek and Latin and imply closed, nestlike, weblike, and netlike structures, respectively. These classifications apply equally well to boranes, heteroboranes, and their metalla analogues, and are intimately connected to a quantity known as the framework, or skeletal, electron count. The partitioning of electrons into framework and exopolyhedral classes allows for predictions of structures in most cases, even though these systematics are not concerned explicitly with the assignments of localized bonds within the polyhedral skeletons of these molecules. That is, the lines depicting the skeletons of the structures illustrated are not electron-pair, or "electron precise", bonds. The lines merely serve to join nearest neighbors and illustrate cluster geometries. However, exopolyhedral lines do represent the usual electron precise bonds.

Proposal of a structure from Figure 2 for a given borane or heteroborane proceeds by (1) selecting the row that corresponds to the number of framework

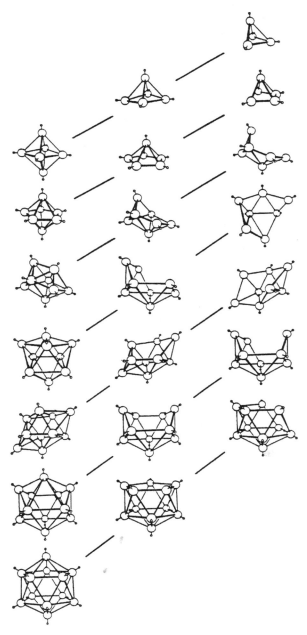

Fig. 2.   Idealized deltahedra and deltahedral fragments for *closo, nido,* and *arachno* boranes and heteroboranes. From left to right, the vertical columns give generic *closo, nido,* and *arachno* frameworks; bridge hydrogens and BH$_2$ groups are not shown, but when appropriate they are placed around the open (nontriangular) face of the framework (see text).

atoms, or polyhedral vertices, n; and (2) determining the number of electrons that can reasonably be assigned to bonding within the polyhedral skeleton as opposed to exopolyhedral bonds. According to what are known as the Williams-Wade rules, framework electron counts of $2n + 2$, $2n + 4$, $2n + 6$, and $2n + 8$ correspond to the cluster classifications *closo, nido, arachno,* and *hypho*, respectively. For *closo, nido,* and *arachno* electron counts, a structure in the appropriate column of Figure 2 is suggested. These systematics emphasize the oxidation–reduction relationship of *closo–nido–arachno–hypho* interconversions for clusters having the same number of vertices. Boranes of the hypho class are relatively rare and are not included in Figure 2. The term *klado* is recommended by the IUPAC to designate the equally rare class of compounds having a $2n+10$ framework electron count. The term *capo* has been proposed to describe boranes, such as $B_{10}H_{10}$, having a $2n$ framework electron count. Compounds having an apparent *closo* structure and possess $<2n+2$ skeletal electrons are referred to as *hypercloso*. Other empirical rules refer to the preferred placement of heteroatoms and so-called extra hydrogen atoms in these clusters. The correlation of skeletal electron count with structure is generally applicable to the metallaborane and metallacarborane derivatives as well as to other non-boron types of cluster molecules such as carbocations (12) and metal clusters (11,13,14).

Note that the eight-vertex *nido* cage shown in Figure 2 is more open than would be expected for a cluster resulting from the simple removal of one vertex from the nine-vertex *closo* cage. However, this is the observed geometry for known eight-vertex boranes and carboranes. This apparent anomaly reflects a pattern that exists for *nido* cages containing only boron and/or carbon framework atoms in which, once cages become large enough, there is an alternation of five- and six-membered open faces (15). That is, 6-, 8-, 10-, and 12-vertex *nido* boranes and carboranes have six-membered open faces and 7-, 9-, and 11-vertex boranes and carboranes have five-membered open faces. This observed pattern can be rationalized in terms of charge density of atoms at the open face.

The assignment of valence electrons and the factoring out of those electrons involved in exopolyhedral bonds provides $2n$ framework electrons for a B_nH_n molecule, two electrons short of the $2n + 2$ *closo* count. In fact, stable neutral B_nH_n molecules are not known; however, the $[B_nH_n]^{2-}$ anions ($n = 6\text{–}12$) and the neutral isoelectronic $C_2B_{n-2}H_n$ carboranes ($n = 5\text{–}12$) are very stable. These are the best known examples of *closo* molecules. Thus, in these respective cases, the double negative charge and the two C–H groups that donate one more valence electron each than a B–H group, furnish the two electrons required to achieve the $2n + 2$ framework electron count. In general, substitution of other moieties for a B–H vertex alters the number of framework electrons contributed by a particular vertex, but as long as the total remains $2n + 2$, the molecule is classified as *closo*. The number of electrons contributed to the framework electron count by a specific vertex group has been generalized in the form of an equation (14). For main group elements the number of framework electrons contributed is equal to $(v + x - 2)$ where v is the number of valence shell electrons of that element, and x is the number of electrons from ligands, eg, for H, $x = 1$, and for Lewis bases, $x = 2$. Examples of $2n + 2$ electron count boranes and heteroboranes, and the number of framework electrons contributed by their skeletal atoms, are given in Table 1.

Table 1. **Electron Counting for 2n+2, 2n+, and 2n+6 Systems**[a]

	Framework electron contribution						
Compound	Extra boron[b]	Carbon[b]	Heteroatom[b]	Hydrogens	Charge	Total	Reference
Closo(2n + 2)e⁻ systems							
$C_2B_3H_5$	3(2)	2(3)			0	12	16
$[B_6H_6]^{2-}$	6(2)				2	14	17
$(CH_3)GaC_2B_4H_6{}^c$	4(2)	2(3)	1(2)		0	16	18
$[B_8H_8]^{2-}$	8(2)				2	18	16
$C_2B_7H_9$	7(2)	2(3)			0	20	19
$[CB_9H_9]^-$	9(2)	1(3)			1	22	20
$C_2B_9H_{11}$	9(2)	2(3)			1	24	21
$SB_{11}H_{11}$	11(2)		1(4)		0	26	22
$SnC_2B_9H_{11}$	9(2)	2(3)	1(2)		0	26	23
$CPB_{10}H_{11}$	10(2)	1(3)	1(3)		0	26	24
Nido(2n + 4)e⁻ systems							
$C_2B_3H_7$	3(2)	2(3)		2	0	14	25
$C_3B_3H_7$	3(2)	3(3)		1	0	16	26
$C_4B_2H_6$	2(2)	4(3)			0	16	27
SB_8H_{10}	8(2)		1(4)	2	0	22	22
SB_9H_{11}	9(2)		1(4)	2	0	24	22
$[CPB_9H_{10}]^{2-}$				9(2)	1(3)	1(3)	
	2	26	24				
Arachno(2n + 6)e⁻ systems							
B_5H_{11}	5(2)			6	0	16	28
$C_2B_7H_{13}$	7(2)	2(3)		4	0	24	29
$[SB_9H_{13}]^-$	9(2)		1(4)	3	1	26	22

[a] Where n is the number of non-hydrogen atoms in the cage structure.
[b] Number of atoms multiplied by, in parentheses, the number of electrons contributed to the framework gives the total electron contribution for the element.
[c] The CH_3 groups are outside the cage.

3.1. *Nido* Clusters (2n + 4 Systems). Many *closo* boranes and heteroboranes add two electrons and undergo a concomitant structural transformation from a deltahedron to a deltahedral fragment. For example, *closo*-2,6-$C_2B_9H_{11}$ [17764-89-0] ($2n+2=24$ e⁻) is readily reduced to [*nido*-7,9-$C_2B_9H_{11}$]²⁻ [39469-99-3] ($2n+4=26$ e⁻) and conversely [*nido*-7,9-$C_2B_9H_{11}$]²⁻, may be oxidized to the *closo* cage (30). An effective reduction also occurs upon addition of donors to these molecules, eg, the molecules *closo*-$C_2B_4H_6$ and *closo*-$C_2B_9H_{11}$ open upon addition of amines to give *nido*-$C_2B_4H_6 \cdot NR_3$ (31) and *nido*-$C_2B_9H_{11} \cdot NR_3$ (32), where R is an alkyl group. Such additions of donor groups can formally be regarded as equivalent to additions of H⁻, ie, *nido*-$C_2B_4H_6 \cdot L$ and *nido*-$C_2B_9H_{11} \cdot L$, where L is a Lewis base, are analogous to [$C_2B_4H_7$]⁻ and [$C_2B_9H_{12}$]⁻, respectively. Other examples of *nido* molecules are given in Table 1.

In molecules such as $C_2B_3H_7$ it can be recognized that there are extra hydrogens, extra in the sense that there are more hydrogens than necessary for each vertex atom to have one exopolyhedral hydrogen atom. Extra hydrogens are generally regarded as contributing to the framework electron count. Usually extra hydrogens are found at open, or nontrigonal, faces of deltahedral fragments in the form of bridging hydrogens in B—H—B groups or as second hydrogen atoms in BH_2 groups. Both of these types of hydrogen locations are reminiscent of framework positions in that the bridge positions usually reside close to a spheroidal extension of the polyhedral surface, and in that one hydrogen of the BH_2 group is usually *endo* (close to a framework extension) and the other *exo*. In addition, extra hydrogens are often acidic and can be removed using bases to give anions, frequently without substantially altering framework geometry. In this respect, extra hydrogens may be regarded conceptually as protonated framework electrons. This concept suggests that the addition of a lone-pair donor, such as a hydride ion H^-, to a polyhedral framework adds two electrons and changes the molecule's classification accordingly.

3.2. *Arachno* Clusters (2n + 6 Systems).   In comparison to the number of known *closo* and *nido* boranes and heteroboranes, there are relatively fewer *arachno* species. Partly because of the lack of a large number of structures on which to base empirical rules, *arachno* structures appear to be less predictable than their *closo* and *nido* counterparts. For example, there are two isomeric forms of B_9H_{15}, one with the *arachno* [19465-30-6] framework shown in Figure 2 (33), the other with a framework more reminiscent of that shown for the nine-atom *nido* classification (34). Structures of *arachno* molecules involve the presence of even more extra hydrogens or other electron-donating heteroatoms than *nido* molecules. Typical examples are given in Table 1.

3.3. *Hypho* Clusters (2n + 8 Systems).   *Hypho* molecules are even more electron rich. Members of the *hypho* class are fairly rare and, therefore, examples of their structures have not been included in Figure 2. The compounds $B_5H_6[P(CH_3)_3]_2$ [39661-74-0], $[B_5H_{12}]^-$ [11056-98-7], and $B_6H_{10}[P(CH_3)_3]_2$ [57034-29-4], three molecules that contain $2n+8$ framework electrons and that represent members of the *hypho* class of boron hydrides, have been prepared and structurally characterized (35). As expected, this class adopts structures that are more open than their *arachno* and *nido* counterparts. The *hypho* molecule $B_5H_9[P(CH_3)_3]_2$, illustrated in Figure 3, contains two nonbonding basal B—B distances (those not bridged by hydrogen). In *arachno*-B_5H_{11} [18433-84-6] there is only one nonbonding basal B—B distance and in *nido*-B_5H_9 [19624-22-7] all basal distances of the pyramid are bonding (see the $n = 5$ horizontal row of Fig. 2). Another example of a *hypho* molecule is $B_4H_6[P(CH_3)_3]_2$ [66750-83-2] (36).

3.4. Metalla Derivatives.   Compounds formed from main group metals and borane or heteroborane cages can be treated using the electron-counting systematics described. For example, the metallacarborane complexes *closo*-$MC_2B_9H_{11}$ (M = Ge [27071-59-9], Sn [23151-46-4], Pb [27071-51-8]) may be regarded as tricarbaborane analogues in which the Group 14 (IVA) metals are present as bare vertices (37). The metals in these clusters in theory possess a nonbonding lone pair of electrons (38) and contribute their remaining two valence electrons to the framework to give a 26-electron *closo* icosahedron.

In addition to satisfying the framework electron requirements of the cage, transition-metal metallaboranes and metallaheteroboranes also generally adhere to the 18 electron rule, and therefore require a somewhat different electron accounting treatment. Assuming that the metal vertex uses only three

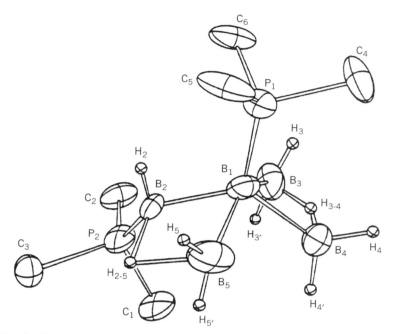

Fig. 3.   The *hypho* molecule $B_5H_9[P(CH_3)_3]_2$. Courtesy of the American Chemical Society.

orbitals in cluster bonding, then 12 of the 18 valence electrons available at a metal vertex are not involved in cluster bonding. Thus the metal d-electrons may effectively be treated as not included as framework electrons. These premises have been generalized to give the number of skeletal electrons per metal vertex as $(v + x\text{-}12)$, where v is the number of valence electrons of the metal and x is the number of electrons donated by exopolyhedral substituents and ligands (39). In this formalism, moieties such as $(CO)_3Fe$ and $(\eta^5\text{-}C_5H_5)Co$ can be regarded as donating two electrons to cage bonding and are analogous to a B–H vertex. In the same way, the $(\eta^5\text{-}C_5H_5)Ni$ moiety functions as a three-electron donor vertex analogous to a C–H vertex. Other examples of common vertex groups and their electron contributions to framework bonding are given in Table 2 (14) and Figure 4. The extension of these principles to organometallics

Table 2. Framework Electron Contributions for Metal Moieties[a]

Metal (M)	v	$M(CO)_2\, x = 4$	$M(\eta^5\text{-}C_5H_5)\, x = 5$	$M(CO)_3\, x = 6$
Cr	6	(−2)	(−1)	0
Mn	7	(−1)	0	1
Fe	8	0	1	2
Co	9	1	2	3
Ni	10	2	3	4

[a] The general contribution for a metal with v valence electrons and exopolyhedral ligands donating x electrons is $v + x - 12$ framework electrons (see text).

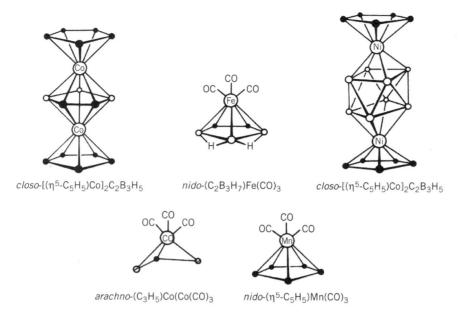

closo-[(η⁵-C₅H₅)Co]₂C₂B₃H₅   nido-(C₂B₃H₇)Fe(CO)₃   closo-[(η⁵-C₅H₅)Co]₂C₂B₃H₅

arachno-(C₃H₅)Co(Co(CO)₃   nido-(η⁵-C₅H₅)Mn(CO)₃

Fig. 4. Metallaboranes and organometallics of the *closo, nido,* and *arachno* classification.

is straightforward: $(\eta^5\text{-}C_5H_5)Mn(CO)_3$ and $(C_3H_5)Co(CO)_3$ may be considered to be *nido* and *arachno* species, respectively, as indicated in Figure 4.

3.5. Placement of Heteroatoms. Many of the deltahedra and deltahedral fragments of Figure 2 have two or more nonequivalent vertices. Nonequivalent vertices are recognized as having a different order; ie, a different number of nearest neighbor vertices within the framework. Heteroatoms generally exhibit a positional preference based on the order of the polyhedral vertex and the electron richness of the heteroatom relative to boron. Electron-rich heteroatom groupings contribute more framework electrons than a : electrondots B–H moiety, which has two framework electrons, and generally prefer low order vertices, ie, those having fewer neighbors (40). For example, two of the three isomeric forms of $C_2B_8H_{10}$ can be isomerized thermally to the 1,10-isomer [13653-23-8] (19), the molecule with the carbons at the lowest order vertices. The pyrolysis of *nido*-6-SB_9H_{11} [59120-72-8] gives *closo*-1-SB_9H_9 [41646-56-4], with sulfur at the lowest order vertex (22) (see Fig. 1 for numbering conventions). When the heteroatom is in the same group as boron it preferably adopts a high order vertex; eg, $CH_3GaC_2B_4H_6$ [36607-02-2] (18). The transition metal moieties occur predominantly at high order vertices. The carborane *nido*-1,2-dicarbapentaborane [26249-71-8], $C_2B_3H_7$, presents a notable exception to the predilection of the carbon for low order vertices. It has been suggested that this exception is related to the placement of bridge hydrogens on the open face (10).

3.6. Placement of Extra Hydrogens. The placement of extra hydrogens plays a crucial role in determining the structures adopted by boranes and carboranes. However, the exact position of extra hydrogens sometimes depends on the physical state of the molecule, eg, the tridecahydrodecaborate(1−) anion,

$[B_{10}H_{13}]^-$ [36928-50-4] exhibits different bridge hydrogen placements in the crystal (41) and in solution (42) as can be inferred from experimental evidence, but the solution data are also consistent with a dynamic process of bridge hydrogen tautomerism.

A well-documented example of fluxionality for bridge hydrogens is provided by B_6H_{10} (43). In spite of the controversy regarding hydrogen placement in certain boranes, some empirical rules are evident: (1) bridging hydrogens generally occur only between two adjacent boron atoms at an open nontriangular face of the skeleton, and only occasionally bridging a triangular array of borons (44); (2) when possible, the bridge termini are the low order vertices of the open face; and (3) there is only one bridging hydrogen per edge. Generally, BH_2 groups may be postulated as tautomeric intermediates in fluxional *nido* boranes, but they occur as ground state moieties in *arachno* molecules and then at vertices of order three or lower. In the metallaboranes, hydrogen atoms often bridge between boron and a metal.

The placement of bridge hydrogens may be the most important variable in the determination of relative isomer stabilities, outranking placement of heteroatoms (10,40). A number of cases exist where heteroatoms adopt high order vertices in deference to bridge hydrogen placement at low order vertices, eg, in *nido*-1,2-$C_2B_3H_7$ one of the carbon atoms is at an unanticipated high order vertex, apparently because of bridge hydrogen atom placement (45,46).

3.7. M–H–B Bridges. Numerous metallaboranes and metallaheteroboranes are known to contain hydrogens bridging between a metal atom and a skeletal boron atom, but complexes containing covalently bound tetrahydroborate(1−) [16971-29-2], $[BH_4]^-$, constitute the prototypical class (47). Metal tetrahydroborates have been reviewed (48). Polyboranes coordinate through bridge hydrogens in a variety of ways as shown in Figure 5. Although the bonds utilize B–H hydrogens in bridging to the metal just as for the tetrahydroborates, the presence of the cage atoms affords distinctly different structural possibilities. The B–H can be regarded as a two-electron donor to metals and often plays a role similar to that of CO in metal carbonyls to stabilize clusters.

3.8. Exceptions to Structural Systematics. When strong electron-donating or electron-withdrawing groups are present as substituents attached

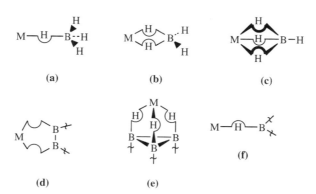

Fig. 5. Modes of M–H–B bonding where M–H–B represents a three-center hydrogen bridge bond for (**a**), (**b**), (**c**) tetrahydroborates and for (**d**), (**e**), (**f**) polyhedral boranes.

to boron in polyboranes, there is the possibility of structural anomalies. In some cases, electron deficiency of boron apparently can be ameliorated by back-donation instead of by the multicenter bonding afforded in a cage framework. Thus it has been suggested that exceptions to the electron-counting paradigms may occur where back-donation from the substituent to a cluster boron is possible. For example, although $C_4B_2H_6$ [12403-20-2] has a pyramidal *nido* geometry, the structural evidence for $C_4B_2F_2H_4$ [20534-09-2] favors a planar form where B—H has been replaced by B—F (49).

Some metallacarboranes present anomalies to the electron-counting formalisms. Symmetrical [*commo*-3,1,2-M^n + $(C_2B_9H_{11})_2]^{(n-4)}$ — sandwich complexes of the d^6 metals, where M is Fe^{2+} [51868-94-1] and Co^{3+} [11078-84-5] fit the paradigm nicely (50). However, the corresponding d^8 complexes, where M is, eg, Cu^{3+} [15721-63-8] or Ni^{2+} [36733-09-2], might be expected to show an unsymmetrically distorted partially open structure. However, symmetrically "slipped" sandwich structures are observed, suggestive of electron delocalization. The slipped structure can be explained in terms of a reduction of the *closo* molecule with a concomitant distortion as observed for *closo* carboranes (49–51). In accord with these ideas, the d^9 copper(II) complex is opened slightly more than the d^8 complex (50) and this distortion can be rationalized in terms of Jahn-Teller arguments (52). Alternatively, it is sometimes more satisfactory to view borane and carborane cages in their metal complexes as donor ligands that coordinate to metals in an appropriate fashion to contribute the number of electrons required by the specific metal center to produce a filled shell. For example, the [*nido*-7,8-$C_2B_9H_{11}]^{2-}$ cage can generally be regarded as a two-, four-, or six-electron donor when bound to transition metals in the η^1, η^3, or η^5 bonding modes, respectively. The η^3 π-bonding mode, which can be considered to occur in the d^8 metal complexes, involves the B_3 set of the five-membered carborane cage face and produces complexes that may be considered analogous to the well-known metal allyl complexes of traditional organometallic chemistry.

Because the electron-counting paradigm incorporates the 18-electron rule when applied to transition-metal complexes, exceptions can be expected as found for classical coordination complexes. Relatively minor exceptions are found in $(\eta^5$-$C_5H_5)_2Fe_2C_2B_6H_8$ [54854-86-3] (53) and $[Ni(B_{10}H_{12})_2]^{2-}$ [11141-32-5] (54). The former ($2n$ electrons) is noticeably distorted from an idealized structure, and the latter is reminiscent of the d^8 and d^9 complexes discussed above. An extremely deficient electron count is obtained for complexes such as $[Cr(C_2B_9H_{11})_2]^-$ [37036-06-9], which have essentially undistorted *closo* structures (55). Exceptional cases occur for certain other metallacarboranes that contain electron-rich d^8, d^9, and d^{10} metals. For example, electron-counting formalisms predict *closo* structures for 3-[$(C_2H_5)_2NCS_2$]-3,1,2-$AuC_2B_9H_{11}$ [62572-50-3] (56) and 8,8-[$(CH_3)_3P]_2$-7,8,10-Pt-$C_2B_8H_{10}$ [58348-40-6] (57), but *nido* structures are observed by X-ray crystallography. In some cases, ambiguities arise because bridging hydrogen atoms have not been observed in X-ray crystallographic studies.

4. Bonding

4.1. Localized Bonds.

Because boron has more valence orbitals than valence electrons, its compounds are often called electron deficient. This electron deficiency is partly responsible for the great interest surrounding boron hydride chemistry and molecular structure. The structure of even the simplest boron hydride, diborane(6) [19287-45-7], B_2H_6, was sufficiently challenging that it was debated for years before finally being resolved (58) as the hydrogen bridged structure shown.

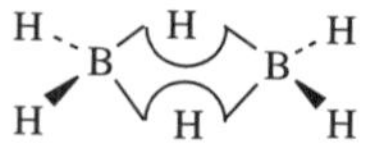

The elucidation of the structure of diborane(6) led to the description of a new bond type, the three-center two-electron bond, in which one electron pair is shared by three atomic centers (59). The delocalization of a bonding pair over a three-center bond allows for the utilization of all the available orbitals in an electron-deficient system. This key point led to the formulation of a valence bond description of the bonding in boron hydrides, sometimes termed a topological description (60). The valence structures of this topological approach give localized bonding descriptions that include delocalized three-center bonds in the basis set of bond types. In addition to the B—H—B three-center bridge bond depicted, a B—B—B three-center bond was introduced to describe bonding in the framework.

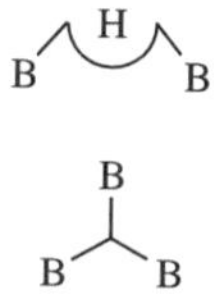

The valence theory (4) includes both types of three-center bonds shown as well as normal two-center, B—B and B—H, bonds. For example, one resonance structure of pentaborane(9) is given in projection in Figure 6. An octet of electrons about each boron atom is attained only if three-center bonds are used in addition to

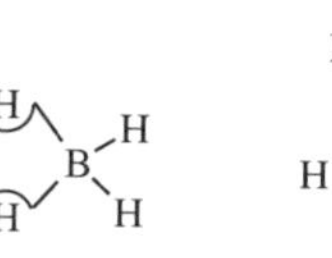

4120 B_5H_9

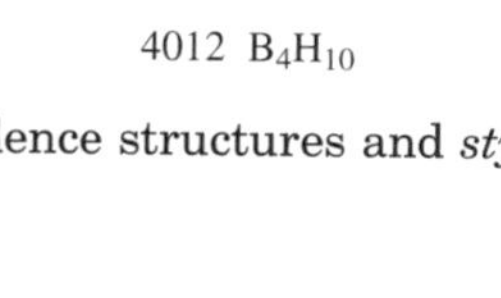

4012 B_4H_{10}

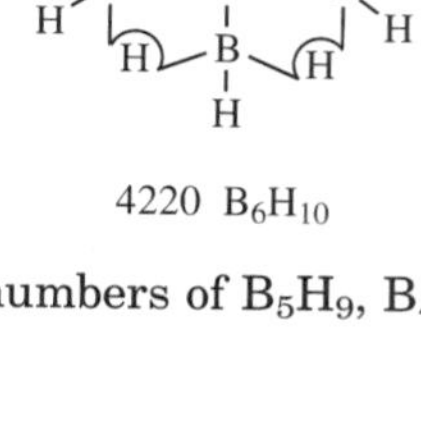

4220 B_6H_{10}

Fig. 6. The valence structures and *styx* numbers of B_5H_9, B_4H_{10}, and B_6H_{10}.

two-center bonds. In many cases involving boron hydrides the valence structure can be deduced. First, the total number of orbitals and valence electrons available for bonding are determined. Next, the B—H and B—H—B bonds are accounted for. Finally, the remaining orbitals and valence electrons are used in framework bonding. Alternative placements of hydrogen atoms require different valence structures.

The possible number of valence structures for a given boron hydride has been defined exactly using three general equations of balance. For a borane of composition $[B_p H_{q + p + c}]^c$, where c is the charge, the equations are

$$s + x = q + c \tag{1}$$

$$s + t = p + c \tag{2}$$

$$t + y = p - c - \frac{1}{2} q \tag{3}$$

where p is the number of terminal B—H units, s is the number of B—H—B three-center bonds, t is the number of B—B—B three-center bonds, y is the number of B—B bonds, and x is the number of BH_2 groups. There are usually several possible solutions to the equations of balance differentiated by a so-called *styx* number, a four-digit number that gives the respective values of s, t, y, and x. The 2D representation of $B_5 H_9$ shown in Figure 6 has the *styx* number 4120. Representations of $B_4 H_{10}$ and $B_6 H_{10}$, also shown in Figure 6, have *styx* numbers of 4012 and 4220, respectively. The *styx* formalism is equally applicable to carboranes and carbocations.

The correct *styx* number reflects the true molecular geometry, but the other *styx* structures may be transition states for fluxional molecules. Hexaborane(10) provides an example of bridge hydrogen tautomerism that is detectable by nmr (43). The tautomerism proceeds through a 3311 valence structure in which a bridge hydrogen of 4220 ground state has been converted to a BH_2 group. Note that the sum of the digits of the *styx* number gives one-half of the electrons involved in framework bonding, ie, the $2n + 2$, $2n + 4$, and $2n + 6$ framework electrons of the electron-counting formalism (10,11) and the topological descriptions are intimately related. Therefore, after using the electron-counting formalism to arrive at a framework structure, a valence bond description of the localized bonds can be determined from the *styx* formalism (4). The simultaneous application of these two formalisms allows the prediction of rearrangement during certain reactions and the prediction of transition state structure. Significant skeletal rearrangement would not be anticipated as long as the number of skeletal electrons remained unchanged, as would be the case for both associative and dissociative electrophilic mechanisms, because H^+ is the model electrophile. For the model nucleophile H^-, associative and dissociative nucleophilic mechanisms increase and decrease the framework count, respectively, and framework rearrangement would be expected during the course of the reaction (61).

Although the *styx* numbers quantify the various structural features present in a compound, molecules that are closely related may have very different *styx* numbers. For example, $[C_6 H_6]^{2+}$ and $B_6 H_{10}$ have similar structures, but *styx* numbers of 0260 and 4220, respectively. A simplified styx system has been introduced that solves this problem (15). In this system, the s and y of *styx* are added together and identified as S of *Stx*. This number has been termed the *Chop-Stx* number. It can be readily seen that compounds structurally related to B_6H_{10} have the *Chop-Stx* number 620. Although at first sight it appears that information regarding the number of B–H–B groups is lost in converting styx to *Chop-Stx*, it turns out that nothing is actually lost. This is because the styx number is always used in conjunction with an empirical formula revealing the number of B–H–B groups. For example, the formula B_6H_{10} indicates that there are four more hydrogen atoms than boron atoms and the *Stx* number 620 indicates that there are no ($x = 0$) BH_2 groups, therefore all extra hydrogens are present as B–H–B groups. The *Chop-Stx* numbers are very useful for making structural correlations. The cataloging of all boron hydride compounds using a 10-digit number, which includes the number of framework electrons over $2n$ (n = number of vertices), the number of boron atoms, the *Chop-Stx* number, the size of the largest open face, and the number of bridging hydrogens has also been proposed (15).

4.2. Molecular Orbital Descriptions.

In addition to the localized bond descriptions, molecular orbital (MO) descriptions of bonding in boranes and carboranes have been developed (4). Early work on boranes helped develop one of the most widely applicable approximate methods, the extended Hückel method. Molecular orbital descriptions are particularly useful for *closo* molecules where localized bond descriptions become cumbersome because of the large number of resonance structures that do not accurately reflect molecular symmetry. Such descriptions show that the highest occupied molecular orbital (HOMO) is degenerate for most deltahedral B_nH_n molecules, but that a closed shell is obtained for the corresponding $[B_nH_n]^{2-}$ anions. After accounting for the electrons in exopolyhedral bonds, $2n + 2$ electrons remain for framework bonding giving some theoretical justification for the electron-counting formalisms (10,11). Symmetry considerations (62) also give justification for the opening of *closo* boranes (63) or carboranes (45) upon addition of electrons.

The *ab initio* MO calculations for smaller boranes and carboranes have progressed to the point where definitive answers to structural problems can be provided in many cases. The correlation of experimental boron nmr chemical shifts with those calculated ab initio using the method known as individual gauge for localized molecular orbitals (IGLO) (64) has proven to be a valuable technique for solving structural problems. This method is especially useful when calculated chemical shifts are sensitive to geometrical changes. For example, comparison of the observed and calculated ^{11}B nmr chemical shifts for *nido*-$C_2B_6H_{10}$ strongly favors a symmetrical structure having a six-membered open face out of the three candidate structures for this compound (63). Also, the proposed existence of three separate structures for $[closo-B_8H_8]^{2-}$, one in the crystal, D_{2d}, and two in solution, C_{2v} and D_{4d}, has been reinvestigated using experimental–theoretical ^{11}B nmr chemical shift comparisons (65). These studies indicated that the higher temperature solution structure of $[closo-B_8H_8]^{2-}$ is D_{2d}, the same as seen in the crystal, and is fluxional involving a somewhat less stable C_{2v} configuration intermediate having a square open face. The structure of the third isomer is unknown. Ab initio calculations of the relative energies of isomeric structures of boranes and carboranes provide quantitative measures that reinforce, extend, and provide a basis for the formerly qualitative rules governing structural features and carbon and bridge-hydrogen placement (40). These rules now allow precise predictions of borane and carborane geometries and structural features.

5. Boranes

5.1. Nido and Arachno Boranes.
These boranes are generally more reactive and less stable thermally than the corresponding *closo* boranes. The most extensively studied boranes include diborane(6), B_2H_6, tetraborane(10), B_4H_{10}, pentaborane(9), B_5H_9, and decaborane(14), $B_{10}H_{14}$. This subject has been reviewed (58). A great deal of early work in this area was associated with the government-sponsored high energy fuels programs. Some of this work is summarized (3). The *nido* and *arachno* boranes smaller than $B_{10}H_{14}$ are quite reactive toward oxygen and water. The properties of selected boranes are given in Table 3.

Diborane(6). This compound is manufactured by Callery Chemical Co. in Pennsylvania and Voltaix Inc. in New Jersey. Laboratory-scale preparations are given in equations 4–6, of which the last may be the most convenient method. Diborane is an important starting material for many other boron hydrides.

$$3\,NaBH_4 + 4\,(C_2H_5)_2 \cdot BF_3 \xrightarrow{\text{diglyme}} 2\,B_2H_6 + 3\,NaBF_4 + 4\,(C_2H_5)_2O \quad (4)$$

$$2\,NaBH_4 + H_2SO_4 \longrightarrow B_2H_6 + 2\,H_2 + Na_2SO_4 \quad (5)$$

$$2\,NaBH_4 + I_2 \xrightarrow{\text{diglyme}} B_2O_6 + 2\,NaI + 2\,H_2 \quad (6)$$

It is a spontaneously flammable gas having an extremely high heat of combustion.

$$B_2H_6 + 3\,O_2 \longrightarrow B_2O_6 + 3\,H_2O \quad \Delta H = -2165\ kJ/mol\ (-517.4\ kcal/mol) \quad (7)$$

Only H_2, BeH_2, and $Be(BH_4)_2$ have higher heats of combustion. When diborane is pyrolyzed >100°C in a sealed tube, it is decomposed to higher boron

Table 3. Physical Properties of Boranes

Borane	CAS Registry Number	Molecular formula	Mp, °C	ΔS°_{298}, Bp, °C	ΔH°_f, kJ/ mol^a	ΔG°_f, kJ/ mol^a	J/ (K·mol)^a
diborane(6)	[19287-45-7]	B_2H_6	−164.9	−92.6	35.5	86.6	232.0
tetraborane(10)	[18283-93-7]	B_4H_{10}	−120	18	66.1		
pentaborane(9)	[19624-22-7]	B_5H_9	−46.6	48	73.2	174	275.8
pentaborane(11)	[18433-84-6]	B_5H_{11}	−123	63	103.0		
hexaborane(10)	[23777-80-2]	B_6H_{10}	−62.3	108	94.6		
decaborane(14)	[17702-41-9]	$B_{10}H_{14}$	99.7	213	31.5	216.1	353.0

^{a} To convert J to cal, divide by 4.184.

hydrides and hydrogen gas in a complex sequence of reactions. This reaction has been investigated in considerable detail (66).

Octahydrotriborate(1−). The octahydrotriborate(1−) anion [12429-74-2], $[B_3H_8]^-$, commonly referred to as the triborohydride ion, is produced by the reduction of diborane with sodium amalgam. Large quantities can be prepared more conveniently by the reaction of sodium tetrahydroborate and iodine.

$$3\,NaBH_4 + I_2 \xrightarrow[100^\circ C]{diglyme} Na[B_3H_8] + 2\,H_2 + 2\,NaI \tag{8}$$

The tetraalkylammonium salts of $[B_3H_8]^-$ formed by ion-exchange reactions are useful synthetic reagents because of their thermal and air stabilities. The structure of the $[B_3H_8]^-$ ion has been determined by an X-ray study (67) and shown to have the 2013 *styx* structure, C_{2v} symmetry. Mechanisms for the formation of this ion have been proposed (68). Tetraborane(10) can be easily obtained from salts of $[B_3H_8]^-$ (eq. 9).

$$4\,HCl + 4\,Na[B_3H_8] \longrightarrow 2\,B_4H_{10} + 4\,NaCl + 3\,H_2 \tag{9}$$

Pentaborane(9). Pentaborane(9) and $B_{10}H_{14}$ can be prepared by gas-phase pyrolysis of B_2H_6 under different conditions. Pentaborane(9) is a low boiling, highly flammable, and extremely toxic material. It is not commercially available in any significant quantities. Callery Chemical Co., which produced pentaborane in the 1950s for the military, issued a safety bulletin in 1998 describing the neurotoxic properties and hazards of handling and disposing of this material. Despite its hazardous nature, pentaborane is a useful reagent for the preparation of other less dangerous boron hydride compounds. In addition to metalla derivatives, pentaborane(9) can be used to selectively prepare a number of higher boranes, such as $B_9H_{13}\cdot O(C_2H_5)_2$, $K[B_{11}H_{14}]^-$, and carboranes (69) (eqs. 10–13). In most cases, these reactions can be carried out as one-pot procedures, such as by combining the reactions shown in equations 10 and 11 with other reactions.

$$1.8\,B_5H_9 + NaH \xrightarrow{diglyme} Na[B_9H_{14}] + H_2 + minor\ products \tag{10}$$

$$Na[B_9H_{14}] + HCl + (C_2H_5)_2O \longrightarrow B_9H_{14}\cdot O(C_2H_5)_2 + NaCl + H_2 \tag{11}$$

$$Na[B_9H_{14}] + BCl_3 + [(CH_3)_4N]Cl \xrightarrow[BCl_3]{diglyme}$$

$$B_{10}H_{14} + O(C_2H_5)_2 + [(CH_3)_4N][BCl_3H] + NaCl \tag{12}$$

$$B_9H_{13}\cdot O(C_2H_5)_2 + 4\,HC{\equiv}CH \longrightarrow$$

$$5,6 - C_2B_8H_{12} + (C_2H_5)_2O\cdot B(CH{=}CH_2)_3 \tag{13}$$

Decaborane(14). As one of the most important and intensely studied of the polyhedral boron hydrides, this colorless, flammable, crystalline solid has been produced in large quantities for military use from the 1940s to the 1970s. It can be handled in air and purified by sublimation. Decaborane(14) can be prepared on a laboratory scale by the pyrolysis of B_2H_6 at 100–200°C in the presence of a catalytic amount of a Lewis base such as dimethylether, $(CH_3)_2O$. In addition to the gas-phase pyrolysis of diborane, $B_{10}H_{14}$ can be prepared by a solution-phase process developed at Union Carbide Corp. Decaborane is a key intermediate in the preparation of many carboranes and metalla derivatives. This important compound is not currently manufactured on a large scale and is only available in laboratory quantities. Prices for decaborane are in the $10–15,000/kg range.

5.2. Reactions of Boranes with Lewis Bases.   Boranes that contain a BH_2 moiety, eg, B_2H_6, B_4H_{10}, B_5H_{11}, hexaborane (12) [28375-94-2], B_6H_{12}, and Nonaborane (15) [19465-30-6], B_9H_{15}, can generally be cleaved by nucleophiles in two ways termed symmetrical and unsymmetrical bridge cleavage (70). By using neutral bases, the two modes of cleavage lead to molecular and ionic fragments, respectively, as illustrated in equations 14–19.

$$B_2H_6 + 2\,NH_3 \longrightarrow \left[H_2B(NH_3)_2\right]^+ (BH_4)^- \tag{14}$$

$$B_2H_6 + 2(CH_3)_2S \longrightarrow 2(CH_3)_2S \cdot BH_3 \tag{15}$$

$$B_4H_{10} + 2\,NH_3 \longrightarrow \left[H_2B(NH_3)_2\right]^+ [B_3H_8]^- \tag{16}$$

$$B_4H_{10} + 2\,N(CH_3)_3 \longrightarrow (CH_3)_2N \cdot BH_3 + (CH_3)_3N \cdot B_3H_7 \tag{17}$$

$$B_5H_{11} + 2\,NH_3 \longrightarrow \left[H_2B(NH_3)_2\right]^+ [B_4H_9]^- \tag{18}$$

$$B_5H_{11} + 2\,CO \longrightarrow BH_3(CO) + B_4H_8(CO) \tag{19}$$

Certain base adducts of borane, BH_3, such as triethylamine borane [1722-26-5], $(C_2H_5)_3N \cdot BH_3$, dimethylsulfide borane [13292-87-0], $(CH_3)_2S \cdot BH_3$, and tetrahydrofuran borane [14044-65-6], $C_4H_8O \cdot BH_3$, are more easily and safely handled than B_2H_6 and are commercially available. These compounds find wide use as reducing agents and in hydroboration reactions (57). Base displacement reactions can be used to convert one adduct to another. The relative stabilities of BH_3 adducts as a function of Groups 15 (V A) and 16 (VI A) donor atoms are $P > N$ and $S > O$. This order has sparked controversy because the trend opposes the normal order established by BF_3. In the case of anionic nucleophiles, base displacement leads to ionic hydroborate adducts (eqs. 20 and 21).

$$(C_2H_5)_2O \cdot BH_3 + KF \longrightarrow K^+[BH_3F]^- + (C_2H_5)_2O \tag{20}$$

$$(C_2H_5)_2O \cdot BH_3 + Na(SCN) \longrightarrow Na^+(BH_3SCN)^- + (C_2H_5)_2O \tag{21}$$

Unsymmetrical cleavage of B_2H_6 by metal hydrides gives metal tetrahydroborate salts, also called metal borohydrides or hydroborates.

$$2\,MH + B_2H_6 \longrightarrow 2\,MBH_4 \qquad (M = Li,\ Na,\ K) \tag{22}$$

Decaborane is a multifunctional species that simultaneously acts as a Brønsted acid and a Lewis acid. Weak bases fail to directly deprotonate decaborane but do react resulting in the evolution of H_2 and the formation of species that contain ligands coordinated at the six- and nine-positions of the decaborane skeleton (see Fig. 1).

$$B_{10}H_{14} + 2\,L \longrightarrow B_{10}H_{12}L_2 + H_2 \qquad (23)$$

Base displacement reactions (71) have been used to establish the relative basicities of a number of ligands toward $B_{10}H_{12}$ to be as follows: $(C_6H_5)_3P >$ pyridine $> (C_2H_5)_3N > CH_3CON(CH_3)_2 > HCON(CH_3)_2 > (C_2H_5)_2NCN > CH_3CN > (CH_3)_2S$. The $B_{10}H_{12}L_2$ species are important intermediates in the synthesis of two-key *closo* species, $[B_{10}H_{10}]^{2-}$ and $1,2\text{-}C_2B_{10}H_{12}$.

In addition to borane adducts with Lewis bases, the organoboron products resulting from the hydroboration of olefins are widely used in commercial organic syntheses to carry out a variety of highly selective reductions. These are especially important in the pharmaceutical industry. These include 9-borabicyclo[3.3.1]nonane (9-BBN), obtained by hydroboration of 1,5-cyclooctene, and diispinocamphenylborane (DIP_2BH), produced by hydroboration of α-pinene. The latter, being optically active, is useful for asymmetric syntheses. A wide variety of borane reducing agents and hydroborating agents is available from Aldrich Chemical Co., Inc. and Callery Chemical Co.

5.3. Proton Abstraction.   Although the exopolyhedral hydrogens of *nido* and *arachno* boranes are generally considered hydridic, the bridge hydrogens are acidic as first demonstrated by titration of $B_{10}H_{14}$ and deuterium exchange (72). Some typical reactions are

$$B_{10}H_{14} + NaOH \longrightarrow Na[B_{10}H_{13}] + H_2O \qquad (24)$$

$$B_{10}H_{14} + NaH \longrightarrow Na[B_{10}H_{13}] + H_2 \qquad (25)$$

$$B_5H_9 + n - C_4H_9Li \longrightarrow Li[B_5H_8] + C_4H_{10} \qquad (26)$$

$$B_4H_{10} + NaH \longrightarrow Na[B_4H_9] + H_2 \qquad (27)$$

$$B_6H_{10} + KH \longrightarrow K[B_6H_9] + H_2 \qquad (28)$$

The deprotonation of $B_{10}H_{14}$ at a $B-H-B$ bridge position produces the yellow species $[B_{10}H_{13}]^-$ (73–76). Reaction of the $[B_{10}H_{13}]^-$ anion and an electron-pair donor L, produces $[B_{10}H_{13}L]^-$ (77). Hydration of $B_{10}H_{14}$ results in the acidic species $B_{10}H_{14}OH_2$, which ionizes to form the colorless $[B_{10}H_{14}OH]^-$ anion (76). Both $B_{10}H_{14}OH_2$ and $[B_{10}H_{14}OH]^-$ are isoelectronic with the $[B_{10}H_{15}]^-$ anion (78). The hydropolyborate ions formed by proton abstraction from decaborane are useful intermediates for the preparation of metallaboranes and heteroboranes.

5.4. Polyhedral Expansion.   The term polyhedral expansion is used to describe a host of reactions in which the size of the polyhedron is increased by the addition of new vertex atoms whether boron, heteroelements, or metals. In the case of the boranes, the pyrolysis of B_2H_6 has been used to obtain B_5H_9 and

$B_{10}H_{14}$ industrially. Although a subject of much study, the mechanism of such pyrolytic expansions is not well understood.

Expansion of $B_{10}H_{14}$ to $[B_{11}H_{14}]^-$ is brought about by a reaction with $[BH_4]^-$ (79).

$$[BH_4]^- + B_{10}H_{14} \longrightarrow [B_{11}H_{14}]^- + 2\,H_2 \tag{29}$$

Other expansion reactions between diborane and borane anions with a $B{-}B$ edge bond have been reported (80), eg,

$$2[B_4H_9]^- + B_2H_6 \longrightarrow 2[B_5H_{12}]^- \tag{30}$$

$$2[B_5H_8]^- + B_2H_6 \longrightarrow 2[B_6H_{11}]^- \tag{31}$$

Boron halides have also been shown to insert into $B{-}B$ bonds to give initial products with the new boryl moiety in a bridge position (81).

$$[B_5H_8]^- + (CH_3)_2BCl \longrightarrow \mu - [(CH_3)_2B] \cdot B_5H_8 + Cl^- \tag{32}$$

$$\mu - [(CH_3)_2B] \cdot B_5H_8 \xrightarrow{(C_2H_5)_2O} (CH_3)_2B_6H_8 \tag{33}$$

5.5. Electrophilic Attack. A variety of boranes, heteroboranes, and metallaboranes undergo electrophilic substitution. Susceptibility of boranes to electrophilic attack is often detected by deuteron–proton exchange experiments. For example, electrophilic hydrogen–deuterium exchange of $B_{10}H_{14}$ occurs at the 1-,2-,3-, and 4-positions when exposed to DCl in the presence of $AlCl_3$ (82,83). The trend to increasing positive sites in $B_{10}H_{14}$ is 2,4 < 1,3 < 5,7,8,10 < 6,9. Initial halogenation and alkylation of $B_{10}H_{14}$ also occurs at the 2,4-positions (82,84). Electrophilic substitution of pentaborane(9) has also been observed to give halogenation, alkylation, and deuteration. The more negative apical site (1-position) is substituted preferentially, but the 1-isomer can be catalytically converted to the basally substituted isomer (2-position) (85) (Fig. 1a). The basal $B{-}B$ bond of B_6H_{10} can be protonated to give the isolable polyhedral borane cation, $[B_6H_{11}]^+$ (86) (Fig. 1d).

5.6. Closo Borane Anions. This group contains a homologous series of very stable polyhedral anions, $[closo\text{-}B_nH_n]^{2-}$, $n = 6{-}12$. Just as the previously known boron hydrides might be considered as analogues of aliphatic hydrocarbons, the *closo* borane anions are analogues of aromatic hydrocarbons. The stability of the *closo* anions is attributable to electron delocalization in a unique 3D aromaticity. Unlike their *nido* and *arachno* counterparts with bridging hydrogens, proton abstraction does not, for practical purposes, occur in *closo* borane chemistry. Instead, acid catalysis is important in their substitution chemistry. The best known members of this series, $[closo\text{-}B_{10}H_{10}]^{2-}$ [12356-12-6] and $[closo\text{-}B_{12}H_{12}]^{2-}$ [12356-13-7], were first reported in 1959 and 1960 (87) and were the subject of detailed studies (88).

In aqueous solution, *closo* borane anions are very stable as their conjugate acids, which possess acidity similar to sulfuric acid, yet their chemistry is remarkably different. Large unipositive cations, such as Tl^+, Cs^+, Rb^+,

$[(CH_3)_4N]^+$, and $[(CH_3)_3S]^+$, yield water-insoluble salts of $[B_{12}H_{12}]^{2-}$ and $[B_{10}H_{10}]^{2-}$ (89). Small unipositive cations and most dipositive cations, such as Ba^{2+} and Ca^{2+}, form water-soluble salts that are strong electrolytes and give hydrates on evaporation. The divalent transition and rare-earth elements also give soluble salts and hydrates, but solubilities decrease when the water of the coordination sphere is replaced by ligands such as NH_3. Polarizable cations, such as Ag^+, Cu^+, Tl^+, and Hg^{2+} form water-insoluble salts. The latter compounds contain $M-H-B$ interactions in the solid state. Salts of cations, which are not readily reducible, display exceptional thermal stabilities. Thus $Cs_2[B_{12}H_{12}]$ and $Cs_2[B_{10}H_{10}]$ can be heated to 810 and 600°C, respectively, in a sealed, evacuated tube and recovered unchanged.

Salts of $[B_6H_6]^{2-}$ [12429-97-9], $[B_7H_7]^{2-}$ [12430-07-8], $[B_8H_8]^{2-}$ [12430-13-6], $[B_9H_9]^{2-}$ [12430-00-0], and $[B_{11}H_{11}]^{2-}$ [12430-44-3] appear to exhibit similar behavior, but less definitive data are available. Although silver salts of $[B_6H_6]^{2-}$, $[B_9H_9]^{2-}$, and $[B_{11}H_{11}]^{2-}$ have been isolated, they are shock-, heat-, and light-sensitive (90,91). Anhydrous $Cs_2[B_6H_6]$, $Cs_2[B_8H_8]$, and $Cs_2[B_9H_9]$ are thermally stable to 600°C, but $Cs_2[B_{11}H_{11}]$ disproportionates to $Cs_2[B_{10}H_{10}]$ and $Cs_2[B_{12}H_{12}]$ at temperatures >400°C (17).

The base-promoted closure of Decaborane(14) yields salts of the $[B_{10}H_{10}]^{2-}$ anion (eq. 34). Relatively strong Lewis bases, such as trialkylamines, are required to accomplish this reaction as weaker bases, such as diethyl sulfide and acetonitrile, form stable $6,9\text{-}L_2B_{10}H_{12}$ species where $L = (C_2H_5)_2S$, H_3CCN, etc, which are important synthetic intermediates (92).

$$B_{10}H_{14} + 2\,N(C_2H_5)_3 \xrightarrow{-H_2} 6,9-[(C_2H_5)_2N]_2\,[B_{10}H_{12}] \longrightarrow$$

$$6,9-[(C_2H_5)_2NH]_2\,[B_{10}H_{10}] \tag{34}$$

The reaction of B_2H_6 with $NaBH_4$ or $(C_2H_5)_3N\cdot BH_3$ at 180°C in $N(C_2H_5)_3$ at high pressure gives $[B_{12}H_{12}]^{2-}$ in nearly quantitative yield (93). In diglyme at 162°C, the same reactants give a 5–10% yield of $[B_6H_6]^{2-}$ at even lower (85°C) temperatures and at atmospheric pressure $Na[B_3H_8]$ [12429-74-2] is obtained (91,94).

$$2\,NaBH_4 + 5\,B_2H_6 \longrightarrow Na_2[B_{12}H_{12}] + 13\,H_2 \tag{35}$$

$$2\big[(C_2H_5)_3N\big]\cdot BH_3 + 5\,B_2H_6 \longrightarrow [(C_2H_5)_3NH][B_{12}H_{12}] + 11\,H_2 \tag{36}$$

Pyrolysis of $Cs[B_3H_8]$ at 230°C gives $Cs_2[B_9H_9]$ (60%) along with some $Cs_2[B_{10}H_{10}]$, $Cs_2[B_{12}H_{12}]$, and $CsBH_4$ (95). The sensitivity of polyhedral expansion reactions to solvent, temperature, and pressure is further exemplified by the results in dioxane at 120°C under pressure. To obtain the *closo* borane, $Na[B_{11}H_{14}]$ is first converted to $Cs_2[B_{11}H_{13}]$, which can be pyrolyzed to give $Cs_2[B_{11}H_{11}]$ (91).

$$NaBH_4 + 5\,B_2H_6 \xrightarrow{120°C} Na[B_{11}H_{14}] + 10\,H_2 \tag{37}$$

Pyrolysis of $[(C_2H_5)_4N][BH_4]$ at $185°C$ gives a 90% yield of $[(C_2H_5)_4N]_2$-$[B_{10}H_{10}]$ (96). The $[B_{12}H_{12}]^{2-}$ anion can be prepared by the reaction of $(C_2H_5)_3N \cdot BH_3$ with decaborane in an inert high boiling hydrocarbon solvent at $190°C$ (94).

$$2[(C_2H_5)_3N \cdot BH_3 + B_{10}H_{14} \xrightarrow{190°C} [(C_2H_5)_3NH]_2 [B_{12}H_{12}] + 3 H_2 \qquad (38)$$

The $[B_6H_6]^{2-}$, $[B_7H_7]^{2-}$, $[B_8H_8]^{2-}$, $[B_9H_9]^{2-}$, and $[B_{11}H_{11}]^{2-}$, *closo* anions are hydrolytically less stable than the $[B_{10}H_{10}]^{2-}$ and $[B_{12}H_{12}]^{2-}$ *closo* anions. All of these anions are more stable in basic than in acidic solution. The $[B_7H_7]^{2-}$ ion is the least stable hydrolytically and is degraded even in basic media. The $[B_6H_6]^{2-}$, $[B_8H_8]^{2-}$, and $[B_9H_9]^{2-}$ *closo* anions are stable in neutral and alkaline solutions but react rapidly with aqueous acid. Strongly acidic solutions ($>2N$ HCl) are necessary for the hydrolysis of $[B_{11}H_{11}]^{2-}$. The $[B_{12}H_{12}]^{2-}$ anion is the most hydrolytically stable borane anion, withstanding even $3\ N$ HCl at $95°C$, conditions that slowly degrade $[B_{10}H_{10}]^{2-}$ (17,90). A salt of the $[B_{12}H_{12}]^{2-}$ anion, produced by Callery Chemical Co., currently finds commercial application in the fuse of the passenger-side automotive airbag.

Much work has been done on the functionalization of *closo* polyborane anions. Both nucleophilic and electrophilic substitution routes have been explored. Most acid-catalyzed nucleophilic substitutions of $[B_{10}H_{10}]^{2-}$ with, eg, amides, ethers, and sulfones, give products having equatorial substituents (97).

$$(B_nH_n)^{2-} + 2\ R_2O^+ \longrightarrow [B_nH_{n-2}(OR)_2]^{2-} + RH \qquad (39)$$

$$[B_nH_n]^{2-} + H^+ + HCON(CH_3)_2 \longrightarrow [B_nH_{n-1}(OCH{=}N(CH_3)_2)]^- + H_2 \qquad (40)$$

$$[B_nH_n]^{2-} + H^+ + R_2SO_2 \longrightarrow [B_nH_{n-1}OS(O)R_2]^- + H_2 \qquad (41)$$

These O-bonded substituents are easily cleaved with hydroxide ion to give the corresponding hydroxyl derivative, $[B_nH_n\text{-}1(OH)]^{2-}$ or $[B_nH_n\text{-}2(OH)_2]^{2-}$, $n = 10,12$. It was initially shown that reaction of $[B_{12}H_{12}]^{2-}$ with H_2S leads to thiolation of the dodecaborate cage (98). The product $[HSB_{12}H_{11}]^{2-}$, known as "BSH", was found to be a promising reagent for boron neutron capture therapy (BNCT) of cancer, discussed below. The acid-catalyzed nucleophilic substitution of $[B_{12}H_{12}]^{2-}$ using N-methylthiopyrrolidone or N-methylbenzothiazole provides thioethers that undergo alkaline hydrolysis to give the BSH thiolate (99). Sulfur can also be introduced to the dodecaborate cage by electrophilic substitution using acetlysulfenyl chloride (100). The isomeric inner sulfonium salts $[Me_2SB_{12}H_{11}]^-$ and $(Me_2S)_2B_{12}H_{10}$ have been prepared and can be converted to the $[MeSB_{12}H_{11}]^-$ and $[(MeS)_2B_{12}H_{10}]^{2-}$ thioether anions by reduction with excess sodium or potassium (101). Further reduction of these methyl thioethers using lithium in methylamine results in cleavage to BSH (102).

Highly or persubstituted *closo* polyhedral borane anions have been dubbed "camouflaged" boranes since substantial subtitution of vertex hydrogens by larger groups results in a substituent shell enclosing and chemically concealing the boron polyhedron. The persubstituted compounds technically are not boron hydrides but fit decidedly within this class. The $[closo\text{-}B_{12}H_{12}]^{2-}$ anion can be perhydroxylated to $[closo\text{-}B_{12}(OH)_{12}]^{2-}$ by treatment with hydrogen peroxide(103,104). This compound serves as a useful reagent for the further functionalization of borane cages. For example, reaction of a $[closo\text{-}B_{12}(OH)_{12}]^{2-}$ salt with benyl chloride in the presence of ethyldiisopropylamine results in the per-O-benzylated compound $[closo\text{-}B_{12}(OCH_2Ph)_{12}]$ (105). This large spherical anion undergoes reversible one-electron oxidation to give the radical anion $[hyper\text{-}closo\text{-}B_{12}(OCH_2Ph)_{12}]^{-\cdot}$, where the *hypercloso* designation is used for compounds having an apparent *closo* structure that possess $<2n+2$ skeletal electrons. Although the subject of computational studies, the parent *hypercloso*-$B_{12}H_{12}$ has not been synthesized.

The $[closo\text{-}B_{12}H_{12}]^{2-}$ anion can also be peralkylated. For example, prolonged reaction of tetraalkylammonium salts of $[closo\text{-}B_{12}H_{12}]^{2-}$ with methyl iodide in trimethylaluminum solvent yields the corresponding $[closo\text{-}B_{12}(CH_3)_{12}]^{2-}$ salts, which can be cation exchanged to give salts of a variety of other cations (106). The $[closo\text{-}B_{12}(CH_3)_{12}]^{2-}$ anion is a unique large symmetrical and hydrophobic anion that, like the O-benzylated B_{12} compound above, undergoes reversible one-electron oxidation to give a paramagnetic radical anion.

The reaction of $[B_{10}H_{10}]^{2-}$ with excess nitrous acid gives an explosive intermediate that can be reduced to the nonexplosive bis inner-diazonium salt 1,10-$(N_2)_2B_{10}H_8$[66750-86-5] (eq. 42). This diazonium species is a useful synthetic intermediate.

$$[B_{10}H_{10}]^{2-} \xrightarrow{HNO_2} [\text{explosive intermediate}] \xrightarrow{NaBH_4} 1,10-(N_2)_2B_{10}H_8 \qquad (42)$$

Unfortunately, $[B_{12}H_{12}]^{2-}$ does not undergo the corresponding reaction. The N_2 group is the only moiety that can be displaced readily from the B_{10}-cluster by a variety of nucleophiles (107).

$$1,10-(N_2)_2B_{10}H_8 + 2\,L \longrightarrow 1,10-L_2B_{10}H_8 + 2\,N_2 \qquad (43)$$

where L = ammonia, amines, nitriles, hydrogen sulfide, azide ion, hydroxide ion, and carbon monoxide.

The dicarbonyl [12539-66-1] available from 1,10-$(N_2)_2B_{10}H_8$ is another important species because of the scope of its chemistry. Carbonyls of $[B_{12}H_{12}]^{2-}$ can be formed from CO and the conjugate acid of $[B_{12}H_{12}]^{2-}$. The B_{10}- and B_{12}-carbonyls exhibit very similar reactivity (108). The carbonyls can be considered anhydrides of carboxylic acids and accordingly react with alcohols and amines to give esters and amides:

$$B_{10}H_{10}(CO)_2 \begin{cases} \xrightarrow{H_2O} H_2[B_{10}H_8(COOH)_2] & (44) \\ \xrightarrow{ROH} H_2[B_{10}H_8(COOR)_2] & (45) \\ \xrightarrow{R_2NH} [R_2NH_2]_2[B_{10}H_8(CONR_2)_2] & (46) \end{cases}$$

Halogenation of *closo*-boranes has been studied extensively. The exhaustive fluorination of *closo*-boranes leads to compounds such as $[closo\text{-}B_{12}F_{12}]^{2-}$ (109). Halogenation of $[B_{12}H_{12}]^{2-}$ and $[B_{10}H_{10}]^{2-}$ occurs using elemental halogen in solvents such as water, alcohols, or tetrachloroethane. Initial rates are extremely high in all cases with $[B_{10}H_{10}]^{2-} > ([B_{12}H_{12}]^{2-}$. The kinetic order is $F \geq Cl > Br > I$ (110). Typical products are $[B_{10}Cl_{10}]^{2-}$ [12430-33-0],$[B_{10}H_3Br_7]^{2-}$ [12360-16-6][$B_{10}I_{10}]^{2-}$ [12430-43-2], $[B_{12}Cl_6H_6]^{2-}$ [12430-46-5], $[B_{12}H_3Br_6Cl_3]^{2-}$ [12536-79-7], and $[B_{12}I_{12}]^{2-}$ [12587-25-6]. In general, the alkali and alkaline earth metal salts of the B_{10}- and B_{12}-halogenated derivatives have excellent thermal, oxidative, and hydrolytic stabilities.

Due to their extreme kinetic stabilities, oxidative degradation of $[B_{10}H_{10}]^{2-}$ and $[B_{12}H_{12}]^{2-}$ to boric acid is extremely difficult and requires Kjeldahl digestion or neutral permanganate. Degradation is catalyzed by crown ethers under milder conditions. The heat of reaction obtained from the permanganate degradation leads to a calculated heat of formation for $[B_{10}H_{10}]^{2-}$ (aq) of 92.5 ± 21.1 kJ/mol (22.1 ± 5.0 kcal/mol) (108).

The oxidative coupling of both $[B_{10}H_{10}]^{2-}$ and $[B_{12}H_{12}]^{2-}$ has been studied in some detail (112–116). The $[B_{10}H_{10}]^{2-}$ anion can be oxidized chemically or electrochemically to give $[B_{20}H_{19}]^{3-}$ and $[B_{20}H_{18}]^{2-}$[59724-35-5] (eqs.47–49).

$$(B_{10}H_{10})^{2-} \rightleftharpoons (B_{10}H_{10})^- + e^- \tag{47}$$

$$2[B_{10}H_{10}]^{2-} \longrightarrow [B_{20}H_{19}]^{3-} + H^+ \tag{48}$$

$$[B_{20}H_{19}]^{3-} \longrightarrow [B_{20}H_{18}]^{2-} + H^+ + 2e^- \tag{49}$$

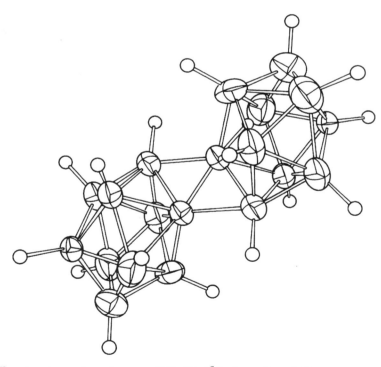

Fig. 7. The structure of one isomer of $[B_{20}H_{18}]^{2-}$ where ○ and ⌀ represents BH and B. Courtesy of the American Chemical Society (114).

These are conjuncto-boranes consisting of two B_{10}-clusters linked via a double three-center B–B–B bond as originally proposed (112) and later confirmed by X-ray crystallography (113). The oxidation of $[B_{10}H_{10}]^{2-}$ with aqueous ferric ion at room temperature was more recently shown to yield $[cis\text{-}B_{20}H_{18}]^{2-}$, an isomer in the which the two B_{10}-cages are bound together in a cisoid configuration (114). The structures of the isomeric $[cis\text{-}B_{20}H_{18}]^{2-}$ and $[trans\text{-}B_{20}H_{18}]^{2-}$ anions are shown in Figure 7. Reduction of $[B_{20}H_{18}]^{2-}$ by sodium metal in liquid ammonia gives $[B_{20}H_{18}]^{4-}$ [59724-36-6], in which the two B_{10}-clusters are joined by a single two-center B–B bond through the apical or equatorial positions. The nondestructive electrochemical oxidation of $[B_{12}H_{12}]^{2-}$ leads to the $[B_{24}H_{23}]^{2-}$ anion (115).

5.7. Tetrahydroborates. The tetrahydroborates constitute the most commercially important group of boron hydride compounds. Tetrahydroborates of most of the metals have been characterized and their preparations have been reviewed (46). The most important commercial tetrahydroborates are those of the alkali metals. Some properties are given in Table 4.

Sodium tetrahydroborate, also called sodium borohydride, is the single most important industrial boron hydride material in terms of tonnage produced. It is manufactured by Rohm & Haas and Finnish Chemicals (Nokia) using boric acid as the source of boron. Treatment of trimethyl borate with a metal hydride, eg, NaH, in the absence of a solvent yields sodium hydrotrimethoxyborate [16940-17-3], $Na[HB(OCH_3)_3]$, (eq. 50) which disproportionates in the presence of solvents such as tetrahydrofuran (THF) at 60–70°C (eq. 51) (123).

$$MH + B(OCH_3)_3 \longrightarrow M[BH(OCH_3)_3] \quad M = Li,\ Na,\ K \tag{50}$$

$$4\,M[BH(OCH_3)_3] \longrightarrow MBH_4 + 3\,M[B(OCH_3)_4] \tag{51}$$

Addition of diborane (eq. 52) under the latter conditions renders the production of MBH_4 essentially continuous until consumption of the metal hydride is complete because trimethyl borate is regenerated.

$$3\,M[BH(OCH_3)_4] + 2\,B_2H_6 \longrightarrow 3\,MBH_4 + 4\,B(OCH_3)_3 \tag{52}$$

Variations on this method are used for the commercial production of $NaBH_4$, but are less satisfactory for the manufacture of $LiBH_4$ and KBH_4. Some metathetical conversions are shown in equations 53–58.

$$TiNO_3 + KBH_4 \xrightarrow{\ H_2O\ } TiBH_4 + KNO_3 \tag{53}$$

$$NaBH_4 + KOH \xrightarrow{\ H_2O\ } KBH_4 + NaOH \tag{54}$$

$$[(C_6H_5)_4P]F + KBH_4 \xrightarrow{\ H_2O\ } [(C_6H_5)_4P][BH_4] + KF \tag{55}$$

$$NaBH_4 + LiCl \xrightarrow{\ isopropylamine\ } LiBH_4 + 2\,LiCl \tag{56}$$

$$MgCl_2 + 2\,NaBH_4 \longrightarrow Mg[BH_4]_2 + 2\,NaCl \tag{57}$$

$$NH_4F + NaBH_4 \longrightarrow NH_4[BH_4] + NaF \tag{58}$$

Table 4. **Properties of Alkali Metal Tetrahydroborates**

	Compound					
Property	LiBH$_4$	NaBH$_4$	KBH$_4$	RbBH$_4$	CsBH$_4$	References
CAS Registry Number	[16949-15-8]	[16940-66-2]	[13762-51-1]	[20346-99-0]	[19193-36-3]	
mp, °C	268	505	585			(116–118)
decomp. temp., °C	380	315	584	600	600	(116–118)
density, g/mL	0.68	1.08	1.17	1.71	2.40	(117, 119, 120)
refractive index	1.547	1.490	1.487	1.498	122	
lattice energy, kJ/mol[a]	792.0	697.5	657	648	630.1	117
ΔH_f°, kJ/mol[a]	−184	−183	−243	−246	−264	120, 121
ΔS_{298}°, J/(mol·K)[a]	−128.7	−126.3	−161	−179	−192	122

[a] To convert J to cal, divide by 4.184.

194

There is considerable interest in sodium borohydride for the safe and efficient storage and transportation of energy. Although concentrated aqueous solutions of sodium borohydride with the addition of $\sim$1% sodium hydroxide are stable and nonflammable, they are readily decomposed catalytically by metals such as ruthenium, nickel, cobalt, and platinum with the evolution of substantial quantities of H_2 gas. This

$$NaBH_4 + 4\,H_2O \xrightarrow{\text{catalyst}} NaB(OH)_4 + 4\,H_2 \qquad (59)$$

provides a method of producing "hydrogen on demand" for fuel cells or combustion engines. It is notable that half of the hydrogen generated in this reaction comes from water. Thus the deliverable hydrogen content of a concentrated sodium borohydride solution is quite high, and competitive with other known technologies for mobile H_2 storage applications in terms of energy content per unit weight requirement.

Sodium borohydride can also be used in "direct" fuel cells (essentially batteries) since H_2 can be generated directly at a platinum anode. In this case the decomposition of $NaBH_4$ can be expressed by the following anode half reaction, where each mole of $NaBH_4$ provides eight electrons.

$$NaBH_4 + 8\,OH^- \longrightarrow NaB(OH)_4 + 4\,H_2O + 8\,e^- \qquad (60)$$

As a result, 20 mg of $NaBH_4$ is sufficient to operate a cell at 100 mA for $\sim$1 h. Millennium Cell Corp. in the United States is developing a variety of commercial power sources that utilize sodium borohydride as fuel, including borohydride powered automobiles that run on either H_2-powered internal combustion engines or fuel cell powered electric motors. Prototype borohydride vehicles have demonstrated similar range and performance characteristics to conventional automobiles. The current barrier to the commercial use sodium borohydride in such large-scale power applications is the need for an efficient method of recycling the by-product sodium metaborate back to sodium borohydride.

With a hydrogen economy predicted for the future of energy management it is likely that sodium borohydide will play an increasingly important role in the storage and transport of hydrogen, if not for mass transportion then at least for small-scale or portable power sources.

The physical and chemical properties of the tetrahydroborates show more contrasts than the salts of nearly any other anion, ranging from highly ionic compounds such as $CsBH_4$ and $Ba(BH_4)_2$ to covalent volatile species such as $Al(BH_4)_2$ and $Zr(BH_4)_4$. The alkali metal salts are the most stable. In dry air, $NaBH_4$ is stable to 300°C and in vacuo to 400°C with only partial decomposition. In contrast, several tetrahydroborates, including the titanium, thallium, gallium, copper, and silver salts, are unstable at or slightly above ambient temperatures. The chemical and physical properties of the tetrahydroborates are closely related to molecular structure. Sodium tetrahydroborate, which is typical of the alkali metal tetrahydroborates except for the lithium salt, has a face-centered cubic (fcc) crystal lattice that is essentially ionic and contains the tetrahedral $[BH_4]^-$ anion. The tetrahydroborates of the polyvalent metals are in many cases the

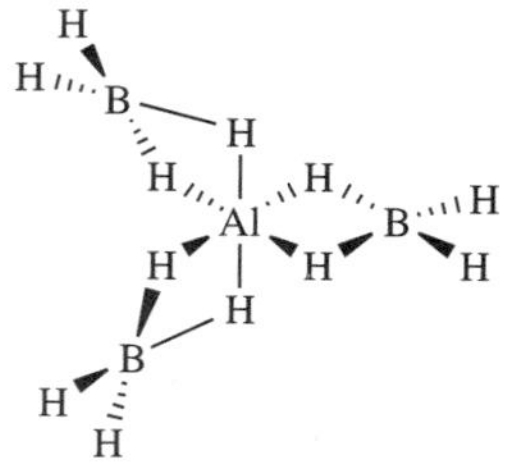

Fig. 8.   The structure of Al(BH$_4$)$_3$.

most volatile derivatives of these metals known. Aluminum tris(tetrahydroborate)[16963-07-5], Al[BH$_4$]$_3$, has a normal boiling point of 44.5°C and uranium bis(tetrahydroborate) [33725-14-3], U[BH$_4$]$_2$, has a vapor pressure of 530 Pa (4 torr) at 61°C. Other covalent tetrahydroborates include Be[BH$_4$]$_2$, Zr[BH$_4$]$_4$, Hf[BH$_4$]$_4$, and U[BH$_4$]$_4$. These compounds contain M–H–B-type bonds. The structure of Al[BH$_4$]$_3$ is shown in Figure 8. The alkali metal tetrahydroborates are stable in dry air, and the sodium and potassium salts can be crystallized from aqueous solution. Alternatively, Al[BH$_4$]$_3$ is hydrolyzed explosively and is pyrophoric in air.

Sodium tetrahydroborate is quite soluble in liquid ammonia and soluble to some extent in a variety of other solvents. It is appreciably soluble only in polar solvents of high dielectric constant and those that can solvate the metal ion, such as water, amines, N,N-diethylformamide (DMF), and glyme ethers. The rate of hydrolysis of NaBH$_4$ in water is increased by either lowering the pH or by increasing the temperature. It can be recrystallized from alkaline solutions. Dissolution in water results in a slow hydrolysis until the solution becomes alkaline. A 0.01 M solution of NaBH$_4$ gives initial pH 9.6. Only very slow hydrolysis occurs at pH values >12.9. Thus aq NaBH$_4$ solutions can be rendered stable by addition of ~1% NaOH.

The tetrahydroborates have been used as reducing agents for a variety of inorganic reductions. Many metal cations are reduced by borohydrides in protic or aprotic solvents. The products of these reductions may be lower valent cations, free elements, volatile hydrides, or metal borides. For example, Sn, Ge, As, Sb, and Bi salts or oxides can be reduced to SnH$_4$, GeH$_4$, AsH$_3$, SbH$_3$, and BiH$_3$. Several of these reactions are utilized in quantitative analytical procedures. Reactive metal powder can be prepared by borohydride reduction of metal compounds such as cobalt chloride, chromium oxide, tungsten oxide, and molybdenum oxide or chloride. Sodium tetrahydroborate, as well as amine boranes, are used in electroless plating (qv), particularly of nickel, palladium, and platinum, on both metallic and nonmetallic substrates. Many transition metal hydride complexes have been prepared by reactions utilizing borohydrides. Borohydrides also find use in the bleaching of paper pulp and clays, purification of organic chemicals and pharmaceuticals, the recovery of valuable metals, and the treatment of wastewater from industrial process streams. Covalent metal tetrahydroborates derived from Ti, Zr, Co, Ni, and Rh have shown catalytic activity in hydrogenation, polymerization, and isomerization reactions (47). An important industrial use of sodium tetrahydroborate is in the production of the dithionite anion by reduction of the bisulfite anion (124).

The use of tetrahydroborates, as well as the boranes and organoboranes, for organic transformations has proven to be even more significant because these reduction reactions are highly selective and nearly quantitative (125). The reducing characteristics of borohydrides may be varied by changing the associated cation and by changing the solvent. Borohydrides are often the reagents of choice for the reduction of aldehydes and ketones to the corresponding alcohols, especially when selective reduction in the presence of other functional groups is required. Many other functional groups, such as acid chlorides, imines, and peroxides, can also be reduced using borohydrides.

6. Heteroboranes

Heteroboranes contain heteroelements classified as nonmetals. The heteroatoms known to form part of a borane polyhedron include C, N, O, Si, P, Ge, As, S, Se, Sb, and Te either alone or in combination (126). In principle, heteroboranes containing a variety of heteroatoms could have a wide range of skeletal sizes. Of these, the carboranes have by far the greatest demonstrated scope of chemistry.

6.1. Carboranes.   The term carborane is widely used as a contraction of the IUPAC approved nomenclature carbaborane. The first carboranes, including isomers of $C_2B_3H_5$, $C_2B_4H_6$, and $C_2B_5H_7$, were prepared in the mid-1950s. These carboranes were obtained as mixtures in low yield from the reaction of smaller boranes such as pentaborane(9) with acetylene in a silent electric discharge. The discovery of the icosahedral *closo*-1,2-dicarbadodecaborane(12) [16872-09-6], 1,2-$C_2B_{10}H_{12}$, came soon after and led to a rapid development of carborane chemistry (127). This latter carborane, usually called *ortho*-carborane, is prepared in good yields by the reaction of acetylene and one of the 6,9-$L_2B_{10}H_{12}$ species such as $[(C_2H_5)_2S]_2B_{10}H_{12}$[28377-92-6] or 6,9-$(H_3CCN)_2B_{10}H_{12}$, where L is a Lewis base (Fig. 9). A variety of *C*-substituted *ortho*-carboranes isobtained by utilizing substituted acetylenes, RC≡CR′. The symbols

$$HC\underset{\underset{B_{10}H_{10}}{\diagdown O \diagup}}{}CH$$

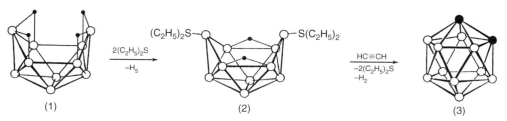

Fig. 9.   The synthesis of *closo*-1,2-$C_2B_{10}H_{12}$ (3) from *nido*-$B_{10}H_{14}$ (1) via *arachno*-$[(C_2H_5)_2S]_2B_{10}H_{12}$ (2) where ○ represents BH; ● CH; and ● H.

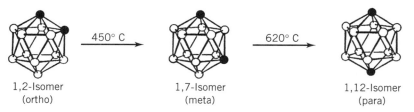

1,2-Isomer   1,7-Isomer   1,12-Isomer
(ortho)   (meta)   (para)

Fig. 10. Thermal rearrangement of *closo*-$C_2B_{10}H_{12}$, where $\bigcirc$ represents BH; $\bullet$ CH.

$RCB_{10}H_{10}CR'$, and

$$\overline{HCB_{10}H_{10}CH}$$

are used in the literature to represent substituted *closo*-carboranes. The carbons-apart isomers, called *meta*-carborane [16986-21-6], and *para*-carborane [20644-12-6], respectively, are obtained by thermal isomerization (Fig. 10) of 1,12-$C_2B_{10}H_{12}$ (see Fig. 10 for numbering conventions). The C–H vertices in the carboranes are acidic and *C*-substituted carboranes can be obtained conveniently through the intermediacy of lithium reagents such as 1,2-dilithium-*o*-carborane [22220-85-5], 1,2-Li_2-1,2-$C_2B_{10}H_{10}$, and 1,7-dilithium-*meta*-carborane [17217-89-9], 1,7-Li_2-1,7-$C_2B_{10}H_{10}$, which are readily prepared by treatment of *o*- or *m*-carborane with *n*-C_4H_9Li. Furthermore, it has been shown (128) that 1,2-dehydro-*o*-carborane, $C_2B_{10}H_{10}$, which can be generated through loss of lithium bromide from the lithium salt of the 2-bromo-*o*-carboranyl anion, $C_2B_{10}H_{11}Br$, exhibits chemistry similar to 1,2-dehydrobenzene [462-80-1] (benzyne), C_6H_4. This reactivity was demonstrated through 2 + 2, 4 + 2, and related cycloaddition reactions with dienes, leading to *C*-substituted organocyclic carboranes.

The discovery (129) of the base-promoted degradation of the isomeric *closo*-$C_2B_{10}H_{12}$ cages provided one of the most important carborane anion systems, the isomeric [*nido*-$C_2B_9H_{12}$]$^-$ anions,

$$closo - C_2B_{10}H_{12} + RO^- + 2\,ROH \longrightarrow [nido - C_2B_9H_{12}]^- + B(OR)_3 + H_2 \quad (61)$$

where R = CH_3, C_2H_5, etc. The [*nido*-$C_2B_8H_{12}$]$^-$ cages, and their *C*-substituted derivatives, are commonly referred to as dicarbollide ions, derived from the Spanish *olla*, meaning a bowl. Note that the rules governing the numbering of heteroborane cages calls for different systems for *closo* and *nido* cages (7). Thus the base-promoted degradation of 1,2-, 1,7-, and 1,12-$C_2B_{10}H_{12}$ cages leads to 7,8-, 7,9-, and 2,9-*nido* 11-vertex cages, respectively. The [*nido*-7,8-$C_2B_9H_{12}$]$^-$ features a five-membered C_2B_3 open face having a perpendicular plane of symmetry passing between the two adjacent carbon atoms and through the unique boron atom of the open face. In addition to 11 terminal B–H and C–H hydrogens in [*nido*-7,8-$C_2B_9H_{12}$]$^-$ there is a twelfth hydrogen atom associated with the open face, that is referred to as an *endo*-hydrogen. Although the subject of some controversy, the position of this hydrogen atom was shown to reside in the solid-state structures of at least some of its salts in an unsymmetrical bridging position between the unique boron of the open face (B10) and an adjacent boron atom (B9/B11) (130–132). Deprotonation of this extra hydrogen of $[nido\text{-}C_2B_9H_{12}]^-$ leads to the $[nido\text{-}C_2B_9H_{11}]^{2-}$ dianion. Protonation of the $[nido\text{-}7,8\text{-}C_2B_9H_{12}]^-$ anion with strong acids leads to the neutral highly acidic $C_2B_9H_{13}$ molecule.

Aside from their extensive use in metallacarborane chemistry, the dicarbollide anions are important intermediates in the synthesis of other carborane compounds. For example, aqueous ferric chloride oxidation of the $[7,8\text{-}C_2B_9H_{11}]^-$ anion results in the 10-vertex cage $nido\text{-}5,6\text{-}C_2B_8H_{12}$ (133) and the aqueous chromic acid oxidation of $[7,9\text{-}C_2B_9H_{11}]^-$ yields $arachno\text{-}1,3\text{-}C_2B_7H_{13}$ [17653-38-2] (29).

Nonicosahedral carboranes can be prepared from the icosahedral species by similar degradation procedures or by reactions between boranes such as B_4H_{10} and B_5H_9 with acetylenes. The degradative reactions for intermediate $C_2B_nH^{n+2}$ species ($n = 6$–9) have been described in detail (134). The small $closo\text{-}C_2B_nH_n + 2$ species ($n = 3$–5 are obtained by the direct thermal reaction (500–600°C) of B_5H_9 using acetylene in a continuous-flow system. The combined yields approach 70% and the product distribution is $\sim$5:5:1 of $2,4\text{-}C_2B_5H_7$ [20693-69-0] to $1,6\text{-}C_2B_4H_6$ [20693-67-8] to $1,5\text{-}C_2B_3H_5$ [20693-66-7] (135). A similar reaction (eq. 62) employing base catalysts, such as 2,6-dimethylpyridine, at ambient temperature gives $nido\text{-}2,3\text{-}C_2B_4H_8$ [21445-77-2] (136). The C-substituted derivatives, $nido\text{-}(CR)_2C_2B_4H_6$, can be prepared conveniently. The trimethylsilyl-substituted derivatives, $nido\text{-}2,3\text{-}[(CH_3)_3Si]_2\text{-}2,3\text{-}C_2B_4H_6$, with carbon atoms adjacent in the five-membered open face, and $nido\text{-}2,4\text{-}[(CH_3)_3Si]_2\text{-}2,4\text{-}C_2B_4H_6$, with carbon atoms apart in the open face, have been used extensively in research studies (137). The $[nido\text{-}C_2B_4H_7]^-$ anion can be prepared by deprotonation of $closo\text{-}C_2B_4H_8$.

$$B_5H_9 + HC\equiv CH \xrightarrow{\text{L}} 2,3 - C_2B_4H_8 + L\cdot BH_3 \quad (L = \text{base}) \qquad (62)$$

The $arachno$ carboranes $1,3\text{-}C_2B_7H_{13}$ (29) and $6,9\text{-}C_2B_8H_{14}$ [38670-58-5] (138) are unusual in that two of the extra hydrogens occur in CH_2 groups. The other two extra hydrogens are present as B–H–B bridges. The compound $arachno\text{-}CB_8H_{14}$ contains one CH_2 group and four bridging hydrogens. The CH_2 groups in these $arachno$ carboranes are quite acidic and can be deprotonated readily to give the corresponding carborane mono- and dianions, where are useful synthetic reagents.

As with the simple boranes, the $closo$ carboranes are generally more thermally stable than the corresponding $nido$ and $arachno$ species. Thermal decomposition of $nido$ and $arachno$ carboranes often leads to one or more $closo$ carborane. For example, pyrolysis of $2,3\text{-}C_2B_4H_8$ is another route to $2,3\text{-}C_2B_5H_7$ [30347-95-6], $1,2\text{-}C_2B_4H_6$ [20693-68-9], $1,6\text{-}C_2B_4H_6$ [20693-67-8], and $1,5\text{-}C_2B_3H_5$ [20693-66-7] (139).

A readily accessible carborane is $nido\text{-}7\text{-}(NH_3)\text{-}7\text{-}CB_{10}H_{12}$ [12539-44-5], a zwitterionic species formally derived from $[CB_{10}H_{13}]^-$ by replacement of a H$^-$ by NH$_3$. It has been shown (140) that this monocarbaborane can be obtained in excellent yield by treatment of $B_{10}H_{14}$ with CN$^-$ followed by passage through an acidic ion-exchange column (eq. 63).

$$B_{10}H_{14} + 2NaCN \xrightarrow[-HCN]{H_2O} Na_2[B_{10}H_{13}CN] \xrightarrow{H_2} CB_{10}H_{12}(NH_3) \tag{63}$$

The related mono-N-alkylated carboranes, 7-(NH$_2$R)-7-CB$_{10}$H$_{12}$, can be prepared by treatment of decaborane(14) with alkyl isocyanides (141).

$$B_{10}H_{14} + RNC \longrightarrow (H_2NR)\, CB_{10}H_{12} \tag{64}$$

The nitrogen of these aminocarboranes can be alkylated to give, eg, 7-[N(CH$_3$)$_3$]-7-CB$_{10}$H$_{12}$ [31117-16-5]. These compounds give [$closo$-2-CB$_{10}$H$_{11}$]$^-$ [38102-45-0] upon treatment with Na in THF followed by iodine oxidation (eq. 65) (142).

$$[CB_{10}H_9(Nr_3)] \xrightarrow[THF]{Na} [CB_{10}H_{11}]^{3-} \xrightarrow{I_2} [CB_{10}H_{11}]^- + 2\,I^- \tag{65}$$

Other large monocarbaboranes include $nido$-6-(NR$_3$)-6-CB$_9$H$_{11}$[$closo$-1-CB$_9$H$_{10}$]$^-$; [38192-43-7] and [$closo$-CB$_{11}$H$_{12}$]$^-$; [39102-46-0]. The $closo$ monocarbaboranes can be functionalized at carbon via lithiation using reagents such as n-butyl lithium in a manner similar to the two-carbon carboranes. The small monocarbaboranes $closo$-1-CB$_5$H$_7$ [25301-90-0], $nido$-2-CB$_5$H$_9$ [12385-35-2], and a variety of their alkylated derivatives are also known (143,144).

Synthetic methods have been developed to prepare a variety of tricarbaboranes that contain three carbons in their polyhedral skeletons. The isomeric 11-vertex $nido$-C$_3$B$_8$H$_{12}$ and its C-substituted derivatives form a family analogous to the dicarbollides discussed above, with the [$nido$-C$_3$B$_8$H$_{12}$]$^-$ anion behaving as a strong base having no extra hydrogens about its open five-membered face (145). Other tricarbaboranes include $nido$-C$_3$B$_7$H$_{10}$, and $arachno$-C$_3$B$_7$H$_{12}$, for which various isomers and C-substituted derivatives are known (146). Tetracarbaboranes, containing four carbon atoms in a single polyhedral skeleton, were rare until the discovery (147) of the metallacarborane-mediated synthesis of (CH$_3$)$_4$C$_4$B$_8$H$_8$ [58815-26-2].

$$[2,3-(CH_3)_2-2,3-C_2B_4H_5]^- \xrightarrow{CoCl} [(CH_3)_2C_2B_4H_4]_2CoH \xrightarrow{[O]} (CH_3)_4C_4B_8H_8 \tag{66}$$

As the C$_4$B$_n$−4H$_n$ series of tetracarbaboranes is classified in the electron-counting formalism as $nido$, these molecules can have more open structures even though extra hydrogens are absent, as shown for the isomers of 2,3,4,5-C$_4$B$_2$H$_6$ [28323-17-3] and (CH$_3$)$_4$C$_4$B$_8$H$_8$, at least one isomer of the latter having an open nonicosahedal structure (148). Other tetracarbaboranes include isomers of $nido$-C$_4$B$_8$H$_{10}$, $nido$-C$_4$B$_7$H$_{11}$, and $arachno$-8-[CH$_3$OC(O)]-7,8,9,10-C$_4$B$_8$H$_{13}$ (149).

Cage rearrangements in polyhedral carboranes are well known. Although most carborane cages are stable at room temperature, many undergo rearrangements at elevated temperatures. Carborane isomers obtained by conventional synthetic routes are often kinetic products and not the thermodynamically most stable isomers. When subjected to elevated temperatures below the ultimate decomposition temperatures, carboranes often undergo rearrangements to the more stable isomers. This process may involve the sequential formation of a series of successively more stable isomers. Isomerization of the $closo$-1,2-$C_2B_{n-2}H_n$ $(5 \leq n \leq 12)$ carboranes and their C-substituted derivatives has attracted considerable interest. Perhaps most intensely studied is the rearrangement of $closo$-$C_2B_{10}H_{12}$ (see Fig. 10). This rearrangement reflects a progression toward greater stability as a result of increasing carbon atom separation within the cage. The mechanisms for this cage rearrangement, as well as the rearrangements of other carboranes, heterocarboranes, and metallacarboranes, has been a topic of much controversy (150). Several mechanisms have been proposed including most notably the diamond–square–diamond, modified diamond–square–diamond, triangular face rotation, pentagonal rotation, and opening closure processes, among others. In addition to thermal cage rearrangements, a number of carborane species are believed to undergo reversible rearrangements in solution at or near room temperature. For example, ^{11}B nmr spectral data indicates that the $[CB_{10}H_{11}]^-$ ion (151) and $closo$-$C_2B_6H_8$ (152) may be fluxional in solution.

Considerable effort has been devoted to the synthesis of novel materials using carborane building blocks that may find application as components for nano technology (153,154). Icosahedral carboranes can be linked together by direct C–C bond formation to produce concatenated compounds having geometrically distinct structures that may depend on the isomeric carbon configuration of the carborane cages. Carborane molecules can be linked together either by direct C–C bonds or via variety of linking group to form a "carboracycles" or chains. Linking together p-carborane molecules by direct C–C bonds leads to rigid rod-shaped molecules referred to as "carborods" (155). A carborod tetramer is shown in Figure 11. These cylindrical molecules have van der Waals diameters of ~7 Å, which is relatively thick compared to other known rigid-rod molecules, such as staffanes, rodanes, and cubanes. Related carborods can be prepared by connecting the carbon atoms of a series of p-carborane molecules with various linker groups, eg, biphenyl.

As with the $closo$-boranes discussed above, a family of highly substituted (camouflaged) $closo$-carboranes also exists. For example, exhaustive fluorination of icosahedral two-carbon carboranes provides $closo$-1,2-$(H)_2$-1,2-$C_2B_{10}F_{10}$ (156), $closo$-1,7-$C_2B_{10}F_{12}$ (156, 157) and $closo$-1,12-$(H)_2$-1,12-$C_2B_{10}F_{10}$ (156), all of which are air stable. In aqueous media, they hydrolyze to boric acid. The fluorinated one-carbon monoanion 1-$[HCB_{11}F_{12}]^-$ has also been prepared and shown to undergo nucleophilic substitution by OH^- at B–F vertices in basic aqueous media (158). $Closo$-carboranes can also be peralkylated carboranes. For example, 1,12-$C_2B_{10}(CH_3)_{12}$ (159), $[CB_{11}(CH_3)_{12}]^-$ (160), $closo$-1,12-$(H)_2$-1,12-$C_2B_{10}(CH_3)_{10}$ (161) and related compounds have been prepared. Oxidation of $[CB_{11}(CH_3)_{12}]^-$ leads to the stable free radical $[CB_{11}(CH_3)_{12}]^-$ (162), which is a moderately air-stable strong oxidizing agent. Smaller carboranes may be highly substituted as well, as illustrated by the preparation of $[closo$-1-H-$CB_9X_9]^-$, $[closo$-1-NH_2-$CB_9X_9]^-$ $[closo$-1-H-$CB_9X_9]^-$ anions, which X is chlorine, bromine, and iodine (163) and related compounds.

Subsequent halogenation of the alkyl camouflaged carboranes leads to halogenoalkylcarboranes. For example, chlorination of 1,12-$C_2B_{10}(CH_3)_{12}$ provides

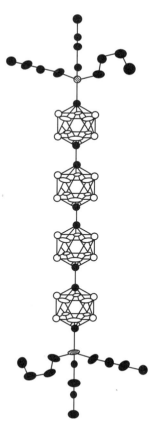

Fig. 11. A carborane rigid rod (carborod) containing four *p*-carborane units terminated by tri-*n*-butylsilyl groups, where ◯, ●, O, and O represents BH; C, CH₂, and CH₃;   (Si, and hydrogen is omitted.)

1,12-C₂B₁₀(CHCl₂)₁₂, where the steric bulk of chloride prevents full complete substitution of methyl hydrogens (164). This spherical molecule is comparable in size the fullerene C₆₀. Fluorination of Cs[CB₁₁(CH₃)₁₂] using F₂ and K₂NiF₆ provides Cs[CB₁₁(CF₃)₁₂]⁻, the anion of which has been described as simultaneously explosive and "inert" (165). Salts of the [CB₁₁(CF₃)₁₂]⁻ anion are flammable and percussion sensitive, exploding upon scraping with a metal spatula. However, this compound is chemically inert and resists attack by 20% KOH in ethanol, concentrated H₂SO₄, anhydrous CF₃SO₃H, and BF₃/HF mixtures. The anion likely remains unprotonated in aqueous solution. Prolonged treatment of *closo*-1,12-(H)₂-1,12-C₂B₁₀(CH₃)₁₀ with F₂ at temperatures >35°C leads to quantitative conversion to *closo*-1,12-(F)₂-1,12-C₂B₁₀(CF₃)₁₀ (166). This air-stable compound is unchanged in acidic solution but is slowly degraded in basic media. It is stable to 300°C, however, mixture of this and related fluoromethylcarborane were reported to detonate upon scraping with a spatula. *Extreme care should be taken when handling any of the fluoroalkylcarboranes.*

6.2. Weakest Anions and Strongest Acids. Carborane anions have received much attention as robust anions that have very low nucleophilicity and weak coordination to metal ions anions. For example, $[closo\text{-}CB_{11}H_{12}]^-$ and $[closo\text{-}CB_9H_{10}]^-$ have been studied as members of a class described as the weakest coordinating anions (167). Such anions are important as counterions in catalytic metal complexes, where their lack of significant coordination results in the stabilization of highly reactive cationic metal centers (168–172). Such catalyst systems are especially useful in stereochemically controlled olefin polymerizations (173). The functionalization of carborane anions can be used to enhance or fine tune their properties for specific applications (174). For example, the halogenated anions $[CB_{11}H_6X_6]$ $(X = Cl, Br)$, are extremely weak nucleophiles and are chemically unreactive. The weakest anion character of these carborane anions has led to a counterpart chemistry, the strongest acids (175). Carboranes such as $closo\text{-}CB_{11}H_{13}$, the conjugate acid of $[closo\text{-}CB_{11}H_{12}]^-$, and especially its halogenated derivative are superacids (those acids stronger than 100% sulfuric acid). For example, the superacids $H[CB_{11}H_6X_6]$ $(X = Cl, Br)$, halogenated at the cage positions farthest from the carbon, are among the strongest acids known, having acid strength strong enough to protonate any arenes. The chlorine-containing derivative is shown in Figure 12. Most strong acids have serious limitations because they are too nucleophilic and have oxidizing capacity. Thus their protonation reactions are usually accompanied by oxidative degradation. Hence, very strong acids are generally associated with highly corrosive liquids. The lack of significant nucleophilicity in the carboranyl anions, however, allows their conjugate acids to participate in pure protonation reactions without the destructive effects usual found with other superacids. Moreover, these are perhaps the only superacidic reagents that are easily weighed and delivered in stoichiometric quantities. The carborane superacids should prove useful in many acid-catalyzed chemical processes, including industrial polymerizations.

A diversity of polyhedral carborane cage-containing polymers has been prepared. The best known of these are elastomeric polycarboranylsiloxanes which were developed by Olin Corp. (176) and Union Carbide Corp. (177) in the 1970s. These are based on m-carborane cages linked by polysiloxane groups with direct C–Si bonds. The properties of these materials can be varied by

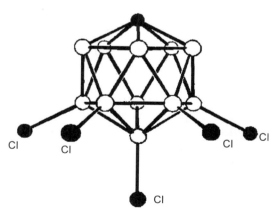

Fig. 12. The highly robust $[CB_{11}H_6Cl_6]^-$ anion is the conjugate base of one of the strongest known superacids, where ○ represents B; ○ C ; ● Cl and hydrogen is omitted.

Table 5. **Heteroboranes**

Heteroborane[a]	CAS Registry Number	References
Group 15 (V) heteroatoms		
arachno-6-NB_8H_{13}	[58920-21-1]	179
nido-10-$C_6H_5CH_2$-7,8,10-$C_2NB_8H_{10}$	[58614-34-9]	179
nido-7-CH_3-7-$PB_{10}H_{12}$	[57108-87-9]	180
closo-$C_6H_5PB_{11}H_{11}$	[57139-68-1]	181
closo-$CPB_{10}H_{11}$	[17398-92-4]	181
[*nido*-7,8-CPB_9H_{11}]$^-$	[52110-38-0]	24
closo-1,2-$As_2B_{10}H_{10}$	[51292-90-1]	182
[*nido*-7-$AsB_{10}H_{12}$]$^-$	[51292-97-8]	182
[*nido*-7,8-$As_2B_9H_{10}$]$^-$	[51358-26-0]	183
[*closo*-$AsB_{11}H_{11}$]$^-$	[51898-88-5]	182
closo-1,2-$CAsB_{10}H_{11}$	[23231-66-5]	184
Group 16 (VI) heteroatoms		
closo-$SB_{11}H_{11}$	[56464-75-6]	22
[*arachno*-6-$SB_{11}H_{12}$]$^-$	[51358-27-1]	185
nido-4-SB_8H_{10}	[59351-07-4]	22
closo-1-SB_9H_9	[41646-56-4]	22
nido-7-$SB_{10}H_{12}$	[58984-44-4]	186
arachno-6,8-$S_2B_7H_9$	[63115-77-5]	187
arachno-6,8-CSB_7H_{11}	[63115-78-6]	187
nido-7-$SeB_{10}H_{12}$	[61649-90-9]	188
nido-7,8-$Se_2B_9H_9$	[61618-06-2]	188
nido-7-$TeB_{10}H_{12}$	[61649-91-0]	188

[a] The *closo, nido, arachno* classifications are given on the basis of framework electron count and not structure.

changing the length and substituents of the polysiloxane linkages as well as their overall molecular weights. Some of these materials have excellent thermal stabilities, chemical resistance, and high temperature elastomeric properties. Polymers of this type, known under the trade name Dexsil, were commercial materials, useful as high temperature stationary phases in gas chromatography among other applications. These compounds, however, have not been produced commercially for many years. The organic and organometallic chemistry of *closo* carborane derivatives has been reviewed (5,178).

6.3. Other Heteroboranes.   Other well-documented families of heteroboranes include the azaboranes, thiaboranes, phosphaboranes, arsenaboranes, stibaboranes, selenaboranes, and telluraboranes (126). Table 5 lists representative examples of heteroboranes from Groups 15 (V) and 16 (VI). The thiaboranes are the most extensively developed class of heteroboranes after the carboranes. The thiaboranes [*arachno*-6-SB_9H_{12}]$^-$ [45979-10-0] and *nido*-6-SB_6H_{11} [59120-72-8] (185) can be converted (eq. 67) to 1-SB_9H_9 [41646-56-4], which has a nonicosahedral fragment structure (22).

$$B_{10}H_{14} \xrightarrow[H_2O]{S_x^{2-}} [arachno-6-SB_9H_{12}]^- \xrightarrow[C_6H_6]{I_2}$$

$$nido-6-SB_9H_{11} \xrightarrow{\Delta} closo-1-SB_9H_9 \tag{67}$$

Preparation of the smaller thiaboranes, $4\text{-}SB_8H_{10}$, occurs via the thiaboranes $[arachno\text{-}6\text{-}SB_9H_{12}]^-$ and larger species are attainable by expansion reactions.

$$closo\text{-}1\text{-}SB_9H_9 \xrightarrow[\text{CH}_3\text{OH}]{\text{KOH}} [4\text{-}SB_8H_9]^- \underset{-\text{H}^+}{\overset{+\text{H}^+}{\rightleftharpoons}} 4\text{-}SB_8H_{10} \tag{68}$$

Closo-1-SB_9H_9 and *closo*-$SB_{11}H_{11}$, unlike most other thiaboranes, are resistant to moisture and air oxidation. Other thiaboranes include the nine-vertex clusters *arachno*-$S_2B_7H_9$ (187) and *arachno*-SB_8H_9 (22), which are isostructural with *arachno*-$C_2B_7H_{13}$. These can be prepared in high yield.

$$[6\text{-}SB_9H_{12}]^- \xrightarrow{\Delta, \text{H}^+} 7\text{-}SB_{10}H_{12} \xrightarrow{\text{R}_3\text{NBH}_3} SB_{11}H_{11} \tag{69}$$

7. Metallaboranes

7.1. Transition-Element Metallaboranes.

The transition-metal hydroborate cluster, $HMn_3(CO)_{10}(BH_3)_2$, containing a B_2H_6 moiety, which is multiplied bridging between three manganese carbonyl and manganese carbonyl hydride centers via M–H–B bridges, might be regarded as the first structurally characterized metallaborane cluster (189). This and similar clusters were isolated in the 1960s as by-products in the synthesis of transition-metal carbonyl hydrides by sodium borohydride reduction of metal carbonyls, a standard method for the preparation of transition-metal hydride complexes and clusters since the 1970s (190). Indeed, the $[BH_4]^-$ anion acts as a ligand in a wide variety of metal complexes in which from one to all four hydrogen atoms are involved in bonding to metals (191). However, the chemistry of stable metallaboranes that incorporate metals in vertex positions of polyhedral borane clusters was developed somewhat later. To date a great many metallaborane clusters have been characterized covering a wide range of metals, sizes, and polyhedral fragment geometries.

One of the most extensive classes of metallaboranes are those derived from decaborane, which in most cases produces 11-vertex metallaborane products. The $[B_{10}H_{12}]^-$ [12430-37-4] anion can also be considered as a bidentate ligand that coordinates metals between boron atoms 2,11 and 3,8, the metal at position 7 (Fig. 1**n**) such that the metal in effect occupies the position of a bridge hydrogen of the conjugate acid borane. Situations in which a metal vertex may be regarded as equivalent to an H^+, BH^{2+}, or the BH_2^+ moiety have also been discussed (192,193). In the case of the $[B_{10}H_{12}]^-$ ligand, the bridge hydrogens lie in positions 8,9 and 10,11. Typical complexes containing $[B_{10}H_{12}]^{2-}$ include $[M(B_{10}\text{-}H_{12})_2]^{2-}$, where $M = \text{Zn, Cd, Hg}$ (194), Co, Ni, Pd, Pt (195); $L_2M(B_{10}H_{12})$, where $M = \text{Pd, Pt}$; $L = PR_3$ (195); and $[L_3M(B_{10}H_{12})]^-$ where $M = \text{Co, Rh, Ir}$; $L = \text{CO, PR}_3$ (195). The X-ray structure of $[\text{Ni}(B_{10}H_{12})_2]^{2-}$ [31388-28-0] is shown in Figure 13 (196). If the $[B_{10}H_{12}]^{2-}$ ligand in the Ni complex is considered to be bidentate, coordination about the metal is effectively square planar. The geometry of the metal in analogous complexes varies according to the requirements of the

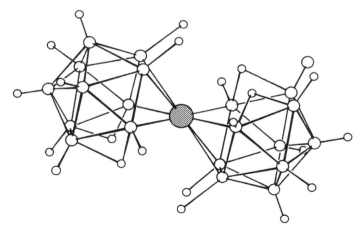

Fig. 13. The structure of $[Ni(B_{10}H_{12})_2]^{2-}$ (196) where ⬤ represents the Ni; ◯ B; and ◯ H.

metal. For example, $[Zn(B_{10}H_{12})_2]^{2-}$ [19154-53-1] is tetrahedral (197). The $[B_{10}H_{13}]^-$ ion [36928-50-4] (198) is especially useful for the synthesis of 11-vertex *nido* metallaboranes. These syntheses are influenced by (1) the cation of the $[B_{10}H_{13}]^-$ salt; (2) the nature of other ligands about the metal; and (3) the availability of a proton trap for the reaction (195,199).

The first *closo* metallaborane complexes prepared (200) were the nickelaboranes $[closo-(\eta^5-C_5H_5)Ni(B_{11}H_{11})]^-$ and $closo-1,2-(\eta^5-C_5H_5)_2-1,2-Ni_2B_{10}H_{10}$ [55266-88-1] (Fig. 14). These species are equivalent to $[closo-CB_{11}H_{12}]^-$ and $closo-C_2B_{10}H_{12}$ by the electron-counting formalism. The mixed-bimetallic anion $[closo-(\eta^5-C_5H_5)_2CoNi(B_{10}H_{10})]^-$ and other related species were reported later (201). These metallaboranes display remarkable hydrolytic, oxidative, and thermal stability.

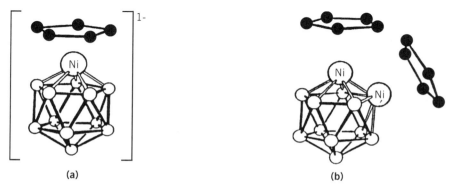

(a)   (b)

Fig. 14. The structures of *closo* metallaboranes where ◯ represents BH; ⬤ CH: (**a**) $[closo-(\eta^5-C_5H_5)Ni(B_{11}H_{11})]^-$; (**b**) $closo-1,2-(\eta^5-C_5H_5)_2-1,2-Ni_2B_{10}H_{10}$.

Closo metallaboranes can also be formed by the direct interaction of polyborane and metal carbonyl clusters. For example,

$$arachno-B_{10}H_{12}[S(C_2H_5)_2]_2 + Co_2(CO)_5 \xrightarrow[reflux°C]{toluene}$$

$$closo-(CO)_5Co_2B_{10}H_8[S(C_2H_5)_2] + 3\ CO + 2\ H_2 \qquad (70)$$

yields two isomers of this 12-vertex *closo* compound in good yield (202). Both air-stable isomers contain a $Co_2(CO)_8$ cobalt carbonyl cluster fragment, which can be considered a four framework electron-donor group. The retention of two $S(C_2H_5)_2$ substituents allows these clusters to comply with electron-counting rules for a *closo* compound. The B–SR$_2$ groups each donate three framework electrons. The structure of one isomer of *closo*-$(CO)_5Co_2B_{10}H_8[S(C_2H_5)_2]_2$ is shown in Figure 15.

A number of novel products have been isolated from the reaction of $[B_5H_8]^-$ [31426-87-6] and $CoCl_2$ and $[C_5H_5]^-$ in THF (203,204). The predominant product is *nido*-2-(CpCo)-B_4H_8 [43061-99-0]. Also obtained are isomeric clusters containing up to four cobalt atoms, eg, $(\eta^5$-$C_5H_5Co)_4B_4H_8$[59370-82-0]. Characterization of these clusters indicates an unusual $2n$ framework electron count having geometries reminiscent of strictly metallic clusters (11,205). Other metallaboranes, particularly those containing early transition metals, such as the series $(Cp^*Re)_2B_nH_n$, where $n = 8$–10, have structural characteristics that are borderline between the classical polyhedral boranes and metal borides (206). These have structures that exhibit nondeltahedral geometries and cross-cluster bonds.

7.2. Main Group Element Metallaboranes. A variety of metallaborane clusters, which incorporate main group metals in vertex positions of polyhedral metallaborane clusters, have been reported. Examples are $(BH_4)BeB_5H_{10}$

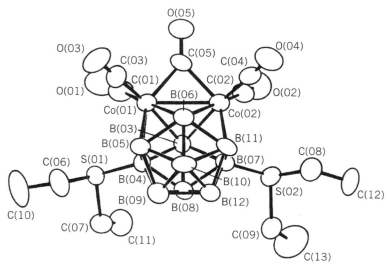

Fig. 15. The structure of one isomer of *closo*-$(CO)_5Co_2B_{10}H_8[S(C_2H_5)_2]_2$ without the hydrogens. Courtesy of the American Chemical Society (202).

(207), $MgB_{10}H_{12} \cdot 2O(C_2H_5)_2$ (208), $[(CH_3)HgB_{10}H_{12}]^-$ (209), $[AlB_{10}H_{14} \cdot 2O(C_2H_5)_2]^-$ (210), $[(CH_3)AlB_{11}H_{11}]^{2-}$ (211), $(CH_3)InB_{10}H_{12}$ (209), $[(CH_3)_2TlB_{10}H_{12}]^-$ (209), and $(CH_3)_2MB_{10}H_{12}$, where, $M = Si$ (212), Ge (213), or Sn (213). A number of main group metal complexes have been reported that incorporate borane moieties via M−H−B bridges, such as those found in $Al(BH_4)_3$ (Fig. 8). Examples of other compounds in this class include the octahydrotriborate complexes $Be(B_3H_8)_2$ (214), $(C_5H_5)BeB_3H_8$ (214), $[(CH_3)BeB_3H_8]_2$ (214), $(CH_3)_2AlB_3H_8$ (215), $H_2GaB_3H_8$ (216), $(CH_3)_2GaB_3H_8$ (216), $Mg(B_3H_8)_2 \cdot 6THF$ (217), and $(BH_4)MgB_3H_8 \cdot 5THF$ (217). Complexes containing the $[B_3H_8]$ unit are generally fluxional in solution. The low temperature static structure of $Be(B_3H_8)_2$ exhibits C_{2v} symmetry and features a tetrahedral beryllium center covalently linked via four Be−H−B bonds to two B_3H_8 units (218). The nmr spectral data for $[(CH_3)Be(B_3H_8)]_2$ indicates a methyl bridged dimer structure.

Several complexes containing the $B_5H_{10}Be$ unit, such as $(BH_4)BeB_5H_{10}$ and $Be(B_5H_{10})_2$ (207) have been characterized. X-ray diffraction studies (219) of these compounds show pentagonal pyramid structures having the beryllium atom in a basal position with three of the five bridging hydrogen atoms at the open face taking part in Be−H−B bonds. In $(BH_4)BeB_5H_{10}$, the BH_4 moiety is linked to the beryllium center by two Be−H−B bonds. The structure of $Be(B_5H_{10})_2$ consists of two pyramidal $B_5H_{10}Be$ units linked at a common basal position beryllium atom.

The volatile, air-sensitive liquid species $(CH_3)_2AlB_3H_8$ and $(CH_3)_2GaB_3H_8$ are prepared by the direct reaction of the corresponding main group metal halide and salts of the $[B_3H_8]^-$ ion, in the absence of solvent (220). The reaction of $(CH_3)_2AlB_3H_8$ and $Al(BH_4)_3$ results in the species $(BH_4)_2AlB_3H_8$. These small metallaboranes are fluxional in solution and have limited thermal stability at room temperature. The high yield preparation of the *closo*-aluminaborane anion $[(CH_3)AlB_{11}H_{11}]^{2-}$ has been described (211).

$$Na_2[B_{11}H_{13}] + Al(CH_3)_3 \longrightarrow Na_2[(CH_3)AlB_{11}H_{11}] + 2\,CH_4 > 90\% \qquad (71)$$

Similar synthetic strategies involving the elimination of alkyl groups from organometallic reagents and acidic B−H−B groups have been used to prepare a number of other metallaboranes and metallacarboranes. The $[(CH_3)AlB_{11}H_{11}]^{2-}$ anion is isostructural with *closo*-$[B_{12}H_{12}]^{2-}$. The methyl group is attached to aluminum projecting radially from the icosahedral AlB_{11} cage.

The silacarborane analogue of C,C-dimethyl-*ortho*-carborane, *closo*-1,2-$(CH_3)_2$-1,2-$Si_2B_9H_{11}$ [128270-48-4] has been reported (221). This *o*-silacarborane, which has an icosahedral framework much like *o*-carborane and is reported to be stable to air and moisture, was obtained in low yield from the reaction of decaborane and bis(dimethylamino)methylsilane in refluxing benzene.

$$B_{10}H_4 + CH_3(H)Si[N(CH_3)_2]_2 \longrightarrow closo-1,2-(CH_3)_2-1,2-Si_2B_9H_{11}$$

$$+\, 6,9-[(CH_3)_2NH]_2B_{10}H_{12} + \text{other products} \qquad (72)$$

7.3. Exopolyhedral Metallaboranes. Polyboranes may bind exopolyhedral metals in a variety of ways. Most commonly metals are bound via

M–H–B interactions. In other cases, metals may formally replace bridging hydrogen atoms at edge positions to give B–M–B interactions. Metals may also be attached to polyborane cages by direct M–B σ-bonds. The M–H–B bonds are found in $Cu[B_{10}H_{10}]$ (222) and in $\{[(C_6H_5)_3P]_2Cu\}_2$-μ-$B_{10}H_{10}$ [54020-26-7] (193). In both cases, the metal centers are bound through a bidentate interaction with two adjacent B–H groups.

A series of divalent lanthanide metal metallaborane derivatives has been prepared by the redox reaction of metallic lanthanides and boron hydrides and by the metathesis reaction of boron hydride salts with $LnCl_2$ where Ln = Sm, Eu, Yb (223,224). The species $(CH_3CN)_6Yb[(\mu\text{-}H)_2B_{10}H_{12}]$, $(CH_3CN)_4Yb[(\mu\text{-}H)_3BH]_2$, and $(C_6H_5N)_4Yb[(\mu\text{-}H)_3BH_4]_2$ have been structurally characterized by X-ray crystallography and shown to contain ytterbium to boron hydride Yb–H–B linkages. Thermal decomposition of lanthanaboranes can be used to generate lanthanide metal borides.

Metallaboranes containing M–B σ-bonds can be prepared by nucleophilic displacement reactions (225) and oxidative addition (226) of B–H and B–Br bonds to metal centers. For example, the reaction of $IrCl(CO)[P(CH_3)_3]_2$ and 1- or 2-BrB_5H_8 results in 2-$[IrBr_2(CO)\text{-}[P(CH_3)_3]_2B_5H_8]$ in which the B_5H_8 polyhedron serves as ligand for the metal.

Boranes also form derivatives in which main group elements occupy a bridging position between two boron atoms, rather than a polyhedral vertex. An extensively studied system is μ-$R_3MB_5H_8$, where R = H, CH_3, C_2H_5, halogen, and M = Si, Ge, Sn, Pb (227). The structure of 1-Br-μ-$[(CH_3)_3Si]$-B_5H_7 [28323-19-5] is shown in Figure 16 (228).

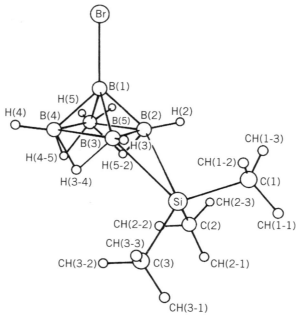

Fig. 16. Structure of 1-Br-μ-$[(CH_3)_3Si]B_5H_7$. Courtesy of the American Chemical Society (228).

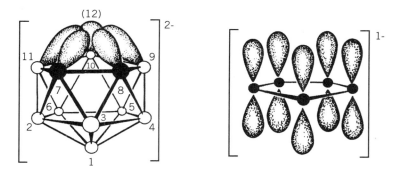

Fig. 17.   Structural analogy between the $[7,8\text{-}C_2B_9H_{11}]^{2-}$ dicarbollide dianion and the $[C_5H_5]^-$ anion (6), where ◯ represents BH; ● CH; showing p orbital lobes.

7.4. Metallacarboranes.   In the early 1960s, M. F. Hawthorne recognized (6,299) the bonding similarities between the pentagonal face of the isomeric $[nido\text{-}C_2B_9H_{11}]^{2-}$ ions and the well-known cyclopentadienide ion $[C_5H_5]^-$ (Cp$^-$) (Fig. 17). The isomeric $[nido\text{-}C_2B_9H_{11}]^{2-}$ anions and their C-substituted derivative, which are commonly known as dicarbollide ions, form stable complexes with most of the metallic elements. An extensive chemistry also exists for the isomeric C-substituted derivatives of the smaller carborane anion $[nido\text{-}R_2C_2B_4H_6]^{2-}$. Many other carboranes are also known to form metal complexes. Indeed nearly all metals can be combined with polyboron hydride clusters to produce an apparently limitless variety of cluster compounds. The chemistry of metallacarboranes incorporating d-block metals has been reviewed (230).

7.5. Transition-Metal Metallacarboranes.   The first demonstration of the insertion of a transition metal into an open face of a carborane cage was with the iron sandwich compound $[commo\text{-}Fe(C_2B_9H_{11})_{12}]^{2-}$ [12541-50-3] (231). This product is readily air oxidized to $[commo\text{-}(C_2B_9H_{11})_2Fe]^-$ [12547-76-1], a complex containing a formal Fe^{3+} center. These complexes, as well as those formed from a variety of formally d^3, d^5, d^6, and d^7 transition metals (Table 6),

Table 6. **Structure of $[(C_2B_9H_{11})_2M^{n+}]^{n-4}$ Complexes as a Function of Electronic Configuration**

		Electronic configuration			
d^3	d^5	d^6	d^7	d^8	d^9
		Unslipped sandwich[a]			
Cr^{3+}	Fe^{3+}	Fe^{2+}	Co^{2+}		
		Co^{2+}	Ni^{3+}		
		Ni^{4+}	Pd^{2+}		
		Pd^{2+}			
		Slipped sandwich[b]			
				Cu^{3+}	Cu^{2+}
				Ni^{2+}	Au^{2+}
				Pd^{2+}	
				Au^{3+}	

[a] See Figure 18**a**.
[b] See Figure 18**b**.

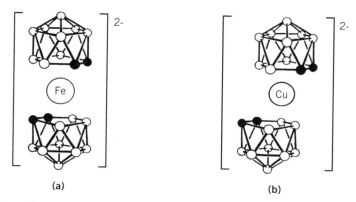

Fig. 18. Exemplary structures of (**a**) unslipped and (**b**) slipped metallacarborane dicarbollide sandwich derivatives where ○ represents BH; ● CH.

have symmetrical sandwich structures (232) of the type shown in Figure 18**a**. By contrast, d^8 and d^9 metals form slipped sandwich structures as shown in Figure 18**b** (6,233). A great many icosahedral complexes have been produced by incorporating transition metals into clusters based on the dicarbollide cage. (233,234). Examples are

$$[nido\text{-}C_2B_9H_{11}]^{2-} + [C_5H_5]^- + M^{2+} \longrightarrow [closo\text{-}\eta^5\text{-}C_5H_5)MC_2B_9H_{11}]^-$$

$$\downarrow \; [O] \qquad\qquad (73)$$

$$closo\text{-}(\eta^5\text{-}C_5H_5)MC_2B_9H_{11}$$

where M = Fe, Co, Ni, Cr.

Metallacarborane dicarbollide complexes are generally more robust than the corresponding cyclopentadiene complexes. The bis(dicarbollide) sandwich complexes of general formula $[M(C_2B_9H_{11})_2]^-$, where M is Fe^{3+}, Co^{3+}, and Ni^{3+}, exhibit great thermal, chemical, redox, and radiolytic stability. These species are also unusual in that they are extremely hydrophobic anions that form very strong conjugate acids. This unique combination of features leads to a number of potential uses such as the extraction of organic compounds from extremely dilute solutions and the isolation of metal cations, including the quantitative separation of radionuclides, eg, ^{137}Cs (235).

Representative icosahedral metallacarborane carbonyl complexes are prepared as shown (236).

$$[nido-C_2B_9H_{11}]^{2-} + M(CO)_6 \xrightarrow[\text{M=Cr, Mo, W}]{h\nu} [closo-(C_2B_9H_{11})M(CO)_3]^- + 3\,CO$$

$$(74)$$

$$[nido-C_2B_9H_{11}]^{2-} + BrM(CO)_5 \xrightarrow[\text{M=Mn, Re}]{\Delta} [closo-(C_2B_9H_{11})M(CO)_3]^-$$

$$+\,Br^- + 2\,CO \qquad (75)$$

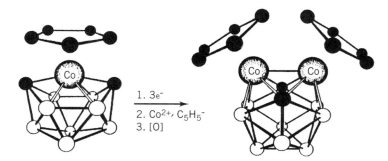

Fig. 19. A characteristic polyhedral expansion reaction leading to isomeric $[(\eta^5\text{-}$ $C_5H_5)Co]_2C_2B_8H_{10}$ clusters where represents $\bigcirc$ BH; $\bullet$ CH. The 2,9-$(\eta^5\text{-}C_5H_5)$-2,9-Co_2-1,12-$C_2B_8H_{10}$ isomer is shown. Courtesy of the American Chemical Society (245).

Fundamental methodologies for the synthesis and transformation of metallacarboranes(6) include polyhedral expansion (237–242), contraction (243), subrogation (244), and related reactions. Many carboranes are readily reduced from *closo* to *nido* molecules with a concomitant opening of the cage structures (see Fig. 2). The open, nontriangular faces of the resulting *nido* anions are generally capable of coordination to metals. The polyhedral expansion reaction (237–242), also called metal insertion, is most general for two-carbon carboranes.

$$closo - C_2B_nH_{n+2} + 2e^- \longrightarrow [nido - C_2B_nH_{n+2}]^{2-} \tag{76}$$

$$M^x \longrightarrow [commo - M(C_2B_nH_{n+2})_2]^{n-4} \tag{77}$$

Multiple polyhedral expansion reactions, carried out either simultaneously or sequentially, have been used to prepare metallacarboranes having multiple metal centers (245). An example is given in Figure 19.

In most cases, the prolonged treatment of a *closo* carborane or borane with strong base results in the removal of a single boron vertex to yield a *nido* cluster, inert to further degradation. This principle is exploited in the polyhedral contraction and subrogation synthetic strategies. In the prototypical case, the polyhedral contraction reaction (243) involves the degradative removal of a formal $[BH]^{2+}$ vertex from a *closo* n vertex metallacarborane followed by a two-electron oxidation of the *nido* intermediate to produce the corresponding *closo* n-1 vertex metallacarborane. An example of the polyhedral contraction reaction is given in Figure 20 (243). Polyhedral subrogation (244) is similar to the polyhedral contraction except that instead of oxidation of the *nido* metallacarborane intermediate, this intermediate is trapped by insertion of a metal atom to provide a bimetallacarborane framework. In practice, this procedure often leads to polymetallacarboranes as shown in Figure 21 (245).

Application of the polyhedral expansion methodology to $C_2B_{10}H_{12}$ leads to supraicosahedral metallacarboranes such as *closo*-$(\eta^5\text{-}C_5H_5)CoC_2B_{10}H_{12}$ [33340-90-8] (237–242). Further expansion of 13-vertex species or thermal metal transfer reactions leads to the 14-vertex cluster $[(\eta^5\text{-}C_5H_5)Co]_2C_2B_{10}H_{12}$ [52649-56-6] and [52649-57-7] (242). Similar 14-vertex species have been obtained from

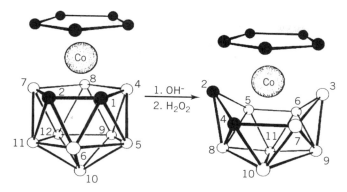

Fig. 20. A characteristic polyhedral contraction reaction leading to $(\eta^5\text{-}C_5H_5)CoC_2B_8H_{10}$ where represents BH; CH. Courtesy of the American Chemical Society (243).

tetracarbaboranes (246) and show unusual structures. The isomeric bimetallic cobaltacarborane complexes $closo\text{-}(\eta^5\text{-}CpCo)_2C_2B_8H_{10}$ can be formed by either polyhedral expansion or contraction reactions. Six isomers of this cluster are formed in the thermally induced intermolecular metal transfer and polyhedral expansion of the 11-vertex $closo\text{-}(\eta^5\text{-}Cp)CoC_2B_8H_{10}$.

Metallacarboranes are subject to thermal rearrangement reactions similar to those of carboranes and heteroboranes (247–249). In a study of the *closo* metallacarboranes $(\eta^5\text{-}C_5H_5)CoC_2B_nH_{n+2}$, $n=6\text{-}10$ (202), it was concluded that the empirical rules governing thermal rearrangements are (1) the transition-metal atom preferentially occupies the highest order vertex; (2) the carbon atoms do not decrease their mutual separation; (3) carbon atoms migrate to the lowest order vertices; and (4) carbon atoms migrate away from the transition metal providing rules (2) and (3) are not violated. Exceptions have been found, but at least some exceptions result from kinetic rather than thermodynamic control of the rearrangement (246).

The highly robust nickelacarborane system $[commo\text{-}3,3'\text{-}Ni(3,1,2\text{-}C_2B_9H_{12})_2]^{n-}$ $(n = 0 - 2)$ is particularly illustrative of the electronic effects associated with metallacarboranes (50,250). The monoanion, containing a formal Ni^{3+} center, is a symmetrical sandwich compound with the metal bound in η^5-fashion to

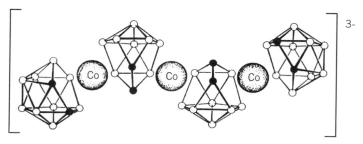

Fig. 21. A polymetallacarborane anion resulting from the polyhedral subrogation reaction where represents BH; CH Courtesy of the American Chemical Society (245).

each dicarbollide cage face and the carbon of the two dicarbollide cage of opposite sides in a transoid configuration. Reduction of this monoanion results in a dianionic sandwich compound containing a formal Ni^{2+} center in which the two dicarbollide cages are symmetrically slipped such that the metal is closer to the three boron atoms of each cage face. Alternatively, oxidation of the monoanion gives an unslipped, but distorted, neutral sandwich having a cisoid configuration with the carbon atoms of the dicarollide cages lying on the same side. This bright yellow, air-stable compound is a rare example of a complex containing a formal Ni^{4+} center, and illustrates the ability of the dicarbollide cage to stabilize high or otherwise unusual oxidation states of metals. This neutral nickelacarborane sandwich [*commo*-3,3'-Ni(3,1,2-$C_2B_9H_{12}$)$_2$], forms intensely colored charge-transfer complexes in the presence of excess electron donor species such as *N,N*-dimethylaniline, naphthalene, and pyrene, accompanied by dramatic color changes (251).

Another example of the relationship between metallacarborane structure and the electronics of the metal center is found in the "pinwheel cluster" shown in Figure 22 (252). In this trimeric cupracarborane three *nido* dicarbollide ligands, each charge compensated by substitution of an H:$^-$ by methyl nicotinate, are η^3-bonded to a Cu^+ ion and to a neighboring copper center via B–H–Cu bonds. This cluster, which has crystallographic C_3 symmetry and three

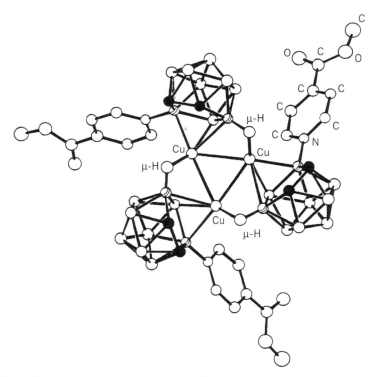

Fig. 22.   Molecular structure of the metallacarborane pinwheel cupracarborane complex $Cu_3(\mu\text{-H})_3C_2B_9H_9[4\text{-}(C_5H_4N)CO_2CH_3]_3$ where ⊘ within the cage structure represents ○ B; ○ BH; and ● carboranyl CH. Courtesy of the American Chemical Society (252).

equivalent Cu—Cu bonds, is stabilized by M—H—B bonding in much the same way that metal carbonyl clusters are stabilized by bridging CO groups.

Metallacarboranes containing early transition metals including Ti, V, Cr, Mn, Zr, and Hf in a variety of oxidation states are especially unique (253). For example, the titanium 2+, 3+, and 4+ compounds $[commo\text{-}Ti(C_2B_{10}H_{12})_2]^{2-}$, $closo\text{-}(\eta^5\text{-}C_5H_5)TiC_2B_{10}H_{12}$, $closo\text{-}(C_8H_8)TiC_2B_{10}H_{12}$, and $closo\text{-}(C_8H_8)TiC_2B_9H_{11}$ have no counterparts in traditional organometallic chemistry (254). These compounds exhibit remarkable thermal stability.

A precursor to several $closo$ 10-vertex cobaltacarboranes is $arachno\text{-}[C_2B_7H_{11}]^{2-}$ [42319-46-0], which is obtained by the deprotonation of 6,8-$C_2B_7H_{13}$ [17653-38-2]. When treated with excess $CoCl_2$ and cyclopentadienide ion [12127-83-2], $[C_2B_7H_{11}]^{2-}$ gives $closo\text{-}(\eta^5\text{-}C_5H_5)Co(C_2B_7H_9)$ [37381-23-0] and [51539-00-5] (eq. 78), which occurs as two isomers, $closo\text{-}2\text{-}[(\eta^5\text{-}C_5H_5)Co]\text{-}1,6\text{-}C_2B_7H_9$ [41348-11-2], and $closo\text{-}2\text{-}[(\eta^5\text{-}C_5H_5)Co]\text{-}1,10\text{-}C_2B_7H_9$ [42808-86-6] (29).

$$[C_2B_7H_{11}]^{2-} + [C_5H_5]^- + \frac{3}{2}Co^{2+} \longrightarrow H_2 + \frac{1}{2}Co + (C_5H_5)Co(C_2B_7H_9) \qquad (78)$$

An interesting $closo$ 10-vertex monocarbon metallacarborane species is $closo\text{-}10\text{-}[(C_5H_5)Ni]\text{-}1\text{-}CB_8H_9$ [52540-76-8] (255), which contains a metal atom bound to a B_4-face. $Nido$ 10-vertex metallaboranes frequently have boatlike frameworks where the metal is at the 6- or 9-position; however, the nature of the bonding depends on the metal. Examples of nine-vertex metallacarboranes include $[closo\text{-}2\text{-}(CO)_3\text{-}2,1,6\text{-}MnC_2B_6H_8]^-$ [41267-49-6] (256), which has the expected tricapped trigonal prism structure, and the clusters $[(C_5H_5)Ni]_2C_2B_5H_7$ [51108-05-5] (257) and 6,8-$(CH_3)_2$-1,1-$[P(CH_3)_3]_2$-1,6,8-$PtC_2B_6H_6$ (258).

Many structures are possible for the smaller metallacarboranes, and various synthetic strategies are available. Especially noteworthy is the occurrence of triple- and tetradecker sandwich compounds (259,260). The polyhedral expansion synthetic strategy can also be used with small carboranes (257). For example, the small metallacarborane $closo\text{-}1,1,1\text{-}(CO)_3\text{-}1,2,3\text{-}FeC_2B_4H_6$ [32761-40-3] (257) is obtained from $C_2B_4H_8$ upon treatment with $Fe(CO)_5$.

7.6. Exopolyhedral Metallacarboranes. Many metallacarboranes are known that exhibit exopolyhedral bonding to metals. Most commonly in these compounds metals are bound via M—H—B interactions in which the B—H group can be regarded as a two-electron donor to the metal center. In other cases, M—B, M—C, or M—M bonding may be involved. For example, electron-rich transition-metal complexes are capable of activating carboranyl B—H bonds leading to B-metalated metallacarboranes. Thus the d^8 Ir^+ complex $[Ir(C_8H_{14})_2Cl]_2$ reacts with 1,2-, 1,7-, and 1,12-$C_2B_{10}H_{12}$ carboranes in the presence of triphenylphosphine to give regiospecific B-metalated oxidative addition products of the type $[(C_6H_5)_3P]_2IrHCl\text{-}closo\text{-}C_2B_{10}H_{11}$ (261,262). Similarly, the C-substituted phosphinacarborane 1-$[(CH_3)_3P]\text{-}closo\text{-}C_2B_{10}H_{11}$ reacts with $[Ir(C_8H_{14})_2Cl]_2$ to give a metallacycle containing an Ir—B bond (262, 263). An example of exopolyhedral metal binding by a combination of M—M and M—H—B bonding is given in Figure 23. This complex is formed in the reaction of copper(I) chloride, $(C_6H_5)_3P$, and $[closo\text{-}3,1,2\text{-}TlC_2B_9H_{11}]^-$ (264). A variety of

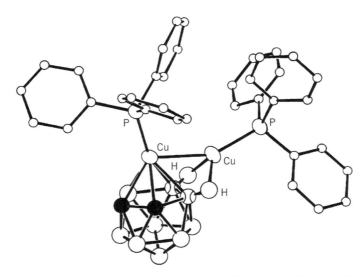

Fig. 23. The molecular structure of [*closo-exo*-4,8-((-H)$_2$Cu[P(C$_6$H$_5$)$_3$]-3-[(C$_6$H$_5$)$_3$P]-3,1,2-CuC$_2$B$_9$H$_9$, where ⊘ within the cage structure represents B; ○ BH; and ● carboranyl CH. Courtesy of the American Chemical Society (264).

metallacarborane complexes containing metal–metal bonds between cage framework metal centers and exopolyhedral metal centers have also been prepared (265).

Carborane cages can also be B-metalated by direct mercuration. The strongly electrophilic reagent (CF$_3$CO$_2$)$_2$Hg in CF$_3$CO$_2$H reacts *closo*-C$_2$B$_9$H$_{12}$ to give the B-mercuricarborane, C$_2$B$_{10}$H$_{11}$-Hg-O$_2$CCF$_3$. Disproportionate of this compound leads to the bis(B-carbonyl)mercury compound, Hg(C$_2$B$_{10}$H$_{11}$)$_2$, containing B–Hg σ-bonds (266). Mercuration primarily occurs at the 9-position, which is most remote from the carbon atoms of *o*- and *m*- carborane. These compounds have significant preparative value since they can be used as reagents for the synthesis of many B-substituted carborane derivatives. For example, reaction of Hg(C$_2$B$_{10}$H$_{11}$)$_2$ with MX$_3$, where M is Ga, In, Tl, P, As, or Sb, and X is a halide, provides a range of B-substituted carboranes of the type C$_2$B$_8$H$_{11}$MX$_2$. Mercuration can also be done on carborane cages of metallacarboranes (267).

Metallation of carborane cages at carbon also provides useful synthetic agents. Lithiation at carbon using butyllithium is often employed to form intermediates to *C*-functionalized carboranes. Also, the copper *C*-metallated compound Cu$_2$C$_2$B$_{10}$H$_{10}$ is a convenient reagent, particularly for the C-arylation to the carborane cage, since this compound reacts with aryl halides in the presence of pyridine (268).

7.7. Host–Guest Chemistry-Carborane Anticrowns. An extensive literature exists for compounds that complex cations, including the crown ethers and more complex host–guest chemistries. However, compounds that display selective anion complexation are more unusual. Anion complexation has received increasing attention recently because of its importance to biology and analytical chemistry (269). A class of macrocyclic metallacarboranes has been developed that act as unique multidentate Lewis acid hosts (270). Macrocyclic

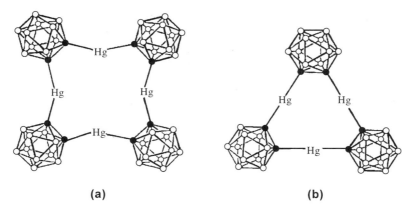

(a)   **(b)**

Fig. 24. Macrocyclic mercuracarborands, [9]mercurand-3 (**a**) and [12]mercurand-4 (**b**), where ○ represents BH; ● C.

mercuracarboranes have been prepared involving three or four *closo*-1,2-$C_2B_{10}H_{10}$ cages linked into cycles by an equal number of bridging mercury(II) atoms that are σ-bonded to the carborane carbons. These cycles are termed mercurands, where a numerical suffix indicates the cycle size and prefix in brackets indicated the total number of atoms completing the cycle, ie, [9] mercurand-3 (Fig. 24**a**) and [12] mercurand-4 (Fig. 24**b**). The mercury atoms in these compounds contain two empty p orbitals and exhibit Lewis acidity toward a variety of nucleophiles. For example, the trimer [9] mercurand-3 will sequester I^-, Br^-, Cl^-, SCN^-, NO_3^-, and ClO_4^-, in which the anions lie at the center of the cycles and are coordinated symmetrically to the mercury atoms. The [9] mercurand-3 cycle has also been incorporated as the active component in chloride-selective membrane electrodes. The tetrameric cyclic [12] mercurand-4 can sequester I^-, Br^-, Cl^-, CN^-, NO_3^-, and O_2^{2-}, $CH_3CO_2^-$, $C_6H_5S^-$, OH^-, and *closo*-$B_{10}H_{10}^{2-}$. The formation constant for [12] mercurand-4 can sequester Cl^- has been estimated at 10^7.

7.8. Metallacarboranes in Catalysis.   Perhaps the most intensely studied of all metallacarborane complexes is the exopolyhedral metallacarborane *closo*-3,3-[P(C$_6$H$_5$)$_3$]$_2$-3-H-3,1,2-RhC$_2$B$_9$H$_{11}$[61250-52-0], shown in Figure 25**a**,

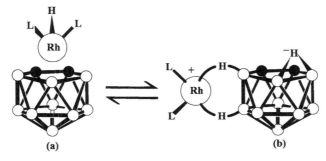

(a)   **(b)**

Fig. 25.   (**a**) The structure of *closo*-3,3-[(C$_6$H$_5$)$_3$P]$_2$-3-H-3,1,2-RhC$_2$B$_9$H$_{11}$, and (**b**) one isomer of its tautomer *exo–nido*-(L$_2$Rh)$_2$-(μ-H)$_2$-7,8-C$_2$B$_9$H$_{12}$, where L is (C$_6$H$_5$)$_3$P.

and its cage C-substituted derivatives. The three available isomers of closo-$[P(C_6H_5)_3]_2(H)Rh-C_2B_9H_{11}$ are synthesized in high yield by the oxidative addition of $[P(C_6H_5)_3]_2RhCl$ with the appropriate $[nido\text{-}C_2B_9H_{12}]^-$ ion (271), which may also be made chiral by the attachment of a single-alkyl or -aryl group at a carbon position (272). The resulting hydridorhodacarboranes are quite robust and catalyze a number of reactions including the isomerization and hydrogenation of olefins, the deuteration of B–H groups, and the hydrosilanolysis of alkenyl acetates. These species function as homogeneous catalyst precursors for the isomerization and hydrogenation of olefins as well as other reactions (273). Extensive investigations of rhodacarborane catalysts and the mechanisms responsible for their activity revealed the novel closo–nido tautomerism illustrated in Figure 25**a** and **b**. When alkyl or aryl substituents are present at the carborane carbon atoms, the usual rhodacarborane complex isolated is not the closo species, but rather an exo–nido complex in which the $[P(C_6H_5)_3]_3$ Rh^+ center is bonded to the $[nido\text{-}7,8\text{-}(R)_2\text{-}7,8\text{-}C_2B_9H_{12}]^-$, (R = alkyl, aryl), ions by a pair of Rh–H–B three-center, two-electron bridge bonds (274). The formation of exo–nido tautomers in this rhodacarborane system has been attributed primarily to steric factors. The isolation of closo structures for C-substituted isomers, which carry their substituent steric bulk removed from the metal center, such as closo-2,2-$[P(C_6H_5)_3]$-2-H-1-(CH_3)-7-(C_6H_5)-2,1,7-$RhC_2B_9H_{11}$, supports this contention.

The generality of closo–exo–nido redox equilibria in solution for these rhodacarborane species was demonstrated by labeling studies using derivatives of the nido-carborane precursor anion having a bridging B–D–B group. The deuterium is specifically transferred to the rhodium vertex of the closo tautomer (275). Retention of the D-label at the rhodium center during catalytic reactions proves that the H- (or D-) ligand attached to the $[P(C_6H_5)_3]_2RhH$ vertex does not participate in the catalytic processes, but is sequestered in the B–D–B bridge of the catalytically active exo–nido tautomer or related structure. The contention that the rhodacarborane exo–nido tautomer is the actual catalyst in solution is further supported by the observation that the catalytic hydrogenation and isomerization of alkenes exhibits the same characteristics and reaction rate law regardless of whether these reactions are conducted using a closo species or one of its exo–nido counterparts (276). In addition, the occurrence of facile carborane cage exchange reactions coupled with rate data obtained for such processes also implicates a reactive exo–nido intermediate (277).

Rhodacarborane catalysts have been immobilized by attachment to polystyrene beads with appreciable retention of catalytic activity (278). A 13-vertex closo-hydridorhodacarborane has also been synthesized and demonstrated to possess catalytic activity similar to that of the icosahedral species (279). Air-oxidation of closo-3,3-$[P(C_6H_5)_3]_2$-3-H-3,1,2-$RhC_2B_9H_{11}$ results in a brilliant purple dimer. This compound contains two formal Rh^{2+} centers linked by a sigma bond and a pair of Rh–H–B bridge bonds. A number of similar dimer complexes have been characterized, and the mechanism of dimer formation in these rhodacarborane clusters has been studied in detail (280).

The exopolyhedral metallacarborane complex $Ti(C_2B_{10}H_{11})_4$, which is prepared by the reaction of $TiCl_4$ and 1-Li-1,2-$C_2B_{10}H_{11}$, has also been reported to be an active heterogeneous catalyst for the polymerization of olefins when supported on alumina and in the presence of $(C_2H_5)_2AlCl$ cocatalyst (281).

7.9. Main Group Element Carborane Derivatives.

Main group element carborane derivatives have been reviewed (282). Although Group 1 (IA) metallacarboranes were prepared before other metallacarboranes, representatives of this class were only studied in detail in the past decade. For example, degradation of o-carborane (closo-$C_2B_{10}H_{12}$) in alcoholic KOH leads to the potassium salt of the dicarbollide monoanion ([nido-$C_2B_9H_{11}$]$^-$) in high yield. This monoanion salt is either used as a synthetic reagent or further deprotonated with potassium or sodium hydride or n-butyllithium to give salts of the dicarbollide dianion ([nido-$C_2B_9H_{11}$]$^{2-}$), which are often used in the preparation of other metallacarboranes. Similar methods are used in the preparation of other Group 1 (IA) metallacarborane derivatives, which are generally treated as intermediate synthetic reagents. The carborane nido-$C_2B_4H_8$, which has two bridging hydrogens associated with its pentagonal open face, can be deprotonated with NaH to give Na[$C_2B_4H_7$]. Further deprotonation to the dianion does not occur even in the presence of excess NaH (283), but can be effected using solution bases such as n-butyllithium. This behavior was perplexing until the X-ray crystal structure was determined for the THF solvated sodium salt of the C-trimethylsilyl-substituted carborane monoanion {2,3-[(CH$_3$)$_3$Si]-2,3-$C_2B_4H_5$}$^-$ (284). This compound, which exists as a dimer in the solid state, has a THF solvated sodium atom capping its open face and apparently shielding the extra-hydrogen from heterogeneous reaction with solid bases. An analogous compound was prepared having tetramethylenediamine (TMEDA) solvent in place of THF and shown to have a similar structure and reactivity. However, a more extensively solvated variation with two TMEDA molecules per sodium atom has an exo–nido structure with a (TMEDA)$_2$Na moiety is not capping but bound to the carborane cage via a pair of Na–H–B bridges (285). This compound is readily deprotonated by NaH. Structures have now been determined for a number of other Group 1 (IA) metallacarboranes, which often contain metals complexed by the carborane cage and thus participating in the polyhedral framework (282).

Structures have been determined for TMEDA solvated dilithium salts of isomericdianions {2,3-[(CH$_3$)$_3$Si]-2,3-$C_2B_4H_4$}$^{2-}$ and {2,4-[(CH$_3$)$_3$Si]-2,4-$C_2B_4H_4$}$^{2-}$. These contain one TMEDA solvated lithium atom capping the pentagonal open face and another TMED–Li moiety bound in exo-polyhedral fashion via a pair of Li–H–B bridges to two borons of the five-membered face (286). Solution nmr data shows little or no exchange of the lithiums in solution indicating that the solution structure is similar to that seen in the solid state. This type of structure is probably typical of many Group 1 (IA) salts of carboranyl dianions. A TMEDA solvated monolithium salt of the {2,3-[(CH$_3$)$_3$Si]-2,3-$C_2B_4H_5$}$^-$ anion has also been prepared (287). Slow vacuum sublimation of this compound at 160–170°C results in formation of the lithiacarborane anionic sandwich {commo-1-1'-Li-2,3-[(CH$_3$)$_3$Si]$_2$-2,3-$C_2B_4H_5$}$^-$ (288). The lithium atom in this complex is displaced toward the carbon atoms and one of the face boron atoms due to the presence of the exo-hydrogen atom bridging between the remaining two boron atoms. It is unclear if the Group 1 (IA) metals exhibit covalent bond character in their interactions with the open faces of carborane cages or if bonding to the cage faces is merely electrostatically preferred.

A number of alkaline earth element metallacarborane derivatives have been characterized. The icosahedral beryllacarborane, closo-3-[(CH$_3$)$_3$N]-3,1,

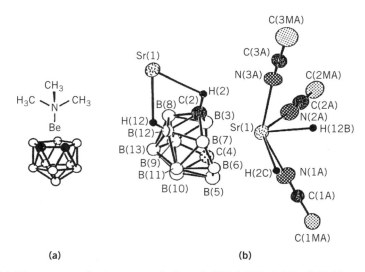

Fig. 26. (**a**) The proposed structure of *closo*-3-(CH$_3$)$_3$N-3,1,2-BeC$_2$B$_9$H$_{11}$, where ◯ represents BH; ● CH; and (**b**) the X-ray structure of the polymeric [*closo*-1,1,1-(CH$_3$CN)$_3$-1,2,4-SrC$_2$B$_{10}$H$_{12}$]$_n$ showing the metal-to-carborane interaction with the metal of the adjacent monomer unit. A indicates an atom on the CH$_3$CN ligand; MA is the methyl group; H(12B) is the hydrogen bound to the boron at position 12; H(2C) is the hydrogen bound to the carbon at position 2. Courtesy of the American Chemical Society (289,291).

2-BeC$_2$B$_9$H$_{11}$, shown in Figure 26**a**, was prepared via the reaction of *nido*-7,8-C$_2$B$_9$H$_{13}$ and Be(CH$_3$)$_2$·[O(C$_2$H$_5$)$_2$]$_2$ followed by reaction of the diethyletherate product and trimethylamine (289).

The reaction of calcium iodide and strontium iodide and the [*nido*-C$_2$B$_{10}$H$_{12}$]$^{2-}$ ion in THF followed by treatment with acetonitrile provides the 13-vertex metallacarboranes [*closo*-1,1,1,1-(CH$_3$CN)$_4$-1,2,4-CaC$_2$B$_{10}$H$_{12}$ (290) and [*closo*-1,1,1-(CH$_3$CN)$_3$-1,2,4-SrC$_2$B$_{10}$H$_{12}$]$_n$ (291), respectively. Both of these highly air-sensitive compounds have been structurally characterized by X-ray crystallography. The calcium complex contains a [Ca(CH$_3$CN)$_4$]$^{2+}$ moiety that caps the hexagonal C$_2$B$_4$ face of the [*nido*-C$_2$B$_{10}$H$_{12}$]$^{2-}$ cage to complete the 13-vertex polyhedron. The strontium compound, shown in Figure 26**b**, is polymeric and features a Sr(CH$_3$CN)$_3$H$_2$ capping moiety, the hydrogen atoms of which are involved in Sr–H–C and Sr–H–B bridging interactions with terminal hydrogen atoms of the carborane cage of adjacent strontium carborane repeating units, producing a spiral polymer chain. Bridging M–H–C interactions are rare in metallacarborane chemistry because B–H groups are generally more basic than C–H groups. A barium carborane complex was reported that contains a barium center solvated by two tetrahydrofuran molecules and bonded to three 2,3-bis(trimethylsilyl)-2,3-*nido*-C$_2$B$_4$H$_5$ cages. Two of the carborane cages coordinate to barium via a pair of Ba–H–B three-center bonds involving two boron atoms of the C$_2$B$_3$ pentagonal face and, the third cage is coordinated by two carbons of the open face and a bridging hydrogen lying between two of the borons of open face to form a four-center BaHB$_2$ bond (292).

The cogener relationship between boron and aluminum has prompted considerable interest in the aluminacarboranes. The Lewis acid–base adduct

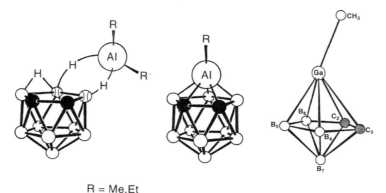

R = Me, Et

Fig. 27. The molecular structures of (**a**) 3-R-3,1,2-AlC$_2$B$_9$H$_{11}$, and (**b**) *exo–nido*-9,10-(μ-AlR$_2$)(μ-H)$_2$-7,8-C$_2$B$_9$H$_{10}$ where ◯ represents BH; ● CH; ⬮ B; and R = CH$_3$, C$_2$H$_5$, and (**c**) 1-CH$_3$-1,2,3-GaC$_2$B$_4$H$_6$. Courtesy of the American Chemical Society (296).

3-(C$_2$H$_5$)-3,1,2-AlC$_2$B$_9$H$_{11}$·2THF was first prepared by the reaction of Na$_2$[7,8-C$_2$B$_9$H$_{11}$] and (C$_2$H$_5$)AlCl$_2$ in THF solution (293). The reaction of the acidic species *nido*-C$_2$B$_9$H$_{13}$ and trialkylaluminum reagents (294) results in loss of 1 equiv of alkane and formation of *nido* aluminacarborane species of the type *exo–nido*-9,10-(μ-AlR$_2$)(μ-H)$_2$-7,8-C$_2$B$_9$H$_{10}$ (Fig. 27**b**). Upon heating these species, a second equivalent of alkane is lost with the formation of 12-vertex *closo* aluminacarboranes of the type *closo*-3-R-3,1,2-AlC$_2$B$_9$H$_{11}$ (Fig. 27**a**). These two steps can be combined to produce the *closo* aluminacarboranes in a single step:

$$nido - 7,8 - C_2B_9H_{13} + MR_3 \longrightarrow closo - 3 - R - 3,1,2 - MC_2B_9H_{11} \qquad (79)$$

for M = Al; R = CH$_3$, C$_2$H$_5$; for M = Ga; R = C$_2$H$_5$. The same methodology can be applied to the preparation of the corresponding gallacarborane (294). In the absence of Lewis bases, these clusters contain a main group element bound in η^5-fashion to the five-membered face of the dicarbollide cage. The X-ray crystal structure (295) of *closo*-3-(C$_2$H$_5$)-3,1,2-AlC$_2$B$_9$H$_{11}$ reveals an undistorted icosahedral cluster; however, the ethyl group on aluminum is tilted away from the normal to the plane of the cage bonding face in the direction of the carbon atoms at an angle of 19.4°. This aluminacarborane and its *C*-substituted derivatives form adducts with Lewis bases in which the aluminum atom is slipped dramatically with respect to the dicarbollide cage face in a direction away from the two-carbon atoms (296).

The *closo*-3-(C$_2$H$_5$)-3,1,2-AlC$_2$B$_9$H$_{11}$ complex is especially interesting because in aromatic solvents it exists in temperature-dependent equilibrium with a unique highly fluxional dimer molecule consisting of the [*commo*-Al(C$_2$B$_9$H$_{11}$)$_2$]$^-$ sandwich ion complexed with a [(C$_2$H$_5$)$_2$Al]$^+$ moiety via two Al–H–B interactions (296). The free [Al(C$_2$B$_9$H$_{11}$)$_2$]$^-$ sandwich ion (Fig. 28**a**), in which the aluminum atom is bound in η^5-fashion to the pentagonal faces of two dicarbollide cages, has been prepared independently and structurally

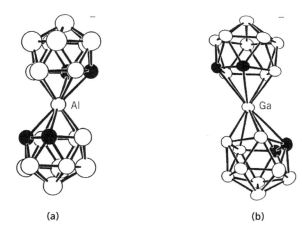

(a)   (b)

Fig. 28. The X-ray structures of (**a**) [*commo*-3,3′-Al(3,1,2-AlC$_2$B$_9$H$_{11}$)$_2$]$^-$, and (**b**) [*commo*-3,3′-Ga(3,1,2-GaC$_2$B$_9$H$_{11}$)$_2$]$^-$. Courtesy of the American Chemical Society (296).

characterized (296,297). The analogous gallacarborane sandwich anion [Ga(C$_2$B$_9$H$_{11}$)$_2$]$^-$ (Fig. 28**b**) has also been characterized and found to possess a significantly distorted structure (296,298). Various aluminacarboranes have been prepared based on the [C$_2$B$_4$H$_6$]$^{2-}$ anion and its *C*-substituted derivatives (299,300), which contain aluminum atoms that either participate in the polyhedral framework of carborane cages or bridge between boron atoms in the polyhedral frameworks. In addition, the aluminacarboranes [Al(C$_2$B$_6$H$_8$)$_2$]$^-$, C$_2$H$_5$AlC$_2$B$_8$H$_{10}$, and [Al(C$_2$B$_8$H$_{10}$)$_2$]$^-$ (296) contain aluminum atoms that are σ-bound to carbon atoms in the carborane frameworks.

Other smaller Group 13 (III A) element metallacarboranes include the isomeric seven-vertex gallacarborane cluster *closo*-CH$_3$GaC$_2$B$_4$H$_6$ (18), shown in Figure 27c, which exhibits a tilt distortion of the Ga—CH$_3$ group similar to that observed for *closo*-3-(C$_2$H$_5$)-3,1,2-AlC$_2$B$_9$H$_{11}$. The alkylaluminum and alkylgallium *closo*-metallacarboranes react with Lewis bases to give product having two two-electron donor bases coordinated to the metal accompanied by slippage of the metal toward the side or edge of the pentagonal face (301). A novel gallacarborane complex of the *C*-substituted 2,4-[Si(CH$_3$)$_3$]$_2$-2,4-*nido*-C$_2$B$_4$H$_4$ cage was prepared under conditions permitting loss of the sustituent at gallium to provide a seven-vertex *closo*-gallacarborane dimer (Fig. 31) in which the cage-capping gallium atoms link two seven-vertex clusters together via a gallium—gallium bond, as shown in Figure 29 (302). This complex exhibits an unusually short Ga—Ga bond (2.340 Å).

The thallacarborane anion [*closo*-3,1,2-TlC$_2$B$_9$H$_{11}$]$^-$ and its *C*-methyl derivatives are often used as synthetic reagents since they are air-stable and have excellent shelf life (303). The parent anion is prepared by the reaction of [*nido*-C$_2$B$_9$H$_{12}$]$^-$ and thallium(I) acetate [563-68-8] in aqueous base to give the salt Tl[TlC$_2$B$_9$H$_{11}$], in which one thallium atom caps the dicarbollide cage and the second thallium atom serves as a counterion to the 12-vertex anion. This compound, a yellow, air-stable, insoluble powder, is a convenient synthetic reagent for the preparation of other metallacarboranes. A variety of other thallacarborane

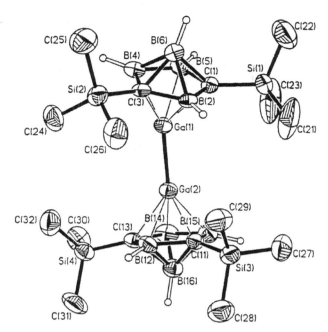

Fig. 29. The X-ray structure of [*closo*-Ga-2,4-[Si(CH$_3$)$_3$]$_2$-2,4-C$_2$B$_4$H$_4$]$_2$, containing a Ga$-$Ga bond.

salts, eg, [(C$_6$H$_5$)$_3$PCH$_3$][3,1,2-TlC$_2$B$_9$H$_{11}$] and [PPN][3,1,2-TlC$_2$B$_9$H$_{11}$], where PPN=[(C$_6$H$_5$)$_3$P]$_2$N$^+$, can be prepared by cation-exchange reactions using the dithallium compound. The [PPN]$^+$ salt is a useful reagent because of its enhanced solubility in organic solvents. Several of these salts have been characterized by X-ray crystallography. The thallium atom of [*closo*-3,1,2-TlC$_2$B$_9$H$_{11}$]$^-$ is positioned at the open five-membered face of the dicarbollide cage and is slipped slightly toward the unique boron atom of the cage face. Metal–cage bonding in this anion and its *C*-substituted derivatives has been the subject of some controversy. On one hand the thallium cage atom distances are relatively long, suggesting an ionic interaction. On the other hand, ^{1}H and ^{11}B nmr spectra show strong coupling between thallium and the cage atoms, indicative of a covalent interaction. The yellow color of the *closo*-[3,1,2-TlC$_2$B$_9$H$_{11}$]$^-$ anion has been attributed to metal–cage charge transfer. In some salts, including that of the [(C$_6$H$_5$)$_3$P]$_2$N$^+$ cation, the solid-state structure of this anion is that of a dimer linked by four Tl–H–B bridging interactions (304).

A series of compounds of the type *closo*-MC$_2$B$_9$H$_{11}$ in which M is divalent Ge, Sn, and Pb, have been prepared (305).

$$[nido-7,8-C_2B_9H_{11}]^{2-} + MX_2 \longrightarrow closo-3,1,2-MC_2B_9H_{11} + 2X^- \qquad (80)$$

As the B–H group, the 2+ main group atoms act as two-electron donors to the *closo*-MC$_2$B$_9$H$_{11}$ cage system. Bonding considerations suggest that these compounds should possess an unshared lone pair of electrons available for bonding at the metal center. However, the complexes do not exhibit Lewis base

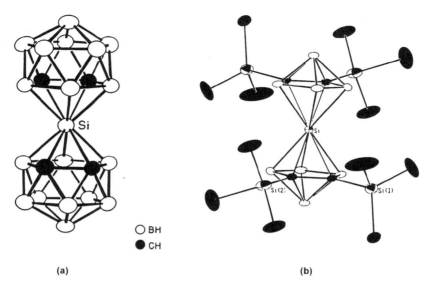

○ BH
● CH

(a)   (b)

Fig. 30. The X-ray structures (**a**) *commo*-3,3′-Si(3,1,2-SiC$_2$B$_9$H$_{11}$)$_2$, where ○ represents BH; ● CH, and (**b**) *commo*-2,2(,3,3′-[(CH$_3$)$_2$Si]$_4$-*commo*-1,1′-Si(1,2,3-C$_2$B$_4$H$_4$). Courtesy of the American Chemical Society (308).

properties and actually act primarily as Lewis acids, forming donor–acceptor complexes with a variety of donor molecules. The analogous seven-vertex cluster *closo*-SnC$_2$B$_4$H$_6$ and its *C*-substituted derivatives form similar donor complexes with Lewis bases. The cluster *closo*-CH$_3$GeCB$_{10}$H$_{11}$ (306) has been described. In this compound a Ge–CH$_3$ moiety, which may be regarded as a three-electron donor, caps a monocarbon carborane cage, to give a 12-vertex, 26-electron *closo* system. The methyl group on germanium in this compound is reversibly removed by reaction with pyridine to yield the [GeCB$_{10}$H$_{11}$]$^-$ anion.

The X-ray structure of the bis(dicarbollide) sandwich compound *commo*-3,3′-Si(3,1,2-SiC$_2$B$_9$H$_{11}$)$_2$ (Fig. 30**a**) contains a silicon atom in a highly unusual bonding mode (307). This compound, which has good thermal stablity, is prepared in good yield.

$$2\,\mathrm{Li}_2[7,8-\mathrm{C}_2\mathrm{B}_9\mathrm{H}_{11}] + \mathrm{SiCl}_4 \longrightarrow commo-3,1,2-\mathrm{SiC}_2\mathrm{B}_9\mathrm{H}_{11_2} + 4\,\mathrm{LiCl} \quad 80\%$$

(81)

It is isoelectronic and isostructural with the aluminum sandwich ion [*commo*-3,3′-Al(3,1,2-AlC$_2$B$_9$H$_{11}$)$_2$]$^-$ shown in Figure 28**a**. The silicon is η^5-bonded in unslipped fashion to the C$_2$B$_3$ faces of two dicarbollide cages. This bis (dicarbollide) silicon sandwich also forms adducts of a variety of structural types with Lewis bases such as pyridine and trimethylphosphine.

A series of sandwich compounds of the type *commo*-[(CH$_3$)$_3$Si(R)-C$_2$B$_4$H$_4$]$_2$M, where M is Si, Ge, Sn, and Pb, and R is variously (CH$_3$)$_3$Si, CH$_3$, and H, have also been prepared (308). These compounds, formed by reactions between salts of the [(CH$_3$)$_3$Si(R)C$_2$B$_4$H$_4$]$^{2-}$ and [(CH$_3$)$_3$Si(R)C$_2$B$_4$H$_5$]$^-$ ions and

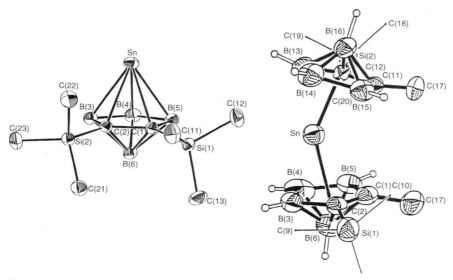

Fig. 31. The X-ray structure of (**a**) 2,3-[(CH$_3$)$_2$Si]$_2$-1,2,3-SnC$_2$B$_4$H$_4$ and (**b**) *commo*-1,1'-Sn-{2-[Si(CH$_3$)$_3$]-3-(CH$_3$)-1,2,3-SnC$_2$B$_4$H$_4$}$_2$. Courtesy of the American Chemical Society (18,310312).

appropriate main group element halides, have structures containing central main group elements in the 4+ oxidation states similar to the bis(dicarbollide) silicon sandwich compound. The structure of the silicon sandwich compound *commo*-[(CH$_3$)$_3$Si$_2$C$_2$B$_4$H$_4$]$_2$Si is shown in Figure 30**b**.

In compounds of the type *closo*-MC$_2$B$_4$H$_6$, where M is Ge, Sn, and Pb, the main group elements complete one apex of a pentagonal bipyramid (309). As in the case of the analogous icosahedral clusters, the lone pair of electrons at the divalent main group element center has no tendency to form donor–acceptor complexes with Lewis acids. A number of C-substituted compounds of the type *closo*-MC$_2$(R)$_2$B$_4$H$_4$, where M is Si, Ge, Sn, Pb and R is variously H, CH$_3$, Si(CH$_3$)$_3$, have also been characterized (310). The X-ray crystal structure of one of the tin-containing members of this series is shown in Figure 31. The compounds in this group react with Lewis bases to form adducts having a variety of intriguing structures, a number of which have been characterized by X-ray crystallography (311). The formal Sn^{4+} sandwich compound *commo*-1,1'-Sn-{2-[Si(CH$_3$)$_3$]-3-(CH$_3$)-1,2,3-SnC$_2$B$_4$H$_4$}$_2$, shown in Figure 31b, has been prepared and shown to exhibit a bent configuration (312). The (ring centroid 1)-Sn-(ring centroid 2) angle is142.5°, which is similar to those found for (η^5-C$_5$H$_5$)$_2$Sn and [η_5-C$_5$(CH$_3$)$_5$]$_2$Sn, 145.8° and 144.1°, respectively (313).

A number of related clusters has been prepared based on other heterocarborane cages with faces analogous to [C$_5$H$_5$]$^-$. Some of these include the [*nido*-ECB$_9$H$_{10}$]$^{2-}$ anions where E = P, As, which form metal complexes with many transition and main group elements. *Closo*-GePCB$_9$H$_{10}$ and *closo*-GeAsCB$_9$H$_{10}$ are examples (314).

7.10. f-Block Element Metallacarborane Derivatives. The chemistry of metallacarboranes containing *f*-block elements has been reviewed (230). The first actinide metallacarborane complex, *commo*-$(C_2B_9H_{11})UCl_2$, was prepared in 1977 (315). The coordination geometry of this complex can be described as distorted tetrahedral with the four positions occupied by two η^5-bound [7,8-$C_2B_9H_{11}$]$^-$ cages and two chloride ions. Complexes of this type are often referred to as bent sandwiches because of the configuration of the two-dicarbollide cage about the metal center, which is analogous to the corresponding pentamethyl cyclopentadiene–metal complexes.

The synthesis of lanthanacarboranes has been described (316). The metathetical reaction of [*nido*-$C_2B_9H_{11}$]$^{2-}$ salts with LnI_2, where Ln = Sm or Yb, in THF leads to Ln^{2+} complexes of the type $(C_2B_9H_{11})Ln(thf)_4$. These complexes, which are stable to temperatures >200°C, undergo ligand exchange reactions in a variety of donor solvents. The reaction in DMF (solvent) gives the corresponding $(C_2B_9H_{11})Ln(dmf)_4$ complexes. The structure of the ytterbium complex has been determined to be a *closo* icosahedral cluster consisting of a $C_2B_9H_{11}$ cage capped by an ytterbium atom to which four dmf ligands are coordinated via oxygen. These reactions are similar to those for calcium and strontium metallacarboranes. The reaction of $(C_2B_9H_{11})Sm(thf)_4$ in THF (solvent) with a soluble salt of the [*closo*-3,1,2-$TlC_2B_9H_{11}$]$^-$ anion affords the bent samarium sandwich anion, [*commo*-$(C_2B_9H_{11})_2Sm(thf)_2$]$^-$ (Fig. 32). The two η^5-dicarbollide cages and two thf ligands are in the coordination sphere of samarium and the average dicarbollide–Sm–dicarbollide angle is 132°. The metal–dicarbollide cage distances found in structurally characterized lanthanacarboranes are similar to the metal–carbon distances observed for the corresponding lanthanum–pentamethylcyclopentadiene complexes. Neutral monocage complexes of Sm and Yb have also been prepared using the 12-vertex [*nido*-$C_2B_{10}H_{12}$]$^{2-}$ cage.

7.11. Boron Neutron Capture Therapy.   There is great interest in the use of polyboron hydride compounds for boron neutron capture therapy (BNCT) for the treatment of cancers (315–320) and other diseases. Boron-10 is unique

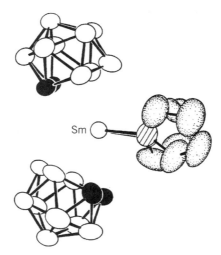

Fig. 32.   The X-ray structure of [3,3-$(C_4H_8O)_2$-*commo*-3,3′-Sm(3,1,2-$SmC_2B_9H_{11})_2$]$^-$, where ○ represents BH; ● CH; ▨ O ; ◉ CH_2. Courtesy of the American Chemical Society (316).

among the light elements in that it possesses an unusually high neutron capture nuclear cross-section ($3.8 \times (10^{-25}$ m^2), 0.02–0.05 eV neutron). The nuclear reaction between ^{10}B and low energy thermal neutrons yields alpha particles and recoiling lithium-7 nuclei:

$$
^{10}_{5}\text{B} + {}^{1}n_{th} \longrightarrow [{}^{11}_{5}\text{B}^*]
\begin{array}{l}
\xrightarrow{94\%}\ {}^{4}_{2}\text{He} \quad + \quad {}^{7}_{3}\text{Li} \quad + \quad \gamma \\
\qquad\quad 1.47\ \text{Mev} \quad\ \ 0.84\ \text{Mev} \quad\ \ 0.48\ \text{Mev} \\
\\
\xrightarrow{6\%}\ {}^{4}_{2}\text{He} \quad + \quad {}^{7}_{3}\text{Li} \\
\qquad\quad 1.78\ \text{Mev} \quad\ \ 1.01\ \text{Mev}
\end{array}
\tag{82}
$$

2.28 M eV is released as kinetic energy.

Because the cytotoxic effects of the energetic lithium-7 and α particles are spacially limited to a range of only about one-cell diameter, the destructive effects are confined to only one or two cells near the site of the event. Thus BNCT involves the selective delivery of sufficiently high concentrations of ^{10}B-containing compounds to tumor sites followed by the irradiation of these sites with a beam of relatively nondestructive thermal neutrons. The resulting cytotoxic reaction can then in theory destroy the tumor cells that are intimately associated with ^{10}B target. The great advantage of BNCT is its cell level selectivity for destroying tumor cells without significant damage to healthy tissue. This attribute makes BNCT particularly valuable in the treatment of tumors that are difficult or impossible to remove by surgical methods, including malignant brain tumors such as glioblastoma and astrocyctoma.

It has been estimated that using available neutron intensities, such as 10^{10} neutrons/cm$^2 \cdot$s, concentrations of ^{10}B from 10–30 µg/g of tumor with a tumor cell to normal cell selectivity of at least five are necessary for BNCT to be practical. Hence, the challenge of BNCT lies in the development of practical means for the selective delivery of approximately 10^9 ^{10}B atoms to each tumor cell for effective therapy using short neutron irradiation times. Indeed, many experimental ^{10}B delivery systems have been developed (315–317). Derivatives of ^{10}B-enriched *closo*-borane anions and carboranes appear to be especially suitable for BNCT because of their high concentration of ^{10}B and favorable hydrolytic stabilities under physiological conditions. To date, the most extensively studied polyboron hydride compounds in BNCT research have been the icosahedral mercaptoborane derivatives $Na_2[B_{12}H_{11}SH]$ (called BSH) and $Na_4[(B_{12}H_{11}S)_2]$, which have been used in human trials, particularly in Japan, with some success. New generations of tumor-localizing boronated compounds are being developed. The dose-selectivity problem of BNCT has been approached using boron hydride compounds in combination with a variety of delivery vehicles including boronated polyclonal and monoclonal antibodies, porphyrins, amino acids, nucleotides, carbohydrates, and liposomes. BNCT has also been proposed as a treatment for other diseases such as arthritis.

7.12. Boranes as Pharmacophores. The unique properties of polyhedral boranes, such as hydrophobicity, steric bulk, stability under physiological conditions, and ease of functionalization, have been exploited in the design of new pharmaceutical agents. For example, the dicarba-*closo*-dodocaborane moiety has been studied as hydrophobic pharmacophores in the modulation of tumor necrosis factor (321) and the design of estrogen antagonists (322) Carboranes have also been used as pharmacophores in retinoid antagonists (323) and other biologically active molecules of therapeutic interest.

A related potential medical application of metallacarboranes is based on the highly favorable kinetic stability of many metallacarborane complexes under physiological conditions and ability to form stable complexes with a wide range of metals. These features make certain functionalized metallacarboranes containing radiometals ideal choices for use as medical imaging reagents. The use of antibody-conjugated bridged dicarbollide metallacarborane (venus fly trap) chelate complexes incorporating γ-emitting $^{57}Co^{3+}$ in the imaging tumors has been reported (324,325).

8. Economic Aspects

Despite the fact that many boron hydride compounds possess unique chemical and physical properties, very few of these compounds have yet undergone significant commercial exploitation. This is largely owing to the extremely high cost of most boron hydride materials, which has discouraged development of all but the most exotic applications. Nevertheless, considerable commercial potential is foreseen for boron hydride materials if and when economical and reliable sources become available. Only the simplest of boron hydride compounds, most notably sodium tetrahydroborate, $NaBH_4$, diborane(6), B_2H_6, and some of the borane adducts, eg, amine boranes, are now produced in significant commercial quantities.

8.1. Sodium Tetrahydroborate, Na[BH$_4$]. This air-stable white powder, commonly referred to as sodium borohydride, is the most important commercial boron hydride material. It is used in a variety of industrial processes including bleaching of paper pulp and clays, preparation and purification of organic chemicals and pharmaceuticals, textile dye reduction, recovery of valuable metals, wastewater treatment, and production of dithionite compounds. Sodium borohydride is produced by Rohm & Haas Co., in manufacturing plants in the Netherlands and the state of Washington, and in Scandinavia by Finnish Chemicals (Nokia), with Rohm & Haas supplying most of the market. More than 6 million lb of this material supplied as powder, pellets, and aqueous solution (called Borol solution), were produced in 2001.

8.2. Diborane(6), B$_2$H$_6$. This spontaneously flammable gas is consumed primarily by the electronics industry as a dopant in the production of silicon wafers for use in semiconductors. It is also used to produce amine boranes and the higher boron hydrides. Callery Chemical Co., a division of Mine Safety Appliances Co., and Voltaix, Inc., are the main U.S. producers of this substance. Several hundred tons were manufactured worldwide in 2001.

8.3. Borane Adducts. Trialkylamine and dialkylamine boranes, such as tri-*tert*-butylamine borane and dimethylamine borane, are mainly used in electroless plating processes. Other borane adducts, such as THF–borane and dimethyl sulfide–borane are used for specialized reduction reactions. These are produced by Callery Chemical Co. and Aldrich Chemical Co.

8.4. Organoboron Hydrides. A variety of organoboron hydrides produced by hydroboration of olefins is commercially available. These are widely used in organic synthesis and the manufacture of pharmaceuticals. Examples include 9-borabicyclo[3.3.1]nonane (9-BBN), obtained by hydroboration of 1,5-cyclooctene, and diispinocampheylborane (DIP$_2$BH), produced by hydroboration of α-pinene. The latter is useful for asymmetric syntheses. A wide variety of borane reducing agents and hydroborating agents is available from Aldrich Chemical Co., Inc. and Callery Chemical Co.

8.5. Polyhedral Boron Hydrides. Although relatively large quantities of polyhedral boron hydrides and carboranes have been produced under various government contracts, these materials are not currently produced on large-scale commercial basis. Decaborane(14) (*nido*-B$_{10}$H$_{14}$), and carborane isomers, such as the *o*, *m*-, and *p*-carborane (*closo*-1,2-, 1,7-, and 1,12-C$_2$B$_{10}$H$_{12}$), as well as various other derivatives, are available in experimental quantities. Prices for these compounds range, according to purity, composition, and quantity purchased, from $30 to 500 or more per gram. Callery Chemical Co. has the capability to produce some of these materials on demand.

Demonstrated areas of potential commercial applications for other boron hydride-based materials are listed according to the classes of these materials.

Polyhedral Boron Hydrides. These are used as experimental agents in neutron capture therapy of cancers (317–320), and as burn rate modifiers (accelerants) in gun and rocket propellant compositions. A salt of the [B$_{12}$H$_{12}$]$^{2-}$ anion is used in the fuse system of the passenger-side automotive airbag.

Carboranes. These are used as experimental agents in neutron capture therapy (317–320), and as burn rate modifiers in gun and rocket propellants. They have been used as high temperature elastomers and high temperature gas–liquid chromatography stationary phases and have potential for use in other unique materials, optical switching devices (326), and gasoline additives (327).

Metallacarboranes. These have potential for use in homogeneous catalysis (273), including hydrogenation, hydrosilylation, isomerization, hydrosilanolysis, phase-transfer, burn rate modifiers in gun and rocket propellants, neutron capture therapy (317–320), medical imaging (324,325), processing of radioactive waste (327), analytical reagents, and as ceramic precursors.

BIBLIOGRAPHY

"Boron Hydrides" under "Boron Compounds" in *ECT* 1st ed., Vol. 2, pp. 593–600, by S. H. Bauer, Cornell University; "Boron Hydrides and Related Compounds" under "Boron Compounds," Suppl. 1, pp. 103–130, by S. H. Bauer, Cornell University; "Diborane and Higher Boron Hydrides" under "Boron Compounds," Suppl. 2, pp. 109–113, by W. J. Shepherd and E. B. Ayres, Callery Chemical Company; "Boron Hydrides" under "Boron Compounds" in *ECT* 2nd ed., Vol. 3, pp. 684–706, by G. W. Campbell, Jr., U.S. Borax Research Corporation; "Boron Hydrides and their Metalla Derivatives" under "Boron Compounds," in *ECT* 3rd ed., Vol. 4, pp. 135–183, by R. W. Rudolph, The University of Michigan; "Boron Hydrides and their Metalla Derivatives" under "Boron Compounds," in *ECT* 4th ed., Vol. 4, pp. 439–500, by D. M. Schubert, U.S. Borax Inc.; "Boron Hydrides, Heteroboranes, and their Metalla Derivatives" in *ECT* (online), posting date: December 4, 2000, by David M. Schubert, U.S. Borax Inc.

CITED PUBLICATIONS

1. A. Stock, *Hydrides of Boron and Silicon*, Cornell University Press, Ithaca, N. Y., 1933.
2. H. C. Brown and B. C. Subba Rao, *J. Am. Chem. Soc.* **78**, 5694 (1956); H. C. Brown, *Hydroboration*, W. A. Benjamin, Inc., New York, 1962; S. E. Thomas, *Organic Synthesis, The role of Boron and Silicon*, Oxford University Press, Oxford, 1991; H. C. Brown and P. V. Ramachandran, in W. Siebert, ed., *Advances in Boron Chemistry*, The Royal Society of Chemistry, Cambridge, U.K., 1997, p. 151.
3. R. T. Holtzmann, ed., *Production of Boranes and Related Research*, Academic Press, New York, 1967.
4. W. N. Lipscomb, *Science* **196**, 1047 (1977).
5. T. L. Heying, J. W. Ager, S. L. Clark, D. J. Mangold, H. L. Goldstein, M. Hillman, R. J. Polak, and J. W. Szymanski, *Inorg. Chem.* **2**, 1089 (1963); M. M. Fein, J. Bobinski, N. Meyers, N. Schartz, and M. S. Cohen, *Inorg. Chem.* **2**, 111 (1963); L. I. Zakharkin, V. I. Stanko, V. A. Brattsev, Y. A. Struchkov, *Izw. Akad. Nauk. SSSR, Ser. Khim.* 2069 (1963).
6. M. F. Hawthorne, *Acc. Chem. Res.* **1**, 281 (1968); M. F. Hawthorne, *Pure Appl. Chem.* **29**, 547 (1972); **33**, 475 (1973); K. P. Callahan and M. F. Hawthorne, *Pure Appl. Chem.* **39**, 475 (1974); R. N. Grimes, ed., *Metal Interactions with Boron Clusters*, Plenum Press, New York, 1982.
7. G. L. Leigh, ed., *Nomenclature of Inorganic Chemistry, Recommendations 1990*, IUPAC, Oxford, UK, 1990, Chapt. I-11.
8. J. B. Casey, W. J. Evans, W. H. Powell, and T. E. Sloan, "A Structural Definitive Descriptor and Numbering System for Cluster Compound Nomenclature," Chemical Abstract Service, presented at the *198th National Meeting of the American Chemical Society*, Miami Beach, Fl., Sept. 1989.
9. R. E. Williams, *Inorg. Chem.* **10**, 210 (1971); K. Wade, *J. Chem. Soc. Chem. Commun.* 792 (1971); R. W. Rudolph and W. R. Pretzer, *Inorg. Chem.* **11**, 1974 (1971).
10. R. E. Williams, *Adv. Inorg. Chem. Radiochem.* **18**, 67 (1976).
11. R. W. Rudolph, *Acc. Chem. Res.* **9**, 446 (1976).
12. H. Hogenveen and P. W. Kwant, *Acc. Chem. Res.* **8**, 413 (1975).
13. E. L. Muetterties, T. N. Rhodin, E. Band, C. F. Brucker, and W. R. Pretzer, *Chem. Rev.* **79**, 91 (1979).
14. K. Wade, *Adv. Inorg. Chem. Radiochem.* **18**, 1 (1976).
15. R. E. Williams, in G. A. Olah, R. E. Williams, and K. Wade, eds., *Electron Deficient Compounds of Boron and Carbon*, John Wiley & Sons, Inc., New York, 1991, Chapt. 2.
16. I. Shapiro, C. D. Good, and R. E. Williams, *J. Am. Chem. Soc.* **84**, 3837 (1962).
17. F. Klanberg, D. R. Eaton, L. J. Guggenberger, and E. L. Muetterties, *Inorg. Chem.* **6**, 1271 (1967).
18. R. N. Grimes and W. J. Rademaker, *J. Am. Chem. Soc.* **91**, 6498 (1969).
19. P. M. Garrett, J. C. Smart, G. S. Ditta, and M. F. Hawthorne, *Inorg. Chem.* **8**, 1907 (1969).
20. W. H. Knoth, *J. Am. Chem. Soc.* **89**, 1274 (1967).
21. F. N. Tebbe, P. M. Barret, and M. F. Hawthorne, *J. Am. Chem. Soc.* **86**, 4222 (1964).

22. W. R. Pretzer and R. W. Rudolph, *J. Am. Chem. Soc.* **98**, 1441 (1976).
23. R. L. Voorhees and R. W. Rudolph, *J. Am. Chem. Soc.* **91**, 2173 (1969).
24. L. J. Todd, J. L. Little, and H. T. Silverstein, *Inorg. Chem.* **8**, 1698 (1969).
25. D. A. Franz and R. N. Grimes, *J. Am. Chem. Soc.* **92**, 1438 (1970).
26. R. N. Grimes, C. L. Bramlett, and R. L. Vance, *Inorg. Chem.* **7**, 1066 (1968).
27. P. Binger, *Tetrahedron Lett.*, 2675 (1966).
28. K. Borer, A. B. Littlewood, and C. S. G. Phillips, *J. Inorg. Nucl. Chem.* **15**, 316 (1960).
29. P. M. Garrett, T. A. George, and M. F. Hawthorne, *Inorg. Chem.* **8**, 2008 (1969).
30. W. J. Evans, G. B. Dunks, and M. F. Hawthorne, *J. Am. Chem. Soc.* **95**, 4565 (1973); V. Chowdhry and co-workers, *J. Am. Chem. Soc.* **95**, 4560 (1973).
31. B. Lockman and T. Onak, *J. Am. Chem. Soc.* **94**, 7923 (1972).
32. T. E. Berry, F. N. Tebbe, and M. F. Hawthorne, *Tetrahedron Lett.* **12**, 715 (1965).
33. W. N. Lipscomb and co-workers, *J. Chem. Phys.* **27**, 200 (1957).
34. P. C. Keller, *Inorg. Chem.* **9**, 75 (1970).
35. M. Mangion and co-workers, *J. Am. Chem. Soc.* **98**, 449 (1976); A. V. Fratini, G. W. Sullivan, M. L. Denniston, R. K. Hertz, and S. G. Shore, *J. Am. Chem. Soc.* **96**, 3013 (1974).
36. G. Kodama and A. R. Dodds, paper presented at meeting, *Imeboron III*, Ettal, Germany, July 6, 1976.
37. R. W. Rudolph, R. L. Voorhees, and R. E. Cochoy, *J. Am. Chem. Soc.* **92**, 3351 (1970).
38. R. W. Rudolph and V. Chowdhry, *Inorg. Chem.* **13**, 248 (1974).
39. D. M. P. Mingos, *Nature Phys. Sci.* **236**, 99 (1972).
40. M. Hofmann, M. A. Fox, R. Greatrex, P. v. R. Schleyer, and R. E. Williams, *Inorg. Chem.* **40**, 1790 (2001).
41. L. G. Sneddon, J. C. Huffman, R. O. Schaeffer, and W. E. Streib, *J. Chem. Soc., Chem. Commun.* 474 (1972).
42. A. R. Siedel, G. M. Bodner, and L. J. Todd, *J. Inorg. Nucl. Chem.* **33**, 3671 (1971).
43. V. T. Brice, H. D. Johnson II, and S. G. Shore, *J. Am. Chem. Soc.* **95**, 6629 (1973).
44. G. L. McKnown, B. P. Don, R. A. Beaudet, P. J. Vergamini, and L. H. Jones, *J. Chem. Soc., Chem. Commun.*, 765 (1974).
45. D. A. Franz, V. R. Miller, and R. N. Grimes, *J. Am. Chem. Soc.* **94**, 412 (1972).
46. B. J. Meneghelli and R. W. Rudolph, *Inorg. Chem.* **14**, 1429 (1975).
47. M. G. H. Wallbridge, *Prog. Inorg. Chem.* **11**, 99 (1970).
48. T. J. Marks and J. R. Kolb, *Chem. Rev.* **77**, 263 (1977).
49. P. L. Timms, *J. Am. Chem. Soc.* **90**, 4585 (1968); N. S. Hosmane, in W. Siebert, ed., *Advances in Boron Chemistry*, The Royal Society of Chemistry, Cambridge, U.K. 1997, p. 350.
50. M. F. Hawthorne and co-workers, *J. Am. Chem. Soc.* **90**, 879 (1968); R. M. Wing, *J. Am. Chem. Soc.* **90**, 4828 (1968).
51. L. F. Warren, Jr., and M. F. Hawthorne, *J. Am. Chem. Soc.* **90**, 4823 (1968).
52. C. Glidewell, *J. Organometal. Chem.* **102**, 339 (1975).
53. K. P. Callahan, W. J. Evans, F. Y. Lo, C. E. Strouse, and M. F. Hawthorne, *J. Am. Chem. Soc.* **97**, 296 (1975).
54. L. J. Guggenberger, *J. Am. Chem. Soc.* **94**, 114 (1972).
55. H. E. Ruhle and M. F. Hawthorne, *Inorg. Chem.* **7**, 2279 (1968); D. St. Clair, A. Zalkin, and D. H. Templeton, *Inorg. Chem.* **10**, 2587 (1971).
56. H. M. Colquhoun, T. J. Greenhough, and M. G. H. Wallbridge, *J. Chem. Soc., Chem. Commun.*, 1019 (1976).
57. G. K. Barker and co-workers, *J. Chem. Soc., Chem. Commun.*, 804 (1975).
58. S. G. Shore, in E. L. Muetterties, ed., *Boron Hydride Chemistry*, Academic Press, New York, 1975, Chapt. 3.; E. L. Muetterties and W. H. Knoth, *Polyhedral Boranes*, Marcel Dekker, Inc., New York, 1968.

59. H. C. Longuet-Higgins, *Q. Rev. Chem. Soc.* **11**, 121 (1957).

60. W. H. Eberhardt, B. L. Crawford Jr., and W. N. Lipscomb, *J. Chem. Phys.* **22**, 989 (1954).

61. R. W. Rudolph and D. A. Thompson, *Inorg. Chem.* **13**, 2779 (1974).

62. R. G. Pearson, *J. Am. Chem. Soc.* **91**, 4947 (1969).

63. R. W. Rudolph and W. R. Pretzer, *Inorg. Chem.* **11**, 1974 (1972).

64. W. Kutzelnigg, *Isr. J. Chem.* **19**, 193 (1980); W. Kutzelnigg, *J. Phys. Chem.* **76**, 1919 (1982).

65. J. W. Bausch, G. K. Prakash, and R. E. Williams, paper presented at the 2nd Boron USA Workshop (BUSA II), Research Triangle Park, N. C., June 1990.

66. N. N. Greenwood and R. Greatrex, *Pure Appl. Chem.* **59**, 857 (1987), and references cited therein.

67. C. R. Peters and C. E. Nordman, *J. Am. Chem. Soc.* **82**, 5757 (1960).

68. S. Hermanek and J. Plesek, *Coll. Czech. Chem. Commun.* **31**, 177 (1966); R. E. Enrione and R. Schaeffer, *J. Inorg. Nucl. Chem.* **18**, 103 (1961).

69. M. A. Toft, J. B. Leach, F. L. Himpsl, and S. G. Shore, *Inorg. Chem.* **21**, 1952 (1982); *J. Am. Chem. Soc.* **103**, 988 (1981); S. H. Lawrence and co-workers, *Inorg. Chem.* **25**, 367 (1986); N. S. Hosmane, J. R. Wermer, Z. Hong, T. D. Getman, and S. G. Shore, *Inorg. Chem.* **26**, 3638 (1987), and references cited therein.

70. R. W. Parry and L. J. Edwards, *J. Am. Chem. Soc.* **81**, 3554 (1959).

71. R. J. Pace, J. Williams, and R. L. Williams, *J. Chem. Soc.*, 2196 (1961).

72. G. A. Guter and G. W. Schaeffer, *J. Am. Chem. Soc.* **78**, 3546 (1956); M. F. Hawthorne, A. R. Pitochelli, R. D. Strahm, and J. J. Miller, *J. Am. Chem. Soc.* **82**, 1825 (1960).

73. M. F. Hawthorne and J. J. Miller, *J. Am. Chem. Soc.* **80**, 754 (1958).

74. M. F. Hawthorne, *J. Am. Chem. Soc.* **80**, 3480 (1958).

75. J. J. Miller and M. F. Hawthorne, *J. Am. Chem. Soc.* **81**, 4501 (1959).

76. M. F. Hawthorne, A. R. Pitochelli, R. D. Strahm, and J. J. Miller, *J. Am. Chem. Soc.* **82**, 1825 (1960).

77. B. M. Graybill, A. R. Pitochelli, and M. F. Hawthorne, *Inorg. Chem.* **1**, 622 (1966).

78. J. A. DuPont and M. F. Hawthorne, *Chem. Ind.*, 405 (1962).

79. V. D. Aftandilian, H. C. Miller, G. W. Parshall, and E. L. Muetterties, *Inorg. Chem.* **1**, 734 (1962).

80. H. D. Johnson, II and S. G. Shore, *J. Am. Chem. Soc.* **93**, 3798 (1971).

81. D. Gaines and T. V. Iorns, *J. Am. Chem. Soc.* **92**, 4571 (1970).

82. J. A. DuPont and M. F. Hawthorne, *J. Am. Chem. Soc.* **84**, 1804 (1962).

83. D. F. Gaines, H. Beall *Inorg. Chem.* **39**, 1812 (2000).

84. L. I. Zakharkin and V. N. Kalinin, *J. Gen. Chem. USSR* **36**, 2154 (1966); R. L. Williams, I. Dunstan, and N. J. Blay, *J. Chem. Soc.*, 5006 (1960).

85. D. Gaines and J. A. Martens, *Inorg. Chem.* **7**, 704 (1968).

86. H. D. Johnson, II, V. T. Brice, G. L. Brubaker, and S. G. Shore, *J. Am. Chem. Soc.* **94**, 6711 (1972).

87. M. F. Hawthorne and A. R. Pitochelli, *J. Am. Chem. Soc.* **81**, 5519 (1959); A. R. Pitochelli and M. F. Hawthorne, *J. Am. Chem. Soc.* **82**, 3228 (1960).

88. W. H. Knoth and co-workers, *J. Am. Chem. Soc.* **84**, 1056 (1962).

89. E. L. Muetterties, J. H. Balthis, Y. T. Chia, W. H. Knoth, and H. C. Miller, *Inorg. Chem.* **3**, 444 (1964).

90. F. Klanberg and E. L. Muetterties, *Inorg. Chem.* **5**, 1955 (1966).

91. J. L. Boone, *J. Am. Chem. Soc.* **86**, 5036 (1964).

92. S. Y. Tyree, Jr., ed., *Inorg. Synth.* **9**, 16 (1967).

93. H. C. Miller, M. E. Miller, and E. L. Muetterties, *Inorg. Chem.* **3**, 1456 (1964).

94. E. L. Muetterties, *Inorg. Synth.* **10**, 82 (1967).

95. W. L. Jolly, *Inorg. Synth.* **11**, 27 (1968).

96. J. M. Makhlouf, W. V. Hough, and G. T. Hefferan, *Inorg. Chem.* **6**, 1196 (1967).

97. E. L. Muetterties and W. H. Knoth, Polyhedral Boranes, Marcel Dekker, Inc. New York, 1968, p. 1

98. W. H. Knoth, J. C. Sauer, D. C. English, W. R. Hertler, E. L. Muetterties, *J. Am. Chem. Soc.* **86**, 3973 (1964).

99. E. I. Tolpin, G. R. Wellum, S. A. Berley, *Inorg. Chem.* **24**, 2934, (1976); T. Nakagawa, T. Nagai, *Chem. Pharm. Bull.* **98**, 1515 (1998).

100. E. I. Tolpin, G. R. Wellum, and S. A. Berley, *Inorg. Chem.* **17**, 2867, (1978).

101. E. J. M. Hamilton, G. T. Jordan IV, E. A. Meyers, and S. G. Shore, *Inorg. Chem.* **35**, 5335 (1996); R. G. Kultyshev, J. Meyers, and S. G. Shore, *Inorg. Chem.* **38**, 4913 (1999); R. G. Kultyshev, J. Meyers, and S. G. Shore, *Inorg. Chem.* **39**, 3333 (2000).

102. R. G. Kultyshev, S. Liu, and S. G. Shore, *Inorg. Chem.* **39**, 6094 (2000).

103. T. Peymann, A. Herzog, C. B. Knobler, and M. F. Hawthorne *Angew. Chem., Int. Ed. Engl.* **38**, 1062 (1999).

104. M. Andreas, C. B. Knobler, and M. F. Hawthorne *Angew. Chem., Int. Ed. Engl.* **40**(9), 1662 (2001).

105. T. Peymann, C. B. Knobler, S. I. Khan, and M. F. Hawthorne, *Angew. Chem. Int. Ed. Engl.* **40**, 1664 (2001).

106. T. Peymann, C. B. Knobler, S. I. Khan, and M. F. Hawthorne, *Inorg. Chem.* **40**, 1291 (2001).

107. W. H. Knoth, *J. Am. Chem. Soc.* **88**, 935 (1966).

108. A. Kaczmarczyk, W. C. Nichols, W. H. Stockmayer, and T. B. Eames, *Inorg. Chem.* **7**, 1057 (1968).

109. W. H. Knoth, J. C. Sauer, J. H. Balthis, and H. C. Miller, *J. Am. Chem. Soc.* **89**, 4842 (1967).

110. W. H. Knoth, H. C. Miller, D. C. England, G. W. Parshall, and E. L. Muetterties, *J. Am. Chem. Soc.* **84**, 1056 (1962).

111. W. H. Knoth, J. C. Sauer, J. H. Balthis, Y. T. Chia, and E. L. Muetterties, *Inorg. Chem.* **3**, 159 (1964).

112. M. F. Hawthorne, R. L. Pilling, P. F. Stokely, and P. M. Garrett, *J. Am. Chem. Soc.* **85**, 3704 (1963); M. F. Hawthorne, R. L. Pilling, and P. F. Stokely, *J. Am. Chem. Soc.* **87**, 1893 (1965).

113. R. L. Pilling, M. F. Hawthorne, and E. A. Pier, *J. Am. Chem. Soc.* **86**, 3568 (1964).

114. C. H. Schwalbe and W. N. Lipscomb, *Inorg. Chem.* **10**, 151 (1971).

115. F. Li, K. Shelly, C. B. Knobler, and M. F. Hawthorne, *Angew. Chem. Int. Ed. Engl.* **37**, 1868 (1998).

116. R. J. Wiersema and R. L. Middaugh, *J. Am. Chem. Soc.* **85**, 3704 (1963); R. L. Middaugh and F. Farha, Jr., *J. Am. Chem. Soc.* **88**, 4147 (1966).

117. E. M. Fedneva, V. I. Alpatova, and V. I. Mikheeva, *Russ. J. Inorg. Chem.* **9**, 826 (1964).

118. V. I. Mikheeva and S. M. Arkhipev, *Russ. J. Inorg. Chem.* **11**, 805 (1966).

119. V. I. Mikheeva, M. S. Selivokhina, and G. N. Kryukova, *Russ. J. Inorg. Chem.* **1**, 838 (1962).

120. S. C. Abrahams and J. Kalnajs, *J. Chem. Phys.* **22**, 434 (1954).

121. W. D. Davis, L. S. Mason, and G. Stegman, *J. Am. Chem. Soc.* **71**, 2775 (1949).

122. M. D. Banus, R. W. Bragdon, and A. A. Hinckley, *J. Am. Chem. Soc.* **76**, 3848 (1954).

123. A. P. Altschuler, *J. Am. Chem. Soc.* **77**, 5455 (1955).

124. M. B. Smith and G. E. Bass, *J. Chem. Eng. Data* **8**, 342 (1963).

125. H. I. Schlesinger and co-workers, *J. Am. Chem. Soc.* **75**, 199 (1953).

126. G. S. Panson and C. E. Weill, *J. Inorg. Nucl. Chem.* **15**, 184 (1960).

127. A. Pelter, K. Smith, and H. C. Brown, *Borane Reagents*, Academic Press, London, 1988.

128. J. Plesek, S. Hermanek, J. C. Huffman, P. Ragatz, and R. Schaeffer, *J. Chem. Soc. Commun.* 935 (1975); B. Stibr, J. Plesek, and S. Hermanek, in J. F. Liebman, A. Greenberg, and R. E. Williams, eds., *Molecular Structure and Energetics, Vol. 5, Advances in Boron and Boranes*, VCH Publishers, New York, p. 35 (1988). P. Paetzold, J. Muller, F. Meyer, H.-P. Hansen, and L. Schneider, p. 337; J. Holub, T. Jelinek, B. Stibr, J. D. Kennedy, and M. Thornton-Pett, p. 359, A. Ouassas, B. Fenet, H. Mongeot, and B. Frange, in G. W. Kalbalka, ed., *Current Topics in The Chemistry of Boron*, The Royal Society of Chemistry, Cambridge, 1994, p. 363; A. E. Willie, K. Su, P. J. Carroll, and L. G. Sneddon.

129. R. N. Grimes, *Carboranes*, Academic Press, New York, 1970.

130. H. L. Gingrich, T. Ghosh, Q. Huang, and M. Jones, Jr., *J. Am. Chem. Soc.* **112**, 4082 (1990).

131. M. F. Hawthorne and R. A. Wiesboeck, *J. Am. Chem. Soc.* **86**, 1642 (1964); M. F. Hawthorne and co-workers, *J. Am. Chem. Soc.* **90**, 862 (1968).

132. J. Buchanan, E. J. M. Hamilton, D. Reed, and A. J. Welch, *J. Chem. Soc. Dalton Trans.*, 677 (1990).

133. M. G. Davidson, M. A. Fox, T. G. Hibbert, J. A. K. Howard, A. MacKinnon, I. S. Neretin, K. Wade *Chem. Commun.*, 1649 (1999).

134. M. A. Fox, A. E. Goeta, J. A. K. Howard, A. K. Hughes, A. L. Johnson, D. A. Keen, K. Wade, C. C. Wilson, *Inorg. Chem.* **40**, 173 (2001).

135. J. Plesek and S. Hermanek, *Collect. Czech. Chem. Commun.* **38**, 338 (1973).

136. G. B. Dunks and M. F. Hawthorne, *Acc. Chem. Res.* **6**, 124 (1973).

137. J. F. Ditter, E. B. Klusmann, J. D. Oates, and R. E. Williams, *Inorg. Chem.* **9**, 889 (1970).

138. T. Onak, F. J. Gerhart, and R. E. Williams, *J. Am. Chem. Soc.* **85**, 3378 (1963).

139. U. Siriwardane, M. S. Uslam, T. A. West, N. S. Hosmane, J. A. Maguire, A. H. Cowley, *J. Am. Chem. Soc.* **109**, 4600 (1987).

140. B. Stibr, J. Plesek, and S. Hermanek, *Coll. Czech. Chem. Commun.* **39**, 1805 (1974).

141. T. Onak, R. P. Drake, and G. B. Dunks, *Inorg. Chem.* **3**, 1686 (1964).

142. W. H. Knoth, *Inorg. Chem.* **10**, 598 (1971).

143. D. E. Hyatt, F. R. Scholer, L. J. Todd, and J. L. Warner, *Inorg. Chem.* **6**, 2229 (1967).

144. R. J. Wiersema and M. F. Hawthorne, *Inorg. Chem.* **12**, 785 (1973).

145. G. B. Dunks and M. F. Hawthorne, *Inorg. Chem.* **8**, 2667 (1969).

146. E. Groszek, J. R. Leach, G. T. F. Wong, C. Ungermann, and T. Onak, *Inorg. Chem.* **10**, 2770 (1971).

147. B. Stibr, J. Holub, and F. Teixidor, in W. Siebert, ed., *Advances in Boron Chemistry*, The Royal Society of Chemistry, Cambridge, 1997 p. 331.

148. K. Su, B. Barnum, P. J. Carroll, and L. G. Sneddon, *J. Am. Chem. Soc.* **114**, 2730 (1992); K. Su, P. J. Carroll, and L. G. Sneddon, *J. Am. Chem. Soc.*, **115**, 10004 (1993).

149. W. M. Maxwell, V. R. Miller, and R. N. Grimes, *J. Am. Chem. Soc.* **96**, 7116 (1974); K. K. Fonda and R. N. Grimes, *Polyhedron* **9**, 949 (1990), and references cited therein.

150. J. P. Pasinski and R. A. Beaudet, *J. Chem. Phys.* **61**, 683 (1974); D. P. Freyberg, R. Weiss, E. Sinn, and R. N. Grimes, *Inorg. Chem.* **16**, 1847 (1977).

151. B. Stibr, T. Jelinek, E. Drdakova, S. Hermanek, and J. Plesek, *Polyhedron* **7**, 669 (1988); K. Su, P. J. Carroll, and L. G. Sneddon, *J. Am. Chem. Soc.* **115**, 10004 (1993).

152. S. Wu and M. Jones Jr., *J. Am. Chem. Soc.* **111**, 5373 (1989), and references cited therein; B. M. Gimarc and J. J. Ott, *Main Group Met. Chem.* **12**, 77 (1989).

153. R. J. Wiersema and M. F. Hawthorne, *Inorg. Chem.* **12**, 785 (1973).

154. M. L. Thompson and R. N. Grimes, *J. Am. Chem. Soc.* **93**, 6677 (1971).

155. M. F. Hawthorne, in W. Siebert, ed., *Advances in Boron Chemistry*, The Royal Society of Chemistry, Cambridge, U.K. 1997, p. 261.

156. P. Kaszynski, D. Lipiak, K. A. Bairamov, E. Brady, M. K. Patel, and J. Laska, in W. Siebert, ed., *Advances in Boron Chemistry*, The Royal Society of Chemistry, Cambridge, U.K. 1997, p. 508.

157. X. Yang, W. Jiang, C. B. Knobler, and M. F. Hawthorne, *J. Am. Chem. Soc.* **114**, 9719 (1992); J. Muller, K. Base, T. F. Magnera, and J. Michl, *J. Am. Chem. Soc.* **114**, 9721 (1992).

158. S. Kongpricha, and H. Schroeder *Inorg. Chem.* **8**, 2449 (1969).

159. R. J. Lagow, and J. L. Margave, *J. Inorg. Nucl. Chem.* **35**, 2084 (1973).

160. S. V. Ivanov, J. J. Rockwell, O. G. Polyakov, C. M. Gaudinski, O. P. Anderson, K. A. Solntsev, and S. H. Strauss, *Angew. J. Am. Chem. Soc.* **120**, 4224 (1998).

161. W. Jiang, D. E. Harwell, M. D. Mortimer, C. B. Knobler, and M. F. Hawthorne, *Inorg, Chem.* **25**, 4355 (1996).

162. B. T. King, Z. Janousek, B. Gruner, M. Trammell, B. C. Noll, J. Michl, *J. Am. Chem. Soc.* **118**, 3313 (1996); I. Zharov, B. T. King, Z. Havlas, A. Pardi, and J. Michl, *J. Am. Chem. Soc.* **122**, 10253 (2000).

163. W. Jiang, C. B. Knobler, M. D. Mortimer, and M. F. Hawthorne, *Angew. Chem.* **107**, 1470 (1995).

164. B. T. King, B. C. Noll, A. J. McKinley, and J. Michl, *J. Am. Chem. Soc.* **118**, 10902 (1996).

165. C.-W. Tsang, Q. Yang, E. T.-P. Sze, T. C. W. Mak, D. T. W. Chan, and Z. Xie, *Inorg. Chem.* **39**, 3582 (2000); C.-W. Tsang, Q. Yang, E. T.-P. Sze, T. C. W. Mak, D. T. W. Chan, and Z. Xie, *Inorg. Chem.* **39**, 5851 (2000).

166. W. Jiang, C. B. Knobler, and M. F. Hawthorne, *Angew. Chem. Int. Ed. Engl.* **35**, 2536 (1996); W. Jiang, C. B. Knobler, and M. F. Hawthorne, *Angew. Chem. Int. Ed. Engl.* **108**, 2653 (1996).

167. B. T. King, J. Michl, *J. Am. Chem. Soc.* **122**, 10255 (2000).

168. A. Herzog, R. P. Callahan, C. L. B. Macdonald, V. M. Lynch, M. F. Hawthorne, and R. J. Lagow, *Angew. Chem. Int. Ed. Engl.* **40**, 2121 (2001).

169. C. A. Reed, *Acc. Chem. Res.* **31**, 133 (1998).

170. S. M. Ivanova, S. V. Ivanova, S. M. Miller, O. P. Andersen, K. A. Solntsev, S. H. Strauss, *Inorg. Chem.* **38**, 3756 (1999).

171. A. J. Lupinetti, M. D. Havighurst, S. M. Miller, O. P. Andersen, S. H. Strauss *J. Am. Chem. Soc.* **121**, 11920 (1999).

172. C. A. Reed, Z. Bau, R. Bau, A. Bernesi, *Science*, **262**, 402 (1993).

173. Z. Xie, R. Bau, C. A. Reed *Angew. Chem., Int. Ed. Engl.* **33**, 2433 (1994).

174. Z. Xie, R. Bau, C. A. Reed *Inorg. Chem.* **34**, 5403 (1995).

175. C. A. Reed, N. L. P. Fackler, K.-C. Kim, D. Stasko, D. R. Evans *J. Am. Chem. Soc.* **121**, 8466 (1999)

176. Eur. Pat. Appl. 88,300,698, 88,300,699 (1998), H. W. Turner.

177. C. A. Reed, K-C Kim, R. D. Bolskar, and L. J. Mueller, *Science*, **289**, 101 (2000).

178. S. Papetti, B. B. Schaeffer, A. P. Gray, and T. L. Heying, *J. Polymer. Sci.* **4** (A-1), 1623 (1966); U. S. Pats. 3,388,090 (1968); 3,388,091 (1968); 3,388,092 (1968), T. L. Heying, S. Papetti, and O. G. Shaffling.

179. E. Hedaya and co-workers, *J. Polym. Sci., Polym. Chem. Ed.* **15**, 2229 (1977), and references cited therein.

180. S. Bresadola, in R. N. Grimes, ed., *Metal Interactions with Boron Clusters*, Plenum Press, New York, 1982.

181. K. Base and co-workers, *J. Chem. Soc., Chem. Commun.*, 934 (1975).

182. J. L. Little and A. C. Wong, *J. Am. Chem. Soc.* **93**, 522 (1971).

183. J. L. Little, J. T. Moran, and L. J. Todd, *J. Am. Chem. Soc.* **89**, 5495 (1967).

184. J. L. Little, S. S. Pao, and K. K. Sugathan, *Inorg. Chem.* **13**, 1752 (1974).
185. J. L. Little, *Inorg. Chem.* **18**, 1598 (1979).
186. L. J. Todd, A. R. Burke, H. T. Silverstein, J. L. Little, and G. S. Wilkholm, *J. Am. Chem. Soc.* **91**, 3376 (1969).
187. W. R. Hertler, F. Klanberg, and E. L. Muetterties, *Inorg. Chem.* **6**, 1696 (1967).
188. W. R. Pretzer and R. W. Rudolph, *J. Am. Chem. Soc.* **95**, 931 (1973).
189. J. Plesek, S. Hermanek, and Z. Janousek, *Coll. Czech. Chem. Commun.* **42**, 785 (1977).
190. J. L. Little, G. D. Friesen, and L. J. Todd, *Inorg. Chem.* **16**, 869 (1977).
191. H. D. Kaesz, W. Fellmann, G. R. Wilkes, and L. F. Dahl, *J. Am. Chem. Soc.* **87**, 2753 (1965).
192. H. D. Kaesz and R. B. Saillant, *Chem. Rev.* **72**, 344 (1972).
193. H. Nöth, M. Thomann, M. Bremer, and G. Wagner, in G. W. Kabalka, ed., *Advances in Boron Chemistry*, The Royal Society of Chemistry, Cambridge, U.K. 1994, p. 387.
194. P. Wegner, in E. L. Muetterties, ed., *Boron Hydride Chemistry*, Academic Press, New York, 1975, Chapt. 12.
195. J. T. Gill and S. Lippard, *Inorg. Chem.* **14**, 751 (1975).
196. N. N. Greenwood and N. F. Travers, *J. Chem. Soc. A*, 3257 (1971).
197. F. Klanberg, P. A. Wegner, G. W. Parshall, and E. L. Muetterties, *Inorg. Chem.* **7**, 2072 (1968).
198. L. J. Guggenberger, *J. Am. Chem. Soc.* **94**, 114 (1972).
199. N. N. Greenwood, J. A. McGinnety, and J. D. Owen, *J. Chem. Soc. A*, 809 (1971).
200. B. M. Graybill, A. R. Pitochelli, and M. F. Hawthorne, *Inorg. Chem.* **1**, 622 (1962).
201. N. N. Greenwood and D. N. Sharrocks, *J. Chem. Soc. A*, 2334 (1969).
202. B. P. Sullivan, R. N. Leyden, and M. F. Hawthorne, *J. Am. Chem. Soc.* **97**, 455 (1975).
203. R. N. Leyden, B. P. Sullivan, R. T. Baker, and M. F. Hawthorne, *J. Am. Chem. Soc.* **100**, 3758 (1978); R. N. Leyden and M. F. Hawthorne, *J. Chem. Soc., Chem. Commun.*, 310 (1975); C. G. Salentine, R. R. Rietz, and M. F. Hawthorne, *Inorg. Chem.* **12**, 3025 (1975).
204. D. M. Schubert, C. B. Knobler, P. A. Wegner, and M. F. Hawthorne, *J. Am. Chem. Soc.* **110**, 5219 (1988).
205. V. R. Miller and R. N. Grimes, *J. Am. Chem. Soc.* **95**, 5078 (1973).
206. *Ibid.* **98** 1600 (1976).
207. D. F. Gaines and J. L. Walsh, *Inorg. Chem.* **17**, 1238 (1978).
208. S. Ghosh, M. Shang, Y. Li, and T. P. Fehlner, *Angew. Chem. Int. Ed. Engl.* **40**, 1125 (2001).
209. D. L. Denton, W. R. Clayton, M. Mangion, S. G. Shore, and E. A. Meyers, *Inorg. Chem.* **15**, 541 (1976).
210. N. N. Greenwood and N. F. Travers, *J. Chem. Soc. A*, 15 (1968).
211. N. N. Greenwood, B. S. Thomas, and D. W. Waite, *J. Chem. Soc., Dalton Trans.*, 299 (1975).
212. N. N. Greenwood and J. A. McGinnety, *J. Chem. Soc. A*, 1090 (1966).
213. T. D. Getman and S. G. Shore, *Inorg. Chem.* **27**, 3439 (1988).
214. R. E. Loffredo and A. D. Norman, *Inorg. Nucl. Chem. Lett.* **13**, 599 (1977).
215. R. E. Loffredo and A. D. Norman, *J. Am. Chem. Soc.* **93**, 5587 (1971).
216. D. F. Gaines, J. Morris, D. Hillenbrand, and J. L. Walsh, *Inorg. Chem.* **17**, 1516 (1978).
217. J. Borlin and D. F. Gaines, *J. Am. Chem. Soc.* **94**, 1367 (1972).
218. S. Hermanek and J. Plesek, *Coll. Czech. Chem. Commun.* **31**, 177 (1966).
219. W. V. Hough, L. J. Edwards, and A. J. McElroy, *J. Am. Chem. Soc.* **80**, 1828 (1958).
220. J. C. Calabrese, D. F. Gaines, S. J. Hildebrandt, and J. H. Morris, *J. Am. Chem. Soc.* **98**, 5489 (1976).

221. D. F. Gaines and J. L. Walsh, *Chem. Commun.*, 482 (1976); D. F. Gaines, J. L. Walsh, and J. C. Calabrese, *Inorg. Chem.* **17**, 1242 (1978).

222. J. Borlin and D. F. Gaines, *J. Am. Chem. Soc.* **94**, 1367 (1972).

223. D. Seyferth, K. Büchner, W. S. Rees, Jr., and W. M. Davis, *Angew. Chem. Int. Ed. Engl.* **29**, 918 (1990).

224. T. E. Paxson, M. F. Hawthorne, L. D. Brown, and W. N. Lipscomb, *Inorg. Chem.* **13**, 2772 (1974).

225. J. P. White, III, H.-B. Deng, and S. G. Shore, *J. Am. Chem. Soc.* **111**, 8946 (1989).

226. J. P. White, III, and S. G. Shore, paper presented at the the *2nd Boron USA Workshop (BUSA II)b*, Research Triangle Park, N. C., June 1990.

227. D. F. Gaines and T. V. Iorns, *Inorg. Chem.* **7**, 1041 (1968).

228. M. R. Churchill, J. J. Hackbarth, A. Davison, D. D. Traficante, and S. S. Wreford, *J. Am. Chem. Soc.* **96**, 4041 (1974).

229. D. F. Gaines and T. V. Iorns, *J. Am. Chem. Soc.* **90**, 6617 (1968).

230. J. C. Calabrese and L. F. Dahl, *J. Am. Chem. Soc.* **93**, 6042 (1971).

231. M. F. Hawthorne, D. C. Young, and P. A. Wegner, *J. Am. Chem. Soc.* **87**, 1818 (1965); M. F. Hawthorne, D. C. Young, and co-workers, *J. Am. Chem. Soc.* **90**, 879 (1968).

232. A. K. Saxena and N. S. Hosmane, *Chem. Rev.* **93**, 1081 (1993).

233. M. F. Hawthorne and T. D. Andrews, *J. Chem. Soc., Chem. Commun.*, 443 (1965); R. J. Wilson, L. F. Warren, Jr., and M. F. Hawthorne, *J. Am. Chem. Soc.* **91**, 758 (1969); L. F. Warren, Jr., and M. F. Hawthorne, *J. Am. Chem. Soc.* **89**, 470 (1967).

234. M. F. Hawthorne and R. L. Pilling, *J. Am. Chem. Soc.* **87**, 3987 (1965).

235. L. F. Warren, Jr., and M. F. Hawthorne, *J. Am. Chem. Soc.* **90**, 4823 (1968).

236. R. J. Wilson, L. F. Warren, Jr., and M. F. Hawthorne, *J. Am. Chem. Soc.* **91**, 758 (1969).

237. J. Plesek and co-workers, *Collec. Czech. Chem. Commun.* **49**, 2776 (1984).

238. M. F. Hawthorne and T. D. Andrews, *J. Am. Chem. Soc.* **87**, 2496 (1965); M. F. Hawthorne and H. W. Ruhle, *Inorg. Chem.* **8**, 176 (1969).

239. G. B. Dunks and M. F. Hawthorne, *J. Am. Chem. Soc.* **92**, 7213 (1970).

240. W. J. Evans and M. F. Hawthorne, *J. Am. Chem. Soc.* **93**, 3063 (1971).

241. D. Dustin, G. B. Dunks, and M. F. Hawthorne, *J. Am. Chem. Soc.* **95**, 1109 (1973).

242. W. J. Evans, G. B. Dunks, and M. F. Hawthorne, *J. Am. Chem. Soc.* **95**, 4565 (1973).

243. W. J. Evans and M. F. Hawthorne, *Inorg. Chem.* **13**, 869 (1974).

244. W. J. Evans and M. F. Hawthorne, *J. Chem. Soc., Chem. Commun.*, 38 (1974).

245. C. J. Jones, J. N. Francis, and M. F. Hawthorne, *J. Am. Chem. Soc.* **94**, 8391 (1972); C. J. Jones, J. N. Francis, and M. F. Hawthorne, *J. Chem. Soc., Chem. Commun.*, 900 (1972); J. N. Francis and M. F. Hawthorne, *Inorg. Chem.* **10**, 863 (1971); D. F. Dustin, W. J. Evans, and M. F. Hawthorne, *J. Chem. Soc., Chem. Commun.*, 805 (1973).

246. D. F. Dustin and M. F. Hawthorne, *J. Am. Chem. Soc.* **96**, 3462 (1974).

247. M. R. Churchill, A. H. Reis, Jr., J. N. Francis, and M. F. Hawthorne, *J. Am. Chem. Soc.* **92**, 4993 (1970).

248. W. M. Maxwell, R. Weiss, E. Sinn, and R. N. Grimes, *J. Am. Chem. Soc.* **99**, 4016 (1977).

249. M. K. Kaloustian, R. J. Wiersema, and M. F. Hawthorne, *J. Am. Chem. Soc.* **94**, 6679 (1972).

250. D. F. Dustin, C. J. Jones, and R. J. Wiersema, *J. Am. Chem. Soc.* **96**, 3085 (1974).

251. W. J. Evans, C. J. Jones, B. Stibr, R. A. Grey, and M. F. Hawthorne, *J. Am. Chem. Soc.* **96**, 7405 (1974); C. G. Salentine and M. F. Hawthorne, *J. Am. Chem. Soc.* **97**, 6382 (1975).

252. R. J. Wilson, L. F. Warren, Jr., M. F. Hawthorne, *J. Am. Chem. Soc.* **92**, 1157 (1970).

253. D. M. Schubert, D. E. Harwell, C. B. Knobler, and M. F. Hawthorne, *Acta Chem. Scand.* **53**, 721 (1999).

254. H. C. Kang, Y. Do, C. B. Knobler, and M. F. Hawthorne, *J. Am. Chem. Soc.* **109**, 6530 (1988); H. C. Kang, Y. Do, C. B. Knobler, and M. F. Hawthorne, *Inorg. Chem.* **27**, 1716 (1988).

255. C. G. Salentine and M. F. Hawthorne, *Inorg. Chem.* **15**, 2872 (1976).

256. C. G. Salentine and M. F. Hawthorne, *J. Chem. Soc., Chem. Commun.*, 848 (1975).

257. W. M. Maxwell, V. R. Miller, and R. N. Grimes, *J. Am. Chem. Soc.* **98**, 4818 (1976).

258. M. F. Hawthorne and A. D. Pitts, *J. Am. Chem. Soc.* **89**, 7115 (1967); A. D. George and M. F. Hawthorne, *Inorg. Chem.* **8**, 1801 (1969); F. J. Hollander, D. H. Templeton, and A. Zalkin, *Inorg. Chem.* **12**, 2262 (1973).

259. R. N. Grimes, D. C. Beer, L. G. Sneddon, V. R. Miller, and R. Weiss, *Inorg. Chem.* **13**, 1138 (1974).

260. A. J. Welch, *J. Chem. Soc. Dalton Trans.*, 225 (1976).

261. R. N. Grimes, *J. Am. Chem. Soc.* **93**, 261 (1971); R. N. Grimes, *Coord. Chem. Rev.* **28**, 47 (1979); A. Fassenbecker, M. D. Attwood, R. N. Grimes, M. Stephan, H. Pritzkow, U. Zenneck, and W. Siebert, *Inorg. Chem.* **29**, 5164 (1990), and references cited therein.

262. R. N. Grimes, *Chem. Rev.* **92**, 251 (1992).

263. E. L. Hoel and M. F. Hawthorne, *J. Am. Chem. Soc.* **96**, 6770 (1974).

264. *Ibid.* **97**, 6388 (1975).

265. *Ibid.* **95**, 2712 (1973).

266. H. C. Kang, Y. Do, C. B. Knobler, and M. F. Hawthorne, *Inorg. Chem.* **27**, 1716 (1988).

267. S. J. Dossett, I. J. Hart, and F. G. A. Stone, *J. Chem. Soc., Dalton Trans.*, 3481 (1990); S. J. Dossett, I. J. Hart, M. U. Pilotti, and F. G. A. Stone, *J. Chem. Soc., Dalton Trans.*, 3489 (1990), and references cited therein.

268. V. I. Bregadze, V. Ts. Kampel, and N. N. Godovikov, *J. Organometal. Chem.* **136**, 281 (1976); V. I. Bregadze, V. Ts. Kampel, A. Ya. Usiatinsky, and N. N. Godovikov, *Pure Appl Chem.* **63**, 835 (1991); V. I. Bregadze, *Chem. Rev.* **92**, 209, (1992).

269. V. I. Bregadze, A. Ya. Usiatinsky, O. B. Zhidlova, O. M. Khitrova, P. V. Petrovskii, F. M. Dolgushin, A. I. Yanovsky, and Yu. T. Struchkov, in W. Siebert, ed., *Advances in Boron Chemistry*, The Royal Society of Chemistry, Cambridge, U.K., 1997, p. 341.

270. W. Clegg, H. M. Colquhoun, R. Coult, M. A. Fox, W. R. Gill, P. L. Herbertson, J. A. H. MacBride, and K. Wade, in G. W. Kabalka, ed., *Advances in Boron Chemistry*, The Royal Society of Chemistry, Cambridge, U.K., 1994, p. 232.

271. A. Muller, H. Reuter, S. Dillinger, *Angew. Chem. Int. Ed. Engl.* **34**, 2328 (1995).

272. M. F. Hawthorne, Z. Zheng, *Acc. Chem. Res.* **30**, 267 (1997).

273. T. E. Paxson and M. F. Hawthorne, *J. Am. Chem. Soc.* **96**, 4674 (1974); D. C. Busby and M. F. Hawthorne, *Inorg. Chem.* **21**, 4101 (1982).

274. R. T. Baker, M. S. Delaney, R. E. King III, C. B. Knobler, J. A. Long, T. B. Marder, T. E. Paxson, R. G. Teller, and M. F. Hawthorne, *J. Am. Chem. Soc.* **106**, 2965 (1984).

275. P. M. Garrett, G. S. Ditta, and M. F. Hawthorne, *Inorg. Chem.* **9**, 1947 (1970); P. M. Garrett, G. S. Ditta, and M. F. Hawthorne, *J. Am. Chem. Soc.* **93**, 1265 (1971).

276. J. A. Long, T. B. Marder, P. E. Behnken, and M. F. Hawthorne, *J. Am. Chem. Soc.* **106**, 2979 (1984); C. B. Knobler and co-workers, *J. Am. Chem. Soc.* **106**, 2990 (1984).

277. J. A. Long, T. B. Marder, P. E. Behnken, and M. F. Hawthorne, *J. Am. Chem. Soc.* **106**, 2979 (1984).

278. P. E. Behnken and co-workers, *J. Am. Chem. Soc.* **106**, 3011 (1984); P. E. Behnken and co-workers, *J. Am. Chem. Soc.* **106**, 7444 (1984).

279. J. A. Long, T. B. Marder, and M. F. Hawthorne, *J. Am. Chem. Soc.* **106**, 3004 (1984).

280. B. A. Sosinsky, W. C. Kalb, R. A. Grey, V. A. Uski, and M. F. Hawthorne, *J. Am. Chem. Soc.* **99**, 6768 (1977).

281. J. D. Hewes, C. B. Knobler, and M. F. Hawthorne, *J. Chem. Soc., Chem. Commun.,* 206 (1981).

282. P. E. Behnken and co-workers, *J. Am. Chem. Soc.* **107**, 932 (1985).

283. *Res. Discl.* **292**, 588 (1988).

284. N. S. Hosmane and J. A. Maguire, *Adv. Organometal. Chem.* **30**, 99 (1990); A. K. Saxena, J. A. Maguire, and N. S. Hosmane, *Chem. Rev.* **97**, 2421 (1997).

285. T. Onak and G. B. Dunks, *Inorg. Chem.* **5**, 439 (1966).

286. N. S. Hosmane, U. Siriwardane, G. Zhang, H. Zhu, and J. A. Maguire, *J. Chem. Soc. Chem. Commun.* 1128 (1989).

287. N. S. Hosmane, L. Jia, Y. Wang, A. K. Saxena, H. Zhang, and J. A. Maguire, *Organometallics* **13**, 4113 (1994).

288. N. S. Hosmane, A. K. Saxena, R. D. Barreto, H. Zhang, J. A. Maguire, L. Jia, Y. Wang, A. R. Oki, K. V. Grover, S. J. Whitten, K. Dawson, M. A. Tolle, U. Siridawardane, U. Demissie, and J. S. Fagner, *Organometallics* **12**, 3001 (1993); H. Zhang, Y. Wang, A. K. Saxena, A. R. Oki, J. A. Maguire, and N. S. Hosmane, *Organometallics* **12**, 3933 (1993).

289. Y. Wang, H. Zhang, J. A. Maguire, and N. S. Hosmane, *Organometallics* **12**, 3781 (1993).

290. N. S. Hosmane, J. Yang, H. Zhang, and J. A. Maguire, *J. Am. Chem. Soc.* **118**, 5150 (1996).

291. R. Khattar, C. B. Knobler, and M. F. Hawthorne, *J. Am. Chem. Soc.* **112**, 4962 (1990).

292. R. Khatter, C. B. Knobler, and M. F. Hawthorne, *Inorg. Chem.* **29**, 2191 (1990).

293. G. Popp and M. F. Hawthorne, *Inorg. Chem.* **10**, 391 (1971).

294. M. Westerhausen, C. Gückel, S. Schneiderbauer, H. Nöth, and N. S. Hosmane, *Angew. Chemie Int. Ed. Engl.* **40**, 1902 (2001).

295. B. M. Mikhailov and T. V. Potapova, *Izv. Akad. Nauk. SSSR, Ser. Khim.* **5**, 1153 (1968).

296. D. A. T. Young, R. J. Wiersema, and M. F. Hawthorne, *J. Am. Chem. Soc.* **93**, 5687 (1971).

297. M. R. Churchill, A. H. Reis, Jr., D. A. T. Young, G. R. Willey, and M. F. Hawthorne, *J. Chem. Soc., Chem. Commun.*, 298 (1971); M. Churchill and A. H. Reis, *J. Chem. Soc., Dalton Trans.*, 1317 (1972).

298. D. M. Schubert, M. A. Bandman, W. S. Rees, Jr., C. B. Knobler, P. Lu, W. Nam, and M. F. Hawthorne, *Organometallics* **9**, 2046 (1990).

299. M. A. Bandman, C. B. Knobler, and M. F. Hawthorne, *Inorg. Chem.* **27**, 2399 (1988).

300. *Ibid.* **28**, 1204 (1989).

301. C. P. Magee, L. G. Sneddon, D. C. Beer, and R. N. Grimes, *J. Organomet. Chem.* **86**, 159 (1975).

302. J. S. Beck, L. G. Sneddon, *J. Am. Chem. Soc.* **110**, 3467 (1988).

303. N. S. Hosmane, A. K. Saxena, K.-J. Lu, J. A. Maguire, H. Zhang, Y. Wang, C. J. Thomas, D. Zhu, B. R. Grover, T. G. Gray, J. F. Eintracht, H. Isom, and A. H. Cowley, *Organometallics* **14**, 5104 (1995).

304. A. K. Saxena, H. Zhang, J. A. Maguire, N. S. Hosmane, and A. H. Cowley, *Angew. Chem. Int. Ed. Engl.* **34**, 332 (1995).

305. J. L. Spencer, M. Green, and F. G. A. Stone, *J. Chem. Soc., Chem. Commun.*, 1178 (1972).

306. M. J. Manning, C. B. Knobler, M. F. Hawthorne, and Y. Do, *Inorg. Chem.* **30**, 3589 (1991).

307. R. L. Voorhees and R. W. Rudolph, *J. Am. Chem. Soc.* **91**, 2173 (1969); R. W. Rudolph, R. L. Voorhees, and R. E. Cochoy, *J. Am. Chem. Soc.* **92**, 3351 (1970);

V. Chowdhry, W. R. Pretzer, D. N. Rai, and R. W. Rudolph, *J. Am. Chem. Soc.* **95**, 4560 (1973).

308. L. J. Todd, A. R. Burke, H. T. Silverstein, J. L. Little, and G. S. Wilkholm, *J. Organomet. Chem.* **91**, 3376 (1969).

309. D. M. Schubert, W. S. Rees, Jr., C. B. Knobler, and M. F. Hawthorne, *Organometallics* **9**, 2938 (1990).

310. U. Siriwardane and co-workers, *J. Am. Chem. Soc.* **109**, 4600 (1987); N. S. Hosmane, P. de Meester, U. Siriwardane, M. S. Islam, and S. S. C. Chu, *J. Chem. Soc., Chem. Commun.*, 1421 (1986).

311. K.-S. Wong and R. N. Grimes, *Inorg. Chem.* **16**, 2053 (1977).

312. N. S. Hosmane, N. N. Sirmokadam, and R. H. Herber, *Organometallics* **3**, 1665 (1984); N. S. Hosmane and co-workers, *Organometallics* **5**, 772 (1986); A. H. Cowley, P. Galow, N. S. Hosmane, P. Jutzi, and N. C. Norman, *J. Chem. Soc., Chem. Commun.*, 1564 (1984).

313. A. K. Saxena, H. Zhang, J. A. McGuire, N. S. Hosmane, and A. H. Cowley, *Angew, Chem. Int. Ed. Engl.* **34**, 332–334 (1995).

314. L. Jia, H. Zhang, N. S. Hosmane, *Organometallic* **11**, 2957 (1992).

315. J. L. Atwood, W. E. Hunter, A. H. Cowley, R. A. Jones, C. A. Stewart, *J. Chem. Soc., Chem. Commun.* 925 (1981); P. Jutzi, *Adv. Organomet. Chem.* **26**, 217 (1986).

316. N. S. Hosmane and co-workers, *Organometallics* **7**, 1893 (1988); N. S. Hosmane, K. J. Lu, U. Siriwardane, and M. S. Shet, *Organometallics* **9**, 2798 (1990).

317. D. C. Beer and L. J. Todd, *Organometal. Chem.* **50**, 93 (1973).

318. F. R. Fronczek, G. W. Halstead, and K. N. Raymond, *J. Am. Chem. Soc.* **99**, 1769 (1977).

319. M. J. Manning, C. B. Knobler, and M. F. Hawthorne, *J. Am. Chem. Soc.* **110**, 4458 (1988).

320. D. A. Feakes, in M. F. Hawthorne, and co–workers, eds, *Frontiers in Neutron Capture Therapy*, Vol. 1, Kluwer Academic/Plenum Publishers, New York, 1998, Chapter 3, p. 23.

321. G. L. Locker, *Am. J. Roentgenol.* **36**, 1 (1936); R. F. Barth, A. H. Soloway, and R. G. Fairchild, *Sci. Am.*, Oct. (1990).

322. F. Alam, R. F. Barth, and A. H. Soloway, *Antibodies, Immunoconjugates, Radiopharmaceut.* **2**, 145 (1989); R. F. Barth, A. H. Soloway, and R. G. Fairchild, *Cancer Res.* **50**, 1061 (1990).

323. Y. Endo, et al. *Biol. Pharm. Bull.* **23**, 513 (2000).

324. Y. Endo, et al. *Chem. Pharm. Bull.* **48**, 314 (2000); Y. Endo and co–workers, *Biomed. Chem. Lett.* **9**, 3701 (1999); *Biomed. Chem. Lett.* **9**, 3313 (1999).

325. Y. Endo, and co-workers *Biomed. Chem. Lett.* **10**, 1733 (2000); Y. Endo, et al. *Chem. Pharm. Bull.*, **47**, 585 (1999).

326. M. F. Hawthorne, A. Varadarajan, C. B. Knobler, J. D. Beatty, and S. Chakrabarti, *J. Am. Chem. Soc.* **112**, 5365 (1990).

327. M. F. Hawthorne and A. Maderna, *Chem. Rev.* **99**, 3421 (1999).

328. U.S. Pat. 3,711,180 (1973), T. J. Klingen and J. R. Wright (to the University of Mississippi).

329. U.S. Pat. 3,539,330 (1970), D. C. Young (to Union Oil Co.).

DAVID M. SCHUBERT
U.S. Borax Inc.

BORON OXIDES, BORIC ACID, AND BORATES

1. Introduction

Boron, the fifth element in the periodic table, does not occur in nature in its elemental form. Rather, boron combines with oxygen as a salt or ester of boric acid. There are more than 200 minerals that contain boric oxide but relatively few that are of commercial significance. In fact, three minerals represent almost 90% of the borates used by industry: borax, a sodium borate; ulexite, a sodium–calcium borate; and colemanite, a calcium borate. These minerals are extracted in California and Turkey, and to a lesser extent in Argentina, Bolivia, Chile, Peru, and China. China and Russia also have some commercial production from magnesium borates and calcium borosilicates. These deposits furnish nearly all of the world's borate supply at this time (1).

According to legend, borates were used by Egyptians in mummification and by the ancient Romans in glassmaking. But the first historically verifiable use of borates was by Arabian gold- and silversmiths who used the minerals as refining and soldering agents in the eighth century A.D. The earliest confirmed use of borates in ceramic glazes was in China in the tenth century A.D. Three hundred years later, regular imports of borates from Tibet to Venice began along trade routes taken by Macro Polo's caravans.

The discovery of significant deposits of borates in both North and South America in the mid-nineteenth century and their subsequent development helped to lower the price and greatly increase global usage, particularly as a key ingredient in soaps and detergents. Early in the twentieth century, larger and higher grade deposits found first in Death Valley and then in the Mojave Desert also helped speed the rate of borate consumption. Another world-class deposit was discovered in Turkey in the late 1960s, making that country a major world producer.

Refined borates are usually sold on the basis of their B_2O_3 content and defined by their water or hydration content; borax pentahydrate or "5 mol," borax decahydrate or "10 mol," and boric acid are the three most commonly sold borate products. United States–based Borax, and Etibank, the Turkish national mineral producer, supply approximately 90% of the borates and refined borate products worldwide.

There is a large global market for borates because of their unique chemical properties and relative cost-effectiveness. Today, borates are used in a wide variety of products and processes; the major markets are for insulation fiberglass, textile fiberglass, detergents and bleaches, enamels and frits, and agricultural products. Borates are essential for imparting strength as well as heat- and impact-resistance to glass and glass fibers. The element boron is also an essential micronutrient for plants. Thousands of everyday items, including contact lens solutions, barbecue charcoal, brake fluid, and kitchenware—all contain borates.

Other developments include the increased usage of ulexite and colemanite in the manufacture of some insulation products, and expanding applications for borates as pest control products, fertilizers, herbicides, wood preservatives, and flame retardants.

Boron, in trace amounts, is an essential micronutrient for plants. Although it has not yet been proved that humans need boron to live, there is almost universal agreement that boron is nutritionally important to maintain optimal human health. Studies indicate that people in a wide variety of cultures consume ~1–3 mg of boron per day through a combination of foods and drinking water in their local diets.

Nevertheless, boron has come under some scrutiny in the latest revisions of drinking-water standards, and there has been some legislation proposed in Europe to limit the amount of borate in detergent formulations as a result.

Table 1. **Oxides and Borates Referred to in Text**

Compound	CAS Registry Number	Molecular Formula
ammonium pentaborate tetrahydrate	[12229-12-8]	$NH_4B_5O_8 \cdot 4H_2O$
barium metaborate hydrate	[13701-59-2]	$BaO \cdot B_2O_3 \cdot xH_2O$
boron dioxide	[13840-88-5]	BO_2
boron monoxide	[12505-77-0]	BO
boron oxide (6:1)	[11056-99-8]	B_6O
boron oxide (7:1)	[12447-73-3]	B_7O
boron oxide (13:2)	[56940-67-1]	$B_{13}O_2$
boron phosphate	[13308-51-5]	BPO_4
boron suboxide	[54723-68-1]	$B_{12}O_2$
boron oxide (12:2)	[54723-68-1]	$B_{12}O_2$
diammonium tetraborate tetrahydrate	[12228-87-4]	$(NH_4)_2O \cdot 2B_2O_3 \cdot 4H_2O$
diboron dioxide	[13766-28-4]	B_2O_2
boron oxide (2:2)	[13766-28-4]	B_2O_2
diboron trioxide	[1303-86-2]	B_2O_3
boron oxide (2:3)	[1303-86-2]	B_2O_3
dicalcium hexaborate pentahydrate	[12291-65-5]	$2CaO \cdot 3B_2O_3 \cdot 5H_2O$
dipotassium tetraborate tetrahydrate	[12045-78-2]	$K_2O \cdot 2B_2O_3 \cdot 4H_2O$
disodium octaborate tetrahydrate	[12280-03-4]	$Na_2O \cdot 4B_2O_3 \cdot 4H_2O$
disodium tetraborate	[1330-43-4]	$Na_2O \cdot 2B_2O_3$
disodium tetraborate decahydrate (borax)	[1303-96-4]	$Na_2O \cdot 2B_2O_3 \cdot 10H_2O$
disodium tetraborate pentahydrate	[12045-88-4]	$Na_2O \cdot 2B_2O_3 \cdot 5H_2O$
disodium tetraborate tetrahydrate	[12045-87-3]	$Na_2O \cdot 2B_2O_3 \cdot 4H_2O$
dizinc hexaborate heptahydrate	[12280-01-2]	$Zn_2B_6O_{11} \cdot 7H_2O$
metaboric acid	[13460-50-9]	HBO_2
orthoboric acid	[10043-35-3]	$B(OH)_3$
potassium pentaborate tetrahydrate	[12229-13-9]	$KB_5O_8 \cdot 4H_2O$
sodium calcium pentaborate octahydrate	[1319-33-1]	$NaCaB_5O_9 \cdot 8H_2O$
sodium calcium pentaborate pentahydrate	[12229-14-0]	$NaCaB_5O_9 \cdot 5H_2O$
sodium metaborate dihydrate	[16800-11-6]	$NaBO_2 \cdot 2H_2O$
sodium metaborate tetrahydrate	[10555-76-7]	$NaBO_2 \cdot 4H_2O$
sodium pentaborate pentahydrate	[12046-75-2]	$NaB_5O_8 \cdot 5H_2O$
sodium perborate tetrahydrate	[10486-00-7]	$NaBO_3 \cdot 4H_2O$
sodium perborate trihydrate	[28962-65-4]	$NaBO_3 \cdot 3H_2O$
sodium perborate monohydrate	[10332-33-9]	$NaBO_3 \cdot H_2O$
zinc salt (1:2), hydrate	[12447-61-9]	$2ZnO \cdot 3B_2O_3 \cdot 3.5H_2O$
zinc diborate dihydrate	[27043-84-1]	$ZnO \cdot B_2O_3 \cdot 2H_2O$
zinc triborate monohydrate	[12429-73-1]	$Zn(B_3O_3(OH)_5) \cdot H_2O$

Possible health benefits and the safe health limits of boron in the human system are still undergoing intensive investigation.

The oxides and oxyacids of boron as well as a variety of hydrated and anhydrous metal borates are discussed herein. An alphabetical list of compounds referred to in the text is given in Table 1.

The confusing and often ambiguous systems of nomenclature encountered in the literature of inorganic borates have been described (2). The accumulation of detailed structural data for many of the crystalline compounds has led to derivation of more complex names and formulas in an effort to convey more precise information; *Chemical Abstracts* has adopted a classification system based on a series of the usually hypothetical boric acids. For example, the compound having empirical formula $Zn_2B_6O_{11} \cdot 7H_2O$ has been called dizinc hexaborate heptahydrate. Applying the resolved oxide system proposed by the IUPAC, the substance becomes $2ZnO \cdot 3B_2O_3 \cdot 7H_2O$, known as zinc (2:3) borate heptahydrate. This latter system has gained wide acceptance and is followed herein. However, knowledge of the crystal structure allows a more precise structural formulation, $Zn(B_3O_3(OH)_5) \cdot H_2O$, ie, zinc triborate monohydrate, which is listed in *Chemical Abstracts* as boric acid, $H_7B_3O_8$, zinc salt [12429-73-1] (2). Because many authors continue to use the older formulations, a second listing has been devised by *Chemical Abstracts* for the same compound, ie, boric acid, $H_4B_6O_{11}$, zinc salt (1:2) heptahydrate [12280-01-2].

The principal borate minerals are listed in Table 2. A much more complete listing is available in the literature (4,5). Crystal structures of known borate compounds have been compiled (6).

Reports have been made concerning the minerals of the Searles Lake (7), the Boron-Kramer, and the Death Valley (7) areas in the United States.

The single largest tonnage use of borates worldwide is in insulation fiberglass, a soda lime borosilicate glass used as thermal and acoustic insulation and filtration media. Borates are essential components that impart a unique combination of melt and product properties (9). They contribute to insulating values of fiberglass by increasing the infrared absorption of the glass. The second

Table 2. Commercial Borate Minerals

Mineral	CAS Registry Number	Empirical Formula	B_2O_3 Content, wt %	Location
borax (tincal)	[1303-46-4]	$Na_2B_4O_7 \cdot 10H_2O$	36.5	USA, Turkey, Argentina
kernite	[12045-87-3]	$Na_2B_4O_7 \cdot 4H_2O$	51.0	USA
Ulexite (cottonball)	[1319-33-1]	$NaCaB_5O_9 \cdot 8H_2O$	43.0	Turkey, South America
Colemanite	[12291-65-5]	$Ca_2B_6O_{11}5H_2O$	50.8	Turkey
Inderite	[12260-26-3]	$Mg_2B_6O_{11} \cdot 15H_2O$	37.3	Kazakhstan
Szaibelyite (aschari-te)[a]	[12447-04-0]	$Mg_2B_2O_5 \cdot H_2O$	41.4	China
Suanite	[36564-04-2]	$Mg_2B_2O_5$	46.3	China
Datolite		$Ca_2B_2Si_2O_9 \cdot H_2O$	21.8	Russia

[a] This material has two CAS Registry Numbers.

largest application is in textile or continuous filament fiberglass, a calcium alu-mino borosilicate glass, used for electrical insulation and for reinforcement of plastics (10). The absence of alkali makes it difficult to melt and fiberize, which is overcome by the addition of boric oxide. Borates are also finding an increasing use in ceramic glaze. Since about 1990, there has been movement away from lead-containing glaze for tiles and tableware, and toward borosilicate glazes (11).

A number of boron-containing products are prepared directly from boric acid. These include synthetic inorganic borate salts: zinc borates, barium meta-borate, boron phosphate, fluoroborates, boron trihalides, boron carbide; and, metal alloys such as ferroborn. There is a growing interest in use of boron in refractory boron compounds such as boron nitrides and boron carbide.

Inorganic boron compounds—namely, boric acid, sodium borates, zinc borates, and ammonium borates — are generally good fire retardants and are particularly effective in reducing flammability for both plastic and cellulosic materials (12). On heating cellulosic materials, formation of borate esters are believed to block the pyrolytic decomposition to volatile flammable products. In both cellulose and plastics, borates also function as a flux, forming glassy vitr-eous char that protects against further pyrolysis.

Boron is an essential micronutrient for plants and is added to boron-deficient soils to increase crop vitality and fruit yield. The element has also been shown to be of nutritional importance in humans and has been classified as "probably essential in animals" (13). In biological systems, it is generally thought that the Lewis acid properties and crosslinking ester formation is important in boron's role. At high concentrations, this interaction can also inhibit metabolism and leads to the biostatic effects utilized in the preservative applications of borates. These include biodeterioration control, particularly the protection of wood and cellulosic products against termites, other insects, and fungi.

Compounds of boron that do not contain oxygen directly bonded to boron are of lesser commercial volume; these include boron halides and boron hydrides (13) (see BORON HALIDES). The only commercially important boron halides are boron trichloride, BCl_3, and to a smaller extent boron tribromide, BBr_3. The boron halides are strong Lewis acids due to the B–X bond being strongly polarized and of high π-bonding character. These boron halides react with Lewis bases having reactive O, S, N, or P moieties. The boron halides are primarily used as Friedel–Craft polymerization catalysts and in chemical vapor deposition (CVD) processes for the production of elemental boron fibers for high strength reinfor-cement applications. Boron hydrides are represented by a unique and diverse class of compounds, but have few commercial applications because of their high cost. Only sodium borohydride, $NaBH_4$, has significant commercial use as a powerful reducing agent used primarily as a bleaching agent to whiten paper and clays. Diborane, B_2H_6, is used in very small amounts as a semiconduc-tor dopant in the electronics industry (see BORON HYDRIDES HETEROBORANES, AND THEIR METALLA DERIVATIVES (14).

The application of crystallographic data to borate geology has been described (15).

2. Occurrence

The known world-class borate deposits are located in southern California and northwestern Anatolia, Turkey. In California, sodium borates are produced from a large deposit located in Boron in the northwestern Mojave Desert, and as a coproduct from brines pumped from shallow depths at Searles Lake west of Death Valley. Minor amounts of colemanite are extracted from a deep underground mine on the edge of Death Valley National Park.

In Turkey, there are three main producing districts: Kirka, which produces borax; the Emet basin, which contains two large colemanite deposits; and the Bigadic basin, which produces both ulexite and colemanite from several deposits.

Portions of four South American countries—northern Argentina, northern Chile, southern Bolivia, and southern Peru—contain numerous ulexite deposits. Northern Argentina also hosts the Tincalayu borax deposit and several ulexite–colemanite deposits. These deposits supply the South American market and export a minor amount to Pacific Rim countries. Eastern China, where a number of relatively small magnesium borate deposits are located, and far eastern Russia, which mines a large borosilicate deposit, supply their local markets and represent most of the remainder of the world's borate production.

Borax, and to a lesser extent kernite, are mined from huge open pits at Boron in California, Kirka in Turkey, and Tincalayu in northern Argentina. These three deposits, plus the brines from Searles Lake, California, furnish most of the sodium borates used by industry. Sodium borates are easily soluble in water, making them preferable for many end uses. All four of these facilities also produce boric acid at nearby refineries.

Colemanite is mined from several deposits in the Emet and Bigadic basins of western Turkey, one Death Valley deposit, and small deposits in northern Argentina. This calcium borate is used in end products that have a low sodium requirement, and in making boric acid.

Ulexite is the common marsh or playa (salar) borate. It is produced commercially from numerous playas in South America and the Provinces of Quinghi and Xizang (Tibet) in western China. Ancient playa deposits 5–20 million years old—are the source of Turkish ulexite, ulexite found in Death Valley, and some of the ore produced in Argentina. Much of this material is ground and used as a crude product in agriculture or to manufacture boric acid.

The less easily refined magnesium borates of China are found in the Liaodong peninsula adjacent to North Korea. These are mined by underground methods and refined in small, local plants to produce sodium borates and boric acid for domestic use. A unique borosilicate deposit near Dalnegorsk in far-eastern Russia is mined to produce boric aicd. Borates from this souce were once shipped by rail to western Russia and eastern Europe when they operated under the Former Soviet Union but now are shipped through local ports to markets in the Far East.

Other borate deposits, some of which were once mined commercially, are reported in a number of other areas as well (Table 2). These include the Inder region of Kazakhstan, the Stassfurt district of eastern Germany, northern Iran, and Yugoslavia. Numerous existing deposits in western United States, western Turkey, and in South America are not currently in production due to various economic reasons (16).

3. Boron Oxides

3.1. Boric Oxide.   Boric oxide, B_2O_3, formula wt 69.62, is the only commercially important oxide. It is also known as diboron trioxide, boric anhydride, or anhydrous boric acid. B_2O_3 is normally encountered in the vitreous state. This colorless, glassy solid has a Mohs' hardness of 4 and is usually prepared by dehydration of boric acid at elevated temperatures. It is mildly hygroscopic at room temperature, and the commercially available material contains ca 1 wt % moisture as a surface layer of boric acid. The reaction with water:

$$B_2O_3 \text{ (glass)} + 3\ H_2O \longrightarrow 2\ B(OH)_3$$

is exothermic, $\Delta H^\circ = -75.94$ kJ/mol (-18.15 kcal/mol) B_2O_3 (17).

Boric oxide is an excellent Lewis acid. It coordinates even weak bases to form four-coordinate borate species. Reaction with sulfuric acid produces $H[B(HSO_4)_4]$ (18). At high (>1000°C) temperatures molten boric oxide dissolves most metal oxides and is thus very corrosive to metals in the presence of oxygen.

Molten boric oxide reacts readily with water vapor above 1000°C to form metaboric acid in the vapor state.

$$B_2O_3 \text{ (glass)} + H_2O(g) \longrightarrow 2\ HBO_2(g)$$

A value of $\Delta H_{298} = -199.2 \pm 8.4$ kJ/mol (-47.61 ± 2.0 kcal/mol) has been calculated for this reaction, which is of considerable economic importance to glass manufacturers because B_2O_3 losses during glass (qv) processing are greatly increased by the presence of water. For this reason anhydrous borates or boric oxide are often preferred over hydrated salts, eg, borax or boric acid for glass manufacture. The presence of MgO has been found to reduce volatilization of B_2O_3 from glass charges (19).

The physical properties of vitreous boric oxide (Table 3) are somewhat dependent on moisture content and thermal history. Much of the older physical data has been revised following development of more reliable techniques for sample preparation (23,24). Many physical properties are sensitive to moisture present as metaboric acid, not as free water. Water can be reduced to 0.17% by heating in air at 1000°C and a level of 10 ppm has been achieved by prolonged heating in a vacuum, 0.13 kPa (1 mm Hg), in a carbon crucible. Removal of residual water causes the density to decrease and softening point at 6×10^6 Pa·s (6×10^7 P) to increase. At 0.28 wt% of water the density of boric oxide is 1.853, softening point 240–275°C; nearly anhydrous B_2O_3, having 20 ppm water has density of 1.829 g/mL and a softening point of 300–325°C (25). Thermal expansion, viscosity, and refractive index are all affected by moisture

Table 3. **Physical Properties of Vitreous Boric Oxide**

Property	Value	Reference
vapor pressure[a], 1331–1808 K	$\log P = 5.849 - \dfrac{16960}{T}$	20
heat of vaporization, ΔH_{vap}, kJ/mol[b]		
1500 K	390.4	21
298 K	431.4	21
boiling point, extrapolated	2316°C	17
viscosity, $\log \eta$, mPa·s(= cP)		
350°C	10.60	22
700°C	4.96	22
1000°C	4.00	22
density, g/mL		
0°C	1.8766	
18–25°C	1.844	
18–25°C[c]	1.81	
500°C[d]	1.648	22
1000°C[d]	1.528	22
index of refraction, 14.4°C	1.463	
heat capacity (specific), J/(kg·K)[b]		
298 K	62.969	17
500 K	87.027	17
700 K	132.63	17
1000 K	131.38	17
heat of formation[e], ΔH_f, kJ,[b] 298.15 K	-1252.2 ± 1.7	17

[a] P is in units of kPa; T is in K. To convert kPa to torr, multiply by 7.5.
[b] To convert J to cal, divide by 4.184.
[c] Well-annealed.
[d] Quenched.
[e] For $2\,B(s) + \frac{3}{2}O_2(g) \longrightarrow B_2O_3$ (glass).

content. Boric oxide becomes pourable on heating to about 500°C. The viscosity of boric oxide with temperature is given below.

Temperature, °C	Viscosity, Pa·s	Temperature, °C	Viscosity, Pa·s
260	6.1×10^{10}	700	8.5×10
300	4.4×10^8	800	2.6×10
400	1.6×10^5	900	1.2×10
500	3.9×10^3	1000	7.4
600	4.8×10^2	1100	4.3

The historical debate over the molecular structure of vitreous and molten boric oxide may never be completely resolved because of its amorphous nature (26). There are only trigonal borons in the solid glass and these are believed to have a branched network of planar boroxol (−BO−)$_3$ rings (**1**). The three exocyclic oxygens, outside the ring, form bridges to neighboring rings or to planar BO$_3$ groups (27,28). This network breaks down as the glass melts, and spectroscopic features attributed to the boroxol group, eg, the strong Raman line at 808 cm^{-1}, decrease as the liquid is heated to 800°C. It has been proposed (24) that above

800°C the liquid consists of discrete, but strongly associated, small molecules, conceivably the same monomeric B_2O_3 units observed in the vapor state (29).

(1)

Two crystalline forms of boric oxide have been prepared, and the structures of both materials have been determined by x-ray diffraction (18). The phase relationships between the liquid and crystalline forms have also been developed (30). The more common hexagonal crystal phase, B_2O_3-I or α-form ($d = 2.46$ g/mL, mp = $455 - 475°C$), is more stable than the vitreous phase. The effect of residual water in crystalline B_2O_3, as in the vitreous phase, is to lower the melting, softening, and freezing points (31). For the transformation B_2O_3–I $\longrightarrow$ B_2O_3 (glass), $\Delta H° = +18.24$ kJ/mol (4.36 kcal/mol) (17). However, vitreous B_2O_3 does not crystallize in the absence of seed crystals or increased pressure. Crystallization of B_2O_3 can be induced by prolonged heating of melt with <18 wt % water below 235°C or in the presence of 5 wt % water and addition of crystalline B_2O_3 seed at 250°C. Crystals do not form at any temperature from melt containing <1 wt % water. Crystalline B_2O_3 can also be made by prolonged heating of boric acid seeded with boron phosphate at 220 to 260°C (32). A second dense monoclinic crystalline phase, B_2O_3-II or β-form ($d = 2.95$ g/mL, mp = 510°C) can be obtained at 400°C and >2.23 GPa (>22,000 atm). The crystal lattice of B_2O_3-II consists of a highly compact network of BO_4 tetrahedra where the four apical oxygens are shared by either two or three boron atoms. The acidic character associated with trigonal BO_3 groups is thus masked in B_2O_3-II. Although this material is thermodynamically unstable under ordinary conditions, it reacts very slowly with Lewis bases such as water and fluoride ion.

In the United States a high (99% B_2O_3) purity grade is produced by fusing refined, granular boric acid in a glass furnace fired by oil or gas. The molten glass is solidified in a continuous ribbon as the melt flows over chill-rolls. The amorphous solid product is crushed, screened, and packed in sacks or drums with moisture-proof liners. The price of this product has increased 11% since the mid-1980s and more than doubled since 1977. The carload (>36 metric tons) price in January 1990 was $2780–2950/t (fob plant) depending on mesh size and packaging (33). Boric oxide is no longer commercially produced by mixing borax and sulfuric acid in a fusion furnace. There is no commercial source of crystalline boric oxide (B_2O_3).

Boric oxide reacts with water to form boric acid, with halogens to form boron trihalides, with halogen salts to form glasses, and with P_2O_5 to form boron phosphate. It also is a powerful Lewis acid solvent for dissolving metal oxides, has a low surface tension, and readily wets metal surfaces. Boric oxide can be used as a solvent for metal reductions such as $2\,CuO + C \longrightarrow CO_2 + 2\,Cu$, for growing crystals of garnet, refractory oxides, and preparation of lead titanate, barium titanate, and calcium zirconates from the corresponding oxides.

The uses of boric oxide relate to its behavior as a flux, an acid catalyst, or a chemical intermediate. The fluxing action of B_2O_3 is important in preparing many types of glass, glazes, frits, ceramic coatings, and porcelain enamels (qv).

Boric oxide is used as a catalyst in many organic reactions. It also serves as an intermediate in the production of boron halides, esters, carbide, nitride, and metallic borides.

3.2. Boron Monoxide and Dioxide.

High temperature vapor phases of BO, B_2O_3, and BO_2 have been the subject of a number of spectroscopic and mass spectrometric studies aimed at developing theories of bonding, electronic structures, and thermochemical data (1,34). Values for the principal thermodynamic functions have been calculated and compiled for these gases (35).

Vibrational emission spectra indicate that the B_2O_2 molecule has a linear O=B−B=O structure. Values of 782 and 502 kJ/mol (187 and 120 kcal/mol) were calculated for the respective B=O and B−B bond energies (36).

Two noncrystalline solid forms of BO have been prepared (1,34). Several polymeric $(BO)_n$ or $(B_2O_2)_n$ structures have been proposed for these materials. Although conclusive structural evidence is unavailable, the presence of B−B bonds appears likely. The low temperature form is a white, water-soluble powder produced at 220°C by vacuum-dehydration of tetrahydroxydiborane(4), [13675-18-8]$B_2(OH)_4$, that can be prepared from tetrakis(dimethylamino)diborane(4)[1630-79-1] (37). This product is irreversibly converted to an insoluble, light brown modification on heating above 500°C. The latter material was also prepared by reduction of B_2O_3 by boron at 1330°C, by carbon, or by boron carbides at 1250°C (38). Both BO polymorphs are strong reducing agents that decompose slowly in water to yield hydrogen gas and boric acid.

$$B_2O_3 + C \longrightarrow B_2O_2 + CO$$

$$5\,B_2O_3 + B_4C \longrightarrow 7\,B_2O_2 + CO$$

3.3. Lower Oxides.

A number of hard, refractory suboxides have been prepared either as by-products of elemental boron production (1) or by the reaction of boron and boric acid at high temperatures and pressures (39). It appears that the various oxides represented as B_6O, B_7O, $B_{12}O_2$, and $B_{13}O_2$ may all be the same material in varying degrees of purity. A representative crystalline substance was determined to be rhombohedral boron suboxide, $B_{12}O_2$, usually mixed with traces of boron or B_2O_3 (39). A study has been made of the mechanical properties of this material, which exhibits a hardness comparable to that of boron carbide (40). At temperatures above 1000°C, $B_{12}O_2$ gradually decomposes to B(s) and B_2O_2(g).

4. Boric Acid

The name boric acid is usually associated with orthoboric acid, which is the only commercially important form of boric acid and is found in nature as the mineral sassolite. Three crystalline modifications of metaboric acid also exist. All these forms of boric acid can be regarded as hydrates of boric oxide and formulated as $B_2O_3 \cdot 3H_2O$ for orthoboric acid and $B_2O_3 \cdot H_2O$ for metaboric acid.

4.1. Forms of Boric Acid.   Orthoboric acid, $B(OH)_3$, formula wt, 61.83, crystallizes from aqueous solutions as white, waxy plates that are triclinic in nature; sp gr^{14}_4, 1.5172. Its normal melting point is 170.9°C, however, when heated slowly it loses water to form metaboric acid, HBO_2, formula wt, 43.82, which may exist in one of three crystal modifications. Orthorhombic HBO_2-III or α-form (d = 1.784 g/mL,mp = 176°C) forms first around 130°C and gradually changes to monoclinic HBO_2-II or β-form (d = 2.045 g/mL,mp = 200.9°C). Water-vapor pressures associated with these decompositions follow. To convert kPa to mm Hg, multiply by 7.5.

Temperature, °C	Vapor pressure of H_2O over $B(OH)_3$	
	and HBO_2-III, kPa	and HBO_2-II, kPa
25	0.048	0.16
100	8.4	16
130	39.9	62.5
150	102	143

At temperatures above 150°C, dehydration continues to yield viscous liquid phases beyond the metaboric acid composition (39). The most stable form of metaboric acid, cubic HBO_2-I or γ-form (d = 2.49 g/mL, mp = 236°C) crystallizes slowly when mixtures of boric acid and HBO_2-III are melted in an evacuated, sealed ampul and held at 180°C for several weeks (41).

The relationships between condensed phases in the B_2O_3–H_2O system are shown in Figure 1 (42). There is no evidence for stable phases other than those shown. B_2O_3 melts and glasses containing less than 50 mol% water have mechanical and spectroscopic properties consistent with mixtures of HBO_2 and vitreous B_2O_3.

Vapor phases in the B_2O_3 system include water vapor and $B(OH)_3(g)$ at temperatures below 160°C. Appreciable losses of boric acid occur when aqueous solutions are concentrated by boiling (43). At high (600–1000°C) temperatures, $HBO_2(g)$ is the principal boron species formed by equilibration of water vapor and molten B_2O_3 (44). At still higher temperatures a trimer $(HBO_2)_3(g)$ (**2**) is formed.

(**2**)

The crystal structure of orthoboric acid consists of planar sheets made up of hydrogen-bonded, triangular $B(OH)_3$ molecules. The stacking pattern of the molecular layers is completely disordered, indicative of relatively weak van der Waals forces between the planes. This accounts for its slippery feel and the ease with which the crystals are cleaved into thin flakes (45). The structures of all three forms of metaboric acid are also known (46). The basic structural unit of

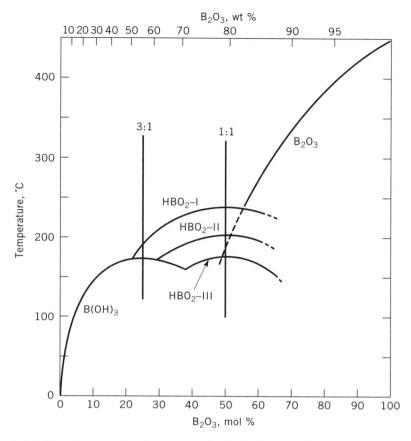

Fig. 1. Solubility diagram for the system $H_2O–B_2O_3$ (42). Courtesy of *The American Journal of Science*.

HBO$_2$-III is the trimeric ring (**2**) and consists of only planar trigonal BO_3 units. Metaborate-I has only tetrahedral BO_4 structural units and HBO$_2$-II contains both trigonal and tetrahedral borons in a ratio of 2:1. The HBO$_2$-III trimer may persist to some extent in the vapor phase, but infrared spectra indicate that monomeric O=B–OH species predominate in gaseous metaboric acid (47).

4.2. Properties. The standard heats of formation of crystalline ortho-boric acid and the three forms of metaboric acid are $\Delta H_f^\circ = -1094.3$ kJ/ mol (-261.54 kcal/mol) for B(OH)$_3$; -804.04 kJ/mol (-192.17 kcal/mol) for HBO$_2$-I; -794.25 kJ/mol (-189.83 kcal/mol) for HBO$_2$-II; and -788.77kcal/mol (188.52 kcal/mol) for HBO$_2$-III (48). Values for the principal thermodynamic functions of B(OH)$_3$ are given in Table 4 (17).

The solubility of boric acid in water (Table 5) increases rapidly with temperature. The heat of solution is somewhat concentration dependent. For solutions having molalities in the range $0.03–0.9$ m, the molar heats of solution fit the empirical relation (49):

$$\Delta H = \left[22062 - 222\, m + 979\, e^{-1230\, m} \right] \text{ kJ/mol}$$

Table 4. **Thermodynamic Properties of Crystalline Boric Acid, B(OH)$_3$**[a]

Temperature, K	C°_p,J/(kg·K)[b]	S°, J/K[b]	$H^{\circ}-H^{\circ}_{298}$, J/mol[b]
0	0	0	−13393
100	35.92	28.98	−11636
200	58.74	61.13	−6866
298	81.34	88.74	0
400	100.21	115.39	9284

[a] Ref. 44.
[b] To convert J to cal, divide by 4.184.

Table 5. **Aqueous Solubility of Boric Acid**

Temperature, °C	B(OH)$_3$, wt %	Temperature, °C	B(OH)$_3$, wt %
−0.76[a]	2.47	60	12.97
0	2.52	70	15.75
10	3.49	80	19.10
20	4.72	90	23.27
30	6.23	100	27.53
40	8.08	103.3[b]	29.27
50	10.27		

[a] Melting point.
[b] Boiling point.

The presence of inorganic salts may enhance or depress the aqueous solubility of boric acid: it is increased by potassium chloride as well as by potassium or sodium sulfate but decreased by lithium and sodium chlorides. Basic anions and other nucleophiles, notably borates and fluoride, greatly increase boric acid solubility by forming polyions (44).

Boric acid is quite soluble in many organic solvents (Table 6). Some of these solvents, eg, pyridine, dioxane, and diols, are known to form boric acid complexes.

Dilute aqueous solutions of boric acid contain predominantly monomeric, undissociated B(OH)$_3$ molecules. The acidic properties of boric acid relate to behavior as a base acceptor, ie, as a Lewis acid, rather than as a proton donor. For the reaction

$$B(OH)_3 + H_2O \rightleftharpoons B(OH)_4^- + H^+$$

an equilibrium constant of 5.80×10^{-10} at 25°C has been reported (50). However, calculated pH values based on this constant deviate considerably from measured ones as the boric acid concentration is increased, as shown in Table 7. The increased acidity has been attributed to secondary equilibria involving condensation reactions between B(OH)$_3$ and B(OH)$_4$, tetrahydroxyborate[15390-83-7], to produce polyborates. A trimeric species B$_3$O$_3$(OH)$_4^-$ [17927-69-4] appears to be the most important of these complex ions (52).

The apparent acid strength of boric acid is increased both by strong electrolytes that modify the structure and activity of the solvent water and by reagents that form complexes with B(OH)$_4^-$ and/or polyborate anions. More than one

Table 6. **Solubility of Boric Acid, Borax Decahydrate, and Borax Pentahydrate in Organic Solvents**

Solvent	Temperature, °C		Solubility, wt %	
			$B(OH)_3$	$Na_2B_4O_7 \cdot 10H_2O$
glycerol, 86.5%	20	21.1	47.1	
glycerol, 98.5%	20	19.9	52.6	
glycerol	25	17.5		
ethylene glycol	25	18.5	41.6	
propylene glycol	25	15.1		31.2
diethylene glycol	25	13.6	18.6	21.9
mannitol, 10%	25	6.62		10.0
methanol	25	173.9[a]	19.9	
ethanol	25	94.4[a]		16.9
n-propanol	25	59.4[a]		
n-butanol	25	42.8[a]		
2-methylbutanol	25	35.3[a]		
isoamyl alcohol	25	2.39		
acetone	25	0.6	0.60	
methyl ethyl ketone	20	0.7		
ethyl acetate	25	1.5	0.14	
diethyl ether	20	0.008		
dioxane	25	ca 14.6[a]		
pyridine	25	ca 70[a]		
aniline	20	0.15		
acetic acid, 100%	30	6.3		

[a] Solubility values are in g/L.

mechanism may be operative when salts of metal ions are involved. In the presence of excess calcium chloride the strength of boric acid becomes comparable to that of carboxylic acids, and such solutions may be titrated using strong base to a sharp phenolphthalein end point. Normally titrations of boric acid are carried out following addition of mannitol or sorbitol, which form stable chelate complexes with $B(OH)_4^-$ in a manner typical of polyhydroxy compounds. Equilibria of the type:

$$B(OH)_4^- \; + \; \begin{matrix} HO \\ HO \end{matrix} R \; \rightleftharpoons \; R \begin{matrix} O \\ O \end{matrix} B \begin{matrix} OH \\ OH \end{matrix} \; + \; 2\,H_2O$$

Table 7. **Observed and Calculated pH Values for Boric Acid**[a]

Concentration, M	pH observed	pH calculated
0.0603	5.23	5.23
0.0904	5.14	5.14
0.1205	5.01	5.08
0.211	4.71	4.96
0.422	4.22	4.80
0.512	4.06	4.76
0.753	3.69	4.54

[a] Ref. 51.

and

$$R^{O}_{O}\overline{B}^{OH}_{OH} + {HO \atop HO}R \ \rightleftharpoons \ R^{O}_{O}\overline{B}^{O}_{O}R \ + \ 2\,H_2O$$

have been exploited in other applications besides analytical determinations of boric acid (53). Ion-exchange resins containing polyols have been developed that are highly specific for removing borates from solution (54). A number of aliphatic and aromatic diols have been patented as extractants for borates and boric acid (55).

Boric acid and fluoride ion react to form a series of fluroborates where OH^-; is displaced by F^-; (see FLUORINE COMPOUNDS, INORGANIC, FLUOROBORIC ACID AND FLUORO-BORATES). Stepwise formation of the ions fluorotrihydroxyborate[32554-53-3], $BF(OH)_3^-$, difluorodihydroxyborate[32554-52-2], $BF_2(OH)_2^-$, and trifluorotrihydroxyborate[18953-00-9], $BF_3(OH)_3^-$, proceeds rapidly in acidic solutions, but tetrafluoroborate[14874-70-5], BF_4^-, forms slowly (56). A fluorosubstituted polyborate, $B_3O_3F_6^{3-}$ [59753-06-9], has also been identified (52).

Alcohols react with boric acid with elimination of water to form borate esters, $B(OR)_3$. A wide variety of borate salts and complexes have been prepared by the reaction of boric acid and inorganic bases, amines, and heavy-metal cations or oxyanions (44,45). Fusion with metal oxides yields anhydrous borates or borate glasses.

4.3. Manufacture. The majority of boric acid is produced by the reaction of inorganic borates with sulfuric acid in an aqueous medium. Sodium borates are the principal raw material in the United States. European manufacturers have generally used partially refined calcium borates, mainly colemanite from Turkey. Turkey uses both colemanite and tincal to make boric acid.

When granulated borax or borax-containing liquors are treated with sulfuric acid, the following reaction ensues:

$$Na_2B_4O_7 \cdot xH_2O + H_2SO_4 \longrightarrow 4\,B(OH)_3 + Na_2SO_4 + (x - 5)\,H_2O$$

In the United States boric acid is produced by United States Borax & Chemical Corp. in a 103,000 B_2O_3 metric ton per year plant by reacting crushed kernite ore with sulfuric acid. Coarse gangue is removed in rake classifiers and fine gangue is removed in thickeners. Boric acid is crystallized from strong liquor, nearly saturated in sodium sulfate, in continuous evaporative crystallizers, and the crystals are washed in a multistage countercurrent wash circuit.

When boric acid is made from colemanite, the ore is ground to a fine powder and stirred vigorously with diluted mother liquor and sulfuric acid at about 90°C. The by-product calcium sulfate [7778-18-9] is removed by settling and filtration, and the boric acid is crystallized by cooling the filtrate.

A unique liquid–liquid extraction process for manufacturing boric acid from sodium borate brines has been operated at Searles Lake, Trona, California, by the North American Chemical Co. since 1962. Both potassium sulfate and sodium sulfate are produced as coproducts in this process.

Boric acid crystals are usually separated from aqueous slurries by centrifugation and dried in rotary driers heated indirectly by warm air. To avoid overdrying, the product temperature should not exceed 50°C. Powdered and impalpable boric acid are produced by milling the crystalline material.

The principal impurities in technical-grade boric acid are the by-product sulfates, <0.1 wt %, and various minor metallic impurities present in the borate ores. A boric acid titer is not an effective measure of purity because overdrying may result in partial conversion to metaboric acid and lead to $B(OH)_3$ assays above 100%. High purity boric acid is prepared by recrystallization of technical-grade material.

4.4. Uses. Boric acid has a surprising variety of applications in both industrial and consumer products (6,57). It serves as a source of B_2O_3 in many fused products, including textile fiber glass, optical and sealing glasses, heat-resistant borosilicate glass, ceramic glazes, and porcelain enamels (see ENAMELS, PORCELAIN AND VITREOUS). It also serves as a component of fluxes for welding and brazing (see SOLDERS AND BRAZING FILLER METALS; WELDING).

A number of boron chemicals are prepared directly from boric acid. These include synthetic inorganic borate salts, boron phosphate, fluoborates, boron trihalides, borate esters, boron carbide, and metal alloys such as ferroboron [11108-67-1].

Boric acid catalyzes the air oxidation of hydrocarbons and increases the yield of alcohols by forming esters that prevent further oxidation of hydroxyl groups to ketones and carboxylic acids (see HYDROCARBON OXIDATION).

The bacteriostatic and fungicidal properties of boric acid have led to its use as a preservative in natural products such as lumber, rubber latex emulsions, leather, and starch products.

NF-grade boric acid serves as a mild, nonirritating antiseptic in mouth-washes, hair rinse, talcum powder, eyewashes, and protective ointments (see DISINFECTANTS AND ANTISEPTICS). Although relatively nontoxic to mammals (58), boric acid powders are quite poisonous to some insects. With the addition of an anticaking agent, they have been used to control cockroaches and to protect wood against insect damage (see INSECT CONTROL TECHNOLOGY).

Inorganic boron compounds are generally good fire retardants (59). Boric acid, alone or in mixtures with sodium borates, is particularly effective in reducing the flammability of cellulosic materials. Applications include treatment of wood products, cellulose insulation, and cotton batting used in mattresses (see FLAME RETARDANTS).

Because boron compounds are good absorbers of thermal neutrons, owing to isotope ^{10}B, the nuclear industry has developed many applications. High purity boric acid is added to the cooling water used in high pressure water reactors (see NUCLEAR REACTORS).

4.5. Analytical Methods. Boric acid is such a weak acid that it cannot be accurately determined by direct alkali titration. Howevr, it can be transformed into a relatively strong acid by adding polyols such as glycerol or mannitol. The resulting ester complexes are much stronger acids and can be titrated accurately (60).

In practice, the boric acid sample is first dissolved in hot water and boiled to remove carbon dioxide, which interfres with the titration. Mannitol (or glycerol) is then added and the resulting solution is titrated with sodium hydroxide (NaOH) solution using phenolphthalein as the indicator. The B_2O_3 content is calculated from the volume and the normality of the NaOH titrant.

5. Solutions of Boric Acid and Borates

5.1. Polyborates and pH Behavior.   Whereas boric acid is essentially monomeric in dilute aqueous solutions, polymeric species may form at concentrations above 0.1 M. The conjugate base of boric acid in aqueous systems is the tetrahydroxyborate [15390-83-7] anion sometimes called the metaborate anion, $B(OH)_4^-$. This species is also the principal anion in solutions of alkali metal (1:1) borates such as sodium metaborate, $Na_2O \cdot B_2O_3 \cdot 4H_2O$ (61). Mixtures of $B(OH)_3$ and $B(OH)_4^-$ appear to form classical buffer systems where the solution pH is governed primarily by the acid:salt ratio, ie, $[H^+] = K_a[B(OH)_3]/[B(OH)_4^-]$. This relationship is nearly correct for solutions of sodium or potassium (1:2) borates, eg, borax, where the ratio $B(OH)_3{:}B(OH)_4^- = 1$, and the pH remains near 9 over a wide range of concentrations. However, for solutions that have pH values much greater or less than 9, the pH changes greatly on dilution as shown in Figure 2 (62).

This anomalous pH behavior results from the presence of polyborates, which dissociate into $B(OH)_3$ and $B(OH)_4^-$ as the solutions are diluted. Below pH of about 9 the solution pH increases on dilution; the inverse is true above pH 9. This is probably because of the combined effects of a shift in the equilibrium concentration of polymeric and monomeric species and their relative acidities. At a $Na_2O{:}B_2O_3$ mol ratio equal to 0.41 at pH 8.91, or $K_2O{:}B_2O_3$ mol ratio equal to 0.405 at pH 9 the pH is independent of concentration. This ratio and the pH associated with it have been termed the isohydric point of borate solutions (63).

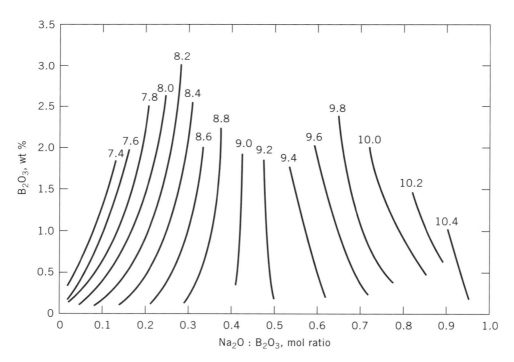

Fig. 2.   Values of pH in the system $Na_2O-B_2O_3-H_2O$ at 25°C (62).

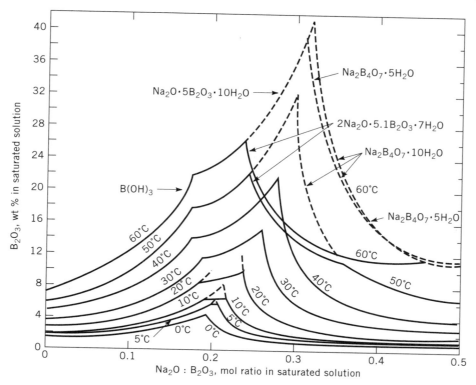

Fig. 3. Solubility isotherms for the system $Na_2B_4O_7$–B_2O_3–H_2O at 0–60°C. The compound $2Na_2O \cdot 5.1B_2O_3 \cdot 7H_2O$ (Suhr's borate) usually does not appear because it crystallizes very slowly in the absence of seed.

The presence of metal salts, particularly those containing alkaline-earth cations and/or halides, cause some shifts in the polyborate equilibria. This may result from direct interaction with the boron–oxygen species, or from changes in the activity of the solvent water (64).

5.2. Solubility Trends. Formation of polyborates greatly enhances the mutual solubilities of boric acid and alkali borates. Solubility isotherms in the system Na_2O–B_2O_3–H_2O are shown in Figure 3. When borax, $Na_2B_4O_7 \cdot 10H_2O$, is added to a saturated boric acid solution or when boric acid is added to a saturated borax solution, the B_2O_3 weight percent in the solution greatly increases. Polymerization decreases the concentrations of $B(OH)_3$ and $B(OH)_4^-$ in equilibrium with the solid phases, thus permitting more borax or boric acid to dissolve.

Sodium borate solutions near the Na_2O:B_2O_3 ratio of maximum solubility can be spray-dried to form an amorphous product with the approximate composition $Na_2O \cdot 4B_2O_3 \cdot 4H_2O$ commonly referred to as sodium octaborate (65). This material dissolves rapidly in water without any decrease in temperature to form supersaturated solutions. Such solutions have found application in treating cellulosic materials to impart fire-retardant and decay-resistant properties (see CELLULOSE).

5.3. The Polyborate Species. From a series of very rigorous pH studies, a series of equilibrium constants involving the species $B(OH)_3$, $B(OH)_4^-$,

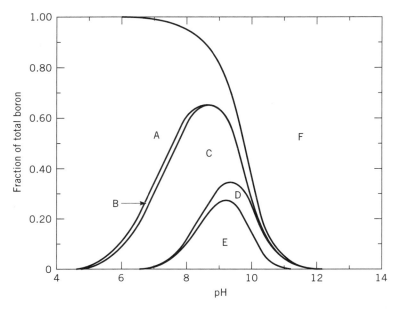

Fig. 4. Distribution of boron in A, $B(OH)_3$; B, $B_5O_6(OH)_4^-$; C, $B_3O_3(OH)_4^-$; D, $B_3O_3(OH)_5^{2-}$; E, $B_4O_5(OH)_4^{2-}$; F, $B(OH)_4^-$; where total B_2O_3 concentration is 13.93 g/L. At a given pH, the fraction of the total boron in a given ion is represented by the portion of a vertical line falling within the corresponding range (66).

and the plyions $B_3O_3(OH)_5^{2-}$ [12344-78-4], $B_3O_3(OH)_4^-$ [12344-77-3], $B_5O_6(OH)_4^-$ [12343-58-7], and $B_4O_5(OH)_4^{2-}$ [12344-83-1] have been calculated (65). The relative populations of these species as functions of pH are shown in Figure 4. It is clear that species containing three, four, and five borons are significant at intermediate pH values. The ratio between the total anionic charge and the number of borons per ion increases with increasing pH.

The polyions postulated in solution all have known structural analogues in crystalline borate salts. Investigations of the Raman (67) and ^{11}B nmr (68) spectra of borate solutions have confirmed the presence of three of these species: the triborate (3), $B_3O_3(OH)_4^-$, tetraborate (4), $[B_4O_5(OH)_4^{2-}]$, and pentaborate (5) $B_5O_6(OH)_4^-$, polyanions. Skeletal structures were assigned based on coincidences between the solution spectra and those solid borates for which definitive structural data are available (52). These same ions have been postulated to be present in alkali metal borate glasses as well. A rapid equilibrium exists among the various polyborate species in aqueous solutions.

6. Borates

6.1. Sources and Supplies.

A limited number of geographic regions on earth contain borate deposits. The two main producers and regions are Borax in California and Etibank in Turkey. Other minor producers exist in the Andes region of northern Chile and Argentina, Bolivia, Peru, California, Russia, and China (69).

Borax. Borax is the world's leading supplier of refined borates, distinguished from other producers by its integrated approach to mining, refining, and distribution. The Borax Group is includes mining and refining facilities in the United States, Argentina, and France; shipping terminals in the United States, the Netherlands, France, and Spain; corporate, sales, or technical offices in the United States, the United Kingdom, Argentina, Belgium, Brazil, Germany, Spain, France, Italy, Japan, and Singapore; and warehouses in Austria, Germany, Russia, and the Ukraine. Borax is owned by world's largest mining company, Rio Tinto. The company's headquarters are in Valencia, California, with regional headquarters in England and Singapore.

Borax operates California's largest open-pit mine in Boron, about 90 m (144.81 km) northeast of Los Angeles. The mine produces approximately one million tons of borates annually, enough to satisfy nearly half the world's demand. At an onsite refinery, the company employs two separate processes to refine ore: the boric acid process and the refined borax process. The boric acid process uses kernite as feedstock: the refined borax process uses a combination of borax ore and kernite ore that has been converted to borax.

This primary deposit consists of borax and kernite. These ores are interbedded with montmorillonite-illite clays, which are overlain with claystones containing ulexite and minor colemanite. The ulexite and colemanite are currently being stockpiled in the ore extraction process. All kernite ore is kept separte from the borax and borax/kernite blended ores.

The company's refinery and shipping terminal in the Port of Los Angeles in Wilimington, California, currently produces special quality grades of borates and boric acid, other enhanced borates, and a range of speciality products that fall under the general categories of flame retardants, agricultural products, pest control products, and wood preservatives.

Under its Borax Argentina division, the company runs mining and refining operations in the Salta Province of northern Argentina. Ores mined include borax from Tincalayu, hydroboracite from Sijes, and ulexite from Salar Diablillos and Cauchari. The annual combined production from these operations combined is less than 100,000 tons of borates. The majority of product mined by Borax Argentina supplies South American markets, and the remainder, mainly hydroboracite, is shipped to Spain for use in ceramic frits and glases.

IMC Chemicals Inc. IMC chemicals Inc. (IMCC), formerly North American Chemical Co. (NACC), is a subsidiary of IMC Global Inc. The IMC chemicals business unit is responsible for the soda ash–boron buisness. IMC is the second largest producer of boron products in North America. The feed for the Searles Lake operations come from two salt horizons saturated with brines extracted by solution mining (continuous and selective crystallization and precipitation of different salts). The brines contain bicarbonate, borate, chloride, carbonate, Potassium,

Table 8. **Scope and Size of Etibank Operations**

Location	Approximate number of mines	Main Boron minerals	Approximate concentrate capacity $\times 10^3$ t
Kirka	1 open-pit	tincal	600
Kestelek	1 open-pit	colemanite	100
Bigadic	3 open-pit	colemanite	400
	2 underground ulexite		120
Emet	2 open-pit	colemanite	500

and sulfate ions. The total B_2O_3 production level is about 60,000 tons per year including borax pentahydrate, borax decahydrate, boric acid, and anhydrous boric acid production.

In addition to the Searles Lake operation, IMCC also has an Italian associate (Societa Chimica Larderello) based in Tuscany, which produces high specification boric acid and speciality products using imported colemanite and ulexite as fedback.

Fort Cady Minerals Corp. Mining operatiions are located in Barstow, California. Borates are mined using a solution mining process where a dilute acid solution is injected into the colemanite ore body to create a high purity calcium borate. The annual capacity of this operation is ~4000 tons/year.

American Borate Co. American Borate Company (ABC) operates an underground mine in Death Valley, California. The colemanite and ulexite ore is extracted by means of a froth flotation and calcining operation to produce 15,000–20,000 tons/year of mineral concentrate. The majority of these concentrates are exported to Asia, primarily for textile fiberglass manufacturing.

Etibank. The second largest supplier of borates in the world and the largest supplier of calcium borates is state-owned Etibank in Turkey. Etibank consists of mines in several locations, including Kirka, Kestelek, Bigadic, and Emet (see Table 8).

Etibank controls an estimated 60% of the discovered borate reserves in the world and currently produces around 400,000 tons/year of B_2O_3. (see Table 9). It is estimated that the majority of the product — colemanite, colemanite concentrates, and some ulexite — is exported. This high export volume accounts for one-quarter of Turkey's overall mineral exports. The generally accepted method of distribution is through various agents.

Argentina. The total borate production in Argentina is estimated at approximately 245×10^3 t. Most producers are small mining companies, which make crude ulexite concentrates. Aside from the Borax operations in argentina,

Table 9. **Feedback, Main Products, Capacity, and B_2O_3 content of Etibank**

Minerals processed	Products	Capacity, $\times 10^3$ t	Grade B_2O_3, %
tincal	tincal concentrate	600	32.0
	borax pentahydrate	320	47.8
	anhydrous borax	60	68.87
colemanite and concentrates	borax decahydrate	50	36.47
	boric acid	135	56.25
	borax pentahydrate	5	47.75

the other main producers are S.R. Minerals and Ulex S.A. It was estimated that S.R. Minerals produced 30,000 t in 1997 from their Loma Blanca mine. Ulex S.A. produces 10,000 t/yr of concentrated colemantic and hydroboracite. The majority of their production is exported.

Chile. There are various producers of borates in Chile, including Quiborax Boroquimica and Sdad Boroquimica, who mine ulexite reserves from salars in the north of the country. Although most production is consumed locally for applications like agriculture, a substantial amount of product is exported. The total borate production in 1995 was estimated at 212×10^3 t; of this, 250 t of borax and 30,100 of boric acid were exported.

Bolivia. Like Chile, Bolivia has a number of relatively small producers who mainly exploit ulexite reserves. Most reserves are located in the Bolivian Altiplano around Salar de Uyuni. The largest producer in this country is Cia Minera Tierra near the Chilean border. In 1996, an estimated 4100 t of ulexite, 4400 t of boric acid, and 200 t of borax were exported. The total production capacity is estimated to be around 13,000 t.

Peru. The main production of borates in Peru originates from Cia Minera Ubinas and Quimica Oquendo, both owned by the Collorabia group in Italy. Like their South American competitors, they mine Ca 180,000 t of ulexite during the nonwinter months. The ore is used to produce both a crude washed ulexite and boric acid.

Former Soviet Union. Although thers are numerous borate deposits in this area, only one is a significant producer of borates. JSC Bor in Russia mines datolite in the city Dalnergorsk, producing datolite concentrate and refined borates, including calcium borates, borax, boric acid and sodium perborate. Bor is the only Russian producer to export borates, mostly to Japan.

China. Like the South American countries, Chinab also has many small producers. Most production originates from the Liaoning province in the Liaodong peninsula. The ores in this area of the world are mainly magnesium borates such as szaibelyte. However, ulexite, pinnoite, hydroboracite, and borax are also mined in Qinghai. The Chinese products are generally a lower quality than the refined products produced by in the United States. The annual Chinese production is estimated at around 100,000 tons/year (70).

7. Sodium Borates

7.1. Disodium Tetraborate Decahydrate (Borax Decahydrate). Disodium tetraborate decahydrate, $Na_2B_4O_7 \cdot 10H_2O$ or $Na_2O \cdot 2B_2O_3 \cdot 10H_2O$, formula wt, 381.36; monoclinic; sp gr, 1.71; specific heat 1.611 kJ/(kg · K) [0.385 kcal/(g°C] at 25–50°C (70); heat of formation, -6.2643 MJ/mol $(-1497.2$ kcal/mol) (71); exists in nature as the mineral borax. Its crystal habit, nucleation, and growth rate are sensitive to inorganic and surface active organic modifiers (72).

The solubility–temperature curves for the $Na_2O–B_2O_3–H_2O$ system are given in Figure 5 (Table 10). The solubility curves of the penta- and decahydrates intersect at 60.6–60.8°C, indicating that the decahydrate, when added to a saturated solution above this temperature, dissolves with crystallization of the pentahydrate and the reverse occurs below this temperature. This transition temperature may be lowered in solutions of inorganic salts, eg, 49.3°C in

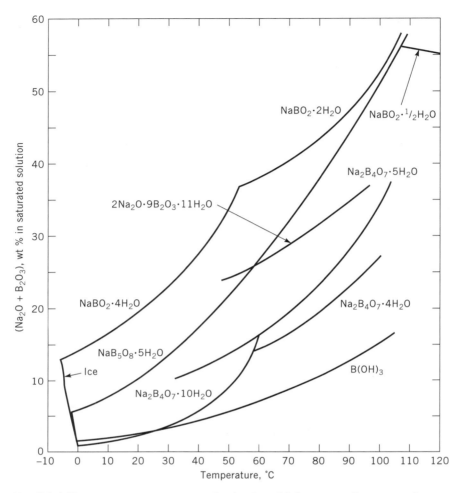

Fig. 5. Solubility–temperature curves for boric acid, borax, sodium pentaborate, and sodium metaborate (73). Courtesy of The American Chemical Society.

solutions saturated with sodium sulfate and 39.6°C with sodium chloride. Heats of solution for borax have been determined (68,75) and the manufacturer quotes a value of about 283 kJ/kg (67.6 kcal/mol) (33).

The pH of a borax solution increases slightly with increasing concentration (Table 11) and drops slightly with increasing temperature. The vapor pressures of aqueous saturated borax solutions at various temperatures are (75,76):

Temperature, °C	Pressure, kPa
57.94	17.25
57.99	17.33
58.23	17.51
58.56	17.74
58.82	17.94
58.91	18.05
59.42	18.42

Table 10. Aqueous Solubilities of Alkali Metal and Ammonium Borates at Various Temperatures

Compound	CAS Registry Number	Solubility, wt% anhydrous salt, at °C											
		0	10	20	25	30	40	50	60	70	80	90	100
$Li_2O \cdot 5B_2O_3 \cdot 10H_2O$ [a]	[37190-10-6]	2.2–2.5	2.55	2.81	2.90	3.01	3.26	3.50	3.76	4.08	4.35	4.75	5.17
$Li_2O \cdot 2B_2O_3 \cdot 4H_2O$	[39291-91-3]												
$Li_2O \cdot B_2O_3 \cdot 16H_2O$ [b]	[41851-38-1]	0.88	1.42	2.51	3.34	4.63	9.40						
$Li_2O \cdot B_2O_3 \cdot 4H_2O$	[15293-74-0]						7.40	7.84	8.43	9.43	10.58, 9.75	11.8, 9.7	13.4, 9.70[c]
$Na_2O \cdot 5B_2O_3 \cdot 10H_2O$	[12046-75-2]	5.77	7.90	10.55	12.20	13.72	17.50	20.88, 21.72	24.34, 26.88	27.98, 32.25	31.79, 38.1	36.2, 44.3	41.2, 51.0
$Na_2O \cdot 2B_2O_3 \cdot 10H_2O$	[1303-96-4]	1.18	1.76	2.58	3.13	3.85	6.00	9.55	15.90				
$Na_2O \cdot 2B_2O_3 \cdot 4.67H_2O$ [d]	[12045-88-4]								16.40	19.49	23.38	28.37	34.63
$Na_2O \cdot 2B_2O_3 \cdot 4H_2O$ [e]	[12045-87-3]								14.82	17.12	19.88	23.31	28.22
$Na_2O \cdot B_2O_3 \cdot 8H_2O$ [f]	[10555-76-7]	14.5	17.0	20.0	21.7	23.6	27.9	34.1					
$Na_2O \cdot B_2O_3 \cdot 4H_2O$	[16800-11-6]								38.3	40.7	43.7	47.4	52.4
$K_2O \cdot 5B_2O_3 \cdot 8H_2O$	[12229-13-9]	1.56	2.11	2.82	3.28	3.80	5.12	6.88	9.05	11.7	14.7	18.3	22.3
$K_2O \cdot 2B_2O_3 \cdot 4H_2O$	[12045-78-2]		9.02	12.1	13.6	15.6	19.4	24.0	28.4	33.3	38.2	43.2	48.4
$K_2O \cdot B_2O_3 \cdot 2.5H_2O$	[27516-44-5]		42.3	43.0	43.3	44.0	45.0	46.1	47.2	48.2	49.3	50.3	
$Rb_2O \cdot 5B_2O_3 \cdot 8H_2O$	[37190-12-8]	1.58	2.0	2.67	3.10	3.58	4.82	6.52	8.69	11.4	14.3	18.1	23.75[g]
$Cs_2O \cdot 5B_2O_3 \cdot 8H_2O$ [h]	[12229-10-6]	1.6	1.85	2.5	2.97	3.52	4.8	6.4	8.31	10.5	13.8	18.0	23.45[i]
$(NH_4)_2O \cdot 2B_2O_3 \cdot 4H_2O$	[10135-84-9]	3.75	5.26	7.63	9.00	10.8	15.8	21.2	27.2	34.4	43.1	52.7	
$(NH_4)_2O \cdot 5B_2O_3 \cdot 8H_2O$	[12229-12-8]	4.00	5.38	7.07	8.03	9.10	11.4	14.4	18.2	22.4	26.4	30.3	

[a] Incongruent solubility below 37.5 or 40.5°C.

[b] Transition point to tetrahydrate, 36.9 or 40°C.

[c] At 101.2°C.

[d] Commonly known as the five hydrate(74), transition point to decahydrate, 60.7°C, 16.6% $Na_2B_4O_7$.

[e] Transition point to decahydrate, 58.2°C, 14.55% $Na_2B_4O_7$.

[f] Transition point to tetrahydrate, 53.6°C, 36.9% $Na_2B_2O_4$.

[g] At 102°C.

[h] Dicesium tetraborate pentahydrate [12228-83-0], $Cs_2O \cdot 2B_2O_3 \cdot 5H_2O$, and dicesium diborate heptahydrate [66634-85-3], $Cs_2O \cdot B_2O_3 \cdot 7H_2O$, also exist. The former has incongruent solubility; the latter has a solubility of 36.8 wt% anhydrous salt at 18°C.

[i] At 101.65°C.

Table 11. **pH of Aqueous Borate Solutions**

Compound	Concentration, wt %						
	0.1	0.5	1.0	2.0	4.0	10.0	15.0
$Na_2B_4O_7 \cdot 10H_2O$	9.2	9.2	9.2	9.2	9.3^a		
$Na_2B_8O_{13} \cdot 4H_2O$		8.5	8.5	8.5	8.1	7.6	7.3
$NaB_5O_8 \cdot 5H_2O$			8.5	8.4	8.1	7.6	7.3
$NaBO_2 \cdot 4H_2O$	10.5	10.8	11.0	11.2	11.4	11.8	11.9
$NaBO_2 \cdot 2H_2O$	10.6	10.9	11.1	11.3	11.5	11.8	12.0
$K_2B_4O_7 \cdot 4H_2O$	9.2	9.1	9.1	9.2	9.3		
$KB_5O_8 \cdot 4H_2O$		8.4	8.4	8.3	7.9	7.6	
$NH_4B_5O_8 \cdot 4H_2O$	8.5	8.4	8.3	8.2	7.8	7.3	

^{a} Saturated solution, 4.71 wt %.

To convert from kPa to mm Hg, multiply by 7.5. Values for the specific heat of aqueous borax solutions as a function of weight percent decahydrate are (75):

Borax decahydrate, wt %	Specific heat, kJ/(kg · K)
1.9	4.13
4.7	4.08
7.2	4.04
9.5	3.99
19.0	3.84
22.8	3.78
26.6	3.71
30.4	3.65
38.0	3.52
45.6	3.57
55.1	3.68

To convert from kJ/(kg · K) to cal/(g · °C), divide by 4.184. The solubilities of borax in organic solvents are given in Table 6.

If borax has been previously warmed to 50°C, it dehydrates reversibly into the pentahydrate and water vapor. The equilibrium vapor pressure for this transition at various temperatures is (76, 77): 15°C, 0.933 kPa (7.0 mm Hg); 19.8°C, 1.33 kPa (10.0 mm Hg); 25°C, 1.87 kPa (14.0 mm Hg); 59°C, 17.7 kPa (133.0 mm Hg). If the decahydrate has not been warmed above 50°C, it develops a vapor pressure of only 0.213 kPa (1.6 mm Hg) at 20°C. In this case, when placed over P_2O_5, it does not form the crystalline pentahydrate but decomposes gradually to form an amorphous product having about 2.4 molecules water content.

Heats of dehydration per mole of water vapor are (76) decahydrate to pentahydrate, 54.149 kJ (12.942 kcal), and decahydrate to tetrahydrate, 54,074 kJ (12.924 kcal). Borax stored over a saturated sucrose–sodium chloride solution maintains exactly 10 moles of water and can thus be used as an analytical standard. Commercial borax tends to lose water of crystallization if stored at high temperature or in dry air.

A single-crystal x-ray diffraction study has shown that the borate ion present in borax has the formula $[B_4O_5(OH)_4]^{2-}$ (4), the sodium ions occupying two unique sites, and eight moles of water of crystallization and two moles of water existing as hydroxyl groups. The formula is best represented as $Na_2[B_4O_5(OH)_4] \cdot 8H_2O$ (78). The same borate ion (4) exists in the pentahydrate, explaining the ready interconversion of the penta- and decahydrates (79). Slow dehydration of borax results in the loss of eight moles of water between 50 and 150°C.

Rapid heating of either borax decahydrate or pentahydrate causes the crystal to dissolve before significant dehydration, and at about 140°C, puffing occurs from rapid vaporization of water to form particles having as high as 90% void volume and very low bulk density (80).

7.2. Disodium Tetraborate Pentahydrate (Borax Pentahydrate).
Although referred to as borax pentahydrate, well-formed crystals actually contain not five but 4.67 moles of water, $Na_2B_4O_7 \cdot 4.67H_2O$ or $Na_2O \cdot 2B_2O_3 \cdot 4.67H_2O$. This structure has been confirmed by an x-ray single crystal analysis that showed that two of the three water of crystallization sites are only partially filled (73). The structural formula is best represented as $Na_2[B_4O_5(OH)_4] \cdot 2.67H_2O$; formula wt, 286.78; trigonal; rhombohedral crystal shape; sp gr, measured 1.880, crystallographic 1.912; specific heat, 1.32 kJ/(kg·K) [0.316 kcal/(g·°C)] (70); heat of formation, -4.7844 MJ/mol (-1143.5 kcal/mol) (71). It is found in nature as a fine-grained mineral, tincalconite, formed by dehydration of borax.

Solubility data in water are given in Figure 5 and in Table 10, solution pH in Table 11, and the solubility in organic solvents is given in Table 6. Heats of solution in water have been determined (70,75). The pentahydrate, in contact with its aqueous solution, is metastable with respect to the tetrahydrate (kernite) at temperatures above 58.2°C and metastable to borax decahydrate below 60.6–60.8°C. Kernite can be slowly crystallized from a near saturate solution heated near the boiling point for several days.

Pentahydrate is reversibly converted to an amorphous dihydrate, at 88°C and 0.26 kPa (2 mm Hg) or by boiling with xylene (75,77). The heat of dehydration for the pentahydrate to tetrahydrate has been calculated to be 53.697 kJ (12.834 kcal) per mole of water (76). Thermogravimetric analyses show that 2.75 moles of water are lost on heating to 140°C. Like borax, pentahydrate puffs when heated rapidly to give a product having a bulk density of 0.042 g/mL (81).

A single-crystal x-ray structure determination has shown that the borate ion in the pentahydrate and in borax are identical (79).

7.3. Disodium Tetraborate Tetrahydrate.
Disodium tetraborate tetrahydrate, $Na_2B_4O_7 \cdot 4H_2O$ or $Na_2O \cdot 2B_2O_3 \cdot 4H_2O$, formula wt, 273.27; monoclinic; sp gr, 1.908; specific heat, ca 1.2 kJ/(kg·K) [0.287 kcal/(g·°C)] (62); heat of formation, -4.4890 MJ/mol (-1072.9 kcal/mol) (71); exists in nature as the mineral kernite and has a structural formula $Na_2[B_4O_6(OH)_2] \cdot 3H_2O$. The crystals have two perfect cleavages and when ground, form elongated splinters.

The water solubility of kernite is shown in Figure 5 and in Table 10. Kernite is the stable phase in contact with its solutions from 58.2°C to ca 95°C (73). Its rate of crystallization is, however, much slower than that of the pentahydrate. Large kernite crystals can be grown slowly by seeding saturated borax solutions.

At relative humidities above 70%, kernite absorbs water irreversibly to form borax. Kernite loses water slowly over P_2O_5 in vacuum or by heating at 100–120°C, forming a crystalline dihydrate, metakernite, which reverts to kernite at 60% relative humidity (76).

The structure of kernite consists of parallel infinite chains of the $[B_4O_6(OH)_2]_n^{2n-}$ ion (**6**) composed of six membered rings (82). The polymeric nature of the anion is consistent with the slow rate of dissolution and crystallization observed for kernite.

(**6**)

7.4. Disodium Tetraborate (Anhydrous Borax). Disodium tetraborate, $Na_2B_4O_7$ or $Na_2O \cdot 2B_2O_3$, formula wt, 201.21; sp gr (glass), 2.367, (α-crystalline form), 2.27; heat of formation (glass), -3.2566 MJ/mol (-778.34 kcal/mol), (α-crystalline form), -3.2767 MJ/mol (-783.2 kcal/mol) (17); exists in several crystalline forms as well as a glassy form (75). The most common α-crystalline form that melts congruently at 742.5°C is obtained by dehydrating borax hydrates and is the stable form above 600–700°C (75). A large amount of heat capacity data has been reported (17,83). Anhydrous borax glass dissolves in water more slowly than the hydrated forms. Heats of solution have been measured (71), and the manufacturer lists a value of -213.8 kJ/kg (-51.1 kcal/kg) (33). The solubilities of finely divided crystalline disodium tetraborate at 25°C expressed as weight percent $Na_2O \cdot 2B_2O_3$ is 16.7% in methanol, 30% in ethylene glycol, and 40.6 g/L in formamide (62).

Crystalline anhydrous borax takes up some water from moist air even at 300°C. It becomes anhydrous near 700°C and melts at 742.5°C. The heat of hydration to borax has been calculated as 161 kJ/mol (38.5 kcal/mol) of $Na_2O \cdot 2B_2O_3$ (75,84). The heat of fusion has been reported as 81.2 kJ/mol (19.4 kcal/mol) (17).

A single-crystal x-ray diffraction study has shown that the borate anion in anhydrous borax is polymeric in nature and is formed via oxygen bridging of triborate and pentaborate groups (85). The chemistry of anhydrous borax has been reviewed (75,86).

7.5. Disodium Octaborate Tetrahydrate. The composition of a commercially available sodium borate hydrate, 66.3 wt% B_2O_3, POLYBOR (65), corresponds quite closely to that of a hypothetical compound, disodium octaborate tetrahydrate, $Na_2B_8O_{13} \cdot 4H_2O$ or $Na_2O \cdot 4B_2O_3 \cdot 4H_2O$. This product dissolves rapidly in water without the temperature decrease, which occurs when the crystalline borates dissolve, and easily forms viscous supersaturated solutions at elevated temperatures. The solution pH decreases as the concentration increases

(Table 11). The solubility of the product is shown compared with that of borax (33):

Temperature, °C	Solubility, POLYBOR, wt %	Concentration, B_2O_3 in saturated solutions, wt %	
		POLYBOR	Borax
0	2.4	1.6	0.7
10	4.5	3.0	1.1
20	9.5	6.3	1.7
30	21.9	14.5	2.6
40	27.8	18.4	4.1
50	32.0	21.2	6.5
60	35.0	23.2	11.1
75	39.3	26.0	14.7
94	45.3	30.0	21.0

7.6. Sodium Pentaborate Pentahydrate. Sodium pentaborate pentahydrate, $NaB_4O_8 \cdot 5H_2O$ or$Na_2O \cdot 5B_2O_3 \cdot 10H_2O$; formula wt, 295.11; monoclinic; sp gr, 1.713; exists in nature as the mineralsborgite[12272-01-4]. Heat capacity, entropy, and other thermal measurements have been made at 15–345 K (87).

Sodium pentaborate can easily be crystallized from a solution having a $Na_2O:B_2O_3$ mol ratio of 0.2. Its water solubility (Fig. 5 and Table 10) exceeds that of borax and boric acid. Its pH decreases with solution concentration (Table 11). It is stable in contact with its own solution between 2 and 59.5°C. When a saturated pentaborate solution is agitated for some time at temperatures near boiling, the compound $2Na_2O \cdot 9B_2O_3 \cdot 11H_2O$, also known as Taylors borate, sp gr, 1.903; crystallizes very slowly if seed is present. Pentaborate pentahydrate, which is metastable to Taylors borate at higher temperatures, readily forms supersaturated solutions and crystallizes as the kinetic product. In the absence of seed crystals, however, the stable phase above 106°C shifts to pentaborates of lower hydration (75).

Crystalline sodium pentaborate pentahydrate is stable in the atmosphere. When heated in vacuum, it is stable to 75°C; however, above 75°C, four of its five H_2O molecules are lost (75).

A single-crystal x-ray diffraction study gives a structural formula of $Na_2[B_5O_6(OH)_4] \cdot 3H_2O$ and contains the pentaborate ion analogous to that found in the corresponding potassium compound (88).

7.7. Sodium Metaborate Tetrahydrate. Sodium metaborate tetrahydrate, $NaBO_2 \cdot 4H_2O$ or $Na_2O \cdot B_2O_3 \cdot 8H_2O$; formula wt, 137.86; triclinic; sp gr, 1.743; is easily formed by cooling a solution containing borax and an amount of sodium hydroxide just in excess of the theoretical value. It is the stable phase in contact with its saturated solution between 11.5 and 53.6°C. At temperatures above 53.6°C, the dihydrate, $NaBO_2 \cdot 2H_2O$, becomes the stable phase. The water solubility of sodium metaborate is given in Figure 5 and in Table 10 and the pH with concentration is given in Table 11.

Heat capacity data for metaborate solutions have been reported (89). The solubility of sodium metaborate tetrahydrate in methanol at 40°C is 26.4 wt% (62).

The relative humidity over a saturated solution of the tetrahydrate at 14–24°C is $90 \pm 1\%$, and the humidity over mixtures of the tetra- and dihydrates is 42% at 22°C; 43% at 24.8°C; 45% at 27.0; and 39% at 91.3°C (90). The heat of hydration for the dihydrate to tetrahydrate conversion has been calculated as 52.51 kJ (12.55 kcal) per mole of water (90). The thermogravimetric curve shows a loss of 0.5 moles of water at 130°C; two moles at 140°C; three moles at 280°C; and the last at temperatures up to 800°C (91).

Sodium metaborate absorbs CO_2 from the atmosphere, forming borax and sodium carbonate. Crystals of the tetrahydrate melt in its water of crystallization at about 54°C. The solid-state structure of the tetrahydrate, $Na[B(OH)_4] \cdot 2H_2O$, consists of discrete tetrahedral $B(OH)^-{}_4$ groups (92).

7.8. Sodium Metaborate Dihydrate. Sodium metaborate dihydrate, $NaBO_2 \cdot 2H_2O$ or $Na_2O \cdot B_2O_3 \cdot 4H_2O$; formula wt, 101.83; triclinic; sp gr, 1.909; can be prepared by heating a slurry of the tetrahydrate above 54°C, by crystallizing metaborate solutions at 54–80°C, or by dehydrating the tetrahydrate in vacuum. Large crystals can be grown by heating the solid in its mother liquor for several days. The dihydrate is the stable phase in contact with its saturated solution between 53.6 and 105°C. At higher temperatures a hemihydrate, $NaBO_2 \cdot 0.5H_2O$, is formed (75).

The water solubility for the dihydrate is shown in Figure 5 and in Table 10 and solution pH with concentration is given in Table 11. The solubility of the dihydrate inethanol is 0.3 wt % at boiling, and in methanol it is 17.8% at 22°C, 19.5% at 40°C, and 24.6% at 60°C (62).

The dihydrate loses water slowly at room temperature. Its heat of dehydration to $NaBO_2 \cdot 0.5H_2O$ has been calculated as 58.1 kJ/mol (13.9 kcal/mol) of H_2O (90). Sodium metaborate dihydrate reacts with atmospheric CO_2 to produce sodium carbonate and borax. The melting point is 90–95°C, compared to 54°C for the tetrahydrate. Some crystallographic work has been done (93).

7.9. Sodium Perborate Hydrates. Peroxyborates are commonly known as perborates, written as if the perborate anion were BO_3^-. X-ray crystal structure has shown that they contain the dimeric anion $[(HO)_2B(O_2)_2B(OH)_2]^{2-}$ (**7**) (94). Three sodium perborate hydrates, $NaBO_3 \cdot xH_2O$ ($x = 1$, 3, and 4), are known. Only the mono- and tetrahydrate are of commercial importance, primarily asbleaching agents (qv) in laundry products.

$$\begin{bmatrix} HO & O-O & OH \\ \diagdown B \diagup & & \diagdown B \diagup \\ HO & O-O & OH \end{bmatrix}^{2-}$$

(**7**)

Sodium perborate tetrahydrate, $NaBO_3 \cdot 4H_2O$ or $Na_2B_2(O_2)_2(OH)_4 \cdot 6H_2O$, is triclinic; heat of formation, -2112 kJ/mol (-504.8 kcal/mol) (crystal), -921 kJ/mol (-220.2 kcal/mol) (1M soln); and contains 10.4 wt % active oxygen. It melts at 63°C by dissolving in its own water of hydration and on heating to 250°C decomposes rapidly and completely to oxygen and sodium metaborate. In water its decomposition, which is important in its use as a bleach, is accelerated by catalysts or elevated temperature. Typical solutions at room temperature are unstable and lose active oxygen unless a stabilizer is present. The rate of decomposition increases with pH. The solubility in water is 2.5 wt% $NaBO_3 \cdot 4H_2O$ at 20°C and 3.6 wt% at 29°C (95). The solubility is enhanced by certain polyhydroxy compounds which form complexes with borates, such as tartaric acid, citric acid, mannitol, glycerol, and most significantly, by alkali polyphosphates. Dilute solutions contain the monoperoxyborate anion, $B(OH)_3(OOH)^-$. More concentrated solutions contain this anion plus $B(OH)_2(OOH)_2^-$, $B(OH)(OOH)_3^-$, $B(OOH)_4^-$, and polyperoxyborate anions (96).

Commercial preparation of sodium perborate tetrahydrate is by reaction of a sodium metaborate solution, from sodium hydroxide and borax pentahydrate, and hydrogen peroxide followed by crystallization of tetrahydrate (97). The trihydrate and monohydrate can be formed by reversible dehydration of the tetrahydrate.

Sodium perborate trihydrate, $NaBO_3 \cdot 3H_2O$ or $Na_2B_2(O_2)_2(OH)_4 \cdot 4H_2O$, triclinic, contains 11.8 wt% active oxygen (98). It has been claimed to have better thermal stability than the tetrahydrate but has not been used commercially. The trihydrate can be made by dehydration of the tetrahydrate or by crystallization from a sodium metaborate and hydrogen peroxide solution in the present of trihydrate seeds. Between 18 and 50°C the trihydrate is more stable but slower to crystallize than the tetrahydrate. Below 15°C the trihydrate is spontaneously converted into the tetrahydrate.

Sodium perborate monohydrate, $NaBO_3 \cdot H_2O$ or $Na_2B_2(O_2)_2(OH)_4$, 16.0 wt% active oxygen, is commercially prepared by dehydration of the tetrahydrate. The monohydrate has the same peroxyborate anion (**7**), as the higher hydrates and is the anhydrous sodium salt of this anion. Further dehydration results in decomposition of the peroxyborate.

7.10. Analysis. The alkali metal and ammonium borates are analyzed for M_2O and B_2O_3 content by dissolving the compound in water, titrating the M_2O content with dilute HCl and determining the B_2O_3 content by complexation with excess mannitol followed by titration with dilute NaOH (99). The B_2O_3 content for calcium borates and other borates of low water solubility is determined by extraction into acid solution followed by mannitol complexation and titration with dilute base. The commercial hydrates are often overdried, leading to apparent B_2O_3 assays over 100%.

Borate reacts with curcumin [458-37-7], $C_{21}H_{20}O_6$, in the presence of a mineral acid to give a colored 1:2 borc acid: curcumin complex that has been used to determine microamounts of boron. Carminic acid [1260-17-9], $C_{22}H_{20}O_{13}$, (100) and azomethine-H (101) also form a colored complex useful for low level detection of borates. Boron compounds give a characteristic green color when burned in a flame.

Methods for analysis of industrial borate chemicals have been reviewed (103).

Crude (Mineral) Borates. The titrimetric determination of B_2O_3 in crude borates — such as tincalconite, kernite, colemanite, or ulexite — is prone to interference. Common interfering elements and compounds are iron. alumina, soluble silica, and manganese. To remove these substances, a procedure called the *barium carbonate method* was developed by Pacific Coast Borax Company (now U.S. Borax Inc.)(102).

The procedure is based on the fact that barium borate, formed by the addition of barium carbonate to boric acid, is quite soluble. It acts as a buffer solution to cause precipitation of the hydroxides of heavy metals.

The sample is first finely ground to <100 mesh. It is digested in HCl and neutralized by NaOH using methyl red as the indicator. Saturated bromine water is added to oxidize Fe(II) to Fe(III). Excess bromine is removed by boiling. Barium carbonate is added. The hydroxides of heavy metals are removed by filtering. The sample is acidified using HCl and boiled to remove carbon dioxide. It is neutralized with NaOH to a boric acid solution using methyl red as the indicator. Mannitol (or glycerol) is then added, and the resulting solution is titrated with NaOH solution using phenolphalein as the indicator.

7.11. Manufacturing, Production and Processing. Both sedimentary and metamorphic borate mineral deposists are exploited, although sedimentary deposits are by far the larger sources. Ore extraction is typically accomplished using conventional surface and underground mining tehniques. Solution mining has also been employed, albeit on a small scale. Sedimentary sodium borate minerals borax and kernite, the sedimentary calcium mineral colemanite, the sedimentary calcium–sodium mineral ulexite, and the metamorphic mineral datolite are the principal ore types exploited around the world.

In nearly all end uses, borates are either dispersed in low concentrations (eg, detergents, agricultural products) or incorporated into products from which borate seperation would be difficult (eg, ceramics, fiberglass). Because borates cannot easily be recycled, nearly all borates used in commerce are obtained from virgin sources located in remote desert regions.

Before mining begins, geologists collect core samples and prepare three-dimensional models of the ore body. These models include informatiom about ore body geometry, ore mineralogy and grade, and impurity types and concentrations. Information gleaned from the models is used to develop an optimal mine plan and to determine precise refining requirements.

Commercial borate ores fall into basic types: sodium borates, which are relatively soluble in water; and calcium or sodium–calcium borate ores, which are relatively insoluble in water. Because of these intrinsic properties, sodium borate ores are often used to produce refined borate products (various forms of sodium borates and boric acid) whereas calcium and sodium–calcium borate ores are often used to supply mineral borates.

Most large-scale borate extraction is performed using conventional surface mining techniques. Overburden material is drilled, blasted, and stripped away using large shovels and haul trucks to expose the ore body. Once the ore is exposed, it is blasted to loosen it. Electric or hydryaulic shovels load the ore into trucks with capacities as large as $\geq$200 tons. The haul trucks carry the ore to crushers for size reduction. mine ore can range in size from a few millimetres to $\geq$2 m in diameter. Crushers reduce the particle size to prepare the ore for sale or refining.

Some commercial uses for borates, such as insulation fiberglass and ceramic glazes, do not typically require high purity feedstocks. For these limited uses, raw or upgraded minerals can provide the borate source. Few ores are of high enough grade to be used directly, but ore is upgraded using gravity or magnetic separation techniques. Gravity separation takes advantage of the differences in specific gravity between the borate salts and the surrounding clay matrix. Magnetic separation takes advantage of the weakly magnetic properties of the clays impurities. Borate recovery from these separation processes is often relatively poor, but the cost of the processing is low.

Most commercial applications of borates require the use of refined borates. The feedstock to the refineries can be raw or upgraded ore. The refining process includes four steps: dissolution, purification, recrystallization, and drying.

Refining processes for sodium borate ores take advantage of their highly soluble nature. In the first step of refining, crushed ore is dissolved in hot water using agitation and steam. Dissolution produces a hot, saturated salt solution containing sodium borates and a suspension of insoluble of insolube matter.

Either sodium- or calcium-containing borate minerals may be used to produce boric acid. The first step in refining is essentially the same; however, sulfuric or some other mineral acid is added to assist dissolution and to convert the borate mineral to soluble boric acid and sodium or calcium sulfate, which is typically discarded because of its low value. In boric acid production, dissolution also results in a hot, saturated salt solution containing boric acid along with a suspension of insoluble matter.

Impurities present in the borate solutions typically consist of insoluble clays or other minerals (and, in the case of boric acid production, either insoluble sodium or calcium sulfate). Impurity removal is accomplished by passing the solution through a series of screens, settling tanks, and filters.

Following particular removal, the saturated borate solution is cooled in vacuum crystallizers to produce a slurry of pure sodium borate or boric acid crystals. As the solutions enter the crystallizers, a vacuum is applied, causing the solution to boil. As the saturated solutions cools, crystals form. The conditions used in the vacuum crystallizers determine the form of refined borate product generated. Sodium borates are typically crystallized into forms containing either 5 or 10 molecules of water of hydration. In the industry, these hydrated forms of sodium borates are classified as pentahydrate, or 5 mol and decahydrate, or 10 mol sodium borates.

The crystal slurry is dewatered in centrifuges or belt filters, generating a moist cake of sodium borate or boric acid crystals. These crystals may be washed to remove soluble impurities and fine particulates. The final step in refining involves drying in steam or gas-fired dryers to drive off remaining moisture. Heated air is passed countercurrent to the flow of borate salt crystals, removing free moisture and producing a dry, flowable granular product.

The finished product is stored in silos or warehouses to await shipment. Because the final product is white, contamination from rust, dirt, or other material is readily apparent. Storage is normally fully enclosed to ensure a contaminant-free product.

High value, speciality borates are commonly produced from 5-mol, 10-mol, or boric acid products. High purity borates are produced by redissolving and recrystallizing the commercial-gradeproducts. Low water or anhydrous forms of borates are produced by dehydrating or melting the commercial-grade products to drive off waters of hydration.

7.12. Shipment. Because there are so few primary sources of borates, and because of the wide diversity of end uses and end-use locations, distribution of finished products is of critical importance. This is as true today as it was centuries ago when traders carried borates to Europe along the same trade routes used by Marco Polo or in the late 1800s when the famous Twenty Mule Team carried borates out of Death Valley to the nearest railhead.

Today, borates are moved by virtually all available forms of transportation. Bulk, intercontinental shipments are made in ocean-going vesels with capacities of $\geq 40,000$ tones. Barges, railcars, and trucks move borate products from sources or shipping ports to customers. Intermediate stock points are located around the world to provide working inventories and to guard aganist supply interruptions. The major producers maintain worldwide networks of sales representatives, agents and distributors to interact with customers and ensure reliable supplies and service.

7.13. Grades, Specifications, and Quality Control. The major mineralization in the borax deposit in borax deposit in Boron, California is ulexite ($Na_2O \cdot 2CaO \cdot 5B_2O_3 \cdot 16H_2O$), kernite ($Na_2O \cdot 2B_2 \cdot O_3 \cdot 4H_2O$), and borax ($Na_2O \cdot 2B_2O_3 \cdot 10H_2O$). Currently only the borax are present in a one-to-one ratio with an average ore grade of 23% B_2O_3.

While both minerals are salts, their aqueous solubility requires different refining methods. Borax at an average ore grade of 23% is fed to the sodium borate plant, where it is dissolved in hot water, clarified, and crystallized (under controlled conditions) to yield Neobor ($Na_2O \cdot 2B_2O_3 \cdot 5H_2O$) and borax. Kernite at an average B_2O_3 content of 29% is either hydrated to form borax and is fed to the sodium borate plant or is processed with sulfuric acid, clarified, filtered, and crystallized to form boric acid. Both Neobor and boric acid are fused to their anhydrous forms, Dehybor ($Na_2B_4O_7$) and boric oxide (B_2O_3) in separate plants.

Quality control starts at the ore mining stage. After an ore block is mapped, samples are taken and tested for mineralization type and selected impurities. On the basis of these determinations, the various ores are blended and crushed in several stages to yield plant feeds.

After crushing and blending, feedstocks are tested for B_2O_3 content and various impurities to ensure a consistent plant feed. Both liquid and solid samples are assayed during processing to ensure optimum recovery and a high quality product. To ensure consistent quality, both of the fusing plants are fed directly from production with minimal testing.

Specifications for the maximum allowable impurity levels for borate products are given in Table 12. Where maximum levels are not set, typical values are given. Typical levels of impurities generally fall well below the maximum specification. Both borax decahydrate and pentahydrate are sometimes over-dried in manufacture and may give higher than theoretical assays.

7.14. Economic Aspects. The pie chart in Figure 6 demonstrates the end-use makeup of the estimated 1.2×10^6 t global market for borates.

7.15. Health and Safety. Cases of industrial intoxication on exposure to inorganic borates have not been reported (104). There is a large body of literature on the toxicology of boric acid and borax (105). Acute oral LD_{50} in the rat is 3000–4000 mg/kg for boric acid and 4500–6000 mg/kg for borax (58). These values are comparable to sodium chloride, LD_{50} 3750 mg/kg. Ingested boric acid is excreted rapidly in the urine with a half-life of 21 h (106). Chronic ingestion studies (high dosage level and repeated exposure) indicates some reproductive toxicity in animals, but adequate evidence for these effects in humans is lacking (107). Studies

Table 12. **Maximum Impurity Specifications for Borates**^{a}, wt %

Chemical	Grade^b	Cl$^-$	SO$_4^{2-}$	Fe$_2$O$_3$	Na$^+$	Ca^{2+}	Heavy metals as Pb	H$_2$O insolubles
Na$_2$O·2B$_2$O$_3$·10H$_2$O	T	0.07	0.06	0.003				0.02
	SQ^c	0.4^d	1.0^d	2.8^d		50^d	10^d	10^d
Na$_2$O·2B$_2$O$_3$·5H$_2$O	T	0.05	0.08	0.004				
NaBO$_2$·4H$_2$O	T	0.1	0.1	0.003		0.002^e	0.0005^e	0.002^e
NaBO$_2$·2H$_2$O	T	0.1	0.1	0.007		0.003^e	0.0005^e	0.002^e
K$_2$O·2B$_2$O$_3$·4H$_2$O	T	0.05	0.05	0.0014	0.10	0.002^e	0.0005^e	0.002^e
KB$_5$O$_8$·4H$_2$O	T	0.05	0.05	0.003	0.10		0.0005^e	
(NH$_4$)$_2$O·2B$_2$O$_3$·4H$_2$O	T	0.05	0.05	0.0014	0.0026^e			
NH$_4$B$_5$O$_8$·4H$_2$O	T	0.05	0.05	0.0014			>0.0001^e	
	SQ	0.4^d	1^d	5^d			2^d	10^d

^{a} Ref. 33.
^{b} T = technical and SQ = special quality.
^{c} Also contains 10 ppm phosphate.
^{d} Values are ppm.
^{e} Maximum values are not set. These are typical values.

indicate no evidence of carcinogenic (108) or mutagenic activity (109). Boric acid and borax are poorly absorbed through healthy skin and do not cause skin irritation. A permissible exposure limit (PEL) of 10 mg/m^3 of sodium borate dust has been adopted by OSHA in the United States (110).

Sodium metaborate hydrates are more alkaline than borax and greater care is required in handling. The metaborate material is harmful to the eyes and can cause skin irritation. Gloves, goggles, and a simple dust mask should be used when handling sodium metaborate powder.

Boron in the form of borate is an essential micronutrient for the healthy growth of plants and is present in the normal daily human diet at an estimated level of 3–40 mg as boron. It is not a proven essential micronutrient for animals (111).

The handling of boric acid and borax is generally not considered dangerous. There are no fire risks associated with the storage or use of inorganic borates, and they are not explosive.

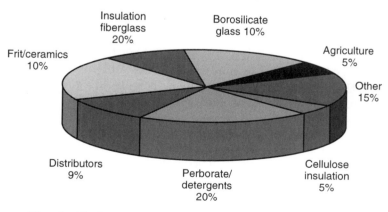

Fig. 6.   End-use makeup of the global market for borates.

7.16. Uses. The single largest end use for borate is insulation fiberglass or glass wool. Insulation fiberglass accounts for 20% of the total world B_2O_3 demand and >30% of the total demand in North America. Of the estimated half million tons of borates used for insulation fiberglass production in 1996. 60% was used in Norh America, 30% in Europe, and 10% in other regions of the world.

The second largest end use is perborate and cleaning products, accounting for just under 20% of world demand. Much of this use category consists of perborate, which is included in the formulations of various soapers. Approximately 95% of the perborate demand originates in Europe with the remainder arising mostly from North America. Conversely, the demand for cleaning products is approximately 90% from North America and the remainder from Europe.

The third largest end use for borates is textile fiberglass, accounting for roughly 15% of world borate demand. More than half of the world's textile fiberglass producers are in North America, another 25% in Europe, and 20% in Asia.

Another 10% of the world's borates end up in ceramics and frits. The leading market is Europe with more than half of that production taking place in Spain and Italy. Producers in Asia and South America each account for another 15% of borate use in ceramics and frits.

The final catchall category, "other uses," accounts for an additional 15% of global demand for borates. Most end uses for these categories are relatively minor applications such as cellulose and adhesives. Therefore, these manufacturers have a small annual tonnage requirement.

Borosilicate glass accounts for 10% of B_2O_3 world demand; >70% of this demand arises from North America and Europe. Cellulose insulation is the smallest of the end use categories and accounts for only 2% of world demand, 80% of which comes from producers in North America.

Disodium Tetraborate Decahydrate. In the United States, nearly all the refined borax is used for household cleaning products. Small amounts are used as fertilizers and herbicides. USP-grade borax is used in cosmetic and toilet goods, in which purity is demanded. Special quality-grade borax is used in electrolytic capacitors, in nuclear applications, and as a laboratory chemical.

Disodium Tetraborate Pentahydrate. Refined pentahydrate consumed in the United States is used in insulation fiber glass, glass, fertilizers, and herbicides. Smaller amounts are used in antifreeze (see ANTIFREEZES AND DEICING FLUIDS), ceramic glazes, and cleaning agents. About 40% of the pentahydrate produced in the United States is exported (112). A large-scale application of this chemical is in the preparation of perborate bleaches.

Disodium Tetraborate. In the United States, anhydrous borax finds most application in the glass industry for enamels, borosilicate glass, and fiber glass insulation. It is also used as an antifreeze additive and as an algicide in industrial water.

Disodium Octaborate Tetrahydrate. Commercially available products, having the approximate composition of a hypothetical disodium octaborate tetrahydrate, have found application in wood (qv) preservatives, fertilizer sprays, insecticides, herbicides, and fire retardants. In many applications the large water solubility of these products is an asset.

Disodium octaborate tetrahydrate (TIM-BOR) is registered for a variety of pests including termites, wood destroying beetles, and carpenter ants. This same compound is also used to control fly larvae in manure piles and is marketed as POLYBOR 3 for this application.

Disodium octaborate tetrahydrate is also used to protect wood from wood destroying fungi and pests. Whereas it has mainly been used for this application in New Zealand, it is being introduced into the United States for this use.

Sodium Metaborate Tetrahydrate and Dihydrate. The sodium metaborates are components in textile finishing, sizing and scouring compositions, adhesives, and detergents. They are also used in many photographic applications. In agriculture they are used in both herbicides and fertilizer sprays. The dihydrate is less affected by heat.

8. Other Alkali Metal and Ammonium Borates

8.1. Dipotassium Tetraborate Tetrahydrate.
Dipotassium tetraborate tetrahydrate, $K_2B_4O_7 \cdot 4H_2O$ or $K_2O \cdot 2B_2O_3 \cdot 4H_2O$, formula wt, 305.49; orthorhombic; sp gr, 1.919; is much more soluble than borax in water. Solubility data are given in Table 10; pH with concentration is given in Table 11.

Phase relationships in the system $K_2O-B_2O_3-H_2O$ have been described and a portion of the phase diagram is given in Figure 7. The tetrahydrate,

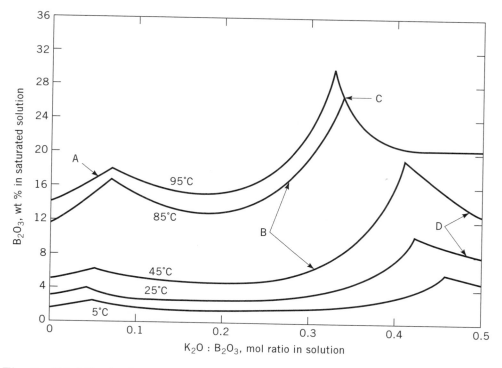

Fig. 7. Solubility isotherms for the $K_2O-B_2O_3-H_2O$ system at temperatures from 5 to 95°C where A, B, C, and D represent the solid phases $B(OH)_3$, $K_2O \cdot 5B_2O_3 \cdot 8H_2O$, $2K_2O \cdot 5B_2O_3 \cdot 5H_2O$, and $K_2O \cdot 2B_2O_3 \cdot 4H_2O$, respectively (113).

which can be dried at 65°C without loss of water of crystallization, begins to dehydrate between 85 and 111°C, depending on the partial pressure of water vapor in the atmosphere. This conversion is reversible and has a heat of dehydration of 86.6 kJ/mol (20.7 kcal/mol) of H_2O. Thermogravimetric curves indicate that two moles of water are lost between 112 and 140°C, one more between 200 and 230°C and the last between 250 and 290°C (114).

Single-crystal x-ray studies have shown that the borate ion in the potassium compound is identical to that found in borax (4) and has the structural formula $K_2[B_4O_5(OH)_4] \cdot 2H_2O$ (115).

8.2. Potassium Pentaborate Tetrahydrate.

Potassium pentaborate tetrahydrate, $KB_5O_8 \cdot 4H_2O$ or $K_2O \cdot 5B_2O_3 \cdot 8H_2O$; formula wt, 293.20; orthorhombic prisms; sp gr, 1.74; heat capacity, 329.0 J/(mol·K) [78.6 cal/(mol·K)] at 296.6 K; is much less soluble than sodium pentaborate (Tables 10 and 11). Heat capacity measurements on the solid have been made over a broad temperature range (87).

The tetrahydrate is stable under normal conditions of storage. Its heat of dehydration has been calculated as 110.8 kJ/mol (26.5 kcal/mol) between 106.5 and 134°C (114). Its thermal stability is highly dependent on the partial pressure of atmospheric water. It is stable when heated in a vaccum up to 105°C; in an atmosphere saturated with water at 90°C, it is stable up to 170°C.

The solid-state structural formula is $K[B_5O_6(OH)_4] \cdot H_2O$ (116), which is analogous to that found in sodium pentaborate (5).

8.3. Diammonium Tetraborate Tetrahydrate.

Diammonium tetraborate tetrahydrate, $(NH_4)_2B_4O_7 \cdot 4H_2O$ or $(NH_4)_2O \cdot 2B_2O_3 \cdot H_2O$; formula wt, 263.37; monoclinic; sp gr, 1.58; is readily soluble in water (Table 10). The pH of solutions of diammonium tetraborate tetrahydrate is 8.8 and independent of concentration. The compound is quite unstable and exhibits an appreciable vapor pressure of ammonia. Phase relationships have been outlined and the x-ray crystal structure formula is $(NH_4)_2[B_4O_5(OH)_4] \cdot 2H_2O$ (117).

8.4. Ammonium Pentaborate Tetrahydrate.

Ammonium pentaborate tetrahydrate, $NH_4B_5O_8 \cdot 4H_2O$ or $(NH_4)_2O \cdot 5B_2O_3 \cdot 8H_2O$; formula wt, 272.13; sp gr, 1.567; heat capacity, 359.4 J/(mol·K) [85.9 cal/(mol·K)] at 301.2 K; exists in two crystalline forms, orthorhombic (α) and monoclinic (β). The α-form, which crystallizes as the kinetic product, is the commercial form of ammonium pentaborate tetrahydrate and the β-form is the thermodynamic product but is slow to crystallize. Its heat capacity has been measured over a broad temperature range (87). Solubility data are given in Table 10 and pH data in Table 11.

Ammonium pentaborate tetrahydrate is very stable in respect to ammonia loss. On heating from 100 to 230°C, it loses 75% of its water content but less than 1% of the ammonia. At 200°C, under reduced pressure, the water content drops to 1.15 mol, but only 2% of the ammonia is lost (62). At still higher temperatures all ammonia and water are expelled to give boric oxide (118).

The pentaborate is shown by x-ray data to contain the $[B_5O_6(OH)_4]^-$ ion (5), analogous to that found in the sodium and potassium compounds. The α-form has the structural formula $NH_4[B_5O_6(OH)_4] \cdot 2H_2O$ and the β-form $NH_4[B_5O_8] \cdot 4H_2O$ (119).

8.5. Lithium Borates.

Two lithium borates are of minor commercial importance, the tetraborate trihydrate and metaborate hydrates.

Dilithium tetraborate trihydrate, $Li_2B_4O_7 \cdot 3H_2O$ or $Li_2O \cdot 2B_2O_3 \cdot 3H_2O$, has a density of 1.88 g/mL. It crystallizes with difficulty from a supersaturated solution of lithium hydroxide and boric acid, which on standing forms a gelatinous deposit that is converted to hydrate crystals after boiling for several hours. The trihydrate is stable up to 180°C, then the compound dehydrates becoming anhydrous up to about 320°C, and fuses at 890°C.

Lithium metaborate octahydrate, $LiBO_2$ $8H_2O$ or $Li_2O \cdot B_2O_3 \cdot 16H_2O$, hexagonal; $d = 1.825$ g/mL; has the structural formula $Li[B(OH)_4 \cdot 6H_2O$ (120). On heating to 70°C six waters are lost; the last two waters are lost between 140 and 280°C (121).

The octahydrate is the stable solid phase in contact with its solution below 36.9°C. Above this temperature lithium metaborate dihydrate, $LiBO_2 \cdot 2H_2O$ or $Li_2O \cdot B_2O_3 \cdot 4H_2O$, becomes the stable solid phase. Dihydrate crystals are orthorhombic having a density of 1.825 g/mL and a structural formula $Li[B(OH)_4]$. In solution above 150°C a hemihydrate, $LiBO_2 \cdot 1/2H_2O$, forms and the anhydrous salt crystallizes above 225°C.

8.6. Manufacture. Potassium tetraborate tetrahydrate may be prepared from an aqueous solution of KOH and boric acid having a $B_2O_3:K_2O$ ratio of about 2 or by separation from a KCl–borax solution (122). Potassium pentaborate is prepared in a manner analogous to that used for the tetraborate, but the strong liquor has a $B_2O_3:K_2O$ ratio near 5.

Ammonium tetraborate tetrahydrate is prepared by crystallization from an aqueous solution of boric acid and ammonia having a $B_2O_3:(NH_4)_2O$ ratio of 1.8:2.1. Ammonium pentaborate is similarly produced from an aqueous solution of boric acid and ammonia having a $B_2O_3:(NH_4)_2O$ ratio of 5. Supersaturated solutions are easily formed and the rate of crystallization is proportional to the extent of supersaturation (123). A process for the production of ammonium pentaborate by precipitation from an aqueous ammonium chloride–borax mixture has been patented (124).

8.7. Economic Aspects. The potassium, lithium, and ammonium borates are low volume products. Annual production figures are in the range of hundreds of metric tons.

8.8. Health and Safety. Little toxicological data are available on borates other than boric acid and borax. Most water-soluble borates have the same toxicological effects as borax when adjusted to account for differences in B_2O_3 content.

8.9. Analytical Methods. *Sodium, Potassium, and Ammonium Borates.* Sodium, Potassium, and ammonium salts of borates are first dissolved in hot water. The solution is neutralized using hydrochloric acid (HCl). An excess amount of HCl is added to the solution. Carbon dioxide is removed by boiling. It is neutralized with NaOH to a boric acid solution using methyl red as the indicator. Mannitol (or glycerol) is then added, and the resulting solution is titrated with NaOH solution using phenolphthalein as the indicator. The purity of the product is calculated by dividing the actual B_2O_3 content by the theoretical B_2o_3 content in the chemical formula.

8.10. Uses. Dipotassium tetraborate tetrahydrate is used to replace borax in applications where an alkali metal borate is needed but sodium salts cannot be used or where a more soluble form is required. The potassium compound is used as a solvent for casein, as a constituent in welding fluxes, and a component in diazotype developer solutions. Potassium pentaborate tetrahydrate is used in fluxes for welding and brazing of stainless steels for nonferrous metals. Diammonium tetraborate tetrahydrate is used when a highly soluble borate is desired but alkali metals cannot be tolerated. It is used mostly as a neutralizing agent in the manufacture of urea–formaldehyde resins and as an ingredient in flameproofing formulations. Ammonium pentaborate tetrahydrate is used as a component of electrolytes for electrolytic capacitors, as an ingredient in flameproofing formulations, and in paper coatings.

9. Calcium-Containing Borates

9.1. Dicalcium Hexaborate Pentahydrate. Dicalcium hexaborate pentahydrate, $Ca_2B_6O_{11} \cdot 5H_2O$ or $2CaO \cdot 3B_2O_3 \cdot 5H_2O$; formula wt, 411.08; monoclinic; sp gr, 2.42; heat of formation, -3.469 kJ/mol (-0.83 kcal/mol)(125); exists in nature as the mineral colemanite. Its solubility in water is about 0.1% at 25°C and 0.38% at 100°C. Heats of solution have been determined in HCl (125). Colemanite is slowly formed on heating saturated solutions of inyoite, $2CaO \cdot 3B_2O_3 \cdot 13H_2O$, or other higher hydrates. Colemanite decrepitates violently at 480°C losing all its water and forming an anhydrous very low bulk density powder (126).

The crystal structure of colemanite has been shown to contain $[B_3O_4\text{-}(OH)_3]^{2n-};_n$ polyanion chains. The structural relationships between colemanite and the other minerals of the series $2CaO \cdot 3B_2O_3 \cdot nH_2O$ ($n = 1, 5, 7, 9, 13$), and structural changes accompanying the ferroelectric transition of colemanite have been outlined (127).

9.2. Sodium Calcium Pentaborate Octahydrate. Sodium calcium pentaborate octahydrate, $NaCaB_5O_9 \cdot 8H_2O$ or $Na_2O \cdot 2CaO \cdot 5B_2O_3 \cdot 16H_2O$; formula wt, 405.23; triclinic; sp gr, 1.95; exists in nature as the mineral ulexite. The compound can be prepared by seeding a solution of 110 g $CaB_2O_4 \cdot 6H_2O$, 40 g boric acid, 100 g borax, 450 g $CaCl_2$ and 2.5 L H_2O (111). Ulexite is slowly converted to $NaCaB_5O_9 \cdot 5H_2O$, probertite, when seed is added to a moistened sample at 80–100°C. When crystals of ulexite are heated, four moles of water are lost at 80–100°C, 8.5 more until 175°C, and the remaining 3.5 on heating to 450°C (128).

The x-ray crystal structure consists of isolated pentaborate polyanions and the structural formula is $NaCa[B_5O_8(OH)_6] \cdot 5H_2O$ (129). Some specimens of ulexite have fiber optic properties with surprisingly good resolution of projected images. The fiber is aligned along the c-axis with index of 1.529. Cladding results from random orientation of crystals about the fiber direction, producing a core-to-cladding index difference ranging from 0 to a maximum of $\gamma - \alpha = 0.038$ (130).

The solubility in water at 25°C is 0.5% as $NaCaB_5O_9$. Calcining at 200–500°C increases its solubility to 9–13 g/L.

9.3. Sodium Calcium Pentaborate Pentahydrate. Sodium calcium pentaborate pentahydrate, $NaCaB_5O_9 \cdot 5H_2O$ or $Na_2O \cdot 2CaO \cdot 5B_2O_3 \cdot 10H_2O$; formula wt 351.19; monoclinic; sp gr, 2.14; exists in nature as the mineral probertite. Probertite can be prepared by heating a mixture of two parts ulexite and one part borax to about 60°C (131) or by heating a borax and calcium metaborate solution at 105°C for eight days (76). The structural formula $NaCa[B_5O_7(OH)_4] \cdot 3H_2O$ has been determined from the crystal structure (132). By thermogravimetric analysis two moles of water are lost at 100°C, four more from 100° to 180°C, and slow loss of the last four up to 400°C.

9.4. Manufacture. The alkaline-earth metal borates of primary commercial importance are colemanite and ulexite. Both of these borates are sold as impure ore concentrates from Turkey, which is the principal world supplier. Colemanite and ulexite mining areas in Turkey are the Bigadic, Emet, and Kestelek regions. In 1986 Etibank run-of-mine production was about 793,000 t/yr of colemanite and 185,000 t/yr of ulexite. The concentrates produced are glass-grade material. The mining is both open-pit and underground.

At Hisarcik, in the Emet District, Etibank operates an open-pit mine and a colemanite concentrating plant. The production from this plant is relatively high in arsenic, about 3500 ppm. The ore consists of colemanite nodules, closely packed with shale. The presence of high concentrations of arsenic sulfides has been indicated. Plant capacity is about 184,000 t/yr as B_2O_3. At Espey, Etibank operates an underground mine, which was to be converted to an open-pit mine, that had a capacity of 25,000 t/yr as B_2O_3. A concentrating plant at Bigadic has a capacity of 132,000 t/yr as B_2O_3 (6,133).

Death Valley, California, has historically been a significant source of both colemanite and ulexite, but mining in the Death Valley National Monument has been forbidden as a result of environmental concerns. In 1986, the American Borate Co. ceased mining in Death Valley, but continues to market ore concentrate from inventory as well as borates and concentrates imported from Turkey.

9.5. Specifications and Shipping. The colemanite, which is to be used in the production of glass fibers, must conform to the purchasers' specifications on Fe and As. Colemanite is available in bags and bulk.

9.6. Uses. Colemanite, $2CaO \cdot 3B_2O_3 \cdot 5H_2O$, is used in the production of boric acid and borax, as well as in several direct applications. It is a highly desirable material for the manufacture of the E-glass used in textile glass fibers and plastic reinforcement (where sodium cannot be tolerated). High As or Fe levels in the ore concentrate can limit its use in this application. Colemanite has seen limited application as a slagging material in steel manufacture. It is also used in some fire retardants and as a precursor to some boron alloys.

Ulexite, $NaCaB_5O_9 \cdot 8H_2O$, and probertite, $NaCaB_5O_9 \cdot 5H_2O$, have found application in the production of insulation fiber glass and borosilicate glass as well as in the manufacture of other borates.

Borate Melts and Glasses. Like silicon oxide and lead oxide, boric oxide (B_2O_3) is a natural glass network-forming oxide having very strong covalent bonds. These glass-forming oxides are capable of existing in the vitreous state either alone or in combination with other oxides. When heated alkali metal oxides, hydroxides, or carbonates fuse with boric acid or hydrated alkali metal borates to form a clear liquid melt. If these liquids are high (M_2O/B_2O_3 from 0 to 2 mol) in boric oxide content, they become viscous on cooling and form glasses.

Most of the interest in alkali metal borate glasses has centered on reports indicating the existence of maxima and minima in some of the physical properties

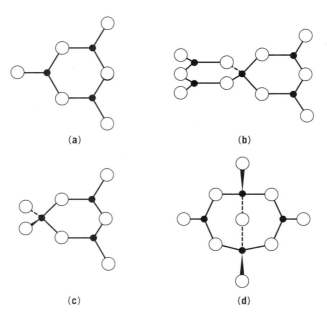

Fig. 8. The borate glass structural groups where • = boron; ○ = oxygen; (**a**) boroxol; (**b**) pentaborate; (**c**) triborate; and (**d**) diborate (135).

of the glasses, such as viscosity, density, and thermal expansion coefficient, that occur with increasing metal oxide content. This phenomenon has been called the boron oxide anomaly (134). Modern theory on borate glass structure, however, indicates that these property changes are not anomalous but are the result of well-defined structural changes in the glass at the molecular level. Four different borate structural groups have been theorized to be present in alkali borate glasses below 34 mol % M_2O (135). These groups are shown in Figure 8. The triborate and pentaborate groups always occur in pairs which are then referred to as tetraborate groups. The argument that pure vitreous B_2O_3 consists of flat BO_3 triangles in the form of boroxol rings connected by BO_3 chains is strongly supported by boron nuclear quadrapole resonance spectroscopy (136). Addition of alkali oxide to boron oxide to form an alkali borate glass results in transformation of BO_3 triangles to BO_4 tetrahedra in a continuous manner. On addition of M_2O, up to 20 mol %, tetraborate groups are formed. Between 20 and 34 mol % M_2O, diborate groups form at the expense of the tetraborate groups. Infrared (135) and laser Raman (28) data on borate glasses and the analogous crystalline anhydrous borates support this reasoning. Changes in the physical properties of the glass with M_2O content represent compromises between the effect of adding more metal ions to the system and the effect of making the borate structural entities more rigid by converting trigonally coordinated borons to tetrahedrally coordinated borons.

Another widely studied phenomenon in alkali borate glasses is the mixed alkali effect, the nonlinear change in glass properties when a second kind of alkali oxide is added into the single-alkali glass. Models have been suggested to explain the mixed alkali effect (137), but a universally accepted model has not been developed as of this writing.

A number of reviews have appeared covering the various aspects of borate glasses. The structure, physical properties, thermochemistry, reactions, phase equilibria, and electrical properties of alkali borate melts and glasses have been presented (75). The application of x-ray diffraction, nmr, Raman scattering, ir spectroscopy, and esr to structural analysis is available (26). Phase-equilibrium diagrams for a large number of anhydrous borate systems are included in a compilation (138), and thermochemical data on the anhydrous alkali metal borates have been compiled (17).

The largest single commercial use of borates is in fiber glass. There are two basic types of glass fibers: insulation (soda lime borosilicate glass) and textile (low alkali lime aluminosilicate glass) grades. Borax pentahydrate is the most common source of B_2O_3 for making insulation fiber glass. Textile or E-glass fiber requires low sodium formulations and for this reason boric acid or colemanite is commonly used. Only borates having low arsenic content are suitable for use in glass making. Smaller amounts of borates are consumed in heat-resisting (Pyrex or low thermal expansion) glass, sealing glass, glazes and enamels (frit), optical glass, nuclear waste storage glass, and in the making of vycor. The typical range of B_2O_3 content in commercial glasses is shown in Table 13. Borates are not generally used in container or flat glass.

Boron oxide, B_2O_3, can be added to a glass formulation from a variety of boron-containing compounds, but because the boron is taken into vitreous solution, it is often immaterial which source of boron is used. The choice of raw material is usually determined by consideration of the price per contained B_2O_3 unit, uniformity of composition, purity, hydration state, and compatibility of the cation in the finished glass. In addition, boron-containing raw materials are usually the only water-containing constituents in a glass batch and because the water must be removed in the melting furnace, the dehydration characteristics are important (137).

Boron oxide is added to borosilicate glass formulations to improve properties both in the finished glass and in the glass-making process (146). The benefits of B_2O_3 use in glass making are (1) creation of a low melting flux to dissolve refractory silica; (2) a lower liquidus temperature and inhibition of devitrification; (3) lower melt viscosity; (4) enhanced melt rate; and (5) improved draw qualities in fiber production. The benefits of B_2O_3 in the finished glass product

Table 13. **Borate Content in Commercial Glass**

Glass type	B_2O_3, wt %
fiber	
textile	6–13
insulation	3–7
heat resisting (pyrex)	12–15
sealing glass	8–30
porcelain enamel (frit)	11–13
vycor	20[a]

[a] This is the percent B_2O_3 in glass prior to acid leaching to form vycor.

are (1) improved chemical durability; (2) lower thermal expansion; (3) increased mechanical strength; (4) decreased devitrification tendency; (5) improved scratching hardness; and (6) enhanced refraction, color, and brilliance.

10. Other Metal Borates

Borate salts or complexes of virtually every metal have been prepared. For most metals, a series of hydrated and anhydrous compounds may be obtained by varying the starting materials and/or reaction conditions. Some have achieved commercial importance.

In general, hydrated borates of heavy metals are prepared by mixing aqueous solutions or suspensions of the metal oxides, sulfates, or halides and boric acid or alkali metal borates such as borax. The precipitates formed from basic solutions are often sparingly-soluble amorphous solids having variable compositions. Crystalline products are generally obtained from slightly acidic solutions.

Anhydrous metal borates may be prepared by heating the hydrated salts to 300–500°C, or by direct fusion of the metal oxide with boric acid or B_2O_3. Many binary and tertiary anhydrous systems containing B_2O_3 form vitreous phases over certain ranges of composition (138).

10.1. Barium Metaborate. Three hydrates of barium metaborate, $BaO \cdot B_2O_3 \cdot xH_2O$, are known. The tetrahydrate (140) and pentahydrate (141) both contain the $B(OH)_4^-$ anion, and are properly formulated as $Ba[B(OH)_4]_2 \cdot xH_2O$, where $x = 0$ or 1. These compounds crystallize when solutions of barium chloride and sodium metaborate are combined at room temperature (142). The higher hydrate is favored when excess sodium metaborate is used. Saturated aqueous solutions contain 13.5 g/L of $BaO \cdot B_2O_3 \cdot 4H_2O$ at 25°C. Both forms dehydrate at temperatures above 140°C (143). Barium metaborate may also be prepared from barium sulfide formed by prior reduction of barium sulfate. The presence of sulfide impurities in the product may render it unsuitable for some applications (62). Crystals of a hydrate, $x = 1.67$ H_2O, form from a boiling solution having the B:Ba molar ratio <2. Dehydration of this hydrate at 300°C gives $BaO \cdot B_2O_3$ in which boron atoms are both triangularly and tetrahedrally coordinated (144).

Barium metaborate is used as an additive to impart fire-retardant and mildew-resistant properties to latex paints, plastics, textiles, and paper products (6).

10.2. Copper, Manganese, and Cobalt Borates. Borate salts of copper, manganese, and cobalt are precipitated when borax is added to aqueous solutions of the metal(II) sulfates or chlorides (145). However, these materials are no longer produced commercially.

10.3. Zinc Borates. A series of hydrated zinc borates have been developed for use as fire-retardant additives in coatings and polymers (59,146). Worldwide consumption of these zinc salts is several thousand metric tons per year. A substantial portion of this total is used in vinyl plastics where zinc borates are added alone or in combination with other fire retardants such as antimony oxide or alumina trihydrate.

Zinc borate $2ZnO \cdot 3B_2O_3 \cdot 7H_2O$ is formed when borax is added to aqueous solutions of soluble zinc salts at temperatures below about 70°C. An x-ray structure determination has indicated that this compound is orthorhombic and has a zinc triborate monohydrate structure, $Zn[B_3O_3(OH)_5] \cdot H_2O$ (2). Zinc borates $2ZnO \cdot 3B_2O_3 \cdot 7H_2O$ and $ZnO \cdot B_2O_3 \cdot 2H_2O$ lose water of hydration when heated from 130 to 250°C (59).

A different crystalline hydrate, $2ZnO \cdot 3B_2O_3 \cdot 3.5H_2O$, equivalent to $4ZnO \cdot 6B_2O_3 \cdot 7H_2O$, is produced when the reaction between zinc oxide and boric acid is carried out at temperatures of 90–100°C (147). This product has also been crystallized from solutions containing borax, zinc chloride, and sodium hydroxide (148). It is marketed by the United States Borax & Chemical Corp. under the trademark FIREBRAKE ZB, BOROGARD ZB, and under AMAX, Inc. as ZB-467. This compound has the unusual property of retaining its water of hydration at temperatures up to 290°C. This thermal stability makes it attractive as a fireretardant additive for plastics and rubbers that require high processing temperatures. It is also used as an anticorrosive pigment in coatings. Zinc borates are also manufactured by Storey (UK) and Waardals (Norway).

Zinc borate $2ZnO \cdot 3B_2O_3 \cdot 3.5H_2O$ has an acute oral toxicity in rats $LD_{50} > 10,000$ mg/kg body weight and acute dermal toxicity in rabbits $LD_{50} > 10,000$ mg/kg body weight. It is not a skin irritant and gives a negative response in the Ames mutagenicity test.

11. Boron Phosphate

Boron phosphate, BPO_4, is a white, infusible solid that vaporizes slowly above 1450°C, without apparent decomposition. It is normally prepared by dehydrating mixtures of boric acid and phosphoric acid at temperatures up to 1200°C.

$$B(OH)_3 + H_3PO_4 \longrightarrow BPO_4 + 3\,H_2O$$

Complete dehydration requires temperatures above 1000°C.

The structure of boron phosphate prepared under normal atmospheric conditions consists of tetragonal bipyrimids analogous to the high cristobalite form of silica. Both the boron and phosphorus are tetrahedrally coordinated by oxygen. Similar silicalike structures are found for $BAsO_4$ and $TaBO_4$ (149). A quartzlike form of boron phosphate can be prepared by heating the common form to 500°C at 5.07 GPa (50,000 atm) (150).

The tri-, tetra-, penta-, and hexahydrates of boron phosphate have been reported. All of these decompose rapidly in water to give solutions of the parent acids. Anhydrous boron phosphate hydrolyzes in a similar fashion, though the reaction proceeds quite slowly for material that has been ignited at high temperatures.

The principal application of boron phosphate has been as a heterogeneous acid catalyst (151).

Although boron phosphate is derived from two of the three most common glass-forming oxides, it exhibits little tendency to form a glass itself. Boron phosphate is a primary phase over a considerable portion of the B_2O_3–SiO_2–P_2O_5 system (152).

12. Environmental Concerns, General

Boron is present everywhere — in soil, water, plants, and animals — in trace amounts. Although scientists refer to "levels of boron" when addressing environmental, health, and safety issues, is important to note that the element boron does not exist by itself in nature. Rather, boron combines with oxygen and other elements to form boric acid, or inorganic salts called borates.

Concentrations of boron compounds large enough to mine are limited to only a few places on the planet. In fact, the vast majority are mined at one of two world-class deposits. Boron compounds are essential ingredients in many industrial processes including the manufacture of glass, ceramics, fiberglass insulation, detergents, flame retardants, and wood protection products.

Despite the millions of tons of industrial borates mined, processed, and distributed around the world every year, a far larger quantity of boron travels around the planet by way of natural forces. Rain, volcanic activity, condensation, and other atmospheric activities move at least twice as much boron as all commercial practices combined.

Boron was identified in the 1920s as one of the seven micronutrients essential to all plants. The element is integral to a plant's reproductive cycle: controlligg flowering, pollen production, germination, and seed and fruit development. Boron also acts as a fuel pump, aiding the transmission of sugars from older leaves yo new growth areas and root systems (153).

In some regions, heavy rainfall, geologic characteristics, or farming practices have leached boron from the soil, leaving too little to support plant and crop survival. Many crop plants — including canola, alfalfa, clover, corn, cotton, peanuts, apples, carrots, celery, table beets, and fruit and nut trees — are regularly treated with foliar or systemic boron supplements to enhance crop quality and yield.

On the other hand, high concentrations of boron can be harmful to some types of plants. Citrus trees, for instance, are particularly susceptible to boron toxicity.

Overall, agriculture relies heavily on boron supplementation to ensure an adequate food supply. Studies indicate that people in a wide variety of cultures consume about 1–3 mg of boron per day through a combination of foods and drinking water in their local diets (154). Although it has yet to be proved that humans need boron to live, there is almost universal agreement that boron is nutritionally important to maintaining optimal human health.

In summary, trace amounts of boron are enviornmentally ubiquitous, essential to plant life and nutritionally important to humans. However, as with any substance, the dose makes the poison. In other words, overexposure to boron compounds can be hazardous to plants, animls, and even people. Concentration of boron in the environment are discussed below (155).

12.1. Boron in Water.   Boron occurs naturally in seawater at an average concentration of 5 mg of boron per liter. Boron generally occurs in freshwater at concentrations of up to 1 mg/L, or one part per million (ppm) (156).

A boron concentration of 1 ppm is believed to be safe for all aquatic creatures. As context, 1 ppm is equivalent to a handful of borax in a body of water the size of an Olympic swimming pool.

Higher concentrations of boron in water — ranging from 10 to 1000 ppm — have been created in laboratory settings in order to measure boron toxicity and its effects. These lab studies show that extremely high levels of boron may have a negative impact on the reproductive systems of aquatic organisms. Zebrafish and rainbow trout – species with a heightened sensitivity to boron in their larval stages — are most often used in this reserch adverse. It is interesting to note that studies using zebra fish have shown that the fish exhibit adverse reproductive and developmental effects when exposed to enviornments where *not enough* boron is present, as well (157). Simply put, not enough boron is as harmful as too much.

12.2. Boron in Soil.   The earth's soil can be categorized as low boron (<10 ppm boron) or high boron (on the order of 100 ppm boron). The average overall concentration of boron in all soil ranges from 10 to 20 ppm.

Because borates dissolve easily, their concentration in soil is greatly dependent on the physical and chemical characteristics of the soil, as well as the availability of water moving through it. As mentioned above, when natural boron levels have been depleted through leaching by rainwater or uptake by plants, soils are regularly treated with borates to encourage crop production.

Extremely high boron soils are rare. In fact, there are only two known deposits of borates on the planet: in California's Mojave Desert and in Turkey. Both deposists were formed over many millions of years as a result of precise geologic conditions that fostered high concentrations of borates becoming encapsulated in nonpermeable layers of clay.

12.3. Boron in Air.   Relatively little information is available about how much boron is present in the air. Boron has been detected in measurable quantities in rain, snow, evaporation from seawater and hot springs, and industrial airborne dust.

In rain and snow, boron has been reported in concentrations ranging between 0.003 and 0.005 mg/L. On a global basis, boron moves through the atmoshere at a rate of 5–7 million tonnes per year.

13. Health and Safety Factors, General

This section introduces some of the effects of both overexposure and deficiency of boron. Boron compounds have relatively low acute toxicity — meaning that ingestion of small amounts is not likely to cause health problems. Laboratory studies show that very large doses of boron compounds can cause reproductive and developmental effects in animals. However, similar effects have *not* been observed in humans.

Although both humans and animals readily absorb boron compounds, the vast majority is excreted rapidly; in humans, 80–100 % of that elimination is through urination. Further, boron is not metabolized by animals or by humans. Boron compounds are poorly absorbed through intact skin in both humans and animals, although they can be absorbed through damaged skin. Boric acid and other boron compounds are used at a concentration of 5% in cosmetics and talc in the United States and Europe. Data also indicate that inhaling boron compounds does not pose a significant risk to animals or humans.

13.1. Effects in Humans.   Early literature reported the toxicity of boric acid when it was used as a medicinal treatment; that use has long since been discontinued. Accidental misuse resulting in poisoning and death occured in hospitals — most notably when a 30% solution of boric acid was mistaken for distilled water and used to make up baby formula for newborns and from repeated use of 100% boric acid powder on large areas of diaper rash.

Adults may display adverse health effects after ingesting ~3 g of boric acid. This amount represents 1000–3000 times more boron than an adult consumes each day in a normal healthy diet (158).

Of 784 cases of acute boric acid ingestion reported to the National Capital Poison Review Center and the Maryland Poison Center between 1981 and 1985, 88.3% were asmyptomatic. Among the remaining 11.7% less frequent symptoms included lethargy, headache, lightheadness, and rash. Although one death has been reported after a person injested about 30 g of boric acid, ≤ 88 g of boric acid has been ingested, by an adult, without undue harm (159).

A study of mine and refinery workers exposed to high levels (≤ 14 mg/m^3) of borate dust measured only minor respiratory effects such as mild nose, eye, and throat irritation. No pulmonary or other health effects were observed in workers exposed to higher than average borate levels over time (160).

The reproductive performance of workers at a large mining – refining facility was also studied. The study measured the rate of births to the wives of workers after prolonged occupational exposure to boron compounds; 72% of the eligible workers took part in this study. The study found that these workers had more offspring than the national average; indicating no adverse effect on human reproduction (161).

13.2. Animal Studies: Acute Toxicity.   Boron compounds have been tested on numerous species, including invertebrates; vertebrates, such as fish and frogs; and mammalian species, such as rats, mice, and dogs. These studies have been conducted to measure the effect of high levels of boron on animals in two ways: *acute*. meaning a single high dose; and *chronic*, meaning repeated exposures over longer periods of time.

The commonly used figure for determining acute toxicity is the LD$_{50}$, which is the threshold at which the administered dose is lethal to at least half of the animals being tested. The oral LD$_{50}$ for boric acid and most borates in rats ranges from 3.5 to 5.0 g per kilogram body weight. Oral LD$_{50}$ refers to test results in which animals ingest such a dosage on a one-time basis (162).

Dermal toxicity for borate products has been determined to be >5 g/kg body weight. The extremely low dermal toxicity is due to the fact that borates do not penetrate intact skin. In fact, the permeability of human skin to borate products has been determined to be <1%.

A wide range of inhalation studies indicate that borates are not toxic by inhalation. Research shows that the toxic inhalation threshold for boron compounds is >2 mg/L for boron compounds. In other words, animals tested at this extremely high air concentration did not die; rather, they displayed the effect of being exposed to a high level of dust, but recovered after a short time outside the inhalation chamber.

In skin and eye studies, boron compounds were found to be mildly to moderately irritating, but insufficiently harmful to classify and label them as irritants. The finding makes sense, as borates and boric acid are used in cosmetics, talc, oral hygiene products, eyewashes, and contact lens solutions for humans. Finally, there is no evidence of skin sensitization in humans or animals as a result of boron exposure.

13.3. Animal Studies: Chronic Toxicity.

Boron compounds have also been tested to determine their long-term or chronic toxic effects. The results of these studies are highlighted in the following paragraphs (162).

Numerous studies testing the effect of boron compounds on both bacteria and mammalian cells prove that they are not mutagenic (ie, do not cause genetic mutation), nor are boron compounds considered carcinogenic, or cancer-causing.

The main treatment-related effects of long-term exposure to high doses of boron compounds among animals are testicular atrophy and reduced fertility. Rats that ingested 58.5 mg boron/kg body weight per day for an extended period exhibited sterility with no viable sperm in males and decreased ovulation in females. In mice, 111 mg boron/kg body weight per day impaired male fertility but did not reduce female fertility.

High doses of boron compounds have also been shown to cause developmental effects in rats, mice, and rabbits. Symptoms of developmental toxicity (eg, decreased fetal body weight, skeletal defects, and circulatory system damage) were noted in pregnant animals at doses ranging from 13.3 to 87 mg boron/kg body weight per day. Maternal toxicity also occured at the higher exposure levels. The definitive rat developmental toxicity study of boric acid accepted today by all health and regulatory authorities demonstrated a no-effect level of 9.6 mg boron/kg body weight per day (163).

13.4. Essentiality.

Another area of boron research concerns essentiality, or what role boron plays in maintaing and promoting health in animals and humans. To demonstrate born's nutritonal essentiality, researchers are working to show that insufficient boron in the diet results in adverse health effects — from those that severely interfere with an organism's quality of life to those that result in death. The World Health Organization has concluded that boron is probably essential to humans (164).

In humans, boron's importance in energy metabolism, bone health and strength, and brain function has been established. In animals, reserch is currently under way to measure boron's impact on reproduction and development. One study involving zebrafish has demonstrated that boron deficiency significantly reduces embryo survival rate, and that embryonic death begins very early in the post-fertilization period in the absence of sufficient boron (165).

Early results of another study indicate that insufficient boron results in pooor-quality embryos — some dead and most developing poorly and dying prematurely — in frogs (166). Although it is more difficult to effectively eliminate dietary boron in mammalian studies, essentiality research is also being conducted with rats and mice.

13.5. Safety.

Boron is an ubiquitous element, essential to plants and a normal component of a healthy human diet. Although scientists create artificial levels of boron in laboratories to study the effects of over- and underexposure its distribution in the natural world poses little or no risk to plants, animals, or people.

Research on exposure to boron compounds proves that too much or too little boron triggers adverse health effects in a variety of plants and animals. These plant and animal studies also form the basis for determining safe levels of boron exposure for people.

13.6. Risk Assessments.   Risk assessments are conducted to pinpoint safe levels of boron in the air, food, water, or other exposure sources. To make this assessment, reults of existing boron toxicity and exposure studies are subjected to complex calculations to produce a generally acceptable guideline for human intake of a substance.

For instance, a particular risk assessment of boron in the the United States determined that 18 mg of boron per day would be an acceptable daily intake from all sources (167). Given that average dietary intake of boron is ~1.0 mg/day, and that the median boron level in United States drinking-water supplies is <0.10 mg boron/L, this and other risk assessments have concluded that boron exposure from all sources are not expected to pose any public health risk.

13.7. Occupational Safety.   Safety standards are most important for the people exposed to higher than usual amounts of boron compounds, including workers at mining and refining facilities, and at industrial plants where borates and boric acid are used to manufacture other products. These standards are also important tools for the cadre of local, national, and international regulators entrusted with protecting human health and the enviornment.

Occupational exposure limits are set by federal and state agencies to control worker exposure to borate dust in industrial settings. These exposure limits range within $1-15$ mg dust/m^3 air. The higher limits are consistent with the default values applied to "nuisance dusts" or materials that are not otherwise classified.

Within some industrial settings, there are areas of particularly high boron dust concentration. In these areas, personal protective equipment is required. The equipment can include respirators or dust masks, eye protection, gloves, and protective clothing.

13.8. Product Safety.   Since the late nineteenth century, borate products have enjoyed an excellent reputation for safety and effectiveness when used as directed. Borates and boric acid are used to manufacture a wide range of products, including fiberglass insulation, heat-resistant glass, flame retardants, cosmetics, fertilizers, and household laundry detergents.

Some products that contain boron compounds — such as fertilizers, pesticides, and pharmaceutical products — are regulated by state and federal agencies. Moreover, most industrial borate and boric acid products are regulated under the federal Occupational Safety and Health Administration (OSHA) hazard communication standard. Therefore, these products require Material Safety Data Sheets (MSDSs) and proper labeling.

MSDs are standardized documents that list a comprehensive range of information, including chemical composition and characteristics, hazard identification, first aid measures, firefighting and accidental release measures, proper handling and storage guidance, exposure controls and personal protection recommendations, detail on toxicologic and ecologic information, disposal considerations, transport information, and regulatory information.

BIBLIOGRAPHY

"Boron Oxides, Boric Acids, and Borates" under "Boron Compounds" in *ECT* 1st ed., Vol. 2, pp. 600–622, by M. H. Pickard, Pacific Coast Borax Co.; "Boron Compounds, Boron Oxides, Boric Acid, and Borates" in *ECT* 2nd ed., Vol. 3, pp. 608–652, by N. P. Nies, U.S. Borax Research Corp.; "Boron Oxides, Boric Acids, and Borates" under "Boron Compounds" in *ECT* 3rd ed., Vol. 4, pp. 67–110, by D. J. Doonan and L. D. Lower, U.S. Borax Research Corp.; "Boron Oxides, Boric Acids, and Borates," under Boron Compounds, in *ECT* 4th ed.; Vol. 4, pp. 365–413 by Robert A. Smith and Robert B. McBroom, U.S. Borax Research Corporation; "Boron Oxides, Boric Acid, and Borates" in *ECT* (online), posting date: November 27, 2000 by Robert A. Smith, Robert B. McBroom, US. Borax Research Corporation.

CITED PUBLICATIONS

1. R. Kister and C. Helvaci in D. Carr, ed. *Industrial Minerals and Rocks*, 6th ed., Littleton, Colo., 1994, pp. 171–185.
2. N. P. Nies and G. W. Campbell, in R. M. Adams, ed., *Boron, Metallo-Boron Compounds, and Boranes*, Interscience Publishers, New York, 1964, pp. 192–194.
3. J. Ozols, I. Tetere, and A. Ievins, *Latv. PSR Zinat. Akad. Vestis Kim. Ser.* (1), 3 (1973).
4. R. Thompson and A. J. E. Welch, eds., *Mellor's* Comprehensive Treatise on Inorganic and Theoretical Chemistry, Vol. V, *Boron*, Part A, *Boron-Oxygen Compounds*, Longman, New York, London, 1980.
5. G. Heller, in K. Niedenzu and K. C. Buschbeck, eds., *Gmelin Handbuch der Anorganischen Chemie, Band 28, Teil 7*, Springer-Verlag, Berlin, 1975, pp. 2–4.
6. G. Heller, in F. L. Boschke, ed., *Topics in Current Chemistry*, Vol. 131, Springer-Verlag, Berlin, Heidelberg, 1986, p. 39.
7. R. Will, Y. Sakuma, and R. Willhalm, "Boron Minerals and Chemicals Report," in *Chemical Economics Handbook*, SRI International, Menlo Park, Calif., Sept. 1990, p. 717.1000A.
8. J. M. Barker and S. J. LeFond, eds., *Borates: Economic, Geology and Production, Proceeding of a Symposium, Oct. 24, 1984*, SME-AIME in Denver, Colo., Society of Mining Engineers of the American Institute of Mining, Metallurgical, and Petroleum Engineering, Inc., New York, 1985.
9. F. L. Foster, J. F. Bauer, H. H. Russell III, and X. Xu in A. C. Wright, S. A. felter, and A. C. Hannon, eds., *Proc. Second International Conference on Boron Crystals and Melts*, Society of Glass Technology, UK. 1997, p. 324.
10. K. Loewenstein, *Manufacturing Technology of Continuous Glass Fibers*, 3rd ed., Elsevier, 1993.
11. S. Baele, *Mining Eng. Mag.*, 19–21 (June 1999).
12. W. G. Woods, in V. M. Bhatnagar, ed., *Fire Retardants*, P. 2, Vol. 3 of *Progress in Fire Retardance Series*, 1973, p. 120.
13. W. G. Woods, *J. Trace Elem. Exp. Med.* **9**, 153–163 (1996).
14. V. Morgan, "Boron Geochemistry," in Ref. 4, section A2, p. 119.
15. C. L. Christ, *J. Geol. Educ.* **20**, 235 (1972).
16. D. Garrett, *Borates, Handbook of Deposits, Processing, Properties, and Uses.* Academic Press, San Diego, 1998.
17. *Thermochemical Tables*, 2nd ed., Nat. Stand. Ref. Data Ser., Nat. Bur. Stand. (U.S.), JANAF, Washington, D.C., 1971.

18. Ref. 4, pp. 7–15.
19. G. S. Bogdanova, S. L. Antonova, and V. I. Kislyak, *Steklo. Keram.*, 13 (1975).
20. J. R. Soulen, P. Sthapitanonda, and J. R. Margrave, *J. Chem. Phys.* **59**, 132 (1955).
21. D. L. Hildebrand, W. F. Hall, and N. D. Potter, *J. Chem. Phys.* **39**, 296 (1963).
22. A. Napolitano, P. B. Macedo, and E. G. Hawkins, *J. Am. Corain. Soc.* **48**, 613 (1965).
23. J. Boow, *Phys. Chem. Glasses* **8**, 45 (1967); K. H. Stern, *J. Res. Nat. Bur. Stand.* **A69**, 281 (1965).
24. L. L. Sperry and J. D. MacKenzie, *Phys. Chem. Glasses* **9**, 91 (1968).
25. W. Poch, *Glastech. Ber.* **37**(12), 533 (1964).
26. D. L. Griscom, in L. D. Pye, V. D. Frechette, and N. J. Kreidl, eds., *Borate Glasses: Structure, Properties, and Application*, Vol. 12, Materials Science Research, Plenum Press, New York, 1979.
27. R. L. Mozzi and B. E. Warren, *J. Appl. Crystallogr.* **3**, 251 (1970); G. E. Jellison, Jr. and P. J. Bray, *Solid-State Commun.* **19**, 517 (1976).
28. W. L. Konijnendijk and J. M. Stevels, *J. Non-Cryst. Solids* **818**, 307 (1975).
29. P. L. Hanst, V. H. Early, and W. Klemperer, *J. Chem. Phys.* **42**, 1097 (1965).
30. J. D. MacKenzie and W. F. Claussen, *J. Am. Ceram. Soc.* **44**, 79 (1961).
31. R. Brueckner, *Glastech. Ber.* **37**(9), 413 (1964).
32. D. Kline, P. J. Bray, and H. M. Kriz, *J. Chem. Phys.* **48**(11), 5277 (1968).
33. *Industrial Products Catalog and Price Schedules*, United States Borax & Chemical Corp., Los Angeles, Calif., 1991.
34. Ref. 4, pp. 5–7.
35. D. D. Wagman and co-workers, *Nat. Bur. Stand. U.S. Tech. Note* **270-2**, 26 (1966).
36. D. White and co-workers, *J. Chem. Phys.* **32**, 481 (1960).
37. A. L. McCloskey, R. J. Brotherton, and J. L. Boone, *J. Am. Chem. Soc.* **83**, 4750 (1961).
38. F. A. Kanda and co-workers, *J. Am. Chem. Soc.* **83**, 1509 (1956).
39. D. R. Petrak, R. Ruh, and B. F. Goosey, *Proc. 5th Mater. Res. Symp. Solid-State Chem. Nat. Bur. Stand. U.S. Special Pub. No. 364*, 605 (1972).
40. D. R. Petrak, R. Ruh, and G. R. Atkins, *Bull. Am. Ceram. Soc.* **53**, 569 (1974).
41. M. V. Kilday and E. J. Prosen, *J. Am. Chem. Soc.* **82**, 5508 (1960).
42. F. C. Kracek, G. W. Morey, and H. E. Merwin, *Am. J. Sci.* **35-A**, 143 (1938).
43. C. Feldman, *Anal. Chem.* **33**, 1916 (1961).
44. R. W. Sprague, "Properties and Reactions of Boric Acid," in Ref. 4, section A6, p. 224.
45. Ref. 1, pp. 67–69.
46. C. R. Peters and M. E. Milberg, *Acta Crystallogr.* **17**(3), 229 (1964); W. H. Zachariasen, *Acta Crystallogr.* **16**, 380 (1963); **16**, 385 (1963).
47. D. White and co-workers, *J. Chem. Phys.* **32**, 488 (1960).
48. Ref. 35, pp. 27–28.
49. J. Smisko and L. S. Mason, *J. Am. Chem. Soc.* **72**, 3679 (1950).
50. H. O. Jenkins, *Trans. Faraday Soc.* **41**, 138 (1945).
51. J. O. Edwards, *J. Am. Chem. Soc.* **75**, 6151 (1953).
52. L. Maya, *Inorg. Chem.* **15**, 2179 (1976).
53. B. R. Sanderson, "Coordination Compounds of Boric Acid" in Ref. 4, section A18, p. 721.
54. R. Kunin and F. Preuss, *Ind. Eng. Chem. Prod. Res. Dev.* **3**, 304 (1964); U.S. Pat. 3,887,460 (June 3, 1975), C. J. Ward, C. A. Morgan, and R. P. Allen (to United States Borax & Chemical Corp.).
55. U.S. Pat. 2,969,275 (Jan. 24, 1961), D. E. Garrett (to American Potash and Chemical Corp.); Ger. Pat. 1,164,997 (Mar. 12, 1964), D. E. Garrett and co-workers (to American Potash and Chemical Corp.); U.S. Pat. 3,479,294 (Nov. 18, 1969), F. J. Weck (to American Potash and Chemical Corp.); U.S. Pat. 3,424,563 (Jan. 28,

1969), R. R. Grinstead (to The Dow Chemical Company); U.S. Pat. 3,493,349 (Feb. 3, 1970), C. A. Schiappa and co-workers (to The Dow Chemical Company).

56. R. E. Mesmer, K. M. Palen, and C. F. Bates, *Inorg. Chem.* **12**, 89 (1973).

57. K. A. L. G. Watt, *World Minerals and Metals' No. 12*, British Sulphur Corp. Ltd., London, UK, 1973, pp. 5–12.

58. R. J. Weir and R. S. Fisher, *Toxicol. Appl. Pharmacol.* **23**, 351 (1972).

59. W. G. Woods, in V. M. Bhatnager, ed., *Advances in Fire Retardants*, Part 2, Technomic, Pa., 1973, 120–153; J. P. Neumeyer, P. A. Koenig, and N. B. Knoepfler, *U.S. Agric. Res. Serv., South Reg., ARS-S-64*, 70 (1975).

60. I. M. Kolthoff, E. B. Sandell, E. J. Meehan, and S. Bruckenstein, *Quantitative Chemical Analysis*, 4th ed., Macmillan, Collier-Macmillan Ltd., 1969, p. 787.

61. J. O. Edwards, G. C. Morrison, and J. W. Schultz, *J. Am. Chem. Soc.* **77**, 266 (1955).

62. E. Colton and R. E. Brooker, *J. Phys. Chem.* **62**, 1595 (1958).

63. Ref. 1, pp. 85–89.

64. L. Ciavatta, G. Nunziata, and M. Vicedomini, *Ric. Sci., Rend., Sez. A* **8**(5), 1096 (1965).

65. *Technical Data Sheets*, United States Borax & Chemical Corp., 1991; U.S. Pat. 2,998,310 (Aug. 29, 1961), P. J. O'Brien and G. A. Connell (to United States Borax & Chemical Corp.).

66. N. Ingri, *Sven. Kem. Tidskr.* **75**(4), 199 (1963).

67. M. Maeda, *J. Inorg. Nucl. Chem.* **41**, 1217 (1979).

68. C. G. Salentine, *Inorg. Chem.* **22**, 3920 (1983).

69. P. Lyday, *Boron, Mineral Industry Surveys*, 1997 Annual Review, U.S. Geological Surey, Dept. of Interior, 1998.

70. T. Kendall, *Ind. Minerals Mag.* 51–66 (Nov. 1997).

71. S. Scholle and M. Szmigielska, *Chem. Prumysl* **15**, 530 (1965); *Nat. Bur. Stand. U.S. Circ.* **500**, 481 (1952).

72. A. D. Randolph and A. D. Puri, *AIChE J.* **27**, 92 (1981).

73. N. P. Nies and R. W. Hulbert, *J. Chem. Eng. Data* **12**, 303 (1967).

74. R. P. Douglas, D. F. Gaines, P. J. Zerella, and R. A. Smith, *Acta Crystallogr.* **C47**, 2279 (1991).

75. N. P. Nies, "Alkali-Metal Borates: Physical and Chemical Properties," in Ref. 4, section A9, p. 343.

76. H. Menzel and H. Schulz, *Z. Anorg. Chem.* **245**, 157 (1940).

77. H. Menzel and co-workers, *Z. Anorg. Chem.* **224**, 1 (1935).

78. N. Morimoto, *Mineral J.* **2**, 1 (1956); H. A. Levy and G. C. Lisensky, *Acta Crystallogr.* **B34**, 3502 (1978).

79. C. Giacovazzo, S. Menchetti, and F. Scordari, *Am. Mineral.* **56**, 523 (1973).

80. U.S. Pat. 4,412,978 (Nov. 1, 1983), R. T. Ertle (to Stokely Van Camp Inc.).

81. U.S. Pat. 3,454,357 (July 8, 1969), R. C. Rhees and H. N. Hammer (to American Potash and Chemical Corp.).

82. R. F. Giese, *Science* **154**, 1453 (1966); W. F. Cooper, F. K. Larsen, P. Coppens, and R. F. Giese, *Am. Mineralogist* **58**, 21 (1973).

83. E. F. Westrum and G. Grenier, *J. Am. Chem. Soc.* **79**, 1799 (1957); E. F. Westrum, *Thermodynamic Transport Properties Gases, Liquids, and Solids*, Papers Symposium, Lafayette, Ind., 1959, p. 275; C. R. Fuget and J. F. Masi, *Thermodynamic Properties for Selected Compounds, CCC-1024, TR-263*, U.S. Atomic Energy Communication, Washington D.C., 1957.

84. A. Predvoditelev, *Z. Phys.* **51**, 136 (1928).

85. J. Krogh-Moe, *Acta Crystallogr.* **B30**, 578 (1974).

86. Ref. 1, p. 176.

87. G. T. Furukawa, M. L. Reilly, and J. H. Piccirelli, *J. Res. Nat. Bur. Stand.* **68A**, 381 (1964).

88. S. Merlino and F. Sartori, *Acta Crystallogr.* **B28**, 3559 (1972).

89. S. N. Sidorova, L. V. Puchkov, and M. F. Federov, *Ah. Prikl. Khim.* **48**, 253 (1975).

90. H. Menzel and H. Schulz, *Z. Anorg. Chem.* **251**, 167 (1943).

91. E. Svares, V. Grundstein, and A. Ievins, *Zh. Neorg. Khim.* **12**, 2017 (1967).

92. S. Block and A. Perloff, *Acta Crystallogr.* **16**, 1233 (1963).

93. J. Krc, *Anal. Chem.* **23**, 806 (1951).

94. M. A. A. F. de C. T. Carrondo and A. C. Skapski, *Acta Crystallogr.* **B34**, 3551 (1978).

95. C. Frances, B. Biscans, and C. Laguerie, *J. Chem. Eng. Data* **35**, 423 (1990).

96. J. Flanagan, W. P. Griffith, R. D. Powell, and A. P. West, *J. Chem. Soc. Dalton Trans.*, 1651 (1989); B. N. Chernyshov, *Russ. J. Inorg. Chem.* **35**(9), 1333 (1990).

97. A. Chianese, A. Contaldi, and B. J. Mazzarotta, *Cryst. Growth* **78**, 279 (1986); U.S. Pat. 4,298,585 (Nov. 3, 1981) (to Air Liquide); U.S. Pat. 3,726,959 (Apr. 10, 1973) (to Kali-Chemie); U.S. Pat. 2,828,183 (Mar. 25, 1958) (to E. I. du Pont de Nemours & Co., Inc.).

98. W. P. Griffith, A. C. Skapski, and A. P. West, *Chem. Ind. (London)* **5**, 185 (1984).

99. I. M. Kolthoff and E. B. Sandell, *Textbook of Quantitative Inorganic Analysis*, MacMillan, New York, 1952, p. 534.

100. D. L. Callicoat and J. D. Wolszom, *Anal. Chem.* **31**, 1434 (1959).

101. F. J. Krug and co-workers, *Anal. Chem. Acta* **125**, 29 (1981).

102. N. H. Furman ed., *Standard Methods of Chemical Analysis*, 6th ed., Van Nostrand, 1962, p. 223.

103. F. D. Snell and C. L. Hilton, eds., *Encyclopedia of Industrial Chemical Analysis*, Wiley-Interscience, New York, 1968, p. 368.

104. L. J. Casarett and J. Doull, eds., *Casarett and Doull's Toxicology*, 2nd ed., MacMillan Publishing Co. Inc., New York, 1980, p. 440.

105. *Safe Use at Work of Borax, Boric Acid, and Other Inorganic Borates*, Technical Service Bulletin 101, Market Services Department, Borax Consolidated Ltd., Borax House, London, 1990; L. Butterwick, N. de Oude, and K. Raymond, *Ecotoxicol. Environ. Safety* **17**, 339 (1989).

106. J. A. Jansen, J. Andersen, and J. S. Schou, *Arch. Toxicol.* **55**, 64 (1984).

107. S. M. Barlow and F. M. Sullivan, *Reproductive Hazards of Industrial Chemicals*, Academic Press, London, 1982, pp. 126–135.

108. *Toxicology and Carcinogenesis Studies of Boric Acid in B6C3F1 Mice*, National Toxicology Program (NTP) Technical Report Series No. 324, U.S. Department of Health and Human Services, Washington D.C., Oct. 1987.

109. W. H. Benson, W. J. Birge, and H. W. Dorough, *Environ. Toxicol. Chem.* **3**, 209 (1984).

110. *Fed. Regist.* **54**(12), 2436, 2451, 2452, 2584, 2590, 2591 (Jan. 19, 1989).

111. J. Ploquin, *Bull. Soc. Sci. Hyg. Aliment. Aliment. Ration.* **55**(103), 70 (1967).

112. P. A. Lyday, in *Boron*, U.S. Department of the Interior, Bureau of Mines, Washington D.C., 1991, p. 8.

113. G. Capreni, *Bull. Soc. Chim. Fr.*, 1327 (1955); G. Carpeni, J. Haladjian, and M. Pilard, *Bull. Soc. Chim. Fr.*, 1634 (1960).

114. J. Haladjian and G. Carpeni, *Bull. Soc. Chim. Fr.*, 1629 (1960).

115. M. Marezio, H. Plettinger, and W. Zachariasen, *Acta Crystallogr.* **16**, 975 (1963).

116. J. P. Ashmore and H. G. Petch, *Can. J. Phys.* **48**, 1091 (1970).

117. R. Janda, G. Heller, and J. Pickardt, *Z. Kristallogr.* **154**, 1 (1981).

118. U.S. Pat. 2,867,502 (Jan. 6, 1959), H. Strange and S. L. Clark (to Olin Mathieson Chemical).

119. X. Solans and co-workers, *J. Appl. Crystallogr.* **16**, 637 (1983).

120. S. Nakamura and H. Hayashi, *J. Ceram. Soc. Japan, Yagyo Kyo Kaishi* **83**, 38 (1975).
121. R. Bouaziz, *Bull. Soc. Chim. Fr.*, 1451 (1962).
122. U.S. Pat. 2,776,186 (Jan. 1, 1957), F. H. May (to American Potash and Chemical Corp.).
123. V. N. Al'ferova, *Acta Univ. Voronegiensis* **11**, (3,7) (1939).
124. U.S. Pat. 2,867,502 (Jan. 6, 1959), H. Strange (to Olin Mathieson Chemical Corp.).
125. V. M. Gurevich and V. A. Sokolov, *Geokhimiya* **3**, 455 (1976).
126. C. Cipriani, *Atti. Soc. Toscana Sci. Nat. Pisa Mem. P.V. Ser. A* **65**, 284 (1958).
127. J. R. Clark, D. E. Appleman, and C. L. Christ, *J. Inorg. Nucl. Chem.* **26**, 73 (1964).
128. Ref. 1, p. 129.
129. S. Ghose, C. Wan, and J. R. Clark, *Am. Mineral.* **63**, 160 (1978).
130. E. J. Weichel-Moore and R. J. Potter, *Nature* **200**, 1163 (1963).
131. C. Palache, H. Berman, and C. Frondell, *Dana's System of Mineralogy*, John Wiley & Sons, Inc., New York, 1957, p. 347.
132. S. Menchetti, C. Sabelli, and R. Trosti-Ferrari, *Acta Crystallogr.* **B38**, 3072 (1982).
133. *The Economics of Boron*, 6th ed., Roskill Information Services, London, UK, 1989, p. 24.
134. H. Doweidar, *J. Mater. Sci.* **25**, 253 (1990).
135. J. Krogh-Moe, *Phys. Chem. Glasses* **6**, 46 (1965).
136. S. J. Gravina, P. J. Bray, and G. L. Peterson, *J. Non-Cryst. Solids* **123**, 165 (1990).
137. G. Tomandl and H. A. Schaeffer, *J. Non-Cryst. Solids* **73**, 179 (1985); J. Zhong and P. J. Bray, *J. Non-Cryst. Solids* **111**, 67 (1989).
138. E. M. Levin, H. F. McMurdie, and F. P. Hall, *Phase Diagrams for Ceramacists*, Part I, 1956; Part II, 1959; Supplement I, 1964; and Supplement II, 1969; The American Ceramic Society, Columbus, Ohio.
139. R. A. Smith, *J. Non-Cryst. Solids* **84**, 421 (1986).
140. N. B. Kravehenko, *Zh. Strukt. Khim.* **6**, 724 (1965).
141. L. Kutschabsky, *Acta Crystallogr.* **B25**, 1181 (1969).
142. S. Vimba, A. Ievins, and J. Ozols, *Zh. Neorg. Khim.* **3**, 325 (1958); **2**, 2423 (1957).
143. Ref. 1, p. 133.
144. H. A. Lehmann, K. Muehmel, and D-F. Sun, *Z. Anorg. Chem.* **355**, 288 (1967).
145. A. Kesans, *Riga: Izdatel. Alcad, Nauk Latv. S.S.R.*, 179 (1955).
146. W. G. Woods and J. G. Bower, *Mod. Plast.* **47**, 140 (1970); K. K. Shen, *J. Therm. Insul.* **3**, 190 (1980); R. W. Sprague and K. K. Shen, *J. Therm. Insul.* **2**, 161 (1979).
147. U.S. Pat. Re 27,424 (July 4, 1972), N. P. Nies and R. W. Hulbert (to United States Borax & Chemical Corp.); U.S. Pat. 3,718,615 (Feb. 27, 1973), W. G. Woods, J. C. Whiten, and N. P. Nies (to U.S. Borax & Chemical Corp.).
148. U.S. Pat. 3,649,172 (Mar. 14, 1972), N. P. Nies and R. W. Hulbert (to United States Borax & Chemical Corp.).
149. A. F. Wells, ed., *Structural Inorganic Chemistry*, 5th ed., Clarendon Press, Oxford, UK, 1984, p. 1078.
150. Ref. 1, 184–186.
151. B. P. Long, "Boron–Oxygen Compounds of Groups V and VI," in Ref. 4, section A13, p. 651.
152. W. J. Englert and F. A. Hummel, *J. Soc. Glass Technol.* **39**, 121T (1955).
153. W. D. Loomis and R. W. Dunst, *BIofactors* **3**: 229–239 (1992).
154. C. Rainy and L. Nyquist, *Biol. Trace Elem. Res.* **66**, 79–86 (1998).
155. P. Arqust, *Bio. Trace Elem. Res.* **66**, 131–143 (1998).
156. J. R. Coughlin, *Biol. Trace Elem. Res.* **66**, 87–100 (1998).
157. R. I. Rowe and co-workers, *Biol. Trace Elem.*
158. B. D. Culver and S. A. Hubbard, *J. Trace Elem. Exp. Med.* **9**, 175–184 (1996).
159. T. L. Litovity and co-workers, *Am. J. Energ. Med.* **6**, 209–213 (1988).

160. D. H. Wegman and co-workers, *Environ. Health Perspect.* **102**(Supp. 17), 119–128 (1994).
161. M. D. Whorton and co-workers, *Occup. Environ. Med.* **51**, 761–767 (1994).
162. S. A. Hubbard and F. M. Sullivan, *J. Trace Elem. Exp. Med.* **9**, 165–173 (1996).
163. C. J. Price and co-workers, *Fund. Appl. Toxicol.* **32**, 179–193 (1996).
164. World Health Organization, *Trace Elements in Human Nutrition and Health*, Geneva, 1996, 175–179.
165. R. I. Rowe and C. D. Eckhert, *J. Exp. Biol.* (in press).
166. D. J. Fort and co-workers, *J. Trace Elem. Exp. Med* **12**(3) (in press).
167. J. A. Moore and an Expert Scientific Committee, *Reprod. Toxicol.* **11**, 123–160 (1997).

GENERAL REFERENCES

Anonymous "Boron," in Grew and Anovitz eds., *Reviews in Mineralogy*, Min. Society of America, 1996.
Anonymous *Mineral Commodity Summaries 1993*; U.S. Dept. Interior, Bureau of Mines, 1993, pp. 36–37.
Anonymous, *The Economics of Boron 1989*, 6th ed., Roskill Information Service, U.K., 1989.
W. Buhler, *Borasit, the Story of the turkish Boron Mines And Their Impact on the Boron Industry*, Imprimeire Chabloz SA, Switzerland, 1996.
Gaines and co-workers, *Dana's New Mineralogy*, 8th ed., Wiley, New york, 1997.
P. Harben, *The Industrial Minerals Handybook*, 2nd ed., Industrial Min. Information Ltd., Industrial Min. Division of metal Bulletin, U.K., 1995.
P. Harben and Bates, *Geology of the Nonmetallics*; R. Hartnoll Ltd., U.K. 1984.
P. Harbenand and Kuzvart, *Industrial Minerals, a Global geology*, Industrial, Min. Information Ltd., Industrial Min. Division of Metal Bulletin, U.K., 1996.
C. Helvaci, *Econ. Geol.* **90**, 1237–1260 (1995).
T. Kendall, *Industrial Minerals*, 1997, pp. 51–69.
R. Kistler and C. Helvaci in D. Carr, ed., *Industrial Minerals and Rocks*, 6th ed., Society of Minning Engineering, New York, 1994, pp. 171–186.
Morgan and Erd, *Minerals of the Kramer Borate District, Calif.*, Min. Information Service, Calif. Division of Mines and Geolog., 1969, Vol. 22, Nos. 9, 10 pp. 143–154, 165–172.
Palmer and C. Helvaci, *Geochim. Cosmochim.* **61**(15), 3161–3169 (1997).
J. Siefke in Mckibben, ed., *The Diversity of mineral and energy resources of Southern California*, SEG Guidebook Series, 1991, Vol. 12, pp. 4–15.
G. Smith, *Subsurface of Stratigraphy and Geochemistry of Later Quaternary Evaporites, Searles Lake, Calif.*, U.S. Geological Survey, 1979, p. 1047.
Swihart and co-workers, *Geochim. Cosmochim. Acta.* **50**, 1297–1301 (1986).
Travis and Cocks, *The Tincal Trail*, Harrao, U.K., 1984.

MICHAEL BRIGGS
U.S. Borax Inc.

BROMINE

1. Introduction

Bromine, Br_2, [7726-95-6], is the only nonmetallic element that is a liquid at standard conditions. The Bromine (Br) atom [10097-32-2] has at no. 35, at wt 79.904, and belongs to Group 17 (VIIA) of the Periodic Table, the halogens. Its electronic configuration is $1s^2 2s^2 2p^6 3s^2 3p^6 3d^{10} 4s^2 4p^5$. The element's known isotopes range in mass number from 74 to 90. Isotopes usable as radioactive tracers are 77, 80, 80m (metastable), and 82. Bromine has two stable isotopes, ^{79}Br and ^{81}Br. The most common valence states are -1 and $+5$, but $+1$, $+3$, and $+7$ are also observed. The covalent radius of bromine is 0.1193 nm. The ionic radius of the bromide ion [24959-67-9], is 0.197 nm and of Br(VII) Br^{+7}, [20681-12-3] is 0.039 nm. The name bromine is derived from the Greek word, *bromos,* meaning stench.

Bromine occurs in the hydrosphere mainly as soluble bromide salts. Its concentration ranges from 65 mg/L in seawater up to 6.5 g/L in the southern basin of the Dead Sea. In 1826, Antoine-Jerome Balard in France published the discovery of bromine, which was isolated by chlorinating seawater bitterns and distilling out bromine. Bromine had been prepared earlier by Joss and Liebig but neither of them recognized it as an element (1). Around 1840 bromine was used in photography. The first mineral to contain bromine was apparently silver bromide, discovered in 1941 by Berthier. Its first medical use was in 1857 when bromides were used for the treatment of epilepsy. The first commercial bromine production from salt brines was in 1846 at Freeport, Pennsylvania. In 1858, potash was discovered in the Stassfurt salt deposits in Germany and bromine was a by-product from the potash production. Herbert Dow invented the "blowing out" process for Midland (Michigan) brines in 1889. The antiknock properties of tetraethyllead, $(CH_3CH_2)_4Pb$, [78-00-2] were discovered in 1921 and soon after ethylene dibromide, $C_2H_4Br_2$, [106-93-4] was found to aid the removal of lead from combustion chambers. At one time, ~80% of all bromine was used to produce ethylene dibromide. Bromine was first commercially extracted from seawater in 1934, using the Dow process. The richest source of Br in the world is found in the Dead Sea brines, which contain up to 12 g/L of Br^-. Its commercial exploitation by DSBG started in 1957. In the 1950s, bromine was discovered in south Arkansas brines, the only significant source of bromine in the United States, with a bromine content of 2–5 g/L (see CHEMICALS FROM BRINE).

2. Physical Properties

Bromine is a dense, dark red, mobile liquid that vaporizes readily at room temperature to give a red vapor that is highly corrosive to many materials and human tissues. Bromine liquid and vapor, up to ~600°C, are diatomic (Br_2). Table 1 summarizes the physical properties of bromine.

Remark: While bromine does not form a true azeotrope with water, bromine saturated with water or mixed with it creates a solution that boils at 54.3°C, compared with 58.8°C for dry bromine. The distillate contains ~2% water.

Table 1. **Physical Properties of Bromine**[a]

Property	Value
stable isotope abundance, %	
^{79}Br	50.54%
^{81}Br	49.46%
mol wt	159.808
freezing point, °C	−7.25
bp, °C	58.8
density, g/mL	
15°C	3.1396
20°C	3.1226
25°C	3.1055
30°C	3.0879
vapor density, g/L, 0°C, 101.3 kPa[b]	7.139
refractive index	
20°C	1.6083
25°C	1.6475
viscosity, mm^2/s (=cSt)	
20°C	3.14×10^{-1}
30°C	2.88×10^{-1}
40°C	2.64×10^{-1}
50°C	2.45×10^{-1}
surface tension, mN/m (=dyn/cm), 25°C	40.9
solubility parameter, 25°C, (J/cm^3)$^{1/2c}$	23.5
critical temperature, °C	311
critical pressure, MPa[d]	10.3
thermal conductivity, W/(m · K)	0.123
specific conductivity, $(\Omega \cdot cm)^{-1}$	9.10×10^{-12}
dielectric constant, 25°C, 10^5 Hz	3.33
electrical resistivity, 25°C, $\Omega \cdot cm$	6.5×10^{10}
expansion coefficient from 20−30°C, per °C	0.0011
compressibility, 20°C from 0−10 MPa[d]	62.5×10^{-6}
heat of vaporization, 50°C, J/g[c]	187
heat of fusion, −7.25°C, J/g[c]	66.11
heat capacity, J/mol[c]	
15 K	7.217
30 K	22.443
60 K	36.33
240 K	57.94
265.9 K	61.64
265.9 K[e]	77.735
288.15 K[f]	78.66
electronegativity	3.0
electron affinity, kJ[c]	330.5

[a] References 2–5.

[b] To convert kPa to mmHg, multiply by 7.50.

[c] To convert J to cal, divide by 4.184.

[d] To convert MPa to bar, multiply by 10.

[e] Solid bromine.

[f] Liquid bromine.

Bromine is moderately soluble in water, 33.6 g/L at 25°C. It gives a crystalline hydrate having a formula of $\sim$Br$_2$ $\leq$7.9H$_2$O (6). The solubilities of bromine in water at several temperatures are given in Table 2. Aqueous bromine solubility increases in the presence of bromides or chlorides because of complex ion

Table 2. **Aqueous Solubility of Bromine**[a]

Temperature, °C	Solubility, g/100 g soln	Temperature, °C	Solubility, g/100 g soln
0	2.31 (4.05)[b]	20	3.41
3	3.08 (3.85)[b]	25	3.35
5	3.54 (3.77)[b]	40	3.33
10	3.60	53.6[c]	3.50

[a] References 7–9.
[b] These solutions are metastable.
[c] This is the boiling point.

formation. This increase in the presence of bromides is illustrated in Figure 1. Equilibrium constants for the formation of the tribromide and pentabromide ions at 25°C have been reported (11).

$$Br_2 + Br^- \rightleftharpoons Br_3^- \qquad K = 16.85 \ M^{-1}$$

$$Br_3^- + Br_2 \rightleftharpoons Br_5^- \qquad K = 1.45 \ M^{-1}$$

Bromine is soluble in nonpolar solvents and in certain polar solvents such as alcohol and sulfuric acid. It is miscible with alcohol, ether, carbon disulfide, and many halogenated solvents. Bromine reacts with some of these solvents under certain conditions.

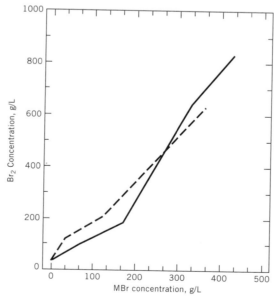

Fig. 1. Solubility of bromine in the presence of (—) NaBr and (– – –) KBr (10).

Bromine can function as a solvent. One of the very few metal bromides that has significant solubility in bromine is cesium bromide, 19.3 g/100 g of solution, thus providing a method of separating cesium bromide from the other alkali bromides (12). Aluminum bromide also is reported to have significant solubility in bromine but the published solubility values are not in good agreement (13). Bromine serves as the solvent in some brominations of organic compounds, such as 1,2-diphenylethane (14).

3. Chemical Properties

One of the central features of the chemistry of the halogens is the tendency to acquire an electron to form either a negative ion, X^-, or a single covalent bond, $-X$, and bromine is no exception. The halogens are electron rich systems having few potential bonding orbitals. Except for helium and neon, all of the elements in the Periodic Table form halides with one or more of the halogens. Halides that are predominately ionic tend to have high conductivities when fused, high boiling points, and if soluble in water, are generally not hydrolyzed. Predominately covalent halides are volatile, nonconductive in the liquid state, and usually readily hydrolyzed (15).

Nonmetal halides are generally hydrolyzed to a hydrogen halide and to an oxy-acid containing the other element. The first row nonmetal halides, eg, CCl_4, resist hydrolysis because the nonmetal element cannot expand its octet of electrons to form a bond to water before its bond to the halide is broken. Hydrolysis requires either an energetic water molecule to strike the halide or ionization of the covalent nonmetal–halide bond, processes that tend to be quite slow (16).

3.1. Reaction with Hydrogen and Metals.   Bromine combines directly with hydrogen at elevated temperatures and this is the basis for the commercial production of hydrogen bromide [10036-10-6]. Heated charcoal and finely divided platinum metals are catalysts for the reaction (17).

Bromine reacts with essentially all metals, except tantalum and niobium, although elevated temperatures are sometimes required, eg, solid sodium does not react with dry bromine but sodium vapor reacts vigorously. Metals such as lead, magnesium, nickel, and silver react with bromine to form a surface coat of bromide that resists further attack. This protective coating allows lead and nickel to be used as linings in bromine containers. Metals tend to be corroded by bromine faster in the presence of moisture than without, probably because of the formation of hydrobromic and hypobromous acids.

Bromine reacts with some metal oxides, eg, thorium oxide, at high temperatures in the presence of reducing agents to form bromides (18). Certain nonhydrated metal halides can be formed by precipitation. These include AgBr, CuBr, AuBr, TlBr, $PbBr_2$, $PtBr_2$, and Hg_2Br_2 (19).

3.2. Reaction with Other Halides.   Bromide ion is oxidized by chlorine to bromine, which is the basic reaction in the production of bromine from brines, seawater, bitterns, or bromine containing wastes.

$$2\,Br^- + Cl_2 \longrightarrow Br_2 + 2\,Cl^-$$

Iodide ion is oxidized by bromine to iodine.

$$2\,I^- + Br_2 \longrightarrow I_2 + 2\,Br^-$$

Among the interhalogen compounds containing bromine are BrF, BrF_3, BrF_5, BrCl, and IBr. Bromine is soluble in chlorine in all proportions, yielding an equilibrium mixture of the interhalogen bromine chloride:

$$Br_2 + Cl_2 \rightarrow 2BrCl \qquad K_{eq} = 2.94 \; at \; 25°C$$

This equilibrium mixture can be separated by distillation.

The interhalogens are characterized by great reactivity. The higher fluorides are quite thermally stable. Bromine pentafluoride [7789-30-2], BrF_5, which is stable up to 460°C, is the most reactive of the higher fluorides and reacts with all of the elements except nitrogen, oxygen, and the noble gases (15). Solid polyhalide salts are known. Examples are NH_4IBr_2, $RbBrCl_2$, and KClIBr (20).

3.3. Reaction with Nonmetals.   Bromine oxidizes sulfur and a number of its compounds.

$$H_2S + Br_2 \longrightarrow S + 2\,HBr$$

$$S + 3\,Br_2 + 4\,H_2O \longrightarrow H_2SO_4 + 6\,HBr$$

$$SO_2 + Br_2 + 2\,H_2O \longrightarrow 2\,HBr + H_2SO_4$$

$$Na_2S + 4\,Br_2 + 8\,NaOH \longrightarrow Na_2SO_4 + 8\,NaBr + 4\,H_2O$$

Bromine also oxides red phosphorus and some phosphorus compounds.

$$2\,P + 3\,Br_2 + 6\,H_2O \longrightarrow 6\,HBr + 2\,H_3PO_3$$

$$H_3PO_3 + Br_2 + H_2O \longrightarrow H_3PO_4 + 2\,HBr$$

$$H_3PO_3 + Br_2 + 3\,NaOH \longrightarrow NaH_2PO_4 + 2\,NaBr + 2\,H_2O$$

Ammonia, hydrazine, nitrites, and azides are oxidized by bromine. Nitrogen is often a product of such reactions.

$$2\,NH_3 + 3\,Br_2 + 6\,NaOH \longrightarrow 6\,NaBr + N_2\uparrow + 6\,H_2O$$

$$8\,NH_4OH + 3\,Br_2 \longrightarrow 6\,NH_4Br + N_2\uparrow + 8\,H_2O$$

$$N_2H_4 + 2\,Br_2 + 4\,NaOH \longrightarrow 4\,NaBr + N_2\uparrow + 4\,H_2O$$

$$NaNO_2 + Br_2 + H_2O \longrightarrow NaNO_3 + 2\,HBr$$

Under certain circumstances bromine reacts with ammonia and amino compounds to form bromamide, NH_2Br, [14519-10-9] bromimide, $NHBr_2$, [14519-03-0], or nitrogen bromide, NBr_3, [15162-90-0]. These compounds can decompose explosively so great care should be exercised any time bromine and ammonia or amino compounds might come in contact with each other (see BROMINE INORGANIC COMPOUNDS).

Bromine oxidizes carbon and reacts with carbon monoxide to form carbonyl bromide [593-95-3].

$$C + Br_2 + H_2O \longrightarrow 2\,HBr + CO\uparrow$$

$$Br_2 + CO \longrightarrow COBr_2$$

3.4. Reactions in Water.   The ionization potential for bromine is 11.8 eV and the electron affinity is 3.78 eV. The heat of dissociation of the Br_2 molecule is 192 kJ (46 kcal). The reduction potentials for bromine and oxybromide anions in aqueous acid solutions at 25°C are(21)

$$Br_2(aq) + 2\,e^- \longrightarrow 2\,Br^- \qquad E^\circ = +1.0873\ V$$

$$Br_2(l) + 2\,e^- \longrightarrow 2\,Br^- \qquad E^\circ = +1.066\ V$$

$$HOBr + H^+ + 2\,e^- \longrightarrow Br^- + H_2O \qquad E^\circ = +1.331\ V$$

$$HOBr + H^+ + e^- \longrightarrow \tfrac{1}{2}Br_2(aq) + H_2O \qquad E^\circ = +1.574\ V$$

$$HOBr + H^+ + e^- \longrightarrow \tfrac{1}{2}Br_2(l) + H_2O \qquad E^\circ = +1.596\ V$$

$$BrO^- + H_2O + 2\,e^- \longrightarrow Br^- + 2\,OH^- \qquad E^\circ = +0.761\ V$$

$$BrO_3^- + 6\,H^+ + 5\,e^- \longrightarrow \tfrac{1}{2}Br_2 + 3\,H_2O \qquad E^\circ = +1.482\ V$$

$$BrO_3^- + 6\,H^+ + 6\,e^- \longrightarrow Br^- + 3\,H_2O \qquad E^\circ = +1.423\ V$$

$$BrO_3^- + 3\,H_2O + 6\,e^- \longrightarrow Br^- + 6\,OH^- \qquad E^\circ = +0.61\ V$$

When bromine dissolves in water, it partially disproportionates.

$$Br_2 + H_2O \;\rightleftharpoons\; HOBr + H^+ + Br^-$$

The equilibrium constant for this reaction at 25°C is $7.2 \times 10^9\ M^2$(22). Light catalyzes the decomposition of hypobromous acid to hydrogen bromide and oxygen.

$$HOBr \xrightarrow{h\nu} HBr + \tfrac{1}{2}O_2$$

In the dark, hypobromous acid decomposes to bromic acid and bromine. Bromic acid is relatively unstable and decomposes slowly to give bromine and oxygen.

$$5\,HOBr \longrightarrow HBrO_3 + 2\,Br_2 + 2\,H_2O$$

$$4\,HBrO_3 \longrightarrow 2\,Br_2 + 5\,O_2 + 2\,H_2O$$

In alkaline solution, bromine reacts rapidly to produce hypobromite.

$$Br_2 + 2\,OH^- \rightleftharpoons Br^- + BrO^- + H_2O \qquad K = 2 \times 10^8$$

It is necessary to maintain this reaction below 0°C to minimize the disproportionation of hypobromite to bromate and bromide.

$$3\,BrO^- \longrightarrow 2\,Br^- + BrO_3^-$$

Because they are unstable, hypobromites are usually prepared just before use for such uses as textile bleaching and desizing. In alkaline solutions at 50–80°C bromine reacts to form bromide and bromate. This reaction is reversed in acidic solutions.

$$3\,Br_2 + 6\,OH^- \longrightarrow 5\,Br^- + BrO_3^- + 3\,H_2O$$

$$5\,Br^- + BrO_3^- + 6\,H^+ \longrightarrow 3\,Br_2 + 3\,H_2O$$

3.5. Reactions with Organic Compounds. The addition of bromine to unsaturated carbon compounds occurs readily.

$$CH_2{=}CH_2 + Br_2 \longrightarrow BrCH_2CH_2Br$$

Conjugated double bond systems usually undergo 1,4-addition.

$$CH_2{=}CH{-}CH{=}CH_2 + Br_2 \longrightarrow BrCH_2CH{=}CHCH_2Br$$

Bromine reacts directly with alkanes but this reaction has little value because mixtures are obtained. However, photochemical bromination of alkyl bromides can be quite selective (23). The bromination of aromatic hydrocarbons can occur either in a side chain or on the ring, depending on conditions. In the presence of sunlight, alkylbenzenes are brominated predominately in the side chain (24).

$$Br_2 \xrightarrow{h\nu} 2\,Br\cdot$$

$$Br\cdot + C_6H_5CH_3 \longrightarrow C_6H_5CH_2\cdot + HBr$$

$$C_6H_5CH_2\cdot + Br_2 \longrightarrow C_6H_5CH_2Br + Br\cdot$$

$$2\,Br\cdot \longrightarrow Br_2$$

In the presence of halogen Lewis acids, such as metal halides or iodine, aromatic hydrocarbons are halogenated on the ring (24). Some polynuclear aromatics, such as anthracene, can be brominated without a catalyst (23).

Phenols and phenol ethers readily undergo mono-, di-, or tribromination in inert solvents depending on the amount of bromine used. In water the main product is the 2,4,6-tribromophenol $C_6H_3Br_3O$, [118-79-6] (23). In water or acetic acid anilines also give the tribrominated product (25). Tribromophenol can be further brominated in buffered acetic acid to give 2,4,4,6-tetrabromo-2,5-cyclo-hexadien-1-one [20244-61-5], a useful brominating agent (26). Heterocyclic compounds range from those, such as furan, which is readily halogenated and tends to give polyhalogenated products, to pyridine, which forms a complex with aluminum chloride that can only be brominated to 50% reaction (23).

Aliphaticketones (qv) are readily brominated in the α position. Mixtures are usually obtained (24).

$$RCH_2COCH_2R + Br_2 \longrightarrow RCHBrCOCH_2R + HBr$$

Bromination of aldehydes (qv) is more complicated because bromination can take place on the aldehyde carbon as well as the α-carbon. Acetals are brominated satisfactorily in cold chloroform solution in the presence of calcium carbonate, which reacts with the hydrogen bromide formed (24).

$$2\,RCH_2CH(OC_2H_5)_2 + 2\,Br_2 + CaCO_3 \longrightarrow$$

$$2\,RCHBrCH(OC_2H_5)_2 + CaBr_2 + H_2O + CO_2\uparrow$$

Acids and esters (see ESTERS, ORGANIC) are less easily brominated than aldehydes or ketones. Acid chlorides and anhydrides are more easily brominated (23).

$$RCH_2COCl + Br_2 \longrightarrow RCHBrCOCl + HBr$$

Bromination of α-chloro ethers proceeds readily and often gives 90–95% yields (24).

$$RCH_2CHClOR' + Br_2 \longrightarrow RCHBrCHBrOR' + HCl$$

Bromine can replace sulfonic acid groups on aromatic rings that also contain activating groups. Phenolic sulfonic acids, for example, are polybrominated (24).

$$4-HOC_6H_4SO_3K + Br_2,\ H_2O \longrightarrow 2,4,6-Br_3C_6H_2OH + KHSO_3$$

Organometallic compounds can react with bromine to give bromides, but because organometallic compounds are frequently made from bromides the reaction with iodine to give iodides is of more synthetic significance (24).

$$RHgX + Br_2 \longrightarrow RBr + HgXBr$$

Bromine reacts with the silver salts of carboxylic acids to give an alkyl bromide containing one less carbon atom than the acid (24).

$$RCOOAg + Br_2 \longrightarrow RBr + CO_2 + AgBr$$

Amides and imides can be N-brominated in the cold by alkali hypobromites (24).

$$RCONH_2 + Br_2 + KOH \longrightarrow RCONHBr + KBr + H_2O$$

During some brominations a hydroxyl group can be converted to a ketone on an adjacent carbon atom (27).

$$(CH_3)_2C(OH)CH_2CH_3 + Br_2 \longrightarrow (CH_3)_2CBrCHBrCH_3 + H_2O$$

$$(CH_3)_2CBrCHBrCH_3 + H_2O \longrightarrow (CH_3)_2CHCOCH_3 + 2\ HBr$$

In the presence of base, bromine reacts with acetylenes to displace a hydrogen (28).

$$C_6H_5C{\equiv}CH + Br_2 + NaOH \longrightarrow C_6H_5C{\equiv}CBr + NaBr + H_2O$$

Hydrazines can be oxidized by bromine (29). Bromine has been used to synthesize organoselenium compounds (30).

In an aqueous acetate buffer at pH 5 bromine oxidizes ethers containing an α-hydrogen (31).

$$(CH_3CH_2CH_2)_2O\ +\ Br_2\ \xrightarrow{H_2O}\ 2\ CH_3CH_2COOH\ +\ 2\ HBr$$

Ca 100%

$$(C_6H_5CH_2)_2O\ +\ Br_2\ \xrightarrow{H_2O}\ 2\ C_6H_5CHO\ +\ 2\ HBr$$

98%

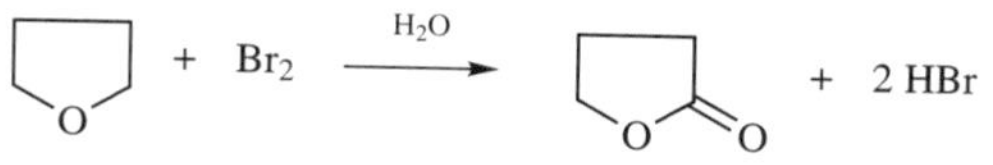

+ Br$_2$ $\xrightarrow{H_2O}$ + 2 HBr

ca 100%

In the presence of a silver salt, bromine reacts with a tertiary alcohol to give a product corresponding to an insertion of oxygen (32). Bromine can oxidize certain tertiary amines to lactams (33). When tertiary alcohols are oxidized with bromine and a silver salt, tetrahydrofuran derivatives result (34).

Bromine has been used to form cyclobutane-1,2-dione [33689-28-0], C$_4$H$_4$O$_2$, when other methods failed (35,36). Regioselective bromination of ketones at the more highly substituted α-position is effected by photocatalytic bromination in the presence of 1,2-epoxycyclohexane (37).

$$CH_3COCH(CH_3)_2 + Br_2 \xrightarrow[\text{1,2-epoxycyclohexane}]{h\nu} CH_3COCBr(CH_3)_2 + HBr$$

Bromine or chlorine dissolved in hexamethylphosphoric triamide (HMPT) [680-31-9], with a base, eg, NaH$_2$PO$_4$, present, oxidizes primary and secondary alcohols to carbonyl compounds in high yield (38). Brominating epoxides in CCl$_4$ under irradiation gives α-bromo ketones (39).

Bromine in a two-phase system, H$_2$O–CH$_2$Cl$_2$, with KHCO$_3$ can convert sulfides to sulfoxides in good yields (40). Aldehydes can be directly converted to esters using bromine in alcohol solvents with sodium bicarbonate buffer (41).

Organic compounds that are easily oxidized are destroyed by bromine.

$$HCHO + 2\,Br_2 + 4\,NaOH \longrightarrow 4\,NaBr + CO_2\uparrow + 3\,H_2O$$

$$HCOOH + Br_2 \xrightarrow{H_2O} CO_2\uparrow + 2\,HBr$$

$$NH_2CONH_2 + 3\,Br_2 + 6\,NaOH \longrightarrow 6\,NaBr + N_2\uparrow + CO_2\uparrow + 5\,H_2O$$

3.6. Charge-Transfer Compounds.   Similar to iodine and chlorine, bromine can form charge-transfer complexes with organic molecules that can serve as Lewis bases. The frequency of the intense ultraviolet (uv) charge transfer absorption band is dependent on the ionization potential of the donor solvent molecule. Electronic charge can be transferred from a π-electron system as in the case of aromatic compounds or from lone pairs of electrons as in ethers and amines.

Charge-transfer compounds can be isolated in the crystalline state, although low temperatures are often required. The bromine–dioxane compound, eg, has a chain structure (42).

4. Occurrence

Bromine is widely distributed in nature but in relatively small amounts. Its abundance in igneous rock is 0.00016% by weight and in seawater is 0.0065% by weight. The only natural minerals that contain bromine are some silver halides, including bromyrite AgBr, [14358-95-3], embolite Ag (Cl, Br), [1301-83-3], and iodobromite, Ag (Cl, Br, I). The biggest source of commercial bromine are the Dead Sea brines. They contain 5 g/L bromine in the open sea, 6.5 g/L in the southern basin, near Ein Bokek, and up to 12 g/L in the end brine of potash production, that is the raw material for production of bromine in Israel. Their quantity is practically unlimited. Other important sources are underground brines in Arkansas, which contain 3–5 g/L bromine, and in China, Russia, and the United Kingdom; bitterns from mined potash in France and Germany; seawater or seawater bitterns in France, India, Italy, Japan, and Spain (43).

An average of ∼7 ppm of bromine is found in terrestrial plants, and edible foods contain up to 20 ppm. Among animals the highest bromide contents are found in sea life, such as fish, sponges, and crustaceans (44). Animal tissues contain 1–9 ppm of bromide and blood 5–15 ppm. The World Health Organization has set a maximum acceptable bromide intake for humans at 1 mg/kg of body weight per day. In adult males, the bromine content in serum has been found to be 3.2–5.6 µg/mL, in urine 0.3–7.0 µg/mL, and in hair 1.1–49.0 µg/mL. Bromine may be an essential trace element as are the other halides (45).

Bromine compounds are found in the atmosphere in small amounts; the sea is a primary source. Rainfall over the Pacific and Indian Oceans has been found to contain 60–80 µg/L of bromine (46). Approximately 15–20 parts per trillion (ppt) (v/v) of bromine is found in the stratosphere (47). It is up from 10 ppt a decade ago (48), maybe due to the use of brominated fire suppressants (CF_3Br, etc). The inorganic forms of Br in the stratosphere are likely involved in ozone destruction processes.

5. Manufacture

5.1. From Natural Sources.
Bromine occurs in the form of bromide in seawater and in natural brine deposits (see CHEMICALS FROM BRINE), always together with chloride. In all current methods of bromine production, chlorine, which has a higher reduction potential than bromine, is used to oxidize bromide to bromine.

$$\frac{1}{2} Cl_2(g) + e^- \longrightarrow Cl^- \qquad E^\circ = +1.356\ V$$

$$\frac{1}{2} Br_2(g) + e^- \longrightarrow Br^- \qquad E^\circ = +1.065\ V$$

$$2\,Br^- + Cl_2 \longrightarrow Br_2 + 2\,Cl^-$$

There are four principal steps in bromine production: (1) oxidation of bromide to bromine; (2) stripping bromine from the aqueous solution; (3) separation of bromine from the vapor; and (4) purification of the bromine. Most of the differences between the various bromine manufacturing processes are in the stripping and purification step.

5.2. Traditional Processes.
The two primary stripping vapors are steam and air. Steam is used when the concentration of bromine in brine is >1000 ppm. The advantage is that bromine can be condensed directly from the steam. Air is used, when seawater is the source of bromine because very large volumes of stripping gas are needed and steam would be too expensive. When air is used the bromine needs to be trapped in an alkaline or reducing solution to concentrate it.

Typical brines received at an Arkansas bromine plant have 3–5 g/L bromide, 200–250 g/L chloride, 0.15–0.20 g/L ammonia, 0.1–0.3 g/L hydrogen sulfide, 0.01–0.02 g/L iodide, and additionally may contain some dissolved organics, including natural gas and crude oil. The bromide-containing brine is first treated to remove natural gas, crude oil, and hydrogen sulfide prior to introduction into the contact tower (48).

The average composition of the liquors left from potash production in the Dead Sea, which are the raw material for production of bromine in Israel (49), was in 2001: 45 g/L calcium; 85 g/L magnesium; 350 g/L chloride; and 12 g/L bromide; L.

The biggest single bromine plant was erected by the Dead Sea in Israel, and is operated by the Dead Sea Bromine Group (DSBG). In 2000, it produced 210,000 metric tons (49).

In the steaming-out process, excess chlorine is used and recycled. The major process conditions that are measured and controlled are temperature, pressure, pH, and oxidation potential.

Materials that come in contact with wet halogens must be corrosion-resistant. Glass, ceramics, tantalum, and fluoropolymers are suitable materials. Granite has been used in steaming-out towers.

In the blowing-out process, used when the source of bromine is seawater, air is used instead of steam to strip bromine from solution. At the pH of seawater, the liberated bromine hydrolyzes to hypobromous acid and bromide. Bromide traps bromine as the tribromide ion and little bromine is released. Before stripping, enough sulfuric acid is added to the seawater to reduce the pH to 3–3.5.

The exiting air containing bromine is absorbed in a sodium carbonate solution.

$$3 \, Na_2CO_3 + 3 \, Br_2 \longrightarrow 5 \, NaBr + NaBrO_3 + 3 \, CO_2\uparrow$$

When the alkalinity of the absorbing solution becomes low it is moved to storage. Acidifying the absorbing solution with sulfuric acid reconstitutes the bromine that can then be steamed out.

$$NaBrO_3 + 5 \, NaBr + 3 \, H_2SO_4 \longrightarrow 3 \, Br_2 + 3 \, Na_2SO_4 + 3 \, H_2O$$

An alternative absorbing solution uses sulfur dioxide.

$$Br_2 + SO_2 + 2 \, H_2O \longrightarrow 2 \, HBr + H_2SO_4$$

The bromine is recovered by oxidizing the bromide with chlorine and steaming it out of solution.

Treatment with sulfuric acid and fractional distillation are the main methods used to purify bromine. It is especially important to reduce the water content to <30 ppm to prevent corrosion of metal transportation and storage containers.

5.3. Newer Process Modifications. Patents describe a single-stage vacuum process (50) and a double-stage vacuum process (48) for recovering bromine from brines. The former is essentially the steaming-out process carried out at subatmospheric pressure. In the double-stage process the tail brines from the first stripping are stripped again under greater vacuum.

According to the patents, Arkansas brines reach the bromine plant at elevated temperatures and in the usual steaming-out process are further heated by steam to the boiling point. Additional steam is required to strip the bromine from the brine. Vacuum is used in the modified process, which, by matching the vapor pressure of the brine eliminates the need for steam to heat the brine. Because of the lower volume of steam used in the vacuum process, the capacity of a given size of contact tower is increased. A further benefit is that at the lower operating temperature of the vacuum process, chlorine undergoes fewer side reactions and

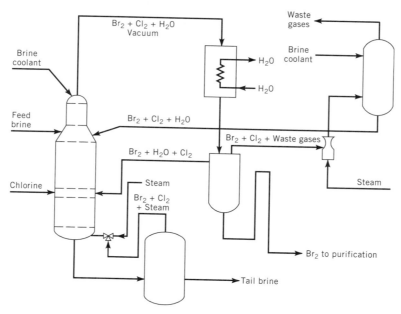

Fig. 2. Schematic of a two-stage vacuum bromine steaming-out process.

less hydrolysis so chlorine use can be reduced. In the two-stage process (Fig. 2), a second steam stripping of the tail brine is done. Other claimed advantages of the vacuum process are a reduction in the amount of lime required to treat the spent brine, lower plant maintenance costs, and decreased waste gases.

DSBG preferred to optimize its atmospheric plant rather than using vacuum. An elaborate network of heat exchangers decreased the steam consumption to 2 ton per ton bromine. DSBG claims that it is more economical than using vacuum.

A potential modification would be to replace the stripping with steam by direct extraction of the Br, resulting from the oxidation. In a process recommended in (51), after oxidation of the brines with chlorine gas, the Br_2 is extracted with CCl_4. The resulting bromine CCl_4 solution can be used for organic brominating reactions without further purification. Another method, recommended in (52) uses bromobenzene or bromotoluene to extract the bromides from natural brines containing ~0.7 g bromides/L and 160 g chlorides/L. The Br^- containing extract can be used for the preparation of elemental Br_2 by oxidation, or utilized directly to make metal bromides, and various Br derivs. No industrial application of either method was reported.

From Wastes. Increasing amount of bromine is manufactured from wastes. The main source is HBr, which is obtained as a by-product in many organic reactions (see BROMINATED INORGANIC COMPOUNDS). Part of the HBr comes from incineration of organic wastes in a BRU–bromine recovery unit (see section Spills and Disposal Procedures below). The flue gases from BRU contain both HBr and Br_2, which are usually separated prior to the processing of the HBr. The basic process is similar to the production of Br_2 from bromides, ie, by oxidation with Cl_2 or other oxidating agent.

Recent researches try to recover bromine from waste hydrogen bromide streams using electrolytic membranes (53) to electrolytically decompose hydrogen bromide into its molecular constituents. These processes are not utilized yet.

Bromine Carriers. The concept of bromine carriers was developed to decrease the hazard of transportation and storage of bromine. The basic idea is to make a salt with a high concentration of bromine from the bromine and to ship it—either as a concentrated solution or as a solid—to the customers. The customer erects in his site a small bromine generator, using the standard method based on chlorine (see above). This may be useful especially for customers who have their own chlorine or at least are located near to a chlorine manufacturer, and who on the other hand are far from a source of bromine and from a port where bromine is available. Another option is to use anionic exchangers to absorb the bromine (54).

The natural choice is $CaBr_2$, which contains 80% weigh bromine, and whose 52% solution is used worldwide as a drilling fluid (see DRILLING FLUIDS). Another option may be $MgBr_2$, which contains 85% bromine.

The drawback in the use of bromine carriers is the extra cost of making the fluid from bromine, the need to prepare the bromine in-situ, and to ship chlorine (unless the customer has his own), which creates greater hazard than the bromine itself. Until now no commercial use of bromine carriers was reported.

6. Economic Aspects

Facilities for manufacturing bromine are primarily located near sources of natural brines or bitterns containing usable levels of bromine. In 1990, the United States had seven bromine plants owned by four companies. Six of the plants are in southern Arkansas and are operated by two U.S. producers: Great Lakes Chemical Corporation and Ethyl Corporation. The biggest single bromine plant is erected by the Dead Sea in Israel, and is operated by the DSBG.

The costs of building and maintaining a bromine plant are high because of the corrosiveness of brine solutions that contain chlorine and bromine and require special materials of construction. The principal operating expenses are for pumping, steam, environmental costs, energy, and chlorine. The plants are very capital intensive.

Figure 3 shows the prices of bromine (excluding shipment) in tank car quantities from 1985 to 1997 (55). The price rose 125% over these years, an aver-

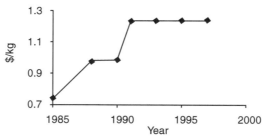

Fig. 3. Price of bromine in the period of 1985–2000.

Table 3. **Annual Bromine Production in Thousands of Metric Tons**[a]

Country	1991	1992	1993	1994	1995	1996	1997	1998	1999	2000
Azerbaijan		5	4	3	2	2	2	2	2	2
China	12.1	16.65	24.6	31.4	32.7	41.4	50.1	40	45	45
France	3	3.2	2.29	2.19	2.26	2.024	1.974	1.95	2	2
India	1.3	1.3	1.4	1.4	1.5	1.5	1.5	1.5	1.5	1.5
Israel	135	135	135	135	130	160	180	185	185	210
Italy	0.4	0.3	0.3	0.3	0.3	0.3	0.3	0.3	0.3	0.3
Japan	15	15	15	15	15	15	20	20	20	20
Spain	0.3	0.25	0.2	0.2	0.2	0.1	0.1	0.1	0.1	0.1
Turkmenistan	0	12	0.1	0.1	0.1	0.102	0.13	0.15	0.1	0.2
Ukraine	24[a]	7	5	3	3.5	3	3	3	3	3
United Kingdom	29.3	29.9	27.4	33.8	26.2	30.6	35.6	30	28	30
United States	170	171	177	195	218	227	247	230	239	229
Total	*390.4*	*396.6*	*392.3*	*420.4*	*431.8*	*483.0*	*541.7*	*514*	*526.0*	*543.1*

[a] Ref. 55.

Source: 1991–1993—Bromine, U.S. Bureau of Mines 1997; 1994–1999—Bromine, U.S. Bureau of Mines 1999; 2000 Bromine, U.S. Bureau of Mines 2001.

age of 7.0% a year. But the average in this case is misleading: between 1990 and 1991 the prices jumped by 25%, and remained stable since then. Even when inflation is taken into account, the price in constant dollars rose significantly. It should be mentioned that in the period of 1976–1985 the price in constant dollars fell slightly (55). The cost of shipment in bromine is high—about 60% of the price of Br, ie, ~30% of the final price of the Br is shipment costs.

Estimates of bromine production around the world are shown in Table 3.

Figure 4 shows the United States and Israel bromine production with respect to bromine production in the rest of the world. In particular, Israel, has increased its production in recent years (55,56). Between the years 1976 and 1990 U.S. production fell from 234,000 to 177,000 metric tons but it has now recovered. In the year 2000 it was back at 229,000 metric tons. Israel's production increased from 23,000 in 1976 to 210,000 metric tons at 2000.

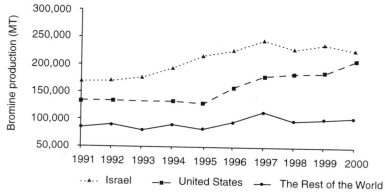

Fig. 4. Annual bromine production by the main manufacturers.

Table 4. **Bromine Specification**

Parameter	ACS requirements	Product specifications
bromine content, wt %[a]	99.5	99.9
specific gravity, 20/15°C[a]		3.1
water, ppm[b]		30
chlorine, ppm[b]	500	100
organic halogen compounds, ppm[a]	[c]	80
nonvolatile matter, ppm[a]	50	30
iodine	10	[c]
sulfur (as S)	10	[c]
heavy metals (as Pb)	2	[c]
nickel	5	[c]

[a] Minimum value.
[b] Maximum value.
[c] Must pass ACS test.

7. Requirements and Specifications

The American Chemical Society (ACS) requirements for bromine used as a reagent chemical (57) are listed in Table 4. Typical specifications for bromine produced in a modern plant (58) generally exceed these requirements.

8. Analytical Methods

To assay liquid bromine, an ampule of bromine is crushed under the surface of an aqueous potassium iodide solution and the resultant iodine ir titrated with standard sodium thiosulfate.

$$2\,I^- + Br_2 \longrightarrow I_2 + 2\,Br^-$$

$$I_2 + 2\,Na_2S_2O_3 \longrightarrow Na_2S_4O_6 + 2\,NaI$$

Bromine vapor can be analyzed by the same procedure. The specific gravity of bromine is determined by hydrometer (54).

Bromine and bromides can be detected qualitatively by a number of methods. In higher concentrations bromine forms colored solutions in solvents such as carbon tetrachloride [56-23-5] and carbon disulfide [75-15-0]. Bromine reacts with yellow disodium fluorescein [518-47-8] to form red disodium tetrabromofluorescein (eosin) $C_{20}H_6Br_4Na_2O_5$, [548-26-5]. As little as 0.3 µg of bromide can be detected and chlorides do not interfere (59). Bromine reacts with platinum sulfate $Pt(SO_4)_2$, [7446-29-9], solution to form red to brown crystals of potassium hexabromoplatinate [16920-93-7], K_2PtBr_6 (60).

Impurities in bromine may be determined quantitatively (61). Weighing the residue after evaporation of a bromine sample yields the total nonvolatile matter. After removing the bromine, chloride ion may be determined by titration with mercuric nitrate, and iodide ion by titration with thiosulfate; water and organic compounds may be detected by infrared ir spectroscopy; sulfur may be determined turbidimetrically as barium sulfate; and heavy metals may be determined either colorimetrically after conversion to sulfides, or by turbidimetry. An individual metal may be detected by the dissolution of the residue and ICP determination.

Quantitative methods for determining bromide include: the Mohr method, using $AgNO_3$ titrant and potassium chromate indicator; the Volhard method using excess $AgNO_3$ titrated with potassium thiocyanate and ferric ammonium sulfate indicator; Fajans method with $AgNO_3$, as titrant, eosin as absorption indicator; silver nitrate titrant with the end point determined potentiometrically using a silver indicator electrode; and a gravimetric method as AgBr. Bromides can be detected in acidic solutions by titrating with mercuric nitrate using sodium nitroprusside indicator. Trace amounts of bromides can be determined quantitatively by (1) the van der Meulen method, which is useful in presence of large amounts of chloride, the bromide is oxidized to bromate and determined iodometrically; (2) by constant-current and constant-potential coulometry, used for fractions of a milligram up to several grams of bromide; (3) by ion chromography, which is useful for detecting bromide in the presence of other ions; (4) by polography, useful for microgram quantities (61); (5) by spectrophotometric methods useful for microgram quantities in the presence of chloride (68); and (6) by activation analysis with thermal neutrons which is useful for nanogram quantities.

Bromine in organic compounds can be determined chemically following oxidation of the organic compounds and reduction of the bromine to bromide. In the Shoniger method a few milligrams of sample is burned inside of a stoppered Erlenmeyer flask filled with oxygen. After ignition by electrical or other means, the combustion products are absorbed and the bromine content is determined acidimetrically (62,63). An alternative method employs a fusion with sodium peroxide in a Paar bomb. Infrared spectroscopy allows the determination of bromine with an accuracy of ~1%. Neutron activation, X-ray fluorescence, ir spectroscopy, and atomic emission spectroscopy are also used to determine bromine in organic materials.

Bromine is used as an analytical reagent to determine the amount of unsaturation in organic compounds because carbon–carbon double bonds add bromine quantitatively, and for phenols that add bromine in the ortho and para positions. Standard bromine is added in excess and the amount unreacted is determined by an indirect iodine titration. Bromine is also used to oxidize several elements, such as Tl(I) to Tl(III). Excess bromine is removed by adding phenol. Bromine plus an acid, such as nitric and/or hydrochloric, provides an oxidizing acid mixture useful in dissolving metal or mineral samples prior to analysis for sulfur. Solutions of bromine in Br^- under acidic or even mildly basic conditions are used for leaching of gold and platinum group elements from ores as a quick analytical method.

9. Health and Safety Aspects and Handling

9.1. Consequences of Exposure.
Bromine has a sharp, penetrating odor. The Occupational Safety and Health Administration (OSHA) threshold

limit value–time-weighted average for an 8-h workday and 40-h workweek is 0.1 ppm in air (64). Monitors are available for determining bromine concentrations in air. Concentrations of ~1 ppm are unpleasant and cause eyes to water; 10 ppm are intolerable. Inhalation of 10 ppm and higher concentrations of bromine causes severe burns to the respiratory tract and is highly toxic. Symptoms of overexposure include coughing, nose bleed, feeling of oppression, dizziness, headache, and possibly delayed abdominal pain and diarrhea. Pneumonia may be a late complication of severe exposure.

Liquid bromine produces a mild cooling sensation on first contact with the skin, which is followed by a sensation of heat. If bromine is not removed immediately by flooding with water, the skin becomes red and finally brown, resulting in a deep burn that heals slowly. Contact with concentrated vapor can also cause burns and blisters. For very small areas of contact in the laboratory, a 10% solution of sodium thiosulfate in water can neutralize bromine and such a solution should be available when working with bromine. Bromine is especially hazardous to the tissues of the eyes where severely painful and destructive burns may result from contact with either liquid or concentrated vapor. Ingestion causes severe burns to the gastrointestinal tract (65,66).

9.2. Detection of Bromine Vapor.

Bromine vapor in air can be monitored by using an oxidant monitor instrument that sounds an alarm when a certain level is reached. An oxidant monitor operates on an amperometric principle. The bromine oxidizes potassium iodide in solution, producing an electrical output by depolarizing one sensor electrode. Detector tubes, useful for determining the level of respiratory protection required, contain o-toluidine which produces a yellow-orange stain when reacted with bromine. These tubes and sample pumps are available through safety supply companies (57). The useful concentration range is 0.2–30 ppm.

9.3. Protective Equipment.

Totally enclosed systems should be used for processes involving bromine. For handling bromine in the laboratory, the minimum safety equipment should include chemical goggles, rubber gloves (Buna-N or neoprene rubber), laboratory coat, and fume hood. For handling bromine in a plant, safety equipment should include hard hat, goggles, neoprene full coverage slicker, Buna-N or neoprene rubber gloves, and neoprene boots. For escaping from an area where a bromine release has occurred, a full face respirator with an organic vapor–acid gas canister is desirable. For emergency work in an area with bromine concentrations >0.1 ppm, a self-contained breathing apparatus can be used until the air supply gets low. For longer term work in elevated bromine concentrations, an air-line respirator is essential.

9.4. Reactivity.

Bromine is nonflammable but may ignite combustibles, such as dry grass, on contact. Handling bromine in a wet atmosphere, extreme heat, and temperatures low enough to cause bromine to solidify (−6°C) should be avoided. Bromine should be stored in a cool, dry area away from heat. Materials that should not be permitted to contact bromine include combustibles, liquid ammonia, aluminum, titanium, mercury, sodium, potassium, and magnesium. Bromine attacks some forms of plastics, rubber, and coatings (66).

9.5. Spills and Disposal Procedures.

If a spill occurs outdoors, personnel should stay upwind of it. If the spill is in a diked area it may be possible to recover much of the bromine, by collecting it into sealable containers.

Otherwise it should be absorbed with appropriate material. Bromine vapors may be neutralized by gaseous ammonia. Small spills may be neutralized with lime water slurry (most common procedure) or 10–30% of K_2CO_3 or Na_2CO_3 solution. No decontaminants other than water should be used on humans. Under the Comprehensive Environmental Response, Compensation, and Liability Act (CERCLA) regulations in effect at the end of 1986, bromine is regulated as a hazardous waste or material. Therefore, it must be disposed of in an approved hazardous waste facility in compliance with the Environmental Protection Agency (EPA) and/or other applicable local, state, and federal regulations and should be handled in a manner acceptable to good waste management practice (66).

The modern procedures of disposal are based on BRU, in which the organic (sometimes mixed with inorganic) wastes are incinerated at high temperature. In one process (67), an organic bromide, eg, ethylene dibromide, is reacted in the gas phase with H_2 at 400–500°C in the presence of a catalyst to prepare HBr. Pelleted Al_2O_3-supported Cr_2O_3 cracking catalyst is suitable. The Br recovery as HBr is ~95%. The process makes it possible to ship organic bromides as sources of HBr. Recent processes operate at very high temperatures (>1000°C), to prevent the creation and survival of brominated dioxins, and the resulting bromine and hydrogen bromide are recovered by absorption in water and release by distillation. The hydrogen bromide may be further oxidized to yield bromine (see section on Bromine Manufacturing from Wastes). The costs of the incineration exceed the commercial value of the recoved bromine values, but as it is the only effective way to get rid of brominated organic wastes, without creating an ecological hazard, they become an integral part of more and more bromine manufacturing plants.

An example of disposal is the recovery of bromine from Br-containing photographic wastewater, which is done by spray incineration of the wastewater with alkali metal hydroxides or salts under an oxidative atmosphere, collection of the ashes formed by the incineration in water, and treatment of the aqueous salt solutions with Cl_2 to recover the bromine (68).

9.6. Materials of Construction. Glass has excellent corrosion-resistance to wet or dry bromine. Lead is very useful for bromine service if water is <70 ppm. The bromine corrosion rate increases with concentrations of water and organics. Tantalum and niobium have excellent corrosion-resistance to wet or dry bromine. Nickel and nickel alloys such as Monel 400 and Hastelloys B and C have useful resistance for dry bromine but is rapidly attacked by wet bromine. Steel and stainless steel materials are not recommended. The fluoropolymers Kynar, Halar, and Teflon are highly resistant to bromine but are somewhat permeable. The rate depends on temperature, pressure, and structure (density) of the fluoropolymer (69). Other polymers are not recommended.

9.7. Storage and Transportation. Bromine in bulk quantities is shipped in the United States in 7570 and 15,140 L lead-lined pressure tank cars or 6435–6813 L nickel-clad pressure tank trailers. The trailers must be filled at least 92% full to prevent inertia effects of the heavy liquid while on the highway. International shipments made by DSBG are in 15.2–23.3 metric ton (volume of 5300–8000 L), lead-lined tanks containers (isotanks). For smaller quantities lead lined tanks ("goslars") of 3.5 metric tons (four tanks packed on one isoframe) and cylinders of 400 kg are used. The relatively high freezing point of bromine ($-7.25°C$) may cause some problems in shipping and storage. If bromine has frozen in a tank car, it is necessary to circulate warm (below $54°C$) water through the heating coils.

Dry nitrogen gas is recommended for use in pressure transferring bromine, although dry ($-40°C$ dew point) air may be used. The gas used to pad the bromine in the storage tank must be absolutely dry or severe corrosion results. Dry bromine picks up water from air having a dew point of $-70°C$. When exposed to a high humidity atmosphere the water content of bromine can exceed 300 ppm. Bromine is nonreactive to lead, monel, and a few other alloys when water content is <30 ppm. If the water content increases above 70 ppm, corrosivity to many metals increases greatly. Fluorinated plastics are widely used in equipment, piping, valves, and gaskets.

The concept of shipping solid bromine was considered on a R&D level and verified in a pilot plant (70), but has not been commercially implemented. The basic idea is to freeze the bromine inside a cooled isotank, and keep it cooled during the ship transport. The main advantage of this method is, that in the case of an accident that results in dropping the tank into the sea, in the port or on the highway, there is enough time for a rescue operation, before thawing of the solid, decreasing the risk of liquid and vapor spillage.

10. Uses

An important use of bromine compounds is in the production of flame retardants (qv). These are of the additive-type, which is physically blended into polymers, and the reactive type, which chemically reacts during the formation of the polymer. Bromine compounds are also used in fire extinguishers. Brominated polymers are used in flame retardant applications and bromine-containing epoxy sealants are used in semiconductor devices.

Bromine has some use in swimming pools and in bleaching. It is also a disinfectant for cooling water and wastewater. Its main use is as a chemical reactant. Bromine compounds are frequently intermediates in the production of other organic chemicals. Bromine is found in certain dyes. Bromides have been used for many years in the pharmaceutical industry as sedatives and as intermediates for drugs. Some therapeutic powers are claimed for certain iodine–bromine spa waters. Alkali bromides are used in the photographic industry. Bromine is used in making some perfumes and certain bromine compounds are disinfectants, eg, bromochlorodimethylhydantoin, which is used in swimming pools and spas.

Zinc–bromine storage batteries (qv) are under development as load-leveling devices in electric utilities (71). Photovoltaic batteries have been made of selenium or boron doped with bromine. Graphite fibers and certain polymers can be made electrically conductive by being doped with bromine. Bromine is used in quartz–halide light bulbs. Bromine is used to etchaluminum, copper, and semi-conductors. Bromine and its salts are known to recover gold and other precious metals from their ores (72). Gold, as a precious metal, is found naturally as an element, at very low concentration . A mixture of bromine and bromide salts

Table 5. **Consumption of Bromine by End Users (Thousands of Metric Tons)**[a,b]

	1965	1970	1975	1980	1985	1990	1993	1996
flame retardants	5.9	7.7	21.8	25.4	48	73	79	89.8
drilling fluids	[d]	[d]	[d]	[d]	18	21.8	12.2	54
brominated agr. chemicals	4.5	7.7	9.1	15.3	17.2	23.3	29.5	28.5
biocides/water treatment	[b]	[b]	[b]	[b]	6.7	9.8	7.7	15.9
gasoline additives	81.6	111.2	98	72.6	32	14	10.9	neg.
others[c]	25	29.5	54.2	54.6	20.3	32.9	32.5	36.5
Total	*117*	*156.1*	*183.1*	*167.9*	*142.2*	*174.8*	*171.8*	*224.7*

[a] Data through 1982 are compiled annually from a voluntary survey. Data are not complete because many smaller bromine consumers are not included. In addition, data may not be reported consistently by different end users.

[b] Reference (56).

[c] Includes bromobutyl rubber, photographic chemicals, dyes, pharmaceuticals and brominated intermediates. Data for 1955–1980 also include drilling fluids, biocides, and water treatment chemicals.

[d] Included in "Others".

may serve as an oxidant and complexant of the gold, according to the overall formula:

$$2 \, Au + 3 \, Br \longrightarrow 2 \, AuBr_4]^- + Br^-$$

The complex is reduced back to a metal, adsorbed on carbon granules, either in CIS (carbon in solution) or CIP (carbon granules) process. This bromine route is much quicker and efficient than the standard cyanide process, and significantly less toxic. In a process developed by Kaljas (73) and tested in Australia in 1989 and 1992 by Reid and Storhok (72), the CIP technique was applied successfully, at ambient temperature, at pH <10, using 0.3% bromine. No industrial application has been reported.

Bromine can be used to desulfurize fine coal (see COAL CONVERSION PROCESSES; CLEANING AND DESULFURIZATION). Table 5 shows estimates of the primary uses of bromine.

BIBLIOGRAPHY

"Bromine" in *ECT* 1st ed., Vol. 2, pp. 629–645, by V. A. Stenger, The Dow Chemical Company; in *ECT* 2nd ed., Vol. 3, pp. 750–766, by V. A. Stenger, The Dow Chemical Company; in *ECT* 3rd ed., Vol. 4, pp. 226–241, by C. E. Reineke, Dow Chemical U.S.A.; "Bromine" in *ECT* 4th ed., Vol. 4, pp. 536–560, by Philip F. Jackisch, Ethyl Corporation; "Bromine" in *ECT* (online), posting date: December 4, 2000, by Philip F. Jackisch, Ethyl Corporation.

CITED PUBLICATIONS

1. M. C. Sneed, J. L. Maynard, and R. C. Brasted, *Comprehensive Inorganic Chemistry*, Vol. **3**, D. Van Nostrand, New York, 1954, p. 70.
2. A. J. Downs and C. J. Adams, in A. F. Trotman-Dickenson, ed., *Comprehensive Inorganic Chemistry*, Pergamon Press, New York, 1973, p. 1107.

3. F. Yaron, in Z. E. Jolles, ed., *Bromine and Its Compounds*, Ernest Benn Ltd., London, 1966, 43–49.

4. V. A. Stenger, *Angew. Chem. Int. Ed.* **5**(3), 280 (1966).

5. *Bromine—Unloading, Storing, Handling*, Form No. 101-2-76, Dow Chemical U.S.A., 1976.

6. G. H. Cady, *J. Phys. Chem.* **89**, 3302 (1985).

7. J. d'Ans and P. Hofer, *Angew Chem.* **47**, 71 (1934).

8. F. H. Rhodes and C. H. Bascom, *Ind. Eng. Chem. Ind. Ed.* **19**, 480 (1927).

9. L. W. Winkler, *Chem. Ztg.* **23**, 687 (1899).

10. W. F. Linke, ed., *Solubilities Inorganic and Metal–Organic Compounds*, 4th ed., Vol. **1**, American Chemical Society, Washington, D.C., 1958, 442–444.

11. A. I. Popov, in V. Guttman, ed., *Halogen Chemistry*, Vol. **1**, Academic Press, New York, 1967, p. 225.

12. U.S. Pat. 2,481,455 (Sept. 6, 1949), W. R. Kramer and V. A. Stenger (to The Dow Chemical Company).

13. Ref. 10, p. 161.

14. U.S. Pat. 5,008,477 (Apr. 16, 1991), S. Hussain (to Ethyl Corp.).

15. J. Kleinberg, W. J. Argersinger, Jr., and E. Griswold, *Inorganic Chemistry*, D. C. Heath, Boston, Mass., 1960, Chapt. 16.

16. E. S. Gould, *Inorganic Reactions and Structure*, Henry Holt, New York, 1955, Chapt. 14.

17. Ref. 1, p. 118.

18. H. S. Booth, ed., *Inorganic Syntheses*, McGraw-Hill, New York, 1939, p. 51.

19. R. B. Heslop and P. L. Robinson, *Inorganic Chemistry*, 3rd ed., Elsevier, Amsterdam, The Netherlands, 1967, Chapt. 28.

20. T. Moeller, *Inorganic Chemistry*, John Wiley & Sons, Inc., New York, 1952, Chapt. 13.

21. R. C. Weast, ed., *CRC Handbook of Chemistry and Physics*, CRC Press, Boca Raton, Fla., 70th ed., 1989, p. D152.

22. A. J. Downs and C. J. Adams, in A. F. Trotman-Dickenson, ed., *Comprehensive Inorganic Chemistry*, Pergamon Press, New York, 1973, p. 1191.

23. C. A. Buehler and D. E. Pearson, *Survey of Organic Syntheses*, Wiley-Interscience, New York, 1970, Chapt. 7.

24. R. B. Wagner and H. D. Zook, *Synthetic Organic Chemistry*, John Wiley & Sons, Inc., New York, 1953, Chapt. 4.

25. F. Yaron, in Z. E. Jolles, ed., *Bromine and its Compounds*, Academic Press, New York, 1966, 71–72.

26. G. J. Fox, G. Hallas, J. D. Hepworth, and K. N. Paskins, *Org. Synth.* **VI**, 181 (1988).

27. F. C. Whitmore, W. L. Evers, and H. S. Rothrock, *Org. Synth.* **II**, 408 (1943).

28. S. I. Miller, G. R. Ziegler, and R. Wieleseck, *Org. Synth.* **V**, 921 (1973).

29. P. A. Wender, M. A. Eissenstat, N. Sapuppo, and F. E. Ziegler, *Org. Synth.* **VI**, 334 (1988).

30. L. Blanco, P. Amice, and J. M. Conia, *Synthesis*, 194 (1976).

31. N. C. Deno and N. H. Potter, *J. Am. Chem. Soc.* **89**, 3550 (1967).

32. R. A. Sneen and N. P. Matheny, *J. Am. Chem. Soc.* **86**, 3905 (1964).

33. A. Picot and X. Lusinchi, *Synthesis*, 109 (1975).

34. N. M. Roscher and E. J. Jedziniak, *Tetrahedron Lett.*, 1049 (1973).

35. H.-G. Heine, *Ber. Dtsch. Chem. Ges.* **104**, 2869 (1971).

36. J. M. Conia and J. M. Denis, *Tetrahedron Lett.*, 2845 (1971).

37. V. Calo, L. Lopez, and G. Pesce, *J. Chem. Soc. Perkin Trans. I*, 501 (1977).

38. M. Al Neirabeyeth, J.-C. Ziegler, and B. Gross, *Synthesis*, 811 (1976).

39. V. Calo, L. Lopez, and D. S. Valentino, *Synthesis*, 139 (1978).

40. J. Drabowicz, W. Midura, and M. Mikolajczyk, *Synthesis*, 39 (1979).

41. D. R. Williams, F. D. Klingler, E. E. Allen, and F. W. Lichtenthaler, *Tetrahedron Lett.* **29**(40), 5087 (1988).
42. F. A. Cotton and G. Wilkinson, *Advanced Inorganic Chemistry*, 2nd ed., Wiley-Interscience, New York, 1966, 563–564.
43. P. A. Lyday, in *Minerals Yearbook 1987*, Vol. **1**, U.S. Bureau of Mines, Washington, D.C., p. 172.
44. P. J. Scheuer, *Chemistry of Marine Nature Products*, Academic Press, New York, 1973.
45. R. E. Cuenca, W. J. Pories, and J. Bray, *Biol. Trace Elem. Res.* **16**(2), 151 (1988).
46. G. D. Supatashvili and G. A. Makharadze, *Soobshch. Adad. Nauk Gruz. SSR* **120**(1), 121 (1985).
47. S. C. Wofsy, M. B. McElroy, and Y. L. Yung, *Proc. Conf. Clim. Impact Assess. Program, 4th 1975*, 286 (1976).
48. U.S. Pat. 4,719,096 (Jan. 12, 1988), K. C. Lesher and H. W. Henry (to Ethyl Corp.).
49.
    DSBG Internal Data
50. U.S. Pat. 4,725,425 (Feb. 16, 1988), K. C. Lesher and H. W. Henry (to Ethyl Corp.).
51. A. P. Krasnov, V. F. Trifonov, O. V. Lebedev, *Khim. Prom-st.* (Moscow) (1986), (**4**), 221
52. C. Yang, Z. Wang, H. Xiao, L. D. X. Nanjing, **23**(5), 466 (1999) (in Chinese)
53. C. N. Wauters, J. Winnick, *AIChE* **44** p. 2144 (10 Oct 1998)
54. Rus. Pat. 2070537 (Dec. 20, 1996) Y. N. Fedulov, L. L. Koroleva, L. N. Pisarenko, N. P. Sokolova, V. Danilov,
55. *Bromine*, U.S. Bureau of Mines, Washington, D.C., annual issues from 1975 to 2001.
56. CEH, 1998 by the Chemicals Economics Handbook - SRI International.
57. *Reagent Chemicals*, 7th ed., American Chemical Society, Washington, D.C., 1986, 158–160.
58. *Bromine*, Ethyl Corp., Baton Rouge, La., Jan. 1985, p. 7.
59. N. D. Cheronis and J. B. Entrikin, *Semimicro Qualitative Organic Analysis*, Wiley-Interscience, New York, 1957, p. 179.
60. G. W. Armstrong, H. H. Gill, and R. F. Rolf, "The Halogens," in *Treatise on Analytical Chemistry*, Part II, Vol. **7**, Wiley-Interscience, New York, 1961.
61. F. Feigl and V. Anger, *Spot Tests in Inorganic Analysis*, Vol. **1**, Elsevier, New York, 1972, p. 143.
62. W. Shoniger, *Mikrochim. Acta*, 123 (1955).
63. *Ibid.*, 869 (1956).
64. *Threshold Limit Values and Biological Exposure Indices for 1989–1990*, American Conference of Governmental Industrial Hygienists, Cincinnati, Ohio, 1989, p. 14.
65. *Material Safety Data Sheet: Bromine*, Ethyl Corp., Baton Rouge, La., Sept. 20, 1991.
66. *Bromine Safe Handling Seminar*, Ethyl Corp., Magnolia, Ark., Oct. 4–6, 1988.
67. U.S. Pat. 3919398 (Nov. 11, 1975), R. A. Davis, (for Dow Chemical Co., USA).
68. Jap. Pat. 07171581 (July 11, 1995), M. Yamada and T. Nakamura, H. Asano, and T. Adachi (for Fuji Photo Film Co. Ltd, Japan).
69. "Chemical Profile: Bromine" in *Chem. Mark. Rep.* (Apr. 16, 1979); (Apr. 22, 1982); (July 22, 1985); (July 11, 1988); and (July 15, 1991).
70. S. Wajc, IMI(TAMI) Internal Report (1997)
71. T. N. Veziroglu, ed., "Alternate Energy Sources," *Proceedings of the Miami International Conference*, Vol. **1**, Hemisphere, Washington, D.C., 1983, 327–333.
72. M. Freiberg, *Rev. Chem. Eng.*, **9**, 3–4, 333, (1993)
73. U.S. Pat. 4684404 (Aug. 4, 1987), Gui I. Z. Kalocsal (for Kaljas Ltd., Australia).

GENERAL REFERENCES

Z. E. Jolles, ed., *Bromine and its Compounds*, Academic Press, New York, 1966.

V. Gutmann, ed., *MTP Int. Rev. Sci.: Inorg. Chem., Ser. One*, **3** (1972).

A. J. Downs and C. J. Adams, "Chlorine, Bromine, Iodine, and Astatine," in J. C. Bailar, Jr. and co-eds., *Comprehensive Inorganic Chemistry*, Vol. **2**, Pergamon Press, Oxford, 1973, 1107–1594.

V. Gutmann, ed., *Halogen Chemistry*, Academic Press, New York, 1967 (three volumes).

BARUCH GRINBAUM
IMI(TAMI) Institute for
Research and Development

MIRA FREIBERG
Dead Sea Bromine Group

BROMINE, INORGANIC COMPOUNDS

1. Introduction

The aim of this article is to provide a brief summary of physical and chemical properties of several classes of inorganic bromine compounds. The main classes of inorganic bromine compounds covered include (a) bromamines, (b) hydrogen bromide and hydrobromic acid, (c) metal bromides, (d) non-metal bromides, (e) bromine halides, (f) bromine oxides, (g) oxygen acids and their salts. Wherever possible, updated uses and applications have been cited for the specific compounds surveyed.

2. Bromamines

The bromamines are highly unstable compounds, having a tendency to explode at low temperatures, if they are isolated. Traces of these compounds can be formed when water containing small amounts of bromide and ammonia is chlorinated.

Monobromamine, NH_2Br, bromamide, [14519-10-9], was first prepared by reacting a dilute solution of bromine in anhydrous ether with a 50% excess of ammonia (in anhydrous ether) at $-60°C$ (1) according to the reaction

$$2\,NH_3 + Br_2 \rightarrow NH_2Br + NH_4Br$$

After the ammonium bromide and excess ammonia are removed, a pale straw-colored product solution remains. If the bromamine is isolated at low temperatures, it decomposes violently when its temperature is allowed to reach $-70°C$ (2). The bromamine reacts in ether with Grignard reagents to produce primary amines, ammonia, and nitrogen.

Dibromamine, $NHBr_2$ (bromimide) [14519-03-0], is prepared by adding an ether solution of ammonia to an excess of bromine (in ether) at a temperature of $-50°C$, until the color of the bromine solution changes from red to yellow (1,3).

$$3 NH_3 + 2 Br_2 \rightarrow NHBr_2 + 2 NH_4Br$$

An ether solution of $NHBr_2$, after the removal of ammonium bromide, is stable for several hours at $-72°C$. When dibromamine reacts with Grignard reagents, primary and secondary amines are formed, together with ammonia and N_2. The disinfecting efficiency of dibromamine in water exceeds that of dichloramine (4).

At $-78°C$, Br_2 and liquid ammonia form pale yellow-to-red solutions containing ammonium bromide and an equilibrium mixture of NH_2Br and violet $NBr_3 \cdot 6NH_3$.

Tribromamine, NBr_3 (nitrogen tribromide) [15162-90-0], can be obtained as an ammoniate, ie, $NBr_3 \cdot 6NH_3$, by the reaction between bromine and ammonia below $-70°C$ (2). The tribromamine is predominant in water at $pH < 8$ when $2-3$ mol Br_2 are added per mole ammonia (5); it may be extracted from water with chloroform.

3. Bromides

3.1. Hydrogen Bromide.
Hydrogen bromide, HBr (hydrobromic acid) [10035-10-6], is a colorless, corrosive gas that fumes strongly in moist air. It is extremely irritating to the eyes, nose, and throat. Some of the physical properties of anhydrous hydrogen bromide gas are summarized in Table 1.

Hydrogen bromide gas is highly soluble in water, forming azeotropic mixtures whose compositions at various pressures have been determined (6). At normal atmospheric pressure, the boiling point of the azeotrope is 124.3°C and the HBr content is 47.63%. It freezes at about $-11°C$ and has a density of 1.482 g/mL at 25°C. At very low temperatures, HBr forms hydrates with 1,2,3, and 4 molecules of water (7).

Table 1. **Physical Characteristics of Hydrogen Bromide**

mp $-86°C$
bp $-67°C$ (101.3 kPa)
liquid density 2.152 g/mL
heat of fusion at mp 29.8 kJ/kg (7.12 kcal/kg)
heat of vaporization at $-66.7°C$, 218 kJ/kg (52 kcal/kg)
heat capacity J/(kg·K) [cal/(kg·K)]
   (i) solid at $-91°C$, 636 [152]
   (ii) liquid 737 [176]
   (iii) gas at 27°C, 356 [85]
critical temperature 89.8°C
critical pressure 8510 kPa (84 atm)

Methods for HBr gas preparation include

1. Reaction of a mixture of excess hydrogen with bromine vapor at a temperature of 500°C, or reaction of hydrogen and bromine at 375°C over platinized silica gel or platinized asbestos catalysts (8).
2. The reaction of bromine with sulfur or phosphorus and water. In the above-mentioned procedures, the HBr vapor that is formed is passed through hot, activated charcoal or iron to remove the free bromine (9), and is then either liquefied by cooling for shipment in cylinders or is absorbed into water.
3. Hydrogen bromide is obtained as a by-product during the bromination of aromatic compounds and separated by distillation. This hydrogen bromide is less pure and is usually used as a raw material for clear drilling fluids.

Laboratory methods for HBr preparation include

1. Distillation of HBr from potassium bromide and dilute sulfuric acid.
2. Reaction of bromine with sulfur dioxide and water followed by distillation.
3. For preparation of a dilute HBr solution, an alkali bromide solution can be passed through the hydrogen form of a cation-exchange resin that is suitable for dilute solutions.

Hydrobromic acid is one of the strongest mineral acids known. It is considered a more effective leaching agent than hydrochloric acid for some mineral ores because of its higher boiling point and stronger reducing action. Certain higher oxides such as ceric oxides are readily dissolved in HBr. The acid forms complexes with the bromides of several metals. Examples of such complexes are hydrogen tetrabromoferrate, $HFeBr_4$, amber [19567-68-1]; hydrogen tribromocuprate, $HCuBr_3$, violet [31415-59-5]. Bromine is highly soluble in concentrated aqueous hydrobromic acid.

Safety and Environmental Considerations. The liquids and vapors of HBr are highly corrosive to human tissue. The threshold limit value for hydrogen bromide gas in an 8-h day is 3 ppm time-weighted average. Inhalation of the vapor (when present at highly hazardous concentrations) is highly irritating to the nose and throat. Symptoms from HBr overexposure include coughing, choking, burning in the throat, wheezing, or asphyxiation. Ingestion of the HBr vapor causes severe burns of the mouth and stomach while skin contact can cause severe burns. In the case of liquid or vapor contact with the eyes, permanent damage may result. It is therefore imperative to employ the proper safety equipment when handling HBr including a safety shower and eye bath.

For 48% hydrobromic acid (which is corrosive to eyes, skin and mucous membranes) the following measures are suggested: (a) Eye contact: Hold eyelids apart, flush eyes promptly with large amounts of flowing water for at least 20 min. (b) Skin contact: Skin should be flooded with water. The skin should be washed thoroughly with mild soap and plenty of water for 15 min. (c) Inhalation:

Remove person to fresh air. Keep person quiet and warm. Apply artificial respiration if needed. (d) Ingestion: If swallowed, wash mouth thoroughly with plenty of water and give water to drink. In all of the above cases, medical attention should be obtained immediately.

If a leak or spill occurs when working with HBr, the exposure to its vapors should be strictly avoided. If a high concentration of the hydrobromic acid is accidentally spilled, it should be diluted immediately with water to reduce its fuming before neutralization with soda ash or lime. Aqueous sodium hydroxide may also be used to neutralize the diluted acid. HBr also reacts with metals to produce highly explosive hydrogen gas.

Most metals, concrete, and other construction materials are corroded by HBr. Suitable materials of construction include some fiber glass-reinforced plastics, some chemically resistant rubbers, polyvinyl chloride (PVC), Teflon, polypropylene and ceramic-, rubber-, and glass-lined steel. Metals that are used include Hastelloy B, Hastelloy C, and titanium. The Hastelloys are only suitable at ambient temperatures. Hydrogen bromide under pressure in glass at or above room temperature can attack the glass, resulting in unexpected shattering.

Technical 48 and 62% acids range from colorless-to-light yellow liquids, which are available in drums or tank trailers and tank car quantities. They are classified under DOT regulations as corrosive materials. Anhydrous HBr is available in cylinders under its vapor pressure ($\sim$2.4 MPa or 350 psi) at 25°C. It is classified as a nonflammable gas.

Uses and Applications

1. Hydrogen bromide is used in the manufacture of inorganic bromides. Metal hydroxides or carbonates are used for the neutralization.

2. Hydrogen bromide is also a raw material in the synthesis of alkyl bromides from alcohols.

3. Hydrobromination of olefins. The addition can take place by an ionic mechanism, usually in a polar solvent according to Markovnikob's rule to yield a secondary alkyl bromide. By using a free-radical catalyst in aprotic, nonpolar solvents, dry HBr reacts with an olefin to produce a primary alkyl bromide as the predominant product. Primary alkyl bromides are used in synthesizing other compounds and are 40–60 times as reactive as the corresponding chlorides (10).

4. Hydrogen bromide adds to acetylene to form vinyl bromide or ethylidene bromide, depending on the stoichiometry.

5. Hydrogen bromide cleaves acyclic and cyclic ethers.

6. Hydrogen bromide adds to the cyclopropane group by ring opening.

7. Addition of hydrogen bromide to quinones produces bromohydroquinones.

8. Hydrogen bromide and aldehydes can be used to introduce bromoalkyl groups into various molecules, eg, bromoethylation of aromatic nuclei (10).

9. In the petroleum industry, HBr serves as an alkylation catalyst.

Table 2. **Physical Properties and Characteristics of Selected Metal Bromides**

Compound	mp, °C	bp, °C	d_4^{25}	Other characteristics
sodium bromide	747	1390	3.26	bitter, salty hygroscopic white solid
potassium bromide	730	1435	2.75	bitter, salty hygroscopic white solid
calcium bromide	730[a]	810	3.35	white, hygroscopic, salty crystal
lithium bromide	550	1265	3.46	eye and skin exposure can irritate
ammonium bromide	452 (subl.)		2.43	white, yellow hygroscopic solid
zinc bromide	394	650	4.20	metallic taste, hygroscopic solid
aluminum bromide	97.5	263[b]	3.0	white, yellow hygroscopic solid
cadmium bromide	568	1136		

[a] Some decomposition occurs.
[b] At a pressure of 99.6 kPa. To convert kPa to mmHg multiply by 7.5.

10. Hydrogen bromide is claimed as a catalyst in the controlled oxidation of aliphatic and alicyclic hydrocarbons to ketones, acids, and peroxides (11,12).

11. Applications of HBr with NH_4Br (13) or with H_2S and HCl (14) as promoters for the dehydrogenation of butene to butadiene.

12. Hydrogen bromide is used in the replacement of aliphatic chlorine by bromine - in the presence of an aluminum catalyst.

13. Hydrogen bromide also finds use in the electronics industry.

14. Hydrogen bromide is used to make a catalyst for PTA production or is used as the catalyst itself.

3.2. Metal and Non-Metal Bromides. *Properties.* Physical characteristics of common metal bromides are summarized in Table 2 (15). Table 3 presents data for the solubilities in water for selected metal bromides as a function of temperature.

Other physical properties have been redetermined in recent years for some of the above metal bromide solutions. Some examples include

Table 3. **Solubilities of Various Inorganic Bromides in Water as Function of Temperature**[a]

Compound	t, °C					
	0	20	40	60	80	100
lithium bromide	58.8	62.7	67.5	69.1	71.0	72.8
sodium bromide	44.2	47.5	51.4	54.1	54.1	54.8
potassium bromide	35.1	39.6	43.2	46.2	48.8	51.2
ammonium bromide	37.3	42.6	47.3	51.2	53.9	57.4
calcium bromide	55.5	58.8	68.1	73.5	74.7	75.0
zinc bromide	79.6	81.6	85.5	86.1	86.6	87.0
magnesium bromide	49.5	50.3	51.6	52.8	—	55.6
strontium bromide	46.0	50.6	55.2	60.0	64.5	69.0
barium bromide	48.5	50.3	52.4	54.3	56.5	58.9
nickel bromide	53.0	56.7	59.1	60.4	60.6	60.8

[a] Solubilities in % by weight

1. Densities and sound velocities were measured (16) for aqueous solutions of zinc bromide, calcium bromide, and sodium bromide in the ionic strength range of 0–0.8 at 25°C.

2. Viscosities and densities of aqueous solutions of LiBr and $ZnBr_2$ were also measured for both single salt (17) and two-salt solutions (18). For the single-salt solutions, the densities were presented in tabular form and as 10-parameter correlations.

3. Densities of 0.1, 0.5 and 1.0 $mol \cdot kg^{-1}$ solutions of sodium and potassium bromide were determined at 1°K temperature intervals from a temperature of 277.15°K to a temperature of 343.15°K (19).

Manufacture. The metal bromides are generally prepared by the reaction of hydrogen bromide with the metals, hydroxides, carbonates, or oxides of the metals. For the specific cases of alkali and alkaline earth bromides, the procedure involves the neutralization of the corresponding hydroxide or carbonate with hydrobromic acid, evaporation of excess water, and crystallization. For example,

$$Ca(OH)_2 + 2\,HBr \rightarrow CaBr_2 + 2\,H_2O$$

$$Mg(OH)_2 + 2\,HBr \rightarrow MgBr_2 + 2\,H_2O$$

Both $CaBr_2$ and $MgBr_2$ may be extracted from concentrated brines, using a suitable solvent. Both solvent extraction processes were developed by IMI (TAMI). $MgBr_2$ is extracted using higher alcohols (20) while $CaBr_2$ is extracted with a composite solvent. The calcium bromide process was tested on a semicommercial scale by DSBG (21) with positive results.

An alternative preparation used is the van der Meulen process where bromine and a reducing agent such as ammonia, formic acid, or carbon is used in place of the hydrobromic acid (22). The reaction using ammonia is given below

$$3\,K_2CO_3 + 3\,Br_2 + 2\,NH_3 \rightarrow 6\,KBr + N_2 + 3\,CO_2 + 3\,H_2O$$

Ammonium bromide [12124-9-9] is prepared on a commercial scale by the reaction of bromine with aqueous ammonia according to the following equation

$$3\,Br_2 + 8\,NH_3 \rightarrow 6\,NH_4Br + N_2$$

Some other routes (23) to metal bromides that have been considered include

1. High temperature bromination of metal oxides, sometimes in the presence of CBr_4 or carbon, to assist in the removal of oxygen. For example, Nb_2O_5 can be converted to $NbBr_5$ using CBr_4 as the brominating agent at a temperature of 370°C. Similarly, Ta_2O_5 can form $TaBr_5$ in the presence of carbon and Br_2 above 460°C.

2. A closely related route is a halogen exchange, usually in the presence of an excess of the halogenation agent

$$FeCl_3 + BBr_3 \text{ (excess)} \rightarrow FeBr_3 + BCl_3$$

$$MCl_3 + 3\,HBr \text{ (excess)} \rightarrow MBr_3 + 3\,HCl \qquad 400-600°C \text{ (M = Ln, Pu)}$$

3. Reductive halogenation of the higher bromide with the parent metal, another metal or hydrogen

$$3\,WBr_5 + Al \rightarrow 3\,WBr_4 + AlBr_3 \qquad \text{Thermal gradient}: 475 \rightarrow 240°C$$

4. Thermal decomposition or disproportionation to yield the lower bromide

$$2\,TaBr_4 \rightarrow TaBr_3 + TaBr_5 \qquad \text{at } 500°C$$

Metal bromide solutions can be prepared on a laboratory scale using the reaction of the appropriate carbonate or sulfate with barium bromide.

A more recent process suggested (24) for the preparation of inorganic bromide salts is the reaction of the respective chloride salt with 1,2-dibromoethane (DBE). The reaction is reportedly carried out under convenient phase-transfer catalytic conditions. The process used to produce calcium bromide is based on mixing aqueous calcium chloride with DBE at 90°C in the presence of the phase-transfer catalyst didecyldimethylammonium bromide (2%).

Other procedures that have appeared more recently in the chemical literature are described here briefly:

1. A recent patent (25) claimed an inexpensive process for the production of lithium bromide having a desired purity level, where LiCl or $LiSO_4$ are used as raw materials. Either of the above-mentioned lithium salts is reacted with NaBr or KBr in an aqueous, semiaqueous or organic medium. The precipitated salts are removed so that a LiBr solution of the desired purity can be achieved. The process claims the avoidance of the use of highly acidic materials, thus reducing the cost of the raw materials and the need for specialized equipment.

2. Another method (26) offered for the production of an ammonium bromide, and optionally, a calcium bromide brine, uses sodium bromide brine as the raw material. The process consists of introducing ammonia and carbon dioxide into a sodium bromide brine to form a product solution of ammonium bromide, ammonium bicarbonate, and a sodium bicarbonate precipitate. Following removal of the sodium bicarbonate solids from the product solution, two options are available: (a) Ammonium bromide can be recovered from the solution by evaporative crystallization. (b) Calcium bromide can be produced by the addition of calcium hydroxide or calcium oxide to the product solution, following removal of the $NaHCO_3$ solids.

3. Zinc by-products (containing low grade zinc oxide from industrial products) are used (27) for the manufacture of clear brine fluids by mixing the zinc feedstock (containing metal impurities) with HBr at   pH 3.5–4.5 to produce an impure zinc bromide solution. Metal impurities are removed by precipitation and filtration of the manganese and iron impurities at pH 3–5, and then the zinc bromide solution is concentrated. The solution is then contacted with elemental zinc (zinc shot) at 40–80°C to cement out nickel, lead, cadmium, copper, mercury, and cobalt under an alkalinity of <1.0% zinc oxide. It is preferable to contact the

zinc bromide solution with the zinc metal in multiple passes through packed column systems.

4. Two additional recent Chinese processes deal with the production of ammonium bromide (28) and potassium bromide (29). The process for ammonium bromide is initiated by the preparation of a solution of stoichiometric amounts of sodium bicarbonate and urea. A stoichiometric amount of bromine is added slowly to the solution, with continuous stirring. After the removal of Fe, filtering, and evaporation, reagent grade ammonium bromide is crystallized. The process for potassium bromide is similar: The reaction of bromine with potassium carbonate and urea is the basis of the process. The first step of the process involves the addition of K_2SO_4 to the potassium carbonate solution, followed by heating to 80°C. After the lead-containing precipitate is removed by filtration, the bromine and urea are added, and the temperature and pH are adjusted to 30°C and 6.0–6.5, respectively. Potassium bromide is recovered by recrystallization after reduction of volume of the reacting solution by evaporation. The sulfate can be removed from the solution by addition of $BaBr_2$.

The production of the nonmetal bromides is accomplished through the reaction of the nonmetal with hydrogen bromide or bromine at an elevated temperature. Uses and physical properties of selected inorganic nonmetal bromides are listed below.

Boron tribromide, BBr_3 [10294-33-4]
Physical characteristics: specific gravity 2.64, mp −46°C, bp 91.3°C. Volatile liquid.
Uses: (1) In diborane and ultrahigh purity boron manufacture. (2) Arylacetone boron tribromide aldol condensation cyclization (30).

Phosphorus pentabromide, PBr_5 [7789-69-7]
Physical characteristics: mp 83.8°C, bp > 106°C. Red-yellow solid.
Uses: (1) Brominating agent: Converts organic acids to acyl bromide. (2) Converts phenols and sec-alcohols to bromides (31,32).

Phosphorus tribromide, PBr_3 [7789-60-8]
Produced by the reaction between bromine and yellow phosphorus in PBr_3 at 130°C.
Physical characteristics: Specific gravity 2.85, mp −40°C, bp 173°C. Colorless liquid.
Uses: (1) Converts alcohols to alkyl bromides (33). (2) Catalyst in specific organic reactions.

Selenium tetrabromide, $SeBr_4$ [7789-65-3]
Physical characteristics: Decomposes at 75°C. Red-brown crystals.
Uses: (1) Dopant for a photoreceptor for electrophotography (34). (2) Additive for a rapid bright silver electroplating bath (35).

Silicon tetrabromide, $SiBr_4$ [7789-66-4]
Produced from reaction of silicon and Br_2 vapor > 600°C.
Physical characteristics: mp 5°C, bp 153°C. Colorless liquid.

Use: Production of high purity silicon (36).

Tellurium tetrabromide, TeBr$_4$ [10031-27-3]

Physical characteristics: Specific gravity 4.31, mp $380 \pm 6°C$, bp $421°C$ (decomposes).

Orange crystals.

Use: Catalyst for synthesis of acids (37,38).

Thionyl bromide, SOBr$_2$ [507-16-4]

Produced by passing HBr into thionyl chloride with cooling, without the introduction of moisture. Physical characteristics: Specific gravity 2.69, mp $-52°C$, bp $39°C$ (2 kPa). Orange-yellow liquid.

Use: Brominating agent in organic chemistry, eg, converts alcohols to alkyl bromides (39).

Tribromosilane, SiHBr$_3$ [7789-57-3]

Produced by reaction of metallurgical grade silicon with SiBr$_4$ (tetrabromosilane) at $600-800°C$ in hydrogen atmosphere. Physical characteristics: mp $-73.5°C$, bp $112°C$.

Colorless liquid.

Use: Production of high purity silicon (36).

Applications. The main applications of inorganic bromides are in drilling fluids, biocides, photography, pharmaceutics, catalysts, and brominating agents.

Drilling Fluids (see also Drilling Fluids). The calcium, zinc, and sodium bromides find extensive applications in the oil and gas drilling industry for high density, clear drilling completion, packer, and workover fluids. They find extensive use in deep, high pressure oil and gas wells. These fluids are advantageous in that they do not plug the formation in workover and completion operations, as conventional drilling muds can. Higher production rates and longer lifetimes are the result. In the United States, calcium bromide is the highest volume bromide salt used for this application, followed closely by zinc bromide and sodium bromide. Sodium bromide finds specialized use where calcium/zinc bromides can cause minimal swelling or in strata containing carbon dioxide. Bromine-based completion fluids are used for offshore and onshore drilling.

Calcium bromide is available for oil fields as a 95% minimum purity solid or as a 52–54% aqueous solution (specific gravity 1.70–1.71 at 15.5°C). Zinc bromide is available as a 75% aqueous brine having a specific gravity of 2.4. Also marketed is a zinc bromide/calcium bromide brine for use when prevailing pressures require densities up to 2.3 g/mL (19.2 lb/US gal). Assay for this mixture is 18–21% CaBr$_2$, 53–58% ZnBr$_2$, and its maximum crystallization point is $-7°C$.

Biocides. Ammonium bromide and sodium bromide are used to control microbial, fungal, and algal growth in industrial water systems such as cooling towers, pulp, and paper mills, once through cooling water and in wastewater disinfection. Sodium bromide, as a biocide (40% solution), is typically applied in conjunction with chlorine gas (Cl$_2$), 12.5% sodium hypochlorite (NaClO), sodium dichloro-*s*-triazinetrione, or sodium trichloro-*s*-triazinetrione to form the active biocidal species hypobromous acid (HBrO), or a predefined mixture of hypochlorous/hypobromous acid (40).

For example,

$$Cl_2 + H_2O + NaBr \rightarrow HBrO + NaCl + HCl$$

$$NaClO + NaBr \rightarrow NaBrO + NaCl$$

At the elevated pH levels commonly found in use in today's cooling water programs, hypobromous acid exhibits much greater biocidal efficacy when compared to chlorine. Hypobromous acid is effective in the presence of ammonia and can help to minimize TRO (total residual oxidant) levels in order to enable the end user to meet stricter environmental discharge regulations. Sodium bromide activated by an oxidizer is used also for pools and spas sanitation.

Concentrated Aqueous Solutions. A number of nondrilling fluid applications have been suggested in recent years for concentrated zinc bromide brines and related bromide solutions. Some of these are referred to briefly.

1. Filaments of polyketones can be formed into various fibers using concentrated aqueous zinc bromide or lithium bromide solutions as solvents for dissolving the polyketones (41).

2. A process (42) was described for the electrophilic substitution of an aromatic compound, where an aromatic compound, a precursor of the desired substituent and an aqueous reagent containing a high concentration of zinc bromide are the constituents of the reacting mixture. In certain instances, a water soluble compound, lithium bromide, is added in order to enhance the effectiveness of the catalysis and to increase the reaction yield.

3. Use of concentrated zinc bromide in water and aqueous polar organic solvents for the separation of alcohols from hydrocarbons and alkyl halides (43).

4. Separation of a nonanediol and nonol mixture in methylene chloride by extraction of a zinc bromide-containing water-methanol system (44).

5. Zinc bromide solutions have been suggested (45) as a component of a multiphase mixture in hydraulic systems.

Information on uses of wide range of metal bromides is summarized below.

Aluminum bromide, AlBr$_3$ [7727-15-3]
Uses: Anhydrous compound used as acid catalyst in Friedel–Crafts reactions, olefin polymerizations, hydrocarbon isomerization, and olefin alkylation of aromatics.

Ammonium bromide, NH$_4$Br [12124-97-9]
Uses: (*1*) A biocide to control microbial, fungal, and algal growth in industrial water systems. (*2*). For light-sensitive photographic emulsions and developing solutions. (*3*) Process engraving and lithography. (*4*) Wood fireproofing. (*5*) Corrosion inhibitor. (*6*) Flame retardant for chipboard. (*7*) Sedative. (*8*) Textile finishing.

Barium bromide, $BaBr_2$[10553-31-8]
Uses: (*1*) Reactant for bromide manufacture. (*2*) X-ray storage phosphor (46).
Beryllium bromide, $BeBr_2$ [7787-46-4]
Use: Adhesive for polyvinyl alcohol (47).
Bismuth bromide, $BiBr_3$ [7787-58-8]
Uses: (*1*) Claimed: catalyst for dehydrating cyclohexanol to cyclohexene (48).
(*2*) Claimed: Part of solid electrolyte in primary lithium batteries (49).
(*3*) Polymerization of methyl methacrylate monomer to yield transparent
resins (50). (*4*) Catalyst for benzylation of aliphatic alcohols. (51).
Calcium bromide, $CaBr_2$ [7789-41-5]
Uses: (*1*) Applications in oil and gas well completion, workover, and packer
brines (52–54% calcium bromide, specific gravity >1.7). (*2*) Water treatment.
(*3*) Photography. (*4*) Gravity separation fluid. (*5*) Electrical conductivity fluid
Cerous bromide, $CeBr_3$ [14457-87-5]
Use: In molten salt bath for reduction of uranium oxide by magnesium
(52).
Cesium bromide, $CsBr$ [7787-69-1]
Uses: (*1*) X-ray fluorescent screens. (*2*) Spectrometer prisms. (*3*) Adsorption
cell windows. (*4*) Claimed: Optical fibers.
Chromous bromide, $CrBr_2$ [10049-25-9]
Use: Chromizing.
Chromic bromide, $CrBr_2$, [10031-25-1]
Use: Catalyst for polymerizing olefins (53).
Cobaltous Bromide, $CoBr_2$ [7789-43-7]
Use: (*1*) In hydrometers. (*2*) $CoMnBr_n$ catalyst for organic reactions (54,55).
Cupric bromide, $CuBr_2$[7789-45-9]
Uses: (*1*) Intensifier in photographic chemicals. (*2*) Brominating agent in
organic syntheses (56,57). (*3*) Catalyst in production of acrylic acid, olefins,
and vinyl chloride. (*4*) Humidity indicator. (*5*) Wood preservative. (*6*) Solid
electrolyte batteries.(*7*) Stabilizer for acetylated polyformaldehyde.
Cuprous bromide, $CuBr$ [7787-70-4]
Uses: (*1*) Catalyst for organic reactions (58,59). (*2*) Water-activated CuBr
battery (60).
Ferric bromide, $FeBr_3$ [10031-26-2]
Uses: (*1*) Catalyst for C_{60} bromination (61). (*2*) Catalyst for organic reactions,
particularly in brominations of aromatics.
Ferrous bromide, $FeBr_2$ [7789-46-0]
Use: Component of metamagnets (62).
Gold tribromide, $AuBr_3$ [10294-28-7]
Use: Claimed: Sensor component for halogenated gases (63).
Lithium bromide, $LiBr$ [7750-35-8]
Uses: (*1*) Absorption-type chillers (55% solution) (64). (*2*) Heat pump (65).
(*3*) Dehumidifier. (*4*) Lithium - bromine battery. (*5*) Sedative.
Magnesium bromide, $MgBr_2$ [7789-48-2]
Uses: (*1*) Used in organic syntheses (66,67). (*2*) Sedative, anticonvulsant.
Manganese bromide, $MnBr_2$ [13446-03-2]
Uses: (*1*) Catalyst: In formation of aromatic aldehydes from alkylbenzenes
(68) and phthalic acids from xylenes (69). (*2*) Claimed: Catalyst in

ammoxidation conversion of (*o*)-xylene to phthalimide (70). (*3*) Preparation of substituted haloalkylopyrazoles (71).

Mercurous bromide, Hg_2Br_2 [10031-18-2]
Use: Reacts with HBr to form hydrogen quantitatively (72).

Nickel bromide, $NiBr_2$ [13462-88-9]
Uses: (*1*) Catalyst: Dimerization of butadiene (73). (*2*) Catalyst: Condensing butadiene onto ring systems and benzyl ketones (74,75). (*3*) Catalyst: oxidation of sec-alcohols to ketones (76). (*4*) Preparation of biaryls from aryl halides (77).

Palladium bromide, $PdBr_2$, [13444-94-5]
Use: Catalyst for various carbonylation reactions (78,79).

Platinum bromide, $PtBr_2$ [13455-12-4]
Use: Dehydrodimerization catalyst for boron hydrides and carboranes (80,81).

Potassium bromide, KBr [7758-02-3]
Uses: (*1*) For light-sensitive photographic emulsions and developing solutions. (*2*) Used in process engraving and lithography. (*3*) Pharmaceutical preparations. (*4*) Heat stabilizer for nylon. (*5*) Controls epileptic seizures in dogs.

Praseodymium bromide, $PrBr_3$ [13536-53-3]
Uses: (*1*) In molten salt bath for reduction of uranium oxide by magnesium (52). (*2*) Light filter for cathode ray tube (82).

Rhodium bromide, $RhBr_3$ [15608-29-4]
Use: Catalyst: Carbonylating methanol to acetic acid (83).

Rubidium bromide, RbBr [7789-39-1]
Uses: (*1*) Component of X-ray intensifier screens (84). (*2*) Antiferromagnets (85).

Silver bromide, AgBr [7785-23-1]
Uses: (*1*) Photography. (*2*) Topical antiinfective. (*3*) Astringent.

Sodium bromide, NaBr [7647-15-6]
Uses: (*1*) Oxidizing biocide used to control microbial, fungal, and algal growth in industrial water systems. (*2*) Oil well completion fluid. (*3*) Bleaching agent (used with oxidizer). (*4*) For light-sensitive photographic emulsions and developing solutions. (*5*) Leaching agent, in combination with Br_2, for gold/silver ores (86). (*6*) Pharmaceutical preparations.

Stannic bromide, $SnBr_4$ [7789-67-5]
Use: In metallurgical separation of minerals (87).

Stannous bromide, $SnBr_2$ [10031-24-0]
Use: Claimed: Catalyst for preparation of lubricant antioxidant (88).

Strontium bromide, $SrBr_2$ [10476-81-0]
Uses: (*1*) Anticonvulsant. (*2*) Chemical heat pump (89).

Thallium bromide, TlBr [7789-40-4]
Use: Claimed: Component in radiographic image conversion panels (90).

Titanium bromide, $TiBr_4$ [7789-68-6]
Use: Catalyst for olefin polymerizations (91).

Tungsten bromide, WBr_6 [13701-86-5]
Use: Catalysts for polymerizing olefins (53).

Zinc bromide, $ZnBr_2$ [7699-45-8]

Uses: (*1*) Used to make silver bromide collodion emulsions (photography). (*2*) Radiation shielding. (*3*) Gravity separation. (*4*) Electrical conductivity fluid. (*5*) Storage battery electrolyte for electric vehicles and load leveling applications (92). (*6*) Used in completion, workover, and packer brines in oil well drilling and maintenance.

4. Bromine Halides

Bromine is capable of forming a number of bromine halide compounds under varying process conditions. In general, these compounds have properties intermediate between those of the parent halogens. The most important of these are described below.

Bromine chloride, BrCl [13863-41-7]: This compound is formed when bromine and chlorine react reversibly in the liquid and vapor phases at room temperature. Where an equimolar mixture of these halogens is reacted, $\sim$60% of the mixed halogens are present as BrCl (93). The bromine chloride is a dark red, fuming, irritating liquid having an mp $-66°C$ and bp $5°C$. BrCl has uses in organic synthesis involving the addition across olefinic double bonds to produce bromochloro compounds, and for aromatic brominations, where an aromatic bromide and HCl are produced. Some other characteristic reactions for BrCl are

$$(1) \qquad BrCl + RH \rightarrow HCl + RH$$
$$(2) \qquad BrCl + H_2O \rightarrow HCl + HOBr$$
$$(3) \qquad BrCl + NH_3 \rightarrow HCl + NH_2Br$$

BrCl also finds use as a disinfectant in wastewater treatment (94) where it enjoys the following advantages over chlorine: activity maintained over a wider pH range, more rapid disinfection, effective at lower residual concentrations, and lower aquatic toxicity (95). Specific uses in water treatment: Effluent streams from sewage plants and cooling water towers. In addition, bromine chloride has applications as a brominating agent in the preparation of fire-retardant chemicals, pharmaceuticals, high density brominated liquids, agricultural chemicals, dyes, and bleaching agents.

The time-weighted average concentration of bromine chloride should not exceed 0.1 ppm for an 8-h day. Suitable materials of construction for shipping and storing BrCl are low carbon steel or nickel, or its alloys, such as Monel (95).

Bromine monofluoride, BrF [13863-59-7]: The monofluoride may be prepared by direct reaction between bromine and fluorine. Because of its high tendency to disproportionate, it has never been prepared in its pure form (96). However, the BrF can be prepared in situ by the reaction of bromine with silver fluoride in benzene (97) or by the reaction of *N*-bromoacetamide and hydrofluoric acid in ether (98). BrF adds to simple alkenes at room temperature to give trans-addition products. BrF also adds to various alkynes to give the family of compounds containing CF_2CBr_2

$$PhCCCOOEt + 2\,BrF \rightarrow PhCF_2CBr_2COOEt.$$

Bromine trifluoride, BrF_3 [7787-71-5]: The trifluoride can be formed by the reaction between gaseous fluorine and liquid bromine at a temperature of 200°C (99):

$$Br_2 + 3F_2 \rightarrow 2BrF_3$$

The BrF_3 can be purified by distillation to give a pale straw-colored liquid.

Bromine trifluoride undergoes vigorous reaction with water, organic compounds, and metals. (It is somewhat less vigorous a fluorinating agent than the corresponding trifluoride of chlorine.) Reaction of BrF_3 with oxides often evolves oxygen quantitatively

$$B_2O_3 + 2BrF_3 \rightarrow 2BF_3 + Br_2 + \tfrac{3}{2}O_2$$

The high specific conductivity of the BrF_3, ie, $> 8 \times 10^{-3}\ \Omega^{-1}\ cm^{-1}$, is due to its autoionization in the liquid state to the $BrF_2^+BrF_4^-$ species. This Lewis acid–base character is thus demonstrated by the ability of BrF_3 to dissolve in a number of halide salts, eg, CsF according to the reaction

$$BrF_3\ (liquid) + CsF\ (solid) \rightarrow CsBrF_4\ (solution)$$

Bromine trifluoride is a useful solvent for those ionic reactions that need to be carried out under highly oxidizing conditions. As a strong fluorinating agent, it finds use in organic syntheses and in the formation of inorganic fluorides. Bromine trifluoride is highly toxic and corrosive to all tissues. Rail shipments of bromine trifluoride require "Oxidizer" and "Poison" labeling.

Bromine Pentafluoride, BrF_5 [13863-59-7]: Bromine pentafluoride can be formed either by (1) the reaction of bromine with excess fluorine above 150°C or by (2) the reaction of BrF_3 vapor and gaseous fluorine at 200°C (99). Small-scale preparation of BrF_5 can be achieved by the fluorination of KBr at a temperature of 25°C. Analogous to the trifluoride, the pentafluoride reacts with alkali fluorides, eg, MF to form $MBrF_6$ salts. BrF_5 is also used as a fluorinating agent in organic syntheses (100–102) and as the oxidizer component of some rocket propellants. It can fluorinate silicates at 450°C as follows

$$KAlSi_3O_8 + 8BrF_5 \rightarrow KF + AlF_3 + 3SiF_4 + 4O_2 + 8BrF_3$$

BrF_5 is highly toxic and corrosive to the skin. It requires the "Oxidizer" label for rail shipments and the "Corrosive" label for air shipments.

The two bromine fluorides—the trifluoride and pentafluoride—are available commercially.

A summary of the physical properties of the Br–F compounds is given in Table 4 below (103).

Table 4. Properties of Bromine–Fluorine Binary Compounds

	mp (°C)	bp (°C)	Specific conductivity at 25°C ($\Omega^{-1}\ cm^{-1}$)	Structure
BrF	−33	20	—	—
BrF_3	9	126	$>8.0 \times 10^{-3}$	planar; distorted T
BrF_5	−60	41	9.1×10^{-8}	square pyramidal

Materials of construction for the bromine fluorides include nickel, Monel metal, or teflon.

Iodine bromide, IBr [7789-33-5], is a black solid (mp 41°C, subl 50°C) that is more stable than bromine monochloride. It is formed by the reaction between bromine and iodine. There is also evidence for the existence of a tribromo species, iodine tribromide IBr$_3$ [7789-58-4]. These iodine–bromine compounds are soluble in carbon tetrachloride and acetic acid and are used as halogenating agents (104,105).

5. Bromine Oxides

None of the bromine oxides are stable at ordinary temperatures and none are considered to be of any practical importance.

Dibromine oxide, bromine monoxide, Br$_2$O [21308-80-5]. This compound is a dark-brown solid that is moderately stable at −60°C (mp −17.5°C with decomposition). Br$_2$O is a bent symmetrical molecule like Cl$_2$O, with a C$_{2v}$ symmetry. It is formed by (1) the low temperature decomposition of BrO$_2$ in vacuum or by (2) the reaction of Br$_2$ vapor with HgO. The latter reaction is carried out in a carbon tetrachloride or fluorotrichloromethane solvent

$$HgO + 2\,Br_2 \rightarrow Br_2O + HgBr_2$$

Excess mercuric oxide and mercuric bromide are then removed by filtration. The solution obtained can be used in brominations. Br$_2$O oxidizes iodine to I$_2$O$_5$, benzene to 1,4-quinone and yields OBr$^-$ in alkaline solutions.

Bromine dioxide: This compound is a yellow solid formed by the oxidation of Br$_2$ in CF$_3$Cl at −78°C with ozone. BrO$_2$ is thermally unstable above −40°C, and decomposes violently at ∼0°C. Slower warming of BrO$_2$, as mentioned above, yields Br$_2$O. Alkaline hydrolysis of BrO$_2$ leads to disproportionation to bromate and bromide. A recent study (106) claims that BrO$_2$ is a largely or totally a mixture of Br$_2$O$_3$ and Br$_2$O$_5$.

6. Oxygen Acids and Salts

The oxygen acids of bromine are unstable strong oxidizing agents, which exist only at ambient temperatures in solution.

Hypobromous acid, HBrO [7486-26-2]. This compound can be prepared by reacting either mercuric oxide or silver nitrate with bromine water according to the following reactions (107):

$$2\,Br_2 + H_2O + HgO \rightarrow 2\,HBrO + HgBr_2$$

$$Br_2 + H_2O + AgNO_3 \rightarrow HBrO + AgBr + HNO_3$$

To obtain more concentrated solutions from the products of the above reactions, the weak HBrO solution is distilled under vacuum following removal of AgBr or

HgBr$_2$ by filtration. An alternative method proposed for obtaining concentrated solutions was to carry out the mercuric oxide reaction in Freon 11, without using water, to yield a solution of bromine monoxide, which was then filtered and hydrolyzed. HBrO is a weak acid that is only slightly ionized. Its dissociation constant is 1.6×10^{-9} at 25°C. Hypobromous acid is used as a strong bactericide and as a water disinfectant.

The salts of hypobromous acid, referred to as hypobromites, undergo slow disproportionation to bromate and bromide

$$3\,NaBrO \rightarrow 2\,NaBr + NaBrO_3$$

Cobalt, nickel, and copper catalyze the disproportionation reaction (108), so that these impurities should be avoided.

Alkali metal hypobromites are generally prepared by reacting bromine with an alkali hydroxide (eg, sodium hydroxide) in aqueous solution at a temperature < 0°C

$$2\,NaOH + Br_2 \rightarrow NaBrO + NaBr + H_2O$$

At temperatures of 50–80°C, quantitative yields of BrO$_3^-$ are obtained according to the reaction

$$6\,NaOH + 3\,Br_2 \rightarrow NaBrO_3 + 5\,NaBr + 3\,H_2O$$

Pure crystalline hydrates such as NaOBr·5H$_2$O [13824-96-9] and KOBr·3H$_2$O [13824-97-0] were not described until the work of Scholder and Kraus (109). Solid alkaline earth hypobromites such as CaBr(OBr) [67530-61-4] have been known since the early twentieth century. The hypobromites find use as bleaching and desizing agents in the textile industry.

Bromous acid, HOBrO [7486-26-2]: In general, halite ions and halous acids do not arise during the hydrolysis of halogens. Thus, the existence of bromous acid is therefore considered to be in doubt (96).

Sodium bromite, NaBrO$_2$ [7486-26-2]: This compound is available in the United States at a solution concentration of ~10%, and is used as a desizing agent in the textile industry. Sodium bromite has been synthesized as 99.6% pure NaBrO$_2$·3H$_2$O, and is available commercially in Europe. This compound is made from a concentrated hypobromite solution at pH 11–12, at a temperature <0°C. After ammonia has been added to the solution to decompose any hypobromite that remains after reaction, a yellow sodium bromite trihydrate is isolated through concentration and filtration. The sodium salt is stable and can be stored for long periods of time.

Barium bromite, Ba(BrO$_2$)$_2$ [14899-01-5]; **potassium bromite**, KBrO$_2$ [76908-17-3]; and **lithium bromite** LiBrO$_2$ [14518-92-4]: These compounds can be prepared in a similar way to that described for the sodium salt under vacuum at 0°C (110). Anhydrous lithium and barium bromite have been prepared by heating bromate salts (111)

$$2\,LiBrO_3 + LiBr \rightarrow 3\,LiBrO_2 \qquad \text{at } 190°C$$

$$Ba(BrO_3)_2 \rightarrow Ba(BrO_2)_2 + O_2 \qquad \text{at } 250°C$$

Such compounds are stable for at least 3 months. The bromites find industrial application as desizing agents for textiles and in water treatment.

Bromic acid, (HBrO$_3$ [7789-31-3]: This compounds can be prepared by (1) the reaction of alkali bromates with dilute sulfuric acid, or (2) by the electrolysis of bromine in bromic acid solution with platinum or lead dioxide anodes (112). Another preparation method is by the passage of an alkali bromate solution ($\sim$0.5 M) through a cation exchanger in the H$^+$-form (113,114). The acid coming from ion exchange can be concentrated (in vacuum) to $\sim$ 50%. The 50% solution begins to decompose at around 40°C while more concentrated solutions are not stable. HBrO$_3$ is a strong acid (pK < 0). It is also a strong oxidant, with a standard potential of 1.47 V for the reaction HBrO$_3 \rightarrow 0.5$ Br$_2$ (in acid medium).

The industrially important bromates—sodium and potassium bromate—are commonly produced by the electrolytic oxidation of the corresponding aqueous bromide solution in the presence dichromate at a temperature of 65–70°C (115). Cathodes may be made of stainless steel or copper while the anode is usually lead dioxide or iron. The bromate formed is then crystallized by cooling the hot solution, after removing it from the cell. The mother liquor is treated with more bromide and recycled.

Another method for producing bromate salts is the reaction of alkali hydroxide or carbonates with bromine at elevated temperatures, as cited previously. In this case, sodium bromide is also produced along with the bromate. The relatively high sodium bromide solubility permits the bromates to be recovered by cooling and filtration. The commercial sodium bromate product is available at a 99.5% minimum assay. An Israeli patent (116) describes the manufacture of bromate–bromide mixtures.

A recent patent for sodium bromate production (117) is comprised of dissolving sodium carbonate in water, passing bromine and chlorine gases into the solution, filtering the hot solution to remove sodium chloride, and cooling the filtrate to 5–10°C to obtain a crude sodium bromate. The crude product is then redissolved in hot water and recrystallized in the temperature range of 5–10°C.

Sodium bromate, NaBrO$_3$ [7789-38-0]: This compound is a strong oxidizing agent, which can cause fires upon contact with organic materials. As the bromates are a source of active oxygen, when either heated, subjected to shock or acidified, they represent potential fire and explosion hazards. The bromates are usually packed in polyethylene-lined filter drums. Metal drums have also been used for packaging. The bromates are considered quite stable in storage.

Some physical characteristics and uses are listed for selected bromates:

1. **Sodium bromate**: Colorless crystal or powder. specific gravity 3.34, mp 381°C (decomposes with oxygen evolution). Water solubility at 25°C is 28.39 g/100 g solution while at 80°C, the solubility is 43.1 g/100 g solution. Uses: (a) Neutralizer–oxidizer in hair wave preparations. (b) Used in mixtures with sodium bromide in gold mining applications. (c) Used in applications in textile bleaching.

2. **Potassium bromate**, KBrO$_3$ [7758-01-2]: White crystals or powder, specific gravity 3.27, decomposes at $\sim$327°C. Water solubility at 25°C is 7.53 g/100-g solution while at 80°C, the solubility is 25.4 g/100-g solution. This bromate is a powerful oxidizing agent, either in the pure state

or when blended with magnesium carbonate. Uses: (a) Flour treatment to improve baking characteristics. (b) Hair wave solutions. (c) Malting of beer. (d) Cheese manufacture. (e) Used as an analytic standard.

3. **Barium bromate**, $Ba(BrO_3)_2 \cdot H_2O$ [10326-26-8]: Exists as white crystals. Specific gravity 4.0. Loses water of crystallization at 180–200°C and decomposes at ~270°C. Produced by reaction between bromine and barium hydroxide. This bromate is used as a corrosion inhibitor.

Some of the characteristic reactions of aqueous bromates include

1. Reaction with hydrogen peroxide to yield O_2.
2. Reaction with PH_3 to yield phosphoric acid.
3. Reaction with HBr to yield Br_2.
4. Reaction with XeF_2, F_2 (alkali) to yield $BrO_4{}^-$.
5. Reaction with OCl^- (slow) to yield $ClO_4{}^-$.
6. Reaction with nitrous acid to yield nitric acid.

Perbromic acid $HBrO_4$ [19445-25-1]: This compound is a strong acid, which is completely dissociated in aqueous solutions. This acid is generally prepared by passing sodium perbromate solution through a cation-exchange resin, which is in the hydrogen form. Perbromic acid solutions, at concentrations above 6 M (55%), are air unstable. They are believed to undergo an autocatalytic decomposition that is catalyzed by some metal ions, eg, Ag^+ and Ce^{4+}. Perbromic acid is very similar to perchloric acid, having little oxidizing activity in dilute solution but being quite capable of reacting violent when concentrated. A 3 M perbromic acid solution attacks stainless steel, a 6 M solution at 100°C rapidly oxidizes Mn^{2+} to $MnO_4{}^-$ and a 12 M solution explodes on contact with cellulose.

The perbromates were first prepared by E. Appelman at the Argonne National Laboratory in 1968 by oxidizing a bromate either electrolytically or with xenon difluoride (118). A fluorination procedure was reported a year later by Appelman (119) to produce larger quantities of the perbromates. The newer method consisted of introducing fluorine into strongly basic solutions of bromates. The normal oxidation potential of the $BrO_4{}^-/BrO_3{}^-$ couple in acid solution is 1.74 V. Appelman and co-workers(120) reported the thermodynamic properties of perbromate and bromate ions, as well as the effects of photolysis and pulse-radiolysis on perbromate in aqueous solutions (121,122). Perbromic acid can also be used to prepare perbromate salts.

Sodium perbromate, $NaBrO_4$ [33497-30-2]: This compound is highly soluble in water while **potassium perbromate**, $KBrO_4$ [22207-96-1], is soluble only to the extent of ~0.2 M at room temperature.

BIBLIOGRAPHY

"Bromine Compounds" in *ECT* 1st ed., Vol. 2, pp. 645–660, by V. A. Stenger and G. J. Atchison, The Dow Chemical Company; in *ECT* 2nd ed., Vol. 3, pp. 766–783, by V. A. Stenger and G. J. Atchison, The Dow Chemical Company; in *ETC* 3rd ed., Vol. 4, pp. 243–263, by N. A. Stenger, Dow Chemical U.S.A.; in *ECT* 4th ed., Vol. 4, pp. 560–589 by Philip F. Jackisch, Ethyl Corporation; "Bromine" in *ECT* (online), posting date: December 4, 2000, by Philip F. Jackish, Ethyl Corporation.

CITED PUBLICATIONS

1. G. H. Coleman, H. Soroos, and C. B. Yeager, *J. Am. Chem. Soc.* **55**, 2075 (1933); **56**, 965 (1934).
2. J. Jander and E. Kurzbach, *Z. Anorg. Allgem. Chem.* **296**, 117 (1958); J. Jander and Lafrenz, *Z. Anorg. Allgem. Chem.* **349**, 57 (1966).
3. G. H. Coleman and G. E. Goheen, in H. Booth, ed. *Inorganic Syntheses*, Vol. 1, McGraw-Hill Book Co., Inc., New York, 1939, p. 62.
4. J. D. Johnson and R. Overby, *J. Sanit. Eng. Div. Am. Soc. Civil Eng.* **97**(SA5), 617 (1971).
5. H. Gal-Gorchev and J. C. Morris, *Inorg. Chem.* **4**, 899 (1965).
6. W. D. Bonner, L. G. Bonner, and F. J. Jurney, *J. Am. Chem. Soc.* **55**, 1406 (1933).
7. J. O. Lundgren and J. Olovsson, *J. Chem. Phys.* **49**, 1068 (1968).
8. C. P. Smyth and C. S. Hitchcock, *J. Am. Chem. Soc.* **55**, 1830 (1933).
9. U.S. Pat. 2,070,263 (Feb. 9, 1937), G. F. Dressel and O. C. Ross (to the Dow Chemical Company).
10. *Hydrobromic Acid*, Ethyl Corp., Nov. 1985, pp. 7–8.
11. U.S. Pats. 2,369,182 (Feb. 13, 1945); 2,380,675 (July 31, 1945); 2,383,919 (Aug. 28, 1945); 2,391,740 (Dec. 25, 1945); F. F. Rush and co-workers (to Shell Development Co.).
12. Ger. Offen. 1,960,558 (June 9, 1971), H. Jenker and O. Rabe (to Chemische Fabrik Kalk GmbH).
13. U.S. Pat. 3,426,093 (Feb. 4, 1969), O. C. Karkalits, Jr., and C. A. Leatherwood (to Petro-Tex Chem. Corp.).
14. M. Vadekar and I. S. Pasternak, *Can. J. Chem. Eng.* **48**, 216 (1970).
15. H. Stephen and T. Stephen, *Solubilities of Inorganic and Organic Compounds.* Volume 1 (Binary Systems) Part 1. Pergamon Press 1963.
16. W. Grzybkowski, and G. Atkinson, *J. Chem. Eng. Data* **31**, 309 (1986).
17. J. M. Wimby and J. Berntsson, *J. Chem. Eng. Data* **39**, 68 (1994).
18. J. M. Wimby and J. Berntsson, *J. Chem. Eng. Data* **39**, 73 (1994).
19. A. Apelblat and E. Manzurola, *Steam, Water, Hydrotherm. Syst.: Proc. Int. Conf. Prop. Water Steam*, 13th, Meeting Date 1999, pp. 183–190. P. R. Tremaine, ed., National Research Council of Canada: Ottawa, Ont. 2000.
20. R. Blumberg in Lo, Bard, Hanson (eds.) *Handbook of Solvent Extraction*, John Wiley & Sons, Inc., New York 1983, p. 832.
21. The Development of the CaBr$_2$-by-Extraction Process, *Israel Chem. Eng.* **19**, 14 (Aug. 1990).
22. U.S. Pat. 1,775,598 (Sept. 8, 1930), J. H. van der Meulen.
23. N. N. Greenwood and A. Earnshaw, *Chemistry of the Elements*, Pergamon Press 1984. Chapter 17: The Halogens: Fluorine, Chlorine, Bromine, Iodine, and Astatine.
24. R. Aizenberg, O. Arrad, and Y. Sasson, *Ind. Eng. Chem. Res.* **31**, 431 (1992).
25. PCT Int. Appl. WO98/13297 (April 2, 1998), V. C. Mehta.
26. U.S. Pat. 5,290,531 (March 1, 1994), R. A. Fisher, K. M. Surendra, and R. T. Swartwout (to Tetra Technologies Inc.).
27. PCT Int. Appl. WO99/31014 (June 24, 1999), R. A. Fisher, D. J. Hanlon, and P. Wayland (to Tetra Technologies Inc.).

28. R. Li, X. Feng, J. Liu, and J. Zhang, *Huaxue Shiji* **16**(4), 255 (1994).
29. M. Wang, *Haihuyan Yu Huagong* **22**(4), 37 (1993).
30. R. DuPont and P. Cotelle, *Synthesis*, **9**, 1651 (1999).
31. C. E. Kaslow and M. M. Marsh, *J. Org. Chem.* **12**, 456 (1947).
32. E. L. Eliel and R. G. Haber, *J. Org. Chem.* **24**, 143 (1959).
33. R. E. Kent and S. M. McElvain, *Org. Synth.* **III**, 493 (1955).
34. Ger. Offen. 2,849,573 (May 17, 1979), T. Teshima, N. Nozaki, M. Koyama, and K. Katoh (to Stanley Electric Co.).
35. Pol. Pat. 102,785 (June 30, 1979), M. Zatorski and S. Daszkiewicz (to Instytut Mechaniki Precyzyznej).
36. Jpn Kokai Tokkyo Koho JP 63,129,011 (June 1, 1988), Y. Ihara and N. Ogawa (to Tosoh).
37. U.S. Pat. 4,124,633 (Nov. 7, 1978), J. J. Leonard and J. Kao (to Atlantic Richfield).
38. U.S. Pat. 4,237,314 (Dec. 2, 1980), M. N. Sheng and J. Kao (to Atlantic Richfield).
39. M. J. Frazer, W. Gerrard, G. Machell, and B. D. Shepherd, *Chem. Ind.*, 931 (1954).
40. U.S. Pat. 5,955,019 (Sept. 21, 1999), C. E. Ash (to Shell Oil Co.).
41. R. D. Bartholomev, *Proceedings of the International Water Conference*, IWC-98-74, 1–30 (1998).
42. PCT Int. Appl. WO 99/19275 (April 22, 1999), A. Ewenson, D. Itzhak, M. Freiberg, A. Shushan, B. Croituro, D. Beneish, and N. Faza (to DSBG).
43. S. M. Leschev, S. F. Furs, and I. Y. Rumyantsev, *Zh. Prikl. Khim. (S.-Peterburg)* **65**(8), 1864 (1992).
44. U.S.S.R. Pat. SU 1,432,046 (Oct. 23, 1988), E. M. Rakhman'ko and co-workers (to Scientific Research Institute of Physical Chemical Problems, Minsk).
45. Brit. UK Pat. Appl. GB 2314341 (Dec. 24, 1997), J. R. Drewe and H. S. Sussman.
46. B. Nensel, P. Thielemann, and G. J. Decker, *Appl. Phys.* **83**(4), 2276 (1998); M. Thoms, S. Bauchau, D. Hausermann, M. Kunz, T. LeBihan, M. Mezouar and D. Strawbridge, *Nucl. Instrum. Methods Phys. Res., Sect. A* **413**(1), 175 (1998).
47. Jpn Pat. 73 20,013 (June 18, 1975), J. Kawada, M. Fujita, and T. Harayama (to Nippon Synthesis Chem. Ind.).
48. Jpn Kokai Tokkyo Koho JP 76 11,736 (Jan. 30, 1976), H. Aizawa, A. Kuroda, M. Minaga, K. Onishi, and S. Matsuhisa (to Toray Ind.).
49. U.S. Pat. 4,288,505 (Sept. 8, 1981), A. V. Joshi and W. P. Sholette (to Rat-O-Vac Corp).
50. J. Smid, *Makromol. Chem. Rapid Commun.* **8**, 543 (1987).
51. B. Boyer, E.-M. Keramane, J.-P. Roque, and A. A. Pavia, *Tetrahedron Lett.* **41**(16), 2891 (2000).
52. U.S. Pat. 4,534,792 (Aug. 13, 1985), G. R. B. Elliott.
53. Jpn. Pat. 77,124,094 (Oct. 18, 1977), K. Sakashita and Y. Nakano (to Mitsui Petrochemical Ind.).
54. D.-B. Kim, S. Park, W. Cha, H.-D. Roh, and K. D. Kwak, *Kongop Hwahak*, **10**(6), 863 (1999).
55. Y. Zhang, *Ranliao Gongye* **36**(6), 22, 33 (1999).
56. S. Lu, *Huaxue Yanjiu* **10**(2), 35 (1999).
57. Czech Rep. Pat. 285486 (Aug. 11, 1999), P. Hradil, R. Gabriel, P. Indrak, P. Slezar, and M. Zatloukal (to Farmak, A. S.).
58. PCT Int. Appl. WO 99/15567 (Apr. 11, 1999), Y. Nakagawa, K. Kitano, M. Fujita, and N. Fujita (to Kaneka Corporation).
59. PCT Int. Appl. WO 2000/46273 (Aug. 10, 2000), A. J. F. M. Braat, H. G. E. Ingelbrecht, and R. Trion (to General Electric Company).
60. K. Vuorilehto and H. Rajantie, *J. Appl. Electrochem.* **29**(8), 903 (1999).

61. A. Djordjevic, M. Vojinovic-Miloradov, N. Petranovic, A. Devecerski, D. Lazar, and B. Ribar, *Fullerene Sci. Technol.* **6**(4), 689 (1998).

62. H. A. Katori, *RIKEN Rev.* **27**, 35 (2000).

63. Jpn. Kokai Tokkyo Koho JP 59 19,303 (May 4, 1984) to Toshiba.

64. T. Berlitz, P. Satzger, F. Summerer, F. Ziegler, and G. Alefeld, *Int. J. Refrig.* **22**(1), 67 (1999); S. B. Riffat, S. E. James, and C. W. Wong, *Int. J. Energy Res.* **22**(12), 1099 (1998).

65. Jpn. Kokai Tokkyo Koho JP 11080978, (Mar. 26, 1999) (to Tokyo Gas Co. Ltd.).

66. V. M. Swamy, S. K. Mandal, and A. Sarkar, *Tetrahedron Lett.* **40**(33), 6061 (1999); A. Matsumoto and S. Nakamura, *J. Appl. Polym. Sci.* **74**(2), 290 (1999).

67. M. N. Shemaneva, L. V. Mel'nik, Y. N. Razina, L. M. Egorova, and S. I. Kryukov, *Neftekhimiya* **39**(3), 215 (1999); J.-W. Huang, C.-D. Chen, and M.-K. Leung, *Tetrahedron Lett.* **40**(49), 8647 (1999).

68. Jpn. Pat. 69 12,727 (June 9, 1969) Y. Yamada, T. Kaneda, and K. Furukawa (to Ube Industries).

69. Jpn. Pat. 69 29,053 (Nov. 27, 1969), O. Morita and T. Kasamaru (to Mitsui Petro-chemical Industries).

70. U.S. Pat. 3,305,561 (Feb. 21, 1967), W. G. Toland (to Chevron Research).

71. U.S. Pat. 5,869,688 (Feb. 9, 1999), B. C. Hamper and M. K. Mao (to Monsanto Co.).

72. Ger. Offen. 2,258,463 (July 19, 1973), G. Schuetz (to Euraton).

73. J. Kiji, K. Masui, and S. Furukawa, *Bull. Chem. Soc. Japan* **44**, 1956 (1971).

74. R. Baker, A. H. Cook and T. N. Smith, *Tetrahedron Lett.*, 503 (1973).

75. S. Yoshikawa, S. Nishimura, and J. Furukawa, *Tetrahedron Lett.*, 3071 (1973).

76. M. P. Doyle, W. J. Patrie, and S. B. Williams, *J. Org. Chem.* **44**, 2955, (1979).

77. K. Takagi, N. Hayama, and S. Inokawa, *Chem. Lett.* 917 (1979).

78. Ger. Offen. 2,827,453 (Jan. 18, 1979), H. S. Kesling, Jr., and L. R. Zehner (to Atlantic Richfield Co.).

79. Ger. Offen. 2,949,936 (July 3, 1980), J. E. Hallgren (to General Electric Co.).

80. E. W. Corcoran, Jr., and L. G. Sneddon, *J. Am. Chem. Soc.* **106**, 7793 (1984).

81. *Ibid.* **107**, 7446 (1985).

82. Jpn. Pat. 84,134,532 (Aug. 2, 1984) to Philips Electronics.

83. U.S.S.R. Pat. SU 1,108,088 (Aug. 15, 1984), E. F. Vainshtein, A. L. Lapidus, and J. Popov.

84. Jpn. Kokai Tokkyo Koho JP 01 18,099 (Jan 20, 1989), K. Nakano, N. Kakamaru, S. Honda, H. Tsuchino, and F. Shimado (to Konica Co.).

85. K. Morishita, T. Kato, K. Iio, T. Mitsui, M. Nasui, T. Tojo, and T. Atake, *Ferroelectrics* **238**(1–4), 669 (2000).

86. M. Freiberg, *Rev. Chem. Eng.* **9**(3–4), 333 (1993); T. McNulty, *Mining Mag.*, 256 (May 2000).

87. J. D. Sullivan, Bureau of Mines Tech Paper No. 381 (1927).

88. U.S.S.R. Patent SU 1,498,773 (Aug. 7, 1989), P. A. Kirpichnikov, N. A. Mukmeneva, V. K. Kadyrova, A. S. Sharifullin, L. M. Volozhin, A. F. Gafarova, and N. A. Abdullina (to Kazan Chem. Tech. Inst.).

89. PCT Int. Appl. WO 2000/031206 (June 2, 2000), S. Jonsson, R. Olsson, and M. Karebring-Olsson (to Suncool AB, Sweden).

90. Jpn Kokai Tokkyo Koho JP 63,179,922 (July 23, 1988), S. Honda, K. Amitani, N. Kakamaru, and F. Shimado (to Konica Co.).

91. U.S. Pat. 3,401,157 (Sept. 10, 1968), W. H. Coover, Jr. (to Eastman Kodak).

92. T. N. Verizoglu, ed., "Alternate Energy Sources", Proceedings of the 6th Miami International Conference, Vol. 1, Hemisphere, Washington D. C. pp. 327–333 (1983).

93. H. G. Vesper and G. K. Rollefson, *J. Am. Chem. Soc.* **56**, 620 (1934).

94. J. F. Mills, *Disinfectants: Water Wastewater* 113 (1975).

95. Bromine Chloride for Treating Cooling Water and Wastewater, Ethyl Corp., Sept. 1984.

96. F. A. Cotton and G. Wilkinson, *Advanced Inorganic Chemistry*, 5th ed., John Wiley & Sons, Inc., New York, 1988, pp. 572–573.

97. L. D. Hall and D. L. Jones, *Can. J. Chem.* **51**, 2902 (1973).

98. L. Eckes and M. Hanack, *Synthesis*, 217 (1978).

99. W. Kawasnik in G. Brauer, ed., *Handbook of Preparative Inorg. Chem.*, 2nd ed., Vol. 1, Academic Press, New York, 1963, pp. 155–157.

100. U.S. Pats. 2,432,997 (Dec. 23, 1947); 2,471,831 (May 31, 1949); 2,480,080 (Aug. 23, 1949); 2,489,969 (Nov. 29, 1949); W. B. Ligett, E. T. McBee, and V. V. Lindgren (to Purdue Research Foundation).

101. Brit. Pat. 1,059,234 (Feb. 15, 1967), R. A. Davis and E. R. Larsen (to the Dow Chemical Company).

102. R. A. Davis and E. R. Larsen, *J. Org. Chem.* **32**, 3478 (1967).

103. D. Shriver and P. Atkins, *Inorganic Chemistry*, 3rd ed., W. H. Freeman & Co., New York 1999.

104. W. Militzer, *J. Am. Chem. Soc.* **60**, 256 (1938).

105. E. P. White and P. W. Robertson, *J. Chem. Soc.* **142**, 1509 (1939).

106. K. Seppelt, *Acc. Chem. Res.* **30**, 111 (1997).

107. A. J. Downs and C. J. Adams in J. C. Bailar, Jr., and co-ed., *Comprehensive Inorganic Chemistry*, Vol. 2, Pergamon, Oxford, U.K. 1973, p. 1400.

108. A. Luneckas and A. Prokopcikas, *Liet. TSR Mokslu Akad. Darbei* **B**(3), 53 (1960).

109. R. Scholder and K. Kraus, *Z. Anorg. Allgem. Chem.* **268**, 279 (1952).

110. A. Massagli, A. Indelli, and F. Pergola, *Inor. Chim. Acta* **4**(4), 593 (1970).

111. Ref. 107, pp. 1419–1420.

112. V. I. Bogatyrev and A. I. Vulikh, *J. Appl. Chem. U.S.S.R.* **36**, 205 (1963).

113. U.S. Pat. 3,187,044 (June 1, 1965), D. Robertson (to Arapahoe Chemicals Inc.).

114. U.S.S.R. Pats. 254,490 (Oct. 17, 1969); 297,576 (March 11, 1971), D. P. Semchenko, V. I. Lyubushkin, and S. S. Khadirov.

115. T. Osuga, K. Sugino, *J. Electrochem. Soc.* **104**(7), 448 (1957).

116. Israeli Pat. 84,830 (Jan. 24, 1995), A. Prager, S. Smilovitch, M. Freiberg, and H. Hariton (to DSBG).

117. Chinese Pat. 1,102,818 (May 24, 1995), N. Zhao, Z. Wu, and G. Sun (to Daqinghe Saltern).

118. E. H. Appelman, *J. Am. Chem. Soc.* **90**, 1900 (1968).

119. E. H. Appelman, *Inorg. Chem.* **8**, 223 (1969).

120. G. K. Johnson, P. N. Smith, E. H. Appelman, and W. N. Hubbard, *Inorg. Chem.* **9**, 119 (1970).

121. V. K. Klaening, K. J. Olsen, and E. H. Appelman, *J. Chem. Soc. Faraday Trans. 1* **71**, 473 (1975).

122. K. J. Olsen, K. Sehested, and E. H. Appelman, *Chem. Phys. Lett.* **19**, 213 (1973).

S.D. UKELES
OB IMI (TAMI)

M. FREIBERG
DSBG

BROMINE, ORGANIC COMPOUNDS

1. Introduction

Organic bromine compounds, which are organic compounds in which the bromine is covalently bonded to carbon or (rarely) to nitrogen and oxygen, are a very important group of organic halogen compounds. Even naturally occurring bromine containing organic compounds produced by marine and terrestrial plants, bacteria, fungi, insects, marine animals, and some higher animals, number nearly 1500 compounds (1). Historically, the organic bromine compound, Tyrian, or Royal Purple, (dibromoindigo [19201-53-7],) extracted from a Mediterranean Sea mollusk, was one of the first used dyes (2). However, far more important are synthetic organic bromine compounds. Organic bromine compounds are the predominant industrial bromine compounds and in terms of bromine consumption, account for ∼80% of bromine production. The industrially produced organic bromine compounds can be divided into two main groups.

(1) Organic bromine compounds in which the bromine atom is retained in the final molecular structure, and where its presence contributes to the properties of the desired products, are the largest segment in terms of consumed volumes. This segment includes mainly flame retardants, biocides, gasoline additives, halons, bromobutyl rubber, pharmaceuticals, agrochemicals, and dyes. Some of the products in this category, such as methyl bromide, ethylene dibromide, and halons are being subjected to environmental restrictions that will lower their consumption, although the overall market of this segment is still forecast to use because of the increasing demand of other products, especially flame retardants and biocides.

(2) Organic bromine compounds have traditionally played an important role as intermediates in the production of agrochemicals, pharmaceuticals and dyes, while new process developments that result in new applications in ultraviolet (uv) sunscreens, high performance polymers, and others, are forecast to increase their market share. The world consumption of bromine for intermediates is dwarfed by the corresponding consumption of chlorine. On the simple grounds of raw material halogen cost, organic chlorine intermediates have dominated in the manufacture of low value, high volume commodity products. Bromine, however, has tended to compete more favorably with chlorine for application as an intermediate in the above mentioned, more specialized, higher value areas (3). The diverse applications of organic bromine intermediates in commercial manufacture provide ample illustrations of the many virtues of bromine chemistry serving to outweigh the penalty of high halogen cost. These virtues are reflected in process benefits for both the production and application of intermediates, summarized below.

(a) Improved selectivity in the production of bromine intermediates.
(b) Improved reactivity of the intermediates in bromine displacement, resulting in (c).

(c) Cleaner processes, reduction of waste, and reduced environmental impact. While (a) and (b) provide obvious improvements in manufacturing economics, the rapidly increasing demand for environmentally clean chemical processes makes (c) a major, and often dominant factor, in the choice of the intermediate and process route for a new production.

Of course, the above division is relative. Several compounds can be used both as final products and as intermediates for the production of other products.

The scope of this article is limited to those organic bromine compounds having industrial application. A short description of the chemical properties and routes for the synthesis of the organic bromine compounds as a specific group, including references to the corresponding monographs and reviews, is given.

2. Chemical Properties

Substitution of bromine by other groups proceeds as a nucleophilic, electrophilic, or radical process (4). Nucleophilic displacement of bromine, both in aliphatic and aromatic molecules, by neutral or anionic nucleophiles, is the leading process in the application of brominated intermediates. As a rule, the reactivity of bromine compounds in the nucleophilic substitution is greater than the corresponding chlorine derivatives, owing to the difference in the bond energies (C–Br 276 kJ/mol vs. C–Cl 328 kJ/mol). This reactivity is the main advantage of using bromine compounds as intermediates. The reaction of aliphatic bromides, either with water or with dilute aqueous solutions of bases, gives alcohols (5), the reaction with alcoholates gives ethers, the reaction with salts of carbonic acids gives esters, and the reaction with sodium cyanide gives nitriles. Interaction with ammonia, both in solution and in the gaseous phase, gives primary, secondary, or tertiary amines and quaternary ammonium salts, depending on the reaction conditions. Aldehydes and ketones are formed by the hydrolysis of dibromides, $RCHBr_2$ or $RCBr_2R'$, respectively, and the hydrolysis of tribromides, $RCBr_3$, gives carbonic acids.

The nucleophilic substitution in the aliphatic series may be accompanied by elimination, and the yield of the target compound depends on a number of factors (structure of the initial compound, presence and nature of solvent and catalyst, etc). Thus, alkyl bromides eliminate HBr under the action of concentrated solutions of bases, forming alkenes:

$$CH_3CH_2Br \rightarrow CH_2{=}CH_2 + HBr \tag{1}$$

Vicinal dibromoalkanes are debrominated by zinc, forming unsaturated compounds.

$$RCHBrCH_2Br + Zn \rightarrow RCH{=}CH_2 + ZnBr_2 \tag{2}$$

When the bromine atoms are located at more remote carbons, a cyclic compound can be formed

$$CH_2BrCH_2CH_2Br + Zn \longrightarrow \underset{\displaystyle CH_2}{H_2C{-}CH_2} + ZnBr_2 \tag{3}$$

The action of Mg or Li on the organic bromides leads to the formation of alkyl magnesium bromides (Grignard reagents) or alkyl lithium reagents, widely used both in laboratory and industrial organic synthesis.

The bromine atoms in organic bromine compounds can be replaced by hydrogen forming hydrocarbons

$$BrCH_2CH_2Br + H_2 \rightarrow CH_3CH_3 \tag{4}$$

Aliphatic bromides are used for the alkylation of aromatic hydrocarbons in the presence of Lewis acids (6)

$$RBr + \bigcirc \xrightarrow{AlCl_3} \bigcirc^R \tag{5}$$

Under the same conditions, these alkyl bromides add to unsaturated hydrocarbons and take part in telomerization reactions (7)

$$CBr_4 + n\, CH_2{=}CH_2 \rightarrow CBr_3(CH_2 - CH_2)_n Br \tag{6}$$

Figure 1 illustrates the various reactivities of organic bromine compounds.

The reactions with Mg, Li, and NaR are also typical for aromatic bromine compounds; the reactions in the above scheme proceed in the aromatic series either under drastic conditions or with aromatic compounds containing activated bromine. Substitution of an aromatic bromine with ammonia, amines, phenols, and other nucleophiles is accelerated by the addition of a copper catalyst (8). The Ullmann ether condensation is a more widely used industrial process, carried out in the presence of copper (9).

The modern metal-catalyzed coupling reactions of aryl and alkenyl bromides are very efficient and reliable methods for the formation of a new carbon–carbon bond. The Heck stereospecific palladium-catalyzed coupling of alkenes with organic bromides lacking an sp^3 hybridized β-hydrogen is one of the most valuable strategies in modern organic synthesis, including industrial

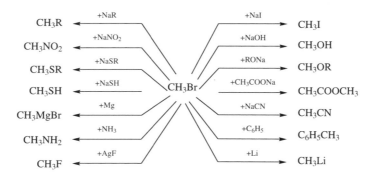

Fig. 1.   The reactivity of alkyl bromides.

processes (10–12). Palladium-catalyzed Suzuki cross-couplings of organic bromides with organoboron derivatives are also powerful methods for C—C bond formation (12,13).

3. Preparation and Production

Organic bromine compounds can be produced by a great number of different chemical reactions (14–23); however, substitution and addition reactions (substitutive and additive bromination) are the most common methods employed in industrial processes.

3.1. Substitutive Bromination. The process of bromination of saturated hydrocarbons (alkanes and alkylarenes) both in the gas and liquid phase, proceeds by a free-radical chain mechanism and requires photolytic or thermal initiation

$$
\begin{aligned}
&\text{Br}_2 \xrightarrow{h\nu} 2\,\text{Br}^{\cdot} \\
&\text{RH} + \text{Br}^{\cdot} \rightarrow \text{R}^{\cdot} + \text{HBr} \\
&\text{R}^{\cdot} + \text{Br}_2 \rightarrow \text{RBr} + \text{Br}^{\cdot} \\
&\text{R} = \text{alkyl or aralkyl}
\end{aligned}
\tag{7}
$$

This substitution reaction is selective in the case of alkylarenes and branched alkanes, giving correspondingly, 1-bromo-1-arylalkanes and tertiary alkyl bromides. In the presence of Lewis acids, the bromination of branched alkanes occurs by an electrophilic mechanism.

The direct bromination of carboxylic acids in the α-position is carried out by an acid -catalyzed process. Various reagents are available for the introduction of a bromine atom into a fatty acid but in the conventional Hell–Volhard–Zelinsky reaction, bromine and red phosphorus are used and α-bromination normally results (24). A number of modified procedures are available, including the use of bromine in the presence of a catalytic amount of phosphorus trichloride or phosphorus tribromide. The substitution of α-hydrogen atoms in aliphatic carbonyl compounds occurs via the addition of bromine to the enol form of the carbonyl compound (25)

$$
\text{CH}_3\text{COR} \;\rightleftharpoons\; \underset{\text{CH}_2=\text{CR}}{\overset{\text{OH}}{|}} \;\xrightarrow{\text{Br}_2}\; \underset{\text{CH}_2-\text{CRBr}}{\overset{\text{Br}\quad\text{OH}}{|\qquad|}} \;\longrightarrow\; \text{CH}_2\text{BrCOR} + \text{HBr}
\tag{8}
$$

Other substitution reactions that are often used in industrial chemical processes for the production of brominated aliphatic compounds include

(1) Substitution of the hydroxyl group of alcohols by bromine using HBr, PBr$_3$, or P + Br$_2$.

(2) Replacement of the chlorine atoms of chlorinated hydrocarbons by bromine using various Lewis acids and a phase-transfer catalyst (26).

(3) Replacement of a carboxyl group by the action of bromine on the salts of carboxylic acids (27,28).

The substitutive nuclear bromination of aromatic and heteroaromatic compounds proceeds by an electrophilic mechanism. No catalyst is required for the bromination of activated aromatics, such as polyalkylbenzenes, naphthalene, and polycyclic aromatic compounds (29). Phenols and anilines readily undergo mono-, di-, or tribromination (30). The bromination of nonactivated and deactivated aromatics is usually carried out in the presence of a Lewis acid catalyst (AlCl$_3$, FeBr$_3$, I$_2$, etc) (31)

$$C_6H_6 + Br_2 + AlCl_3 \longrightarrow \underset{\underset{H \; AlCl_3Br^-}{Br}}{\boxed{\bighexagon{+}}} \longrightarrow C_6H_5Br + HBr + AlCl_3 \qquad (9)$$

Very strongly deactivated aromatic compounds containing two or more electron-withdrawing groups, eg, dinitrobenzene or di(trifluoromethyl)benzene, undergo bromination in the presence of sulfuric acid in combination with nitric acid (32).

The amount and orientation of the bromine substitution of aromatic compounds depends on the effect of other substituents in the aromatic ring, the activity of the catalyst used, the reaction conditions, and other factors. In general, bromine is a less strong and more selective electrophile than chlorine; therefore its application enables the obtaining of more valuable haloaromatic intermediates than with the application of chlorine. For example, it is known that aluminum trihalides and other Lewis acids cause the isomerization of haloaromatics. The migration of chlorine is generally most likely to be intermolecular. In contrast, the migration of bromine may be both intra- and intermolecular. The intramolecular migration of bromine is used as a practical route for the preparation of some meta-substituted and meta-disubstituted bromobenzenes (33,34). The isomerization of a mixture of brominated anilines gives the para-brominated isomer in high yield (35).

The catalytic decomposition of aromatic diazocompounds in the presence of Cu (Gattermann reaction) or of Cu$_2$Br$_2$ (Sandmeyer reaction) is often used for the production of bromoaromatic compounds (36).

Since the use of bromine as a brominating agent is not always easy because of its volatile and toxic character, several bromine derivatives, especially N-bromo derivatives, are used as brominating agents. The more often used reagents are N-bromosuccinimide (NBS) [128-08-5], (37), 1,3-dibromo-5,5-dimethylhydantoin [77-48-5], (38), and stable crystalline organic ammonium tribromides, such as benzyltrimethylammonium tribromide [111865-47-5] (39), pyridinium tribromide [39416-48-3] (40), and phenyltrimethylammonium tribromide [4207-56-1], (41). These compounds are used for the Wohl–Ziegler radical bromination of the methyl or methylene groups of alkenes, alkylarenes, and heteroaromatic compounds (42); electrophilic bromination of activated aromatic compounds— phenols, anilines; bromination of carbonyl compounds, and several other brominations. Bromination using these solid reagents usually proceeds under milder conditions and is more selective than using bromine itself. The application of polymer-supported N-brominating compounds is a recent achievement of the bromination technology (43).

3.2. Additive Bromination. The addition of bromine to aromatic and heteroaromatic compounds proceeds by a radical mechanism under the action of light or heat. The uncatalyzed addition of bromine to a C=C bond is rapid and quantitative. In the case of asymmetrical reagents such as hydrogen bromide (hydrobromination), BrCl, or hypobromite, the reaction proceeds by a radical or electrophilic mechanism (44,45). In the absence of peroxides, the reaction with hydrogen bromide proceeds by an electrophilic mechanism. The obtained product is formed from the most stable intermediate carbocation, generated by the addition of a proton to the double bond (Markovnikov addition) (46). The carbocation stability follows the order tertiary > secondary > primary, and therefore hydrogen adds to the most hydrogenated carbon atom

$$CH_3CH=CH_2 + HBr \rightarrow CH_3CHBrCH_3 \tag{10}$$

Under free-radical conditions, the direction of the addition of hydrogen bromide takes place in a reverse order (anti-Markovnikov addition)

$$CH_3CH=CH_2 + HBr \rightarrow CH_3CH_2CH_2Br \tag{11}$$

The addition of hydrogen bromide to the C=C, conjugated with electron-withdrawing groups, proceeds more slowly than addition to an isolated double bond. The bromine atom adds in the β-position to the electron-withdrawing group

$$CH_2=CHR + HBr \rightarrow BrCH_2CH_2R \quad R = COR', COOR', CN$$
$$R' = H, alkyl, Ar \tag{12}$$

In nucleophilic solvents like water, dimethylsulfoxide (DMSO), dimethylformamide (DMF), carboxylic acids, alcohols, nitriles, and even ethers, the solvent can compete with the bromide ion, leading to incorporation of the latter. These cobromination (mixed bromination) processes have great synthetic potential (47,48).

4. Aliphatic Bromine Compounds

4.1. Methyl Bromide. Methyl bromide, CH_3Br, (bromomethane) [74-83-9], is a colorless liquid or gas with practically no odor. Its physical properties are mp $-93.7°C$; bp $3.56°C$; d^{20}_4 1.6755 kg/m^3; 3.974 kg/m^3; n_D^{20} 1.4218; vapor pressure at 20°C, 189.3 kPa (1420 mmHg); viscosity at -20, 0, and 25°C: 0.475, 0.397, and 0.324 mPa·s, respectively. Heat capacity of the liquid at $-13°C$ and of the vapor at 25°C, 824 (197) and 448 (107) J/kg·K, (cal/kg·K), respectively; heat of vaporization at 3.6°C, 252 J/g (60.2 cal/g); critical temperature (calculated) 194°C; expansion coefficient -15 to 3°C, 0.00163/K; dielectric constant at 0°C and 0.001–0.01 MHz, 9.77; dipole moment gas, 1.81D. Methyl bromide is miscible with most organic solvents and forms a bulky, crystalline hydrate below 4°C. Its solubility in water varies with pressure: at normal pressure, methyl bromide plus water vapor, the solubility is 1.75 g/100-g solution (20°C).

Methyl bromide reacts with several nucleophiles and is a useful methylation agent for the preparation of ethers, sulfides, amines, etc. Tertiary amines are methylated by methyl bromide to form quaternary ammonium bromides. The reactivity of methyl bromide is summarized in Figure 1.

Methyl bromide, when dry (<100 ppm water), is inert toward most materials of construction. Carbon steel is recommended for storage vessels, piping, pumps, valves, and fittings. Copper, brass, nickel, and their alloys are sometimes used. Aluminum, magnesium, zinc, and alloys of these metals should not be used, because under some conditions dangerous pyrophoric Grignard-type compounds may be formed. A severe explosion due to the ignition of a methyl bromide–air mixture by pyrophoric methylaluminum bromides produced by the corrosion of an aluminum fitting has been reported. Nylon and poly(vinyl chloride) (PVC) should also be avoided for handling methyl bromide.

Methyl bromide is nonflammable over a wide range of concentration in air at atmospheric pressure, and offers practically no fire hazards. With an intense source of ignition its explosive limits are from 13.5 to 14.5% by volume.

The commercial manufacture of methyl bromide is based on the reaction of hydrogen bromide with methanol. The hydrogen bromide used could be generated in situ from bromine and a reducing agent. The uses of sulfur (49) or hydrogen sulfide (50) as reducing agents are described, the latter process having the advantage. A new continuous process for the production of methyl bromide from methanol and aqueous HBr in the presence of a silica supported heteropolyacid catalyst has recently been described (51). Methyl bromide can also be coproduced with other organic bromine compounds by the reaction of the methanol solvent with hydrogen bromide formed as a by-product. The processes include coproduction of methyl bromide with bromostyrenes (52), tribromophenol (53), potassium and sodium bromide (54), and especially tetrabromo bisphenol A (55,56).

The major world producers of methyl bromide are as follows (1996): the Dead Sea Bromine Group (DSBG–Israel) supplies ~41% of the world market, two U.S. producers–Great Lakes Chemical Corporation (31%) and Albermarle Corporation (12%), supply ~43% of the world market; Elf Atochem, the dominant producer in Western Europe, had a world market share of ~6%; and five producers in Japan have a collective share of ~10%.

The world consumption of methyl bromide in 1996 was 68.4 thousand metric tons. The United States accounted for nearly 31% of total world consumption, Western Europe accounted for ~23% and Japan accounted for 13%.

Worldwide, methyl bromide is used principally as a space fumigant used for killing soil parasites (nematodes, fungi, weeds, insects, and rodents) in agriculture and for the sanitation of cereal and other crops under storage and before shipment (57,58). Methyl bromide is also used as an intermediate for the manufacture of pharmaceuticals (clidinium bromide [3485-62-9], clobazam [22316-47-8], glycopyrrolate [596-51-0], mepenzolate bromide [25990-43-6], mepivacaine hydrochloride [1722-62-9], methscopolamine bromide [155-41-9], pancuronium bromide [15500-66-0], propantheline bromide [50-34-0], pyridostigmine bromide [101-26-8]),biocides (CTAB [57-09-0]), insecticides (pirimicarb [23103-98-2]), and chemical reagents (methylmagnesium bromide [75-16-1] and tetramethylammonium bromide [64-20-0]). The current world consumption of methyl bromide as an intermediate is ~1000 t/a (~1.5% of total world consumption).

Due to its role in the depletion of the ozone layer, an international agreement has been reached calling for its reduced consumption and complete phasing out in the developed countries. In September 1997 at the Ninth Meeting of the Parties of Montreal Protocol, members agreed to a schedule for a reduction in the use of the fumigant. The final agreements include

(1) For industrial nations, a 25% reduction in use in 1999 followed by a 50% reduction in 2001 and a complete phaseout in 2005 with an allowance for critical exemptions.

(2) For developing nations, a 20% reduction in consumption in 2005 at a level based upon average consumption across 1995–1998 followed by a complete phaseout in 2015.

The nonagricultural uses of methyl bromide (its use in organic synthesis) are not restricted, provided that the compound is destroyed during the reaction.

Methyl bromide is a toxic compound (59–61). Repeated splashes on the skin result in severe skin lesions. In cases of lesser exposure, a severe itching dermatitis can develop. Overexposure to methyl bromide may cause dizziness, nausea, vomiting, headache, drowsiness, dimming of vision, and convulsions in the short term. Repeated and prolonged exposure to lower concentrations (30–100 ppm) causes severe nervous system effects. The time-weighted average limit for daily 8-h exposure to the vapor in air is 5 ppm by volume, or 19 mg/m^3; the short-time exposure limit is 15 ppm (62).

Bromochloromethane, CH_2BrCl, (methylene chlorobromide) [74-97-5], is a colorless liquid with a characteristic sweet odor. Its physical properties are bp 68.1°C; mp − 86.5°C; d^{25}_4 1.9229 kg/m^3; n^{25}_D 1.4808; vapor pressure at 20°C 117 mmHg; heat of vaporization at bp 232 kJ/kg (55.4 kcal/kg). The liquid is completely miscible with common organic solvents. Its solubility in water is 0.9%. For the solubility data, bromochloromethane–water system, see (63).

Common routes for its production involve the partial replacement of chlorine in dichloromethane [75-09-2] by a halogen-exchange reaction using either bromine (64) or hydrogen bromide (65). Both processes are carried out in the presence of aluminum or aluminum trihalide. Other patented processes to produce bromochlorometane include the gas-phase bromination of methyl chloride with a mixture of chlorine and HBr (66) or bromine and chlorine (67), and the liquid-phase displacement reaction of dichloromethane with inorganic bromides (68,69). A mixture of bromochloromethane and dibromomethane is formed in all the above reactions. The compounds are separated by fractional distillation.

The major use of bromochloromethane is as a fire-extinguishing fluid (70); its effectiveness per unit weight makes it suitable for use in aircraft and portable systems. It is also used as an explosion suppression agent, as a solvent, and as an intermediate in the manufacture of some insecticides (chlormephos [24934-91-6]).

The TWA limit or daily 8-h exposure to the vapor of bromochloromethane in air is 200 ppm by volume, or 1050 mg/m^3; the short-time exposure limit is 250 ppm.

Dibromomethane, CH_2Br_2, (methylene bromide) [74-95-3], is a similar liquid, mp $-52.7°C$; bp $96.9°C$; d^{20}_4 2.4956 kg/L; n^{25}_D 1.5419; vapor pressure 34.9 mmHg ($20°C$). Its solubility in water is 1.17 g/100 g at $15°C$. For the solubility data, dibromomethane–water system, see (71). The compound is produced by the same methods as bromochloromethane. The compound is used as high-density solvent (mineral separation, gauge fluid) and as an intermediate (piperonal [120-57-0], methylene bisthiocyanate [6317-18-6]). Dibromomethane is more toxic than bromochloromethane.

Both dibromomethane and bromochloromethane (or mixtures of these two compounds) are used as solvents for bromination reactions, especially for the production of polybrominated aromatic compounds and polymers (72–74).

Tribromomethane, $CHBr_3$, (bromoform), [75-25-2]. The pure liquid has a mp $8.3°C$, bp $149.5°C$, d^{20}_4, 2.8912 kg/L, n^{19}_D 1.5419, vapor pressure 5 mmHg ($20°C$). Its water solubility is $\sim$0.3 g/100 g at $25°C$. For the solubility data, tribromomethane–water system, see (75). Bromoform is prepared by reaction of bromine with acetone or ethanol in the presence of sodium hydroxide (76). Uses have been found as high-density solvent for mineral separation, in gauge fluids and as an intermediate (deltametrin [52918-63-5]). Bromoform is a toxic and irritant compound, TLV 0.5 ppm (skin).

Tetrabromomethane, CBr_4, (carbon tetrabromide), [558-13-4] crystallized in two forms, α-form mp $48.4°C$, β-form mp $92.5°C$, bp $189.5°C$; d^{20}_4 3.240 kg/m^3, $d^{99.5}$ 2.9609, $n^{99.5}_D$ 1.600. It is prepared by the replacement of chlorine in carbon tetrachloride using hydrogen bromide and an aluminum halide catalyst (65) or by action of sodium hypobromite on bromoform by an extension of the haloform reaction (77). Tetrabromomethane is used as intermediate both in ionic (78) and in homolytic reactions and telomerisation processes (7).

Bromotrifluoromethane, $CBrF_3$, [75-63-8], bromochlorodifluoro methane, $CBrClF_2$ [353-59-3], and 1,2-dibromotetrafluoroethane, $CBrF_2CBrF_2$, [124-73-2], are volatile bromine-containing halogenofluorocarbons, known under the technical name "halons". Their physical properties are presented in Table 1.

Halons are fire-extinguishing agents, which replace the more toxic methyl bromide and carbon tetrachloride.

4.2. Ethylene Dibromide. Ethylene dibromide CH_2BrCH_2Br (ethylene bromide, 1,2-dibromoethane), [106-93-4] (commonly abbreviated as EDB) is a clear, colorless liquid with a characteristic sweet odor. Its properties include: mp $9.9°C$; bp $131.4°C$; d^{20}_4 2.1792 kg/L, n^{20}_D 1.5380, vapor pressure 1.13 (8.5),

Table 1. **Physical Properties of Halons**

Halon	Formula	bp °C (1 atm)	mp °C	d Liquid (25°C)	Critical temperature (°C)	Toxicity[a] (% in air)
1301	$CBrF_3$	−57.8	−168	1.539	67	TWA1000
1211	$CBrClF_2$	−3.9	−161	1.83	154	32
2402	$CBrF_2CBrF_2$	47.3	−110	2.16	214	12.6

[a] Approximately lethal concentration for 15-min exposure.

15.98 (119.8), and 38.03 kPa·s (285.2 mmHg) at 20, 75, and 100°C, respectively; viscosity 1.727 mPa·s (20°C); heat capacity of the solid at 15.3°C, 519 J/kg·K (124 cal/kg·K) and of the liquid at 21.3°C, 724 J/kg·K (173 cal/kg·K); heat of fusion at 9.9°C, 53.4 J/g (12.76 cal/g); heat of vaporization at bp 191 J/g (45.7 cal/g); heat of transition at −23.6°C, 10.34 J/g (2.47 cal/g); critical temperature, 309.8°C; critical pressure 7154 kPa (70.6 atm); expansion coefficient at 15–30°C, 0.000958/K; dielectric constant at 20.5°C (0.1 MHz), 4.77. The liquid is completely miscible with carbon tetrachloride, benzene, gasoline, and anhydrous alcohols at 25°C and its solubility in water at 20°C is 0.404 g/100-g solution.

EDB is nonflammable and quite stable under ordinary conditions. Ethylene glycol is produced by its high temperature hydrolysis under pressure. Reaction with metals (zinc, magnesium) yields ethylene, reaction with ammonia proceeds with explosion, yielding ethylenediamine and higher polymers.

EDB is manufactured via uncatalyzed, liquid-phase bromination of ethylene. Gaseous ethylene is brought into contact with bromine by various methods, allowing for dissipation of the heat of the reaction (79–81).

The largest single application of EDB has traditionally been its use as a lead scavenger in leaded gasoline. Since the U.S. Environmental Protection Agency (EPA) mandated a reduction in the lead content in gasoline beginning in 1974, U.S. consumption of EDB in antiknock mixes has declined dramatically, from 60 thousand metric tons in 1978 to <1000 metric tons in 1997.

The second-largest traditional use of EDB was as an insect fumigant and soil nematocide. In 1983, however, the EPA banned the use of EDB in most agricultural applications because of concerns about the chemical's toxicity. As a result, EDB consumption in this market has also dropped. Currently, most EDB in the USA is produced for export.

Other uses of EDB are believed to consume ~1000 metric tons annually (USA, 1999). Major uses in this category are as an intermediate for pharmaceuticals (tetramisole [5036-02-2], theodrenaline [13460-98-5]), herbicides (diquat dibromide [85-00-7]), and dyes, (vat Blue 16), where it provides an "ethylene bridge" in the molecular structure. EDB is used as a nonflammable solvent for resins, gums, and waxes. Additionally, EDB can be used as a raw material in the synthesis of chemicals such as vinyl bromide (a precursor of flame-retardants) and styrenic block copolymers.

EDB is an acutely toxic, severely irritating to skin, mutagenic, and carcinogenic compound. Its current time-weighted average limit is 20 ppm. The toxicology of EDB has been reviewed (82).

5. Industrial Chemical Intermediates

Tables 2–4 list some organic bromine compounds used as intermediates and reagents on the industrial scale, their physical properties, and main applications.

6. Industrial Chemical Products

6.1. Flame Retardants.　Brominated flame retardants are the more significant and voluminous part of all bromine derivatives. Bromine consumption in

Table 2. **Industrial Aliphatic Bromine Compounds**

Compound	CAS Registry Number	Molecular formula	mp, °C	bp, °C[a]	d_4^{20}	n_D^{20}	Derivatives, uses and miscellaneous properties
allyl bromide	[106-95-6]	$CH_2=CHCH_2Br$	−119	71.3	1.398	1.4654	methohexital, nalorhine
n-amyl bromide	[110-53-2]	$CH_3(CH_2)_3CH_2Br$	−95	130	1.218	1.4436	photographic coupling agent intermediate
benzyl bromoacetate	[5437-45-6]	$BrCH_2COOCH_2Ph$		$166^{2.9}$	1.446	1.5440	Merbac-35 (preservative for water-based paints)
bromoacetic acid	[79-08-3]	$BrCH_2COOH$	50	208	1.93		methoprene
bromoacetyl bromide	[598-21-0]	$BrCH_2COBr$		148	2.317		cefotetan, labetalol, sotalol
2-bromobutyric acid	[80-58-0]	$CH_3CH_2CHBrCOOH$	−4	$181^{4.5}$	1.567	1.4720	commercial product is racemate
3-bromo-3-butene-1-ol	[76334-36-6]	$CH_2=CBrCH_2CH_2OH$		$64^{1.2}$	1.522	1.4990	building block for sesquiterpenes
1-bromo-3-chloropropane	[109-70-6]	$BrCH_2CH_2CH_2Cl$		142–5	1.46		fluphenazine, perphenazine, prochlorperazine, trazodone, verapamil
2-bromopropionic acid	[598-72-1]	$CH_3CHBrCOOH$	25	204	1.700	1.4750	naproxene, prilocaine
3-bromopropionic acid	[590-92-1]	$BrCH_2CH_2COOH$	62.5	140^6	1.480		quaternization reagent for biosensors
5-bromovaleric acid	[2067-33-6]	$Br(CH_2)_4COOH$	38–40				allethrolone
trans-bromostyrene	[103-64-0]	$PhCH=CHBr$	7	$110^{2.7}$	1.427	1.6070	fragrance ingredient
n-butyl bromide	[109-65-9]	$CH_3(CH_2)_3Br$	−112	101	1.276	1.4390	bupivacaine, tetracaine, tetra-n-butylammonium bromide
cetyl bromide	[112-82-3]	$CH_3(CH_2)_{15}Br$	17.3	$190^{1.5}$	1.99	CTAB	
cetylpyridinium bromide	[140-72-7]	$CH_3(CH_2)_{15}Py^+\ Br^-$					topical disinfectant
cyclopropyl bromide	[4333-56-6]	$cyclo\text{-}C_3H_5Br$		68–70	1.510	1.4600	biotin, nortriptiline
cyclopropylmethyl bromide	[7051-34-5]	$cyclo\text{-}C_3H_5CH_2Br$		105–107	1.392	1.4570	buprenorphine, naltrexone, prazepam
1,4-dibromobutane	[110-52-1]	$BrCH_2CH_2CH_2CH_2Br$		194–196	1.81	1.5186	pentoxyverine
1,3-dibromopropane	[109-64-8]	$BrCH_2CH_2CH_2Br$	−34.2	166.7	1.9822	1.5232	building block

Compound	CAS Number	Formula	bp, °C	mp, °C	Density	n_D	Uses
2,3-dibromopropan-1-ol	[96-13-9]	$BrCH_2CHBrCH_2OH$	$95^{1.3}$			1.5590	building block for fire retardant
ethyl bromide	[74-96-4]	C_2H_5Br	38	−119	1.45		pentobarbital, phenobarbital, thiopental
ethyl bromoacetate	[105-36-2]	$BrCH_2COOC_2H_5$	159	−14	1.50	1.451	building block for pharmaceuticals
n-hexyl bromide	[111-25-1]	$CH_3(CH_2)_5Br$	154	−85	1.18		building block for fragrances
isobutyl bromide	[78-77-3]	$(CH_3)_2CHCH_2Br$	91.5	−119	1.260	1.4350	building block for fragrances and dyes
isopropyl bromide	[75-26-3]	$(CH_3)_2CHBr$	59	−89	1.310	1.4251	ipratropium bromide, verapamil
octyl bromide	[111-83-1]	$CH_3(CH_2)_7Br$	201	−55	1.118	1.4518	building block for fragrances
n-propyl bromide	[106-94-5]	$CH_3CH_2CH_2Br$	71	−110	1.353	1.4341	penconazole, valproic acid
tetrabromoethane	[79-27-6]	$Br_2CHCHBr_2$	246	0	2.96		high density solvent for mineral separation
vinyl bromide	[593-60-2]	$CH_2=CHBr$	16	−138	1.51		fire retardant comonomer

[a] At 101 kPa, unless otherwise indicated in parentheses; pressure in kilo pascals (kPa). To convert kPa to mmHg, multiply by 7.5.

Table 3. **Industrial Aromatic Bromine Compounds (Benzene Series)**

Compound	CAS Registry Number	Molecular formula	mp, °C	bp, °C	d_4^{20}	n_D^{20}	Derivatives, uses and miscellaneous properties
2-amino-3-bromo-5-nitrobenzonitrile	[17601-94-4]	$R_1 = NH_2$, $R_2 = CN$, $R_4 = NO_2$, $R_3 = R_5 = H$	186				disperse Blue 183
4-bromoaniline	[106-40-1]	$R_3 = NH_2$, $R_1 = R_2 = R_4 = R_5 = H$	64				metobromuron, resorantel
bromobenzene	[108-86-1]	$R_1 = R_2 = R_3 = R_4 = R_5 = H$	−30.6	156.1	1.4951	1.5604	bromopropylate, clobazam, fenoprofen, flubiprofen, phenylmagnesium bromide
4-bromobenzyl cyanide	[16532-79-9]	$R_3 = CH_2CN$, $R_1 = R_2 = R_4 = R_5 = H$	48				brompheramine
2-bromo-4,6-dinitroaniline	[1817-73-8]	$R_1 = NH_2$, $R_2 = R_4 = NO_2$, $R_3 = R_5 = H$	152				disperce Blue 79, dyestuffs intermediate
bromo-2,4-difluorobenzene	[348-57-2]	$R_1 = R_3 = F$, $R_2 = R_4 = R_5 = H$		145–146	1.708	1.5050	fluconazole, saperconazole, trovafloxacin
4-bromofluorobenzene	[460-00-4]	$R_3 = F$, $R_1 = R_2 = R_4 = R_5 = H$	−17	155	1.593	1.5270	flustiazole, flutriafol
tetrabromo-o-cresol	[576-55-6]	$R_1 = R_2 = R_3 = Br$, $R_4 = OH$, $R_5 = CH_3$	206				disinfectant
2,4,6-tribromoaniline	[147-82-0]	$R_1 = NH_2$, $R_2 = R_4 = Br$	120	300			building block for fire-retardants
2,4,6-tribromophenol	[118-79-6]	$R_1 = OH$, $R_2 = R_4 = Br$	95	290			wood preservative

Table 4. **Industrial Organic Bromine Compounds—Derivatives of Anthraquinone**

Compound	CAS Registry Number	Molecular formula	mp °C	Derivatives
2-amino-3-bromo-anthraquinone	[6337-00-4]	$R_1 = R_4 = H$, $R_3 = Br$, $R_2 = NH_2$	235–303	vat Blue 30, vat Red 10
1-amino-2-bromo-4-hydroanthraquinone	[116-82-5]	$R_1 = NH_2$, $R_2 = Br$, $R_3 = H$, $R_4 = OH$		disperse Red 60
1-amino-2-bromo-4-(4′-methylphenylsulfamido)-anthraquinone	[26868-32-6]	$R_1 = NH_2$, $R_2 = Br$, $R_3 = H$, $R_4 = NHSO_2C_6H_4CH_3$		acid Blue 45
1-amino-2,4-dibromoanthraquinone	[81-49-2]	$R_1 = NH_2$, $R_2 = R_4 = Br$, $R_3 = H$	226	acid Blue 96
bromamine acid, 1-amino-4-bromoanthraquinone-2-sulfonic acid	[116-81-4]	$R_1 = Br$, $R_2 = H$, $R_3 = SO_3H$, $R_4 = NH_2$		acid Blue 25, 40, 41, 53, 62, 111, 124, 127, 129, 138, 145, 230, direct Green 28, pigment Red 177, reactive Blue 2, 4, 19, 94
1-bromo-4-methylaminoanthraquinone	[128-93-8]	$R_1 = Br$, $R_2 = H$, $R_3 = SO_3H$, $R_4 = NHCH_3$	194	acid Blue 27, acid Violet 80, basic Blue 22

flame retardants has risen substantially from the early 1990s and at present forms, on average, one-half of organic bromine compounds consumption and ~40% of the total bromine consumption.

Brominated flame retardants can be divided into two groups according to their chemical structure: brominated aliphatic compounds and brominated aromatic compounds. The latter are much more stable and may be used in thermoplastics at fairly high temperatures without the use of stabilizers and at very high temperature with stabilizers.

Brominated flame retardants can also be divided into two general classes according to their relation to polymers—additive flame retardants and reactive flame retardants. Additives are mixed into the polymer in common polymer processing equipment. Reactive flame retardants literally become part of the polymer by either reacting into the polymer backbone or grafting onto it. The characteristics of commercial brominated flame retardants are given in Tables 5–7. The world consumption of brominated flame retardants is presented in Table 8.

The average annual growth rate for brominated compounds is forecast through 2003 at 2.5–3% per year for additive compounds and 4% for reactive compounds. These products will benefit from more exacting fire safety standards in consumer products, building products, automobile and aircraft components. The proliferation of computers and other consumer electronics is boosting demand for plastics that have enhanced flame retardancy characteristics. Because of the effectiveness and performance advantages of brominated flame retardants, smaller amounts can be used, which enhances the cost effectiveness of these products while maintaining the functional characteristics of the host material.

6.2. Pesticides. A list of bromine-containing pesticides is presented in Table 9.

A more significant and expanding group of brominated biocides is for water treatment (83). While chlorine controls the majority of the water treatment market, brominated compounds are becoming increasingly popular. In general, both industrial and consumer segments of the water treatment industry are increasingly replacing chlorine and chlorinated compounds as sanitizers and biocides with bromine-based products. Chlorine is now subject to a wide range of EPA limitations, and although bromine is itself a halogen, no restrictions have yet been placed on brominated biocides. In addition, the greater strength of bromine-based biocides allows treatment of a given amount of water with considerably less biocide. This not only reduces the amount of halogen released into the environment, but can also reduce costs for municipal and industrial water treatment plants. Benefits accrue through reduced chemical costs, avoidance of the dechlorination step common in chlorine water treatment, and reduced corrosion to condensers, tubing and other equipment.

Brominated biocides also perform better than chlorinated biocides in a number of industrial applications because of their higher tolerance to a wide range of pH levels, a concern in cooling towers and process waters. A trend in recent years has been to run cooling water towers at a higher pH to minimize corrosion, but this often leads to a larger formation of algae. Brominated

Table 5. **Brominated Additive Flame Retardants**

Compound	CAS Registry Number	Molecular formula	mp, °C	Bromine content, %	Producers[a]
bis(2-ethylhexyltetrabromophthalate)	[26040-51-7]	$C_{24}H_{34}Br_4O_4$		45	EA, GL
bis(methyl)tetrabromophthalate	[55481-60-2]	$C_{10}H_6Br_4O_4$		63	EA
bis(tribromophenoxy)ethane	[37853-59-1]	$C_{14}H_8Br_6O_2$		68	GL
brominated trimethylphenylindane	[155613-93-7]	$C_{18}H_{12}Br_8$	224	73	DS
decabromobiphenyl	[13654-09-6]	$C_{12}Br_{10}$	240–255	84.5	EA
decabromodiphenyl ether	[1163-19-5]	$C_{12}Br_{10}O$	303–307	83	AL, ASC, DS, GL, ISC, MI, MT, NC, TS, WI
decabromodiphenylethane	[137563-36-1] [84852-53-9]	$C_{14}H_4Br_{10}$	350–356	82.3	AL
dibromoethyldibromocyclohexane	[3322-93-8]	$C_8H_{12}Br_4$	175–185	73	AL, DS, GL, ISC
ethylenebisdibromonorbornanedicar-boximide	[52907-07-0] [41291-34-3]	$C_{20}H_{20}Br_4N_2O_4$	294	45	AL, GL
ethylenebis(tetrabromophthalimide)	[32588-76-4]	$C_{18}H_4Br_8N_2O_4$	456	67.2	AL
hexabromocyclododecane	[3194-55-6]	$C_{12}H_8Br_6$	180	74.7	AL, DS, GL, ISC
octabromodiphenyl ether	[32536-52-0]	$C_{12}H_2Br_8O$	70–150	78	AL, DS, GL, ISC, MI
pentabromotoluene	[87-83-2]	$C_7H_3Br_5$		82	DS
tetrabromobisphenol A bis(2,3-dibromopro-pyl ether)	[21850-44-2]	$C_{23}H_{20}Br_8O_2$	95	67.7	AL, DS, GL
tetrabromobisphenol A	[79-94-7] [6386-73-8]	$C_{15}H_{12}Br_4O_2$	180	58.4	AL, ASC, DS, GL, T, TS
tetradecabromodiphenoxybenzene	[58956-66-5]	$C_{18}Br_{14}O_2$	370	82	AL
tris-dibromopropylisocyanurate	[52434-90-9]	$C_{12}H_{15}Br_6N_3O_3$	106–108	65.8	AC, ASC, T

[a] Company and country are as follows: AC = Akzo Chemicals BV, Netherlands; AL = Albemarle, USA; ASC = Asahi Chemical, Japan; DS = DSBG, Israel; EA = Elf Atochem, France; GL = Great Lakes, USA; ISC = ISC Chemicals, Ltd, UK; MI = Manac Inc., Japan; M = Marubishi, Japan; MT = Mitsui Toatsu Fine Chemicals, Japan; NC = Nippon Chemicals Corp., Japan; T = Teijin, Japan; TS = Tosoh, Japan; WI = Warwick Int., UK; CECA = CECA, SA, France.

Table 6. **Polymeric and Oligomeric Additive Flame Retardants**

Compound	CAS Registry Number	Molecular Formula[a]	mp, °C	Bromine content, %	Producers[b]
brominated polystyrene low molecular weight	[88497-56-7]	$(C_8H_{5.3}Br_{2.7})_n$	130–140	66	KC
brominated polystyrene	[88497-56-7]	$(C_8H_{5.3}Br_{2.7})_n$	195	66	KC
poly(dibromophenylene oxide)	[26023-27-8]	$(C_6H_2Br_2O)_n$	225	62	GL
poly(pentabromobenzylacrylate)	[59447-55-1]	$(C_{10}H_5Br_5O_2)n$	210	71	DS
poly(dibromostyrene)	[62354-98-7]	$(C_6H_6Br_2)_n$	155–165	59	GL
tetrabromobisphenol A carbonate oligomer, phenoxy end capped	[94334-64-2]	$(C_{16}H_{12}Br_4O_3)_n$	210–230	52	GL, MG, T
tetrabromobisphenol A carbonate oligomer, tribromophenoxy end capped	[71342-77-3]	$(C_{16}H_{12}Br_4O_3)_n$	230–260	58	GL
tetrabromobisphenol A epoxy oligomers	[68928-70-1]	$(C_{16}H_{12}Br_4O_3)_n$		52–54	DS, DI, DC, H, MI, SY, TK
tetrabromobisphenol A epoxy oligomers, tribromophenoxy end capped	[400039-93-8]	$(C_{16}H_{12}Br_4O_3)_n$		55–58	DI, H, TK

[a] Formulas for polymeric compounds are for the repeat unit only and ignore the end groups.
[b] Company and country are as follows: DS = DSBG, Israel; DI = Dainippon Ink and Chemical, Japan; DC = Dow Chemical, USA; GL = Great Lakes, USA; H = Hitachi, Japan; KC = Keil Chemical Div., Ferro Corp., USA; MI = Manac Inc., Japan; MG = Mitsubishi Gas Chemical Co, Japan; SY = Sakamoto Yukuhin Co., Japan; T = Teijin, Japan; TK = Tohto Kasei, Japan.

Table 7. **Brominated Reactive Flame Retardants**

Compound	CAS Registry Number	Molecular formula	mp, °C	Bromine content,%	Producers[a]
dibromoneopentyl glycol	[3296-90-0]	$C_5H_{10}Br_2O_2$	109.5	60	AL, DS
pentabromobenzylacrylate	[59447-55-1]	$C_{10}H_5Br_5O_2$		71.0	DS
pentabromobenzyl bromide	[38521-51-6]	$C_7H_2Br_6$		84.8	DS
tetrabromobisphenol A	[79-94-7]	$C_{15}H_{12}Br_4O_2$	181	58.5	AL, ASC, CECA, DS, GL, T, TS
tetrabromobisphenol A bis(allyl ether)	[25327-89-3]	$C_{21}H_{20}Br_4O_2$	119	51.2	DS, GL
tetrabromobisphenol A bis(2-hydroxyethyl ether)	[4162-45-2]	$C_{19}H_{20}Br_4O_4$	116	51.6	DS, GL
tetrabromophthalic anhydride	[632-79-1]	$C_8Br_4O_3$	270	68	AL, CECA, GL
tetrabromophthalic anhydride/diol	[7709807-8]	$C_{15}H_{16}Br_4O_7$	liquid	46	AL, GL
tribromophenylmaleinimide	[59789-51-4]	$C_{10}H_8Br_3NO_2$		57.9	DS
tribromoneopentyl alcohol	[1522-92-5]	$C_5H_9Br_3O$	62–67	73.6	AL, DS
2,4,6-tribromophenol	[118-79-6]	$C_6H_3Br_3O$	93	72.3	DS, GL, MI
vinyl bromide	[593-60-2]	C_2H_3Br	liquid	74.5	AL

[a] Company and country are as follows: AL = Albemarle, USA; ASC = Asahi Chemical, Japan; CECA = CECA, SA, France; DS = DSBG, Israel; GL = Great Lakes, USA; MI = Manac Inc., Japan; TS = Tosoh, Japan.

Table 8. Consumption of Brominated Flame Retardants 1998, $\times 10^3$ t

USA	Western Europe	Japan	Other Asia	Total	Total value (millians of dollars)
68.3	51.5	47.8	97	264.5	790

Table 9. Organic Bromine Compounds Used as Pesticides

Compound	CAS Registry Number	Toxicity LD_{50} mg/kg rats
Acaricide		
bromopropylate	[18181-80-1]	1700
Fungicides		
(2-bromo-1,2-diiodoacryl)ethylcarbonate	[77352-88-6]	500–600
BMPCA, *N*-(4-bromo-2-methylphenyl)-2-chloracetamide	[96686-51-0]	4044
Herbicides		
bromacil	[314-40-9]	
bromobutide	[74712-19-9]	
bromofenoxim	[13181-17-4]	1200
bromoxynil	[1089-84-5]	
bromoxynil octanoate	[1689-99-2]	
chlorbromuron	[13360-45-7]	
diquat dibromide	[85-00-7]	300
metobromuron	[3060-89-7]	
Insecticides		
bromofos (and acaricide)	[2104-96-3]	2800
deltametrin	[52918-63-5]	
leptophos	[21609-90-5]	
naled (and acaricide)	[300-76-5]	430
profenofos	[41198-08-7]	358
Microbicides (water treatment biocides)		
benzyl bromoacetate[a]	[5437-45-6]	
bis-1,2-(bromoacetoxy)ethane[b]	[3785-34-0]	
bis-1,4-(bromoacetoxy)-2-butene[b]	[20679-58-7]	
bromochlorophen[c]	[15435-29-7]	3700–8000
BNP (2-bromo-2-nitropropanol)	[24403-04-1]	~300
BNS (2-bromo-2-nitrostyrene)[b]	[7166-19-0]	
bronidox	[30007-47-7]	590
bronopol	[52-51-7]	325
cetrimonium bromide CTAB	[57-09-0] [77-48-5]	
DBDMH (1,3-dibromo-5,5-dimethylhydantoine)	[77-48-5]	
DBNPA (2,2-dibromo-2-cyanoacetamide)	[10222-01-2]	126
1,2-dibromo-2, 4-dicyanobutane[a]	[35691-67-7]	
dibromohexamidine isethionate	[93856-82-7]	541
disanyl	[87-12-7]	>4000
halobrom (BCDMH, bromochlorodimethylhydantoin)	[126-06-7], [32718-18-6], [107846-11-7]	1700
2-hydroxyethyl 2,3-dibromo-propionate[a]	[68479-77-6]	
trisanyl	[87-12-7]	570

Table 9   (*Continued*)

Compound	CAS Registry Number	Toxicity LD$_{50}$ mg/kg rats
Nematocides		
1,2-dibromo-3-chloropropane	[96-12-8]	173
1,2,3-tribromopropane	[96-11-7]	
Rodenticides		
brodifacum	[56073-10-0]	
bromadiolone	[28772-56-7]	
bromethalin	[63333-35-7]	

[a] Used as preservative for in-can protection of water based paints, adhesives and polishes.
[b] Used as ingredient in nonpersistent slimicides in the paper industry.
[c] Used as a preservative for cosmetics.

Table 10. Bromine Demand in Water Treatment Biocides in USA, ×10^3 t

Year	1985	1989	1995	2000	2005
Water treatment biocides	12.7	22.7	36.3	47.2	61.3

Table 11. Bromine Containing Pharmaceuticals

Compound	CAS Registry Number	Compound	CAS Registry Number
Adrenergic		*Anticoagulant*	
hydroxyamphetamine hydrobromide[a]	[306-21-8]	bromindione	[1146-98-1]
Analgesic		*Anticonvulsant*	
		cinromide	[58473-74-8]
bromadoline	[67579-24-2]		
bromfenac	[91714-94-2]	*Antidepressant*	
phenazocine	[1239-04-9]		
		zimeldine	[56775-88-3]
Anesthetic (inhalation)		*Antihistaminic*	
halothane[a]	[151-67-7]	bromodiphenhy-dramine[a]	[118-23-0]
roflurane	[679-90-3]	brompheniramine[a]	[86-22-6]
teflurane	[124-72-1]	dexbrompheniramine[a]	[132-21-8]
		temelastine	[86181-42-2]
Antiadrenergic		*Antihypertensive*	
bretylium tosylate	[61-75-6]		
		guanisonium	[154-73-4]
Anticholinergic		quinuclium bromide	[35425-83-3]
anisotropine methylbromide	[80-50-2]	*Antiinfective*	
benzilonium bromide	[1050-48-2]		
clidinium bromide[a]	[3485-62-9]	domiphen bromide	[538-71-6]
glycopyrrolate[a]	[596-51-0]		
heteronium bromide	[7247-57-6]	*Antiinflamatory*	
hyoscyamine hydrobromide[a]	[306-03-6]		
mepenzolate bromide	[76-90-4]	broperamole	[33144-79-5]
methantheline bromide	[53-46-3]	halopredone acetate	[57781-14-3]
methscopolanine bromide	[155-41-9]		

Table 11   (*Continued*)

Compound	CAS Registry Number	Compound	CAS Registry Number
penthienate bromide	[60-44-6]	*Antineoplastic*	
pipenzolate bromide	[125-51-9]		
propantheline bromide[a]	[50-34-0]	bropirimine	[56741-95-8]
scopolanine hydrobromide[a]	[114-49-8]	pipobroman[a]	[54-91-1]
thihexinol methylbromide	[7219-91-2]		
valethamate bromide	[90-22-2]	*Antipsychotic*	
Anticholinergic (ophthalmic)		brofoxine	[21440-97-1]
		bromperidol	[10457-90-6]
homatropine hydrobromide[a]	[51-56-9]		
homatropine methylbromide[a]	[80-49-9]		
Antitussive		*Expectorant*	
dextromethorphan		bromohexine	[3572-43-8]
hydrobromide[a]	[125-69-9]		
		Inhibitor aldose reductase	
Antiviral			
		ponalrestat	[72702-95-5]
sorivudine	[77181-69-2]		
		Neuromuscular blocker	
Bronchodilator			
		pancuronium bromide	[15500-66-0]
ipratropium bromide	[22254-24-6]	pipecuronium bromide	[52212-02-9]
rimiterol hydrobromide	[32953-89-2]	rocuronium bromide	[119302-91-9]
		vecuronium bromide	[50700-72-6]
Cholinergic			
		Progestin	
neostigmine bromide[a]	[114-80-7]	haloprogesteron	[3538-57-6]
pyridostigmine bromide[a]	[101-26-8]		
		Sceletal muscle relaxant	
Cholinergic ophtalmic			
		azumolene	[64748-79-4]
demecarium bromide[a]	[56-94-0]	hexafluorenium	
		bromidea	[317-52-2]
Detergent			
		Sedative and hypnotic	
thonzonium bromide	[553-08-2]		
		brotizolam	[57801-81-7]
Diagnostic aid (hepatic function)			
		Tranquilizer minor	
sulfobromophthalein	[71-67-0]		
sodium			
		bromazepam	[1812-30-2]
Diagnostic aid (hepatobiliary function)			
		Vasodilator	
mebrofenina	[78266-06-5]		
		brovincamine	[57475-17-9]
Diuretic		nicergoline	[27848-84-6]
brocrinate	[72481-99-3]	*Veterinary medicine*	
pamabrom	[606-04-2]		
		halofuginone	[55837-20-2]
Enzyme inhibitor (prolactin)			
bromocriptine	[25614-03-3]		
bromocriptine mesylate[a]	[22260-51-1]		

[a] Listed in the *U.S. Pharmacopeia* 24 (2000).

Table 12. **Bromine-Containing Dyes**

Dye	CAS Registry Number	CI Number	Dye used for
acid orange 11, 4′,5′-dibromofluorescein	[596-03-2]	45370	wool, silk, paper
alizarin pure Blue B	[6424-75-5]	1088	wool
Ciba Bordeaux B, 5,5′-dibromothioindigo	[6371-14-8]	1208	cotton
5,5′-dibromoindigo	[19201-53-7]	1183	cotton
disperse Blue 20	[26846-51-5]		
disperse Blue 56	[12217-79-7]	63285	polyester, acetate
disperse Yellow 64 4-bromo-3-hydroxy-quinophthalone	[10319-14-9]	47023	polyester, acrylic, acetate
eosine B, acid Red 91, 4′,5′-dibromo-2′,7′-dinitrofluorescein	[548-24-3]	45400	acetate, polyamide, 4-polyester
eosine Y, Acid Red 87, 2′,4′,5′,7′-tetrabromofluorescein	[17372-87-1]	45380	wool, cotton, paper differential staining
pigment Green 36 brominated phthalocyanine	[14302-13-7]	74265	cosmetics, paper and inks, biological stain
pigment Red 216, tribromopyranthrone	[1324-33-0]	59710	paints, printing inks, plastics
vat Blue 5, 5,5′,7,7′-tetrabromoindigo	[2475-31-2]	73065	paints, plastics
vat Blue 19, dibromodibenzanthrone	[1328-18-3]	59805	cotton, viscose, silk
vat Orange 1, dibromodibenzopyrenequinone	[1324-11-4]	59105	cotton, viscose
vat Orange 2, 4,12-dibromopyranthrone	[1324-35-2]	59705	cotton
vat Orange 3, dibromoanthanthrone	[4378-61-4]	59300	cotton, viscose, silk
			cotton, plastics

361

Table 13. **Bromine-Containing Indicators**

Indicator	CAS Registry Number	Transition range, pH	Color change
bromophenol blue	[115-39-9]	3.0–4.5	yellow–blue violet
bromochlorophenol blue	[2553-71-1]	3.2–4.6	yellow–blue violet
bromocresol green	[76-60-8]	3.8–5.4	blue–green
bromophenol red	[2800-80-8]	5.2–6.8	yellow–purple
bromocresol purple	[115-40-2]	5.2–6.8	yellow–purple
bromoxylenol blue	[40070-59-5]	5.7–7.5	yellow–blue
bromothymol blue	[76-59-5]	6.0–7.6	yellow–blue

products (especially DBDMH and Halobrom) are both more effective at higher pH ranges than chlorine, and are estimated to be three times more effective than chlorine at controlling algae blooms.

Apart from their increasing use in industrial and municipal water treatment, bromine derivatives are also registering gains in the consumer water conditioning market as biocides for spas and hot tubs. In these applications, DBDMH and Halobrom are displacing chlorinated biocides for the same reasons as in large-scale water treatment.

In addition, brominated biocides are more stable than chlorine in higher temperature waters, do not readily decompose on exposure to sunlight and are less irritating to the eyes and mucous membranes. This is particularly important in the recreational segment of the water treatment market. As a result, brominated biocides for water treatment will continue to expand their market share (Table 10).

6.3. Pharmaceuticals.   Organic pharmaceuticals containing bromine can be divided into two groups. The main group includes actual organic bromine compounds, in which bromine is bonded with the carbon atom. The second group includes salts of hydrobromic acid and ammonium organic compounds. Both these groups are presented in Table 11.

6.4. Dyes and Indicators.   The effect of bromine in dye (Table 12) or indicator molecules in place of hydrogen includes a shift of light absorption to longer wavelengths, increased dissociation of phenolic hydroxyl groups, and lower solubility. The first two effects probably result from increased polarization caused by bromine's electronegativity compared to that of hydrogen.

Bromine containing indicators are listed in Table 13.

BIBLIOGRAPHY

"Bromine Compounds" in *ECT* 1st ed., Vol. 2, pp. 645–660, by V. A. Stenger and G. J. Atchison, The Dow Chemical Company; in *ECT* 2nd ed., Vol. 3, pp. 766–786, by V. A. Stenger and G. J. Atchison, The Dow Chemical Company; in *ECT* 3rd ed., Vol. 4, pp. 243–263, by N. A. Stenger, Dow Chemical U.S.A.; in *ECT* 4th ed., Vol. 4, pp. 560–589 by Philip F. Jackisch, Ethyl Corporation; "Bromine Compounds" in *ECT* (online), posting date: April 12, 2000, by Philip F. Jackisch, Ethyl Corporation.

CITED PUBLICATIONS

1. G. W. Gribble, *Verh.- K. Ned. Akad. Wet., Afd Natuurkd., Tweede Reeks* **98**, 1 (1997) (in English); *Chem. Abstr.* **128**, 267–273 (1998).
2. P. Friedlender, *Chem. Ber.* **55**, 1655 (1922).
3. P. Ashworth and J. Chetland, *Ind. Chem. Libr.*, **3** (*Adv. Organobromine Chem.* **1**), 263 (1991).
4. P. B. D. de la Mere and B. E. Swedlund in S. Patai, ed., *The Chemistry of Carbon– Halogen Bond*, Part 1, John Wiley & Sons, Inc., New York, 1973, pp. 407–548.
5. C. Hill, *Activation and Functionalization of Alkanes*, John Wiley and Sons, Inc., New York, 1989.
6. F. A. Drahowzal in G. Olah, ed., *Friedel–Crafts and Related Reactions*, Interscience Publishers, New York, Vol. 2, 1964, pp. 417–475.
7. A. B. Terent'ev and T. T. Vasil'eva, *Russ. Chem. Rev.* **63**, 269 (1994).
8. J. Lindley, *Tetrahedron* **40**, 1433 (1984).
9. A. M. Moroz and M. S. Shvartsberg, *Russ. Chem. Rev.* **43**, 679 (1974).
10. R. F. Heck, *Org. Reactions* **27**, 345 (1982).
11. S. Brase and A. De Meijere in F. Diderich and P. Stang, eds., *Metal-Catalyzed Cross-Coupling Reactions*, Wiley-VCH, Weinheim, 1998, pp. 99–166.
12. R. Franzen, *Can. J. Chem.* **78**, 957 (2000).
13. A. Suzuki, *J. Organomet. Chem.* **576**, 147 (1999).
14. A. Roedig in Houben-Weyl, *Methoden der Organische Chemie*, 4th ed., Thieme Verlag, Stuttgart, V/4, 1960, pp. 13–516.
15. R. C. Larock, *Comprehensive organic transformations*, VCH, New York, 1989, pp. 307–383.
16. P. L. Spargo in A. R. Katritzky, O. Meth-Cohn, and C. W. Rees, eds., *Comprehensive Organic Functional Group Transformations*, Pergamon, Oxford, 1995, Vol. 2, pp. 1–37.
17. C. J. Urch in Ref. 16, pp. 605–635.
18. P. L. Spargo, *Contemp. Org. Synth.* **1**, 113 (1993).
19. P. L. Spargo, *Contemp. Org. Synth.* **2**, 85 (1994).
20. S. P. Marsden, *Contemp. Org. Synth.* **3**, 133 (1995).
21. S. P. Marsden, *Contemp. Org. Synth.* **4**, 118 (1996).
22. S. D. Christie, *J. Chem. Soc. Perkin Trans.* **1**, 1577 (1998).
23. S. D. Christie, *J. Chem. Soc. Perkin Trans.* **1**, 737 (1999).
24. H. J. Harwood, *Chem. Rev.* **62**, 99 (1962).
25. N. De Kimpe and R. Verhe in S. Patai and Z. Rappoport, eds., *The Chemistry of Functional Groups*, Update Vol., John Wiley and Sons, Inc., New York, 1988.
26. Y. Sasson, M. Weiss, and G. Barak in D. Price, B. Iddon, and B. J. Wakefield, eds., *Bromine Compounds Chemistry and Applications*, Elsevier, 1988, pp. 252–271.
27. R. A. Sheldon and J. K. Kochi, *Org. Reactions* **19**, 279 (1972).
28. D. Crich, *Comp. Org. Synth.* **7**, 723 (1991).
29. S. M. Kelley and H. Schad, *Helv. Chim. Acta* **68**, 813 (1985).
30. O. S. Tee, *Ind. Chem. Libr.* **3** (*Adv. Organobromine Chem.* **1**), 99 (1991).
31. H. P. Braendlin and E. T. McBee in G. Olah (ed.), *Friedel–Crafts and Related Reactions*, Interscience Publishers, New York, Vol. 3, (1964), p. 1517.
32. A. M. Andrievskii, M. V. Gorelik, S. V. Avidon and E. S. Altman, *Russ. J. Org. Chem.* **29**, 1519 (1993).
33. G. A. Olah, W. C. Tolguesi, and R. E. Dear, *J. Org. Chem.* **31**, 1262 (1962).
34. U.S. Pat. 4,347,390 (1982), R. Nishiyama and co-workers (to Ishihara Sangyo Kaisha Ltd).
35. D. Ioffe, *Mendeleev Comm.* 16 (1993).

36. H. Zollinger, *Diazo Chemistry I*, VCH, Weinheim, 1994, pp. 230–235.
37. J. S. Pizey, *Synthetic Reagents* **2**, John Wiley and Sons, Inc., New York, 1974, pp. 1–63.
38. M. Zviely, J. Hermolin, and A. Kampf, *Ind. Chem. Libr.* **3** (*Adv. Organobromine Chem.* **1**), 171 (1991); GB 2,175,895 (Apr 01 1985, to Bromine Compounds Ltd.).
39. S. Kajigaeshi and T. Kakinami, *Ind. Chem. Libr.* **7** (*Adv. Organobromine Chem.* **2**, 29 (1995).
40. M. K. Chaudhuri and co-workers, *Tetrahedron Lett.* **39**, 8163 (1998).
41. J. Jacques and A. Marquet, *Org. Synth.* **VI**, 175 (1988).
42. A. Nechvatal, *Adv. Free-Radical Chem.* **4**, 175 (1972).
43. M. Zupan and N. Segatin, *Synthetic Comm.* **24**, 2617 (1994).
44. M. F. Ruasse, *Ind. Chem. Libr.* **7** (*Adv. Organobromine Chem.* **2**), 100 (1995).
45. G. Belluchi, C. Chiappe, and R. Bianchini, *Ind. Chem. Libr.* **7** (*Adv. Organobromine Chem.* **2**), 128 (1995).
46. F. W. Stacey and J. F. Harri, *Org. Reactions* **13**, 150 (1963).
47. L. S. Boguslavskaya, *Russ. Chem. Rev.* **41**, 740 (1972).
48. J. Rodriguez and J. P. Dulcere, *Synthesis* 1177 (1993).
49. GB Pat. 768,893 (Feb. 20, 1957, to Degussa).
50. U.S. Pat. 2,717,911 (1955), L. Hunter, and H. Veith (to Degussa).
51. N. A. Alekar and co-workers, *Indian J. Chem. Technol.* **7**, 79 (2000).
52. Ger. Offen 2,339,612 (Aug 08, 1973), D. Vofsi, M. Levy, S. Daren, and E. Cohen.
53. H. M. Bhavnagary and S. K. Maiumder, *Res. Ind.* **29**, 5 (1984).
54. JP-Kokai 74,108,003 (1974), K. Matsuda, M. Sigino, and S. Kaji (to Nippon Kayaku Co).
55. U.S. Pat. 3,182,088 (1965), H. E. Hennis (to Dow Chemical).
56. U.S. Pat. 5,0177,728 (May 21 1991), B. G. McKinnie and D. A. Wood (to Ethyl Corp).
57. C. Bell, N. Price, and B. Chakrabarti, eds., *The Methyl Bromide Issue. Vol. 1: Agrochemical and Plant Protection*, UK 1996.
58. J. Katan, *J. Plant. Pathol.* **81**, 153 (1999).
59. G. V. Alexeev and W. W. Kilgore, *Residue Rev.* **88**, 101 (1983).
60. R. Yang and co-workers, *Rev. Environ. Contam. Toxicol.* **142**, 65 (1995).
61. D. J. Guth and co-workers, *Inhalation Toxicol.* **6**, 327 (1994).
62. "Threshold Limit Values and Biological Exposure Indices for 1989-1990", American Conference of Governmental Industrial Hygienists, Cincinnati, Ohio, 1989.
63. A. L. Horvath, *Solubility Data, Ser.* **60**, 143 (1995).
64. Ger. Offen 727,690 (Oct. 8, 1942), O. Scherer, F. Dostal, and K. Dachlauer (to I. G. Farbenindustrie A.- G); U.S. Pat. 2,347,000 (Apr. 18, 1944, to Alien Property Custodian).
65. U.S. Pat. 2,553,518 (May 15, 1951), D. E. Lake and A. A. Asadorian (to Dow Chemical).
66. GB 874,062 (May 31, 1959) (to Soc. Chimica Dell'Aniene SpA).
67. DE 1,283,214 (1968), U. Giacopelli and M. Manca (to Solvay et Cie).
68. Fr. 1,441,233 (June 03, 1966), (to Shell International Research).
69. U.S. Pat. 3,923,914 (1975), P. Kobetz and K. L. Lindsay (to Ethyl Corp.).
70. U.S. Pat. 5,207,953 (Nov. 27, 1991), D. Thorssen and D. Loree, (to Trisol Inc).
71. Ref. 63, p. 146.
72. Eur. Pat. Appl. EP 995,733 (Apr. 26, 2000), N. Kornberg, T. Fishler, and S. Antebi, (to Bromine Compounds Ltd).
73. PCT Int. Appl. WO 9,813,396 (Apr. 02, 1998), M. Ao and co-workers (to Albermarle Corp.).
74. PCT Int. Appl. WO 9,504,409 (Nov. 12, 1998), B. Dalgar and co-workers (to Albermarle Corp.).

75. Ref. 63, p. 82.
76. A. Kergomard, *Bull. Soc. Chim.* 2360 (1961).
77. W. H. Hunter and D. E. Edgar, *J. Am. Chem. Soc.* **54**, 2025 (1932).
78. E. Abele and E. Lukevics, *Org. Prep. Proced. Int.* **31**, 359 (1999).
79. U.S. Pat. 2,746,999 (May 22, 1956), A. A. Gunkler, D. E. Lake, and B. C. Potts (to Dow Chemical Company).
80. Brit. Pats. 804,995 and 804,996 (Nov. 26, 1958), W. J. Read and co-workers (to Associated Ethyl Co.).
81. U.S. Pat. 2,921,967 (Jan. 19, 1960), F. Yaron (to Dead Sea Bromine Co.).
82. S. D. Humphreys, H. G. Huw, and P. A. Routledge, *Adverse Drug React. Toxicol. Rev.* **18**, 125 (1999).
83. W. Paulus, *Microbiocides for the Protection of Materials*, Chapman & Hall, London, 1993.

DAVID IOFFE
IMI (TAMI)
ARIEH KAMPF
DSBG

BUTADIENE

1. Introduction

Butadiene, C_4H_6, exists in two isomeric forms: 1,3-butadiene [106-99-0], $CH_2{=}CH{-}CH{=}CH_2$, and 1,2-butadiene [590-19-2], $CH_2{=}C{=}CH{-}CH_3$. 1,3-Butadiene is a commodity product of the petrochemical industry with a 2000 U.S. production of 4.4 billion pounds (2.0×10^9 kg) (1). Although this is not very different from the production in 1971, it represents significant rebound from the low production in the mid-1980s. Elastomers consume the bulk of 1,3-butadiene, led by the manufacture of styrene–butadiene rubber (SBR). 1,3-Butadiene is manufactured primarily as a coproduct of steam cracking to produce ethylene in the United States, Western Europe, and Japan. However, in certain parts of the world it is still produced from ethanol. The earlier manufacturing processes of dehydrogenation of n-butane and oxydehydrogenation of n-butenes have significantly declined in importance and output. Efforts have been made to make butadiene from other feedstocks such as other hydrocarbons, coal (2,3), shale oil (4); and renewable sources like animal and vegetable oil (5), cellulose, hemicellulose, and lignin (6,7), but in the United States none of these have moved beyond the research and development stage.

The other isomer, 1,2-butadiene, a small by-product in 1,3-butadiene production, has no significant current commercial interests. The production of 1,2-butadiene in purities of 85% has been described (8). However, there are a number of publications and patents on its recovery and applications, particularly in the specialty polymer area (9,10) and as a gel inhibitor (11).

2. Properties

1,3-Butadiene is a noncorrosive, colorless, flammable gas at room temperature and atmospheric pressure. It has a mildly aromatic odor. It is sparingly soluble in water, slightly soluble in methanol and ethanol, and soluble in organic solvents like diethyl ether, benzene, and carbon tetrachloride. Its important

Table 1. **Physical Properties of 1,3-Butadiene**[a]

Property	Value
CAS Registry Number	[106-99-0]
RTECS accession number	EI9275000
UN number	1010
molecular formula	C_4H_6
molecular weight	54.092
boiling point at 101.325 kPa[b], °C	−4.411
freezing point, °C	−108.902
critical temperature, °C	152.0
critical pressure, MPa[c]	4.32
critical volume, cm³/mol	221
critical density, g/mL	0.245
density (liquid), g/mL at	
0°C	0.6452
15°C	0.6274
20°C	0.6211
25°C	0.6194
50°C	0.5818
density (gas) (air = 1)	1.9
heat capacity at 25°C, J/(mol·K)[d]	79.538
refractive index, n_D at −25°C	1.4292
solubility in water at 25°C, ppm	735[e]
viscosity (liquid), mPa·s (=cP) at	
−40°C	0.33
0°C	0.25
40°C	0.20
heat of formation, gas, kJ/mol[d]	110.165
heat of formation, liquid, kJ/mol[d]	88.7
free energy of formation, kJ/mol[d]	150.66
heat of vaporization, J/g[d] at	
25°C	389
boiling point	418
flash point, °C	−85
autoignition temperature, °C	417.8
explosion limits in air, vol %	
lower	2.0
upper	11.5
minimum oxygen for combustion (MOC), %v/vO₂	
N₂−air	10
CO₂−air	13
absorption	
λ, cm⁻¹	217
log ε	4.32

[a]Refs. 23–28.
[b]To convert kPa to mm Hg, multiply by 7.5.
[c]To convert MPa to psi, multiply by 145.
[d]To convert J to cal, divide by 4.184.
[e]245 mol ppm.

Table 2. **Physical Properties of 1,2-Butadiene**[a]

Property	Value
CAS Registry Number	[590-19-2]
molecular formula	C_4H_6
molecular weight	54.092
boiling point at 101.325 kPa,[b] °C	10.85
freezing point, °C	−136.19
density (liquid), g/mL at	
0°C	0.676
20°C	0.652
heat of formation at 25°C (gas), kJ/mol[c]	162.21
heat of vaporization at 25°C, kJ/mol[c]	23.426
refractive index at 1.3°C	1.4205

[a]Refs. 24,29.
[b]To convert kPa to mm Hg, multiply by 7.5.
[c]To convert kJ to kcal, divide by 4.184.

physical properties are summarized in Table 1 (see also Ref. 12,13). 1,2-Butadiene is much less studied. It is a flammable gas at ambient conditions. Some of its properties are summarized in Table 2.

1,3-Butadiene, the simplest conjugated diene, has been the subject of intensive theoretical and experimental studies to understand its physical and chemical properties. The conjugation of the double bonds makes it 15 kJ/mol (3.6 kcal/mol) (14) more thermodynamically stable than a molecule with two isolated single bonds. The s-trans isomer, often called the trans form, is more stable than the s-cis form at room temperature. Although there is a 20 kJ/mol (4.8 kcal/mol) rotational barrier (15,16), rapid equilibrium allows reactions to take place with either the s-cis or s-trans form (17,18).

s-cis   s-trans

The double-bond length in 1,3-butadiene is 0.134 nm, and the single-bond, 0.148 nm. Since normal carbon–carbon single bonds are 0.154 nm, this indicates the extent of double-bond character in the middle single bond. Upon complexing with metal carbonyl moieties like $Fe(CO)_3$, the two terminal bonds lengthen to 0.141 nm, and the middle bond shortens even more to 0.145 nm (19).

Solubilities of 1,3-butadiene and many other organic compounds in water have been extensively studied to gauge the impact of discharge of these materials into aquatic systems. Estimates have been advanced by using the UNIFAC derived method (20,21). Similarly, a mathematical model has been developed to calculate the vapor–liquid equilibrium (VLE) for 1,3-butadiene in the presence of steam (8).

3. Reactions

Since the discovery of 1,3-butadiene in the nineteenth century, it has grown into an extremely versatile and important industrial chemical (30). Its conjugated double bonds allow a large number of unique reactions at both the 1,2- and 1,4-positions. Many of these reactions produce large volumes of important industrial materials.

3.1. Addition Reactions.   1,3-Butadiene reacts readily via 1,2- and 1,4- free radical or electrophilic addition reactions (31) to produce 1-butene or 2-butene substituted products, respectively.

$$Y{-}X \; + \; \underset{CH_2 \quad CH_2}{CH{-}CH} \; \longrightarrow \; \underset{Y{-}CH_2 \quad CH_2}{\overset{X}{CH{-}CH}} \qquad \text{1,2-addition product}$$

$$Y{-}X \; + \; \underset{CH_2 \quad CH_2}{CH{-}CH} \; \longrightarrow \; \underset{Y{-}CH_2 \qquad CH_2\text{-}X}{CH{=}CH} \qquad \text{1,4-addition product}$$

The intermediate in these reactions in the case of the addition of YX is consistent with the addition of Y to the 1-position to form an allylic intermediate to which X adds to produce either the 1,2- or 1,4-product.

$$Y{-}X \; + \; \underset{CH_2 \quad CH_2}{CH{-}CH} \; \longrightarrow \; \underset{Y{-}CH_2 \; \underset{\text{or}}{+} \; CH_2}{\overset{\overset{1,2}{\frown}\; X^- \;(\text{or } X{\cdot})}{\underset{1,4}{CH{-}CH}}}$$

The addition of HX, where X is a halogen, has been thoroughly investigated (32,33). Whether 1,2- or 1,4-product dominates depends on reaction conditions. For example, although HCl adds to butadiene at low temperatures to produce 75–80% of the 1,2-addition product, the thermodynamically more stable 1,4-isomer is favored at higher temperatures (34). On the other hand, HI has been shown to add to butadiene in the vapor phase by a pericyclic mechanism to produce the 1,4-product (35).

$$\underset{\underset{CH{-}CH}{\diagdown}}{\overset{\overset{H{-}I}{\diagup}}{CH_2 \quad CH_2}} \; \longrightarrow \; \underset{CH{=}CH}{CH_3 \qquad CH_2{-}I}$$

Addition of water (36) or alcohols (37–39) directly to butadiene at 40–100°C produces the corresponding unsaturated alcohols or ethers. Acidic ion exchangers have been used to catalyze these reactions. The yields for these latter reactions are generally very low because of unfavorable thermodynamics. At 50°C addition of acetic acid to butadiene produces the expected butenyl acetate with 60–100% selectivity at butadiene conversions of 50%. The catalysts are ion-exchange resins modified with quaternary ammonium, quaternary phosphonium, and ammonium substituted ferrocenyl ions (40). Addition of amines yields unsaturated alkyl amines. The reaction can be catalyzed by homogeneous catalysts such as $Rh[P(C_6H_5)_3]_3Cl$ (41) or heterogeneous catalysts such as MgO and other solid bases (42).

The manufacture of hexamethylenediamine [124-09-4], a key comonomer in nylon-6,6 production proceeds by a two-step HCN addition reaction to produce adiponitrile [111-69-3], $NCCH_2CH_2CH_2CH_2CN$. The adiponitrile is then hydrogenated to produce the desired diamine. The other half of nylon-6,6, adipic acid (qv), can also be produced from butadiene by means of either of two similar routes involving the addition of CO. Reaction between the diamine and adipic acid [124-04-5] produces nylon-6,6.

The first CO route to make adipic acid is a BASF process employing CO and methanol in a two-step process producing dimethyl adipate [627-93-0], which is then hydrolyzed to the acid (43–46). Cobalt carbonyl catalysts such as $Co_2(CO)_8$ are used. Palladium catalysts can be used to effect the same reactions at lower pressures (47–49).

The other CO route for adipic acid manufacture involves 1,4-addition of CO and O_2 to butadiene to produce an intermediate, which is subsequently hydrogenated and hydrolyzed to adipic acid (50). This is called the oxycarbonylation process. Both the BASF and the oxycarbonylation processes have been intensively investigated.

Halogenation of butadiene has also attracted a lot of interest. Both 1,2- and 1,4-isomers are formed. Since the *trans*-1,4-isomer was observed from the 1,4-addition product, researchers postulate that the electrophilic X^+ forms a 1,2-cyclic intermediate and not a 1,4-cyclic intermediate that would form the *cis*-1,4-addition product (51,52).

$$H_2C-CH-CH=CH_2$$
$$\underset{X}{\diagdown\diagup}$$
$$+$$

Fluorination with XeF_2 or $C_6H_5IF_2$ gives both the 1,2- and 1,4-difluoro products. This reaction proceeds via the initial electrophilic addition of F^+ to the diene (53).

Chloroprene (qv), 2-chloro-1,3-butadiene, [126-99-8] is produced commercially from butadiene in a three-step process. Butadiene is first chlorinated at 300°C to a 60:40 mixture of the 1,2- and 1,4-dichlorobutene isomers. This mixture is isomerized to the 3,4-dichloro-1-butene with the aid of a $Cu-Cu_2Cl_2$ catalyst followed by dehydrochlorination with base such as NaOH (54).

The 1,4-dichloro-2-butene can also be separated and hydrolyzed with aqueous NaOH to form 1,4-butenediol, which is hydrogenated with Ni catalyst to produce 1,4-butanediol. In 1971, this process was commercialized in Japan (55). The plant is now shut down because of unfavorable economics.

Butadiene also undergoes a 1,4-addition reaction with SO_2 to give sulfolene [77-79-2]. This reaction followed by hydrogenation is commercially used to manufacture sulfolane [126-33-0] (56).

$$\underset{\text{CH}_2}{\overset{\text{CH}_2}{\underset{|}{\overset{|}{\text{CH}-\text{CH}}}}} \quad + \quad SO_2 \quad \longrightarrow \quad \bigcirc SO_2 \quad \overset{H_2}{\longrightarrow} \quad \bigcirc SO_2$$

Formaldehyde also reacts with butadiene via the Prins reaction to produce pentenediols or their derivatives. This reaction is catalyzed by a copper-containing catalyst in a carboxylic acid solution (57) or $RuCl_3$ (58). The addition of hydrogen also proceeds via 1,2- and 1,4-addition.

3.2. Hydrogenation Reactions.   Butadiene can be hydrogenated to n-butanes and n-butane using a large number of heterogeneous (59) and homogeneous (60–64) catalysts. Palladium-containing membranes have also been used to allow the use of permeated hydrogen to effect hydrogenation (65–67). Many catalysts have been developed and used commercially to remove small quantities ($\leq 3\%$) of butadiene from 1-butene streams (68–71). Since 2-butene [107-01-7] is more stable thermodynamically than 1-butene [106-98-9] under mild conditions, catalysts that promote 1,2-addition and do not isomerize 1-butene are essential for getting high 1-butene selectivity. Many of the palladium catalysts require the use of CO to improve 1-butene selectivity (72–74).

Selectivities to various isomers are more difficult to predict when metal oxides are used as catalysts. Zinc oxide (ZnO) preferentially produced 79% 1-butene and several percent of cis-2-butene [624-64-6] (75). Cadmium oxide (CdO) catalyst produced 55% 1-butene and 45% cis-2-butene. It was also reported that while interconversion between 1-butene and cis-2-butene was quite facile on CdO, cis–trans isomerization was slow. This finding was attributed to the presence of a π-allyl anion intermediate (76). High cis-2-butene selectivities were obtained with molybdenum carbonyl encapsulated in zeolites (77). On the other hand, deuteration using ThO_2 catalyst produced predominantly the 1,4-addition product, $trans$-2-butene-d_2 with no isotope scrambling (78).

$$\begin{array}{c} \qquad\qquad CH_2D \\ \qquad\qquad / \\ CH{=}CH \\ / \\ DCH_2 \end{array}$$

Although supported Pd catalysts have been the most extensively studied for butadiene hydrogenation, a number of other catalysts have also been the object of research studies. Some examples are Pd film catalysts, molybdenum sulfide, metal catalysts containing Fe, Co, Ni, Ru, Rh, Os, Ir, Pt, Cu, MgO, $HCo(CN)_5^{3-}$ on supports, and $LaCoC_3$ Perovskite. There are many others (79–85). Studies on the well-characterized Mo(II) monomer and Mo(II) dimer on silica carrier catalysts have shown wide variations not only in catalyst performance, but also of activation energies (86).

Another method to hydrogenate butadiene occurs during an oxidation–reduction reaction in which an alcohol is oxidized and butadiene is reduced. Thus copper–chromia or copper–zinc oxide catalyzes the transfer of hydrogen from 2-butanol or 2-propanol to butadiene at 90–130°C (87,88).

3.3. Oxidation Reactions.   Like all reactions between oxygen and hydrocarbons, complete oxidation of butadiene is controlled by limiting the oxygen and operating at specific temperature ranges. Other ways to control selectivity to specific products involve the use of catalysts and/or conducting the reaction in the presence of other reagents. Some of the many oxidized products are depicted in Figure 1.

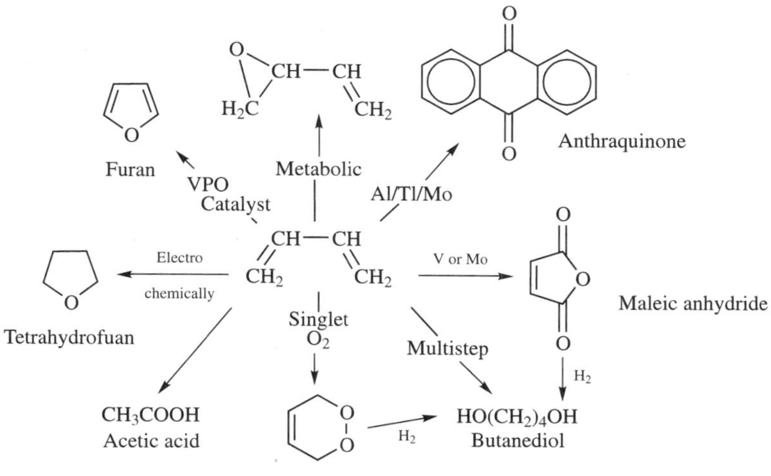

Fig. 1. Oxidative reactions of butadiene.

The vapor-phase oxidation (VPO) of butadiene with air at 200–500°C produces maleic anhydride [108-31-6]. Catalysts used are based on vanadium and molybdenum oxides (89,90). Alternatively, when using a catalyst containing Al, Mo, and Ti, butadiene undergoes a complex series of condensations and oxidations to form anthraquinone at 250°C (91).

Reaction between oxygen and butadiene in the liquid phase produces polymeric peroxides that can be explosive and shock-sensitive when concentrated. Both Ir(I) and Rh(I) complexes have been shown to catalyze this polymerization at 55°C (92). These peroxides, which are formed via 1,2- and 1,4-addition, can be hydrogenated to produce the corresponding 1,2- or 1,4-butanediol [110-63-4] (93). Butadiene can also react with singlet oxygen in a Diels–Alder type reaction to produce a cyclic peroxide that can be hydrogenated to 1,4-butanediol.

Oxygen has also been shown to insert into butadiene over a VPO catalyst, producing furan [110-00-9] (94). Under electrochemical conditions butadiene and oxygen react at 100°C and 0.3 amp and 0.43 V producing tetrahydrofuran (THF) [109-99-9]. The selectivity to THF was 90% at 18% conversion (95) and it can also be made via direct catalytic oxidation of butadiene with oxygen. Active catalysts are based on Pd in conjunction with polyacids (96), Se, Te, and Sb compounds in the presence of Cu_2Cl_2, $LiCl_2$ (97), or Bi–Mo (98).

The oxidation reaction between butadiene and oxygen and water in the presence of CO_2 or SO_2 produces 1,4-butenediol. The catalysts consist of iron acetylacetonate and LiOH (99). The same reaction was also observed at 90°C with groups 8–10 (VIII) transition metals such as Pd in the presence of I_2 or iodides (100). The butenediol can then be hydrogenated to butanediol [110-63-4]. In the presence of copper compounds and at pH 2, hydrogenation leads to furan (101).

Alternatively, butadiene can be oxidized in the presence of acetic acid to produce butenediol diacetate, a precursor to butanediol. The latter process has been commercialized (102–104). This reaction is performed in the liquid phase at 80°C with a Pd–Te–C catalyst. A different catalyst system based on

$PdCl_2(NCC_6H_5)_2$ has been reported (105). Copper- (106) and rhodium- (107) based catalysts have also been studied.

$$CH_2{=}CH{-}CH{=}CH_2 + 2\ CH_3COOH + O_2 \longrightarrow CH_3COOCH_2$$

$$CH{=}CHCH_2OOCCH_3 \longrightarrow H_2\ CH_3COOCH_2CH_2CH_2CH_2OOCCH_3\ ^{H_2O}$$

$$HOCH_2CH_2CH_2CH_2OH + 2\ CH_3COOH$$

Another butadiene oxidation process to produce butanediol is based on the 1,4-addition of *tert*-butyl hydroperoxide to butadiene (108). Cobalt on silica catalyzes the first step. This is followed by hydrogenation of the resulting olefinic diperoxide to produce butanediol and *tert*-butyl alcohol.

Butadiene can also be readily epoxidized with peracids to the monoepoxide or the diepoxide (109,110). These have been proposed as important intermediates in the metabolic cycle of butadiene in the human body (111).

3.4. Diels–Alder Reactions.   The important dimerization between 1,3-dienes and a wide variety of dienophiles to produce cyclohexene derivatives was discovered in 1928 by Otto Diels and Kurt Alder. In 1950, they won the Nobel Prize for their pioneering work. Butadiene has to be in the *s*-cis form in order to participate in these concerted reactions. Typical examples of reaction products from the reaction between butadiene and maleic anhydride (1), or cyclopentadiene (2), or itself (3), are *cis*-1,2,3,6-tetrahydrophthalic anhydride [27813-21-4], 5-vinyl-2-norbornene [3048-64-4], and 4-vinyl-1-cyclohexene [100-40-3], respectively.

(1)   (2)   (3)

Diels–Alder reactions with butadiene are generally thermally reversible and can proceed in both gas and liquid phases. The reactions are exothermic and follow second-order kinetics; first-order with respect to each reactant.

The dienophiles for reaction with butadiene can be alkenes, allenes, and alkynes. Simple alkenes like ethylene are poor dienophiles resulting in sluggish reactions. Substituted olefins, X—C=C—X′, are more reactive when X and/or X′ are C=C, Ar, COOR, COOH, COH, COR, COCl, CN, halogens, and many other electron-withdrawing substituents (112–116). A compilation of the reaction parameters between butadiene and C2–C4 olefins in the temperature range of 510–750°C has been published (117). Other double-bond or triple-bond compounds, such as C=N—, —N=N—, O=C—, —C≡N, and O_2 can also act as dienophiles to give heterocyclic products. These types of concerted reactions have been the subject of extensive orbital symmetry studies (118,119).

Diels–Alder reactions are thermal reactions requiring no catalysts (120). However, over the years both acid- and metal-based homogeneous or heterogeneous catalysts have been developed (121–127). Some catalysts used in

Diels–Alder catalyzed reactions of butadiene are $Fe(NO)_2Cl–(CH_3CH_2)_2AlCl$, $Pd[P(C_6H_5)_3]_4$, Cu(I) exchanged silica–alumina (128,129), large pore zeolites (130), and carbon molecular sieves. An electrochemical process has also been used to catalyze the self-condensation to vinylcyclohexene (131). When the asymmetric Ni catalyst (**4**) was used, specificity to the enantomeric (S)-4-vinylcyclohexene (132,133) was observed (26% enantiomeric excess).

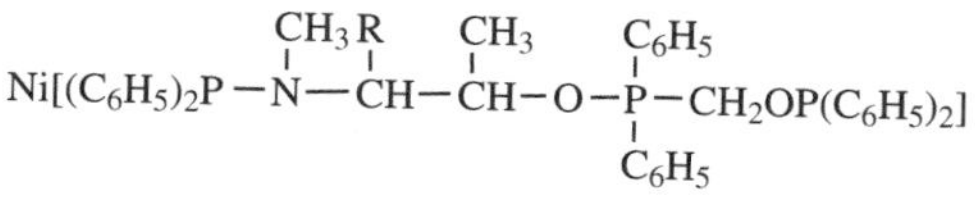

$$Ni[(C_6H_5)_2P-\overset{\overset{\displaystyle CH_3R}{|}}{N}-\overset{\overset{\displaystyle CH_3}{|}}{CH}-\overset{\overset{\displaystyle CH_3}{|}}{CH}-O-\overset{\overset{\displaystyle C_6H_5}{|}}{\underset{\underset{\displaystyle C_6H_5}{|}}{P}}-CH_2OP(C_6H_5)_2]$$

(**4**)

When the Diels–Alder reaction between butadiene and itself is carried out in the presence of alkali metal hydroxide or carbonate (such as KOH, Na_2CO_3, and K_2CO_3 on alumina or magnesia supports) dehydrogenation of the product, vinylcyclohexene, to ethylbenzene can occur at the same time (134). The same reaction can take place on simple metal oxides like ZrO_2, MgO, CaO, SrO, and BaO (135).

The Diels–Alder reaction between 2 mol of butadiene and 1 mol of quinone [106-51-4] produces tetrahydroanthraquinone [28758-94-3] (136).

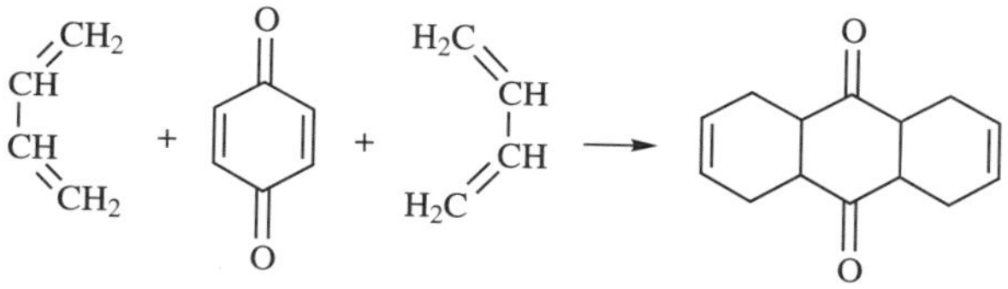

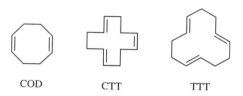

3.5. Dimerization and Oligomerization Reactions.
Besides Diels–Alder type dimerization reactions, butadiene undergoes a number of other dimerization or oligomerization reactions to produce cyclic or linear products. With the proper catalysts these reactions proceed quite selectivity. Noncatalyzed or photo-catalyzed dimerizations produce compounds like divinylcyclobutanes and have been studied in detail (137,138).

A fascinating series of cyclodimerization or cyclotrimerization reactions was first observed in the labs of Wilke to produce 1,5-cyclooctadiene [111-78-4] (COD) and *cis,trans,trans*-1,5,9-cyclododecatriene [2765-29-9] (CTT), or *trans,-trans, -trans*-1,5,9-cyclododecatriene [676-22-2] (TTT).

COD   CTT   TTT

These cyclodimerization and cyclotrimerization reactions are catalyzed by low valent Ziegler-type Ni catalysts (139–144). Large ligands, such as tris-*o*-biphenylyl phosphite on nickel tend to favor COD formation while smaller ligands favor the linear dimer, 1,3,7-octatriene. The dimer yield at 80°C and 101.3 kPa (1 atm) is 96%. The nickel catalyst can also be placed on a support so that it can be recycled (145). Many other type catalysts have been reported for this reaction (146). The linear 1,3,7-octatriene and its 1,3,6 isomer are also obtained by a Pd catalyzed dimerization (147–151). The kinetics of thermally induced dimerization to cod has also been studied (152).

One of the butadiene dimerization products, cod, is commercially manufactured and used as an intermediate in a process called FEAST to produce linear α,φ-dienes (153). 1,5-cyclooctadiene or cyclooctene [931-87-3], obtained from partial hydrogenation, is metathesized with ethylene to produce 1,5-hexadiene [592-42-7] or 1,9-decadiene [1647-16-1], respectively. Many variations to make other diolefins have been demonstrated. Huls AG also metathesized cyclooctene with itself to produce an elastomer useful in rubber blending (154). The cyclic *cis, -trans,trans*-triene described above can be hydrogenated and oxidized to manufacture dodecanedioic acid [693-23-2]. The product was used in the past for the production of the specialty nylon-6,12, Qiana (155,156).

The trimerization to produce *cis,trans,trans*-1,5,9-cyclododecatriene has also been practiced commercially using a Ziegler–Natta catalyst $TiCl_4$–Al_2 $(C_2H_5)_3Cl_3$ (157).

Linear dimerization and oligomerization of butadiene can be achieved by using a number of catalyst systems based on Pd, Ni (158–161), and Fe (162). 1,7-Octadiene can be obtained selectively when the dimerization is carried out in the presence of a reducing agent such as formic acid (163–165) or H_2/CO_2 (166).

Ziegler-type catalysts based upon Co, Ni, and Fe and in the presence of aluminum alkyls codimerize butadiene with olefins such as ethylene, producing 1,4-hexadiene derivatives. A rhodium-based catalyst can be used to produce all the *trans*-1,4-hexadiene used as the termonomer in the commercial EPDM (155,156) (see ELASTOMERS, SYNTHETIC; ETHYLENE–PROPYLENE RUBBER). In contrast, cobalt and iron catalysts are known to give the cis-isomer.

3.6. Telomerization Reactions.

Butadiene can react readily with a number of chain-transfer agents to undergo telomerization reactions. The more often studied reagents are carbon dioxide (167–178), water (179–181), ammonia (182), alcohols (183–185), amines (186), acetic acid (187), water and CO_2 (188), ammonia and CO_2 (189), epoxide and CO_2 (190), mercaptans (191), and other systems (171). These reactions have been widely studied and used in making unsaturated lactones, alcohols, amines, ethers, esters, and many other compounds.

Reaction between butadiene and CO_2 has been extensively studied (171) since the reaction was first demonstrated (167–170). This reaction has been shown to be catalyzed by Pd (172,173), Ni (174), Ru (175), Pt (178), and Rh (172,173) catalysts. Products include gamma (**5**) and delta lactones (**6**), acids (**7,8**), and esters (**9**). Mechanistic studies have shown that butadiene initially forms a dimer (Pd, Ru, Ni) or trimer (Rh) intermediate followed by CO_2 insertion

(171). The fate of these intermediates depends on the metal, the ligands, and the reaction conditions.

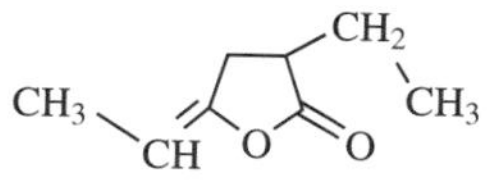

(5) 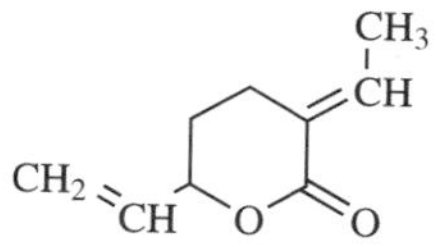   (6) 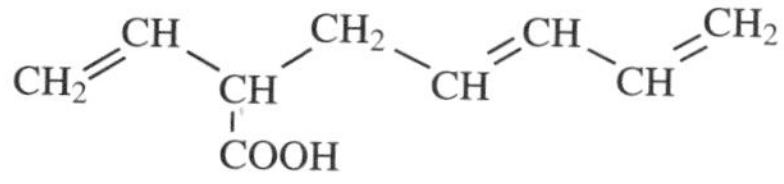   (7)

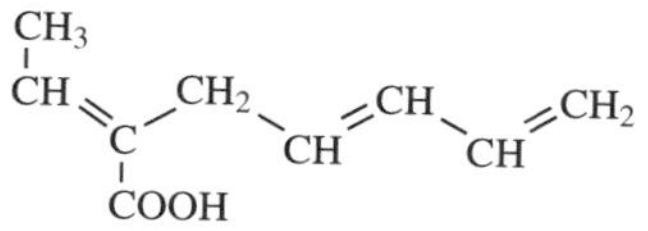

(8) 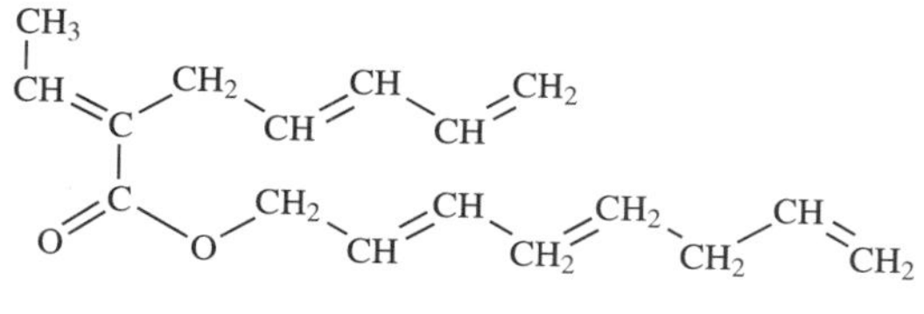   (9)

The delta lactone can be obtained in very high yields when triisopropylphosphine or tricyclohexylphosphine is the ligand along with Pd(acac)$_2$ (acac = acetylacetoacetonate) as the metal source (171).

Coupling of butadiene with CO_2 under electrochemically reducing conditions produces decadienedioic acid, and pentenoic acid, as well as hexenedioic acid (192). A review article on diene telomerization reactions catalyzed by transition metal catalysts has been published (193).

3.7. Polymerization Reactions.

The polymerization of butadiene with itself and with other monomers represents its largest commercial use. The commercially most important polymers are SBR, polybutadiene (BR), styrene–butadiene latex (SBL), acrylonitrile–butadiene–styrene polymer (ABS), and nitrile rubber (NR). The reaction mechanisms are free-radical, anionic, cationic, or coordinate, depending on the nature of the initiators or catalysts (194–196).

Different grades of SBR copolymers are prepared by either free-radical initiated emulsion polymerization or anionic solution polymerization. The technology was developed during World War II to produce a substitute for natural rubber. The properties of these SBRs are so good that natural rubber has never again regained its importance. Today, SBR represents the single largest use of butadiene (see ELASTOMERS, SYNTHETIC, POLYBUTADIENE; STYRENE–BUTADIENE RUBBER).

The original SBR process is carried out at ~50°C and is referred to as hot polymerization. It accounts for only ~5% of all the rubber produced today. The dominant cold polymerization technology today employs more active initiators to effect polymerization at about 5°C. It accounts for ~ 85% of the products manufactured. Typical emulsion polymerization processes incorporate about 75% butadiene. The initiators are based on persulfate in conjunction with mercaptans (197), or organic hydroperoxide in conjunction with ferrous ion (198). The rest of SBR is produced by anionic solution polymerization. The density of unvulcanized SBR is 0.933 (199). The T_g ranges from −59 to −64°C (199).

Homopolymerization of butadiene can proceed via 1,2- or 1,4-additions. The 1,4-addition produces the geometrically distinguishable trans or cis structures with internal double bonds on the polymer chains. 1,2-Addition, on the other

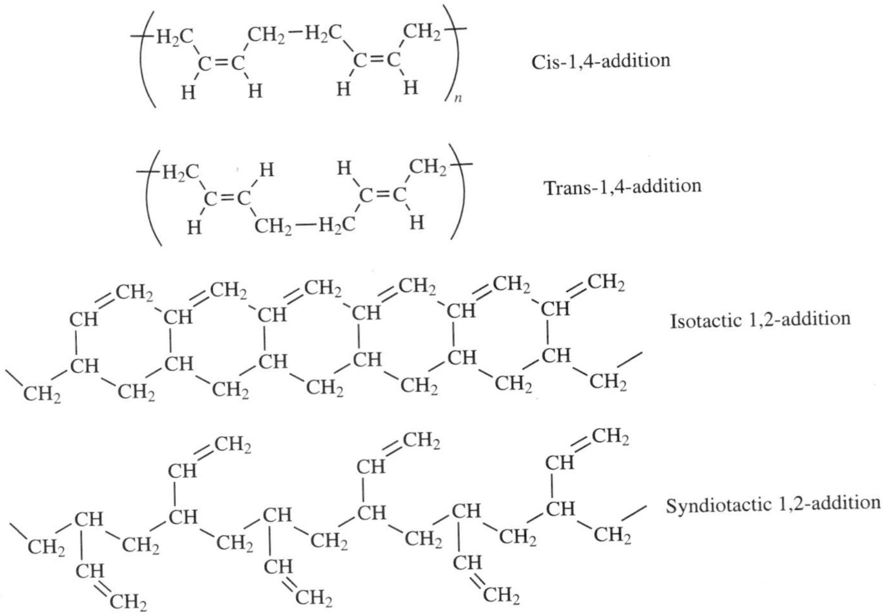

Fig. 2. Modes of addition of butadiene.

hand, yields either atactic, isotactic, or syndiotactic polymer structures with pendent vinyl groups (Fig. 2). Commercial production of these polymers started in 1960 in the United States. Firestone and Goodyear account for >60% of the current production capacity (see ELASTOMERS, SYNTHETIC–POLYBUTADIENE).

Very high cis-1,4-addition (≥80%) imparts more desirable properties for applications like heavy-duty tires (200). Structural properties of some typical polybutadienes are listed in Table 3 (200–202).

Extensive efforts have been made to develop catalyst systems to control the stereochemistry, addition site, and other properties of the final polymers. Among the most prominent ones are transition metal-based catalysts including Ziegler or Ziegler–Natta type catalysts. The metals most frequently studied are Ti (203,204), Mo (205), Co (206–208), Cr (206–208), Ni (209,210), V (205), Nd (211–215), and other lanthanides (216). Of these, Ti, Co, and Ni complexes

Table 3. Properties of Polybutadiene

Type[a]	Unit cell	Density	Melting point, °C	T_g, °C
1,4-cis	monoclinic	1.01	2	−106
1,4-trans-	hexagonal[b]			−107
modification I		0.97	97	
modification II		0.93	145	
1,2-syndiotactic	rhombic	0.96	154	−28
1,2-isotactic	rhombohedral	0.96	120	

[a]See Figure 2.
[b]Up to 60°C.

have been used commercially. It has long been recognized that by varying the catalyst compositions, the trans/cis ratio for 1,4-additions can be controlled quite selectively (204). Catalysts have also been developed to control the ratio of 1,4- to 1,2-additions within the polymers (203).

In situ preparation of polymer blends of 1,4-polybutadiene with polystyrene, or poly(1-butene) has been achieved by using the heterogeneous Ziegler–Natta type catalyst $(C_2H_5)_2AlCl–Ti(OC_4H_9)_4$ in the host polymers (217). Homogeneous catalysts can also be used to catalyze these reactions (218).

Anionic polymerization of butadiene has been intensively investigated over the years. Alkali metals and their alkyl derivatives are most frequently used. The process employing alkyllithium compounds as the initiators was first commercialized in the 1950s. Typical vinyl (1,2-addition) content is ~10–25%, with the balance about evenly divided between cis- and trans-1,4-addition, depending on the alkyllithium catalyst selected (219). The vinyl content can be increased substantially by the addition of polar compounds such as ethers or amines to the reaction mixture (202,220,221). The most common catalyst used commercially appears to be butyllithium. The products are mostly amorphous and provide desirable vulcanized rubber properties (202). By taking advantage of the living polymer nature of the reaction, products containing various end groups, such as —OH, —COOH, —SH, and others, can be prepared by terminating the polymerization with the proper reagent (195).

Another better studied system is the Alfin (alkoxide–olefin) catalyst, which is composed of a sodium salt, sodium alkoxide, and allylsodium (222). Similarly, there are many different modifications of the system to produce polymers with different 1,2- to 1,4-addition ratios as well as other properties (223).

Acrylonitrile–butadiene copolymers (nitrile–butadiene rubber, NBR) are also produced via emulsion polymerization of butadiene with acrylonitrile, $CH_2=CH—CN$ [107-13-1].

$$CH_2=CH-CH=CH_2 + CH_2-CH-CN \longrightarrow \sim\!\!\sim CH_2-CH=CH-CH_2-CH_2-CH\!\!\sim\!\!\sim$$
$$\underset{CN}{|}$$

Acrylonitrile–butadiene–styrene resins are two-phase blends. These are prepared by emulsion polymerization or suspension grafting polymerization. Products from the former process contain 20–22% butadiene; those from the latter, 12–16%.

Butadiene can also be copolymerized with a large number of other olefins (224) and SO_2 (225).

3.8. Other Reactions. Due to the highly reactive conjugated double bonds, butadiene can undergo many reactions with transition metals to form organometallic complexes. For example, iron pentacarbonyl reacts with butadiene to produce the tricarbonyl iron complex (**10**) (226). This and many other organometallic complexes have been covered (227).

$$\begin{array}{c} CH-CH \\ CH_2 \quad CH_2 \end{array} + Fe(CO)_5 \longrightarrow \underset{CO}{\overset{Fe}{\underset{CO}{CO}}} + 2\,CO$$

(**10**)

Side-chain anionic alkylation reactions with aromatic compounds take place when catalyzed with strong basic catalysts, like Na−K (228). The yield is 83% when *o*-xylene reacts with butadiene.

4. Manufacture and Processing

The pattern of commercial production of 1,3-butadiene parallels the overall development of the petrochemical industry. Since its discovery via pyrolysis of various organic materials, butadiene has been manufactured from acetylene as well as ethanol, both via butanediols (1,3- and 1,4-) as intermediates (see ACETYLENE-DERIVED CHEMICALS). On a global basis, the importance of these processes has decreased substantially because of the increasing production of butadiene from petroleum sources. China and India still convert ethanol to butadiene using the two-step process while Poland and the former USSR use a one-step process (229,230). In the past butadiene also was produced by the dehydrogenation of *n*-butane and oxydehydrogenation of *n*-butenes. However, butadiene is now primarily produced as a by-product in the steam cracking of hydrocarbon streams to produce ethylene. Except under market dislocation situations, butadiene is almost exclusively manufactured by this process in the United States, Western Europe, and Japan.

4.1. Steam Cracking.   Steam cracking is a complex, highly endothermic pyrolysis reaction. During the reaction a hydrocarbon feedstock is heated to $\sim 800°C$ and 34 kPa (5 psi) for less than a second during which carbon−carbon and carbon−hydrogen bonds are broken. As a result, a mixture of olefins, aromatics, tar, and gases are formed. These products are cooled and separated into specific boiling range cuts of C_1, C_2, C_3, C_4, etc. The C_4 fraction contains butadiene, isobutylene, 1-butene, 2-butene, and some other minor hydrocarbons. The overall yields of butadiene depend on both process parameters (231) and the composition of feedstocks (Table 4) (232). Generally, heavier steam cracking feedstocks produce greater amounts of butadiene as a by-product. Thus, with heavier feedstocks like light naphtha or virgin gas oil, up to about 5.4 wt% of the total product is butadiene. The processes of separating butadiene from the other C_4 compounds are described later.

4.2. Dehydrogenation of *n*-Butane.   Dehydrogenation of *n*-butane [106-97-8] via the Houdry process is carried out under partial vacuum, 35−75 kPa (5−11 psi), at $\sim 535–650°C$ with a fixed-bed catalyst. The catalyst consists of aluminum oxide and chromium oxide as the principal components. The reaction is endothermic and the cycle life of the catalyst is ~ 10 min because of coke buildup.

Table 4. **Product Distribution from Steam Cracking Various Feedstocks**[a]

Feedstock	Product yield, wt %					
	Ethylene	Propylene	Butadiene	Butenes	BTX	Fuel products
ethane	77.5	2.8	1.9	0.8		17.0
propane	42.0	16.8	3.0	1.3	3.0	33.9
light naphtha	33.7	15.6	4.5	4.2	9.1	32.9
light VGO[b]	20.4	14.1	5.4	6.3	8.5	45.3

[a]Ref. 232.
[b]VGO = vacuum gas oil.

Several parallel reactors are needed in the plant to allow for continuous operation with catalyst regeneration. Thermodynamics limits the conversion to ~30–40% and the ultimate yield is 60–65 wt% (233).

4.3. Oxydehydrogenation of *n*-Butenes. Normal butenes can be oxidatively dehydrogenated to butadiene in the presence of high concentration of steam with fairly high selectivity (234). The conversion is no longer limited by thermodynamics because of the oxidation of hydrogen to water. Reaction temperature is below ~600°C to minimize over oxidation. Pressure is ~34–103 kPa (5–15 psi).

4.4. Separation and Purification. Separation and purification of butadiene from other components is dominated commercially by the extractive distillation process. The most commonly used solvents are acetonitrile and dimethylformamide. Dimethylacetamide, furfural, and N-methyl-2-pyrrolidinone also accomplish the separation. These solvents are aprotic polar compounds that have high complexing affinity toward the more polarizable butadiene than other olefins in the streams. Among the many factors that must be considered in choosing the solvent process are cost, solvency, thermal stability, viscosity, toxicity, corrosivity, heat of vaporization, and fouling. The fact that no single solvent has dominated the process suggests that there are no significant differences in both operation and economics using most of these solvents. 1,3-Butadiene separation from a crude C_4 stream using catalytic extractive distillation has been described (235). There are continuous efforts around the world to search for solvents (236,237) that reduce fouling (238,239) and improve operations (240,241). A typical process schematic is shown in Figure 3. Processes using membranes for butadiene separation have also been patented (242,243). A system for purifying a 1,3-butadiene feedstock by using a separation column has been patented (244).

Another approach to separate butadiene from other hydrocarbons is to use a solution containing cuprous ammonium acetate that forms a weak copper(I) complex with butadiene (245,246). The latter process has been used in a number of plants.

In commercial extraction operations, the C_4 fractions that contain butadiene, isobutene, and 1- and 2-butenes usually first go through a butadiene extraction unit in which the butadiene is removed. This may be followed by isobutylene removal via reaction between isobutylene and methanol to form methyl *tert*-butyl ether [1634-04-4] (MTBE). The butenes are then distilled from the MTBE. 1-Butene may then be separated from 2-butene by distillation.

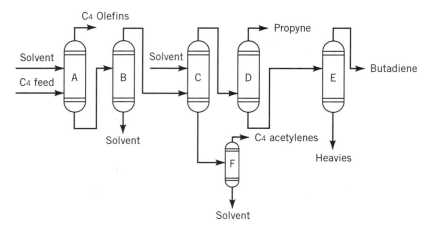

Fig. 3. Separation and purification of butadiene: A, first extraction distillation tower; B, solvent stripper; C, second extraction distillation tower; D, topping tower; E, tailing tower; F, solvent recovery towers.

5. Handling, Storage, and Shipping

Large quantities of butadiene are manufactured, stored, transported, and handled in a safe manner every day. Typical product specifications are listed in Table 5. However, butadiene reacts with a large number of chemicals, has an inherent tendency to dimerize and polymerize, and is toxic. Therefore, specific handling, storage, and shipping procedures must be followed. A review of means to prevent occupational exposure during handling, storage, and shipping has been published (247).

Table 5. **Butadiene Specifications and Test Methods**[a]

Component	Specification	Test method
1,3-butadiene	≥99.5%	ASTM D2593
inhibitor[b], wppm	50–150	ASTM D1157
impurities, wppm, max		
1,2-butadiene	20.0	ASTM D2593
propadiene	10.0	ASTM D2593
acetylenes (methyl, ethyl, vinyl)	20.0	ASTM D2593
dimers	500.0	ASTM D2426
isoprene	10.0	ASTM D2593
C5s	500.0	ASTM D2593 or D2426
sulfur	5.0	ASTM D4045 or D2784
peroxides (calculated as H_2O_2)	5.0	ASTM D1022
ammonia	5.0	water wash and ASTM D1426
water	300.0	Karl Fischer or panametric dew point
carbonyls	10.0	ASTM D4423
nonvolatile residues, wt %	0.05	ASTM D1025
oxygen in gas phase, vol %	0.10	teledyne oxygen analyzer

[a]Courtesy of Exxon Chemical Co.
[b]*tert*-Butylcatechol [27213-78-1] TBC.

Table 6. **Effects of Temperature on Dimerization Rate of Butadiene**

Temperature, °C	Butadiene, % dimerized/h
20	0.00015
40	0.0014
60	0.013
80	0.12
100	1.1

The thermally induced Diels–Alder dimerization reaction producing vinyl-cyclohexene is very difficult to prevent except by lowering the storage temperature (248). Since the reaction rate increases about ninefold for every 20°C increase in temperature (Table 6), care must be taken to keep butadiene at a low temperature.

Butadiene reacts readily with oxygen to form polymeric peroxides, which are not very soluble in liquid butadiene and tend to settle at the bottom of the container because of their higher density. The peroxides are shock sensitive; therefore it is imperative to exclude any source of oxygen from butadiene. Addition of antioxidants like t-butylcatechol (TBC) or butylated hydroxy toluene (BHT) removes free radicals that can cause rapid exothermic polymerizations. Butadiene shipments now routinely contain ~100 ppm TBC. Before use, the inhibitor can easily be removed (249,250). Inert gas, such as nitrogen, can also be used to blanket contained butadiene (251).

Butadiene is also known to form rubbery polymers caused by polymerization initiators like free radicals or oxygen. Addition of antioxidants like TBC and the use of lower storage temperatures can substantially reduce fouling caused by these polymers. Butadiene and other olefins, such as isoprene, styrene, and chloroprene, also form so-called popcorn polymers (252). These popcorn polymers are hard, opaque, and porous. They have been reported to ignite spontaneously on exposure to air and, therefore, are usually kept under water. They grow faster in the presence of seeds, or oxygen and rust. Because popcorn polymers can grow exponentially (253), they can generate tremendous pressure resulting in sudden rupture or plugging of containers, distillation towers, and pipes (254). It is reported that rigorous exclusion of oxygen from the system, metal surface passivation, and removal of popcorn polymer seeds can mitigate most of this problem (253). Addition of antioxidants will also help, but the high boiling points of many of these materials render them effective primarily in the liquid phase. Recently, a comprehensive study of this phenomenon and effective mitigation methods have been investigated, developed, and are available for license (255).

Butadiene is primarily shipped in pressurized containers via railroads or tankers. U.S. shipments of butadiene, which is classified as a flammable compressed gas, are regulated by the Department of Transportation (256). Most other countries have adopted their own regulations (30). Other information on the handling of butadiene is also available (257). As a result of the extensive emphasis on proper and timely responses to chemical spills, a comprehensive handbook from the National Fire Protection Association is available (258).

Table 7. **U.S. Producers of Butadiene and their Capacities**

Producer	Capacity[a] $\times 10^6$ kg (10^6lb)
BP Amoco, Chocolate Bayou, Texas	90.7 (200)
Equistar, Channelview, Texas	281.2 (620)
Equistar, Chocolate Bayou, Texas	68.0 (150)
Equistar, Corpus Christi, Texas	90.7 (200)
Exxon, Baton Rouge, La.	149.7 (330)
Exxon, Baytown, Texas	122.5 (270)
Huntsman, Port Neches, Texas	370.6 (850)
Shell, Deer Park, Texas	136.0 (300)
Shell, Norco, La.	260.8 (575)
Texas Petrochemicals, Houston, Texas	415.0 (915)
Total	*1985 (4,410)*

[a]Per year of finished butadiene capacity.

6. Economic Aspects

World capacity grew at 3.3%/yr in 1994–1999 and reached 10×10^6 t as of January 2000 (259). Most capacity was in Asia, South America, and the Middle East. Asia is the world's leading producer.

United States producers are listed in Table 7 (1). The estimated world demand for 2003 is 5.8 billion lb (2.6 million tons) (1).

During 1994–1999 the price of butadiene was a high of $0.24/lb and a low of $0.13/lb tanks fob Gulf (1).

Since the bulk of butadiene is recovered from steam crackers, its economics is very sensitive to the selection of feedstocks, operating conditions, and demand patterns. Butadiene supply and, ultimately, its price are strongly influenced by the demand for ethylene, the primary product from steam cracking. Currently, there is a worldwide surplus of butadiene. Announcements of a number of new ethylene plants will likely result in additional butadiene production, more than enough to meet worldwide demand for polymers and other chemicals (1). When butadiene is in excess supply, ethylene manufacturers can recycle the butadiene as a feedstock for ethylene manufacturer.

7. Health and Safety Aspects

Butadiene has been used widely in producing many important industrial polymers and other products. Thus over the years the effects on plant workers who have been exposed to different levels of butadiene have been under increasing scrutiny from manufacturers, users, health organizations, as well as government agencies. Short-term exposure to high concentrations of butadiene may cause irritation to the eyes, nose, and throat. Dermatitis and frostbite may result from exposure to the liquid and the evaporating gas (247). Long-term physiological reactions to 1,3-butadiene may vary individually.

Exposure studies have been made using mice and rats (260). These experiments have demonstrated species differences in butadiene toxicity and

Table 8. **Exposure Limits of Butadiene in Selected Countries**

Country	Exposure limit[a]
Belgium	10 ppm
France	10 ppm
Germany	carcinogen
Russia	STEL 100 mg/m^3
The Netherlands	21 ppm
United Kingdom	10 ppm
United States	2 ppm (ACGIH)
	5 ppm for 15 min (OSHA)
	1 ppm 8-h time weighted average (OSHA)

[a]Ref. 263.

carcinogenicity. Butadiene was found to be a potent carcinogen in the mouse, but only a weak carcinogen in the rat. The interpretations have focused on differences in toxification rates and detoxification metabolisms as causative factors (257). The metabolism is believed to proceed through intermediates involving butadiene monoepoxide and butadiene diepoxide (260). A similar mechanism has been proposed for its biodegradation pathway (261).

A retrospective epidemiological study covering a period of 36 years and a population of ~14,000 workers from eight SBR production facilities in the United States and Canada was the largest of several such studies conducted (260–262). The study covered the period between 1943 and 1976 with an update from 1977 to 1985. Despite the difficulty in ascertaining the exposure levels, the authors of that paper suggested there were no statistically significant differences in tumor mortality in total or for any specific cause of death as compared to the general population (260). In several epidemiological studies elevation in mortality were observed for small subgroups and tumor types. See Table 8 for exposure limits in selected countries.

There have been many reviews published on the toxicity of butadiene (264). A summary of environmental health perspectives was presented at the 1988 Symposium on Toxicology, Carcinogenesis, and Human Health Effects of 1,3-Butadiene (265). Detailed comparisons of various personal monitoring devices are available (266), and control of occupational exposure to 1,3-butadiene has been reviewed (267).

8. Uses

Butadiene is used primarily in polymers, including SBR, BR, ABS, SBL, and NR. In 2000 these uses accounted for about 89% of butadiene consumed in the United States (1). Styrene–butadiene rubber, the single largest user of butadiene, consumes 29% of the total. It is followed by polybutadiene rubber at 27%. Consumption for the other polymers, ABS, SBL, polychloroprene, and nitrile rubber are listed in Table 9.

Another significant butadiene use is for manufacturing adiponitrile, NC(CH$_2$)$_4$CN [111-69-3], a precursor for nylon-6,6 production. Other miscellaneous chemical uses, such as for ENB (ethylidene norbornene) production, account for 7% combined (Fig. 4) (271).

Table 9. **Pattern of Butadiene Uses in the United States, 2000**[a]

End use	Percentage of total, %
synthetic elastomers	61
styrene–butadiene rubber (SBR)	29
polybutadiene (BR)	27
polychloroprene (Neoprene)	3
nitrile rubber (NR)	2
polymers and resins	18
acrylonitrile–butadiene–styrene (ABS)	6
styrene–butadiene copolymer (latex)	12
chemicals and other uses	21
adiponitrile	14
others	7

[a]Ref. 1

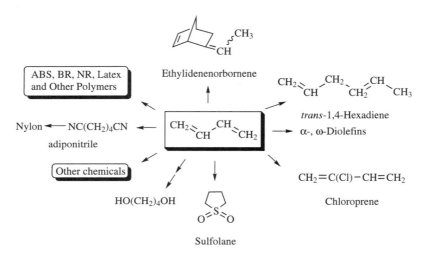

Fig. 4. Commercial uses of butadiene.

New butadiene applications are expected to grow from a small base, particularly liquid hydroxy-terminated homopolymers used in polyurethane for sealants, waterproofing, electrical encapsulation, and adhesive (1). A recent patent describes an epoxy resin composition for semiconductor encapsulation comprising an epoxy resin, phenolic resin, and butadiene rubber particles (268). An invention for improved polybutadiene composition suitable for use in molded golf ball cores has been described (269). A butadiene rubber adhesive composition with specified low syndiotactic butadiene has been patented (270).

BIBLIOGRAPHY

"Butadiene" in *ECT* 1st ed., Vol. 1, pp. 669–674, by A. A. Dolnick and M. Potash, Publicker Industries, Inc.; in *ECT* 2nd ed., Vol. 3, pp. 784–815 by I. Kirshenbaum and

R. P. Cahn, Exxon Research and Engineering Company; "Manufacture of Butadiene" in *ECT* 3rd ed., Vol. 11, pp. 857–870, by C. E. Morrell, Esso Laboratories, Standard Oil Development Company; "Butadiene" in *ECT* 4th ed., Vol. 4, pp. 663–690, by H. N. Sun and J. P. Wristers, Exxon Chemical Company; "Butadiene" in *ECT* (online), posting date: December 4, 2000, by H. N. Sun and J. P. Wristers, Exxon Chemical Company.

CITED PUBLICATIONS

1. "Butadiene", *ChemExpo Chemical Profile*, Chemical News and Data, March 6, 2000, http://www.chemexpo.com/Profile, searched Feb. 14, 2002.
2. S. K. Chatterjee and A. Das, *Chem. Age, India* **34**, 751 (1983).
3. L. M. Stock, *Prepr. Am. Chem. Soc., Div. Fuel Chem.* **32**, 463 (1987).
4. C. G. Rudershausen and J. B. Thompson, *Prepr. Div. Pet. Chem., Am. Chem. Soc.* **23**, 241 (1978).
5. Neth. Pat. 87,00,587 (Mar. 11, 1987), (to Lummus Crest B.V.).
6. R. J. Evans and T. A. Milne, *Energy Fuels* **1**, 123 (1987).
7. T. Funazukuri, R. R. Hudgins, and P. L. Silverston, *J. Anal. Appl. Pyrolysis* **10**, 225 (1987).
8. U.S. Pat. 6,175,049 (Jan. 16, 2001), A. Stuwe, and co-workers (EC Erdolchemie GmbH).
9. Jpn. Pat. 83 25,635 (Feb. 15, 1983), (to Toray Industries, Inc.).
10. M. Ghalamkar-Moazzam and T. L. Jacobs, *J. Polym. Sci. Polym. Chem. Ed.* **16**, 701 (1978).
11. U.S. Pat. 4,239,870 (Dec. 16, 1980), R. L. Smith (to Phillips Petroleum Co.).
12. R. W. Gallant, *Physical Properties of Hydrocarbons*, Gulf Publishing Co., Houston, Tex., 1968.
13. D. R. Stuff, E. F. Westrum, Jr., and G. C. Sinke, *The Chemical Thermodynamics of Organic Compounds*, John Wiley & Sons, Inc., New York, 1969, pp. 330, 331.
14. G. B. Kistiakowsky, J. R. Ruhoff, H. A. Smith, and W. E. Vaughan, *J. Am. Chem. Soc.* **57**, 876 (1935); **58**, 146, 237 (1936).
15. T. R. Durig, W. E. Bucy, and A. R. H. Cole, *Can. J. Phys.* **53**, 1832 (1976).
16. L. A. Carriera, *J. Chem. Phys.* **62**, 3851 (1975).
17. M. E. Squillacote, R. S. Sheridan, O. L. Chapman, and F. A. L. Anet, *J. Am. Chem. Soc.* **101**, 3657 (1979).
18. J. G. Aston, G. Szasz, H. W. Wooley, and F. G. Brickwedde, *J. Chem. Phys.* **14**, 67 (1946).
19. D. J. Marais, N. Shepard, and B. P. Stoicheff, *Tetrahedron* **17**, 163 (1962).
20. A. A. Al-Sahhaf, *J. Environ. Science and Health* **A24**, 49 (1989).
21. S. Banerjee and P. H. Howard, *Environ. Sci. Technol.* **22**, 839 (1988).
22. O. Grotewohl, *Chem. Tech.* **40**, 485 (1988).
23. W. Baker and A. L. Mossman, *Matheson Gas Data Book* **83**, (1980) F. D. Rossini, K. S. Pitzer, R. L. Arnett, R. M. Braun, and G. C. Pimentel, *Selected Values of Physical and Thermodynamic Properties of Hydrocarbons and Related Compounds*, American Petroleum Institute Research Project 44, Carnegie Press, Pittsburgh, Pa., 1953.
24. R. C. Weast, *CRC Handb. Chem. Phys.* **70**, C161 (1990).
25. C. L. Yaws and P.-Y. Chiang, *Chem. Eng.* **95**(13), 81 (1988).
26. N. Gee, K. Shinsaka, J. P. Dodelet, and G. R. Freeman, *J. Chem. Thermodyn.* **18**, 221 (1986).
27. C. L. Yaws, H.-C. Yang, J. R. Hopper, and K. C. Hansen, *Chem. Eng.* **97**(4), 177 (1990).

28. *Standard on Explosion Prevention Systems*, National Fire Protection Association, NFPA 69, Boston, Mass., 1973.
29. C. L. Yaws, H.-C. Yang, and W. A. Cawley, *Hydrocarbon Process.* 87 (June 1990).
30. H. J. Muller and E. Loser, *Ullmanns Encycl. Tech. Chem.* **A4**, 431 (1985).
31. J. Tsuji, *Acct. Chem. Res.* **6**, 8 (1973).
32. U.S. Pat. 2,123,504 (July 12, 1938), H. B. Dykstra (to E. I. du Pont de Nemours & Co., Inc.).
33. H. Kubota, *Rev. Phys. Chem. Jpn.* **37**, 25 (1967); L. M. Mascavage, H. Chi, S. La, and D. R. Dalton, *J. Org. Chem.* **56**, 595 (1991).
34. M. S. Kharasch, J. Kritchevsky, and F. R. Mayo, *J. Org. Chem.* **2**, 489 (1938).
35. P. J. Gorton and R. Walsh, *J. Chem. Soc., Chem. Commun.*, 782 (1972).
36. U.S. Pat. 4,645,863 (Feb. 24, 1987), W. Rebafka and H. Nickels (to BASF A.-G.).
37. U.S. Pat. 4,590,300 (May 20, 1986), M. J. Mullins and P. J. Hamlin (to Dow Chemical Company).
38. Ger. Pat. 2,550,902 (May 26, 1977), K. Janowsky, H.-G. Wegner, and J. Woeliner (to Deutsche Texaco AG).
39. Eur. Pat. Appl. 252,401 (Mar. 18, 1981), E. Drent (to Shell International Research Maatschappij B.V.).
40. U.S. Pat. 4,450,286 and U.S. Pat. 4,450,287 (May 22, 1984), T. E. Paxton (to Shell Oil Co.).
41. R. Baker and D. E. Halliday, *Tetrahedron Lett.* **27**, 2773 (1972).
42. Y. Kakuno and H. Hattovi, *J. Catal.* **85**, 509 (1984).
43. U.S. Pat. 3,876,695 (Apr. 8, 1975), N. von Kutepow (to BASF A.-G.).
44. U.S. Pat. 4,550,195 (Oct. 29, 1985), R. Platz, R. Kummer, H. W. Schneider, and K. Schwirten (to BASF A.-G.).
45. Ger. Pat. 3,638,218 (May 11, 1988), P. Panitz, H. Reitz, G. Schuch, and W. Seyfert (to BASF A.-G.).
46. Ger. Pat. 3,638,219 (May 11, 1988), R. Maerkl and co-workers (to BASF A.-G.).
47. Eur. Pat. Appl. 271,145 (June 15, 1988), E. Drent (to Shell Internationale Research Maatschappij B.V.).
48. Eur. Pat. Appl. 273,489 (July 6, 1988), E. Drent (to Shell Internationale Research Maatschappij B.V.).
49. Eur. Pat. Appl. 284,170 (Sept. 28, 1988), E. Drent and J. Van Gogh (to Shell Internationale Research Maatschappij B.V.).
50. U.S. Pat. 4,171,450 (Oct. 16, 1979), H. S. Kesling and L. R. Zehner (to Atlantic Richfield Co.).
51. K. Mislow and H. M. Hellman, *J. Am. Chem. Soc.* **73**, 244 (1951).
52. K. Mislow, *J. Am. Chem. Soc.* **75**, 2512 (1953).
53. D. F. Shellhamer, R. J. Conner, R. E. Richardson, V. L. Heasley, and E. G. Heasley, *J. Org. Chem.* **49**, 5015 (1984).
54. J. H. Prescott, *Chem. Eng.* **78**, 47 (Feb. 8, 1971).
55. U.S. Pat. 3,720,704 (Apr. 13, 1973), H. Sakomura, H. Kisaki, T. Tada, and S. Mabuchi (to Toyo Soda).
56. D. Craig, *J. Am. Chem. Soc.* **65**, 1006 (1943).
57. Jpn. Pat. 76 115,408 (Oct. 12, 1976) and 76 88,908 (Aug. 4, 1976), W. Funakoshi and T. Urasaki (to Teijin Ltd.).
58. J. Thivolle-Cazat and I. Tkatchenko, *J. Chem. Soc., Chem. Commun.* **19**, 1128 (1982).
59. P. Rylander, *Catalytic Hydrogenation in Organic Synthesis*, Academic Press, Inc., New York, 1979, Chapt. 3.
60. F. J. McQuillin, *Homogeneous Hydrogenation in Organic Chemistry*, D. Reidel Publishing Co., Boston, Mass., 1976, pp. 46–48.

61. B. R. James, *Homogeneous Hydrogenation*, John Wiley & Sons, Inc., New York, 1973, p. 148.
62. J. Halpern and L. Y. Wong, *J. Am. Chem. Soc.* **90**, 6665 (1968).
63. M. G. Burnett and P. J. Conolly, *J. Chem. Soc. A*, 991 (1968).
64. U.S. Pat. 3,009,969 (Nov. 21, 1961), M. S. Spencer and D. A. Dowden (to ICI).
65. H. Inoue, H. Nagamoto, and M. Shinkai, *Asahi Garasu Kogyo Gijutsu Shoreikai Kenkyu Hokoku* **31**, 277 (1977).
66. H. Nagamoto and H. Inoue, *Bull. Chem. Soc. Jpn.* **59**, 3935 (1986).
67. J. Tamaki, S. Nakayama, M. Yamamura, and T. Imanaka, *Shokubai* **29**, 170 (1987).
68. M. Derrien, J.-P. Boitiaux, L. Quicke, and J. W. Andrews, *AIChE Spring Meeting Preprints*, paper 82f (1989).
69. Ger. Pat. 3,417,549 (Nov. 14, 1985), H. Zimmermann and H. J. Wernicke (to Linde A.-G.).
70. J.-P. Boitiaux, J. Cosyns, M. Derrien, and G. Leger, *Hydrocarbon Process.* 51 (Mar. 1985).
71. H. Lauer and E. Kohle, *Erdgas Petrochem.* **36**, 249 (1983).
72. Jpn. Pat. 86 85,333 (Apr. 30, 1986), M. Okamura, M. Yamase, K. Horii, Y. Setoguchi, and Y. To (to Sumitomo Chem. Co., Ltd.).
73. Jpn. Pat. 84 05,127 (Jan. 12, 1984), (to Sumitomo Chemical Co., Ltd.).
74. Y. Furukawa and co-workers, *Bull. Jpn. Petrol. Inst.* **15**, 71 (1973).
75. S. Naito, Y. Sakurai, H. Shimizu, T. Onishi, and K. Tamaru, *Trans. Faraday Soc.* **67**, 1539 (1971).
76. T. Okuhara and K. I. Tanaka, *J. Catal.* **61**, 135 (1980).
77. Y. Okamoto, A. Maezawa, H. Kane, and T. Imanaka, *J. Chem. Soc. Chem. Commun.* 380 (1988).
78. Y. Imizu, K. Tanabe, and H. Hattori, *J. Catal.* **56**, 303 (1979).
79. G. C. Bond and P. B. Wells, *Adv. Catal.* **15**, 92 (1964).
80. G. C. Bond, *Catalysis by Metals*, Academic Press, Inc., New York, 1962, Chapts. 11 and 12.
81. J. Tamaki, S. Nakayama, M. Yamamura, and T. Imanaka, *Shokubai* **29**, 170 (1987).
82. I. Suzuki and A. Shibata, *Utsunomiya Daigaku Kyoikugakubu Kiyo, Dai-2-bu* **34**, 43 (1983).
83. A. M. Rosan, R. K. Crissey, and I. L. Mador, *Tetrahedron Lett.* **25**, 3787 (1984).
84. K. Shimazu and H. Kita, *J. Chem. Soc., Faraday Trans. I* **81**, 175 (1985).
85. J. N. Nudel, B. S. Umansky, R. O. Piagentini, and E. A. Lombardo, *J. Catal.* **89**, 362 (1984).
86. Y. Iwasawa, *Adv. Catal.* **35**, 218 (1987).
87. Brit. Pat. 2,106,129 (Apr. 7, 1983), C. S. John and J. A. Van Broekhoven (to Shell International Research Maatschappij, B.V.).
88. U.S. Pat. 4,421,933 (Dec. 20, 1983), J. A. M. Van Broekhoven and C. S. John (to Shell Oil Co.).
89. D. J. Hucknall, *Selective Oxidation of Hydrocarbons*, Academic Press, Inc., New York, 1973, Chapt. 4.
90. M. Ai, *J. Catal.* **71**, 88 (1981).
91. U.S. Pat. 4,400,324 (Aug. 23, 1983), T. S. Brima and A. B. Cottingham (to National Distillers and Chemical Corp.).
92. F. Mares and R. Tang, *J. Org. Chem.* **43**, 4631 (1978).
93. Jpn. Pat. 78,91,999 (Aug. 12, 1978), S. Mabuchi, S. Kumoi, and M. Sumida (to Toyo Soda Manufacturing Co., Ltd.).
94. G. Centi and F. Trifiro, *J. Mol. Catal.* **35**, 255 (1986).
95. U.S. Pat. 4,409,076 (Oct. 11, 1983), W. C. Seidel and D. N. Staikos (to E. I. du Pont de Nemours & Co., Inc.).

96. U.S. Pat. 4,298,531 (Nov. 3, 1981), R. V. Lindsey, Jr. and W. W. Richard (to E. I. du Pont de Nemours & Co., Inc.).

97. Jpn. Pat. 81,122,370 (Sept. 25, 1981), (to Mitsuibishi Chemical Industries).

98. U.S. Pat. 4,322,358 (Mar. 30, 1982), T. A. Bither and W. R. McCellan (to E. I. du Pont de Nemours & Co., Inc.).

99. U.S. Pat. 4,234,747 (Nov. 18, 1980), C. J. McIndoe (to ICI, Ltd.).

100. Jpn. Pat. 84,84,831 (May 16, 1984), (to Mitsui Taotsu Chemicals, Inc.).

101. Eur. Pat. Appl. 8,457 (Mar. 5, 1980), D. I. Garnett and M. L. Peterson (to E. I. du Pont de Nemours & Co., Inc.); Eur. Pat. Appl. 31,729 (July 8, 1981), R. V. Lindsey, Jr., and W. W. Vernon (to E. I. du Pont de Nemours & Co., Inc.).

102. T. Onoda, *Petrotech (Japan)* **10**, 810 (1987).

103. Y. Tanabe, *Hydrocarbon Process.* **60**, 187 (Sept. 1981).

104. Eur. Pat. Appl. 289,725 (Nov. 9, 1988), J. Haji, I. Yokotake, T. Yamaguchi, M. Sato, and N. Murai (to Mitsubushi Kasei Corp.).

105. A. V. Devekki and M. I. Yakushkin, *Kinet. Katal.* **28**, 1099 (1987).

106. U.S. Pat. 4,044,041 (Aug. 23, 1977), P. R. Stapp (to Phillips Petroleum Co.).

107. I. Takehira, J. A. Trinidad Chena, S. Niwa, T. Hayakawa, and T. Ishikawa, *J. Catal.* **76**, 354 (1982).

108. U.S. Pat. 4,384,146 (May 13, 1983), S. C. Tang (to Shell Oil Co.).

109. P. D. Lawley, in C. E. Searle, ed., *Carcinogenesis by Alkylating Agents, in Chemical Carcinogens ACS Monograph 173*, American Chemical Society, Washington, D.C., 1976, pp. 83–244.

110. E. Malvoisin, G. Lhoest, F. Poncelet, M. Roberfroid, and M. Mercier, *J. Chromatogr.* **178**, 419 (1979).

111. E. Loeser, *Annu. Meet. Proc. Int. Inst. Synth. Rubber Prod.* Paper 13 (1987).

112. H. H. Wasserman, *Diels–Alder Reactions*, Elsevier Science Publishing Co., Inc., New York, 1965.

113. W. von E. Doering, M. Franck-Neumann, D. Hasselmann, and R. L. Kaye, *J. Am. Chem. Soc.* **94**, 3833 (1972).

114. W. Oppolzer, *Angew. Chem. Int. Ed.* **16**, 10 (1977).

115. A. U. Konovalov, *Russ. Chem. Rev.* **52**, 1064 (1983).

116. G. Consiglio and R. M. Waymouth, *Chem. Rev.* **89**, 257 (1989).

117. T. Sakai, in L. F. Albright, B. L. Crynes, and W. H. Corcoran, eds., *Pyrolysis: Theory and Industrial Practice*, Academic Press, Inc., New York, 1983, pp. 100, 101.

118. R. B. Woodward and R. Hoffmann, *The Conservation of Orbital Symmetry*, Academic Press, Inc., New York, 1970.

119. K. Fukui, *Acc. Chem. Res.* **4**, 57 (1971).

120. U.S. Pat. 3,454,665 (July 8, 1969), H. E. Cler (to Esso Research and Engineering Co.).

121. R. Z. Dolor and P. Vogal, *J. Mol. Catal.* **60**, 65 (1990).

122. P. V. Bonnesen, C. L. Puckett, R. V. Honeychuck, and W. H. Hersh, *J. Am. Chem. Soc.* **111**, 6070 (1989).

123. Jpn. Pat. 80 87,728 (July 2, 1980), (to Mitsui Petrochemical Industries, Ltd.).

124. U.S. Pat. 3,767,593 (Oct. 23, 1973), C. L. Meyers (to Phillips Petroleum Company).

125. U.S. Pat. 4,125,483 (Nov. 14, 1978), R. S. Downing, J. van Amstel, and A. H. Joustra (to Shell Oil Co.).

126. A. Goliaszewski and J. Schwartz, *Organometallics* **4**, 415 (1985).

127. U.S. Pat. 4,384,153 (May 17, 1983) and 4,413,154 (Nov. 1, 1983), R. M. Dessau (to Mobil Oil Corp.).

128. I. E. Maxwell, R. S. Downing, and S. A. J. van Langen, *J. Catal.* **61**, 485 (1980).

129. I. E. Maxwell, J. J. de Boer, and R. S. Downing, *J. Catal.* **61**, 493 (1980).

130. R. M. Dessau, *J. Chem. Soc. Chem. Commun.* 1167 (1986).

131. A. Mortreux, J. C. Bavay, and F. Petit, *Nouv. J. Chim.* **4**, 671 (1980).

132. P. Cros and G. Buono, *New J. Chem.* **11**, 573 (1987).
133. W. J. Richter, *J. Mol. Catal.* **13**, 201 (1981).
134. U.S. Pat. 4,367,358 (Jan. 4, 1983), L. G. Wideman, L. A. Bente, and J. A. Kuczkowski (to Goodyear Tire and Rubber Co.).
135. H. Suzuka and H. Hattori, *Appl. Catal.* **47**, L7 (1989); *J. Mol. Catal.* **63**, 371 (1990).
136. K. T. Finley, in S. Patai, *The Chemistry of the Quinonoid Compounds*, Part 2, John Wiley & Sons, Inc., New York, 1974, 986–1068.
137. J. J. Gajewski, *Organic Chemistry Series*, in H. H. Wasserman, ed., *Hydrocarbon Thermal Isomerization*, Vol. 5, Academic Press, Inc., New York, 1981, pp. 271–282.
138. J. A. Berson and P. B. Dervan, *J. Am. Chem. Soc.* **95**, 267 (1973).
139. U. Schuchardt and F. Santos Dias, *J. Mol. Catal.* **29**, 145 (1985).
140. H. Heimbach, P. W. Jolly, and C. Wilke, *Adv. Organomet. Chem.* **8**, 29 (1970).
141. U.S. Pat. 3,219,716 (Nov. 23, 1965), D. Wittenberg, H. Lautenschlager, N. von Kutepow, F. Meier, and H. Seibt (to BASF A.-G.).
142. G. Wilke, ed., *Angew. Chem., Int.* **2**, 105 (1963).
143. G. Wilke and co-workers, *Angew. Chem.* **69**, 397 (1957); and **71**, 574 (1959).
144. P. W. Jolly and G. Wilke, *The Organic Chemistry of Nickel*, Vol. 2, Academic Press, Inc., New York, 1975.
145. D. E. Bergbreiter and R. Chandran, *J. Chem. Soc. Chem. Commun.* 1396 (1985).
146. H. tom Dieck and J. Dietrich, *Angew. Chem. Int. Ed. Engl.* **24**, 781 (1985).
147. A. Goliaszewski and J. Schwartz, *Organometallics* **4**, 415 (1985).
148. Jpn. Pat. 82 03,651 (Jan. 22, 1982), (to Kureha Chemical Industry Co., Ltd.).
149. P. Denis, A. Mortreux, F. Pepit, G. Buono, and G. Peiffer, *J. Org. Chem.* **49**, 5274 (1984).
150. G. Wilke and co-workers, *Angew. Chem. Int. Ed. Engl.* **5**, 151 (1966).
151. A. Musco and A. Silvani, *Organomet. Chem.* **88**, C41 (1975).
152. G. Huybrechts, L. Luyckx, T. Vandenboom, and B. Van Mele, *Int. J. Chem. Kinet.* **9**, 283 (1977).
153. *FEAST Olefins* Technical Bulletin, Shell Oil Co., 1990.
154. K. J. Ivin, *Olefin Metathesis*, Academic Press, Inc., New York, 1983, p. 162.
155. G. W. Parshall, *Homogeneous Catalysis*, Wiley-Interscience, New York, 1980.
156. G. W. Parshall and W. A. Nugent, *Chemtech* **5**, 314 (1988).
157. W. Ring and J. Gaube, *J. Chem. Eng. Tech.* **36**, 1041 (1966).
158. W. Keim, *Angew. Chem. Int. Ed. Engl.* **29**, 235 (1990).
159. W. J. Richter, *J. Mol. Catal.* **34**, 145 (1986).
160. Eur. Pat. 161,979 (Nov. 21, 1985), A. Mortreux, F. Pepit, P. Denis, G. Buono, and G. Peiffer (to Société Chimique des Charbonnages).
161. Ger. Pat. 118,409 (June 12, 1976), W. Lindenlaub, W. Walter, H. Fuellbier, and E. Alder.
162. G. P. Potapov, V. V. Punegov, and U. M. Dzhemilev, *Izv. Akad. Nauk. SSSR, Ser. Khim.* **7**, 1468 (1985).
163. C. U. Pittman, Jr., R. M. Haines, and J. J. Yang, *J. Mol. Catal.* **15**, 377 (1982).
164. Ger. Pat. 3,024,879 (Jan. 28, 1982), B. Schleppinghoff and H. Schoeneberger (to EC Erdoelchemie GmbH).
165. Eur. Pat. Appl. 8,139 (Feb. 20, 1980), K. Nozaki (to Shell International Res. Maatschappij, B. V.).
166. Eur. Pat. Appl. 43,039 (Jan. 6, 1982), B. Schleppinghoff, H. D. Koehler, and H. Schoeneberger (to EC Erdoelchemie GmbH).
167. Y. Sasaki, Y. Inoue, and H. Hashimoto, *J. Chem. Soc. Chem. Commun.* 605 (1976).
168. Y. Inoue, Y. Sasaki, and H. Hashimoto, *Bull. Chem. Soc. Jpn.* **51**, 2375 (1978).
169. A. Musco, C. Prego, and V. Tartiari, *Inor. Chim. Acta* **28**, L147 (1978).

170. U.S. Pat. 4,167,513 (Nov. 9, 1979), A. Musco, R. Santi, and G. P. Chiusoli (to Montedison SpA).
171. P. Braunstein, D. Matt, and D. Nobel, *Chem. Rev.* **88**, 747 (1988).
172. Eur. Pat. Appl. 234,668 (Sept. 2, 1987), E. Drent (to Shell International Research Maatschappij B.V.).
173. A. Behr, K. D. Juszak, and R. He, *Int. Congr. Catal. [Proc]*, 8th ed. **5**, V565 (1984).
174. H. Hoberg, Y. Peres, A. Milchereit, and S. Gross, *J. Organomet. Chem.* **345**, C17 (1988).
175. A. Behr, *Bull. Soc. Chim. Belg.* **94**, 671 (1985).
176. A. Behr, in W. Keim, ed., *Catalysis in C_1 Chemistry*, D. Reidel Publishing Co., Dordrecht, Holland, 1983, pp. 191–194.
177. P. Braunstein, D. Matt, and D. Nobel, *J. Am. Chem. Soc.* **110**, 3207 (1988).
178. U.S. Pat. 4,554,375 (Nov. 19, 1985), J. J. Lin and D. C. Alexander (to Texaco Inc.).
179. Eur. Pat. Appl. 287,066 (Oct. 19, 1988), Y. Tokitoh and N. Yoshimura (to Kuraray Co., Ltd.).
180. J. P. Bianchini, B. Waegell, E. M. Gaydou, H. Rzehak, and W. Keim, *J. Mol. Catal.* **10**, 247 (1981).
181. Fr. Pat. 2,499,978 (Aug. 20, 1982), N. Yoshimura and M. Tamura (to Kuraray Co., Ltd.).
182. J. Tsuji, T. Takahashi, Y. Kobayashi, H. Yamada, and I. Minami, *Kenkyo Hokoku–Asahi Garasu Kogyo Gijutsu Shoreikai* **38**, 301 (1981).
183. K. Kaneda, M. Higuchi, and T. Imanaka, *Chem. Express* **3**, 335 (1988).
184. Jpn. Pat. 87 63,538 (Mar. 20, 1987), (to National Distillers and Chemical Co.).
185. Eur. Pat. Appl. 104,602 (Apr. 4, 1984), A. J. Chalk and K. L. Purszcki (to Givaudan, L., et Cie SA).
186. U. M. Dzhemilev, F. A. Selimov, and G. A. Tolstikov, *Izv. Akad. Nauk SSSR, Ser. Khim.* 348 (1980).
187. J. Tsuji, I. Shimizu, H. Suzuki, and Y. Naito, *J. Am. Chem. Soc.* **101**, 5070 (1979).
188. Jpn. Pat. 86 06,807 (Mar. 1, 1986), T. Onoda, H. Wada, K. Sato, and Y. Kasori (to Mitsubishi Chemical Industry).
189. U. M. Dzhemilev, R. V. Kunakova, and V. V. Sidorova, *Izv. Akad. Nauk SSSR, Ser. Khim.* **2**, 403 (1987).
190. T. Fujinami, T. Suzuki, M. Kamiya, S. Fukuzawa, and S. Sakai, *Chem. Lett.*, 199 (1985).
191. U. M. Dzhemilev, R. V. Kunakova, and N. Z. Baibulatova, *Izv. Akad. Nauk SSSR, Ser. Khim.*, 128 (1986).
192. D. Pletcher and J. T. Girault, *J. Appl. Electrochem.* **16**, 791 (1986).
193. Arno Behr in R. Ugo, ed., *Aspects of Homogeneous Catalysis*, Vol. 5, D. Reidel Publishing Co., Dordrecht, Holland, 1984, pp. 3–73.
194. W. Cooper, *Dev. Polym.* **1**, 103 (1979).
195. D. P. Tate and T. W. Bethea, in J. I. Kroschwitz, ed., *Encyclopedia of Polymer Science and Engineering*, Vol. 2, Wiley-Interscience, New York, 1985, pp. 537–590.
196. D. H. Richards, *Chem. Soc. Rev.* **6**, 235 (1977).
197. Ref. 20, p. 553.
198. E. B. Storey, *Rubber Chem. Tech.* **34**, 1402 (1961).
199. L. A. Wood in J. Brandrup and E. H. Immergut, eds., *Polymer Handbook*, 3rd ed., John Wiley & Sons, Inc., New York, 1989, p. 9.
200. H. F. Mark, in A. Vidal and J. B. Dannet, eds., *Chemistry and Technology of Rubber, Applied Polymer Symposium*, Vol. 39, John Wiley & Sons, Inc., New York, 1982, p. 15.
201. W. S. Bahary, D. I. Sapper, and J. H. Lane, *Rubber Chem. Technol.* **40**, 1529 (1967).
202. H. L. Stephens, in Ref. 197, pp. V1–V5.

203. N. G. Gaylord, T. K. Kwei, and H. F. Mark, *J. Polym. Sci.* **42**, 417 (1960).
204. Belg. Pat. 573,680 (1958), (to Montecatini).
205. G. Natta, *J. Polym. Sci.* **48**, 219 (1960).
206. Y. Lin and co-workers, *Huaxue Xuebao* **43**, 438 (1985).
207. U.S. Pat. 3,094,514 (June 18, 1963), H. Tucker (to Goodrich-Gulf Chemicals).
208. W. Cooper, *Ind. Eng. Chem. Prod. Res. Develop.* **9**, 457 (1970).
209. P. Teyssie, P. Hadjiandreou, M. Julemont, and R. Warin, *NATO ASI Ser., Ser. C.* **215** (1987).
210. U.S. Pat. 4,155,880 (May 22, 1979), M. C. Throckmorton and W. M. Saltman (to Goodyear Fire and Rubber Co.).
211. J. Witte, *Angew. Makromol. Chemie* **94**, 119 (1981).
212. U.S. Pat. 4,260,707 (Apr. 7, 1981), G. Sylvester, J. Witte, and G. Marwede (to Bayer, AG).
213. T. Sun and co-workers, *Yingyong Huaxue* **2**, 47 (1985).
214. Z. M. Sabirov, N. K. Minchenkova, N. A. Vakhrusheva, and Y. B. Monakov, *Dokl. Akad. Nauk SSSR* **312**, 147 (1990).
215. Z. M. Sabirov, Y. B. Monakov, and G. A. Tolstikov, *J. Mol. Catal.* **56**, 194 (1989).
216. Y. B. Monakov, N. G. Marina, and G. A. Tolstikov, *Chem. Stosow.* **32**, 547 (1988).
217. M. E. Gavin and S. A. Heffner, *Macromolecules* **22**, 3307 (1989).
218. F. W. Breitbarth and co-workers, *Acta Polym.* **37**, 508 (1986).
219. G. V. Vinogradov, E. A. Dzyura, A. Y. Malkin, and V. A. Grechanovskii, *J. Polym. Sci.* **A29**, 1153 (1971).
220. H. E. Adams, R. L. Bebb, L. E. Forman, and L. B. Wakefield, *Rubber Chem. Technol.* **45**, 1252 (1972); I. Kuntz and A. Gerber, *J. Polym. Sci.* **42**, 299 (1960).
221. A. W. Langer, Jr., *Polym. Prepr. Div. Polym. Chem. Amer. Chem. Soc.* **7**, 132 (1966).
222. A. A. Morton in N. M. Bikales, ed., *Encyclopedia of Polymer Science and Technology*, Vol. 1, Wiley-Interscience, New York, 1964, p. 629.
223. Ref. 194, p. 563.
224. D. Su, D. Yu, L. Hu, and S. Jiao, *J. Polmer. Sci.* **A27**, 3769 (1989).
225. Z. Florjanczyk, T. Florjanczyk, and B. B. Klopotek, *Makromol. Chem.* **188**, 2811 (1987).
226. R. Pettit and G. F. Emerson, *Adv. Organometallic Chem.* **1**, 1 (1964).
227. A. J. Pearson, *Metallo-Organic Chemistry*, John Wiley & Sons, Inc., New York, 1985.
228. Jpn. Pat. 81 34,571 (Aug. 11, 1981); Neth. Appl. 74 14,746 (May 14, 1976), (to Teiji Ltd.).
229. *Chemical Economics Handbook*, SRI International, Dec. 1990.
230. S. Kvisle, A. Aguero, and R. P. A. Aneeden, *Appl. Catal.* **43**, 117 (1988).
231. A. K. K. Lee and A. M. Aitani, *Oil Gas J.*, 60 (Sept. 10, 1990), and references cited therein.
232. F. P. Wilcher, B. V. Vora, and P. R. Pujado, *De Witt 1990 Petrochemical Review, T-1* (1990).
233. *Hydrocarbon Process.* **94** (Nov. 1989).
234. U.S. Pat. 3,907,918 (Sept. 23, 1975), T. Hutson, Jr., (to Phillips Petroleum Co.).
235. U.S. Pat. 6,040,489 (March 21, 2000), T. Imai (to UOP LLC).
236. Jpn. Pat. 82,120,532, and 82,120,533 (July 27, 1982), (to Nippon Oil Co., Ltd.).
237. U.S. Pat. 4,596,655 (June 24, 1986), A. T. Van Eiji (to The Dow Chemical Company).
238. U.S. Pat. 4,268,361 (May 19, 1981), (to B. F. Goodrich Co.); U.S. Pat. 4,269,668 (May 26, 1981), P. V. Patel (to B. F. Goodrich Co.).
239. Eur. Pat. Appl. 30,673 (June 24, 1981), H. Hiroshi and N. Iwaki (to Nippon Zeon Co., Ltd.).
240. Ger. Pat. 2,724,365 (Nov. 30, 1978), D. Stockburger and co-workers, (to BASF A.-G.).
241. K. Volkamer, K. J. Schneider, A. Lindner, D. Bender, and E. Kohle, *Erdgas, Petrochem.* **34**, 343 (1981).

242. Jpn. Pat. 81,02,811 (Jan. 13, 1981), (to Showa Denko K.K.).
243. Jpn. Pat. 85,90,005 (May 21, 1985), (to Agency of Industrial Sciences and Technology).
244. U.S. Pat. 6,093,285 (July 25, 2000), D. T. Fernald and co-workers (to Phillips Petroleum Co.).
245. U.S. Pat. 3,192,282 (June 29, 1965), C. E. Porter (to Esso Research and Engineering Co.).
246. U.S. Pat. 2,985,687 (Oct. 25, 1961), R. P. Cahn (to Esso Research and Engineering Co.).
247. M. Sittig, *Handbook of Toxic and Hazardous Chemicals and Carcinogens*, Noyes Publications, N.J., 1985, p. 153.
248. Ger. Pat. 2,051,548 (Apr. 4, 1972), (to Erdolchemie GmbH).
249. W. Braker and A. L. Mossman, eds., *Matheson Gas Data Book*, 6th ed., 1980, p. 80.
250. *Technical Bulletin No. AL-154*, Aldrich Chemical Co.
251. M. Nitsche, *Chem.-Anlagen Verfahren* **12**, 86 (1977).
252. M. S. Kharasch, *Ind. Eng., Chem.* **39**, 830 (1947).
253. G. H. Miller and co-workers, *J. Polym. Sci.* **C**, 1109 (1964).
254. L. Bretherick, *Handbook of Reactive Chemical Hazards*, 3rd ed., Butterworths, UK, 1985, p. 419.
255. Exxon Chemical Co., unpublished data, 1989.
256. *Code of Federal Regulations*, Title **49**, 1975.
257. *Properties and Essential Information for Safe Handling and Use of Butadiene*, Pamphlet SD-55, the Chemical Manufacturing Association, Washington, D.C., 1990.
258. *Hazardous Materials Response Handbook*, National Fire Protection Association, 1989.
259. J. Larson, Butadiene, *Chemical Economics Handbook*, SRI, Menlo Park, Calif., Nov. 2000.
260. E. Loeser, *Annu. Meet. Proc.-Int. Inst. Synth. Rubber Prod.* Paper 13 (1988).
261. R. J. Watkinson and H. J. Somerville, *The Microbial Utilization of Butadiene*, Shell Research Ltd., Sittingbourne Research Centre, Kent, UK, 1976.
262. R. K. Hinderer, *Annu. Meet. Proc.- Int. Inst. Synth. Rubber Prod.*, 32 (1988).
263. T. Carreon, "Aliphatic Hydrocarbons" in E. Bingham, B. Cohrrsen, C. H. Powell, eds *Patty*'s Toxicology, Vol. 4, John wiley & Sons, Inc., New York, 2001, Chapt. 49, pp. 110, 111.
264. C. D. Meester, *Mutat. Res.* **195**, 273 (1988).
265. *Environmental Health Perspectives*, U.S. Department of Health and Human Services, Washington, D.C., Vol. 86, 1990.
266. D. W. Gosselink, D. L. Braun, H. E. Mullins, S. T. Rodriguez, and F. W. Snowden, *Chemical Hazards in the Workplace*, American Chemical Society Symposium Series 149, Las Vegas, Nev., 1980, pp. 195–207.
267. E. R. Krishnan and T. K. Corwin, *Proc. APCA Annu. Meet. 80th* **5**, 87/84A.14 (1987).
268. U.S. Pat. 6,288,169 (Sept. 11, 2001), H. Usui and co-workers (to Nitto Denko Corp.).
269. U.S. Pat. 6,277,920 (Aug. 21, 2001), R. D. Nesbitt (to Spalding Sports Worldwide, Inc.).
270. U.S. Pat. 6,187,855 (Feb. 13, 2001), Y. Nagai and K. Yamaguchi (to Bridgestone Corporation).
271. Exxon Chemical Company, unpublished data, 1990.

H. N. SUN
J. P. WRISTERS
Exxon Chemical Company

BUTYL ALCOHOLS

1. Introduction

Butyl alcohols encompass the four structurally isomeric 4-carbon alcohols of empirical formula $C_4H_{10}O$. One of these, 2-butanol, can exist in either the optically active $R(-)$ or $S(+)$ configuration or as a racemic ($\pm$) mixture [15892-23-6].

2. Physical and Chemical Properties

The butanols are all colorless, clear liquids at room temperature and atmospheric pressure with the exception of t-butyl alcohol which is a low melting solid (mp 25.82°C); it also has a substantially higher water miscibility than the other three alcohols. Physical constants (1) of the four butyl alcohols are given in Table 1.

Physical constants (2) for the optically pure stereoisomers of 2-butanol have been reported as follows:

	CAS Registry Number	d^t_4	n^{20}_D	$[\alpha]^{27}_D$
(S)-(+)-2-butanol	[4221-99-2]	0.8025^{27}	1.3954	+13.52
(R)-(-)-2-butanol	[14898-79-4]	0.8042^{25}	1.3970	−13.52

The most common azeotropes (3,4) formed by the butanols are given in Table 2. Butyl alcohol liquid vapor pressure/temperature responses (5,6), which are important parameters in direct solvent applications, are presented in Figure 1. Similarly, viscosity/temperature plots (1) for the four butanols are presented in Figure 2.

The butanols undergo the typical reactions of the simple lower chain aliphatic alcohols. For example, passing the alcohols over various dehydration catalysts at elevated temperatures yields the corresponding butenes. The ease of dehydration increases from primary to tertiary alcohol: t-butyl alcohol undergoes dehydration with dilute sulfuric acid at low temperatures in the liquid phase whereas the other butanols require substantially more stringent conditions.

With the exception of the t-butyl compound, the butyl alcohols are dehydrogenated to the corresponding carbonyl compounds when passed over copper or silver catalysts at temperatures around 300°C. Thus, n- and isobutyl alcohols are dehydrogenated to n- and isobutyraldehyde, respectively, while 2-butanol gives methyl ethyl ketone (2-butanone). Continued or more vigorous oxidation of n- and isobutyl alcohol yield the corresponding carboxylic acids whereas 2-butanol is degraded to acids of shorter chain length.

The butyl alcohols undergo esterification with organic acids in the usual manner in the presence of trace amounts of mineral acid catalysts. Esterification is fastest with t-butyl alcohol and slowest with the primary alcohols although t-butyl alcohol undergoes substantial dehydration in the presence of the typical acid esterification catalysts.

Table 1. **Physical Properties of the Butyl Alcohols (Butanols)**

Common Name CAS Registry Number	n-Butyl alcohol [71-36-3]	Isobutyl alcohol [78-83-1]	sec-Butyl alcohol [78-92-2]	t-Butyl alcohol [75-65-0]
systematic name formula	1-butanol $CH_3(CH_2)_3OH$	2-methyl-1-propanol $(CH_3)_2CHCH_2OH$	2-butanol $CH_3CH(OH)C_2H_5$	2-methyl-2-propanol $(CH_3)_3COH$
critical temperature, °C	289.90	274.63	262.90	233.06
critical pressure, kPa[a]	4423	4300	4179	3973
critical specific volume, m^3/kg mol	0.275	0.273	0.269	0.275
normal boiling point, °C	117.66	107.66	99.55	82.42
melting point, °C	−89.3	−108.0	−114.7	25.82
ideal gas heat of formation at 25°C, kJ/mol[b]	−274.6	−283.2	−292.9	−312.4
heat of fusion, kJ/mol[b]	9.372	6.322	5.971	6.703
heat of vaporization at normal boiling point, kJ/g[b]	43.29	41.83	40.75	39.07
liquid density, kg/m^3 at 25°C	809.7	801.6	806.9	786.6[c]
liquid heat capacity at 25°C, kJ/(mol·K)[b]	0.17706	0.18115	0.19689	0.2198 at mp
refractive index at 25°C	1.3971	1.3938	1.3949	1.3852
flash point, closed cup, °C	28.85	27.85	23.85	11.11
dielectric constant, ε	17.5[25]	17.93[25]	16.56[20]	12.47[30]
dipole moment $\times 10^{30}$, C·m[d]	5.54	5.47	5.54	5.57
solubility parameter, $(MJ/m^3)^{0.5}$ [e] at 25°C	23.354	22.909	22.541	21.603
solubility in water at 30°C, % by weight	7.85	8.58	19.41	miscible
solubility of water in alcohol at 30°C, % by weight	20.06	16.36	36.19	miscible

[a] To convert kPa to mm Hg, multiply by 7.50.

[b] To convert kJ to kcal, divide by 4.184.

[c] For the subcooled liquid below melting point.

[d] To convert C · m to debyes, divide by 3.336×10^{-30}.

[e] To convert $(MJ/m^3)^{0.5}$ to $(cal/cc)^{0.5}$, multiply by $0.239^{0.5}$.

394

Table 2. **Azeotropic Mixtures of the Butyl Alcohols**

Components	Weight %	Boiling point of mixture, °C
Binary azeotropes		
1-butanol		
water	42.4	92.6
n-butyl acetate	32.8	117.6
n-butyl formate	76.3	105.8
methyl isovalerate	67	113
cyclohexane	90	79.8
tetrachloroethylene	68	110.0
ethyl isobutyrate	83	109.2
toluene	68	105.5
isobutyl alcohol		
water	33	90
cyclohexane	86	78.1
benzene	91	79.8
toluene	56	101.1
2-butanol		
water	32	88.5
2-butyl acetate	13.7	99.6
Ternary azeotropes		
1-butanol	10	83.6
n-butyl formate	68.7	
water	21.3	
1-butanol	27.4	89.4
n-butyl acetate	35.3	
water	37.3	

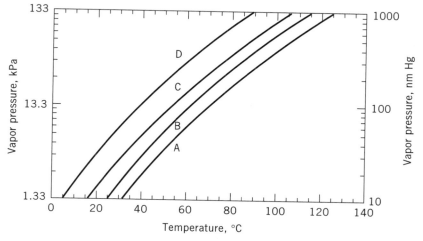

Fig. 1. Vapor pressure of butyl alcohols: A, *n*-butyl; B, isobutyl; C, *sec*-butyl; D, *t*-butyl. To convert kPa to mm Hg, multiply by 7.5.

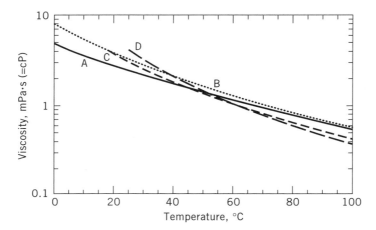

Fig. 2. Liquid viscosity of butyl alcohols: A, 1-butanol; B, isobutyl alcohol; C, 2-butanol; D, *t*-butyl alcohol.

1-Butanol and isobutyl alcohol are aminated with ammonia over alumina at 300–350°C to give the corresponding mono-, di-, and tributylamines.

3. Manufacture

The principal commercial source of 1-butanol is *n*-butyraldehyde [123-72-8], obtained from the Oxo reaction of propylene. A mixture of *n*- and isobutyralde- hyde [78-84-2] is obtained in this process; this mixture is either separated initially and the individual aldehyde isomers hydrogenated, or the mixture of isomeric aldehydes is hydrogenated directly and the *n*- and isobutyl alcohol product mix separated by distillation. Typically, the hydrogenation is carried out in the vapor phase over a heterogeneous catalyst. For example, passing a mixture of *n*- and isobutyraldehyde with 60:40 $H_2:N_2$ over a CuO–ZnO–NiO catalyst at 25–196°C and 0.7 MPa proceeds in 99.95% efficiency to the corre- sponding alcohols at 98.6% conversion (7,8) (see BUTYRALDEHYDES; OXO PROCESS).

In a process which is now largely of historical interest, 1-butanol has been produced from ethanol [64-17-5] via successive dehydrogenation (to acetaldehyde [75-07-0]), condensation (to crotonaldehyde [4170-30-3]) and hydrogenation.

$$CH_3CH_2OH \longrightarrow CH_3CHO \longrightarrow CH_3CH{=}CHCHO \longrightarrow CH_3CH_2CH_2CH_2OH$$

Alternatively, the intermediate acetaldehyde (qv) for this process was obtained from ethylene by the Wacker process (9). A small amount of *n*-butyl alcohol is produced in the United States by the Ziegler-Natta chain growth reaction from ethylene [74-85-1] (10).

The earliest commercial process to 1-butanol, still practiced extensively in many Third World countries, employs fermentation of molasses or corn products with Clostridium acetobutylicum (11–13). Acetone and ethanol are obtained as coproducts.

A fermentation route to 1-butanol based on carbon monoxide employing the anaerobic bacterium, Butyribacterium methylotrophicum has been reported (14,15). In contrast to other commercial catalytic processes for converting synthesis gas to alcohols, the new process is insensitive to sulfur contaminants. Current productivities to butanol are 1 g/L, about 10% of that required for commercial viability. Researchers hope to learn enough about the bacteria's control mechanisms to be able to use recombinant DNA to make the cells produce more butanol.

A novel route to 1-butanol from butadiene has been disclosed in the recent patent literature [36]. Butadiene is converted catalytically initially into a mixture of linear and branched butenyl ethers. These are isomerized to the linear butenyl ether, which is subsequently converted into the linear butyral, which is finally hydrogenated to 1-butanol. Alternatively, the intermediate acetal can be hydrolyzed to n-butyraldehyde. High overall efficiencies to n-butanol or n-butyraldehyde are claimed.

As of December 11, 1997, the total U. S. capacity of 1-butanol was 8,111 t per year. Capacity for an additional 733,000 t per year of n-butyl alcohol is operated in Western Europe, and an additional 320,000 t in Southeast Asia (Japan, China and Korea) (10).

Commercial isobutyl alcohol is made almost exclusively from the hydrogenation of isobutyraldehyde obtained by the hydroformylation of propylene. However, this alcohol is also commonly obtained as a coproduct in the Fischer Tropsch synthesis of methanol (16,17).

2-Butanol is produced commercially by the indirect hydration of n-butenes. However, current trends are towards the employment of inexpensive Raffinate II type feedstocks, ie, C-4 refinery streams containing predominantly n-butenes and saturated C-4s after removal of butadiene and isobutylene. In the traditional indirect hydration process, n-butenes are esterified with liquid sulfuric acid and the intermediate butyl sulfate esters hydrolyzed. DEA Mineraloel (formerly Deutsche Texaco) currently operates a 2-butanol plant employing a direct hydration of n-butenes route (18) with their own proprietary catalyst.

The Arco Propylene Oxide Process produces t-butyl alcohol as a coproduct of propylene oxide [75-56-9] when isobutane is used as a starting material.

$$(CH_3)_3CH + O_2 \longrightarrow (CH_3)_3COOH$$

$$(CH_3)_3COOH + CH_2{=}CHCH_3 \longrightarrow (CH_3)_3COH + CH_3CH\overset{O}{\overset{\diagup\diagdown}{-}}CH_2$$

The process can be modified to give predominantly or solely t-butyl alcohol. Thus, t-butyl hydroperoxide (and t-butyl alcohol) produced by oxidation of isobutane in the first step of the process can be decomposed under controlled, catalytic conditions to give gasoline grade t-butyl alcohol (GTBA) in high selectivity (19–22).

$$(CH_3)_3COOH \xrightarrow{\text{Mo catalyst}} (CH_3)_3COH + \frac{1}{2}O_2$$

The oxygen released is recycled to the isobutane oxidation step. GTBA contains some methanol and acetone coproducts and is used as a blending agent for gasoline.

The other significant industrial route to t-butyl alcohol is the acid catalyzed hydration of isobutylene (24), a process no longer practiced in the United States. Raffinate I, C-4 refinery streams containing isobutylene [115-11-7], n-butenes and saturated C-4s or C-4 fluid catalytic cracker (FCC) feedstocks (23) may be employed.

4. Uses

The largest volume commercial derivatives of 1-butanol in the United States are n-butyl acrylate [141-32-2] and methacrylate [97-88-1] (10). These are used principally in emulsion polymers for latex paints, in textile applications and in impact modifiers for rigid poly(vinyl chloride). The consumption of n-butanol in the United States for acrylate and methacrylate esters was 270,000 t in 1997(10).

Butyl glycol ethers, the largest volume derivatives of n-butyl alcohol used in solvent applications (10), are obtained from the reaction of 1-butanol with ethylene oxide. The most important of these derivatives, 2-butoxyethanol, is used principally in vinyl and acrylic paints as well as in lacquers and varnishes. It is also employed in aqueous cleaners to solubilize organic surfactants. 2-Butoxyethanol [111-76-2] has achieved some growth at the expense of the lower alkoxyethanols (ie, methoxy and ethoxyethanol) because of 2-butoxyethanol's lower toxicity.

1-Butanol is used as a direct solvent in paints and other surface coatings. It acts synergistically with butyl acetate as a latent solvent system for nitrocellulose lacquers and thinners to give a solvent system stronger than either solvent alone. Other direct solvent applications for n-butyl alcohol are in the formulation of pharmaceuticals, waxes, and resins. Slightly more 1-butanol is consumed in western Europe in direct solvent applications than in the production of butyl acrylate and methacrylate in contrast to the United States where the acrylate and methacrylate esters are the predominent end products.

Butyl acetate [123-86-4], one of the more important derivatives of n-butyl alcohol produced commercially, is employed as a solvent in rapid drying paints and coatings. In some instances, butyl acetate, $C_6H_{12}O_2$, has replaced ethoxyethyl acetate [111-15-9] due to the latter's reported toxicity and teratogenicity. Butyl acetate is used in leather treatment, perfumes, and as a process or reaction solvent and is also used extensively with wood coatings, maintenance coatings, and in coatings for containers and closures.

Additional commercial markets for 1-butanol include plasticizer esters (eg, dibutyl phthalate), butylated melamine–formaldehyde resins, and mono-, di-, and tributylamines.

Historically, isobutyl alcohol was an unwanted by-product of the propylene Oxo reaction. Indeed, isobutyraldehyde the precursor of isobutyl alcohol was occasionally burned for fuel. However, more recently isobutyl alcohol has replaced n-butyl alcohol in some applications where the branched alcohol appears to have preferred properties and structure. However, supplies of isobutyl alcohol have declined relative to overall C-4 alcohols, especially in Europe, with the conversion of many Oxo plants to rhodium based processes which give higher normal to isobutyraldehyde isomer ratios. Further the supply of isobutyl alcohol at any given time can fluctuate greatly, since it is the lowest valued derivative of isobutyraldehyde, after neopentyl glycol, methyl isoamyl ketone and certain condensation products (10).

The principal industrial application for isobutyl alcohol is as a direct solvent replacement for 1-butanol. It is also used as a process solvent in the flavor and fragrance, pharmaceutical, and pesticide industries. The maximum employment of isobutyl alcohol was in the mid-1980s when it had a distinct price advantage over 1-butanol (10). More recently, however, with increased demand for other value added derivatives of isobutyraldehyde, the price differential between isobutyl and n-butyl alcohols has diminished resulting in a switching back by some consumers to 1-butanol.

Some commercially important isobutyl derivatives include isobutyl acetate, employed as a replacement solvent for n-butyl acetate; zinc dialkyldithiophosphate (ZDPP) lube oil additives; isobutyl acrylate [106-62-8] and methacrylate [97-86-9] monomers; isobutylamines; and amino resins (qv).

t-Butyl alcohol, obtained from hydration of Raffinate I, can be dehydrated and subsequently refined to high purity, polymer-grade isobutylene (25). However, the major use of t-butyl acohol is to produce methyl t-butyl ether (MTBE) gasoline additive by initial dehydration of t-butyl alcohol to isobutylene which is subsequently reacted with methanol in the presence of an acid catalyst (see ESTERS ORGANIC).

t-Butyl alcohol is employed as a feedstock in Japan to make methyl methacrylate monomer. In one such process (26), the alcohol is oxidized (in two steps) to acrylic acid, which is then esterified with methanol. In a similar process (27), t-butyl alcohol is oxidized in the presence of ammonia to give methacrylonitrile [126-98-7]. The latter is hydrolyzed to methacrylamide [79-39-0] which then reacts with methanol to yield methyl methacrylate [80-62-6].

The glycol ethers obtained from t-butyl alcohol and propylene oxide, eg, 1-t-butoxy-2-propanol, have lower toxicities than the widely employed 2-butoxyethanol and are used in industrial coatings and to solubilize organic components in aqueous formulations (28).

2-Butanol is employed almost exclusively to make the solvent methyl ethyl ketone [78-93-3], accounting for almost 90% of its global use [10]. In analogy to the primary butanols, 2-butanol is also employed as a direct solvent, as well as a blend with aromatic hydrocarbons, ketones and other alcohols.

5. Quality Specifications and Analysis

With the exception of gasoline grade t-butyl alcohol (GTBA), whose purity is typically 93–94% the butanols are generally marketed in bulk in the pure isomeric form. ASTM specifications (29) for n-, iso- and sec-butyl alcohol are given in Table 3. Butanol specification purity is routinely obtained by gas chromatography (30).

A cosmetics industry specification for t-butyl alcohol (31) is 99.5% alcohol, a maximum 0.002% acidity (as acetic acid), a maximum of 0.1% water, and a maximum of 0.001% nonvolatile matter.

Table 3. **ASTM Specification for Butyl Alcohols**

ASTM standard	n-Butyl D 304-85	Isobutyl D 1719-86	sec-Butyl D 1007-85	Specification method
apparent specific gravity				
20/20°C	0.810–0.813	0.802–0.804	0.809–0.809	D268
25/25°C	0.807–0.810	0.794–0.801	0.804–0.806	
color (Pt–Co), max	10	10	10	D1209
distribution range,[a] °C	117 ± 1.7	107.9 ± 2	98–101	D1078
nonvolatile material, mg/100 mL, max	5	5	5	D1353
water, wt % max	0.1	0.2	0.5	D1364 and 1476
acidity, as acetic acid, wt % max	0.005	0.003	0.002	D1613

[a] At 101.3 kPa = 760 mm Hg.

6. Health, Safety, and Environmental Considerations

All four butanols are thought to have a generally low order of human toxicity (32). However, large dosages of the butanols generally serve as central nervous system depressants and mucous membrane irritants. Animal toxicity and irritancy data (32) are given in Table 4.

The reported odor threshold limits (33) for n-, iso, sec- and t-butyl alcohol are 0.83, 1.6, 2.6, and 47 ppm, respectively.

Flammability characteristics (1) of the four butanols are given in Table 5.

Table 4. **Animal Toxicity and Irritancy Data for Butanols**

	LD_{50} rats, mg/kg, oral	LD_{50} rabbits, g/kg, intravenous	Inhalation	Eye injury to rabbits
1-butanol	4.36	5.3	8,000 ppm for 4 h (all rats survived)	5 µL (severe corneal damage)
isobutyl alcohol	3.4	0.6	10,600 ppm for 6 h (all mice survived)	one drop caused moderate to severe irritation
2-butanol	6.5	0.8	16,000 ppm for 4 h (5 of 6 rats died)	severe corneal injury (rabbit)
t-butyl alcohol	3.6	1.5 (mouse)		

Table 5. **Flammability Characteristics of the Butanols**

			Explosive limits, vol %	
	Flash point, °C	Autoignition temperature, °C	lower	upper
n-butyl	2–3	342.85	1.4	11.2
isobutyl	28	426.85	1.7	10.9
sec-butyl	21	405.85	1.7	9.8
t-butyl	11	477.85	2.4	8.0

All four butanols are registered in the United States on the Environmental Protection Agency Toxic Substances Control Act (TSCA) Inventory, a prerequisite for the manufacture or importation for commercial sale of any chemical substance or mixture in quantities greater than a 1000 pounds (454 kg). Additionally, the manufacture and distribution of the butanols in the United States are regulated under the Superfund Amendments and Reauthorization Act (SARA), Section 313, which requires that anyone handling at least 10,000 pounds (4545 kg) a year of a chemical substance report to both the EPA and the state any release of that substance to the environment.

7. Storage and Handling

The C-4 alcohols are preferably stored in baked phenolic-lined steel tanks. However, plain steel tanks can also be employed provided a fine porosity filler is installed to remove any contaminating rust (34).

Storage under dry nitrogen is also recommended since it limits flammability hazards as well as minimizing water pickup. There is a report of an explosion occurring during distillation of a sample of aged 2-butanol (35), suggesting that dangerous levels of peroxides can form in 2-butanol on storage in air.

Piping and pumps used for transfer of the butanols can be made of the same metal as tanks. Centrifugal pumps with explosion-proof electric motor drives are recommended (34).

BIBLIOGRAPHY

"Butyl Alcohols" in *ECT* 1st ed., Vol. 2, pp. 674–680, by C. L. Gabriel and A. A. Dolnick, Publicker Industries, Inc.; in *ECT* 2nd ed., Vol. 3, pp. 822–830;in *ECT* 3rd ed., Vol. 4, pp. 338–345, by P. D. Sherman, Jr., Union Carbide Corporation;"Butyl Alcohols" in *ECT* 4th ed., Vol. 4, pp. 691–700; "Butyl Alcohols" in *ECT*(online), posting date: December 4, 2000, by Ernst Billig, Union Carbide Chemicals and Plastics Company Inc.

CITED PUBLICATIONS

1. *AIChE Design Institute for Physical Property Data*, Project 801 Source File Tape, Revision Dates Aug. 1989 and 1990.
2. *Merck Index* **11**, 1541 (1989).
3. L. H. Horsley, *Anal. Chem.* **19**, 588 (1947).
4. L. H. Horsley, *Azeotropic Data—III*, American Chemical Society, Washington, D.C., 1973.
5. *TRC* Table 23-2-1-(1.1020)-k, Pg. 1, June 30, 1965, Pg. k-5030.
6. *TRC* Table 23-2-1-(1.1020)-k, Pg. 1, June 30, 1966, Pg. d-5030.
7. Ger. Offen. 3737277 (May 19, 1988), J. L. Logsdon, R. A. Loke, J. S. Merriam, and R. W. Voight (to Union Carbide Corporation).
8. J. B. Cropley, L. M. Burgess, and R. A. Loke, *Chemtech.* **14**, 374–380 (1984).
9. R. H. Crabtree, *The Organometallic Chemistry of the Transition Metals*, John Wiley & Sons, Inc., New York, 1988, 173–176.

10. *Chemical Economics Handbook*, SRI Consulting, Menlo Park, Calif.
11. S. Donmez, F. Ozcelik, and H. H. Pamir, *Doga: Turk Tarim Ormancilik Derg* **14**, 71–81 (1990).
12. Chin. Pat. 1040824 (Mar. 28, 1990), Y. Wang, S. Liang, and K. Wu (to People's Republic of China).
13. G. S. Kwon, B. H. Kim, and A. S. H. Ong, *Elaeis* **1**, 91–102 (1989).
14. *Chem. Wk.*, 17 (Nov. 22, 1989).
15. *Chem. Mark. Rep.*, 5 (Nov. 20, 1989).
16. W. Keim, *J. Organomet. Chem.* **372**, 15–23 (1989).
17. K. Klier and co-workers, *Report, DOE/PC/70021-T1-Rev. 1;* Order No. DE89003390, 305 pp., 1988.
18. *Eur. Chem. News.*, 11 (Jan. 16, 1984).
19. U.S. Pat. 4,296,263 (Oct. 20, 1981), (to Atlantic Richfield).
20. U.S. Pat. 4,239,926 (Dec. 16, 1980), (to Atlantic Richfield).
21. U.S. Pat. 4,294,999 (Oct. 13, 1981), (to Atlantic Richfield).
22. U.S. Pat. 4,296,262 (Oct. 20, 1981), (to Atlantic Richfield).
23. F. Nierlich, *Erdoel, Erdgas, Kohle* **103**, 486–489 (1987).
24. *Fed. Regist.* **48**(85), 19779–19780 (May 2, 1983).
25. *Chem. Eng. News*, 7 (Feb. 19, 1979).
26. *Jpn. Chem. Week*, 3 (Dec. 17, 1987).
27. *Jpn. Chem. Week*, 2–3 (Feb. 2, 1984).
28. R. A. Heckman, *Mod. Paint Coat.* **76**(6), 36–42 (1986).
29. *1990 Annual Book of ASTM Standards*, Philadelphia, Section 6, Vol. **06.03**.
30. *J. Am. Coll. Toxicol.* **8**(4), 627–641 (1989).
31. *Reagent Chemicals—American Chemical Society Specification*, American Chemical Society, Washington, D.C., 1986, 165–168.
32. F. E. C. George and D. Clayton, eds., *Patty's Industrial Hygiene and Toxicology*, Vol. **2C**, John Wiley & Sons, Inc., New York, 1982, 4571–4586.
33. X. Rousselin and M. Falcy, *Cah. Notes Doc.* **124**, 331–344 (1986).
34. *Brochure F 42379C, 10/79*, Union Carbide Chemical and Plastics, Inc., 1979.
35. V. F. Pozdnev, A. I. Tochilkin, and S. I. Kirrillova, *Zh. Org. Khim.* **13**, 456–457 (1977).

ERNST BILLIG
Union Carbide Chemicals
and Plastics Company Inc.

BUTYLENES

1. Introduction

Butylenes are C_4H_8 mono-olefin isomers: 1-butene, *cis*-2-butene, *trans*-2-butene, and isobutylene (2-methylpropene). These isomers are usually coproduced as a mixture and are commonly referred to as the C_4 fraction. These C_4 fractions are usually obtained as by-products from petroleum refinery and petrochemical complexes that crack petroleum fractions and natural gas liquids. Since the C_4 fractions almost always contain butanes, it is also known as the B–B stream. The linear isomers are referred to as butenes.

2. Physical Properties

For any industrial process involving vapors and liquids, the most important physical property is the vapor pressure. Table 1 presents values for the constants for a vapor-pressure equation and the temperature range over which the equation is valid for each butylene.

$$\ln P = A + B/T + C * \ln T + DT * N \tag{1}$$

P is in Pa and T is in K

where P is in Pa and T is in KA screening technique (1) was used to select the experimental data (2) used in the regressions. Large deviations often occur at low temperatures because of the inability of the equation to model the data over the entire temperature range accurately. In order to ensure that the equations are of practical value, the regressions are performed so that emphasis is placed on conditions of industrial importance; ie, data at subatmospheric conditions are weighed much less than those at pressures exceeding 101.3 kPa (1 atm).

Figure 1 presents the ratio of the vapor pressure of a compound to the vapor pressure of n-butane at the same temperature. For the chemically similar species included in this figure, this ratio is a first approximation of the relative volatility of the compound to n-butane. Whenever the ratios of two compounds approach one another, it becomes increasingly difficult to separate the compounds by simple distillation. Since the butylenes are usually present in mixtures containing the butanes, the butylenes, and the butadienes, Figure 1 shows the ratios for all these species. Figure 1 implies that separating either isobutylene, 1-butene and 1,3-butadiene, or n-butane, trans-2-butene and cis-2-butene, by conventional distillation would be very difficult, if not impossible. In fact, some binary mixtures containing these components form homogeneous azeotropes. The difficulty of these separations greatly influences the design of all industrial processes involving these compounds. If it is necessary to isolate one of these species from the others, it can be expected that the separation process will be expensive.

Table 1. Vapor-Pressure Equation Constants for the Butanes, Butylenes, and Butadienes[a]

	A	B	C	D	N	Temperature range, K
n-butane	61.5623	−4259.90	−6.20315	3.07575×10^{-7}	2.5	135–423
isobutane	66.7163	−4237.62	−7.08156	4.00506×10^{-7}	2.5	129–408
1–butene	78.8760	−4713.65	−9.05743	1.28654×10^{-5}	2.0	126–416
cis–2–butene	71.9534	−4681.34	−7.87527	1.00237×10^{-5}	2.0	203–358
trans–2–butene	74.3950	−4648.45	−8.33977	1.20897×10^{-5}	2.0	195–358
isobutylene	83.8683	−4822.95	−9.90214	1.51060×10^{-5}	2.0	194–359
1,2-butadiene	49.5031	−4021.95	−4.28893	5.13547×10^{-6}	2.0	200–284
1,3-butadiene	73.0016	−4547.77	−8.11105	1.14037×10^{-5}	2.0	164–425

[a] See equation 1.

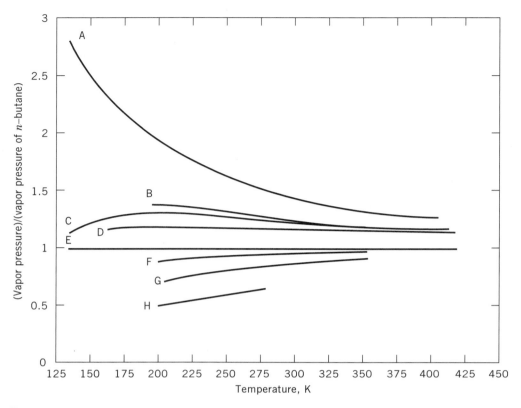

Fig. 1. Vapor-pressure ratios of the C_4 alkanes, alkenes, and dienes with respect to *n*-butane: A, isobutane; B, isobutylene; C, 1-butene; D, 1,3-butadiene; E, *n*-butane; F, *trans*-2-butene, G, *cis*-2-butene; and H, 1,2-butadiene.

Table 2 presents other important physical properties for the butylenes (3). Thermodynamic and transport properties can also be obtained from other sources (4).

3. Chemical Properties

The carbon–carbon double bond is the distinguishing feature of the butylenes and as such, controls their chemistry. This bond is formed by sp^2 orbitals (a sigma bond and a weaker pi bond). The two carbon atoms plus the four atoms in the alpha positions therefore lie in a plane. The pi bond which lies over the plane of the atoms acts as a source of electrons in addition reactions at the double bond. The carbon–carbon bond, acting as a substitute, affects the reactivity of the carbon atoms at the alpha positions through the formation of the allylic resonance structure. This structure can stabilize both positive and negative charges. Thus allylic carbons are more reactive to substitution and addition reactions than alkane carbons (5). Therefore, reactions of butylenes can be

Table 2. **Physical Properties of the Butylenes**

Property	Values			
	1-Butene	cis-2-Butene	trans-2-Butene	Isobutylene
CAS Registry Number	[106-98-9]	[590-18-1]	[624-64-6]	[115-11-7]
molecular weight	56.11	56.11	56.11	56.11
melting point, K	87.80	134.23	167.62	132.79
boiling point, K	266.89	276.87	274.03	266.25
critical temperature, K	419.60	435.58	428.63	417.91
critical pressure, MPa[a]	4.023	4.205	4.104	4.000
critical volume, L/mol	0.240	0.234	0.238	0.239
critical compressibility factor	0.277	0.272	0.274	0.275
acentric factor	0.1914	0.2018	0.2186	0.1984
Ideal gas properties[b] *at* 298.15 K				
Hf, kJ/mol	−0.126	−6.99	−11.18	−16.91
Gf, kJ/mol	71.34	65.90	63.01	58.11
Cp, J/mol · K	85.8	79.4	88.3	90.2
$Hvap$, kJ/mol	20.31	22.17	21.37	20.27
$Hcomb$, kJ/mol	−2719	−2712	−2708	−2724
Saturated vapor at 298.15 K				
viscosity, mPa · s(= cP)	0.00776	0.00782	0.00763	0.00816
thermal conductivity, W/(m · K)	0.0151	0.0135	0.0144	0.0158
Saturated liquid				
density at 298.15 K, mol/L	10.47	11.00	10.69	10.49
surface tension at 298.15 K, mN/m(= dyn /cm)	0.0121	0.0140	0.0132	0.0117
Cp at 266 K, J/mol · K[b]	121.6	118.8	121.8	123.3
viscosity at 266 K, mPa · s(= cP)	0.186	0.214	0.214	0.228
thermal conductivity at 266 K, W/(m · K)	0.120	0.124	0.121	0.117
Flammability limits, vol % in air				
lower limit	1.6	1.6	1.8	1.8
upper limit	9.3	9.7	9.7	8.8
autoignition temperature, K	657	598	597	738

[a] To convert MPa to atm, multiply by 9.869.
[b] To convert kJ to kcal, divide by 4.184.

divided into two broad categories: (*1*) those that take place at the double bond itself, destroying the double bond; and (*2*) those that take place at alpha carbons.

Differences in reactivity of the double bond among the four isomers are controlled by substitution pattern and geometry. Inductive effects imply that the carbons labeled B in Table 3 should have less electron density than the A carbons. ^{13}C nmr shift data, a measure of electron density, confirm this.

The electron-rich carbon−carbon double bond reacts with reagents that are deficient in electrons, eg, with electrophilic reagents in electrophilic addition (6,7), free radicals in free-radical addition (8,9), and under acidic conditions with another butylene (cation) in dimerization.

3.1. Electrophile Addition. The addition of electrophilic (acidic) reagents HZ involves two steps: the slow transfer of hydrogen ion from :Z to

Table 3. ^{13}C Nmr Shifts and Hydration Activation Energies
for Butylenes

Butylene	^{13}C Nmr^a ppm shift at		E_a for hydration, kJ/mol^b at	
	A	B	A	B
$C_A{=}C_B{-}C{-}C$	80.4	53.5	188	145
$C_A{=}C_B$ with two C groups	82.0^c	50.4^c	711	108
$C_A{=}_BC$ with two C groups	69.1	69.1	203	203
$C_A{=}_BC$ with C groups	67.7	67.7	203	203

^{a} Relative to external CS$_2$ reference.
^{b} To convert kJ to kcal, divide by 4.184.
^{c} These values obtained by interpolation.

the butylene to form a carbocation; and, a rapid combination of the carbocation
with the base :Z.

$$\underset{/}{\overset{\backslash}{C}}{=}\underset{\backslash}{\overset{/}{C}} + H{:}Z \xrightarrow{\text{slow}} {-}\underset{\overset{|}{H}}{\overset{|}{C}}{-}\overset{|}{\underset{+}{C}}{-} + {:}Z \xrightarrow{\text{fast}} {-}\underset{\overset{|}{H}}{\overset{|}{C}}{-}\underset{\overset{|}{Z}}{\overset{|}{C}}{-}$$

where HZ = HCl, HBr, HI, H$_2$SO$_4$, H$_3$O$^+$

The rate of addition depends on the concentration of both the butylene and
the reagent HZ. The addition requires an acidic reagent and the orientation of
the addition is regioselective (Markovnikov). The relative reactivities of the isomers are related to the relative stability of the intermediate carbocation and are
isobutylene $\gg$ 1 $-$ butene $>$ 2 $-$ butenes. Addition to the 1-butene is less hindered than to the 2-butenes. For hydrogen bromide addition, the preferred orientation of the addition can be altered from Markovnikov to anti-Markovnikov by
the presence of peroxides involving a free-radical mechanism.

Sulfuric acid is about one thousand times more reactive with isobutylene
than with the 1- and 2-butenes, and is thereby very useful in separating isobutylene as *tert*-butyl alcohol from the other butenes. The reaction is simply carried
out by bubbling or stirring the butylenes into 45–60% H$_2$SO$_4$. This results in
the formation of *tert*-butyl hydrogen sulfate. Dilution with water followed by
heat hydrolyzes the sulfate to form *tert*-butyl alcohol and sulfuric acid. The
Markovnikov addition implies that isobutyl alcohol is not formed. The hydration
of butylenes is most important for isobutylene, either directly or via the butyl
hydrogen sulfate.

Certain oxidizing agents convert butylenes into 1,2-diols. Of the numerous
oxidizing agents that bring about hydroxylation, two of the most commonly used

are cold alkaline potassium permanganate, $KMnO_4$, and peroxy acids such as peroxyformic acid, HCO_2OH. Aqueous hydrolysis of the intermediate epoxide is required. $KMnO_4$ gives syn-addition whereas peroxyformic acid gives antiaddition.

Bromine and chlorine convert the 1- and 2-butenes to compounds containing two atoms of halogens attached to adjacent carbons (vicinal dihalides). Iodine fails to react. In this two-step addition mechanism the first step involves the formation of a cation. The halonium ion formed (a three-membered ring) requires antiaddition by the anion.

$$\diagdown\!\!C\!\!=\!\!C\diagup + Br_2 \longrightarrow \overset{Br^+}{-\overset{|}{C}-\overset{|}{C}-} + Br^- \longrightarrow \overset{Br}{\underset{Br}{-\overset{|}{C}-\overset{|}{C}-}}$$

bromonium ion

Addition to *cis*- and *trans*-2-butene therefore yields different optical isomers (10,11). The failure of chlorine to attack isobutylene is attributed to the high degree of steric hindrance to approach by the anion. The reaction intermediate stabilizes itself by the loss of a proton, resulting in a very rapid reaction even at ambient temperature (12).

$$CH_3-\underset{CH_3}{\overset{|}{C}}=CH_2 + Cl_2 \longrightarrow CH_3-\underset{CH_3}{\overset{Cl^+}{\overset{|}{C}}}-CH_2 + Cl^- \longrightarrow CH_2=\underset{CH_3}{\overset{|}{C}}-CH_2Cl + HCl$$

Addition of chlorine or bromine in the presence of water can yield compounds containing halide and hydroxyl on adjacent carbon atoms (haloalcohols or halohydrins). The same products can be obtained in the presence of methanol (13) or acetic acid (14). As expected from the halonium ion intermediate, the addition is anti. As expected from Markovnikov's rule, the positive halogen goes to the same carbon that the hydrogen of a protic reagent would.

Butylenes can be catalytically hydrogenated in the presence of Pt, Pd, or Ni in an exothermic reaction. In the absence of a catalyst, this reaction proceeds at a negligible rate, even at elevated temperatures. Heats of hydrogenation in kJ/mol are as follows: 1-butene, −126.8; isobutylene, −118.8; *cis*-2-butene, −119.7; and *trans*-2-butene, −115.5.

Since a carbocation can add to an alkene to form a larger cation, under acidic conditions isobutylene can dimerize to form 2,4,4-trimethyl-1-pentene [107-39-1] and 2,4,4-trimethyl-2-pentene [107-40-4], which can then be hydrogenated in the presence of nickel to form isooctane [540-84-1]. This reaction is no longer of commercial significance.

Alkylation of isobutylene and isobutane in the presence of an acidic catalyst yields isooctane. This reaction proceeds through the same mechanism as dimerization except that during the last step, a proton is transferred from a surrounding alkane instead of one being abstracted by a base. The cation thus formed bonds with the base. Alkylation of aromatics with butylenes is another addition reaction and follows the same general rules with regard to relative rates and product structure. Thus 1- and 2-butenes yield *sec*-butyl derivatives and isobutylene yields *tert*-butyl derivatives.

Two other reactions of interest are oxymercuration–demercuration and hydroboration–oxidation. Both reactions amount to hydration of the double bond to the alcohol. The former gives Markovnikov addition whereas the latter yields anti-Markovikov addition. In the first reaction the butylene reacts in aqueous mercuric acetate to add $-OH$ and $-HgOOCCH_3$, to the double bond. Then the $—HgOOCCH_3$ is replaced by $—H$ from sodium borohydride. This reaction is very fast and proceeds with 90% yield. A mercurinium ion (in analogy with a halonium ion) is invoked to explain the addition products. In hydroboration, hydrogen and boron from BHR_2 add to the double bond, then the boron is displaced by hydrogen peroxide in alkaline solution. The intermediate here is a four-centered transition state. As boron gains the pi electrons it becomes increasingly willing to release the hydrogen (see HYDROBORATION).

$$\underset{\delta^+\;\;\;\;\delta^-}{CH_3CH_2CH}{=}CH_2 \; + \; \underset{\delta^-\;\;\delta^+}{H}{-}\underset{\diagdown}{B} \quad \longrightarrow \quad \underset{\underset{H{-}B{-}\delta^-}{|\;\;\;\;\;|}}{\overset{\delta^+}{CH_3CH_2CH}{=}CH_2} \quad \longrightarrow \quad CH_3CH_2CH_2CH_2B\overset{\diagup}{\diagdown}$$

Butylene isomers also can be expected to show significant differences in reaction rates for metallation reactions such as hydroboration and hydroformylation (addition of $HCo(CO)_4$). For example, the rate of addition of di(*sec*-isoamyl)borane to *cis*-2-butene is about six times that for addition to *trans*-2-butene (15). For hydroformylation of typical 1-olefins, 2-olefins, and 2-methyl-1-olefins, specific rate constants are in the ratio $100:31:1$, respectively.

The composition of the products of reactions involving intermediates formed by metallation depends on whether the measured composition results from kinetic control or from thermodynamic control. Thus the addition of diborane to 2-butene initially yields tri-*sec*-butylborane. If heated and allowed to react further, this product isomerizes about 93% to the tributylborane, the product initially obtained from 1-butene (15). Similar effects are observed during hydroformylation reactions; however, interpretation is more complicated because the relative rates of isomerization and of carbonylation of the reaction intermediate depend on temperature and on hydrogen and carbon monoxide pressures (16).

These reactions are also quite sensitive to steric factors, as shown by the fact that if 1-butene reacts with di(*sec*-isoamyl)borane the initially formed product is 99% substituted in the 1-position (15) compared to 93% for unsubstituted borane. Similarly, the product obtained from hydroformylation of isobutylene is about 97% isoamyl alcohol and 3% neopentyl alcohol (17). Reaction of isobutylene with aluminum hydride yields only triisobutylaluminum.

Selectivity among butylene isomers also occurs in vapor-phase heterogeneous catalysis, at least in the case of dehydrogenation of butenes to butadiene, where maximum yields can be obtained by employing slightly different conditions for each isomer (18). In practice, mixtures of isomers are used and an average set of conditions is employed.

3.2. Free-Radical Addition.   Free-radical attack on a butylene occurs so that the most stable radical carbon structure forms. Thus, in peroxide-catalyzed addition of hydrogen halides, the addition is anti-Markovnikov.

$$\ce{>C=C< + Y:Z -> >C-C< + Y. -> >C-C-}$$

This reaction proceeds through a chain mechanism. Free-radical additions to 1-butene, as in the case of HBr, RSH, and H_2S to other olefins (19–21), can be expected to yield terminally substituted derivatives. Some polymerization reactions are also free-radical reactions.

3.3. Polymerization. Polymerization reactions, which are addition reactions, are used to produce the principal products formed directly from butlylenes: butyl elastomers; polybutylenes; and polyisobutylene (see ELASTOMERS, SYNTHETIC; OLEFIN POLYMERS).

3.4. Substitution Reactions. The chemistry at alpha positions hinges on the fact that an allylic hydrogen is easy to abstract because of the resonance structures that can be established with the neighboring double bond. The allylic proton is easier to abstract than one on a tertiary carbon; these reactions are important in the formation of alkoxybutenes (ethers).

3.5. Isomerization. Isomerization of any of the butylene isomers to increase supply of another isomer is not practiced commercially. However, their isomerization has been studied extensively because: formation and isomerization accompany many refinery processes; maximization of 2-butene content maximizes octane number when isobutane is alkylated with butene streams using HF as catalyst; and isomerization of high concentrations of 1-butene to 2-butene in mixtures with isobutylene could simplify subsequent separations (22). One plant (Phillips) is now being operated for this latter purpose (23,24). The general topic of isomerization has been covered in detail (25–27). Isomer distribution at thermodynamic equilibrium in the range 300–1000 K is summarized in Table 4 (25).

The three isomerizations, *cis*-2-butene ⇌ *trans*-2-butene, 1-butene ⇌ 2-butene, and butenes ⇌ isobutylene, require increasingly severe reaction conditions. When the position of the double bond is shifted, cis-trans isomerization

Table 4. Equilibrium Butylenes Distribution, Ideal Gas[a,b]

Temperature, K	Mol %				
	1-Butene	*cis*-2-Butene	*trans*-2-Butene	Total butenes	Isobutylene
300	0.4	3.8	11.8	16.0	84.0
400	1.9	8.3	18.0	28.2	71.8
500	4.5	11.9	21.6	38.0	62.0
600	7.5	14.4	23.4	45.3	54.7
700	10.8	15.8	24.2	50.8	49.2
800	14.0	16.6	24.5	55.1	44.9
900	16.9	17.0	24.6	58.6	41.4
1000	19.6	17.2	24.4	61.2	38.8

[a] At 101.3 kPa = 1 atm.
[b] Ref. 25.

Table 5. **Isomerization of Butylenes**

Catalyst	Temperature, °C	References
cis-2-*Butene* ⇌ trans-2-*butene*		
^{60}Co γ-rays	ambient	28
Na mordenite and porous Vycor	24.5	28
bis(acetylacetonato)Pd–SiO$_2$	61.5	29
1-*Butene* ⇌ -2-*butene*		
Pt–SiO$_2$–Al$_2$O$_3$	−10	30
RhCl$_3$–SnCl$_2$–CH$_3$OH	ambient	31
Cl$_2$[(C$_4$H$_9$)$_3$P]$_2$Ni–(C$_2$H$_5$)$_2$AlCl–SiO$_2$	ambient	32
BF$_3$–Al$_2$O$_3$	ambient	33
H$_3$PO$_4$, 85%	73	34
Ga$_2$O$_3$	190–330	35
iodine	200–250	36
ZnCrFeO$_4$	465	37
Butene ⇌ *isobutylene*		
Pt–Al$_2$O$_3$–SiO$_2$	475	38
SiO$_2$	520	39

also occurs, and mixtures of butenes result when the carbon skeleton is rear-ranged. However, during isomerization of 1-butene to 2-butene, with solid cata-lysts, the cis isomer is preferentially formed initially even though it is the thermodynamically less favored isomer.

An extremely wide variety of catalysts, Lewis acids, Brønsted acids, metal oxides, molecular sieves, dispersed sodium and potassium, and light, are effec-tive (Table 5). Generally, acidic catalysts are required for skeletal isomerization and reaction is accompanied by polymerization, cracking, and hydrogen transfer, typical of carbenium ion intermediates. Double-bond shift is accomplished with high selectivity by the basic and metallic catalysts.

4. Manufacture

The C$_4$ isomers are almost always produced commercially as by-products in a petroleum refinery/petrochemical process as shown in Figure 2. Environmental regulations mandated by recent changes in the laws of the United States to reduce the aromatic content in gasoline will have an impact on butylene produc-tion in this country. As petroleum refiners search for alternative routes to replace the aromatics in the gasoline pool, oxygenated hydrocarbons will become increasingly attractive, not only for regaining the lost octane value in the gaso-line but also for producing a clean burning fuel. Among this class of oxygenates, methyl-*tert*-butyl ether (MTBE) produced from isobutylene appears to be a lead-ing contender (see ETHERS). Free-standing facilities at the gas well head to produce an isobutylene-rich B–B stream cannot be ruled out in the future.

There are two important sources for the commercial production of buty-lenes: catalytic or thermal cracking, and steam cracking. In these two processes, butylenes are always produced as by-products. A catalytic cracking process is

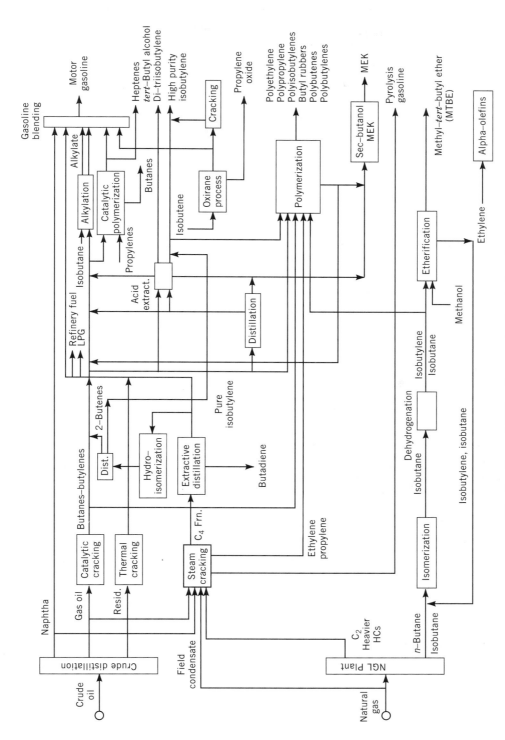

Fig. 2. General U.S. refinery-petrochemical system.

411

always associated with a petroleum refining complex that upgrades high boiling fractions of hydrocarbons to high octane gasoline. Steam cracking converts a variety of hydrocarbon feedstocks that range from natural gas liquids to heavy petroleum fractions to produce light olefins. As demand for butylenes picks up in the future, a third important source for the commercial production of these products is likely to be the dehydrogenation of butanes.

There are other commercial processes available for the production of butylenes. However, these are site or manufacturer specific, eg, the Oxirane process for the production of propylene oxide; the disproportionation of higher olefins; and the oligomerization of ethylene. Any of these processes can become an important source in the future. More recently, the Coastal Isobutane process began commercialization to produce isobutylene from butanes for meeting the expected demand for methyl-*tert*-butyl ether (40).

Table 6 shows the yield distribution of the C_4 isomers from different feedstocks with specific processing schemes. The largest yield of butylenes comes from the refineries processing middle distillates and from olefins plants cracking naphtha. The refinery product contains 35 to 65% butanes; olefins plants, 3 to 5%. Catalyst type and operating severity determine the selectivity of the C_4 isomer distribution (41) in the refinery process stream. Processes that parallel fluid catalytic cracking to produce butylenes and propylene from heavy crude oil fractions are under development (42).

4.1. Steam Cracking. Steam cracking is a nonselective process that produces many products from a variety of feedstocks by free-radical reactions. An excellent treatise on the fundamentals of manufacturing ethylene has been given (44). Feedstocks range from ethane on the light end to heavy vacuum gas oil on the heavy end. All produce the same product slate but in different amounts depending on the feedstock.

Significant products from a typical steam cracker are ethylene, propylene, butadiene, and pyrolysis gasoline. Typical wt % yields for butylenes from a steam cracker for different feedstocks are ethane, 0.3; propane, 1.2; 50% ethane/50% propane mixture, 0.8; butane, 2.8; full-range naphtha, 7.3; light gas oil, 4.3. A typical steam cracking plant cracks a mixture of feedstocks that results in butylenes yields of about 1% to 4%. These yields can be increased by almost 50% if cracking severity is lowered to maximize propylene production instead of ethylene.

Cracking conditions and feed slate are usually selected to maximize production of light olefins. Selectivity to light olefins depends on the temperature and pressure profiles in the pyrolysis reactor coil, and thus the residence time. These profiles are unique for a given reactor coil, so a great deal of attention goes into the selection of the reactor. Older plants that have a residence time of about 1 s have since been modernized to under 0.4 s by replacing the reactor coil. Newer plants have reactor coil designs that give residence times of 0.1–0.2 s.

Typically, cracking is done at a weight ratio of steam to hydrocarbon that ranges from 0.2 to 1. The high end of this ratio is used for heavy feeds such as vacuum gas oil. Desired cracking severity is achieved at 780–875°C at the reactor coil outlet and at slightly above 130 kPa (19 psi) pressure. Hot furnace effluent from the reactors is quenched rapidly to stop undesirable secondary

Table 6. Typical Yields and % Compositions of C4 Fractions from Cracking Operations[a]

Butylene yield	Catalytic cracking — Gas oil	Catalytic cracking — Residue		Thermal cracking of residue — Delayed coking		Thermal cracking of residue — Flexicoking		Steam cracking of naphtha and light gas oil	
	Total	Olefin	Total	Total	Olefin	Total	Olefin	Total	Olefin
wt % on crude	0.5–5		0.1–0.6	0.15–0.8				0.4–0.5	
wt % on feed	3–10		1–1.5	1.5–2					
C4 composition									
butane	7–13		7	47		14–23		2–5	
isobutane	28–52		18–14	12		5		1.5–0.6	
isobutylene	26–8	40–23	79	16	40	13	20–18	27.4–22.0	48
1-butene	8–7	12–20	79	13	31	17	26–24	16.0–14.0	30
cis-2-butene	31–20	48–57	75–79	5	12	35–42	54–58	5.5–4.8	10
trans-2-butene	31–20	48–57	70	7	17	35–42	54–58	6.5–5.8	12
1,3-butadiene	0.1–0.5			0.5		7–9		37.0–47.5	

[a] Ref. 43.

413

reactions. Effluent streams are cooled quickly in heat exchangers to slightly above their dew point, about 120–370°C depending on the feedstock.

Figure 3 shows a typical arrangement of a steam cracker in the United States. Furnace effluent from a steam cracker consists of three phases at ambient temperatures including aqueous liquid, hydrocarbon liquid, and hydrocarbon gas. Effluent from the heat exchangers are further cooled in oil and water wash towers. The oil wash essentially removes the heavy fuel oil fraction and also limits the end point of pyrolysis gasoline in the overhead of this tower. The water wash condenses the dilution steam and the pyrolysis gasoline. The overhead from the water wash tower contains mostly C_4 and lighter fractions. Several fractionation sequences to separate high purity products are available commercially. The choice of sequence depends on the feed slate and economics. Figure 3 shows a front-end demethanizer scheme, which is usually used in steam crackers producing significant amounts of butylenes. Whatever the fractionation scheme used, the C_4 fraction is removed as overhead from the debutanizer. References 45 and 46 give an overview of the ethylene manufacturing process.

The C_4 stream from steam crackers, unlike its counterpart from a refinery, contains about 45% butadiene by weight. Steam crackers that process significant amounts of liquid feedstocks have satellite facilities to recover butadiene from the C_4 stream. Conventional distillation techniques are not feasible because the relative volatility of the chemicals in this stream is very close. Butadiene and butylenes are separated by extractive distillation using polar solvents.

The selection of solvent is dictated by the process used. Strongly dipolar, aprotic solvents alone or mixed with a second solvent to improve separation selectivity are used. The second solvent is usually water, and good solubility in water is an advantage. Toxicity is also an important consideration. Reference 47 is valuable in the selection of solvents. Several extractive distillation technologies are available commercially. Three technologies are used widely including the Nippon-Zein process using dimethylformamide [68-12-2] (48,49), the Shell process using acetonitrile [75-05-8] (50), and the BASF process using N-methyl-pyrrolidinone [872-50-4] (51). All these processes produce polymer-grade 1,3-butadiene and a B–B stream. C_4 Acetylenes and 1,2-butadiene in the B–B product are hydrogenated to produce a clean B–B stream.

4.2. Catalytic Cracking. This is a refinery process that produces a mixture of butylenes and butanes with very small amounts of butadiene. The specific composition of the C_4 mixture depends on the catalyst and process conditions. Most catalytic cracking processes employ temperatures about 450–650°C at pressures of about 250–400 kPa (36–58 psi). The two types of catalysts, the amorphous silica–alumina (52) and the crystalline aluminosilicates called molecular sieves or zeolites (53), exhibit strong carbonium ion activity. Although there are natural zeolites, over 100 synthetic zeolites have been synthesized and characterized (54). Many of these synthetic zeolites have replaced alumina with other metal oxides to vary catalyst acidity to effect different type catalytic reactions, for example, isomerization. Zeolite catalysts strongly promote carbonium ion cracking along with isomerization, disproportionation, cyclization, and proton transfer reactions. Because butylene yields depend on the catalyst and process conditions, Table 6 shows only approximations.

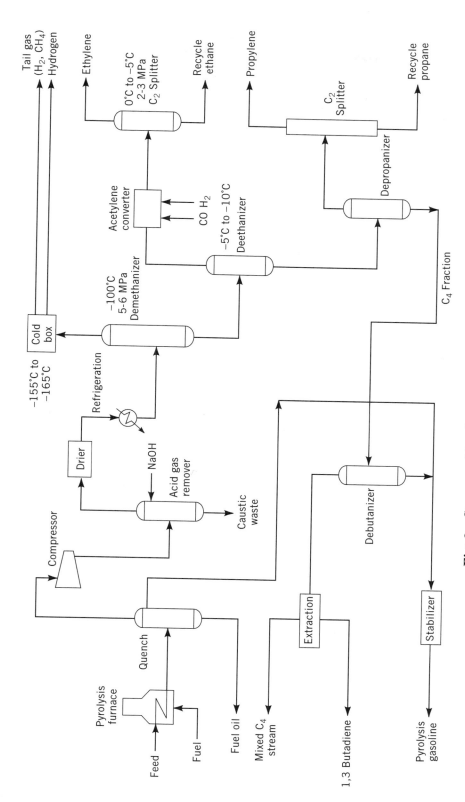

Fig. 3. Steam cracking front-end demethanizer scheme.

415

Zeolites have largely replaced the silica–alumina catalysts. In addition, the catalytic property is further improved by changing the silica and alumina ratio and by introducing cations such as hydrogen and sodium to impart specific catalytic properties. The most significant advance is in improved selectivity to gasoline range products and not in increased activity. Detailed information on the chemistry of catalytic cracking is available (55).

The most dominant catalytic process in the United States is the fluid catalytic cracking process. In this process, partially vaporized medium-cut petroleum fractions called gas oils are brought in contact with a hot, moving, freshly regenerated catalyst stream for a short period of time at process conditions noted above. Spent catalyst moves continuously into a regenerator where deposited coke on the catalyst is burnt off. The hot, freshly regenerated catalyst moves back to the reactor to contact the hot gas oil (see CATALYSTS, REGENERATION).

4.3. Thermal Cracking. Heavy petroleum fractions such as resid are thermally cracked in delayed cokers or flexicokers (44,56,57). The main products from the process are petroleum coke and off-gas which contain light olefins and butylenes. This stream also contains a considerable amount of butane. Process conditions for the flexicoker are more severe than for the delayed coker, about 550°C versus 450°C. Both are operated at low pressures, around 300–600 kPa (43–87 psi). Flexicokers produce much more linear butenes, particularly 2-butene, than delayed cokers and about half the amount of isobutylene (Table 6). This is attributed to high severity of operation for the flexicoker (43).

4.4. Oxirane Process. In Arco's Oxirane process, *tert*-butyl alcohol is a by-product in the production of propylene oxide from a propylene–isobutane mixture. Polymer-grade isobutylene can be obtained by dehydration of the alcohol. *tert*-Butyl alcohol [75-65-0] competes directly with methyl-*tert*-butyl ether as a gasoline additive, but its potential is limited by its partial miscibility with gasoline.

4.5. Disproportionation of Olefins. Disproportionation or the metathesis reaction offers an opportunity to convert surplus olefins to other desirable olefins. Phillips Petroleum and Institut Français du Petrôle have pioneered this technology for the dimerization of light olefins. The original metathesis reaction of Phillips Petroleum was intended to convert propylene to 2-butene and ethylene (58). The reverse reaction that converts 2-butene in the presence of excess ethylene to propylene has also been demonstrated (59). In this process, ethylene is first dimerized to 2-butene followed by metathesis to yield propylene. Since this is a two-stage process, 2-butene can be produced from the first stage, if needed. In the dimerization step, about 95% purity of 2-butene is achieved at 90% ethylene conversion.

In the Institut Français du Petrôle process (60), ethylene is dimerized into polymer-grade 1-butene (99.5% purity) suitable for the manufacture of linear low density polyethylene. It uses a homogeneous catalyst system that eliminates some of the drawbacks of heterogeneous catalysts. It also inhibits the isomerization of 1-butene to 2-butene, thus eliminating the need for superfractionation of the product (61,62). The process also uses low operating temperatures, 50–60°C, and pressures (63).

Many heterogeneous catalysts have been commercialized to dimerize ethylene to selectively yield 1-butene or 2-butene (64–68). Since ethylene is generally priced higher than butylenes, economics favor the production of butylenes from steam crackers, not from ethylene. An excellent review on metathesis is available (69).

4.6. Oligomerization of Ethylene.

1-Butene is a small by-product in the production of linear alpha-olefins by oligomerization of ethylene. Linear alpha-olefins have one double bond at the terminal position and comprise the homologous series of compounds with carbon atoms between 4 and 19. The primary use of alpha-olefins is in the detergent industry.

4.7. Other Technology.

Several technologies are emerging for the production of isobutylene to meet the expected demand for isobutylene: (1) deep catalytic cracking; (2) superflex catalytic cracking; (3) dehydrogenation of butanes; and (4) the Coastal process of thermal dehydrogenation of butanes. Of these, both the dehydrogenation technology and the high pressure thermal pyrolysis technology (the Coastal process) have been around for a long time. These technologies were not economical since inexpensive sources of butylenes were available from petroleum refineries and steam crackers. During the 1960s isobutane was in plentiful supply, and the first commercial unit using the Coastal process was built in 1969 at Corpus Christi, Texas, with a capacity of about 150 million t/yr (40). The dehydrogenation technology was also in use where there was a surplus of inexpensive isobutane.A process for producing isobutylene and methanol by the decomposition of methyl-*tert*-butyl ether has been reported (70).

Deep Catalytic Cracking. This process is a variation of fluid catalytic cracking. It uses heavy petroleum fractions, such as heavy vacuum gas oil, to produce propylene- and butylene-rich gaseous products and an aromatic-rich liquid product. The liquid product contains predominantly benzene, toluene, and xylene (see BTX PROCESSING). This process was developed by SINOPEC in China (42,71).

Superflex Catalytic Cracking. A new process called Superflex is being commercialized to produce predominantly propylene and butylenes from low valued hydrocarbon streams from an olefins complex (72). In this process, raffinates (from the aromatics recovery unit and the B–B stream after the recovery of isobutylene) and pyrolysis gasoline (after the removal of the C_6–C_8 aromatics fraction) are catalytically cracked to produce propylene, isobutylene, and a crude C_6–C_8 aromatics fraction. All other by-products are recycled to extinction.

Dehydrogenation of Butanes. These processes are based on the propane dehydrogenation technology commercialized about 35 years ago. Thermodynamics dictate that the operation be carried out at high temperatures and low pressures to improve selectivity. In the dehydrogenation process, conversion of feedstock is equilibrium limited, and thus conversions are low relative to steam cracking. Work has been carried out by Air Products, UOP, Shell, Ashland, ICI, Monsanto, Phillips, and Petrotex. Among these, five distinct technologies are available for converting isobutane to isobutylene. These technologies include Oleflex from UOP (73), Catofin from CDTECH (74), fluidized-bed from Snamprogetti (75), STAR from Phillips Petroleum (76), and Coastal Thermal Cracking from Foster-Wheeler (40,77).

The UOP Oleflex process uses a proprietary platinum catalyst. Dehydrogenation of isobutane to isobutylene is endothermic, and optimum catalyst activity is maintained by supplying the heat of reaction through interheaters. The catalyst system employs UOP's Continuous Catalyst Regeneration (CCR) technology. The bed of catalyst slowly flows concurrently with the reactants and is removed from the last reactor and regenerated in a separate section. The reconditioned catalyst is then returned to the top of the first reactor. The CCR process is usually applied in the reforming of naphtha to aromatics. When supply is limited, n-butane can be isomerized to isobutane using UOP's Butamer process (78). The Butamer process is a fixed-bed, vapor-phase catalytic process that uses organic halides as promoters.

The Catofin process, which was formerly the property of Air Products (Houdry Division), uses a proprietary chromium catalyst in a fixed-bed reactor operating under vacuum. There are actually multiple reactors operating in cyclic fashion. In sequence, these reactors process feed for about nine minutes and are then regenerated for nine minutes. The chromium catalyst is reduced from Cr^{6+} to Cr^{3+} during the regeneration cycle.

The Snamprogetti fluidized-bed process uses a chromium catalyst in equipment that is similar to a refinery catalytic cracker (1960s cat cracker technology). The dehydrogenation reaction takes place in one vessel with active catalyst; deactivated catalyst flows to a second vessel, which is used for regeneration. This process has been commercialized in Russia for over 25 years in the production of butenes, isobutylene, and isopentenes.

The Phillips Steam Active Reforming (STAR) process catalytically converts isobutane to isobutylene. The reaction is carried out with steam in tubes that are packed with catalyst and located in a furnace. The catalyst is a solid, particulate noble metal. The presence of steam diluent reduces the partial pressure of the hydrocarbons and hydrogen present, thus shifting the equilibrium conditions for this system toward greater conversions.

The Coastal process uses steam pyrolysis of isobutane to produce propylene and isobutylene (as well as other cracked products). It has been suggested that the reaction be carried out at high pressure, >1480 kPa (∼15 atm), to facilitate product separation. This process was commercialized in the late 1960s at Coastal's Corpus Christi refinery.

These processes are all characterized by low isobutane conversion to achieve high isobutylene selectivity. The catalytic processes operate at conversions of 45–55% for isobutane. The Coastal process also operates at 45–55% isobutane conversion to minimize the production of light ends. This results in significant raw material recycle rates and imposing product separation sections.

Dehydrogenation of isobutane to isobutylene is highly endothermic and the reactions are conducted at high temperatures (535–650°C) so the fuel consumption is sizeable. For the catalytic processes, the product separation section requires a compressor to facilitate the separation of hydrogen, methane, and other light hydrocarbons from-the paraffinic raw material and the olefinic product. An excellent overview of butylenes is available (79).

4.8. Separation and Purification of C_4 Isomers.

1-Butene and isobutylene cannot be economically separated into pure components by conventional distillation because they are close boiling isomers (see Table 1 and Fig. 1). 2-Butene can be separated from the other two isomers by simple distillation. There are four types of separation methods available: (1) selective removal of isobutylene by polymerization and separation of 1-butene; (2) use of addition reactions with alcohol, acids, or water to selectively produce pure isobutylene and 1-butene; (*3*) selective extraction of isobutylene with a liquid solvent, usually an acid; and (*4*) physical separation of isobutylene from 1-butene by absorbents. The first two methods take advantage of the reactivity of isobutylene. For example, isobutylene reacts about 1000 times faster than 1-butene. Some 1-butene also reacts and gets separated with isobutylene, but recovery of high purity is possible. The choice of a particular method depends on the product slate requirements of the manufacturer. In any case, 2-butene is first separated from the other two isomers by simple distillation.

There are currently three important processes for the production of isobutylene: (*1*) the extraction process using an acid to separate isobutylene; (*2*) the dehydration of *tert*-butyl alcohol, formed in the Arco's Oxirane process; and (*3*) the cracking of MTBE. The expected demand for MTBE will preclude the third route for isobutylene production. Since MTBE is likely to replace *tert*-butyl alcohol as a gasoline additive, the second route could become an important source for isobutylene. Nevertheless, its availability will be limited by the demand for propylene oxide, since it is only a coproduct. An alternative process has emerged that consists of catalytically hydroisomerizing 1-butene to 2-butenes (*80*). In this process, trace quantities of butadienes are also hydrogenated to yield feedstocks rich in isobutylene which can then be easily separated from 2-butenes by simple distillation.

The acid extraction process uses strong mineral acids, such as sulfuric, hydrochloric, and phosphoric. These acids selectively remove isobutylene from mixed butylenes. The Exxon, BASF, and the Compagnie Française de Raffinage (CFR) processes have been commercialized. The Nippon Petrochemical extraction process uses hydrochloric acid along with a heavy-metal catalyst. Reportedly, the selectivity to isobutylene hydration is significantly higher than in the sulfuric acid process. However, because of corrosion problems of stainless steel, this process requires titanium and palladium alloys.

The CRF, Exxon, and BASF processes use sulfuric acid as the extraction medium. The BASF process is the dominant process in Europe. It uses the dilutest acid of any commercial process. This permits selective reaction even in the presence of butadiene. The BASF process uses vapor–liquid extraction unlike the Exxon and CFR processes which are of the liquid–liquid type.

The desired extraction process is the exothermic proton-catalyzed hydrolysis of isobutylene to *tert*-butyl alcohol. This alcohol is further dehydrated to yield pure isobutylene. At low concentrations the hydrolysis reaction is favored:

$$(CH_3)_2C{=}CH_2 + H_2O \xrightarrow{\;H^+\;} (CH_3)_3COH$$

At high concentrations, there is a tendency to form an organic hydrosulfate:

$$(CH_3)_2C{=}CH_2 + H_2SO_4 \longrightarrow (CH_3)_3CHSO_4$$

The main differences between these processes are the acid concentration and the extraction temperature to effect selective removal of isobutylene. The acid concentration range is 45–65%. Figure 4 shows a simplified flow diagram of the CFR process.

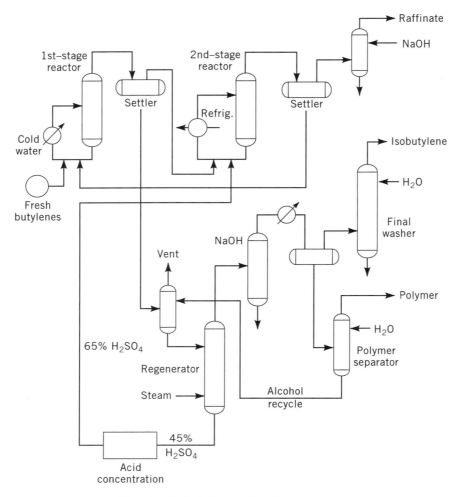

Fig. 4. Isobutylene extraction process.

In the physical separation process, a molecular sieve adsorbent is used as in the Union Carbide Olefins Siv process (81–83). Linear butenes are selectively adsorbed, and the isobutylene effluent is distilled to obtain a polymer-grade product. The adsorbent is a synthetic zeolite, Type 5A in the calcium cation exchanged form (84). UOP also offers an adsorption process, the Sorbutene process (85). The UOP process utilizes a liquid B–B stream, and uses a proprietary rotary valve containing multiple ports, which direct the flow of liquid to various sections of the adsorber (86,87). The cis- and trans-isomers are alkylated and used in the gasoline blending pool.

1-Butene can be separated from 2-butenes by simple distillation. If the B–B streams contain dienes, these must be hydrogenated prior to the separation of the linear butenes. If not hydrogenated, these contaminants tend to divide themselves between the purified isomers. Trace quantities of acetylinic compounds and butadiene are hydrogenated selectively using a noble metal catalyst.

Hydrogenation after separation is not desirable as the catalyst used for hydrogenation isomerizes butenes, which affects product purity. If butanes are also present, as they are in the refinery streams, they also distribute themselves in the purified products. If pure isomers are required, butanes can be separated by extractive distillation, and the residual C_4 isomers can be isomerized. These all increase the cost of the separation process. There is a balance between the purity sought and the cost associated in achieving it.

5. Shipment and Handling

5.1. Storage and Transportation.
Handling requirements are similar to liquefied petroleum gas (LPG). Storage conditions are much milder. Butylenes are stored as liquids at temperatures ranging from 0 to 40°C and at pressures from 100 to 400 kPa (1–4 atm). These conditions are much lower than those required for LPG. Their transportation is also similar to LPG; they are shipped in tank cars, transported in pipelines, or barged.

6. Economic Aspects

The United States, Western Europe, and Japan account for 90% of the butylenes market. In 2001, 32×10^6 t of butylenes were consumed in these three regions (88).

Fuel markets comprise 90% of the total use of butylenes. The fuel market includes the manufacture of gasoline, blending of components, gasoline alkylates, polymer gasoline, and dimersol. Isobutylene is a raw material for oxygenate, methyl-*tert*-butyl ether, and ethyl-*tert*-butyl ether. Butylenes are directly blended into gasoline for volatility control. Another market is with propane and butane as liquefied petroleum gas.

The other 10% of the market consists of *n*-butenes as precursors for *sec*-butyl alcohols, butadiene, and 1-butene. Chemical markets worldwide are growing slowly. Isobutylene has grown slowly in the U.S., Western Europe, and Japan, especially in the U.S. Other isobutylene derivatives, such as butyl rubber and polybutylenes are mature markets.

7. Analytical Methods

Butenes are best characterized by their property of decolorizing both a solution of bromine in carbon tetrachloride and a cold, dilute, neutral permanganate solution (the Baeyer test). A solution of bromine in carbon tetrachloride is red; the dihalide, like the butenes, are colorless. Decoloration of the bromine solution is rapid. In the Baeyer test, a purple color is replaced by brown manganese oxide (a precipitate) and a colorless diol. These tests apply to all alkenes.

Identification of C_4 isomers is now routinely performed by gas chromatography. Advances in column technology permit rapid analysis with good accuracy in capillary columns (89). Pure isomers require quantification of contaminants, usually in parts per million. Gas–liquid chromatography and mass spectroscopy are the most commonly used analytical tools. ASTM standards are excellent sources of analytical procedures for hydrocarbon mixtures. Al_2O_3–KCl PLOT capillary gas chromatographic columns provide very good and fast separation in the following elution order: *trans*-2-butene, 1-butene, isobutylene, and *cis*-2-butene.

8. Health and Safety Factors

The effect of long-term exposure is not known, hence, they should be handled with care. Reference 90 lists air and water pollution factors and biological effects. They are volatile and asphyxiants. 1-Butene, *cis* 2-butene, and *trans* 2-butene are very dangerous fire hazards when exposed to heat or flame. *cis* 2-Butene and *trans*-2-butene are very likely to explode. 1-Butene is moderately explosive. 2-Butene emits acrid smoke and irritating vapors when heated to decomposition (91). Care should be taken to avoid spills because they are extremely flammable. Physical handling requires adequate ventilation to prevent high concentrations of butylenes in the air. Explosive limits in air are 1.6 to 9.7% of butylenes. Their flash points range from −80 to −73°C. Their autoignition is around 324 to 465°C (Table 2). Water and carbon dioxide extinguishers can be used in case of fire.

9. Uses

Among the butylenes, isobutylene has become one of the important starting materials for the manufacture of polymers and chemicals. There are many patents that describe the use of isobutylene or its derivatives to produce insecticides, antioxidants, elastomers, additives for lubricating oils, adhesives, sealants, and caulking compounds.

9.1. Fuels. *Alkylate.* Alkylation means the chemical combination of isobutane with any one or a combination of propylene, butylenes, and amylenes to produce a mixture of highly branched paraffins that have high antiknock properties with good stability. These reactions are catalyzed by strong acids such as sulfuric or hydrofluoric acid and have been studied extensively (45,92–96). In the United States mostly butylenes and propylene are used as the olefins. The alkylate contains a mixture of isoparaffins, ranging from pentanes to decanes and higher, regardless of the olefins used. The dominant paraffin in the product is 2,2,4-trimethylpentane, also called isooctane. The reaction involves methide-ion transfer and carbenium-ion chain reaction, which is catalyzed by strong acid.

The C_7 and C_8 paraffins comprise about 90% of the alkylate; C_8 accounts for over 60%. Over 70% of the commercial alkylation processes employ sulfuric acid as the catalyst. Among the butylenes, 2-butene is superior to 1-butene. The C_3-C_4 fraction from the catalytic crackers is considered to be a superior feedstock to the alkylation unit.

Polymer Gasoline. Refinery trends tend to favor alkylation over polymerization. Unlike the alkylation process, polymerization does not require isobutane. The catalyst is usually phosphoric acid impregnated on kieselghur pellets. Polymerization of butylenes is not an attractive alternative to alkylation unless isobutane is unavailable. The motor octane number of polymer gasoline is also low, and there is considerable shrinkage in product volume. The only commercial unit to be built in recent years is at Sasol in South Africa. The commercial process was developed by UOP in the 1940s (97).

Gasoline Blending and LPG. Direct blending of butylenes into gasoline has the highest value since there is no shrinkage in product volume or conversion cost. The amount of butylenes that can be blended is limited by vapor-pressure specifications, amounting to only 8 to 10% of the gasoline pool. More butylenes could be used in winter to increase volatility for easy starting. In warm seasons, the butylenes have to be reduced to prevent vapor lock in the motor carburetor. Environmental concern in recent years has reduced the amount of butylenes, which could be blended into gasoline.

Since the heating values are similar to LPG, butylenes may be blended with LPG for bottle gas (98,99). In Europe, because LPG is unavailable, it is common to use butylenes as fuel. In the United States, butylenes have a higher value as an alkylate feed. LPG, which is readily available, is used as fuel instead.

9.2. Chemicals. Although the amount of butylenes produced in the United States is roughly equal to the amounts of ethylene and propylene produced, the amount consumed for chemical use is considerably less. The utilization of either ethylene or propylene for each of at least five principal chemical derivatives is about the same or greater than the utilization of butenes for butadiene, their main use. This production is only about one-third of the total; the two-thirds is derived directly from butane. The underlying reasons are poorer price–performance compared to derivatives of ethylene and propylene and the lack of applications of butylene derivatives. Some of the C_4 products are more easily derived from 1-, 2-, and 3-carbon atom species, eg, butanol, 1,4-butanediol, and isobutyl alcohol (see ACETYLENE-DERIVED CHEMICALS; BUTYL ALCOHOLS).

$$2\ CH_2{=}CH_2 \xrightarrow{[O]} CH_3CHO \longrightarrow CH_3CH{=}CHCHO \xrightarrow{H_2} CH_3CH_2CH_2CH_2OH$$

$$HC{\equiv}CH + 2\ CH_2O \xrightarrow{H_2} HOCH_2CH_2CH_2CH_2OH$$

$$CH_3CH{=}CH_2 \xrightarrow[H_2]{CO} CH_3CH_2CH_2CH_2OH + (CH_3)_2CHCH_2OH$$

The price of butanes and butylenes fluctuates seasonally depending on the demand for gasoline in the United States. Since much chemical-product usage is determined by price–performance basis, a shift to development of butylene-based technology may occur. Among the butylenes, demand for isobutylene is likely to increase (and so its price) as more derivatives such as methyl methacrylate and methacrylic acid are produced from isobutylene instead of the conventional acetone cyanohydrin process.

Butadiene. Most butadiene [106-99-0] is produced by extraction of C_4 streams from olefins units. Only about 10% of butadiene was produced from butenes in the United States (100). In Western Europe and Japan hardly any butenes are used to produce butadiene. Butadiene requirements in these countries are entirely met by the extraction units, since they crack only naphthas and heavier hydrocarbon feedstocks. The recent trend in the United States is to rely on the extraction units for its requirements (see BUTADIENE).

sec-Butyl Alcohol. *sec*-Butyl alcohol [78-92-2] is produced entirely from butenes using indirect hydration with sulfuric acid. Nearly all of the *sec*-butyl alcohol is converted to methyl ethyl ketone (MEK) [78-93-3] by catalytic dehydrogenation. MEK is an outstanding solvent for a wide variety of coating resins. *sec*-Butyl alcohol growth rate is closely tied in with the demand for MEK.

A typical feed to a commercial process is a refinery stream or a steam cracker B–B stream (a stream from which butadiene has been removed by extraction and isobutylene by chemical reaction). The B–B stream is a mixture of 1-butene, 2-butene, butane, and isobutane. This feed is extracted with 75–85% sulfuric acid at 35–50°C to yield butyl hydrogen sulfate. This ester is diluted with water and stripped with steam to yield the alcohol. Both 1-butene and 2-butene give *sec*-butyl alcohol. The sulfuric acid is generally concentrated and recycled (101) (see BUTYL ALCOHOLS).

$$CH_3CH_2CH{=}CH_2 \text{ or } CH_3CH{=}CHCH_3 \xrightarrow{H_2SO_4} \underset{\underset{OSO_3H}{|}}{CH_3CH_2CHCH_3} \xrightarrow{H_2O} \underset{\underset{OH}{|}}{CH_3CH_2CHCH_3} + H_2SO_4$$

Di- and Triisobutylenes. Diisobutylene [18923-87-0] and triisobutylenes are prepared by heating the sulfuric acid extract of isobutylene from a separation process to about 90°C. A 90% yield containing 80% dimers and 20% trimers results. Use centers on the dimer, C_8H_{16}, a mixture of 2,4,4-trimethylpentene-1 and -2. Most of the dimer-trimer mixture is added to the gasoline pool as an octane improver. The balance is used for alkylation of phenols to yield octylphenol, which in turn is ethoxylated or condensed with formaldehyde. The water-soluble ethoxylated phenols are used as surface-active agents in textiles, paints, caulks, and sealants (see ALKYLPHENOLS).

$$C_8H_{16} + C_6H_5OH \longrightarrow (C_8H_{17})C_6H_4OH \xrightarrow{n-C_2H_4O}$$

$$(C_8H_{17}){-}C_6H_4{-}(OCH_2CH_2)_nOH$$

The octylphenol condensate is used as an additive to lubricating oils and surface-active agents. Other uses of dimer are amination to octylamine and octyldiphenylamine, used in rubber processing; hydroformylation to nonyl alcohol for phthalate production; and carboxylation via Koch synthesis to yield acids in formulating paint driers (see DRYING).

Butylated Phenols and Cresols.   Butylated phenols and cresols, used primarily as oxidation inhibitors and chain terminators, are manufactured by direct alkylation of the phenol using a wide variety of conditions and acid catalysts, including sulfuric acid, *p*-toluenesulfonic acid, and sulfonic acid ion-exchange resins (102,103). By use of a small amount of catalyst and short residence times, the first-formed, ortho-alkylated products can be made to predominate. For the preparation of the 2,6-substituted products, aluminum phenoxides generated *in situ* from the phenol being alkylated are used as catalyst. Reaction conditions are controlled to minimize formation of the thermodynamically favored 4-substituted products (see ALKYLPHENOLS). The most commonly used is *p-tert*-butylphenol [98-54-4] for manufacture of phenolic resins. The *tert*-butyl group leaves only two rather than three active sites for condensation with formaldehyde and thus modifies the characteristics of the resin.

2,6-Di-*tert*-butyl-4-methylphenol [25377-21-3] (di-*tert*-butyl-*p*-cresol) or butylated hydroxytoluene (BHT)) is most commonly used as an antioxidant in plastics and rubber. Use in food is decreasing because of legislation and it is being replaced by butylated hydroxy anisole (BHA) (see ANTIOXIDANTS; FOOD ADDITIVES).

Heptenes.   Heptenes, C_7H_{14}, are used for the preparation of isooctyl alcohol [26952-21-6] by hydroformylation (see OXO PROCESS). The heptenes are prepared by very carefully controlled fractionation of polymer gasoline. Specifications generally call for >99.9% C_7 content (including some paraffin that is also formed) to simplify processing.

tert-Butylamine.   *tert*-Butylamine [75-64-9] is used as an intermediate in the manufacture of lubricating oil additives and miscellaneous chemicals. It is manufactured using the Ritter reaction. Isobutylene first reacts with sulfuric acid and then HCN to yield *tert*-butylamine.

$$(CH_3)_2C{=}CH_2 \xrightarrow{H_2SO_4} (CH_3)_3COSO_3H \xrightarrow{HCN} (CH_3)_3CN{=}CHOSO_3H \xrightarrow{H_2O}$$

$$(CH_3)_3CNHCHO \xrightarrow{OH^-} (CH_3)_3CNH_2$$

tert-Butyl Alcohol.   *tert*-Butyl alcohol [75-65-0] is an intermediate in the separation of isobutylene from a mixed butane–butylene stream. It is manufactured by the hydrolysis of the acid extract in the isobutylene separation process. A small amount is used as a solvent. It is also a significant coproduct in Arco's (now BP) Oxirane process (104) which produces propylene oxide [75-56-9]. *tert*-Butyl alcohol is derived from isobutane, which is the oxygen carrier for the process:

$$(CH_3)_3CH \xrightarrow{[O]} (CH_3)_3COOH$$

$$(CH_3)_3COOH + CH_3CH{=}CH_2 \longrightarrow CH_3CH{-}CH_2 + (CH_3)_3COH$$

It competes directly with MTBE as an octane enhancer in the gasoline pool. Since MTBE is more desirable than *tert*-butyl alcohol because of its total miscibility with gasoline, *tert*-butyl alcohol will be an important source of isobutylene used in the manufacture of MTBE. High purity isobutylene, C_4H_8, can be obtained by dehydration of *tert*-butyl alcohol, $C_4H_{10}O$.

tert-Butyl Mercaptan.   *tert*-Butyl mercaptan [75-66-1] is used primarily as an odorant at <30 ppm for natural gas so that leaks can be easily detected. It is manufactured by the reaction of isobutylene and hydrogen sulfide in the presence of acid catalysts (105).

Primary Amyl Alcohols.   Primary amyl alcohols (qv) are manufactured by hydroformylation of mixed butenes, followed by dehydrogenation (106). Both 1-butene and 2-butene yield the same product though in slightly different ratios depending on the catalyst and conditions. Some catalyst and conditions produce the alcohols in a single step. By modifying the catalyst, typically a cobalt carbonyl, with phosphorus derivatives, such as tri(*n*-butyl)phosphine, the linear alcohol can be the principal product from 1-butene.

$$
\begin{array}{ccc}
& & \text{CHO} & & \text{CH}_2\text{OH} \\
& & | & & | \\
\text{CH}_3\text{CH}=\text{CHCH}_3 & \xrightarrow[\text{catalyst}]{\text{H}_2-\text{CO}} & \text{CH}_3\text{CH}_2\text{CHCH}_3 & \xrightarrow{\text{H}_2} & \text{CH}_3\text{CH}_2\text{CHCH}_3 \\
\text{or} & & + & & + \\
\text{CH}_3\text{CH}_2\text{CH}=\text{CH}_2 & & \text{CH}_3\text{CH}_2\text{CH}_2\text{CH}_2\text{CHO} & & \text{CH}_3\text{CH}_2\text{CH}_2\text{CH}_2\text{CH}_2\text{OH}
\end{array}
$$

The main use of the amyl alcohols is as esters such as acetates for solvents.

Di- and Triisobutylaluminums.   Triisobutylaluminum [100-99-2] is prepared by reaction of isobutylene with aluminum at 80°C and 20.3 MPa (200 atm) of hydrogen (107). It is used as a catalyst for ethylene oligomerization to prepare even numbered, linear 1-olefins. Use of stochiometric quantities of triisobutylaluminum followed by oxidation of the resulting mixture of long-chain aluminum alkyls yields even numbered, terminal primary alcohols in the plasticizer and detergent range (108). Oxychem uses this process in the United States to manufacture plasticizer (C_6-C_{10}) and detergent ($C_{16}-C_{22}$) range alcohols (see ALCOHOLS, HIGHER ALIPHATIC).

Triisobutylaluminum is converted to diisobutylaluminum chloride [1779-25-5] and diisobutylaluminum hydroxide [1191-15-7], which are used as cocatalysts for Ziegler polymerization systems. Corresponding ethyl compounds are prepared via the reaction of triisobutylaluminum with ethylene.

Butylene Oxide.   Butylene oxides are prepared on a small scale by Dow by chlorohydrin technology. There appears to be no technical reason why they could not be prepared by the propylene oxide Oxirane process (see CHLOROHYDRINS).

A significant use of butylene oxide [26249-20-7] is as an acid scavenger for chlorine-containing materials such as trichloroethylene. Inclusion of about 0.25–0.5% of butylene oxide, based on the solvent weight, during preparation of vinyl chloride and copolymer resin solutions minimizes container corrosion which may be detrimental to resin color and properties.

p-tert-Butyltoluene.   *p-tert*-Butyltoluene [98-51-1], prepared by acid catalyzed alkylation of toluene with isobutylene under mild conditions (109,110), is an intermediate in the production of *p-tert*-butylbenzoic acid [98-73-7]. This acid is used as a chain-length control agent in the preparation of unsaturated polyester resins. Solubility characteristics offer some advantage over benzoic acid.

Neopentanoic (Pivalic) Acid. Neopentanoic acid [75-98-9] is prepared using the Koch technology in which isobutylene reacts with carbon monoxide in the presence of strong acids such as H_2SO_4, HF, and $BF_3 \cdot H_2O$ (111–114). General reaction conditions are 2–10 MPa (about 20–100 atm) of CO and 40–150°C.

$$(CH_3)_2C{=}CH_2 \xrightarrow{H_2SO_4} (CH_3)_3C^+OSO_3H^- \xrightarrow{CO} (CH_3)_3C\overset{\overset{O}{\|}}{C}OSO_3H$$

$$\xrightarrow{H_2O} (CH_3)_3C\overset{\overset{O}{\|}}{C}OH + H_2SO_4$$

The acids are converted to peroxy esters for use as polymerization initiators. The metal salts are used as driers in paint formulations (see CARBOXYLIC ACIDS, TRIALKYLACETIC ACIDS).

Methylallyl Chloride. Methallyl chloride [563-47-3] is the principal product when isobutylene and chlorine react over a wide range of temperatures (115). Very little addition takes place.

$$\underset{\underset{CH_3}{|}}{CH_3C}{=}CH_2 + Cl_2 \longrightarrow \underset{\underset{CH_3}{|}}{CH_2{=}CCH_2Cl}$$

This allylic chloride is a chemical intermediate for various specialty products, but it has no single significant commercial use (see CHLOROCARBONS AND CHLOROHYDRO-CARBONS, ALLYL CHLORIDE).

Butylated Hydroxy Anisole (BHA). This material is an oxidation inhibitor and has been accepted for use in foods where the use of butylated hydroxytoluene (BHT) is restricted (see FOOD ADDITIVES). It is manufactured by the alkylation of 4-hydroxyanisole [150-76-5] with isobutylene that yields a mixture of 2- and 3-*tert*-butyl isomers as products (116).

BHA

Methyl Methacrylate and Methacrylic Acid. The traditional production of methyl methacrylate [80-62-6] and methacrylic acid [79-41-4] involves the reaction of acetone with HCN and subsequent conversion to methyl ester and by-product ammonium bisulfate.

The handling of toxic materials and disposal of ammonium bisulfate have led to the development of alternative methods to produce this acid and the methyl ester. There are two technologies for production from isobutylene now available: ammoxidation to methyl methacrylate (the Sohio process), which is then solvolyzed, similar to acetone cyanohydrin, to methyl methacrylate; and direct oxidation of isobutylene in two stages via methacrolein [78-85-3] to methacrylic acid, which is then esterified (117). Since direct oxidation avoids the need for HCN and NH_3, and thus toxic wastes, all new plants have elected to use this technology.

Methyl tert-Butyl Ether (MTBE).   Methyl *tert*-butyl ether [1634-04-4] is made by the etherification of isobutylane with methanol, and there are six commercially proven technologies available. These technologies have been developed by Arco, IFP, CDTECH, Phillips, Snamprogetti, and Hüls (licensed jointly with UOP). The catalyst in all cases is an acidic ion-exchange resin. The United States has been showing considerable interest in this product. Western Europe has been manufacturing it since 1973 (ANIC in Italy and Hüls in Germany). Production of MTBE in the U.S. was reported recently at 2 59,190 barrels/day (118).

The etherification reaction is equilibrium limited, requiring an excess of methanol to drive the reaction. Conversion is favored by low temperature whereas the reaction kinetics are favored by high temperature. A compromise on temperature must be made in order to obtain an economic design. The etherification reaction is exothermic, and these technologies differ primarily in the type of reactor employed and the method for removing heat of reaction. In these processes, the reaction is carried out in two reactors connected in series to facilitate heat removal and also for economic reactor design. Typically, isobutylene conversion is about 95%, with most of the conversion taking place in the first reactor. Units can also be designed to obtain greater than 99% conversion.

The first reactor in series in the Arco, IFP, and Phillips processes is adiabatic (vessel filled with catalyst). The exothermic heat of reaction is removed in a pump-around loop where a portion of the reactor contents are taken from the reactor, pumped through an external exchanger, cooled, and returned to the reactor.

The Snamprogetti process utilizes a tubular isothermal reactor (tubes filled with catalyst) for the first reactor with cooling water on the shell side to control temperature. The Hüls process uses either an adiabatic or isothermal reactor for the first reactor.

In the CDTECH process (formerly CR&L technology), the first reactor is adiabatic. The heat of reaction is removed partly by vaporization of the reaction mix. The operating temperature is controlled by adjusting the operating pressure.

The reactor combinations for the two reactors in series consist of two fixed-beds for the Arco process; an expanded bed followed by a catalytic distillation reactor for IFP; a fixed-bed followed by a catalytic distillation reactor for CDTECH; and two fixed-beds for Phillips. The Hüls process uses an adiabatic reactor for the second reactor.

The various sources of isobutylene are C_4 streams from fluid catalytic crackers, olefin steam crackers, isobutane dehydrogenation units, and isobutylene produced by Arco as a coproduct with propylene oxide. Isobutylene concentrations (weight basis) are 12 to 15% from fluid catalytic crackers, 45% from olefin steam crackers, 45 to 55% from isobutane dehydrogenation, and high purity isobutylene coproduced with propylene oxide. The etherification unit should be designed for the specific C_4 feedstock that will be processed.

Polymers. Polymers account for about 3–4% of the total butylene consumption and about 30% of nonfuels use. Homopolymerization of butylene isomers is relatively unimportant commercially. Only stereoregular poly (1-butene) [9003-29-6] and a small volume of polyisobutylene [25038-49-7] are produced in this manner. High molecular weight polyisobutylenes have found limited use because they cannot be vulcanized. To overcome this deficiency a butyl rubber copolymer of isobutylene with isoprene has been developed. Low molecular weight viscous liquid polymers of isobutylene are not manufactured because of the high price of purified isobutylene. Copolymerization from relatively inexpensive refinery butane–butylene fractions containing all the butylene isomers yields a range of viscous polymers that satisfy most commercial needs (see OLEFIN POLYMERS; ELASTOMERS, SYNTHETIC-BUTYL RUBBER).

BIBLIOGRAPHY

"Butylenes" in *ECT* 2nd ed., Vol. 3, pp. 830–865, by C. E. Morrell, Esso Research and Engineering Company; in *ECT* 3rd ed., Vol. 4, pp. 346–375, by M. C. Hoff, UK Im, W. F. Hauschildt, and I. Puskas, Amoco Chemicals Company; in *ECT* 4th ed., Vol. 4, pp. 701–735, by N. Calamur, M. E. Carrera, and R. A. Wilsak, Amoco Chemical Company; "Butylenes" in *ECT* (online), posting date: December 4, 2000, by Narasimhan Calamur, Martin E. Carrera, and Richard A. Wilsak, Amoco Chemical Company.

CITED PUBLICATIONS

1. R. A. Wilsak and G. Thodos, *Ind. Eng. Chem. Fundam.* **23**, 75 (1984).
2. *DIPPR, Project 801, Data Compilation* (July 1990) and earlier references cited therein; R. D. Goodwin, *NBSIR 79-1621*, Thermophysical Properties Division, National Engineering Laboratory, National Bureau of Standards, Boulder, Colo. 1979; G. B. Kistiakowsky, J. R. Ruhoff, H. A. Smith, and W. E. Vaughan, *J. Am. Chem. Soc.* **57**, 876 (1935); *ibid.* **58**, 146 (1936); P. B. Ayscough, K. J. Ivin, and J. H. O'Donnell, *Trans. Faraday Soc.* **61**, 1601 (1965); J. B. Garner, *Petrol. Refiner* **24** (2), 99 (1945); J. G. Aston and G. J. Szasz, *J. Am. Chem. Soc.* **69**, 3108 (1947); D. G. Laird and C. S. Howat, *Fluid Phase Equilibria* **60**, 173 (1990).
3. B. D. Smith and R. Srivastava, *Thermodynamic Data for Pure Compounds*, Part A, Elsevier Science Publishers, Amsterdam, The Netherlands, 1986; *DIPPR, Project 801, Data Compilation* (July 1990); R. C. Reid, J. M. Prausnitz, and B. E. Poling, *The Properties of Gases and Liquids*, 4th ed., McGraw-Hill Book Co., New York, 1987; *Beilstein Online*, Beilstein Institute for Organic Chemistry, Springer-Verlag, Heidelberg, F. R. Germany, 1990.
4. *American Petroleum Institute Research Project 44*, Thermodynamics Research Center, Texas Engineering Experiment Station, Texas A & M University, College Station, Tex., 1976; *Technical Data Book—Petroleum Refining*, Refining Department, 4th ed., American Petroleum Institute, 1983; *Physical Property Data for the Design Engineer*, C. F. Beaton and G. F. Hewitt, eds., Hemisphere Publishing Corp., New York, 1989; E. W. Flick, ed., *Industrial Solvents Handbook*, 3rd ed., Noyes Data Corp., Park Ridge, N.J., 1985.
5. R. T. Morrison and R. N. Boyd, *Organic Chemistry*, 4th ed., Allyn and Bacon, Boston, Mass., 1983; D. E. Dorman, M. Jantelot, and J. D. Roberts, *J. Org. Chem.* **36**, 2157 1971; I. Hirana, O. Kikuchi, and K. Suzuki, *Bull. Chem. Soc. Jpn.* **49**, 3321 (1976).

6. J. Hine, *Physical Organic Chemistry*, McGraw-Hill Book Co., New York, 1962, Chapt. 9.

7. P. B. D. de la Mare and R. Bolton, *Electrophilic Addition to Unsaturated Systems*, Elsevier Science Publishing Co., New York, 1966.

8. D. C. Nonhebel, J. M. Tedder, and J. C. Walton, *Radicals*, Cambridge University Press, 1979.

9. W. A. Pryer, *Introduction to Free Radical Chemistry*, Prentice-Hall, Englewood Cliffs, N.J., 1965.

10. J. B. Hendrickson, O. J. Cram, and G. S. Hammond, *Organic Chemistry*, McGraw-Hill Book Co., New York, 1970, p. 614.

11. M. C. Hoff, K. W. Greenlee, and C. E. Boord, *J. Am. Chem. Soc.* **73**, 3329 (1951).

12. J. Burgin and co-workers, *Ind. Eng. Chem.* **13**, 1413 (1939).

13. J. Chetron, M. Henant, and G. Marinier, *Bull. Chim. Soc. Fr.*, 1966 (1969).

14. J. H. Polstar and K. Yates, *J. Am. Chem. Soc.* **91**, 1469 (1969).

15. H. C. Brown, *Hydroboration*, W. A. Benjamin Inc., New York, 1962, pp. 114, 192, 200.

16. I. Wender and co-workers, *J. Am. Chem. Soc.* **78**, 5401 (1956).

17. *Ibid*, **77**, 5760 (1955).

18. U.S. Pat. 2,555,054 (May 22, 1951), J. Owen (to Phillips Petroleum Co.).

19. L. F. Fieser and M. Fieser, *Advanced Organic Chemistry*, Reinhold Publishing Co., New York, 1961, p. 162.

20. U.S. Pat. 2,392,294 (Jan. 1, 1946), W. E. Vaughn and F. W. Rust (to Shell Development Co.).

21. U.S. Pat. 2,925,443 (Feb. 1, 1960), W. L. Wash (to Gulf Research and Development Co.).

22. V. J. Guerico, *Oil Gas J.*, 68 (Feb. 21, 1977).

23. *Chem. Eng.*, 62 (Feb. 13, 1978).

24. *Chem. Week*, 49 (Nov. 16, 1977).

25. J. E. Germain, *Catalytic Conversion of Hydrocarbons*, Academic Press, Inc., New York, 1969.

26. T. Brooks and co-workers, *The Chemistry of Petroleum Hydrocarbons*, Reinhold Publishing Corp., New York, 1955.

27. G. Egloff, G. Hulla, and V. I. Komarewski, *Isomerization of Pure Hydrocarbons*, Reinhold Publishing Corp., New York, 1942; D. R. Stuhl, E. F. Westrum, and G. C. Sinke, *The Chemical Thermodynamics of Organic Compounds*, John Wiley & Sons, Inc., New York, 1969.

28. W. G. Burns and co-workers, *Trans. Faraday Soc.* **64**, 129 (1968); K. Otsuka and A. Morikawa, *Chem. Commun.* (6),218 (1975).

29. Y. Misono, Y. Saito, and Y. Yoneda, *J. Catal.* **10**(2), 200 (1968).

30. R. Nicolova and co-workers, *C. R. Acad. Sci. Ser. C* **265**(8), 468 (1967).

31. K. Tanaka, *Sci. Pap. Inst. Phys. Chem. Res. Jpn.* **69**(2), 50 (1975).

32. U.S. Pat. 3,641,184 (Feb. 8, 1972), C. E. Smith and B. J. White (to Phillips Petroleum Co.).

33. K. Matsura, A. Suzuki, and M. Itoh, *J. Catal.* **23**, 396 (1971).

34. W. B. Smith and W. Y. Watson, *J. Am. Chem. Soc.* **84**, 3174 (1962).

35. T. A. Gilmore and J. J. Rooney, *Chem. Commun.*, 219 (1975).

36. K. W. Egger and S. W. Benson, *J. Am. Chem. Soc.* **88**, 236 (1966).

37. U.S. Pat. 3,527,834 (Sept. 8, 1970), W. L. Kehl, R. J. Rennard, and H. E. Swift (to Gulf Research and Development Co.).

38. J. Dubien, L. DeMourgues, and Y. Trambouze, *Bull. Soc. Chem. Fr.* (1), 108 (1967).

39. U.S. Pat. 3,479,415 (Nov. 18, 1969), E. Shull (to Air Products and Chemicals, Inc.).

40. M. J. McGrath and M. Soudek, *1990 Summer National Meeting, AIChE*, San Diego, Calif., Aug. 19–22, 1990.

41. E. G. Wollaston and co-workers, *Hydrocarbon Process.* **54**(9), 93 (1975).
42. Eur. Pat. 305,720 A2 (July 22, 1988), L. Zelting, L. Shunhua, and G. Xingpin (to Research Institute of Petroleum Processing, China).
43. D. F. Blaser, in J. J. McKetta, ed., *Encyclopedia of Chemical Processing and Design,* Vol. 10, Marcel Dekker, Inc., New York, 1979, p. 13.
44. S. B. Zodnik, E. J. Green, and L. P. Halle, *Manufacturing Ethylene*, The Petroleum Publishing Co., Tulsa, Okla., 1972.
45. D. M. Considine, ed., *Chemical and Process Technology Encyclopedia*, McGraw-Hill Book Co., Inc., New York, 1974, p. 429.
46. R. E. Haney, *High. Polm.* **24**(2), 577 (1971).
47. C. Marsden, *Solvent Guide,* 2nd ed., Interscience Publishers, New York, 1963.
48. S. Tukao, *Hydrocarbon Process.* **45**(11), 151 (1966).
49. U.S. Pat. 3,001,608 (Sept. 26, 1961), L. Lorenz, and co-workers (to BASF).
50. H. D. Evans and D. H. Sarno, *Seventh World Pet. Congr.* **5**, 259 (1967).
51. H. Klein and H. M. Weitz, *Hydrocarbon Process.* **47**(11), 135 (1968).
52. L. B. Ryland, M. W. Tamele, and J. N. Wilson, in P. H. Emmett, ed., *Catalysis,* Vol. VI, Reinhold Publishing Corp., New York, 1968, p. 1.
53. R. M. Barrer, *Chem. Ind. London*, 1203 (1968).
54. D. W. Breck, *Zeolite Molecular Sieves*, John Wiley & Sons, Inc., New York, 1974.
55. B. C. Gates, J. R. Katzer, and G. C. Schuit, *Chemistry of Catalytic Process*, McGraw-Hill Book Co., Inc., New York, 1979, p. 112–180.
56. *Oil Gas J.* **73**, 53 (Mar. 10, 1975).
57. Ref. 45, p. 1089.
58. R. L. Banks and G. C. Bailey, *Ind. Eng. Chem., Prod. Res. Dev.* **3**, 170 (1964).
59. R. L. Banks, *J. Mol. Catal.* **8**, 269 (1980).
60. J. F. Boucher and co-workers *Oil Gas J.* (Mar. 1982).
61. Y. Chauvin, J. Gillard, B. Juguin, L. Leonard, and M. Derrien, *1987 Spring National Meeting*, AIChE, Houston, Tex., Apr. 1987.
62. *Hydrocarbon Process.*, 73–75 (Mar. 1990).
63. *Hydrocarbon Process.*, 18–120 (Nov. 1984); 26 (Nov. 1985).
64. U.S. Pat. 3,341,620 (Sept. 12, 1967), (to Phillips Petroleum Co.).
65. Brit. Pat. 1,064,829 (Apr. 12, 1967), (to British Petroleum Co.).
66. U.S. Pat. 3,915,897 (Oct. 28, 1975), (to Phillips Petroleum Co.).
67. U.S. Pat. 4,242,531 (Dec. 30, 1980), (to Phillips Petroleum Co.).
68. U.S. Pat. 4,487,847 (Dec. 11, 1984), (to Phillips Petroleum Co.).
69. E. D. Cooper and R. L. Banks, *1985 Spring National Meeting*, AIChE, Houston, Tex., 1985.
70. U.S. Pat. Appl. 20030013933 (Jan. 16, 2003), M. Yamase and Y. Suguki (to Sumitomo Chemical Co., Ltd.).
71. L. Ziating, J. Fukang, and M. Enze, paper presented at the *1990 Annual Meeting of National Petroleum Refiners Association*, San Antonio, Tex., Mar. 25–27, 1990, paper no. AM-90-40.
72. *Chem. Eng. News* **68**(41) 12 (Oct. 8, 1990).
73. F. P. Wilcher, B. V. Vora, and P. R. Pujado, *Dewitt 1990 Petrochemical Review,* paper S-1, Houston, Tex., Mar. 27–29, 1990.
74. R. G. Craig, T. J. Delaney, and J. M. Dufallo, in Ref. 75, paper T-1.
75. G. Fusco, D. Sanflippo, and F. Buonomo, in Ref. 75, paper W-1.
76. R. O. Dunn, in Ref. 75, paper U-1.
77. J. J. Lacatena, in Ref. 75, paper V-1.
78. *Hydrocarbon Process.*, 104 (Apr. 1986).
79. J. P. Kennedy and I. Kirshenbaum, *High Polym.* **24**(2), 691 (1971); Ref. 65, p. 677.

80. Y. Chauvin and co-workers, "Dimersol B: A Key to Olefin Interconversion," *1987 Spring Meeting AIChE, Apr. 1, 1987, Session 70, Paper No. 70; Hydrocarbon Process.*, 106 (Apr. 1986).

81. J. R. Barber, J. J. Collins, and T. C. Sayer, paper presented at *68th National Meeting*, AIChE, Houston, Tex., Feb. 1971.

82. *Hydrocarbon Process.* **55**(9), 226, (1976).

83. M. S. Adler and co-workers, paper presented at the *85th National Meeting*, AIChE, June 1978.

84. U.S. Pat. 3,721,064 (Mar. 20, 1973), (to Union Carbide Corp.).

85. A. J. DeRosset and co-workers, paper presented at the *Symposium on Recent Advances in the Production and Utilization of Light Olefins*, American Chemical Society, Anaheim, Calif., Mar. 12–17, 1978; ACS Preprint **23**(2) 766–774 (1978).

86. U.S. Pat. 2,985,589 (May 23, 1961) (to UOP).

87. *Chem. Eng. News*, 48 (May 12, 1978).

88. T. Lacson, T. Kaelin, and T. Sasano, *Chemical Economics Handbook*, SRI, Menlo Park, Calif., April 2002.

89. H. M. McNair and E. J. Bonelli, *Basic Gas Chromatography*, Varian Instrument Division, Calif., Mar. 1969.

90. K. Verscheren, ed., *Handbook of Environmental Data on Organic Chemicals*, Van Nostrand Reinhold Co., New York, 1983, pp. 304, 317.

91. R. J. Lewis, *Sax's Dangerous Properties of Industrial Materials*, 10th ed., Vol. 2, John Wiley & Sons, Inc., New York, 2000.

92. C. R. Cupit, J. E. Gwyn, and E. C. Jernigan, *Petr. Manage.* **33**(12), 203 (Dec. 1961); **34**(1), 207 (Jan. 1962).

93. I. H. Gary and G. E. Handwerk, *Petroleum Refining*, Marcel Dekker, Inc., New York, 1975, p. 142.

94. J. P. Kennedy and I. Kirshenbaum, *High. Polym.* **24**(2), 701 (1971).

95. Ref. 55, p. 15.

96. C. L. Thomas and E. J. McNeils, *Seventh World Pet. Congr.*, Review Paper **12** (Apr. 1967).

97. P. C. Weinert and G. Egolff, *Pet. Process.* **3**, 585 (1948).

98. Ref. 93, p. 7.

99. G. D. Hobson and W. Pohl, *Modern Petroleum Technology,* 5th ed., John Wiley & Sons, Inc., New York, 1984, p. 517.

100. "Butadiene Chemical Profile", *Chemical Market Reporter*, March 25, 2002.

101. A. L. Waddams, *Chemicals from Petroleum,* 3rd ed., John Murray, London, England, 1973.

102. G. A. Olah, *Friedel-Crafts and Related Reactions,* Vol. II, Part 1, Interscience Publishers, New York, 1964.

103. P. Wiseman, *Industrial Organic Chemistry*, Wiley-Interscience, New York, 1970, p. 169.

104. *Ibid.*, p. 93.

105. A. V. Hahn, *The Petroleum Industry*, McGraw-Hill Book Co., Inc., New York, 1975, p. 591.

106. Ref. 111, p. 219.

107. C. E. Coates, M. L. H. Green, and K. Wade, *Organometallic Compounds,* Vol. I, Mathew & Co., Ltd., London, 1968, p. 299.

108. Ref. 111, p. 245.

109. Ref. 111, p. 153.

110. Ref. 113, p. 526.

111. *Chem. Eng. News*, 46 (Aug. 1963).

112. U.S. Pat. 3,296,286 (1967), (to Esso Research and Engineering).

113. Brit. Pat. 1,174,209 (Dec. 17, 1969), A. Kiwantes and B. Stouthamer (to Shekk Internationale Research Maalschppij N.V.).

114. Jpn. Pat. 73 23, 413 (July 13, 1973), Y. Komatsu, T. Tamura, and H. Okayama (to Maruzen Oil Co.).

115. J. Burgin and co-workers, *Ind. Eng. Chem.* **31**, 1413 (1939).

116. Ref. 102, p. 93.

117. Y. Oda and co-workers, *Hydrocarbon Process.* (10), 115 (1975).

118. MTBE, Chemical Profile, *Chemical Market Reporter*, Jan. 6, 2003.

NARASIMHAN CALAMUR
MARTIN E. CARRERA
RICHARD A. WILSAK
Amoco Chemical Company

BUTYL RUBBER

1. Introduction and Background

Isobutylene has been of interest since the early days of synthetic polymer research when Friedel Crafts catalysts were used to prepare elastic materials. Isobutylene polymers of commercial importance include homopolymers and copolymers containing small amounts of isoprene or *p*-methylstyrene. Currently, chlorinated and brominated derivatives of butyl(poly[isobutylene-*co*-isoprene]) have the highest sales volume.

Isobutylene was first observed to polymerize in 1873 but high molecular weight polymers were not made until work at I. G. Farben in Germany established that molecular weight increases with decreasing polymerization temperature. Polyisobutylene was synthesized at $-75°C$ using BF_3 as a catalyst. However, as the polymer was saturated, it could not be cross-linked into a rubbery network and no commercial uses were found (1).

Poly[isobutylene-*co*-isoprene] or butyl rubber was synthesized in 1937 at the Standard Oil Development Co., forerunner of ExxonMobil Chemical Co. (2). The first sulfur-curable copolymer was prepared in ethyl chloride over an aluminum chloride catalyst with 1,3-butadiene as the comonomer; however, it was soon found that isoprene was a better comonomer and that methyl chloride was a better polymerization diluent. During World War II, the natural rubber supply to the United States was drastically curtailed, boosting the production of synthetic rubber. The commercial production of butyl rubber in 1943 was an enormous scientific and engineering achievement given the very early state of the art and complexity of this technology.

The discovery of butyl rubber was, in fact, the discovery of the limited-functionality elastomers. Unlike natural rubber and polybutadiene, which have reactive sites on every monomer unit, the unsaturation in butyl rubber is widely spaced along a saturated, flexible hydrocarbon chain. The principle of limited functionality has been subsequently used in other elastomers, eg, ethylene–propylene terpolymers and chlorosulfonated polyethylene. Halogenated butyl rubber was first synthesized at Goodrich (3). A brominated butyl rubber was commercialized via a small-scale batch method in 1954 starting from N-bromosuccinimide and butyl rubber. The product was withdrawn in 1969. Following the withdrawal of Goodrich bromobutyls from the market, Polymer Co. of Canada developed the commercial process using elemental bromine, which is the currently used commercial process. Chlorinated butyl rubber was developed at ExxonMobil and commercialized in 1961. It is made by the continuous chlorination of a solution of butyl rubber (4). Currently, brominated butyl rubber is also manufactured by a similar continuous-solution process.

The first use for butyl rubber was as inner tubes, whose air-retention characteristics contributed significantly to the safety and convenience of tires. Good weathering, ozone resistance, and oxidative stability have led to applications in mechanical goods and elastomeric sheeting. Automobile tires were manufactured for a brief period from butyl rubber, but poor abrasion resistance curtailed this development.

Halogenated butyl rubber greatly extended the usefulness of butyl rubber by providing much higher vulcanization rates and improving the compatibility with highly unsaturated elastomers, such as natural rubber and styrene–butadiene rubber (SBR). These properties permitted the production of tubeless tires with chlorinated or brominated butyl innerliners. The retention of air pressure (5) and low intercarcass pressure (6) extended tire durability.

Polyisobutylene is produced in a number of molecular weight grades and each has found a variety of uses. The low molecular weight liquid polybutenes have applications as adhesives, sealants, coatings, lubricants, and plasticizers, and for the impregnation of electrical cables (7). Moderate molecular weight polyisobutylene was one of the first viscosity-index modifiers for lubricants (8). High molecular weight polyisobutylene is used to make uncured rubbery compounds and as an impact additive for thermoplastics.

2. Process Chemistry

Butyl rubber is prepared from 2-methylpropene [115-11-7] (isobutylene) and 2-methyl-1,3-butadiene [78-79-5] (isoprene). Isobutylene with a purity of >99.5 wt% and isoprene with a purity of >98 wt% are used to prepare high molecular weight butyl rubber. Water and oxygenated organic compounds are minimized by feed purification systems because these impurities interfere with the cationic polymerization mechanism. Copolymers of isobutylene can also be prepared from mixed C_4 olefin containing streams that contain n-butene. These copolymers are generically known as polybutenes.

2.1. Isobutylene Polymerization Mechanism. The carbocationic polymerization of isobutylene and its copolymerization with viable comonomers like isoprene and p-methylstyrene is mechanistically complex (9–11). The initiating system is typically composed of two components: an initiator and a Lewis acid coinitiator. Typical Lewis acid coinitiators include $AlCl_3$, (alkyl)$AlCl_2$, BF_3, $SnCl_4$, $TiCl_4$, etc. More recently uncommon Lewis acid such as methylaluminoxane

(MAO) (12,13) and specifically designed weakly coordinating Lewis acids such as $B(C_6F_5)_3$ (14,15) have been used as Lewis acids in initiating systems for isobutylene polymerization. Common initiators include Brønsted acids such as HCl, RCOOH, H_2O, alkyl halides, eg, $(CH_3)_3CCl$, $C_6H_5(CH_3)_2Cl$, esters, ethers, peroxides, and epoxides. More recently, transition-metal complexes, such as metallocenes and other single-site catalyst systems, when activated with weakly coordinating Lewis acids or Lewis acid salts, have been used to initiate isobutylene polymerization (16). In the initiation step, isobutylene reacts with the Lewis acid coinitiator–initiator pair to produce a carbenium ion. Additional monomer units add to the formed carbenium ion in the propagation step. Temperature, solvent polarity, and counterions affect the chemistry of propagation. These reactions are fast and highly exothermic. The propagation rate constant has been determined to be around 108 L/(mol·s), essentially diffusion-limited (17,18). Polymerizations at low temperature give extremely high polymerization rates in either hydrocarbon or halogenated hydrocarbon solvents. Isoprene is copolymerized mainly by trans-1,4-addition (>90%), and to a lesser extent by either 1,2-addition (19,20) or as a branched 1,4-addition product (21). The propagation proceeds until chain transfer or termination occurs.

In the chain-transfer step, the carbenium ion chain end reacts with isobutylene, isoprene, or a species with an unshared electron pair, ie, RX, solvents, counterion, and olefins. Reaction with these species terminates the growth of this macromolecule and permits the formation of a new chain. The activation energy of chain transfer is larger than propagation, thus the molecular weight of the polymer is strongly influenced by the polymerization temperature. Lower temperatures lead to higher molecular weight polymer. As comonomers exhibit their own chain-transfer characteristics, the presence of comonomer can also influence the final molecular weight of a copolymer. Higher isoprene contents typically lower the molecular weight of prepared butyl rubber (22,23). Termination results from the irreversible destruction of the propagating carbenium ion and discontinuance of the kinetic chain. Termination reactions include the collapse of the carbenium ion–counterion pair, hydride abstraction from comonomer, formation of dormant or stable allylic carbenium ions, or by reaction of the carbenium ion with nucleophiles, eg alcohols or amines. The reactivity ratios are strongly affected by the polymerization conditions (Table 1). A laboratory procedure for the preparation of butyl rubber is described in Reference 34

Fundamental rate constants for the initiation, propagation, chain-transfer, and termination steps in the polymerization process are difficult to measure because of the rapid rate of reaction. However, recent work (35) has shown that the propagation rate constant is of the order of 6×108 L/(mol·s).

Studies of the living cationic polymerization of isobutylene and copolymerization with isoprene have begun (36,37). The living copolymerization of isobutylene and isoprene has so far produced a random copolymer with narrow molecular weight distribution and a well-defined structure. For example, the BCl_3/cumyl acetate polymerization system in methyl chloride or methylene chloride at −30°C provides for copolymers with 1–8 mol% trans-1,4-isoprene units and M_n between 2000 and 12,000 with a M_w/M_n of under 1.8. The advent of living polymerization of isobutylene has brought about the preparation of a large number of new isobutylene-based materials.

Table 1. **Copolymerization Reactivity Ratios of the Isobutylene/Isoprene System**[a]

Initiating system	Solvent	$T,°C$	r_1	r_2	Reference
AlCl₃	CH₃Cl	−103	2.5 ± 0.5	0.4 ± 0.1	24
EtAlCl₂	CH₃Cl	−100	2.17	0.5	25
AlCl₃	CH₃Cl	−90	2.3		26
EtAlCl₂ + Cl₂	CH₃Cl	−35	2.5		27
AlCl₃	EtCl	−95	2.26	0.38	28,29
EtAlCl₂	EtCl	−90	2.27	0.44	28,29
EtAlCl₂	CH₃Cl 88%/hexane 12%	−80	2.15	1.03	30
EtAlCl₂	CH₃Cl 50%/hexane 50%	−80	1.90	1.05	30
EtAlCl₂	CH₃Cl 12%/hexane 88%	−80	1.17	1.08	30
EtAlCl₂	Hexane 100%	−80	0.80	1.28	30
AlCl₃	C₅/CH₂Cl₂	−70	1.56 ± 0.19	0.95 ± 0.17	31,32
AlCl₃	CH₃Cl	−80	1.6		33

[a] Ref. 24.

High molecular weight copolymers of isobutylene and isoprene have been prepared at temperatures 40–50°C higher than is commercially practiced using metallocene and single-site initiators (12–16). Newer Lewis acids like methylaluminoxane and weakly coordinating anions or their salts like $B(C_6F_5)_3$ or $R^+[B(C_6F_5)_4]^-$ also prepare high molecular weight copolymers at these higher temperatures (12–16).

2.2. Modification of Butyl Rubbers. *Halobutyls.*

Chloro- and bromo-butyls are commercially the most important derivatives. The halogenation reaction is carried out in hydrocarbon solution using elemental chloride and bromine (equimolar with the enchained isoprene). The halogenation is fast and proceeds mainly by an ionic mechanism. The structures that may form include the following:

$$\begin{array}{c} CH_3 \\ | \\ {+}CH_2{-}C{=}CH{-}CH_2{+} \end{array} + \; X_2 \longrightarrow \begin{array}{c} CH_3 \\ | \\ {+}CH_2{-}\underset{(R)}{C}{\overset{\oplus}{\cdots}}CH{-}CH_2{+} \\ X \end{array} + \; X^{\ominus} \longrightarrow$$

$$\begin{array}{c} CH_2 \\ \| \\ {+}CH_2{-}C{-}CH{-}CH_2{+} \\ | \\ X \end{array} + \begin{array}{c} CH_2X \\ | \\ {+}CH_2{-}C{=}CH{-}CH_2{+} \end{array} + \begin{array}{c} CH_3 \\ | \\ {+}CH{=}C{-}CH{-}CH_2{+} \\ | \\ X \end{array}$$

$$\qquad\qquad 1 \qquad\qquad\qquad\qquad 2 \qquad\qquad\qquad\qquad 3$$

Normally structure **1** is dominant (>80%) (38,39). More than one halogen atom per isoprene unit can also be introduced. However, the reaction rates for excess halogens are lower and the reaction is complicated by chain fragmentation (40).

Other Derivatives. Various other derivatives have appeared on the market or reached the market development stage. Conjugated-diene butyl is obtained by the controlled dehydrohalogenation of halogenated butyl rubber

(41). The product can be cross-linked with peroxide or exposure to radiation. Free-radical grafting with vinyl monomers, eg styrene, can be used in a graft cure, leading to a transparent rubber with a T_g of about $-59°C$. Carboxy-terminated polyisobutylene useful in forming networks with epoxies or aziridine has been prepared from high molecular weight butyl rubber or a poly(isobutylene-*co*-piperylene) [26335-67-1] copolymer (42). The resulting carboxy-terminated polymers are viscous liquids. High molecular weight isobutylene–cyclopentadiene rubbers containing up to 40% cyclopentadiene were produced and developed by ExxonMobil (43) and were recently reexamined by Daelim (43). Highly branched polyisobutylene are prepared by use of an appropriately designed initiator that is also a comonomer in the polymerization (44).

Isobutylene–Isoprene–Divinylbenzene Terpolymers. A partially cross-linked terpolymer of isobutylene, isoprene, and partially reacted divinyl benzene is commercially available from Rubber Division, Bayer Inc., Canada. The residual vinyl functionality may be cross-linked with peroxides, a treatment that would normally degrade conventional butyl rubbers. This material is used primarily in the manufacture of sealant tapes and caulking compounds (45).

Liquid Butyl Rubber. Degradation of high molecular weight butyl rubber by extrusion at high shear rates and temperatures produces a liquid rubber with a viscosity average molecular weight (M_v) between 20,000 and 30,000. The relatively low viscosity aids in formulating high solids compounds for use in sealants, caulks, potting compounds, and coatings. Resulting compounds can be poured, sprayed, and painted.

2.3. New Materials. *Star-Branched (SB) Butyl.* Butyl rubbers have unique processing characteristics because of their viscoelastic properties and lack of crystallization of compounds on extension. They exhibit both low green strength and low creep resistance as a consequence of high molecular weight between entanglements. To enhance the strength of uncured traditional butyl rubber a relatively high molecular weight is required. Increasing molecular weight also causes an increase in relaxation time along with high viscosity. In such situations it is usually helpful to broaden the molecular weight distribution, but this is difficult to accomplish in conventional butyl rubber polymerization. Physical blending of low and high molecular weight polymer can also provide broader molecular weight distributions, but it results in other processing problems such as high extrudate swelling in flow-through shaping dies.

SB butyl has a bimodal molecular weight distribution with a high molecular weight branched mode and a low molecular weight linear component. The polymer is prepared by a conventional carbocationic copolymerization of isobutylene and isoprene at low temperature, but in the presence of a polymeric branching agent. The high molecular weight branched molecules are formed during the polymerization via a graft-from or a graft-onto mechanism. A graft-from reaction takes place when a macroinitiator/macrotransfer reagent, such as hydrochlorinated poly(styrene-*co*-isoprene) or chlorinated polystyrene, is used. A graft-onto reaction takes place when a multifunctional terminating agent, eg, poly(styrene-*co*-butadiene), is employed as the branching agent. In general, the SB butyl has 10–20% high molecular weight branched molecules, which have a random comb-like structure with 20–40 butyl branches. Although this is not a true star topology, it approaches a star structure since the branching agent is relatively short and the branching density is relatively high, ie, the molecular weight between branching points is low compared to the segment length of the butyl branches. SB butyl rubbers offer a unique balance of viscoelastic properties, resulting in significant processability improvements. Dispersion in mixing and mixing rates are improved. Compound extrusion rates are higher, die swell is lower, shrinkage is reduced, and surface quality is improved. The balance between green strength and stress relaxation at ambient temperature is improved, making shaping operations such as tire building easier. Several grades of ExxonMobil SB butyl polymers including copolymer, chlorinated, and brominated copolymers are commercially available.

Brominated Poly(isobutylene-co-p-methylstyrene).   *para*-Methylstyrene [622- 97-9] (PMS) can be readily copolymerized with isobutylene via classical carbocationic copolymerization using a strong Lewis acid, eg, $AlCl_3$ or alkyl aluminum in methyl chloride, at low temperature. The copolymer composition is very similar to the feed monomer ratio because of the similar copolymerization reactivity ratios, ie, $r_1 = 1$ and $r_2 = 1.4$, under commercial polymerization conditions. These new high molecular weight copolymers encompass an enormous range of properties, from polyisobutylene-like elastomers to poly(*p*-methylstyrene)-like tough, hard plastic materials with T_g's above $100°C$, depending on monomer ratio. A highly reactive and versatile benzyl bromide functionality, $C_6H_5CH_2Br$ can be introduced by the selective free-radical bromination of the benzyl group in the copolymer. The brominated copolymer can be cross-linked with a variety of cross-linking systems. This new functionalized copolymer preserves polyisobutylene properties, low permeability, and unique dynamic response, while adding the behavior of inertness to ozone, a property similar to ethylene–propylene rubbers. Copolymers with PMS below 10 mol% are most useful for elastomeric applications because the T_g's are near $-60°C$. Several grades of the brominated copolymer (Exxpro^{TM} Specialty Polymers) are available from ExxonMobil (46,47).

The benzyl bromide in the brominated copolymer can also be easily converted by nucleophilic substitution reactions to a variety of other functional groups and graft copolymers as desired for specific properties and applications (47). Ionomers (48), grafted copolymers (49), radiation-cured rubbers (50), and rubber-toughened nylons (51) are a few examples of the derivatives and functions that modification of brominated poly(isobutylene-*co-p*-methylstyrene) can offer.

Thermoplastic Elastomers.   With the structural control inherent in living polymerization, new block copolymers containing polyisobutylene are possible (36). ABA triblock copolymers (A = polymethyl vinyl ether, B = polyisobutylene) that provide morphologies capable of exhibiting properties of an elastomer at use temperature, while processing like a thermoplastic, have been made from polyisobutylene and several styrenic derivatives (52–58). As thermoplastic elastomers, these materials offer other advantages owing to the intrinsic properties of polyisobutylene, namely low permeability and low dynamic modulus. More recently, star block copolymers of the A_2B_2 type have been prepared and exhibit unique physical properties (59). Many more architecturally designed polyisobutylenes are possible through living polymerization techniques. Kuraray Inc., Japan, commercially manufactured polyisobutylene–polystyrene block copolymers. Also, Kaneka Inc., Japan, produce Epion, a functionalized polyisobutylene.

3. Manufacturing

Most of the butyl polymers made commercially are produced by copolymerizing isobutylene and isoprene in precipitation processes that use methyl chloride as the diluent and a catalyst system comprising a Lewis Acid and an alkyl halide. The Lewis acid used in many of the commercial butyl rubber plants is aluminum chloride, which is low cost, a solid, and soluble in methyl chloride. Aluminum alkyls are now becoming popular because they simplify catalyst preparation and have been shown to increase monomer conversion.

The manufacture of butyl rubber, poly(isobutylene-*co-p*-methylstyrene) (Exxpro backbone) and high molecular weight polyisobutylene (Vistanex) requires a complex manufacturing process consisting of feed purification, feed blending, polymerization, slurry stripping, and finishing. A schematic flow diagram (Fig. 1) shows the major units in a butyl plant.

An alternative solution process, developed in Russia, uses a C_5–C_7 hydrocarbon as solvent and an aluminum alkyl halide as the initiator. The polymerization is conducted in scraped surface reactors at −90 to −50°C. The solution process avoids the use of methyl chloride, which is an advantage when butyl rubber is to be halogenated. However, the energy costs are higher than for the slurry

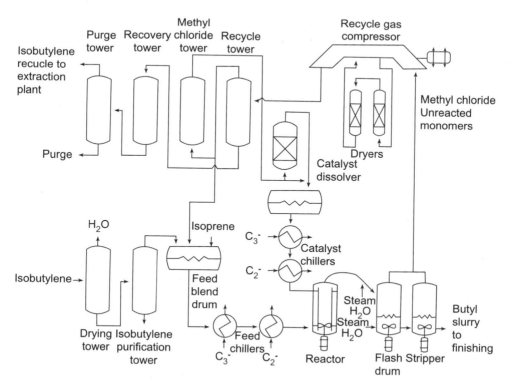

Fig. 1.   Butyl plant flow plan.

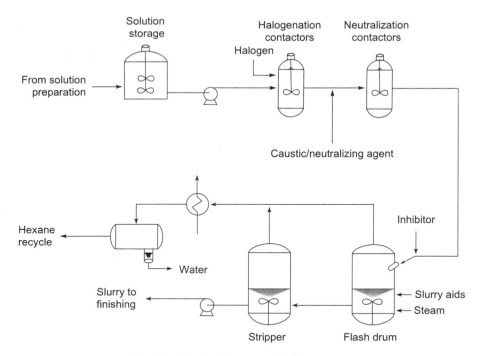

Fig. 2. Block diagram of halogenation.

process because of the higher viscosity of the polymer solution. Consequently, it is unlikely that the well-established slurry process will be displaced.

The manufacture of halobutyl rubbers such as Bromobutyl, Chlorobutyl, and Exxpro [bromopoly(isobutylene-*co*-*p*-methylstyrene)] requires a second chemical reaction: the halogenation of the polymer backbone. This can be achieved in two ways, the finished polymer produced in the butyl plant can be dissolved in a hydrocarbon solvent such as hexane or pentane, or a solvent replacement process can be used to dissolve the polymer from the slurry leaving the reactor. A schematic flow diagram of the halogenation process is shown by Figure 2.

3.1. Monomer Purification. To make high molecular weight polymers with good isobutylene conversion and good reactor service factor, the feed must be pure and dry. We must start with high quality isobutylene (>99%), dry it, and remove other olefins, eg butene-1, butene-2, propene, and oxygenated hydrocarbons such as dimethyl ether and methanol. A number of commercial processes are available for production of the required high purity isobutylene. An extraction process based on sulfuric acid has been developed by several companies (60,61) and is used extensively. Significant quantities of isobutylene are also produced by dehydration of *tert*-butyl alcohol (62). The highest purity isobutylene is produced by MTBE (methyl-*t*-butyl ether, [1634-04-4]) decomposition plants. This process starts with the selective reaction of dilute isobutylene in a C_4 stream with methanol over an acid ion-exchange resin, eg, Amberlyst 15, to form MTBE. This ether is produced mainly as a high octane blending component for low lead gasoline. Catalytic decomposition at 170–200°C and 600 kPa

(5.9 atm) over a fixed-bed acid catalyst, eg, $SiO_2Al_2O_3$ or Amberlyst 15, produces high purity isobutylene (63–65).

The isobutylene is then dried by azeotropic distillation and purified in a super-fractionating distillation column to reduce the butenes to less than 1000 ppm. Note that if water is not removed it will cause icing in the feed chillers and this will lead to a poor reactor service factor.

The purified isobutylene is then blended with a recycled methyl chloride stream containing a low level of isobutylene (~5%). Finally, the comonomer, isoprene or p-methylstyrene, is added. In this blending process, control of the ratio of comonomer to isobutylene is very important. This is because it has a significant impact on the composition of the polymer produced, the conversion of monomer, and the stability of reactor operation. For these reasons, a combination of both an analyzer and a mass balance control can be used to maintain the composition of the feed blend. The feed blend contains 20–40 wt% of isobutylene and 0.4–1.4 wt% of isoprene or 1–2 wt% of p-methylstyrene, depending on the grade of butyl rubber to be produced; the remainder is methyl chloride.

3.2. Polymerization. Catalyst solution is produced by passing pure methyl chloride through packed beds of granular aluminum chloride at 45°C. The concentrated solution formed is diluted with additional methyl chloride to which a catalyst activator is added and the solution later stored. The feed blend and catalyst solutions are chilled to −100 to −90°C in a series of heat exchangers before entry to the reactor.

The cold feed and catalyst are introduced continuously to a reactor comprising a central vertical draft tube surrounded by concentric rows of cooling tubes. The reactors have an aspect ratio of 28 (length) to 8 (diameter) and contain an axial flow pump located at the bottom of the draft tube that circulates the slurry through the cooling tubes. The reactor is constructed with 3.5 or 9 wt% nickel steel, or alloys that have adequate impact strength at the low temperature of the polymerization reaction. The production of high molecular weight butyl requires a polymerization temperature below −135°F (−90°C) and the reaction is exothermic, generating 0.82 MJ/kg (350 Btu/lb) of polymer. This requires a two-stage refrigeration system that uses boiling propylene or propane and ethylene as the refrigerants. In some plants ammonia is used in the first stage of the refrigeration process.

The reactors are the epicenter of all butyl plants and are mechanically complex. The operation of the remainder of the plant is dictated by the reactors. Essentially, the reactors are stirred tanks that contain a heat exchanger to keep them cold.

In the reactor, polymer chains are initiated by the catalyst and propagate in solution. Chain propagation occurs in microseconds but the overall reaction rate is controlled by the slower initiation sequence. As individual polymer molecules are formed they precipitate to produce a fine milky slurry of sub–micron-size particles. These particles grow in the reactor by accreting new polymer chains and by agglomeration.

The polymer slurry circulates through the reactor tubes, and boiling liquid refrigerant in the reactor jackets keeps the reactor contents cold. The butyl polymerization process is complex because of the combination of low temperature operation, polymer slurry formation, and difficulty in directly measuring polymer quality. To achieve high molecular weight, the reaction temperature must be kept low to reduce the amount of chain transfer. In the slurry process, the viscosity is relatively low, and good heat transfer can be achieved. To maintain a stable slurry, the temperature must be kept below the glass-transition temperature, T_g, of polyisobutylene ($-68°C$) and, as the slurry is shear-thinning, there must be a high level of shear. The axial flow pump that provides a high degree of circulation in the reactor achieves this.

One of the common theories to explain fouling in a butyl reactor is that rubber particles in the reactor slurry are not completely glassy because of diluent and monomers in the surface layer. This causes them to be sticky as well as agglomerating, and the particles will stick to the heat-transfer surfaces in the reactor. This causes the heat-transfer resistance to increase and the slurry temperature to warm. Eventually, this warming can cause the slurry to destabilize and the viscosity to increase rapidly. Ultimately, this could lead to plugging, which requires extended solvent washing to clean the reactor. Therefore, production in a reactor is stopped every 2–4 days for solvent washing. This forces a cyclical multireactor operation in which some reactors are producing polymer while others are being washed to remove rubber fouling. Typical runs are from 18 to 60 h, depending on feed purity, slurry solids concentration, and production rate.

The production rate is 2.0–4.5 t/h, depending on feed rate, monomer concentration in the feed, and monomer conversion. The conversion of isobutylene and isoprene typically ranges from 75–95% and 45–85%, respectively, depending on the grade of butyl rubber being produced. The molecular weight of the polymer produced is set by the ratio of chain-making to chain-terminating processes. In commercial plants the molecular weight and composition of the polymer formed are controlled by the concentration of monomers in the reactor liquid phase and the amount of terminating or transfer species present to interrupt chain growth. The slurry composition depends on the monomer content of the feed stream and the extent of monomer conversion. In practice, the flow rate and the composition of catalyst to the reactor is the principal operating variable; reactor residence time is often in the range 30–60 min.

The original reactor design, known as the draft tube reactor, has been used commercially since the initial development of butyl in the 1940s. An improved design in which the draft tube is replaced by additional tubes and the circulation pump is redesigned has recently been proposed (66).

In addition to these changes to the mechanical design of the reactor, improvements to the polymerization chemistry and diluent have been investigated in the last decade. Examples are the use of supercritical carbon dioxide (67) and the use of aluminum alkyls in the catalyst system (68).

3.3. Halogenation. Chlorinated and brominated butyl rubber can be produced in the same plant in blocked operation. However, there are some differences in equipment and reaction conditions. A longer reactor residence time is required for bromination because of the slower reaction rate compared to chlorination. Separate facilities are needed to store and meter the individual halogens to the reactor. Additional facilities are required because of the complexity of stabilizing bromobutyl rubber.

The halogenation process begins with the preparation of a hexane solution of butyl rubber with the desired molecular weight and unsaturation. Slurry from a butyl polymerization reactor is dissolved (69–71) by transferring it into a drum containing hot liquid hexane that rapidly dissolves the fine slurry particles. Hexane vapor is added to flash methyl chloride and unreacted monomers overhead for recovery and recycle. The solution of butyl in hexane passes to a stripping column where the final traces of methyl chloride and monomers are removed. The hot solution is brought to the desired concentration for halogenation, typically 20–25 wt% in an adiabatic flash step.

Alternatively, bales of finished butyl rubber are chopped or ground to small pieces and conveyed to a series of agitated dissolving vessels or to a large vessel divided into multiple stages. Solutions containing 15–20% polymer can be prepared in 1–4 h depending upon temperature, particle size, and agitation. This method has the advantage of being independent from the butyl polymerization process but requires storage and careful inventory control between the two stages of the process. This process also requires two finishing operations: one to produce dry butyl backbone, the second to finish the halobutyl product. Investment costs for the two dissolving processes are similar, but energy costs for the dissolving process can be higher.

Halogenation is the second major chemical reaction in the production of halobutyl polymers. It is usually carried out by adding bromine liquid or chlorine vapor to a solution of rubber in hydrocarbon solvent (hexane or pentane), which is often referred to as cement. The cement must be essentially free of monomers, or low molecular weight toxic species will be formed during the chlorine or bromine reactions. The halogenation of the butyl backbone is an ionic-substitution reaction in which the halogen is added to the cement stream in a well-mixed reactor (eq. 1). This reaction is unusual for polymers where it's more typical for the bromine to add across the double bond.

Insufficient mixing will lead to poor distribution of the halogen on the polymer, which can cause the product cure characteristics to be unsatisfactory. Another potential problem is the vigorous reaction of liquid chlorine with hydrocarbon. This leads to complete breakdown of some of the polymer, producing carbon and HX. This causes the polymer to be gray rather than the normal off-white and the product properties to be adversely affected.

For both chloro- and bromobutyl rubber two isomers are formed as shown in equation 1. The ratio of these isomers must be carefully controlled in order to keep consistent product properties.

The by-product from this reaction, HX, is normally reacted with aqueous caustic solution to give a soluble salt. Incomplete neutralization will leave HX in the rubber, and subsequent reactions during the drying process can destabilize the polymer. The key to good neutralization is good mixing of the halogen and cement. Further improvements to the neutralization process can be achieved by the addition of a surfactant (72).

Because of the generation of HCl or HBr, the maximum efficiency for halogen usage in this process is 50%. An improvement in efficiency can be achieved by adding hydrogen peroxide to the process to convert the acid back to halogen (73).

The bromination reaction required to produce Exxpro requires a free-radical reaction because there is no unsaturation in the backbone. The bromine and a free-radical initiator are added to the cement, mixed well, and then passed through a series of stirred tank reactors where the chain reaction shown by equation 2 occurs.

$$I_2 \xrightarrow{\text{Heat}} 2I^{\bullet} \xrightarrow{\text{Br}_2} \text{IBr} + \text{Br}^{\bullet}$$

Chain reaction

An alternative process is bulk-phase halogenation (74–76). Dry butyl rubber is fed into a specially designed extruder reactor and contacted in the melt phase with chlorine or bromine vapor. By-product halogen acids are vented directly, avoiding the need for a separate neutralization step. Halogenated rubbers comparable in composition and properties to commercial products can be obtained.

3.4. Finishing.   The halogenated rubber solution then passes into a vertical drum where the solvent is flashed and stripped by steam and hot water. Calcium stearate is added to the slurry in this drum to prevent polymer agglomeration. A second vessel in series provides additional residence time for the solvent to diffuse from the rubber and be vaporized. The final solvent content and the steam usage for solvent removal depend on the conditions in each vessel. Typically, the lead flash drum is operated at 105–120°C and 200–300 kPa (29–43.5 psi). Conditions in the final stripping stage are 101°C and 105 kPa (15 psi). The hexane can be reduced to 0.5–1.0 wt% with a steam usage of 2.0–2.5 kg/kg rubber.

The resultant polymer/water slurry is kept agitated and then screened to separate the bulk water from the rubber. The polymer is then dried in a series of extrusion dewatering and drying steps to a final moisture content of <0.3 wt%. Antioxidants and stabilizers, eg, BHT and epoxidized soybean oil, protect against dehydrohalogenation and are added in the final extrusion step. Fluid-bed conveyors and/or airvey systems are used to cool the product to an acceptable packaging temperature. The resultant dried product is in the form of small "crumbs," which are subsequently weighed and compressed into 75-lb bales for wrapping in polyethylene film and packaged into wooden or metal crates or cardboard containers.

4. Structure, Properties, and Product Applications of Isobutylene-Based Polymers

4.1. Molecular Structure.   *Polyisobutylene.*   Isobutylene polymerizes in a regular head-to-tail sequence to produce a polymer having no asymmetric carbon atoms. The two pendant methyl groups bonded to alternative chain atoms in polyisobutylene (PIB) produce steric crowding, for which a partial relief is achieved by a distortion of the methylene carbon to 124° as compared to 110° for the tetrahedral carbon (73,74). Methyl crowding is further alleviated by distorting the dihedral angle, leading to a splitting of the trans-rotational isomer by about 25° (73–75). The splitting of the trans isomer in two equivalent states is a special feature of PIB, not observed in its vinyl analogue, and has significant impacts on its local chain-segment dynamics (76). The additional fast trans-to-trans skeletal relaxation process in PIB results in a faster segmental motion and a lower glass-transition temperature despite the fact that PIB is a sterically strained and crowded macromolecule (76). The glass-transition temperature of PIB is about −70°C (77). It is an amorphous elastomer in the unstrained state. On extension at room temperature, PIB of high molecular weight crystallizes into crystals of an 8/3 helix with an alternating distorted trans-gauche structure (78). One chain end of PIB is typically unsaturated because of chain-transfer and termination mechanisms. Molecular weights could range from several hundred to several millions. There is no long-chain branching unless special synthesis methods are employed. The molecular weight distribution is commonly the most probable with $M_w/M_n = 2.0$.

Polybutenes. Copolymerization of mixed isobutylene and 1-butene containing streams with a Lewis acid catalyst system yields low molecular weight copolymers, from several hundreds to a few thousand, that are clear, colorless, and viscous liquids. The chain ends are unsaturated and are often chemically modified to provide a certain functionality (79,80).

Butyl Rubber. In butyl rubber, isoprene is enchained predominantly (90–95%) by 1,4-addition and head-to-tail arrangement (81–84). The remaining minor isoprene species may be interpreted spectroscopically either as 1,2-enchained (83,84) or as branched species from 1,4-addition (85). Depending on the grade, the unsaturation in butyl rubber is between 0.5 and 3 mol%. The glass-transition temperature of butyl rubber is about −65°C (86). With the low content of isoprene and a near-unity reactivity ratio between isoprene and isobutylene (87), a random distribution of enchained isoprene monomer in butyl rubber is achieved. The molecular weight distribution in butyl rubber is mostly with M_w/M_n of 3–5.

Halogenated Butyl Rubber. Halogenation at the isoprene site in butyl rubber proceeds by a halonium ion mechanism leading to the formation of a predominated exomethylene alkyl halide structure in both chlorinated and brominated rubbers (88,89). Upon heating, the exo allylic halide rearranges to give an equilibrium distribution of exo and endo structures (90–92). Halogenation of the unsaturation has no apparent effects on the butyl backbone and on the butyl glass-transition temperature. However, cross-linked halobutyl rubbers do not crystallize on extension as a result of the backbone irregularities introduced by the halogenated isoprene units.

Star-Branched (SB) Butyl. Introduction of a branching agent of styrene–butadiene–styrene (SBS) block copolymer during cationic polymerization of butyl leads to a SB butyl that, in general, contain 10–15% of star polymers, with remaining linear butyl chains (93,94). SB butyl is a reactor blend of linear polymers and star polymers, where the star molecules were synthesized during polymerization by cationic grafting of propagating linear butyl chains onto the branching agent (95). In the solid state, these star polymers aggregate to form ~30-nm-diameter spherical domains separated by about 90 nm and covered by about 6 vol% (96,97). It was proposed that these domains consist of predominantly the styrene–butadiene–styrene branching agent (96). A much broader molecular weight distribution is achieved in SB butyl with M_w/M_n greater than 8. Halogenation of SB butyl results in the same halogenated structure in the linear butyl chains and in the linear butyl chain arms of the star fraction as that in the halogenated butyl. Unreacted butadiene blocks of the SBS branching agent were 100% brominated and 60% chlorinated after halogenation (98).

Brominated Poly(isobutylene-co-p-methylstyrene). Copolymerization of isobutylene with *para*-methylstyrene (PMS) produces a saturated copolymer backbone with randomly distributed pendant *para*-methyl-substituted aromatic rings. During radical bromination after polymerization, some of the substituted *para*-methyls are converted to bromomethyls for vulcanization and functionization (99,100). Brominated poly(isobutylene-*co-p*-methylstyrene) polymers are saturated terpolymers containing isobutylene, 1–6 mol% PMS, and 0.5–1.5 mol% BrPMS (brominated *para*-methylstyrene). Their T_gs increase with increasing PMS content and is around −60°C. The molecular weight distribution of brominated poly(isobutylene-*co-p*-methylstyrene) is narrow, with M_w/M_n less than 3 (99).

4.2. Physical Properties. *Permeability.*

Primary uses of polyisobutylene and isobutylene copolymers of butyl, halobutyl, SB butyl, and brominated poly(isobutylene-*co-p*-methylstyrene) in elastomeric vulcanized compounds rely on their properties of low air permeability and high damping. In comparison with many other common elastomers, polyisobutylene and its copolymers are notable for their low permeability to small-molecule diffusants as a result of their efficient intermolecular packing (101–103) as evidenced by their relative high density (density of 0.917 g/cm3). This efficient packing in isobutylene polymers leads to their low fractional free volumes and low diffusion coefficients for penetrants (101). In combination with the low solubilities of small-molecule diffusants in isobutylene polymers, low permeability values for small molecules, such as He, H_2, O_2, N_2, CO_2, and others, are observed in isobutylene polymers as compared with other elastomers (104).

As shown in Figure 3, diffusion coefficients of nitrogen in both various diene rubbers and in butyl rubber increase with increasing differences between the measurement temperature and the corresponding rubber's glass-transition temperature. However, although the rate of increase in diffusion coefficient with $T-T_g$ is about the same between diene rubbers and butyl rubber, the absolute values of diffusion coefficient in butyl rubber are significantly less than that of diene rubbers. Considering isobutylene, copolymers discussed contain only small amounts of comonomers besides isobutylene, their temperature-dependent permeability values follow the same curve in Figure 3 for butyl rubber. With brominated poly(isobutylene-*co-p*-methylstyrene) having the highest T_g among isobutylene copolymers, it has the corresponding lowest permeability value at room temperature among isobutylene copolymers (105).

Dynamic Damping. Polyisobutylene and isobutylene copolymers are high damping at 25°C, with loss tangents covering more than eight decades of

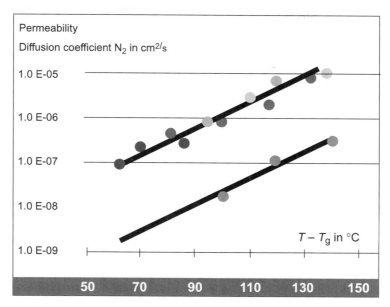

Fig. 3. Diffusion coefficients of nitrogen in diene rubbers and in butyl rubber as a function of $T - T_g$ (Data from Ref. 104). ● Chloroprene rubber (neoprene); ● styrene butadiene rubber; ● natural rubber; ● nitrile butadiene rubber; ● butyl rubber.

frequencies even though their T_gs are less than $-60°C$ (106,107). This broad dispersion in PIB's dynamic mechanical loss modulus is unique among flexible-chain polymers and is related to its broad glass–rubber transition (108). The narrowness of the glass–rubber transition, as defined by the steepness index, for PIB is 0.65, which is much smaller than most polymers. In addition, PIB has the most symmetric and compact monomer structure among amorphous polymers, which minimizes the intermolecular interactions and contributes to its unique viscoelastic properties (109,110). As a result, a separation in time scale between the segmental motion and the Rouse modes is broader in glass–rubber transition, leading to the appearance of the sub-Rouse mode (108,109). In Rouse's theory, a polymer chain is represented as a sequence of equivalent Gaussian submolecules. The Rouse mode refers to motions of these submolecules, and sub-Rouse mode involves motions of chain units smaller than these submolecules. Considering the differences in temperature dependencies of these motions, the glass transitions of PIB and its copolymers are thermo-rheological complexes and they do not follow time–temperature superposition (108,111). PIB and its copolymers have high entanglement molecular weights (112) and corresponding low plateau moduli which, in turn, contribute to their high tack or self-adhesion in the uncross-linked state. Entanglement molecular weight of PIB is about 9000 as compared with 1300 for polyethylene, 1700 for ethylene–propylene copolymer, 2000 for 1,4-polybutadiene, 3000 for styrene–butadiene copolymer, and 6100 for hevea rubber.

Chemical Properties.   PIB is a saturated hydrocarbon where the unsaturated chain ends could undergo reactions typical of a hindered olefin. These end groups are used, particularly in low molecular weight materials where plenty of chain ends are available, as a route to functionization, such as the introduction of amine groups to PIB for producing dispersants for lubricating oils. As for butyl rubbers, their in-chain unsaturations could be attacked by atmospheric-ozone-leading degradation and, hence, require protection by antioxidants (113). Chlorobutyls are thermally more stable than bromobutyls. Upon thermal exposure up to 150°C, no noticeable decomposition took place in chlorobutyl except for some allylic chlorine rearrangement, whereas the elimination of HBr occurred in bromobutyl concurrently with isomerization to produce conjugated dienes that subsequently degrade (91,92). Brominated poly(isobutylene-*co*-*p*-methylstyrene), without any unsaturation, is the most thermally stable isobutylene copolymer. In addition, the strong reactivity of the benzylic bromine functionality in brominated poly(isobutylene-*co*-*p*-methylstyrene) with nucleophiles allows functionization and grafting in addition to its uses for vulcanization (99,100). Polyisobutylene and its copolymers, including butyl, halobutyl, and brominated poly(isobutylene-*co*-*p*-methylstyrene), are readily soluble in nonpolar solvents where cyclohexane is an excellent solvent, benzene is a moderate solvent, and dioxane is a nonsolvent (114).

4.3. Elastomeric Vulcanizates.   As with almost all rubbers, the applications of isobutylene copolymers in rubber goods require compounding and subsequent vulcanization, or cross-linking. During compounding, various fillers, processing aids, plasticizers, tackifiers, curatives, and antidegradants are added.

Fillers.   Filler addition is a common practice in rubber industries to stiffen and strengthen amorphous rubbers. In addition to the reinforcement, fine and particulate fillers, most notably carbon black, suppress elasticity of gum rubbers and render better processability, such as less die swell, less shrinkage, less melt fracture, and less nerve (or less elastic spring-back in uncured state) (115). Physical interactions between carbon blacks and rubber are strong but can yield upon stressing. The effects of carbon black on properties of butyl, halobutyl, and brominated poly(isobutylene-*co*-*p*-methylstyrene) rubbers are similar to that of other elastomers. As particle size decreases with a corresponding increase in specific surface area, the strength, modulus, hardness, viscosity, and damping of carbon-black-filled vulcanized rubbers increase (116,117). Tensile strength, in general, is raised to a maximum with carbon black content and then decreases at higher loading. For butyl rubber, this maximum tensile strength is obtained at 50–60 phr (part per hundred of rubber) of carbon black.

Beyond filler addition in rubbers, elastomeric behavior can be tailored by proper blending of filled elastomers. Major reasons for the usage of blends are to reduce compound cost, to improve processability, and to enhance final product performance (118). But, in this case, selective filler partition may alter the blend morphology and produce significant changes in blend properties. Elastic and fracture properties, such as resilience, tear strength, and fatigue resistance, of rubber blends have been found to depend strongly on filler phase distribution (119,120). Carbon aggregates were known to preferentially reside in the polymer with higher unsaturation, lower viscosity, and higher polarity, generally in that order of significance (121). Hence, a preferential filler partition into the diene rubber phase, such as BR and SBR, in diene rubber blends with butyl rubber was found (122,123). However, with the strong interactions between allylic bromine or benzylic bromine in bromobutyl and in brominated poly(isobutylene-*co-p*-methylstyrene), respectively, and carbon blacks (124), the opposite in terms of preferential filler partition into the bromobutyl or brominated poly (isobutylene-*co-p*-methylstyrene) phases in their blends with high diene rubbers was indicated (125,126).

Mineral fillers are used for light-colored rubber compounds for cost reduction and physical improvement. In butyl and halobutyl, alkaline fillers, such as calcium silicates, and hygroscopic fillers can strongly retard cure, while acid clays can accelerate cure. Silicas provide maximum reinforcement among mineral fillers but they promote compound stiffness and strongly affect cure rates by their absorption of curatives. Clays are less reinforcing and calcined clay is particularly suited for pharmaceutical stopper applications for its purity, low water content, and particle size uniformity. Talc is semireinforcing and improves resistance to fragmentation during needle penetration and withdrawal in pharmaceutical closures. Nonreinforcing fillers, such as calcium carbonate, are added to lower compound cost without affecting cure. Strong reductions in permeability in black or light-colored compounds of isobutylene copolymers could be achieved by adding small percents (less than 5 wt%) of exfoliated high-aspect-ratio atomic-thickness sheets of layered mineral, such as organonanoclays, without significant compromises in flexibility (127–129). The alignment of the nanoclay leads to a nematic-phase morphology and introduces torturosity for the diffusants, resulting in the lowering in permeability (130).

Plasticizers and Processing Aids.   Petroleum-based oils are commonly used as plasticizers to lower compound viscosity and to improve processability and low temperature properties but with an increase in compound air permeability. Plasticizers are selected for their compatibility with isobutylene polymers and for their low temperature properties. Butyl rubber has a solubility parameter that is similar in value to that of polybutene, and paraffinic and naphthenic oils for which they are preferred. Although not to be considered as solvating agents, paraffin waxes and low molecular weight polyethylene are added for improved processing. For an improvement in low temperature flexibility, ester-type plasticizers, such as adipates and sebacates, can be used. In addition to functioning as a plasticizer, Struktol 40 MS can enhance filler dispersion and cure adhesion to high diene rubber substrates. Materials such as mineral rubber and asphaltic pitch have also been used for this dual function.

Other Ingredients.   Other compound ingredients include tackifiers, flame retarders, odorants, and lubricants. Hydrocarbon tackifier resins improve the tack of unvulcanized compounds without any side reactions but are less effective than the phenolic type. However, phenol-formaldehyde tackifier resins react with halobutyl through their methylol groups, leading to premature cure. Hindered phenolic antioxidants added during manufacture of butyl and halobutyl prevent oxidation during their finishing, storage, and usage. These antioxidants also improve the resistance of vulcanized butyl and halobutyl compounds to ozone cracking. Additional stabilization against dehydrohalogenation at elevated temperature is required for halobutyls (26,131). Calcium stearate is used in chlorobutyl and a combination of calcium stearate and an epoxidized soybean oil is applied in bromobutyl to prevent dehydrohalogenation. No antioxidants or other stabilizers are added in brominated poly(isobutylene-*co-p*-methylstyrene).

4.4. Vulcanization. Vulcanization or curing in isobutylene polymers introduces chemical cross-links via reactions involving either allylic hydrogen or allylic halogen in butyl or halobutyl, respectively, or benzylic bromine in brominated poly(isobutylene-*co-p*-methylstyrene) to form a polymer network. In butyl and halobutyl, most vulcanizates have about $1 \times 10-4$ mol of cross-links per cm^3, or about 250 backbone carbon atoms between cross-links.

Butyl. The basic curing agent for butyl rubber is elemental sulfur, with the formation of sulfide cross-links. Because of the low unsaturation in butyl rubbers, sulfur accelerators, such as thiurams or dithiocarbamates, are required to obtain acceptable rates and states. To activate accelerators and to stabilize cross-links already formed, butyl compounds require zinc oxide. Sulfur cross-links rearrange at elevated temperatures as a result of their low bonding energy that could lead to creep and permanent set in strained butyl at high temperatures. The resin cure in butyl rubber is based on the reaction of the methylol groups in the phenol-formaladehyde resin with allylic hydrogen in butyl, usually with a Lewis acid catalyst, to yield carbon−carbon cross-links that are thermally stable. Resin cure is preferred in butyl tire-curing bladders for their required thermal fatigue stability.

Halobutyl. The basic curing agent for halobutyl is zinc oxide, with the formation of carbon−carbon cross-links by alkylation catalyzed by a zinc halide produced after dehydrohalogenation of halobutyl (90,91,132). The ease of halogen elimination is the main difference between chlorobutyl and bromobutyl during curing where the bromobutyl cures faster (133). Esterification occurs at the halogen site in the presence of stearic acid. As a result, stearic acid accelerates cure rate and lowers the cure state. The zinc oxide cure is sensitive to acidic and basic compounding ingredients. In general, acids accelerate cure while bases retard. Water is a strong retarder because it complexes with reactive intermediates. In some applications, such as special pharmaceutical stoppers, the absence of zinc is mandatory. On the basis of the strong reactivity between amine and allylic halogen (134), diamines can be used as the curative. While regular butyl undergoes molecular weight breakdown in the presence of peroxide, halobutyl can be cross-linked with organic peroxides. The best cure rate and optimal properties are achieved with a suitable co-agent, such as *m*-phenylene bismaleimide, at high temperatures.

Brominated Poly(isobutylene-co-p-methylstyrene). Using mainly zinc oxide and stearic acid, the cross-linking of brominated poly(isobutylene-*co-p*-methylstyrene) involves the formation of carbon−carbon bonds through Friedel−Crafts alkylation coordinated by a zinc complex (135,136). Unlike halobutyls, brominated poly(isobutylene-*co-p*-methylstyrene) cures extremely slowly with zinc oxide alone. A nucleophilic agent, such as zinc stearate, is needed to promote rapid displacement of benzyl bromide. The zinc stearate can be added directly or can be formed during vulcanization through the reaction of zinc oxide and stearic acid. In cases where faster cure is desired, dinucleophilic substitution using dinucleophiles, such as diamines, is the method of choice.

Table 2. **Annual Butyl and Halogenated Butyl Rubber Production Capacity (YE 2002)**

Producer	Products	Capacity, 10^3 t
ExxonMobil Chemical Co.		
Baton Rouge, La.	Halobutyl rubbers	143
Baytown, Tex.	Butyl rubber	27
	Halobutyl rubbers and Exxpro	75
Notre Dame de Gravenchon, France	Butyl rubber	54
Fawley, U.K.	Halobutyl rubbers	97
Japan Butyl Co., Ltd.[a]		
Kawasaki, Japan	Butyl rubber	78
Kashima, Japan	Halobutyl rubbers	52
Bayer (152)		
Sarnia, Ontario, Canada	Butyl, halobutyl rubbers	140
Zwijndrecht, Belgium	Butyl, halobutyl rubbers	115
CIS (152)		
Nizhnekamsk, Russia	Butyl rubbers	60
Togliati, Russia	Butyl rubbers	35
Sinopec (152)		
Yanshan, P.R. China	Butyl rubbers	30

[a]Jointly owned with Japan Synthetic Rubber Co., Ltd.

5. Economic Aspects

Table 2 shows the manufacturing capacity for the world's major butyl and halobutyl polymers manufacturers. In addition to ExxonMobil Chemical and Bayer, during the last 10 years, butyl rubber produced at both Nizhnekamsk and Togliati has been sold throughout the world. Also, in the last 5 years, Sinopec has started to produce butyl polymers at its Yanshan plant in the People's Republic of China. All of the world's butyl plants use the more efficient slurry process except for the Togliati plant, which uses the solution process.

Manufacturing capacity has continued to expand in the last decade to keep up with the increasing demand due to expanding car sales and the growth in popularity of sports–utility vehicles, particularly in the United States. Apart from the economic downturns in 1992 and 2001, Figure 4 shows the worldwide sales of halobutyl polymers increasing steadily at approximately 4% per year. The widespread use of halobutyl innerliners in the tires of both passenger cars and light trucks accounts for most of these sales, although there is a small market for pharmaceutical stoppers. For most of the decade, halobutyl manufacturing plants have been running well above 85% capacity.

In the 1990s, sales of regular butyl polymers have not seen the same rate of growth as halobutyl. The majority of sales are to inner tube and tire bladder applications and, except for trucks and in some developing countries, these markets have been overtaken by the more popular tubeless tire. The introduction of the Russian and Chinese manufacturers has also dampened demand and not until the early 2000s has there been significant growth. Figure 5 shows the sales of regular butyl polymers during the past decade.

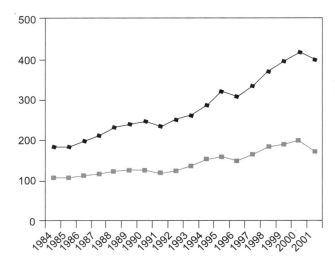

Fig. 4. Worldwide halobutyl sales. —◆— Halobutyl, worldwide; —■— Halobutyl, Americas.

6. Health and Safety Factors

Polyisobutylene, isobutylene–isoprene copolymers, and isobutylene-p-methyl-styrene copolymers are considered to have no chronic hazards associated with exposure under normal industrial use. Some grades can be used in chewing-gum base and are regulated by the FDA in 21 CFR 172.615. Vulcanized products prepared from butyl rubber or halogenated butyl rubber may contain small

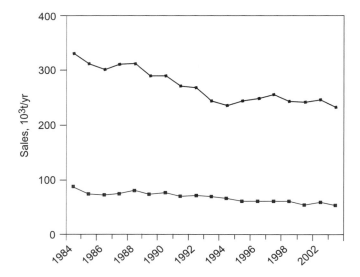

Fig. 5. Worldwide butyl sales. —◆— Regular butyl, worldwide; —■— Regular butyl, Americas.

amounts of toxic materials as a result of the particular vulcanization chemistry. Although many vulcanizates are inert, eg, zinc oxide cured chlorobutyl is used extensively in pharmaceutical stoppers, specific recommendations should be sought from suppliers.

7. Uses

The polyisobutylene portion of the isobutylene copolymers, such as butyl, halobutyl, and brominated poly(isobutylene-co-p-methylstyrene), imparts chemical and physical characteristics that make them highly useful in a wide variety of applications. The low degree of permeability to gases accounts for the largest uses of butyl and halobutyl rubbers, namely as inner tubes and tire innerliners. These same properties are also of importance in air cushions, pneumatic springs, air bellows, accumulator bags, and pharmaceutical closures. The thermal stability of butyl rubber makes it ideal for rubber tire-curing bladders, high temperature service hoses, and conveyor belts for hot material handling. With the added thermal stability in brominated poly(isobutylene-co-p-methylstyrene), it has made inroads in applications of bladders and curing envelopes.

Isobutylene-based polymers exhibit high damping and have uniquely broad damping and shock absorption ranges in both temperature and frequency. Molded rubber parts from butyl and halobutyl find wide applications in automobile suspension bumpers, exhaust hangers, and body mounts.

Blends of halobutyl or brominated poly(isobutylene-co-p-methylstyrene) with high diene rubbers are used in tire sidewalls and tread compounds (137–139). In sidewalls, ozone resistance, crack cut growth, and appearance are critical to their performance. Properly formulated blends with high diene rubbers that exhibit phase cocontinuity yield excellent sidewalls (140). The property balance for tire tread compounds can be enhanced by the incorporation of a more damping halobutyl or brominated poly(isobutylene-co-p-methylstyrene) rubber phase (141). Improvements in wet-, snow-, and ice-skid resistances and in dry traction without compromises in abrasion resistance and rolling resistance for high performance tires can be accomplished by using bromobutyl or brominated poly(isobutylene-co-p-methylstyrene) up to 30 phr in tread compounds (142).

Blends of isobutylene polymers with thermoplastic resins are used for toughening of these compounds (143,144). High-density polyethylene and isotactic polypropylene are often modified with 5–30 wt% of PIB. In some cases, a halobutyl phase is cross-linked as it is dispersed in the plastic phase to produce a highly elastic compound that is processable in thermoplastic molding equipment (145,146). The strong reactivity of the benzylic bromine group in brominated poly(isobutylene-co-p-methylstyrene) with amines during blending of brominated poly(isobutylene-co-p-methylstyrene) with polyamides results in surface-reactive compatibilization that suppresses coalescence and leads to extremely fine dispersions in these blends (147). Combining the low permeability values of both polyamides and brominated poly(isobutylene-co-p-methylstyrene), dynamic vulcanized thermoplastic blends of polyamides with brominated poly(isobutylene-co-p-methylstyrene) have been demonstrated to be suitable for tire innerliner applications (148–151).

Polybutenes enjoy extensive uses as adhesives, caulks, sealants, and glazing compounds. They are also used as plasticizers in rubber formulations with butyl, SBR, and natural rubber. In linear low density polyethylene (LLDPE) blends, they induce cling to stretch-wrap films. Polybutenes when modified at their chain-end unsaturations with polar functionality are widely employed in lubricants as dispersants. Blends of polybutene with polyolefins produce semisolid gels that can be used as potting and electrical-cable filling materials.

It should be noted that proper compounding and formulation are critical to the successful uses of all elastomer materials. It is recommended that the suppliers be contacted for information.

BIBLIOGRAPHY

"Rubber, Halogenated Butyl" in *ECT* 1st ed., 2nd Suppl. Vol., pp. 716–734 by F. P. Baldwin and I. Kuntz, Esso Research and Engineering Co.; "Elastomers, Synthetic" in *ECT* 2nd ed., Vol. 7, pp. 676–705 by W. M. Saltman, The Goodyear Tire & Rubber Co.; "Elastomers Synthetic (Butyl Rubber)" in *ECT* 3rd ed., Vol. 8, pp. 470–484 by F. P. Baldwin and R. H. Schatz, Exxon Chemical Co.; "Butyl Rubber" in *ECT* (online), posting date: December 4, 2000, by Edward Kresge, H.-C. Wang, Exxon Chemical Company; "Butyl Rubber" in *ECT* (online), posting date: December 4, 2000, by Edward Kresge, H.-C. Wang, Exxon Chemical Company.

CITED PUBLICATION

1. E. G. M. Tornqvist, in J. P. Kennedy and E. G. M. Tornqvist, eds., *Polymer Chemistry of Synthetic Elastomers*, Part 1, John Wiley & Sons, Inc., New York, 1968, p. 76.
2. R. M. Thomas, I. E. Lightbown, W. J. Sperks, P. K. Frolich, and E. V. Murphree, *Ind. Eng. Chem.* **32**, 1283–1292 (1940).
3. R. T. Morrissey, *Rubber World* **138**, 725 (1955).
4. F. P. Baldwin, *Rubber Chem. Technol.* **52**, 677 (1979).
5. D. Coddington, *Rubber Chem. Technol.* **59**, 905 (1979).
6. S. A. Banks, F. Brzenk, J. A. Rae, and C. S. Hua, *Rubber Chem. Technol.* **38**, 158 (1965).
7. Bulletin 12-K, Amoco Chemical Corp. Chicago, Ill.
8. Ref. 2, p. 209.
9. K. Matyjaszewski, ed., *Cationic Polymerizations*, Marcel Dekker, Inc., New York, 1996.
10. J. P. Kennedy and M. Marechal, *Carbocationic Polymerization*, John Wiley & Sons, Inc., New York, 1982.
11. P. H. Plesch and A. Gandini, *The Chemistry of Polymerization Process*, Monograph 20, Society of Chemical Industry, London, 1966.
12. A. Lisovskii, E. Nelkenbaum, V. Volkis, R. Semiat, and M. S. Eisen, *Inorg. Chim. Acta* **334**, 243 (2000).
13. U.S. Pat. 5,668,232 (Sept. 16, 1997), G. Langstein, D. Freitag, M. Lanzendörfer, and K. Weiss (to Bayer, AG).
14. T. D. Shaffer and J. R. Ashbaugh, *J. Polym. Sci., Part A: Polym. Chem.* **35**, 329 (1997).
15. S. Jacob, P. Zhenglie, and J. P. Kennedy, *Polym. Bull.* **41**, 503 (1998).
16. M. C. Baird, *Chem. Rev.* **100**, 1471 (2000).
17. M. Roth and H. Mayr, *Macromolecules* **29**, 6104 (1996).
18. H. Schlaad, Y. Kwon, L. Sipos, R. Faust, and B. Charluex, *Macromolecules* **33**, 8225 (2000).

19. D. M. Cheng, I. J. Gardner, H.-C. Wang, C. B. Fedrick, and A. H. Dekmezian, *Rubber Chem. Technol.* **63**, 265 (1990).
20. J. E. Puskas and C. Wilds, *Rubber Chem. Technol.* **67**, 329 (1994).
21. J. L. White, T. D. Shaffer, C. J. Ruff, and J. P. Cross, *Macromolecules* **28**, 3290 (1995).
22. J. P. Kennedy and R. M. Thomas, *Polymerization and Polycondensation Processes* (Advances in Chemistry Series 34), American Chemical Society, Washington, D.C., 1962, p. 111.
23. Ref. 22, p. 326
24. U.S. Pat. 2,356,128 (1944), R. M. Thomas and W. J. Sparks.
25. J. P. Kennedy and N. H. Canter, *J. Polym. Sci., Part A1* **5**, 2455 (1967).
26. S. Cesca, P. Giusti, P. L. Magagnini, and A. Priola, *Makromol. Chem.* **176**, 2319 (1975).
27. S. Cesca, A. Priola, M. Bruzzone, G. Ferraris, and P. Giusti, *Markromol. Chem.* **176**, 2339 (1975).
28. J. Bandrupt and E. H. Immergut, eds., *Polymer Handbook*, 3nd ed., John Wiley & Sons, Inc., New York, 1989.
29. Polymer Corp., private communication.
30. W. A. Thaler and J. D. Buckley Jr., *Rubber Chem. Technol.* **49**, 960 (1976).
31. C. Corno, A. Priola, G. Spallanzani, and S. Cesca, *Macromolecules* **17**, 37 (1984).
32. J. P. Kennedy and I. Krishenbaum in E. C. Leonard, ed., *Vinyl and Diene Monomers, Part 2*, New York, 1971, Chapt. 3, p. 691.
33. J. E. Puskas, M. Verhelst, P. Collart, and J. B. Schmidt, *Kautsch. Gummi Kunstst.* **48**, 866 (1995).
34. E. N. Kresge, R. H. Schatz, and H.-C. Wang, in J. I. Kroschwitz, ed., *Encyclopedia of Polymer Science and Engineering*, Vol. 8, 2nd ed., John Wiley & Sons, Inc., New York, 1987, pp. 423–448.
35. M. Roth and H. Mayr, *Macromolecules* **29**, 6104–6109 (1996).
36. B. Ivan and J. P. Kennedy, *Designed Polymers by Carbocationic Macromolecular Engineering: Theory and Practice*, Hanser Publishers, Munich, 1991.
37. R. Faust, A. Fehervari, and J. P. Kennedy, *Br. Polym. J.* **19**, 379 (1987).
38. R. Vukov, *Rubber Chem. Technol.* **57**, 275 (1984).
39. A. Van Tongerloo and R. Vukov, in *Proceedings of the International Rubber Conference*, Venice, 1979, p. 70.
40. F. P. Baldwin and co-workers, *Rubber Plast. Age* **42**, 500 (1960).
41. F. P. Baldwin and I. J. Gardner, *Chemistry and Properties of Cross-Linked Polymers*, Academic Press, Inc., New York, 1977, p. 273.
42. F. P. Baldwin and co-workers, *Adv. Chem. Ser.* **91**, 448 (1969).
43. U.S. Pat. 5,883,207 (Mar. 16, 1999), H. K. Youn, B. W. Hwang, C. S. Chu, Y. S. Chung, G. S. Han, Y. H. Yeom, C. J. Lee, M. K. Jun, and K. D. Park (to Daelim Industrial Co., Ltd.).
44. U.S. Pat. 6,156,859 (Dec. 5, 2000), G. Langstein, W. Obrecht, J. E. Puskas, O. Nuyken, and K. Weiss (to Bayer AG).
45. J. Walker, G. J. Wilson, and K. J. Kumbhani, *J. Inst. Rubber Ind.* **8**, 64 (1974).
46. U.S. Pat. 5,162,445 (Nov. 10, 1992), K. W. Powers and H.-C. Wang (to Exxon Chemical Co.).
47. H.-C. Wang and K. W. Powers, *Elastomerics* **124**, 14 and 22 (1992).
48. P. Arjunan, H.-C. Wang, and J. A. Olkusz, *Functional Polymers*, (ACS Symposium Series 704), American Chemical Society, Washington, D.C., 1998, pp. 199–216.
49. U.S. Pat. 5,548,029 (Aug. 20, 1996), K. W. Powers, H.-C. Wang, T.-C. Chung, A. J. Dias, and J. A. Olkusz (to Exxon Chemical Patents, Inc.).
50. U.S. Pat. 5,824,717 (Oct. 20, 1998), N. A. Merrill, H.-C. Wang, and A. J. Dias (to Exxon Chemical Patents, Inc.).

51. T. Yu, H.-C. Wang, K. W. Powers, A. F. Yee, and D. Li, *Polym. Prepr. (Am. Chem. Soc., Polym. Chem. Div.)* **33**, 622 (1992).
52. U.S. Pat. 4,946,899 (Aug. 7, 1990), J. P. Kennedy, J. E. Puskas, G. Kaszas, and W. G. Hager (to The University of Akron).
53. G. Kaszas, J. E. Pusaks, J. P. Kennedy, and W. G. Hager, *J. Polym. Sci., Polym. Chem. Ed. A* **29**, 427 (1991).
54. M. Gyor, Z. Fodor, and H.-C. Wang, *J. Macromol. Sci., A: Pure Appl. Chem.* **31B**, 2055 (1994).
55. X. Cao, L. Sipos, and R. Faust, *Polym. Bull.* **45**, 121 (2000).
56. D. Li and R. Faust, *Macromolecules* **28**, 4893 (1995).
57. J. E. Puskas, G. Kaszas, J. P. Kennedy, and W. G. Hager, *J. Polym. Sci., Polym. Chem. Ed. A* **30**, 41 (1992).
58. J. P. Kennedy, S. Midha, and Y. Tsunogae, *Macromolecules* **26**, 429 (1993).
59. Y. C. Bae and R. Faust, *Macromolecules* **31B**, 2480 (1998).
60. G. T. Baumann and M. R. Smith, *Hydrocarbon Process. Pet. Refiner.* **33**(5), 156 (1954).
61. A. M. Valet and co-workers, *Hydrocarbon Process. Pet. Refiner.* **41**(5), 119 (1962).
62. H. Kroper, K. Schlomer, and H. M. Weitz, *Hydrocarbon Process. Pet. Refiner.* **48**(9), 195 (1969).
63. U.S. Pat. 3,665,048 (Jan. 14, 1970), H. R. Grane and I. E. Katz (to Atlantic Richfield Co.).
64. V. Fattore and co-workers, *Hydrocarbon Process.* **60**(8), 101 (1981).
65. Belg. Pat. 882,387 (July 16, 1980), L. A. Smith Jr. (to Chemicals Research and Licensing Co.); Ger. Pat. 2,534,544 (Feb. 12, 1976), R. Tesei, V. Fattore, and F. Buonomo (to SNAM Piogetti SpA).
66. U.S. Pat. 5,417,930 (May 23, 1995), M. F. McDonald, D. J. Lawrence, and D. A. Williams (to Exxon Chemical Inc.).
67. T. Pernecker, and J. P. Kennedy, *Polym. Bull. (Berlin)* **33**(1), 13–19 (1994); *Polym. Bull. (Berlin)* **32**(5/6), 537–543 (1994).
68. Eur. Pat. 0 344 021 A2 (Jan. 22, 1992), K. W. Powers and H.-C. Wang (to Exxon Chemical).
69. U.S. Pat. 3,023,191 (Feb. 27, 1962), B. R. Tegge, F. P. Baldwin, and G. E. Serniuk (to Esso L Research & Engineering Co.).
70. U.S. Pat. 2,940,960 (June 14, 1960), B. R. Tegge and co-workers (to Esso Research & Engineering Co.).
71. U.S. Pat. 3,257,349 (June 21, 1966), J. A. Johnson Jr. and E. D. Luallin (to Esso Research & Engineering Co.).
72. U.S. Pat. 5,286,804 (Feb. 15, 1994), R. N. Webb, K. W. Powers, H.-C. Wang, M. F. McDonald, J. V. Fusco, and H. F. VanBrackle.
73. U.S. Pat. 5,681,901 (Oct. 28,1997), N. F. Newman.
74. U.S. Pat. 4,384,072 (May 17, 1983), N. F. Newman and R. C. Kowalski (to Exxon Research & Engineering Co.).
75. U.S. Pat. 4,573,116 (Apr. 23, 1985), R. C. Kowalski, W. M. Davis, and L. Erwin (to Exxon Research & Engineering Co.).
76. U.S. Pat. 4,548,995 (Oct. 22, 1985), R. C. Kowalski and co-workers (to Exxon Research & Engineering Co.).
77. L. A. Wood, *Rubber Chem. Technol.* **49**, 189 (1976).
78. G. Allegra, E. Benedetti, and C. Pedone, *Macromolecules* **3**, 727 (1970).
79. Bulletin 12-M, Amoco Chemical Corp., Chicago, Ill., 1990.
80. I. Puskas, E. M. Banas, and A. G. Nerhein, *J. Polym. Sci., Polym. Symp.* **56**, 191 (1976).
81. C. Y. Chu and R. Vukov, *Macromolecules* **18**, 1423 (1985).
82. I. Kuntz and K. D. Rose, *J. Polym. Sci., Part A: Polym. Chem.* **27**, 107 (1989).
83. D. M. Cheng, I. J. Gardener, H.-C. Wang, C. B. Frederick, and A. H. Dekmezian, *Rubber Chem. Technol.* **63**, 265 (1990).

84. J. E. Puskas and C. Wilds, *Rubber Chem. Technol.* **67**, 329 (1994).
85. J. L. White, T. D. Shaffer, C. J. Ruff, and J. P. Cross, *Macromolecules* **28**, 3290 (1995).
86. H.-C. Wang, unpublished data, 1990.
87. J. P. Kennedy and E. Marechal, *Carbocationic Polymerization*, John Wiley & Sons, Inc., New York, 1982.
88. U.S. Pat. 5,162,445 (Nov. 10, 1992), K. W. Powers, H.-C. Wang, T.-C. Chung, A. J. Dias, and J. A. Olkusz (to Exxon Chemical).
89. R. M. Thomas and W. J. Sparks, in G. S. Whitby, ed., *Synthetic Rubber*, John Wiley & Sons, Inc., New York, 1954, Chapt. 24.
90. C. C. Chu and R. Vukov, *Macromolecules* **18**, 1423 (1985).
91. R. Vukov, *Rubber Chem. Technol.* **57**, 284 (1984).
92. J. S. Parent, D. J. Thom, G. White, R. A. Whitney, and W. Hopkins, *J. Polym. Sci., Part A: Polym. Chem.* **39**, 2019 (2001).
93. U.S. Pat. 5,071,913 (Dec. 10, 1991), K. W. Powers, H.-C. Wang, D. C. Handy, and J. V. Fusco (to Exxon Chemical).
94. H.-C. Wang, K. W. Powers, and J. V. Fusco, in *ACS Rubber Division Meeting*, Mexico City, Mexico, May, 1989.
95. I. Duvdevani, L. Gursky, and I. J. Gardener, in *ACS Rubber Division Meeting*, Mexico City, Mexico, May, 1989.
96. L. L. Ban, I. Duvdevani, and H.-C. Wang, in *ACS Rubber Division Meeting*, Las Vegas, May, 1990.
97. A. H. Tsou, unpublished data, 2000.
98. H.-C. Wang and I. Duvdevani, unpublished data, 1990.
99. H.-C. Wang and K. W. Powers, in *ACS Rubber Division Meeting*, Toronto, Canada, May, 1991.
100. U.S. Pat. 5,162,445 (Nov. 10, 1992), K. W. Powers, H.-C. Wang, T.-C. Chung, A. J. Dias, and J. A. Olkusz (to Exxon Chemical).
101. R. H. Boyd and P. V. Krishna Pant, *Macromolecules* **24**, 6325 (1991).
102. P. V. Krishna Pant and R. H. Boyd, *Macromolecules* **25**, 494 (1992).
103. P. V. Krishna Pant and R. H. Boyd, *Macromolecules* **26**, 679 (1993).
104. G. J. van Amerongen, *Rubber Chem. Technol.* **37**, 1065 (1964).
105. G. E. Jones, unpublished data, 2001.
106. E. R. Fitzgerald, L. D. Grandine Jr., and J. D. Ferry, *J. Appl. Phys.* **24**, 650 (1953).
107. J. D. Ferry, L. D. Grandine Jr., and E. R. Fitzgerald, *J. Appl. Phys.* **24**, 911 (1953).
108. D. J. Plazek, I.-C. Chay, K. L. Ngai, and C. M. Roland, *Macromolecules* **28**, 6432 (1995).
109. D. J. Plazek and K. L. Ngai, *Macromolecules* **24**, 1222 (1991).
110. A. K. Rizos, K. L. Ngai, and D. J. Plazek, *Polymer* **38**, 6103 (1997).
111. K. L. Ngai and D. J. Plazek, *Rubber Chem. Technol.* **68**, 376 (1995).
112. J. D. Ferry, *Viscoelastic Properties of Polymers*, 3rd ed., John Wiley & Sons, Inc., New York, 1980.
113. J. F. S. Yu, J. L. Zakin, and G. K. Patterson, *J. Appl. Polym. Sci.* **23**, 2493 (1979).
114. R. A. Orwoll, *Rubber Chem. Technol.* **50**, 451 (1977).
115. J.-B. Donnet, R. C. Bansal, and M. J. Wang, eds., *Carbon Black*, 2nd ed., Marcel Dekker, Inc., New York, 1993.
116. A. R. Payne, *J. Appl. Polym. Sci.* **7**, 873 (1963).
117. M. Morton, ed., *Rubber Technology*, 3rd ed., Chapman and Hall, London, 1995.
118. E. T. McDonel, K. C. Baranwal, and J. C. Andreis, in D. R. Paul and S. Newman, eds., *Polymer Blends*, Vol. II, Academic Press, New York, 1978, Chap. 19.
119. W. M. Hess and V. E. Chirico, *Rubber Chem. Technol.* **50**, 301 (1977).
120. W. M. Hess, P. C. Vegvari, and R. A. Swor, *Rubber Chem. Technol.* **58**, 350 (1985).
121. J. E. Callan, W. M. Hess, and C. E. Scott, *Rubber Chem. Technol.* **44**, 814 (1971).
122. E. M. Dannenberg, *Rubber Chem. Technol.* **48**, 410 (1975).

123. A. M. Gessler, *Rubber Chem. Technol.* **41**, 1494 (1969).
124. K. Yurekli, R. Krishnamoorti, M. F. Tse, K. O. McElrath, A. H. Tsou, and H.-C. Wang, *J. Polym. Sci., Part B: Polym. Phys.* **39**, 256 (2001).
125. A. H. Tsou and W. H. Waddell, *Kautschuk Gummi Kunststoffe* **55**, 382 (2002).
126. A. H. Tsou, G. E. Zheng, and M. C. Boyce, *J. Polym. Sci., Part B: Polym. Phys.* (submitted).
127. E. P. Giannelis, *Adv. Mater.* **8**, 29 (1996).
128. U.S. Pat. 5,576,372 (Nov. 19, 1996), E. N. Kresge and D. J. Lohse (to Exxon Chemical).
129. U.S. Pat. 5,883,173 (Mar. 16, 1998), C. W. Elspass, D. G. Peiffer, E. N. Kresge, D.-I. Hsieh, J. J. Chludzinski, and K. S. Liang (to Exxon Research & Engineering).
130. A. A. Gusev and H. R. Lusti, *Adv. Mater.* **13**, 1641 (2001).
131. F. P. Baldwin, D. J. Buckley, I. Kuntz, and S. B. Robinson, *Rubber Plas. Age* **42**, 500 (1961).
132. I. Kuntz, R. L. Zapp, and R. J. Pancirov, *Rubber Chem. Technol.* **57**, 813 (1984).
133. R. L. Zapp and A. A. Oswald, *Rubber Chem. Technol.* **48**, 860 (1978).
134. J. S. Parent, G. D. F. White, R. A. Whitney, and W. Hopkins, *Macromolecules* **35**, 3374 (2002).
135. J. M. Frechet, R. Bielski, H.-C. Wang, J. V. Fusco, and K. W. Powers, *Rubber Chem. Technol.* **66**, 98 (1993).
136. R. Bielski, J. M. Frechet, J. V. Fusco, K. W. Powers, and H.-C. Wang, *J. Polym. Sci., Part A* **31**, 755 (1993).
137. E. T. McDonel, K. C. Baranwal, and J. C. Andries, in D. R. Paul and S. Newman, eds., *Polymer Blends*, Vol. 2, Academic Press, Inc., New York, 1978, Chapt. 19.
138. J. E. Rogers and W. H. Waddell, *Rubber World* **219**, 24 (1999).
139. K. O. McElrath and A. L. Tissler, in *ACS Rubber Division Meeting*, Anaheim, May, 1997.
140. A. H. Tsou, I. Duvdevani, and K. O. McElrath, *Rubber Chem. Technol.* (in press).
141. W. H. Waddell, J. H. Kuhr, and R. R. Poulter, in *ACS Rubber Division Meeting*, Cleveland, October, 2001.
142. W. H. Waddell, R. C. Napier, and R. R. Poulter, *Kautschuk Gummi Kunststoffe* **54**, 1 (2001).
143. J. M. Willis, B. D. Favis, and J. Lunt, *Polym. Eng. Sci.* **30**, 1073 (1990).
144. E. Kresge, *Rubber Chem. Technol.* **64**, 469 (1991).
145. R. C. Puydak and D. R. Hazelton, *Plast. Eng.* **44**, 27 (1988).
146. U.S. Pat. 5,051,477 (Sept. 24, 1991), B. Wenderoth, F. Schuetz, S. Brand, H. Sauter, E. Ammerman, and G. Lorezn (to Exxon Chemical).
147. U.S. Pat. 6,013,727 (Jan. 11, 2000), R. Dharmarajan, R. C. Puydak, H.-C. Wang, K. W. Powers, T. C.-C. Chi, and D. R. Hazelton (to Exxon Chemical).
148. EP Pat. 0857761A1 (Aug. 12, 1998), J. Watanabe (to Yokohama Rubber Co.).
149. EP Pat. 0722850B1 (May 6, 1999), H. Takeyama (to Yokohama Rubber Co.).
150. EP Pat. 0969039A1 (Jan. 5, 2000), J. Watanabe (to Yokohama Rubber Co.).
151. U.S. Pat. 6,359,071 (Mar. 19, 2002), J. Watanabe (to Yokohama Rubber Co.).
152. International Institute of Synthetic Rubber Producers, Inc., *Worldwide Rubber Statistics*, 2001.

ROBERT N. WEBB
ExxonMobil

TIMOTHY D. SHAFFER
ExxonMobil

ADDY H. TSOU
ExxonMobil

BUTYRALDEHYDES

1. Introduction

The two isomeric butanals, n- and isobutyraldehyde, C_4H_8O, are produced commercially almost exclusively by the Oxo Reaction of propylene. They also occur naturally in trace amounts in tea leaves, certain oils, coffee aroma, and tobacco smoke.

2. Physical Properties

The butanals are highly flammable, colorless liquids of pungent odor. Their physical properties are shown in Table 1.

These aldehydes are miscible with most organic solvents, eg, acetone, ether, ethanol, and toluene, but are only slightly soluble in water. Some azeotropes of n-butyraldehyde are given in Table 2.

Table 1. **Physical Properties of C-4 Aldehydes**

	n-Butyraldehyde	Isobutyraldehyde
formula	$CH_3CH_2CH_2CHO$	$(CH_3)_2CHCHO$
CAS Registry Number	[123-72-8]	[78-84-2]
systematic name	butanal	2-methylpropanal
critical temperature, °C	263.95	233.85
critical pressure, kPa[a]	4000	4100
critical specific volume, m^3/(kg-mol)[b]	0.258	0.263
melting point, °C	−96.4	−65.0
normal boiling point, °C	74.8	64.1
coefficient of expansion at 20°C	0.00114	
refractive index at 25°C	1.3766	1.3698
liquid density at 20°C, kg/m^{3c}	801.6	789.1
liquid heat capacity at 25°C, kJ/(mol·K)[d]	0.16333	0.15581
heat of vaporization at normal boiling point, kJ/mol[d]	30.72	31.23
ideal gas heat of formation at 25°C, kJ/mol[d]	−204.8	−215.8
heat of fusion, kJ/mol[d]	11.1	12.0
dipole moment, C·m[e]	9.07×10^{-30}	9.0×10^{-30}
dielectric constant ε at °C	13.426	
solubility parameter, at 25°C, (MJ/m^3)$^{0.5f}$	18.666	18.446
solubility in water at 25°C, wt %	8.36	6.47
solubility of water in at 25°C, wt %	3.45	2.60

[a] To convert kPa to mm Hg, multiply by 7.50.
[b] To convert m^3/kg-mol to mL/mol, multiply by 1000.
[c] To convert m^3/kg-mol to mL/mol, multiply by 1000.
[d] To convert kJ to kcal, divide by 4.184.
[e] To convert C·m to Debye, D, divide by 3.36×10^{-30}.
[f] To convert (MJ/m^3)$^{0.5}$ to (cal/cc)$^{0.5}$, multiply by $0.239^{0.5}$.

Table 2. **Azeotropes of *n*-Butyraldehyde**

Other component(s)	Wt %	Bp, °C
Homogeneous binary azeotropes		
ethanol	60.6	70.7
methanol	51	62.6
hexane (commercial)	74	60
Heterogeneous binary azeotropes		
water	12	68[a]
Heterogeneous ternary azeotropes		
ethanol	11	67.2[b]
water	9	

[a] The upper layer (94 vol %) contains 3.5 wt % water. The lower layer (6 vol %) contains 91.8% water.

[b] The upper layer (97.8 vol %) contains 11 wt % ethanol and 7 wt % water.

3. Chemical Properties

The reactions of *n*- and isobutyraldehyde are characteristic aldehyde reactions of oxidation, reduction, and condensation.

Aldehydes (qv) are intermediate in the sequence:

$$RCH_2OH \underset{oxidation}{\overset{reduction}{\rightleftharpoons}} RCHO \underset{oxidation}{\overset{reduction}{\rightleftharpoons}} RCOOH$$
$$\text{alcohol} \qquad\qquad \text{aldehyde} \qquad\qquad \text{acid}$$

Thus, *n*-butyl [71-36-3] [71-36-3] and isobutyl alcohol [78-83-1] [78-83-1] are obtained by hydrogenation of their respective aldehydes and butyric and isobutyric acid are produced by oxidation.

Hydrogenation of *n*- and isobutyraldehyde to the corresponding alcohols, $C_4H_{10}O$, can be carried out in high yield over various heterogeneous catalysts. Particularly effective in this application are $NiO-SiO_2-Al_2O_3$, Raney copper, and Raney nickel (1,2). Quantitative hydrogenation of butyraldehyde to butanol has also been effected with a homogeneous catalyst, $IrH_3[P(C_6H_5)_3]_2$, in acetic acid (3).

Oxidation of butyraldehyde to butyric acid [107-92-6]is most commonly carried out employing air or oxygen as the oxidant. Alternatively, organic oxidants, eg, cumene hydroperoxide, can also be employed effectively to give high yields of butyric acid, $C_4H_8O_2$ (4).

Catalytic oxidation of isobutyraldehyde with air at 30–50°C gives isobutyric acid [79-31-2] in 95% yield (5). Certain enzymes, such as horseradish peroxidase, catalyze the reaction of isobutyraldehyde with molecular oxygen to form triplet-state acetone and formic acid with simultaneous chemiluminescence (6).

Several species of bacteria under suitable conditions cause *n*-butyraldehyde to undergo the Cannizzaro reaction (simultaneous oxidation and reduction to butyric acid and butanol, respectively); this reaction can also be catalyzed by Raney nickel (7). The direct formation of butyl butyrate [109-21-7] or isobutyl isobutyrate [97-85-8](Tishchenko reaction) from the corresponding aldehyde takes place rapidly with aluminum ethylate or aluminum butyrate as catalyst (8). An essentially quantitative yield of butyl butyrate, $C_8H_{16}O_2$, from butyraldehyde has been reported using a ruthenium catalyst, $RuH_2[P(C_6H_5)_3]_4$ (9).

Hydrogen chloride or a few drops of hydrochloric acid catalyze the conversion of *n*-butyraldehyde into the trimer, parabutyraldehyde, $C_{12}H_{24}O_3$, (2,4,6-tripropyl-1,3,5-trioxane [56769-26-7], (1). The reaction is reversed by heating the parabutyraldehyde in the presence of acid. Anhydrous hydrogen chloride at $-40°C$ converts *n*-butyraldehyde into 1,1′-dichlorodibutyl ether, (2) in 70–75% yield (10).

$$(1) \qquad\qquad (2)$$

In the presence of dilute sodium or potassium hydroxide, *n*-butyraldehyde undergoes the aldol reaction to form 2-ethyl-3-hydroxyhexanal [496-03-7] which, on continued heating, is converted into 2-ethyl-2-hexenal [26266-68-2]. Hydrogenation of the latter gives 2-ethyl-1-hexanol [104-76-7], a principal plasticizer alcohol.

Many commercially important isobutyraldehyde derivatives are prepared through aldol and/or Tischenko condensation reactions. For example, isobutyraldehyde undergoes the aldol reaction to form isobutyraldol (2,2,4-trimethyl-3-hydroxypentanal [918-79-6]) which, when hydrogenated, gives 2,2,4-trimethyl-1,3-pentanediol (TMPD) [144-19-4].

Isobutyraldehyde also undergoes consecutive aldol and Tischenko condensations to give 2,2,4-trimethyl-1,3-pentanediol monoisobutyrate [25265-77-4] (Texanol, Filmer IBT), alternatively prepared by the esterification of TMPD with isobutyric acid.

Neopentyl glycol (2,2-dimethyl-1-propanol [126-30-7]), another important industrial derivative of isobutyraldehyde, is obtained from the aldol reaction product of isobutyraldehyde with formaldehyde followed by hydrogenation.

Crossed aldol reactions of butyraldehyde with other aldehydes also occur though these are generally not useful because they produce too many products. However, the commercially important solvent, methyl amyl ketone [110-43-0] is derived from the crossed aldol addition of acetone [67-64-1] with butyraldehyde. Similarly, formaldehyde [50-00-00] reacts cleanly with *n*-butyraldehyde to give 2,2′-(dihydroxymethyl)butanal [41966-25-0] (11), which can be hydrogenated to trimethylolpropane, 2-ethyl-2-(hydroxymethyl)-1,3-propanediol (TMP) [77-99-6]. Alternatively, formaldehyde and *n*-butyraldehyde react under hydrogen (6 MPa for 4 h at 70–120°C) over a $CuO-Al_2O_3$ catalyst in the presence of triethylamine, to give TMP in a single step in 85% yield (12). The reaction of formaldehyde with butyraldehyde, when carried out in the vapor phase (275–300°C) over a tungsten oxide on silica catalyst, gives 2-ethylacrolein [922-63-4] in 95% selectivity at 50% conversion (13).

n-Butyraldehyde undergoes stereoselective crossed aldol addition with diethyl ketone [96-22-0] in the presence of a stannous triflate catalyst (14) to give a predominantly erythro product (3). Other stereoselective crossed aldol reactions of n-butyraldehyde have been reported (15).

$$CH_3\overset{\diagdown}{}\underset{CH_2}{}\overset{O}{\underset{}{\overset{\|}{C}}}\overset{\diagup}{}\underset{CH}{}\overset{OH}{\underset{}{\overset{\blacktriangledown}{CH}}}\overset{\diagup}{}\underset{CH_2}{}\overset{CH_2}{}\overset{\diagup}{}CH_3$$

(3)

When the butanals are treated with alcohols in the presence of a mineral acid or calcium chloride, an acetal of the carbonyl group is produced.

$$CH_3CH_2CH_2CHO + 2ROH \xrightarrow{\text{H}^+} CH_3CH_2CH_2CH(OR)_2 + H_2O$$

Reaction of poly(vinyl alcohol) [9002-89-5] with n-butyraldehyde yields poly(vinyl butyral) [63148-65-2] (PVB), a commercially important resin.

Various alkyl-substituted pyridine derivatives are formed from the condensation of butyraldehyde with ammonia at high temperatures. For example, cocondensation of n-butyraldehyde with acrolein [107-02-8] and ammonia at 400°C over a borosilicate zeolite gives 3-ethylpyridine [536-78-7] in 70% yield (16). Similarly, condensation of n-butyraldehyde with ammonia over Co₃Al₂-(PO₄)₃ at 350°C gives a 74% yield of 3,5-diethyl-2-propylpyridine [4808-75-7] (17).

Isobutyraldehyde reacts with aqueous ammonia at 0–10°C to give hexahydro-2,4,6-triisopropyl-s-triazine, (4) (18), whereas under refluxing conditions the eneazomethine [5339-41-3], (5), is formed (19). Isobutyraldehyde condenses with two mole equivalents of urea in the presence of an acid catalyst to give isobutylidenediurea [2224-20-6] (IBDU), (6) a slow release fertilizer (20).

(4)   (5)   (6)

Butyraldehyde undergoes facile acyloin condensation via a novel thiazolium salt catalyzed procedure to give butyroin [496-77-5], (7), in 71–74% yield (21).

The butanals undergo some interesting homologation reactions employing certain quaternary triazolium or sulfonium salts as alkylating (homologating) agents. Isobutyraldehyde, for example, is converted into isovaleraldehyde [590-86-3], (8), in 72% yield employing 3-methylthio-1,4-diphenyl-1,2,4-triazolium chloride (22). Similarly, n-butyraldehyde undergoes consecutive homologation and isomerization to 1,2-epoxypentane [1003-14-1], (9), in 75% yield in the presence of dimethylhexylsulfonium methylsulfate (23).

(7)   (8)   (9)

4. Manufacture

The earliest commercial route to *n*-butyraldehyde was a multistep process starting with ethanol, which was consecutively dehydrogenated to acetaldehyde, condensed to crotonaldehyde, and reduced to butyraldehyde. In the late 1960s, production of *n*-butyraldehyde (and isobutyraldehyde) in Europe and the United States switched over largely to the Oxo reaction of propylene.

$$CH_3CH{=}CH_2 + CO + H_2 \xrightarrow{\text{Co or Rh}} CH_3CH_2CH_2CHO + (CH_3)_2CHCHO$$

The earliest modification of the Oxo process (qv) employed cobalt hydrocarbonyl, $HCo(CO)_4$, as catalyst. The reaction was carried out in the liquid phase at 130–160°C and 10–20 MPa (1450–2900 psi) to give a ratio of *n*- to isobutyraldehyde of between 2:1 to 4:1. *n*-Butyraldehyde, the straight-chain isomer and the precursor of 2-ethylhexanol, was the more valuable product so that a high isomer ratio of *n*- to isobutyraldehyde was obviously advantageous.

In the mid-1970s, a process employing a rhodium complex catalyst, $HRhCO[P(C_6H_5)_3]_3$, was commercialized by Union Carbide. This technology (24,25), subsequently licensed worldwide by Union Carbide and Davy McKee, operates at low temperatures 80–120°C and low pressure, 0.7–3 MPa (100–450 psi) and gives an isomer ratio of *n*- to isobutyraldehyde of 8:1 to 12:1. The advantages of the rhodium process for making butanals besides the lower temperatures and operating pressures, include a higher efficiency to the more valuable normal isomer and less by-product formation. The product butanals are separated continuously by vaporization from the nonvolatile catalyst, a distinct procedural advantage over the unmodified high pressure cobalt Oxo reaction. The latter requires a continuous separation and regeneration of the volatile cobalt hydrocarbonyl catalyst, which codistills with the butanal product. The rhodium catalyst, which is almost one thousand times more reactive than the cobalt hydrocarbonyl catalyst, requires relatively minute amounts of rhodium to achieve commercial rates of reaction.

In the mid-1980s, Ruhrchemie (now Hoechst) converted its oxo capacity to a proprietary water soluble rhodium catalyzed process (26,27), a technology developed jointly with Rhône-Poulenc. Product separation in this process is by decantation. Isomer ratios of *n*- to isobutyraldehyde of about 20:1 are obtained.

Mitsubishi Chemical uses a proprietary medium pressure rhodium-catalyzed process (28) in some plants which operate at 90–120°C and 5–10 MPa (725–1450 psi), and gives isomer ratios of about 4:1.

5. Economic Aspects

In 2001, 8.4×10^6 t of oxo chemicals were produced. The major portion (90%) of the butyraldehyde produced went into the production of 2-ethylhxanol. n-Butryaldehyde is also used to produce n-butanol.

The biggest markets for C_4–C_5 alcohols are solvents and coatings.

Oxo chemicals are expected to grow at the rate of 1.4% through the years 2002–2007. Production is expected to shift to Southeast Asia (29).

2-Ethylhexanol is used in the production of acrylate and methacrylic esters. Their principal markets are as acrylic emulsion polymers for pressure sensitive adhesives, textiles, and surface coatings. Demand for water-based acrylic products has increased because of the stringent air-emission regulations on solvent-based products (30).

The largest use of n-butanol is in the production of n-butyl acrylate and n-butyl methacrylate. These compounds are use in emulsified and solution polymers that are used in latex surface coatings and enamels (31).

6. Specifications

Some standard industrial quality specifications for n- and isobutyraldehyde are given in Table 3. Many times, however, specification limits are tailored to individual customer requirements.

7. Analytical Methods

The butanals form the conventional aldehyde hydrazone, semicarbazone, and dimedone-type derivatives. In the absence of other aldehydes and ketones, n-butyraldehyde can be determined by addition of sodium bisulfite and the excess bisulfite determined with iodine or thiosulfate (36).

Analysis for the butanals is most conveniently carried out by gas chromatography. Trace quantities of n-butyraldehyde (18 ppb) in exhaust gases have been determined employing a combination of capillary gas chromatography with thermionic detection (37). Similarly, trace amounts of n-butyraldehyde in cigarette smoke and coffee aroma have been determined by various capillary gc techniques (38,39).

The infrared carbonyl stretching frequencies of n- and isobutyraldehyde in the condensed phase occur at 1727.6 and 1738.0 cm^{-1}, respectively (40). The proton nmr spectra of both aldehydes are well-known (41).

8. Storage and Handling

Stainless steel, baked phenolic lined steel, or aluminum are often used for storage and handling of n- and isobutyraldehyde. The butanals are flammable and reactive, are easily oxidized on exposure to air, and in contact with acid, bases, or certain metal ions (eg, iron), will undergo exothermic condensation

Table 3. **Quality Specifications for the Butanals**

	n-Butyraldehyde[a]	Isobutyraldehyde[b]	Method[c]
aldehyde, wt %, min	98.5	99.5	capillary gc
distillation[d]	ibp 72.0°C min	ibp 62.5°C min	standard ASTM
	95 mL 80°C max	97 mL 67.0°C max	distillation for lacquer solvents
acidity	0.5 wt %, max[e]	0.2 wt %, max[e]	titration to phenolphthalein end point
water	0.3 wt %, max	0.07 wt %, max	Karl-Fischer titration
color	15 Pt–Co max	10 Pt–Co max	colorimetry
specific gravity at 20/20°C	0.801 to 0.806	0.788 to 0.793	densitometry
suspended matter	substantially free	substantially free	visual
iron	20 ppm max	20 ppm max	atomic absorption
copper		10 ppm max	atomic absorption
nickel		10 ppm max	atomic absorption
alcohols		0.3 wt %, max	capillary gc

[a] Ref. 32.
[b] Ref. 33.
[c] Refs. (34,35).
[d] At 101.3 kPa = 1 atm.
[e] Calculated as butyric acid.

465

Table 4. **Animal Toxicity and Irritancy Data for Butanals**

	n-Butyraldehyde[a]	Isobutyraldehyde[b]
LD_{50}, oral, rats	5.9 mL/kg	1.6–3.7 mL/kg
LD_{50}, dermal, rabbits	1.26 mL/kg	7.13 mL/kg
inhalation, rats	4 h LC_{50} = 16,400 (10,600 to 25,300) ppm	8,000 ppm killed 1 in 6 in 4 h
primary irritation, rabbits	grade 2 (no irritation on 3 rabbits, moderate capillary injection on 2)	grade 1 (no irritation)
eye injury, rabbits	grade 5 (severe corneal injury with iritis from 0.02 mL, moderate corneal injury from 0.005 mL)	grade 5 (severe corneal injury from 0.02 mL, minor from 0.005 mL)

[a] Ref. 44.
[b] Ref. 45.

reactions. Storage of the aldehydes under nitrogen will avoid these problems and preserve the integrity of the material (42). There is some evidence that water stabilizes aldehydes against certain types of exothermic condensation reactions, possibly by precipitating any soluble iron species as hydrous iron oxides.

9. Health, Safety, and Environmental Factors

n-Butyraldehyde is moderately toxic by ingestion, inhalation skin contact, intraperitoneal, and subcutaneous routes. It is a severe skin and eye irritant. It is a highly flammable liquid. When heated to decomposition, it emits acid smoke and fumes (43). No threshold limit value has been assigned for either butyraldehyde or isobutyraldehyde. Isobutyraldehyde is moderately toxic by ingestion, skin contact and inhalation. It is a severe skin and eye irritant. It is a highly flammable liquid and can react vigorously with reducing materials. When heated to decomposition, it emits smoke and fumes (43). Because of the ease of oxidation of the butanals to the corresponding butyric acids, precautions associated with these carboxylic acids must also be noted. Reported animal toxicity and irritancy values for the butanals are given in Table 4.

The biological oxygen demand (BOD) in aqueous streams for both butanals is 1.62 wt/wt for five days (46). The NFPA Hazard classification (46) for both aldehydes are health (blue) 2; flammability (red) 3; and reactivity (yellow) 0.

The flammability characteristics of the butanals are given in Table 5. The flash points for the butanals are well below room temperature. Thus, precautions must be taken to avoid heat, sparks, or open flame.

Table 5. **Flammability Characteristics of Butanals**

	n-Butyraldehyde	Isobutyraldehyde
flash point (closed cup), °C	−11	−40
autoignition temperature, °C	230	254
explosive limits in air, %	2.5–12.5	1.6–10.6

Both butanals are on the United States Toxic Substances Control Act (TSCA) Inventory, a prerequisite for the manufacture or importation for commercial sale of any chemical substance or mixture in quantities greater than one thousand pounds (455 kg). Additionally, the manufacture and distribution of the butanals in the United States are regulated under the Superfund Amendments and Reauthorization Act (SARA), Section 313, which requires that anyone handling at least ten thousand pounds (4550 kg) a year of a chemical substance report to both the EPA and the state any release of that substance to the environment.

10. Uses

The majority (90% in 2001) of the butyraldehyde produced in the United States is converted into 1-butanol and 2-ethylhexanol (2-EH). 2-EH is most widely used as the di(2-ethylhexyl) phthalate [117-81-7] ester for the plasticization of flexible PVC. Other uses for 2-EH include production of intermediates for acrylic surface coatings, diesel fuel, and lube oil additives (29).

The remaining n-butyraldehyde production of the United States goes into (in decreasing order): poly(vinyl butyral), 2-ethylhexanal, trimethylolpropane, methyl amyl ketone, and butyric acid.

Poly(vinyl butyral) is employed most widely as the adhesive interlayer in laminated automobile safety glass; it is also employed in architectural applications such as skylights, atriums, and glazing of office buildings. There are many grades of PVB. In general, 18–23% of the alcohol groups in the poly(vinyl alcohol) backbone remain unreacted, and there may be 1–3% residual ester groups from precursor poly(vinyl acetate).

2-Ethylhexanal, the reduced aldol condensation product of n-butyraldehyde, is converted into 2-ethylhexanoic acid [149-57-5], which is converted primarily into salts or metal soaps. These are used as paint driers and heat stabilizers for poly(vinyl chloride).

Trimethylolpropane (TMP), the reduced crossed aldol condensation product of n-butyraldehyde and formaldehyde, competes in many of the same markets as glycerol (qv) and pentaerythritol. The largest market for TMP is as a precursor in unsaturated polyester resins, short-oil alkyds, and urethanes for surface coatings (see ALKYD RESINS).

Methyl amyl ketone, derived from the crossed aldol condensation of n-butyraldehyde and acetone, is used predominantly as a high solids coatings solvent. It is also employed as a replacement for the very toxic 2-ethoxyethyl acetate [111-15-9].

Butyric acid, the simple oxidation product of n-butyraldehyde, is used chiefly in the production of cellulose acetate butyrate [9004-36-8]. Sheets of cellulose acetate butyrate are used for thermoformed sign faces, blister packaging, goggles, and face shields.

Most isobutyraldehyde goes into the production of isobutyl alcohol and isobutyraldehyde condensation and esterification products. The other principal isobutyraldehyde derivative markets are neopentyl glycol; isobutyl acetate; isobutyric acid; isobutylidene diurea; and methyl isoamyl ketone.

2,2,4-Trimethyl-1,3-pentanediol (TMPD), the hydrogenated aldol condensation product of isobutyraldehyde, is a modifying agent in alkyd resins (qv), high solids coatings, and moisture-set inks.

The monoisobutyrate ester of TMPD, Texanol, or Filmer IBT, formally an isobutyraldehyde trimer, is prepared in a single step from isobutyraldehyde or, alternatively, by the esterification of TMPD with isobutyric acid. This monoester is most commonly employed as a coalescing agent for latex-based paints and water-based ink formulations.

The diisobutyrate ester [6846-50-0] of TMPD, Kodaflex TXIB, is used as a viscosity control agent in various plastisol, rotomolding, and rotocasting operations. Kodaflex TXIB is also used in the production of rolled sheet vinyl flooring where a high percentage of fugitive plasticizer flashes off during the fusion of the PVC resin to impart a harder flooring surface.

The principal markets for neopentyl glycol (NPG), the hydrogenated, crossed aldol condensation product of isobutyraldehyde and formaldehyde, are in water-borne and alkyd-surface coatings.

Isobutyl isobutyrate, the Tischenko condensation product of two molecules of isobutyraldehyde, is a slow evaporating ester solvent that has been promoted as a replacement for ethoxyethyl acetate. Although produced primarily by the acetylation of isobutyl alcohol, some isobutyl acetate is produced commercially by the crossed Tischenko condensation of isobutyraldehyde and acetaldehyde. Isobutyl acetate [110-19-0] is employed mainly as a solvent, particularly for nitrocellulose coatings.

Isobutyric acid, the simple oxidation product of isobutyraldehyde, is employed in the esterification of TMPD to form the mono- and diesters of TMPD. Some isobutyric acid is also used in the production of isobutyronitrile, an organo-phosphate pesticide precursor.

Isobutylidene diurea (IBDU), a slow release fertilizer, is formed from isobutyraldehyde and two moles of urea.

Methyl isoamyl ketone (MIAK), a product derived from the aldol condensation of isobutyraldehyde and acetone, is used principally as a solvent for lacquers, cellulosics, and epoxies.

BIBLIOGRAPHY

"Butyraldehydes" in *ECT* 1st ed., Vol. 2, pp. 684–693, by M. S. W. Small, Shawinigan Chemicals, Ltd., and P. R. Rector, Carbide and Carbon Chemicals Corp.; in *ECT* 2nd ed., Vol. 3, pp. 865–877, by A. P. Lurie, Eastman Kodak Co.; in *ECT* 3rd ed., Vol. 4, pp. 376–385, by P. D. Sherman, Jr., Union Carbide Corp.; in *ECT* 4th ed., Vol. 4, pp. 736–747, by Ernst Billig, Union Carbide Chemicals and Plastics Co., Inc.; "Butyraldehydes" in *ECT* (online), posting date: December 4, 2000, by Ernst Billig, Union Carbide Chemicals and Plastics Company, Inc.

CITED PUBLICATIONS

1. Ger. Pat. 1,115,232 (July 18, 1958), W. Rottig (to Ruhrchemie AG).
2. J. Jadot and R. Braine, *Bull. Soc. Roy. Sci. Liege* **25**, 62 (1956).

3. R. S. Coffey, *J. Chem. Soc. Chem. Comm.*, 923 (1967).
4. E. G. Hawkins, *J. Chem. Soc.*, 2169 (1950).
5. H. G. Hagemeyer and G. C. DeCroes, *The Chemistry of Isobutyraldehyde and its Derivatives*, Eastman Kodak Co., 1953.
6. C. Bohne, I. D. MacDonald, and H. B. Dunford, *J. Biol. Chem.* **262**(8), 3527 (1987).
7. C. Neuber and F. Windisch, *Biochem. Z.* **166**, 454 (1925).
8. H. S. Kulpinski and F. F. Nord, *J. Org. Chem.* **8**, 256 (1943).
9. T. Itoh and co-workers, *Bull. Chem. Soc. Jap.* **55**(2), 504 (1982).
10. L. Brandsma and J. F. Arens, *Rec. Trav. Chem. Pays Bas.* **81**, 33 (1962).
11. O. Neunhoeffer and H. Neunhoeffer, *Ber.* **95**, 102 (1962).
12. Eur. Pat. 142,090 (May 22, 1985), F. Merger and co-workers (to BASF AG).
13. G. Albanesi and P. Moggi, *Appl. Catal.* **37** (1–2), 315–322 (1988).
14. T. Mukaiyama, R. W. Stevens, and N. Iwasawa, *Chem. Lett.* 353 (1982).
15. J. R. Green and co-workers, *Tetrahedron Lett.* **27** (5), 535 (1986).
16. Eur. Pat. 263,464 (Apr. 13, 1988), W. Hoelderich, N. Goetz, and G. Fouquet (to BASF AG).
17. C. R. Adams and J. Falbe, *Brennstoff Chem.* **47**(6), 184 (1966).
18. A. T. Nielsen and co-workers, *J. Org. Chem.* **38**, 3288 (1973).
19. U.S. Pat. 4,477,674 (Oct. 16, 1984), B. A. O. Alink(to Petrolite Corp.).
20. H. Hamamoto, *New Fert. Mat.* 28–37 (1968).
21. W. Tagaki and H. Hara, *J. Chem. Soc. Chem. Comm.* 891 (1973).
22. G. Doleschall, *Tetrahedron Lett.* **21**(43), 4183 (1980).
23. Eur. Pat. 205,400 (Dec. 17, 1986), K. J. Coers and P. Radimarski (to Ciba Geigy AG).
24. U.S. Pat. 3,527,809 (Sept. 8, 1970), R. L. Pruett and J. A. Smith (to Union Carbide Corp.).
25. U.S. Pat. 4,593,127 (June 3, 1986), D. L. Bunning and M. A. Blessing (to Union Carbide Corp.).
26. Ger. Pat. 3,234,701 (Aug. 18, 1982), B. Cornils and co-workers (to Ruhrchemie AG).
27. Ger. Pat. 3,341,035 (Nov. 12, 1983), H. Kalbfell and co-workers (to Ruhrchemie AG).
28. Jpn. Pat. 7,480,888 (July 15, 1974), (to Mitsubishi Chem. Ind. KK).
29. S. Bizzani, R. Gubler, and A. Kishi, *Chemical Economics Hand book*, SRI International, Menlo Park, Calif., Nov. 2002.
30. "2-Ethylhexanol, " Chemical Profile, *Chemical market Reporter*, 27 (Nov. 4, 2002).
31. "n-Butanol," Chemical Profile, *Chemical market Reporter*, (Oct. 28, 2002).
32. *Union Carbide Specification (Butyraldehyde)* 1-4A4-1m, Aug. 2, 1989.
33. *Union Carbide Specification (Isobutyraldehyde)* 1-4B1-1n, Oct. 5, 1989.
34. *Union Carbide Specification Method (Butyraldehyde)* 1B-4A4-1i, June 1, 1990.
35. *Union Carbide Specification Method (Isobutyraldehyde)* 1B-4B1-1h, May 6, 1988.
36. A. E. Parkinson and E. C. Wagner, *Ind. Eng. Chem. Anal. Ed.* **18**, 433 (1934).
37. H. Nishikawa and co-workers, *Bunseki Kagaku* **36**(6), 381 (1987).
38. Y. Wang and co-workers, *Huaxue Tongbao* (11), 43–45 (1986).
39. R. Liardon and J. C. Spadone, *Colloq. Sci. Int. Cafe* [C.R.], **11**, 181–196, 1985.
40. C. J. Pouchert, *The Aldrich Library of FT-IR Spectra*, 1st ed., Vol. 1, Milwaukee, Wis., 1985.
41. *Standard Proton NMR Collection*, Sadtler Research Laboratories, Division of Bio-Rad Laboratories, Inc., Philadelphia, Pa., 1980.
42. *Aldehydes*, Union Carbide Corp., New York, 1974.
43. R. J. Lewis, Sr., *Dangerous Properties of Industrial Materials*, 10th ed., John Wiley & Sons, Inc., New York, 2000.

44. *Union Carbide Toxicity Report 40–73, Butyradlehyde,* 1977.
45. *Union Carbide Toxicity Report 15–55, Isobutyraldehyde,* 1952.
46. *Hazardous Chemicals Data Book*, G. Weiss, ed., Noyes Data Corp., 1980, pp. 119, 200.

ERNST BILLIG
Union Carbide Chemicals
and Plastics Company Inc.

C

CADMIUM AND CADMIUM ALLOYS

1. Introduction

Cadmium (Cd) [7440-43-9] a name derived from the Latin *cadmia* and Greek *kadmeia*, which are ancient names for calamine or zinc oxide, is a member, along with zinc and mercury, of group 12 (IIB) of the periodic table of elements. It is generally characterized as a soft, ductile, silver-white or bluish-white metal, and was discovered almost simultaneously in 1817 by Strohmeyer and Hermann, both in Germany, as an impurity in zinc carbonate (1). Germany produced the first commercial cadmium metal in the late nineteenth century and was the only important producer of cadmium metal and cadmium sulfide [1306-23-6] pigments until World War I, recovering cadmium from the smelting of zinc ores from Upper Silesia. Production in the United States began in 1907 at the Grasselli Chemical Company in Cleveland, Ohio, where metallic cadmium was recovered as a byproduct of zinc smelting. Initial production in 1907 in the United States was 6.3 mt, which expanded to its peak production level of 5736 tonnes in 1969 but has declined to approximately 1200 tonnes today (2).

Cadmium metal may be recovered as a byproduct of zinc, lead, or copper smelting operations and is utilized commercially as a corrosion-resistant coating on steel, aluminum, and other nonferrous metals (see Zinc and zinc alloys; Copper; Lead). Cadmium is also added to some nonferrous alloys to improve properties such as strength, hardness, wear resistance, castability, and electrochemical behavior or to other nonferrous alloys to obtain lower melting temperatures and improved brazing and soldering characteristics. Cadmium compounds are employed mainly as the negative electrode materials in NiCd batteries, as ultraviolet light and weathering stabilizers for polyvinyl chloride (PVC), and as red, orange, and yellow pigments in plastics, ceramics, glasses, enamels, and artists' colors.

2. Occurrence

Cadmium is a naturally occurring metallic element, one of the minerals present in minor amounts in the earth's crust and its oceans, and present everywhere in our environment. The crustal abundance of cadmium has variously been reported between 0.1 and 0.5 ppm, but much higher and much lower values have also been cited depending on a large number of factors. Igneous and metamorphic rocks tend to show lower values, 0.02–0.2 ppm, whereas sedimentary rocks have much higher values, from 0.1–25 ppm. Naturally, zinc, lead, and copper ores, which are mainly their sulfides and oxides, contain even higher levels of cadmium, 200–14,000 ppm for zinc ores and ~500 ppm for typical lead and copper ores. The raw material for iron and steel production contain approximately 0.1–5.0 ppm, while those for cement production contain about 2 ppm. Fossil fuels contain 0.5–1.5 ppm cadmium, but phosphate fertilizers may contain levels from 10 to as high as 200 ppm cadmium (3).

A number of cadmium-containing minerals have been identified, and are summarized in Table 1.

The most common cadmium-containing mineral is greenockite, which is not found in any isolated deposits, but is nearly always associated with the zinc ore, sphalerite. Approximately 80% of current cadmium production arises from primary zinc production; the remaining 20% is a result of lead and copper byproduct production and cadmium product recycling.

The average cadmium content in the world's oceans has variously been reported as low as <5 ng/L (4) and 5–20 ng/L (5,6) to as high as 110 ng/L (7) 100 ng/L (3) and 10–100 ng/L (8). Higher levels have been noted around certain coastal areas (8) and variations of cadmium concentration with the ocean depth, presumably due to patterns of nutrient concentrations, have also been measured (4,5). Even greater variations are quoted for the cadmium contents of rainwater, freshwater, and surface water in urban and industrialized areas. Levels from of 10–4000 ng/L have been quoted in the literature depending on specific location and whether total cadmium or dissolved cadmium is measured (3,4,7).

Ambient-air cadmium concentrations have generally been estimated to range within 0.1–5 ng/m^3 in rural areas, 2–15 ng/m^3 in urban areas, and 15–150 ng/m^3 in industrialized areas (4,5,8) although some much lower values have been noted in extremely remote areas and some much higher values have been recorded in the past near uncontrolled industrial sources. There are generally little or no differences noted in cadmium levels between indoor and outdoor

Table 1. **Cadmium-Containing Minerals**

Name	Formula	CAS registry number	Cadmium, %
sphalerite	ZnS	[1314-98-3]	0.02–1.4
wurtzite	ZnS	[1314-98-3]	0.02–1.4
galena	PbS	[1314-87-0]	500 ppm
chalopyrite	CuFeS$_2$	[1308-56-1]	500 ppm
otavite	Zn(Cd)CO$_3$	[513-78-0]	<1.2
greenockite	CdS	[1306-23-6]	77.8
monteponite	CdO	[1306-19-0]	87.5
cadmoselite	CdSe	[1306-24-7]	58.7

Table 2. **World Production of Refined Cadmium Metal**

Year	Production, t
1991	21,268
1992	20,197
1993	19,497
1994	18,411
1995	19,478
1996	19,108
1997	19,946
1998	19,851
1999	18,767
2000 (estimated)	18,065

air in nonsmoking environments. The presence of tobacco smoke, however, may substantially affect indoor ambient-air cadmium concentrations.

3. Sources and Supplies

World production of cadmium from 1990 through 1999 varied between 18,000 and 21,000 t/yr according to the World Bureau of Metal Statistics, and is summarized in Table 2. Historically, the Western world cadmium production trend growth has been 0.6% per year from 1970 through 1982 but increased to 0.8% per year during 1982–1995. This production trend could increase further as zinc production capacity increases, recycling of cadmium products increases, and production from countries such as China, Kazakhstan, Russia, and Canada increases. Overall, refined primary cadmium metal production has shown decreases in recent years as secondary or recycled cadmium production has increased.

The world production of cadmium continues to originate from the three principal producing areas—Europe, the Americas, and Asia—with lesser but still significant production from Australia and only very small production from Africa. The relative cadmium productions of these five areas in 1997 are summarized in Table 3.

Even though the cadmium production figures from countries in Asia such as China, Russia, Kazakhstan, and North Korea are often difficult to obtain and veriify, there is still no doubt that significant amounts of cadmium are being produced in Asia. Japan remains one of the largest cadmium producers in the world.

Table 3. **Geographic Summary of 1997 World Cadmium Production**

Area	Production, t
Asia	6,986
Europe	5,928
Americas	5,815
Australia	632
Africa	52

Table 4. **Leading Producers of Refined Cadmium Metal, 1994–1997**

Country	Production, t			
	1994	1995	1996	1997
Canada	2,167.8	2,342.6	2,832.3	3,082.6
Japan	2,614.1	2,628.5	2,356.7	2,373.3
China	1,282.0	1,471.0	1,300.0	1,300.0
Belgium	1,556.1	1,710.4	1,579.2	1,420.0
United States	1,125.7	1,361.8	1,238.4	1,179.1
Germany	1,145.0	1,144.8	1,144.8	1,144.8
South Korea	909.0	908.4	930.0	930.0
Kazakhstan	601.0	600.2	593.0	819.0
Netherlands	306.7	704.0	603.2	752.9
Russia	582.3	670.0	716.0	716.0
Mexico	646.0	689.0	784.0	784.0
Australia	909.5	838.1	638.9	631.5
Peru	507.0	560.0	405.0	562.0
Finland	548.0	540.0	540.0	540.0
United Kingdom	469.5	548.6	537.2	454.6
Italy	475.0	308.0	296.0	287.0

Countries which are leading producers of cadmium are listed in Table 4 along with their refined cadmium metal production according to the World Bureau of Metal Statistics.

The total amounts of secondary or recycled cadmium produced cannot accurately be estimated because in some recycling operations such as the processing of baghouse dusts from lead and copper smelters, the cadmium recovered subsequently enters primary cadmium production circuits at zinc refining operations and is assessed in the production statistics for primary cadmium metal. However, estimated figures are available from the recycling of spent NiCd batteries and manufacturing wastes based on plant capacities and tonnages of batteries and wastes processed, and these figures give some indication of the levels of cadmium recovered. Table 5 summarizes the estimated amounts of cadmium recovered in 1996 and 1997 at NiCd battery recycling plants around the world (10).

Although the estimates listed in Table 5 are very approximate and include NiCd battery manufacturing wastes which may be lower in cadmium content than the nominal level (12–15%) in spent NiCd batteries, it indicates that

Table 5. **Estimates of Cadmium Production from NiCd Recycling**

Company	Country	Cadmium recycled, mt	
		1996	1997
INMETCO	USA	338	395
SNAM	France	650	975
SAFT	Sweden	225	225
BAJ Program[a]	Japan	600	613
Total		*1813*	*2208*

[a]Battery Association of Japan Collection and Recycling Program.

Table 6. **Summary of U.S. DLA Cadmium Disposals**

Fiscal year	Authorized limit	Cadmium sales, t
1992/1993	227.3	148.9
1993/1994	340.9	339.8
1994/1995	340.9	360.9
1995/1996	443.2	0
1996/1997	544.3	141.7
1997/1998	544.3	118.8
1998/1999	544.3	544.3
1999/2000	544.3	329.2

recycling of NiCd batteries is adding more than 2200 t or more than 11% of current total production. If recycling of other manufacturing wastes such as filter cakes, electroplating sludges, electric-are furnace (EAF) dusts, and spent anodes is added to the NiCd battery recycling numbers, then cadmium recycling is probably 15–20% of the total supply rather than the 10–15% level currently suggested. Cadmium recycling rates are expected to continue to grow in the future.

Another factor in the cadmium supply situation in recent years has been the decision by the U.S. Defense Logistics Agency (DLA) in 1991 to dispose of its entire cadmium stockpile of 2877 t as excess material. As of September 2000, 1983.6 t of DLA stockpile cadmium has been sold into the market, mostly to traders and much of it at prices substantially below the existing market prices published in *Metal Bulletin*. In addition, the DLA has raised the authorized annual sales limit for any one fiscal year, which normally runs from October 1 to September 30, from the original level of 500,000 lb (227.3 t) to 1,200,000 lb (544.3 t). The status of DLA disposals and authorized disposal limits for fiscal years 1993–1999 are summarized in Table 6 (10).

Although the total amount of cadmium sold from the DLA stockpile is relatively small compared to total production and consumption, the continual availability of more than 500 t/yr of surplus cadmium does not help the cadmium market recover, especially in times of weak prices and general oversupply.

3.1. Stocks. World stocks of cadmium built significantly during 1996 following a trend begun in 1994 after decreases during 1992–1994. In 1997, cadmium stocks decreased by ~14%, but still remained at >3800 t at the end of the year. The world stock situation by geographic area for 1993–1997 is summarized in Table 7. These figures do not include the DLA stockpile amounts and reflect mainly producer inventories rather than consumer inventories that have returned to more normal levels.

Table 7. **World Cadmium Stocks by Geographic Area**

Area	Stocks, t				
	1993	1994	1995	1996	1997
Europe	1,424.5	1,142.4	1,401.9	1,746.5	1,370.2
Americas	920.0	733.4	1,029.2	1,855.7	1,686.5
Other	556.3	417.8	746.9	856.8	784.9
Total	*2900.8*	*2293.6*	*3178.0*	*4459.0*	*3841.6*

Table 8. **Natural Isotopes of Cadmium**

Mass	Relative abundance, atomic %
106	1.25
108	0.89
110	12.49
111	12.80
112	24.13
113	12.22
114	28.73
116	7.49

4. Properties

The electronic structure of elemental cadmium is $1s^2 2s^2 2p^6 3s^2 3p^6 3d^{10} 4s^2 4p^6 4d^{10} 5s^2$, and its normal oxidation state in almost all of its compounds is +2, although the +1 oxidation state has been reported in rare instances. There are eight natural isotopes of cadmium and their mass and relative abundance are summarized in Table 8.

The standard electrochemical reducion potential for the reaction

$$Cd^{2+} + 2e^- \longrightarrow Cd$$

is −0.403 V at 25°C (7). Its electrochemical equivalent is 582.4 Mg/C. Because of its position in the electromotive series of elements, cadmium is displaced from solution by more electropositive metals such as zinc and aluminum but is more galvanically active than steel and thus protects it in a sacrificial manner.

The physical, thermal, electrical, magnetic, optical, and nuclear properties of cadmium metal are summarized in Table 9. The chemical properties of cadmium metal in general resemble those of zinc, especially under reducing conditions and in covalent compounds. In oxides, fluorides and carbonates, and under oxidizing conditions, cadmium may behave similarly to calcium. It also forms a relatively large number of complex ions with other ligand species such as ammonia, cyanide, and chloride. Cadmium is a fairly reactive metal. It dissolves slowly in dilute hydrochloric or sulfuric acids, but dissolves rapidly in hot dilute nitric acid. All of the halogens, phosphorus, sulfur, selenium, and tellurum also react readily with cadmium at elevated temperatures.

Unlike zinc, cadmium is not markedly amphoteric, and cadmium hydroxide is virtually insoluble in alkaline media. Like zinc, however, cadmium forms a protective oxide film that reduces its corrosion/oxidation rate in atmospheric service. Both metals exhibit low corrosion rates over the range of pH ∼5−10. (1,11) In more acidic and alkaline environments, their corrosion rates increase dramatically. Cadmium is generally preferred for marine or alkaline service, whereas zinc is often as good or better in heavy industrial exposures containing sulfur or ammonia.

5. Manufacturing and Processing

5.1. Production of Cadmium Metal.
Cadmium metal is produced mainly as a byproduct of the beneficiation and refining of zinc sulfide ore concentrates,

Table 9. **Physical Properties of Cadmium**

Property	Value
atomic weight	112.40
melting point, °C	321.1
boiling point, °C	767
latent heat of fusion, kJ/mol[a]	6.2
latent heat of vaporization, kJ/mol[a]	99.7
specific heat, J/(mol · K)[a]	
20°C	25.9
321–700°C	29.7
coefficient of linear expansion at 20°C, μm/(cm.°C)	0.313
electrical resistivity, μω · cm	
22°C	7.27
400°C	34.1
600°C	34.8
700°C	35.8
electrical conductivity, % IACS[b]	25
density, kg/m^3	
26°C	8642
330°C (liq>)	8020
400°C	7930
600°C	7720
volume change on fusion, % increase	4.74
thermal conductivity, W/(m · K)	
273 K	98
373 K	95
573 K	89
vapor pressure, kPa[c]	
382°C	0.1013
473°C	1.013
595°C	10.13
767°C	101.3
surface tension, mN/m (= dyn/cm)	
330°C	564
420°C	598
450°C	611
viscosity, mPa·s (= cP)	
340°C	2.37
400°C	2.16
500°C	1.84
600°C	1.54
molar magnetic susceptibility, cm^3/mol (= emu/mol)	-19.8×10^{-6}
Brinell hardness, kg/mm^2	16–23
tensile strength, MPa[d]	71
elongation, %	50
Poisson's ratio	0.33
modulus of elasticity, GPa[e]	49.9
shear modulus, GPa[e]	19.2
thermal neutron capture cross-section at 2200 m/s, m^2/atom	$2450 \pm 50 \times 10^{-28}$

[a]To convert J to cal, divide by 4.184.
[b]IACS = International Annealed Copper Standard.
[c]To convert kPa to mm Hg, multiply by 7.5.
[d]To convert MPa to psi, multiply by 145.
[e]To convert GPa to psi, multiply by 145,000.

and, to a lesser degree, from the processing of complex zinc, lead, and copper ores and their concentrates. Cadmium is also increasingly being recovered through the recycling of nickel–cadmium batteries. Mined zinc ores, which may contain 0.02–1.4% cadmium (3), are first crushed and ground to liberate Zn(Cd)S particles from the host rock. Differential flotation techniques are employed to separate the sulfide particles from waste rock yielding a high grade zinc concentrate that normally contains 0.3–0.5% cadmium (2). It is estimated that 90–98% of the cadmium present in zinc ores is recovered in the mining and beneficiation stages of the extraction process. A schematic flowsheet of the mining and beneficiation processes for a typical lead–zinc ore is shown in Figure 1.

Refining of zinc/cadmium concentrates to separate and purify the two metals can be accomplished by either hydrometallurgical (electrolytic) or pyrometallurgical (high temperature) techniques. Today, most zinc is produced by electrolytic production techniques, but considerable cadmium may also be generated from cadmium fumes and dust from the processing of complex zinc–lead–copper concentrates. In both cases, the concentrate is first converted from a

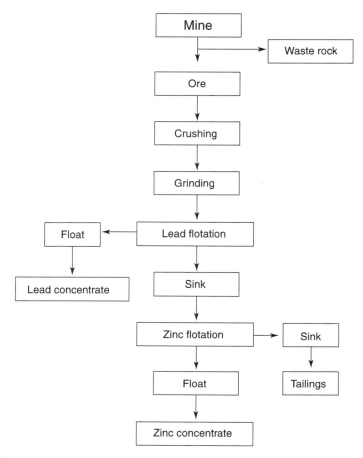

Fig. 1. Schematic flow sheet of the mining and beneficiation processes for a typical lead–zinc ore (2)

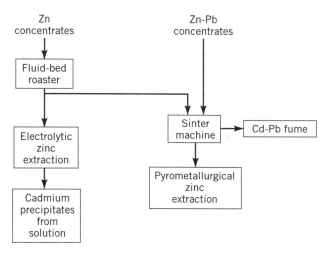

Fig. 2. Preliminary cadmium roasting processes.

zinc/cadmium sulfide to a zinc/cadmium oxide by roasting in a fluidized-bed roaster in an excess of oxygen usually at temperatures below 1000°C, which removes sulfur in the form of sulfur dioxide (SO_2). The SO_2 offgas is stripped of all entrapped dust and other impurities, which are recycled, and then converted to sulfuric acid in an acid plant. The preliminary roasting processes for zinc and zinc–lead concentrates results in cadmium recovery either as precipitates from solution or as cadmium–lead fume, respectively, as shown in Figure 2.

Precipitates are treated by a hydrometallurgical process and cadmium recovered by electrolytic extraction, while the cadmium present as fumes may be treated by several different methods to recover the lead and cadmium.

In the hydrometallurgical process, the crude zinc calcine is dissolved in the sulfuric acid leach of the roasted zinc ore to form an impure zinc sulfate solution, and then neutralized to precipitate any iron. Copper and cadmium are the most common impurities in these solutions, which are subjected to electrolysis for zinc recovery. Both may be precipitated from the solution by successive zinc dust additions in controlled amounts to produce a cadmium cake containing about 25% cadmium, 50% zinc, and minor amounts of copper and lead. The cadmium cake is redissolved in sulfuric acid with only zinc and cadmium going into solution. Two additional acid dissolution and zinc dust precipitation stages ultimately produce a relatively pure cadmium sponge, which is finally dissolved to produce a high purity cadmium solution from which high purity cadmium metal may be electrodeposited. After the electrodeposition process, cadmium metal is stripped from the cathodes and melted and cast into slabs, ingots, sticks, or balls generally of 99.95–99.99% minimum purity. Vacuum distillation techniques may be further employed to produce 99.999 and 99.9999% purity metals that are utilized in semiconductor and photovoltaic applications. A detailed flowchart of the electrolytic cadmium production process is shown in Figure 3.

In the pyrometallurgical or high temperature process, zinc concentrate is roasted as above under oxidizing conditions to remove sulfur and produce a granular sinter for the Imperial Smelting Furnace (ISF) lead–zinc blast furnace

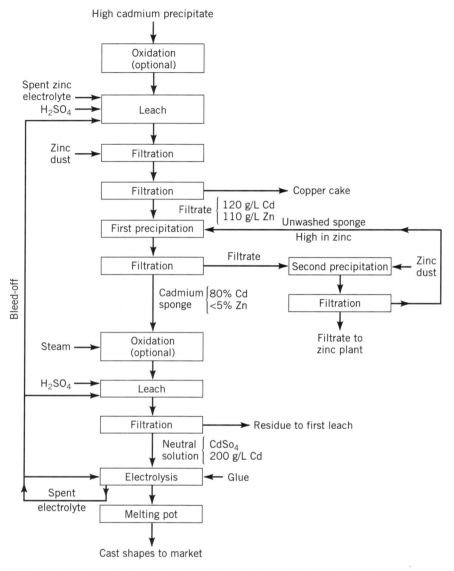

Fig. 3. Schematic flow sheet for electrolytic production of cadmium (12).

process. In the sintering operation, ≤70% of the cadmium content of the concentrate will volatilize and be collected as dust and fume. The collected material is leached with sulfuric acid to precipitate lead as lead sulfate and then treated with zinc dust to precipitate cadmium as a metallic sponge that is subsequently dried and refined by distillation to cadmium metal. In all pyrometallurgic processes, the calcine containing oxidized zinc and cadmium is heated to ~1100–1350°C, reduced by carbonaceous material, and the zinc and cadmium are volatilized. Most of the cadmium collects with zinc metal and may be removed by fractional distillation since the boiling points to the three metals (cadmium 767°C, zinc 906°C, lead 1750°C) are well separated. The dusts, powder, and alloy are repeatedly redistilled under reducing conditions to produce a pure metal. A flowchart for cadmium recovery from cadmium-bearing fumes is shown in Figure 4(12) illustrating several possible alternatives during the processing of these materials for cadmium recovery.

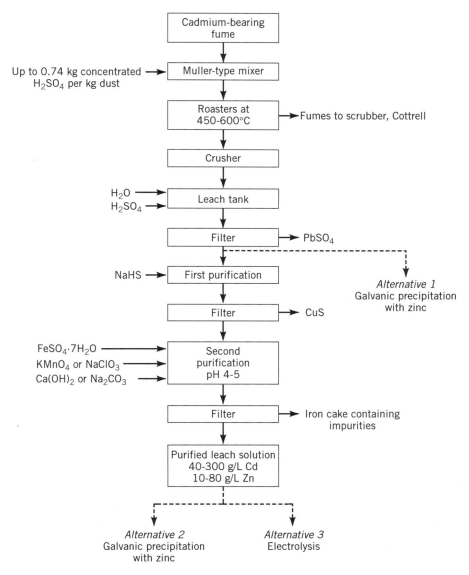

Fig. 4. Cadmium recovery from cadmium-bearing fumes (12).

Table 10. **World Consumption of Refined Cadmium Metal**

Year	consumption, t
1991	20,283
1992	17,870
1993	19,165
1994	18,149
1995	18,847
1996	17,726
1997	18,506
1998	18,104
1999	18,936
2000 (estimated)	19,180

6. Economic Aspects

6.1. Consumption. Accurate cadmium consumption figures are more difficult to establish than those for cadmium production because of two important confounding factors: (1) the conversion of cadmium metal into cadmium oxide and other cadmium compounds and (2) shipments of cadmium-containing residues to zinc smelters from recycling operations. Once cadmium metal is produced, it may be converted to an intermediate product in one country, incorporated into batteries, pigments, stabilizers, or coatings in another country, and finally manufactured and sold in a product in a third country. It is difficult to establish consumption patterns accurately on a geographic basis, and the figures reported in Table 10 by the World Bureau of Metal Statistics (WBMS) refer to *apparent* consumption of refined metal and include the conversion of refined metal into compounds such as the oxide, hydroxide, sulfide, or carboxylate as would be utilized respectively in batteries and coatings, pigments, and stabilizers. Table 10 shows that refined cadmium metal consumption has decreased substantially since 1996, and that cadmium metal production has exceeded cadmium metal consumption, especially during 1997 and 1998.

The world's leading consumers of refined cadmium metal, according to the World Bureau of Metal Statistics, are summarized in Table 11.

Table 11. **World's Leading Consumers of Refined Cadmium Metal**

Country	Apparent consumption			
	1994	1995	1996	1997
Japan	6,615.3	8,363.5	6,527.2	7,247.0
United States	2,236.6	2,007.6	1,700.5	1,355.9
Belgium	2,944.0	2,017.0	2,017.1	2,017.2
France	1,969.0	1,968.0	1,476.0	1,476.0
Germany	750.0	750.0	750.0	750.0
China	600.0	600.0	600.0	600.0
United Kingdom	663.5	587.4	617.7	631.2
India	411.0	446.4	446.4	446.4
South Korea	380.4	380.4	380.0	380.4
Sweden	293.0	393.0	259.0	300.0

With regard to the data shown in Table 11, Belgium is not a large cadmium consumer producing cadmium-containing products, but it does convert an enormous amount of cadmium metal to cadmium oxide, which is then shipped to the NiCd battery manufacturers around the world. Japan is the largest cadmium consumer in the world for cadmium products because the world's two largest NiCd battery producers, Sanyo and Panasonic, manufacture NiCd battery electrode material there and then ship it to battery assembly plants all over the world. NiCd battery production accounts for over 90% of Japan's cadmium consumption.

6.2. Prices.   Cadmium has always shown a high price volatility as a byproduct metal with an inelastic supply dependent primarily on zinc production and as a heavily regulated "heavy metal" subject to many proposed product and production restrictions, especially in the European Union. Another problem with cadmium pricing is that the published price very largely reflects trader spot sales, which make up only a relatively small part of total cadmium sales volume. Thus, cadmium prices have varied from as high as $10/lb in the late 1980s to as low as $0.20/lb, their levels in 1998 and 1999. Cadmium prices began at $0.70– 0.85/lb in early 1997 for 99.95 and 99.99% purity grades as published in *Metal Bulletin*. Prices decreased steadily throughout the year, and by the end of the year stood at $0.35–0.45/lb, close to the all-time lows recorded in 1993. Cadmium metal prices were as high at $2.50/lb in late 1994 and late 1995, but declined to < $0.50/lb for all of 1998. The published *Metal Bulletin* prices for 99.99% Cd from 1993 through September 1998 are shown in Figure 5.

Prices for cadmium chemicals and for high-purity cadmium metal command a substantial premium over the base metal prices, depending on the specific chemical required, its purity and particle size, lot size, shipping arrangements, and

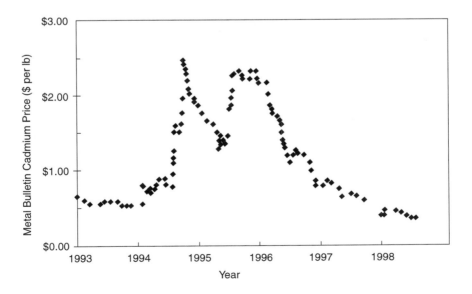

Fig. 5.   Metal Bulletin Prices for 99.99% cadmium, 1993–1998 (10).

Table 12. **Specifications for Cadmium and Its Alloys, Compounds, and Products**^{a}

Number	Title
	ASTM Specifications
B201-89	*Testing Chromate Coatings on Zinc and Cadmium*
B440-00	*Cadmium*
B635-91	*Coatings of Cadmium-Tin Mechanically Deposited*
B696-91	*Coatings of Cadmium Mechanically Deposited*
B699-93	*Coatings of Cadmium Vacuum Deposited*
B766-93	*Electrodeposited Coatings of Cadmium*
B774-95	*Low Melting Point Alloys*
B781	*Silver-Cadmium Oxide Contact Material Guide*
B816-91	*Coatings of Cadmium-Zinc Mechanically Deposited*
F326	*Electronic Hydrogen Embrittlement Test for Cadmium Electroplating Processes*
F1135-93	*Cadmium or Zinc Chromate Organic Corrosion Protective Coating for Fasteners*
	ISO Specifications
ISO 2082-86	*Electroplated Coatings of Cadmium on Iron or Steel*
ISO 3613-80	*Chromate Conversion Coatings on Zinc and Cadmium—Test Methods*
ISO 4520-81	*Chromate Conversion Coatings on Electroplated Zinc and Cadmium Coatings*
	National and Military Specifications
A-A-50800(1)	*Cadmium Oxide (U.S.A)*
A-A-51126A	*Cadmium Anodes (U.S.A)*
ANSI C18.2-1984	*Sealed Rechargeable Nickel-Cadmium Cylindrical Bare Cells (U.S.A.)*
BS 2868:1968	*Cadmium Anodes and Cadmium Oxide for Electroplating (U.K.)*
MIL-C-8837B(1)	*Cadmium Coating, Vacuum Deposited (U.S.A)*
MIL-C-81562B(1)	*Mechanically Deposited Cadmium, Cadmium Tin and Zinc Coating*
MIL-F-14072D	*Finish for Ground Signal Equipment (U.S.A)*
MIL-P-23408B(1)	*Plating, Tin-Cadmium (Electrodeposited) (U.S.A)*
MIL-S-5002D(1)	*Surface Treatments and Inorganic Coatings for Metal Surfaces of Weapons Systems (U.S.A)*
MIL-STD-171E	*Finishing of Metal and Wood Surfaces (U.S.A)*
MIL-STD-870BB	*Electrodeposited Cadmium Plating, Low Embrittlement (U.S.A)*
MIL-STD-889B(3)	*Dissimilar Metals (U.S.A)*
MIL-STD-1500	*Electrodeposited Cadmium-Titanium Plating, Low Embrittlement (U.S.A)*
NAS672	*National Aerospace Standard for Electrodeposited Cadmium (U.S.A)*
QQ-P-416F	*Electrodeposited Cadmium Plating (U.S.A)*
	Commercial Specifications
AMS/SAE 2400S	*Cadmium Plating*
AMS/SAE 2401D	*Cadmium Plating, Low Hydrogen Content Deposit*
AMS/SAE 2416G	*Diffused Nickel-Cadmium Plating*
AMS/SAE 2419A	*Cadmium-Titanium Alloy Plating*
AMS/SAE 2426B	*Cadmium Plating, Vacuum Deposition*

^{a}*Nomenclature:* ASTM = American Society for Testing and Materials; ISO = International Standards Organization; ANSI = American National Standards Institute; BS = British Standard; NAS = National Aerospace Standard; AMS = Aerospace Materials Specification; SAE = Society of Automotive and Aerospace Engineers.

many other commercial factors. For example, in May 1999, when the price of cadmium metal was published in *Metal Bulletin* at approximately \$0.19–0.24/ lb, the corresponding prices for various cadmium chemicals were published in the *Chemical Market Report* ranging from \$6.75/lb pound for cadmium nitrate to \$25.24/lb for chemically pure, dark red cadmium sulfide used for pigment applications. Cadmium/zinc sulfide yellows are priced at approximately \$12– 13/lb, while the cadmium selenide lithopones are priced in the \$5–12/lb range. Depending on specific purity and particle size requirements, cadmium oxide and cadmium hydroxide for NiCd battery production generally may be priced at several times the base price for the high purity (99.99% Cd) metal. The 1999 prices for high purity cadmium powder were \$5.00–7.00/lb, compared to base metal prices of \$0.25/lb for cadmium sticks and balls.

7. Specifications for Cadmium and Cadmium Compounds

There are a number of specifications for cadmium, cadmium compounds, and cadmium products, some of which are listed in Table 12.

In addition, there are many specific specifications for cadmium products prepared by various materials and products manufacturers. For example, each NiCd battery producer has a specific requirement for the cadmium compounds such as cadmium oxide, cadmium hydroxide, or cadmium nitrate, which they may utilize in the production of their NiCd batteries. Similarly, pigment and stabilizer producers have specific compositional, particle size, and other requirements for the cadmium compounds used in their formulations. Many different zinc-, lead-, and copper-based products also have maximum cadmium contents levels.

8. Analytical Methods

Analytical methods for cadmium are broadly classified into techniques for determining the impurity levels of antimony, arsenic, copper, lead, silver, thallium, tin, and zinc in cadmium metal and cadmium alloys, and methods for establishing cadmium impurities in other base metals such as aluminium and aluminum alloys, nickel, uranium oxide, and zinc and zinc alloys. The ASTM specifications for determining the impurity levels of certain metals in cadmium and of cadmium impurities in certain other systems are summarized in Table 13.

Excellent, if somewhat dated, reviews of the analytical techniques used for the determination of cadmium have been presented by Farnsworth (1) and Fernando and Freizer (13). Farnsworth, for example, summarizes the various analytical techniques available and their sensitivity and accuracy, and her summary is shown in Table 14.

More information (13) indicates that the sensitivity of a number of these techniques has improved greatly and that it is now possible to detect cadmium ion concentrations down to 0.01 parts per billion (ppb) (see CHROMIUM COMPOUNDS).

Table 13. **ASTM Specifications for Analysis of Cadmium**

Number	Title
E34-94	*Chemical Analysis of Aluminum and Aluminum-Base Alloys*
E227-90	*Optical Emission Spectrometric Analysis of Aluminum and Aluminum Alloys by the Point-to-Plane Technique*
E396-98	*Standard Test Methods for Chemical Analysis of Cadmium*
E402-95	*Standard Test Method for Spectrographic Analysis of Uranium Oxide (U_3O_8) by Gallium Oxide-Carrier Technique*
E536-98	*Standard Test Methods for Chemical Analysis of Zinc and Zinc Alloys*
E607-90	*Standard Test Method for Optical Emission Spectrometric Analysis of Aluminum and Aluminum Alloys by the Point-to-Plane Technique, Nitrogen Atmosphere*
E1277-96	*Standard Test Method for Chemical Analysis of Zinc-5% Aluminum-Mischmetal Alloys by ICP Emission Spectrometry*
E1587-94	*Standard Test Methods for Chemical Analysis of Refined Nickel*

A 1986 summary of analytical techniques for cadmium (14) estimates the following detection limits for cadmium ions by various analytical techniques:

Method	Detection limit, ppb
Atomic absorption spectroscopy (graphite furnace)	0.008
Polarography (square-wave)	10
X-ray fluorescence (energy-dispersive)	5,000
Neutron activation analysis	1
Isotope dilution	10
Inductively coupled plasma emission spectra	1

At higher concentrations, cadmium may be estimated gravimetrically following precipitation with sulfide, β-naphthoquinoline, or after plating from a cyanide-containing solution onto a stationary platinum cathode. Volumetric

Table 14. **Analytic Techniques for Cadmium**

Technique	Sensitivity ppb	Accuracy, %
spectrophotometry		
dithizone	50	5–10
electrochemistry		
d-c polarography	1000	5–10
pulse polarography	10	10
anodic stripping voltammetry	1	20
spectroscopy		
arc/spark emission	20,000	30
atomic emission	50	5
atomic absorption	10	5
atomic absorption with extraction	1	10
atomic absorption with furnace	1	10
atomic fluorescence	1	10
neutron activation	1–10	20
mass spectroscopy		
spark source	10–100	20
isotope dillution	1	5

Source: Ref. 1.

procedures rely on preliminary precipitation of the sulfide that is purified and then dissolved in acid whereupon the liberated H_2S may be titrated with iodine. An alternative, should zinc be a likely contaminant, is to precipitate with diethyl-dithiocarbamate and then to redissolve in acid and titrate with sodium ethylene-triaminetetraacetate (EDTA) using Eriochrome Black T as indicator (see CHROMIUM COMPOUNDS).

9. Recycling

There has been increasing emphasis on the collection and recycling of spent cadmium products to reduce any risk to human health and the environment from the disposal of cadmium products. Since NiCd batteries account for the vast majority (75%) of cadmium consumption, the main efforts have centered on the collection and recycling of NiCd batteries. However, cadmium-coated products, cadmium alloys, and CdTe photovoltaic modules are also recyclable and efforts have been made to collect and recycle these products as well. For example, cadmium coatings may be recycled through the treatment of electric-arc furnace (EAF) dust, which is generated by the remelting of large quantities of scrap steel, some of which may be cadmium-coated or contain cadmium as an impurity in galvanized coatings. In this case, EAF dust is processed mainly to recover zinc and lead, but could be processed to recover cadmium as well if the economics were more favorable. Similarly, certain types of cadmium-containing brazing and soldering alloys are recycled, mainly to recover more valuable metals such as silver and nickel, but in the process cadmium can be recovered as well. More recently, First Solar LLC. in the United States has developed successful techniques for recycling both cadmium and tellurium from CdTe manufacturing scraps and spent solar panels. However, the large volumes recycled and the massive programs that the cadmium and NiCd battery industries have developed arise from the recycling of nickel–cadmium batteries.

Nickel–cadmium batteries, as well as other cadmium-containing products, are relatively easy to recycle and may be recycled by either pyrometallurgical (high temperature) or hydrometallurgical (wet chemical) processes. There are also processes that include both hydrometallurgical and pyrometallurgical components in their overall flowcharts. The principal materials present in NiCd batteries are iron, nickel, cadmium, potassium hydroxide electrolyte, some copper contact materials, plastic (polypropylene or polyethylene) casings, and organic separator materials. In the recycling of large industrial NiCd batteries, the electrolyte is normally drained off, and plastic casings and separator materials are diverted from the metallic waste stream of mainly iron, nickel, and cadmium and their oxides or hydroxides. The objective of most recycling operations is to convert the metal oxide/hydroxide back to a metal with a purity, quality, and form that can subsequently be utilized for the manufacture of new batteries.

Since nickel, iron, and cadmium all have relatively low (<200 kJ) negative free energies of formation of their respective oxides, they are easier to reduce than are the oxides of other battery metals such as zinc, manganese, chromium, titanium, vanadium, aluminum, zirconium, and the rare-earth elements, all of which have higher (>400 kJ) negative free energies of formation of their

Table 15. **World's Major NiCd Battery and NiCd Scrap Recycling Plants**

Company	Location	Type	Capicity, tonnes
INMETCO	USA	Stainless Steel	3,000
Hanil Metal Recycle Co.	Korea	NiCd Recycler	3,000
Mitsui Mining & Smelting	Japan	Zinc Refinery	1,800
Toho Zinc Co., Ltd.	Japan	Zinc Refinery	1,700
Kansai Catalyst	Japan	Zinc Refinery	500
Hydrometal S.A.	Belgium	Hydrometallurgical	1,300
SAFT	Sweden	NiCd Recycler	1,500
SNAM	France	NiCd Recycler	1,000
ACCUREC	Germany	NiCd Recycler	1,0000

respective oxides. Therefore, NiCd and NiFe batteries will require less energy to recycle than do alkaline manganese and zinc–carbon batteries, which, in turn, require less energy to recycle than NiMH and lithium-ion batteries. Cadmium may be recycled indefinitely from spent NiCd batteries and manufacturing scrap, and continually produced to the same specifications as cadmium produced from zinc ore and meeting all the requirements of NiCd battery manufacturers. Some of the details of the various types of pyrometallurgical and hydrometallurgical NiCd battery processes are summarized below, including a listing of the world's major NiCd battery recycling plants and detailed process flowcharts. More complete details, including an extensive bibliography on NiCd battery recycling, are available in an OECD document (15).

Nickel–cadmium battery and manufacturing scrap recycling systems are usually based on pyrometallurgical (high temperature) or hydrometallurgical (wet chemical) processes, and the pyrometallurgical processes are further divided into three basic types: stainless-steel recyclers, dedicated NiCd battery recyclers, and primary zinc refineries. A summary of the world's NiCd battery and battery manufacturing scrap recycling facilities indicating their basic type and recycling capacity is shown in Table 15. In the case of the three hydrometallurgical recyclers listed, these are not commercially processing NiCd batteries or scraps at the present time and are included to illustrate both types of recycling processes.

Recycling of cadmium coatings and cadmium alloys is accomplished primarily by pyrometallurgical or high temperature techniques, while processing of CdTe manufacturing scraps and spent modules is carried out utilizing a hydrometallurical process. Details of all of the processes described above are available in the OECD *Guidance Document* (15) on nickel–cadmium battery recycling.

10. Environmental Concerns

10.1. Cadmium Emissions.

Cadmium emissions arise from two principal source categories: natural sources and man-made or anthropogenic sources. Emissions occur to the three major compartments of the environment—air, water, and soil—but there may be considerable transfer between the three compartments after initial deposition. Emissions to air are considered more mobile than those to water, which in turn are considered more mobile than those to soils.

Natural Cadmium Emissions. Even though the average cadmium concentration in the earth's crust is generally placed between 0.1 and 0.5 ppm, much higher levels may accumulate in sedimentary rocks, and marine phosphates and phosphorites have been reported to contain levels as high as 500 ppm (3,4). Weathering and erosion of parent rocks result in the river transport of up to 15,000 t per annum of cadmium to the world's oceans (4,5). Volcanic activity is also a major natural source of cadmium release to the atmosphere, and estimates have been placed as high as 820 t per year (4,5,16,17). Forest fires contribute to cadmium air emissions, with estimates of 1–70 t emitted to the atmosphere each year (16).

Anthropogenic Cadmium Emissions. Man-made cadmium emissions arise either from the manufacture, use and disposal of products intentionally utilizing cadmium, or the presence of cadmium as a natural but not functional impurity in noncadmium-containing products. The former category of cadmium-containing products includes.

- Nickel–cadmium batteries
- Cadmium-pigmented plastics, ceramics, glasses, paints, and enamels
- Cadmium-stabilized PVC products
- Cadmium-coated ferrous and nonferrous products
- Cadmium alloys
- Cadmium electronic compounds

The latter category of noncadmium-containing products includeds:

- Nonferrous metals and alloys of zinc, lead, and copper
- Iron and steel
- Fossil fuels (coal, oil, gas, peat, and wood)
- Cement   Phosphate fertilizers

Anthropogenic cadmium emissions have declined dramatically since the late 1960s and are still declining today. Studies in Europe, Japan, and the United States have all shown significant decreases in cadmium emissions in the time period from approximately 1960 to 1995 (13,18–21). Considerable progress is now also being made in reduction of diffuse contamination from cadmium products through collection and recyling programs of cadmium-containing products (3,9,19,22). Cadmium emissions from products where cadmium is present as an impurity have not been reduced as significantly (3,18,21) and appear to be the one remaining area where additional reductions might be achieved.

Cadmium in Municipal Solid Waste. There are many sources of cadmium in municipal solid waste, not just products to which cadmium has intentionally been added. As noted above, other sources include products such as iron and steel; gypsum, cement, and other construction debris; nonferrous (Zn, Pb, Cu) metal and alloy products; fossil fuels and fossil fuel residues; and natural substances such as grass, plants, and foods, all of which may contain cadmium (23). Some studies have determined that incineration accounts for only ~1% of the total sources of human cadmium exposure (5,21,24–26). The human exposure resulting from the disposal of cadmium-containing products by incineration is even less and is generally considered insignificant for all of these reasons (24–26).

Partitioning of Cadmium Emissions to Compartments. The vast majority of cadmium emissions, approximately 80–90%, partition initially to soils. Although some transfer does occur from soils back to the air or water compartments, the net flux into the soil is generally regarded as positive since there is deposition from both air and water onto soils. Thus, most cadmium emissions eventually return to soil. In soils, cadmium is largely bound to the nonexchangeable fraction, such as on clays and manganese and iron oxides. For this reason, its mobility and transfer into the animal and human food chain is limited. The remaining 10–20% of anthropogenic cadmium emissions partition between air and water and depend largely on the emission source. For example, cadmium electroplating operations result in no air emissions but, in the past, did result in water emissions. Today, water effluent regulations ensure that even cadmium electroplating water emissions are negligible. Any cadmium waste from cadmium electroplating is now contained in electroplating sludge from the pollution control equipment, which may be recycled to recover valuable metals.

Sources of Cadmium Emissions to Air, Water, and Soil. Cadmium emissions to air arise, in decreasing order of importance, from the combustion of fossil fuels, iron and steel production, nonferrous metals production, and municipal solid-waste combustion (3,21,27,28).Cadmium emissions to water arise, in decreasing order of importance, from phosphate fertilizers, nonferrous metals production, and the iron–steel industry (15,21,24).

Cadmium emissions to soils must be considered in three distinct categories: as inputs to agricultural soils, as emissions to nonagricultural soils, and as depositions in controlled landfills. In the first case, the main inputs to agricultural soils that are of primary relevance to human exposure to cadmium arise from atmospheric deposition, sewage sludge application, and phosphate fertilizer application (6,21). In the second case, inputs to nonagricultural soils arise mainly from the iron–steel industry, nonferrous metals production, fossil-fuel combustion, and cement manufacture (5,24,27).

In the case of cadmium present in controlled landfills, these amounts can arise from disposal of spent cadmium-containing products, products with cadmium impurities, and wastes such as grass, food, and soil with naturally occurring trace levels of cadmium (23). Cadmium input to agricultural soils is of far greater relevance to human health than cadmium input to nonagricultural soils. Input to controlled landfills is of even less importance because cadmium is largely immobilized in controlled landfills. For example, numerous studies of the leachate from municipal solid-waste landfills have conclusively demonstrated that, even after long periods of time, the leachate from these landfills meets the WHO cadmium drinking-water standards of 3 µg/L (29,30).

Cadmium Emissions vs. Natural Cadmium Emissions. Earlier estimates of anthropogenic versus natural cadmium emissions had indicated approximately 8000–10,000 t/yr for anthropogenic emissions compared to 800–1000 tonnes per year for natural cadmium emissions (4,16,17). Anthropogenic cadmium emissions, however, have decreased by $\leq 98\%$. Thus, natural cadmium emissions, expressed in terms of contained cadmium, which in the past have been stated to be only about 10% of the level of anthropogenic cadmium emissions, are more nearly equal to synthetic emissions today. This parity is becoming increasingly likely because of the rapidly decreasing anthropogenic cadmium emissions to the environment compared to the relatively stable natural cadmium emissions.

11. Levels of Cadmium in the Enviroment

11.1. Cadmium in Air.   Three distinct categories may be recognized with respect to cadmium-in-air concentrations: cadmium in ambient air, cadmium air levels in occupational exposure situations, and cadmium in air from the smoking of tobacco. Cadmium in ambient air represents, by far, the majority of total airborne cadmium. Inputs from all three categories may affect human cadmium intake and human health, but the levels and the transfer mechanisms to humans are substantially different for the three. Whereas cadmium emissions from occupational environments and from cigarette smoke are transferred directly to humans, cadmium in ambient air is generally deposited onto waters or soils, then eventually transferred to plants and animals, and finally enters the human body through the food chain. Thus the amount of cadmium transferred through occupational exposure or cigarette smoking will be much higher than the amount transferred through atmospheric deposition.

Cadmium in Ambient Air.   Ambient-air cadmium concentrations have generally been estimated to range from 0.1 to 5 ng/m^3 in rural areas, from 2 to 15 ng/m^3 in urban areas, and from 15 to 150 ng/m^3 in industrialized areas (4,5,8), although some much lower values have been noted in extremely remote areas and some much higher values have been recorded in the past near uncontrolled industrial sources. There are generally little or no differences noted in cadmium levels between indoor and outdoor air in nonsmoking environments. Smoking, however, may substantially affect indoor ambient-air cadmium concentrations.

Cadmium in Occupational Enviroments.   Cadmium air concentrations may be elevated in certain industrial settings, but these exposures are closely controlled today by national occupational exposure standards. Historically, average exposure levels and regulatory permissible exposure limits (PELs) have decreased markedly in since late 1950s in recognition of the importance of cadmium inhalation to human health and with the significant improvements in air-pollution-control technology over that period (4,8). Occupational exposure standards that were formerly set at 100–200 μg/m^3 are now specified at 2–50 μg/m^3 along with requirements to maintain biological indicators such as cadmium-in-blood and cadmium-in-urine below certain levels to assure no adverse human health effects from cadmium occupational exposure (31–33).

Cadmium in Tobaco Smoke.   Tobacco leaves naturally accumulate and concentrate relatively high levels of cadmium, and therefore smoking of tobacco is an important source of air cadmium exposure for smokers. It has been reported that one cigarette contains about 0.5–2 μg of cadmium and that ~10% of the cadmium content is inhaled when the cigarette is smoked (4,8). Smokers generally exhibit significantly higher cadmium body burdens than do nonsmokers.

11.2. Cadmium in Water.   The average cadmium content in the world's oceans has variously been reported as low as <5 ng/L and as high as 110 ng/L (3–8). Higher levels have been noted around certain coastal areas and variations of cadmium concentration with the ocean depth, presumably due to patterns of nutrient concentrations, have also been measured (4,5). Even greater variations are quoted for the cadmium contents of rainwater, freshwater, and surface water in urban and industrialized areas. Levels of 10–4000 ng/L have been quoted in the literature depending on specific location and whether total cadmium or dissolved cadmium is measured (4,5,8).

Cadmium is a natural, usually minor constituent of surface and groundwaters. It may exist in water as the hydrated ion; as inorganic complexes such as carbonates, hydroxides, chlorides, or sulfates; or as organic complexes with humic acids (5). Much of the cadmium entering freshwater may be rapidly adsorbed by particulate matter, and thus sediment may be a significant sink for cadmium emitted to the aquatic environment (4). Some data shows that recent sediments in lakes and streams range from 0.2 to 0.9 ppm in contrast to the levels of generally less than 0.1 ppm cited above for fresh waters (3). Partitioning of cadmium between the adsorbed-in-sediment state and dissolved-in-water state is therefore an important factor in whether cadmium emitted to waters is or is not available to enter the food chain and affect human health.

Rivers containing excess cadmium can contaminate surrounding land, either through irrigation for agricultural purposes, dumping of dredged sediments or flooding. It has also been demonstrated that rivers can transport cadmium for considerable distances, up to 50 km, from the source (4). Nonetheless, studies of cadmium contamination in major river systems in the late twentieth century conclusively demonstrated that cadmium levels in these rivers decreased significantly since the 1960s and 1970s (13,18,19,21). For example, studies on the Rhine River Basin during 1973–1987 indicated that the point source cadmium discharges to the Rhine River decreased from 130 to 11 tonnes/year over that 14-year timespan, a reduction of over 90% (18). Similarly, data on total cadmium and dissolved cadmium at the Dutch/German border during 1971–1987 have shown comparable reductions (34).

11.3. Cadmium in Soil.   *Sources.*   Cadmium is much less mobile in soils than in air and water. The major factors governing cadmium speciation, adsorption and distribution in soils are pH, soluble organic matter content, hydrous metal oxide content, clay content and type, presence of organic and inorganic ligands, and competition from other metal ions (5). The use of cadmium-containing fertilizers and sewage sludge is most often quoted as the primary reason for the increase in the cadmium content of soils in Europe (6). Atmospheric cadmium emissions deposition onto soils has generally decreased significantly over that same time period (13,19). Indeed, studies in Europe have documented that atmospheric emissions do not have a significant impact on the cadmium content of soils (35).

Cadmium in Agricultural Soils. Numerous agencies have focused on the presence of cadmium in agricultural soils, the means by which agricultural soils may be enriched by cadmium, the degree to which cadmium is taken up by foodstuffs, and the subsequent transfer of cadmium to humans via foodstuffs. Because cadmium is a naturally occuring component of all soils, all foodstuffs contain some cadmium and therefore all humans are exposed to natural levels of cadmium. Although much attention has been focused on the cadmium content of agricultural soils, it is important to recognize that the cadmium content of food items varies more as a function of the nature of the crop grown and the agricultural practices followed. Except in cases of extreme contamination, the concentration of cadmium in soils is not the primary determinant of cadmium in the human diet. For example, leafy vegetables and potato tubers naturally accumulate higher levels of cadmium than do fruits and cereals (36). Moreover, tillage and crop rotation practices similarly have a greater impact on the cadmium content of food than does the concentration of cadmium in soils (46). Cadmium absorption may also depend on other factors as well, as described below.

Cadmium Levels in Foodstuffs. Cadmium levels can vary widely in various types of foodstuffs. Leafy vegetables such as lettuce and spinach and certain staples such as potatoes and grain foods exhibit relatively high values from 30 to 150 ppb. Peanuts, soybeans, and sunflower seeds also exhibit naturally high values of cadmium with seemingly no adverse health effects. Meat and fish normally contain lower cadmium contents, from 5 to 40 ppb. Animal offal such as kidney and liver can exhibit extraordinarily high cadmium values, up to 1000 ppb, as these are the organs in animals where cadmium concentrates (4,37). The cadmium contents of foodstuffs may vary widely with the agricultural practices utilized in the particular areas such as phosphate fertilizer, sewage sludge and manure application, the types of crops grown, and atmospheric cadmium deposition from natural or anthropoenic sources. Since various studies have shown that human cadmium intake, as least for nonsmokers, comes principally (~95%) from the ingestion of foods rather than from inhalation of cadmium in air, it is the cadmium levels of foods which most affect the general population. There are strong indications that cadmium levels in foodstuffs have substantially decreased with the progressive control of cadmium emissions to the environment (38–40). More recent studies have further documented that the cadmium content of food crops in Europe and many other countries are now stable and not increasing with time (41).

Cadmium Contamination of Agricultural Soils. In the past, there were examples of marked cadmium contamination in areas where food has been grown. This was particularly so for rice crops in Japan in the 1950s and 1960s where cadmium concentrations of 200–2000 ppb were found (8). In general, soils that have been historically contaminated with cadmium from industrial operations are no longer used for agricultural purposes. In those cases where old industrial installations that are cadmium-contaminated are subsequently employed for growing crops, suitable remediation techniques do exist to immobilize the cadmium present in the soil and thus to control the risk to human health. There is, however, no doubt that old sites that are so contaminated do require proper risk management and control by cleaning up or immobilizing the existing excess cadmium in the soil.

12. Health and Safety Factors

It has been well established that excess cadmium exposure produces adverse health effects on human beings. For virtually all chemicals, adverse health effects are noted at sufficiently high total exposures. For certain elements such as copper and zinc, which are essential to human life, a deficiency as well as an excess can cause adverse health effects. Cadmium is not regarded as essential to human life. The relevant questions with regard to cadmium exposure are the total exposure levels and the principal factors that determine the levels of cadmium exposure and the adsorption rate of the ingested or inhaled cadmium by the individual, in other words, the pathways by which cadmium enters the food chain, which is the principal pathway of cadmium exposure for most human beings.

13. Principal Factors that Determine Levels of Human Exposure

Human normally absorb cadmium into the body by either ingestion or inhalation. Dermal exposure (uptake through the skin) is generally not regarded to be of signifigance (42). It is widely accepted (14,37) that approximately 2–6% of the cadmium ingested is actually taken up into the body. Factors influencing cadmium absorption are the form in which cadmium is present in the food, and the iron status of the exposed individual. In contrast, 30–64% of inhaled cadmium is absorbed by the body, with some variation as a function of chemical form, solubility, and particle size of the material inhaled. Thus, a greater proportion of inhaled cadmium is retained by the body than when cadmium is taken in by ingestion. For the non–occupationally exposed individual, inhalation exposure to cadmium does not usually contribute significantly to overall body burden. The exception to this generalization is the cigarette smoker. One model for human cadmium intake (26) has estimated that ingestion accounts for 95% of total cadmium intake in a nonsmoker. For a smoker, this model estimates that roughly 50% of their cadmium intake arises from cigarettes with the balance due to ingestion and the low levels of cadmium naturally present in ambient air. In the past, occupational exposure was also a significant contributor to total cadmium intake, but with very stringent occupational standards in place today, occupational cadmium intake is much less of a consideration than it was in the late 1970s. Thus, the principal determinants of human cadmium exposure today are smoking habits, diet, and, to a lesser extent, occupational exposure.

13.1. Human Intake of Cadmium. *Ingestion.* Much of the cadmium that enters the body by ingestion comes from terrestrial foods, or from plants grown in soil or meat from animals that have ingested plants grown in soil. Thus, directly or indirectly, it is the cadmium present in soil and the transfer of that cadmium to food plants together with the cadmium deposited from the atmosphere on edible plant parts that establishes the vast majority of human cadmium intake. Some have estimated that 98% of the ingested cadmium comes from terrestrial foods, whereas only 1% comes from aquatic foods such as fish and shellfish, and 1% arises from cadmium in drinking water (27).

Cadmium Intake from Foods.   Many studies have attempted to establish the average daily cadmium intake resulting from foods. In general, these studies show that the average daily diet for nonsmokers living in uncontaminated areas is at present at the low end of the range of 10–25 µg of cadmium (5,8,37). This general trend is confirmed by decreasing blood cadmium levels in the general population in several countries during this time period (43,44). In another evaluation, the International Programme on Chemical Safety (IPCS) assessed the average daily intake at the lower end of this range (4).

The World Health Organization (WHO) has established a provisional tolerable weekly intake (PTWI) for cadmium at 7 µg/kg of body weight. This PTWI weekly value corresponds to a daily tolerable intake level of 70 µg of cadmium for the average 70-kg man and 60 µg of cadmium per day for the average 60-kg woman (4). Clearly, the daily cadmium intake for the general population from food, which is by far the dominant source of cadmium, is well below the guidelines established by the World Health Organization. The average daily cadmium intake for the general population in the Western world has shown a distinct downward trend from 1970 through 1992 (31), a reduction presumed to be due to the marked decreases in direct atmospheric deposition of cadmium onto crops and soils. Other studies have suggested that, over the timeframe of 1980–1985, levels of cadmium intake have been relatively constant (5). At an absorption rate of 5% from ingestion, the average person is believed to retain about 0.5–1.0 µg of cadmium per day from food.

There is considerable information in the literature regarding the cadmium contents of foods grown in contaminated areas (4,5,8). Detailed studies have indicated that only a small percentage of these contaminated areas were actually utilized for growing foods that were subsequently consumed with the exception of rice fields in Japan, where considerable cadmium did find its way into the average person's diet through rice grown on contaminated rice beds (8). In specific cases, management measures to reduce the transfer of cadmium from historically contaminated soils into the local food chain have proved successful (45).

Inhalation.   Cadmium inhalation is a far smaller contributor to total cadmium body burden except, as previously noted, in the cases of smokers or some highly exposed workers of the past. Today, the inhalation route is well controlled in the occupational setting, and is well controlled from point sources such as those which directly pertain to the nonferrous, cadmium, or cadmium products industries. Ambient-air emissions from fossil-fuel power generation plants, the iron–steel industry, and other major industries where cadmium may be present as a low concentration impurity, on the other hand, may be substantial because the volumes of the waste gases generated are substantial.

Cadmium Intake from Cigarette Smoking.   Smokers absorb amounts of cadmium comparable to those from food, about 1–3 µg of cadmium per day, from cigarettes smoking. It has been reported that one cigarette contains ~1–2 µg of cadmium and that ~10% of the cadmium content is inhaled when the cigarette is smoked (4). Cigarette construction, the use of filters, and variations in the cadmium contents of tobaccos could decrease cadmium exposure by this route, but in general cigarette smoking is a habit that can more than double the average person's daily cadmium intake. Cigarette smokers who are also occupationally exposed may increase their total cadmium intake even further.

Cadmium Intake from Occupational Exposure. Before the 1960s, elevated cadmium in air exposure levels were measured in some workplaces, sometimes as high as 1 mg/m^3. Since that time, workplace exposures and standards have decreased markedly so that most occupational exposure standards today are in the range of 2–50 µg/m^3. The result has been that occupational exposures today are generally below 5 µg/m^3, and most cadmium workers are exposed at levels considered safe (37). In rare cases where cadmium air levels are higher, the use of personal protective equipment is obligatory. Extensive preventative hygiene programs and medical follow-up programs have been developed to control the risk related to cadmium exposure at the workplace (32,33,42,46). Considering present levels of occupational exposure cadmium intake, general dietary intake, and cigarette smoking intake, it still would appear, however, that today's average daily cadmium intake is well below the values recommended by the World Health Organization.

Human Health Effects of Cadmium. The kidney is the critical target organ for the general population as well as for occupationally exposed populations. Cadmium is known to accumulate in the human kidney for a relatively long time, from 20 to 30 years, and, at high doses, is also known to produce health effects on the respiratory system and has been associated with bone disease. Most of the available epidemiological information on cadmium has been obtained from workers occupationally exposed at high levels or on Japanese populations in highly contaminated areas. Most studies have centered on the detection of early signs of kidney dysfunction and lung impairment in the occupational setting, and, in Japan, on the detection and screening for bone disease in general populations exposed to cadmium-contaminated rice. More recently, the possible role of cadmium in human carcinogenesis has also been studied in some detail.

General Population. The actual levels of cadmium intake resulting from food ingestion varies as a function of multiple factors. For example, certain crops (eg, sunflowers) and shellfish contain naturally elevated amounts of cadmium. Individuals who consume large amounts of these materials might thus at first seem to be at increased risk. However, recent studies have demonstrated that foods which are naturally enriched in cadmium are also enriched in substances that inhibit the uptake of cadmium into the body. Thus, individuals who ingest large amounts of sunflower seeds may ingest up to 100 µg cadmium per day, yet these individuals do not have levels of cadmium in blood or urine higher than those of individuals with far lower levels of cadmium intake (47). Similarly, consumption of a diet rich in shellfish can double the intake of dietary cadmium without producing significant impacts on blood cadmium (48). These studies illustrate that the cadmium content of food is just one of a number of factors that determine the actual uptake of cadmium into the body. Indeed, more recent studies (48) have suggested that overall nutritional status is a more important determinant of cadmium uptake into the body than is the actual amount of cadmium ingested. For example, women subsisting on a vegetarian diet and with reduced iron stores have increased uptake of ingested cadmium. For these women, iron deficiency is a more important determinant of cadmium uptake than is the actual amount of cadmium ingested.

The present levels of cadmium intake in most European countries are far below the PTWI recommended by WHO. Indeed, as a result of numerous public health policies implemented during the late twentieth century, the cadmium body burden of the general population appears to be rapidly declining (49). Present exposure levels in many European countries are now comparable to, or lower than, those that characterize "unacculturated populations" residing in the jungles of South America (50).

Present levels of general population exposure to cadmium have no known adverse health consequences. Existing standards such as the PTWI are based on biological models that associate cadmium exposure and increased urinary excretion of low molecular weight proteins. This has been estimated to occur in humans with a lifelong daily intake of approximately 200 µg. Only in highly contaminated agricultural areas, and only if the cadmium levels in the food grown there were significantly increased, can levels of exposure be sufficient so as to produce kidney dysfunction. Such a situation did occur in the 1950s and 1960s in Japan, where heavy cadmium contamination of rice fields, along with nutritional deficiencies for iron, zinc, and other minerals, led to renal impairment and bone disease (Itai Itai disease) in exposed populations. Studies utilizing very sophisticated biomarkers have detected mild alterations in kidney functions at lower levels of exposure (51). However, more recent studies in the occupational setting have suggested that such alterations have no actual clinical consequences (52).

Occupationally Exposed Populations. Occupational exposure to cadmium is mainly by inhalation but also may include additional intakes through food, tobacco, and poor personal hygiene practices. In the past, the total cadmium in air level in the workplace has varied according to the type of industry, type of workplace, and industrial hygiene programs in place. Depending on the level and duration of cadmium exposure, a wide variety of effects have been observed in occupationally exposed groups. For acute exposure by ingestion, the principal effects are gastrointestinal disturbances such as nausea, vomiting, abdominal cramps, and diarrhea. Acute poisoning by inhalation may lead to respiratory manifestations such as severe bronchial and pulmonary irritation, subacute pneumonitis, and lung emphysema, and, in the most severe situations, death from pulmonary edema may occur (42).

Chronic obstructive airway disease has been associated with long-term high level occupational exposure by inhalation (4,5). Although there have been suggestions in past studies that such exposure may cause lung or prostate cancer, more recent epidemiologic analyses of cadmium-exposed cohorts have dismissed the prostate cancer association and indicate that arsenic rather cadmium may be responsible for the observed increase in lung cancer mortality rates (53,54). In addition, most of the other data cited as evidence for the carcinogenicity of cadmium (43) relies on studies that are confounded by the presence of other carcinogenic substances such as nickel, asbestos, or tobacco smoke as well as arsenic.

For chronic cadmium exposure, effects occur mainly on the kidneys, lungs, and bones. A relationship has been established between cadmium air exposure and proteinuria (an increase in the presence of low molecular weight proteins in the urine and an indication of kidney dysfunction (4,5). Cadmium is known to accumulate in the renal cortex, and there is evidence that the level of cadmium in the renal cortex associated with increased urinary excretion is about 200–250 µg/g (wet weight). Depending on exposure level and other sources of cadmium, this level might be reached after 20 years of occupational exposure.

However, more recent work has demonstrated that these effects are reversible at low exposure levels once the cadmium exposure has been removed or reduced (52). With today's low occupational exposure standards, coupled with required biological monitoring of cadmium exposure levels (eg, cadmium-in-blood and cadmium-in-urine) and kidney funcion parameters (eg, β_2-microglobulin, a low molecular weight protein), there is every assurance that kidney dysfunction or other effects will not develop in occupationally exposed workers as they did in the past.

13.2. Sources of Human Cadmium Exposure. Although sources of cadmium emissions to the environment have been listed in some detail in this report and others (3–6,17,21,24,27,28,55), there have been very few attempts to quantitatively partition human cadmium exposure to its various sources. Van Assche (21,26) has developed a model for cadmium exposure for human beings and allocated this exposure to its various sources. Some of the assumptions and the data inputs for the model have been based in large part on actual data from Belgium and the European Community, and, in particular, on the Environmental Resources Limited report on the sources of human and environmental contamination in Europe (24) and the updated data on cadmium emissions contained in the OECD Monograph on Cadmium (5).

The analysis acknowledges that, for the general population, most human cadmium exposure comes from ingestion of food, and most of that arises from the uptake of cadmium by plants from fertilizers, sewage sludge, manure, and atmospheric deposition. Specifically, the model estimated that the relative importance of various cadmium sources to human exposure is as follows (26):

Phosphate fertilizers	41.3%
Fossil-fuel combustion	22.0%
Iron and steel production	16.7%
Natural sources	8.0%
Nonferrous metals	6.3%
Cement production	2.5%
Cadmium products	2.5%
Incineration	1.0%

Clearly, of the anthropogenic sources of cadmium, phosphate fertilizers, fossil-fuel combustion, and some industrial activities contribute far more to human cadmium exposure than do production, use, and disposal of cadmium products and incineration of all cadmium-containing materials. However, as shown earlier, this observation should be put in the perspective of average daily cadmium intakes well below those necessary to ensure human health. Thus no action seems required to reduce cadmium from its existing levels.

14. Uses

Cadmium metal and cadmium compounds are utilized in five principal product areas that include NiCd batteries, pigments, stabilizers, coatings, and alloys and other miscellaneous products. Definitive consumption figures for these markets do not exist, but the International Cadmium Association has for some years

Table 16. **Western World Cadmium Consumption Patterns**

Market Segment	Percent of Total Cadmium Consumption				
	1993	1994	1995	1996	1997
Batteries	63	65	67	69	70
Pigments	15	15	14	13	13
Stabilizers	11	10	9	8	7
Coatings	8	8	8	8	8
Alloys etc	3	2	2	2	2

made estimates of cadmium consumption patterns by various end-use categories, which are summarized in Table 16 (10).

Approximately 80% of the cadmium consumed in the NiCd battery market is for the small consumer cells utilized in power tools, cellular telephones, camcorders, portable household appliances, laptop computers, and other cordless devices, while the remaining 20% is utilized in the large industrial NiCd batteries for railroad, aerospace, and other industrial applications. The total NiCd battery market, particularly its consumer cell portion, is continuing to grow at the expense of some of the traditional cadmium markets such as pigments, stabilizers, and alloys. Cadmium coatings, which accounted for more than 50% of cadmium consumption in the late 1960s, have decreased substantially but now appear to have stabilized at about 8% of total cadmium consumption. At present the total NiCd battery market, which is growing at a rate of about 4% per year, is expected to account for more than 75% of all cadmium consumption.

14.1. Batteries. Nickel–cadmium (NiCd) batteries are characterized by higher energy and power densities and longer cycle lives than are rechargeable lead acid batteries, but are generally more expensive in initial cost and less expensive in lifecycle cost. NiCd batteries also exhibit excellent high and low temperature performance, marked resistance to mechanical and electrical abuse, and rapid charge and discharge characteristics. Because of their high discharge capability, they are the only battery chemistry presently suitable for portable power tools and other high drain rate applications. They are also utilized extensively in cellular and cordless telephones and pagers, emergency lighting and security, household appliances, and hobby and toy uses. Their earlier usage in portable computers and camcorders is being replaced by higher energy desity batteries such as the lithium-ion battery. A breakdown of the Japanese consumer NiCd battery market, which accounts for approximately 70% of the world's consumer NiCd batteries, is summarized in Table 17 for 1994–1997.

Applications for industrial NiCd batteries are fairly evenly divided between aviation starting and emergency power applications and railroad/subway emergency power, switching, and signaling uses. Some production of industrial NiCd batteries for electric vehicles has begun, and future opportunities exist as starters for small engines and in telecommunications uses. (see BATTERIES, LEAD-ACID SECONDARY CELLS)

14.2. Pigments. Cadmium sulfide–based pigments are bright yellow, orange, and red colors used in plastics, ceramics, glasses, enamels, and artists'

Table 17. **Japanese NiCd Consumer Battery Market, 1994–1997**

	Market Share by Year, (Percent of Total)			
Application	1994	1995	1996	1997
Home appliances	27	31	28	25
Power tools	15	17	22	24
Communications	34	25	19	18
Emergency power	7	8	11	13
Miscellaneous	5	7	7	9
Office equipment	10	9	7	6
Retail sales	2	3	6	5

colors. A wide spectrum of colors from golden yellow to maroon can be produced by substituting zinc or mercury for cadmium and selenium for sulfur in the CdS crystal lattice. These pigments are characterized by extremely low solubility and high stability, and resistance to high temperature and high pressure, which makes them excellent pigments in applications where high temperature or high stress processing are employed. The approximate breakdown of cadmium pigments usage by applications is summarized in Table 18 (56–58).

Cadmium pigments are widely used to color polymers such as acrylonitrile butadiene styrene (ABS) and other high impact polystyrenes, polyamides (nylon), polycarbonates, high density polyethylene (HDPE), high density polypropylene (HDPP), and silicone resins. It has been very difficult to substitute for cadmium pigments in these polymers, although some substitution has occurred in other polymers processed at lower temperatures.

14.3. Coatings.   Cadmium coatings are generally applied by electroplating, mechanical plating, or vacuum coating onto iron and steel, aluminum, and titanium surfaces to obtain a superior combination of high corrosion resistance, particularly in saltwater or alkaline media, plus either low coefficient of friction, low electrical resistivity, or good brazing and soldering characteristics (see CORROSION AND CORROSION INHIBITORS). No other galvanically protective coating on steel, is able to match cadmium's combination of these properties. Estimates of U.S. cadmium coatings applications were made by the U.S. Department of Commerce in 1989 and by Morrow in 1992 and are summarized in Table 19 (59). Cadmium coatings are usually deposited from alkaline cyanide, acid sulfate, neutral chloride, or acid fluoborate baths; the cyanide process is the most widely utilized because of its ability to produce bright, uniform deposits. At one time, cadmium coatings accounted for almost half of total cadmium consumption, but this amount has decreased substantially in the face of product restrictions in Europe. The usage of cadmium coatings in the United States remains much

Table 18. **Applications for Cadmium Pigments**

Application	Percent of total
plastics	80
ceramics	8
paints	7
others	5

Table 19. **U.S. Cadmium Coatings Markets (1989 vs 1992)**

	Percent of total	
Category	1989	1992
automotive	30.0	15.0
communications	22.5	31.0
fasteners	17.5	20.0
aircraft	12.5	15.0
ammunition	6.0	10.0
shipbuilding	5.0	5.0
railroads	2.5	2.5
hardware	2.5	1.0
appliances	1.5	0.5

higher than in Europe, and environmental and human health concerns are managed by regulation and pollution control technology as well as increased recycling of cadmium products.

14.4. Stabilizers. Cadmium-bearing stabilizers are utilized to retard the degradation processes in PVC on exposure to heat and ultraviolet light (sunlight). These stabilizers normally consist of mixtures of lead, cadmium, and barium organic salts, which may include cadmium laurate or cadmium stearate, and that are incorporated into the PVC during processing to arrest any degradation processes. They ensure that the PVC develops good initial color and clarity, allows the use of high temperature processing, and promotes a longer service life. The most common type of cadmium-bearing stabilizers are the barium/cadmium stabilizers, which may be either liquid stabilizers containing octoates, phenolates, neodecanoates, naphthenates, and benzoates or the solid barium/cadmium stabilizers, which contain laurates and stearates.

Barium/cadmium-stabilized PVC may be prepared by three different processing techniques: extrusion, calendaring or dispersion coated resins. Typical applications of cadmium-stabilized extruded PVC include pipe for water and for sewage, electrical conduits, shaped profiles, hoses, electrical insulation, credit card stock, automotive exterior trim, belts, fencing, and PVC film and sheet. Applications for cadmium-stabilized calendered PVC include decals, waterbeds, swimming pool liners, adhesive and magnetic tape, laminating materials, ceiling tile, shower curtains, tarps, Band-Aids, and roofing membranes. Barium/cadmium-stabilized products are also formulated into dispersion coated resins used for vinyl flooring, raincoats, coated cloth, and automotive undercoating. There has also been a tendency to replace barium/cadmium stabilizers where feasible with barium/zinc, calcium/zinc or organotin type stabilizers.

14.5. Alloys. Cadmium-containing alloys are generally considered in two groups: those where small amounts of cadmium are utilized to improve some feature of the base alloy and those where cadmium is part of a complex low melting point alloy where the presence of cadmium helps to lower the melting point (60). Examples in the former category include

- Copper alloys where additions of 0.8–1.2% Cd double the mechanical strength and wear resistance of copper but result in only a 10% loss in conductivity. Major uses include telephone wires, contact wire, catenary strand for railway overhead electrification, and automotive radiators and fittings.
- Zinc alloys containing small amounts of cadmium exhibit improved mechanical properties such as strength, drawability, and extrudability, and may be used in applications such as the battery can material in carbon–zinc and alkaline–manganese primary batteries. Zinc alloys containing 0.025–0.15% Cd are used as sacrificial anodes for the corrosion protection of structural steel immersed in seawater.
- Lead alloys with ≤0.075% Cd are sometimes used as sheaths for cables subject to cyclic stress.
- Tin-bearing alloys with ≤1% Cd have improved tensile and fatigue strength for use in marine engines and gearboxes.
- Precious-metal alloys incorporate cadmium to improve strength and hardness. Levels of ≤5% Cd in gold–silver–copper alloys make Greek gold, a greenish-tinged gold.
- Silver electrical contact alloys utilize 10–15% cadmium or cadmium oxide for heavy-duty electrical applications such as relays, switches, and thermostats. Cadmium suppresses electrical arcing and improves resistance to electric erosion and material transfer.

Examples and applications of the low melting joining and fusible alloys include

- Lead–tin–bismuth–cadmium alloy ("Woods Metal") is used in the bonding of metallized ceramic and glass components to metal frames and chassis where higher soldering temperatures are not possible. Other cadmium-containing solders are used in light electrical and electronic assemblies and when it is necessary to make a lower temperature second joint adjacent to a conventional tin–lead solder joint.
- Intermediate temperature soldering alloys include alloys of silver, zinc and/ or tin alloyed with cadmium. These alloys have higher tensile strengths than do most common solders in this temperature range, and are used in applications where temperature sensitivity prohibits the use of silver solders with higher tensile strength. Zinc–cadmium alloys are used to solder aluminum, while cadmium–zinc–tin alloys are employed for soldering magnesium.
- Silver brazing alloys are quaternary alloys of silver, copper, zinc, and cadmium, which are distinguished by their ability to produce high strength joints at working temperatures more than 100°C below the melting points of the ternary silver–copper–zinc brazing alloys.
- Fusible alloys include a wide variety of very low melting and rapidly solidifying materials that can be utilized as heat-sensitive fusible links in fire safety devices or as control devices in kilns and ovens. Woods metal is utilized in water sprinkler valves to automatically activate water flow when

the temperature exceeds 70°C. These alloys are also employed to mount firmly complex or delicate parts during machining or grinding operations.

14.6. Minor Applications. Cadmium forms a number of compounds that exhibit semiconducting or electronic behavior that makes them useful for a wide variety of applications. The amounts utilized are usually quite small, but these minor uses often have major technological and social importance. For example, cadmium–telluride (CdTe) and cadmium–sulfide [1306-23-6] (CdS) are both used in photovoltaic or solar cells that convert sunlight directly into energy. Cadmium sulfide photoconductive cells are used in photographic exposure meters. Cadmium sulfide is also used as a highly sensitive photoreceptor in electrophotographic systems in photocopiers. Other cadmium compounds, including those containing sulfide, tungstate, borate, and silicate, are essential in the preparation of light-emitting phosphors that are activated by electron beams. These phosphors are utilized in color displays, X-ray instruments, luminescent dials, fluorescent lamps, and cathode-ray tubes.

Cadmium mercury telluride is utilized in infrared imaging systems, and cadmium selenides are used in thin-film transistors for switching applications.

Cadmium salts of organic acids are also used as polymerization catalysts in the production of a wide variety of organic compounds, as ethylating agents and as antiwear additives. Diethyl cadmium is a polymerization catalyst for production of vinyl chloride, vinyl acetate, and methylmethacrylate. Admixed with titanium tetrachloride, it catalyzes the polymerization of the respective monomers to form polyethylene and a highly crystalline polypropylene suitable for filaments, textiles, glues, and coatings.

Silver–indium–cadmium alloys are utilized as control rods in some pressurized-water nuclear reactors, and cadmium sheet is used for nuclear shielding applications because of its high neutron absorption characteristics (61).

14.7. Future Applications. Applications that show promise for the future include NiCd battery–powered electric vehicles, hybrid electric vehicles and electric buses. NiCd batteries offer an excellent compromise between low price–low performance lead acid batteries and high price–high performance NiMH or lithium-ion technologies. Renault and PSA Peugeot in France have already selected NiCd batteries for their initial electric vehicle production. Another area of promise for future cadmium applications is CdTe solar cells and NiCd batteries for energy storage of the output from those solar cells. This area has become one of increasing interest, especially in Third World nations in tropical zones, where solar power is a viable option and extensive power infrastructures are not available.

Whatever new applications are developed for cadmium for the future, they will have to be recyclable. Today, batteries, coatings and alloys are all recyclable, and major industry efforts have been and are being undertaken to ensure that these products are recycled. Likewise, the CdTe solar cell industry and the electric vehicle NiCd battery industry are making sure that recycling is part of the lifecycle of their respective products. In addition, the plastics industry has undertaken research to determine the feasibility of recycling cadmium pigments and stabilizers from plastics.

BIBLIOGRAPHY

"Cadmium and Cadmium Alloys" in *ECT* 1st ed., Vol. 2, pp. 716-732, by S. J. Dickinson, American Smelting and Refining Co.; in *ECT* 2nd ed., Vol. 3, pp. 884-899, by H. E. Howe, American Smelting and Refining Co.; in *ECT* 3rd ed., Vol. 4, pp. 387-396, by M. L. Hollander and S. C. Carapella, Jr., ASARCO Inc.; "Cadmium and Cadmium Alloys," in *ECT* 4th ed., Vol. 4, pp. 748–760, by D. S. Carr, International lead zinc Research Organization; "Cadmium and Cadmium Alloys" in *ECT* (online), posting date: December 4, 2000, by D. S. Carr, International Lead Zinc Research Organization, Inc.

CITED PUBLICATIONS

1. M. Farnsworth, *Cadmium Chemicals*, Internal Lead Zinc Research Organization, Inc., New York, 1980.
2. T. O. Llewellyn, *Cadmium (Materials Flow)*, Bureau of Mines Information Circular IC 9380, U.S. Dept. Interior, Washington, D.C., 1994.
3. M. E. Cook, and H. Morrow, "Anthropogenic Sources of Cadmium in Canada," *National Workshop on Cadmium Transport into Plants*, Canadian Network of Toxicology Centres, Ottawa, Ontario, Canada, June 20–21, 1995.
4. World Health Organization (WHO), *Environmental Health Criteria 134—Cadmium*, International Programme on Chemical Safety (IPCS) Monograph, 1992.
5. Organisation for Economic Co-operation and Development (OECD), *Risk Reduction Monograph No. 5: Cadmium* OECD Environment Directorate, Paris, 1994.
6. A. Jensen, and F. Bro-Rasmussen, *Rev. Environ. Contam. Toxicol.* **125**. 101–181 (1992).
7. *CRC Handbook of Chemistry and Physics*, 77th ed. CRC Press, Boca Raton, Fla., 1996.
8. C.-G. Elinder, "Cadmium: Uses, Occurrence, and Intake," in *Cadmium and Health: A Toxicological and Epidemiological Appraisal*, CRC Press, Boca Raton, Fla., 1985.
9. H. Morrow, "Cadmium," in *Metals & Minerals Annual Review—1997*, The Mining Journal Ltd., London, 1997.
10. H. Morrow, "Cadmium," in *Metals & Minerals Annual Review—1998*, The Mining Journal Ltd., London, 1998; H. Morrow, "Cadmium," Ryan's Notes Minor Metals Conf. Sept. 17, 1998, London, 1998.
11. T. K. Christman, and co-workers, *Zinc: Its Corrosion Resistance*, 2nd ed., International Lead Zinc Research Organization, Inc., New York, 1983.
12. D. S. Carr, "Cadmium and Cadmium Alloys" in *Encyclopedia of Chemical*, 4th ed., Vol. **4**, Wiley, New York, 1992, pp. 748–760.
13. Q. Fernando, and H. Freizer, in *Treatise on Analytical Chemistry*, I. M. Kolthoff and P. J. Elving, eds., Interscience, New York, 1963, Vol. **3**, pt. II, pp. 171–229.
14. K. Matsumoto, and K. Fuwa, in E. C. Foulkes, ed., *Handbook of Experimental Pharmacology*, Vol. **80**, Springer-Verlag, Berlin and Heidelberg, 1986.
15. H. Morrow, *Draft OECD Guidance Document for the Development of Successful Systems for the Collection and Recycling of Nickel-Cadmium Batteries*, Risk Management Programme, Environment Directorate, Joint Meeting of the Chemicals Committee and the Working Party on Chemicals, June 8–11, 1999, Paris, 1999.
16. J. O. Nriagu, in *Cadmium in the Environment*, Part I, *Ecological Cycling*, Wiley, New York, 1980, pp. 71–114.
17. J. O. Nriagu, *Nature* **338**. 47–49 (1989).

18. F. Elgersma, B. S. Anderberg, and W. M. Stigliani, "Emission Factors for Aqueous Industrial Cadmium Emissions in the Rhine River Basin; A Historical Reconstruction for the Period 1970–1988," in *Proc. 7th Int. Cadmium Conf.*, New Orleans, Cadmium Assoc. (London), Cadmium Council (Reston, Va.), International Lead Zinc Research Organization (Research Triangle Park N.C.), 1992.

19. J. Mukunoki, and K. Fujimoto, "Collection and Recycling of Used Ni-Cd Batteries in Japan," *Sources of Cadmium in the Environment*, Inter-Organization Programme for the Sound Management of Chemicals (IOMC), Organisation for Economic Co-Operation and Development (OECD), Paris, 1996, 1996.

20. United States Environmental Protection Agency (USEPA), *1994 Toxics Release Inventory—Public Data Release*, USEPA Office of Pollution Prevention and Toxics, Washington, D.C., 1996.

21. F. J. Van Assche, and P. Ciarletta, "Cadmium in the Environment: Levels, Trends and Critical Pathways, *Proc. 7th Int. Cadmium Conf.*, New Orleans, Cadmium Assoc. (London), Cadmium Council (Reston VA.), and International Lead Zinc Research Organization. (Research Triangle Park, N.C.), 1992.

22. H. Morrow, and J. Keating, "Overview Paper for OECD Workshop on the Effective Collection and Recycling of Nickel-Cadmium Batteries," OECD Workshop on the Effective Collection and Recycling of Nickel-Cadmium Batteries, Lyon, France, Sept. 23–25, 1997.

23. A. J. Chandler, "Characterizing Cadmium in Municipal Solid Waste," *Sources of Cadmium in the Environment*, Inter-Organization Programme for the Sound Management of Chemicals (IOMC), Organisation for Economic Co-Operation and Development (OECD), Paris.

24. Environmental Resources Limited (ERL), *Evaluation of the Sources of Human and Environmental Contamination by Cadmium*, report prepared for the Commission of the European Community, Directorate General for Environment, Consumer Protection and Nuclear Safety, London, Feb. 1990.

25. J. Owens, *The Environmental Impacts of Disposed Nickel-Cadmium Rechargeable Batteries in the Nation*'s Municipal Solid Waste Stream, U.S. Bureau of Mines, July 1994.

26. F. J. Van Assche, "A Stepwise Model to Quantify the Relative. Contribution of Different Environmental Sources to Human Cadmium Exposure," *Proc. 8th Int. Nickel-Cadmium Battery Conf.*, Prague, Czech Republic, Sept. 21–22, 1998.

27. T. Jackson, and A. MacGillivray, *Accounting for Cadmium*, Stockholm Environment Institute, London, 1993.

28. R. Jones, T. Lapp, and D. Wallace, *Locating and Estimating Air Emissions from Sources of Cadmium and Cadmium Compounds*, report prepared by Midwest Research Institute for the U.S. Environmental Protection Agency, Office of Air and Radiation, Report EPA-453/R-93-040, Sept. 1993.

29. U. Eggenberger, and H. N. Waber, "Cadmium in Seepage Waters of Landfills: A Statistical and Geochemical Evaluation," report of Nov. 20, 1997 for the OECD Advisory Group on Risk Management Meeting, Feb. 9–10, 1998, Paris, 1998.

30. NUS Corp. *Characterization of Municipal Waste Combustor Ashes and Leachates from Municipal Solid Waste Landfills, Monofills, and Codisposal Sites*, report prepared for the U.S. Environmental Protection Agency, Office of Solid Waste, R-33-6-7-1, Washington, D.C., 1987.

31. International Labour Organisation, *Occupational Exposure Limits Jor Airborne Toxic Substances*, 3rd ed., Occupational Safety and Health Series No. 37, 1991.

32. American Conference of Governmental Industrial Hygienists (ACGIH), *1996 TLVs and BEIs, Threshold Limit Values and Biological Exposure Indices for Chemical Substances and Physical Agents*, 1996.

33. Occupational Safety and Health Administration (OSHA), *Occupational Exposure to Cadmium; Final Rules*, U.S. Dept. Labor, 29 CFR Part 1910.1027, Sept. 14, 1992.

34. G. Van Urk, and J. M. Marquenie, in *Heavy Metals in the Environment*, Geneva, Vol. **II**, 1989, pp. 456–459.

35. J. Bak, and co-workers, *Sci. Total Environ.* **207**. 179–186 (1997).

36. M. J. Mench, *Agricult. Ecosyst. Environ.* **67**. 175–187 (1998).

37. Agency for Toxic Substances and Disease Registry (ATSDR), *Draft Toxicological Profile for Cadmium*, Public Health Service, U.S. Department of Health & Human Services, Atlanta, GA., 1997.

38. F. J. Van Assche, and P. Ciarletta, in *Heavy Metals in the Environment*, Toronto, Sept. 1993, Vol. **I**, pp. 34–37.

39. T. Watanable, and co-workers, *Int. Arch. Occup. Environ. Health* **65**. S205–S208 (1993).

40. T. Watanabe, and co-workers, *Bull. Environ. Contam. Toxicol.* **52**. 196–202 (1994).

41. A. M. Chaudri, and co-workers, *J. Environ. Qual.* **24**. 850–855 (1995).

42. R. R. Lauwerys, *Health Maintenance of Workers Exposed to Cadmium*, The Cadmium Council, Inc., New York, 1980.

43. G. Ducoffre, F. Claeys, and F. Sartor, *Arch. Environ. Health.* **91**. 157 (1992).

44. Minister fur Umwelt, Raumordnung und Landwirtschaft (MURL), *Luftreinhaltun in Nord Rhein-Westfalen: Eine Erfolgungsbilanz der Luftreinhalteplanenung* (1975–1988), Dusseldorf, 1989, p. 215.

45. J. Staessen, and co-workers, *Transfer of Cadmium from a Sandy Acidic Soil to Man: A Population Study*, Hypertension and Cardiovascular Rehabilitation Unit, Dept. Pathophysiology, Univ. Leuven, Belgium, 1991.

46. G. L. Leone, B. W. Lewis, and D. Morikawa, *Using Cadmium Safely: A Guide for the Workplace*, The Cadmium Council, Inc., New York, 1986.

47. P. G. Reeves, and R. A. Vanderpool, *Environ. Health Perspect.* **105**. 1098–1104 (1997).

48. M. Vahter, co-workers, *Toxicol. Appl. Pharmacol.* **136**. 332–341 (1996).

49. L. Friis, L. Petersson, and C. Edling, *Environ. Health Perspect.* **106**. 175–178 (1998).

50. L. H. Hecker, co-workers. *Arch. Environ. Health* **29**. 181–185 (1974).

51. J.-P. Buchet, and co-workers, *Lancet* **336**. 699–702 (1990).

52. H. A. Roels, co-workers, *Am. J. Ind. Med.* **31**. 645–652 (1997).

53. T. Sorahan and co-workers, *Occup. Environ. Med.* **52**(12), 804–812 (Dec. 1995).

54. T. Sorahan, and R. J. Lancashire, *Occup. Environ. Med.* **54**(3), 194–201 (March 1997).

55. J. O. Nriagu, and J. M. Pacyna, *Nature* **333**. 134–139 (1988).

56. R. F. Lynch, *Plastics Eng.* (April 1985).

57. J. Dickenson, "Cadmium Pigments in the 1990s," *Proc. 7th Int. Cadmium Conf.*, "Cadmium 92," April 6–8, 1992, New Orleans, Cadmium Assoc. (London), Cadmium Council (Reston, V.A.), and International Lead Zinc Research Organization, Inc. (Research Triangle Park, N.C.), 1992.

58. M. E. Cook, "Cadmium Pigments: When Should I Use Them?" *Inorganic Pigments: Environmental Issues and Technological Opportunities*, Industrial Inorganic Chemicals Group, Royal Society of Chemistry, London, Jan. 12, 1994.

59. H. Morrow, "The Future of Cadmium," *Proc. 1st Annual Cadmium Alternatives Conf.*, National Defense Center for Environmental Excellence, Johnstown, Pa., July 7, 1995, 1995.

60. Cadmium Association, London and Cadmium Council, New York, *Technical Notes on Cadmium: Cadmium Coatings, Technical Notes on Cadmium: Cadmium in Stabilizers for Plastics, Technical Notes on Cadmium: Cadmium Pigments, Technical Notes on Cadmium: Cadmium in Alloys, Technical Notes on Cadmium: Cadmium in Batteries*, 1978.

61. Cadmium Association, London, and Cadmium Council, Greenwich, Conn. *Technical Notes on Cadmium: Cadmium Production, Properties and Uses*, 1991.

GENERAL REFERENCES

H. Morrow, "Cadmium (CD)," in *Metals Handbook*, 10th ed. Vol. **2**, ASM International, Metals Park, Ohio, 1990.

Organisation for Economic Co-operation and Development (OECD), report from Session F, "Sources of Cadmium in Waste," *Chairman's* Report of the Cadmium Workshop, ENV/ MC/CHEM/RD(96)1, Stockholm, Sweden, Oct. 1995.

D. A. Temple, and D. N. Wilson, "Cadmium—Markets and Trends," *Proc. 4th Int. Cadmium Conf.* Munich, Cadmium Assoc. (London), Cadmium Council (New York), and International Lead Zinc Research Organization (New York), 1983.

HUGH MORROW
International Cadmium Association

CADMIUM COMPOUNDS

1. Introduction

Naturally occurring cadmium compounds are limited to the rare minerals, greenockite [1317-58-4], CdS, and otavite (1), an oxycarbonate, but neither is an economically important source of cadmium metal or its compounds. Instead, cadmium compounds are more usually derived from metallic cadmium [7440-43-9] which is produced as a by-product of lead–zinc smelting or electrolysis (see CADMIUM AND CADMIUM ALLOYS). Typically, this cadmium metal is burnt as a vapor, to produce the brown-black cadmium oxide [1306-19-0], CdO, which then acts as a convenient starting material for most of the economically important compounds.

2. Properties

Cadmium is a member of Group 12 (Zn, Cd, Hg) of the Periodic Table, having a filled d shell of electrons $4d^{10}5s^2$ which dictates the usual valence state of +2. In rare instances the +1 oxidation state may be produced in the form of dimeric Cd_2^{2+} species [59458-73-0], eg, as dark red melts of Cd^0 dissolved in molten cadmium halides or as diamagnetic yellow solids such as $(Cd_2)^{2+}$ $(AlCl^-_4)_2$ [79110-87-5] (2). The Cd^{2+}_2 species is unstable in water or other donor solvents, immediately disproportionating to Cd^{2+} and Cd. In general, cadmium compounds exhibit properties similar to the corresponding zinc compounds. Compounds and properties are listed in Table 1. Cadmium(II) [22537-48-0] tends to favor tetrahedral coordination in its compounds, particularly in solution as complexes, eg, tetraamminecadmium(II) [18373-05-2], $Cd(NH_3)^{2+}_4$. However, solid-state cadmium-containing oxide or halide materials frequently exhibit octahedral coordination at the Cd^{2+} ion, eg, the rock-salt structure found for CdO.

Table 1. Physical and Chemical Properties of Selected Cadmium Compounds[a]

Compound	CAS Registry Number	ΔH°_{f298},[b] kJ/mol	ΔG°_{f298},[b] kJ/mol	S°_{298},[b] J/mol·K	Density, g/mL	C°_p, J/mol·K	Mp, °C	ΔH_{fus}, kJ/mol	Aqueous solubility, g/100 g H_2O[c]	Crystal[d] structure	Unit cell dimensions, nm
cadmium anti-monide CdSb	[12050-27-0]	−14.4	−13.0	92.9	6.92		452	32.05		ortho-rhomb	$a = 0.6471$ $b = 0.8253$ $c = 8.526$
cadmium bromide CdBr$_2$	[7789-42-6]	−316	−296	137.2	5.192	76.7	568	20.92	95_{18}	hex	$a = 0.395$ $c = 1.867$
cadmium carbonate CdCO$_3$	[513-78-0]	−751	−669	92.5	4.26		332 dec		2.8×10^{-6}	rhomb	$a = 0.61306$
cadmium chloride CdCl$_2$	[10108-64-2]	−391	−344	115.3	4.05	74.7	568	22.176	128.6_{30}	hex	$a = 0.3854$ $c = 1.746$
cadmium fluoride CdF$_2$	[7790-79-6]	−700	−648	77.4	6.39		1048	22.594	3.45_{25}	cubic	$a = 0.53880$
cadmium hydroxide Cd(OH)$_2$	[21041-95-2]	−561	−474	96.2	4.79		150 dec		2.6×10^{-4}	hex	$a = 0.3475$ $c = 0.467$
cadmium iodide CdI$_2$, α-form	[7790-80-9]	−203	−201	161.1	5.67	80.0	387	33.472	86_{25}	hex (hex)[e]	$a = 0.424$ $c = 0.684$ $(c = 1.367)$[e]
cadmium nitrate Cd(NO$_3$)$_2$	[10325-94-7]	−456	−255	197.9			350		109_0	cubic	$a = 0.756$
cadmium nitrate tetrahydrate Cd(NO$_3$)$_2$·4H$_2$O	[10022-68-1]	−1649			2.455		59.4	32.636	215_0	ortho-rhomb	$a = 0.583$ $b = 2.575$ $c = 1.099$
cadmium oxide CdO	[1306-19-0]	−258	−228	54.8	8.2	43.4	1540 sub	243.509 sub	9.6×10^{-4}		

cadmium selenide CdSe, α-form; cubic; [1306-24-7]	−136	−100	96.2		$a = 0.46953$; 5.81	1350 dissoc	305.307 dissoc	hex; (cubic)e	$a = 0.4309$; $c = 0.7021$; $(a = 0.605)$e
cadmium m-silicate CdSiO₃; [13477-19-5]	−1189	−1105	97.5	88.6	4.928	1242		monocli-nic	$a = 1.504$
cadmium sulfate CdSO₄; [10124-36-4]	−933	−823	123.0	99.6	4.691	1000	20.084 76.6_{20}	ortho-rhomb	$b = 0.710$; $c = 0.696$; $a = 0.4717$
cadmium sulfate hydrate CdSO₄·H₂O; [13477-20-8]	−1240	−1069	154.0	134.6	3.79	105 trans		monocli-nic	$b = 0.6559$; $c = 0.4701$; $a = 7.607$
cadmium sulfate hydrate 3CdSO₄·8H₂O; [7790-89-3]	−1729	−1465	229.6	213.3	3.09	80 trans	113_0	monocli-nic	$b = 0.7541$; $c = 8.186$; $a = 0.947$; $b = 1.184$; $c = 1.635$
cadmium sulfide CdS, α-form; [1306-23-6]	−162	−156	64.9		4.82 (4.50)e	980 sub in N₂	201.669 subl	hex; (cubic)e	$a = 0.41348$; $c = 0.6749$; $1.3 \times 10^{-4}_{18}$
cadmium telluride CdTe; [1306-25-8]	−92	−92	100.4		6.20	1045		(cubic)e; hex; (cubic)e	$(a = 0.5818)$e; $a = 0.457$; $c = 0.747$; $(a = 0.6480)$e

[a] Refs. 3–12.
[b] To convert J to cal, divide by 4.184.
[c] Subscript denotes temperature in °C.
[d] Ortho–rhomb is orthorhombic and hex is hexagonal.
[e] β-form.

3. Inorganic Compounds

3.1. Cadmium Arsenides, Antimonides, and Phosphides.

Cadmium arsenide [12511-93-2], CdAs, cadmium diarsenide [12044-40-5], $CdAs_2$, and tricadmium diarsenide [12006-15-4], Cd_3As_2, are known. Cd_3As_2 is prepared as grey tetragonal crystals, $a = 0.8945$ nm, $c = 1.265$ nm; $d = 6.25$ g/mL; mp 721°C (4, 7); $\Delta H_{f,298}^{\circ} = -41.84$ kJ/mol ($-$ kcal/mol) (3) by heating stoichiometric amounts of the elements to fusion in an argon atmosphere. It is an n-type semiconductor having high electron mobility (10, 000 cm^2/V·s), electron concentration of 3×10^{18}/cm^3 and a band gap energy of 0.13 eV. It may also be prepared by wet chemical methods involving passage of arsine gas through a weakly ammoniacal solution of cadmium sulfate. The fine black precipitate liberates arsine when treated with acid and inflames upon treatment with oxidants. Thin films (qv) of Cd_3As_2 find application as ultrasonic multipliers, photodetectors, thermodetectors, and Hall generators (13). $CdAs_2$ is prepared by slow heating, to 700°C, of stoichiometric amounts of the elements in a sealed, evacuated, quartz ampul. The grey tetragonal crystals, $a = 0.465$ nm; $c = 0.793$ nm; $d = 5.80$ g/mL; mp 621°C; $\Delta H_{f,298}^{\circ} = -17.5$ kJ/mol (-4.18 kcal/mol) (3), decompose upon heating to give arsenic and Cd_3As_2. The additional, less well characterized phase, CdAs, has been prepared by decomposition of $CdAs_2$ at high pressure to give orthorhombic grey crystals, $a = 0.5993$ nm, $b = 0.7819$ nm, $c = 0.8011$ nm; $d = 6.63$ g/mL; which decomposes instead of melting (14).

Antimonides of formulas CdSb and Cd_3Sb_2 have been reported. Both are usually prepared by direct union of the elements, the former is a hole-type semiconductor (9), with properties shown in Table 1, and finds use as a thermoelectric generator. Reagent-grade material costs $2.00/g in small lots. The band gap energy is 0.46 eV (2.70 μm) (15); ΔH_{vap} is 138 kJ/mol (33.0 kcal/mol). Dicadmium triantimonide [12014-29-8], Cd_2Sb_3, is a metastable, white crystalline compound of monoclinic symmetry: $a = 0.72$ nm, $b = 1.351$ nm, $c = 0.616$ nm, $\beta = 100°14'$; $d = 7.014$ g/mL; mp 423°C (7).

The phosphides tricadmium diphosphide [12014-28-7], Cd_3P_2, cadmium diphosphide [12133-44-7], CdP_2, and cadmium tetraphosphide [12050-26-9], CdP_4, may all be prepared by indirect fusion of the elements, usually by passing phosphorus vapors, in a nitrogen or hydrogen carrier gas, over heated cadmium. Cd_3P_2 forms grey metallic needles of tetragonal symmetry, $a = 0.894$ nm, $c = 1.228$ nm; $d = 5.95$ g/mL; mp 700°C (4,7). It may also be prepared by passage of phosphine gas through a solution of cadmium ion and, if a surfactant such as polyphosphate is present, may be maintained in solution as a size-quantized semiconductor colloid the color of which ranges from pale yellow to black, dependent on particle size (16). It is an n-type semiconductor with near metallic mobilities (15) and was the first $A_3(II)B_2$ (15) compound to show laser action (13). CdP_2 may be prepared by heating a mixture of ammonium phosphate, cadmium carbonate, and carbon black and has a structure consisting of P chains where each P atom is bonded to two other P atoms and two Cd atoms. CdP_4 has the Cd atoms in octahedral coordination sandwiched between layers of P atoms (17).

3.2. Cadmium Borates.

Of general formula nCdO · mB$_2$O$_3$, the cadmium borates are prepared from CdO–B_2O_3 melts (7) and are used as phosphors. Materials $n = 1$, $m = 3$ [20571-45-3]; $n = 2$, $m = 3$, and $n = 3$, $m = 1$, all show green cathodoluminescence. Mn-doped material with $n = 2$, $m = 1$ luminesces strongly orange in electromagnetic radiation. Cadmium borotungstate [1306-26-9], $2CdO \cdot B_2O_3 \cdot WO_3 \cdot 18H_2O$, solutions can have densities up to 3.28 g/mL and have utility as flotation (qv) media for mineral separations. Cadmium fluoroborate [14486-19-2], $Cd(BF_4)_2$, is very hygroscopic and is used to prepare electroplating baths for high strength steels where the normal cyanide baths cause problems of hydrogen embrittlement.

3.3. Cadmium Carbonate. Pure cadmium carbonate, as the hemihydrate, $CdCO_3 \cdot \frac{1}{2}H_2O$, is obtained only when ammonium carbonate is used to precipitate the white, prismatic crystals from cadmium ion solutions; alkali carbonates precipitate the oxycarbonate. The carbonate is also produced by heating an acidified solution of cadmium chloride and urea in a sealed tube at 200°C (7). Cadmium oxide also slowly absorbs carbon dioxide to form the normal carbonate. The decomposition $CdCO_3 \longrightarrow CdO + CO_2$ gives a CO_2 partial pressure of 101 kPa (1 atm) at 357°C. The carbonate eliminates CO_2 in acids and acts as a convenient source of other Cd compounds in this type of reaction. It is soluble, because of complex ion formation, in ammonium ion- or cyanide ion-containing solutions.

3.4. Cadmium Complexes. Aqueous cadmium ion complexes are listed in Table 2. Cadmium binds four or less anionic ligands in solution generally resulting in colorless complexes. Many organic ligands form complexes with cadmium ion, the more common being methylamine, thiourea, oxalic acid, tartaric acid, dimethylglyoxime, pyridine, acetic acid, ethylenediamine tetraacetic acid, thiols, and glycolic acid. Stability constants are provided in the literature (18).

Cadmium complexes are of importance to the electroplating industry because baths containing complexed cadmium have excellent covering power and yield dense, fine grains. The complexed cation permits high cathodic overvoltages and this changed deposition potential allows codeposition of other metals.

Table 2. Thermodynamic and Stability Constant Data for Selected Aqueous Cadmium Complexes[a,b]

Complex ion	CAS Registry Number	$\Delta H^\circ_{f,298}$ kJ/mol[c]	$\Delta G^\circ_{f,298}$ kJ/mol[c]	Stability constant
$CdCl^+$	[14457-58-0]	−240.5	−224.4	$\log K_1 = 1.32$
$CdCl^-_3$	[21439-35-0]	−561.0	−487.0	$\log K_3 = 0.09$
$Cd(CN)^{2-}_4$	[16041-14-8]	428.0	507.5	$\log K_4 = 3.58$
$Cd(NH_3)^{2+}_2$	[47942-20-1]	−266.1	−159.0	$\log K_2 = 2.24$
$Cd(NH_3)^{2+}_4$	[18373-05-2]	−450.2	−226.4	$\log K_4 = 1.18$
$CdBr^+$	[15691-37-9]	−200.8	−193.9	$\log K_1 = 1.97$
$CdBr^-_3$	[21439-36-1]		−407.5	$\log K_3 = 0.24$
CdI^+	[15691-38-0]	−141.0	−141.4	$\log K_1 = 2.08$
CdI^-_3	[15691-42-6]		−259.4	$\log K_3 = 2.09$
CdI^{-2}_4	[15975-72-1]	−341.8	−315.9	$\log K_4 = 1.59$
$CdSCN^+$	[18194-99-5]		7.5	$\log K_1 = 1.90$
$Cd(SCN)^{-2}_4$	[19438-35-8]			$\log K_4 = ca\ 0.1$
$Cd(N_3)^{-2}_4$	[16408-27-8]		1,295.0	$\log K_4 = 0.76$

[a] Standard state $M = 1$.
[b] Refs. 3 and 18.
[c] To convert kJ to kcal, divide by 4.184.

In commercial operation the cyanide-plating bath, containing the complex ion $Cd(CN)^{2-}_4$ and formed from CdO (24 g/L), Cd metal (25 g/L), and NaCN (105 g/L), is the bath of choice. The exception is for high strength steels, which suffer embrittlement as a result of hydrogen incorporation, or in areas where cyanide effluent control is particularly difficult. In such cases the fluoroborate bath is the usual alternative although other electrolytes such as sulfate, sulfamate, chloride, and pyrophosphate have been used.

3.5. Cadmium Halides. Cadmium halides show a steadily increasing covalency of the metal–halide bond proceeding from fluoride through to iodide. Bond lengths increase through the series: F, 0.197 nm; Cl, 0.221 nm; Br, 0.237 nm; I, 0.255 nm. The fluoride is much less soluble in water than the others (see Table 1) and the Cl, Br, and I compounds dissolve to a significant extent in alcohols, ethers, acetone, and liquid ammonia. Boiling points and corresponding ΔH_{vap}'s are CdF_2, 1747°C, 234 kJ/mol (55.9 kcal/mol); $CdCl_2$, 960°C, 125 kJ/mol (29.9 kcal/mol); $CdBr_2$, 963°C, 113 kJ/mol (27.0 kcal/mol); and CdI_2, 787°C, 106 kJ/mol (25.3 kcal/mol).

Aqueous solutions have low conductivities resulting from extensive complex ion formation. The halides, along with the chalcogenides, are sometimes used in pyrotechnics to give blue flames and as catalysts for a number of organic reactions.

Cadmium Fluoride. Elemental fluorine reacts with cadmium metal as well as the oxide, sulfide, and chloride to give CdF_2 [7790-79-6]. Alternatively, treatment of $CdCO_3$ with 40% HF yields a solution of CdF_2, which may be evaporated to recover efflorescent crystals of the dihydrate. CdF_2 has been used in phosphors, glass manufacture, nuclear reactor controls, and electric brushes.

Cadmium Chloride. Data for anhydrous material are listed in Table 1 but cadmium chloride also exists as hydrates having 1, 2, 2.5, and 4 molecules of water per formula unit, all of which are efflorescent in dry air. The pentahydrate $2CdCl_25H_2O$ [7790-78-5] is the most normal commercial form of the chloride and exists as colorless crystals of $d = 3.33$ g/mL; $\Delta H_{f,298} = -1132$ kJ/mol (−270.6 kcal/mol) (3). It may be prepared by dissolving the metal, the carbonate, oxide, sulfide, or hydroxide in hydrochloric acid and evaporating the solution. Anhydrous material may be derived from this by heating in a stream of dry HCl. It has been used in photography, photocopying, dyeing, and calico printing (with thiosulfate), vacuum tube manufacture, cadmium pigment manufacture, galvanoplasty, lubricants, ice-nucleation agents, and in the manufacture of special mirrors.

Cadmium Bromide. The hydrated bromide is prepared by dissolution of cadmium carbonate, oxide, sulfide, or hydroxide in hydrobromic acid. The white crystalline material is cadmium bromide tetrahydrate [13464-92-1], $CdBr_2 \cdot 4H_2O$, $\Delta H_{f,298} = -1492.55$ kJ/mol (−356.73 kcal/mol) (3) which dehydrates to the monohydrate at 36°C and to the yellow, anhydrous material $CdBr_2$ at 200°C. The anhydrous material may be prepared directly from the elements at elevated temperature or from anhydrous cadmium acetate [543-90-8], $Cd(CH_3COO)_2$, mixed with glacial acetic acid and acetyl bromide. Uses include photography, process engraving, and lithography.

Cadmium Iodide. Two crystal morphologies exist for CdI_2 [7790-80-9], the white α-form (see Table 1) and the brown β-form. The latter crystallizes from fused-salt mixtures. The more common α-form has a saltlike layered crystal structure where individual layer sheets contact each other through van der Waals' interactions of the outer iodide ions (17). This structure dictates the highly lamellar and easily cleaved nature of CdI_2 crystals. CdI_2 is prepared either by direct combination of the elements in the absence of oxygen or by the dissolution of cadmium metal, the oxide, carbonate, hydroxide, or sulfide in hydroiodic acid. Precipitation of CdI_2 from a solution of the sulfate using KI also yields the hexagonal, lamellar, lustrous crystals of the α-form. The iodide is used in electro-deposition of Cd, as a nematocide, in phosphors, lubricants, photoconductors, in photography, process engraving, and lithography (qv).

3.6. Cadmium Hydroxide. $Cd(OH)_2$ [21041-95-2] is best prepared by addition of cadmium nitrate solution to a boiling solution of sodium or potassium hydroxide. The crystals adopt the layered structure of CdI_2: there is contact between hydroxide ions of adjacent layers. $Cd(OH)_2$ can be dehydrated to the oxide by gentle heating to 200°C; it absorbs CO_2 from the air forming the basic carbonate. It is soluble in dilute acids and solutions of ammonium ions, ferric chloride, alkali halides, cyanides, and thiocyanates forming complex ions.

$Cd(OH)_2$ is much more basic than $Zn(OH)_2$ and is soluble in 5 N NaOH at 1.3 g/L as the anionic complex tetrahydroxocadmate [26214-93-7], $Cd(OH)_4^{2-}$. Its most important utility is as the active anode in rechargeable Ni–Cd and Ag–Cd storage batteries. The chemical reaction responsible for the charge–discharge of the batteries is (19):

$$Cd(OH)_2 + 2\ Ni(OH)_2 \ \underset{\text{discharge}}{\overset{\text{charge}}{\rightleftharpoons}} \ Cd + 2\ NiOOH + 2\ H_2O$$

These batteries accounted for 75% of 2001 U.S. annual consumption of cadmium and are used in heavy-duty, long-life applications such as rechargeable tools, appliances, instruments, and electronics (20). Several new applications could include use in electric and hybrid electric vehicles, remote area power storage systems, and solar cells (21).

3.7. Cadmium Nitrate. Anhydrous $Cd(NO_3)_2$ [10325-94-7] is obtained by action of nitric acid on the carbonate to give cadmium nitrate tetrahydrate [10022-68-1] by crystallization; this may then be dried by careful exposure to concentrated nitric acid at 20°C. The tetrahydrate, bp 132°C, is soluble in alcohols, acetone, and ethyl acetate and most polar organic solvents.

Cadmium nitrate is the preferred starting material for $Cd(OH)_2$ for use as the anode in alkaline batteries. The sintered anode matrix of such batteries is saturated with cadmium nitrate (480–500 g/L Cd) and cadmium hydroxide is formed therein by standardized electrolysis and drying (22). Other uses include photographic emulsions and as a colorant in glass and ceramics.

3.8. Cadmium Oxide. Cadmium vapor burns in air to produce the dark brown oxide CdO [1306-19-0] and the commercial process for its production is as follows. Pure Cd metal is melted and then vaporized whereupon air is blown through the hot vapor, oxidizing the cadmium and carrying the product to a baghouse. The resultant oxide (particle size controlled by the air–Cd ratio) is calcined at 550°C to ensure uniform properties. Other preparative approaches include calcination of the carbonate, nitrate, sulfate, or hydroxide in air;

oxidation of the sulfide by heating in air and by pyrolysis of cadmium formate [4464-23-7] or cadmium oxalate [814-88-0]. These last two methods give very finely divided material of high activity. Oxide smokes of extreme toxicity may be produced by spontaneous combustion of the cadmium alkyls, eg, $Cd(CH_3)_2$, in air. CdO has the rock-salt crystal structure with octahedral coordination at the Cd and O ions. Cadmium peroxide [12139-22-9], CdO_2, has been reported and is also cubic having $a = 0.5313$ nm and $d = 6.396$ g/mL.

CdO is soluble in dilute acids but not in water or alkalies and forms a variety of soluble complexes, the most important being with sodium cyanide in the bath used in electroplating. CdO is an n-type semiconductor of band gap energy 222 kJ/mol (53.1 kcal/mol). It may be reduced to the metal with hydrogen, carbon, or carbon monoxide at 600°C. It finds uses as a starting material for PVC heat stabilizers and other inorganic Cd compounds and as the cadmium source in cyanide plating baths. Ag–CdO contacts are used in electrical devices whereas high purity CdO is used as a second depolarizer, along with Ag_2O, in Ag–Zn storage batteries (23). CdO is also used in nitrile rubbers and plastics such as Teflon where it improves high temperature properties and heat resistance. Its use as a high temperature resistor material takes advantage of its low specific resistivity and low negative temperature coefficient of resistivity. Other uses include in phosphors, as a glass colorant, as a nematocide, as an ascaricide or anthelmintic in swine, and as a catalyst of a variety of organic chemical reactions.

3.9. Cadmium Phosphates.   Cadmium phosphate [13477-17-3], $Cd_3(PO_4)_2$, is prepared by reaction of cadmium nitrate with potassium dihydrogen phosphate in the presence of sodium hydroxide to neutralize the KH_2PO_4 (7). This material is both a catalyst and a phosphor. Cadmium dihydrogen phosphate [17695-54-4], $Cd(H_2PO_4)_2$, is prepared by adding phosphoric acid to a slurry of cadmium carbonate in water and acts as a catalyst for the polymerization of gaseous olefins (24). Roasting the dihydrogen compound to 300°C leads to cadmium metaphosphate [14466-83-2], $Cd(PO_3)_2$, which finds utility as an exceptionally bright and stable phosphor.

3.10. Cadmium Selenide and Telluride.   Both materials are n-type semiconductors having band gap energies of 1.74 eV (712 nm) for CdSe [1306-24-7] and 1.45 eV (855 nm) for CdTe [1306-25-8] (25). They are best prepared by direct reaction of the elements at elevated temperatures in evacuated, sealed quartz tubes. However, the materials may also be prepared by simple wet chemical techniques whereby a solution of cadmium ion is exposed to H_2Se or H_2Te gases in an inert atmosphere or to solutions of the alkali metal selenide or telluride. The heavy precipitates can vary in color from yellow-orange to deep red-black, depending on the particle size of the compound, and this is a manifestation of the size-quantization phenomenon (16) which is particularly well demonstrated by the cadmium chalcogenides. Colloidal suspensions of the semiconductors may be maintained by addition of a surfactant (polyphosphate or micelle-forming agent) to the precipitating solution and such solutions of nanoparticulates have been under intense investigation as nonlinear optical media (16) (see NONLINEAR OPTICAL MATERIALS).

CdSe forms solid solutions with CdS which are used as pigments ranging in color from orange to deep maroon and are called cadmium sulfoselenides. Other uses are in photocells, rectifiers, luminous paints, and as a ruby colorant for glass manufacture.

CdTe is used in infrared optics (26), phosphors, electroluminescent devices, photocells, and as a detector for nuclear radiation (27).

3.11. Cadmium Silicates.

Cadmium orthosilicate [15857-59-2], Cd_2SiO_4, (mp 1246°C; $d = 5.83$ g/mL) and cadmium metasilicate [13477-19-5], $CdSiO_3$, are both prepared by direct reaction of CdO and SiO_2 at 390°C under 30.4 MPa (300 atm) or at 900°C and atmospheric pressure in steam. The materials are phosphors when activated with Mn(II) ion and are both fluorescent and phosphorescent.

3.12. Cadmium Sulfate.

$CdSO_4 \cdot \frac{8}{3}H_2O$ [7790-84-3] is the normal form of cadmium sulfate and is prepared by crystallization of solutions made by dissolving cadmium metal, oxide, sulfide, hydroxide, or carbonate in sulfuric acid. Alternatively, the $\frac{8}{3}$ hydrate is precipitated from such solutions with alcohol. The monohydrate [13577-20-8] may be prepared by dehydrating this material by heating to 80°C. Anhydrous cadmium sulfate [10124-36-4] is prepared by oxidation of the sulfide or sulfite under carefully controlled oxidizing atmospheres at high temperature. An alternative method involves treatment of powdered cadmium nitrate, halides, oxide, or carbonate with dimethyl sulfate (7).

Cadmium sulfate solutions are used in the standard Weston cell and as electrolytes in electroplating as alternatives to cyanide baths. Other uses include phosphors and as a nematocide.

3.13. Cadmium Sulfide.

CdS [1306-23-6] is dimorphic and exists in the sphalerite (cubic) and wurtzite (hexagonal) crystal structures (25). At very high pressures it may exist also as a rock-salt structure type. It is oxidized to the sulfate, basic sulfate, and eventually the oxide on heating in air to 700°C, especially in the presence of moisture (9).

CdS may be prepared ranging in color from white to deep orange-red depending on the preparative method and resultant particle size of the material. The smaller the particle size the lighter the coloration (16) and glass colored with cadmium sulfide is colorless when first cast from the melt where the particle size of the CdS is less than 2 nm. Upon annealing (striking) at 700°C, yellow-orange color develops as the particle size increases as a result of aggregation in the glass matrix. Direct reaction between H_2S and Cd vapor or between sulfur and cadmium metal or its oxide at high temperature produces CdS. However, a simple method involves treatment of an acidic or neutral cadmium ion solution with H_2S or Na_2S and collection of the dense yellow precipitate is straightforward. At room temperature this method gives yellow solids whereas from boiling solutions one obtains yellow solids at neutral pH but reddish solids at low pH. Acidified cadmium acetate solutions precipitate yellow CdS; ammoniacal solutions give a red modification.

CdS finds its main use as a pigment, particularly in the glass and plastics industry.

Increased concerns over the toxicity and environmental impact of cadmium materials and increased imports of Cd pigments from Europe contributed to the decrease in production in the latter 1980s and 1990s. Pure yellow cadmium sulfides are formulated with red cadmium selenides to form solid solutions called C.P. toners ranging from yellows and oranges (low selenium content) to reds and maroons (high selenide). Such pigments are manufactured on an industrial scale in three steps: (1) cadmium sticks are dissolved in sulfuric acid to give a sulfate solution; (2) sodium sulfide and selenide are mixed in the desired proportion and added to the cadmium sulfate solution; and (3) the precipitated pigment is filtered, washed, dried, and calcined at 700°C in an inert atmosphere to obtain a uniform product.

The cadmium lithopones, ZnS–CdS/Se–BaSO$_4$, are additional cadmium sulfide-based pigments prepared by adding barium sulfide to zinc–cadmium sulfate mixtures. Again the colors range from yellow to deep red-maroon depending on additive content. Cadmium pigments in general are very resistant to H$_2$S, SO$_2$, light, heat, and other atmospheric conditions and are dense, heavy colorants having good covering power and bright deep shades.

CdS colorants find use in plastics, paints, soaps, rubber, paper, glass, printing inks, ceramic glazes, textiles, and fireworks. Luminescent pigments based on CdS–ZnS are also produced.

Other uses of CdS take advantage of its semiconducting properties. It is an n-type semiconductor with a band gap (wurtzite phase) at 2.58 eV (480 nm). It is used as a thin-film cell to convert solar energy to electrical power, is a photoconductor, and is electroluminescent (25). These properties have found use in phosphors, photomultipliers, radiation detectors, thin-film transistors, diodes and rectifiers, electron-beam pumped lasers, and smoke detectors (19). CdS, when shock fractured by pressure release from the cubic phase, has been reported to be a high temperature superconductor having a critical temperature for conversion to the superconductive state in excess of 190 K (28). Finally the nonlinear optical properties of colloidal suspensions of CdS in glass or polymer matrices have been explored as possible light transistors for optical computing applications (16).

3.14. Cadmium Tungstate. Cadmium tungstate [7790-85-4], CdWO$_4$, forms white or yellow monoclinic crystals which are highly luminescent, $a = 0.5029$ nm, $b = 0.5859$ nm, $c = 0.5074$ nm; $d = 8.033$ g/mL. It is prepared by the action of tungstic acid on cadmium oxide in a little water (7) and finds use in x-ray screens, scintillation counters, phosphors, and as a catalyst in organic reactions (29).

4. Organic Compounds

Many organocadmium compounds are known but few have been of commercial importance. Wanklyn first isolated diethylcadmium in 1856. The properties of this and other dialkylcadmiums are listed in Table 3. In general, these materials are prepared by reaction of an anhydrous cadmium halide with a Grignard or alkyllithium reagent followed by distillation of the volatile material in an inert atmosphere or *in vacuo*. Only the liquid dimethyl compound is reasonably stable and then only when stored in a sealed tube. Dimethylcadmium is mildly pyrophoric in air and produces dense clouds of white, then brown, cadmium oxide smoke, which is highly toxic if breathed (30). When dropped into water, the liquid sinks in large droplets that decompose with a series of small explosive jerks and pops. For this reason, and particularly because of the low thermal stability, most dialkylcadmium materials are prepared and used *in situ* without separation, eg, in the conversion of acid chlorides to speciality ketones (qv):

$$2 \text{ RCOCl} + \text{CdR}'_2 \longrightarrow 2 \text{ RCOR}' + \text{CdCl}_2$$

Table 3. **Properties of Dialkyl Cadmium Compounds**

Compound	CAS Registry Number	Molecular formula	Mp, °C	Bp, °C[a]	Density, g/mL
dimethylcadmium	[506-82-1]	$(CH_3)_2Cd$	−4.5	105.5 (101.3)	1.9846
diethylcadmium	[592-02-9]	$(C_2H_5)_2Cd$	−21	64(2.6)	1.6564
dipropylcadmium	[5905-48-6]	$(C_3H_7)_2Cd$	−83	84(2.8)	1.4184
dibutylcadmium	[3431-67-2]	$(C_4H_9)_2Cd$	−48	103.5 (1.6)	1.3054
diisobutylcadmium	[3431-67-2]	$(C_4H_9)_2Cd$	−37	90.5 (2.6)	1.2674
diisoamylcadmium	[35061-27-9]	$(C_5H_{11})_2Cd$	−115	121.5 (2.0)	1.2184

[a] Pressure of bp determination is given in kPa in parentheses. To convert kPa to mm Hg, multiply by 7.5.

Dimethylcadmium has a linear C–Cd–C core with a Cd–C bond length of 0.211 nm (17).

Many dialkyl and diaryl cadmium compounds have found use as polymerization catalysts. For example, the diethyl compound catalyzes polymerization of vinyl chloride, vinyl acetate, and methyl methacrylate (30), and when mixed with $TiCl_4$, can be used to produce polyethylene and crystalline polypropylene for filaments, textiles, glues, and coatings (30). With >50% $TiCl_4$, diethyl cadmium polymerizes dienes. Diethyl cadmium may be used as an intermediate ethylating agent in the production of tetraethyllead. The diaryl compounds such as diphenylcadmium [2674-04-6], $(C_6H_5)_2Cd$, (mp 174°C) are also polymerization catalysts. These compounds are also prepared using Grignard or aryllithium reagents in tetrahydrofuran (THF) solvent but may be prepared by direct metal substitution reactions such as:

$$(C_6H_5)_2Hg + Cd \longrightarrow (C_6H_5)_2Cd + Hg$$

Dimethylcadmium has found use as a volatile source of Cd for metal organic chemical vapor deposition (MOCVD) production of cadmium-containing semiconductor thin films (qv) such as CdS, $Cd_{1-x}Hg_xTe$, or $Cd_{1-x}Mn_xTe$, as multiple quantum well species (16).

Cadmium alkyl and aryl halides, RCdX, as well as cadmium allyls have been prepared by Grignard reactions but, as yet, have not realized any commercially important uses despite reactivity toward a number of organic and inorganic materials.

4.1. Cadmium Acetate. Cadmium acetate [543-90-8], $Cd(CH_3COO)_2 \cdot n$ H_2O, can exist as the anhydrous salt ($n = 0$) mp 256°C, $d = 2.341$ g/mL or as one of a series of hydrates ($n = 1 - 3$). The anhydrous material may be prepared by treating cadmium nitrate with acetic anhydride or by very careful heating and drying the dihydrate at ~130°C. The cadmium acetate dihydrate [5743-04-4], $d = 2.01$ g/mL, is obtained by dissolving cadmium metal or its oxide, hydroxide, or carbonate in acetic acid and crystallizing. Cadmium acetate monohydrate [543-90-8] may be obtained from the dihydrate by careful drying. All acetates are very soluble in water and alcohols.

Cadmium acetate is a colorant for glass and textiles, a glaze for ceramics where it produces iridescent effects, a starting material for preparation of the cadmium halides, and is an alternative to the cyanide bath for cadmium electro-plating.

4.2. Organocadmium Soaps.   Other salts of organic acids, apart from the acetate, have found wide usage as heat and light stabilizers for plastics, especially flexible PVC. During the molding process, unstabilized PVC begins to lose HCl at 95°C resulting in marked yellowing of the molded article. In addition, the free HCl acts as a catalyst for further degradation of the polymer and the unsaturated polymer chains left behind upon HCl elimination absorb more uv radiation which also breaks down the polymer chains leading to embrittlement. Cadmium salts of long-chain fatty acids such as laurate and stearate, cadmium soaps, are acid acceptors that react with HCl to give the weak organic acid and $CdCl_2$, and so their incorporation into the plastic article prevents early discoloration. However, cadmium chloride is a strong Lewis acid, capable of initiating polymer degradation itself. For this reason, barium compounds of the same fatty acids are used in conjunction with the cadmium stabilizers because there is a rapid exchange of chloride from cadmium to barium. A phosphite chelator is typically added to the mixture to produce an almost complete stabilizer package (19).

The solid soaps are prepared from cadmium chloride solution by precipitation with sodium salts of the fatty acids. Cadmium laurate [2605-44-9], $Cd(C_{12}H_{24}O_2)_2$, cadmium stearate [2223-93-0], $Cd(C_{18}H_{36}O_2)_2$, cadmium palmitate [6427-86-7], $Cd(C_{16}H_{32}O_2)_2$, and cadmium myristate [10196-67-5], $Cd(C_{14}H_{28}O_2)_2$, are of this type. Liquid stabilizers such as cadmium octoate [2191-10-8], $Cd(C_8H_{16}O_2)_2$, cadmium phenolate [18991-05-4], $Cd(C_6H_6O)_2$, cadmium decanoate [2847-16-7], $Cd(C_{10}H_{20}O_2)_2$, cadmium benzoate [3026-22-0], $Cd(C_7H_6O_2)_2$, and cadmium naphthenate are more versatile and economical in use and are prepared from CdO and the organic acid in an inert solvent. The water by-product is driven off as the reaction proceeds and the clear solution of cadmium soap in the organic solvent is used directly in plastics manufacture. FDA regulations have decreed that plastics that contact foodstuffs may not contain Cd–Ba stabilizers. Overall environmental concerns have led most plastics manufacturers to move away from heavy metal-based stabilizers, toward alternatives such as Ca–Zn, Ba–Zn or organotin materials (21,31). There has therefore been a decrease in consumption of Cd compounds in the stabilizer field that began in the late 1970s.

5. Economic Aspects

Compared with 2000, both production and consumption of cadmium in the U.S. declined during 2001. Domestic production of cadmium metal declined by 64% and production of cadmium compounds, including cadmium sulfide, declined by more than 93%. World consumption was 16,000 t in 2001, 17% less than in 2000 (21).

In 2000 the United States was the net exporter of cadmium sulfide, most of which was exported to the Philippines (59%), Germany (26%), and Saudi Arabia (10%) (21). Table 4 gives U.S. production of cadmium compounds.

One of the most promising applications is the use of cadmium telluride solar cells to convert sunlight into electricity and the use of NiCd batteries to store

Table 4. **U.S. Production of Cadmium Compounds, t, Cd content**[a]

Year	Cadmium sulfide[b]	Other cadmium compounds[c]
2000	42	417
2001	31	0

[a] Ref. 21.
[b] Includes cadmium lithopone and cadmium sulfoselenide.
[c] Includes oxide and plating salts (acetate, carbonate, nitrate, sulfate, etc.).

that energy (21). Critical to the cadmium market is the industry's ability to recycle cadmium products.

6. Analytical Methods

Because of the increasing emphasis on monitoring of environmental cadmium the determination of extremely low concentrations of cadmium ion has been developed. Table 5 lists the most prevalent analytical techniques and the detection limits. In general, for soluble cadmium species, atomic absorption is the method of choice for detection of very low concentrations. Mobile prompt gamma *in vivo* activation analysis has been developed for the nondestructive sampling of cadmium in biological samples (32).

At higher levels, cadmium may be estimated gravimetrically following precipitation with sulfide (34), β-naphthoquinoline (35), or after plating from a cyanide-containing solution onto a stationary platinum cathode. Volumetric procedures rely on preliminary precipitation of the sulfide that is purified and then dissolved in acid whereupon the liberated H_2S may be titrated with iodine. An alternative, should zinc be a likely contaminant, is to precipitate with diethyldithiocarbamate and then to redissolve in acid and titrate with sodium ethylenetriaminetetraacetate (EDTA) using Eriochrome Black T as indicator (36).

7. Health and Safety Factors

Cadmium, both as the free metal and in its compounds, is highly toxic and has been designated one of the 100 most hazardous substances under Section 110 of

Table 5. **Analytical Methods and Detection Limits for Cadmium Ion**[a]

Method	Detection limit, ppb
atomic absorption spectroscopy (graphite furnace)	0.008
polarography (square wave)	10
x-ray fluorescence (energy dispersive)	5000
neutron activation analysis	1
isotope dilution	10
inductively coupled plasma emission spectra	1

[a] Ref. 33.

the Superfund Amendments and Reauthorization Act of 1986 (37). Poisoning may occur either via inhalation or ingestion. Only about 6% of the estimated 40–50 µg/d of ingested cadmium is absorbed by the body, whereas 25–50% of the 2–10 µg/d of cadmium in inhaled dust is absorbed (33,38). Cadmium is present in cigarette smoke and as much as 0.1–0.2 micrograms per cigarette may be absorbed by the lungs (39).

Chemical pneumonitis or pulmonary edema may result from acute exposure to cadmium fumes, as oxide or chloride aerosols, at a dose of 5 mg/m^3 over an 8-h period. One mg/m^3 inhaled over the same time period gives rise to clinically evident symptoms in sensitive individuals. Deaths from acute cadmium poisoning have resulted from inhalation of cadmium oxide smokes and fumes, usually from welding operations on cadmium plated steels in poorly ventilated areas. Acute ingestion of cadmium concentrations above ∼15 ppm (0.1–1.0 mg/(kg · d)) produce symptoms of nausea, vomiting, abdominal cramps, and headache (40). Possible sources of such poisoning have been traced to cadmium-plated cooking utensils, cadmium solders in water coolers, or from acid juices stored in ceramic pots glazed using cadmium-containing compounds. Table 6 lists exposure limits for cadmium and cadmium compounds as Cd.

Cadmium is efficiently scavenged by the body and biological half-times for cadmium excretion are on the order of 10–30 years. The kidneys and liver appear to be the organs of concentration for cadmium; kidney damage leading to proteinuria is probably the most common manifestation of chronic cadmium exposure. The combination of dietary deficiency and high cadmium exposure resulted in the most infamous example of suspected cadmium poisoning on record where a disease known as itai-itai afflicted elderly Japanese women from the Zinzu river basin after World War II. Whereas the role of cadmium in the disease is still controversial, it seems clear that the severe weakening of bone tissue associated with the disease was a result of demineralization induced by cadmium (33). Although there is some evidence that cadmium may be carcinogenic in animals under certain exposure conditions, the association between cadmium exposure and cancer in humans remains tenuous. The increased risk of prostatic cancer in workers exposed to cadmium dusts and fumes has been reported to be significant, but the number of cases reported so far is very small (33) and the conclusions have been questioned. More recent data have described a possible link

Table 6. **Exposure Limits for Cadmium and Cadmium Compounds as Cd**[a]

Exposure Limits	OSHA PEL	NIOSH Exposure Limit	ACGIH TLV
time-weighted average	0.005 mg/m^3	lowest feasible conc.	0.01 mg/m^3 0.002 mg/m^3 (respirable fraction)
short-term exposure limit			
ceiling limit			
biological limits (if available)	3 µg/g creatinine (urine) 5 µg/L (blood)		5 µg/g creatinine (urine) 5 µg/L (blood)

[a] Ref. 41.

between cadmium exposure and lung cancer in humans (40). In 1989 the EPA denied a petition requesting removal of CdS and CdSe from their list of toxic chemicals, citing available cancer data on CdS and other cadmium compounds (42). EPA maintains that cadmium is a probable human carcinogen (Group B1) but only by the inhalation route.

8. Environmental Concerns

Cadmium discharges to air and water are decreasing as primary zinc producers have largely converted to electrolytic processing and as more efficient pollution control technologies take effect (43). Most of the cadmium released to the environment is now in the form of solid wastes such as coal ash, sewage sludge (5–20 ppm), flue dust, and fertilizers (2–20 ppm). Effluent limits of all sources have been strictly regulated in recent years and cadmium emissions are controlled by the best available technology including membrane filtration (see EXHAUST CONTROL, INDUSTRIAL). Recycling programs have been instituted by several battery manufacturers, eg, in France, Belgium, Japan, Sweden, and Korea (37), aimed at reducing cadmium pollution from spent Ni–Cd batteries. U.S. and Canadian collection and recycling programs for small rechargeable batteries are expected to expand. About 1,700 t of rechargeable batteries were recycled in 2000 (21).

9. Uses

The principal areas of cadmium usage in terms of U.S. consumption at the end of 2001 were batteries (75%); pigments (12%); coatings and plating (8%); plastic stabilizers (4%); nonferrous alloys and other uses (1%).

Cadmium hydroxide is the anode material of Ag–Cd and Ni–Cd rechargeable storage batteries (see BATTERIES, SECONDARY CELLS). Cadmium sulfide, selenide, and especially telluride find utility in solar cells (see SOLAR ENERGY). Cadmium sulfide, lithopone, and sulfoselenide are used as colorants (orange, yellow, red) for plastics, glass, glazes, rubber, and fireworks.

A cadmium sulfide interface layer can improve III-V semiconductor device performance (44).

In flexible PVC, cadmium salts of long-chain organic acids, such as stearate and laurate, are used in combination with similary Ba^{2+} salts as heat and light stabilizers (see HEAT STABILIZERS) but these uses are in decline since these compounds can now be replaced by less toxic compounds. Cadmium cyanide, acetate, fluoroborate, or sulfate is used as an electrolyte in coating a thin cadmium layer, ie, electroplating (qv), onto other metals thereby imparting enhanced corrosion protection. Cadmium protective overlayers are also deposited by mechanical plating or vapor deposition (see METALLIC COATINGS).

The cadmium chalcogenide semiconductors (qv) have found numerous applications ranging from rectifiers to photoconductive detectors in smoke alarms. Many Cd compounds, eg, sulfide, tungstate, selenide, telluride, and oxide, are used as phosphors in luminescent screens and scintillation counters.

Glass colored with cadmium sulfoselenides is used as a color filter in spectroscopy and has recently attracted attention as a third-order, nonlinear optical switching material (see NONLINEAR OPTICAL MATERIALS). Dialkylcadmium compounds are polymerization catalysts for production of poly(vinyl chloride) (PVC), poly(vinyl acetate) (PVA), and poly(methyl methacrylate) (PMMA). Mixed with $TiCl_4$, they catalyze the polymerization of ethylene and propylene.

Dimethyl cadmium is toxic and can be replaced with a more stable cadmium oxide to form nanocrystals for use in electronics and optoelectronic devices (45). At the University of California, Berkeley, new forms of cadmium sulfide crystals have been developed for use in the manufacture of solar panels. The new crystals facilitate electron flow and could boost efficiency by more than 20% (46).

BIBLIOGRAPHY

"Cadmium Compounds" in *ECT* 1st ed., Vol. 2, pp. 732–738, by G. U. Greene, Fenn College; in *ECT* 2nd ed., Vol. 3, pp. 899–911, by G. U. Greene, New Mexico Institute of Mining and Technology; in *ECT* 3rd ed., Vol. 4, pp. 397–411, by P. D. Parker, AMAX Base Metals Research & Development, Inc.; in *ECT* 4th ed., Vol. 4, pp. 760–776, by Norman Herron, E. I. du Pont de Nemours & Co., Inc.; "Cadmium Compounds" in *ECT* (online), posting date: December 4, 2000, by Norman Herron, E. I. du Pont de Nemours & Co., Inc.

CITED PUBLICATIONS

1. U.S. Bureau of Mines, *Mineral Facts and Problems*, U.S. Dept. of the Interior, Washington, D.C., 1970, p. 516.
2. R. A. Potts, R. D. Barnes, and J. D. Corbett, *Inorg. Chem.* **7**, 2558 (1968).
3. D. D. Wagman and co-workers, *Selected Values of Chemical Thermodynamic Properties*, National Bureau of Standards Technical Note No. 270-3, Washington, D.C., 1968, p. 248.
4. B. J. Aylett, in J. C. Bailar and A. F. Trotman-Dickenson, eds., *Comprehensive Inorganic Chemistry*, Pergamon Press, Oxford, UK, 1973, pp. 187, 190, pp. 258–272.
5. K. E. Almin, *Acta. Chem. Scand.* **2**, 400 (1948).
6. F. D. Rossini and co-workers, *Selected Values of Chemical Thermodynamic Properties*, National Bureau of Standards Circular No. 500, Washington, D.C., 1952.
7. H. M. Cyr, in M. C. Sneed and R. C. Brasted, eds., *Comprehensive Inorganic Chemistry*, Vol. IV, Van Nostrand, Princeton, N.J., 1955, pp. 71–90.
8. R. W. G. Wyckoff, *Crystal Structures*, 2nd ed., Vols. I–V, Wiley-Interscience, Inc., New York, 1963–1965.
9. D. M. Chizhikov, *Cadmium*, trans. D. E. Hayler, Pergamon Press, Oxford, UK, 1966, pp. 10–48, 61, 63, 68–70.
10. P. Goldfinger and H. Jeunnehomme, *Trans. Faraday Soc.* **59**, 2851 (1963).
11. O. Kubaschewski and E. L. Evans, *Metallurgical Thermochemistry*, Academic Press, New York, 1951, p. 268.
12. M. Farnsworth, *Cadmium Chemicals*, International Lead Zinc Research Org., New York, 1980.
13. W. Zdanowicz and L. Zdanowicz, *Ann. Rev. Mater. Sci.* **5**, 301 (1975).
14. J. Clark and J. K. Range, *Z. Naturforsch., B: Anorg. Chem., Org. Chem.* **31**, 58 (1976).

15. N. A. Goryunova, in J. C. Anderson, ed., *Chemistry of Diamond-Like Semiconductors*, The M.I.T. Press, Massachusetts Institute of Technology, Cambridge, Mass., 1965.

16. Y. Wang and N. Herron, *J. Phys. Chem.* **95**, 525 (1991).

17. A. F. Wells, *Structural Inorganic Chemistry*, 4th ed., Clarendon Press, Oxford, UK, 1975.

18. L. G. Sillen and A. E. Martell, eds., *Stability Constants of Metal–Ion Complexes*, Chemical Society, No. 17, Suppl. 1, London, 1971.

19. S. F. Radtke, *Proceedings of the 1st International Cadmium Conference*, Metal Bulletin Ltd., London, UK, 1978, p. 41.

20. *Nickel Cadmium Battery Update 90*, Cadmium Association, London, 1990.

21. J. Plachy, "Cadmium," *Minerals Yearbook*, U.S. Geological Survey, Reston, VA, 2001.

22. S. U. Falk and A. J. Salkind, *Alkaline Storage Batteries*, John Wiley & Sons, Inc., New York, 1969, p. 132.

23. L. Hadju and J. Zahoran, *Acta Tech. (Budapest)* **73**(1–2), 117 (1972).

24. U.S. Pats. 2,128,126 (Aug. 23, 1938), A. Dunstan (to Iranian Oil Co.); 2,206,227 and 2,206,226 (July 2, 1940), W. Grommbridge and T. Dee (to Celanese Co.).

25. B. Ray, *II–VI Compounds*, Pergamon Press, Oxford, UK, 1969.

26. C. L. Gupta and R. C. Tyagi, *Def. Sci. J.* **24**(2), 71 (1974).

27. A. J. Strauss, *Proceedings of the International Symposium on Cadmium Telluride as a Material for Gamma-ray Detectors*, 1972, p. I-1.

28. T. C. Collins, *Ferroelectrics* **73**, 469 (1987).

29. A. Karl, *Compt. Rend.* **196**, 1403 (1933).

30. J. H. Harwood, *Industrial Applications of the Organometallic Compounds*, Reinhold Publishing Co., New York, 1963, p. 59.

31. R. Monks, *Plastics Tech.* **36**, 48 (1990); **36**, 113 (1990).

32. D. Vartsky, K. J. Ellis, N. S. Chen, and S. H. Cohn, *Phys. Med. Biol.* **22**, 1085 (1977).

33. K. Matsumoto and K. Fuwa, "Cadmium," in E. C. Foulkes, ed., *Handbook of Experimental Pharmacology*, Vol. 80, Springer-Verlag, Berlin, Heidelberg, 1986.

34. W. W. Scott, in N. H. Furman, ed., *Standard Methods of Chemical Analysis*, Vol. I, 5th ed., Van Nostrand, Princeton, N.J., 1945, pp. 197–204.

35. ASTM Standards, *Part 12—Chemical Analysis of Metals and Metal Bearing Ores*, Std. E40-58, 1974, pp. 192–193.

36. *Ibid.*, Std. E56-63, p. 295.

37. U.S. Bureau of Mines, *Mineral Industry Surveys*, Cadmium, U.S. Dept. of the Interior, Washington, D.C., 1989 and 1990.

38. H. A. Schroeder and J. J. Balassa, *J. Chronic. Dis.* **14**, 236 (1961).

39. C. G. Elinder and co-workers, *Environ. Res.* **32**, 220 (1983).

40. *Toxicological Profile for Cadmium*, U.S. Dept. Commerce, NTIS, Washington, D.C., Mar. 1989.

41. M. Jakubowski, in E. Bingham, B. Cohrssen, and C. H. Powell, eds., *Patty's Toxicology*, 5th ed., Vol. 2, John Wiley & Sons, Inc., New York, 2001, p. 309.

42. *Cadmium Sulfide Cadmium Selenide; Tox. Chem. Release Rep.; Comm. Right-To-Know* **54**(201), 42962 (1989).

43. Versar Inc., *Technical and Microeconomic Aspects of Cadmium and its Compounds*, Final Report EPA Contract 68-01-2926, Task 1, EPA, Washington, D.C., 1976, p. 56.

44. U.S. Pat. 5,689,125 (Nov. 11, 1997), K. Vaccaro (to the USA as represented by the Secretary of Air).

45. *Business Week* **3722**, 30B (March 5, 2001).

46. *Business Week* **3722**, 100 (March 5, 2001).

NORMAN HERRON
E. I. du Pont de Nemours & Co., Inc.

CALCIUM AND CALCIUM ALLOYS

1. Introduction

Calcium [7440-70-2], Ca, a member of Group 2 (IIA) of the Periodic Table between magnesium and strontium, is classified, together with barium and strontium, as an alkaline-earth metal and is the lightest of the three. Calcium metal does not occur free in nature; however, in the form of numerous compounds, it is the fifth most abundant element, constituting 3.63% of the earth's crust.

The word calcium is derived from calx, the Latin word for lime. The Romans used large quantities of calcium oxide or lime as mortar in construction (see LIME AND LIMESTONE). Because calcium compounds are very stable, elemental calcium was not produced until 1808, when a mercury amalgam resulted from electrolysis of calcium chloride in the presence of a mercury cathode. However, attempts to isolate the pure metal by distilling the mercury were only marginally successful.

Calcium metal was produced in 1855 by electrolysis of a mixture of calcium, strontium, and ammonium chlorides, but the product was highly contaminated with chlorides (1). By 1904 fairly large quantities of calcium were obtained by the electrolysis of molten calcium chloride held at a temperature above the melting point of the salt but below the melting point of calcium metal. An iron cathode just touched the surface of the bath and was raised slowly as the relatively chloride-free calcium solidified on the end. This process became the basis for commercial production of calcium metal until World War II.

Prior to 1939, calcium was manufactured exclusively in France and Germany. However, with the outbreak of World War II, an electrolytic calcium plant was constructed in the United States at Sault Ste. Marie, Michigan, by the Electro Metallurgical Corp. Large amounts of calcium were required as the reducing agent for uranium production (see URANIUM AND URANIUM COMPOUNDS). In addition, calcium was used to produce calcium hydride, which could easily be transported to remote areas and used as a source of hydrogen for meteorological balloons.

Calcium is mainly used as a reducing agent for many reactive, less common metals; to remove bismuth from lead (qv); as a desulfurizer and deoxidizer for ferrous metals and alloys; and as an alloying agent for aluminum, silicon, and lead. Small amounts are used as a dehydrating agent for organic solvents and as a purifying agent for removal of nitrogen and other impurities from argon and other rare gases (see HELIUM-GROUP GASES).

2. Physical Properties

Pure calcium is a bright silvery white metal, although under normal atmospheric conditions freshly exposed surfaces of calcium quickly become covered with an oxide layer. The metal is extremely soft and ductile, having a hardness between

Table 1. **Physical Properties of Calcium**[a]

Property	Value		
atomic weight	40.08		
electron configuration	$1s^2 2s^2 2p^6 3s^2 3p^6 4s^2$		
stable isotopes			
atomic weight 40 42 43	44	46	48
natural abundance, % 96.947 0.646 0.135 0.18	2.083	0.186	
specific gravity at 20°C, kg/m^3	1.55×10^3		
melting point, °C	839 ± 2		
boiling point, °C	1484		
heat of fusion, ΔH_{fus}, kJ/mol[b]	9.2		
heat of vaporization, ΔH_{vap}, kJ/mol[b]	161.5		
heat of combustion, kJ/mol[b]	634.3		
vapor pressure			
pressure, kPa[c] 0.133 1.33 101.3	13.3	53.3	
temperature, °C 800 970	1200	1390	1484
specific heat at 25°C, J/(g·K)[b]	0.653		
coefficient of thermal expansion, 0–400°C, m/(m·K)	22.3×10^{-6}		
electrical resistivity at 0°C, $\mu\Omega\cdot$cm	3.91		
electron work function, eV	2.24		
tensile strength (annealed), MPa[c]	48		
yield strength (annealed), MPa[c]	13.7		
modulus of elasticity, GPa[c]	22.1–26.2		
hardness (as cast)			
HB[d]	16–18		
HR B[e]	36–40		

that of sodium and aluminum. It can be work-hardened to some degree by mechanical processing. Although its density is low, calcium's usefulness as a structural material is limited by its low tensile strength and high chemical reactivity (2).

Calcium has a face-centered cubic crystal structure ($a = 0.5582$ nm) at room temperature but transforms into a body-centered cubic ($a = 0.4477$ nm) form at 428 ± 2°C (3). Some of the more important physical properties of calcium are given in Table 1. For additional physical properties, see Refs. (7–12). Measurements of the physical properties of calcium are usually somewhat uncertain owing to the effects that small levels of impurities can exert.

3. Chemical Properties

Calcium has a valence electron configuration of $4s^2$ and characteristically forms divalent compounds. It is very reactive and reacts vigorously with water, liberating hydrogen and forming calcium hydroxide, $Ca(OH)_2$. Calcium does not readily oxidize in dry air at room temperature but is quickly oxidized in moist air or in dry oxygen at about 300°C. The oxide layer is nonprotective, and complete oxidation of a massive piece of calcium eventually occurs. Calcium reacts with fluorine at room temperature and with the other halogens at 400°C. When heated to 900°C, calcium reacts with nitrogen to form calcium nitride [12013-82-0], Ca_3N_2 (see CALCIUM COMPOUNDS, SURVEY). The metal becomes incandescent when heated to 400–500°C in an atmosphere of hydrogen with the formation of calcium hydride [7789-78-8], CaH_2, which reacts with water to give hydrogen:

$$CaH_2 + 2\,H_2O \longrightarrow Ca(OH)_2 + 2\,H_2$$

Thus the hydride is a very efficient carrier of hydrogen. Upon heating, calcium reacts with boron, sulfur, carbon, and phosphorus to form the corresponding binary compounds and with carbon dioxide to form calcium carbide [75-20-7], CaC_2, and calcium oxide [1305-78-8], CaO.

Calcium is an excellent reducing agent and is widely used for this purpose. At elevated temperatures, it reacts with the oxides or halides of almost all metallic elements to form the corresponding metal. It also combines with many metals, forming a wide range of alloys and intermetallic compounds. Among the phase systems that have been better characterized are those with Ag, Al, Au, Bi, Cd, Co, Cu, Hg, Li, Na, Ni, Pb, Sb, Si, Sn, Tl, Zn, and the other Group 2 (IIA) metals (13).

Commercially produced calcium metal is analyzed for metallic impurities by emission spectroscopy. Carbon content is determined by combustion, whereas nitrogen is measured by Kjeldahl determination.

4. Manufacture

4.1. Electrolysis. Although in Western countries the aluminothermic process has now completely replaced the electrolytic method, electrolysis is believed to be the method used for calcium production in the People's Republic of China and the Commonwealth of Independent States (CIS). This process likely involves the production of a calcium–copper alloy, which is then redistilled to give calcium metal.

4.2. Aluminothermal Method. Calcium metal is produced by high-temperature vacuum reduction of calcium oxide in the aluminothermal process. This process, in which aluminum [7429-90-5] metal serves as the reducing agent, was commercialized in the 1940s. The reactions, which are thermodynamically unfavorable at temperatures below 2000°C, have been summarized as follows:

$$6\,CaO + 2\,Al \rightleftharpoons 3CaO \cdot Al_2O_3 + 3\,Ca(g)$$

$$33\,CaO + 14\,Al \rightleftharpoons 12CaO \cdot 7Al_2O_3 + 21\,Ca(g)$$

$$4\,CaO + 2\,Al \rightleftharpoons CaO \cdot Al_2O_3 + 3\,Ca(g)$$

In the range of 1000–1200°C a small but finite equilibrium pressure of calcium vapor is established. The calcium vapor is then transferred using a vacuum pump to a cooled region of the reactor where condensation takes place, shifting the equilibrium at the reaction site and allowing more calcium vapor to be formed.

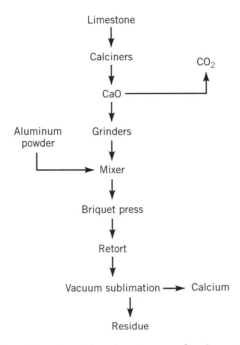

Fig. 1. Flow sheet for aluminum reduction process.

A typical flow sheet for the process is given in Figure 1. High calcium limestone, $CaCO_3$, is quarried and calcined to form calcium oxide. The calcium oxide is ground to a small particle size and dry-blended with the desired amount of finely divided aluminum. This mixture is then compacted into briquettes to ensure good contact of reactants. The briquettes are placed in horizontal metal tubes, ie, retorts made of heat-resistant steel and heated in a furnace to 1100–1200°C. The open ends of the retorts protrude from the furnace and are cooled by water jackets to condense the calcium vapor. The retorts are then sealed and evacuated to a pressure of less than 13 Pa (0.1 mm Hg). After the reaction has been allowed to proceed for about 24 h, the vacuum is broken with argon and the condensed blocks of ca. 99% pure calcium metal, known as crowns, and calcium aluminate [12042-78-3] residue are removed. Large amounts of energy are required by this method, partially because of the high temperatures of the process and partially because of the energy-intensive raw materials employed, ie, the calcined CaO and electrolytically produced aluminum.

The calcium crowns can be sold as such for certain applications. However, further processing may be required, and the crowns can be reduced in size to pieces of about 25 cm or nodules of about 3 mm. They can also be melted under a protective atmosphere of argon and cast into billets or ingots. Calcium wire can be made by extrusion, and calcium turnings are produced as lathe cuttings from cast billets. Technologies have also been developed to manufacture calcium metal particulates and powders by atomization, comminution, and grinding processes.

4.3. Redistillation. For certain applications, especially those involving reduction of other metal compounds, better than 99% purity is required. This can be achieved by redistillation. In one method, crude calcium is placed in the bottom of a large vertical retort made of heat-resistant steel equipped with a water-cooled condenser at the top. The retort is sealed and evacuated to a pressure of less than 6.6 Pa (0.05 mm Hg) while the bottom is heated to 900–925°C. Under these conditions, calcium quickly distills to the condensing section, leaving behind the bulk of the less volatile impurities. Variations of this method have been used for commercial production. Subsequent processing must take place under exclusion of moisture to avoid oxidation.

Redistillation does not greatly reduce the impurity level of volatile materials such as magnesium. Volatile alkali metals can be separated from calcium by passing the vapors over refractory oxides such as TiO_2, ZrO_2, or Cr_2O_3 to form the nonvolatile Na_2O and K_2O (14). Purification techniques include reactive distillation (15), growth of crystals from the melt (16), and combined crystal growth and distillation techniques (17).

5. Shipment

Because of its extreme chemical reactivity, calcium metal must be carefully packaged for shipment and storage. The metal is packaged in sealed argon-filled containers. Calcium is classed as a flammable solid and is nonmailable. Sealed quantities of calcium should be stored in a dry, well-ventilated area so as to remove any hydrogen formed by reaction with moisture.

6. Economic Aspects

Calcium is produced in the United States by Pfizer Incorporated in Canaan, Conn., in Canada by Timminco Ltd., and in France by Société Electrometallurgique du Planet (S.E.M.P.) (18). The world capacity is 5000 t/yr. The United States accounts for production of over 50% of the calcium used worldwide (19).

Domestic consumption of calcium metal increased in 1991. Table 2 lists the amounts and values of the United States imports of calcium from 1987 to 1991 (recent data were not able to be found as this has no longer been collected). Imports increased by more than 26% compared with imports from the previous year. The main countries calcium is imported from are, in decreasing order, China, France, and Canada.

Table 2. **United States Imports for Consumption of Calcium**

Year	Quantity, kg	Value, $
1987	352,089	1,918,099
1988	664,419	3,243,663
1989	679,603	3,210,216
1990	651,000	3,779,410
1991	821,457	5,237,589

Calcium consumption is primarily for the production of maintenance-free and sealed lead–acid batteries, in the steel industry, and for permanent magnet manufacture. These markets are fairly stable and strong markets. Use as a reducing agent of rare-earth oxides for permanent magnet manufacture is expected to increase due to an expansion of this market.

7. Grades and Specifications

Calcium is usually sold as crowns, broken crown pieces, nodules, or billets. The purity of these forms is at least 98%. If a higher quality of the metal is required, it can be redistilled to remove additional impurities. There are three different grades of calcium metal: commercial, melted, and redistilled. In both commercial- and melted-grade calcium, the minimum calcium content is 98.8%. In the redistilled grade, the minimum calcium content is 99.5%. The difference between the commercial and melted grades is in the percents of impurities in the product. Impurities include magnesium, nitrogen, aluminum, iron, manganese, and copper (only in the commercial grade). The redistilled grade has a much lower level of these impurities with the addition of 150 ppm of carbon being present (19).

8. Health and Safety Factors

Inhalation of calcium metal produces damaging effects on the mucous membranes and upper respiratory tract. Symptoms may include irritation of the nose and throat and difficulty breathing. It may also cause lung edema, which is a medical emergency. If breathing is difficult, give oxygen and get medical attention (20). If ingested, caustic lime will form due to reaction with moisture. Large amounts can have a corrosive effect. Abdominal pain, nausea, vomiting, and diarrhea are symptoms. If swallowed, do not induce vomiting but give large amounts of water. As calcium metal is corrosive, contact with the skin can cause pain, redness, and a severe burn. Contact with the eyes will cause redness and pain with possible burns and damage to the eye tissues. If calcium metal comes in contact with the skin or eyes, flush with water for at least 15 min and get medical attention. Protective clothing should include boots, gloves, lab coat, apron, or coveralls in addition to chemical safety goggles. As calcium metal is a very water-reactive flammable sold, reaction with water, steam, and acids to release flammable/explosive hydrogen gas should be avoided. Water must not be used to extinguish a fire with calcium metal.

Calcium metal should be stored in a tightly closed contained in a cool, dry, ventilated area under nitrogen. It should be kept away from water or locations where water may be needed for a fire. This material should be handled as a hazardous waste for diposal purposes.

In the case of a spill, the material should be collected quickly and transferred to a container of kerosene, light oil, or similar hydrocarbon fluid for recovery. Exposure to air should be minimized. Do not use water on the metal or where it spilled if significant quantities still remain. Waste calcium should be packaged under hydrocarbon fluid and sent to an approved waste disposal facility.

9. Calcium Alloys

Calcium alloys can be produced by various techniques. However, direct alloying of the pure metals is normally used in the production of 80% calcium–magnesium, 70% magnesium–calcium, and 75% calcium–aluminum alloys.

Lead alloys containing small amounts of calcium are formed by plunging a basket containing a 77 or 75% calcium–23–25% Al alloy into a molten lead bath or by stirring the Ca–Al alloy into a vortex created by a mixing impellor (21).

Alloys of calcium with silicon are used in ferrous metallurgy (qv) and are generally produced in an electric furnace from CaO (or CaC_2), SiO_2, and a carbonaceous reducing agent (22). The resulting alloy, calcium disilicide [12013-56-8], is nominally of composition $CaSi_2$ and has a typical wt % analysis of 30–33% Ca, 60–65% Si, 1.5–3% Fe (23). Proprietary Ca–Si alloys containing other elements such as Ba, Al, Ti, or Mn are sometimes produced by a combination of carbothermic ore reduction followed by direct alloying. In general, the chemical reactivity of calcium is greatly reduced when it is present in an alloyed state.

10. Uses

The most significant use of calcium is for improvement of steel (qv). It is used as an aid in removing bismuth in lead refining and as a desulfurizer and deoxidizer in steel refining (24–27). Addition of calcium causes inclusions in the steel to float out by modifying the melting point of these inclusions. Any remaining inclusions will be morphologically altered, making them spherical in shape and very small. This results in an overall improvement of the quality of steel and its properties.

Calcium metal is also used in the manufacture of maintenance-free and sealed lead–acid batteries (28). Use of calcium improves the electrical performance and battery life (see BATTERIES). The calcium improves the conductivity and current capacity of the cell.

Calcium has multiple other uses: reducing oxides of the rare-earth neodymium and boron for alloying with metallic iron for use in neodymium–iron–boron permanent magnets and use as a reducing agent to recover hafnium (29), plutonium (30), thorium (31), tungsten (32), uranium (33,34), vanadium (35), and the rare-earth (36) metals from their oxides and fluorides.Calciumatomic weight 40.08 electron configuration $1s^2 2s^2 2p^6 3s^2 3p^6 4s^2$ atomic weight natural abundance 40 96.947% atomic weight natural abundance 42 0.646 % atomic weight natural abundance 43 0.135% atomic weight natural abundance 44 2.083% atomic weight natural abundance 46 0.186% atomic weight natural abundance 48 0.18% specific gravity at 20°C 1.55×10^3 kg/m^3 melting point 839 ± 2 °C boiling point 1484 °C heat of fusion, ΔH_{fus}, 9.2 kJ/mol heat of vaporization, ΔH_{vap} 161.5 kJ/mol heat of combustion 634.3 kJ/mol temperature 0.133 800 °C temperature 1.33 970 °C temperature 13.3 1200 °C temperature 53.3 1390 °C temperature 101.3 1484 °C specific heat at 25°C 0.653 J/(g · K) coefficient of thermal expansion, 0–400°C 22.3×10^{-6} m/(m · K) electrical resistivity at 0°C 3.91 $\mu\Omega \cdot$ cm electron work function 2.24 eV tensile strength (annealed) 48 MPa yield strength (annealed) 13.7 MPa modulus of elasticity 22.1–26.2 GPa hardness (as cast) HB 16–18 hardness (as cast) HR B 36–40

BIBLIOGRAPHY

"Calcium and Calcium Alloys" under "Alkaline-Earth Metals and Alkaline-Earth Metal Alloys," in *ECT* 1st ed., Vol. 1, pp. 458–463, by C. L. Mantell, Consulting Chemical Engineer; "Calcium and Calcium Alloys" in *ECT* 2nd ed., Vol. 3, pp. 917–927, by O. N. Carlson and J. A. Haffling, Ames Laboratory, United States Atomic Energy Commission; in *ECT* 3rd ed., Vol. 4, pp. 412–421, by C. J. Kunesh, Pfizer Inc. "Calcium and Calcium Alloys", in *ECT* 4th ed., Vol. 4, pp. 777–786, by Stephen G. Hibbins, Timminco Metals; "Calcium and Calcium Alloys", in *ECT* (online), posting date: November 27, 2000, by Stephen G. Hibbins, Timminco Metals.

CITED PUBLICATIONS

1. J. W. Mellor, *Comprehensive Treatise on Inorganic and Theoretical Chemistry*, Vol. **3**, Longmans, Green & Co., Inc., New York, 1923, 619–631.
2. W. Hodge, R. I. Jafee, and B. W. Gonser, *RAND Corp. Report R-123*, Battelle Memorial Institute, Santa Monica, Calif., Jan. 1, 1949.
3. J. Katerberg and co-workers, *J. Phys. F.* **5**, L74 (1975).
4. R. C. Weast, ed., *Handbook of Chemistry and Physics*, 57th ed., CRC Press, Cleveland, Ohio, 1976, pp. B1, B3, B12, B276, D62, D165, D211, E81, F170.
5. F. Emley, in D. M. Considine, ed., *Chemical and Process Technology Encyclopedia*, McGraw-Hill Book Co., Inc., New York, 1974, p. 192.
6. C. L. Mantell, in C. A. Hampel, ed., *Rare Metals Handbook*, 2nd ed., Reinhold Publishing Corp., London, UK, 1961, 20–21.
7. F. X. Kayser and S. D. Sonderquist, *J. Phys. Chem. Solids* **28**, 2343 (1967).
8. G. De Maria and V. Piacente, *J. Chem. Thermodyn.* **6**, 1 (1974).
9. K. L. Agarwal and J. O. Betterton, Jr., *J. Low Temp. Phys.* **17**, 509 (1974).
10. E. Schuermann, P. Fünders, and H. Litterscheidt, *Arch. Eisenhuettenwes.* **45**, 433 (1974).
11. J. G. Cook, M. J. Laubitz, and M. P. van der Meer, *Can. J. Phys.* **53**, 486 (1975).
12. E. Schurman and R. Schmid, *Arch. Eisenhuettenwes.* **46**, 773 (1975).
13. M. Hansen, *Constitution of Binary Alloys*, McGraw-Hill Book Co., Inc., New York, 1958, 11–13, 75–77, 190, 302–303, 394–414.
14. U.S. Pat. 2,375,198 (May 8, 1945), P. P. Alexander (to Metal Hydrides, Inc.).
15. J. Evans and co-workers, *J. Less Common Met.* **30**, 83 (1973).
16. A. V. Vakhobov, V. G. Khudaiberdiev, and M. K. Nasyrova, *Izv. Akad. Nauk SSSR Met.* (4), 162 (1974).
17. A. V. Vakhobov, V. N. Vigdorovich, and V. G. Khudaiberdiev, *Izv. Vyssh. Uchebn Zavad. Tsvetn. Metall.* (4), 115 (1973).
18. In W. Gerhartz, ed., *Ullmann's Encyclopedia of Industrial Chemistry*, 4th ed., Vol. **A4**, Weinheim, Germany; Deerfield Beach, Fla., 1995, p. 515.
19. M. Miller, "Calcium and Calcium Compounds," *Minerals Yearbook*, Vol. **1**, U.S. Bureau of Mines, U.S. Department of the Interior, Washington, D.C., 1991, pp. 317–324.
20. *Calcium Metal MSDS*, Mallinckrodt Baker, Inc., Phillipsburg, N.J., 2000.
21. U.S. Pat. 4,439,398 (Mar. 27, 1984), R. D. Prengaman (to RSR Corp.).

22. H. Walter, in G. Volkert, ed., *Metallurgy of Ferroalloys*, Springer, Berlin, Germany, 1972, 570–581.
23. C. L. Mantell, in C. A. Hampel, ed., *The Encyclopedia of the Chemical Elements*, Reinhold Publishing Corp., New York, 1968, p. 102.
24. U.S. Pat. 5,041,160 (Aug. 20, 1991), D. Zuliani and B. Closset (to Timminco Ltd.).
25. U.S. Pat. 1,428,041 (Sept. 11, 1922), W. Kroll.
26. T. R. A. Davey, *Lead–Zinc–Tin '80*, AIME, 477–507 (1980).
27. M. V. Rose and J. A. Young, *5th International Lead Conference, Paris, France, Nov. 1974*, preprints, 1974.
28. R. P. Clarke and K. R. Grothaus, *J. Electrochem. Soc.* **118**, 1680 (1971).
29. O. N. Carlson, F. A. Schmidt, and H. A. Wilhelm, *J. Electrochem. Soc.* **104**, 51 (1957).
30. W. Z. Wade and T. Wolf, *J. Nucl. Sci. Technol.* **6**, 402 (1969).
31. H. A. Wilhelm, *The Metal Thorium*, American Society for Metals, Novelty, Ohio, 1958, 78–103.
32. U.S. Pat. 2,763,542 (Sept. 18, 1956), (to E. I. du Pont de Nemours & Co., Inc.).
33. H. A. Wilhelm, *J. Chem. Ed.* **37**, 56 (1960).
34. T. Oki and J. Tanikawa, *Nippon Kinzoku Gakkaishi* **31**, 1048 (1967).
35. R. K. McKechnie and A. U. Seybolt, *J. Electrochem. Soc.* **97**, 311 (1950).
36. F. H. Spedding and A. H. Daane, *Met. Rev.* **5**(19), 297 (1960).

Lisa M. Vrana
Consultant

CALCIUM CARBIDE

1. Introduction

Chemically pure calcium carbide [75-20-7], is a colorless solid; however, the pure material can be prepared only by very special techniques. China is the world's largest producer of calcium carbide. Commercial calcium carbide is composed of calcium carbide, calcium oxide [1305-78-8], CaO, and other impurities present in the raw materials. The commercial product's calcium carbide content varies and is sold on the basis of the acetylene yield. Industrial-grade carbide contains about 80% as CaC_2, 15% CaO, and 5% other impurities.

Commercial production by the electric furnace method was developed about 1892 by Henri Moissan in France and, independently by T. L. Willson and J. T. Morehead in the United States. The first commercial plant was built ay Spray, North Carolina in 1895. Development of the carbide industry for the generation of acetylene expanded rapidly after that.

Originally acetylene [74-86-2], was used as an illuminant since it burned with a brightness 10–12 times that of coal gas. It later found a use as a fuel gas in oxyacetylene welding and cutting since it developed flame temperatures of 3000°C compared to 1900°C for other gases available at the time. After World War I acetylene became the starting material in the synthesis of a host of organic material such as solvents, plastics, and synthetic rubber. Increased chemical use of acetylene increased demand for calcium carbide (See ACETYLENE-DERIVED CHEMICALS; ACETYLENE). The manufacture of calcium cyanamide from carbide also increased demand. Annual worldwide production of calcium carbide reached a peak of 8000 10^3 t. in the 1960s and has declined steadily since then to about 4700 10^3 t today (at the time of writing). The principal reason for decline was the substitution of acetylene from petrochemical sources (from by-product ethylene production and thermal cracking of hydrocarbons). Other factors include the use of alternative fuel gases replacing acetylene, development of electric welding techniques, and reduced production of calcium cyanamide [156-62-7]. Calcium carbide is, however, used extensively as a desulfurizing reagent in steel and ductile iron production, which allows these manufacturers to use high sulfur coke without the penalty of excessive sulfur in the resultant products. Many countries produce calcium carbide; the largest producer is China.

2. Properties

Table 1 list the more important physical properties of calcium carbide. Additional properties are given in the literature (1). Figure 1 gives the phase diagram calcium carbide–calcium oxide for pure and technical grades.

Table 1. **Physical Properties of Calcium Carbide**

Property	Value
mol wt	64.10
mp, °C	2300
crystal structure	
phase I, 25–447°C	face centered tetragonal
phase II,	triclinic
phase III[a]	monoclinic
phase IV, >450°C	fcc
commercial	grain structure, 7–120 μm
specific gravity, commercial-grade	
at 15°C	2.34
2000°C[b]	1.84
electrical conductivity, technical-grade, $(ohm-cm)^{-1}$	
at 25°C	3,000–10,000
1000°C	200–1,000
1700°C[b]	0.36–0.47
1900°C[b]	0.075–0.078
viscosity at 1900°C, MPa · s (=CP)	
50% CaC_2	6000
87% CaC_2	1700
specific heat, 0–2000°C, J/mol · K[c]	74.9
heat of formation, H_f,298, kJ/mol[c]	-59 ± 8
latent heat of fusion, ΔH_{fus}, kJ/mol[c]	32

[a] Phase III is metastable.
[b] Material is a liquid.
[c] To convert from J to cal, divide by 4.184.

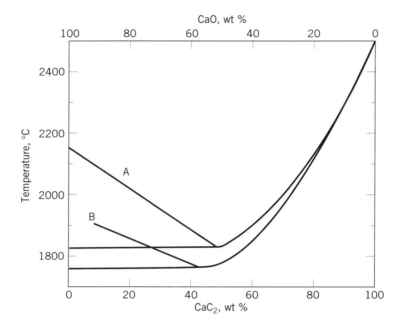

Fig. 1. Calcium carbide–calcium oxide phase diagram using A, pure CaC_2, and B, technical-grade CaC_2.

2.1. Reaction With Water. The following exothermic reaction of calcium carbide and water yielding acetylene forms the basis of the most important industrial use of calcium carbide:

$$CaC_2 + 2\,H_2O \longrightarrow C_2H_2 + Ca(OH)_2 \qquad (H = -130\ kJ/mol) \qquad (1)$$

where H_r = heat produced in the reaction. Wet and dry processes are in use for generating acetylene from calcium carbide. In the wet process, carbide is added to excess water in a generator. Water absorbs the heat of reaction, maintaining a temperature of less than 100°C. A 10% lime sludge slurry is a coproduct. In the dry process, an equal amount of water is added to carbide at a rate such that the heat of reaction evaporates the excess water. The coproduct, $Ca(OH)_2$, containing 2–3% water is a free-flowing powder that can be sold for acid neutralization or soil stabilization. The wet slurry can be used for flue-gas desulfurization at utilities. Both processes yield acetylene of 99.6% purity.

2.2. Reaction With Sulfur. An important use of calcium carbide has developed in the iron and steel industy, where it was found to be a very effective desulfurizing agent for blast-furnace iron. Calcium carbide reacts with sulfur present in the molten metal as follows:

$$CaC_2 + S \longrightarrow CaS + 2\,C \qquad (2)$$

Sulfur was controlled in the past in the iron and steel industry by careful selection of raw materials. Since the availability of high-grade raw materials has declined, and in an attempt to maximize production rates, producers have shifted toward external desulfurization using additives such as calcium carbide in a separate step in the reduction process. Desulfurization by using nitrogen gas to inject calcium carbide powders into a ladle or torpedo car is both fast and efficient. Injection using a refractory lance below the metal surface can easily reduce the sulfur content from 0.1% to 0.01%. Other ingredients, such as calcium carbonate or asphaltites, which acts a gas-generating agent; or graphite, which assists in the injection process, are generally added to the desulfurizing reagent. The most modern process employs calcium carbide in combination with magnesium, either as a blend or in a sequential coinjection process. (see SULFUR REMOVAL AND RECOVERY).

2.3. Reaction With Nitrogen. Calcium cyanamide is produced from calcium carbide

$$CaC_2 + N_2 \longrightarrow CaCN_2 + C \qquad (H_r = -295\,kJ/mol) \qquad (3)$$

The reaction is carried out in a refractory oven by passing nitrogen gas through finely pulverized carbide at a temperature of 1000–1200°C. To initiate the reaction, the carbide is heated electrically using a graphite electrode located at the center of the charge. Since the reaction is strongly exothermic, it proceeds autogenously. In addition to the batch process, some European and Japanese producers have developed continuous nitrogenation furnaces based on a rotary kiln design. Powdered carbide and nitrogen gas is fed to the kiln and the cyanamide product removed in a granular form.

3. Manufacture and Production

Calcium carbide is produced commercially by reaction of high purity quicklime and a reducing agent such as metallurigical or petroleum coke in an electric furnace at 2000–2200°C.

$$CaO + 3\,C \longrightarrow CaC_2 + CO \qquad (H_r = 466\ kJ/mol\ ) \qquad (4)$$

Commercial calcium carbide, containing about 80% CaC_2, is formed in the liquid state. Impurities are mainly CaO and impurities present in raw materials. CO is usually collected and used as a fuel in lime production or drying of the coke used in the process. The liquid calcium carbide is tapped from the furnace into cooling molds.

3.1. Raw Materials. The carbon reducing agent can be metallurgical coke, petroleum coke, anthracite, or charcoal, depending on price and impurities. Metallurgical coke is the most common, because of it's availability. This coke typically contains 85–88% fixed carbon, 9–11% ash, and 2% volatiles. Lime is obtained by calcining limestone in a rotary or shaft kiln. Limestone should contain a minimum of 95–97% $CaCO_3$. Both furnace raw materials should not contain excessive amounts of dust or fines as this will result in furnace eruptions and unstable operation. In many cases, raw materials are prescreened to remove fines. The collected fines can be reintroduced to the furnace through a hollow electrode feed system.

3.2. Furnace Design.   Commercial calcium carbide furnaces have capacities ranging from 25,000 t/y (12 MW) to 130,000 tons/year (50 MW). The majority of furnaces are operated with three-phase current. The electrode arrangement can be either inline, or of a triangular symmetrical configuration. The inline arrangement, due to unsymmetrical arrangement of electrical conductors, causes an inductive effect, which results in a live phase (high power) and a dead phase (low power). The symmetrical arrangement has balanced power to all electrodes.

A cross section of a covered 40-MW furnace is shown in Fig. 2. The shell diameter is 9 m. A taphole to withdraw liquid carbide is located on top of carbon

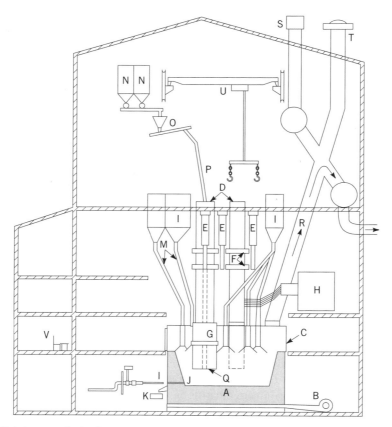

Fig. 2.   Calcium carbide furnace. The furnace crucible (a) is constructed of brick sidewalls, a carbon lined bottom; the bottom is cooled by a fan (b). The entire furnace is closed by a cover (c). The Soderberg electrodes (d) are supported and can be moved vertically by hydraulic cylinders (e). They are fitted with a slipping device (f). Contact plates (g) provide power connection to the electrodes from a single-phase transformer (h). A tapping electrode (i) assists the flow of carbide from the taphole (j) to a chill car (k). Raw material is stored in bins (l) and introduced to the furnace through charging chutes (m). Dust raw material is stored in separate bins (n) and fed to the furnace via screw conveyor (o), flexible connector (p), through the hollow electrode (q). Furnace gas is removed through a duct (r). A crane (u) charges fresh electrode paste. An operator monitors operation from a control room (v).

blocks that line the furnace bottom. The furnace cavity is covered with a relatively gastight cover to facilitate CO gas collection. Preweighed raw materials enter the furnace through water-cooled charge tubes. Three electrodes, in a triangular arrangement, enter the furnace through water-cooled sealing devices in the cover. The electrodes are usually of the continuous self-baking Soderberg type with a diameter of 1100–1600 mm. Electric current is bought from transformers by water cooled bus tubes, connecting at the electrode through water-cooled copper contact plates. Each electrode assembly is suspended or supported by hydraulic cylinders, allowing for independent movement and control of electrical current.

Furnaces may be classed as open or closed. Earlier furnaces were open, referring to the fact that the CO gas was allowed to burn open on the surface of the raw-material mix burden. A closed furnace has a sealed cover that allows for 100% collection of the CO gas, which can then be used as a fuel for a lime kiln or drying the carbon reducing agent. Air pollution control requirements are less since only the dust-laden CO gas requires cleaning, compared to the large volume of draft gases from the open furnace. Most modern furnaces incorporate the closed design.

3.3. Electrodes. Furnace electrodes are carbon materials, due to the excellent electrical and thermal conductivity properties of the carbon. Early electrodes were constructed of prebaked carbon pieces fabricated into the electrode shape. Today these have been largely replaced with the Soderberg self-baking electrode (see Fig. 3). This electrode consists of a sheet steel cylinder with vertical ribs filled with Soderberg paste, a formulated mixture of electrically calcined anthracite and coal tar pitch. Heat from the electrical current bakes the green electrode material to form a strong solid baked electrode. As the electrode is consumed in the furnace process, electrode is replaced in a process called "slipping". This is carried out by a pair of hydraulically tightened slipping bands mounted one above the other on the electrode column. The electrode is held by the lower set of bands, allowing the upper set to be raised to a new position. On release of the lower bands, the electrode can be lowered safely using the hydraulic cylinder provided for the upper bands. Normally the electrode is held in the stable position by the action of both bands. New sections of the sheet steel cylinder and electrode paste are added as required.

3.4. Hollow Electrode. Most modern furnaces utilize the hollow electrode feed system. Coke and lime fines, conveyed in a stream of recycle furnace CO gas or nitrogen, can be fed to the furnace through a 10–15-cm pipe channel at the center of each Soderberg electrode. Although the steel pipe melts and disappears in the current carrying zone of the electrode, it remains intact long enough to maintain a continuous opening. Fines delivered to this zone react quickly and provide a valuable tool for adjusting carbide grade and maintaining proper electrical load balance. In addition to the economic gain of utilizing a waste material and eliminating a disposal problem, there is a 30% reduction in usage of electrode carbon. Approximately 15–20% of the total raw material charge can be introduced through the hollow electrodes. It is also possible to recycle calcium carbide dust, which is generated in the carbide crushing process.

3.5. Energy Requirements. Approximately 865 kg of lime and 495 kg of metallurgical coke are required to produce a metric ton of 80% purity calcium

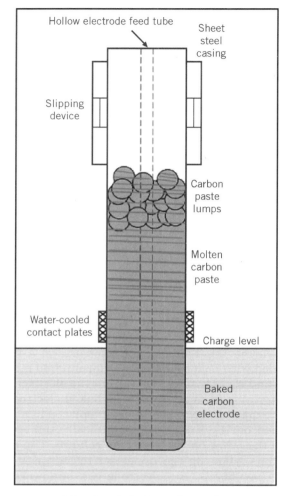

Fig. 3. Soderberg electrode.

carbide. Theoretical energy requirement per metric ton is about 2200 kWh, but because of heat losses, 2900–3300 kWh is required. For every ton of 80% purity carbide produced, about 280 m^3 (15°C) of furnace gas is evolved, which analyzes as 75–85% CO, 5–8% H$_2$, and the balance as N$_2$, CO$_2$, and CH$_4$. Electrode consumption varies from 10–27 kg/ton of carbide.

3.6. Furnace Operation.   In the vicinity of the electrode tip the mix is hot enough to allow the lime to melt. The coke does not melt and the liquid lime percolates downward through the relatively fixed bed of coke forming calcium carbide, which is liquid at this temperature. Both liquids erode coke particles as they flow downward. The weak carbide first formed is converted to richer material by continued contact and reaction with coke particles. The carbon monoxide gas produced in this area must be released by flowing back up through the charge. The process continues down to the taphole level. Material in this area consists solid coke wetted in a pool of liquid calcium carbide at the furnace bottom. The ease with which carbon monoxide can escape from the reaction zone has a bearing on smooth operation of the furnace. The normal furnace charge, consisting of large particles, has a good porosity, allowing a reasonably constant gas flow. Extremely fine material or crusts of condensed impurities impede the escape of gas from the reaction area, allowing pressure to develop and eventually results in "blows" in which hot mix and liquid carbide can be explosively ejected.

Metallurgical coke gives rise to ferrosilicon, which in the liquid phase is more dense than calcium carbide and tends to settle and penetrate the bottom of the furnace. After a lengthy operating period it may extend 30 cm or more below the taphole, eventually reaching the furnace shell, where it causes hot spots requiring repair and replacement of the furnace refractory.

An evenly operating furnace is essential for an efficient process as indicated by steady electrode position, regular descent of mix through all charging chutes, and regular tapping of carbide equivalent to power input to the furnace. These conditions are attained by maintaining standard operating procedures, which include frequent and adequate tapping of carbide constant electrical conditions and hence power input and a constant coke:lime ratio in the mix charge.

3.7. Computer Control. The use of computer control systems to control the operations of submerged arc furnaces, including calcium carbide, has been successfully demonstrated in the United States. Operations directly under control are mix batching, electrode position and slip control, carbide gas yield, power control, and cooling-water systems. Improvements in energy usage, operating time, and product quality are obtained.

3.8. Tapping System. Carbide is tapped from the furnace in a liquid stream at a temperature of 1900–2100°C. Its low thermal conductivity makes it possible to tap directly into cast iron chill molds even though the melting point of the cast iron is lower than that of the carbide. Carbide furnaces can have either a single taphole at one electrode or three tapholes leading to the center of each electrode. With three tapholes, each electrode is tapped in turns at predetermined intervals. Burning through the taphole with a tapping electrode opens the taphole channel. A pneumatic ram bar is also used to poke the taphole at regular intervals and promote the flow of liquid carbide. The liquid carbide is cast into iron chill cars weighing 1–4.5 tons. Following cooling for 24–36 h, the carbide is crushed and screened according to the preestablished sizes that are desired.

4. Grade Specifications

Contracts for acetylene-grade carbide are usually based on size and gas yield specification, with penalties for carbide that fails to meet specified gas yield. In general gas yields range from 280 to 300 L/kg, depending on the screen size of the carbide.

The most important size ranges specified in individual standards are (in millimeters) 0.4–1.7 or 14 ND, 2–4, 4–7, 7–15, 15–25, 25–50, and 50–80. Phosphorus, sulfur, and arsenic levels are determined by analyzing phosphine, hydrogen sulfide, and arsine content of the evolved acetylene according to specified procedures. Gas impurities are typically 0.05% phosphine, 0.15% hydrogen sulfide, and 0.001% arsine.

5. Analytical Methods

Many countries have standards, that specify methods of analysis for gas yield, sieve analysis, and acetylene gas impurities such as phosphine hydrogen sulfide and arsine:

United States	Federal Specification 0-C-101a
Australia	Standards Association of Australia, Publication No. K 49
Europe	Federal Republic of Germany DIN 53922
Great Britain	British Standard 642
India	Indian Standard IS1040
Israel	Israel Standard 267 T (250 × 4) SJ 446
Japan	Japanese Industrial Standard JIS K 1901
Philippines	Philippine standard PNS No. 1991
Mexico	DIN 53922 and Norme Official K 23
South America	Mercosul Standard NM 109:97-Calcium Carbide

Typical analysis of carbide is presented in Table. 2.

6. Shipment and Transportation

Since calcium carbide produces flammable acetylene gas in the presence of water, it is classified as a hazardous material. Each country has developed regulations governing its transport. In the United States these regulations fall under the Department of Transport. In general, calcium carbide is shipped in metal containers, either in bulk bins of up to 4-ton capacity, or in drums ranging in size from 50 to 230 kg. Under certain conditions shipment is allowed in reinforced bulk bags of up to 1-ton capacity. These bags have an advantage of reduced weight of shipping containers. Calcium carbide containers are marked "Flammable solid, dangerous when wet" with the United Nations designation UN 1402.

Calcium carbide for desulfurization is usually sold on the basis of a specified minimum CaC_2 content, together with minimum levels of various additives, and a specified particle size distribution. This can vary considerably according to the

Table 2. **Typical Analysis**

	Calcium carbide
CaC_2	79.0
CaO	13.2
C	0.7
CaS	0.9
mineral impurities	
(Al_2O_3, SiO_2, MgO, FeSi)	6.2

reagent formulation. Domestic shipments are either in steel tote bins of 2–4-ton capacity, bulk railcars, or trucks with pneumatic unloading capabilities. Containers are usually pressurized slightly with an inert gas such as dry nitrogen to avoid reagent contact with atmospheric moisture and generation of acetylene.

6.1. Environmental Concerns. The major environmental problem in carbide production is the prevention of particulate dust emission. Normally cloth filtration equipment is used in handling of raw materials, at furnace tapping, and product crushing and screening. The carbide furnace CO gas steam can be treated by high temperature cloth filtration, ceramic filters, or wet scrubbers. The dust collected by either wet or dry methods may contain trace cyanide that must be treated before disposal.

The hollow electrode system allows for the use of small coke and lime particles, which could not otherwise be used in the furnace in the main furnace charge. A covered furnace collects all the CO gas generated during carbide manufacture. This fuel gas can provide energy for a lime kiln operation or drying carbon raw materials.

Acetylene generators produce a calcium hydroxide coproduct during acetylene manufacture. In the dry process the powder can be used for acid neutralization or soil stabilization. In the wet process the slurry can be used for flue-gas desulfurization at utilities. If the wet process is used the hydrate is collected in settling lagoons for disposal. The dry process produces a powder hydrate that can be used for flue gas desulfurization or soil stabilization.

7. Health and Safety Factors

The usual precautions must be observed around high-tension electrical equipment supplying power. The carbon monoxide formed, if collected in closed furnaces, is usually handled through blowers, and scrubbers, and thence to a pipe transmission system. As calcium carbide exposed to water readily generates acetylene, the numerous cooling sections required must be constantly monitored for leaks. When acetylene is generated, proper precautions must be taken because of the potential explosiveness of air–acetylene mixtures over a wide range of concentrations (from 2.5%–82% acetylene by volume).

Although acetylene is considered to be a material with a very low toxicity, a threshold value (TLV) of 2500 ppm has been established by NIOSH. In the presence of small amounts of water, calcium carbide may become incandescent and ignite the evolved air–acetylene mixture. Non sparking tools should be used when working in the area of acetylene-generating equipment.

Calcium carbide dust, in the presence of hydrated, lime is a skin irritant. Full garment protection (long sleeves, gloves, long underwear, etc) is usually practiced.

8. Economic Aspects

Generally worldwide production of calcium carbide and calcium cyanamide have declined steadily since reaching peaks in the 1960s. Calcium carbide has fallen

Table 3. **Historical Trends of Calcium Carbide Production, 10³**

Region	1936	1962	1972	1982	1997[a]
USA, Mexico	145	982	447	350	270
Canada	209	318	73	100	
Europe	1335	4740	4710	3260	1320
Asia	477	1800	2000	2420	2962
Africa	15	60	80	250	90
South America		100	190	220	70
Total	*2181*	*8000*	*7500*	*6600*	*4712*

[a] Estimated

from about 8000 to 4700 10³ mt/y in 1997 (Table 3). The major reason for decline was due to the substitution of acetylene from petrochemical sources (from by-product ethylene production and thermal cracking of hydrocarbons). This substitution occurred in the late 1960s and early 1970s. Other factors include the use of alternative fuel gases replacing acetylene, development of electric welding techniques, and reduced production of calcium cyanamide. (Prices listed in Table 4).

8.1. North America. There are currently three producers of calcium carbide in North America: The Carbide Graphite Group Inc., Elkem-American Carbide Company in the United States, and Metaloides SA de CV in Mexico (Table 5). There is no production in Canada following the shutdown of Shawbec in Shawinigan, Quebec in 1988 and Cyanamid Canada's carbide plant in Niagara Falls, Ontario in 1992.

Elkem-American and Carbide Graphite supply the domestic U.S. market and exported 33,479 t. in 1995, (3) mainly to Canada. Imports to the United States are relatively low, at 1479 t. in 1995. Metaloides SA de CV supplies the market in Mexico as well as exporting carbide to Central America, South America, and the Caribbean.

8.2. Western Europe. Calcium carbide production has declined from about 2200×10^3 t/yr in the 1970s to about 450×10^3 t/yr today (at the time of writing) (Table 6). The uses are split between calcium cyanamide manufacture, acetylene, and desulfurization reagent.

Table 4. **Estimated Price of Calcium Carbide (FOB plant)[a]**

Country	Price (local currency) per ton	Exchange rate Local Currency per U.S. $	U.S. $ per ton
United States	$500		500
Germany	850 DM	1.80	472
Japan	93,800 yen	125	750
South America			460
South Africa			550
India	21,000 rupees	40	525
Asia	57,500 yen	125	460

[a] Estimated.

Table 5. **Calcium Carbide Producers and Capacity, 1997**[a]

Region	Producers	Estimated annual capacity 10^3 mt	
North America	The Carbide Graphite Group Inc.		
	Calvert City, KY, USA	80	
	Louisville, KY, USA	110	
	Elkem American Carbide Co.		
	Ashtabula, Ohio, USA	110	
	Pryor, OK, USA	35	
	Metaloides SA, Puebla, Mexico	40	
Western Europe	Donau Chemie, Landeck, Austria	40	
	Pechiney Electrometallurgie, Bellegarde, France	60	
	SKW, Hart, Germany	320	
	Odda Smelteverk, Odda, Norway	130	
	Carburos del Cinca, Monzon, Spain	40	
	CEDIE SA , Orense, Spain	30	
	Casco Nobel AB, Sundsvall, Sweden	50	
Eastern Europe	Zakledy Azotowa, Chorzow, Poland	140	
	Bicapa, Tirnaveni, Romania	220	
	Novacke Chemicke, Preividza, Slovak Republic	110	
	Jugorom, Tetovo, Yugoslavia (former)	34	
	Etibank Antalya, Antalya, Turkey	20	
	Russia (several locations)	600	
South Africa	Karbochem, Newcastle	120	
South America	Carbometal SA, Mendoza, Argentina	56	Closed
	Electrometalurgica Andina, San Juan, Argentina	50	
	Companhia Brasileria Carbureto de Calcio, Brazil	60	Closed
	SA White Martins, Iguatama, Brazil	65	
	Columbiana de Carburo y Derivados, Cajica, Columbia		Closed
	Hornos Electros de Venezuela CA	5	
India	Birla Jute \amp Industries Ltd, Calcutta	16	
	Panyam Cements \amp Mineral Industries, Hagari	15	Closed
	Shriram Alkali \amp Chemicals, Jhagadiya	55	
	Tecil Chemicals \amp Hydro Power Ltd.	40	
	Industrial Chemicals \amp Monomers Ltd., Madrass	10	
	Small scale units in other places in India	20	
	Bhutan Carbide \amp Chemical	25	
Asia — Japan	Denki Kagaku Kogyo		
	Omi, Niigate	300	
	Omata, Fukuoka	110	
	Toyo Denka Kogyo, Kochi	60	
Asia—China	Many producersbSee Table 5 for individual producers.	3450	

Table 5   (*Continued*)

Region	Producers	Estimated annual capacity 10^3 mt
Asia—other countries	Emdeki Utama PT, Gresik, East Java	25
	Tae Kyung Industrial Co., Chongson, Korea	60
	MCCI, Iligan City, Philippines	30
	Formosa Plastics, Ilan, Tiawan	70
	MCB Industries, Perak, Malaysia	25
	M Thai Industrial, Samut Sakorn Thailand	30

[a] Estimated.

Calcium carbide-derived acetylene has declined considerably, much for the same reasons as in North America: oxyacetylene welding techniques replaced by electric welding and plasma techniques and less expensive acetylene obtained from ethylene. This decline is expected to continue at about 5% annually.

Calcium carbide–based desulfurization reagents declined in the early 1990s as magnesium-based reagents gained increased use. In the late 1990s years this market stabilized as users realized the benefits of using carbide–magnesium combined processes. Calcium carbide use in this application is expected to continue at the current level.

Exports of calcium carbide are in the order of $10-15,000^2$ t/yr, mainly to North Africa, with some going to Australia.

8.3. Eastern Europe. All countries except Hungary produce calcium carbide. Overall, production has fallen from about 2500×10^3 t. annually in the 1970s to about 870×10^3 t today (Table 6). The reasons for decline are similar to those of Western Europe and North America.

Carbide is used primarily for acetylene generation for industrial applications in oxyacetylene welding and cutting. Acetylene is also used to produce

Table 6. Calcium Carbide Production by Region, 1997, 10^3 t

Region	Total production	Calcium Carbide Consumption[a]		Desulfurization	Calcium cyanamide
		Acetylene			
		Chemical	Industrial		
North America	270	68	135	68	
South America	70		60	10	
Western Europe	450		140	130	150
Eastern Europe	870	675	180	15	
South Africa	90	15	32	23	20
India	100	22	68	10	
Japan	250b	107	45	15	83
China	2512				
East Asia, others	100	25	71	4	
Total	*4712*	*912*	*731*	*275*	*253*

[a] Ref. 2, other values are estimated.

chemicals such as vinyl chloride monomer for PVC. The steel industries are just starting to use calcium carbide for desulfurization of hot metal. It is expected that the use of calcium carbide for chemical manufacture will decline, but its use as a desulfurizing reagent is expected to expand quite rapidly from a low base.

8.4. Japan. There are only two producers of calcium carbide:Denki Kagaku, which also produces calcium cyanamide, and Toyo Denka. As in other regions of the world, production has declined.

Chemical uses of acetylene (eg, PVC, chloroprenes, and acetylene black) account for about 42% of carbide consumption. However, the increased use of olefins such as ethylene and propylene instead of acetylene for various derivatives has weakened the demand for carbide. Industrial uses in cutting and welding have also declined because of changing welding techniques.

Calcium carbide desulfurizing reagent accounts for about 6% of carbide consumption, and use of these reagents has declined gradually with replacement by other less expensive reagents.

8.5. China. China is the worlds largest producer of calcium carbide, with a volume of 2512×10^3 t in 1995 (Table 6) or about 53% of the world's production. Unlike Western countries, where carbide production has declined, China's has increased from a level of $2200^3 \times 10^3$ t in 1990 (estimated). Carbide is used to produce acetylene for industrial and chemical applications (PVC, acetylene black), calcium cyanamide, and dicyandiamide; the majority is probably used in industrial and chemical applications. In the current 9th Five Year Plan energy conservation is a main goal for the Ministry of Chemical Industry (4). For the calcium carbide industry a reduction of electrical energy consumption from 3550 to 3360 kWh/tonne carbide is targeted. To achieve this, new furnace installations must be rated greater than 45,000 t/yr. Also, energy-saving features should be used in new and expanded units. These include technologies such as closed furnace design, hollow electrode system, furnace dry gas cleaning, and furnace gas-fired lime kilns. The major calcium carbide producers in China are listed in Table 7.

8.6. South Africa. The calcium carbide business in South Africa has undergone some restructuring, with Karbochem in Newcastle the only remaining producer in Africa. The AECI carbide plant at Sasolburg was purchased by Samancor in 1993. This plant is currently mothballed and may be converted to ferroalloy production in the future. Another AECI carbide plant at Bellengeich was taken over by Karbochem. The Bellengeich furnaces have been sold to Siltech, who use them for producing ferrosilicon. Carbide from Sasolburg was used extensively to produce acetylene-based chemicals (PVC); however, these are now produced from ethylene sources. Acetylene black for dry batteries continues to be produced from carbide. South Africa exports an estimated 20×10^3 t annually to neighboring countries.

8.7. South America. Only three carbide producers remain active in South America (Table 6) with Carbometal SA, Mendoza, Argentina, Companhia Brasileria Carbureto de Calcio, Brazil, and Columbiana de Carburo y Derivados, Cajica, now closed. White Martins and Andina are the only suppliers. The present level of carbide demand is expected to continue in both desulfurization and acetylene uses.

Table 7. **Calcium Carbide Producers, China**

Producer	City	Province	Capacity[a]
Yunnan Chemical Engineering Factory	Kumming	Anning	20
Wuhu Chemical Plant	Wuhu	Anhui	
Beijing Chemical Industry Group	Beijing	Beijing	
Fujian Sanming Chemical Industry Complex[b]	Sanming	Fujian	
Gansu Calcium Carbide Plant	Yongdeng	Gansu	106
Hebei Tangshan Dongkuang Chemical Factory	Dongkuang	Hebei	
Ganchan Fuyang Chemical Group Corp.	Ganchan	Hebei	
Hebei Jiheng Chemical Factory	Hengsui	Hebei	
Hebei Luquan Tianyuan Dianhua Factory[b]	Luquan	Hebei	
Wuhan Gehua Industrial Group	Wuhan	Hebei	
Zhangjiakou Xiahuayuan Calcium Carbide Plant	Zhangjiakou	Hebei	105
Harbin Petrochemical Industry Corp.	Harbin	Heilongjiang	
Mudonjiang Petroleum Chemical Co.	Mudonjiang	Heilongjiang	80
Huangshi Calcium Carbide Factory	Huangshi	Hubei	
Hubei Dayan Chemical Fertilizer Plant	Huangshi	Hubei	
Hubei Jinmen Calcium Carbide Works	Jinmen	Hubei	
Hubei Zhongtian Group Company Xinhuo Chemical Corp.	Jinmen	Hubei	
Xitujiazhu Chemical Plant	Sien	Hubei	
Hubei Changyi Resin Plant No.7	Yichang	Hubei	
Hubei Changyli Chemical Plant	Zijiang	Hubei	
Zhuzhou City Xiangdong Chemical Plant	Zhuzhou	Hunan	
Nanjing Lanye Group Corp.	Jiangnin	Jiangsu	
Shanhai Railway Industry Co.	Nanjing	Jiangsu	
Jiangxi Gannan Chemical Factory[b]	Ganzhou	Jiangxi	
Nanton Roudon Chemical Plant	Ludon	Jiansu	
Jilin Chemical Industry Group	Jilin	Jilin	60
Sinochem Siping Chemical Complex	Siping	Jilin	

Table 7　(*Continued*)

Producer	City	Province	Capacity[a]
Fushun Organic Chemical Factory	Fushun	Liaonin	
Jinzhou Organic Chemical Plant	Jinzhou	Liaoning	
Baotou No. 2 Chemical Factory	Baotou	Nei Mongol	80
Huhehote Chemical Factory	Hohhot	Nei Mongol	
Huanghe Chemical Factory	Wuhai	Nei Mongol	
Wuda Calcium Carbide Factory[b]	Wuhai	Nei Mongol	
Wuhai Calcium Carbide Factory	Wuhai	Nei Mongol	
Ningxia Darong Dicyandiamide Co[b]	Shizuishan	Ningxia Hui	
Ningxia Shizuishan National Chemical Co[b]	Shizuishan	Ningxia Hui	
Qingdao Honqui Chemical Works	Qingdao	Shandong	
Zibo Organic Chemical Factory	Zibo	Shandong	
Shanghai Wusong Chemical Factory Calcium Carbide	Baosan	Shanghai	
Chemical Group of Changzhi City	Changzhi	Shanxi	
Yang Quan Chemical Material Plant	Yang Quan	Shanxi	10
Shanxi Chemical Industry Factory	Datong	Shanxi	
Niang Zhi Guan Asia Pacific Calcium Carbide Co.[b]	Ping Ding	Shanxi	11
Taiyuan Chemical Industry Group	Taiyuan	Shanxi	
Tianjin Bohai Chemical Group Co.	Tianjin	Tianjin	
JianDe Genglou Chemical Plant	Hangzhou	Zhejiang	
Huzhou Chemical Factory[b]	Huzhou	Zhejiang	
Juhua Group Corp.[b]	Quozhou	Zhejiang	
Quzhou Chemical Group Corp.[b]	Quzhou	Zhejiang	
Hangzhou Electrochemical Co.[b]	Xiaoshan	Zhejiang	

[a] Probable capacity less than 100×10^3 t per plant.
[b] Also produces calcium cyanamide.

8.8. Asia. In other regions of Asia, such as Korea, the Philippines, Indonesia, and Thailand, carbide is used mainly for industrial uses as a fuel gas, with a small amount for chemicals (acetylene black).

9. Uses

Calcium carbide has three primary applications today. It is used to produce acetylene gas for heating, oxyacetylene welding, metal cutting, acetylene black, and acetylene-derived chemicals such as vinyl ethers and acetylenic chemicals and alcohols. It is also used as a desulfurizing reagent for iron and steel, and as an intermediate for calcium cyanamide manufacture.

9.1. Acetylene. Acetylene production accounts for about 75% of calcium carbide consumption (Table 6). There are two main areas of use: industrial applications, which include heating, welding, and cutting; and chemical uses, which include calcium carbide–derived acetylene for various chemicals.

Acetylene is generated by the chemical reaction between calcium carbide [75-20-7] and water. Most carbide acetylene processes are wet processes from which hydrated lime, $Ca(OH)_2$, is a by-product. The hydrated lime slurry is allowed to settle in a pond or tank after which the supernatant lime-water can be decanted and reused in the generator. Federal, state, and local legislation restrict the methods of storage and disposal of carbide lime hydrate and it has become increasingly important to find consumers for the by-product. The thickened hydrated lime is marketed for industrial wastewater treatment, neutralization of spent pickling acids, as a soil conditioner in road construction, and in the production of sand-lime bricks.

The purity of carbide acetylene depends largely on the quality of carbide employed and, to a much lesser degree, on the type of generator and its operation. Carbide quality in turn is affected by the impurities in the raw materials used in carbide production, specifically, the purity of the metallurgical coke and the limestone from which the lime is produced.

All North American suppliers produce carbide for industrial applications. Carbide is sold to companies such as BOC Gases, Air Liquide America Corp., Air Products and Chemicals Inc., and Liquid Carbonic Industries, which produce acetylene for use as an industrial gas. The use of acetylene for welding and cutting has declined in recent years because of replacement with alternative fuel gases, such as propane and propylene based gas mixtures, and the use of electric welding machines. This tend may continue at an average rate of 2–5% per year. The Carbide Graphite Group is the only company producing calcium carbide derived acetylene for chemical use. It is shipped via pipeline to International Specialty Products Inc. and Air Products and Chemicals Inc. (to produce vinyl esters, acetylenic chemicals and alcohols) and to Du pont for vinyl fluoride.

Since calcium carbide-derived acetylene competes with less expensive acetylene derived from the oxidation of natural gas and acetylene produced as a by-product from the manufacture of ethylene, it is not used on large scale for chemical production.

9.2. Calcium Cyanamide. Calcium cyanamide can be manufactured by either the Frank–Caro batch oven process or a continuous processes such as a rotary furnace developed in Trotsberg, Germany. In both processes the carbide is first ground to a powder in a ball or rod mill. The charge is further prepared by the addition of 1–2% fluorspar and recycled calcium cyanamide. The cyanamide dilutes the carbide, thereby preventing a temperature rise that would tend to decompose cyanamide.

The most important use of calcium cyanamide is as fertilizer, but it is also effective as a herbicide and defoliant. It was used as a starting material for ammonia until it was displaced by the Haber process. Several derivatives of calcium cyanamide have also been developed: hydrogen cyanamide (intermediate for insecticides, pharmaceuticals, soil sterilants), dicyandiamide (intermediate for flame and fire retardants, viscosity reducers for glues and adhesives, nitrification inhibitors in fertilizers), melamine (thermoset plastic), and calcium cyanide (gold extraction and fumigant). Calcium cyanamide production has fallen from a 1960s peak of 1300 to about 320×10^3 t/y today (excluding China). The use of cheaper ammonia-based fertilizers is the main reason for decline, as well as melamine, which is now being produced from urea sources instead of from cyanamide. Producing countries include Norway, Germany, China, and Japan.

Generally worldwide production of *calcium cyanamide* has declined since reaching peaks in the 1960s. The major producers of Calcium cyanamide worldwide are listed in Table 8.

Calcium cyanamide production has fallen from a 1960s peak of 1300 to about 320×10^3 t/yr at the time of writing. Cheaper ammonia based fertilizers is the main reason for decline, as well as melamine (trimer of cyanamide) now being produced from urea sources instead of from cyanamide.

9.3. Desulfurizing Reagent.   Calcium carbide for metallurgical applications accounts for about 25% of its use (Table 6). Carbide for desulfurization in steel mills began in the 1970s and has allowed steel mills to use to use high-sulfur coke in the blast furnace without the penalty of excessive sulfur in the resultant steel. The reagent generally consists of commercial carbide that has

Table 8. **Calcium Cyanamide Producers**[a]

Producer	Cyanamide products produced			
	Cyanamide fertilizer	Dicyandiamide	Cyanamide solutions	Calcium cyanide
SKW, Trotsberg, Germany	Yes	Yes	Yes	
Odda smeltverk AS, Odda, Norway	Yes	Yes		
Polifin, Witbank, South Africa				Yes
Shin-Etsu Kasei, Japan	Yes			
Denki Kagaku Kogyo, Japan	Yes			
Nippon Carbide, Japan	Yes	Yes	Yes	
China	Yes	Yes		

[a] Author's estimate.

been ground to a powder, or the pulverized carbide mixed with other ingredients such as lime, limestone, graphite, coal, solid hydrocarbons, and silicone. The powdered regent is injected into molten iron using an inert-gas carrier. Calcium carbide competes with magnesium based desulfurizing reagents, and switching costs between these reagents are low. Calcium carbide has the advantage of being less expensive; however, magnesium has a higher affinity for sulfur.

Calcium carbide and magnesium are also used as simple blends or in coinjection processes. With coinjection, the two reagents can be injected simultaneously or sequentially. The less expensive carbide removes 90% of the sulfur, and then the remaining sulfur is removed by magnesium injection.

For the foreseeable future it is expected that both reagents will be used in steel mills and demand will change in line with steel production. Carbide Graphite, Elkem-American, and Metaloides all produce powdered carbide for desulfurization. All producers market carbide of a granular size for desulfurizing ductile iron in foundries. This market is expected to be stable.

10. Acknowledgements

The author is grateful to the following companies that supplied information on which "authors estimates" are based. BOC Gases, United Kingdom Carbide Graphite Group, United States Metaloides SA, Mexico Karbochem, South Africa SA White Martins, Iguatama, Brazil Birla Jute & Industries Ltd., Calcutta, India MCCI, Iligan City, Philippines.

BIBLIOGRAPHY

"Carbides" under "Calcium," in *ECT* 1st ed., Vol. 2, pp. 834–846, by A. J. Abbott, Shawinigan Chemicals Ltd.; in *ECT* 2nd ed., Vol. 4, pp. 100–114, by A. G. Scobie, Shawinigan Chemicals Ltd.; "Carbides Calcium" under "Carbides," in *ECT* 3rd ed., Vol. 4, pp. 505–519, by N. B. Shine, Shawinigan Products Dept.; "Calcium carbides" under "Carbides," in *ECT* 4th ed., Vol. 4, pp. 848–891 by William L. Cameron, Cyanamid Canada, Inc; "Calcium Carbide" in *ECT* (online), posting date: December 4, 2000, by William L. Cameron, William Cameron Consulting.

CITED PUBLICATIONS

1. I. R. Juza and H. V. Schuster, *Z. Anorg. Chem.* **311**, 62 (1961).
2. *Chem. Eng. News* (Dec. 15, 1997).
3. *Current Industrial Report*, Fuorgaric Chemicals Series MA28A, U.S. Dept. of Commerce, Bureau of Census.
4. *China Chemical Reporter* (**32**), 163 (Oct. 16, 1996)

GENERAL REFERENCES

D. W. K. Hardie, *Acetylene, Manufacture and Uses*, Oxford University Press, London, UK, 1965.
M. Haley, *Chemical Economics Handbook, Calcium Carbide, United States*, SRI, Menlo Park, Calif., Dec. 1989.

F. W. Kampmann and W. Portz, *Chemicals from Coal via the Carbide Route*, Hoechst A. G., Heurth-Knapsack D-5030, Germany, 1991; *Crit. Rep. Appl. Chem.* **14**, 32–44 (1987).

C. J. Macedo, E. A. O. d'Avila, and J. G. Brosnan, *Startup of a Closed Carbide Furnace Using Charcoal as a Reducing Agent*, Vol. 43, *Electric Furnace Proceeding*, Atlanta, Ga., 1985.

G. E. Healy, *Why a Carbide Furnace Erupts, Electric Furnace Proceeding*, Pennsylvania State University, University Park, Pa., 1965.

U.S. Pat. 1,372,073 (1971), R. A. Casciani and co-workers (to Union Carbide Corp.).

U.S. Pat. 4,491,568 (1985), J. F. Bortnik and co-workers (to Elkem Metals Co.).

H. A. Corver and W. Gmohling, "Hot Metal Desulphurization—North American Experience with CaD," *Iron Steel Eng.* (May 1980).

W. G. Wilson and A. McLean, "Desulphurization of Iron and Steel and Sulfide Shape Control," The Iron and Steel Society of AIME, Warrendale, Pa., 1980.

WILLIAM L. CAMERON
William Cameron Consulting

CALCIUM CARBONATE

1. Introduction

Calcium carbonate [471-34-1], $CaCO_3$, mol wt 100.09, occurs naturally as the principal constituent of limestone, marble, and chalk. Powdered calcium carbonate is produced by two methods on the industrial scale. It is quarried and ground from naturally occurring deposits and in some cases beneficiated. It is also made by precipitation from dissolved calcium hydroxide and carbon dioxide. The natural ground calcium carbonate and the precipitated material compete industrially based primarily on particle size and the characteristics imparted to a product.

Natural ground calcium carbonate has been used for years as the primary constituent of putty. Since 1945, the processing of natural calcium carbonate has seen the introduction of beneficiation by flotation (qv) to remove impurities and the development of grinding processes to manufacture finer products. Precipitated calcium carbonate was first introduced in England in 1850; commercial production started in the United States in about 1913.

Calcium carbonate is one of the most versatile mineral fillers (qv) and is consumed in a wide range of products including paper (qv), paint (qv), plastics, rubber, textiles (qv), caulks, sealants (qv), and printing inks (qv). High purity grades of both natural and precipitated calcium carbonate meet the requirements of the *Food Chemicals Codex* and the *United States Pharmacopeia* and are used in dentifrices (qv), cosmetics (qv), foods, and pharmaceuticals (qv).

2. Properties

Calcium carbonate occurs naturally in three crystal structures: calcite [13397-26-7], aragonite [14791-73-2], and, although rarely, vaterite. Calcite is

Table 1. **Properties of Calcium Carbonate**

Property[a]	Calcite	Aragonite
specific gravity	2.60–2.75	2.92–2.94
hardness, Mohs'	3.0	3.5–4.0
solubility at 18°C, g/100 g H_2O	0.0013	0.0019
melting point, °C	1339[b]	[c]
	dec 900	
index of refraction[d]		
α		1.530
β		1.680
γ		1.685
ω	1.658	
ε	1.486	

[a] Ref. 3.
[b] At 10.38 MPa (102.5 atm).
[c] Decomposes to calcite at temperatures >400°C.
[d] Ref. 2.

thermodynamically stable, aragonite is metastable and irreversibly changes to calcite when heated in dry air to about 400°C. Vaterite is metastable to calcite and aragonite under geological conditions but is found during the high temperature precipitation of calcium carbonate (1). The crystal forms of calcite are in the hexagonal system with $\bar{3}2/m$ symmetry; the crystals are varied in habit and over 300 different forms have been described. Aragonite is orthorhombic with 2/m2/m2/m symmetry and three crystal habits are common: acicular pyramidal, tabular, and pseudohexagonal (2).

The commercial grades of calcium carbonate from natural sources are either calcite, aragonite, or sedimentary chalk. In most precipitated grades aragonite is the predominant crystal structure. The essential properties of the two common crystal structures are shown in Table 1.

3. Manufacturing and Processing

3.1. Natural Calcium Carbonate. The production of natural ground calcium carbonate starts with the quarrying of a deposit of chalk, limestone, or marble. The best deposits for most industrial applications are those having a high (>90% $CaCO_3$) purity and high brightness. Most calcium carbonate quarries are of the open-pit type but there are underground operations. The ore is taken to a primary crusher for size reduction and then into the processing plant. The plant process is dependent on the grade of material being made. Typically, coarse products that do not require high purity, 90–98% $CaCO_3$, go to secondary crushing. This may be a cone- or jaw-type crusher that produces material minus 4 cm. Final grinding for products down to approximately 5 μm median particle size can be done in a roller mill or ball mill. Products finer than 10 μm often involve additional processing, usually in a dry ball mill circuit with air classification.

For those grades requiring high purity or finer material the process is different. Ideally, the secondary crushing step should reduce the ore to the point where mineral impurities are liberated, typically <100 μm, without producing an excess of fines. The material may then be beneficiated through a mineral flotation process in which impurities are floated out. The flotation process produces a higher brightness material that is typically >98% calcium carbonate. Commonly, the flotation product is further ground in a ball mill to produce a product in the 2–50 μm particle range. Products having a median particle size less than 2 μm are usually wet ground in media or sand mills, the final product being a slurry that can be shipped after stabilizers and biocides are added, or dried for powdered products.

3.2. Precipitated Calcium Carbonate.

Precipitated calcium carbonate can be produced by several methods but only the carbonation process is commercially used in the United States. Limestone is calcined in a kiln to obtain carbon dioxide and quicklime. The quicklime is mixed with water to produce a milk-of-lime. Dry hydrated lime can also be used as a feedstock. Carbon dioxide gas is bubbled through the milk-of-lime in a reactor known as a carbonator. Gassing continues until the calcium hydroxide has been converted to the carbonate. The end point can be monitored chemically or by pH measurements. Reaction conditions determine the type of crystal, the size of particles, and the size distribution produced.

The reactions in this method are

calcination	$CaCO_3 \rightarrow CaO + CO_2$
hydration or slaking	$CaO + H_2O \rightarrow Ca(OH)_2$
carbonation	$Ca(OH)_2 + CO_2 \rightarrow CaCO_3 + H_2O$

Following carbonation, the product can be further purified by screening. This screening, also used to control the maximum size of the product, is followed by dewatering (qv). Rotary vacuum filters, pressure filters, or centrifuges are used in the mechanical removal of water. Final drying is accomplished as with natural calcium carbonate in either a rotary, spray, or flash dryer. Products having mean particle sizes from submicrometers (~0.03 μm) to several micrometers are available.

New processes for preparing calcium carbonate with useful properties as paper filters have been described (4,5).

Both natural ground or precipitated calcium carbonate are available as dry products shipped in 22.7 kg multiwall bags, supersacks, or in bulk via truck and railcar. Calcium carbonate slurry, primarily used by the paper industry, is shipped by truck and rail. The solids content of these slurries is typically >70% by weight for ground products and 20–50% for precipitated. In the 1980s small precipitation plants were built at the site of large North American papermills.

Some grades of calcium carbonate are surface coated to improve handling properties and dispersability in plastics. Treatments used are fatty acids, resins, and wetting agents. Coating reduces the surface energy, thereby facilitating dispersion in organic binders.

4. Economic Aspects

Consumption of fine ground calcium carbonate is six times that of the precipitated calcium carbonate. In 1999, the global market for fine ground exceeded 18×10^6 tons and the market for precipitated calcium carbonate was 2.7×10^6 tons. The United States market for fine ground calcium carbonate is expected to grow at the rate of 2% through 2004. In Western Europe the consumption of fine ground $CaCO_3$ is used in the paper industry and exceeds the use of precipitated $CaCO_3$. In 1999, Japan consumed 2×10^6 t of fine ground, which was used mainly in the paper industry. Consumption in Japan grew at a rate of 4.3% between 1991 and 1999 and this trend was expected to continue (13).

5. Specifications, Standards, and Quality Control

The most comprehensive set of test methods for calcium carbonate has been assembled by the Pulverized Limestone Division of the National Stone Association. Methods for particle size, brightness, +325 mesh (44 µm), and percentage of calcium carbonate have been published; standards are available and have been well characterized (7). The Technical Association of the Pulp and Paper Industry (TAPPI) has published methods for calcium carbonate used in the paper industry (8).

Food and pharmaceutical grades of calcium carbonate are covered by the *Food Chemicals Codex* (9) and the *United States Pharmacopeia* (10) and subject to U.S. Food and Drug Administration Good Manufacturing Practices (11). Both purity requirements and test methods are available (9,10). Calcium carbonate is listed in the *U.S. Code of Federal Regulation* as a food additive, and is authorized for use in both paper and plastic food contact applications.

6. Health and Safety Factors

Calcium carbonate is listed as a food additive (9) and not considered a toxic material. The exposure to dust is regulated and a Threshold Limit Value–Time-Weighted Average (TLV–TWA) of 10 mg/m^3 is set (12). OSHA DEL for total dust is 15 mg/m^3 and for respirable fraction, it is 5 mg/m^3 (13). Both natural ground and precipitated calcium carbonates can contain low levels of impurities that are regulated. The impurities depend on the source of material, processing, and the final grade; impurities are typically trace metals and naturally occurring minerals.

7. Uses

The use of calcium carbonate in paint, paper, and plastics make up the principal part of the market. In the paper industry calcium carbonate products find two uses: as a filler in the papermaking process and as a part of the coating on paper.

The benefits of calcium carbonate in papermaking are brighter paper, greater resistance to yellowing and aging, and the economic advantage of substituting inexpensive calcium carbonate for expensive pulp (qv). Depending on paper grade and applications, calcium carbonate can be 25% or more of the sheet. Both ground natural and precipitated calcium carbonate are used as paper fillers depending on the application. Blends of ground and precipitated calcium carbonate have found use in an effort to optimize the properties of both products (see PAPERMAKING MATERIALS AND ADDITIVES) (14).

The other significant market for calcium carbonate in paper is as the pigment in paper coatings. Paper is coated to improve its brightness, opacity, printability, ink receptivity, and smoothness. Ultra fine (< 1 μm mean particle size) ground calcium carbonate, in addition to providing these properties, improves the rheology of coating formulations applied at coater speeds up to 1600 m/min. Calcium carbonate may be the sole pigment in the formulation or may be used in combination with other fillers (qv) such as kaolin (see CLAYS). In coating applications the use of ground natural calcium carbonate far exceeds that of precipitated material.

The plastics industry is a primary consumer of calcium carbonate products. Flexible and rigid PVC, polyolefins, thermosets, and elastomers (qv), including rubber, utilize a wide variety of coated and uncoated grades. Each of these plastics categories benefit by calcium carbonate's lower cost in relation to the polymer. In addition to cost savings, the use of calcium carbonate provides improvements in modulus, heat resistance, hardness, shrinkage reduction, and color fastness. Increases in impact strength and improvements in stability are also benefits, especially with the use of coated grades.

Increased loadings of calcium carbonate in thermosets reduce cost and provide better surface characteristics.

Calcium carbonate is one of the most common filler/extenders used in the paint and coatings industry. Consumer and contractor paint formulas can include products from submicrometer size to coarse mesh sizes. The main function of calcium carbonate in paint is as a low cost extender. It is also used to improve brightness, application properties, stability, and exposure resistance. Coarse products help to lower gloss and sheen or even provide textured finishes. The selection of product type and particle size is determined by the desired performance and cost of the coating.

Calcium carbonate is also used in industrial finishes and powder coatings. These paints typically include finer products; the primary purpose is rheological and gloss control. Calcium carbonate is also used in paints to extend and enhance the use of titanium dioxide. This is accomplished by using the finest of natural ground products or precipitated grades.

Calcium carbonate continues to be used in its original application, putty, as well as caulks, sealants (qv), adhesives (qv), and printing inks (qv). Large volumes are used in carpet backing and in joint cements. It is used to improve body, reinforcement, and other properties.

Calcium carbonate is used in flue gas desulfurization. This application by a variety of engineering processes traps the sulfur–oxygen compounds produced in the combustion of coal (qv) (see COAL CONVERSION PROCESS; EXHAUST CONTROL, INDUSTRIAL; SULFUR REMOVAL AND RECOVERY).

Calcium carbonate is used in food and pharmaceutical applications for both its chemical and physical properties. It is used as an antacid, as a calcium supplement in foods, as a mild abrasive in toothpaste, and in chewing gum to name only a few (see FOOD ADDITIVES). Calcium carbonate can be used as builder in detergent compositions (15).

BIBLIOGRAPHY

"Calcium Carbonate" under "Calcium Compounds" in *ECT* 1st ed., Vol. 2, pp. 750–759, by R. H. Buckie, West Virginia Pulp and Paper Co.; in *ECT* 2nd ed., Vol. 4, pp. 7–11, by R. F. Armstrong, Diamond Alkali Co.; in *ECT* 3rd ed., Vol. 4, pp. 427–432, by R. H. Lepley, Pfizer Inc.; in *ECT* 4th ed., Vol. 4, pp. 796–801, by F. Patrick Carr, David K. Frederick, OMYA, Inc.

CITED PUBLICATIONS

1. R. J. Reeder, ed., *Carbonates, Mineralogy and Chemistry*, Mineralogical Society of America, Washington, D.C., 1990, p. 191.
2. C. Klein and C. S. Hurlbut, Jr., *Manual of Mineralogy*, John Wiley & Sons, Inc., New York, 1985, pp. 328, 335.
3. H. S. Katz and J. V. Milewski, *Handbook of Fillers for Plastics*, Van Nostrand Reinhold Co., New York,1987, p. 123.
4. U.S. Pat. Appl. 20030059362 (March 27, 2003), T. Kazuto and co-workers.
5. U.S. Pat. Appl. 20030049194 (March 13, 2003), T. Y. Nanri, H. Konno.
6. S. Macash, T. Kael, and M. Yoneyama, *Chemical Economics Handbook*, SRI, Menlo Park, CA, July 2000.
7. *Pulverized Limestone Division Test Methods*, National Stone Association, Washington, D.C., 1991.
8. *Tappi Test Methods 1991*, Tappi Press, Atlanta, Ga., 1990, Methods T534, T667, and T671.
9. *Food Chemicals Codex*, 3rd ed., National Academy of Science, Washington, D.C., 1981, p. 46.
10. *United States Pharmacopeia*, 22nd revision, United States Pharmacopeial Convention, Inc., Rockville, Md., 1990, p. 208.
11. *U.S. Code of Federal Regulations*, 21§ 172.5, Government Printing Office, Washington, D.C., 1989.
12. *Threshold Limit Values and Biological Exposure Indices, American Conference of Governmental Industrial Hygienists*, Cincinnati, Ohio, 1989.
13. R. J. Lewis, Sr., *Sax's* Dangerous Properties of Industrial Materials, 10th ed., Vol. 2, John Wiley & Sons, Inc., New York, 2000.
14. M. D. Strutz, P. A. Duncan, and J. C. Pflieger, *1988 Papermakers Conference*, TAPPI Press, Atlanta, Ga., 1988, 55–60.
15. U.S. Pat. Appl. 2002004476 (Jan. 10, 2002), E. J. Pancheri (to Procter and Gamble).

GENERAL REFERENCE

M. D. Strutz and C. T. Sweeney, *Natural Ground Calcium Carbonate, Proceedings Tappi Neutral/Alkaline Short Course October 1990*, Tappi Press, Atlanta, Ga., 1990.

F. PATRICK CARR
DAVID K. FREDERICK
OMYA, Inc.

CALCIUM CHLORIDE

1. Introduction

Calcium chloride [10043-52-4], $CaCl_2$, is a white, crystalline salt that is very soluble in water. In its anhydrous form it is 36.11% calcium and 63.89% chlorine. It forms mono-, di-, tetra-, and hexahydrates. Calcium chloride is found in small quantities, along with other salts, in seawater and in many mineral springs. It also occurs as a constituent of some natural mineral deposits. Natural brines account for 70–75% of the United States $CaCl_2$ production.

Calcium chloride was discovered in the 15th century but received little attention or study until the latter part of the 18th century. All of the early work was done with laboratory prepared samples, since it was not produced on a commercial scale until after the ammonia–soda process for manufacture of soda ash was in operation. It was actually considered a waste product until its uses were discovered.

2. Properties

The properties of calcium chloride and its hydrates are summarized in Table 1. Accurate data are now available for the heats of fusion of the hexahydrate, the incongruent fusion of the tetrahydrate, and the molar heat capacities of the hexahydrate, tetrahydrate, and dihydrate (1). These data are important when considering the calcium chloride hydrates as thermal storage media. A reevaluation and extension of the phase relationships of the calcium chloride hydrates has led

Table 1. **Properties of Calcium Chloride Hydrates**

Property	$CaCl_2 \cdot 6H_2O$	$CaCl_2 \cdot 4H_2O$	$CaCl_2 \cdot 2H_2O$	$CaCl_2 \cdot H_2O$	$CaCl_2$
CAS Registry Number	[7774-34-7]	[25094-02-4]	[10035-04-8]	[22691-02-7]	[10043-52-4]
mol wt	219.09	183.05	147.02	129.00	110.99
composition, wt % $CaCl_2$	50.66	60.63	75.49	86.03	100.00
mp, °C	30.08	45.13	176	187	772
sp gravity, d_4^{25}	1.71	1.83	1.85	2.24	2.16
heat of fusion or transition, kJ/mol[a]	43.4[b]	30.6b	12.9	17.3	28.5
heat of solution in water[c], kJ/mol[a]	15.8	−10.8	−44.05[d]	−52.16d	−81.85[d]
heat of formation, at 25°C, kJ/mola	−2608	−2010	−1403	−1109	−795.4
heat capacity, at 25°C, J/(g·°C)a	1.66[b]	1.35b	1.17b	0.84	0.67

[a] To convert J to cal, divide by 4.184.
[b] Ref. 1.
[c] To infinite dilution.
[d] Ref. 2.

to new values for the heats of infinite dilution for the dihydrate, monohydrate, 0.33-hydrate, and pure calcium chloride (1).

A study on the solubility of calcium chloride hydrates (3) has generated polymonials relating the weight percent of anhydrous salt in a saturated solution to temperature (°C). For $9.33 < °C < 28.16$

$$wt\% \, CaCl_2 = 1.783 + 28.93 \, t^{0.5} - 7.70 \, t + 0.73 \,^{1.5}$$

for $33.54 < t°C < 44.81$

$$wt\% \, CaCl_2 = -238.3 + 146.6 \, t^{0.5} - 25.47 \, t + 1.51 \, t^{1.5}$$

for $49.37 < t°C < 97.65$

$$wt\% \, CaCl_2 = 39.17 + 5.28 \, t^{0.5} - 0.624 \, t + 0.03 \, t^{1.5}$$

These three equations represent saturation with respect to the hexahydrate, tetrahydrate, and dihydrate in the temperature ranges indicated. The phase relationships among calcium chloride, its hydrates, and a saturated solution are illustrated in the diagram in Figure 1.

2.1. Calcium Chloride Solutions.   Because of its high solubility in water, calcium chloride is used to obtain solutions having relatively high densities. For example, densities as high as 1430 kg/m^3 are achieved at 20°C and as high as 1570 kg/m^3 at 80°C. The oil- and gas-drilling industries frequently

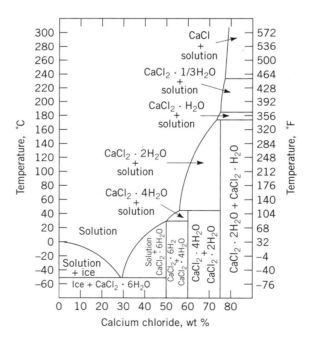

Fig. 1.   The phase relationships among calcium chloride, its hydrates, and a saturated solution.

exploit these high densities when completing or reworking wells. Density, or specific gravity, can also be used to determine the molal concentration, c, of calcium chloride in water (4).

$$c = 30.8 - 129.6\,d + 180.8\,d^2 - 106.8\,d^3 + 24.89\,d^4$$

where c is in units of moles of calcium chloride per kg of water and d is the specific gravity of solution relative to water at 25°C. The densities of calcium chloride solution at various wt% $CaCl_2$ values and different temperatures have been listed (5). Densities and apparent molar volumes of aqueous calcium chloride solutions at temperatures from 323 K (50°C) to 600 K (327°C) and at pressures up to 40 MPa (395 atm) have also been reported (6).

Viscosity is an important property of calcium chloride solutions in terms of engineering design and in application of such solutions to flow through porous media. Data and equations for estimating viscosities of calcium chloride solutions over the temperature range of 20–50°C are available (4). For example, at 25°C and in the concentration range from 0.27 to 5.1 molal (2.87–36.1 wt%) $CaCl_2$, the viscosity increases from 0.96 to 5.10 mPa·s (= cP).

Numerous studies on the thermodynamics of calcium chloride solutions were published in the 1980s. Many of these were oriented toward verifying and expanding the Pitzer equations for determination of activity coefficients and other parameters in electrolyte solutions of high ionic strength. A review article covering much of this work is available (7). Application of Pitzer equations to the modeling of brine density as a function of composition, temperature, and pressure has been successfully carried out (8).

3. Manufacture and Production

Calcium chloride is produced in commercial amounts using many different procedures: (1) refining of natural brines, (2) reaction of calcium hydroxide with ammonium chloride in Solvay soda and production, and (3) reaction of hydrochloric acid with calcium carbonate. The first two processes account for over 90% of the total calcium chloride production (9).

In the United States, the primary route for making calcium chloride is by the evaporation of underground brines. An integrated process is used to extract various brine components. Calcium chloride is derived from the brine left over after processing magnesium chloride into magnesium hydroxide. This brine is ca. 25% $CaCl_2$. A 32–45% solution is produced after being processed through a double- or triple-effect evaporator (9). Unwanted alkali chlorides are precipitated and removed. The brine is then further evaporated, attaining a 78–94% calcium chloride concentration.

Production involves removal of other chlorides (primarily magnesium) by precipitation and filtration followed by concentration of the calcium chloride solution, either for ultimate sale or for evaporation to dry product. Commercial dry products vary by the amount of water removed and by the nature of the drying equipment used. Production and capacity figures for the United States are indicated in Table 2.

Table 2. **Calcium Chloride Statistics**[a]

	Year				
	1987	1988	1989	1990	1991
no. of operations	10	10	10	9	9
production[b], t	597,000	663,000	727,000	626,000	584,000
value[c]	87,400	86,700	133,000	102,000	101,000
exports[d], t	31,496	16,974	20,316	23,300	30,568
imports for consumption, t	208,620	201,328	119,296	145,534	124,094
consumption[e]	774,124	847,354	825,950	748,980	677,526

[a] Ref. 9.

[b] Estimated based on the response of a survey of all but two operations where production from previous years, published plant capacity, and contacts within industry were used to estimate production.

[c] A large percentage of the total was estimated due to producers that did not provide this information.

[d] Bureau of Census.

[e] Consumption = Production + Imprts − Exports.

Mixed chloride brines not yet passing specifications containing calcium chloride, magnesium chloride, sodium chloride, and minor quantities of other salts are also produced in the United States in considerable quantities from natural brines and are marketed mainly in the form of brines of various concentrations.

The Solvay Process Company introduced the ammonia–soda process, which originated in Belgium in 1861, to the United States by 1881. Through a series of purchases and sales, this company now operates under the name of General Chemical with United States headquarters in Parsippany, NJ, and Canadian headquarters in Mississauga, Ontario. About 95% of synthetic calcium chloride produced is recovered from this process. The sole producer of calcium chloride in the United States by that route closed operation in the 1890s and consolidated production with Canada.

In 1896, brine was concentrated further to produce a solid containing 73–75% actual calcium chloride. In 1900, a granular product was produced, which greatly facilitated handling and dissolving. Later the granular product was replaced by flaked material containing 77–80% actual calcium chloride that, through high-temperature drying, acquires a superficial anhydrous coating, thus preventing caking.

The ammonia–soda process involves the reaction of sodium chloride (ordinary salt) with calcium carbonate (limestone) using ammonia as a catalyst to form sodium carbonate (soda ash) and calcium chloride. The process was originally designed to produce soda ash, producing calcium chloride as a waste product. However, the importance of calcium chloride has grown such that calcium chloride is now considered a co-product rather than a by-product.

Additional commercial material is available by action of hydrochloric acid on limestone. Typically the hydrochloric acid is a by-product of some other commercial process and the conversion to calcium chloride is motivated by waste avoidance.

$$CaCO_3 + 2\,HCl \longrightarrow CaCl_2 + H_2O$$

Significant quantities of calcium chloride are produced in the United States, Canada, Mexico, Germany, Belgium, Sweden, Finland, Norway, and Japan. In 1989 there were 10 producers of calcium chloride in the United States. In 1990 this decreased to nine, Table 2 (9). The Dow Chemical Company and Wilkinson Corporation recover calcium chloride from brines in Mason and Lapeer Counties, MI. Calcium chloride pellets, flake, and liquid were produced by Dow's Ludington plant. Wilkinson markets calcium chloride solutions. National Chloride Company of America, Cargill's Leslie Salt Company, and Hills Brothers Chemical Company also produces calcium chloride from dry-lake brine wells in San Bernadino County, CA. Hills Brothers also produced calcium chloride from an operation near Cadiz Lake, CA, and marketed calcium chloride that resulted as a by-product of magnesium production in Rowly, UT, produced by Magnesium Corporation of America.

Calcium chloride was synthesized by Tetra Chemicals at a plant near Lake Charles, LA., and from its liquid plant at Norco, LA. Calcium chloride was recovered as a by-product of the reaction of hydrochloric acid and limestone and marketed by Allied Signal (now Honeywell) Incorporated. Occidental Chemical Corporation also manufactured calcium chloride from this process in Tacoma, WI. Solution production is centered around Michigan (brines), California and Utah (brines), and Louisiana (by-product acid). Michigan is the leading state in natural calcium chloride production with California second.

Calcium chloride is odorless, colorless, and not flammable. Therefore, under current regulations by the DOT, it is not subject to specific handling regulations. Bulk rail cars, bulk trucks, box rail cars, and van and flatbed trucks transport it.

4. Economic Aspects

Calcium chloride consumption is very dependent on the weather. The deicing, dust control, and road stabilization markets are, thus, effected by these conditions. In 1990 and 1991 the winter was mild, thus hurting the deicing market. However, in 1996, production delays, a harsh winter in 1995, and early signs of another one for 1996 created some snugness in the dry calcium chloride market. As for liquid calcium chloride, which is used largely in dust control and oil-well completion, a rainy summer in 1996 reduced demand for calcium chloride as a dust-control product. Aside from yearly changes in precipitation, these markets remain fairly stable. However, oil-drilling activity has declined, slowing the expected increase in the growth of this market. Use as a concrete set accelerator should see an increase as the construction industry continues to soar. About 12% of calcium chloride goes into cement manufacture and concrete accelerating. Industry sources believe that the use of calcium chloride as a growth-enhancing macronutrient may be a future market in the agricultural sector.

There is currently an excess of capacity in the calcium chloride industry, which is only expected to become more actue as additional synthetic or by-product capacity increases. Calcium chloride production is used as a solution to the oversupply of hydrochloric acid. As processes convert from using caustic soda to using hydrated lime for propylene oxide production, an additional 225,000 tons of calcium chloride by-product has the potential for being generated

Table 3. **United States Exports of Calcium Chloride**

Country	1990		1991	
	Quantity, t	Value,[a] $	Quantity, t	Value,[a] $
Canada	16,463	3,214,953	25,006	4,467,701
Egypt	1,014	259,847	336	325,203
Mexico	250	81,370	338	217,005
Taiwan	1	3,684	1,053	263,234
Trinidad	7	8,145	546	187,094
United Arab Emirates	2,428	659,282	238	104,404
United Kingdom	561	189,082	716	211,830
other	2,576	2,197,298	2,335	2,253,710
Total	*23,300*	*6,614,664*	*30,568*	*8,030,181*

[a] U.S. Customs declared value.

(9). With capacity outweighing demand, new niche markets are being developed for using the product: mining, water treatment, fertilizer, pulp and paper, agriculture, and food-grade applications.

Table 2 gives a summary of the calcium chloride statistics for production, value, exports, imports for consumption, and consumption (9). Exports for 1990 and 1991 totaled 23,300 and 30,568 metric tons, respectively. Of this, 16,463 (70%) and 25,006 (81%) metric tons were exported to Canada. Statistics for the United States exports by country are given in Table 3. Calcium chloride canvassing discontinued beginning 1993. However, export statistics were gathered for 1999: 66,197 metric tons were exported with a value of $18,319,470 (10). Of this total, 52.5% ($9,920,969) was to Canada, 2% ($511,167) was to Mexico, 4% ($714,340) was to Trinidad, 4.78% ($631,990) was to Venezuela, 6% ($731,333) was to Italy, 4.5% ($466,411) was to the Netherlands, and 14.9% ($2,420,665) was to United Arab Emirates. The rest was in small amounts experted to other countries (10).

Estimated imports of calcium chloride increased more than ten-fold between 1984 and 1988, from 10,000 to 139,700 metric tons on a 100 wt% basis (9). The United States imports most of its calcium chloride from Canada (1990, 109,880 metric tons; 1991, 92,838 metric tons). The location of production facilities close to the United States/Canada border make this a particularly inviting country to export from because of calcium chloride's use as a deicing material. The other countries the United States imports calcium chloride from are Mexico (1989, 17,800 metric tons), the former Federal Republic of Germany (1989, 6,900 metric tons), and Sweden (1989, 4,800 metric tons). Table 4 lists the United States imports for consumption of calcium by country for 1990 and 1991. As stated previously, calcium chloride canvassing discontinued beginning 1993. However, import statistics were gathered for 1999: 219,249 metric tons were imported with a valve of $26,810,352 (10). Of this total, 78% ($14,347, 984) was from Canada, 16% ($7,160,676) was from Mexico, and 4.7% ($2,342, 994) was from Finland, with the rest being in small amounts from other countries (10).

Table 4. **United States Imports for Consuption of Crude Calcium Chloride**

Country	1990		1991	
	Quantity, t	Value,[a] $	Quantity, t	Value,[a] $
Canada	109,880	13,276,894	92,838	11,276,937
Former Federal Republic of Germany	9,471	2,264,418	13,639	2,901,992
Mexico	18,251	4,143,726	13,000	3,157,007
Sweden	2,517	291,482	3,950	612,256
other	5,415	1,364,858	667	661,960
Total	*145,534*	*21,341,378*	*124,094*	*18,610,152*

[a] U.S. Customs, insurance, freight.

5. Grades and Specifications

Most solution calcium chloride is sold as 38 or 45 wt% concentration; however, different uses require concentrations ranging from 28 to 45 wt%. The principal uses (deicing and dust control) do not require high-purity calcium chloride. However, it must be free of chemicals harmful to the environment. Producers ship the most concentrated form, and the distributors make final adjustments in concentration.

The majority of dry calcium chloride comes in one of two forms: flake or pellet. Lesser amounts are sold as minipellets, powders, or briquettes. Six agencies grade calcium chloride, Table 5. For a product containing 90.5% calcium chloride, the American Society of Testing Materials (ASTM) and the American Association of State Highway and Transportation Officials (AASHTO) has set up standards for calcium chloride content (assay), total alkali chlorides (<8.0% as NaCl), total magnesium (<0.5% as magnesium chloride), and other impurities (<1.0% after accounting for sodium, calcium, potassium, and magnesium chlorides, water, and calcium hydroxide) (9). There are three grades of commercial calcium chloride: Grade 1, 77 wt% $CaCl_2$ minimum; Grade 2, 90 wt% $CaCl_2$; and Grade 3, 94 wt% $CaCl_2$. Adjusted standards exist for all grades (17,18). Calcium chloride manufactured in the United States routinely meets these standards. Table 6 summarized sieve analysis for key commercial grades.

Calcium chloride meeting the Food Chemical Codex (FCC) specifications is used as a food additive. The specifications for this grade of anhydrous calcium chloride are as follows: assay, not less than 93.0%; arsenic (as As), <3 ppm; fluoride, <0.004%; heavy metals (as Pb), <0.002%; lead, <10 ppm; magnesium and alkali salts, <5%; acid-insoluble material, <0.02%; and no particles of sample greater than 2 mm in any dimension.

6. Environmental Concerns

Calcium chloride is not considered to be harmful to the environment. Calcium is essential for all organisms. At concentrations above 1000 ppm, calcium

Table 5. **Calcium Chloride Specifications**

Specification	Uses	Products specified	Reference
ASTM D98–87, AASHTO[a] M144–86	road conditioning/ maintenance; curing concrete	solution: unspecified concentrations; dry: three grades based on 77%, 90%, and 94% minimum assay	11, 12
AWWA[b]	treatment of municipal and industrial water supplies	dry: as flake, pellet, or granular powder or briquette	13
FCC third ed.	sequesterant in food, cross-linker, firming agent in canning, multipurpose food additive	solution: unspecified concentrations; dry: dihydrate 99–107% of formula weights; anhydrous: 93.0% $CaCl_2$ minimum	14
ACS reagent chemicals	reagent-grade desiccant-grade	dihydrate: 74–78% $CaCl_2$ not less than 96% $CaCl_2$	15
USP XXII reagent specifications	dihydrate and anhydrous for drying	use ACS specifications for dihydrate and anhydrous	16
USP XXII	general	dihydrate: 99–107% $CaCl_2 \cdot 2H_2O$; calcium chloride for injections, sterile solution in water, 95–105% of labeled $CaCl_2 \cdot 2H_2O$	

[a] AASHTO = American Association of State Highway and Transportation Officials.
[b] FOOTNOTE ID="CALCREID.A01-T005FNB">AWWA = American Water Works Association.

Table 6. **Sieve Analysis for $CaCl_2$ Commercial Grades, Mass %, Passing**[a]

	Sieve size, mm[b]					
Class, solid form	31.5	9.5	4.75 (No. 4)	2.36 (No. 8)	0.830 (No. 20)	0.600 (No. 30)
Grade 1, 77 wt% $CaCl_2$ min						
A, flake		100	80–100			0–5
B, granular		100	0–80			0–5
Grade 2, 90 wt% $CaCl_2$ min						
A, flake		100	80–100			0–5
B, pellets		100	80–100		0–10	0–5
C, granular	100		0–5			
D, powder			100	80–100		0–65
Grade 3, 94 wt% $CaCl_2$ min						
A, flake		100	80–100			0–5
B, pellets		100	80–100		0–10	0–5
C, granular	100		0–5			
D, powder			100	80–100		0–65

[a] ASTM specifications.
[b] Mesh number appears in parentheses.

chloride has been found to retard plant growth and can damage plant foliage. These effects are most likely caused by excess chloride ion as calcium is a nutrient for plants. In testing United States water supplies, high chloride concentrations are rarely found, even in areas of high salt usage for ice and dust control (19).

Calcium chloride is found in the marine environment. Many organisms and aquatic species are tolerant of the concentrations of calcium and chloride ions in seawater (400 ppm calcium, 18,900 ppm chloride ions). Toxicity arises when possible toxic doses of calcium chloride from spills, surface runoff, or underground percolation get into typically freshwater streams or aquifers. Various agencies have guidelines for calcium and chloride in potable water (20). The European Economic Community (EEC) is the only agency to have a minimum specification for calcium in softened water.

The ability of plants to take up calcium chloride (ion selectivity) and the toxicity of calcium in plants and soils varies widely. Studies of herbaceous crop species, where water defect is not a constraint, point to low levels of chloride ion as being responsible for inhibiting growth (21). However, deicing salts can be toxic to roadside vegetation. The use of both calcium chloride and sodium chloride as deicing salts and the effects on various grasses, shrubs, and trees has been studied. As calcium chloride use with sodium chloride is more effective at deicing roads, thus less is used, the overall chloride ion content is lower than with rock salt alone. From studies in Europe, calcium chloride in blends of deicing salts can have beneficial effects on the regulation of sodium, and of potassium over sodium, in spruce trees (22). Recommendations for calcium chloride tolerant species are available (23,24). Concentrations of 10,000–20,000 ppm in water have been shown to be hazardous to animals and fish. The effects vary widely, ranging from reduced growth rate and impaired reproduction to death. Both calcium chloride (35% solution) and oil-field brine received the lowest toxicity ratings in a study, indicating the environmental advantages of using these products.

7. Health and Safety Factors

In general, calcium chloride is not considered to be toxic. Because calcium chloride is hygroscopic, common safety precautions should be used: wearing gloves, long-sleeved clothing, shoes, and safety glasses. Contact with skin may cause mild irritation on dry skin. Strong solutions or solid in contact with moist skin may cause severe irritation and possibly burns (25). Calcium chloride can irritate and burn eyes from the heat of hydrolysis and chloride irritation. Inhalation may irritate the lungs, nose, and throat with symptoms of coughing and shortness of breath. Ingestion may cause irritation to the mucous membrane due to the heat of hydrolysis. Large amounts can cause gastrointestinal upset, vomiting, and abdominal pain.

Dry bulk calcium chloride can be stored in construction-grade bins. Care should be taken to minimize moisture. It should be kept in a tightly closed container, stored in a cool, dry, ventilated area.

Table 7. **CaCl$_2$ Use in the United States**

Use	Percent
deicing	30
dust control, road stabilization	25
industrial (refrigerant, coal thawing, etc)	15
oil and gas drilling fluids	10
concrete	5
tire ballast	4
miscellaneous	11

8. Uses

Calcium chloride, manufactured for over 100 years, has been used for a variety of purposes. The primary CaCl$_2$ markets have not changed since the 1950s. A breakdown of the United States consumption by percent is given in Table 7 (26). All markets and uses are summarized in Table 8. Significant markets in the United States are for deicing during winter, roadbed stabilization, and as a dust palliative during the summer. Use as an accelerator in the ready-mix concrete industry is sizable, but there is still concern about chlorine usage because of possible corrosion of steel in highways and buildings. Calcium chloride is also used in oil- and gas-well drilling.

8.1. Deicing. The largest market for calcium chloride is for deicing roads, sidewalks, and parking lots (30%). It is more effective than rock salt at lower temperatures. Calcium chloride melts ice at temperatures as low as −51°C (−60 °F). Because it liberates heat upon exposure to moisture, ice melts quickly after application. Anhydrous calcium chloride, 94−97 wt% calcium chloride pellets, and 77−80 wt% calcium chloride flakes are used for highway deicing and in institutional and consumer markets. Under normal conditions, when temperatures drop below −9°C, untreated road salt loses its ability to generate quickly the heat necessary for melting snow and ice. Calcium chloride solutions (28−32 wt%) are used with rock salt or abrasives such as sand or cinders before spreading on highways to enhance their effectiveness. The result is more efficient utilization of road salt and safer roads. Instead of watching the road salt bounce off the roads during colder days, it will be actively melting the ice and snow. Solutions of 42−45 wt% concentration are also used to pretreat stockpiles of these materials. Calcium chloride is the deicer of choice for use at temperatures <−6.7°C (27−29).

8.2. Roadbed Stabilization/Dust Control. One of the earliest uses of calcium chloride was for dust control and roadbed stabilization of unpaved gravel roads. Dust control accounts for ca. 25% of calcium chloride production. Because calcium chloride is hygroscopic and deliquescent, it absorbs moisture from the atmosphere and forms a solution, binding the dust particles and keeping the surface damp. Calcium chloride in dry and solution forms are used both typically and mixed with the aggregate. If aggregate is mixed with dry calcium chloride

Table 8. **Markets and Uses for Calcium Chloride**

Market	Use	Reason for use
building maintenance	freeze-proofing water in fire pails	lowers freezing point
chemical manufacture	production of calcium salts	source of calcium
construction	cold-weather concrete additive	accelerates set
	soil solidification	solidifies loose, sandy soils when injected together with sodium silicate
	tractor tire weighting	lowers freezing temperature of water put in tires to improve traction by incresing weight
drying air and gases	direct drying compound	removes moisture by hygroscopicity
highway construction	shoulder and base stabilization	retains moisture which improves compaction of soil
highway maintenance	dust laying	moisture absorbed from the air prevents dust formation
	snow and ice control	melts ice
mining	dust-profing and freeze-resisting ore and coal	freezing point of residual water
paper manufacture	increases web strength of corrugating media	provides artificial water hardness that allows web to drain better
	improves dye retention	provides artificial water hardness that helps set paper dyes
petroleum	additive to oil well completion fluids	increases density
	cementing finished oil wells	accelerates set of cement
	drilling mud additive	reduces shale swelling
	drying petroleum fractions	absorbs water
portland cement manufacture	additive to klin feed or fuel	reduce alkali content of cement to eliminate expansive reactions in f inished concrete
railroad right-of-way maintenance	weed-killer additive	prevents weed killer from drying out and becoming flammable
refrigeration	brine ingredient	calcium chloride brines have low freezing points
rubber manufacture	coagulating latex emulsions	coagulant
steel and pig iron manufacture	treatment of pelletized ore and blast furnace additive	elimination of alkalies that attack furnace refractory
waterwaste treatment	removal of fluorides	precipitant
	treatment of oily wastes	breaks oil emulsions
	removal of silicates	densifies floc

or a calcium chloride solution and then compacted, the presence of calcium chloride draws in moisture to bind the fine particles in the aggregate matrix. This process leads to well compacted, maximum density gravel road. Due to its low vapor pressure, calcium chloride is slow to evaporate; thus, this dust-free condition is retained over a long period of time.

8.3. Oilfield Uses.   Calcium chloride has two uses in the oil field: as a primary ingredient in completion fluids and as the brine phase in an invert emulsion oil mud. An excellent review of oil-well drilling fluids is available (30).

8.4. Accelerator in Ready-Mix Concrete.   Calcium chloride has been used in concrete since 1885 and finds application mainly in cold weather, when it allows the strength gain to approach that of concrete cured under normal curing temperatures. In normal conditions, calcium chloride is used to speed up the setting and hardening process for earlier finishing or mold turnaround.

Effects of calcium chloride on concrete properties are also widely studied and quantified. Aside from affecting setting time, calcium chloride has a minor effect on fresh concrete properties. It has been observed that addition of $CaCl_2$ slightly increases the workability, reduces the amount of water required to produce a given slump, and reduces bleeding. Using calcium chloride significantly reduces initial and final setting times of concrete. The total effect of adding calcium chloride depends on dosage, type of cement used, and temperature of the mix.

Addition of as little as 1–2% calcium chloride accelerates the set time of concrete, giving it a high early strength development. It is not an antifreeze, but by using it during cold weather, it can offset problems associated with lower temperatures (31). Reviews of the concerns and possible remedies of calcium chloride corrosion problems in concrete are available (32,33). There is no consensus on what the safe levels of calcium chloride in concrete are.

8.5. Food.   Calcium chloride is used in the food industry to increase firmness of fruits and vegetables, such as tomatoes, cucumbers, and jalapenos, and prevent spoilage during processing. Food-grade calcium chloride is used in cheese making to aid in rennet coagulation and to replace calcium lost in pasteurization. It also is used in the brewing industry both to control the mineral salt characteristics of the water and as a basic component of certain beers.

BIBLIOGRAPHY

"Calcium Compounds (Halides)," in *ECT* 1st ed., Vol. 2, pp. 759–761, by G. H. Kimber; "Calcium Compounds (Calcium Chloride)," in *ECT* 2nd ed., Vol. 4, pp. 11–14, by R. F. Armstrong, Diamond Alkali Co.; in *ECT* 3rd ed., Vol. 4, pp. 432–436, by W. L. Shearer, Dow Chemical, U.S.A.; "Calcium Compounds, Calcium Chloride" in *ECT* 4th ed., Vol. 4, pp. 801–812, by Kenneth I. G. Reid, Roger Kust, Tetra Chemicals; "Calcium Chloride" in *ECT* (online), posting date: November 27, 2000, by Kenneth I. G. Reid, Roger Kust, Tetra Chemicals.

CITED PUBLICATIONS

1. K. K. Meissingset and F. Gronvold, *J. Chem. Thermodynam.* (18), 159–173 (1986).
2. G. C. Sinke, E. H. Mossner, and J. L. Curnutt, *J. Chem. Thermodynam.* (17), 893–899 (1985).
3. R. W. Potter and M. A. Clynne, *J. Res. U.S. Geol. Surv.* **6**(6), 701–705 (1978).
4. F. A. Goncalves and J. Kestin, *Ber. Bunsenges. Phys. Chem.* **82**(1), 24–27 (1979).

5. O. Sohnel and P. Novotny, *Densities of Aqueous Solutions of Inorganic Substances*, Elsevier, Amsterdam, The Netherlands, 1985, p. 78.

6. J. A. Gates and R. H. Wood, *J. Chem. Eng. Data*, (34), 53–56 (1989).

7. J. Ananthaswamy and G. Atkinson, *J. Chem. Eng. Data*, (30), 120–128 (1985).

8. N. P. Kemp, D. C. Thomas, G. Atkinson, and B. L. Atkinson, *SPE Prod. Eng.*, 394–400 (Nov. 1989).

9. M. Miller, "Calcium and Calcium Compounds,"*Minerals Yearbook*, Vol. **1**, U.S. Bureau of Mines, U.S. Department of the Interior, Washington, D.C., 1991, pp. 317–324.

10. Statistics for 1999 collected by M. Michael Miller, U.S. Geological Survey, 2000.

11. *ASTM Standard D98-87 Standard Specifications for Calcium Chloride*, Issue 90-06, American Society of Testing Materials, Information Handling Services, Englewood, Colo., Dec. 1990–Jan. 1991.

12. *Standard Specifications for Calcium Chloride*, American Association of State Highway and Transportation Officials, 14th ed., Part I, M144-86, Washington, D.C., 1986, 209–211.

13. *Calcium Chloride*, 2nd ed., Standard B550-90, American Water Works Association, Denver, Colo., 1990.

14. *Food Chemicals Codex*, 3rd ed., National Research Council, National Academy Press, Washington, D.C., 1981, 47–49.

15. Committee on Analytic Reagents, *American Chemical Society, Specifications*, 6th ed., American Chemical Society, Washington, D.C., 1981.

16. *United States Pharmacopeia XXII*, United States Pharmacopeial Convention, Inc., Rockville, Md., 1989.

17. *ASTM Standard D98-87: Standard Specification for Calcium Chloride*, Issue 90-06, American Society of Testing Materials, Information Handling Services, Englewood, Colio., Dec. 1990–Jan. 1991.

18. *Standard Specifications for Calcium Chloride*, American Association of State Highway and Transporation Officials, 14th ed., Part 1, M144-86, Washington, D.C., 1978.

19. F. E. Hutchinson, *The Influence of Salts Applied to Highways on the Levels of Sodium Chloride Ion Present in Water and Soil Samples*, U.S. Department of the Interior, Project No. A-007-ME,U.S. Government Printing Office, Washington, D.C., June 1969.

20. F. W. Pontius, ed.,*Water Quality and Treatment; A Handbook of Community Water Supplies*, American Water Works Association, 4th ed., McGraw-Hill, Inc., New York, 1990, pp. 8–59.

21. H. Greenway and R. Munns, *Annu. Rev. Plant Physiol.* **31**, 149–190 (1980).

22. J. Boegemans, L. Neimnckx, and J. M. Stassart, *Plant Soil* **113**, 3–11 (1989).

23. R. E. Hanes and co-workers, in *National Copperative Highway Research Program Report 170*, Transportation Research Board, National Research Council, Washington, D.C., 1976, p. 88.

24. R. Paul, M. Rocher, and R. Impens, *Sci. Total Environ.* **59**, 277–282 (1987).

25. *Calcium Chloride MSDS*, Mallinckrodt Baker, Inc., Phillipsburg, NH, 2000.

26. In W. Gerhartz, ed., *Ullmann's Encyclopedia of Industrial Chemistry*, 4th ed., Vol. **A4**, Weinheim, Germany; Deerfield Beach, FL, 1995, pp. 547–553.

27. G. C. Sinke and E. H. Mossner, *Transportation Research Record, 598*, Transportation Research Board, National Research Council, Washington, D.C., 1976, pp. 54–57.

28. A. D. McElroy and co-workers, *Transportation Research Record, 1157*, Transportation Research Board, National Research Council, Washington, D.C., 1988, pp. 1–11.

29. *Public Works* **119**, 80–81 (July 1988).

30. G. R. Gray and H. C. H. Darley, *Composition and Properties of Oil Well Drilling Fluids*, 4th ed., Gulf Publishing, Houston, Texas, 1980.

31. National Ready Mixed Concrete Association; *Cold Weather Ready Mixed Concrete*, Silver Spring, Md., 1968, pp. 11–12.
32. ACI Committee 222, *Corrosion of Metal in Concrete*, ACI 222R-85 (89), American Concrete Institute, Detroit, Mich., 1989, p. 30.
33. F. W. Gibon, ed., *Corrosion, Concrete and Chlorides*, ACI SP-102, American Concrete Institute, Detroit, Mich., 1987.

LISA M. VRANA
Consultant

CALCIUM FLUORIDE

1. Introduction

Fluorine chemistry began with observations by Georgius Agricola as early as 1529 that fluorspar lowers the melting point of minerals and reduces the viscosity of slags. This property of fluxing (Latin *fluoere*, to flow) is the origin of the name fluorine. The term fluorspar correctly describes ores containing substantial amounts of the mineral fluorite [14542-23-5], CaF_2, but the word fluorspar is often used interchangeably with fluorite and calcium fluoride.

Calcium fluoride [7789-75-5] has the formula CaF_2 and a molecular weight of 78.07, and it is 51.33% calcium and 48.67% fluorine. Calcium fluoride occurs in nature as the mineral fluorite or fluorspar. It is prepared from the reaction of $CaCO_3$ and HF.

2. Occurrence

Significant mining of fluorspar began in England about 1775 and in the United States after 1820. Substantial use of fluorspar began about 1880 in the basic open-hearth process for making steel (qv). Large increases in demand came with the need for fluorides in the aluminum industry, starting about 1900. A large fluorine chemicals industry based on hydrogen fluoride made from fluorspar followed in production of refrigerants (see REFRIGERATION AND REFRIGERANTS) (1930), alkylation (qv) catalysts for gasoline (1942), materials for nuclear energy (ca 1942), aerosol propellants (see AEROSOLS) (ca 1942), fluoroplastics (ca 1942), and fluorocarbons for soil-repellant surface treatments (early 1950s). Fluorspar is used directly in the manufacture and finishing of glass (qv), in ceramics (qv) and welding (qv) fluxes, and in the extraction and processing of nonferrous metals (see METALLURGY, EXTRACTIVE).

In the geochemistry of fluorine, the close match in the ionic radii of fluoride (0.136 nm), hydroxide (0.140 nm), and oxide ion (0.140 nm) allows a sequential replacement of oxygen by fluorine in a wide variety of minerals. This accounts for the wide dissemination of the element in nature. The ready formation of volatile silicon tetrafluoride, the pyrohydrolysis of fluorides to hydrogen fluoride, and the low solubility of calcium fluoride and of calcium fluorophosphates have provided a geochemical cycle in which fluorine may be stripped from solution by limestone and by apatite to form the deposits of fluorspar and of phosphate rock (fluoroa-patite [1306-01-0]), approximately $CaF_2 \cdot 3Ca_3(PO_4)_2$, which are the world's main resources of fluorine (1).

On average, fluorine is about as abundant as chlorine in the accessible surface of the earth including oceans. The continental crust averages about 650 ppm fluorine. Igneous, metamorphic, and sedimentary rocks all show abundances in the range of 200–1000 ppm. Fluorspar is still the principal source of fluorine for industry.

Fluorspar deposits are commonly epigenetic, ie, the elements moved from elsewhere into the rock. For this reason, fluorine mineral deposits are closely associated with fault zones. In the United States, significant fluorspar deposits occur in the Appalachian Mountains and in the mountainous regions of the West but the only reported commercial production in 1993 was from the faulted carbonate rocks of Illinois.

Worldwide, large deposits of fluorspar are found in China, Mongolia, France, Morocco, Mexico, Spain, South Africa, and countries of the former Soviet Union.

2.1. Supply. Soon after World War II, stockpiling of fluorspar began upon recommendation by the Strategic Material Committee and the Army and Navy Munitions Board. It was decided that this stockpile must be sufficient to sustain United States consumption for at least three years in the case of a national emergency. Domestic output of fluorspar is entirely dependent on sales of material from this stockpile. During the 2002, fiscal year, there were no fluorspar sales authorized to be sold from the national stockpile (2). For fiscal year 2003, 54,400 tons of metallurgical-grade and 10,900 tons of acid-grade are authorized to be sold (2). A global-scale shortage of calcium fluoride could be prevented by (1) use of fluorspar reserves currently considered not to be economically workable, (2) prospecting and discovery of new deposits, and (3) production from hydrofluoric acid and fluorosilicic acid. Processes for synthesis have been proposed, but none have been utilized industrially. In 1997, imports of fluorspar increased almost 4% compared to 1996. From 1995 to 1998, the United States imported approximately 66% of its fluorspar, typically 96.5–97.5% CaF_2, from China, 23% from South Africa, and 11% from Mexico (3).

In 1993, Illinois was the only state reporting production of acid-grade fluorspar, typically 96.5–97.5% CaF_2, and accounted for 100% of all reported shipments. Ozark-Mahoning Company, a wholly owned subsidiary of Elf Atochem North America, Incorporated, operated three deep mines and a flotation mill in Hardin Country, Ill.. A limited amount of metallurgical-grade gravel was produced by Hastie Mining, also located in Hardin County, Ill. (11). Since 1997, there has been no domestic mine production of fluorspar (3). Since 1997, there has been no domestic mine production of fluorspar (3). World mine production, reserves, and reserve base for 2001–2002 are shown in Table 1.

Table 1. **World Mine Production, Reserves, and Reserve Base of Fluorspar,** $\times 10^3$ t^a

Country	Mine production		Reserves[c]	Reserve base[c]
	2001	2002[b]		
United States	NA[d]	NA[d]	NA[d]	6,000
China	2,450	2,450	21,000	110,000
France	110	110	10,000	14,000
Italy	45	50	6,000	7,000
Kenya	108	95	2,000	3,000
Mexico	635	640	32,000	40,000
Mongolia	200	200	12,000	16,000
Morocco	75	95	NA[d]	NA[d]
Namibia	83[e]	86[e]	3,000	5,000
Russia	190	190	moderate	18,000
South Africa	286	240	41,000	80,000
Spain	130	130	6,000	8,000
other countries	220	240	100,000	170,000
World total (may be rounded)	*4,530*	*4,530*	*230,000*	*480,000*

[a] From Ref. 3.
[b] Estimated.
[c] Measured as 100% calcium fluoride.
[d] Not available.
[e] Data are reported in wet tons.

3. Properties

Some of the important physical properties of calcium fluoride are listed in Table 2. Pure calcium fluoride is without color. However, natural fluorite can vary from transparent and colorless to translucent and white, wine-yellow, green, greenish blue, violet-blue, and sometimes blue, deep purple, bluish black, and brown. These color variations are produced by impurities and by radiation damage (color centers). The color of fluorite is often lost upon heating, sometimes with luminescence. Mineral specimens are usually strongly fluorescent, and the mineral thus gives its name to this phenomenon. Specimens vary from well-formed crystals (optical grade) to massive or granular forms.

The crystal structure of fluorite gives its name to the fluorite crystal type. The lattice is face-centered cubic (fcc), where each calcium ion is surrounded by eight fluoride ions situated at the corners of a cube, and each fluoride ion lies within a tetrahedron defined by four calcium ions (23). The bonding is ionic. The unit cell (space group O_h^5) can be pictured as made up of eight small cubes, each containing a fluoride ion, and the eight forming a cube with a calcium ion on each corner and one in the center of each face (Fig. 1). The lattice constant is 0.54626 nm at 25°C (24). The habit is usually cubic, less frequently octahedral, rarely dodecahedral. Cleavage on the [111] planes is perfect. The crystals are brittle with flat-conchoidal or splintery fracture. Luster is vitreous, becoming dull in massive varieties.

Systems of metal oxides with calcium fluoride usually have a simple freezing point composition diagram, commonly exhibiting a eutectic point and no

Table 2. **Physical Properties of Calcium Fluoride**

Property	Value	Reference
formula weight	78.08	
composition, wt %		
Ca	51.33	
F	48.67	
melting point, °C	1402	4
boiling point, °C	2513	5
heat of fusion, kJ/mol[a]	23.0	6
heat of vaporization at bp, kJ/mola	335	7
vapor pressure at 2100°C, Pa[b]	1013	7
heat capacity, C_p, kJ/(mol$\cdot$K)[a]		
solid at 25°C	67.03	8
solid at mp	126	9
liquid at mp	100	9
entropy at 25°C, kJ/(mol$\cdot$K)[a]	68.87	8
heat of formation, solid at 25°C, kJ/mol[a]	−1220	8
free energy of formation, solid at 25°C, kJ/mol[a]	−1167	8
thermal conductivity, crystal at 25°C, W/(m$\cdot$K)	10.96	10
density, g/mL		
solid at 25°C	3.181	11
liquid at mp	2.52	12
thermal expansion, average 25 to 300°C, K^{-1}	22.3×10^{-6}	13
compressibility, at 25°C and 101.3 kPa (=1 atm)	1.22×10^{-8}	14
hardness		
Mohs' scale	4	
Knoop, 500-g load	158	15
solubility in water, g/L at 25°C	0.146	16
refractive index at 24°C, 589.3 nm	1.43382	17
dielectric constant at 30°C	6.64	18
electrical conductivity of solid, ($\Omega\cdot$cm^{-1})		
at 20°C	1.3×10^{-18}	19
at 650°C	6×10^{-5}	20
at mp	3.45	21
optical transmission range, nm	150 to 8000	22

[a] To convert J to cal, divide by 4.184.
[b] To convert Pa to mm Hg, multiply by 7.5×10^{-3}.

abnormal lowering of the melting point (25). When silicates are present, the systems become more complicated and a striking decrease in the viscosity of the glassy melts is observed. The viscosity most likely decreases because of depolymerization of chains or networks of SiO_4 tetrahedra via the replacement of oxide ion by the singly charged fluoride ion, which is close in both size and electronegativity to oxide ion (1). The benefits of calcium fluoride as a metallurgical flux result from both the freezing point depression and the decrease in slag viscosity.

Although stable at ambient temperature, calcium fluoride is slowly hydrolyzed by moist air at about 1200°C, presumably to CaO and HF. Calcium fluoride is not attacked by alkalies or by reactive fluorine compounds but is decomposed by hot, high-boiling acids, as in the reaction with concentrated sulfuric acid, which is the process used to produce hydrogen fluoride. Calcium fluoride is slightly soluble in cold dilute acids and somewhat more soluble in solutions of aluminum halides.

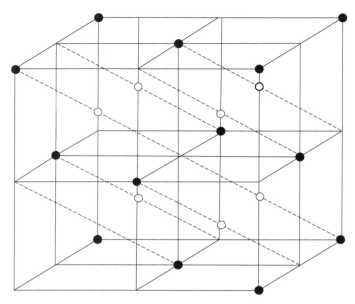

Fig. 1.   Structure of fluorite where • is Ca and ○ is F (3).

4. Mining

For a deposit to be economically workable, the CaF_2 content must be, on average, 20%. Underground mining procedures are used for deep fluorspar deposits, and open-pit mines are used for shallow deposits or where conditions do not support underground mining techniques (see MINERAL RECOVERY AND PROCESSING).

Fluorspar occurs in two distinct types of formation in the fluorspar district of southern Illinois and Kentucky: in vertical fissure veins and in horizontal bedded replacement deposits. A 61-m bed of sandstone and shale serves as a cap rock for ascending fluorine-containing solutions and gases. Mineralizing solutions come up the faults and form vein ore bodies where the larger faults are plugged by shale. Bedded deposits occur under the thick sandstone and shale roofs. Other elements of value associated with fluorspar ore bodies are zinc, lead, cadmium, silver, germanium, iron, and thorium. Ore has been mined as deep as 300 m in this district.

4.1. Beneficiation.   Most fluorspar ores as mined must be concentrated or beneficiated to remove waste. Metallurgical-grade fluorspar is sometimes produced by hand-sorting lumps of high grade ore. In most cases the ore is beneficiated by gravity concentration with fluorspar and the waste minerals, having specific gravity values of >3 and <2.8, respectively. In preconcentration of fluorspar, barite and valuable sulfide minerals are separated from waste as the higher density valuable minerals sink while the waste floats and is discarded. This preconcentration can enrich ores as low as 14% to a concentration of 40%. Multistage froth flotation (qv) is used to take this preconcentrate and produce acid and ceramic grades of fluorspar as well as zinc and lead sulfides. In this process air bubbles are forced through a suspension of pulverized ore, which float the ore into a froth that is continuously skimmed off. After flotation, the fluorspar products are filtered and dried in rotary kilns.

5. Preparation

CaF_2 is manufactured by the interaction of H_2SiF_6 with an aqueous carbonate suspension (26–29); by the reaction of $CaSO_4$ with NH_4F (30); by the reaction of HF with $CaCO_3$ in the presence of NH_4F (31); by reaction of $CaCO_3$ and NH_4F at 300–350°C followed by calcining at 700–800°C (32); by reaction of NH_4F and $CaCO_3$ (33–38); and from the thermal decomposition of calcium trifluoroacetate (39).

High purity CaF_2 is obtained from micro- and ultrafiltration (qv) of raw materials and then crystallization of CaF_2 (40) from the reaction of $CaCO_3$ and the product of H_2SiF_6 and NH_3 (41). High purity CaF_2 having particles of 0.0005–0.5 mm is produced from the reaction of NaF, KF, or NH_4F and $CaCO_3$ with a particle size distribution of 0.01–0.05 nm (42). High purity CaF_2 is also prepared from the reaction of $Ca(NO_3)_2 \cdot 4H_2O$ and NH_4F or a mixture of NH_4HF_2 and NH_4F (43) and obtained by heating impure CaF_2 with 10–15% HCl at 95–100°C (44). Very pure calcium fluoride for the manufacture of special glasses is made by the reaction of hydrofluoric acid with precipitated calcium carbonate. Acicular (whisker-form) CaF_2 particles have been manufactured by continuous feeding of an aqueous $Ca(OH)_2$ solution into water containing CO_2 and subsequent reaction with HF (45). Coarse grain CaF_2 crystals can be prepared by several routes (46–59). CaF_2 can be is crystallized from a wastewater containing fluoride by adding $CaCl_2$ (60). Calcium fluoride can be produced from waste H_2SiF_6 from phosphate product operations and from treating fluoride solution from industrial wastewater with KOH and then with lime (61).

The large amount of fluorine values released from phosphate rock in the manufacture of fertilizers (qv) gives a strong impetus to develop fluorine chemicals production from this source (see PHOSPHORIC ACID AND THE PHOSPHATES). Additional incentive comes from the need to control the emission of fluorine-containing gases. Most of the fluorine values are scrubbed out as fluorosilicic acid, H_2SiF_6, which has limited usefulness. A procedure to convert fluorosilicic acid to calcium fluoride is available (62).

6. Shipment

Truck, rail, barge, and ship are all methods used to transport fluorspar. The different grades are shipped in different forms: metallurgical grade as a lump or gravel; acid grade as a damp filtercake containing 7–10% moisture to facilitate handling and reduce dust.

7. Economic Aspects

In looking at trends for fluorspar use over the past ca 25 years, domestic production has declined since 1976. Pertinent statistics on the United States production

Table 3. **U.S. Pertinent Statistics for Fluorspar,** $\times 10^3$ t[a]

Statistics–United States	1998	1999	2000	2001	2002[b]
production:					
finished, all grades[c]			NA[d]	NA[d]	NA[d]
fluorspar equivalent from phosphate rock	118	122	119	104	126
imports for consumption					
acid grade	462	419	484	495	466
metallurgical grade	41	59	39	27	36
total fluorspar imports	503	478	523	522	502
fluorspar equivalent from hydrofuoric acid plus cryolite	204	192	208	176	189
exports[e]	24	55	40	21	25
shipments from government stockpile	110	131	106	65	13
consumption					
apparent[f]	591	615	601	543	472
reported	538	514	512	536	562
stocks, yearend, consumer and dealer[g]	468	373	289	221	240
employment, mine and mill, number			5	5	5
net import reliance as a percentage of apparent consumption	100	100	100	100	100

[a] From ref. 3.
[b] Estimated.
[c] Shipments.
[d] Not available.
[e] Exports are all general imports re-exported or National Defense Stockpile material exported.
[f] Excludes fluorspar equivalent of flurosilicic acid, hydrofluoric acid, and cryolite.
[g] Industry stocks for three largest consumers, fluorspar distributors, and committed National Defense Stockpile material.

and consumption of fluorspar are given in Table 3 (2,3). Thus, the United States has relied on imports for more than 80% of its fluorspar needs. However, foreign sources are more than adequate. The principal sources are China, Mexico, and South Africa. Foreign producers are able to offer fluorspar at a lower price than that produced domestically because of lower operating costs. Imports from Mexico have declined in part because Mexican export regulations favor domestic conversion of fluorspar to hydrogen fluoride for export to the United States. Statistics for the United States exports by country are given in Table 4 (2).

The large amount of fluorine values released from phosphate rock in the manufacture of fertilizers gives a strong impetus to develop fluorine chemicals production from this source. Additional incentive comes from the need to control the emission of fluorine-containing gases. Most of the fluorine values are scrubbed out as a fluorosilicic acid, H_2SiF_6, which has limited usefulness. A procedure to convert fluorosilicic acid to calcium fluoride is available (63).

8. Grades and Specifications

Ceramic-grade and acid-grade fluorspars (ceramic-spar and acid-spar) have the typical analysis shown in Table 5. Both types are usually finely ground, the bulk

Table 4. **U.S. Exports of Fluorspar**[a]

Country	2000 Quantity, t	2000 Value, $[b]	2001 Quantity, t	2001 Value, $[b]
Canada	18,100	2,930,000	15,800	2,410,000
Dominican Republic	62	9,090		
Italy	13,200	1,210,000		
Mexico	4,520	441,000	3	2,510
Taiwan	3,310	592,000	5,020	733,000
other[c]	647	146,000	374	101,000
Total	*39,800*	*5,330,000*	*21,200*	*3,240,000*

[a] From ref. 2, data are rounded to no more than three significant digits; may not add to totals shown.
[b] Free alongside ship values at U.S. ports.
[c] Includes Australia, Belgium, France, India, Israel, Japan, the Netherlands, Saudi Arabia, Switzerland, and Venezuela.
Source: U.S. Census Bureau.

of the powder passing a 0.23 mm (65 mesh) screen with 22–81% held on a 44 μm (325 mesh) screen. In acid-spar, at least 85–97% calcium fluoride is present with impurities being limited to 1–1.5% silica and 0.03–0.10% sulfide or free sulfur. Ceramic-spar is marketed in different levels: 95–97% CaF_2 (No. 1), 93–95% CaF_2, and 85–93% CaF_2 (No. 2). Specifications on impurities vary but allow a maximum amount of 2.3–3% silica, 1–1.5% calcite, and 0.12% ferric oxide and trace amounts of lead.

Metallurgical fluorspar (met-spar) is sold as gravel, lump, or briquettes. It contains 60–85% effective calcium fluoride. In the United States, the effective value is usually quoted, determined by subtracting 2.5 times the SiO_2 content of the ore from the calcium fluoride content (1). This allows for the fluorine-consuming reaction:

$$2\,CaF_2 + SiO_2 \longrightarrow SiF_2 + 2\,CaO$$

Expressed as effective CaF_2, the specification is usually 70% minimum. Impurities are usually limited to 0.3% sulfilde or free sulfur, 0.25–0.50% lead, and minor amounts of phosphorus.

Table 5. **Analyses of Ceramic- and Acid-Grade Fluorspar, wt %**[a]

Assay	Ceramic	Acid
CaF_2	90.0–95.5	96.5–97.5
SiO_2	1.2–3.0	1.0
$CaCO_3$	1.5–3.4	1.0–1.5
MgO		0.15
B		0.02
Zn		0.02
Fe_2O_3	0.10	0.10
P_2O_5		0.03
$BaSO_4$		0.2–1.3
R_2O_3[b]	0.15–0.25	0.1–0.3

[a] Refs. 64 and 65.
[b] R_2O_3 is any trivalent metal oxide, eg, Al_2O_3.

In steel making, the preferred form of fluorspar flux is a washed gravel, 0.6–5 cm in diameter, containing less than 3% water and assaying 60–80% effective CaF_2 units. The higher CaF_2 ranges are hard to supply in large amounts from some sources of fluorspar. The use of fluorspar briquettes and pellets in the steel industry has declined, but these are still preferred by some producers. The briquettes contain 25–90% CaF_2, are frequently made to customer specifications, and may include fluxing agents and recycled steel mill wastes. Binders used include molasses, lime, and sodium silicate.

Optical-grade calcium fluoride, for special glasses and for growing single crystals, is also supplied in purities up to 99.99% CaF_2. This grade is especially low in transition elements.

9. Health and Safety Factors

Fluorite is not classified as a hazardous material. However, every precaution should be taken to prevent contact of calcium fluoride with an acid since formation of hydrofluoric acid will result. Because of the low solubility of calcium fluoride, the potential problem of fluoride-related toxicity is reduced. However, ingesting large amounts may cause vomiting, abdominal pain, and diarrhea. Water saturated with calcium fluoride has a fluoride concentration of 8.1 ppm as compared to the recommended water fluoridation level of 1 ppm fluoride ion. However, because the solubility of calcium fluoride in stomach acid is higher, continued oral ingestion of calcium fluoride could produce symptoms of fluorosis. The adopted TWA limit for fluorides as F is 2.5 mg/m^3 (66).

Beneficiation facilities require air and water pollution control systems, including efficient control of dust emissions, treatment of process water, and proper disposal of tailings. In handling finished fluorspar, operators must avoid breathing fluorspar dust and contacting fluorspar with acids. Inhalation may cause irritation to the respiratory tract. Symptoms may include coughing and shortness of breath. Upon contact with the skin a mild irritation or redness may result. Protective gloves, safety goggles, and full-body covering are recommended. Proper disposal of spills and the use of respirators and other personal protective equipment must be observed. Spills should be placed in a suitable container for reclamation or disposal, using a method that does not generate dust.

Shipping and storage containers, when empty, can be hazardous as they will contain residues. Calcium fluoride should be kept in a tightly closed container, stored in a cool, dry, ventilated area.

10. Environmental Concerns

Plants sensitive to fluorides have been shown to show signs of injury at concentrations of 1.0–4.0 µg of fluoride/m^3 over a 24-h period of exposure or at 0.5–1.0 µg of fluoride over a one month period of exposure. Cattle consuming a diet containing 40 ppm of fluoride or more developed symptoms of fluorosis: dental defects, bone lesions, lameness, and reduced appetite. The results were weight loss and diminished milk yield (67).

Naturally occurring fluorides are not considered to be health hazards. However, fluorides used in pesticides may cause severe illness or death if ingested. Fluorocarbons are very stable but do pose a health hazard: burning can result in the release of phosgene gas, which is toxic; some fluorocarbons as gases can replace the normal air supply in confined spaces if more dense than air, resulting in suffocation.

11. Uses

Fluorspar is considered to be a commodity of strategic and commercial importance as the United States import reliance is high and fluorspar is necessary in steel and aluminum production. Fluorspar is also the primary source of fluorine and its compounds. Table 5 shows the United States imports for consumption of calcium fluoride by country (2). Over 80% of reported fluorspar consumption is used in the manufacture of hydrofluoric acid and 20% in the steel and iron industry. Fluorspar is the starting material for the production of HF. See Table 6 for U.S. consumption by end use.

An estimated 90% of reported fluorspar consumption went to the manufacture of hydrofluoric acid and aluminum fluoride. The remaining 10% of the reported fluorspar consumption was used as a flux in steel making, in iron and steel foundries, for primary aluminum production, glass manufacture, enamels, welding rod coatings, and other uses or products.

Fluorspar is marketed in several grades. The three principal grades of fluorspar are acid, ceramic, and metallurgical. The specifications (discussed under Grades and Specifications) for the different grades are fairly well defined,

Table 6. **U.S. Reported Consumption of Fluorspar, by End Use, t**[a,b,c]

End use or product	Containing more than 97% calcium fluoride		Containing not more than 97% calcium fluoride		Total	
	2000	2001	2000	2001	2000	2001
hydrofluoric acid and aluminum fluoride	474,000	429,000	16	1,100	474,000	430,000
metallurgical[d]	W	21,300	15,900	43,700	15,900	65,000
other[e]	13,600	23,700	8,650	17,000	22,200	40,700
Total	*487,000*	*474,000*	*24,500*	*61,800*	*512,000*	*536,000*
stocks (consumer), December 31[f]	48,300	71,100	25,700	NA	73,900	NA

[a] Ref. 2.
[b] NA Not available. W Withheld to avoid disclosing company proprietary data.
[c] Data are rounded to no more than three significant digits; may not add to totals shown.
[d] Data for 2000 include consumption for basic oxygen and electric arc furnaces; 2001 data include consumption for all metallurgical uses.
[e] Includes acid grade used in enamel, glass and fiberglass, steel castings, welding rod coatings, and data represented by symbol W.
[f] Because of a change in survey methodology, the 2001 stocks data are only available for hydrofluoric acid and aluminum fluoride.

although some variation is allowed. There has been a general movement toward the use of higher quality fluorspar by many of the consuming industries.

CaF_2 is used for prisms in spectrometers and for cell windows (especially for aqueous solutions), where transparency to visible and ultraviolet radiation is a requirement.

Acid-grade fluorspar is used primarily as a feedstock in the manufacture of hydrofluoric acid and to produce aluminum fluoride. HF is the primary feedstock for the manufacture of most organic and inorganic chemicals containing fluorine. Two companies reported calcium chloride consumption for the production of HF. The largest use of HF is for the production of various fluorocarbon chemicals: hydrofluorocarbons (HFC), hydrochlorofluorocarbons (HCFC), and fluoropolymers. Most of the AlF_3 produced is used in aluminum reduction cells using acid-grade fluorspar.

Ceramic-grade fluorspar is used in the production of glass and enamel (68–70) to make welding rod coatings and as a flux in the steel industry (71).

Metallurgical-grade fluorspar is primarily used as a fluxing agent by the steel industry (71). Calcium fluoride is added to slag to make it more reactive, increasing the fluidity of the slag (by reducing its melting point) and thus increasing the reactivity. Reducing the melting point of the slag brings lime and other fluxes into solution allowing the absorption of impurities. Methods for producing large-volume crystals from calcium chloride for use in photolithography, optics, excimer lasers, wafers, and computer chips have been reported (72,73).

BIBLIOGRAPHY

"Calcium Fluoride" under "Fluorine Compounds, Inorganic," in *ECT* 1st ed., Vol. 6, pp. 689–692, by H. C. Miller, Pennsylvania Salt Manufacturing Co.; "Calcium Fluoride" under "Fluorine Compounds, Inorganic," in *ECT* 2nd ed., Vol. 9, pp. 573–582, by J. F. Gall, Pennsalt Chemicals Corp.; "Calcium" under "Fluorine Compounds, Inorganic," in *ECT* 3rd ed., Vol. 10, pp. 707–717, by J. F. Gall, Philadelphia College of Textiles and Science; "Calcium Fluoride" under "Fluorine Compounds, Inorganic," in *ECT* 4th ed., Vol. 11, pp. 323–335, by Tariq Mahmood and Charles B. Lindahl, Elf Atochem North America, Inc.; "Fluorine Compounds, Inorganic, Calcium" in *ECT* (online), posting date: December 4, 2000, by Tariq Mahmood, Charles B. Lindahl, Elf Atochem North America, Inc.

CITED PUBLICATIONS

1. D. R. Shawe, ed., *Geology and Resources of Fluorine in the United States*, U.S. Geological Survey Professional Paper 933, Washington, D.C., 1976, 1–5, 18, 19, 82–87.
2. M. Miller, "Flurospar," *Minerals Yearbook*, U.S. Geological Survey, Washington, D.C., 2001.
3. M. Miller, "Flurospar," *Mineral Commodity Summaries*, U.S. Geological Survey, U.S. Department of the Interior, Washington, D.C., Jan. 2003.
4. B. Porter and E. A. Brown, *J. Am. Ceram. Soc.* **45**, 49 (1962).
5. D. A. Schulz and A. W. Searcy, *J. Phys. Chem.* **67**, 103 (1963).
6. G. Petit and A. Cremieu, *Compt. Rend.* **243**, 360 (1956).

7. O. Ruff and L. Leboucher, *Z. Anorg, Allg. Chem.* **219**, 376 (1934).
8. *National Bureau of Standards Technical Notes*, Washington, D.C., 1971, 270–276.
9. B. F. Naylor, *J. Am. Chem. Soc.* **67**, 150 (1945).
10. K. A. McCarthy and S. S. Ballard, *J. Appl. Phys.* **36**, 1410 (1960).
11. Ref. 3, p. 69.
12. A. V. Grosse and C. S. Stokes, U.S. Department of Commerce, Office of Technical Service, PB Report 161460, Washington, D.C., 1960.
13. O. J. Whittemore, Jr. and N. N. Ault, *J. Ceram. Soc.* **39**, 443 (1956).
14. E. W. Washburn, ed., *International Critical Tables*, Vol. 3, McGraw-Hill Book Co., Inc., New York, 1929, p. 50.
15. S. S. Ballard, L. S. Combes, and K. A. McCarthy, *J. Opt. Soc. Am.* **42**, 684 (1952).
16. D. W. Brown and C. E. Roberson, *J. Res. U.S. Geol. Surv.* **5**, 509 (1977).
17. *Natl. Bur. Stand. U.S., Tech. News Bull.* **47**, 91 (1963).
18. J. L. Pauley and H. Chessin, *J. Am. Chem. Soc.* **76**, 3888 (1954).
19. E. W. Washburn, ed., *International Critical Tables*, Vol. 6, McGraw-Hill Book Co., Inc., New York, 1929, p. 154.
20. R. W. Ure, Jr., *J. Chem. Phys.* **26**, 1365 (1957).
21. T. Baak, *J. Chem. Phys.* **29**, 1195 (1958).
22. Data sheet, *IR Transmission Materials*, Barnes Engineering Co., Instrument Division, Stamford, Conn., 1992.
23. A. F. Wells, *Structural Inorganic Chemistry*, 3rd ed., Clardon Press, Oxford, U.K., 1962, p. 77.
24. H. E. Swanson and E. Tatge, *Natl. Bur. Stand. U.S. Circ. 539* **1**, 69 (1953).
25. H. Krainer, *Radex Rundsch.*, 19 (1949).
26. V. V. Babkin, V. V. Koryakov, T. A. Sokolova, and N. K. Petrova, *Khim. Prom-St (Moscow)* **3**, 1963–164.25 (1992).
27. H. Gabryel, L. Kacalski, and U. Glabisz, *Chem. Stosow* **33**(4), 673–678 (1989).
28. SU 1286520 A1, (Jan. 30, 1987), I. A. Elizarov and co-workers.
29. U.S. Pat. 4,264,563 (Apr. 28, 1981)
30. SU 1708762 A1 (Jan. 30, 1992), I. G. Saiko, A. A. Perebeinos, L. M. Pupyshevea, N. A. Orel.
31. SU 1699922 A1 (Dec. 23, 1991), M. E. Rakhimov, D. D. Ikrami, L. F. Mansurhodzhaeva, and Sh. A. Khalimov.
32. SU 998352 A1 (Feb. 23, 1983), A. A. Luginina, L. A. Ol'Khovaya, V. A. Reiterov, and D. D. Ikrami.
33. SU 802185 (Feb. 7, 1981), V. I. Rodin and co-workers.
34. A. A. Luginina and co-workers, *Zh. Neorg. Khim* **26**(2) 332–336 (1981).
35. V. V. Tumanov and co-workers, *Khim. Prom-St (Moscow)* (9), 668–671 (1989).
36. SU 83-3558912 (Mar. 2, 1983), M. I. Lyapunov, V. V. Tumonov, L. P. Belova, and G. H. Alekseeva.
37. V. V. Tumanov, L. P. Belova, and G. N. Alekseeva, *Prom-St (Moscow)* **9**, 551–553 (1983).
38. PL 104419 (Nov. 30, 1979), W. Augustyn, M. Dziegielewska, and A. Kossuth.
39. C. Russell, *J. Mater. Sci. Lett.* **11**(3), 152–154 (1992).
40. T. N. Naumova and co-workers, *Zh. Priki. Khim (Leningrad)* **64**(3), 480–484 (1991).
41. W. Augustyn and co-workers, *Prezm. Chem.* **68**(4), 153–155 (1989).
42. PL 106787 (Jan. 31, 1980), W. Augustyn, M. Dziegielewska, and A. Kossuth.
43. RO 88593 B1 (Mar. 31, 1986), H. Glieb, E. Apostol, and C. Dan.
44. SU 983052 A1 (Dec. 23, 1982), V. K. Fomin, N. I. Varlamova, O. V. Leleedev, and A. P. Krasnov.
45. Jpn. Pat. 01083514 A2 (Mar. 29, 1989), Y. Oata, N. Goto, I. Motoyama, T. Iwashita, and K. Nomura.

46. PL 85616 (Sept. 15, 1976), W. Augustyn and co-workers.

47. U.S. Pat. 77,810,047 (June 27, 1977), W. C. Warneke.

48. U. Glabisz and co-workers, *Prezm. Chem.* **68**(1), 20, 29–30 (1989).

49. U.S. Pat. 685,100, AO (Aug. 1, 1986), A. B. Kreuzmann and D. A. Palmer.

50. V. S. Sakharov and co-workers, *Khim Prom-St (Moscow)* (1), 257 (1982).

51. SU 79-2829664 (July 30, 1978), G. A. Loptkina, V. I. Chernykh, and O. D. Fedorova.

52. SU 709537 (Jan. 15, 1980), G. A. Lopatking and V. I. Chernykh.

53. EP 210937 A1 (Feb. 4, 1987), L. Siegneurin.

54. M. S. Nesterova and T. Yu Magda, *Tekhnal. Obogashch. Polezn. Iskop. Sredni.* **A3** 3, 96–99 (1981).

55. R. V. Chernov and D. L. Dyubova, *Zh. Priki Khim. (Leningrad)* **56**(5), 1133–1135 (1983).

56. U. Glabisz, H. Gabryel, L. Kacalski, and B. Kic, *Pr. Nauk. Akad. Ekon. Im. Oskara Langego Wroclawiu* **338**, 165–169 (1986).

57. SU 1224263 A1 (Apr. 15, 1986), V. K. Fomin, N. I. Varlamova, V. P. Kozma, and M. N. Esin.

58. V. V. Pechkovskii, E. D. Dzyuba, and L. P. Valyu, *Zh. Priki (Leningrad)* **53**(5), 961–965 (1980).

59. A. A. Opalovskii and co-workers, *Zh. Neorg. Khim.* **20**(5), 1179–1183 (1975).

60. EP 476773 A1 (Mar. 25, 1992), J. Dijkhorst.

61. U.S. Pat. 82,406,420 A (Nov. 8, 1982), J. P. Harrison.

62. R. C. Kirby and A. S. Prokopovitsh, *Science* **191**, 717 (Feb. 1976).

63. R. C. Kirby and A. S. Prokopvitsh, *Science* **191**, 717 (Feb. 1976).

64. Data sheet, *Fluorspar*, Reynolds Chemicals, Richmond, Va., Mar. 1978.

65. P. L. Braekner, *Allied Chemical*, Industrial Chemicals Division, Morristown, N.J., private communication, Nov. 1978.

66. *Calcium Fluoride MSDS*, Mallinckrodt Baker, Inc., Phillipsburg, N.J., 2001.

67. J. J. McKetta, ed., *Encyclopedia of Chemical Processing and Design*, Vol. 23, Marcel Dekker, Inc., New York, 1985, pp. 270–295.

68. H. Hu, F. Lin, and Y. Yhan, *J. Feng. Mater. Scien. Forum*, 67–68; (Halide Glasses VI) 239–243 (1991).

69. H. Hu and F. Lin, *J. Feng. Guisuanyan Xuebau* **18**(6), 501–505 (1990).

70. V. D. Khalilev and co-workers, *Fiz. Khim. Stekla* **17**(5), 740–743 (1991).

71. Jpn. Pat. 03291324 A2 (Dec. 20, 1991), K. Masame and T. Matsuo (to Heisei).

72. U.S. Pat Appl. 20020038625 (April 4, 2002), S. Sakume and co-workers.

73. U.S. Pat Appl. 20010025598 (Oct. 4, 2001), J. Staeblein and co-workers.

Lisa M. Vrana
Consultant

CALCIUM SULFATE

1. Introduction

Calcium sulfate [7778-18-9], $CaSO_4$, has several forms, ie, calcium sulfate dihydrate (commercially known as gypsum), calcium sulfate anhydrous (anhydrite), calcium sulfate hemihydrate, present in two different structures, α-hemihydrate

Table 1. **Gypsum Forms and Composition**

Common name	CAS Registry Number	Molecular formula	Composition, wt %		
			CaO	SO$_3$	Combined H$_2$O
anhydrite	[7778-18-9]	CaSO$_4$	41.2	58.8	
gypsum	[10101-41-4]	CaSO$_4 \cdot 2H_2O$	32.6	46.5	20.9
stucco	[10034-76-1]	CaSO$_4 \cdot {}^1/_2 H_2O$	38.6	55.2	6.2

and β-hemihydrate (commercial name of β-form: stucco or plaster of Paris [26499-65-0]). In natural deposits, the main form of calcium sulfate is the dihydrate. Some anhydrite is also present in most areas, although to a lesser extent. Mineral composition can be found in Table 1.

Stucco [10034-76-1] has the greatest commercial significance of these materials. Indeed, stucco is the primary constituent used to produce boards and plasters as the primary wall cladding materials in modern building construction and in formulated plasters used in job- or shop-site applications. Other uses of stucco are in Portland cement (qv) set regulation and agricultural soil conditioning. The hemihydrate is normally produced by heat conversion of the dihydrate from which $\frac{3}{2}$ H$_2$O is removed as vapor. Methods for control of the set (hydration time) of hemihydrate conversion to dihydrate were developed by the end of the nineteenth century.

About 20–25 million metric tons of calcium sulfate are consumed annually. About 80% is processed into the commercially usable hemihydrate. Gypsum [10101-41-4] and its dehydrated form have been used by builders and artists in ornamental and structural applications for >5000 years, as evidenced by artifacts from the ancient Egyptian and Greek cultures. Processing of gypsum to the hemihydrate in the United States began ~1835. The ore used was imported from the Canadian Maritime Provinces.

2. Properties

Gypsum mineral has several names that are widely used in the mineral trade.

Selenite is the colorless and transparent variety that shows a pearl-like luster and has been described as having a moon-like glow. The word selenite comes from the Greek for moon and means moon rock. Another variety is a compact fibrous aggregate called *satin spar*. This variety has a very satin-like look that gives a play of light up and down the fibrous crystals. A fine grained massive material is called *alabaster* and is an ornamental stone that has been used in fine carvings for centuries. Among other gypsum names, *rock gypsum* and *gypsite* are worth noting. Figures 1–3 show particularly fine gypsum crystals.

Color is usually white, colorless, or gray, but can also be shades of red, brown, and yellow.

Luster is vitreous to pearly, especially on cleavage surfaces; crystals are transparent to translucent.

Fig. 1.   Five centimeter pencil-sized acicular crystals of gypsum. Specimen from Naica, Chihuahua, S.A., Mexico.

Fig. 2.   Six centimeter bladed rosettes of gypsum. Locality unknown.

Fig. 3.   Gypsum rose Red River Floodway, Winnipeg, 6-cm sphere.

Table 2. **Physical Properties of Calcium Sulfate**

Property	Dihydrate	Hemihydrate	Anhydrite
mol wt	172.17	145.15	136.14
transition point, °C	128[a]	163[b]	
mp[c], °C	1450	1450	1450
specific gravity	2.32		2.96
solubility at 25°C, g/100 g H_2O	0.24	0.30	0.20
hardness, Mohs'	1.5–2.0		3.0–3.5

[a] Hemihydrate is formed.
[b] Anhydrous material is formed.
[c] Compound decomposes.

Crystal system is monoclinic; crystal habits include the tabular, bladed, or blocky crystals with a slanted parallelogram outline; the pinacoid faces dominate with jutting prism faces on the edges of the tabular crystals; long thin crystals show bends and some specimens bend into spirals called Ram's Horn Selenite; two types of twinning are common and one produces a spear head twin or swallowtail twin while the other type produces a fishtail twin.

Cleavage is good in one direction and distinct in two others.

Fracture is uneven but rarely seen. Hardness is 2 and can be scratched by a fingernail.

Specific gravity is ~2.3 (light).

Among other characteristics, thin crystals are flexible but not elastic, meaning that they can be bent but will not bend back on their own. Some samples are also fluorescent. Gypsum has a very low *thermal conductivity* (hence, its use in drywall as an insulating filler), therefore a crystal of gypsum will feel noticeably warmer than a like crystal of quartz (1).

Table 2 lists the physical properties of calcium sulfate in its different forms.

3. Sources

The natural, or mineral, form of calcium sulfate is most widely extracted by mining or quarrying and is commercially used. Natural gypsum is rarely found in pure form. The dihydrate and anhydrous forms are commonly found together. Impurities in gypsum deposits typically include calcium and magnesium carbonates, oxide(s) of silicon, clays, and small amounts of various soluble salts. The last two items generally have the most undesirable effect on commercial processing and production of prefabricated products. In some cases, the crude ore is beneficiated to provide a commercial feedstock in which the percentage of functional dihydrate has been increased. Most commercial gypsum has a purity level of 80% or higher.

The natural ore is quarried or mined in many areas of North America and Europe. Leading regions include Canada, Mexico, and the United States. In

Europe, the countries of France, Spain, Italy, the United Kingdom, and Russia have significant deposits of natural gypsum, as does Germany.

In addition to occurring naturally, calcium sulfate can be obtained by precipitation. In particular, calcium sulfate may crystallize as gypsum, calcium sulfate hemihydrate, and anhydrite.

Gypsum is also obtained as a by-product of various chemical processes. The main sources are from processes involving scrubbing gases evolved in burning fuels that contain sulfur (see SULFUR REMOVAL AND RECOVERY), such as coal (qv) used in electrical power generating plants, and the chemical synthesis of chemicals, such as sulfuric acid, phosphoric acid, titanium dioxide, citric acid, and organic polymers. The ability to market by-product gypsum, mainly for use in wallboard or cement, depends on the supply-demand situation and the gypsum quality, which in turns depends on the type of process and the operating conditions. In general, the added capital investment and processing costs associated with rendering by-product gypsums suitable as feedstocks for the gypsum board and plaster industry have tended to deter their use where good quality and relatively low cost natural gypsums are readily available. However, high gypsum purity makes by-product sources attractive, especially in regions where natural gypsum is scarce. A notable example of this has been Japan, wherein large tonnages of by-product gypsum from its phosphoric acid industry and flue gas desulfurization plants have been used (see PHOSPHORIC ACID AND THE PHOSPHATES). In North America and Europe, the major reason for by-product gypsum production, mainly from flue gas desulfurization (FGD) processes, is to make the disposal of the solid waste less difficult and expensive.

4. Thermodynamic and Kinetics of Gypsum Formation–Decomposition

The thermodynamic properties of gypsum formation by precipitation can be evaluated considering the liquid–solid equilibrium between calcium and sulfate ions in solution and solid $CaSO_4 \cdot 2H_2O$, as described by equation 1.

$$Ca^{2+} + SO_4^{2-} + 2H_2O = CaSO_4 \cdot 2H_2O \qquad (1)$$

The driving force for the formation of calcium sulfate dihydrate is the change in the Gibbs free energy, ΔG, while going from the supersaturated solution to equilibrium. The average energy per ion is given by

$$\Delta G = -\frac{RT}{2} \ln \frac{a_{Ca^{2+}} a_{SO_4^{2-}}}{K_{ps}} \qquad (2)$$

where a_I is the activity expressed as the product of the molality (m_I) and the activity coefficient (γ_I) of the I species ($I = Ca^{2+}$, SO_4^{2-} and H_2O), R and T are the gas constant and the absolute temperature, respectively, and K_{ps} is the thermodynamic solubility product of gypsum. The supersaturation ratio is defined as

$$\sigma = \frac{a_{Ca^{2+}} a_{SO_4^{2-}} a_w^2}{K_{ps}} \qquad (3)$$

The value of K_{ps} can be calculated as a function of temperature by means of the following relationship (2):

$$\ln(K_{ps}) = 390.96 - 152.62 \log T - 12{,}545.62/T + 0.08T \qquad (4)$$

An important parameter in crystal nucleation kinetics is the time that elapses between the onset of supersaturation and the formation of critical nuclei, or embryos [clusters of loosely aggregated molecules that have the same probability of growing (to become crystals) or dissolving (to disappear into the mother solution)]. This time parameter, defined as the true induction period (t^*), primarily depends on solution supersaturation, temperature, and hydrodynamics. However, t^* cannot be experimentally measured, since it is not possible to detect the formation of critical nuclei; rather, in order to perform the measurements, it is necessary to let such nuclei grow until they reach a detectable size. Consequently, it is possible to experimentally evaluate only a time referred to simply as the induction period (t_{ind}), with $t_{ind} \geq t^*$ defined as the time elapsed between the onset of supersaturation and the first changes in the physical properties of the system due to the formation of a solid phase. An analysis of experimentally determined values of t_{ind} gives some important information about the mechanisms of solid-phase formation and the growth process, which leads from critical nuclei to detectable crystals. If the process that takes place is truly homogeneous nucleation, ie, it occurs in a clear solution under the effect of supersaturation alone, t_{ind} is inversely proportional to the nucleation rate, defined as the number of nuclei formed in solution per unit time and volume. Therefore, it is possible to use the experimental knowledge of the induction period to estimate two characteristic thermodynamic quantities: viz, the dependence of t_{ind} on temperature allows us to evaluate the activation energy for nucleation (E_{act}), while its dependence on supersaturation allows us to determine the interfacial tension (γ_s) between crystals and the surrounding solution (3,4). Activation energy and interfacial tension values found in the literature for calcium sulfate dihydrate are reported in Table 3. It is well recognized from the data dispersion, and in particular for γ_s, that a good agreement among researchers has not been found, especially concerning the dependency of γ_s on temperature. The dependence of γ_s on T for $CaSO_4 \cdot 2H_2O$ is a controversial matter: According to some researchers,

Table 3. Values for Gypsum Interfacial Tension and Activation Energy

$T(°C)$	E_{act}(kJ/mol)	γ_s (mJ/m^2)	Reference
25		41.1	6
25–90	52.6		6
25		39.3	6[a]
70		50.6	6[a]
25–50–70	49.7	37.8	8[a]
25		18.0	9
30		76.0	10
25		95.0	11
30		23.2	12
25–50–70	30.0	37.0	13

[a] With NaCl in solution.

surface tension decreases with increasing temperature (5) while according to others, it increases with increasing temperature either linearly (6) or exponentially (7).

The thermodynamic properties of gypsum decomposition involve two distinct steps,

$$CaSO_4 \cdot 2H_2O = CaSO_4 \cdot \frac{1}{2}H_2O + 1\frac{1}{2}H_2O \qquad (5)$$

$$CaSO_4 \cdot \frac{1}{2}H_2O = CaSO_4 + \frac{1}{2}H_2O \qquad (6)$$

which have been the subject of much theoretical and practical study. Two forms of the hemihydrate, α and β, have been identified (14).

The terms α and β are often used to differentiate two generally accepted, yet controversial forms of hemihydrate. The β-hemihydrate has higher energy content and a higher solubility than the α-hemihydrate.

Anhydrite also has three common classifications. Anhydrite I designates the natural rock form. Anhydrite II identifies a relatively insoluble form of $CaSO_4$, which has an orthorhombic lattice. Anhydrite III is a relatively soluble form made by lower temperature decomposition of dihydrate, which has the same crystal lattice as the hemihydrate phase.

5. Manufacture

5.1. Natural Gypsum.   Gypsum rock from the mine or quarry is crushed and sized to meet the requirements of future processing or extracted for direct marketing of the dihydrate as a cement retarder. Once subjected to a secondary crusher, calcining, and drying, the product is fine ground. Fine-ground dihydrate is commonly called land plaster, regardless of its intended use. The degree of fine grinding is dictated by the ultimate use. The majority of fine-ground dihydrate is used as feed to calcination processes for conversion to hemihydrate.

5.2. β-Hemihydrate.   The dehydration of gypsum commonly referred to as calcination in the gypsum industry, is used to prepare hemihydrate, or anhydrite. The β-form is obtained when the dihydrate is partly dehydrated in a vacuum at 100°C or under conditions lacking a nearly saturated steam atmosphere.

Kettle calcination continues to be the most commonly used method of producing β-hemihydrate. The kettle can be operated on either a batch or continuous basis. Its construction is shown in Figure 4. The kettle is a cylindrical steel vessel enclosed in a refractory shell with a plenum in between. The steel vessel is suspended above a fire box from which heated air flows up and into the plenum surrounding the steel vessel and through multiple horizontal flues that completely penetrate the vessel. The plenum and flues provide heat to the kettle contents before the heated air is exhausted. An agitator with horizontal arms penetrates the depth of the kettle and is driven from above. Land plaster, usually ground 85–95% through 100 mesh (149 μm) is fed from the top. In batch operation, using an 18.1 metric ton capacity kettle, filling takes 20–30 min. Another

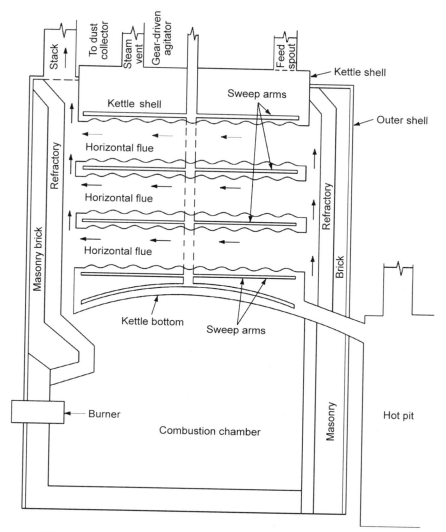

Fig. 4. Generalized vertical cross-section of a calcining kettle.

90–120 min are usually required to convert the dihydrate to hemihydrate. The steam released from the dehydration reaction is vented from the kettle top. When conversion to hemihydrate is complete (usually determined by temperature measurement of the kettle contents), the stucco is discharged by gravity through the quick-opening gate located at the periphery and bottom of the steel vessel. A typical temperature pattern for the kettle contents is shown in Figure 5. Approximately 1 GJ/t of hemihydrate is required in a well-designed kettle.

During the fill portion of a kettle cycle, the firing rate is usually controlled to maintain the kettle contents at a temperature of ∼104°C. When the fill is complete, the firing rate is increased to a level dictated by the desired stucco properties. The mass boils at a temperature of 115–120°C. The boil or drag

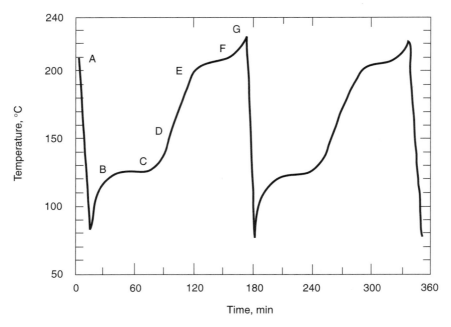

Fig. 5. Time–temperature profile for kettle calcination. Points A–B represent the fill period; B–C, is the boil or drag; C–D is the falling rate or cook-off; D is the discharge for hemihydrate. Points D–E show firing rate to second boil; E–F is the second boil; F–G is the second cook-off; G is the second-settle discharge.

continues for ~1 h, then subsides. Heating continues for a short time period to allow moisture release and the mass temperature increases to ~150–155°C if the hemihydrate form is desired, after which firing is reduced and the contents are dumped. In practice, owing to the inability to heat all particles of gypsum adequately, the discharged mass often contains small percentages of dihydrate, soluble anhydrite, and at times insoluble anhydrite.

If soluble anhydrite is desired, firing is maintained until a second boil occurs accompanied by a second temperature plateau at ~190°C. Virtually all the water of crystallization has been removed at 215°C. Soluble salts are impurities that increase the vapor pressure within the kettle. Aridized stucco refers to kettle-calcined hemihydrate that has been made with the intentional addition of 0.55–1.1 kg NaCl or $CaCl_2$ per metric ton of land plaster. The stucco characteristic of lower water demand permits higher density and higher strength casts. The hygroscopic nature of the chlorides prevents the use of aridized stucco for some applications.

5.3. α-Hemihydrate. The α-form of hemihydrate is prepared by dehydration of gypsum in water at temperatures >97°C and by dissociation in an atmosphere of saturated steam. Three processing methods are used for the production of α-hemihydrate. One, developed in the 1930s, involves charging lump gypsum rock 1.3–5 cm in size into a vertical retort, sealing it, and applying steam at a pressure of 117 kPa (17 psi) and a temperature of ~123°C (15). After calcination under these conditions for 5–7 h, the hot moist rock is quickly dried and pulverized.

Another method (16), first reported in the 1950s, has lower water demand. The dihydrate is heated in a water solution containing a metallic salt, such as $CaCl_2$, at pressures not exceeding atmospheric. A third method (17), developed in 1967, prepares very low water demand α-hemihydrate by autoclaving powdered gypsum in a slurry. A crystal-modifying substance such as succinic acid or malic acid is added to the slurry in the autoclave to produce large squat crystals.

Anhydrite. In addition to kettle calcination (Fig. 4), soluble anhydrite is commercially manufactured in a variety of forms, from fine powders to granules 4.76 mm (4 mesh) in size, by low temperature dehydration of gypsum.

Insoluble anhydrite is manufactured commercially by several methods. Where large rock gypsum is the starting material, beehive kilns are used and 24-h processing times are not unusual. Rotary calciners or traveling grates are often used for small rock feed. Fine-ground gypsum is calcined to the insoluble form in flash calciners. Temperature control is somewhat critical in all methods; low temperatures result in soluble anhydrite being present and too high temperatures dissociate the $CaSO_4$ into CaO and oxides of sulfur.

5.4. By-Product Calcium Sulfate.

There are many industrial chemical processes that produce by-product calcium sulfate in one of its forms. Whereas the most common is the neutralization of spent sulfuric acid, many of those processes do not produce a commercially useful by-product because of contaminants, particle size, or volume produced. There are, however, six chemical processes that do produce sufficient volume to have potential commercial value. Each is named after its chemical process.

The two names commonly given to the by-product gypsum are desulfogypsum or FGD–gypsum which are produced by scrubbing sulfur dioxide out of flue gases (see SULFUR REMOVAL AND RECOVERY). There are three general types of scrubbing processes that produce by-product gypsum: limestone, lime, and dual or double alkali.

The process for limestone scrubbing can be generally described by

Absorption

$$SO_2 + H_2O = H_2SO_3 \tag{7}$$

$$H_2SO_3 = HSO_3^- + H^+ \tag{8}$$

$$HSO_3^- = SO_3^{2-} + H^+ \tag{9}$$

Limestone Dissolution

$$2H^+ + CaCO_3 = Ca^{2+} + CO_2 + H_2O \tag{10}$$

Bisulfite–Sulfite Oxidation

$$HSO_3^- + \frac{1}{2}O_2 = HSO_4^- \tag{11}$$

$$SO_3^{2-} + \frac{1}{2}O_2 = SO_4^{2-} \tag{12}$$

Crystallization

$$Ca^{2+} + SO_4^{2-} + 2H_2O = CaSO_4 \cdot 2H_2O_{(S)} \tag{13}$$

There are several lime-scrubbing processes being marketed. The generalized process is described by

Absorption

$$2\,Ca(OH)_2 + 2\,SO_2 = 2\,CaSO_3 \cdot \frac{1}{2}\,H_2O + H_2O \tag{14}$$

Oxidation / crystallization

$$2\,CaSO_3 \cdot \frac{1}{2}\,H_2O + O_2 + 3\,H_2O = 2\,CaSO_4 \cdot 2H_2O_{(s)} \tag{15}$$

In the dual or double alkali process, an alkali salt that is considerably more soluble in water than limestone is used. The alkali salt is then regenerated using a second alkali, $CaCO_3$. There are several alkalis used in the absorber; the most common are magnesium sulfite, sodium sulfite, and ammonium sulfite. A typical process using magnesium sulfite is

Absorption

$$MgSO_3 + H_2O + SO_2 = Mg^{+2} + 2\,HSO_3^- \tag{16}$$

Oxidation–crystallization

$$2\,HSO_3^- + 2\,CaCO_3 = 2\,CaSO_3 \cdot \frac{1}{2}\,H_2O + 2\,CO_2 + \frac{1}{2}\,O_2 \tag{17}$$

$$2\,CaSO_3 \cdot \frac{1}{2}\,H_2O + O_2 + 3\,H_2O = 2\,CaSO_4 \cdot 2H_2O_{(s)} \tag{18}$$

Note that the dual alkali process is not competitive with respect to the limestone–gypsum process, since it requires a double step not needed in the limestone process. Moreover the limestone–gypsum process gives excellent results for SO_2 removal, and for this reason is the most widespread worldwide. As an example, in the United States the Northen Indiana Public Service Company (NIPSCO) R.M. Schahfer Station converted two FGD systems from dual alkali process to limestone forced oxidation process during the winter 1996–1997 (18). Since this conversion, the FGD systems have consistently produced high quality, wallboard grade gypsum.

Moreover, of all the by-product gypsums from chemical processes, desulfogypsum from coal-fired electric power utility plants has the greatest commercial potential because electric power plants are numerous and many are located near large population centers where there would be a ready market for by-product gypsum wallboard products (see COAL CONVERSION PROCESSES; POWER GENERATION). Utilization of gypsum is dependent on economically removing deleterious chemicals, viz, excess chlorides, water-soluble sodium and magnesium, and unoxidized calcium sulfite.

By-product gypsum made by neutralizing waste sulfuric acid from the sulfate process used to manufacture titanium oxide pigment is called titanogypsum (see TITANIUM COMPOUNDS, INORGANIC). This is commonly a two-industry process in that iron-rich ilmenite ore is first processed to obtain iron and the resulting slag is sold to the TiO$_2$ producers. There are a few locations where titanogypsum is produced in large enough quantities to be considered for commercial use. Limitations are the iron compound contaminants and their average particle size. Titanogypsum has become the second most important source of commercial by-product gypsum after desulfogypsum in the United States.

Phosphogypsum [13397-24-5] is the name given to the by-product gypsum residue when phosphate ore is acidulated to extract phosphoric acid. There are several processes commercially used. All of them digest or acidulate tricalcium phosphate.

$$Ca_3(PO_4)_2 + 3\,H_2SO_4 + 6\,H_2O = 2\,H_3PO_4 + 3\,CaSO_4 \cdot 2H_2O_{(s)} \qquad (19)$$

In the United States, environmental considerations render by-product gypsum from all of the processes inappropriate for the building material industry. Radon and daughter radionuclides are retained in the by-product residue after acidulation as is the heavy metal cadmium (see HELIUM-GROUP GASES; CADMIUM AND CADMIUM ALLOYS). Phosphogypsum's commercial use in the gypsum wallboard industry in Europe and Japan has diminished as desulfogypsum has become more available.

Fluorogypsum is the name ascribed to by-product gypsum from fluorspar acidulation to produce hydrofluoric acid. The chemical reaction

$$CaF_2 + H_2SO_4 \longrightarrow CaSO_4 + 2\,HF(g) \qquad (20)$$

produces anhydrite. Over time, the anhydrite converts to gypsum. Contaminants in fluorogypsum, especially the heavy metal beryllium, render fluorogypsum a better road metal, ie, roadbed material, for which it is used, than as a building material product.

Citrogypsum and borogypsum are named after the respective processes and produce sizeable quantities of by-product gypsum in certain locations. However, contaminants preclude commercial use in the gypsum wallboard industry.

6. Scale Formation

Precipitation fouling or scale formation represents a problem frequently encountered in a variety of industrial applications; deposits are usually formed from those compounds whose solubility decreases with increasing temperature, such as calcium sulfates. Scale can be described as the deposition of a salt from aqueous solution onto surfaces. Scale occurs when an electrolyte solution, in this specific case containing calcium and sulfate ions, is concentrated as, eg, by evaporation; the salts (calcium sulfate anhydrous, hemihydrate, or dihydrate) will precipitate in a certain order depending on the concentration of the various ions and on the temperature. In particular, the solubility of all calcium sulfate

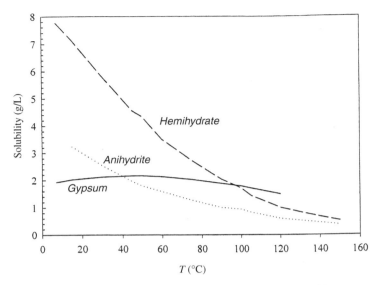

Fig. 6. Solubility of different forms of calcium sulfate.

forms decreases with increasing temperature starting from 40°C, as reported in Figure 6, a fact that is responsible for the formation of scale mostly constituted of a mixture of calcium sulfate dihydrate (gypsum), and calcium sulfate anhydrous (anhydrite). In this view, the comprehension of nucleation and crystal growth mechanisms that regulate the gypsum precipitation is fundamental to understanding processes in which gypsum formation is an unwanted occurrence. Water desalination (19), industrial water recovery in cooling tower technology, water distillation (20), and phosphoric acid production (21) are examples of calcium sulfate scale formation.

The best technique to inhibit gypsum scale formation is the addition of additives in solution, which retard calcium sulfate formation, ie, inhibit the nucleation kinetic mechanism.

Many substances, organic as well as inorganic, have been tested as additives for their capability of retarding the unwanted gypsum precipitation process. Regarding inorganic additives, aluminum ions are the most active among those studied (22,23), enhancing the degree of agglomeration of gypsum crystals and reducing the overall growth and nucleation rate. Moreover, experimental work has showed that chloride salts (NaCl and CaCl$_2$) influence gypsum nucleation by retarding nucleation kinetics (8). Among the organic species whose effect on gypsum precipitation was tested, polymeric additives were the most frequently used, eg, mainly polyelectrolytes (24,25), polyphosphates, and phosphonates (9,26,27). These authors agree that these polymeric species are effective in retarding the kinetics of gypsum crystal nucleation and growth, and their action depends on pH and polymer concentration. Recently, citric acid has proven to have a strong inhibiting effect on gypsum nucleation (28).

An industrially reliable method for calcium sulfate scale removal is the use of EDTA (ethylenediamine tetraacetic acid) and NTA (nitriloacetic acid), which form stable complexes with Ca^{2+} cations: The chelanting agents react with and dissolve solid deposits. As a matter of fact, the more insoluble calcium complex results and is disintegrated into a soft, pumpable slush (29). Moreover, various inorganic alkali solutions such as sodium hydroxide, potassium hydroxide, and ammonium and sodium bicarbonate have been effectively used to remove calcium sulfate, but they are usually referred to as converting solutions, since calcium sulfate is converted to calcium carbonate or calcium hydroxide, and these are in turn removed with acid (29).

7. Shipment

Gypsum and gypsum products are bulky and relatively low in cost. In North America, factors of varying regional supply and demand for building products not withstanding, the normal economic overland shipping range is ~500 km. For overland shipments, there has been a steady shift, starting in the 1950s, from rail-to-motor transport. In some cases, truck shipments are made from plants directly to building construction sites. For continental coastal and lake region markets, crude gypsum is most often transported in specially designed, rapid unloading ships that deliver from quarries to plant sites where the gypsum is then processed into finished products. During the 1980s, there were reports of increased intercontinental trade in both crude gypsum ore and manufactured goods.

As for synthetic gypsum, if the power stations are located on the coast so that economical loading facilities are available for ships, export of FGD gypsum via sea routes to neighboring states can be more advantageous than transportation to the domestic gypsum factories. This situation is encountered in Europe, where almost all power stations have harbor docking facilities (30).

8. Economic Aspects

Crude gypsum is the principal form of calcium sulfate shipped in international trade, although the 1980s saw an increase in the volume of fabricated products moved across international borders.

Figure 7 shows the gypsum usage by source, natural or synthetic, in the United States, showing that the production of synthetic gypsum has raised from 0 to 8.10 million tons from 1985 to the end of 1999 (31).

The quantity of FGD gypsum produced depends on the sulfur content of the coal and on the degree of desulfurization by the FGD plant. For example, a coal-fired power station with an output of 750 MW, an FGD plant efficiency of 95% and a sulfur content in the coal of 0.6–1% produces ~9–14 t of FGD gypsum hourly during full-load operation (32). In the year 2000, flue gas desulfurization plants with FGD gypsum as a by-product are in operation in several countries in Western and Eastern Europe (see Table 4). It has been estimated that in the year 2000, ~94,000 MW of power station capacity have been equipped with wet FGD

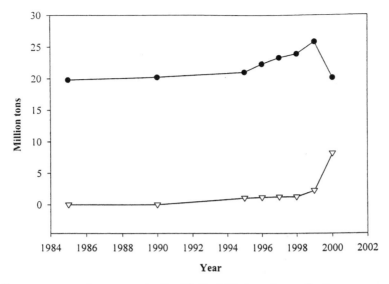

Fig. 7. Gypsum usage by source in the United States; △ synthetic gypsum; ● natural gypsum; (31).

systems and that a total of 15.8 million tons of FGD gypsum is obtained from them (30). It is expected that additional quantities of FGD will come after the installation of further desulfurization plants both in Western and in Eastern Europe, so that an increase of 22% more is expected in the quantity of FGD

Table 4. **FGD Gypsum Volumes in Europe**[a]

Country	Wet FGD systems installed (MW)	FGD gypsum volumes (tons)	
		year 2000 ($\times 10^{-3}$)	year 2005 ($\times 10^{-3}$)
Austria	1,500	100	100
Belgium	500	30	50
Croatia	210	30	110
Czech Republic	5,710	1,900	1,500
Denmark	4,200	330	400
Finland	1,800	170	170
France	1,200	60	90
Germany	51,000	6,200	6,500
Greece	450	120	190
Hungary	600	60	320
Italy	4,180	700	1,100
Netherlands	4,900	340	340
Poland	6,900	1,360	1,830
Slovakia	400	50	50
Slovenia	275	140	320
Spain	1,750	490	490
Turkey	2,670	3,160	4,700
United Kingdom	5,960	600	1,070
Total Europe	*94,205*	*15,840*	*19,330*

[a] Ref. 30.

Table 5. **Quality Specifications Set by the Gypsum Industry for FGD Gypsum Compared With Those of Natural Gypsum**[a]

	Units	Value for FGD gypsum	Value for natural gypsum
moisture content	%	<10	1
purity	%	>95	85
pH value		5–8	7
color (whiteness)	%	>80	
odor	neutral		
average particle size (sieve 32 μm)	%	>60	
minor constituents	%	<5%	
MgO	%	<0.10	0.010
Na$_2$O	%	<0.06	0.005
Cl	%	<0.01	0.003
CaSO$_3$ ½H$_2$O	%	<0.50	
soot	%	<0.10	0.06
Al$_2$O$_3$	%	<0.30	0.30
Fe$_2$O$_3$	%	<0.15	0.20
SiO$_2$	%	<2.50	2.0
CaCO$_3$ + MgCO$_3$	%	<1.50	12
K$_2$O	%	<0.06	0.005
NH$_3$ + NO$_3$	%	0	

[a] Ref. 32.

gypsum up to the year 2005 as a result of retrofitting existing power stations. Therefore FGD gypsum quantity in the year 2005 would reach 20 million tons (30).

The properties of FGD gypsum have been drawn up on a European basis by EUROGYPSUM, the umbrella organization or the European gypsum industry, and has been adopted by all the interested European organizations. In the European view, FGD gypsum is a product and is identical to and of equal value with natural gypsum. FGD gypsum has now been legally accepted as a product in the European and International legislature, so for this reason FGD gypsum is no longer listed as waste in the waste catalogues (32). (See Table 5.)

9. Specifications

Formulated plasters utilizing specially processed calcined gypsum are packaged in multiply paper bags having moisture vapor–resistant liners. This type of packaging protects the contents from airborne moisture keeping the plaster more stable with respect to setting time and mixing water demand over longer periods of warehousing. Manufactured board products are most often bundled, two pieces face to face, stacked in units for transport to dealers' yards, and reshipped to individual job sites as construction schedules dictate. Specialized,

Table 6. **ASTM Gypsum and Gypsum Product Specifications**

ASTM method	Materials
Gypsum and gypsum plasters	
C22-91	gypsum
C28-91	gypsum plasters
C35-89a	inorganic aggregates for use in gypsum plaster
C59-91	gypsum casting and molding plaster
C61-91	gypsum Keene's cement
C317-91	gypsum concrete
C587-91	gypsum veneer plaster
Test methods	
C265-91	calcium sulfate in hydrated Portland cement
C471-91	chemical analysis of gypsum and gypsum products
C472-90a	physical testing of gypsum plasters, etc
Gypsum board products	
C36-91	wallboard (general)
C37-91	lath (base for plaster)
C79-91	sheathing
C442-91	backing board and coreboard
C588-91	base for veneer plasters
C630-91	water-resistant backing board
C931-91	exterior soffit board
C960-91	predecorated board
Test method	
C473-87a	physical testing of gypsum board products

labor saving, power driven handling equipment has been developed for stocking boards on construction sites. The ASTM specifications for gypsum and gypsum products are given in Table 6.

10. Uses

10.1. Uncalcined Gypsum. Calcium sulfate, generally in the form of gypsum, is added to Portland cement (qv) clinker to stop the rapid reaction of calcium aluminates (flash set) [see ALUMINUM COMPOUNDS, ALUMINUM OXIDE (ALUMINA)]. Also, gypsum accelerates strength development. For this reason, gypsum is more properly termed a set regulator, rather than a retarder, for Portland cement. When it is used in proper amounts, it also minimizes volume change. Normal gypsum addition to clinker is 5–6 wt%. Another notable use of uncalcined gypsum is in agricultural soil treatment, wherein it is commonly called land plaster. For this use it is finely ground.

10.2. FGD Gypsum. In countries with many years of gypsum tradition and with a well-developed market for gypsum-based building materials, such as in the United States, Japan, Austria, Belgium, Denmark, Finland, France, Germany, The Netherlands, and The United Kingdom, the gypsum industry

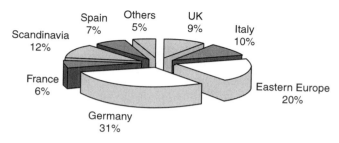

Fig. 8. Estimate of the extra capacity of gypsum required in Europe between 1998 and 2009 (33).

uses more raw gypsum than the amount produced by the FGD plants. This enables complete utilization of FGD gypsum by substitution of a part of the natural raw gypsum that occurs, eg, in Germany, or by substitution of a part of the imported natural raw gypsum, as in the case of The United Kingdom. Figure 8 shows an estimate of the extra capacity of gypsum required in Europe between 1998 and 2009, as estimated by Rumph (33) that could be satisfied by using FGD gypsum.

10.3. Calcined Anhydrite. Soluble anhydrite has physical properties similar to those of gypsum plaster. It hydrates to the dihydrate rapidly in water. Its outstanding property is its extreme affinity for any moisture, which makes it a very efficient drying agent (see DESICCANTS). In ambient moisture-laden air, it readily hydrates to hemihydrate. Soluble anhydrite, under the trade name Drierite, is widely used as a desiccant in the laboratory and in industry. A small amount is also used as an insecticide carrier. Small amounts of soluble anhydrite are unintentionally produced in most commercial calciners during hemihydrate production.

Keenes cement is produced from calcined anhydrite (dead-burned), finely ground and intermixed with special accelerator(s). Although the volume of its use has declined greatly since the 1960s, it is available for job-site mixing with hydrated lime as a composite, hand-finished plaster applied generally over an aggregated, gypsum-base (conventional) plaster.

10.4. Hemihydrate. The ability of plaster of Paris to readily revert to the dihydrate form and harden when mixed with water is the basis for its many uses. Of equal significance, is the ability to control the time of rehydration in the range of 2 min to >8 h through additions of retarders, accelerators, and/or stabilizers. Other favorable properties include its fire resistance, excellent thermal and hydrometric dimensional stability, good compressive strength, and neutral pH.

Upon setting, gypsum expands slightly and this property can be used to reproduce the finest detail, down to ~1 μm, as is done in certain dental and jewelry castings employing the lost wax process. Normal linear expansion upon setting of gypsum plaster is 0.2–0.3%, but by using additives, expansion may be controlled for special uses in the range of from 0.03 to 1.2%.

The calcination procedures and processing techniques produce a family of base stuccos best described by the amount of water, in wt%, of the plaster, which must be added when mixing to obtain standard fluidity. The range of fluidity permits casting neat plaster in the dry range of specific gravity of ~0.85–1.8 and consequent dry compressive strength of ~3.5–70 MPa (35–700 atm). Frequently, these stuccos are formulated with set and expansion control additives as well as many other materials to meet the needs of a particular application. Properties that limit gypsum plaster usage include plastic flow under-load, which is increased under humid conditions, strength loss in a humid atmosphere, and dissolution and erosion in water. Thus gypsum is not normally used for permanent performance structurally or in exposed, exterior locations. To prevent long-term calcination, gypsum products should not be used where temperatures exceed 45°C.

The largest single use of calcined gypsum in North America is in the production of gypsum board. Gypsum wallboard replaced plaster in the United States during the 1960s as the main wall cladding material. During that same time period, new veneer plaster systems were developed as an alternative to gypsum board (drywall) and the classic plastering systems, all of which are specified in building construction (see BUILDING MATERIALS, SURVEY). The veneer plasters are highly proprietary and specially formulated composites that provide good wear-resistant interior wall and ceiling surfaces. They are applied on the construction site either by a one- or two-coat procedure at thicknesses of ~0.19–0.32 cm. As reported in Figure 9, in terms of volume and value, wallboard is the dominant market for gypsum not only in North America but also in those markets where wallboard is well established, such as Western Europe and Japan. Cement is the other main source of demand for gypsum; since the cement production was ~1.6 billion tons in 2000, the cement industry consumed ~5–6% of that quantity, or ~100 million tons of gypsum (33).

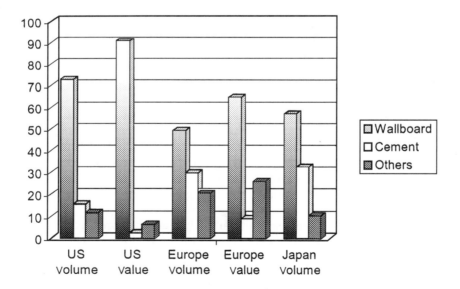

Fig. 9. Gypsum market (billion tons) and value in the United States, Europe, and Japan; (33).

Molding plasters have been used for centuries to form cornices, columns, decorative moldings, and other interior building features. Molding plaster is a good utility plaster where expansion control, high hardness, and strength are not needed. Its miscellaneous uses are numerous. Art plasters are essentially molding plasters modified to increase surface hardness, chip resistance, and to reduce paint absorption of casts made from this material. Orthopedic plasters are used by hospitals and clinics for all types of orthopedic cast work.

A moderate amount of plaster is used in making impressions and casting molds for bridges, and by dental laboratories. Both α- and β-plasters are used by the dental trade (see DENTAL MATERIALS). The α-plasters are also tailored to meet the needs of modern industrial tooling, where they are used for master patterns, models, mock-ups, working patterns, match plates, etc. They are the accepted material because their use results in great time and labor saving costs, as well as the excellent accuracy and stability of cast dimensions. Also, the material is adaptable to intricate, irregular shapes, complex intersections, and quick modification.

BIBLIOGRAPHY

"Calcium Compounds (Calcium Sulfate)," in *ECT* 1st ed., Vol. 2, pp. 767–779, by W. A. Hammond, W. A. Hammond Drierite Co.; in *ECT* 2nd ed., Vol. 4, pp. 14–27, by W. A. Hammond, W. A. Hammond Drierite Co.; in *ECT* 3rd ed., Vol. 4, pp. 437–448, by R. J. Wenk and P. L. Henkels, United States Gypsum Co.; in *ECT* 4th ed., Vol. 4, pp. 812–826, by D. J. Petersen, N. W. Kaleta, L. W. Kingston, National Gypsum; "Calcium Sulfate" in *ECT* (online), posting date: December 4, 2000, by D. J. Petersen, N. W. Kaleta, L. W. Kingston, National Gypsum.

CITED PUBLICATIONS

1. *Web site* "The Mineral Gallery", http://209.51.193.54//minerals/sulfates/gypsum/gypsum.htm
2. W. L. Marshall and R. Slusher, *J. Phys. Chem.* **70**, 4015 (1966).
3. J. W. Mullin, *Crystallization*, 3th ed., Butterworth-Heinemann Ltd, Oxford, (1993).
4. O. Söhnel, and J. Garside, *Precipitation*, Butterworth-Heinemann Ltd, Oxford, (1992).
5. G. R. Wood and A. G. Walton, *J. Appl. Phys.* **41**, 3027(1970).
6. S. He, J. E. Oddo, and M. B. Tomson, *J. Coll. Interf. Sci.* **162**, 297 (1994).
7. D. H. Rasmussen, and A. P. MacKenzie, *J. Chem. Phys.* **59**, 5003 (1973).
8. M. Prisciandaro, A. Lancia, and D. Musmarra, *Ind. Eng. Chem. Res.* **40**, 2335 (2001).
9. P. G. Klepetsanis,E. Dalas, and P. G. Koutsoukos, *Langmuir* **15**, 1534 (1999).
10. V. B. Ratinov, and O. M. Todes, *Dokl. AN SSSR* **132**, 402 (1960).
11. A. E. Nielsen,*Kinetics of Precipitation*, Pergamon, Oxford, 1964.
12. D. M. Keller, R. E. Massey, and O. E. Hileman, Jr., *Can. J. Chem.* **56**, 831 (1978).
13. A. Lancia, M. Prisciandaro, and D. Musmarra, *AIChE J.* **45**, 390 (1999).
14. K. K. Kelly, J. C. Southard, and C. T. Anderson, *U.S. Bureau of Mines Technical Papers*, Technical Paper 625, 1941.

15. U.S. Pats. 1,979,704 (Nov. 6, 1934); 2,074,937 (Mar. 23, 1937), W. S. Randel, M. C. Dailey, and W. M. McNeil (to United States Gypsum Co.).
16. U.S. Pat. 2,616,789 (Nov. 4, 1952), G. A. Hoggatt (to Certain-Teed Products Corp.).
17. Brit. Pat. 1,079,502 (Aug. 16, 1967), G. W. Cafferata (to BPD Industries, Ldt.).U.S. Pat. 3,956,456 (May 11, 1976); Can. Pat. 986, 145 (Mar. 23, 1976), J. A. Keller and R. T. Spitz (to National Gypsum Co.).
18. B. C. Skinner, and T. Cirbo, *Proceedings of Gypsum 2000*, 6th International Conference of Natural and Synthetic Gypsum, May 16–19, 2000, Toronto, Canada, 2000.
19. W. Stumm, and J. J. Morgan, *Aquatic Chemistry*, 3rd ed., John Wiley & Sons, Inc., New York,1996.
20. *BETZ Handbook of Industrial Water Conditioning*, 8th Ed., Betz Lab. Inc. Trevose Pa, 1982,p. 202.
21. S. K. Sikdar, F. Orè, and J. H. Moore, *AIChE Symp. Ser.* **193**, 76,82 (1980).
22. S. Sarig, and J. W. Mullin, *J. Chem. Tech. Biotechnol.* **32**, 525 (1982).
23. J. Budz, A. G. Jones, and J. W. Mullin, *J. Chem. Biothechnol.* **36**, 153 (1986).
24. B. R. Smith, and A. E. Alexander, *J. Coll. Int. Sci.* **34**, 81 (1970).
25. Z. Amjad, and J. Hooley,*J. Coll. Int. Sci.* **111**, 496 (1986).
26. S. T. Liu and G. H. Nancollas, *J. Coll. Int. Sci.* **52**, 593 (1975).
27. M. P. Wilson,A. L. Rohl,A. J. McKinnon, and J. D. Gale, "An experimental and molecular modelling investigation into the inhibition of gypsum crystallisation by phosphonate additives," *Proceedings of Industrial Crystallization*, Cambridge, UK, September 12–16, 1999.
28. E. Badens, S. Veesler, and R. Boistelle, *J. Crystal Growth* **198/199**, 704 (1999).
29. J. Cowan, and D. J. Weintritt, *Water-Formed Scale Deposits.*Gulf Publishing Company-Houston, Tex., 1976.
30. H. Hamm, and R. Hüller, *Proceedings of Gypsum 2000*, 6th International Conference of Natural and Synthetic Gypsum, May 16–19, 2000, Toronto, Canada, 2000.
31. B. Bruce, *Proceedings of Gypsum 2000*, 6th International Conference of Natural and Synthetic Gypsum, May 16–19, 2000, Toronto, Canada. 2000.
32. T. Mallon, *ZKG Int.* **4**, 220 (1998).
33. K. Rumph, *Proceedings of Gypsum 2000*, 6th International Conference of Natural and Synthetic Gypsum, May 16–19, 2000, Toronto, Canada, 2000.

AMEDEO LANCIA
DINO MUSMARRA
MARINA PRISCIANDARO
University of Napoli

CAPILLARY SEPARATIONS

1. Introduction

Capillary scale separations, most typically implemented in chromatography and electrophoresis, generally offer higher performance indices than their larger scale counterparts. Capillary chromatography and electrophoresis are typically used for the analyses of complex samples, containing a large number of analytes. In chromatography, analytes are separated based on their differential partitioning between an immobile stationary phase and a mobile fluid phase. In electrophoresis, different charged analytes are separated based on the different velocities with which they move in an electric field. Micellar electrokinetic chromatography (MEKC) and capillary electrochromatography (CEC) are hybrid techniques where separation by differential partitioning and movement under the influence of an electric field are both involved.

Though chromatography and electrophoresis differ in how separation is achieved, many common principles apply to both. In both techniques, the sample is introduced at the head of the column as a narrow zone and the analytes separate as they move to the exit end. The effluent analyte bands are detected as peaks using suitable detectors. The parameters of the separation process, such as elution/migration time, efficiency (how narrow peak is while eluting at a given time, often given as the number of theoretical plates), resolution between adjacent peaks, etc. are all defined similarly. Some aspects are different between them, however; both similarities and differences are further discussed later in this article.

2. Basic Chromatography Principles

As the analyte is partitioned between the mobile and the stationary phase, an equilibrium or distribution constant K can be defined

$$K = \frac{C_S}{C_M} \tag{1}$$

where C_S and C_M are the concentration of the analyte in the stationary and the mobile phase, respectively. The longer time the analyte spends in the stationary phase, the longer is its retention time, t_r, the total time the analyte spends in the separation process. The mobile phase moves through the column with a constant velocity; all unretained analytes move at the same velocity, eluting at the so-called dead time, t_0. All other analytes, which are retained to any extent by the stationary phase, elute later, the time increasing with increasing K. Figure 1 shows a model chromatogram with two analyte peaks with corresponding retention times t_{r1} and t_{r2} and an unretained molecule (eluting at t_0). Since t_r is a function of the mobile phase flow rate, the retention volume, V_r, the volume of the mobile phase necessary to elute the analyte, which is independent of the flow rate, is sometimes used instead of t_r:

$$V_r = t_r Q \tag{2}$$

where Q is the volumetric flow rate of the mobile phase. Similarly, the dead volume or holdup volume for an unretained analyte, V_0, can be defined from equation (2) by substituting t_0 for t_r.

The retention factor, k', used to describe the movement of the analytes through the column, is also commonly used to describe retention and it is readily

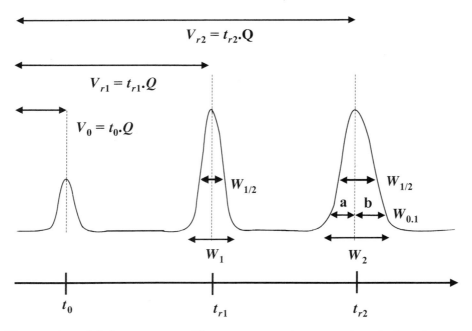

Fig. 1. A model chromatogram. The parameters shown are described in the text.

computed as follows:

$$k' = \frac{t_r - t_0}{t_0} = \frac{V_r - V_0}{V_0} \tag{3}$$

Parameters to compute k' can be easily obtained from a chromatogram such as Figure 1. For practical reasons, best chromatograms result with retention factors in the range of 2–20. If k' values are too low, analytes are not well separated; they all elute near the dead time; if k' values are too high, analysis time increases and the ability to detect an analyte also suffers because a broad wide peak of the same total area as an otherwise sharp narrow peak results.

The difference in retention factors or retention times is not the only factor that governs how well two or more analytes separate. Band broadening occurs during the separation process. This phenomenon was originally explained by Martin and Synge (1) by the "plate theory". Subsequently, the "rate theory" of chromatography advanced by van Deemter and co-workers (2) became more commonly used. This theory describes analyte band broadening in terms of the peak variance, σ^2.

In chromatography, several processes contribute to the final zone width, such as resistance to mass transfer, eddy diffusion (in packed columns), longitudinal diffusion, and extra-column effects. The overall peak variance is the sum of the individual variances:

$$\sigma_{tot}^2 = \sigma_{res}^2 + \sigma_{eddy}^2 + \sigma_{long}^2 + \sigma_{extra}^2 \tag{4}$$

For ideal (Gaussian) analyte peaks, the width of the peak is directly related to the variance, σ^2. The variance per unit column length is the plate height, H:

$$H = \frac{\sigma^2}{L} \tag{5}$$

where L is the column length. Capillary separation systems excel in their separation abilities relative to their larger counterparts in that small plate heights, typically expressed in micrometers (μm), are observed. The total number of theoretical plates in a column, N, which ultimately governs the quality of the separation, is hence given by

$$N = \frac{L}{H} \tag{6}$$

For idealized Gaussian peaks, experimentally N can be readily calculated from the width at half height of the peak, $w_{\frac{1}{2}}$:

$$N = 5.54 \left(\frac{t_r}{w_{1/2}} \right)^2 \tag{7}$$

For capillary gas chromatography (GC), the most mature and perhaps the most widely practiced of all separation techniques, the number of theoretical plates achieved in a typical separation system can be in millions.

True Gaussian peaks, are, however, rarely observed in practice. Foley and Dorsey (3) describe a more realistic approach to calculate N that is applicable to asymmetric peaks. Here the ratio, b/a, of the front and the tail half-widths, $w_{0.1}$, of a peak measured at 10% of the peak height, is used. For a symmetric peak, the value of b/a is unity, for a tailing peak the value will be >1, for a fronting peak the value will be <1. The parameter N is then calculated as:

$$N = \frac{41.7(t_r/w_{0.1})^2}{1.25 + (b/a)} \tag{8}$$

As band broadening increases, plate height, H, increases. In the classical van Deemter equation, H, is described as a function of mobile phase flow velocity:

$$H = A + B/v + Cv \tag{9}$$

where A, B, and C are coefficients pertaining to eddy diffusion, longitudinal diffusion and resistance to mass transfer, respectively, and v is the velocity of the mobile phase.

The eddy diffusion term, of importance only in a packed column, arises from flow inhomogeneities in a packed-bed structure. The coefficient A is directly proportional to the diameter of the packed particles, d_p, and the quality of the packing, λ:

$$A = 2\lambda d_p \tag{10}$$

This term is zero in open tubular columns, as used in capillary GC. Longitudinal diffusion is related to diffusive broadening of analyte zones in the mobile phase and the B-term in the van Deemter equation is directly related to the diffusion coefficient of the analyte in the mobile phase, D_m:

$$B = 2\gamma D_m \tag{11}$$

where γ is related to the packing structure. The shorter t_r is, the less opportunity there is for this type of diffusive broadening; thus the second term in equation (9) is inversely proportional to the mobile phase velocity. This term applies to both packed and open tubular columns; in the latter case, γ is unity.

The third term in equation (9) relates to the rate of transport of the analyte molecules between the stationary and mobile phase. As the mobile phase velocity through the column is increased, it has less and less time to equilibrate with the stationary phase. Some molecules are left behind in the stationary phase that would not have been if equilibrium were achieved. This nonequilibrium condition persists along the column length and results in band broadening that increases with increasing mobile-phase velocity. The coefficient C is proportional to the square of diameter of the packed particles in packed columns or conversely, square of the column diameter for open tubular capillaries and the effective thickness of the active layer on the stationary phase. It is also inversely proportional to the diffusion coefficient of the analyte, both in the mobile and stationary phases.

2.1. Resolution.   Resolution, R_s, provides a quantitative measure of separation between two analytes and is given by equation:

$$R_S = \frac{2(t_{r2} - t_{r1})}{w_1 + w_2} \tag{12}$$

where t_r and w are the retention times and base widths of respective analyte peaks. Sometimes it is useful to relate R_s to the retention factors k'_1 and k'_2, selectivity factor α, defined as $\alpha = k'_2/k'_1$, and N:

$$R_S = \left(\frac{\sqrt{N}}{4}\right)\left(\frac{\alpha - 1}{\alpha}\right)\left(\frac{k'_2}{1 + k'_2}\right) \tag{13}$$

R_S increases with increasing N and α. Using a longer column, decreasing the particle diameter in packed columns, and optimizing the mobile phase velocity can all increase N. All of these are more easily changed in a capillary-based system relative to their large bore counterparts. The parameter R_s is also advantageously increased by increasing the selectivity factor, α, through changing the stationary phase, or by changing the composition/nature of the mobile phase in liquid chromatography (LC).

3. Basic Capillary Electrophoresis (CE) Principles

Whereas chromatography is carried out both in larger bore columns and in the capillary format, for a variety of reasons there is no larger bore equivalent of CE.

In CE, analytes are carried along a capillary in which an electric field is present along the length of the capillary. The analytes separate because they exhibit different mobilities (speeds) due to differences in their size and charge. Every charged particle (ion) in an electric field, E, experiences an electric force, which causes it to accelerate, countered by the friction generated between the particle/ion and the surrounding medium. As a result, the particle/ion moves through the capillary with a constant velocity, v_i, proportional to the applied electric field E:

$$v_i = \mu_i E \tag{14}$$

where μ_i is the electrophoretic mobility of the particle. For spherical particles, a shape approximated by most solvated ions, μ_i can be estimated from

$$\mu_i = \frac{z_i \cdot e}{6\pi \cdot \eta \cdot r_i} \tag{15}$$

where z_i is the formal integer charge on the ion, e is the electron charge in coulombs, η is the viscosity of the medium and r_i is the radius of the solvated ion.

In chromatography, the movement of the sample constituents takes place due to the movement of the mobile phase, in turn accomplished by a pressure gradient. In CE, bulk flow of the separation medium (often called the background electrolyte, BGE) can occur due to electroosmosis (see below) but is not necessarily essential for the separation of the analyte components, because these can move without bulk electroosmotic flow. However, in most cases, electroosmosis is present and plays an important role in the separation process.

CE is most often conducted in fused silica capillaries. The capillary wall contains silanol groups, $-SiOH$, which are deprotonated when the pH of the electrolyte is >2, thereby inducing a net negative charge on the inner capillary surface. Typically, solvated cations such as those of alkali metals are present in the double layer near the capillary wall. As an electric field is applied, these positively charged ions move to the negative end of the capillary. The viscous coupling of the water molecules surrounding the ions and those in the bulk solvent in a narrow bore capillary causes the entire liquid to move; this is termed electroosmosis or electroosmotic flow (EOF). In large bore tubes, this phenomenon does not occur because rather than moving in a unified front, the liquid simply moves along the wall and recirculates by moving in the opposite direction in the center. Additionally, as the tube bore increases, the electrical resistance decreases and extensive joule heating prevents the application of a sustained high electric field.

The sign and magnitude of the EOF is obviously dependent on the sign and magnitude of the charge on the capillary wall, usually expressed as the zeta potential, ξ. As depicted in Figure 2, the flow profile in EOF is flat, significantly different from the parabolic profile common to pressure induced flow. The flat flow profile minimizes axial dispersion and allows the high separation efficiencies observed in CE. The magnitude and direction of the EOF can be derived from the Smoluchovski equation:

$$V_{\text{eof}} = \mu_{\text{eof}} E = \frac{\varepsilon \cdot \xi \cdot E}{\eta} \tag{16}$$

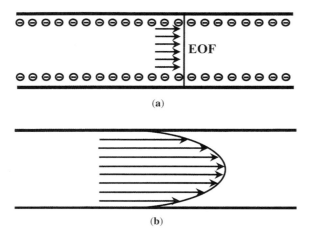

Fig. 2. Flow profiles generated in (**a**) capillary electrophoresis, and (**b**) chromatography.

where ε is the dielectric constant of the electrolyte, ξ is the zeta potential, η is the electrolyte viscosity and E is the electric field applied. The magnitude and direction of the EOF can be modified by adjustment of the electrolyte composition, addition of solvents that change ε or η, or by changing the zeta potential, ξ. The native charge of the fused silica capillary is negative and the EOF is therefore directed from the positive to the negative electrode. However, modifying the wall by any of a variety of techniques can be used to reduce, eliminate, or even reverse the EOF.

The overall velocity of an ion in the fused silica capillary is a combined function of its electrophoretic mobility, μ_i, and the electroosmotic mobility, μ_{eof}. The separation is similar to a boat race along a river where individual analytes represent the individual boats and their speed relative to the river is controlled by μ_i and the bulk flow of the river is represented by μ_{eof}. One can have a race either along or against the river flow, which will control the overall time necessary for the competitors to go from the start to the finish point. In CE, the EOF can be either in the same direction as the electrophoretic movement of the analytes or against it, it can increase or decrease their overall progress to the detection point. When the EOF is zero (much like a completely still river, true zero EOF is difficult to attain), the effective velocity of an analyte is simply given by equation (14).

Note that the uncharged analytes have no electrophoretic mobility and only move along with the EOF. This is similar to the behavior of an unretained analyte in chromatography. However, in CE, a neutral analyte may or may not arrive at the detector, depending on the direction of the EOF. Neutral analytes can be separated by a CE variant, MEKC, described later.

As in chromatography, the plate height, H, or the number of theoretical plates, N, is a measure of the efficiency of the electrophoretic separation. Here, H is defined as

$$H = \frac{\sigma^2}{L_{\text{ef}}} = \frac{2Dt}{L_{\text{ef}}} = \frac{2DL_t}{\mu_i V} \qquad (17)$$

where L_{ef} is the effective capillary length (the length from the injection to the detection), L_t is the total capillary length, V is the voltage applied and D is the diffusion coefficient. The van Deemter equation is also applicable. But there is no stationary phase or packing in CE, only the longitudinal diffusion term (B-term) is relevant. However, other processes, not occurring in chromatography, contribute to the total zone variance:

$$\sigma_{tot}^2 = \sigma_{dif}^2 + \sigma_{joul}^2 + \sigma_{inj}^2 + \sigma_{int}^2 \tag{18}$$

where σ_{dif}^2 is the variance due to the diffusion (B-term), σ_{joul}^2 is the variance caused by joule heating in the capillary, σ_{inj}^2 is the variance caused by processes associated with the injection of the sample plug and σ_{int}^2 is the variance caused by the analyte interaction with the capillary wall (unwanted chromatography). Extensive description of all phenomena contributing to the zone broadening in CE is given in the volume by Foret and co-workers.

Concepts such as number of theoretical plates, N, or resolution, R_s, apply to CE in much the same as chromatography.

4. Sample Preparation for Chromatography and Capillary Electrophoresis

Sample preparation or extraction prior to the actual separation and analysis is most often required in complex samples. Sample preparation is sometimes more challenging than separation/analysis and is often tedious and time consuming. It is particularly important in trace analysis applications (pesticides in fruit/vegetables, polychlorinated biphenyls in soil or sediments, to name a few) where the analyte must first be extracted/separated from the matrix (see TRACE AND RESIDUE ANALYSIS). Currently, several methods are used for sample treatment prior to chromatographic analysis, such as soxhlet extraction, liquid–liquid extraction, supercritical fluid extraction (SFE), accelerated solvent extraction (ASE), solid phase extraction (SPE), solid phase micro extraction (SPME), headspace analysis, and purge-and-trap methods. For air samples, a variety of sample collection devices including adsorbent tubes, filters or denuders are used.

In liquid–liquid extraction (see EXTRACTION, LIQUID–LIQUID) the analyte(s) of interest are extracted from the sample (usually aqueous) into an immiscible extractant phase. Liquid–liquid extraction or soxhlet extraction are the most commonly prescribed analyte extraction method in most United States Environmental Protection Agency (US EPA) and Association of Official Analytical Chemists (AOAC) methods. The main disadvantage of liquid–liquid extraction is the significant consumption of organic solvents (often toxic). SFE (see SUPERCRITICAL FLUIDS), ASE and microwave assisted sample extraction have become popular replacements. Reviews that describe comparison of different extraction procedures as well as description of specialized applications are available (4–8).

For extraction of liquid samples, SPE has become popular. A sample containing the analytes is passed through the solid packing material contained in a SPE-tube or embedded in a disk. In the more common mode of using SPE, analytes of interest are retained on the packing while most matrix components are unretained. Analyte preconcentration factors as high as 100 or more can be obtained. A prewash is used sometimes to remove some unwanted retained material, the analytes of interest are then selectively washed from the SPE phase using a suitable eluting solvent. The eluate is analyzed by any compatible method. In the other approach, unwanted matrix components are retained while the analytes pass through. No analyte preconcentration is possible in this mode. SPE is a convenient, inexpensive, and time saving sample preparation mode for many liquid samples; it significantly reduces the volume of organic solvents used. For complex samples, SPE is often used after SFE or ASE.

Solid phase microextraction (SPME) is a miniaturized form of SPE, particularly useful for the analysis of gaseous samples. A fused silica fiber is coated with an absorbent such as carbowax, polydimethylsiloxane, etc. The fiber is exposed to the sample for a fixed period (usually 2–30 min, with stirring for liquid samples) during which the analytes of interest are taken up by the coating. Commonly, analysis by capillary GC is used; analytes are desorbed from the fiber thermally in the GC injection port. A specially designed LC or CE interface can be used for coupling to these instruments. SPME requires no solvent, the fiber is reusable and can be used for concentration of a variety of analytes and is thus particularly attractive.

Headspace and purge-trap methods are used for analyzing volatile analytes from a liquid, often an aqueous sample, most typically for analysis by capillary GC. Headspace sampling is a static method for the collection of a volatile analyte. The liquid sample is placed in a vial enclosed by a septum. After the equilibrium between the phases has been reached, the vapor in the headspace is withdrawn into a syringe for injection into a GC. The gaseous phase contains analytes in proportion to their concentration in the liquid phase. An SPME fiber is also often used to sample the headspace.

In a purge-and-trap procedure, an inert gas is bubbled through the (typically aqueous) sample causing the removal of the volatile compounds into the gaseous phase. The gas is made to flow through a sorbent trap that retains the analytes. Subsequently, the analytes are desorbed from the trap (typically by heating) and injected into a GC.

5. Capillary Gas Chromatography (GC)

In GC, the sample is vaporized and injected onto the chromatographic column. Flowing gaseous mobile phase (carrier gas) transports the sample through the column and separation between analytes takes place based on the partitioning of the analytes between this mobile phase and the immobilized stationary phase on the capillary wall. The first paper on chromatographic principles by Martin and Synge (1) and experimental demonstration of GC (9) started the phenomenal growth of this technique. Originally, packed columns were used, however, as early as in 1957, Golay (10) realized that higher efficiencies can be achieved with capillary open tubular columns. Further practical development awaited practical capillary columns, based on fused silica capillaries (11). Presently, an estimated 80% of GC analyses are carried out in open tubular capillaries. The high efficiency of open tubular columns is due to the absence of the

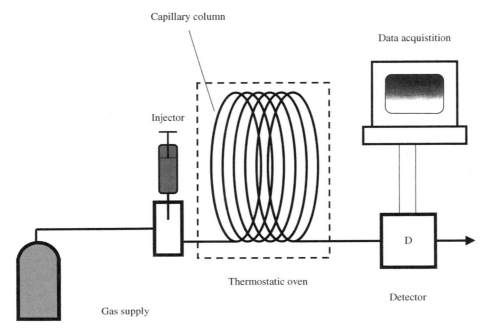

Fig. 3. Schematic diagram of a typical GC system.

column packing and the short radial travel distance to the wall. Capillary columns have much less flow resistance, and hence it is practical to use very long capillary columns. This permits a 2–3 order of magnitude increase in theoretical plate numbers relative to packed columns. On the other hand, due to the limited amount of the stationary phase on the inner-capillary wall, the capacity of the open tubular columns is considerably lower than that of packed columns. They cannot be used for analysis of large sample volumes (ie, <500 or 1 μL for gas and liquid volume, respectively).

5.1. Instrumentation.   A typical GC is depicted in Figure 3. It consists of a pressurized source of a carrier gas (typically a gas cylinder), an injector, a capillary column, a controlled temperature oven and a detector. A typical GC will also contain several pressure regulators and/or flow controllers, since it is vital to control the carrier gas flow rate.

5.2. Carrier Gas.   The carrier gas transports the sample through the column. It does not interact with the analytes. Occasionally, the choice of the carrier gas is dictated by the detector used. Common carrier gases used are He, N_2, and H_2; sometimes Ar, CO_2, and air (especially in portable instruments) are also used. Carrier gas flow rates typically range from 1 to 25 mL/min in capillary GC.

5.3. Sample Injection.   *Microsyringe Injection.*   Direct injection of a liquid sample with a microsyringe, although commonly used with packed columns, is less commonly used in capillary GC since overloading often occurs. The sample is injected with a microsyringe through an injection port that is heated above the boiling point of the least volatile analyte in the sample.

Split Injection. Split injection is one of the most common injection techniques in capillary GC, since it allows injection of samples independent of the solvent choice, at any column temperature and causes minimal band broadening. The vaporized sample is mixed with the carrier gas and a portion of the sample is introduced into the column, while the majority of the sample is vented out of the system. The ratio of the amount of the sample injected to that vented (split ratio) usually range between 1 and 10 and 1 and 1000. Split injections can exhibit poorer quantitative reproducibilities compared to other alternatives due to: (*1*) a pressure pulse is associated with sample injection that can alter the flow of the gas in an irreproducible way, (*2*) discrimination against less volatile analytes can occur. Advances in instrumentation and the use of automated injectors in recent years have ameliorated much of these problems.

Splitless Injection. This injection mode was devised to allow trace analysis, and allows injection of relatively large sample volumes (1–10 µL) on the columns with a bore ≥ 300 µm. The sample is introduced onto the column over a relatively long period of time (several seconds) creating a broad initial zone and then this zone is focused. This focusing relies on *cold trapping* or *solvent effects*. In *cold trapping*, the column inlet is maintained at a temperature of ~80°C lower than the elution temperature of any analyte and not <15°C below the solvent boiling point. The analytes then focus at the cooled column inlet due to its increased retention ability. Focusing due to *solvent effect* occurs when the column inlet is cooled to a temperature of at least 20–30°C lower than the boiling point of the solvent causing it to condense at the column inlet. Provided that the solvent liquid film has sufficient retention power to delay migration of the analytes and sufficient wetting of the stationary phase, broadening of the sample zone is minimized.

In *retention gap injection mode*, a significant length of uncoated and deactivated fused silica capillary is situated prior to the analytical column. This is referred to as the *retention gap*. The temperature during injection is sufficiently low so that the analytes that are not retained in the retention gap are trapped at the beginning of the analytical column. After the injection is accomplished, the temperature is raised and analytes start to migrate through the column. The retention gap technique is used also for injection of large sample volumes. It can be used advantageously with highly reproducible loop type injectors.

Programmed Temperature Vaporization (PTV). A PTV based injector can be used in different modes. In *split injection* with a hot or cold vaporization chamber, the sample is introduced as a liquid into the cold vaporization chamber, which is subsequently rapidly heated to the temperature of the highest boiling analyte. This minimizes both the pressure wave and discrimination against less volatile analytes. In *solvent elimination mode*, the sample is introduced to the cold injector at a temperature close to the boiling point of the solvent, which is vented out of the system by the carrier gas. Thereafter, the injector is heated to the sample vaporization temperature and the sample is introduced onto the column. This technique can be used to inject large sample volumes if the analyte is much less volatile than the solvent. The *cold solvent splitless mode* is used in a similar way as hot splitless injection and relies on *cold trapping* and *solvent effect*. The precision and accuracy attainable by a *PTV* based injector is significantly superior to that attainable in classical split and splitless injection modes.

Cold On-Column Injection. In this mode, a sample is introduced into the column directly as a liquid and is subsequently vaporized. The technique requires a special narrow bore syringe terminus, which fits the inner diameter of the capillary column. This technique eliminates discrimination of less volatile analytes and sample decomposition. It is easy to implement and automate; small sample volumes, 0.2–2 μL, are typically used. One caveat is that since the sample is injected directly on the column, nonvolatile residues in the sample can contaminate the column inlet, causing problems with subsequent analyses.

5.4. Controlled Temperature Oven.
A thermostated oven is used in GC, which requires exact control of the column temperature. Modern gas chromatographs are typically equipped with a programmable temperature oven that quickly and reproducibly changes the temperature as a function of time. The upper temperature limit is at least 350 °C.

5.5. Detectors for Gas Chromatography.
The detector is one of the most important components of a separation system; it may or may not be integrated with the rest of the instrument. An accurate and precise detection of analyte peaks is required for both qualitative and quantitative analysis with any separation system. Ideally the detector should be sensitive, have linear response to the analytes over a wide range, be stable and reproducible, and have a short response time. No detector meets all of the ideal criteria above. In reality, a detector is chosen based on the particular analytical problem. GC detectors, almost all of which are available for use with capillary systems, can be broadly subdivided into ionization detectors, detectors measuring bulk physical properties, optical detectors, and electrochemical detectors.

Ionization Detectors. *Flame Ionization Detector (FID).* In this detector, the effluent from the chromatographic column is made to burn in a hydrogen-air flame. As organic compounds burn in such a flame, ions and electrons are produced that in turn generate a current between electrodes (placed in the flame) with a potential applied across them. The FID detector responds approximately linearly to the number of carbon atoms the analyte molecule contains and in this sense, the detector is more or less a mass sensitive detector for most organic analytes. It is not sensitive to the gases such as CO_2, SO_2, NO_x or H_2O, etc. and is useful for the analysis of samples containing organic compounds. The FID is the most widely used detector in GC because of its good sensitivity, a wide linear response to a variety of organic compounds and affordability. Unfortunately, it does destroy the sample.

Thermionic Ionization Detector (TID). The flame thermionic detector (FTD), thermionic detector (TD), nitrogen/phosphorus detector (NPD) all refer to the same device generally similar in design to the FID. An electrically heated (600–800°C) thermionic bead (a ceramic or glass bead typically doped with Rb) catalytically decomposes N and P containing organic compounds in the presence of an H_2/air plasma. The resulting ions are collected by electrodes, as in the FID, generating a current. The detector does not respond to hydrocarbons. In one mode, no hydrogen is used and the detector responds only to nitro compounds and chlorophenols. The detection mechanism remains poorly understood. It is a destructive detector.

Photoionization Detector (PID). In a PID, the molecules in the effluent stream are ionized by intense ultraviolet (uv) radiation from a discharge lamp (typically 10.3 eV). The effluent passes between two electrodes with applied potential and the electric current resulting from the passage of ionized molecules is measured. The detector is easy to operate and provides higher sensitivity for alkenes and aromatic compounds than the FID. The choice of the discharge wavelength governs what compounds will respond. It is nondestructive and often used in series with an FID.

Electron Capture Detector (ECD). The ECD is a structure-selective ionization detector, in which a stable background current is set up between a radioactive source emitting electrons (a β-source, typically ^{63}Ni or ^{3}H) and a positively charged collector electrode. As the column effluent passes in between the electron source and the collector electrode, compounds containing atoms with high electron affinity, notably halogens, capture some of the electrons. The decrease in the background current is measured. The ECD is highly sensitive and often used for trace analysis of halogenated pesticides or polychlorinated biphenyls, drugs, etc. The linear response range of the ECD is limited. The detector requires ~5% CH_4 for proper operation, which is typically added at the column exit. The response does depend on the temperature of the detector chamber, which can be heated as high as ~400°C.

Pulsed Discharge Detector (PDD). The PDD utilizes a pulsed DC discharge in helium as an ionization source. In the electron capture mode, the PDD is similar in sensitivity and response to a conventional radioactive ECD, and can be operated at temperatures up to 400°C. For operation in this mode, He and CH_4 are introduced just upstream from the column exit. In the helium photoionization mode, the PDD is a universal, nondestructive, high sensitivity detector. The response to both inorganic and organic compounds is linear over a wide range. Response to fixed gases is positive (increase in standing current), with an MDQ in the low ppb range. In this mode, the PDD is a serious competitor to the FID especially when a flame and use of hydrogen are unwanted.

Detectors Measuring Bulk Physical Properties. *Thermal Conductivity Detector (TCD).* The TCD is the only common GC detector based on a bulk property measurement. It is nondestructive and very affordable but not very sensitive, even with He or H_2 as carrier gas, which provides the best response (the thermal conductivity of these gases are 6–10 times higher than most organic vapors). The detector responds to the difference in thermal conductivity of the carrier gas and an effluent analyte, as both flow though a thermostated chamber containing paired sensing elements, typically heated metal filaments.

Optical Detectors (see SPECTROSCOPY, OPTICAL**).** *The Flame Photometric Detector (FPD).* is used for the detection of phosphorous- and sulfur-containing species. Analyte molecules are decomposed and excited by a hydrogen-rich flame. The excited species subsequently emit characteristic band spectra centered at 526 nm for P and 392 nm for S.

In the *Atomic emission detector,* the column effluent is introduced into a microwave-energized helium plasma or discharge. The analyte molecules are destroyed and their atoms are excited by the energy of the plasma. The light that is emitted by the excited atoms have lines characteristic of the individual elements. The emitted light is measured with an array-type spectrometer.

Multiple element detection is possible because several wavelengths can be monitored. The detector is expensive but it is universally applicable and allows simplification of chromatograms containing large amounts of peaks.

In *Chemiluminescence detectors*, the analyte molecules are decomposed in flame to form NO or SO_2. These react with ozone to make products in an excited state; as they return to the ground state, they luminescence.

Electrochemical Detectors. Among the electrochemical detectors, the *Hall electrolytic conductivity detector (HECD)* is the most prominent and is available for use with a capillary GC. Halogenated compounds in the column effluent are catalytically reduced to haloacids by mixing them with hydrogen in a heated nickel reaction tube. The reduced compounds flow into a electrolytic conductivity cell, where they dissolve in a solvent (typically *n*-propanol). The conductivity of the latter is monitored.

5.6. Columns. Columns represent perhaps the heart of a chromatographic system; after all, the actual separation takes place on the column. Columns used in capillary GC today are fused silica or glass capillaries ranging in inner diameter from 50 to 700 μm. The use of polyimide coating on the outside of capillaries ensures their durability, and allows coiling in a reasonably small diameter to place in the chromatographic oven. Other external protective coatings can be used; aluminum coated columns are sometimes used in very high temperature work. There are several types of open tubular capillary columns, such as wall coated open tubular (WCOT), porous layer open tubular (PLOT), or support coated open tubular (SCOT) columns. WCOT columns are capillary tubes coated internally with a thin layer of stationary phase that functions as a virtual liquid. In PLOT columns, the column inner surface consists of a porous layer that is formed either by chemical treatment of the capillary interior, or by coating with a layer of porous particles. In SCOT columns, the inner surface of the capillary is covered with a thin layer of a support material on which the stationary phase, a virtual liquid, is attached. The efficiency of these capillary columns decreases with increasing stationary phase thickness and increasing column diameter. Thicker the stationary phase and larger the bore, greater is the sample capacity. SCOT columns in particular allow the largest amount of samples to be injected.

5.7. Stationary Phases. Stationary phases provide means for the analyte separation. The requirement for a suitable stationary phase is low volatility and thermal stability, chemical inertness and suitable values of k' and α for the analyte mixture to be separated. In the early years of GC, a large number of stationary phases were developed. To compare stationary phases, regardless of some variations in operating conditions and instruments, Kovats (12) developed a scheme of retention indices, *RI*, that is still used and bears his name. Retention indices of a series of *n*-alkanes are used as standards for calculation of the retention index of the analyte according to

$$RI = 100 N_C + 100 \left[\frac{\log t_r'() - \log t_r'(N_C)}{\log t_r'(N_C + 1) - \log t_r'(N_C)} \right] \tag{19}$$

where t_r' is the adjusted retention time ($t_r' = t_r - t_{\text{air}}$), N_C is the carbon number of the *n*-alkane eluting immediately before and $N_C + 1$ is the carbon number of the *n*-alkane eluting immediately after the analyte.

The *RI* of any analyte can be calculated from its retention data, and can be presented on a uniform scale, being generally independent of small instrumental variations. Further, the RI system has the advantage of being based upon readily available standards, which cover a wide boiling range.

To evaluate the performance of capillary columns, Grob and co-workers (13,14) developed a standard test method for capillary columns. This test provides quantitative information about several important parameters of the tested column, such as separation efficiency, adsorptive activity, acid–base character and the stationary phase thickness. Manufacturers and end-users have both used this test.

However, the number of stationary phases in common use has decreased rapidly over the years. Presently, only a limited number of stationary phases is used for the vast majority of applications. The selection of suitable stationary phase is important for the success of the particular application and the "like dissolves like" principle is often applied. The polarity of the stationary phase is the most important parameter in this respect. The most common stationary phases in use are based on polydimethylsiloxane (PDMS). The PDMS itself is a general purpose, nonpolar stationary phase. As polar functional groups such as –CN, –OH. –phenyl, etc, substitute the methyl groups in PDMS, the stationary phase becomes more polar. Table 1 shows the most commonly used stationary phases and their structures.

5.8. Hyphenated Techniques. *GC–MS.* Coupled or hyphenated techniques are among the most powerful in analytical chemistry. Capillary GC is often coupled to a selective detector such as a mass spectrometer (MS, see MASS SPECTROMETRY; ANALYTICAL METHODS, HYPHENATED INSTRUMENTS) or Fourier transform infrared spectroscopy (FTIR), that can provide additional, especially structural, information. Direct introduction of the effluent from a capillary GC into a MS without special interfaces is possible. The effluent from the GC is fed directly to the ionization chamber of the MS. The ions created from the analyte molecules in the ionization chamber are passed through the mass analyzer and detected. The mass spectrum for each chromatographic peak can be compared with the MS spectra library to aid identification. Most often, an electron impact ionization source is used to obtain reproducible spectra for MS library search. The MS unit should be sufficiently fast to provide mass spectra of all peaks eluting from the chromatographic column with sufficient resolution. Mostly quadrupole and magnetic sector mass spectrometers are used in GC–MS systems. Such instruments have been used for the identification and quantitation of thousands of myriad components present in natural and biological systems, including odor and flavor components, water pollutants, or drug metabolites and pharmaceuticals. Extensive reviews on recent developments on GC–MS are available (15–19).

GC–FTIR. Like GC–MS, GC–FTIR provides a potent means for separation and identification of analytes in complicated mixtures. In this technique, the end of the GC column is connected to a narrow bore gold-coated light pipe in which the infrared (ir) absorption spectrum is acquired. Alternatively, the analyte can be trapped on a disc that has been cryogenically cooled and the absorption spectrum then acquired on the solid sample. The ir spectral data (see INFRARED TECHNOLOGY; SPECTROSCOPY, OPTICAL) are stored in memory and compared

Table 1. **The Commonly Used Stationary Phases**

Stationary phase	Applications	Structure
polydimethylsiloxane	general purposes, amines, phenols, solvents, waxes.	
poly[diphenyl(5%) dimethyl(95%)] siloxane	alcohols, alkaloids, halogenated compounds,	
poly[diphenyl(50%) dimethyl(50%)] siloxane	alcohols, drugs, steroids, pesticides, sugars, nitroaromatics	
poly[cyanopropylmethyl(14%)dimethyl(86%)]siloxane	alcohol, alcohol acetates, drugs, fragrances, pesticides	
poly[trifluoropropyl(50%)dimethyl(50%)]siloxane	drugs, environmental samples, chlorinated compounds nitroaromatics	
poly(ethylene glycol)	free acids, flavors, fragrances	$H\!-\!\!\left[O\!-\!CH_2CH_2\right]_n\!\!-\!OH$

to an on-board spectral library as in MS. Several extensive reviews on recent developments of GC–FTIR are available (17–22).

Multidimensional GC. In two-dimensional (2-D) GC or multidimensional GC, one or more peaks eluting from a first GC column is separated on a second GC column offering a different selectivity, resulting in very high resolving power. Typically, the effluent from the primary column is injected in pulses to the secondary column. This switching can be performed using a high speed switching valve or be based on differential pressure in the interface between the columns.

In comprehensive 2-D chromatography, the eluent from the primary column is periodically injected onto a high speed secondary column. The separation on the second column need not be totally complete before a subsequent injection. Since no peaks elute naturally during the dead time for a given run, the dead time is used for elution of peaks from the previous injection. This uses column two run time more effectively. Data can be adjusted accordingly prior to plotting. This technique resembles the fast scanning mass spectra in GC–MS. It is particularly useful in cases where a number of structurally analogous compounds or isomers coexist that mass spectrometry cannot differentiate. Multidimensional GC has been widely applied in the analysis of petroleum products that often contain closely structurally related isomers, in the analysis of traces of structurally analogous chlorinated substances in the environment and the analysis of food and fragrances. Reviews on applications of multidimensional GC are available (21–24).

5.9. Applications.   Capillary GC is and has been applied for the analysis and characterization of an enormous number of compounds in various areas, with the primary focus on volatile, volatilizable and semivolatile compounds. The main application areas include (1) industrial chemical analysis, such as analysis of mixtures of acids, alcohols, aldehydes, amines, esters, phenols, and other aromatics (25–29); (2) analysis of industrial solvents and their mixtures as well as analysis of permanent gases and hydrocarbons in various industries, natural and synthetic gases, refinery gases, sulfur containing gases, permanent and noble gases and freons, gasoline and naphtha, etc (30–34); (3) analysis of flavors, fragrances, antioxidants and preservatives in food, beverages, perfumes, and cosmetic products (35–38); (4) analysis of pharmaceutical formulations including antidepressants, antiepileptics, anticonvulsants, opiates, barbiturates, alkaloids, etc (39–46). Capillary GC is particularly useful for the analysis of volatile and semivolatile analytes in environmental samples (47–54); many US EPA approved methods for the analysis of aromatics, halocarbons, PAHs, organochlorine pesticides, fungicides and various other pollutants in water, soil, air and waste are based on capillary GC (55).

Figure 4 shows some selected applications of gas chromatography. Specific application notes in these and other areas are available from vendors of chromatographic equipment (56–59). Further information on recent advances on GC can be obtained from reviews, eg, the biennial reviews in *Analytical Chemistry* (60–62).

6. Liquid Chromatography

As the name implies, in LC, a liquid mobile phase is used. In contrast to GC, the choice of the mobile phase affects the analyte retention. There are also many more stationary phases available, permitting much greater latitude in developing a separation method. However, a greater choice can also sometimes mean greater complexity.

Tswett (63) described the first LC separation at the turn of the century: Various plant pigments were separated on a glass column packed with finely ground chalk. It was not until 1960s, when the technology for production of uniform spherical particles of diameters ≤10 μm became available that LC became a high efficiency separation method and high-performance liquid chromatography (HPLC) was born. In the intervening years, it has evolved into a separation method of such importance that pharmaceutical or biotechnology industries will cease functioning without it.

The introduction of microcolumn LC is generally attributed to Horvath and co-workers (64,65) in 1967. However, later several research groups published papers that are considered key publications in this area (66–74). The term

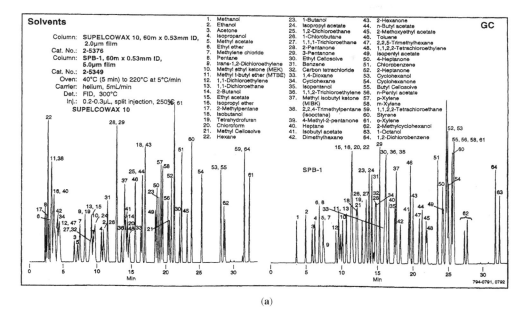

(a)

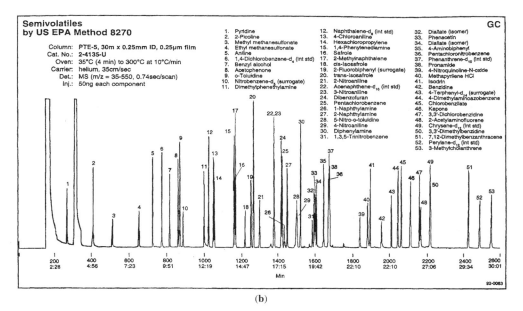

(b)

Fig. 4. Selected chromatograms showing some application areas of gas chromatography, conditions are described in text. (**a**) Separation of organic solvents. (**b**) Separation of semivolatiles by US EPA Method 8270. Reproduced by permission of Supelco, Bellefonte, PA.

microcolumn LC is sometimes used interchangeably with capillary LC. Some commercial instrument makers also use another term, "microbore" LC, to refer to 1- or 2-mm bore columns. The distinction as to where normal bore LC ends and capillary LC begins with "microbore" LC somewhere in between, is rather vague. Most users would not consider 1-mm bore columns as belonging to a capillary format.

As in GC, capillary LC offers better separation performance than its larger bore counterparts. However, it has not yet become as widely used as capillary GC. While capillary GC relies only on open tubular columns, there are few practitioners of open tubular capillary LC (OTLC). While OTLC was reported early on (75,76) and studies continue (77), the big difference is the orders of magnitude lower D_m value in the liquid phase. Either D_m must be increased by increasing temperature (78), or the column bore must be reduced to single digit micrometers (79).

For the most part, capillary LC has established a niche for itself where the ability to analyze minute sample volumes, limited consumption of mobile phase, and high resolution are important. It is also the capillary format of LC that is best coupled with mass spectrometric detectors, although current interfaces allow easy connection of conventional HPLC to MS as well.

The popularity of LC stems from its ability to accomplish almost any separation, given appropriate sample preparation. This is especially true for non-volatile thermally unstable species, such as amino acids, proteins, carbohydrates, drugs, antibiotics, pesticides as well as myriad inorganic and ionic species.

6.1. Instrumentation. A typical liquid chromatograph is depicted in Figure 5. The mobile phase(s) is(are) contained in glass or other inert reservoir(s)

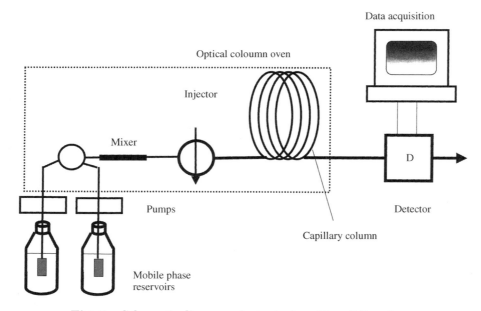

Fig. 5. Schematic diagram of a typical capillary LC system.

and are typically sparged/degassed and filtered through on-line filters prior to entering the pump. With reciprocating pumps, if more than one solvent is used, typically they are blended by a proportioning valve system before entering the pump inlet. The chromatographic pump (see PUMPS) must be capable of pumping at the desired flow rate at very high pressures. Operation at pressures of 1000–2000 psi are common in conventional HPLC. In capillary LC systems, the operating pressure is typically at least as high. The current trend for both conventional and capillary system is to use smaller diameter packing particles and higher pressures to improve sample throughput. The pumped eluent then passes through an injector to the column and to the detector. A short "guard column" is sometimes used ahead of the analytical separation column to protect the latter from particles in the sample or very strongly retained components that do not readily elute and thus poison the column. In capillary LC, it is essential that band dispersion be kept to a minimum. As such, connections associated with the insertion of a guard column produce unwanted dispersion and such a column is often not used. When a column is poisoned, this occurs at the head of the column. One great advantage of a fused silica separation column is that it is always possible to cut and discard a little portion of the column at the top.

6.2. Mobile-Phase Delivery. A detailed review on types of pumps can be found elsewhere (see PUMPS). Capillary LC is sometimes conducted with conventional scale pumps and injectors, most of the flow is discarded or recycled using a splitter. This arrangement cannot take advantage of some of the principal benefits a capillary LC system has to offer. Syringe pumps of 1–10-mL capacity are most commonly used in capillary LC. Use of multiple solvents for gradient elution, etc require multiple syringes. An exception is capillary ion chromatography (IC, a subclass of LC) where eluents can be electrodialytically generated in-line; the syringe is merely used to pump pure water (80). The pump will normally be equipped with a three-way valve arrangement (ruby and sapphire ball and seat check valves are typically used) that allow it to pump at high pressure through the column and then refill from an eluent reservoir. Low-flow single- or multiple-head reciprocating pumps (most commonly used in conventional bore HPLC), with similar check valves at the inlet and the outlet are also used in microcolumn LC, however, considerable innovation and ingenuity goes into making pumps operating in this mode to be pulse free (81).

6.3. Injectors. In early years, injectors that can inject volumes small enough for capillary LC were not available. With the advent of such injectors, approaches of the yesteryears, such as moving injection technique (82,83), static split (84) or pressure pulse driven stopped flow injection (85) have fallen by the wayside. Virtually all injectors in use with microcolumn LC today are two-position rotary-type injectors. For injection volumes less than 1 μL, internal loop injectors (consisting of a slot-in-a-disk inside the injector) can be used and are available down to injector volumes of 20 nL. It is inconvenient to change injection volumes with these valves (it must be disassembled), however, it is possible to use commercially available low internal volume six-port injectors containing external loop for injection volumes ≥1 μL. For smaller injection volumes, the concept of partial loop filling injection can be used with these valves. This technique conserves sample and provides flexibility in the choice of sample volumes to be loaded into the injection valve without changing loops. The precision of either type of injector is usually better than 1%, which allows for accurate and reproducible sample injection and quantitation.

6.4. Detectors. All detectors that are used in capillary scale LC can also be used in CE, which will not be separately discussed. With the exception of refractive index and conductivity detectors, bulk property detectors are not commonly used. Many detection principles allow on-column detection and thus eliminate extraneous dispersion in transit to the detector. Some GC detectors can be used with capillary LC, but the practice is not common.

Absorbance Detectors. The uv-vis absorbance detector is the workhorse in both capillary and conventional HPLC (see Spectroscopy, optical; Photodetectors) and is available in fixed and variable wavelength, fast-scanning, and photodiode array (PDA) forms. Typically a narrow beam of monochromatic light (white light for a PDA) is focused on a window made in the capillary immediately after the column frit (on-packing absorbance detection has also been reported but is not common). Light passes radially through the tube and is then focused or dispersed on the detector. The radial path is short and alternatives such Z-cells and light-pipe based cells of longer path are commercially available.

Fluorescence Detectors. Either monochromatic light from a powerful continuum source (Hg- and Xe-lamps) or increasingly, a laser (see Lasers), is used for excitation. A window is made in the capillary as for absorbance detection above. The light emitted by a fluorescing analyte is measured usually perpendicular to the excitation beam. Filters or monochromators are used to filter the desired excitation and emission wavelengths. Fluorescence detectors, especially Laser-induced fluorescence (LIF) detectors are very sensitive, with concentration LODs in the range of $10^{-9}-10^{-12} M$. The disadvantage is that the analyte must be fluorescent; otherwise a derivatization with fluorogenic agent must be carried out in advance. Such derivatizations are often carried out postcolumn in conventional scale HPLC but is more difficult to implement in capillary LC or CE because of strict requirements to limit band dispersion.

Electrochemical Detectors. Electrochemical detectors rely on measurements of electrochemical properties of the analytes (see Electroanalytical techniques). They can be subdivided into several groups, such as amperometric, polarographic, coulometric, or conductometric detectors. Amperometry at a single graphite fiber was one of the earliest techniques used for detection in capillary IC (76). Electrochemical detectors permit high sensitivity (often to $10^{-9} M$) and respond to several groups of organic compounds typically analyzed by HPLC. However, careful selection of the mobile phase is necessary and electrodes can be easily poisoned from the analytes or the sample matrix. In a conductivity detector, electrical resistivity is measured; strictly speaking, it is not an electrochemical detector because no chemistry is carried out. These detectors are easy to miniaturize, are particularly robust in behavior and are discussed further below.

Conductometric Detectors. Conductivity detectors are mainly used with IC in which ionic analytes are separated by ion exchange. Although a conductometric detector is in principle a bulk property detector, typically IC is used in a "suppression" mode where all the ionic species originally present in the mobile phase are selectively removed, making the detector essentially selective to ionized analytes. Typical sensitivity is similar to the uv absorbance detectors ($\sim10^{-7} M$). Conductometric detectors of many types have been described

(86,87), the current trend is toward contactless conductivity detectors, see Zemann for a review of capillary scale conductometric detectors (88).

Refractive Index (RI) Detectors. The RI detector measures the difference in refractive index of the analyte zones and the mobile phase. The advantage of this detector is its universal response to virtually any compound. There are few RI detectors for use with capillaries. A laser-based microinterferometric back-scatter detector developed by Bornhop (89) is an exception, permitting detection of a RI change of the order of 10^{-7}. The RI detectors must be carefully tempera-ture stabilized to attain good sensitivity. For recent developments in this type of detector and a general review of detectors used in capillary LC/CE, see Swinney and Bornhop (90).

Evaporative Light Scattering Detector. In this detector, the column efflu-ent is nebulized and the mobile phase is evaporated. Nonvolatile analyte parti-cles are detected by laser light scattering. The detector provides universal response to all nonvolatile compounds and is sensitive. Commercially available detectors can be adapted for use with capillaries and can also be used with super-critical fluid chromatography (SFC). Nonvolatile (eg, buffer) components cannot be present in the mobile phase. In one variation of this, the evaporated particles are used as condensation nuclei and single molecules can be detected (91).

6.5. Columns for LC. Although many different styles of columns, eg, drawn packed capillaries (pack a large bore glass capillary with particles and subsequently draw it to the desired diameter using a glass drawing machine—incompatible with thermally sensitive packing), wall-functionalized open tubular columns as in GC, etc have been described, the mainstay today is packed fused silica capillaries with the most popular column bore being ~300 µm, albeit packed columns in 50–500-µm bore capillaries are commercially available. Packed microcolumns are prepared by slurry packing. A frit is made at one end of the capillary to retain the packing material. This frit is usually made from glass wool, metal or porous silica. The particles are suspended in a suitable solvent, having a density similar to that of the particles to avoid particle sedi-mentation. The reservoir with the slurry is agitated or sonicated to achieve a homogenous mixture and the capillary is packed under pressure or vacuum. The particle size, uniformity, and the homogeneity of the packed bed are the major factors contributing to the column performance. Hundreds of column pack-ings are commercially available for conventional HPLC; most major types are available as capillary columns, at least on a semicustom basis. The one exception is capillary IC, no commercial columns are available at this time.

6.6. Stationary and Mobile Phases. In LC both the stationary and mobile phase play an important role in the analyte partitioning mechanism. In virtually all present-day columns, the stationary phase consists of spherical sup-port particles, ranging in diameter of 1–10 µm, which are often chemically mod-ified by covalently bonded functionalities. The particles themselves are composed of silica, alumina, zirconia, carbon, and poly(styrenedivinylbenzene) (PSDVB) or related polymers. In the past 5 years, a tremendous amount of effort was devoted to making organic-modified silica solid supports to overcome temperature and pH problems.

The type of mobile phase depends on the chromatographic mode used. In "normal-phase" LC, a nonpolar solvent such as hexane or isopropyl ether are used, in "reversed phase" (RP) LC, by far the most common mode in which LC is currently practiced, aqueous eluents containing methanol or acetonitrile (occasionally, tetrahydrofuran (THF)) as well as buffering salts/acids/bases are used. Since RPLC is widely used for ionized or ionizable compounds, a proper selection of pH of the mobile phase is important. Phosphate and Tris-based buffers are often used and are optically transparent in most of the wavelength range of interest. In IC, the mobile phase is typically dilute aqueous solution of a strong acid or base.

6.7. Normal Phase LC.   In normal phase LC a column is packed with a polar stationary phase, the mobile phase is of a nonpolar or moderately polar nature. The polarity of the analytes plays an important and crucial role in their retention on the stationary phase. The stationary phase contains either unmodified silica or alumina or a chemically bonded stationary phase, typical functional groups being cyano, amino, dihydroxypropyl modified polysiloxane. A wide variety of organic solvents can and has been used for normal-phase chromatography with polarity ranging from fluorocarbons to methanol. In normal-phase chromatography, the mobile phase is typically less polar than any of the analytes. Usually, a binary mixture of two solvents of different polarity, the less polar solvent being most commonly hexane, and a more polar solvent, are blended to achieve the required elution strength; a gradient (eluent changing composition with time) is sometimes used. Normal-phase LC mode is used in <20% of all LC applications.

6.8. Reversed Phase LC.   RPLC accounts for the majority of present LC separations. In reversed-phase chromatography, the elution order of the analytes is opposite to the normal-phase chromatography, ie, the most polar analyte elutes first. A nonpolar stationary phase and a polar mobile phase is used. Most commonly the stationary phase is porous silica to which polysiloxane, typically modified with alkyl chains, such as C8 (*n*-octyl) or C18 (*n*-octadecyl, this phase is also called octadecyl silica or ODS and is the most commonly used) are bonded. The hydrocarbon chains are aligned parallel to one another and perpendicular to the particle surface making the surface hydrophobic. The PSDVB and porous graphitic carbon (eg, Hypecarb) are alternatives to silica-based stationary phases and can be used in extremes of pH. Zirconia-based phases have extraordinary temperature stability. Separation of aromatic hydrocarbons with pure water as eluent on polybutadiene or carbon modified zirconia phases have been reported at a temperature of $370°C$ with sufficient backpressure to keep the water in the liquid state (92). Hypercarb is more retentive than silica based stationary phases (C18) and has been popular in capillary LC. Its high retaining ability contributes to on-column focusing effects and allows for injection of larger sample volume. Aside from the solvent composition of the mobile phase, appropriate buffering and the selection of the optimum buffer, buffer pH is also very important for optimum separation of many compounds, especially those with weakly acidic or basic character.

6.9. LC–MS.   The great potential in coupling liquid chromatography with mass spectrometry (LC–MS) has been recognized especially in recent years, because conventional GC–MS methods are not adequate for variety of applications due to limited volatility and thermal lability (see ANALYTICAL METHODS, HYPHENATED INSTRUMENTS; MASS SPECTROMETRY). In developing a successful coupling between a LC and a MS, one needs to deal with the fact that a significant amount of solvent comes out as the column effluent and all of this must be evaporated off. Many early MS units also had serious problems with nonvolatile components in the eluent. Capillary LC is much more directly compatible with MS than large bore LC in this regard since the total volumetric flow rates are much lower. The LC–MS interfaces in current use include continuous flow fast atom bombardment (CF–FAB), thermospray, particle beam, electrospray and atmospheric pressure chemical ionization (APCI) interface.

In a CF–FAB interface, a liquid stream is mixed with FAB matrix solvent (typ. glycerol) and ions are generated by bombardment of this liquid by accelerated atoms or ions (Cs^+ or Xe).

In a thermospray interface, small droplets of eluent are generated in a heated vaporizer tube. The ionization of analytes is accomplished by solvent-mediated chemical ionization and ion evaporation processes. This interface is compatible with substantial liquid flow rates, up to 2 mL/min, and has been widely used in many areas such as analysis of drugs, their metabolites, natural products, and in environmental analysis.

In the particle beam interface, the column effluent is nebulized pneumatically or by thermospray into a desolvation chamber, where the high mass analytes are separated from the low mass solvent molecules. The ionization of analytes takes place by a conventional electron impact or chemical ionization. This interface is able to generate electron-impact spectra (but limited to MW <1000), which makes it useful in pharmaceutical and environmental analysis.

In the electrospray interface, a high voltage (typ. 3 kV) is applied between the eluent emerging from a capillary needle attached to the column outlet and a counter-electrode, causing the formation of small droplets of the eluent. Solvent is evaporated and analyte molecules are ionized prior to entering the MS proper.

In an APCI interface, the eluent from the column is nebulized by pneumatic or thermospray nebulization and ionized in an atmospheric pressure ion source.

Coupling of LC to MS provides an important tool in qualitative and quantitative analysis of various samples and is useful in several areas such as analysis of pharmaceuticals, biochemical and environmental samples (93–102).

6.10. Applications of Capillary LC. LC is the most popular and widely used analytical separation method today. The application area is enormous and thousands of LC based analyses are in routine use. Major areas for LC application include analysis of pharmaceutical formulations and drugs, food, and consumer products, large molecules of biological importance such as DNA, DNA-fragments or peptides, and proteins (see BIOSEPARATIONS). The LC analysis of mixtures of nonvolatile or semivolatile pollutants supplement the available GC methodologies; approved methods are listed in the EPA list of methods (55), (see also WATER, ANALYSIS).

Virtually all methods developed for conventional LC with large bore columns can be readily implemented to the capillary scale. While capillary LC is not at the moment poised to displace conventional scale LC (as happened with GC), its future is increasingly brighter. It has already found a niche in bioanalysis (103–105), neuroscience, in vivo measurements, protein/peptide research (106–109) including proteomics and chiral separations (110–112), where the sample availability, cost of reagents and compatibility with MS detection benefit from the capillary scale. Developments in microcolumn LC are summarized in several recent reviews (113–119). In Figure 6, selected chromatograms of micro-column LC applications are shown.

6.11. Ion Chromatography (IC).

IC is an important subclass of LC and is used primarily to determine inorganic and organic anions and cations. In IC,

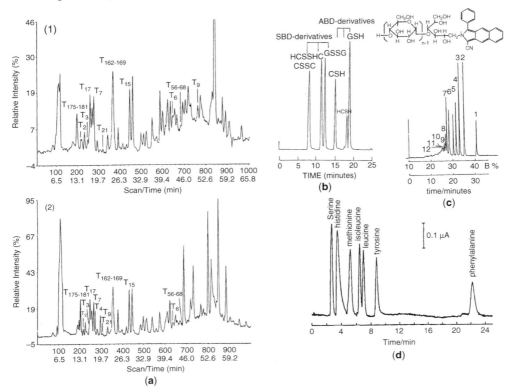

Fig. 6. Selected chromatograms showing some application areas of capillary LC. (**a**) Comparative peptide mapping of enzymatic digests of bovine ribonuclease B with micro-LC-electrospray MS (1) TIC trace of tryptic digest of reduced and (S)-carboxymethylated–ribonuclease B, (2) TIC trace of digested reduced and (S)-carboxymethylated–ribonuclease B with trypsin and peptide-N-glycosidase F. Conditions: capillary column: 33 cm × 0.25 mm id, packed with C_{18}. Mobile phase–Solvent A = 0.1% TFA in water, solvent B = solvent A/acetonitrile (20:80), gradient elution from 0 to 60% solvent B over 120 min. Reproduced from J. Liu, K. J. Volk, E. H. Kerns, S. E. Klohr, M. S. Lee, I. E. Rosenberg, *J. Chromatogr.* **632**, 45 (1993); by permission of Elsevier Science. (**b**) Chromatogram of mixture of thiols and disulfides: Conditions: capillary column 25 cm × 0.32 mm, packed with 5 μm Rosil C_{18}. Mobile phase 150 mM H_3PO_4:acetonitrile, gradient elution from 92:8 to 70:30 in 15 min, 70:30 for 5 min. Peaks: cysteine (CSH), homocysteine (HCSH), glutathione (GSH), cystine (CSSC), homocystine (HCSSC), and oxidized glutathione (GSSG). Reproduced from B. L. Ling, C. Dewaele, W. R. G. Baeyens, *J. Chromatogr.* **553**, 433 (1991); by permission of Elsevier Science. (**c**) Chromatogram of oligosaccharide derivatives: Individual 3-benzoyl1-2-naphtaldehyde-derivatized components of Dextrin 15. Conditions: Capillary column 90 cm × 0.3 mm id, packed with Capcell-C_{18}. Linear gradient elution with water:acetonitrile. Reproduced from J. Liu, O. Shirota, M. Novotny, *J. Chromatogr.* **559**, 223 (1991); by permission of Elsevier Science. (**d**) Chromatogram of underivatized amino acids. Conditions: Capillary column 15 cm × 0.3 mm id, packed with ODS-3 (3 μm). Mobile phase 50 mM phosphate:10 mM KBr, pH 7.5. Amperometric detection. Reproduced from K. Sato, J.-Y. Jin, T. Takeuchi, T. Miwa, Y. Takekoshi, S. Kanno, S. Kawase, *Analyst* **125**, 1041 (2000); by permission of the Royal Society of Chemistry.

the separation is based on ion exchange (see ION EXCHANGE; CHROMATOGRAPHY) between the analytes from the sample and stationary ion exchange sites on a column and displacement by eluent ions. The column capacity is significantly lower than conventional ion exchange resins (eg, as used for water softening) to allow reasonable retention times. The columns are typically based on PSDVB based polymeric beads that have been modified to contain strongly acidic ($SO_3^-H^+$), strongly basic ($NR_3^+OH^-$), weakly acidic (COO^-H^+, $HPO_3^-H^+$), or weakly basic ($-NH_2$, $-NRH$, $-NR_2$) functional groups. Chelating functional groups are present in columns used for the preconcentration of transition metals. One of the most interesting classes of columns used uniquely in IC utilizes a surface-agglomerated packing. If a PSDVB bead is mildly sulfonated, negatively charged sulfonate groups are formed on the surface. If one now passes a suspension of very fine (submicrometer size) quaternized, positively charged latex particles atop the surface sulfonated beads, the positively charged latex microbeads agglomerate electrostatically on the surface of the negatively charged beads. This provides for a highly efficient anion exchanger where ions can readily exchange in the top layer.

Conductometric detection is the mainstay in IC; all ions share the property of electrical mobility. In so-called single column IC, the signal arises from the difference in equivalent conductance between the sample ion and the analyte ion. For example, if dilute HCl is being used as the eluent, the detector background is the conductivity of the eluent. As various cations, eg, Li^+, Na^+, K^+, etc elute from the column, the H^+ concentration decreases in a proportionate manner because the metals replace H^+ in the effluent. Because the metal ions have substantially less electrical mobility than H^+, negative peaks in the conductance signal are observed with the elution of each metal. To limit the background conductance, the eluent concentration needs to be limited. To make this work, the column capacity must be limited. These limitations have largely relegated this technique presently to historical importance only. The detection principle is the same as indirect photometric detection commonly used in CE.

The major mode in which IC is used today is the so-called "suppressed" mode. The eluent is a strong base (eg, KOH) or a very weak acid salt of a strong base (Na_2CO_3) for anion separations on an anion exchanger column or a strong acid (eg, HCl) with a cation exchange column for cation separations. In the anion separation system, the separation column is followed by a "suppression" device and then the conductivity detector. The suppressor device above exchanges all cations for H^+. The eluent KOH is thus converted to water (Na_2CO_3 will be converted to H_2CO_3), a very weakly conducting background. In contrast, when Cl^-, NO_3^-, SO_4^{2-}, etc elute from the separation column and pass through the suppressor, they emerge as strongly conducting HCl, HNO_3, H_2SO_4, etc permitting direct detection. Hence, the suppressor "suppresses" the conductivity of the eluent while amplifying the conductance of the analyte. Similarly, in cation separation system, the suppressor is a OH^- exchanger that converts an HCl or HNO_3 eluent to water while analytes Li^+, Na^+, K^+, reach the detector as LiOH, NaOH, KOH, etc. This technique, the mainstay of ionic analysis today, was invented by Small and co-workers in 1975 (120). The original suppressor device was a periodically regenerated resin column; it is still used in this form in a capillary format but the majority of suppressor devices today are based on a chemically or electrically regenerated membrane or resin-packed device (or a hybrid) (see MEMBRANE TECHNOLOGY).

In an electrically operated membrane suppressor, the eluent ions are removed electrodialytically. The exact reverse of this technique, electrodialytic eluent generation, allows the electrical production of high purity eluents on demand and at specified concentration. Recently, Small and Riviello (121) described "Ion reflux". As applied to IC, this is a new ion-exchange technique where an electrically polarized ion-exchange bed becomes the source of eluent as well as its means of suppression. Using water as the pumped phase, such polarized beds enable the "perpetual" generation and suppression of eluent with little intervention by the user. In one embodiment of ion reflux, continuous eluent generation, ion separation, and continuous suppression are accomplished within a single bed. In another case, where separation is uncoupled from the other two functions, the ion reflux device may be used with existing separators. Indeed, the latter technique holds much promise for capillary IC since eluent generation and suppression can be carried out on a different scales.

Currently, capillary IC instrumentation is commercially unavailable. This makes widespread use impossible. The power of capillary IC is apparent from the chromatogram in Figure 7. Field portable electrical eluent generation based capillary IC has also been demonstrated (122).

6.12. Applications of IC. IC is applied for separation and analysis of small charged analytes in various samples. It is used particularly for the analysis of different types of water samples, such as tap, drinking, rain or wastewater from industrial processes (see WATER, ANALYSIS). Several industries, such as pulp and paper, chemical, pharmaceutical, power generation and the semiconductor industry depend on IC analysis (123–132). Among the compounds mainly analyzed by IC are inorganic anions and cations, small organic molecules such as carboxylic acids or amines, short/long chain surfactants or metalcyano complexes (133–142). Heavy metal speciation by post column reaction detection or element specific detection (124,135,143,144) is an important area. Literature is available for further reading (145).

7. Supercritical Fluid Chromatography

In supercritical chromatography, the sample is injected onto the column and carried through it by a mobile phase consisting of a supercritical fluid, a fluid that is being maintained in the separation system at a temperature higher than its critical temperature T_c and critical pressure P_c. (see SUPERCRITICAL FLUIDS). Properties of supercritical fluids such as density, viscosity, and diffusion coefficient are intermediate between those of liquids and gases. For example, density of the supercritical fluid is high compared to gases; this promotes dissolution of large nonvolatile compounds. On the other hand, its viscosity is close to that of gases, with a proportionately high diffusion coefficient. This makes SFC an interesting alternative to both GC and LC (146). SFC can be advantageously used for compounds not easily analyzed by GC, eg, nonvolatile, thermally labile, or polymeric compounds or compounds lacking chromophores or electrochemical properties that are not so easily detected by conventional LC detectors.

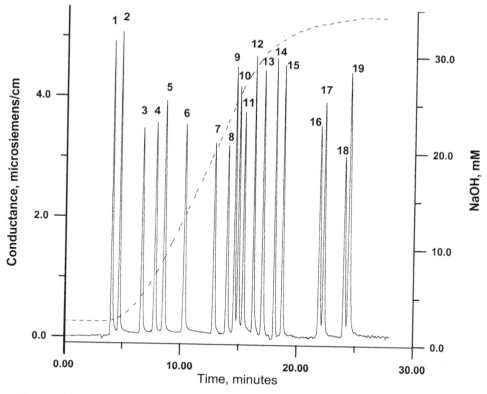

Fig. 7. Capillary ion chromatogram : 56 cm × 0.18 mm column packed with Dionex AS-11 packing. The right ordinate shows the electrodialytically generated gradient (dashed trace) in mM NaOH. Peak identities: 1, Fluoride; 2, Formate; 3, Monochloroacetate; 4, Bromate; 5, Chloride; 6, Nitrite; 7, Trifluoroacetate, 8, Dichloroacetate; 9, Bromide; 10, Nitrate; 11, Chlorate; 12, Selenite; 13, Tartrate; 14, Sulfate; 15, Selenate; 16, Phthalate; 17, Phosphate; 18, Arsenate; 19, Citrate. All compounds injected were 50 µM except selenate and sulfate that were 25 µM. Reproduced from Ref. 122 by permission of the American Chemical Society.

Both packed column SFC and open tubular capillary SFC are in use. The open tubular columns usually provide higher efficiencies, longer analysis times, and low sample capacity, while packed columns exhibit lower efficiencies and high sample capacity, analysis time being shorter. The van Deemter equation (eq. 9) is applicable to SFC. The theoretical plot of plate height versus flow rate for SFC yields a curve with broader minima, shifted toward the higher flow rates, compared to LC. In LC, the typical operating flow velocities are usually well above the optimum values, otherwise, analysis time is too long. In SFC, the operating flow velocities closely match the theoretical optimum values and along with the much larger D_m value, this permits higher efficiencies than in LC. For similar reasons, for comparable separations, SFC usually provides shorter analysis times than LC with efficiencies and analysis times approaching those of GC.

7.1. Instrumentation. Figure 8 depicts the scheme of a typical instrument for capillary SFC. It resembles the instrumentation for GC. The chromato-

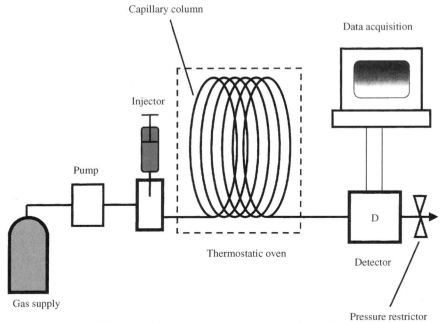

Fig. 8. Schematic diagram of a typical SFC.

graph consists of a gas tank (usually containing CO_2, with or without added modifiers, as the mobile phase), a syringe pump, an injector, a column placed in the thermostatic oven, a pressure restrictor and a detector. The instrumentation for capillary SFC differs slightly from the packed column SFC due to the different pressures generated in the two systems.

7.2. Mobile Phase.   As in LC, the mobile phase affects the retention of the analytes. By changing the composition of the mobile phase the selectivity factor, α, can be changed and the selectivity of the separation altered. Carbon dioxide is the most popular and widely used carrier fluid for SFC due to its favorable critical temperature (T_c) and critical pressure (P_c) values, low cost, and low toxicity. Another advantage of CO_2 as a mobile phase is that it does not elicit a response in typical GC detectors such as the FID. Thus, SFC with CO_2 as a mobile phase can be used together with a FID detector, providing sensitive detection for most organic compounds. However, the polarity of CO_2 is quite low, close to that of pentane or hexane, and it does not dissolve more polar compounds. Binary or ternary mixtures of CO_2 and polar additives, such as methanol, water, formic acid or tetrahydrofuran can be used to enhance the applicability of CO_2 -containing supercritical mobile phases to the analysis of polar compounds. Other solvents used for SFC are listed elsewhere (see Table 1 in SUPERCRITICAL FLUIDS).

7.3. Injectors.   The type of injector and the size of the injected sample depend primarily on the columns used. For capillary columns, with inner diameter of several tens of micrometers, the sample volume should be sufficiently small so that the column overloading is minimized. Sample splitting techniques described in the GC section are widely used.

7.4. Detectors. SFC benefits from being an intermediate methodology between GC and LC, typically detection can be accomplished by either GC or LC -detectors. The GC detectors, such as the FID, TID, FPD, or ECD can be used with simple supercritical fluids and capillary columns. On the other hand, if analyte contain chromophores or is electroactive, uv–vis absorption or electrochemical detectors for LC can be used, with the modification that the detection cell can withstand high pressure and is placed before the pressure restrictor. In many hyphenated techniques, such as SFC–MS or SFC–FTIR, the two halves are more easily coupled than their LC counterparts (147).

7.5. Columns. The use of packed or capillary columns in SFC is dictated by the needs of the particular analysis. Capillary columns are usually made of fused silica, having an inner diameter from 25 to 100 μm and the length of 1–35 m. The column temperature is maintained using a GC type oven, at temperatures usually not exceeding 200°C. The temperature is not the only parameter, which needs to be kept constant or precisely controlled. Pressure on the column dictates the density of the mobile phase and thus its retention power or retention factor, k', as a consequence of the increase of density of the supercritical fluid with increase of the pressure. An increase in pressure generates greater elution strength of the mobile phase and decreases the analyte retention. This effect can be compared to the temperature effect in GC separations or effect of polarity change of the mobile phase in LC. The pressure can be programmed to increase linearly or asymptotically thus improving the analysis time and peak shapes.

7.6. Stationary Phases. Stationary phases used for open tubular SFC are those routinely used in GC (see Table 1), which are further cross-linked to increase their stability. Polysiloxanes with various percentages of more polar groups are used to coat the inner wall of the fused silica capillaries. Thickness of the coating is usually between 0.1 and 3 μm.

7.7. Applications. Supercritical fluid chromatography (often following supercritical fluid extraction, SFE) has been used, eg, for measuring caffeine in coffee or nicotine in tobacco. SFC has been used for the analysis of fats, oils and food related samples, natural products such as terpenoids or lipids, low molecular weight polymers, fossil fuels and synthetic lubricants, thermally labile large molecules, including biomolecules and synthetic polymers. Figure 9 depicts a few selected SFC applications. Several reviews are available addressing recent instrumental developments and application areas of SFC (148–154). The interesting properties of the mobile phase makes SFC a very attractive technique in principle; however, thus far SFC has not enjoyed the degree of commercial success as originally hoped for.

8. Capillary Electrophoresis

Capillary electrophoresis differs in several important ways from the three chromatographic methods described above. The mode of injection, the mode of sample transport, and the mechanism of analyte separation are all different. CE evolved from the planar electrophoretic methods (see ELECTROSEPARATIONS, ELECTROPHORESIS) performed on paper or a thin gel. First electrophoresis conducted in a capillary was reported by Hjerten (155). Later Virtanen (156) used 0.2–0.5-mm bore

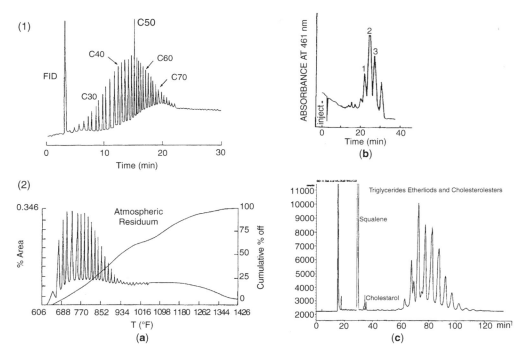

Fig. 9.   Selected supercritical fluid chromatograms. (**a**) (1) Chromatogram of polyethylene wax PE740. Conditions: capillary column 10 m × 0.05 mm, statinary phase 0.2 µm film of 5% diphenyl–95% dimethyl polysiloxane. Mobile phase:CO_2 at 100°C, pressure from 2000 to 5000 psi in 20 min, 5500 psi 10 min. (2) Atmospheric residuum of crude oil obtained by removing low boiling fractions, same conditions as (a). Reproduced from H.E. Schwartz and R.G. Brownlee, *U.S. Pat. 4,971,915.* (**b**) Chromatogram of lycopene (2), alpha-(1) and beta-carotene (3) from tomato extract Conditions: capillary column 10 m × 0.05 mm, fused silica SB-phenyl/50, coated with stationary phase 50% phenyl–50% polymethylsiloxane. CO_2 at 45°C with an asymptotic density rise program with a 1/2 rise time constant of 30 min, initial density 0.66 g/mL. Reproduced from H. H. Schmitz, W. E. Artz, C. L. Poor, J. M. Dietz, J. W. Erdman, *J. Chromatogr.* **479**, 261 (1989); by permission of Elsevier Science. (**c**) Chromatogram of untreated shark (*Centroscymnus coelolepsis*) liver oil. Conditions: capillary column 25 m × 0.1 mm, fused silica DB-5, film thickness 0.1 µm. Mobile phase: CO_2 at 170°C. Reproduced from C. Borch-Jensen, J. Mollerup, *Chromatographia* **42**, 252 (1996); by permission of Vieweg Publishing.

glass tubes for separation of alkali metals. The use of fused silica capillaries by Mikkers and co-workers(157) and Jorgenson and Lukacs(158) in the early 1980s was a real breakthrough and in particular the latter authors showed the extremely high efficiencies attainable by CE. Since then the use of CE as a research tool has grown rapidly.

In CE, a fused silica capillary is filled with a buffered background electrolyte (BGE) solution, which serves as the separation medium. Unlike the chromatographic methods, in CE, there are no pumps or pressurized fluid sources. The two ends of the capillary are immersed in the reservoirs containing the BGE. Two noble metal electrodes, usually Pt, are placed in each vial as well. The movement of the electrolyte and sample through the capillary is accomplished by applying a high voltage (HV) (typ. 10–30 kV across a 40–75 cm long, 25–100-μm bore capillary) across the electrodes. Typical electrolytes used are 1–10 mM phosphate or borate; with the resulting current being in microamperes.

Capillary electrophoresis is characterized by the very high efficiency due to the plug-like flow profile of the electrophoretic flow (see Fig. 2). Low to tens of nanoliters of sample are typically injected (contents of a single cell have been analyzed, however, in most cases actual sample needed is several μl) and the total consumption of BGE per run is a few microliters (μl) (although in practice a greater amount of BGE is needed to prevent gross alteration of the BGE composition by electrolysis during the run). There are several modes of CE operation and these, together with the instrumental aspects, are discussed below.

8.1. Basic Instrumentation in CE.   A simple CE apparatus is depicted in Figure 10. It consists of a HV supply capable of providing a potential difference in the range of ±30 kV, two platinum electrodes, placed in the two electrolyte vials, a fused silica capillary, and a detector.

8.2. Injection Techniques in CE.   Although valve-based injection has been used in CE, this is not common. The two most used injection techniques are the hydrodynamic (HD) and the electrokinetic (EK) injection methods. While both differ from injection methods in chromatography systems, both can be fully automated.

Hydrodynamic Injection.   In the HD injection mode, the electrolyte vial at the head of the capillary (the tail end is the destination or the detector end) is briefly replaced by the sample vial, and a small portion of the sample is forced into the capillary by applying pressure (or vacuum at the tail end), or by gravity

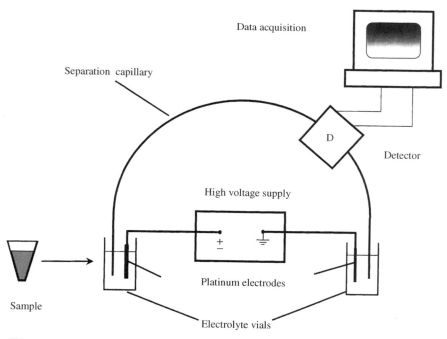

Fig. 10. Schematic diagram of a typical capillary electrophoresis system.

(raise the sample vial with respect to the destination vial). Prior to separation, the sample vial is again exchanged for a BGE vial.

The hydrodynamic injection mode is preferred in CE since the peak areas of the analytes are linearly related to the concentration over a wide concentration range and the injected amount is relatively independent of the sample matrix composition. However, any difference in the viscosity of the sample and the standard will cause a quantitation error. If viscosity difference is significant, the standard addition method (see CHEMOMETRICS) is preferred. In the standard addition method, a small amount of a standard solution of known concentration is added directly to the sample matrix and measurements are made before and after such addition. Most commercial instruments permit good quantitative precision ($\leq 3\%$ RSD) in the HD injection mode.

Electrokinetic Injection. In the EK injection mode, HV of appropriate polarity is applied across the capillary placed in the sample vial and ions from the sample migrate into the capillary. The amount of each analyte introduced is a function of both the mobility of the analyte and the EOF in the capillary.

The EK injection mode is instrumentally simple and allows preconcentration by stacking the analyte ions into a narrow zone (159–161): however, the method has several problems as well. The amount of a particular analyte ion injected in the EK mode increases with increasing electrophoretic mobility, μ_i, therefore ions of larger mobility are injected to a greater extent. This phenomenon, "mobility-induced bias", can in principle be corrected for by correcting peak areas with mobility information that are readily derived from the migration times. A second, more insidious problem arises from differing conductivities between samples and standards. The amount of an analyte introduced is related to the electric field experienced by the sample. In the typical EK injection mode, the capillary is filled with BGE and represents a fixed resistance. If the samples and standards have very different conductivities, the effective electric field experienced by the sample and standard will differ. Under otherwise identical conditions, even when a given analyte is present in equal concentration in the sample and the standard, the actual amount of the analyte introduced in the two cases will differ. Several methods have been proposed to overcome these shortcomings—none are universally applicable.

Flow Injection Approaches. A flow-injection (FI) based sample introduction approach permits repetitive sample injection in CE in a simple manner. The sample is injected in a FI system, typically using the BGE as the carrier and flies by the CE system inlet where a portion of it is injected into the capillary either by hydrodynamic or more typically, electrokinetic, means. The separation of the previously injected sample is not physically disrupted during the subsequent EK injection. Air sensitive or hazardous samples can be injected and the system is easily automated. All limitations of HD and EK injections remain, nevertheless, and sample and BGE consumption are much higher (162,163).

8.3. Detection in CE. As previously stated, detection methods and detectors in CE are essentially identical to those used in capillary LC with one caveat that modifications may need to be made for the high voltage present in the capillaries. In the following, only the special issues unique to adaptation of the detection method in CE are discussed.

Absorbance detection is the most commonly used detection mode in CE. For nonabsorbing analytes, an absorbing electrolyte in the BGE is commonly used. In this indirect detection mode, similar to conductivity detection described for single column IC, analyte elution is detected as negative peaks. Direct absorbance detection is, of course, preferred whenever possible and a large variety of organic substances are generally detected this way using optically transparent phosphate or borate based BGEs. In CE, the detector is always placed near the grounded end of the capillary, to avoid high electric field induced problems.

Conductivity detection is completely analogous to conductivity detection in single column IC, the analyte signal arises from the difference in equivalent conductance between the analyte and the BGE ion of the same charge type. Suppressed conductivity detection has also been demonstrated, in this case, a tubular membrane suppressor is externally grounded electrically. When conductivity measurements must be made within the electric field, either high frequency excitation is used to permit contactless measurement; else, the measurement circuitry is galvanically isolated, eg, by using a transformer.

The grounding and stray current issues also appear in amperometric detection. Detection is normally just at the column outlet. For larger bore capillaries, the current flowing through the capillary itself makes it problematic to detect the much smaller analyte induced current at the working probe electrode. Exact positioning of the working electrode at the outlet of the capillary affects results and reproducible positioning is not trivial. The current flowing through the capillary itself due to the applied HV is inversely proportional to the square of the capillary bore and as such decreases dramatically with decreasing capillary bore. This in turn makes amperometric detection much less of a problem when conducted with smaller bore ($\leq 25 \ \mu m$) capillaries.

8.4. Counter versus Co-current Electrophoretic versus Electroosmotic Flow.

Manipulation of the EOF can be used to achieve additional resolution of closely migrating peaks. CZE can be performed in co- and counter electroosmotic flow modes. In the first, the analytes electrophoretic migration is in the same direction as the electroosmotic flow and the overall velocity is thus additive. Separation times under 5 min are not unusual in this mode. This is often used for analysis of small ions and molecules. In the counterelectroosmotic flow mode, the analytes move electrophoretically opposite to the EOF, but are eventually swept to the detector, because in magnitude the EOF is greater and the vector sum of the EOF and the electrophoretic velocity still has the same direction as the EOF. In a system consisting of bare silica capillary at a neutral to alkaline pH, the EOF is from the anode to the cathode, cationic analytes are moving coelectroosmotically and the anions are moving counterelectroosmotically. Counterelectroosmotic electrophoresis is often used for difficult to resolve molecules allowing them greater separation time. In the extreme case, where two analytes move with almost the same electrophoretic mobility, conditions may be adjusted to attain an EOF that is just greater than the higher mobility analyte (in the opposite direction). Under these conditions, even isotopic separation has been attained.

8.5. Factors Affecting the Separation Selectivity in CE.

Separation selectivity in CE is mainly achieved by modification of the BGE. Several factors as described below affect separation efficiency and migration time and are discussed below.

Electrolyte Composition.   The BGE composition, especially the BGE ion that is of the same charge type as the analytes, is of vital importance. Analyte peaks in CE be fronting, tailing, or highly symmetrical, based on if the mobility of the analyte is less, greater, or equal relative to the BGE coion. The separation of closely migrating analytes will be best if the mobility of the BGE coion is close to the analyte mobilities; thus a proper selection of the electrolyte coion is important.

pH.   Many analytes separated by CE have weak acid–base character and the extent of their dissociation depends on the pH. The effective electrophoretic mobility of an analyte is the product of the mobility of the fully dissociated ion times the fractional extent of ionization. Indeed, the variation of migration time with pH can be used to measure pKa. The choice of pH also affects the EOF. By proper selection of the pH, the separation can be optimized.

Complexation.   Complexation equilibria are also commonly used in CE to tailor the separation. In this case, a complexing agent is added to the BGE. The complexing agent interacts with the analytes based on their association constants. A separation of lanthanides, eg, is commonly achieved with a BGE containing α-hydroxyisobutyric acid. Addition of inorganic ions, such as Zn(II) or Cu(II) can be used to tailor the separation of peptides. Guest–host complexation, especially with cyclodextrins, is also frequently used in CE, especially in the separation of chiral isomers, as cyclodextrins are chiral selectors. A variety of compounds, notably amino acids and drugs have been separated (see CHIRAL SEPARATIONS).

Other ways, such as addition of organic solvents, change in the ionic strength of the electrolyte coating the capillary to modify EOF etc, are all used for tuning a separation.

8.6.   Separation Modes in CE. Related Techniques.   CE is typically carried out in (and what has been described above corresponds to) the zone injection and separation mode, which is therefore often referred to as capillary zone electrophoresis (CZE). There are other modes of capillary electrophoretic separations such as isotachophoresis where the capillary is filled with a high mobility electrolyte (leading electrolyte), which also fills the anode reservoir. The sample is then introduced. A low mobility electrolyte (trailing electrolyte) is now put in; the cathode reservoir contains this electrolyte. As HV is applied, the leading electrolyte moves rapidly toward the anode. The highest mobility component in the sample is drawn toward the leading electrolyte to fill in the conductivity gap left by the quickly migrating leading electrolyte. Progressively, each ion in the sample follows at lower conductivity, just behind an ion of higher mobility and just ahead of an ion of lower mobility. Finally, the "trailing" electrolyte, of lowest mobility, terminates the separation. In this separation mode, the length of each separated band is proportional to their concentration. In capillary gel electrophoresis (CGE), widely used in DNA sequencing, the capillary is filled with a gel (typically polyacrylamide, often referred to as a sieving gel). For this and other related modes, see ELECTROSEPARATIONS, ELECTROPHORESIS.

8.7.   Electrophoresis in Sieving Media. DNA Analysis.   Electrophoresis in sieving media provides a method for separation of high molecular weight molecules, such as proteins, nucleic acids, and their fragments based on their size (see also ELECTROSEPARATIONS, ELECTROPHORESIS). While proteins can usually be separated based on the differences in charge and size using CE, nucleic acids, synthetic nucleotides, DNA restriction fragments and higher DNA strains posses very similar charge densities and their separation by CE is difficult. Two modes are in use: capillary gel electrophoresis (CGE) and dynamic size-sieving CE.

In CGE, mostly polyacrylamide gels (PAG) are used. The size of the pores and cross-linking can be varied during the polymerization phase by adjusting the concentration of monomer and cross-linking agent. The PAG polymer is chemically bonded to the capillary wall.

In dynamic size-sieving CE, the capillary is filled with a low viscosity polymer solution such as linear polyacrylamide, agarose, hydroxyethylcellulose, or hydroxypropylmethylcellulose. In these sieving matrices, pores are created by physical interactions and composition can be varied over a large range. The capillary content can be replaced between the runs, thus enhancing the reproducibility and decreasing the sample carry-over between analyses.

In CGE, the separation of the molecules is based on the differences in both charge and size. To separate the molecules based only on their size, a charged ligand, such as sodium dodecyl sulfate (SDS) can be added both to the sample and the separation electrolyte. SDS binds strongly to protein molecules, giving it a constant net charge per mass unit, and the electrophoretic mobility of the molecule under sieving conditions becomes dependent only on its size or molecular weight.

Electrophoresis in sieving media is an indispensable tool for DNA sequencing and the analysis of nucleic acids, their fragments, and synthetic oligonucleotides.

DNA molecules are unique to each living organism and the knowledge of the nucleotide sequence provides valuable information for genetics, clinical biochemistry and molecular biology. In 1990, the Human Genome Project begun (164). The goals of this project were to identify ~30,000 genes in human DNA and determine the sequence of 3 billion base pairs that make up the human DNA. Knowledge about the effects of DNA variations among individuals can lead to revolutionary new ways to diagnose, treat, and someday prevent the thousands of disorders that affect mankind. Fragments almost a 1000 base long, which differ by a single base, have to be resolved for high throughput analysis. It is the high resolving power of high-speed CE that made this possible. Multiplexing of capillaries and separations on a microchip increase the throughput. For instance, multicapillary array electrophoresis systems have been commercially developed and are routinely used for DNA sequencing. Dye-tagged DNA fragments are introduced into the capillary array. Individual base termini are tagged with differently fluorescent dyes, each with their characteristic fluorescence wavelengths. As the separated fragments near the exit, they are irradiated with a spatially scanning laser. Fluorescence in the four colors is individually measured with dedicated photomultiplier tubes or a multichannel CCD detector. The working draft on the sequencing and analysis of the human genome was published in February 2001 (165).

8.8. Micellar Electrokinetic Chromatography. CE methods discussed thus far can only separate charged analytes. In MEKC, first described by Terabe and co-workers (166), the BGE consists of a surfactant above its critical micelle concentration (CMC, see SURFACTANTS). Surfactants form micelles, spherical aggregates of several molecules of surfactants, above their CMC. In a micelle, the hydrophobic tails (usually alkyl chains) face the interior of the sphere, while the ionic groups extend into the surrounding media. Most MEKC applications use typical anionic surfactants like SDS or cationic surfactants like CTAB. The charged micelles move at a rate different from that of the bulk liquid. Neutral analytes can remain in the bulk liquid or partition to the micelle, the "pseudostationary phase". Differences in this partition behavior for different analytes lead to different effective velocities.

Thus, the differential partitioning of the nonpolar, uncharged, analytes between the BGE and the hydrophobic micellar interior is the principle of separation by MEKC. This is analogous to reversed-phase LC, where nonpolar analytes are differentially retained on the nonpolar stationary phase. Typically, the electrophoretic velocity of the micelles is opposite in direction and lower in magnitude compared to the EOF. The total elution window ranges from that of an unretained analyte [ethanol, dimethyl sulfoxide(DMSO)], t_{eo}, and a fully retained analyte that elute at the same time as the micelles, t_{MC}, (usually marked by a hydrophobic dye, wholly retained within the micelle). All other neutral analytes will elute within t_0 to t_{MC}. The capacity factor of an analyte eluting at t_i can thus be defined as:

$$\bar{k} = \frac{t_i - t_{eo}}{t_{eo}\left(1 - \frac{t_i}{t_{MC}}\right)} \tag{20}$$

The correspondence to the definition of retention factor in liquid chromatography (eq. 3) is evident. When t_{MC} approaches infinite value, eg, the pseudostationary phase is really stationary, the equation becomes equivalent to the equation 3 for liquid chromatography.

The presence of ionic species brings another dimension to the MEKC separations. Since these species possess their own electrophoretic mobilities and their retention times cover much wider range than that of the neutrals, the simultaneous separation of neutral, anionic and even cationic analytes is possible in MEKC. The type of surfactant, its concentration, temperature, electrolyte pH, and the presence of other additives significantly influence the abilities of the MEKC method to separate target analytes. For example, the use of bile salts, such as sodium cholate or sodium deoxycholate instead of SDS, extends the applicability to the more hydrophobic compounds. The same effect can also be achieved by buffer additives such as urea. The addition of organic solvents alters the retention mechanism by altering the polarity of the aqueous phase and generally increases the span of the separation window. The BGE pH influences the EOF and changes the extent of ionization for weak acids and bases; this affects their effective electrophoretic mobility and therefore alters retention.

MEKC extends the separation range of CE-like methods to uncharged molecules. MEKC can be applied for the analysis of variety of analytes such as amino acids and polypeptides, DNA adducts, flavonoids and steroids, drugs and environmental samples (167–175); a selected example is shown in Figure 11.

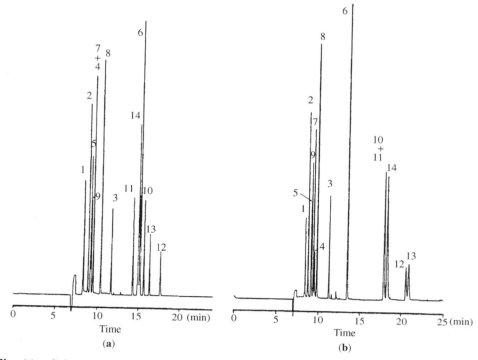

Fig. 11. Selected micellar electrokinetic chromatogram. Separation of 14 active ingredients from a Cold medicine. Conditions: 20 m*M* phosphate-borate, pH 9.0 with (**a**) 100 m*M* sodium cholate, (**b**) 50 m*M* sodium deoxycholate, 20-kV separation voltage, uv detection at 210 nm. Peaks: (*1*) caffeine, (*2*) acetaminophen, (*3*) sulphyrin, (*4*) trimetoquinol, (*5*) guaifenesin, (*6*) naproxen, (*7*) ethenzamide, (*8*) phenacetin, (*9*) isopropylantipyrine, (*10*) noscapine, (*11*) chlorpheniramine, (*12*) tipepidine, (*13*) dibucaine, (*14*) triprolidine. Reproduced from H. Nishi, T. Fukuyama, M. Matsuo and S. Terabe, *J. Chromatogr.* **498**, 313 (1990); by permission of Elsevier Science.

8.9. Capillary Electrochromatography.

Capillary electrochromatography combines the high selectivity and versatility of the stationary phases from LC with the high efficiency of electrically driven flow in CE. The column is essentially the same as a packed LC column (although some aficionados believe that CEC columns need to be packed electrokinetically). Since no pumps are involved for the mobile phase delivery, the column can be packed with the particles as small as 0.5 μm, thus providing very high column efficiencies. The eluent is driven through the column electrokinetically, by applying voltage. Extraordinarily high separation efficiencies (≥1 million plates) have been reported.

Will CEC take up some of the applications of LC? Or will the fate of CEC be more like SFC, which is yet to fulfill its originally envisioned market potential? The technique is still not fully mature and reproducibility is often limited. The EOF in CEC, the flow source, depends on several parameters, such as characteristics of the stationary phase and capillary wall, pH, electrolyte composition, ionic strength, temperature, etc. Bubble formation is another vexing problem to which a multitude of solutions have been devised. CEC in monolithic column

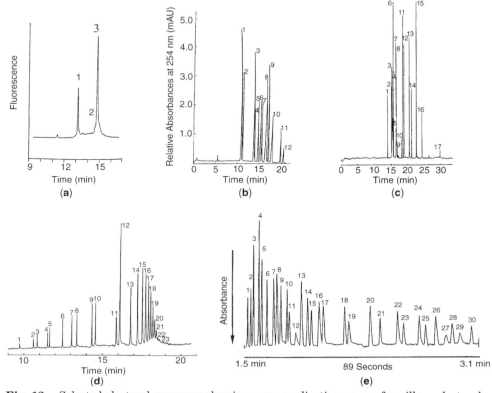

Fig. 12. Selected electropherograms showing some application areas of capillary electrophoresis. (**a**) Electropherogram of major proteins in a single erythrocyte. Conditions: 50 mM Na$_2$B$_4$O$_7$ electrolyte at pH 9.1 Peaks: (*1*) carbonic anhydrase, (*2*) methemoglobin, (*3*) hemoglobin. Reproduced from T. T. Lee, E. S. Yeung, *Anal. Chem.* **64**, 3045 (1992); by permission of the American Chemical Society. (**b**) Electropherogram of twelve 5'-ribonucleotides. Conditions: 30-mM sodium carbonate/bicarbonate electrolyte, pH 9.5, separation voltage 18 kV. Peaks: (*1*) AMP, (*2*) CMP, (*3*) ADP, (*4*) GMP, (*5*) CDP, (*6*) ATP, (*7*) UMP, (*8*) CTP, (*9*) GDP, (*10*) GTP, (*11*) UDP, (*12*) UTP. Reproduced from S. E. Geldart, P. R. Brown, *J. Chromatogr. A* **828**, 317 (1998); by permission of Elsevier Science. (**c**) Electropherogram of 3-(4-carboxybenzoyl)-2quinolinecarboxyaldehyde tagged aminoacids in lysozyme hydrolyzate. Conditions: 50-mM 2-[(*N*)-[tris (hydroxymethyl)methyl]amino] ethane-sulfonic acid, 50-mM SDS, pH 7.02. Separation voltage 25 kV.Peaks: (*1*) Arg, (*2*) Trp, (*3*) Tyr, (*4*) His, (*5*) Met, (*6*) Ile, (*7*) Gln, (*8*)Asn, (*9*) Thr, (*10*) Phe, (*11*) Leu, (*12*) Val, (*13*) Ser, (*14*) Ala, (*15*) Gly, (*16*) Glu, (*17*) Asp. Reproduced from J. Liu, Y. Z. Hsieh, D. Wiesler and M. Novotny, *Anal. Chem.* **63**, 408 (1991); by permission of the American Chemical Society. (**d**) electropherogram of 1000/base pair DNA ladder on a 3% T, 0.5% C polyacrylamide capillary. Conditions: 100-mM TRIS-borate (pH 8.3), 2 mM EDTA. Peaks: (*1*) 75, (*2*) 142, (*3*) 154, (*4*) 200, (*5*) 220, (*6*) 298, (*7*) 344, (*8*) 394, (*9*) 506, (*10*) 516, (*11*) 1018, (*12*) 1635, (*13*) 2036, (*14*) 3054, (*15*) 4072, (*16*) 5090, (*17*) 6108, (*18*) 7126, (*19*) 8144, (*20*) 9162, (*21*) 10180, (*22*) 11198, (*23*) 12216 base pairs. Reproduced from D. N. Heiger, S. A. Cohen and B. L. Karger, *J. Chromatogr.* **516**, 33 (1990); by permission of Elsevier Science. (**e**) Electropherogram of 30 inorganic anions. Conditions: 5 mM chromate, 0.5 mM OFM BT, pH 8. Separation voltage 30 kV. Indirect uv detection at 254 nm. Peaks: (*1*) thiosulfate, (*2*) bromide, (*3*) chloride, (*4*) sulfate, (*5*) nitrite, (*6*) nitrate, (*7*) molybdate, (*8*) azide, (*9*) tungstate, (*10*) monofluorophosphate, (*11*) chlorate, (*12*) citrate, (*13*) fluoride, (*14*) formate, (*15*) phosphate, (*16*) phosphite, (*17*) chlorite, (*18*) galactarate, (*19*) carbonate, (*20*) ethanesulfonate, (*21*) acetate, (*22*) propionate, (*23*) propanesulfonate, (*24*) butyrate, (*25*) butanesulfonate, (*26*) valerate, (*27*) benzoate, (*28*) L-glutarate, (*29*) pentanesulfonate, (*30*) D-gluconate. Reproduced from W. R. Jones and P. Jandik, *J. Chromatogr.* **546**, 445 (1991); by permission of Elsevier Science.

beds is being investigated also as a solution to this problem. The technique is still in its infancy and only time will prove its value. Further reading on CEC is available (176–183).

8.10. CE–MS. Like LC–MS, CE–MS provides identification and structural information (see MASS SPECTROMETRY; ANALYTICAL METHODS, HYPHENATED INSTRUMENTS). Compatibility of the CE BGE to MS, flow rate matching, termination of HV, etc. are important issues in a successful interface.

Sheath-flow, sheathless flow and liquid junction CE–MS interfaces have been developed. The ground (usually cathodic) end of the separation capillary need not to be inserted in an electrolyte vial as long as it is properly biased relative to the HV end. The outlet end of the separation capillary is inserted directly into the interface and the electrical ground of the separation capillary is accomplished either by an externally metallized capillary end, a metallic needle connected to the capillary that functions as the electrospray needle, or via liquid contact with a grounding electrode in the liquid junction interface.

The electrospray ionization mode (ESI) is likely the most popular mode of ionization currently used in CE–MS (and LC–MS). The eluent from the separation capillary is mixed with the hot gas (usually N_2) at the atmospheric pressure. The solvent is evaporated and molecules are ionized prior to entering the MS. FAB–MS, also used in LC–MS, has already been mentioned. Recently matrix assisted laser desorption/ionization (MALDI) has been used off-line for CE separated analytes. Recent reviews on CE–MS provide additional information on instrumentation development and application of CE–MS (184–190).

8.11. Applications. Due to its high resolving power, short analysis times and minute sample consumption, capillary electrophoresis plays an important role in analysis of small samples such as in the life sciences, notably analysis of single cells, human and plant tissues and for monitoring of in vivo processes (191–198). CE can be applied for variety of samples and the range of analytes overlaps application areas of IC and LC. However, it has not significantly displaced either of these for routine quantitative analysis. Despite the advent of MEKC, CE is not competitive in the marketplace with LC or GC for the analysis of neutral analytes. Main application areas of capillary electrophoresis and related techniques are DNA sequencing, analysis of amino acids, bases, nucleosides and nucleotides, proteins and peptides (199–203). High separation efficiency and little need for pretreatment makes it an excellent method for analysis of various enzymatic digests, peptide mapping, analysis of fermentation broths, biological fluids and food samples (204,205).

Another important area of CE is chiral separations, due to the decreased costs of the chiral selectors needed for CE-separation (206–212) (see CHIRAL SEPARATIONS). Further application areas include analysis of drugs, carbohydrates, small organic molecules and inorganic ions (213–226). Figure 12 shows some typical applications of capillary electrophoresis.

BIBLIOGRAPHY

1. A. J. P. Martin and R. L. M. Synge, *Biochem. J. (London)* **35**, 1358 (1941).
2. J. J. van Deemter, F. J. Zuiderweg, and A. Klinkenberg, *Chem. Eng. Sci.* **5**, 271 (1956).

3. J. P. Foley and J. G. Dorsey, *Anal. Chem.* **55**, 730 (1983).
4. B. Kaufmann and P. Christen, *Phytochem. Anal.* **13**, 105 (2002).
5. E. Bjorklund, C. von Holst, and E. Anklam, *Trends Anal. Chem.* **21**, 39 (2002).
6. I. Ferrer and E. T. Furlong, *Anal. Chem.* **74**, 1275 (2002).
7. A. Hubert, K. D. Wenzel, W. Engelwald, and G. Schuurmann, *Rev. Anal. Chem.* **20**, 101 (2001).
8. I. Windal, D. J. Miller, E. De Pauw, and S. B. Hawthorne, *Anal. Chem.* **72**, 3916 (2000).
9. A. T. James and A. J. P. Martin, *Biochem. J.* **50**, 679 (1952).
10. M. J. E. Golay, *Gas Chromatography, Ed. D. H. Desty*, Butterworths, London, 1958, pp. 36–55.
11. R. Dandenneau and E. Zerenner, *J. High Resolut. Chromatogr. Chromatogr. Commun.* **2**, 351 (1979).
12. E. Kovats, *Helv. Chim. Acta* **41**, 1915 (1958).
13. K. Grob, G. Grob, and K. Grob, *J. Chromatogr.* **156**, 1 (1978).
14. K. Grob, G. Grob, and K. Grob, *J. Chromatogr.* **219**, 13 (1978).
15. J. Abian, *J. Mass. Spectrom.* **34**, 157 (1999).
16. V. Navale, *Rev. Anal. Chem.* **18**, 193 (1999).
17. R. L. White, *Appl. Spectrosc. Rev.* **23**, 165 (1987).
18. N. Ragunathan, K. A. Krock, C. Klawun, T. A. Sasaki, and C. L. Wilkins, *J Chromatogr. A.* **703**, 335 (1995).
19. T. A. Sasaki and C. L. Wilkins, *J. Chromatogr. A* **842**, 341 (1999).
20. N. Ragunathan, K. A. Krock, C. Klawun, T. A. Sasaki, and C. L. Wilkins, *J. Chromatogr. A* **856**, 349 (1999).
21. K. Himberg, E. Sippola, and M. Riekkola, *J. Microcolumn Sep.* **1**, 271 (1989).
22. K. A. Krock and C. L. Wilkins, *Trends Anal. Chem.* **13**, 13 (1994).
23. W. Bertsch, *J. High Resolut. Chromatogr.* **22**, 647 (1999).
24. W. Bertsch, *J. High Resolut. Chromatogr.* **23**, 167 (2000).
25. L. G. Blomberg, *J.High Resolut. Chromatogr.* **7**, 232 (1984).
26. G. Spiteller, W. Kern, and P. Spiteller, *J. Chromatogr. A* **843**, 29 (1999).
27. H. Nishikawa and T. Sakai, *J. Chromatogr. A* **710**, 159 (1995).
28. P. I. Demyanov, *Zh. Anal. Khim.* **47**, 1942 (1992).
29. M. Vogel, A. Buldt, and U. Karst, *Fresenius J. Anal. Chem.* **366**, 781 (2000).
30. J. Beens and U. A. Th. Brinkman, *Trends Anal. Chem.* **19**, 260 (2000).
31. Z. Wang and M. Fingas, *J. Chromatogr. A* **774**, 51 (1997).
32. H. P. Tuan, H. G. Janssen, C. A. Cramers, A. L. C. Smit, and E. M. Van-Loo, *J. High Resolut. Chromatogr.* **17**, 373 (1994).
33. R. P. Philp, *J. High. Resolut. Chromatogr.* **17**, 398 (1994).
34. L. Cox, *Proc Int. Sch. Hydrocarbon Meas.* **65**, 427 (1990).
35. T. Cserhati, E. Forgacs, M. H. Morais, and T. Mota, *LC-GC Eur.* **13**, 254 (2000).
36. G. Lercker and M. T. Rodriguez-Estrada, *J. Chromatogr. A* **881**, 105 (2000).
37. A. Mosandl, *J. Chromatogr.* **624**, 267 (1992).
38. D. Juchelka, T. BeckT. U. Hener, F. Dettmar, and A. Mosandl, *J. High. Resolut. Chromatogr.* **21**, 145 (1998).
39. Y. Gaillard and G. Pepin, *J. Chromatogr. B* **733**, 231 (1999).
40. H. H. Maurer, *J. Chromatogr. B* **733**, 3 (1999).
41. B. A. Rudenko, S. A. Savchuk, and E. S. Brodskii, *Zh. Anal. Khim.* **51**, 182 (1996).
42. D. M. Higton and J. M. Oxford, *Appl. Spectrosc. Rev.* **30**, 81 (1995).
43. C. Staub, *Forensic Sci. Int.* **70**, 111 (1995).
44. D. Both, *Chromatogr. Sci.* **49**, 107 (1990).
45. R. L. Barnes, *Chromatogr. Sci.* **49**, 149 (1990).
46. A. S. Chawla and T. R. Bhardwaj, *Indian Drugs*, **23**, 188 (1986).

47. S. Moret and L. S. Conte, *J. Chromatogr. A* **882**, 245 (2000).

48. E. Manoli and C. Samara, *Trends Anal. Chem.* **18**, 417 (1999).

49. I. G. Zenkevich, B. N. Maksimov, and A. A. Rodin, *Zh. Anal. Khim.* **50**, 118 (1995).

50. S. B. Singh and G. Kulshrestha, *J. Chromatogr. A* **774**, 97 (1997).

51. P. Aragon, J. Atienza, and M. D. Climent, *Crit. Rev. Anal. Chem.* **30**, 121 (2000).

52. N. A. Klyuev, *Zh. Anal. Khim.* **51**, 163 (1996).

53. K. Peltonen and T. Kuljukka, *J. Chromatogr. A* **710**, 93 (1995).

54. K. G. Furton, E. Jolly, and G. Pentzke, *J. Chromatogr.* **642**, 33 (1993).

55. http://www.epa.gov

56. http://www.sigmaaldrich.com

57. http://www.chem.agilent.com

58. http://www.altechweb.com

59. http://www.tekmar.com/appnotes/

60. G. A. Eiceman, B. Davani, and J. Gardea-Torresday, *Anal. Chem.* **68**, 291R (1996).

61. G. A. Eiceman, H. H. Hill, Jr., and J. Gardea-Torresday, *Anal. Chem.* **70**, 321R (1998).

62. G. A. Eiceman, H. H. Hill, Jr., and J. Gardea-Torresday, *Anal. Chem.* **72**, 137R (2000).

63. M. Tswett, *Ber. Dtch. Bot. Geo.* **24**, 385 (1906).

64. C. G. Horvath, B. A. Preiss, and S. R. Lipsky, *Anal. Chem.* **39**, 1422 (1967).

65. C. G. Horvath and S. R. Lipsky, *Anal. Chem.* **41**, 1227 (1969).

66. D. Ishii, K. Asai, K. Hibi, T. Jonokuchi, and M. Nagaya, *J. Chromatogr.* **144**, 157 (1977).

67. T. Tsuda and M. Novotny, *Anal. Chem.* **50**, 271 (1979).

68. T. Tsuda and M. Novotny, *Anal. Chem.* **50**, 623 (1979).

69. R. P. W. Scott and P. Kucera, *J. Chromatogr.* **169**, 51 (1979).

70. R. P. W. Scott, P. Kucera, and M. Munroe, *J. Chromatogr.* **186**, 475 (1979).

71. R. P. W. Scott, *J. Chromatogr. Sci* **18**, 49 (1980).

72. C. E. Reese and R. P. W. Scott, *J. Chromatogr. Sci* **18**, 479 (1980).

73. P. Kucera, *J. Chromatogr.* **198**, 93 (1980).

74. F. J. Yang, *J. Chromatogr.* **236**, 265 (1982).

75. M. P. Maskarinec, J. D. Vargo, and M. J. Sepaniak, *J. Chromatogr.* **261**, 245 (1983).

76. L. A. Knecht, E. J. Guthrie and J. W. Jorgenson, *Anal. Chem.* **56**, 479 (1984).

77. G. Desmet and G. V. Baron, *J. Chromatogr. A* **867**, 23 (2000).

78. D. Pyo, P. K. Dasgupta, and L. S. Yengoyan, *Anal. Sci.* **13**(Suppl), 185 (1997).

79. T. Tsuda and G. Nakagawa, *J. Chromatogr. A* **268**, 369 (1983).

80. C. B. Boring, P. K. Dasgupta, and A. Sjögren. *J. Chromatogr. A* **804**, 45 (1998).

81. http://www.microlc.com/PS/Products1.htm

82. M. C. Harvey, S. D. Stearns, and J. P. Averette, *LC Liq. Chromatogr. HPLC Mag.* **3**, 5 (1980).

83. H. A. Claessens, A. Burcinova, C. A. Cramers, P. Mussche, and C. E. van Tilburg, *J. Microcolumn Sep.* **2**, 132 (1990).

84. J. W. Jorgenson and E. J. Guthrie, *J. Chromatogr.* **255**, 335 (1983).

85. A. Manz and W. Simon, *J. Chromatogr.* **387**, 187 (1987).

86. P. K. Dasgupta and L. Y. Bao, *Anal. Chem.* **65**, 1003 (1993).

87. S. Kar, P. K. Dasgupta, H. Liu, and H. Hwang, *Anal. Chem.*, **66**, 2537 (1994).

88. A. J. Zemann, *Trends Anal. Chem.* **20**, 346 (2001).

89. D. J. Bornhop, *Appl. Opt.* **34**, 3234 (1995).

90. K. Swinney and D. J. Bornhop, *Electrophoresis* **21**, 1239 (2000).

91. K. C. Lewis, D. M. Dohmeier, J. W. Jorgenson, S. L. Kaufman, F. Zarrin, and F. D. Dorman, *Anal. Chem.* **66**, 2285 (1994).

92. T. S. Kephart and P. K. Dasgupta, *Talanta* **56**, 977 (2002).

93. D. Barcelo, Ed, *Applications of LC-MS in environmental chemistry*, Elsevier, Amsterdam, The Netherlands, 1996.

94. W. M. A. Niessen, *Liquid chromatography-Mass spectrometry*, Marcel-Dekker Inc, New York, 1999.

95. W. M. A. Niessen, *J. Chromatogr. A* **856**, 179 (1999).

96. W. M. A. Niessen, *J. Chromatogr. A* **812**, 53 (1998).

97. E. L. Esmans, D. Broes, I. Hoes, F. Lemiere, and K. Vanhoutte, *J. Chromatogr. A* **794**, 109 (1998).

98. J. Leonil, V. Gagnaire, D. Molle, S. Pezennec, and S. Bouhallab, *J. Chromatogr. A* **881**, 1 (2000).

99. J. Slobodnik, B. L. M. van-Baar, and U. A. T. Brinkman, *J. Chromatogr. A* **703**, 81 (1995).

100. M. Careri, A. Mangia, and M. Musci, *J. Chromatogr. A* **727**, 153 (1996).

101. M. Careri, P. Manini, and M. Maspero, *Ann. Chim. (Rome)* **84**, 475 (1994).

102. A. Di Corcia, *J. Chromatogr. A* **794**, 165 (1998).

103. D. Rindgen, R. J. Turesky, and P. Vorous, *Chem. Res. Toxicol.* **7**, 82 (1995).

104. R. Straub, M. Lindner, and R. D. Voyksner, *Anal. Chem.* **66**, 3651 (1994).

105. E. J. Caliguri and I. N. Mefford, *Brain Res.* **296**, 156 (1984).

106. W. J. Henzel, C. Grimely, J. H. Bourell, T. M. Billeci, S. C. Wong, and J. T. Stults, *Methods Comparison Enzymol* **6**, 239 (1994).

107. J. R. Yates III, J. K. Eng, and A. L. McCormack, *Anal. Chem.* **67**, 3203 (1995).

108. C. Elicone, M. Lui, S. Geromanos, H. Erdjument-Bromage, and P. Tempst, *J. Chromatogr.* **676**, 121 (1994).

109. D. B. Kassel, B. Sushan, T. Sakuma, and J. P. Salzmann, *Anal. Chem.* **66**, 236 (1994).

110. C. E. Kientz, J. Langenberg, G. J. De Jong, and U. A. Th. Brinkman, *J. High. Resolut. Chromatogr.* **14**, 460 (1991).

111. H. J. Cortes and L. W. Nicholson, *J. Microcol. Sep.* **6**, 257 (1994).

112. R. Hu, T. Takeuchi, J.-Y. Jin, and T. Miwa, *Anal. Chim. Acta* **295**, 173 (1994).

113. J. P. C. Vissers, *J. Chromatogr. A* **856**, 117 (1999).

114. J. P. C. Vissers, H. A. Claessens, and C. A. Cramers, *J. Chromatogr. A* **779**, 1 (1997).

115. J. G. Dorsey, W. T. Cooper, B. A. Siles, J. P. Foley, and H. G. Barth, *Anal. Chem.* **68**, 515R (1996).

116. J. G. Dorsey, W. T. Cooper, B. A. Siles, J. P. Foley, and H. G. Barth, *Anal. Chem.* **70**, 591R (1998).

117. L. D. Rothman, *Anal. Chem.* **68**, 587R (1996).

118. W. R. LaCourse and C. O. Dasenbrock, *Anal. Chem.* **70**, 37R (1998).

119. W. R. LaCourse, *Anal. Chem.* **72**, 37R (2000).

120. H. Small, T. S. Stevens, and W. C. Bauman, *Anal. Chem.* **57**, 1809 (1975).

121. H. Small and J. Riviello, *Anal. Chem.* **70**, 2205 (1998).

122. A. Sjögren, C. B. Boring, P. K. Dasgupta, and J. N. Alexander, IV, *Anal. Chem.* **69**, 1385 (1997).

123. M. Betti, *J. Chromatogr. A* **789**, 369 (1997).

124. W. Frankenberger, H. C. Mehra, and D. T. Gjerde, *J. Chromatogr.* **504**, 211 (1990).

125. H. Klein and R. Leubolt, *J. Chromatogr.* **640**, 259 (1993).

126. A. Henshall, *Cereal Foods World* **42**, 414 (1997).

127. P. L. Buldini, S. Cavalli, and A. Trifiro, *J. Chromatogr. A* **789**, 529 (1997).

128. M. R. Marshall, R. H. Schmidt, and B. L. Walker, *Food Sci. Technol.* **45**, 39 (1991).

129. M. D. H. Amey, D, and A. Bridle, *J. Chromatogr.* **640**, 323 (1993).

130. V. Tusset and J. Hancart, *Steel. Res.* **60**, 241 (1989).

131. J. Weiss, *Galvanotechnik* **77**, 2675 (1986).

132. K. Haak, *Plat. Surf. Finish.* **70**, 34 (1983).

133. B. Lopez-Ruiz, *J. Chromatogr. A* **881**, 607 (2000).

134. R. P. Singh, N. M. Abbas, and S. A. Smesko, *J. Chromatogr. A* **733**, 73 (1996).
135. C. Sarzanini, *J. Chromatogr. A* **850**, 213 (1999).
136. L. M. Thienpont, J. E. Van Nuwenborg, and D. Stoeckl, *J. Chromatogr. A* **789**, 557 (1997).
137. J. Dugay, A. Jardy, and M. Doury-Berthod, *Analusis* **23**, 183 (1995).
138. J. Dugay, A. Jardy, and M. Doury-Berthod, *Analusis* **23**, 196 (1995).
139. E. O. Otu, J. J. Byerley, and C. W. Robinson, *Int. J. Environ. Anal. Chem.* **63**, 81 (1996).
140. M. B. Masters, *Anal. Proc. (London)* **22**, 146 (1985).
141. E. Dabek-Zlotorzynska and M. McGrath, *Fresenius J. Anal. Chem.* **367**, 507 (2000).
142. P. Hajos and L. Nagy, *J. Chromatogr. B* **717**, 27 (1998).
143. T. Guerin, A. Astruc, and M. Astruc, *Talanta* **50**, 1 (1999).
144. C. Sarzanini and E. Mentasti, *J. Chromatogr. A* **789**, 301 (1997).
145. R. E. Smith, *Ion Chromatography Applications*, 1988, CRC Press, Boca Raton, Fla.
146. D. R. Gere, *Science* **222**, 253 (1083).
147. M. Kaplan, G. Davidson, and M. Poliakoff, *J. Chromatogr. A* **673**, 231 (1994).
148. T. A. Becker, *J. Chromatogr. A* **785**, 3 (1997).
149. T. L. Chester, J. D. Pinkton, and D. E. Raynie, *Anal. Chem.* **68**, 487R (1996).
150. T. L. Chester, J. D. Pinkton, and D. E. Raynie, *Anal. Chem.* **70**, 301R (1998).
151. T. L. Chester and J. D. Pinkton, *Anal. Chem.* **72**, 129R (2000).
152. M. Luebke, *Analusis* **19**, 323 (1991).
153. E. B. Hoving, *J. Chromatogr. B* **671**, 341 (1995).
154. W. W. Christie, *Analusis.* **26**, M34 (1998).
155. S. Hjerten, *Ark. Kemi* **13**, 151 (1958).
156. R. Virtanen, *Acta Polytech. Scand.* **123**, 1 (1974).
157. F. E. P. Mikkers, F. M. Everaerts, and T. M. E. P. Verheggen, *J. Chromatogr.* **169**, 1 (1979).
158. J. W. Jorgenson and K. D. Lukacs, *Anal. Chem.* **53**, 1298 (1981).
159. D. S. Burgi and R. L. Chien, *Anal. Chem.* **63**, 2042 (1991).
160. R. L. Chien and D. S. Burgi, *Anal. Chem.* **64**, 1046 (1992).
161. D. S. Burgi., *Anal.Chem.* **65**, 3726 (1993).
162. P. Kuban, A. Engström, J. C. Olsson, G. Thorsen, R. Tryzell, and B. Karlberg, *Anal. Chim. Acta* **337**, 117 (1997).
163. Z. L. Fang, Z. S. Liu, and Q. Shen, *Anal. Chim. Acta* **346**, 135 (1997).
164. http://www.ornl.gov/hgmis
165. The International Human Genome Consortium, *Nature* **409**, 934 (2001).
166. S. Terabe, K. Otsuka, and T. Ando, *Anal. Chem.* **57**, 834 (1985).
167. H. Nishi and S. Terabe, *J. Chromatogr. A* **735**, 3 (1996).
168. H. J. Issaq and K. C. Chan, *Electrophoresis* **16**, 467 (1995).
169. M. A. Strege and A. L. Lagu, *J. Chromatogr. A* **780**, 285 (1997).
170. E. Szoko, *Electrophoresis* **18**, 74 (1997).
171. H. Nishi, *J. Chromatogr. A* **780**, 243 (1997).
172. L. G. Song, Z. H. Xu, J. W. Kang, and J. K. Cheng, *J. Chromatogr. A* **780**, 297 (1997).
173. T. Watanabe and S. Terabe, *J. Chromatogr. A* **880**, 295 (2000).
174. K. Otsuka and S. Terabe, *J. Chromatogr. A* **875**, 163 (2000).
175. S. K. Wiedmer and M. L. Riekkola, *Rev. Anal. Chem.* **18**, 67 (1999).
176. K. D. Bartle and P. Myers Eds., *Capillary Electrochromatography*, Royal Society of Chemistry, Cambridge, U.K., 2001.
177. I. S. Krull, A. Sebag, and R. Stevenson, *J. Chromatogr. A* **887**, 137 (2000).
178. K. D. Altria, *J. Chromatogr. A* **856**, 443 (1999).
179. F. Svec, E. C. Peters, D. Sykora, and J. M. J. Frechet, *J. Chromatogr.* **887**, 3 (2000).
180. A. Karcher and Z. El-Rassi, *Electrophoresis* **20**, 3280 (1999).

181. A. Dermaux and P. Sandra, *Electrophoresis* **20**, 3027 (1999).
182. M. G. Cikalo, K. D. Bartle, M. M. Robson, P. Myers, and M. R. Euerby, *Analyst (Cambridge, UK)* **123**, 87R (1998).
183. C. G. Edmonds, J. A. Loo, C. J. Barinaga, H. R. Udseth, and R. D. Smith, *J. Chromatogr.* **474**, 21 (1989).
184. W. M. A. Niessen, U. R. Tjaden, and J. van der Greef, *J. Chromatogr.* **636**, 3 (1993).
185. J. Cai and J. Henion, *J. Chromatogr. A* **703**, 667 (1995).
186. J. F. Banks, *Electrophoresis* **18**, 2255 (1997).
187. W. F. Smyth, *Trends Anal. Chem.* **18**, 335 (1999).
188. R. D. Suessmuth and G. Jung, *J. Chromatogr. B* **725**, 49 (1999).
189. A. I. Gusev, *Fresenius J. Anal. Chem.* **366**, 691 (2000).
190. G. Guetens, K. van Cauwenberghe, G. de Boeck, R. Maes, U. R. Tjaden, J. van der Greef, M. Highley, A. T. van Oosterom, and E. de Bruijn, *J. Chromatogr. B* **739**, 139 (2000).
191. S. Chen and Y. Chen, *Electrophoresis* **20**, 3259 (1999).
192. H. J. Issaq, *Electrophoresis* **18**, 2438 (1997).
193. H. J. Issaq, *Electrophoresis* **20**, 3190 (1999).
194. A. L. Lagu, *Electrophoresis* **20**, 3145 (1999).
195. E. S. Yeung, *J. Chromatogr. A* **830**, 243 (1999).
196. S. D. Gilman and A. G. Ewing, *J. Capillary Electrophor.* **2**, 1 (1995).
197. M. I. Davies, *Anal. Chim. Acta* **379**, 227 (1999).
198. D. T. Eash and R. J. Bushway, *J. Chromatogr. A* **880**, 281 (2000).
199. W. Thormann, A. B. Wey, I. S. Lurie, H. Gerber, C. Byland, N. Malik, M. Hochmeister, and C. Gehrig, *Electrophoresis* **20**, 3203 (1999).
200. Z. K. Shihabi, *J. Liq. Chromatogr. Relat. Technol.* **23**, 79 (2000).
201. S. N. Krylov and N. J. Dovichi, *Anal. Chem.* **72**, 11R (2000).
202. S. E. Geldart and P. R. Brown, *J. Chromatogr. A* **828**, 317 (1998).
203. S. McWhorter and S. A. Soper, *Electrophoresis* **21**, 1267 (2000).
204. M. A. Strege and A. L. Lagu, *Electrophoresis* **18**, 2343 (1997).
205. H. W. Lahm and H. Langen, *Electrophoresis* **21**, 2105 (2000).
206. R. Vespalec and P. Bocek, *Electrophoresis* **15**, 755 (1994).
207. R. Vespalec and P. Bocek, *Electrophoresis* **18**, 843 (1997).
208. R. Vespalec and P. Bocek, *Electrophoresis* **20**, 2579 (1999).
209. Z. Deyl, I. Miksik, and F. Tagliaro, *Forensic Sci. Int.* **92**, 89 (1998).
210. M. R. Hadley, P. Camilleri, and A. J. Hutt, *Electrophoresis* **21**, 1953 (2000).
211. T. J. Ward, *Anal. Chem.* **72**, 4521 (2000).
212. K. Otsuka and S. Terabe, *J. Chromatogr. A* **875**, 163 (2000).
213. H. Nishi, *Electrophoresis* **20**, 3237 (1999).
214. C. L. Flurer, *Electrophoresis* **20**, 3269 (1999).
215. Y. Nakahara, *J. Chromatogr. B* **733**, 161 (1999).
216. H. H. Maurer, *J. Chromatogr. B* **733**, 3 (1999).
217. Z. E. Rassi, *Electrophoresis* **20**, 3134 (1999).
218. S. Suzuki and S. Honda, *Electrophoresis*, **19**, 2539 (1998).
219. P. E. Jackson and P. R. Haddad, *Trends Anal. Chem.* **12**, 231 (1993).
220. A. R. Timerbaev, *J. Chromatogr. A* **792**, 495 (1997).
221. B. F. Liu, L. B. Liu, and J. K. Cheng, *J. Chromatogr. A* **834**, 277 (1999).
222. D. Kaniansky, M. Masar, J. Marak, and R. Bodor, *J. Chromatogr. A* **834**, 133 (1999).
223. P. Doble and P. R. Haddad, *J. Chromatogr. A* **834**, 189 (1999).
224. V. Pacakova, P. Coufal, and K. Stulik, *J. Chromatogr. A* **834**, 257 (1999).
225. S. M. Valsecchi and S. Polesello, *J. Chromatogr. A* **834**, 363 (1999).
226. J. Sadecka and J. Polonsky, *J. Chromatogr. A* **834**, 401 (1999).

GENERAL REFERENCES

R. A. Meyers, *Encyclopedia of Analytical Chemistry*, Wiley-Interscience, John Wiley & Sons, Ltd., Chichester, U.K., 2000 Volumes 12 and 13.

D. A. Skoog, F. J. Holler, and T. A. Nieman, *Principles of Instrumental Analysis-5th ed.*, Harcourt Brace College Publishers, Philadephia, 1998.

C. F. Poole and S. K. Poole, *Chromatography Today*, Elsevier, Amsterdam, The Netherlands, 1991.

P. R. Haddad and P. E. Jackson, *Ion Chromatography: Principles and Applications*, Elsevier, Amsterdam, The Netherlands, 1990.

H. Small, *Ion Chromatography*, Plenum Press, New York, 1989.

M. Caude, D. Thiebaut, *Practical Supercritical Fluid Chromatography and Extraction*, Harwood Academic Publishers, Amsterdam, The Netherlands, 1999.

F. Foret, L. Krivankova, and P. Bocek, *Capillary Zone Electrophoresis*, VCH, Weinheim, 1993.

A. Chrambach, M. J. Dunn, and B. J. Radola, *Advances in Electrophoresis*, Vols. 1–7, VCH-Weinheim, Germany, 1987–1994.

S. F. Y. Li *Capillary Electrophoresis*, Elsevier, Amsterdam, The Netherlands, 1992.

R. Kuhn and S. Hoffstetter-Kuhn, *Capillary Electrophoresis: Principles and Practice*, Springer Verlag, Berlin, 1993.

PETR KUBAN
PUMENDU K. DASGUPTA
Texas Tech University

CARBIDES, (SURVEY)

1. Introduction

Carbon (qv) reacts with most elements of the Periodic Table to form a diverse group of compounds known as carbides, some of which are extremely important in technology. For example, calcium carbide, CaC_2, is a source of acetylene; silicon carbide, SiC, and boron carbide, B_4C, are used as abrasives (qv); tungsten carbide, WC, titanium carbide, TiC, and tantalum (niobium) carbide, TaC(NbC) find use as structural materials at extremely high temperatures or in corrosive atmospheres. Cementite, Fe_3C, and the multimetallic complexes $(Co,W)_6C$, $(Cr,Fe,Mo)_{23}C_6$, and $(Cr,Fe)_7C_3$ are the components in tool steels and Stellite-type alloys responsible for their hardness, wear resistance, and excellent cutting performance. The general properties and applications of the carbides, including emerging uses of these materials as catalysts, have been described (1). (See TUNGSTEN AND TUNGSTEN ALLOYS; TANTALUM AND TANTALUM COMPOUNDS; NIOBIUM AND NIOBIUM COMPOUNDS; TITANIUM AND TITANIUM COMPOUNDS).

Figure 1 provides a survey of the most important and well-known binary compounds of carbon, according to their position in the periodic system. These may be divided into four main groups: the saltlike, metallic, diamondlike, and volatile compounds of carbon, which have ionic, metallic, semiconductor, or

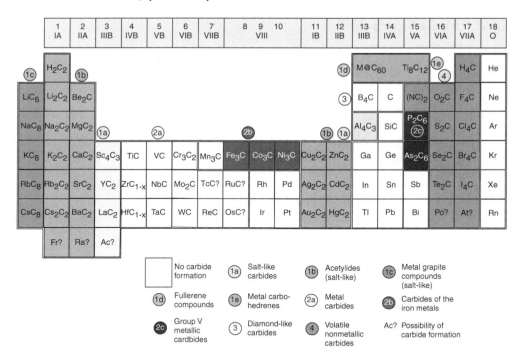

Fig. 1. Principal binary compounds of carbon.

covalent character, respectively. These and further subdivisions, used to characterize the compounds, are not rigid and there are a number of transitional cases. Whereas the members of Groups 2 (IIA) and 3 (IIIB) are classified as saltlike carbides, some of their properties, eg, Be_2C, with a very high degree of hardness, correspond to diamondlike carbides. Conversely, some monocarbides of Group 3 (IIIB), eg, scandium carbide, ScC, and uranium carbide, UC, as well as thorium carbide, ThC, have pronounced metallic characteristics.

Carbides are generally stable at high temperatures and thus can be prepared by the direct reaction of carbon with metals or metal-like materials at high temperatures. This does not apply to the acetylides and the alkali metal–graphite compounds, which although being carbon compounds, fit only marginally into the category of carbides. Similarly, the large class of coordination compounds, ML_y, known as organometallics are not typically considered carbides, even though the ligands, L, are attached to the metal center by metal–carbon bonds. The same applies to compounds such as M_x @ C_{60} or M_x @ C_{70}, formed from diverse elements by association with fullerene structures (see Fullerenes) (2) and metallo-carbohedrene clusters such as Ti_8C_{12}, in which metals are part of the polyhedral cage (3). The volatile compounds of carbon are also excluded.

Table 1 contains an alphabetical listing of carbides referred to in the text.

2. Saltlike Carbides

This group comprises almost all carbides of Groups 1–3 (IA, IIA, and III B) of the Periodic Table. Beryllium carbide and Al_4C_3 may be considered as derivatives of

Table 1. **Carbide Compounds**[a]

Carbide	CAS Registry Number	Formula
aluminum carbide (4:3)	[1299-86-1]	Al_4C_3
arsenic carbide (2:6)		As_2C_6
beryllium carbide	[57788-94-0]	Be_2C
boron carbide (4:1)	[12069-32-8]	B_4C
calcium carbide (2:1)	[75-20-7]	CaC_2
chromium carbide	[12011-60-8]	CrC
chromium carbide (3:2)	[12012-35-0]	Cr_3C_2
chromium carbide (4:1)	[12075-40-7]	Cr_4C
chromium carbide (7:3)	[12075-40-0]	Cr_7C_3
chromium carbide (23:6)	[12105-81-6]	$Cr_{23}C_6$
cobalt carbide (3:1)	[12011-59-5]	Co_3C
cobalt tungsten carbide (6:6:1)	[12538-07-7]	Co_6W_6C
hafnium carbide	[12069-85-1]	HfC
iron carbide	[12069-60-2]	FeC
iron carbide (2:1)	[12011-66-4]	Fe_2C
iron carbide (3:1)	[12011-67-5]	Fe_3C
	[12169-32-3]	
iron carbide (5:2)	[12127-45-6]	Fe_5C_2
iron carbide (7:3)	[12075-42-2]	Fe_7C_3
iron carbide (23:6)	[12012-72-5]	$Fe_{23}C_6$
lanthanum carbide (1:2)	[12071-15-7]	LaC_2
manganese carbide (3:1)	[12121-90-3]	Mn_3C
manganese carbide (23:6)	[12266-65-8]	$Mn_{23}C_6$
magnesium carbide (1:2)	[12122-46-2]	MgC_2
magnesium carbide (2:3)	[12151-74-5]	Mg_2C_3
molybdenum carbide	[12011-97-1]	MoC
molybdenum carbide (2:1)	[12069-89-5]	Mo_2C
molybdenum carbide (23:6)	[12152-15-7]	$Mo_{23}C_6$
nickel carbide	[12167-08-7]	NiC
nickel carbide (3:1)	[12012-02-1]	Ni_3C
niobium carbide	[12069-94-2]	NbC
niobium carbide (2:1)	[12011-99-3]	Nb_2C
plutonium carbide	[12070-03-0]	PuC
plutonium carbide (2:3)	[12076-56-1]	Pu_2C_3
phosphorus carbide (2:6)		P_2C_6
scandium carbide	[12012-14-5]	ScS
silicon carbide	[409-21-2]	SiC
tantalum carbide	[12070-06-3]	TaC
tantalum carbide (2:1)	[12070-07-4]	Ta_2C
thorium carbide	[12012-16-6]	ThC
thorium carbide (1:2)	[12071-31-7]	ThC_2
titanium carbide	[12070-08-5]	TiC
tungsten carbide	[12070-12-1]	WC
tungsten carbide (2:1)	[12070-13-2]	W_2C
uranium carbide	[12170-09-6]	UC
uranium carbide (1:2)	[12071-33-9]	UC_2
uranium carbide (2:3)	[12076-62-9]	U_2C_3
vanadium carbide	[12070-10-9]	VC
vanadium carbide (2:1)	[12012-17-8]	V_2C
zirconium carbide	[12020-14-3]	ZrC

[a] Stoichiometry indicated in paranthesis

methane (C^{4-} anion) and most carbides with C_2 groups (chiefly C_2^{2-} anions) as derivatives of acetylene. This finding is supported to some extent by the following hydrolysis reactions:

$$Be_2C + 4\,H_2O \rightarrow 2\,Be(OH)_2 + CH_4 \tag{1}$$

$$Al_4C_3 + 12\,H_2O \rightarrow 4\,Al(OH)_3 + 3\,CH_4 \tag{2}$$

$$CaC_2 + 2\,H_2O \rightarrow Ca(OH)_2 + C_2H_2 \tag{3}$$

Propyne is obtained from Mg_2C_3 (probably C_3^{4-} anions), formed by thermal decomposition of MgC_2, with separation of graphite:

$$Mg_2C_3 + 4\,H_2O \rightarrow 2\,Mg(OH)_2 + CH_3C{\equiv}CH \tag{4}$$

In their pure state, the carbides of Groups 1 (IA) and 2 (IIA) are characterized by their transparency and lack of conductivity. The carbides of Group 3 (IIIB) (Sc, Y, the lanthanides, and the actinides) are opaque. Some of them, depending on their composition, show metallic luster and electroconductivity. The M^{2+} cation may exist in the MC_2 phases of this group, and the remaining valence electron apparently imparts partly metallic character to these compounds. Methane, ethylene, and hydrogen, as well as acetylene, are formed during the hydrolysis of the Group 3 (IIIB) carbides of varying composition, M_3C, MC, M_2C_3, and MC_2.

Formally, the term acetylides applies to carbides precipitated from aqueous solutions or from solutions in aqueous ammonia by reaction of metal ions with acetylene (4). These include compounds of Cu, Ag, Au, Na, K, Rb, Cs, Zn, Cd, Hg, Pd, Os, Ce, Al, Mg, etc. Because they require additives such as H_2, H_2O, NH_3, C_2H_2, and metal salts for stabilization, it is doubtful whether they can be considered as pure metal–carbon compounds and described as carbides.

The alkali metal–graphite compounds formed by graphite absorption of the fused metals Na, K, Rb, and Cs, represent a special type of metal–carbon compound (5). These are intercalation compounds with formulas such as MC_8, MC_{16}, and MC_{60}. They become strongly graphitic as the carbon content increases, with their color changing from brown to gray to black.

3. Metallic Carbides

This class of compounds comprises the interstitial carbides of the transition metals of Groups 4–6 (IVB–VIB) (see CARBIDES, INDUSTRIAL HARD) and the carbides of metals of Groups 7–10 (VIIB–VIII). The metalloconductive carbides P_2C_6 and As_2C_6 are also included, as well as some members of the lanthanide and actinide series. Their standard heat of formation and free energy of formation are reported in Figure 2. It is seen that the thermodynamic stability of the compounds is high toward the left in the Periodic Table and decreases in moving to the right.

3.1. Group 4–6 (IVB–VIB) Metals.   The early transition elements possess relatively large atomic radii and carbon resides in the interstitial cavities between metal atoms. In these interstitial carbides, the metal atoms form

4 IVB	5 VB	6 VIB	7 VIIB	4 VIII
TiC -184.5/-181.0	VC -101.7/-99.1	Cr_3C_2 -85.4/-102.1	Mn_3C -16.1/-33.2	Fe_3C +25.0/+3.7
ZrC -207.1/-203.7	NbC -138.9/-136.9	Mo_2C -49.4/-58.8		
HfC -225.9/-223.0	TaC -144.1/-142.6	WC -40.6/-39.5		

Fig. 2. Thermodynamic properties of metallic carbides ($\Delta H°298/\Delta G°298$). Values in kJ/mol of metal.

relatively simple structures commonly found among pure metals: hexagonal closed-packed (hcp), face-centered cubic (fcc), and simple hexagonal (hex) structures. The major structure not adopted by carbides is the body-centered cubic (bcc) structure, because this lattice does not accommodate a large interstitial site. In Groups 4 and 5 (IVB and VB), and some of the rare earth carbides the preferred structure is the B1 NaCl structure, where carbon occupies every octahedral site in an fcc arrangement of metal atoms. The resulting stoichiometry is MC, exemplified by TiC, ZrC, HfC, VC, NbC, TaC, UC, and PuC. There is often some carbon deficiency in these substances, which may be disordered, as in the MC_{1-x} phases, or ordered, as in the compounds, V_8C_7, Nb_4C_3, etc. In Group 5 (V B), the stoichiometry M_2C appears, eg, V_2C, Nb_2C, and Ta_2C, which becomes most favorable for Group 6 (VIB), eg, Mo_2C, and W_2C. In this structure, carbon randomly occupies one-half of the octahedral sites in an hcp arrangement of metal atoms. Again, substoichiometry may occur, as well as the formation of ordered vacancies. At high temperatures MoC_{1-x} and WC_{1-x} ($x \sim 0.5$) phases having an fcc arrangement of metal atoms are observed. Group 6 (VIB) also shows retention of the MC stoichiometry, MoC, WC, and (Mo,W)C. However, here carbon does not occupy the octahedral holes of the B1 structure, but rather the more spacious trigonal prismatic sites in a hex arrangement of metal atoms.

The Group 4–6 (IVB–VIB) carbides are thermodynamically very stable, exhibiting high heats of formation, great hardness, elevated melting points, and resistance to hydrolysis by weak acids. At the same time, they have values of electrical conductivity, Hall coefficients, magnetic susceptibility, and heat capacity in the range of metals (6).

3.2. Group 7–10 (VIIB, VIII) Metals.
Carbides of these metals are generally less stable than those of the earlier transition metals. The noble metals of the second and third row (Rh, Pd, Ir, Pt) do not form carbides at all. In moving to

the right in the Periodic Table the size of the atoms decreases, and the metal lattice is no longer able to accomodate interstitial carbon atoms while maintaining close-packed or near close-packed metal atoms. Thus, complex structures arise, eg, Mn_8C_3 is triclinic, Mn_4C is tetragonal, while Mn_3C, Fe_3C, and Co_3C are orthorhombic. The chromium carbides occupy an intermediate position in this respect, eg, Cr_3C_2, is also orthorhombic. Group 7–10 (VI B–VIII) compounds also differ markedly in chemical and physical behavior from the interstitial carbides. Hardness values and melting points are lower and chemical stability to mineral acids is also no longer apparent. However, the materials are still robust; as mentioned earlier, they are of technological importance as components in high speed tool steels and advanced alloys.

3.3. Properties and Nature of Bonding in the Metallic Carbides.

The metallic carbides are interesting materials that combine the physical properties of ceramics (qv) with the electronic nature of metals. Thus, they are hard and strong, but at the same time, good conductors of heat and electricity.

The crystal structure and stoichiometry of the materials are determined from two contributions, geometric and electronic. The geometric factor is derived from the empirical observation (7) that simple interstitial carbides, nitrides, borides, and hydrides are formed for small ratios of nonmetal (X) to metal radii, $r_X/r_M < 0.59$, ie, the Hägg rule. When this ratio is >0.59, as in the Group 7–10 (VI B–VIII) metals, the structure becomes more complex to compensate for the loss of metal–metal interactions. Although there are minor exceptions, the Hägg rule provides a useful basis for predicting structure.

There is also an electronic factor that contributes to the bonding properties of the materials (8). The materials behave electronically as though their electron density increased by alloying. In forming carbide alloys, carbon combines its valence sp electrons with the metal spd band. Engel-Brewer theory may be used to predict the crystal structures of the materials based on the total number of valence sp electrons. Thus, with increasing sp electrons, crystal structure is predicted to change from bcc → hcp → fcc, as observed in the series Nb → Nb_2C → MoC_{1-x} or Ta → Ta_2C → WC_{1-x}. An increase in band occupancy, first to bonding levels and then to antibonding levels, is also indicated by the occurrence of a maximum in melting point at Group 5 (VB) for the carbides, rather than at Group 6 (VIB) for the pure metals. Furthermore, increased metal/nonmetal ratios (MC → M_2C → M_3C) in moving to the right in the Periodic Table suggests rejection of carbon by the metal. This occurs from increased filling of the antibonding portion of the electronic band.

Ideas on bonding have evolved considerably over the years (9). Early models considered bonding to arise from resonance between different covalent canonical forms. The hardness and brittleness of the compounds was attributed to the presence of directed covalent bonds, and the electrical conductivity to the existence of resonance structures. An opposing view, based on the high electronegativity of carbon compared to metals, held that the bonding was ionic in nature. This explained the refractory nature and the B1 NaCl structure of many of the compounds, but did not account for the instability of carbides beyond Group 6 (VIB). The instability was explained by band structure models suggesting that in compounds of the late transition metals, filling of the antibonding portion of the d-band occurred. This finding was consistent with electron donation from carbon

Table 2. **Physical Properties of Diamondlike Carbides and Nonmetallic Hard Materials**

Compound	Molecular formula	Density, g/mL	mp, °C	Micro hardness[a]	Transverse rupture strength, N/mm^{2b}	Compression strength, N/mm^{2b}	Modulus of elasticity, N/mm^{2b}	Heat conductivity, W/(cm·K)	Coefficient of thermal expansion, $\beta \times 10^{-6}$	Electrical resistivity, μΩ·cm
diamond	C	3.52	3800 dec	7600	~300	~2000	~900,000	1.14	0.9–1.18	10^{18}
boron carbide	B$_4$C	2.52	2450	2940	500	1800	450,000	0.27	6.0	10^{4}
silicon carbide	SiC	3.2	2300 dec	2580	<400^b	1400	480,000	0.15	5.7	10^{3}
beryllium carbide	Be$_2$C	2.42	2300	2690		740	350,000	0.21	7.4	10^{3}
alumina, sintered	Al$_2$O$_3$	3.9	2050	2080	<700^c		400,000	0.19	7.8	10^{18}
boron nitride, cubic	BN	3.45	2730^d	4700		3000	600,000			10^{16}
aluminum nitride	AlN	3.26	2250	1230			350,000			10^{13}
silicon nitride	Si$_3$N$_4$	3.44	1900	1700	<750^c		210,000	0.18	2.4	10^{16}
silicon boride	SiB$_6$	2.43	1950	2300	~100		330,000		6.3	10^{5}

[a] See HARDNESS.
[b] To convert N/mm^2 (MPa) to psi, multiply by 145.
[c] Hot-pressed.
[d] Transition from cubic to hexagonal ~1650°C.

to metal. The direction of electron transfer has been controversial, both theoretically and experimentally (10). Rigid-band models and tight-binding linear combination of atomic orbitals (LCAO) models suggest electron transfer from carbon to metal, which is supported by the known chemical behavior of the compounds and similarities in the shapes of the valence bands. However, augmented plane wave (APW) and related calculations suggest electron donation in the opposite direction, and this is confirmed experimentally by X-ray photoelectron spectroscopy and near-edge X-ray absorption fine structure spectroscopy (11). These difficulties may be resolved by considering a narrowing of the d-band in forming the later transition metal carbides (12). This results in greater occupation of the band, as deduced chemically, with simultaneous transfer of electrons from metal to nonmetal, as observed physically.

4. Diamondlike Carbides

This group of materials include the main group compounds silicon carbide and boron carbide (see SILICON CARBIDE BORON COMPOUNDS. Beryllium carbide, which has a high degree of hardness, can also be included (see BERYLLIUM COMPOUNDS). These materials have electrical resistivity in the range of semiconductors, and the bonding is largely covalent. Diamond itself may be considered a carbide of carbon because of its chemical structure, although its conductivity is low.

Table 2 summarizes the properties of the so-called nonmetallic hard materials, including diamond, the diamondlike carbides B_4C, SiC, and Be_2C. Also included in this category are corundum, Al_2O_3, cubic boron nitride, BN, aluminum nitride, AlN, silicon nitride, Si_3N_4, and silicon boride, SiB_6 (13).

BIBLIOGRAPHY

"Survey" under "Carbides" in *ECT* 3rd ed., Vol. 4, pp. 476–482, by R. Kieffer, Technical University of Vienna, and F. Benesovsky, Metallwerk Plansee A. G., Reulte, Austria; in *ECT* 4th, ed., Vol. 4, pp. 841–848, by S.Ted Oyama, Clarkson University and Richard Kieffer, Technical University of Vienna;"Carbides, Survey" in *ECT* (online), posting date: December 4, 2000, by S. Ted Oyama, Clarkson University and Richard Kieffer, Technical University of Vienna.

CITED PUBLICATIONS

1. S. T. Oyama, ed., *The Chemistry of Transition Metal Carbides and Nitrides*, Blackie Academic and Scientific, London, 1996.
2. H. W. Kroto, J. R. Heath, S. C. O'Brien, R. F. Curl, and R. E. Smalley, *Nature (London)* **318**, 162(1985).
3. B. C. Guo, K. P. Kearns, and A. W. Castleman, Jr., *Science* **255**, 1411(1992).
4. W. Reppe, *Acetylenchemie*, Verlag Chemie, Weinheim, 1951.
5. K. Fredenhagen and G. Cadenbach, *Z. Anorg. Allgem. Chem.* **158**, 249, 263(1926).
6. L. Toth, *Transition Metal Carbides and Nitrides*, Academic Press, New York, 1971.
7. G. Hägg, *Z. Physik. Chem.* **B6**, 221(1929); **B12**, 33(1931).

8. S. T. Oyama, *J. Solid State Chem.* **96**, 1(1992).

9. V. A. Gubanov, A. L. Ianovsky, and V. P. Zhukov, *Electronic Structure of Refractory Carbides and Nitrides*, Cambridge University Press, Cambridge, 1994.

10. S. T. Oyama and G. L. Haller, Vol. 5, in G. C. Bond and G. Webb, eds., *Catalysis, Specialist Periodical Reports*, The Royal Society of Chemistry, London, p. 333, 1982.

11. J. G. Chen, *Surf. Sci, Rep.* **30**, 1(1997).

12. C. D. Gelatt, Jr., A. R. Williams, and V. L. Moruzzi, *Conference on the Physics of Transition Metals*, Leeds, U.K. 1980.

13. F. Binder, *Radex Rdsch* 531(1975).

S. TED OYAMA
Virginia Polytechnic Institute
and State University

RICHARD KIEFFER
Technical University of Vienna

CARBIDES, CEMENTED

1. Introduction

Cemented carbides belong to a class of hard, wear-resistant, refractory materials in which the hard carbides of Group 4–6 (IVB-VIB) metals are bound together or cemented by a soft and ductile metal binder, usually cobalt or nickel. Although the term cemented carbide is widely used in the United States, these materials are better known internationally as hard metals (see also REFRACTORIES; REFRACTORY COATINGS; REFRACTORY FIBERS).

Cemented carbides were first developed in Germany in the early 1920s. The first cemented carbide to be produced was tungsten carbide [12070-12-1], WC, having a cobalt [7440-48-4], Co, binder (1). A number of scientific and technological advances provided impetus to development (2): (*1*) discovery of the high hardness of cast WC; (*2*) production of fine particles of WC, by reaction of the elements or by carburizing with hydrocarbons (qv); (*3*) application of sintering technology to the carbides; (*4*) lowering of the high sintering temperature of pure carbides by the use of a liquid phase of eutectic alloys of the iron-group metals; and (*5*) discovery of a unique combination of properties of the WC–Co alloys, including high compressive strength, high elastic modulus, abrasion resistance, toughness, and thermal shock resistance.

Over the years, the basic WC–Co material has been modified to produce a variety of cemented carbides containing WC–TiC, WC–TiC–TaC, WC–TiC–(Ta,Nb)C, WC–Mo_2C–TiC, and other solid solution carbides, which cover a wide range of applications including metalcutting, mining, construction, rock drilling, metal-forming, structural components, and wear parts. About 50% of all cemented carbide production is used for metal-cutting applications. Efforts to replace cobalt completely by nickel or iron in WC-based compositions have not been very successful, although partial replacement with nickel has been shown to offer benefits in certain applications (3).

Attempts to produce WC-free compositions for metal-cutting applications were made in the 1930s with the development of TaC—Ni, TiC—Mo$_2$C—Ni, and TiC—VC—Ni—Fe—Co alloys. But these alloys could not compete with the stronger WC—Co based cutting tools. However, in the 1950s, an understanding of the role of molybdenum in improving the wettability of titanium carbide [12070-08-5], to Ni binder brought the TiC—Mo$_2$C—Ni alloys closer in performance to WC—Co based tools in finish machining of steels. Further improvements in tool performance were obtained by additions of other carbides such as tantalum carbide [12070-06-3], TaC, and niobium carbide [12069-94-2], NbC, to the TiC—Mo$_2$C—Ni Ni alloys.

The first carbonitride alloys based on Ti(C,N)—Ni—Mo were introduced in 1970 followed by (Ti, Mo) (C,N) based compositions having fine microstructures that provided a balance of wear resistance and toughness (4). Continued research on the titanium carbonitride alloys in the 1980s led to the development of complex cermets having a variety of additives such as molybdenum carbide (2:1) [12069-89-5], Mo$_2$C, TaC, NbC, zirconium carbide [12020-14-3], ZrC, hafnium carbide [12069-85-1], HfC, WC, vanadium carbide [12070-10-9], VC, chromium carbide (3:2) [12012-35-0], Cr$_3$C$_2$, and aluminum, Al (5). Various mixes of these additives impart different combinations of wear resistance, thermal shock resistance, and toughness and allow tools to be tailored for a wide range of machining applications.

The binder metal, cobalt or nickel, is obtained as very fine powder and is blended with the carbide powders in ball mills, vibratory mills, or attritors using carbide balls. The mills are lined with carbide, low carbon steel, or stainless-steel sleeves. Intensive milling is necessary to deagglomerate the carbide particles, break up the initial carbide crystallites, and disperse the cobalt among the carbide particles to enhance wetting by cobalt during sintering. Milling is usually performed under an organic liquid such as alcohol, hexane, heptane, or acetone to minimize heating of the powder and to prevent its oxidation. In the milling process, a solid lubricant such as paraffin wax or polyethylene glycol/is added to the powder blend to impart strength to the pressed or consolidated powder mix. The lubricant provides a protective coating to the carbide particles and greatly reduces the oxidation of the powder. After milling, the organic liquid is removed by drying. In a spray-drying process, commonly employed in the cemented carbide industry, a hot inert gas such as nitrogen impinges on a stream of carbide slurry droplets and produces free-flowing spherical powder aggregates (see Fig. 1).

The milled and dried grade powders are pressed to desired shapes in hydraulic or mechanical presses. Special shapes may require a presintering operation followed by machining or grinding to the final form. Cold isostatic pressing, followed by green forming, is also common in the manufacture of wear-resistant components and metal-forming tools. Rods and wires are formed by the extrusion process. Complex parts can be formed by injection molding.

The pressed compacts are normally set on graphite trays and sintered. The vacuum sintering process consists of lubricant removal (400–500°C), oxide reduction (500–1000°C), rapid densification (1300–1350°C), and WC grain growth (1300–1500°C). During the final sintering operation, the cobalt–carbon eutectic melts and draws the carbide particles together, shrinking the compact by 17–25% on a linear scale and producing a virtually pore-free, fully dense product. In addition, the carbide grains coarsen due to solution and reprecipitation of the finer carbide particles in the eutectic melt.

In the 1970s, the cemented carbide industry adapted hot isostatic pressing (HIP) technology to remove any residual internal porosity, pits, or flaws from the sintered product (6). The HIP process involves reheating vacuum sintered material to 25–50°C less than the sintering temperature under an inert gas pressure of 100–150 MPa (14,500–21,750 psi). The sinter-HIP process (7), developed in the early 1980s, employs low pressure HIP, up to 7 MPa (1015 psi), after vacuum sintering. The pressure is applied at the sintering temperature when the metallic binder is still molten, resulting in void-free products. After sintering, cemented carbide products that require shaping to meet surface finish, tolerance, or geometry requirements undergo grinding with metal-bonded diamond wheels or lapping with diamond-containing slurries.

Separate cemented carbide product classifications exist for metal-cutting applications and wear parts (8). Classifications that are generally accepted by producers and users are available (9).

Inhalation of extremely fine carbide, cobalt, and nickel powders should be avoided. Efficient exhaust devices, dust filters, and protective masks are essential when handling these powders.

2. Recycling of Scrap

Recycling of cemented carbide scrap is of growing importance. The nonsintered scrap is reused in the milling of grade powders. The sintered cemented carbide scrap is recycled by several different processes. In one recycling method, the sintered scrap is heated to 1700–1800°C in a vacuum furnace to vaporize some of the cobalt and embrittle the material. After removal from the furnace the material is crushed and screened. In chemical recycling, the cobalt is removed by leaching, leaving carbide particles intact. In the zinc reclaim process, commercialized in the late 1970s, the cleaned scrap is heated with molten zinc in an electric furnace under inert gas at ∼800°C. The zinc reacts with the cobalt binder and the carbide pieces swell to more than twice their original volume. The zinc is distilled off in vacuum at 700–950°C and reclaimed. The treated carbide pieces are pulverized and screened to produce a fine powder. The cobalt is still present and there is no change in grain size from the original sintered scrap. The coldstream reclaim method employs a high velocity airstream to accelerate cemented carbide particles against a target surface with sufficient energy to fracture the particles. The coldstream process, so called because the air cools as it expands from the nozzles, is employed in combination with the zinc reclaim process.

3. Tool Failure Modes

Cutting tools are subjected to various failure modes during use. A primary failure mode is abrasive wear, which can be explained in terms of relative hardness

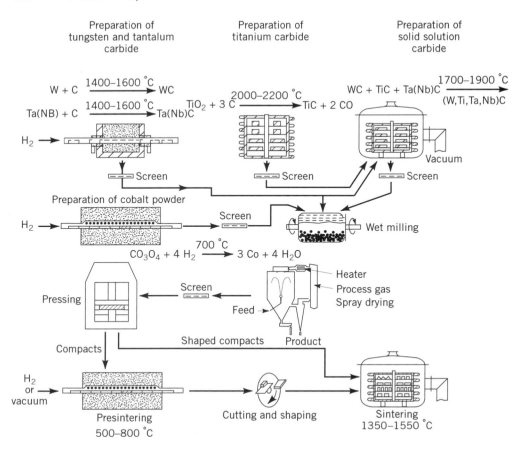

Fig. 1.   Preparation of cemented carbides where cobalt serves as the binding metal.

of the cutting tool and the workpiece material that is machined. In simple terms, a cutting tool must be harder than the material being cut. In addition to abrasive wear, other failure modes can occur due to high-localized tool tip temperatures ($\sim$1000°C) and high stresses ($\sim$700 MPa or $\sim$101,500 psi). High temperatures and stresses can cause blunting of the tool tip from plastic deformation, whereas high stresses can lead to catastrophic fracture. The workpiece may chemically interact with the tool material. The tool may also experience repeated impact loads during interrupted cuts. The useful life of the tool would depend on its response to the conditions existing at the tool tip. Figure 2 presents examples of tool failures often observed in metal-cutting operations.

3.1. Crater Wear.   Crater wear (Fig. 2**a**), observed on the rake or top face of cutting tools, generally occurs during machining of steels and ductile irons at high speeds. It is primarily caused by a chemical interaction between the rake face of a metal-cutting insert and the hot metal chip flowing over the tool. This interaction may involve diffusion or dissolution of the tool material into the chip.

3.2. Flank Wear.   Flank or abrasive wear (Fig. 2**b**) is observed on the flank or clearance face of a metalcutting insert or at the working end of wear

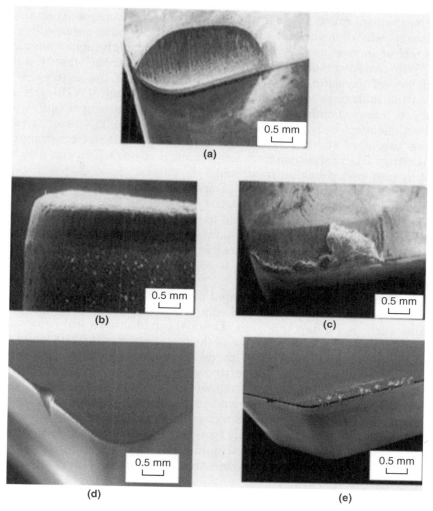

Fig. 2. Tool failure modes: (**a**) Crater wear on a cemented carbide tool produced during machining plain carbon steel. (**b**) Abrasive wear on the flank face of a cemented carbide tool produced during machining gray cast iron. (**c**) Built-up edge produced during low speed machining of a nickel-based alloy. (**d**) Depth-of-cut notching produced in nickel-based alloy machining, and (**e**) Thermal cracks on tool edge produced during face milling of alloy steel.

parts or mining tools and is related to the hard microstructural phases of the workpiece. Harder tool materials provide greater flank and abrasive wear resistance.

3.3. Built-up Edge. Under certain conditions, when ductile materials such as low carbon steels or aluminum are machined, the workpiece may weld to the tool tip as built-up edge, BUE (Fig. 2c). Workpiece build-up can lead to poor part finish and premature tool failure. During machining, the built-up material may also break off carrying with it small fragments of tool material. BUE formation can arise from a mechanical or chemical effect. Built-up edge from a mechanical cause refers to the adherence of the workpiece metal at relatively low speeds when the tool surface is rough. It can be suppressed by polishing the tool or increasing the tool tip temperature through higher metalcutting speed. With coated cemented carbide tools, BUE may occur when the coating layer flakes off exposing the carbide substrate. Even if the coating is intact, some coating materials such as chemical vapor deposition (CVD)–TiN show a tendency to react chemically with steel workpieces causing a built-up edge. BUE from chemical causes can be suppressed if the top TiN layer is removed to expose the more stable underlying Al_2O_3 during postcoat treatment. With rough coatings BUE can occur from a mechanical effect, which can also be suppressed by post-coat polishing treatment.

3.4. Depth-of-Cut Notching. Depth-of-cut notching (DOCN) is a localized tool wear process that occurs at the depth-of-cut line (Fig. 2**d**). Notching is a problem with workpieces that tend to work harden and generate high tool-tip temperatures, such as austenitic stainless steels or high temperature alloys. Notching is attributed to the chemical reaction of the tool material and the atmosphere, or to abrasion by the hard, sawtooth outer edge of the chip. DOCN can lead to tool fracture.

3.5. Thermal Cracks. Cemented carbide tools sometimes exhibit a series of cracks perpendicular to the tool edge when applied in interrupted cutting conditions such as milling (Fig. 2**e**). These thermal cracks are caused by the alternating expansion and contraction of the tool surface as it heats while cutting and then cools outside the cut. With prolonged intermittent cutting, lateral cracks may appear parallel to the cutting edge. The thermal and lateral cracks may join together and cause small fragments of tool material to break away.

Fracture is the least desirable mode of tool failure because it is unpredictable and catastrophic. The preferred tool failure mode is flank wear, because it progresses gradually and can be easily monitored for tool-changing protocol. Tool material development work is focused on minimizing flank wear and preventing unwanted tool failure modes such as catastrophic fracture, gross plastic deformation, BUE, DOCN, and crater wear.

4. Evaluation of Properties

In addition to chemical analysis, a number of physical and mechanical property evaluations are required to determine cemented carbide quality. Standard test methods employed by the industry for abrasive wear resistance, apparent grain size, apparent porosity, coercive force, compressive strength, density, fracture toughness, hardness, linear thermal expansion, magnetic permeability, microstructure, Poisson's ratio, transverse rupture strength, and Young's modulus are set forth by ASTM/ANSI and the ISO.

Among the physical properties, density is very sensitive to composition and porosity of the cemented carbide and is widely used as a quality control test. Magnetic properties most often measured are magnetic saturation and coercive force. Magnetic saturation provides an accurate measure of the carbon content in the cemented carbide alloy and is also used as a quality control test. The carbon content must be controlled within narrow limits to prevent the formation of a

(a)   (b)

Fig. 3. Microstructures of cemented carbides: (**a**) 90% WC—10% Co alloy, coarse grain. (**b**) 86% WC—7%(Ta,Ti,Nb)C—7% Co alloy, medium grain size. The gray angular particles are WC and the dark gray rounded particles are solid solution carbides. The white areas are cobalt binder.

brittle eta-phase, of composition Co_3W_3C or Co_6W_6C, at low carbon levels, or free graphite where carbon levels are high. For a WC-6 wt% Co alloy, the preferred carbon window is 0.09 wt%. Coercive force may vary considerably as sintering temperature increases and indicates the structural changes that take place during sintering. For a given cobalt and carbon content, the coercive force provides a measure of the size and distribution of the carbide phase in the microstructure.

The properties and performance of cemented carbide tools depend not only on the type and amount of carbide but also on carbide grain size and the amount of binder metal. Information on porosity, grain size and distribution of WC, solid solution cubic carbides, and the metallic binder phase is obtained from metallographically polished samples.

Optical microscopy and scanning and transmission electron microscopy are employed for microstructural evaluation. Figure 3 shows typical microstructures of cemented carbides.

Hardness, which determines the resistance of a material to abrasion and deformation, is affected not only by composition but also by porosity and microstructure. Higher cobalt content and larger carbide grain size reduce hardness and abrasion resistance but increase the toughness of cemented carbides. The trade-off between abrasion resistance and toughness enables the cemented carbide manufacturer to tailor these materials to a wide variety of metal-cutting and nonmetal-cutting applications.

Hardness is measured by the Rockwell A-scale diamond cone indentation test (HRA) or by the Vickers diamond pyramid indentation test (HV). Although the Rockwell scale has been used for decades in the carbide industry as a measure of hardness, a true indication of the resistance of the tool to deformation in metal-cutting operations can be obtained only by measuring hardness at elevated temperatures. Hot-hardness tests are performed using Vickers diamond pyramid indentors. The hardness of cemented carbides decreases monotonically with increasing temperatures.

Cemented carbides possess high compressive strength but low ductility at room temperature, but at temperatures associated with metal cutting these materials exhibit a small but finite amount of ductility. Measurement of yield strength is therefore more appropriate at higher temperatures. Like hardness, the compressive yield strength of cemented carbide decreases monotonically with increasing temperatures.

The most common method of determining the fracture strength of cemented carbides is the transverse rupture strength (TRS) test. A disadvantage of this test is the large scatter in the experimental data resulting from surface defects in the test specimens. Nevertheless, TRS is an excellent quality control test and it is particularly useful for large carbide components. A better measure of the intrinsic strength of the cemented carbide is the fracture toughness parameter, K_{Ic}, which indicates the resistance of a material to fracture in the presence of a sharp crack (10). The fracture toughness of carbide materials increases with cobalt content and carbide grain size but decreases with additions of cubic carbides.

Resistance to thermal shock is another important property that determines tool performance in milling. No laboratory test has yet been developed that can consistently predict the resistance to thermal shock of a tool. However, empirical parameters have been suggested that can be used to evaluate tool materials for their thermal shock resistance (11). A commonly used parameter is $\sigma k/E\alpha$, where σ is the transverse rupture strength, k is the thermal conductivity, E is the Young's modulus, and α is the coefficient of thermal expansion. In general, the higher the value of this parameter, the higher is the thermal shock resistance of the tool material.

Although hardness is a measure of abrasion resistance of cemented carbides, tool manufacturers use a wet-sand abrasion test to measure abrasion resistance directly. In this test, a sample is held against a rotating wheel for a fixed number of revolutions in a water slurry containing aluminum oxide particles. Abrasion resistance is reported as a relative ranking based on the reciprocal of volume loss of the material. Note that the abrasion resistance measured at room temperature does not correlate directly with high temperature wear resistance in metal-cutting operations.

5. Metal-Cutting Applications

5.1. Tools and Toolholding. Early carbide metal-cutting tools consisted of carbide blanks brazed to steel holders. When the brazed tool wore out, the carbides were reground and used again. Indexable inserts were introduced in the 1950s. In this configuration the so-called throwaway carbide insert is secured in the holder pocket by a clamp or some other holding device instead of a braze. When a cutting edge wears, a fresh edge is rotated or indexed into place. The main advantages of indexable inserts are ease of replacement, consistent positioning of the cutting edge relative to the workpiece, elimination of regrinding, and the ability to coat the inserts with the chemical vapor deposition (CVD) or the physical vapor deposition (PVD) processes. Indexable inserts also feature chipbreaker grooves, which not only control chip formation but also reduce the cutting forces. The edges of both brazed tools and indexable inserts are often modified, slightly rounded (honed) or chamfered, to prevent chipping and premature failure of a too sharp and therefore a weak cutting edge.

5.2. Compositions. For machining purposes, alloys having 5–12 wt% Co and carbide grain sizes from 0.5 to >5 µm are commonly used. By controlling the amount of Co and WC grain size, different combinations of wear resistance and toughness can be obtained. Powder metallurgical processing allows the manufacture of cemented carbides with any combination of raw materials to obtain the desired properties. The straight WC–Co alloys have excellent resistance to simple abrasive wear and are widely used in machining materials that produce short chips, eg, gray cast-iron, nonferrous alloys, high temperature alloys, etc and in a broad spectrum of industrial applications, including metalworking, metal and coal mining, transportation and construction industries. WC–Co alloys having submicrometer carbide grain sizes have been developed for applications requiring more toughness or tool edge strength. Such applications include indexable metal-cutting inserts and a wide variety of solid carbide drilling and milling tools. Grain refinement in these alloys is obtained by small (0.25–3.0 wt%) additions of TaC, NbC, VC, or Cr_3C_2 during carburization of tungsten or later in the powder blend.

5.3. Ultrafine-Grained WC—Co Alloys. Since the late 1980s there has been a tendency toward ultrafine-grained cemented carbides (WC grain sizes <0.5 µm) for woodworking tools, printed circuit board drills, and endmills. A number of factors have contributed to this development (12): (1) Observation that both hardness (abrasion resistance) and strength (transverse rupture strength) can be substantially improved with ultrafine grained carbide without seriously impacting bulk fracture toughness. (2) The hardness advantage over conventional cemented carbide is maintained up to 800–900°C. (3) Advent of sinter-HIP technology has allowed manufacture of ultrafine-grained carbides without pits and flaws. (4) The decrease in thermal conductivity with decreasing carbide grain size is more than compensated with increase in strength, so that the thermal shock resistance is not adversely impacted. (5) Advances in powder milling, granulation, and pressing technology have allowed the manufacture of ultrafine-grained cemented carbide possible. (6) Availability of advanced PVD and CVD coatings has enhanced tool performance and improved productivity. The use of ultrafine-grained carbides is expected to grow not only for woodworking tools, wear parts, and printed circuit board microdrills, but in metal cutting insert applications (milling of Al-Mg alloys, cast iron engine blocks, etc).

5.4. Cemented Carbides for Steel Machining. Straight WC–Co tools are not suitable for machining steels that produce long chips because they undergo crater wear from diffusion of WC into the steel chip surface. However, additions of cubic carbides such as TiC, TaC, and NbC impart chemical stability and crater wear resistance to WC–Co alloys. A disadvantage of TiC additions to WC–Co alloys is reduction in thermal shock resistance. However, TaC additions to WC–TiC–Co alloys can mitigate the deleterious effect of the TiC. Steel cutting compositions thus typically contain WC–TiC–(Ta,Nb)C–Co. It has been shown that cubic carbides are more effective in resisting crater wear and exhibit greater strength and hardness when they are added as preformed solid solutions such as WC–TiC (50:50) than when added as mixtures of WC and TiC (13). Other preformed solid solutions that are added to steel cutting compositions include TaC–NbC, WC–TiC–TaC, and WC–TaC(NbC). Tantalum carbide is often added as (Ta,Nb)C because the chemical similarity between TaC and NbC makes their separation expensive.

5.5. Coated Carbide Tools. Chemical vapor deposited TiC coatings were used in the late 1960s to combat wear on steel watch parts and cases. When applied to cutting tools, the relatively thin ($\sim$5 µm) TiC coatings extended tool life in steel and cast iron machining by a factor of two to three by suppressing the crater wear and flank wear (14). Hard coatings reduce frictional forces at the chip–tool interface, which in turn reduce the heat generated in the tool resulting in lower tool tip temperatures. Coatings also permit the use of higher cutting speeds boosting machining productivity. Currently, coated carbides account for nearly 75% of all indexable metal-cutting inserts used in the United States and CVD accounts for $\sim$70% of coated carbides. The rest are coated with PVD process.

In the CVD coating process, tools are heated in a sealed reactor with gaseous hydrogen at atmospheric or lower pressure; volatile compounds are added to the hydrogen to supply the metallic and nonmetallic constituents of the coating. For example, TiC coatings are produced by reaction of $TiCl_4$ vapors with methane (CH_4) and hydrogen (H_2) at 970–1000°C. The reaction is

$$TiCl_4 \ (g) + CH_4 \ (g) + H_2 \ (g) \longrightarrow TiC \ (s) + 4 \ HCl \ (g) + H_2 \ (g)$$

During the initial stage of the TiC deposition, a secondary reaction often occurs in which carbon is taken from the cemented carbide substrate. The resulting surface decarburization leads to the formation of a brittle eta phase and associated microporosity at the coating/substrate interface. The eta phase, in turn, can produce premature tool failure resulting from excessive chipping and reduced edge strength (15). However, improvements in CVD coating technology have resulted in coatings with greater thickness uniformity, more adherence, and more consistent morphology and microstructure having minimum interfacial eta phase and associated porosity (16).

Modern CVD coatings feature multiple layers involving various combinations of titanium carbonitride [12347-09-0], TiCN, titanium carbide [25583-20-4], TiC, titanium nitride [25583-20-4], TiN, zirconium cabronitride [25583-20-4], ZrCN, and alumina [25583-20-4], Al_2O_3, with controlled crystal structures (alpha-or kappa-alumina) and grain sizes optimized for various machining applications (Fig. 4). With better control of the CVD process, tool manufacturers can now offer coatings in a wide range of thickness (5–20 µm) with consistency and reproducibility. Multilayer CVD coatings with individual layers as thin as 0.2 µm are also available for machining tough workpiece materials (17).

The high temperatures employed during CVD coating generally ensure good interdiffusion bonding between the substrate and the coating. However, during cooling, the thermal expansion mismatch between the substrate and the coating can cause stresses that adversely affect coating adhesion. In certain cases coating stresses may be relieved by cracks that form in the coating. Because the thermal expansion coefficients of the coating materials (TiC, TiCN, TiN, and Al_2O_3) are higher than those of the WC–Co based substrates, CVD coatings are in residual tension at room temperature. Residual tensile

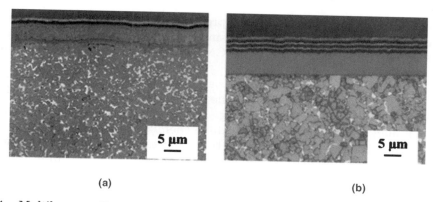

(a)   (b)

Fig. 4. Multilayer coatings on cemented carbide substrates: (**a**) 92% WC—8% Co with a TiN–TiCN–TiC–TiCN–Al$_2$O$_3$–TiN coating ~9 μm thick. (**b**) 86% WC—8%(Ta,Ti,Nb)C—6% Co with TiCN coating supporting multiple alternating coating layers of Al$_2$O$_3$ and TiN.

stresses are most severe at tool corners. To minimize stress, CVD-coated tools are honed at the cutting edge before coating.

In the mid-1980s a new CVD process for depositing TiCN was commercialized. Using a mixture of TiCl$_4$, H$_2$, and an organic compound such as acetonitrile, it was shown that TiCN can be deposited at moderate temperatures (800–900°C) at a faster deposition rate than the conventional CVD process (18). The reduced process temperature and faster deposition rate minimize the formation of the embrittling eta phase at the substrate-coating interface and reduce the thermally induced tensile cracks that are common to higher temperature CVD coatings (19). Tool manufacturers are currently offering TiCN and ZrCN coatings deposited by the medium temperature CVD (MTCVD) process. MTCVD TiCN coating has become an integral part of multilayer CVD coatings on carbide substrates (Fig. 5).

In another modification of the CVD technology called the plasma assisted CVD (PACVD) process, the deposition temperature is further reduced to

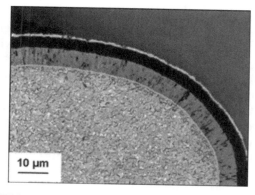

Fig. 5. Multilayer TiN–MTCVD–TiCN/Al$_2$O$_3$/TiN coating (~10 μm total thickness) >90% WC-4%Tac-6% Co substrate.

~600°C but the chemical reaction that produces the hard coating compounds is activated by the use of a plasma, comprising argon and reactive gases (20). Use of plasma assisted CVD coatings (TiN, TiCN, and Al_2O_3) on carbide substrates is still not widespread.

The metal-cutting performance of coated carbide tools is dependent not only on the properties of the substrate and the coating but on the interfacial bonding between the substrate and the coating. In recent years, significant advances have been made in this area. These include chemical as well as abrasive treatment of the substrate prior to the coating process. These precoat treatments improve the surface integrity and smoothness and enhance the adhesion of the coating. Significant advances have also been made in improving the interlayer adhesion through the use of special bonding layers. These new technologies have allowed tool manufacturers to offer thicker coatings with more functional layers.

A recent development in the area of hard coatings for cemented carbide tools is the low pressure synthesis of pure diamond by carbonaceous gas decomposition in presence of hydrogen atoms (21). This process is an alternative to the high pressure synthesis of bulk diamond, which is used to fabricate PCD (polycrystalline diamond) tips that are brazed on to cemented carbide tool inserts. The CVD diamond products are available either as thin films (<30 μm) deposited on carbide substrates or as thick sheets (typically >350 μm) brazed on to carbide substrates. Both products are used in machining abrasive, non-ferrous and nonmetallic workpiece materials and in particular, aluminum silicon alloys and aluminum metal matrix composites (low silicon aluminum alloys reinforced with SiC or Al_2O_3 particles), which are used primarily in automotive (brake rotors, drive shafts, etc) and aerospace (landing gear components) industries. Special pretreatments of the carbide substrate are required to ensure good adhesion of the diamond film. While the earlier diamond films featured coarser, faceted diamond surface, process advances now enable the tools to be produced with smoother diamond morphology with improved workpiece surface finish capabilities.

5.6. Coated Carbides with Functionally Gradient Substrate Microstructures.   A breakthrough in coated carbide cutting tools occurred in the late 1970s when a peripherally cobalt-enriched substrate (2–3 times higher cobalt at the tool insert periphery than in the bulk) was developed for a TiC–TiCN–TiN coated tool (Fig. 6a). The cobalt-enriched periphery was also slightly depleted in cubic carbides. The bulk of the tool insert had lower cobalt and higher level of cubic carbides. The combination provided superior edge strength while maintaining the edge and crater wear resistance of the coating layers (22). The high edge strength of this tool was attributed not only to the higher level of cobalt at the tool periphery but also to the striated morphology of the cobalt binder that prevented crack propagation and catastrophic fracture. This development permitted users to make heavy interrupted machining cuts such as those encountered in scaled forgings and castings at lower speeds. Further refinements to the cobalt-enrichment concept (1.5–2 times higher cobalt and complete depletion of cubic carbides in the tool periphery compared to the bulk and non-striated distribution of cobalt), in combination with Al_2O_3 coating (Fig. 6b), expanded the application range of this type of tool to higher speeds (23,24). Multilayer CVD coated tools with cobalt-enriched substrates can handle

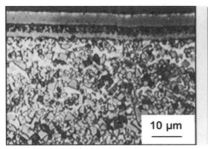 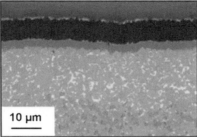

Fig. 6. Microstructure of a cemented carbide alloy: (**a**) 86% WC–8%(Ti,Ta,Nb)C–6% Co, with a cobalt-enriched periphery (first generation) and TiC–TiCN–TiN coating, and (**b**) 86% WC—8%(Ta,Ti,Nb)C—6% Co, with a second generation cobalt-enriched and solid solution carbide depleted periphery and TiCN–Al$_2$O$_3$–TiN coating.

medium to heavy roughing to semifinishing and even finishing operations on a wide range of workpiece materials such as cast irons, carbon, alloy, and stainless steels. The broad application range of cobalt-enriched tools enables them to cover a large percentage of the metalcutting operations of a plant giving the user an added level of tool edge security and performance consistency.

5.7. Postcoat Surface Treatments. One of the major problems in the performance of coated carbide tools is the susceptibility of coatings to flake in certain machining applications. This results in microchipping of the cutting edge and accumulation of workpiece build-up leading to poor part finish and premature tool failure. This problem can be largely suppressed by postcoat polishing treatments that either polish the insert edge or the entire tool surface. The top TiN coating layer is removed exposing the alumina layer underneath at the cutting edge (Fig. 7) or over the entire insert surface.

5.8. Physical Vapor Deposited Coatings. In the mid-1980s (PVD) emerged as a commercially viable process for applying hard TiN coatings onto cemented carbide tools. The coating is deposited in a vacuum, sustaining an

Fig. 7. Microstructure of an 87% WC—6%(Ta,Ti,Nb)C—7% Co cemented carbide substrate with a TiCN–Al$_2$O$_3$–TiN coating and postcoat treatment.

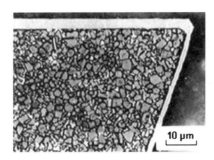

Fig. 8.   An example of PVD TiN coating on a sharp cemented carbide tool.

argon gas plasma, by condensation from a flux of neutral or ionized atoms where the metal species are derived from a variety of sources, including electron-beam evaporation, magnetron sputtering, and arc evaporation (25,26). The evaporated or sputtered metal or cation species (titanium, aluminum, chromium, etc) are made to react with the anions from gaseous species (N_2, CH_4, etc) introduced into the vacuum chamber. Since PVD coatings are deposited at low pressures (10^{-3}–10^{-2} torr), the atoms and molecules have long mean free paths and undergo fewer collisions, making PVD a line-of-sight deposition technique. This requires tool fixture rotation during the deposition process to ensure coating thickness uniformity on the faces of tool inserts. A number of factors make PVD process attractive for use with cemented carbide tools: (1) lower deposition temperature ($<550°C$) prevents eta-phase formation and produces finer grain sizes in the coating layer; (2) PVD coatings are usually crack-free; (3) depending on the deposition technique, compressive residual stresses, which are beneficial in resisting crack propagation, may be introduced in the coating (27); (4) PVD coating preserves the transverse rupture strength of the carbide substrate, whereas the CVD process generally reduces the TRS by as much as 30% (28); and (5) PVD coatings can be applied uniformly over sharp cutting edges (Fig. 8). This is desirable because it leads to lower cutting forces, reduced tool tip temperatures, and finer workpiece finishes. PVD coated tools are thus successfully employed in operations where sharp edges are most beneficial, including milling, turning, boring, drilling, threading, and grooving. Typical workpieces include carbon, alloy, and stainless steels, hardened steels, cast irons, nonferrous materials, and high temperature nickel-base alloys and titanium alloys. Newer PVD coatings are rapidly becoming commercially available (Fig. 9). These include TiCN, titanium aluminum nitride, which ranges in composition from Ti_2AlN [60317-94-4] to TiAlN, chromium carbide [12011-60-8], Cr_3C_2, chromium nitride [24094-93-7], CrN, titanium diboride, TiB_2, multiple alternating layers of TiN–TiAlN, and aluminum-rich AlTiN. Crystalline PVD Al_2O_3 coatings are subject to extensive R&D effort, but are not yet commercially available. Nano-layered PVD coatings (alternating very thin layers ~20 nm each, eg, TiN/AlN) have been recently introduced into the market. The properties of these coatings are dependent on the choice of the individual layers, including the thickness and crystallinity of each layer. It is possible to increase the hardness of the composite by a factor of 2 or more at the optimum nanolayer spacing compared to the hardness of either consituent alone

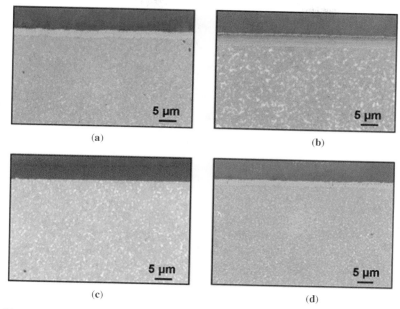

Fig. 9. Four generations of PVD coatings on cemented carbide substrates: (**a**) TiN, (**b**) TiN/TiCN/TiN, (**c**) TiAlN, and (**d**) TiB$_2$.

(29). Nano-composite thin films comprising two or more materials [eg, TiN or TiAlN nano-crystals, typically <4 nm, embedded in a matrix of amorphous Si$_3$N$_4$] with microhardness exceeding 40 GPa are also being developed (30).

As noted earlier, the wear behavior of coated tools and their metalcutting performance can be improved by precoat treatments. For PVD coatings, it has been demonstrated by indentation test that microblasting the tool surface with fine grit alumina or water peening can significantly improve the bonding of PVD coating to the substrate (31).

5.9. Solid Lubricant Coatings. More recent developments in hard coatings for cemented carbide cutting tools include solid lubricant coatings (hard coatings with a low coefficient of friction), eg, amorphous metal–carbon, Me–C:H and soft coatings (eg, MoS$_2$, pure graphite, or WC/C) deposited on a hard coating layer such as PVD TiAlN (31) (see Fig. 10). These coatings provide enhanced metalcutting performance, notably in drilling and tapping of steels and aluminum alloys by resisting chip adhesion to the tool and aiding chip evacuation.

5.10. CVD–PVD Combination Coatings. The late 1980s saw the development of a coating technology, in which an outer layer of PVD TiN was combined with CVD TiN/TiCN inner layers. The development of this technology was based on the observation that the inner CVD layers provide wear resistance and excellent adhesion to the substrate, and the outer PVD layer offers a hard, fine-grained, crack-free, smooth surface endowed with compressive residual stresses. When combined with a cobalt-enriched substrate with good bulk

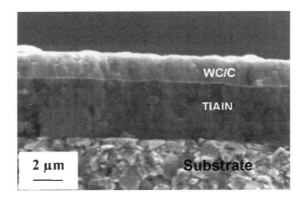

Fig. 10. TiAlN–WC/C coating on carbide substrate (Courtesy: Balzers Inc.)

deformation resistance, the CVD–PVD combination coating has been found to provide improved tool performance in milling of steel workpieces at relatively high speeds (32). Other CVD–PVD combinations involving PVD TiAlN are now commercially available.

5.11. Metalcutting Tool Design. Modern metalcutting tool insert development requires a systems approach that involves (1) substrate design for optimum deformation resistance, fracture resistance, and thermal shock resistance, (2) coating design for wear resistance, lubricity, reduced frictional heat, and workpiece surface finish requirements, (3) macro-geometry or chip-groove design for chip control, and (4) micro-geometry or edge sharpness to control cutting force and surface finish, and to minimize edge chipping.

6. Nonmetal-Cutting Applications

Today almost one half of the total production of cemented carbides is used for nonmetal-cutting applications such as mining, oil and gas drilling, transportation and construction, metalforming, structural and fluid-handling components, and forestry tools. The majority of compositions used in these applications comprise straight WC–Co grades. In general, cobalt contents vary from 5 to 30 wt% and WC grain sizes range from <1 to <8 μm and sometimes up to 30 μm. Extensive discussion of hard metals employed in nonmachining applications is available (33–36).

Metal-forming applications include drawing dies, rolls for hot and cold forming of strips and bars, cold heading dies, forward and back extrusion punches, swaging hammers and mandrels, and can body punches and dies. Applications requiring high impact strength employ grades with 11–25 wt% Co. When wear resistance is of paramount importance, grades having lower cobalt content and finer grain size are suitable choices. When gall resistance, ie, resistance to metal pickup on the tool, is needed, alloy carbides such as (W,Ti)C and (Ta,Nb)C are used. Corrosion resistance applications typically employ grades having finer WC grain sizes and lower cobalt contents or combinations of nickel, cobalt, and chromium. In metal-forming applications, hard coatings, particularly by PVD, can enhance the wear resistance of carbide tools further although the use of coatings is not as widespread as in metal-cutting applications.

The impetus for the synthesis of WC and subsequent development of cemented carbides came from the wire drawing industry where the hard metals are still used. The most commonly used grade is WC—6 wt% Co with medium grain size (1–2 μm). Compositions having higher cobalt content are used in drawing tubes, rods, and bars.

Alloys having 15–30 wt% Co and very coarse WC grain sizes (up to 20 μm) have replaced steel rolls in the production of hot-rolled steel rods. Carbide rolls are also well suited for the cold reduction and finishing of strip products in Sendzimir mills where rigidity and dimensional stability are particularly important. The compositions used in these applications have a 5.5 wt% Co and medium WC grain size (1–2 μm).

The high abrasion resistance and edge strength of carbides make them ideal for use as slitter knives for trimming steel cans and stainless and carbon steel strips, cutting abrasive materials in the paper, cellophane and plastic industries, and for slitting magnetic tapes for audio, video, and computer applications. Carbides having submicrometer grain sizes and 6–10 wt% Co offer sharp cutting edges, good surface finish, and high edge strength in these applications.

Cold-forming equipment such as extrusion or heading punches and dies are made from cemented carbides to produce a variety of parts such as wrist pins, spark plug shells, bearing retainer cups, and propeller shaft ends. Generally, WC—12 wt% Co alloy is used for back extrusion punches and a WC—16 wt% Co grade is recommended for extrusion dies. Submicrometer carbides may also be used for punches. In cold-heading applications involving the manufacture of nuts, bolts, and screws the dies have to withstand considerable stress and repeated impacts and must therefore possess good fatigue strength. Alloys with 20–30 wt% Co or Co—Ni—Fe are required.

The high elastic modulus, compressive strength, and wear resistance of cemented carbides make them ideal candidates for use in boring bars, long shafts, and plungers, where reduction in deflection, chatter, and vibration are concerns. Metal, ceramic, and carbide powder-compacting dies and punches are generally made of 6 and 11 wt% Co alloys, respectively. Another application area for carbides is the synthetic diamond industry where carbides are used for dies and pistons.

The rigidity, hardness, and dimensional stability of cemented carbides, coupled with their resistance to abrasion, corrosion, and extreme temperatures, provide superior performance in fluid-handling components such as seal rings, bearings, valve stems and seats, and nozzles. In the transportation and construction industry, steel tools having cemented carbide cutting tips are used for road planing, soil stabilization, asphalt reclamation, vertical and horizontal drilling, trenching, dredging, tunnel boring, forestry, and for snowplow blades, tire studs, and street sweeper skids.

Cemented carbides play a crucial role in the recovery of metallic ores and nonmetals by underground or open-pit mining practices, recovery of minerals such as coal, potash, and trona, and drilling for oil and gas. The methods of excavation can be broadly classified into three types: rotary drilling, roto-percussive drilling, and flat-seam underground mining. In the oil and gas drilling industry tungsten carbide buttons, having 10–15 wt% Co, are used in steel drill bodies for deep penetration of metamorphic, igneous, and sedimentary rocks.

Cemented tungsten carbides also find use as a support for PCD cutting tips, or as a matrix alloy with cobalt, nickel, copper, and iron, in which diamond particles are embedded. These tools are employed in a variety of industries including mineral exploration and development; oil and gas exploration and production; and concrete, asphalt, and dimension stone cutting.

7. Economic Aspects

Cemented carbide inserts and tools for metal cutting and metal working have traditionally accounted for the largest percentage of carbide industry sales. However, carbide tool consumption in nonmetal-working fields, notably in the construction and transportation industries, has grown rapidly. On the other hand, the demand for primary materials has been somewhat reduced by use of recycled cemented carbide scrap.

There are >200 cemented carbide producers in the world. A majority of hard metal production can be attributed to Cerametal Sarl, Iscar Ltd., Kennametal Inc., Mitsubishi Materials Corporation, Plansee Tizit GmbH, Sandvik AB, Sumitomo Electric Industries Ltd., Allegheny Technologies Inc., Toshiba Tungaloy Co. Ltd., and Zhuzhou Cemented Carbide Industry Company. Many of the smaller producers have narrow manufacturing capabilities and a limited range of product offerings.

Developments in materials, coatings, and insert geometries have claimed an increasing share of research and development budgets in the cemented carbide industry. Important economic benefits of these effects have been an increase in tool performance and significant increase in metal-cutting productivity. Continuing developments in computer numerically controlled machining systems have placed a heavy emphasis on tool reliability and consistency, which in turn puts pressure on the industry to invest increasing amounts of capital in developing new materials and processes.

BIBLIOGRAPHY

"Cemented Carbides" under "Carbides" in *ECT* 2nd ed., Vol. 4, pp. 92–100, by R. Kieffer, University of Vienna, and F. Benesovsky, Metallwerke Plansee A.G.; in *ECT* 3rd ed., Vol. 4, pp. 483–489, by R. Kieffer, Technical University, Vienna, and F. Benesovsky, Metallwerke Plansee A.G.; in *ECT* 4th ed., Vol. 4, pp. 848–860, by A. T. Santhanam, Kennametal, Inc.; "Carbides, Cemented" in *ECT* (online), posting date: December 4, 2000, by A. T. Santhanam, Kennametal Inc.

CITED PUBLICATIONS

1. U.S. Pat. 1,549,615 (Aug. 11, 1925), K. Schroter (to General Electric Co.).
2. R. Kieffer, N. Reiter, and D. Fister, *BISRA–ISI Conference on Materials for Metalcutting*, Scarborough, U.K., 1970, p. 126.

3. E. A. Almond and B. Roebuck, *Mater. Sci. Eng.* **A105**/106, 237 (1988).
4. E. Rudy, S. Worcester, and W. Elkington, *High Temp. High Pressures* **6**, 447 (1974).
5. H. Doi, "Science of Hard Materials," *Proceedings of the 2nd International Conference on the Science of Hard Materials*, Rhodes, Sept. 23–28, 1984, Ser. No. 75, Adam Hilger Ltd., Bristol, U.K., 1986, pp. 489–523.
6. E. Lardner and D. J. Bettle, *Metals Mater.* **7**, 540 (1973).
7. R. C. Lueth, *Refract. Hard Metal J.* **4**, 87 (1985).
8. K. J. A. Brookes, *World Directory and Handbook of Hardmetals and Hard Materials*, 6th ed., International Carbide Data, Pub. East Barnet Hertfordshire, U.K., 1996.
9. ISO Recommendation R-513, *Application of Carbides for Machining by Chip Removal*, 1st ed., International Organization for Standardization, Nov. 1966.
10. J. R. Pickens and J. Gurland, *Mater. Sci. Eng.* **33**, 135 (1978).
11. W. D. Kingery, H. K. Bowen, and D. R. Uhlmann, *Introduction to Ceramics*, 2nd ed., John Wiley & Sons, Inc., New York, 1960, p. 828.
12. G. Gille, B. Szesny, K. Dreyer, H. van den Berg, J. Schmidt, T. Gestrich, and G. Leitner, *Intl. J. Refractory Metals Hard Mater.* **20**, 3 (2002).
13. U.S. Pat. 2,113,353 (Apr. 5, 1938), Philip M. McKenna.
14. U.S. Pat. 3,832,221 (Aug. 27, 1974), C. S. Ekmar.
15. W. Schintlmeister, O. Pacher, K. Pfaffinger, and T. Raine, *J. Electrochem. Soc.* **123**, 924 (1976).
16. V. K. Sarin and J. N. Lindstrom, *J. Electrochem. Soc.* **126**, 1281 (1979).
17. K. Narasimhan and W. C. Russell, *Proceedings of the 14th Plansee Seminar*, Metallwerk Plansee A.G., Reutte/Tyrol, 1997, pp. 290–303.
18. M. Bonetti-Lang, R. Bonetti, H.E. Hintermann, and D. Lohmann, *Int. J. Refractory Hard Metals* **1**, 161 (1982).
19. R. S. Bonetti, H. Wiprachtiger, and E. Mohn, *Metal Powder Rep.* **45**, 837 (1990).
20. R. Tabersky, H. van den Berg, and U. Konig, in E. Broszeit, ed., *Plasma Surface Engineering*, Vol. 1, DGM, Germany, 1989, p. 133.
21. Haubner and B. Lux, *Diamond Related Mater.* **2**, 1277 (1993).
22. B. J. Nemeth, A. T. Santhanam, and G. P. Grab, *Proceedings of the Tenth Plansee Seminar*, Metallwerk Plansee A.G., Reutte/Tyrol, 1981, pp. 613–627.
23. A. T. Santhanam, G. P. Grab, G. A. Rolka, and P. Tierney, *Proceedings of the Conference on High Productivity Machining—Materials and Processes*, New Orleans, La., American Society for Metals, 1985, pp. 113–121.
24. U.S. Pat. 4,610,931 (Sept. 9, 1986), B. J. Nemeth and G. P. Grab.
25. R. F. Bunshah, ed., *Deposition Technologies for Films and Coatings: Developments and Applications*, Noyes Publications, 1982.
26. W.D. Sproul, *Cutting Tool Engineering*, CTE Publications, Inc., 1994, p. 52.
27. D. T. Quinto, A. T. Santhanam, and P. C. Jindal, *Mater. Sci. Eng.* **A105**/106, 443 (1988).
28. G. J. Wolfe, C. J. Petrosky, and D. T. Quinto, *J. Vac. Sci. Technol.* **A4**(6), 2747 (1986).
29. X. Chu, A. Barnett, M. S. Wong, and W. D. Sproul, *Surface and Coatings Tech.* **57**, 13 (1993).
30. S. Veprek, *Surface and Coatings Tech.* **97**, 15 (1997).
31. F. Klocke and T. Krieg, *Coated Tools for Metalcutting—Features and Applications*, ANNALS-CIRP **48**, 515 (1999).
32. U.S. Pat. 5,250,367 (Oct. 5, 1993) A. T. Santhanam, R. V. Godse, G. P. Grab, D. T. Quinto, K. E. Undercoffer, and P. C. Jindal.
33. J. Larsen-Basse, *Powder Metall.* **16**(31), 1 (1973).
34. G. E. Spriggs and D. J. Bettle, *Powder Metall.* **18**(35), 53 (1975).
35. E. Lardner, *Powder Metall.* **21**(2), 65 (1978).
36. W. E. Jamison, in M. B. Peterson and W. O. Winer, eds., *Wear Control Handbook*, American Society of Mechanical Engineers, New York, 1980, pp. 859–998.

GENERAL REFERENCES

H. E. Exner, *Int. Met. Rev.* **24**(4), 149–173 (1979).

ASM Committee on Tooling Materials, *Superhard Tool Materials, Metals Handbook*, Vol. 3, 9th ed., 1980, pp. 448–465.

K. J. A. Brookes, *World Directory and Handbook of Hardmetals and Hard Materials*, 6th ed., International Carbide Data, East Barnet Hertfordshire, U.K., 1996.

B. North, "Indexable Metalcutting Inserts—A Review of Recent Developments," *Proceedings of the 1st International Conference on the Behavior of Material in Machining*, Nov. 8–10, 1988, Stratford-upon-Avon, The Institute of Metals, Paper 35, 1988.

G. Schneider, Jr., *Principles of Tungsten Carbide Engineering*, 2nd ed., Society of Carbide and Tool Engineers, ASM International, Materials Park, Ohio, 1989.

A. T. Santhanam, P. Tierney, and J. L. Hunt, *Metals Handbook, Properties and Selection*, Vol. 2, 10th ed., 1990, pp. 950–977.

E. M. Trent, *Metal Cutting*, 3rd ed., Butterworth-Heinemann Ltd., Oxford, U.K., 1991.

A. T. Santhanam and D. T. Quinto, *ASM Handbook*, Vol. 5, Surface Engineering, 1994, pp. 900–908.

A. T. Santhanam
Kennametal Inc.

CARBIDES, INDUSTRIAL HARD

1. Introduction

The four most important carbides for the production of cemented carbides (a.k.a. hard metals) are tungsten carbide [12070-12-1] (WC), titanium carbide [12070-08-5] (TiC), tantalum carbide [12070-06-3] (TaC), and niobium carbide [12069-94-2] (NbC). The binary and ternary solid solutions of these carbides such as WC–TiC and WC–TiC–TaC (NbC) are also of great importance. Chromium carbide (3:2) [12012-35-0] (Cr_3C_2), molybdenum carbide [12011-97-1] (MoC), and molybdenum carbide (2:1) [12069-89-5] (Mo_2C), vanadium carbide [12070-10-9] (VC), hafnium carbide [12069-85-1] (HfC), and zirconium carbide [12020-14-3] (ZrC), have minor significance. Carbides and their solid solutions (the latter also herein referred to as mixed crystals) are generally combined with cobalt and used in the form of cemented carbides.

2. Preparation

In general, the carbides of metals of Groups 4–6 (IVB–VIB) are prepared by reaction of elementary carbon or hydrocarbons and metals and metal compounds at high temperatures. The process may be carried out in the presence of a protective gas, under vacuum, or in the presence of an auxiliary metal (menstruum).

2.1. Carburization by Fusion. This method is used for the production of cast tungsten carbide eutectic alloy, chiefly for deposition by welding onto many types of machinery wear-surfaces, eg, on rotary rock-drill bits used in the petroleum, mining, and construction industries. Using rapid induction heating of tungsten metal powder and carbon under hydrogen in graphite crucibles, a fused eutectic (~4%C) is cast and chill-cooled and the mass crushed to produce preferred granular size ranges for use as filler for weldable hardfacing products.

$$3\,W + 2\,C \xrightarrow[\text{H}_2\text{C}]{2800^\circ\text{C}} W_2C - WC$$

2.2. Carburization by Thermal Diffusion.

Carburization of chemically processed metal or metal-compound powders is carried out through solid-state, thermal diffusion processes, either in protective gas or vacuum. Carbide solid solutions are prepared by the same methods. Most carbides are made by these processes, using loose or compacted mixtures of carbon and metal or metal-oxide powders. Halides of Group 5 (VB) metals recovered from ores by chlorination are similarly carburized.

$$W + C \xrightarrow[\text{H}_2]{1400-1600^\circ\text{C}} WC \tag{1}$$

$$Ta(H) + C \xrightarrow[\text{H}_2,\text{vacuum}]{1400-1500^\circ\text{C}} TaC + (H)$$

$$W + CH_4 \xrightarrow[\text{H}_2]{1400-1600^\circ\text{C}} WC + 2\,H_2$$

$$WO_3 + 4\,C \xrightarrow[\text{H}_2,\text{CO}]{1400-1600^\circ\text{C}} WC + 3\,CO \tag{2}$$

$$TiO_2 + 3\,C \xrightarrow[\text{H}_2,\text{CO,vacuum}]{1800-2000^\circ\text{C}} TiC + 2\,CO$$

$$Nb_2O_5 + 6\,C + CH_4 \xrightarrow[\text{H}_2]{1400-1700^\circ\text{C}} 2\,NbC + 5\,CO + 2\,H_2$$

2.3. Carburization by Menstruum Process.

The P. M. McKenna method of carburization (1) involves the use of mineral concentrates such as wolframite [1332-08-7] [Fe(Mn)WO$_4$] and microlite [12173-96-5] (Ca$_2$Ta$_2$O$_7$), ferroalloys such as iron tungstide (FeW) or high purity scrap metals in a high-temperature melt of auxiliary (menstruum) metal or metals, with carbon. Upon cooling, carbide crystals are dispersed throughout the metallic mass. The mass is then crushed and the carbide crystals isolated by dissolving the menstruum alloy in mineral acids. Further purification by elutriation follows.

$$Fe(Mn)WO_4 + 5\,C + Fe \xrightarrow{2300-2600^\circ\text{C}} WC + Fe(Mn) + Fe + 4\,Co + C$$

$$FeW + 2\,C \xrightarrow{2300-2600^\circ\text{C}} WC + Fe + C$$

$$TiO_2 + Fe + 4\,C \xrightarrow{2300-2600^\circ\text{C}} TiC + Fe + 2\,CO + C$$

$$FeW + Ti + Fe + 2\,C \xrightarrow{2300-2600^\circ\text{C}} (W,Ti)C + 2\,Fe + C$$

$$Ca_2Ta_2O_7 + Fe + 13\,C \xrightarrow{2300-2600^\circ\text{C}} 2\,TaC + Fe + 7\,CO + 2\,CaC_2 + C$$

2.4. Carburization by Exothermic Thermochemical Reaction. Carburization of tungsten contained in tungsten mineral concentrates is accomplished by means of a simultaneous aluminothermic reduction and carburization (2,3). The reaction produces, upon cooling, a metallic mass in which tungsten carbide crystals are dispersed in a menstruum alloy. After crushing, WC crystals in the mass are isolated by dissolving the menstruum either in acidic ferric chloride solutions or mineral acids, followed by elutriation. The process has, since the 1960s, become an important source of primary WC for the manufacture of cemented hard carbides and other metallurgical products.

$$6\,Fe(Mn)WO_4 + 22\,Al + 3CaC_2 + 3\,Fe_3O_4 \xrightarrow{2500-3000°C}$$

$$6\,WC + 6\,Fe(Mn) + 9\,Fe + 3\,CaO + 11\,Al_2O_3$$

Titanium carbide may be prepared by a thermochemical reaction between finely divided carbon and titanium metal powder. The reaction proceeds exothermically.

$$Ti + C \xrightarrow{1700-2200°C} TiC$$

Additionally, titanium carbonitride (TiCN) for use in cermets may be produced exothermically in a wide range of TiC–TiN solid solutions by reacting TiC and Ti powder-blends or powder compacts in low concentrations of nitrogen in vacuum. For example,

$$Ti + N \xrightarrow[N_2,vacuum]{1500-1700°C} TiN$$

$$TiC + Ti + N \xrightarrow[N_2,vacuum]{1500-1700°C} TiCN$$

The use of titanium carbonitrides in Ti-based cermets marks a significant improvement in metal-cutting performance over TiC.

2.5. Reduction. Reduction of halides using hydrogen–hydrocarbon mixtures is sometimes done in the presence of a graphite carrier or using metals possessing high melting points, ie, the van Arkel gas deposition method (4). If a plasma gun is employed, finely powdered (<1 μm) carbides are obtained (5) (see PLASMA TECHNOLOGY).

A number of hard materials, including TiC, Ti(C,N), Zr(C,N), TiN, and alumina, (Al$_2$O$_3$), are produced by reduction from metal-halide precursors as thin coatings (3–20 μm) directly onto metalcutting insert surfaces by chemical vapor deposition (CVD), in gaseous atmospheres containing various combinations of nitrogen, hydrocarbon gases, eg, methane, and oxygen, at ~1000°C. These coatings reduce cutting temperatures and the various abrasive and chemical wear processes, increase tool life, and allow higher machining speeds, thereby increasing metalcutting productivity. Titanium nitrides, TiN, titanium carbonitrides, TiCN, titanium aluminum nitrides, TiAlN, titanium diborides, TiB$_2$, and chromium nitrides, CrN are also deposited, with tool life and metalcutting productivity benefits, by the physical vapor deposition (PVD) process. The following typify the CVD process:

$$TiCl_4 + CH_4 \xrightarrow[H_2]{1000°C} TiC + 4\ HCl$$

$$TiCl_4 + CH_4 + 1-2\ N_2 \xrightarrow[H_2]{1000°C} TiCN + 4\ HCl$$

$$TiCl_4 + 1-2\ N_2 + 2\ H_2 \xrightarrow[H_2]{1000°C} TiN + 4\ HCl$$

$$2\ AlCl_3 + 3\ H_2O \xrightarrow[H_2]{1000°C} Al_2O_3 + 6\ HCl$$

3. Tungsten Carbide

Traditionally, tungsten ore is chemically processed to ammonium paratungstate [1311-93-9], [$(NH_4)_{10}W_{12}O_{41} \cdot 5H_2O$] and tungsten oxides ($W_xO$). These compounds are then hydrogen-reduced to tungsten [7440-33-7] metal powder (see TUNGSTEN AND TUNGSTEN ALLOYS; TUNGSTEN COMPOUNDS). The fine tungsten powders are blended with carbon and heated in a hydrogen atmosphere between 1400 and 1500°C to produce tungsten carbide particles having sizes varying from 0.5 to 30 μm. Each particle is composed of numerous tungsten carbide crystals. Small amounts of vanadium, chromium, or tantalum are sometimes added to tungsten and carbon powders before carburization to produce very fine (<1 μm) WC powders.

The characteristics of WC, especially grain size, are determined by purity, particle shape and grain size of the starting material, and the conditions employed for reduction and carburization. The course of the reaction $WO_3 \rightarrow W \rightarrow WC$ is dependent on temperature, gas-flow rates, water-vapor concentration in the gas, and the depth of the powder bed. All these factors affect the coarsening of the grain.

Selection of suitable tungsten-containing raw materials and modification of the reduction and carburization conditions permit the preparation of the WC powder in various grain sizes. The following examples are illustrative: (1) very pure tungstic acid of fine (<0.1 μm) grain size is reduced in dry hydrogen at ~800°C, and the fine (<0.5 μm) tungsten powder obtained gives very fine (<1 μm) grained WC on carburization at 1350–1400°C; (2) calcined coarse (<2 μm) WO_3, after reduction with hydrogen in the presence of water vapor at 900–950°C, provides tungsten powder (<6 μm) from which coarse (<10 μm) crystalline WC powder can be prepared by carburization at 1600°C.

The WC leaving the furnace is light gray with a bluish tinge. It is generally caked and must be broken up, milled, and screened before use. It should contain ~6.1–6.25 wt% total C, of which 0.03–0.15 wt% is in the free, unbound state. The theoretical C-content is 6.13 wt%. The great bulk of WC powder used in the manufacture of cemented carbides exceeds ~2 μm in average particle size before conversion to powder grades by milling. During the sintering process, which follows milling and compacting, WC crystal growth occurs in the semiliquid binder alloy, during which the largest crystals may reach 10 μm or more, while the underlying ground mass may range from 1 to 10 μm, depending on the grade composition and the intended application. At the high particle size end of this broad spectrum lie coarser initial WC particle sizes, used in the manufacture of coarser sintered microstructures for mining, construction, and transportation tools. At the lower particle-size end, finer initial WC particle sizes, generally 1–2 μm, enter the picture, yielding finer sintered microstructures (eg, ∼1–6 μm WC crystals), designed to provide combinations of both high strength and high flank-wear resistance, especially for metalcutting and metalforming applications.

In recent years, efforts to achieve higher levels of both sintered strength and wear resistance, using an initial submicron WC particle size typically in the <1-μm range, and sometimes in "ultrafine" or "nanosized" ranges, typically <0.4-μm average particle size, have resulted in much finer sintered microstructures together with increases in transverse rupture strengths. Special WC powder processing controls necessary for the production of initial WC in submicron sizes are demanding. Many control parameters in WC-powder processes influence initial WC particle size and particle size range. They include the characteristics of tungsten-precurser materials, such as ammonium paratungstate and its reduction parameters to form tungsten metal powder, the purity and stream velocity of hydrogen or other gases used in reduction or carburizing furnaces, temperature profiles used in both, the gas–solid contact efficiency in processing furnaces, and choices of processing equipment. Equipment and process departures made by some manufacturers to produce submicron particle sizes include, inter alia, a rotary carburizing furnace with a carbon-rich atmosphere, gaseous

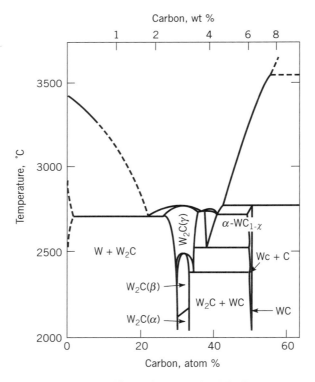

Fig. 1. Phase diagram for W–C.

fluid-bed carburizers, a chemical vapor reaction (CVR) process, fluid-bed gas carburization of dry-powder mixtures of cobalt and tungsten, and a continuous interconnected tungsten reduction and carburization complex. At the same time, enormous increases in the specific surface of WC powder, which must accompany reductions in particle size into the submicron ranges, introduce problems related to powder flow, compacting, sensitivity to adsorption of ambient gases, and other surface related behavior.

The P. M. McKenna aluminothermic process proceeds from tungsten–mineral concentrates, scheelite, wolframite or ferberite, or mixtures of them, to produce tungsten carbide crystals and crystal fragments ranging in size from ~840–44 µm. Macrocrystalline WC, as grown in a menstruum alloy, forms well-developed, angular crystals having a triangular habit. The crystals always contain a perfectly stoichiometric bound-carbon content and are monocrystalline. Both in initially coarse form, and after size-reduction by milling, macrocrystalline WC has comparatively low specific surface and is entirely free of W_2C. Powders are prepared over a wide range of particle sizes, from granular screen-sized ranges to micron-sized powders.

Both macrocrystalline WC and the fused $WC–W_2C$ eutectic (Fig. 1) are important in the manufacture of diamond drill bits used in the mining, oil and gas, and construction industries, and in hardfacing rods and electrodes. The properties of WC are listed in Table 1.

Table 1. **Physical Properties of Primary Carbides**

Property	WC	TiC	TaC	NbC
mol wt	195.87	59.91	192.96	104.92
carbon, wt%	6.13	20.05	6.23	11.45
crystal structure	hex, Bh	fcc, B1	fcc, B1	fcc, B1
lattice constants, nm	$a = 0.29065$ $c = 0.28366$	0.43305	0.4454	0.4470
density, g/cm^3	15.7	4.93	14.48	7.78
mp, °C	2720	2940	3825	3613
microhardness, kg/mm^2	1200–2500	3000	1800	2000
modulus of elasticity, N/mm^{2a}	696,000	451,000	285,000	338,000
transverse rupture strength, N/mm^{2b}	550–600	240–400	350–450	300–400
coefficient of thermal expansion, K^{-1}	$a = 5.2 \times 10^{-6}$ $c = 7.3 \times 10^{-6}$	7.74×10^{-6}	6.29×10^{-6}	6.65×10^{-6}
thermal conductivity, $W/(m \cdot k)$	121	21	22	14
heat of formation, $\Delta H_{f, 298}$, kJ/mol^c	−40.2	−183.4	−146.5	−140.7
specific heat, $J/(mol \cdot)c$	39.8	47.7	36.4	36.8
electrical resistivity, $\mu\Omega \cdot cm$	19	68	25	35
superconducting temperature, $<K$	1.28	1.15	9.7	11.1
Hall constant, $cm^3/(A \cdot s)$	-21.8×10^{-4}	-15.0×10^{-4}	-1.1×10^{-4}	-1.3×10^{-4}
magnetic susceptibility	+10	+6.7	+9.3	+15.3

^{a} Face-centered cubic = fcc and hexagonal = hex.

^{b} To convert N/mm^2 (MPa) to psi, multiply by 145.

^{c} To convert J to cal, divide by 4.184.

4. Titanium Carbide

On an industrial scale, TiC is produced most often through the reaction of TiO_2 with carbon black (see TITANIUM AND TITANIUM ALLOYS; TITANIUM COMPOUNDS).

In industrial production of titanium carbide, pure (99.8%, with minor impurities of Si, Fe, S, P, and alkalies) titanium oxide [13463-67-7] (TiO_2), in the dry or wet state is mixed in 68.5/31.5 ratio with carbon black or finely milled low ash graphite. The dry mixture is pressed into blocks that are heated in a horizontal or vertical carbon-tube furnace at 1900–2300°C; hydrogen that is free of oxygen and nitrogen serves as protective gas. In the vertical push-type furnaces, the liberated CO itself provides protection.

Titanium carbide is generally obtained in the form of gray, well-sintered lumps that are broken up in jaw-crushers and fine-milled in ball mills. Technical-grade TiC contains 0.5–1.5 wt% graphite, in addition to 0.5–1 wt% oxygen and nitrogen and 0.1 wt% impurities, such as Fe, Si, S, and P. The oxygen and nitrogen content may be reduced to 0.1–0.3 wt% by heating the impure carbides under high vacuum for several hours at 2000–2500°C, or less expensively, by formation of solid WC solutions, with or without the use of Mo_2C or Cr_3C_2.

Titanium carbide is also prepared by the menstruum method (6), starting with ferrotitanium (TiO_2), or titanium [7440-32-6] (Ti), metal. Comparatively low levels of oxygen and nitrogen, 0.1 and 0.3 wt%, respectively, are achieved without double processing because of strong outgassing under high temperature liquid menstruum alloy. Menstruum-made TiC strongly reflects the cubic crystal form, initially ranging in size from ~149 down through 44 μm. Among TiC produced by various processes, the combined carbon content of mestruum-process unmilled crystals at 19.7% (theoretical 20.05 wt%) is comparatively high, while free carbon content at 0.1 wt% is comparatively low. Oxygen content of TiC made by the carbothermic process is slightly higher than that for menstruum-made TiC, while free carbon at ~0.6 wt% is higher and combined carbon at 19.0 wt% minimum is lower. As is true with all of the major primary monocarbides and notably in the case of TiC, the closeness of combined carbon content to theoretical carbon content is a useful single-value surrogate indicator of purity and hardness. Properties of titanium carbide are listed in Table 1; the Ti–C phase diagram is shown in Figure 2.

The relatively high hardness and the cubic lattice of titanium carbide, the latter providing strong solvent power for forming solid solutions (mixed crystals) with WC (as well as with NbC and TaC), give it high importance in cemented carbide metal-cutting compositions. With WC as solute, TiC forms a continuous mixed-crystal series up to a maximum of ~25 TiC–75 WC wt%, which is a saturated end member sometimes called $WTiC_2$. In steel machining applications TiC imparts crater wear resistance, generally entering such compositions in the form of a WC–TiC mixed crystal rather than as TiC. As TiC added in any form to cemented carbide compositions decreases transverse rupture strength and thermal shock resistance, its use is generally limited to a maximum of ~10 wt%. After the advent of hard coatings, the modern cemented carbide substrates generally have a maximum of ~2 wt% TiC. Titanium carbide, together with titanium carbonitride, TiCN, are the dominant hard phases in uncoated

Carbon, wt %

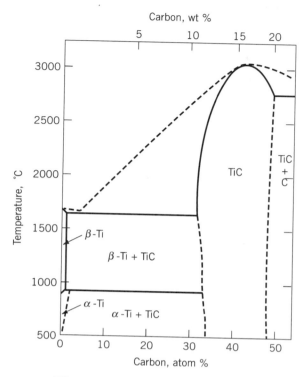

Fig. 2. Phase diagram for Ti–C.

cermet alloys, added for their high chemical stability in metalcutting tools and corrosion-resistant wear and heat-resistant parts.

5. Tantalum Carbide

On an industrial scale, TaC is prepared from tantalum [7440-25-7] (Ta) metal or tantalum hydride [12026-09-4], (Ta_2H, powder), tantalum pentoxide [1314-61-0], Ta_2O_5, high purity scrap obtained in the preparation of ductile Ta, or from ferrotantalum–niobium. The chemical and metallurgical industry produces refined Ta_2O_5 from tantalite mineral concentrate and sometimes from columbite mineral concentrate, and to a lesser extent from microlite mineral concentrate, the latter mineral from the higher tantalum end of the calcium tantalate–calcium niobate solid-solution comprising the microlite–pyrochlore solid-solution mineral series. Until the mid-1980s and the collapse in tin prices, accumulations of Ta- and Nb-containing slags from Southeast Asian tin smelters were a significant resource for Ta (and Nb) oxide recovery, however are now of little importance. The ores are decomposed and separated by fractional crystallization of the double fluorides, fractional distillation of the pentachlorides, or by solvent extraction of the HF-containing solutions. The metals are prepared mainly by alkali metal reduction or fusion electrolysis of the halides.

Tantalum carbide is produced by carburization of the element or the oxide with carbon, in a manner similar to the preparation of WC or TiC. Final carburization in a vacuum gives a golden yellow carbide, free of oxygen and nitrogen, that contains 6.1–6.3 wt% C and 0–0.2 wt% graphite.

The McKenna menstruum carburization process (7) for preparation of TaC parallels that for menstrum-made WC. The golden colored crystals produced have a specific gravity of ~14.2 and a bound-carbon content of 6.2–6.3 wt%. Starting material may be low niobium tantalite or microlite concentrates or high purity tantalum scrap. Low levels of niobium carbide in solid solution with TaC occur when tantalite is used and is of economic benefit, as NbC is widely used as a partial substitute for TaC in cemented carbide manufacture. The properties of TaC are given in Table 1; the Ta–C phase diagram is shown in Figure 3.

WC–TiC–Co grades originally developed for machining long-chip materials have been largely replaced by WC–TiC–TaC(NbC)–Co grades that show higher hot-hardness and better thermal-shock resistance. This, as well as the addition of small (0.5–2 wt%) quantities of TaC to straight WC–Co alloys to prevent grain growth, brought TaC to a position of importance (among the hard carbides), particularly in thermal shock resistant compositions for steel machining. The major area of tantalum metal consumption is electronic components, notably capacitors, and its uses are many and are growing.

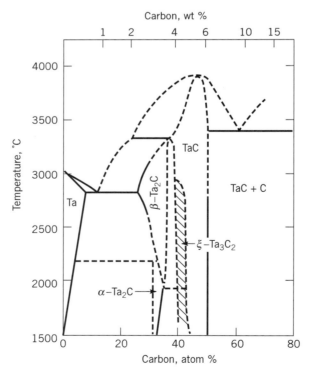

Fig. 3. Phase diagram for Ta–C.

6. Niobium Carbide

The preparation of niobium [7440-03-1], Nb, metal and niobium pentoxide [1313-96-8], Nb_2O_5, is very similar to that of tantalum (see NIOBIUM AND NIOBIUM COMPOUNDS). Niobium carbide is prepared most often by the carburization of Nb_2O_5 with carbon black, and less frequently by reaction of niobium and carbon. The preparation of NbC has special importance because it is used on an industrial scale as a reducing agent to prepare niobium metal. In this process, finely divided mixtures of Nb_2O_5 and carbon black are pressed into cylindrical blocks in large hydraulic presses and converted to NbC in high frequency furnaces in the presence of hydrogen or under vacuum at 1600–1800°C. The carbide may then be mixed with further quantities of the oxide in induction-heated vacuum furnaces and processed to technical-grade niobium metal containing 1–3 wt% O_2, 0.5–1 wt% free graphite.

The menstruum method for niobium carbide (7) may use mineral concentrates of columbite, a niobium-dominant member of the continuous mineral solid-solution series of iron–manganese niobate, or columbite, and iron–manganese tantalate, or tantalite. Since pure end members of the series are rare, the carbide made from columbite is a mixed crystal rich in NbC with low levels of TaC. The production of NbC essentially free of TaC may use ferroniobium as a starting material, of which large quantities are produced in Brazil from calcium niobate, i.e. the mineral pyrochlore, for the production of niobium steels. Pyrochlore lies at the niobium-rich end of a continuous mineral solid-solution series of calcium niobate, or pyrochlore, and calcium tantalate, or microlite. Mixed crystals of NbC carrying subordinate levels of TaC are used in steel-cutting compositions as partial replacements of TaC or TaC-rich TaC–NbC mixed crystals. Such partial replacements of TaC by NbC may result in savings in raw material costs as well as in prior chemical separation of Ta and Nb oxides.

7. Auxiliary Carbides

7.1. Chromium Carbide. Cr_3C_2, the chromium carbide having the highest carbon level, is used as an additive in the preparation of cobalt or nickel-cemented WC or TiC-based carbide alloys designed for corrosion-resistant applications (see CORROSION AND CORROSION CONTROL). Lower carbon forms, eg, chromium carbide (7:3) [12075-40-0], (Cr_7C_3), are not suitable for these purposes. However, Cr_3C_2 is unstable in cobalt or nickel cemented alloys, tending to react with binder metals to produce brittle binary carbides. Lower carbon chromium carbides, however, are useful in reducing binder attack in corrosive applications. Chromium carbide is also used in cemented alloys as a grain-growth inhibitor.

Chromium carbide can be best prepared from pure chromic oxide [1308-38-9], (Cr_2O_3) (see CHROMIUM COMPOUNDS). Compacts containing 74 wt% Cr_2O_3 and 26 wt% carbon black can be heated in carbon-tube furnaces at 1600°C in the presence of hydrogen, giving a carbide containing 13–13.3 wt% total C and 0.1–0.3 wt% free C.

Table 2. **Physical Properties of Auxiliary Carbides**

Property	Cr_3C_2	β-Mo_2C	η-MoC	VC	HfC	ZrC
mol wt	180.05	203.91	107.96	62.96	190.51	103.23
carbon, wt%	13.33	5.89	11.3	19.08	6.30	11.64
crystal structure[a]	rhom, $D5_{10}$	hex, L'3	hex, L'3	fcc, B1	fcc, B1	fcc, B1
lattice constants, nm	$a = 1.147$ $b = 0.554$ $c = 0.283$	$a = 0.3$ $c = 0.4734$	$a = 0.298$ $c = 0.281$	0.4165	0.4648	0.4698
density, g/cm^3	6.68	9.18	9.15	5.36	12.3	6.46
microhardness, kg/mm^2	1350	1500	2200	2900	2600	2700
modulus of elasticity, N/mm^{2b}	373,000	533,000		422,000	352,000	348,000
mp, °C	1810	2520	2600	2684	3820	3420
coefficient of thermal expansion, K^{-1}	10.3×10^{-6}	7.8×10^{-6}		7.2×10^{-6}	6.59×10^{-6}	6.73×10^{-6}
heat of formation $\Delta H_{f, 298}$, kJ/mol^c	−94.2	−49		−124.8	−230.3	−196.8
specific heat, $J/(mol \cdot K)c$	32.7	30.3		32.3	37.4	37.8
electrical resistivity, $\mu\Omega \cdot cm$	75	71		60	37	42
superconducting temperature, <K	<1.2	2.78		<1.2	<1.2	<1.2
Hall constant, $cm^3/(A \cdot s)$	-0.47×10^{-4}	-0.85×10^{-4}		-0.48×10^{-4}	-12.4×10^{-4}	-9.42×10^{-4}
magnetic susceptibility				+28	−25.2	−23

[a] rhom = rhombohedral, hex = hexagonal, fcc = face-centered cubic.
[b] To convert N/mm^2 (MPa) to psi, multiply by 145.
[c] To convert J to cal, divide by 4.184.

Chromium carbide is important in powder preparations designed for thermal spray applications of corrosion and wear-resistant coatings on tool and machine parts. Lower carbon carbides of chromium are important in hard-facing rods and electrodes for weld-applied overlays on machine wear surfaces. However, these carbides are usually formed *in situ* from Cr and C in the rod and not added as preformed carbides. The properties of Cr_3C_2 are listed in Table 2.

7.2. Molybdenum Carbide. Mo_2C can be prepared by the carburization of molybdenum trioxide [1313-27-5] (MoO_3), and molybdenum dioxide [18868-43-4] (MoO_2), with carbon black or, more conveniently, by the reaction of molybdenum [7439-98-7], Mo, powder (93.4 wt%) and carbon black or charcoal (~6.6 wt%) at 1350–1500°C, in the presence of hydrogen (see MOLYBDENUM AND MOLYBDENUM ALLOYS). The carbide formed contains 5.9–6.1 wt% total C and 0.05–0.25 wt% free C. The physical properties are listed in Table 2. There are two molybdenum carbides having higher carbon contents, ie, the cubic MoC_{1-x}, a high-temperature phase often described as Mo_3C_{2-x}, and the hexagonal η-molybdenum carbide [12011-97-1] (MoC), a low temperature phase. Both carbides tend to decompose to graphite and β-molybdenum carbide (2:1) [12069-89-5], (Mo_2C). The latter can be stabilized by W (8), N (9), or W + N (10) so that (Mo,W)C (11) or (Mo,W)(C,N) (10) mixed crystals are formed.

Ferromolybdenum additives to the menstruum method for the production of WC crystals stabilize molybdenum monocarbide, MoC, in a (W,Mo)C mixed crystal. In much the same manner, but with ferrotitanium additives to the menstruum process for TiC, MoC is stabilized in a (Ti,Mo)C mixed crystal. The lower carbon molybdenum carbide, Mo_2C, finds its chief use as an additive to nickel or Ni–Co bonded TiC and TiCN based cermet compositions. Mo_2C, unstable in Ni or Ni–Co binder, reacts with the carbide phase to increase the bond strength between the binder and carbide or carbonitride phases.

7.3. Vanadium Carbide. Vanadium pentoxide [1314-62-1] (V_2O_5), or vanadium trioxide [1314-34-7] (VO_3), are the most satisfactory oxides for the preparation of VC. Vanadium pentoxide is best prepared by igniting chemically pure ammonium vanadate [7803-55-6] (NH_4VO_3), in the presence of moist oxygen to avoid reaction with nitrogen; V_2O_3 is obtained by reduction of V_2O_5 with hydrogen (see VANADIUM COMPOUNDS).

Vanadium carbide is prepared by the reaction of the elements under vacuum. In this process V_2O_5 is reduced at a high temperature with calcium [7440-70-2] and the product is melted in an arc furnace in the presence of argon producing a 99.9% pure material. Vanadium powder of equal purity may be prepared by hydriding and crushing vanadium metal turnings. Vacuum carburization removes nitrogen and oxygen, which are generally present up to 0.5–1 wt% in carbides obtained from vanadium oxides. The properties of VC are given in Table 2.

Although VC is very hard, it is very brittle and has, therefore, been used only in special cemented carbides. For example, submicrometer straight WC–Co alloys are prepared using ~0.5 wt% VC as a grain–growth inhibitor. TiC–VC–Ni–Fe cemented carbides were used in Germany during World War II as tungsten carbide–cobalt-free cutting tool alloys. Small quantities of VC are sometimes used in TiCN–Ni based cermets.

7.4. Hafnium Carbide. The minerals, chiefly zircon, ie, zirconium silicate, and baddeleyite, ie, zirconium oxide, are the sole industrial resources of hafnium metal, a minor constituent of both minerals at a ratio of 1 part hafnium to ~50 parts zirconium. The carbide of hafnium is mutually miscible with the carbides of Zr and Ti, the metal's Group 4 (IVB) comembers. A similar solid-solution relationship between Zr and Hf applies in the mineral forms noted above, placing any possibility of mineral separation at the preliminary ore-dressing stage beyond reach. The need of pure zirconium [7440-67-7] for nuclear reactors prompted the large-scale separation of hafnium [7440-58-6] from zirconium. This in turn made sufficient quantities of hafnium dioxide [12055-23-1] (HfO_2), or Hf metal sponge available for production of HfC for use in cemented carbides (see HAFNIUM AND HAFNIUM COMPOUNDS).

Hafnium carbide can be prepared industrially from hydrided hafnium sponge at 1500–1700°C or from HfO_2 at 2000–2200°C by carburization in vacuum in the presence of hydrogen. The resulting carbide contains almost the theoretical quantity of carbon, 6.30 wt% C, of which a maximum of 0.1 wt% is unbound.

The properties of HfC are listed in Table 2. Addition of HfC as NbC–HfC or TaC–HfC solid solutions to WC–TiC–Co alloys has been shown to improve the hardness (12).

7.5. Zirconium Carbide. ZrC may be prepared by igniting a mixture of 78.8 wt% annealed zirconium oxide [1314-23-4] (ZrO_2) and 21.2 wt% charcoal under hydrogen in a graphite crucible in a carbon-tube furnace at 2400°C. The carbide obtained has the following composition: 11.3 wt% chemically bound C, traces of free C, 88.3 wt% Zr, and 0.3 wt% ($O_2 + N_2$). Alternatively, a pressed mixture of ZrO_2 and carbon black may be induction heated in a graphite crucible in the presence of H_2 at 1800°C, and then comminuted and annealed at 1700–1900°C, under vacuum, after addition of 1–2 wt% carbon black. The product contains 11.8 wt% C (~0.5 wt% free C).

The physical properties of ZrC are listed in Table 2. While zirconium carbide (the "red carbide"), as well as zirconium carbonitride, could effectively replace titanium carbide and carbonitride in steel-cutting compositions, their use in cemented carbides has so far not been wide.

7.6. Solid Solutions of the Major Hard Carbides. Essentially straight WC–Co alloys are of paramount importance in manufacturing industries for machining short-chipping materials, as well as for a host of tools and wear parts across a broad spectrum of industrial applications, including metalworking, metal and coal mining, transportation and construction industries. At the same time, the cubic carbides TiC, TaC, and NbC are relied upon to impart thermal stability and crater resistance to cemented carbide metalcutting tools, most effectively when added to metalcutting compositions as preformed solid solutions (mixed crystals). However, WC, among all carbides, supplies the preeminent platform for a wide variety of tools, owing to its superior bonding strength with cobalt and Co–Ni binders, its comparatively good fracture toughness among the carbides, and its superior thermal conductivity. However, WC is less refractory than the cubic carbides, additions of which, in the form of preformed solid solutions of preferred combinations, make up for the deficiency in machining applications that generate high tool tip temperatures. While the use of preformed solid solutions does not entirely replace random reactions between the carbides, which occur in any case during sintering, it greatly reduces dependence on *in-situ*, intercarbide reactions, thus carrying the desired intercarbide reactions closer to completion than would otherwise be the case.

Preformed solid solutions of the major carbides used for metalcutting applications fall mainly into two categories: those between the solute, WC, and one or more of the solvent cubic carbides, and a number of preferred combinations of two or more cubic carbides, without WC. The former case often prevails, as TiC, the strongest solute among the cubic carbides for WC, forms the most widely used solid solution series for steel machining applications, ranging mainly from WC–TiC 50:50 to WC–TiC 75:25. TaC–NbC mixed crystals are commonly preformed, as also TaC–NbC–TiC formulations to preferred rations, as well as WC–TiC–TaC(NbC), WC–TaC(NbC), and (W,Mo)CN. In addition to providing closer approaches to compositional equilibrium in the form of sintered carbides, these solid solutions carry the additional advantages over monocarbides of improved strength and hardness.

Solid solutions are produced by thermal diffusion of blends of the primary carbides, which themselves have been formed initially either by metal–carbon thermal diffusion or by reduction-diffusion reactions from pure oxides. In a third stage, the resulting preformed solid solution powders are then annealed in vacuum furnaces, with hydrogen, in order to attain acceptable levels of intercarbide reactions as well as to achieve low levels of residual (ie, unreacted or free) carbon, oxygen, and nitrogen. Low levels of cobalt or other binder metals are sometimes added to such preformed solid solution powders prior to annealing, as a means of promoting the diffusion of residual carbon and of lowering oxygen and nitrogen.

Alternatively, preformed carbide solid solutions produced by the menstruum method proceed directly from blends of mineral concentrates, or from ferroalloys or from blends of these, to macrocrystalline mixed crystals. By this method, blends of mineral concentrates containing natural association of Ta and Nb oxides with usually minor Ti, and, if WC is desired in the crystal, tungsten mineral concentrates, are the precursor materials, without prior chemical separation of the primary metal oxides or the subsequent formation of separate monocarbides. The same general procedure applies to the production of members of the TiC–WC solid solution series, in which high-temperature menstruum melts provide complete solid solutioning and low levels of oxygen, nitrogen, and residual free carbon in a single operation. Precursor materials for the menstruum-made WC–TiC solid solution series may be blends of tungsten mineral concentrates and rutile (TiO_2) mineral concentrates. Carbide solid solutions of metals in Groups 4 (IVB), 5 (VB), and 6 (VIB) are thus produced by both the thermal diffusion and menstruum methods.

Carbonitride solid solutioning is illustrated by TiC–cermet history. Cermet metalcutting tools and corrosion-resistant components were for some years based on Ni-bonded TiC, strengthened by additions of Mo, either as metal powder or as molybdenum carbide, Mo_2C, which during sintering reacted with TiC to form Ti(Mo)C at the surfaces of TiC crystals, thereby strengthening the TiC–Ni bond. Subsequent development of titanium nitride, TiN, led to both partial and total replacement of TiC by Ti(C,N), reinforced by additions of Mo_2C, which yielded a sintered microstructural refinement and high hardness, attributed both to nitrogen and solid solutioning effects. Wider use of cermets followed Ti carbonitride developments. Production is by diffusion annealing of TiC and TiN in vacuum furnaces or by vacuum heating of TiC and Ti with nitrogen.

Carburization.   Metal oxide mixtures with carbon black having additives such as Co, Ni, Fe, or Cr(0.5–1%) to promote diffusion, may undergo carburization.

$$WO_3 + TiO_2 + C \xrightarrow[\text{H}_2,\text{CO}]{1600-1800^\circ\text{C}} (W, Ti)C + CO$$

$$WO_3 + TiO_2 + Ta_2O_5 + C \xrightarrow[\text{H}_2]{1600-1800^\circ\text{C}} (W, Ti, Ta)C + CO$$

$$MoO_3 + WO_3 + C + (Fe, Ni, Co) \xrightarrow[\text{H}_2,\text{CO,CH}_4]{1000-1200^\circ\text{C}} (Mo, W)C + CO + (Fe, Ni, Co)$$

Metal powder mixtures with carbon black may also undergo carburization.

$$W + Ta + C \xrightarrow[\text{H}_2,\text{vacuum}]{1500-1600^\circ\text{C}} (W, Ta)C$$

$$Hf + Ta + C \xrightarrow[\text{H}_2,\text{vacuum}]{1600-1800^\circ\text{C}} (Hf, Ti)C$$

Diffusion Annealing.   Mixed preformed carbides can be diffusion annealed at temperatures giving solid solutions. Additives, such as Co, Ni, Fe, or Cr (0.5–1%) promote diffusion.

$$WC + TiC \xrightarrow[\text{H}_2,\text{vacuum}]{1600-1900^\circ\text{C}} (W, Ti)C$$

$$WC + TiC + TaC(NbC) \xrightarrow[\text{H}_2,\text{vacuum}]{1600-1800^\circ\text{C}} (W, Ti, Ta, Nb)C$$

$$HfC + TiC + WC \xrightarrow[\text{H}_2,\text{vacuum}]{1600-1900^\circ\text{C}} (Hf, Ti, W)C$$

$$TiN + TiC \xrightarrow[\text{vacuum}]{1600-1800^\circ\text{C}} Ti(C, N)$$

Menstruum Carburization.   Mineral concentrates, ferroalloys, primary metals, or high purity scrap may also be carburized (13).

$$Fe(Mn)WO_4 + TiO_2 + Fe + C \xrightarrow{2300-2600^\circ\text{C}} (W, Ti)C + Fe(Mn) + Fe + CO + C$$

$$WFe + TiFe + C \xrightarrow{2300-2600^\circ\text{C}} (W, Ti)C + Fe + C$$

$$(Fe, Mn)(Ta, Nb)_2O_6 + Fe + C \xrightarrow{2300-2600^\circ\text{C}} (Ta, Nb)C + Fe(Mn) + Fe + CO + C$$

The monocarbides of Groups 4 (IVB) and 5 (VB) metals are completely miscible except for ZrC–VC and HfC–VC (Fig. 4). At 1400–1500°C, WC is soluble in the carbides of Groups 4 (IVB) and 5 (VB) 25–60 wt%, and at higher (1800–2400°C) temperatures even up to 90 wt% (14,15). WC itself, like the other

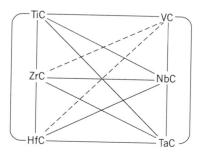

Fig. 4. Schematic illustrating the solid solubility between carbides. Solid line = complete solubility; dashed line = limited solubility.

carbides of Group 6 (VIB), has practically no solubility for face-centered cubic (fcc) carbides. The cubic solid solutions, which are saturated at higher temperatures and contain a high percentage of WC, are very stable. The hexagonal WC forms complete series of solid solutions with the hexagonal MoC.

The pseudoternary system WC–TiC–TaC is especially important in the metallurgy of cemented carbides (16). Figure 5 shows the phase distribution and thus the solubility ratios at 1450°C, which is the typical cemented carbide sintering temperature; at 2200°C, which is the preferred temperature for the formation of pure solid solutions having high WC content; and at 2500°C showing a hypothetical curve, maximum solubility. It is obvious that TiC is a better solvent for WC than TaC, ie, TaC additives in ternary solid solutions reduce the solubility for WC. The more spherical TiC–TaC–WC solid-solution grains can be readily distinguished from the angular WC in the microstructure.

7.7. Carbides of the Actinides, Uranium, and Thorium. The carbides of uranium and thorium are used as nuclear fuels and breeder materials for gas-cooled, graphite-moderated reactors (see NUCLEAR REACTORS). The actinide

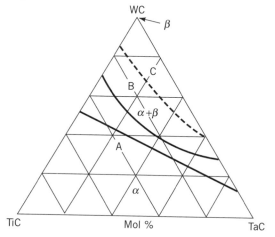

Fig. 5. Phase distribution in the system WC–TiC–TaC at A, 1450°C; B, 2200°C; and C, 2500°C. Dashed line = hypothetical solubility; solid line = experimental solubility.

Table 3. **Physical Properties of Uranium and Thorium Carbides**

Property	UC	UC_2	ThC	ThC_2
mol wt	250.08	262.09	244.06	256.07
carbon, wt%	4.8	9.16	4.92	9.37
crystal structure[a]	fcc, B1	tetr, C11a	fcc, B1	mon
lattice constants, nm	0.49597	$a = 0.3524$	0.5346	$a = 0.6691$
		$c = 0.5996$		$b = 0.4231$
				$c = 0.6744$
				$\beta = 103°50'$
density, g/cm^3	13.63	11.86	10.64	8.65
microhardness, kg/mm^2	920	620	850	600
mp, °C	2560	ca 2500	2625	2655
heat of formation, $\Delta H_{f,298}$, kJ/mol[b]	−97.1	−96.3	−29.3	−125.2
specific heat, J/(mol·K)[b]	50.2	58.6		
electrical resistivity, $\mu\Omega \cdot cm$	40	90	25	30
magnetic susceptibility	+3.15	+3.40		

[a] tetr = tetragonal; mon = monoclinic cubic.
[b] To convert J to cal, divide by 4.184.

carbides are prepared by the reaction of metal or metal hydride powders with carbon or preferably by the reduction of the oxides uranium dioxide [1344-57-6] (UO_2), triuranium octaoxide [1344-59-8] (U_3O_8), or thorium dioxide [1314-20-1] (ThO_2) at 1800–2200°C in carbon-tube furnaces in the presence of hydrogen or in vacuum furnaces. Hot pressing and arc melting are very suitable methods for the preparation of homogeneous compacts, especially if followed by heat treatment in a tungsten-tube furnace in the presence of argon.

The properties of the uranium carbide [12070-09-6] (UC), uranium carbide (1:2) [12071-33-9], thorium carbide [12012-16-7] (ThC), and thorium carbide (1:2) [12071-31-7] (ThC_2), are given in Table 3; the phase diagram for the system U−C is shown in Figure 6. Coefficient of thermal expansion for UC is $9.1 \times 10^{-6} K^{-1}$, and its thermal conductivity is 25 W/(m · K).

Uranium carbide (UC) is comparatively stable whereas UC_2, especially in powder form, hydrolyzes rapidly in moist air. The latter is used in the form of pellets or annealed spherical particles, coated with pyrographite, as a nuclear fuel for high temperature reactors. For breeder reactors using thorium [7440-29-1], it should be noted that UC and ThC form a continuous series of solid solutions, whereas UC_2 and ThC_2 have limited mutual solubility. Furthermore, UC can be stabilized with regard to its carbon content, even at high temperatures, by the formation of solid solutions with ZrC, HfC, NbC, or TaC so that no higher carbides are formed (17). Uranium carbides and plutonium carbides show a high degree of mutual miscibility (see ACTINIDES AND TRANSACTINIDES).

7.8. Carbides of the Iron Group Metals. The carbides of iron, nickel, cobalt, and manganese have lower melting points, lower hardness, and different structures than the hard metallic materials. Nonetheless, these carbides, particularly iron carbide and the double carbides with other transition metals, are of great technical importance as hardening components of alloy steels and cast iron.

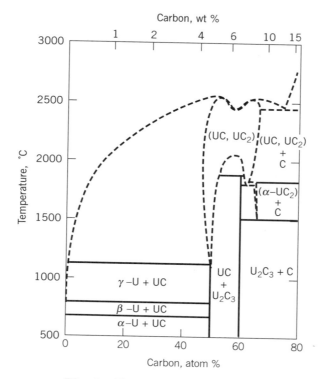

Fig. 6.   Phase diagram for U–C.

The iron–carbon system contains the orthorhombic iron carbide (3:1) [12011-67-5] (Fe₃C), which melts congruently and represents the cementite in steel metallurgy. The existence of other carbides, eg, iron carbide (2:1) [12011-66-4] (Fe₂C), iron carbide (5:2) [12127-45-6] (Fe₅C₂), and iron carbide (7:3) [12075-42-2] (Fe₇C₃), are doubtful (see STEEL).

Iron carbide (3:1) (Fe₃C); mol wt 179.56; carbon 6.69 wt%; density 7.64 g/cm³; mp 1650°C; is obtained from high carbon iron melts as a dark gray air-sensitive powder by anodic isolation with hydrochloric acid. In the microstructure of steels, cementite appears in the form of etch-resistant grain borders, needles, or lamellae. Fe₃C powder cannot be sintered with binder metals to produce cemented carbides because Fe₃C reacts with the binder phase. The hard components in alloy steels, such as chromium steels, are double carbides of the formulas (Cr,Fe)₂₃C₆, (Fe,Cr)₇C₃, or (Fe,Cr)₃C₂, that derive from the binary chromium carbides, and can also contain tungsten or molybdenum. These double carbides are related to η-carbides, ternary compounds of the general formula M₃M′₃C where M = iron metal; M′ = refractory transition metal.

The complex iron carbonitride is the hard component in steels that have been annealed with ammonia (nitrided steels). Complex carbonitrides with iron metals are also present in superalloys in the form of precipitates.

In the nickel–carbon and cobalt–carbon systems, the nickel carbide (3:1) [12012-02-1] (Ni₃C) and cobalt carbide (3:1) [12011-59-5] (Co₃C) are isomorphous with Fe_3C and exist only at low temperatures. The manganese–carbon system contains manganese carbide (3:1) [12121-90-3] (Mn_3C), isomorphous with Fe_3C, and manganese carbide (23:6) [12266-65-8] ($Mn_{23}C_6$), isomorphous with chromium carbide (23:6) [12105-81-6] ($Cr_{23}C_6$). These binary carbides occur frequently in the form of carbide solid solutions or double carbides with other transition metals in alloy steels, superalloys, and special hard metals.

7.9. Complex Carbides. Complex carbides are ternary or quaternary intermetallic phases containing carbon and two or more metals. One metal can be a refractory transition metal; the second may be a metal from the iron or A groups. Nonmetals can also be incorporated.

Complex carbides are very numerous. Many newer compounds of this class have been discovered and their structures elucidated (18). The octahedron M_6C is typical where the metals arrange around a central carbon atom. The octahedra may be connected via corners, edges, or faces. Trigonal prismatic polyhedra also occur. Defining T as transition metal and M as metal or main group nonmetal, the complex carbides can be classified as: (1) T_3M_2C, which has a filled β-manganese structure, eg, Mo_3Al_2C, W_3Re_2C; (2) T_2MC, H-phases, eg, Cr_2AlC, Ta_2GaC, Ti_2SC; (3) T_3MC, perovskite carbides which have the filled Cu_3Au-structure, eg, Ti_3AlC, VRu_3C, URh_3C; (4) T_3M_3C, T_2M_4C, η-carbides that have the filled $Ti2Ni$ structure, eg, Co_3W_3C, Ni_2Mo_4C; and (5) κ-carbides, eg, $W_9Co_3C_4$, $Mo_{12}Cu_3Al_{11}C_6$.

The preferred method for synthesis of complex carbides is the powder metallurgy technique. Hot-pressed powder mixtures must be subjected to prolonged annealing treatments. If low melting or volatile components are present, autoclaves are used.

The η-carbides are not specifically synthesized, but are of technical importance, occurring in alloy steels, stellites, or as embrittling phases in cemented carbides. Other complex carbides in the form of precipitates may form in multicomponent alloys or in high temperature reactor fuels by reaction between the fission products and the moderator graphite, ie, pyrographite-coated fuel kernels.

8. Quality Control Methods for Primary Carbides

Analytical control of primary hard carbides for cemented carbides is normally done on a production-lot basis, with powder samples of uniformly blended lots being issued routinely to control laboratories, which, among major carbide producers commonly have widely diverse analytical capabilities, while more numerous smaller carbide producers rely on independent laboratories. The rapid growth of semi-automatic analytical equipment over the past decades has resulted in large cost savings as well as faster delivery of results.

At the primary carbide stage, control of WC and the principal cubic carbides, TiC, TaC, and NbC, as well as the solid solutions of these, accounts for the bulk of control-lab volume. Combined carbon, among all determinations being the most indicative of purity and hardness, is determined by automated combustion with oxygen and the measurement of resultant CO_2 by infrared detection or by thermal conductivity, as typified by the Leco apparatus. Free (or residual) carbon is determined by decomposition in HNO_3-HF and subsequent filtration to separate carbon. Average particle size is also routine, for which a common method is measurement of air flow through a compacted powder bed, as typified by the Fisher Sub-Sieve particle size analyzer. In the case of submicron or ultrafine WC powder, particle size measurement is based on a determination of the surface area of the powder sample as measured by the volume of gas absorbed under liquid nitrogen and subsequently desorbed, which is then converted to particle size. As in the case of combined carbon, specific gravity is a fundamental indicator of purity among the major hard-metal monocarbides, although unlike the carbon analysis this determination may enter the picture mainly in combination with other tests. For average particle sizes exceeding ~ 2 μm, the pycnometer method with sample in water is used for specific gravity determination. For submicron or ultrafine WC powders, a problem of thorough wetting of very high surface area powders is met by the use of a helium pycnometer, replacing water with helium. The primary solid-solution carbides are of course subject to all of the tests noted above, and are also analyzed by X-ray diffraction for composition and purity.

The abrasive nature of carbide powders carries a potential for low-level, accumulative pick up of impurity metals during contact with processing equipment, notably when powders are in the coarser, primary form. This effect is lessened by a selection of special alloys for critical machine components, accumulations of which are of least effect on ultimate properties, eg, the use of cemented tungsten carbide parts for selected crushing and milling machinery components, as well as of high-hardness ferrous alloys elsewhere. Given normalized acceptable pick-up levels, analyses for selected iron-group metals as well as other metals may be made at intervals, relying on the atomic absorption method for trace metals, inductively coupled plasma method for trace and compositional levels, and on X-ray fluorescence for higher concentrations (19).

Also in the important but nonroutine analytical category, surface examinations may be made of samples in either powder form or in diamond-polished, cemented-alloy form at high magnifications by scanning electron microscope. Such observations include crystal or particle forms, particle surface condition, degree of particle attachment, and particle or crystal size ranges. Qualitative determinations by X-ray line scans may also be made across selected phase areas in polished samples. Observations on a coaser basis are also made by binocular microscope.

9. Economic Aspects

Three categories of metallurgical refining support manufacturers of cemented carbides. The first involves extraction from ore concentrates of tungsten minerals to produce tungstic oxide, WO_3, or ammonium paratungstate (APT), and the extraction from ore concentrates of Ti, Ta, and Nb minerals to produce refined oxides of titanium, tantalum, or niobium. The second step converts WO_3, APT, and oxides of titanium, tantalum, and niobium to the respective primary carbides or solid solutions of these metals. A third category of refining combines the first two steps to convert mineral concentrates of these metals directly into primary carbides. These refining categories also apply in general to the production of the auxiliary carbides.

Some refining, traditionally the conversion of the metal oxides to carbides, is carried out by the cemented carbide manufacturers themselves, chiefly by larger, more vertically integrated companies, while smaller cemented carbide manufacturers obtain their supplies either from independent refiners specializing in primary metal powders, carbides, nitrides, carbonitrides, and many lower volume accessory materials used in cemented carbide production or from the more integrated cemented carbide manufacturers. Consolidation within the cemented carbide industry is however an important ongoing evolution, one aspect of which is growth in cross-supply arrangements between larger companies of primary carbides and other precursor materials, in a mutual search for cost savings.

Recently, many of the advances in the preparation of primary carbides have favored metalcutting and metalworking components in the manufacturing industries rather than the resource industries, eg, the mining and processing of coal and hard minerals, exploration for and development of oil and gas fields, and agriculture. That unalloyed WC has been virtually the sole carbide used in these latter fields suggests that great tool opportunities may yet lie ahead in these fields. For metalcutting and metalworking fields, an ongoing development of production methods for carbides and carbonitrides in the form of chemical vapor- and physical vapor-deposited coatings on cemented carbide tool components stand out as a major leap forward for the cemented carbide tool industry, increasing productivity and profitability. A further important advance in carbide metallurgy has been seen in the development of submicron and ultrafine WC powders, which, in cemented carbide form, reach combinations of high strength and high hardness. Ongoing developments have motivated new independent enterprises specializing in this field, thus creating a potential for a new line of independent suppliers to the industry.

BIBLIOGRAPHY

"Heavy-Metal Carbides" under "Carbides" in *ECT* 1st ed., Vol. 2, pp. 846–854, by P. M. McKenna and J. C. Redmond, Kennametal Inc.; "Industrial Heavy-Metal Carbides" under "Carbides" in *ECT* 2nd ed., Vol. 4, pp. 75–92, by R. Kieffer, University of Vienna, and F. Benesovsky, Metallwerk Plansee A.G.; in *ECT* 3rd ed., Vol. 4, pp. 490–505, by R. Kieffer, Technical University, Vienna, and F. Benesovsky, Metallwerke Plansee A. G., Reutte, Tyrol;"Carbides, Industrial Hard Carbides" in *ECT* 4th ed., Vol. 4, pp. 861–878, by W. M. Stoll, Consulting Metallurgist and A. T. Santhanam, Kennametal Inc.;"Carbides, Industrial Hard" in *ECT* (online), posting date: December 4, 2000, by W. M. Stoll, Consulting Metallurgist and A. T. Santhanam, Kennametal Inc.

CITED PUBLICATIONS

1. U.S. Pat. 2,529,778 (1950), P. M. McKenna.
2. U.S. Pat. 3,379,503 (1968), P. M. McKenna.
3. U.S. Pat. 4,834,963 (1989), C. J. Terry and J. D. Frank.

4. A. E. van Arkel and J. H. de Boer, *Physica* **4**, 286 (1924); *Z. Anorg. Allg. Chem.* **148**, 345 (1925).
5. E. Neuenschwander, *J. Less Common Met.* **11**, 365(1966).
6. U.S. Pat. 2,515,463 (1950), P. M. McKenna.
7. U.S. Pat. 2,124,509 (1939), P. M. McKenna.
8. W. Dawihl, *Z. Anorg. Chem.* **262**, 212 (1950).
9. R. Kieffer and co-workers, *Monatsh. Chem.* **101**, 65 (1970); P. Ettmayer, *Monatsh. Chem.* **101**, 1720 (1970).
10. R. Kieffer, P. Ettmayer, and B. Lux, *Conference on Recent Advances in Hard Metals Production*, Paper 33, Loughborough University, U.K., 1979.
11. J. Schuster, E. Rudy, and H. Nowotny, *Monatsh. Chem.* **107**, 1167 (1976).
12. R. Kieffer, N. Reiter, and D. Fister, *BISRA-ISI Conference on Materials for Metalcutting*, Scarborough, U.K., 1970, p. 126; Fr. Pat. 2,064,842 (1970), R. Kieffer (to Ugine-Carbone).
13. U.S. Pats. 2,113,353 and 2,113,354 (1938), P. M. McKenna.
14. R. Kieffer and H. Nowotny, *Metallforschung* **2**, 257 (1947).
15. J. Norton and A. L. Mowry, *Trans. Am. Inst. Min. Met. Eng.* **185**, 133(1949).
16. H. Nowotny, R. Kieffer, and O. Knotek, *Berg. Huttenmann, Monatsh.* **96**, 6(1951).
17. F. Benesovsky and E. Rudy, *Planseeber. Pulvermet.* **9**, 65(1961); *Monatsh. Chem.* **94**, 204 (1963).
18. H. Nowotny, *Ang. Chem.* **84**, 973 (1972).
19. J. W. Kratofil, personal communications, 2002.

GENERAL REFERENCES

B. Kieffer and F. Benesovsky, *Hartstoffe*, Springer-Verlag, Vienna, Austria, 1963, 1965.

E. K. Storms, *The Refractory Carbides*, Academic Press, New York, 1967.

H. J. Goldschmidt, *Interstitial Alloys*, Butterworth, London, U.K., 1967.

L. Toth, *Transition Metal Carbides and Nitrides*, Plenum Press, New York, 1971.

T. J. Kosolapova, *Carbides: Properties, Production and Applications*, Plenum Press, New York, 1971.

E. Fromm and E. Gebhardt, eds., *Gase und Kohlenstoff in Metallen*, Springer-Verlag, Berlin, Heidelberg, New York, 1976.

K. J. A. Brookes, *World Directory and Handbook of Hardmetals and Hard Materials*, 6th ed., International Carbide Data, East Barnet, Hertfordshire, U.K., 1996.

T. E. Chung, D. S. Coleman, A. G. Dowson, and B. Williams, eds., *Proceedings of the Recent Advances in Hardmetal Production*, Loughborough University of Technology, published by Metal Powder Report, September 17–19, 1979.

R. Eck, "Powder Metallurgy of Refractory Metals and Applications," *Int. J. Powder Metall. Powder Technol.* **17** (3), 201 (1981).

M. MacInnis and T. Kim, in T. C. Lo, M. H. I. Baird, and C. Hanson, eds., *Handbook for Solvent Extraction*, John Wiley & Sons, Inc., New York, 1983.

T. Wilken, C. Wert, J. Woodhouse, and W. Morcom, in H. H. Hausner and P. V. Taubenblat, eds., *Modern Developments in Powder Metallurgy*, Vol. 9, Plenum Press, New York, 1977, pp. 161–169.

W. M. STOLL
Consulting Metallurgist

CARBOHYDRATES

1. Introduction

Carbohydrates are found in all plant and animal cells. They are the most abundant of the organic compounds, so abundant that it is estimated that well over one-half of the organic carbon on earth exists in the form of carbohydrates. Most carbohydrates are produced and found in plants. Carbohydrate molecules make up about three-fourths of the dry weight of plants; most is found in cell walls as structural components. Carbohydrates also constitute important energy reserves in plants; one carbohydrate, starch, provides about three-fourths of the calories in the average human diet on a worldwide basis. But the nutritional aspects are only a part of the story of carbohydrates. They have many important commercial uses in such diverse areas as adhesive, agricultural chemicals, fermentation, food, pharmaceutical, textile and paper and related products, and in petroleum production. Because the basic carbohydrate molecule is functionalized at every carbon atom, and because carbohydrates seldom occur as simple sugars but rather combined with each other or other compounds, the variety of carbohydrates in nature is large, and the number of theoretical possibilities is almost limitless.

2. Classification

The basic carbohydrate molecule possesses an aldehyde or ketone group and a hydroxyl group on every carbon atom except the one of the carbonyl group. As a result, carbohydrates are defined as aldehyde or ketone derivatives of polyhydroxy alcohols and their reaction products. The formula for glucose and related sugars ($C_6H_{12}O_6$) contains hydrogen and oxygen atoms in the ratio in which they are found in water. The name carbohydrate (hydrate of carbon) is derived from the fact that the basic carbohydrate molecule has the formula $C_n(H_2O)_n$.

Monosaccharides, commonly referred to as the simple sugars, are carbohydrates that cannot be broken down by hydrolysis (1,2). They are classified both according to the kind of carbonyl group and according to the number of carbon atoms contained in the molecule. An aldose is a polyhydroxy aldehyde, ie, an aldehyde that has a hydroxyl group on every carbon atom except the carbonyl carbon atom. A ketose is a polyhydroxy ketone. Numical prefixes designating the number of carbon atoms are tri-, tetra-, penta-, hexa-, hepta-, etc. In systemic nomenclature, the suffix for the names of aldehyde sugars is -ose and for ketone sugars-ulose. The two classification systems can be joined in a single-word description. For example, a three-carbon aldose is an aldotriose and a six-carbon ketose is a ketohexose (or hexosulose). Common names are frequently used, creating exceptions to systematic nomenclature.

$$
\begin{array}{cc}
 & CH_2OH \\
 & | \\
HC=O & C=O \\
| & | \\
(CHOH)_n & (CHOH)_n \\
| & | \\
CH_2OH & CH_2OH \\
\\
\text{aldose} & \text{ketose}
\end{array}
$$

Monosaccharides are most often joined together in chains. Oligosaccharides are carbohydrate chains that yield 2–10 monosaccharide molecules upon hydrolysis (2,3). Oligosaccharides are classified according to the number of monosaccharide units in them, eg, di-, tri-, tetra-, pentasaccharides, etc. Polysaccharides are carbohydrate chains that yield at least 35 monosaccharide molecules upon hydrolysis (4–10). Polysaccharides may be linear (unbranched) or branched. They may contain a single kind of monosaccharide unit (homopolysaccharides) or two or more different monosaccharide units (heteropolysaccharides). The generic term for polysaccharides is glycan; therefore, these two groups of polysaccharides are also termed homoglycans and heteroglycans.

Most carbohydrates exist in the form of polysaccharides. Polysaccharides give structure to the cell walls of land plants (cellulose), seaweeds, and some microorganisms and store energy (starch in plants, glycogen in animals). They are important in the human diet and in many commercial applications.

3. Representations

D-Glucose is an aldohexose. Four of the six carbon atoms (C-2, C-3, C-4, C-5) are chiral carbon atoms. To compare the arrangements of atoms in common, simple monosaccharides, structural formulas are written using the convention that all bonds connecting carbon atoms are vertical and project into the plane of the page away from the viewer and the bond to the hydrogen atom and the hydroxyl group on each chiral carbon atom projects out of the plane of the page towards the viewer. Horizontal bonds are often omitted. The acyclic structural formula of D-glucose shown is known as the open-chain, or Fischer, formula. If the hydroxyl group on the most distant chiral carbon atom from the top end (the penultimate carbon atom of the structures in Fig. 1; C-5 of D-glucose) is on the right when the carbon chain of an aldose or ketose is written using this convention, the sugar is said to have the D configuration; if that hydroxyl group is on the left, the sugar belongs to the family of L sugars. Most naturally occurring monosaccharides have the D configuration. An exception is arabinose, which most often occurs as L-arabinose [5328-37-0]. All possible structures of the three-, four-, five-, and six-carbon atom aldoses with the D configuration are given in Figure 1.

Because an aldohexose contains four chiral carbon atoms, there are $2^4 =$ 16 different possible arrangements of the hydroxyl groups in space, ie, there are 16 different stereoisomers of an aldohexose. The structures of one-half of these, the eight D isomers, are shown in Figure 1. Only three of these 16 stereoisomers are commonly found in Nature: D-glucose [50-99-7], D-galactose [59-23-4], and D-mannose [3458-28-4].

4. Chemistry of Saccharides

Most carbohydrates have two kinds of reactive groups: the carbonyl group and primary and secondary hydroxyl groups.

4.1. Reactions of The Carbonyl Group. *Ring Forms.* Aldehydes and ketones react with compounds containing a hydroxyl group (alcohols) to form first hemiacetals and then acetals. Because aldose and ketose molecules have a carbonyl group and hydroxyl groups on the same carbon chain, they

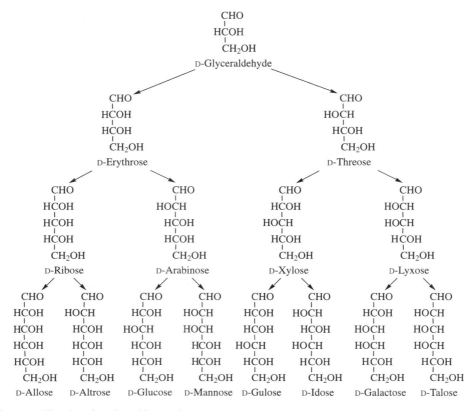

Fig. 1. The family of D-aldoses derive from D-glyceraldehyde by chain extension at the carbonyl carbon atom.

can form hemiacetal structures intramolecularly. Such an intramolecular reaction forms a ring. The most common rings are the six-membered pyranose ring, a cyclic structure composed of five carbon atoms and one oxygen atom, and the five-membered furanose ring, a cyclic structure composed of four carbon atoms and one oxygen atom (Fig. 2).

It is easy to picture the formation of a ring from an open-chain structure if it is remembered that the carbon chain is curving into the plane of the page. When this structure is laid on its side, it is naturally curved into almost the correct shape, with the ends almost touching. However, in order to close the ring, ie, to form the hemiacetal between the hydroxyl group on C-5 and the aldehyde group, C-5 must rotate to bring the hydroxyl group closer to the aldehyde group. The result of this rotation is that the −CH₂OH group sticks up in this (the Haworth) representation of the pyranose ring of the D sugars and the −CH₂OH group projects down in representations of the L sugars. Most, but not all, naturally occurring sugars are D sugars. For C-2, C-3, and C-4, the carbon atoms that are not involved in ring formation, the hydroxyl groups that are on the right in the Fischer projection project down in the Haworth ring form and the hydroxyl groups that are on the left in the Fischer projection stick up in the Haworth ring form.

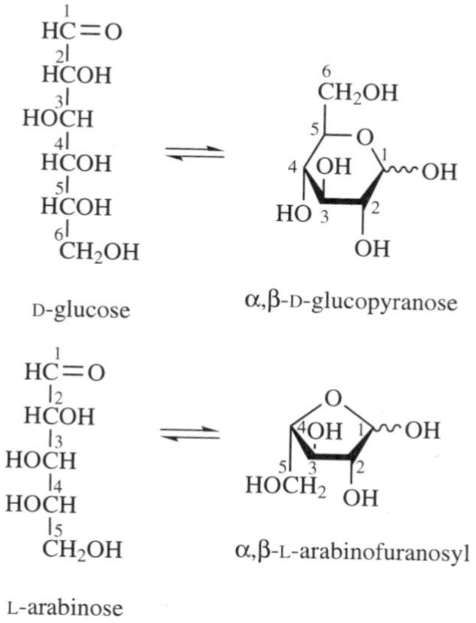

Fig. 2. Conversion from open-chain to ring forms of aldoses (aldehydo sugars). When ring closure occurs, a new chiral center is formed and two C-1 configurational isomers, called anomers, are formed.

When the pyranose (six-membered) ring is formed, a new chiral carbon atom is formed from C-1. Thus there can be two forms of the pyranose ring. D Sugars with the hydroxyl group at C-1 in the up position are said to be in the beta (β) configuration; D sugars with the hydroxyl group at C-1 projecting down are said to be in the alpha (α) configuration. α-D-Glucopyranose [492-62-6] and β-D-glucopyranose [492-61-5] are anomers of each other. In β-L-pyranoses, the hydroxyl group on C-1, termed the anomeric carbon atom, projects down in the Haworth representation. Thus, eg, β-D- and β-L-glucopyranose [39281-65-7] are complete mirror images of each other.

Most free pentoses, hexoses, and heptoses occur primarily in less-strained pyranose rings, but the furanose ring is also quite important. The furanose ring is formed in the same way as the pyranose ring and also occurs in α and β forms. This is demonstrated with L-arabinose, which is commonly found in polysaccharides in the form of α-L-arabinofuranosyl units (see Fig. 2).

Whereas furanose rings are almost, but not quite, flat, pyranose rings are not, thus Haworth representations do not show the actual molecular shape. Pyranose rings assume one of two chair forms designated the $^4C_1$ chair because C-4 is up and C-1 is down (with respect to the plane of O-5, C-2, C-3, and C-5) and the $^1C_4$ form. The $^4C_1$ chair is by far the most prevalent shape of the β-D-glucopyranose molecule because all the bulky groups (the hydroxyl groups at C-1, C-2, C-3, and C-4 and the hydroxymethyl group at C-5) are in equatorial positions which minimizes nonbonded (steric) interactions.

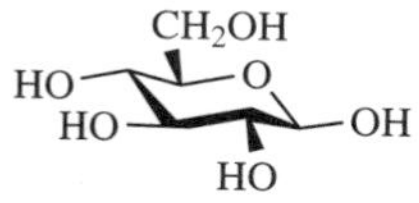

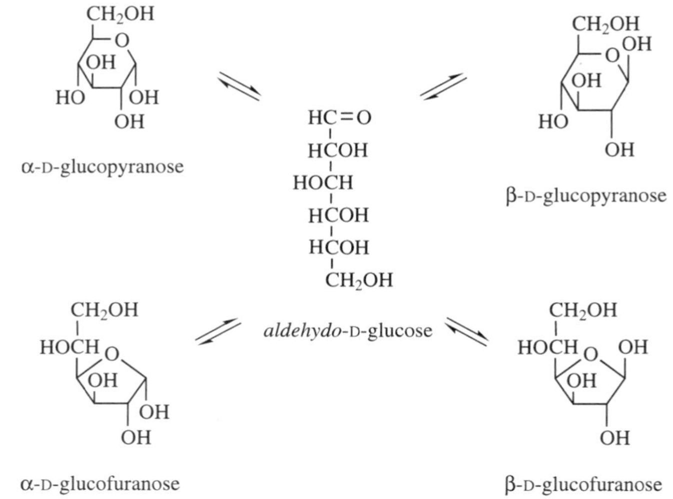

Fig. 3. Equilibrium mixture of D-glucose forms in solution. Pyranose ring forms predominate.

Solutions of aldoses and ketoses reach an equilibrium among various forms, as shown for D-glucose in Figure 3. The exact composition of an equilibrium mixture depends on the temperature and the specific sugar; for D-glucose the approximate composition of a solution at room temperature is <0.01% aldehydo form, 36.2% α-D-glucopyranose, 63.8% β-D-glucopyranose, and traces of the furanose ring forms. The process of conversion among forms is called mutarotation (1) because, when crystals of α-D-glucopyranose are dissolved in water, the initial specific optical rotation ($[\alpha]_D$ at 20°C) of +112° gradually decreases to the equilibrium value of +52.7°. Likewise, when crystals of β-D-glucopyranose are dissolved in water, the initial specific optical rotation of +18.7° gradually increases to the equilibrium value of +52.7°. Mutarotation is both acid and base catalyzed.

The structure of monosaccharides is often written in the acyclic form although only very minor amounts of it ever occur in that form. Because the interconversions are rapid, the carbonyl groups of sugars can and do react both as if they are free and as if they are in a hemiacetal ring form.

Glycosides, Oligosaccharides, and Polysaccharides. Few monosaccharides are found free in nature, and these few are usually present in only small amounts. Most monosaccharides occur in combinations, most often with either more of the same sugar or different sugars in the form of polymers (polysaccharides). Less frequently, except in the case of sucrose, they are joined together in oligosaccharide chains. Mono- and oligosaccharides may also be linked to nonsugar organic compounds. These combined forms of sugars are known as glycosides.

Pyranose and furanose ring forms of carbohydrate molecules are hemiacetals and can react with an alcohol to form glycosides, which are acetals of sugars. Hydrolysis of a glycoside in an acidic solution releases the monosaccharide and the alcohol (1). This forward and reverse process is shown for the reaction of

D-glucose with ethanol to form ethyl α-D-glucopyranoside [19467-01-7] and ethyl β-D-glucopyranoside [3198-49-0].

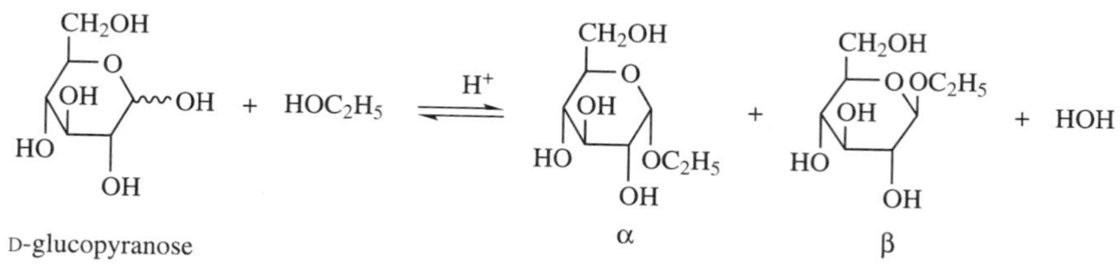

Glycosides, particularly of phenolic compounds, are widely distributed in plant tissues (11,12). Glycosides of anthocyanidins, flavones, flavanols, flavanones, flavanonols, stilbenes and saponins, gallic acid derivatives, and condensed tannins are all common.

In the body, detoxification of drugs and poisonous compounds often involves converting the substance into a more water-soluble compound, which is then excreted in urine. The most common conversion reactions are hydroxylations, oxidations, reductions, and conjugations. Acetaminophen [103-90-2], an analgesic used as an aspirin substitute, contains a hydroxyl group which is combined with the monosaccharide D-glucuronic acid to form the water-soluble β-D-pyranoside. After deacetylation, aspirin [50-78-2] may be conjugated with either D-glucuronic acid or the amino acid glycine (uronic acids are monosaccharides in which the terminal primary alcohol group has been oxidized to a carboxylic acid group).

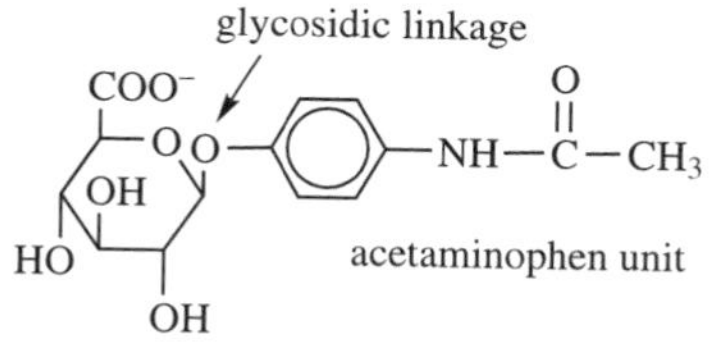

Frequently, the alcohol that forms a glycoside with a sugar is a hydroxyl group of another sugar. The formation of a glycoside between two sugar units joins them, forming a disaccharide, eg, two D-glucopyranosyl units may be linked to form the disaccharide maltose [69-79-4].

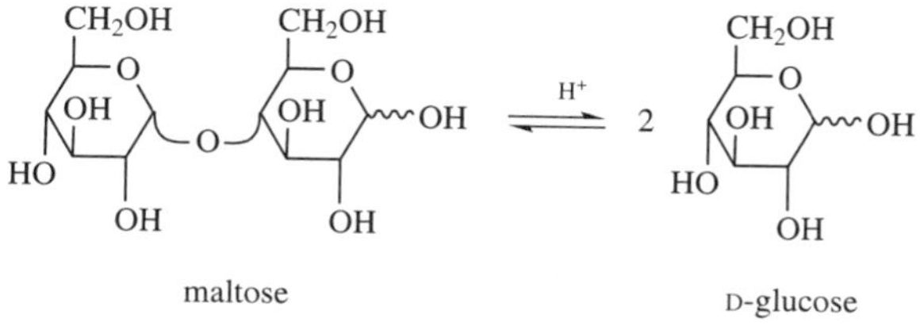

In Nature, however, monosaccharide units are not joined together by such a simple acid-catalyzed condensation as that shown above, which is why the reaction is written as a hydrolytic reaction. For chemical synthesis of di- and higher saccharides, activation of the anomeric carbon atom and blocking of hydroxyl groups not involved in the linkage are required and employed. An exception is found in the manufacture of polydextrose [68424-04-4], which is made by heating D-glucose under dehydrating conditions (vacuum) in the presence of sorbitol (D-glucitol), and citric acid (catalyst) to make a highly branched, low molecular weight polymer.

For the most part, low molecular weight carbohydrates of commerce are made by depolymerization of polysaccharides via enzyme- or acid-catalyzed hydrolysis. Only sucrose and, to a very much lesser extent, lactose (both disaccharides) are commercial low molecular weight carbohydrates not obtained in this way.

Oligo- and polysaccharides have reducing and nonreducing ends. A reducing sugar is a carbohydrate that contains an aldehyde or ketone group, either free or in a hemiacetal form, which in aqueous solution is always in equilibrium with the free form. The aldehyde group (and the ketone group, after isomerization to an aldehyde group under basic conditions) can be oxidized to a carboxyl group, ie, act as a reducing agent. The reducing end of an oligo- or polysaccharide is the one end not involved in a glycosidic linkage and can, therefore, react as an aldehyde or ketone. The sugar units constituting all other ends are attached through glycosidic (acetal) bonds and are, therefore, nonreducing ends. Reducing and nonreducing ends can be demonstrated with the structure of lactose [63-42-3] (β-D-galactopyranosyl-α-D-glucopyranose), the reducing disaccharide of milk.

Additional sugar units added to either end of disaccharides form higher oligosaccharides. For example, if one α-D-glucopyranosyl unit is added to the disaccharide maltose in a $(1 \rightarrow 4)$ linkage, the trisaccharide maltotriose [1109-28-0] is obtained. Another unit extends to the tetrasaccharide maltotetraose [34612-38-9] and yet another to the pentasaccharide maltopentaose [1668-09-3], etc (malto- is a prefix indicating a product originating from depolymerization of starch molecules). When many sugar units are joined together by glycosidic linkages (the acetal bonds connecting sugar units), the structure is that of a polysaccharide.

Polysaccharides are naturally occurring polymers of monosaccharide (sugar) units (4–10). In precise chemical nomenclature, polysaccharides are glycans and are described as being composed of glycosyl units. Polysaccharides, like oligosaccharides, have ends that can be distinguished from each other because the individual monomer units are joined in a specific head-to-tail fashion. Polysaccharides have one reducing end (free or potential aldehyde or ketone group, although ketoses are uncommon constituents of polysaccharides) and one or

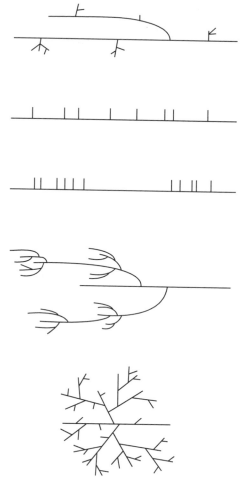

Fig. 4. Branching patterns in polysaccharides.

many nonreducing ends. Polysaccharide molecules can be linear or branched in any of several different ways (Fig. 4). They may be composed of a single type of glycosyl unit (a homoglycan) or from two to six different glycosyl units (a heteroglycan). They generally contain from hundreds to tens of thousands of glycosyl units; some may be larger. Of the heteroglycans, only the bacterial polysaccharides have regular repeating-unit structures, as opposed to plant and animal polysaccharides, because of a different pathway of biosynthesis.

The polysaccharides of starch [9005-25-8] (13) are of great interest and importance. Because starch is the carbohydrate storage material of many plants and the principal component of the seeds of cereal grains (corn, wheat, rice, oat, barley, rye, sorghum, etc) and some edible roots and tubers (potato, cassava, taro, etc.), it is the principal source of carbohydrate in our diet. Starch is also widely used to make D-glucose (often called dextrose in commerce), syrups, and related products. It also has important nonfood applications.

Starch occurs in the form of granules composed of two polysaccharides. Both contain only α-D-glucopyranosyl units and are, therefore, glucans. Amylose [9005-82-7] is an essentially linear polysaccharide of (1 → 4)-linked α-D-glucopyranosyl units. Fine structure analysis has revealed that at least some, especially the larger molecules, are slightly branched. The basic structure of amylose, which is essentially an extension of the maltose structure shown previously, shows that there can be only one sugar unit in any polysaccharide that is not joined to another through a glycosidic bond involving its carbonyl group and only one reducing end. However, branching is possible in polysaccharides because of the multitude of OH groups (Fig. 4). The second polysaccharide in starch granules, amylopectin [9037-22-3], is a highly branched molecule.

Glycogen [9005-79-2], the principal carbohydrate food-reserve substance in animals, has a structure similar to that of amylopectin, ie, it is a branched polymer consisting of linear segments of (1→4)-linked α-D-glucopyranosyl units joined by the (1→6) glycosidic linkages that constitute the branch points. All cells of higher animals may contain some glycogen. Because it is in a dynamic state, glycogen is polymolecular with the range of molecular weights depending on the metabolic state of the tissue. The weight-average molecular weight of rabbit liver glycogen has been reported to be 2.7×10^8, with a range of 6×10^6 to 1.6×10^9. It contains 0.35% of covalently bound protein, a molecule that served as the primer upon which glycogenesis began. The average degree of polymerization (DP, the average number of monosaccharide units in a polysaccharide) of the chain segments depends on the source, but the majority of values are within the range DP 11–14. Glycogen is an amorphous polymer. It is highly soluble and exhibits a fairly ideal hydrodynamic behavior.

In commerce, hydrolysis of glycosidic bonds is far more important than is condensation of sugars with alcohols or other sugar units to form glycosidic bonds. Glycosidic bonds are formed in nature via biosynthetic reactions, and compounds containing them are isolated and used as starting materials for various transformations. Hydrolysis, whether catalyzed by acids or enzymes, follows the same general mechanism (1).

Synthetic Methods.   Although mono- and oligosaccharides are most often made by depolymerization of polysaccharides, oligosaccharides and other compounds with glycosidic bonds can be made synthetically. The classic and still widely used reaction is the Koenigs-Knorr reaction; many modifications of it are known (14–17). The reaction in its simplest form is that given below. Modifications involve the nature of the promoter and blocking groups and have been developed to influence the anomeric configuration, ie, the stereochemistry, of the product.

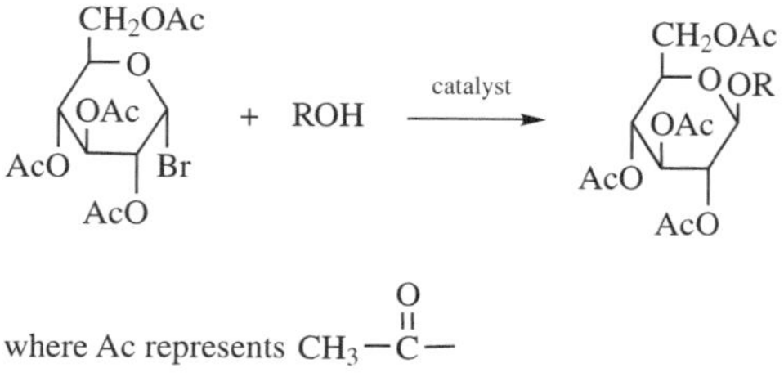

The activated sugar used in this reaction is 2,3,4,6-tetra-*O*-acetyl-α-D-gluco-pyranosyl bromide [572-09-8], commonly called acetobromoglucose. A phenomenon known as the anomeric effect requires that this compound always has the α-D configuration, ie, that the halogen atom be in the axial position. Glycosyl chlorides and glycosyl fluorides are also used in the preparation of glycosides. The excocylic oxygen atom in the product can be replaced with a sulfur atom or an amino group (14,18).

Nucleosides (15,19) and nucleotides are examples of the latter (glycosyla-mine) type.

The reaction to form an acetal can be an intramolecular reaction. The best known example is 1,6-anhydro-β-D-glucopyranose [498-07-7], commonly called levoglucosan.

C-Glycosyl compounds have a carbon atom in place of the exocyclic oxygen atom of the acetal group and, therefore, are branched cyclic ethers. An example is the naturally occurring anthraquinone dye, carminic acid [1260-17-9] (CI Natural Red 4).

Glycoconjugates.   Another class of carbohydrates is the glycoconjugates, which is composed of glycoproteins, proteoglycans, peptidoglycans, and glycoli-pids.

Glycoproteins.   (4,20,21) are molecules containing saccharide chains, often but not always oligosaccharide chains, covalently attached to a polypeptide chain. The saccharide chains may be attached via a glycosidic linkage to a hydro-xyl group of a seryl, threonyl, hydroxylysyl, or hydroxyprolyl unit or via a glyco-sylamine linkage to the amide group of an asparaginyl unit. The percentage of carbohydrate in glycoproteins, which includes the majority of all proteins, varies from <1 to >80%. The surface of cells is covered with a complex mosaic of carbohydrates, called the glycocalyx, many of which are the saccharide units of glycoproteins. Some of those on erythrocytes control blood group antigenicity (see BLOOD, COAGULANTS AND ANTICOAGULANTS).

Proteoglycans (20,21) are components of connective tissue. They have specific polysaccharide chains covalently attached to a polypeptide chain. The specific polysaccharides are glycosaminoglycans (4,20,21), commonly called mucopolysaccharides. As the name suggests, these polysaccharides contain amino sugars (2-amino-2-deoxysugars). All except keratan sulfate contain uronic acid units; all except hyaluronic acid contain sulfate half-ester groups. The polysaccharides found as components of proteoglycans are chondroitin 4-sulfate [24967-93-9], chondroitin 6-sulfate [25322-46-7], dermatan sulfate [24967-94-0], and keratan sulfate [9056-36-4]. The following structure is that of chondroitin 6-sulfate. In dermatan sulfate the −COOH group points down in the Haworth representation, ie, the uronic acid unit is an L-iduronic acid [2073-35-0] unit, and the sulfate group is at C-4.

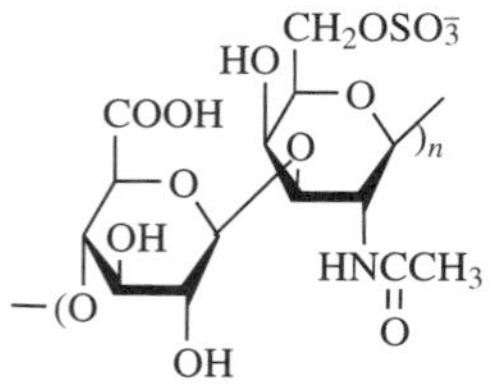

There are other glycosaminoglycans. Hyaluronic acid [9004-61-9] occurs both free and in noncovalent association with proteoglycan molecules. Heparin [9005-49-6] and heparan sulfate [39403-40-2], also known as heparitin sulfate [9050-30-0], occur in mast cells and in the aorta, liver, and lungs.

Compounds with similar structures, ie, polysaccharide chains covalently attached to polypeptide chains, but where the polysaccharides are not glycosaminoglycans, are found commonly in plants and are known as protein-polysaccharides.

Peptidoglycans (4,10) are the primary components of bacterial cell walls. They consist of a heteropolysaccharide called murein cross-linked with short peptide chains.

Glycolipids (4,20) are primarily glycosphingolipids, molecules that have oligosaccharide groups attached to ceramide [104404-17-3]. They are present, at least in small amounts, in the membranes of most, if not all, animal tissues. They too, like cell-membrane glycoproteins, are recognition determinants.

Lipopolysaccharides (10) are cell-wall components of gram-negative bacteria.

Teichoic acids (10) are bacterial polymers in which alditols (glycerol or ribitol) are joined through the primary hydroxyl groups via phosphate diester linkages.

Phosphonomannans (6) are bacterial polymers in which manno-oligosaccharides are joined by phosphate diester linkages. Phosphonogalactans are present in certain fungi.

Sucrose and Derivatives of Sucrose. By far the most abundant of the naturally occurring oligosaccharides is the disaccharide sucrose [57-50-1] (22), ordinary table sugar from sugar cane or sugar beets (see SUGAR). The two monosaccharide units in sucrose are α-D-glucopyranosyl and β-D-fructofuranosyl units.

In sucrose, the ketohexose, D-fructose [30237-26-4], exists as a five-membered furanose ring (β-D-fructofuranose [470-23-5]) formed by reaction between the carbonyl group at C-2 and the hydroxyl group on C-5. Sucrose is unique in that the two glycosyl units are linked head-to-head via an acetal bond rather than head-to-tail. Thus the molecule has no hemiacetal group and no reducing end, and is, therefore, classified as a nonreducing sugar.

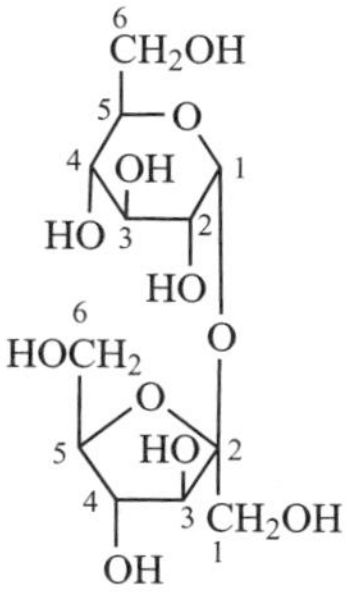

sucrose

The trisaccharide raffinose [512-69-6] consists of a sucrose molecule with an α-D-galactopyranosyl unit linked $1 \rightarrow 6$ to its D-glucosyl unit. Raffinose is the second most abundant oligosaccharide and, like sucrose, may be ubiquitous in the plant kingdom. However, it is present in only minor amounts as compared to sucrose.

The tetrasaccharide stachyose [470-55-3], which contains an additional $(1 \rightarrow 6)$-linked α-D-galactopyranosyl unit, is almost as widely distributed as raffinose, but is present in even lower concentrations. Although raffinose and stachyose occur in all parts of plants, they are concentrated in storage tissues, eg, in sugar beets and beans, and leaves for the most part.

Structures of raffinose and stachyose are given below using official short-hand designations. In this system, the D or L designation is not used if the sugars are D; if a glycosyl unit is from an L sugar, an L is placed before the three-letter abbreviation. Subscripts p and f refer to pyranosyl and furanosyl rings, respectively.

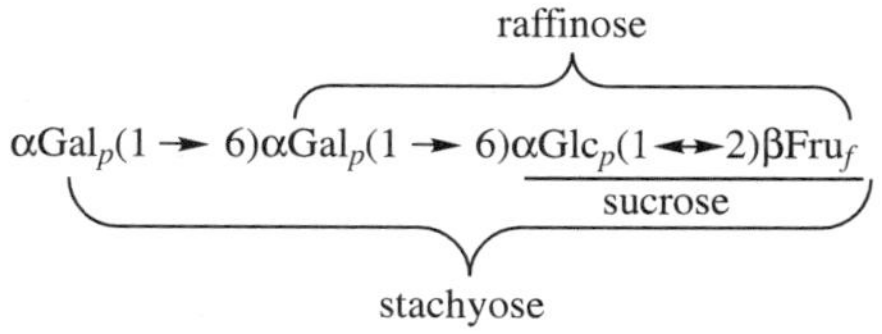

$\alpha\text{Gal}_p(1 \rightarrow 6)\alpha\text{Gal}_p(1 \rightarrow 6)\alpha\text{Glc}_p(1 \leftrightarrow 2)\beta\text{Fru}_f$

raffinose

sucrose

stachyose

Oxidation to Sugar Acids and Lactones. When the aldehyde group of an aldose is oxidized, the resulting compound is an aldonic acid (salt form = aldonate) (14). Some aldonic acids are products of carbohydrate metabolism.

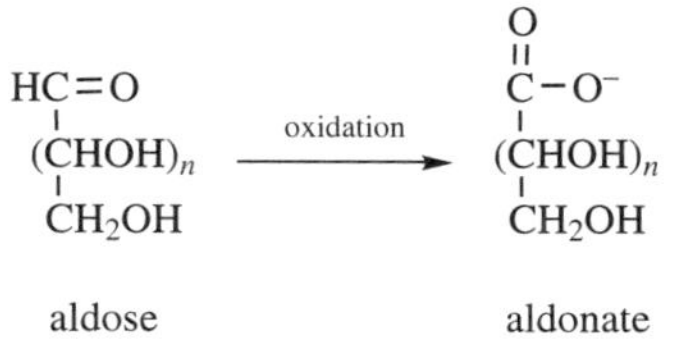

aldose   aldonate

Oxidation of the aldehyde group of an aldose to form a carboxylic acid or carboxylate ion is often used analytically to determine the amount of reducing sugar. The Benedict and Fehling methods measure the amount of reducing sugar present in a fluid. In these reactions, the oxidant, Cu^{2+}, is reduced to Cu^+. The Cu^+ ion precipitates as Cu_2O, which can be measured in a variety of ways. In the Tollens test, Ag^+ is reduced to Ag^0.

Monosaccharides in the pyranose or furanose ring forms can also be oxidized, forming an internal ester, a lactone, that can subsequently open to the acyclic form. The amount of D-glucose is often determined by this kind of oxidation catalyzed by the enzyme glucose oxidase [9001-37-0]. Glucose oxidase-catalyzed oxidation of D-glucose is also used in the commercial production of D-glucono-1,5-lactone (D-glucono-δ-lactone) [90-80-2], which is used for slow acidification, especially as a chemical leavening agent (see BAKERY PROCESSES, CHEMICAL LEAVENING AGENTS).

D-gluconate

Preparation of another widely used aldonate, whose common name is sodium glucoheptonate [31138-65-5], and which is an epimeric mixture of heptonates, involves reaction of D-glucose with cyanide ion. The cyanohydrin is then hydrolyzed to the heptonic acid salt. Both sodium D-gluconate [527-07-1] and sodium D-glucoheptonate are used as components of washing compounds because of their ability to sequester divalent cations, in agriculture to carry trace minerals, and in concrete (largest use). Both can be, and are, produced from glucose syrups as well as from pure crystalline D-glucose.

D-glucose   cyanohydrin   sodium glucoheptonate

Oxidation of the carbon atoms at both ends of the carbon chain produces an aldaric acid. That made from D-galactose is galactaric acid [526-99-8], a meso compound commonly known as mucic acid.

$$
\begin{array}{c}
\text{COOH} \\
| \\
\text{HCOH} \\
| \\
\text{HOCH} \\
| \\
\text{HOCH} \\
| \\
\text{HCOH} \\
| \\
\text{COOH}
\end{array}
$$

galactaric acid

Reduction.   Mono- and oligosaccharides can be reduced to polyols (polyhydroxy alcohols) termed alditols (1,14) (see SUGAR ALCOHOLS). Common examples of compounds in this class are D-glucitol (sorbitol) [50-70-4] made by reduction of D-glucose, and xylitol [87-99-0] made by reduction of D-xylose. Glycerol [56-87-5] is also an alditol. Reduction of D-fructose produces a mixture of D-glucitol and D-mannitol [69-65-8].

$$
\begin{array}{c}
\text{CH}_2\text{OH} \\
| \\
\text{C}=\text{O} \\
| \\
\text{HOCH} \\
| \\
\text{HCOH} \\
| \\
\text{HCOH} \\
| \\
\text{CH}_2\text{OH}
\end{array}
\xrightarrow{\text{reduction}}
\begin{array}{c}
\text{CH}_2\text{OH} \\
| \\
\text{HCOH} \\
| \\
\text{HOCH} \\
| \\
\text{HCOH} \\
| \\
\text{HCOH} \\
| \\
\text{CH}_2\text{OH}
\end{array}
\;+\;
\begin{array}{c}
\text{CH}_2\text{OH} \\
| \\
\text{HOCH} \\
| \\
\text{HOCH} \\
| \\
\text{HCOH} \\
| \\
\text{HCOH} \\
| \\
\text{CH}_2\text{OH}
\end{array}
$$

D-fructose   D-glucitol   D-mannitol

Sorbitol and mannitol are generally recognized as safe (GRAS) by the U.S. Food and Drug Administration. An important use of sorbitol is as a humectant. It can extend shelf life in confections and bakery products. Like other alditols, and unlike reducing sugars, it will not undergo Maillard browning and caramelization. Mannitol can be used as a dusting agent because of its low hygroscopicity. Most food applications of alditols are in dietetic products.

Alditols are sweet. Xylitol has essentially the same sweetness as sucrose; sorbitol is about one-half as sweet as sucrose. In chewing gum, polyols provide texture, sweetness, and mouthfeel and reduce the incidence of dental caries.

Reduction of oligomeric chains of monosaccharides results in the same oligosaccharide terminated at the reducing end with an alditol unit. Products made by hydrogenation of various corn syrups are viscous, hygroscopic, noncariogenic, and sweet, depending on the amounts of sorbitol and maltitol present. Their physical properties are generally similar to those of the syrup from which they are made, usually a high maltose syrup, but they exhibit a greatly decreased tendency to brown, a decreased tendency to crystallize, reduced fermentability, and slower conversion to D-glucose. The latter property makes these products of potential use as carbohydrate sources in diets for diabetics.

A series of sorbitol-based nonionic surfactants is used in foods as water-in-oil emulsifiers and defoamers. They are produced by reaction of fatty acids with sorbitol. During reaction, cyclic dehydration as well as esterification (primary hydroxyl group) occurs so that the hydrophilic portion is not only sorbitol but also its mono- and dianhydride. The product known as sorbitan monostearate [1338-41-6], for example, is a mixture of partial stearic and palmitic acid esters (sorbitan monopalmitate [26266-57-9]) of sorbitol, 1,5-anhydro-D-glucitol [154-58-8], 1,4-sorbitan [27299-12-3], and isosorbide [652-67-5]. Sorbitan esters, such as the foregoing and also sorbitan monolaurate [1338-39-2] and sorbitan monooleate [1338-43-8], can be further modified by reaction with ethylene oxide to produce ethoxylated sorbitan esters, also nonionic detergents approved for food use by the U.S. Food and Drug Administration.

1,5-anhydro-D-glucitol	1,4-anhydro-D-glucitol (1,4-sorbitan)	1,4:3,6-dianhydro-D-glucitol (isosorbide)

Isosorbide 5-nitrate and 2,5-dinitrate relax vascular smooth muscle and are prescribed as vasodilators for prevention of angina pectoris.

Cyclitols. Cyclitols are polyhydroxycycloalkanes and -alkenes (1,23). They are widely distributed in Nature, though never in large quantities. The most abundant of these carbocyclic compounds are the hexahydroxycyclohexanes, commonly called inositols, and their methyl ethers. The nomenclature of cyclitols is problematic; several systems have been proposed and used. *myo*-Inositol [87-89-8] is so common in plants that it is generally regarded as being ubiquitous. It is most often found as ester, ether, and/or glycoside derivatives. Phytic acid [83-86-3], the hexakisphosphate monoester of *myo*-inositol occurs in most, if not all, higher plants. It is present in relatively large amounts in cereal grains and may be recovered as phytin, its mixed calcium–magnesium salt.

myo-inositol (a cyclitol)

4.2. Reactions of Hydroxyl Groups. *Reduction and Oxidation.* Hydroxyl groups can be both oxidized to carbonyl groups (2,14,17,18,24) and removed by reduction. Sugars that have the hydroxyl group missing from one or more of the carbon atoms are called deoxy sugars. The sugar known by the common name 2-deoxy-D-ribose (2-deoxy-D-*erythro*-pentose) [533-67-5], a component of DNA (deoxyribonucleic acid), is so designated because the hydroxyl group on C-2 of D-ribose is missing. A common component of polysaccharides is L-rhamnose (6-deoxy-L-mannose [3615-41-6]) as α-L-rhamnopyranosyl units.

2-deoxy-β-D-ribofuranose   α-L-rhamnopyranose

Oxidation of hydroxyl groups to carbonyl groups can form molecules with two aldehyde groups (dialdoses), two ketone groups (diuloses), or an aldehyde and a ketone group (osuloses). Keto acids are known as ulosonic acids.

Uronic acids are monosaccharides in which the terminal primary alcohol group is oxidized to a carboxylic acid functional group, eg, D-glucuronic acid [6556-12-3] (4,18).

D-glucuronic acid

Esterification. The hydroxyl groups of sugars can react with organic and inorganic acids just as other alcohols do. Both natural and synthetic carbohydrate esters are important in various applications (1,14,17,24). Phosphate monoesters of sugars are important in metabolic reactions. An example is the enzyme-catalyzed, reversible aldol addition of dihydroxyacetone phosphate [57-04-5] and D-glyceraldehyde 3-phosphate [591-57-1] to form D-fructose 1,6-bisphosphate [488-69-7].

Naturally occurring ester groups also occur on polysaccharides. They include phosphate, sulfate, acetate, glycolate, and succinate ester groups. Mono-, oligo-, and polysaccharides are often chemically acylated to give them desirable functional properties. Examples are the fatty acid esters of sorbitol and 1,4-sorbitan already mentioned, fatty acid esters of sucrose used as edible surfactants and biodegradable detergents, highly esterified cellulose acetate (acetate rayon), and starches with low degrees of phosphorylation (see CELLULOSE ESTERS).

Etherification. Carbohydrates are involved in ether formation, both intramolecularly and intermolecularly (1,14,17,24). The cyclic ether, 1,4-sorbitan, an 1,4-anhydroalditol, has already been mentioned. 3,6-Anhydro-α-D-galactopyranosyl units are principal monomer units of the carrageenans. Methyl, ethyl, carboxymethyl, hydroxyethyl, and hydroxypropyl ethers of cellulose (qv) are all commercial materials. The principal starch ethers are the hydroxyethyl and hydroxypropylethers (see CELLULOSE ETHERS; STARCH).

Acetalation.   As polyhydroxy compounds, carbohydrates react with aldehydes and ketones to form cyclic acetals (1,14,17,24). Examples are the reaction of D-glucose with acetone in the presence of a protic or Lewis acid catalyst to form 1,2:5,6-di-*O*-isopropylidene-α-D-glucofuranose [582-52-5] and its reaction with benzaldehyde to form 4,6-*O*-benzylidene-D-glucopyranose [25152-90-3]. The 4,6-*O*-(1-carboxyethylidine) group (related to pyruvic acid) occurs naturally in some polysaccharides, viz, xanthan.

Ester, ether, and cyclic acetal groups are used as blocking groups to allow regiospecific reactions to take place, ie, reaction at specific unblocked hydroxyl groups.

Replacement of Hydroxyl Groups.   Replacement of a hydroxyl group with an amino group at any position produces an aminodeoxysugar (2,24). If a primary or secondary amino group is on the carbon atom delta from the carbonyl group, a six-membered ring containing −NH− or −NR− will form. Thiosugars are ones in which a thiol group has replaced a hydroxyl group (4,24). When the thiol group is on the carbon atom delta to the carbonyl group, a six-membered ring containing −S− will form. Replacement of one or more hydroxyl groups with halogen atoms forms deoxyhalogenosugars (1).

Isomerization.   Both the carbonyl group and the adjacent hydroxyl group are involved in isomerization of monosaccharides. This reaction can be catalyzed by either a base or an enzyme. By this reaction, an aldose is converted into another aldose and a ketose, and a ketose is converted into two aldoses. It is for this reason that ketoses are reducing sugars. They cannot act as reducing agents because they cannot be oxidized to acids, but under alkaline conditions, they can be isomerized to aldoses that are reducing agents.

When this isomerization reaction is catalyzed by alkali, it is termed the Lobry de Bruyn-Alberda van Ekenstein reaction. By it, D-glucose, D-mannose, and D-fructose can be interconverted. The isomerizations involve a common intermediate, the 1,2-enediol.

$$
\begin{array}{ccc}
\text{HC}=\text{O} & & \text{HC}=\text{O} \\
| & & | \\
\text{HCOH} & \rightleftharpoons & \text{HOCH} \\
| & & | \\
\text{R} & & \text{R}
\end{array}
\qquad
\begin{array}{l}
\text{R}=\text{HOCH} \\
\quad\ | \\
\quad \text{HCOH} \\
\quad\ | \\
\quad \text{HCOH} \\
\quad\ | \\
\quad \text{CH}_2\text{OH}
\end{array}
$$

$$
\begin{array}{c}
\text{CH}_2\text{OH} \\
| \\
\text{C}=\text{O} \\
| \\
\text{R}
\end{array}
$$

isomerization

Enzymes are specific. For example, starch is depolymerized to D-glucose (dextrose) using a combination of enzymes, all categorized as amylases. The solution of glucose is then treated with glucose isomerase [9055-00-9] to give D-fructose in about 42% yield. No D-mannose is formed. Addition of isolated D-fructose to this solution gives the common 55% high fructose syrup (HFS) so widely used in soft drinks in the United States. HFS is about 1.5 times as sweet as sucrose.

4.3. Modifications of The Carbon Chain.

Branched-chain sugars are found in nature. For example, cladinose (2,6-dideoxy-3-*C*-methyl-3-*O*-methyl-L-*ribo*-hexose [3758-45-0]) is a component of erythromycin.

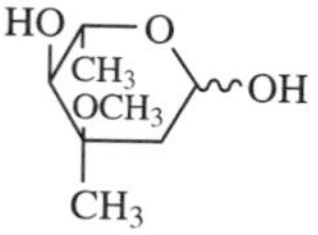

cladinose

Unsaturated sugars are useful synthetic intermediates (14,15,18,25). The most commonly used are the so-called glycals (1,5- or 1,4-anhydroalditol-1-enes). In the presence of a Lewis acid catalyst, 3,4,6-tri-*O*-acetyl-1,5-anhydro-2-deoxy-D-*arabino*-hex-1-enitol [2873-29-2], commonly called D-glucal triacetate, adds nucleophiles in both kinetically controlled and thermodynamically controlled (soft bases predominately at C-3 and hard bases primarily at C-1) reactions (14,15,18,24,25).

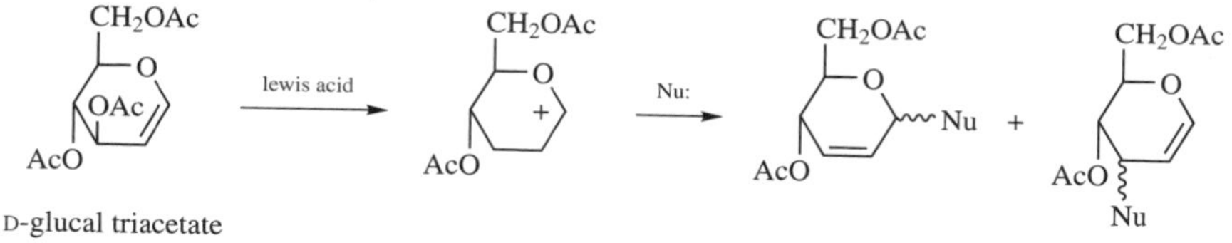

D-glucal triacetate

where Ac represents $CH_3-\overset{\overset{\displaystyle O}{\|}}{C}-$

5. Uses

Carbohydrates have widespread utilization, both as low cost, high volume commodities and as low volume specialty chemicals. Significant uses in terms of volume are surveyed here. Not covered are the lower volume uses involving carbohydrates either in the native state or in modified form; these are mainly pharmaceutical applications involving antibiotics, antigens, and synthetic drugs (2,4,20,21,25,26). In the case of drugs, monosaccharides are important as chiral synthons (chirons) as well as being used more directly to make products such as the nucleoside analogues AraA [9-(β-D-arabinofuranosyl)-9*H*-purin-6-amine] [5536-17-4], an antineoplastic and antiviral compound known by a number of trade names, and AZT (3′-azido-3′-deoxythymidine [30516-87-1]), an antiviral compound also known by a variety of trade names (see ANTIVIRAL AGENTS).

AraA   AZT

Neither are the considerable uses of carbohydrates as carbon sources for various fermentations or the uses of unrefined carbohydrates, flours, eg, described here (see FERMENTATION).

5.1. Monosaccharides.   D-Glucose is produced by complete depolymerization of starch with enzymes that catalyze the hydrolysis of both its $(1 \to 4)$ and $(1 \to 6)$ linkages. In commerce, crystalline α-D-glucopyranose is generally known as dextrose. Glucose is also isomerized to D-fructose to produce high-fructose syrup (HFS). Crystalline D-fructose also finds use in the food industry. The annual consumption, in the United States, of HFS is about 12,000,000 tons.

5.2. Oligosaccharides.   Sucrose is widely used in the food industry to sweeten, control water activity, add body or bulk, provide crispness, give surface glaze or frost, form a glass, provide viscosity, and impart desirable texture. It is used in a wide variety of products from bread to medicinal syrups (22).

Lactose occurs in milk, mainly free, but to a small extent as a component of higher oligosaccharides. Cow and goat milks contain ~4.5% lactose; human milk contains ~7.0%. Lactose is obtained from whey, a by-product of cheese manufacture. It is used as an excipient in tablets to provide bulk and rapid disintegration. It is also used in some food products where it contributes body with only ~40% the sweetness of sucrose and enhances colors and flavors.

Oligo- and higher saccharides are produced extensively by acid- and/or enzyme-catalyzed hydrolysis of starch, generally in the form of syrups of mixtures (13,27,28). These products are classified by their dextrose equivalency (DE), which is inversely proportional to their molecular size and is a measure of their reducing power, with the DE value of anhydrous D-glucose defined as 100.

Maltodextrins [9050-36-6] are mixtures of saccharides with average DE values of <20 (13,27,28). They are rather soluble, have a bland taste, and are widely used in foods. A dextrin is a product obtained by depolymerization of a polysaccharide; malto- is a prefix used with products derived from starch.

Syrup solids are also dry products, have a smaller average size, and are comparatively sweeter (13,27,28). Both maltodextrins and syrup solids are used to prevent caking; enhance dispersibility and solubility; provide body or bulk; impart desirable texture; bind, carry, and protect flavors; control extrusion expansion; provide viscosity; form films and coatings; provide an oxygen barrier; inhibit crystallization; control sweetness; improve sheen; improve organoleptic characteristics; slow meltdown; and improve freeze–thaw stability.

Specifically prepared low DE starch products in the maltodextrin class, especially those from tapioca and potato starches, mimic a fatty mouthfeel and are used as fat replacers and/or sparers (see FAT REPLACERS).

Another class of products are the cyclodextrins or cycloamyloses, a family of cyclic oligosaccharides containing α-D-glucopyranosyl units, most commonly seven (β-cyclodextrin [7585-39-9], cycloheptaamylose, cyclomaltoheptaose) (29,30). All members of this class of compounds are made by action of a specific enzyme, cyclodextrin glycosyltransferase [9030-09-5], on starch. In all, the gluco-syl units are joined by (1→4) glycosidic linkages to form a ring, the cavity of which is especially useful for forming inclusion complexes with hydrophobic guest molecules. These stable complexes are useful in the food industry to pro-vide stable flavors and fragrances in dry powder form, in the pharmaceutical industry, and in other applications where increased chemical and/or physical sta-bility, solubility control, or controlled release is desired, eg, with agricultural chemicals (see INCLUSION COMPOUNDS).

More extensive depolymerization of starch yields syrups. Syrups are puri-fied, concentrated, aqueous solutions of saccharides with an average DE value of >20. Enzymes are most often used to make syrups, although combinations of acid- and enzyme-catalyzed hydrolyses and complete acid conversion may be used. Syrups are grouped into subclasses. Some contain as little as 35% of mal-tooligosaccharides. The maltooligosaccharides are both linear and branched, the branched structures arising from amylopectin. Products with progressively higher concentrations of lower molecular weight products are progressively sweeter and less viscous. By using proper conditions, syrups with specific defined compositions, eg, high maltose syrups, are prepared. The annual consumption in the United States of crystalline dextrose and syrups from corn starch (other than high fructose syrups) is ∼ 5,000,000 tons.

5.3. Polysaccharides.   It has been estimated that > 90% of the carbohy-drate mass in nature is in the form of polysaccharides. In living organisms, carbohydrates play important roles. In terms of mass, the greatest amounts by far are structural components and food reserve materials, in that order and both in plants. However, carbohydrate molecules also serve as structural and energy storage substances in animals and serve a variety of other essential roles in both plants and animals.

Since polysaccharides are the most abundant of the carbohydrates, it is not surprising that they comprise the greatest part of industrial utilization. Most of the low molecular weight carbohydrates of commerce are produced by depolymerization of starch. Polysaccharide materials of commerce can be thought of as falling into three classes: cellulose, a water-insoluble material; starches, which are not water-soluble until cooked; and water-soluble gums.

Cellulose.   Cellulose [9004-34-6] (qv) is the principal cell-wall component of higher plants and the most abundant polysaccharide. Approximately one-half the mass of perennial plants and one-third the mass of annual plants is cellulose. It is a high molecular weight, linear, insoluble polymer of repeating β-D-glucopyr-anosyl units joined by (1→4) glycosidic linkages. Because of their linearity and stereoregular nature, cellulose molecules associate in extended regions, forming polycrystalline, fibrous bundles (31,32).

cellulose

High quality cellulose can be obtained from wood through pulping (delignification) and subsequent purification. The measure of the quality of cellulose is its content of alpha-cellulose, that portion insoluble in 18% alkali, by far the largest fraction. Beta-cellulose is that portion that dissolves in 18% alkali, but precipitates when the solution is neutralized. Gamma-cellulose remains soluble after neutralization of the 18% alkali solution. The greatest amount of cellulose used is the purified, but not highly purified, wood pulp that is used in the manufacture of paper (qv), associated products, absorbants, rayons, and nonwovens. A number of derivatives of cellulose are also commercial entities. The water-soluble ones are covered later.

Every polysaccharide contains glycosyl units with unsubstituted hydroxyl groups available for esterification or etherification. Polysaccharide derivatives are described by their degree of substitution (DS), which is the average number of substituent groups per glycosyl unit. Because each monomeric unit of a cellulose molecule has free hydroxyl groups at C-2, C-3, and C-6, the maximum DS is 3.0 for cellulose and all polysaccharides composed exclusively of neutral hexosyl units, the majority of polysaccharides.

Several cellulose esters (qv) are prepared commercially. Cellulose xanthate [9032-37-5] is made by reaction of cellulose swollen in 8.5–12% sodium hydroxide solution (called alkali cellulose [9081-58-7]) with carbon disulfide and is soluble in the alkaline solution in which it is made. When such a solution, termed viscose, is introduced into an acid bath, the cellulose xanthate decomposes to regenerate cellulose as rayon fibers or cellophane sheets (see FIBERS, REGENERATED CELLULOSICS).

Cellulose acetate [9004-35-7], prepared by reaction of cellulose with acetic anhydride and acetic acid in the presence of sulfuric acid, is spun into acetate rayon fibers by dissolving it in acetone and spinning the solution into a column of warm air that evaporates the acetone. Cellulose acetate is also shaped into a variety of plastic products, and its solutions are used as coating dopes. Cellulose acetate butyrate [9004-36-8], made from cellulose, acetic anhydride, and butyric anhydride in the presence of sulfuric acid, is a shock-resistant plastic.

Cellulose nitrate (pyroxylin) [9004-70-0], made from cellulose and a mixture of nitric and sulfuric acids, is called gun cotton and is used in explosives. Nitrates of lower DS find application in some coatings and adhesives.

Ethylcellulose [9004-57-3] is a cellulose ether (qv). As prepared commercially, ie, of high DS, ethylcellulose is thermoplastic and has a low density. It forms films of good thermostability and excellent flexibility and toughness. Ethylcellulose is used in lacquers, inks, and adhesives and is combined with waxes and resins in the preparation of hot-melt plastics.

Treatment of cellulose with acids results in preferential hydrolysis in the more accessible amorphous regions and produces a product known as microcrystalline cellulose (MCC). MCC is used to prepare fat-free or reduced-fat food products, to strengthen and stabilize food foams, as a pharmaceutical tableting aid, and as a noncaloric bulking agent for dietetic foods. It has GRAS status.

Hemicelluloses and Related Polysaccharides. Hemicelluloses [9034-32-6] are a large group of polysaccharides that are associated with cellulose in the primary and secondary cell walls of all higher plants, but otherwise have no relationship to cellulose (5). They are also present in some other plants.

Hemicelluloses (qv) are heteroglycans. They do not comprise a distinct class of chemical structures. Constituent monosaccharides are D-xylose, D-mannose, D-glucose, D-galactose, L-galactose [15572-79-9], L-arabinose, D-glucuronic acid, 4-O-methyl-D-glucuronic acid [4120-73-4], D-galacturonic acid [685-73-4], and to a lesser extent L-rhamnose, L-fucose, and various methyl ethers of neutral sugars, with a limit of perhaps six different glycosyl units per polysaccharide molecule. Both woody and nonwoody tissues contain 20–35% hemicelluloses. Some hemicelluloses are neutral polymers, but most are acidic. The most abundant have a xylan backbone, ie, a chain of (1→4)-linked β-D-xylopyranosyl units. The chain may be linear, but is often branched, usually containing short side chains and, therefore, being basically linear. The most common acidic hemicelluloses are O-acetylated (4-O-methyl-D-glucurono)xylans [9062-57-1] and L-arabino-(4-O-methyl-D-glucurono)xylans [69865-67-4, 9040-28-2, 98913-73-6], both often containing minor amounts of other sugar units as well. In the former, which are the preponderant hemicelluloses of woody angiosperms, the 4-O-methyl-α-D-glucopyranosyluronic acid units are most often joined to D-xylopyranosyl main chain units by (1→2) linkages. Some hemicelluloses contain unmethylated D-glucopyranosyluronic acid units, both the methylated and unmethylated forms in the same molecule being common. The number of uronic acid units varies considerably. Most hardwood xylans have approximately one uronic acid unit per 10 D-xylosyl units distribulted nonuniformly. Acetyl groups occur to the extent of 3–17%, with the greatest number being present in hardwood hemicelluloses.

The L-arabino-(4-O-methyl-D-glucurono)xylans are found in softwoods and annual plants. The L-arabinose is present primarily as α-L-arabinofuranosyl units, although β-L-arabinopyranosyl units may also be present. In either case, the arabinosyl units are often, but not always, present as single-unit branches, as are the uronic acid units.

Cell walls of woods contain other subgroups of hemicelluloses, in particular those composed primarily of D-mannopyranosyl or D-galactopyranosyl units. Glucomannans [11078-31-2] comprise 3–5% of the wood of angiosperms and 3–12% of the wood of gymnosperms. Galactoglucomannans [9040-29-3] are also common.

Arabinogalactans [9036-66-2] appear to be ubiquitous in plant materials. They form a family of branched polysaccharides with backbones made up predominately of (1→3)-linked β-D-galactopyranosyl units with varying amounts of (1→6)-linked β-D-galactopyranosyl units. The L-arabinose is present primarily as L-arabinofuranosyl units. Some are attached to the backbone as single units; others may be in short chains. Nonreducing ends may be terminated with β-L-arabinopyranosyl units. Other units that may be present in arabinogalactans are L-rhamnopyranosyl (up to 11%), D-mannopyranosyl (up to 16%), D-xylopyranosyl (up to 7%), D-glucopyranosyl (up to 4%), D-glucopyranosyluronic acid and/or 4-*O*-methyl-D-glucopyranosyluronic acid (up to 28%), and D-galacto-pyranosyluronic acid and/or 4-*O*-methyl-D-galactopyranosyluronic acid (up to 26%) units. Not all arabinogalactans are acidic. Water-extractable arabinogalactans are abundant in the wood of larches. The fact that they are water-extractable indicates that they are not associated with lignin (qv) through chemical linkages or physical interactions and not involved in the construction of secondary cell walls. Therefore, larch arabinogalactans [37320-79-9] are probably not properly hemicelluloses.

Some hemicelluloses are partially extractable with water, but they are more completely extracted with alkaline solutions after removal of lipids and lignin. Delignified plant material is termed, holocellulose. Neutralization of the alkaline extract effects precipitation of the more linear and less acidic hemicelluloses, termed the hemicellulose A [63100-39-0] fraction. The more acidic and more branched material, termed hemicellulose B [63100-40-3], is precipitated with ethanol (70%) from the remaining neutral solution. Hemicellulose B types are usually water soluble after extraction.

Certain cereal grains, especially wheat and rye, contain hemicellulose-like arabinoxylans [9040-27-1], commonly called pentosans. Wheat flour pentosans are divided into two types: water-soluble and water-insoluble arabinoxylans, which respectively constitute ~1.1–1.6% and 0.4–0.7% of the total flour. These polysaccharides have functional roles in dough development and baking performance. The water-soluble wheat-flour arabinoxylans consist of a $(1 \rightarrow 4)$-linked chain of β-D-xylopyranosyl units substituted at O-2 and/or O-3 with single-unit α-L-arabinofuranosyl units. Preparations from each source consist of a family of molecules of various molecular weights and xylose/arabinose ratios.

Starches.   Starch (qv) occurs in the form of granules, which must be cooked before they will release their water-soluble molecules. It is common to speak of solutions of polysaccharides, but in general, they do not form true solutions because of their molecular sizes and intermolecular interactions; rather they form molecular dispersions. The general rheological properties of polysaccharides like the starch polysaccharides are described below under the discussion of polysaccharides as water-soluble gums. Starch use is widespread and permeates the entire economy because it (corn starch in particular) is abundantly available, inexpensive, and occurs in the form of granules that can be easily handled and reacted.

All green plants package and store carbohydrate (D-glucose) in the form of starch granules. Starch granules are quasi-crystalline, dense, insoluble in cold water, and only partially hydrated. The sizes and shapes of granules are specific for the plant of origin. Granules can be easily isolated from suspensions by filtration or centrifugation, resuspended, reacted, and recovered (13,33).

Normal corn starch is composed of ~ 28% of the linear polysaccharide amylose and ~72% of the branched polysaccharide amylopectin. Amylose is a linear polysaccharide composed of $(1 \rightarrow 4)$-linked α-D-glucopyranosyl units. Its degree of polymerization (DP, the number of monosaccharide units it contains) is 200–22,000 (mol wt 32,000–3,600,000), depending on the source and method of preparation. Amylose can have several conformations. In the solid state, it probably exists most often as a left-handed, six-fold helix. In solution, it is a loosely wound and extended helix that behaves as a random coil.

Amylopectin has a branch-on-branch structure. Amylopectin molecules are composed of chains of α-D-glucopyranosyl units joined by $(1 \rightarrow 4)$ linkages; branches are formed by joining these chains with α-D-$(1 \rightarrow 6)$ linkages. The average chain length is 20–30, although branch points are not equally spaced. In the currently accepted model of an amylopectin molecule, the branches occur in clusters. The molecular weight of amylopectin has been measured as $5 \times 10^7 - 2 \times 10^8$ (DP $3 \times 10^5 - 2.5 \times 10^6$), depending on the source and method of preparation. Granules of the so-called waxy types of starch contain only amylopectin molecules. Potato starch amylopectin occurs as a natural phosphate ester (13).

Through genetic manipulation, corn cultivars with altered starch compositions have been developed. Various modified and derivatized starches are produced by treating a slurry of starch granules with chemicals or enzymes (13,33). After treatment, the products are again recovered, washed, and dried. Although these modifications and derivatizations are done to effect significant improvements in physical properties, the amount of chemical change required to effect functional property alteration is usually only very slight.

General Properties of Starches. Undamaged starch granules are not soluble in room temperature water. Heating a starch in water causes the granules to gelatinize. Gelatinization is the disruption of molecular order within starch granules, which occurs as they are heated in the presence of water. Loss of organized structure results in irreversible granule swelling and loss of crystallinity. Continued heating of starch granules in excess water effects pasting. Paste formation is a result of further granule swelling and leaching of soluble components, primarily amylose. If shear is applied at this stage, granules are disrupted. A starch paste is a viscous mass consisting of a continuous phase of dissolved starch polymer molecules and a discontinuous phase of granule framents and retrograded starch polymer molecules. Cooling of a hot paste usually produces a firm, viscoelastic gel.

The viscosity obtained by cooking a suspension of starch is determined by the starch type, derivatization and/or modification, solids concentration, pH, amount of agitation during heating, rate of heating, maximum temperature reached, time held at that temperature, agitation during holding, and the presence of other ingredients.

An aqueous dispersion of an unmodified starch containing amylose will gradually form a precipitate through association of linear segments. This process is called retrogradation or set-back.

The properties of starches are a reflection of the properties of their constituent amylose and amylopectin molecules. For example, high amylose starches are difficult to gelatinize because of the extra energy needed to disassociate and hydrate the aggregates of amylose; form firm, opaque gels; and can be used to make strong, tough films. Their solutions and gels will undergo retrogradation. Waxy maize starches, even when they are underivatized, gelatinize more easily and yield viscous, almost transparent solutions that will not form firm gels.

In general, derivatization (etherification or esterification, see below) increases solution and gel clarity, reduces the tendency to gel, improves water binding, increases freeze–thaw stability, reduces the gelatinization temperature, increases peak viscosity, and reduces the tendency to retrograde.

Combinations of substitutions are used to obtain desired properties for specific applications.

In general, all starches can be digested in the human small intestine, and the absorbed D-glucose is used for energy and a source of carbon. Cooked (pasted) starch is much, usually to a high degree, more digestible than is raw starch, and there are nondigestible (resistant) forms of starch.

Oxidized Starches.   Alkaline hypochlorite treatment introduces carboxyl and carbonyl groups, effects some depolymerization, and produces whiter (bleached) products that produce softer, clearer gels (33). Ammonium persulfate is used in some paper mills with continuous thermal cookers to prepare *in situ* high solids, low-viscosity dispersions. Much of the hypochlorite-oxidized starch and all the ammonium persulfate-oxidized starch is used in the paper industry. The low solution viscosity at high solids and good binding and adhesive properties of oxidized starches make them especially effective as textile and paper sizes.

Dextrins.   Dextrins [9004-53-9], like oxidized starches, are in the class of so-called converted starches (33). Dextrins are produced by dry heating starch with or without a catalyst (acidic or alkaline). Because there are a number of variables in the process, a wide range of dextrins with widely varying properties can be produced. All are characterized by higher solubility, lower viscosity, film-forming ability, and loss of the ability to gel. High-solids solutions of some of the more highly converted dextrins produce the tacky, quick-setting adhesives used in paper products.

Acid-Modified Starches.   Acid-modified starches are prepared by treating a suspension of starch granules with dilute mineral acid. In this process, a small amount of glycosidic bond hydrolysis occurs, resulting in products that produce much less viscosity (33). A concurrent weakening of the granule structure occurs. The result is that there is less granule swelling and more granule disintegration when acid-modified starches are heated in water; and although they have reduced viscosity-imparting power, they form gels with improved clarity and increased strength. These acid-modified starches, also called thin-boiling and acid-thinned starches, are used in large quantities as textile warp sizes.

Starch Ethers.   A large number of starch ethers has been prepared and patented; only a few are manufactured and used commercially (13,33). Commercially available starch ethers are the hydroxyalkyl ethers, hydroxyethylstarch [9005-27-0] and hydroxypropylstarch [9049-76-7], and cationic starches.

Essentially all starch derivatives are made by adding the required reagent (in the case of starch hydroxyalkyl ethers, ethylene oxide, or propylene oxide) to an agitated, alkaline (pH 7–12), aqueous starch suspension (35–45% solids) at a slightly elevated temperature. After the required reaction time, the derivatized granules are recovered, washed, and dried. The majority of starch derivatives have degrees of substitution of <0.1. Monofunctional starch derivatives are made to increase starch paste stability. Increased stability results from the introduction of substituent groups that interfere with intermolecular associations.

Hydroxyethylstarch is widely used with synthetic latexes in the surface sizing of paper and as a coating binder. For these uses, the hydroxyethylstarch is acid-thinned, oxidized, or dextrinized. Hydroxypropylstarch is used in foods to provide viscosity stability and to ensure water-holding during low-temperature storage.

Cationic starches are used as wet-end additives in the manufacture of paper (see later).

Starch Esters.   As with the starch ethers, a large number of starch esters have been prepared and patented, but only a few are manufactured and used commercially. Both inorganic and organic acid esters can, and have been, made. The latter are prepared by the same general procedure used to make starch ethers.

Starch acetates [9045-28-7] are made by reaction of starch with acetic anhydride under alkaline conditions (13,33). Starch acetates are used in foods to provide paste clarity and viscosity stability at low temperatures. A waxy maize starch acetate is most commonly used. Waxy maize starch acetates for food use are often also cross-linked. Acetylated starches are also used in warp sizing of textiles.

Starch succinates [39316-70-6] are also used as thickening agents in foods. The 1-octenylsuccinate half-ester [52906-93-1], sold as its sodium salt [66829-29-6], has surface active (emulsifying) properties.

Starch sodium phosphate monoesters [11120-02-8] are prepared by heating mixtures of 10% moisture starch and sodium monohydrogen and dihydrogen phosphates or sodium tripolyphosphate (13,33). Starch phosphate monoesters are used primarily in foods as pudding starches and in oil-in-water emulsions.

Cross-linked Starches.   The polymer chains in starch granules can be cross-linked with difunctional reagents that form diethers or diesters (13,33). The properties imparted to the starch by such cross-linking are unique and, therefore, these derivatives are considered separately. Diphosphate ester cross-links can be introduced by reaction of starch with phosphoryl chloride or sodium trimetaphosphate. Glycerol diethers of starch are made by reaction of starch with epichlorohydrin, although this reaction is no longer used in the United States to prepare modified food starch. A small amount of cross-linking, eg, 1 cross-link per 1000 D-glucopyranosyl units, greatly reduces both the rate and the degree of granule swelling and the sensitivity of starch slurries to processing conditions.

Cross-linking is employed when a stable, high viscosity starch paste is needed and particularly when the dispersion is to be subjected to high temperature, high shear, and/or low pH. Food starches, especially those made from waxy maize, potato, and tapioca starch, are usually both cross-linked and phosphorylated, acetylated, or hydroxypropylated to provide appropriate cooking, viscosity, and textural properties. Examples of their application are their use in canned foods that are to be retort-sterilized and in the preparation of spoonable salad dressings, where products stable to high shear at low pH are required. Starch products that do not gelatinize, even under autoclave conditions, can be made by introducing higher degreees of cross-linking. Starch products with properties of lightly cross-linked starches are made by dry heating under slightly alkaline conditions.

Cationic Starches.   Commercial cationic starches are starch ethers that contain a tertiary amino or quaternary ammonium group, eg, the diethylaminoethyl ether of starch or the 2-hydroxy-3-(trimethylammonio)propyl ether of starch [9063-45-0], sold as its chloride salt [56780-58-6] (13,33).

Cationic starches are used in papermaking. When they are used as a wet-end additive, affinity between the cationic starch and cellulose fibers, which have a negative charge, results in almost complete and irreversible adsorption of the starch. Cationic starches are also used in surface sizing of paper and as coating binders. Amphoteric starches made by introducing anionic groups, such as phosphate monoester or sulfosuccinate ester groups or carboxyl groups produced by oxidation, to cationic starches perform better in some applications.

Pregelatinized Starches.   Suspensions of starches and starch derivatives can be pasted/cooked and dried to yield a variety of products that can be dispersed in cold water to yield pastes comparable to those obtained by cooking granular starch products. These products are made for convenience of use.

Starch Graft Copolymers.   Graft copolymers can be made by forming radicals on a chain of a starch or a modified starch, particularly hydroxyethyl-starch, most commonly with cerium(III) ions, then introducing a monomer (13). Commercial products that have been made in this way are starch–graft–styrene–butadiene latex copolymer and starch–graft–polyacrylonitrile copolymer, which was subsequently treated with alkali to convert the nitrile groups to a mixture of carbamoyl and carboxylate groups.

Cold-Water Swelling Starches.   Special physical treatment produces starch granules that will swell in water without heating. Molecular dispersions can be formed by application of shear to the swollen granules.

Water-Soluble Gums/Hydrocolloids.   Gums (qv) are polymeric substances that, in an appropriate solvent or swelling agent, form highly viscous dispersions or gels at low dry-substance content. Commonly, the term industrial gums refers to water-soluble polysaccharides (glycans in official carbohydrate nomenclature) or polysaccharide derivatives used industrially (8,34). They are classified both by structure (Table 1) and by source (Table 2). Particularly in the food industry, the term hydrocolloid is often used interchangeably with gum.

The usefulness of such industrial gums is based on their physical properties, in particular their capacity to thicken and/or gel aqueous solutions and otherwise to control water. Because all gums modify or control the flow of aqueous solutions, dispersions, and suspensions, the choice of which gum to use for a particular application often depends on its secondary characteristics. These secondary characteristics are responsible for their utilization as adhesives, binders, bodying agents, bulking agents, crystallization inhibitors, clarifying agents, cloud agents, emulsifying agents, emulsification stabilizers, encapsulating agents, film formers, flocculating agents, foam stabilizers, gelling materials, mold release agents, protective colloids, suspending agents, suspension stabilizers, swelling agents, syneresis inhibitors, texturing agents, and whipping agents, in coatings, and for water absorption and binding.

Gums are tasteless, odorless, colorless, and nontoxic. None, except the starches and starch derivatives, are broken down by human digestive enzymes, but all are subject to microbiological attack. All can be depolymerized by acid- and enzyme-catalyzed hydrolysis of the glycosidic (acetal) linkages joining the monomeric (saccharide) units.

All native and modified polysaccharides have a range of molecular weights. The average composition and distribution of molecular weights in a gum sample can vary with the source, the conditions used for isolation or preparation, and

Table 1. **Classification of Selected, Native Polysaccharides by Structure**

	Examples
Classification by shape[a]	
linear	agarose, algins, amyloses, carrageenans, cellulose, chondroitins, chitins, colominic acid [poly(N-acetyl-neuraminic acid], curdlan, dermatan sulfate, furcellaran, gellan, glucomannans, heparin, hyaluronic acid, inulin, keratin sulfate, laminarans[b], mannans, nigeran, pectic acids/pectates, pullulan
branched	
short branches on an essentially linear backbone	arabinans[c], arabinogalactans, galactoglucomannans, galactomannans (guar gum, locust bean gum), konjac glucomannan, psyllium seed gum, rhamsan, scleroglucan, succinoglycan, welan, xanthan, xylans, xyloglucans
branch-on-branch structures	amylopectins, arabinoxylans, flaxseed polysaccharide (acidic), glycogens, gum arabics, gum ghatti, gum karaya, gum tragacanth (tragacanthin), okra gum, rhamnogalacturonans I and II
Classification by monomeric units[d]	
homoglycans	amylopectins, amyloses, arabinans, cellulose, chitins, colominic acid, curdlan, glycogens, laminaransb, mannans, nigeran, pullulan, scleroglucan
diheteroglycans	algins, arabinogalactans, carrageenans, chondroitins, furcellarans, galactomannans, glucomannans, hyaluronic acid, inulin, keratan sulfate, konjac mannan, pectic acids/pectates, succinoglucan, xylans
triheteroglycans	arabinoxylans, dermatan sulfates, galactoglucomannans, gellan, gum karaya, heparin, rhamsan, xanthan
tetraheteroglycans	flaxseed polysaccharide (acidic), gum arabics, okra gum, psyllium seed gum, welan, xyloglucans
pentaheteroglycans	gum ghatti, gum tragacanth (tragacanthin)
octaheteroglycan	rhamnogalacturonan I
nonaheteroglycan	rhamnogalacturonan II
Classification by charge	
neutral	agarose, amylopectins, amyloses, arabinans, arabinogalactans, cellulose, chitins, curdlan, galactoglucomannans, galactomannans, glucomannans, glycogens, inulin, laminarans, mannans, konjac mannan, nigeran, pullulan, scleroglucan, xyloglucans
anionic (acidic)[e]	algins, arabinoxylans, carrageenans, chondroitins, colominic acid, dermatan sulfates, flaxseed polysaccharide, furcellarans, gellan, gum arabics, gum ghatti, gum karaya, gum tragacanth (tragacanthin), heparin, hyaluronic acid, keratan sulfate, okra gum, pectic acids/pectates, pectins, rhamnogalacturonans I and II, psyllium seed gum, rhamsan, succinoglycan, welan, xanthan, xylans
cationic	chitosans (not native)

[a] Primary examples. For example, arabinoxylans occur in different architectures, compositions, and charges.

[b] Contains a few long-chain branches. Some chains are terminated at the reducing end with a second type of unit.

[c] The predominate structure.

[d] Considers only the basic monosaccharide units. A derivatized monosaccharide unit, such as a D-galactopyranosyl 6-sulfate unit, is not considered as a unit separate from a D-galactopyranosyl unit, for example.

[e] From the presence of uronic acid, sulfate half-ester, pyruvyl cyclic acetal, or succinate half-ester groups.

Table 2. **Classification of Commercial Polysaccharides by Source**

Class	Examples
algae (seaweeds)	agars, algins, carrageenans, furcellarans, laminarans
higher plants	
insoluble	cellulose
extract	pectins
seeds	corn starches, rice starches, wheat starches, guar gum, locust bean gum, psyllium seed gum
tubers and roots	potato starch, tapioca/cassava starch, konjac glucomannan
exudates	gum arabics, gum karaya, gum tragacanth
microorganisms (fermentation gums)	curdlan, dextrans, gellan, pullulan, scleroglucan, welan, xanthans
animal	chitins/chitosans (also a cell-wall constituent of some fungi)
derived	
from cellulose	carboxymethylcelluloses, cellulose acetates, cellulose acetate butyrates, cellulose nitrates, ethylcellulose, hydroxyalkylcelluloses, hydroxyalkylalkylcelluloses, methylcelluloses
from starches[a]	starch acetates, starch adipates, starch 1-octenylsuccinates, starch phosphates, starch succinates, carboxymethylstarches, hydroxyethylstarches, hydroxypropylstarches, cationic starches, oxidized starches, dextrins
from guar gum	carboxymethylguar gum, carboxymethyl(hydroxypropyl)-guar gum, hydroxyethylguar gum, hydroxypropylguar gum, cationic guar gum
synthetic	polydextrose

[a] It is common for a commercial modified starch to have undergone two or more different reactions or treatments.

any subsequent treatment(s). In all except bacterial polysaccharides, the percentage of individual monomeric unit types varies from molecule to molecule and from sample to sample. Because both molecular size and structure determine physical properties, various functional types of a given gum are produced by controlling the source and isolation procedure (in the case of natural gums) or derivatization method (in the case of derived gums) and subsequent treatment(s).

In general, gums do not form true solutions. Rather, because of their molecular weights and intermolecular interactions, they form dispersions, where the particles may be dispersed molecules and/or aggregated clusters of molecules. The rheology or flow characteristics and gel properties of gum solutions is a function of particle solvation, particle size, particle shape, particle flexibility and ease of deformation, and the presence and magnitude of charges. In general, the rheology of gum solutions is pseudoplastic or thixotropic, ie, they exhibit shear thinning. Most gums are available in a range of viscosity grades.

Polysaccharide gels in general are composed of 99.0–99.5% water and 0.5–1.0% gum. Important characteristics of gels are means of gelation (chemical gelation, thermogelation), reversibility, texture (brittle, elastic, plastic), rigidity (rigid or firm, soft or mushy), tendency for syneresis, and cutable or spreadable. Gels are composed of interconnected fringed micelles (junction zones).

Algins. Algins are salts (generally sodium [9005-38-3], ammonium [9005-34-9], or potassium [9005-36-1]) or esters (propylene glycol) of alginic acid.

Alginic acid [9005-32-7] is a generic term for polymers of D-mannuronic acid and L-guluronic acid. Alginic acid molecules contain at least three different types of polymer segments: poly(β-D-mannopyranosyluronic acid) segments, poly(α-L-gulopyranosyluronic acid) segments, and segments with alternating sugar units. The ratios of the constituent monomers and the chain segments vary with the source and determine the specific properties of the preparation. All linkages are $1 \rightarrow 4$, making alginates linear polymers. The shapes of the poly(D-mannopyranosyluronic acid) and the poly(L-gulopyranosyluronic acid) segments are quite different because the β-D-mannopyranosyluronic acid units are in the $^4C_1$ conformation and diequatorially linked, whereas the α-L-gulopyranosyluronic acid units are in the $^1C_4$ conformation and diaxially linked. The different conformations make the former segments flat and the latter buckled.

a poly(β-D-mannuronosyl) segment

a poly(α-L-guluronosyl) segment

Algins are extracted from brown algae (Phaeophyceae). The primary U.S. source is the beds of giant kelp (*Macrocystis pyrifera*) that grow off the coast of southern California. The polymer is extracted by treating the seaweed with a sodium carbonate solution. It is recovered from the extract by precipitation as alginic acid or as the calcium salt [9005-35-0], which is then washed with acid to convert it into alginic acid. The alginic acid is then treated with a base to convert it into the desired salt, or partially neutralized alginic acid is treated with propylene oxide to make the propylene glycol ester [9005-37-2].

An important and useful property of alginates is their ability to form gels by reaction with calcium ions. Different types of gels are formed with alginates from different sources. Alginates with a higher percentage of polyguluronate segments form the more rigid, more brittle gels that tend to undergo syneresis. Alginates with the higher percentage of polymannuronate segments form the more elastic, more deformable gels with a lesser tendency to undergo syneresis.

Carrageenans, Agars, and Furcellarans. Carrageenan is a generic term applied to polysaccharides extracted from a number of closely related species of red seaweeds. Agar [9002-18-0] and furcellaran [9000-21-9] are also red seaweed extracts and are members of the same larger family. All polysaccharides in this family are derivatives of linear galactans. All have alternating monosaccharide units and linkages. In all members of the family, one sugar unit is a β-D-galactopyranosyl unit with a glycosidic linkage to O-3. In all except agar, the other unit is a 3,6-anhydro-α-D-galactopyranosyl unit with a glycosidic linkage to O-4. In agar, the other unit is a 3,6-anhydro-α-L-galactopyranosyl unit.

Commercial carrageenans are composed primarily of three types of polymers: kappa-, iota-, and lambda-carrageenan.

kappa-carrageenan, R = H
iota-carrageenan, R = SO_3^2

lambda-carrageenan

The molecular weights of the carrageenans, agars, and furcellaran average ~250,000. The half-ester sulfate contents are 0–3% in agarose [9012-36-6], more properly termed agaran, the linear component of agar, 12–16% in furcellaran, ~25% in kappa-carrageenan [1114-20-8], ~32% in iota-carrageenan [9062-07-1], and ~35% in lambda-carrageenan [9064-57-1]. Each polymer is heterogeneous. Each commercial gum is believed to be generated from native precursor polysaccharides during production.

Kappa- and iota-carrageenans exist as right-handed, threefold helices that form double helices reversibly. The double-helical segments of Kappa- and iota-carrageenans can then interact to form a three-dimensional gel network. The conformation of lambda-carrageenan, a nongelling gum, has been described as a zigzagging ribbon.

Carrageenans and agars are structural polysaccharides of the *Rhodophyceae* (red algae). Carrageenans are extracted primarily from *Chondrus* and *Gigartina* species. Furcellaran is obtained primarily from *Furcellaria* species. Agars are obtained primarily from *Gelidium* and *Gracilaria* species.

A useful property of the red seaweed extracts is their ability to form gels with water and milk. Kappa-carrageenan reacts with milk protein micelles, particularly Kappa-casein micelles. The thickening effect of Kappa-carrageenan in milk is 5–10 times greater than it is in water; at a concentration of 0.025% in milk, a weak thixotropic gel is formed.

Agars are the least soluble of this class of polysaccharides; they can be dispersed only at temperatures >100°C. When agar dispersions are cooled, strong, brittle, turbid gels form. Agar gels remelt when heated, synerese, and are unstable to freeze–thaw cycles. By far the greatest use of agar in the United States is in the preparation of microbiological culture media. Agar is also used in bakery icings. Agar and agarose are used in making gels for electrophoresis, in gel-filtration chromatography, and in several applications in biotechnology.

Guar and Locust Bean Gums. Guaran, the purified polysaccharide from guar gum [9000-30-0], is a galactomannan [11078-30-1]. It has a mannan backbone, a linear chain of (1 → 4)-linked β-D-mannopyranosyl units with, on the average, one of every 1.8 mannosyl units substituted with a (1 → 6)-linked α-D-galactopyranosyl unit. The mannan chain is rather evenly substituted with D-galactopyranosyl units, but still contains some unsubstituted or smooth regions. Its molecular weight is 220,000 ± 20,000 (DP1360 ± 125).

Like guaran, and the endosperm polysaccharides of other legumes, locust bean (carob) gum [9000-40-2] is also a galactomannan. Like guaran, it has a linear backbone of $(1 \rightarrow 4)$-linked β-D-mannopyranosyl units. However, in locust bean gum, approximately only one of every 3.9 β-D-mannopyranosyl units, on the average, is substituted with an α-D-galactopyranosyl unit attached at O-6.

Commercial guar gum is not purified guaran but the ground endosperm of guar seeds. Guar gum forms very high viscosity, pseudoplastic solutions at low concentrations. Guar endosperm preparations can be derivatized with the same reagents and catalysts used to modify starch and cellulose. The following products are prepared by proprietary processes: hydroxypropyl- [39421-75-5], hydroxyethyl- [39465-11-7], sodium carboxymethyl- [51190-15-3], sodium carboxymethyl(hydroxypropyl)- [39454-79-0], and 2-hydroxy-3-(trimethylammonio) propyl - [67034-33-7] (made as its chloride salt [65497-29-2]) guar gums. Derivatives are made to control the rate of hydration, peak viscosity, ash content, insoluble material, heat stability, and compatibility with other materials.

Polymer chains of guar gum and its derivatives, in fact of all galactomannans, are readily cross-linked with borate and titanium ions. Gels formed in this way are rubbery in Nature.

Commercial locust bean gum is the ground endosperm of the seeds of the locust bean (carob) tree. The general properties of locust bean gum are similar to those of guar gum. Differences are its low cold-water solubility and its synergistic gelation with Kappa-carrageenan, furcellaran, and xanthan [11138-66-2].

Gum Arabic. Of the gums of ancient commerce, which were dried exudations collected by hand from various trees and shrubs, only gum arabic [9000-01-5], also called gum acacia and acacia gum, is still in significant use. Gum arabic preparations are mixtures of highly branched, branch-on-branch, acidic polysaccharides. The polysaccharides have a branched main chain of β-D-galactopyranosyl units. Attached to this backbone are side chains containing L-arabinofuranosyl, L-rhamnopyranosyl, D-galactopyranosyl, and D-glucopyranosyluronic acid units in varying amounts depending on the source. Generally accepted values for number average and weight average molecular weights are 250,000 and 580,000, respectively; these values correspond to DPs of 155 and 3600.

Gum arabic comes from various species of *Acacia*. The gum exudes through cracks, injuries, and incisions in the bark and is collected by hand as dried tears. Gum arabic is unique among gums because of its high solubility and the low viscosity and Newtonian flow of its solutions. While other gums form highly viscous solutions at 1–2% concentration, 20% solutions of gum arabic resemble a thin sugar syrup in body and flow properties.

Pectins. Pectic acids [9046-40-6] are galacturonans [poly(α-D-galactopyranosyluronic acids), galacturonoglycans] [9046-38-2, 84149-03-1, 25249-06-3] without, or with only a negligible content of, methyl ester groups. Pectic acids have various degrees of neutralization. Salts of pectic acids are pectates. Pectinic acids are galacturonans with various, but greater than negligible, contents of methyl ester groups. Pectinic acids may have varying degrees of neutralization. Salts of pectinic acids are pectinates. Pectins [9000-69-5, 16048-08-1, 58128-44-2] are mixtures of polysaccharides that originate from plants, contain pectinic acids as primary components, are water-soluble, and whose solutions will gel under suitable conditions. The term pectin is often used in a generic sense to designate those water-soluble galacturonans of varying methyl ester content and degree of neutralization that are capable of forming gels. Commercial pectins are formed during extraction. They are essentially homogalacturonans, but do contain some L-rhamnopyranosyl units in the galacturonan chain.

Commercial pectins are subdivided according to their degree of esterification (DE), a designation of the percent of carboxyl groups esterified with methanol. Pectins with DE >50% are high-methoxyl pectins (HM pectins) [65546-99-8]; those with DE <50% are low-methoxyl pectins (LM pectins) [9049-34-7]. The degree of amidation (DA) indicates the percent of carboxyl groups in the amide form.

The key feature of all pectin molecules is a linear chain of (1→4)-linked α-D-galactopyranosyluronic acid units, making it an α-D-galacturonan [a poly (α-D-galactopyranosyluronic acid) or an α-D-galacturonoglycan] [9046-38-2, 84149-03-1, 25249-06-3]. In all natural pectins, some of the carboxyl groups are in the methyl ester form. Depending on the isolation conditions, the remaining free carboxylic acid groups may be partly or fully neutralized. The DE strongly influences the solubility, gel-forming ability, conditions required for gelation, gelling temperature, and gel properties of the preparation.

Inserted L-rhamnopyranosyl units may provide the necessary irregularities (kinks) in the structure required to limit the size of the junction zones and produce a gel. The presence of side chains composed of D-xylosyl units may also be a factor that limits the extent of chain association. Junction zones are formed between regular, unbranched pectin chains when the negative charges on the carboxylate groups are removed (addition of acid), hydration of the molecules is reduced (addition of a cosolute, usually sugar, to a solution of HM pectin), and/or pectinic acid polymer chains are bridged by multivalent, eg, calcium, cations.

Sodium and calcium pectates, pectic acid, and pectinic acid all occur in the solid state as right-handed helices. In solid pectinic acid, the polymer molecules pack so that the chains are parallel to each other; the pectates pack as corrugated sheets of antiparallel chains. Junction zones in pectinic acid (HM pectin plus sucrose) gels are believed to be formed by a columnar stacking of methyl ester groups to form cylindrical hydrophobic areas parallel to the helix axes. LM pectin [9049-34-7] gels only in the presence of divalent cations. Two models for the formation of junction zones in calcium pectate [12672-40-1, 40022-66-0] gels have been proposed. One suggests an aggregation of chains by a cross-linking of carboxylate groups with calcium ions to form a structure similar to that of the corrugated sheets of antiparallel helices (3–6 chains in an average junction zone) found in solid calcium pectate. The other is the "egg box" model used to describe the formation of calcium alginate [9005-35-0] gels.

Xanthan.   Xanthan, known commercially as xanthan gum [11138-66-2], has a main chain of (1→4)-linked β-D-glucopyranosyl units; therefore, the chemical structure of the main chain is identical to the structure of cellulose [9004-34-6]. However, in xanthan, every other β-D-glucopyranosyl unit in the main chain is substituted on O-3 with a trisaccharide unit. The trisaccharide side chain consists of (reading from the terminal, nonreducing end in towards the main chain) a β-D-mannopyranosyl unit linked (1→4) to a β-D-glucopyranosyluronic acid unit linked $(1 \rightarrow 2)$ to a 6-O-acetyl-α-D-mannopyranosyl unit. About one-half of the terminal β-D-mannopyranosyl units carry a pyruvic acid group as a 4,6-di-O-acetal. The molecular weight is probably on the order of 2×10^6, although much higher figures have been reported.

The unusual properties of xanthan undoubtedly result from its structural rigidity, which in turn is a consequence of its linear, cellulosic backbone that is stiffened and shielded by the trisaccharide side chains. The conformation of xanthan in solution is a matter of debate. It does appear that the conformation changes with conditions.

Xanthan is the extracellular (exocellular) polysaccharide produced by *Xanthomonas campestris*. As with other microbial polysaccharides, the characteristics (polymer structure, molecular weight, solution properties) of xanthan preparations are constant and reproducible when a particular strain of the organism is grown under specified conditions, as is done commercially. The characteristics vary, however, with variations in the strain of the organism, the sources of nitrogen and carbon, degree of medium oxygenation, temperature, pH, and concentrations of various mineral elements.

Xanthan solutions are extremely pseudoplastic and have high yield values. These properties make xanthan almost ideal for the stabilization of aqueous dispersions, suspensions, and emulsions. Whereas other polysaccharide solutions decrease in viscosity when they are heated, xanthan solutions containing a small amount of salt (0.1%) change little in viscosity over the temperature range 0–95°C. Although xanthan is anionic, pH has almost no effect on the viscosity of its solutions over the range pH 1–12. A synergistic viscosity increase results from the interaction of xanthan with galactomannans and with methylcellulose. A combination of xanthan and locust bean gum forms a thermally reversible gel when a hot solution of these two polysaccharides is cooled.

Cellulose Derivatives. Cellulose can be derivatized to make both water-soluble gums and hydrophobic polymers (8,31,32). Preparation of the hydrophobic cellulose esters (qv), cellulose acetates and cellulose nitrates, has already been mentioned. The water-soluble cellulose derivatives are cellulose ethers (qv).

Carboxymethylcelluloses (CMC). Carboxymethylcellulose [9004-42-6] (CMC) is the carboxymethyl ether of cellulose. To prepare CMC, cellulose is steeped in sodium hydroxide solution, and the so-called alkali cellulose is treated under controlled conditions with sodium monochloroacetate to form the sodium salt of carboxymethylcellulose and sodium chloride. Therefore, the CMC of commerce is actually sodium carboxymethylcellulose [9004-32-4].

The physical properties (solution characteristics) of CMC, and all other linear polysaccharides, whether synthetic or natural, are determined by the average chain length or degree of polymerization (DP), the degree of substitution (DS), and the uniformity of substitution. The DS of different CMC types generally ranges from 0.4 to 0.8; some products may approach a DS of 1.5. The most widely used types have a DS of 0.7 or an average of 7 carboxymethyl groups per 10 β-D-glucopyranosyl units.

CMC hydrates rapidly and forms clear solutions. Viscosity building is the single most important property of CMC. Dilute solutions of CMC exhibit stable viscosity because each polymer chain is hydrated, extended, and independent. The sodium carboxylate groups are highly hydrated, and the cellulose molecule itself is hydrated. The cellulose molecule is linear, and conversion of it into a polyanion (polycarboxylate) tends to keep it in an extended form by reason of Coulombic repulsion. This same Coulombic repulsion between the carboxylate anions prevents aggregation of the polymer chains. Solutions of CMC are either pseudoplastic or thixotropic, depending on the type.

Hydroxyethyl- and Hydroxypropylcelluloses. Hydroxyalkylcelluloses are cellulose ethers prepared by reaction of alkali cellulose with ethylene oxide, to prepare hydroxyethylcellulose (HEC) [9004-62-0], or propylene oxide, to prepare hydroxypropylcellulose (HPC) [9004-64-2].

$$\text{Cell}-\text{O}^-\,\text{Na}^+ \; + \; \underset{\substack{\diagdown\;\diagup\\ \text{O}}}{\text{CH}_2-\text{CH}-\text{R}} \; \longrightarrow \; \text{Cell}-\text{O}\!\!\left(\!\text{CH}_2-\overset{\overset{\displaystyle R}{|}}{\text{CH}}-\text{O}\!\right)_{\!n}\!\!\text{H}$$

alkali cellulose

These products are characterized in terms of moles of substitution (MS) rather than DS. MS is used because the reaction of an ethylene oxide or propylene oxide molecule with cellulose leads to the formation of a new hydroxyl group with which another alkylene oxide molecule can react to form an oligomeric side chain. Therefore, theoretically, there is no limit to the moles of substituent that can be added to each D-glucopyranosyl unit. MS denotes the average number of moles of alkylene oxide that has reacted per D-glucopyranosyl unit. Because starch is usually derivatized to a considerably lesser degree than is cellulose, formation of substituent poly(alkylene oxide) chains does not usually occur when starch is hydroxyalkylated and DS = MS.

In general, the MS controls the solubility of both HEC and HPC. For example, water-soluble grades of hydroxyethylcellulose have MS values of 1.6–3.0; those with MS 0.3–1.0 are soluble in aqueous alkali. Higher MS types of hydroxypropylcellulose become soluble in organic solvents, first polar, then nonpolar solvents.

Clear, water-soluble, oil-and grease-resistant films of moderate strength can be cast from hydroxyethylcellulose solutions. HEC is used in joint cements for wallboard, in high salt driling muds and to control the rheology and set time of oil-well cements, in latex paints, and as itself and in several modifications in other applications. Flexible, nontacky, heat-sealable packaging films and sheets can be produced from hydroxypropylcellulose by conventional extrusion techniques. Both gums can be used in the formulation of coatings. HPC can be used to form edible films and coatings.

Methylcelluloses and Hydroxyalkylmethylcelluloses. Methylcellulose [9004-67-5] contains methoxyl groups in place of some of the hydroxyl groups along the cellulose molecule. The primary hydroxyl group of cellulose is somewhat more reactive so, as with other cellulose derivatives, there is a somewhat higher degree of substitution at O-6. The next most acidic hydroxyl group is O-2. Hydroxyalkylmethylcelluloses contain, in addition to methoxyl groups, hydroxyalkoxyl groups in place of some of the hydroxyl groups. As with all other polysaccharide derivatives, the properties of methyl- and hydroxyalkylmethylcelluloses are a function of the type(s) of derivatization, the amount of each type of substituent group, the molecular weight distribution, and to some extent, the physical nature of the product, eg, fibrous vs powdered, granulation size, and surface treatment. Because these variables can be controlled to some degree, the members of this family, like other starch and cellulose derivatives, are a group of tailor-made products.

Methylcellulose is made by reaction of alkali cellulose with methyl chloride until the DS reaches 1.1–2.2. Hydroxypropylmethylcellulose [9004-65-3], the most common member of this family of products, is made by using propylene oxide in addition to methyl chloride in the reaction; MS values of the hydroxypropyl group in commercial products are 0.02–0.3. Both the true methylcelluloses and the hydroxypropylmethylcelluloses are often referred to simply as methylcelluloses. Use of 1,2-butylene oxide in the alkylation reaction mixture gives hydroxybutyl-methylcellulose [9041-56-9, 37228-15-2] (MS 0.04–0.11). Hydroxyethylmethyl-cellulose [9032-42-2] is made with ethylene oxide in the reaction mixture.

Conversion of some of the hydroxyl groups of cellulose molecules into methyl ether groups increases the water solubility of the cellulose molecule and reduces its ability to aggregate, ie, reduces intermolecular interactions. Solubility is increased even more when hydroxyalkyl groups are added to methylcellulose. Solutions of all these products behave somewhat like those of guar and locust bean gums, ie, as linear polysaccharides with short side chains that give stable solutions of high viscosity. As substituent groups are added, the solubility of the products changes from insoluble to soluble in aqueous alkali, to soluble in water, to soluble in various polar organic solvents, such as water–alcohol solutions, alcohols, and alcohol–hydrocarbon solutions.

The most interesting property of these nonionic products is thermal gelation. Solutions of members of this family of gums that are soluble in cold water, like solutions of other polysaccharides, decrease in viscosity when heated. However, unlike most other gums, when a certain temperature is reached, depending on the specific product, the solution viscosity increases rapidly and the solution gels. Gelation can occur from ~45° to ~90°C, depending on the product type. The thermal gelation is reversible.

Methylcelluloses reduce surface and interfacial tension. They form high strength films and sheets that are clear, water-soluble, oil- and grease-resistant, and have low oxygen and moisture vapor transmission rates (see BARRIER POLYMERS). They are used in a variety of applications ranging from shampoo and hair-conditioning compositions to construction applications, their largest volume use. Particularly important uses include formulation of tape joint compounds, gypsum spray plasters, and ceramic tile adhesives, grouts, and mortars.

5.4. Analysis. See references 35 and 36.

BIBLIOGRAPHY

"Carbohydrates" in *ECT* 1st ed., Vol. 2, pp. 867–881, by C. D. Hurd, Northwestern University; in *ECT* 2nd ed., Vol. 4, pp. 132–148, by C. D. Hurd, Northwestern University; in *ECT* 3rd ed., Vol. 4, pp. 535–555, by R. L. Whistler and J. R. Zysk, Purdue University; in *ECT* 4th ed., Vol. 4, pp. 911–948, by James N. BeMiller, Purdue University; "Carbohydrates" in *ECT* (online), posting date: December 4, 2000, by James N. BeMiller, Purdue University.

CITED PUBLICATIONS

1. W. Pigman and D. Horton, eds., *The Carbohydrates: Chemistry and Biochemistry*, Vol. IA, 2nd ed., Academic Press, New York, 1972.

2. B. Fraser-Reid, K. Tatsuta, and J. Thiem, eds., *Glycoscience: Chemistry and Chemical Biology*, Vol. II. Springer, 2001.

3. A. Lipták, P. Fügedi, Z. Szurmai, and J. Harangi, *CRC Handbook of Oligosaccharides*, Vol. 1, *Disaccharides*, 1990; Vol. 2, *Trisaccharides*, 1990, CRC Press, Inc., Boca Raton, Fla.

4. Ref. 2, Vol. III, 2001.

5. G. O. Aspinall, ed., *The Polysaccharides*, Vol. 1, Academic Press, New York/Orlando, 1982.

6. *Ibid.*, Vol. 2, 1983.

7. *Ibid.*, Vol. 3, 1985.

8. R. L. Whistler and J. N. BeMiller, eds., *Industrial Gums*, 3rd ed., Academic Press, San Diego, Calif., 1992.

9. S. Dumitriu, ed., *Polysaccharides*, Marcel Dekker, New York, 1998.

10. E. J. Vandamme, S. DeBaets, and A. Steinbüchel, eds., *Biopolymers*, Vols. 5 and 6, *Polysaccharides I and II*, Wiley-VCH, Weinheim, Germany, 2002.

11. J. W. Rowe, ed., *Natural Products of Woody Plants I. Chemicals Extraneous to the Lignocellulosic Cell Wall*, Springer-Verlag, Berlin, 1989.

12. R. Ikan, ed., *Naturally Occurring Glycosides*, John Wiley & Sons, Inc., Chichester, England, 1999.

13. J. N. BeMiller and R. L. Whistler, eds., *Starch: Chemistry and Technology*, 3rd ed., Academic Press, Orlando, Fla., 2003.

14. Ref. 2, Vol. I, 2001.

15. B. Ernst, G. W. Hart, and P. Sinaÿ, eds., *Carbohydrates in Chemistry and Biology*, Vol. 1, Wiley-VCH, Weinheim, Germany, 2000.

16. *Ibid.*, Vol. 2, 2000.

17. R. V. Stick, *Carbohydrates: The Sweet Molecules of Life*, Academic Press, San Diego, 2001.

18. Ref. 1, Vol. IB, 1980.

19. Ref. 1, Vol. IIA, 1970.

20. A. Varki, R. Cummings, J. Esko, H. Freeze, G. Hart, and J. Marth, eds., *Essentials of Glycobiology*, Cold Spring Harbor Laboratory Press, Cold Spring Harbor, New York, 1999.

21. Ref. 15, Vol. 4, 2000.

22. M. Mathouthi and P. Reiser, eds., *Sucrose. Properties and Applications*, Blackie Academic and Professional, London, 1995.

23. W. W. Wells and F. Eisenberg, Jr., eds., *Cyclitols and Phosphoinositides*, Academic Press, New York, 1978.

24. R. R. Binkley, *Modern Carbohydrate Chemistry*, Marcel Dekker, Inc., New York, 1988.

25. J. Lehmann, *Carbohydrates. Structure and Biology*, Thieme, Stuttgart, Germany, 1998.

26. B. S. Paulsen, ed., *Bioactive Carbohydrate Polymers*, Kluwer Academic Publ., Dordecht, The Netherlands, 2000.

27. G. M. A. Van Beynum and J. A. Roels, eds., *Starch Conversion Technology*, Marcel Dekker, New York, 1985.

28. F. A. Schenck and R. E. Hebeda, eds., *Starch Hydrolysis Products*, VCH Publ., New York, 1992.

29. F. Stoddart, *Cyclodextrins*, Royal Society of Chemistry, London, 1989.

30. R. B. Friedman, ed., *Biotechnology of Amylodextrin Oligosaccharides*, Symposium Series, Vol. 458, American Chemical Society, Washington, D.C., 1991.

31. J. F. Kennedy, G. O. Phillips, and P. O. Williams, and L. Piculell, eds., *Cellulose and Cellulose Derivatives: Physico-chemical Aspects and Industrial Applications*, Woodhead Publ., Cambridge, 1995.

32. D. Kiemm, B. Philipp, T. Heinze, U. Heinze, and W. Wagenknecht, eds., *Comprehensive Cellulose Chemistry*, Vols. 1 and 2, Wiley-VCH, Weinheim, Germany, 1998.
33. O. B. Wurzburg, ed., *Modified Starches: Properties and Uses*, CRC Press, Inc., Boca Raton, Fla., 1986.
34. A. M. Stephen, ed., *Food Polysaccharides and Their Applications*, Marcel Dekker, New York, 1995.
35. H. Scherz and G. Bonn, *Analytical Chemistry of Carbohydrates*, Georg Thieme Verlag, Stuttgart, 1998.
36. J. N. BeMiller, *Carbohydrate Analysis, Food Analysis*, 3rd ed., S. S. Nielsen, ed., Kluwer Academic/Plenum Publishers, New York, 2003, Chapt. 10.

GENERAL REFERENCES

Advances in Carbohydrate Chemistry and Biochemistry, Academic Press, Inc., San Diego, Calif.
J. N. BeMiller and Whistler, *Carbohydrate Chemistry for Food Scientists*, 2nd ed., American Association of Cereal Chemists, St. Paul, Minn., 2003.
Eliasson, ed., *Carbohydrates in Food*, Marcel Dekker, New York, 1996.
El Khadem, *Carbohydrate Chemistry*, Academic Press, Inc., San Diego, Calif., 1988.
References 1,18, and 19.
References 2,4, and 14.
References 15,16, and 21.
Reference 17.
Reference 25.

JAMES N. BeMILLER
Purdue University

CARBON

1. Introduction

Elemental carbon [7440-44-0], atomic number six in the periodic table, at wt 12.011, occurs naturally throughout the world in either its crystalline, more ordered, or amorphous, less ordered, form. Carbonaceous materials such as soot or charcoal are examples of the amorphous form whereas graphite and diamond are crystalline. Activated carbon is a predominately amorphous solid that has an extraordinary surface area and same volume. See the article CARBON ACTIVATED for a detailed discussion. Carbon atoms bond with other carbon atoms as well as with other elements, principally hydrogen, nitrogen, oxygen, and sulfur, to form carbon compounds, which are the subject of organic chemistry. In its many varying manufactured forms, carbon and graphite can exhibit a wide range of electrical, thermal, and chemical properties that are controlled by the selection of raw materials and thermal processing during manufacture (1) (See GRAPHITE, ARTIFICIAL; GRAPHITE, NATURAL.)

There are two allotropes of carbon: diamond and graphite. The diamond, or isotropic form, has a crystal structure that is face-centered cubic with interatomic distances of 0.154 nm. Each atom is covalently bonded to four other carbon atoms in the form of a tetrahedron. The nature of the bonding explains the differences in properties of the two allotropic forms. The hardness of diamond is derived from the regular three-dimensional network of σ-bonds (see DIAMOND, NATURAL and DIAMOND, SYNTHETIC). Graphite, or the anisotropic form, has a structure that is composed of infinite layers of carbon atoms arranged in the form of hexagons lying in planes. The electronic ground state of carbon is $1s^2, 2s^2, 2p^2$. In diamond, the $2s$ and $2p$ electrons mix to form four equivalent covalent σ-bonds. In graphite, three of the four electrons form strong covalent π-bonds with the adjacent in-plane carbon atoms. The fourth electron forms a less strong bond between the planes. A wide variety and range of bulk carbon forms are available within the industry. In general, commercial forms are loosely characterized as carbon or graphite, but they are distinctly different. The term *manufactured carbon* refers to a bonded granular carbon body whose matrix has been subjected to a temperature typically between 900 and 2400°C. *Manufactured graphite* refers to a bonded granular carbon body whose matrix has been subjected to a temperature typically in excess of 2400°C. Natural graphite has been known since the Middle Ages, but carbon was first fabricated by H. Davy in his experiments on the electric arc in the early 1800s. The manufacture of artificial graphite came about only at the end of the nineteenth century, preceded by developments in the fabrication of electrodes. The electric resistance furnace enabled the reaching of approximately 3000°C, the temperature necessary for graphitization. A new application for graphite, its use by E. Fermi in the first self-sustaining nuclear reaction, was followed by new fields of research and new markets opened by development of the aerospace industries, including the use of carbon and graphite fibers in composite materials.

2. Crystallographic Structure

There are two allotropes of carbon: diamond [7782-40-3] and graphite [7782-42-5]. The diamond, or isotropic form, has a crystal structure that is face-centered cubic with interatomic distances of 0.154 nm. Each atom is covalently bonded to four other carbon atoms in the form of a tetrahedron. The nature of the bonding explains the differences in properties of the two allotropic forms. The hardness of diamond is derived from the regular three-dimensional network of σ-bonds; the low electrical conductivity results from fixed-bonding electrons between atoms within the diamond lattice (2). Graphite, or the anisotropic form, has a structure that is composed of infinite layers of carbon atoms arranged in the form of hexagons lying in planes. This structure was first proposed in 1924 (3). The stacking arrangement is ABAB with atoms in alternate planes aligning with each other. Interlayer spacing is 0.3354 nm and interatomic distance within the planes 0.1415 nm. The crystal density is 2.25 g/cm³ compared to 3.51 g/cm³ for diamond.

A rhombohedral form, which occurs in small proportions with the hexagonal form, has a stacking arrangement of ABCABC. Being less stable, it begins to convert to the hexagonal form above 1300°C (4).

In 1990, a third form of solid carbon was confirmed and designated "buck-minsterfullerenes." These 60-carbon (and 78-C) clusters are described as having the shape of a geodesic dome or soccer ball and hence are also known as "bucky balls" (5). See the article, FULLERENES, for a fuel discussion.

The electronic ground state of carbon is $1s^2, 2s^2, 2p^2$, ie, there are four electrons in the outer shell available for chemical bonding. In diamond, the $2s$ and $2p$ electrons mix to form four equivalent covalent σ-bonds. In graphite, three of the four electrons form strong covalent π-bonds with the adjacent in-plane carbon atoms. The fourth electron forms a less strong bond of the van der Waals type between the planes. Bond energy between planes is 17 kJ/mol (4 kcal/mol) (6) and within planes 477 kJ/mol (114 kcal/mol) (7). The weak forces between planes account for such properties of graphite as good lubricity and the ability to form interstitial compounds, whereas the strong π-bonding within the planes contributes to the high electrical and thermal conductivity.

3. Terminology

A wide variety and range of bulk carbon forms are available within the industry. In general, commercial forms are loosely characterized as carbon or graphite, but they are distinctly different. In the United States, the ASTM has issued definitions of terms that relate to manufactured carbon and graphite including processing and property definitions (8). The term manufactured carbon (sometimes called formed carbon, amorphous carbon, or baked carbon) refers to a bonded granular carbon body whose matrix has been subjected to a temperature typically between 900 and 2400°C (8). The process involves mixing carbonaceous filler materials, such as petroleum coke, carbon blacks, or anthracite coal, with binder materials of coal tar or petroleum pitch, forming these mixtures by molding or extrusion, and baking the mixtures in furnaces at temperatures from 900 to 2400°C. Green carbon refers to formed carbonaceous material that has not been baked.

Manufactured graphite (sometimes called synthetic, artificial graphite, electrographite, or graphitized carbon) refers to a bonded granular carbon body whose matrix has been subjected to a temperature typically in excess of 2400°C and whose matrix is thermally stable below that temperature (8). This higher temperature processing, known as graphitization changes not only the crystallographic structure but the physical and chemical properties as well.

Pyrolytic carbons are carbon materials deposited on a heated graphite substrate, or other material, by chemical vapor deposition (CVD) at 800–2300°C; pyrolytic graphite is the product that results from higher temperature treatment and has a crystallite interlayer spacing similar to that of ideal graphite (9). Carbon or graphite fiber forms are produced principally from polyacrylonitrile (PAN) or pitch and are designated carbon or graphite based on crystallographic structure.

Whereas the foregoing are the forms most commonly found in many applications in industry, there are definitions that are necessary not only for industrial purposes but also for consistency in the study of carbon science. Since 1975, the International Committee for Characterization and Terminology of

Carbon has been working to establish definitions and in 1982 published its 30 tentative definitions followed by periodic issues of further tentative definitions (10).

4. History of the Industry

Natural graphite has been known since the Middle Ages for its use in making clay–graphite crucibles and for its lubricating properties. The first known use was for drawing or writing, and it was because of this attribute that the German mineralogist, A. G. Gerner, named graphite after the greek word "graphein" which means to write (11).

Manufacture of artificial graphite did not come about until the end of the nineteenth century. Its manufacture was preceded by developments mainly in the fabrication and processing of carbon electrodes. H. Davy is credited with using the first fabricated carbon in his experiments on the electric arc in the early 1800s. During the nineteenth century, several researchers received patents on various improvements in carbon electrodes. The invention of the dynamo and its application to electric current production in 1875 in Cleveland, Ohio, by C. F. Brush, provided a market for carbon products in the form of arc-carbons for street lighting. The work of a Frenchman, F. Carre, in the late nineteenth century, established the industrial processes of mixing, forming, and baking necessary for the production of carbon and graphite (12).

A significant development occurred when E. G. Acheson patented an electric resistance furnace capable of reaching approximately 3000°C, the temperature necessary for graphitization (13). This development was the beginning of a new industry in which improved carbon and graphite products were used in the production of alkalies, chlorine, aluminum, calcium and silicon carbide, and for electric furnace production of steel and ferroalloys. In 1942, a new application for graphite was found when it was used as a moderator by E. Fermi in the first self-sustaining nuclear chain reaction (14). This nuclear application and subsequent use in the developing aerospace industries opened new fields of research and new markets for carbon and graphite. Carbon and graphite fibers and their use in composite materials are examples of a new form and a new industry.

5. Other Forms of Carbon and Graphite

The versatility and uniqueness of carbon and graphite attest to its widespread use for a variety of industrial applications. Several other forms of carbon and graphite (15) are not covered in industrial articles are presented here.

5.1. Flexible Graphite.   A useful form of graphite is a flexible sheet or foil. Because of graphite's stability at high temperatures, flexible foil is useful in applications requiring thermal stability in corrosive environments, eg, gaskets and valve packings, and is often used as a replacement for asbestos gaskets. The basic structure of flexible graphite results in both mechanical and sealing characteristics (16) without additives, a decisive advantage over contemporary facing materials which are more dependent on binders and impregnants to generate adequate properties.

A common method of manufacturing flexible graphite (17) involves treating natural graphite flake with an oxidizing agent such as a solution of nitric and sulfuric acid to form an intercalated compound with graphite. Upon heating at high temperature, the intercalants in the graphite crystal form a gas that causes the layers of the graphite to separate and the graphite flakes to expand or exfoliate in an accordionlike fashion in the c-direction, ie, the direction perpendicular to the crystalline planes of the graphite. The expanded flakes are then compressed into sheets which are flexible and can be formed and cut into various shapes. Improvements in the process for reducing material and production costs (18) and a product with enhanced electrical and thermal conductivity properties (19) have been reported.

5.2. Carbon and Graphite Foam. Carbon–graphite foam is a unique material that has yet to find a place among the various types of commercial specialty graphites. Its low thermal conductivity, mechanical stability over a wide range of temperatures from room temperature to 3000°C, and light weight make it a prime candidate for thermal protection of new, emerging carbon–carbon aerospace reentry vehicles.

The open cell structure of carbon foam with its greater than 90% porosity and chemical inertness at temperatures below 500°C suggests its use as a filtration media for corrosive liquids and a dispersant for gases.

The earliest foamed graphite was made from exfoliated small crystals of graphite bound together and compacted to a low density (20–22). This type of foam is structurally weak and will not support loads of even a few newtons per square meter. Carbon and graphite foams have been produced from resinous foams of phenolic or urethane base by careful pyrolysis to preserve the foamed cell structure in the carbonized state. These foams have good structural integrity, eg, a typical foam of 0.25 g/cm^3 apparent density has a compressive strength of 9.3–15 MPa (1350–2180 psi) with thermal conductivity of 0.87 W/(m·K) at 1400°C. These properties make the foam attractive as a high temperature insulating packaging material in the aerospace field and as insulation for high temperature furnaces. Variations of the resinous-based foams include the syntactic foams where cellular polymers or hollow carbon spheres comprise the primary volume of the material bonded and carbonized in a resin matrix. Use and production of a carbon-graphite foam as a temperature regulator in engines has been reported (23).

5.3. Pyrolytic Graphite. Pyrolytic graphite was first produced in the late 1800s for lamp filaments. Today, it is produced in massive shapes, used for missile components, rocket nozzles, and aircraft brakes for advanced high performance aircraft. Pyrolytic graphite coated on surfaces or infiltrated into porous materials is also used in other applications, such as nuclear fuel particles, prosthetic devices, and high temperature thermal insulators.

Of the many forms of carbon and graphite produced commercially, only pyrolytic graphite (24,25) is produced from the gas phase via the pyrolysis of hydrocarbons. The process for making pyrolytic graphite is referred to as the chemical vapor deposition (CVD) process. Deposition occurs on some suitable substrate, usually graphite, that is heated at high temperatures, usually in excess of 1000°C, in the presence of a hydrocarbon, eg, methane, propane, acetylene, or benzene.

The largest quantity of commercial pyrolytic graphite is produced in large, inductively heated furnaces in which natural gas at low pressure is used as the source of carbon. Deposition temperatures usually range from 1800 to 2000°C on a deposition substrate of fine-grain graphite.

The properties of pyrolytic graphite exhibit a high degree of anisotropy. For example, the tensile strength in the *ab* direction is five to ten times greater than that of conventional graphite and the strength in the *c* direction is proportionately lower. Similarly, the thermal conductivity of pyrolytic graphite in the *ab* direction ranks among the highest of elementary materials, whereas in the *c* direction its thermal conductivity is quite low. At room temperature, the thermal conductivity values in the *ab* direction are three hundred times greater than in the *c* direction. Pyrolytic graphite with a density of 2.0–2.1 g/cm^3 is the most dense of the commercially produced graphites, exhibiting low porosity and low permeability.

A special form of pyrolytic graphite is produced by annealing under pressure at temperatures above 3000°C. This pressure-annealed pyrolytic graphite exhibits the theoretical density of single-crystal graphite, and though the material is polycrystalline, the properties of the material are close to single-crystal properties. The highly reflective, flat faces of pressure-annealed pyrolytic graphite have made the material valuable as an x-ray monochromator (see X-RAY TECHNIQUES).

5.4. Glassy Carbon.   Glassy, or vitreous, carbon is a black, shiny, dense, brittle material with a vitreous or glasslike appearance (26,27). It is produced by the controlled pyrolysis of thermosetting resins; phenol–formaldehyde and polyurethanes are among the most common precursors. Unlike conventional artificial graphites, glassy carbon has no filler material. The liquid resin itself becomes the binder.

There is little crystal growth during carbonization, which always occurs in the solid phase. The solid cross-linking that occurs at this time does not lend itself to crystal growth. The glassy carbons are composed of random crystallites of the order of 5.0 nm across and are not significantly altered by ordinary graphitization heat treatment to 2800°C.

The properties of glassy carbon are unlike those of conventional carbon and graphites. Exhibiting a density of 1.4–1.5 gm/cm^3, they have low open porosity and low permeability. The hardness and brittleness of this material is the same as that of ordinary glass. Chemical inertness and low permeability have made glassy carbon a useful material for chemical laboratory crucibles and other vessels. It is used as a container/heater for the epitaxial growth of silicon crystals and as crucibles for the growth of single crystals. This type of carbon is useful for metallurgical crucibles (26), heating elements, heat-resistant tubes, machine parts, and electrical parts. A composite glassy carbon disk substrate for a data storage device has been reported (29).

5.5. Carbon and Graphite Paper.   Carbon and graphite paper is produced from carbon fibers by conventional papermaking methods. The carbon or graphite fibers are cut or chopped to a size suitable for processing, about onefourth inch in length, homogeneously intermixed with water and a starch binder to form an aqueous slurry, and then deposited from the slurry on a substrate to form a sheet. The sheet is then processed by conventional papermaking techniques to produce a carbon or graphite paper.

This form of carbon and graphite has outstanding electrical conductivity, corrosion resistance, and moderately high strength. These properties have promoted its use in electrodes for electrostatic precipitators. Composites made of laminated carbon paper (30) are excellent high temperature thermal insulators, having a thermal conductivity of less than 1.4 W/(m · K) (0.8 (BTU · ft)/(ft^2 · h · °F)) at room temperature. The material is not substantially affected by being subjected to high temperature. The thermal conductivity increases to 0.5 W/(m · K) (0.3 (BTU · ft)/(ft^2 · h · °F)) at 2000°C, but is still significantly low, particularly in view of its low density (0.5 g/mL).

BIBLIOGRAPHY

"Structure, Terminology, and History" under "Carbon (Carbon and Artificial Graphite)" in *ECT* 3rd ed., Vol. 4, pp. 556–560, by J. C. Long, Union Carbide Corp.; in *ECT* 4th ed., Vol. 4, pp. 949–952, by J. C. Long, UCAR Carbon Company Inc.; "Baked and Graphitized Products, Uses" under "Carbon" in *ECT* 2nd ed., Vol. 4, pp. 202–243, by W. M. Gaylord, Union Carbide Corp.; "Other Forms of Carbon and Graphite: Carbon" under "Carbon (Carbon and Artificial Graphite)" in *ECT* 3rd ed., Vol. 4, pp. 628–631, by R. M. Bushong, Union Carbide Corp.; in *ECT* 4th ed., Vol. 4, pp. 1012–1015, by J. M. Criscione, UCAR Carbon Company Inc., "Carbon, Structure, Terminology, and History" in *ECT* (online), posting date: December 4, 2000, by J. M. Criscione, UCAR Carbon Company Inc.

CITED PUBLICATIONS

1. P. L. Walker, Jr., *Chem. Ind. London*, 683 (Sept. 18, 1982).
2. H. Marsh, *Introduction to Carbon Science*, Butterworths, Boston, Mass., 1989, pp. 4, 5.
3. J. D. Bernal, *Proc. R. Soc. London Ser. A* **106**, 749 (1924).
4. G. E. Bacon, *Acta Crystallogr.* **4**, 253 (1952); H. P. Boehm and R. W. Coughlin, *Carbon* **2**, 1, (1964).
5. R. M. Baum, *Chem. Eng. News.* **68**, 22 (Oct. 29, 1990).
6. G. J. Dienes, *J. Appl. Phys.* **23**, 1194 (1952).
7. M. A. Kanter, *Phys. Rev.* **107**, 655 (1957).
8. *Standard Definitions of Terms Relating to Manufacturing Carbon and Graphite*, ASTM Standard C 709-90, Vol. 15.01 American Society for Testing and Materials, Philadelphia, Pa., 1991, p. 189.
9. A. W. Moore, *Chem. Phys. Carbon* **8**, 71 (1973).
10. International Committee for Characterization and Terminology of Carbon, *Carbon* **20**, 445 (1982); *Carbon* **21**, 517 (1983); *Carbon* **23**, 601 (1985); *Carbon* **24**, 246 (1986); *Carbon* **25**, 317 (1987); *Carbon* **25**, 449 (1987).
11. F. Cirkel, *Graphite: Its Properties, Occurrence, Refining and Use*, Vol. 202, Department of Mines, Montreal, Canada, 1906, p. 251.
12. F. Jehl, *The Manufacture of Carbons for Electric Lighting and Other Purposes*, "The Electrician" Printing and Publishing Co., Ltd., London, UK, 1899.
13. U.S. Pat. 568,323 (Sept. 28, 1896), E. G. Acheson.
14. E. Fermi, *Collected Papers of Enrico Fermi*, Vol. 2, University of Chicago Press, Chicago, Ill., 1965.
15. R. W. Cahn and B. Harris, *Nature (London)* **221**, 132 (Jan. 11, 1969).
16. R. A. Mercuri, R. A. Howard, and J. J. McGlamery, *Advanced High Temperature Test Methods for Gasket Materials, Automotive Eng.* **97**, 49–52 (July 1989).

17. U.S. Pat. 3,404,061 (Oct. 10, 1968), J. H. Shane, R. J. Russell, and R. A. Bochman (to Union Carbide Corp.).
18. U.S. Pat. 4,895,713 (Jan. 23, 1990), R. A. Greinke, R. A. Mercuri, and E. J. Beck (to Union Carbide Corp.).
19. U.S. Pat. Appl. 2003002496 (Feb. 6, 2003), R. A. Mercuri, and co-workers (to Graftech Inc.).
20. R. A. Mercuri, T. R. Wessendorf, and J. M. Criscione, *Am. Chem. Soc. Div. Fuel Chem. Prepr.* **12**(4), 103 (1968).
21. C. R. Thomas, *Mater. Sci. Eng.* **12**, 219 (1973).
22. S. T. Benton and C. R. Schmitt, *Carbon* **10**, 185 (1972).
23. U.S. Pat. Appl. 2003000486 (Jan. 2, 2003), R. Ott, A. D. McHelan, and A. Choudhury (to Ut-Batelle UC).
24. J. C. Bokros, *Chem. Phys. Carbon* **5**, 1 (1969).
25. A. W. Moore, *Chem. Phys. Carbon* **11**, 69 (1973).
26. F. C. Cowtard and J. C. Lewis, *J. Mater. Sci.* **2**, 507 (1967).
27. G. M. Jenkins and K. Kawamura, *Polymeric Carbons–Carbon Fiber, Glass, and Char.*, Cambridge University Press, New York, 1976, 178 pp.
28. C. Nakayama and co-workers, *Proceedings of the Carbon Society of Japan*, Annual Meeting, 1975, p. 114; C. Nakayama, M. Okawa, and H. Nageshima, *13th Biennial Conference on Carbon, 1977*, Extended Abstracts and Program, American Carbon Society, Irvine, Calif., 1977, p. 424.
29. U.S. Pat. Appl. 20020192421 (Dec. 19, 2002), T. A. Jennings, D. H. Piltingsrod, and S. F. Starcke.
30. U.S. Pat. 3,844,877 (Oct. 29, 1974), T. R. Wessendorf and J. M. Criscione (to Union Carbide Corp.).

GENERAL REFERENCES

P. Threner, ed., *Carbon*, Pergamon Press, New York, 1963.
Tanso (Carbons), Tanso Zairyo Kenkyukai, Tokyo, Japan, 1949.
Carbon & Graphite Conference, 5th, Society of Chemical Industry, London, UK, 1978.
International Symposium on Carbon, 1982, Carbon Society of Japan, Tokyo, Japan, 1982.
Carbon'84, Centre de Recherche de Chimie Structurale "Paul Pascal," Paris, France, 1984.
Conferences on Carbon, American Carbon Society, St. Marys, Pa. (held every two years since 1953).
International Symposium on Carbon—New Processing and New Applications, Tsukuba, Japan, 1990.
T. Ishikaws, T. Nagaoki, and I. C. Lewis, *Recent Carbon Technology*, IEC Press, Cleveland, Ohio, 1983.
B. T. Kelly, *Physics of Graphite*, Applied Science Publishers, London, UK, 1981.
H. Marsh, *Introduction to Carbon Science*, Butterworths, London, UK, 1989.
S. Otani and H. Sanada, *Fundamentals of Carbonization Engineering*, Ohm, Tokyo, Japan, 1980.
P. A. Thrower, ed., *Chemistry and Physics of Carbon; A Series of Advances*, Marcel Dekker, New York, 1999.
H. O. Pierson, *Handbook of Carbon, Graphite, Diamond and Fullerenes: Processes and Applications*, Noyes Publications, New York, 1994.
S. Yoshimura and R. P. Chang, *Supercarbons: Synthesis, Properties, and Applications*, Springer Verlag, New York, 2000.

N. N. Greenwood and A. Earnshaw, *Chemistry of the Elements*, 2nd ed., Butterworth Heineman Inc. Woburn, Mass., 1997.

A. Swertka and E. Swertka, *A Guide to the Elements*, Oxford University Press, New York, 1998.

J. C. Long
J. M. Criscione
UCAR Carbon Company Inc.

CARBON, ACTIVATED

1. Introduction

Activated carbon is a predominantly amorphous solid that has an extraordinarily large internal surface area and pore volume. These unique characteristics are responsible for its adsorptive properties, which are exploited in many different liquid- and gas-phase applications. Activated carbon is an exceptionally versatile adsorbent because the size and distribution of the pores within the carbon matrix can be controlled to meet the needs of current and emerging markets (1). Engineering requirements of specific applications are satisfied by producing activated carbons in the form of powders, granules, and shaped products. Through choice of precursor, method of activation, and control of processing conditions, the adsorptive properties of products are tailored for applications as diverse as the purification of potable water and the control of gasoline emissions from motor vehicles.

In 1900, two very significant processes in the development and manufacture of activated carbon products were patented (2). The first commercial products were produced in Europe under these patents: Eponite, from wood in 1909, and Norit, from peat in 1911. Activated carbon was first produced in the United States in 1913 by Westvaco Corp. under the name Filtchar, using a by-product of the papermaking process (3). Further milestones in development were reached as a result of World War I. In response to the need for protective gas masks, a hard, granular activated carbon was produced from coconut shell in 1915. Following the war, large-scale commercial use of activated carbon was extended to refining of beet sugar and corn syrup and to purification of municipal water supplies (4). The termination of the supply of coconut char from the Philippines and India during World War II forced the domestic development of granular activated carbon products from coal in 1940 (5). More recent innovations in the manufacture and use of activated carbon products have been driven by the need to recycle resources and to prevent environmental pollution.

2. Physical and Chemical Properties

The structure of activated carbon is best described as a twisted network of defective carbon layer planes, cross-linked by aliphatic bridging groups (6). X-ray diffraction patterns of activated carbon reveal that it is nongraphitic, remaining amorphous because the randomly cross-linked network inhibits reordering of the structure even when heated to 3000°C (7). This property of activated carbon contributes to its most unique feature, namely, the highly developed and accessible internal pore structure. The surface area, dimensions, and distribution of the pores depend on the precursor and on the conditions of carbonization and activation. Pore sizes are classified (8) by the International Union of Pure and Applied Chemistry (IUPAC) as micropores (pore width <2 nm), mesopores (pore width 2–50 nm), and macropores (pore width >50 nm) (see ADSORPTION).

The surface area of activated carbon is usually determined by application of the Brunauer-Emmett-Teller (BET) model of physical adsorption (9,10) using nitrogen as the adsorptive (8). Typical commercial products have specific surface areas in the range 500–2000 m²/g, but values as high as 3500–5000 m²/g have been reported for some activated carbons (11,12). In general, however, the effective surface area of a microporous activated carbon is far smaller because the adsorption of nitrogen in micropores does not occur according to the process assumed in the BET model, which results in unrealistically high values for surface area (10,13). Adsorption isotherms are usually determined for the appropriate adsorptives to assess the effective surface area of a product in a specific application. Adsorption capacity and rate of adsorption depend on the internal surface area and distribution of pore size and shape but are also influenced by the surface chemistry of the activated carbon (14). The macroporosity of the carbon is important for the transfer of adsorbate molecules to adsorption sites within the particle.

Functional groups are formed during activation by interaction of free radicals on the carbon surface with atoms such as oxygen and nitrogen, both from within the precursor and from the atmosphere (15). The functional groups render the surface of activated carbon chemically reactive and influence its adsorptive properties (6). Activated carbon is generally considered to exhibit a low affinity for water, which is an important property with respect to the adsorption of gases in the presence of moisture (16). However, the functional groups on the carbon surface can interact with water, rendering the carbon surface more hydrophilic (15). Surface oxidation, which is an inherent feature of activated carbon production, results in hydroxyl, carbonyl, and carboxylic groups that impart an amphoteric character to the carbon, so that it can be either acidic or basic. The electrokinetic properties of an activated carbon product are, therefore, important with respect to its use as a catalyst support (17). As well as influencing the adsorption of many molecules, surface oxide groups contribute to the reactivity of activated carbons toward certain solvents in solvent recovery applications (18).

In addition to surface area, pore size distribution, and surface chemistry, other important properties of commercial activated carbon products include pore volume, particle size distribution, apparent or bulk density, particle density, abrasion resistance, hardness, and ash content. The range of these and other properties is illustrated in Table 1 together with specific values for selected commercial grades of powdered, granular, and shaped activated carbon products used in liquid- or gas-phase applications (19).

Table 1. Properties of Selected U.S. Activated Carbon Products[a]

Property	Typical range	Gas-phase carbons			Liquid-phase carbons		
Manufacturer		Calgon	Norit	Westvaco	Calgon	Norit	Westvaco
Precursor		Coal	Peat	Wood	Coal	Peat	Wood
Product grade		BPL	B4	WV-A 1100	SGL	SA 3	SA-20
Product form		Granular	Extruded	Granular	Granular	Powdered	Powdered
particle size, U.S. mesh[b,c]	<4	12 × 30	3.8 mm	10 × 25	8 × 30	64%	65 – 85%
apparent density, g/cm^3	0.2–0.6	>0.48	0.43	0.27	0.52	0.46	0.34–0.37
particle density, g/cm^3	0.4–0.9	0.80		0.50	0.80		
hardness number	50–100	>90	99	0.50			
abrasion number					>75		
ash, wt %	1–20		6			6	3–5
BET surface area, N$_2$, m^2/g	500–2500	1050–1150	1100–1200	1750	900–1000	750	1400–1800
total pore volume, cm^3/g	0.5–2.5	0.8	0.9	1.2	0.85		2.2–2.5
CCl$_4$ activity, wt %	35–125	>60					
butane working capacity, g/100 cm^3	4–14			>11.0			
iodine number	500–1200	>1050			>900	800	>1000
decolorizing index							
Westvaco	15–25						>20
molasses number							
Calgon	50–250				>200		
Norit	300–1500					440	
heat capacity at 100°C, J/(g·K)[d]	0.84–1.3	1.05			1.05		
thermal conductivity, W/(m·K)	0.05–0.10						

[a] Specific values shown are those cited in manufacturers' product literature (19). Typical ranges shown are based on values reported in the open literature.
[b] Unless otherwise noted.
[c] Approximate mm corresponding to cited meshes are mesh: mm—4: 4.76; 8: 2.38; 10: 2; 12: 1.68; 25: 0.72; 30: 0.59; 325: 0.04.
[d] To convert J to cal, divide by 4.184.

3. Manufacture and Processing

Commercial activated carbon products are produced from organic materials that are rich in carbon, particularly coal, lignite, wood, nut shells, peat, pitches, and cokes. The choice of precursor is largely dependent on its availability, cost, and purity, but the manufacturing process and intended application of the product are also important considerations. Manufacturing processes fall into two categories, thermal activation and chemical activation. The effective porosity of activated carbon produced by thermal activation is the result of gasification of the carbon at relatively high temperatures (20), but the porosity of chemically activated products is generally created by chemical dehydration reactions occurring at significantly lower temperatures (1,21).

3.1. Thermal Activation Processes.
Thermal activation occurs in two stages: thermal decomposition or carbonization of the precursor and controlled gasification or activation of the crude char. During carbonization, elements such as hydrogen and oxygen are eliminated from the precursor to produce a carbon skeleton possessing a latent pore structure. During gasification, the char is exposed to an oxidizing atmosphere that greatly increases the pore volume and surface area of the product through elimination of volatile pyrolysis products and from carbon burn-off. Carbonization and activation of the char are generally carried out in direct-fired rotary kilns or multiple hearth furnaces, but fluidized-bed reactors have also been used (22). Materials of construction, notably steel and refractories, are designed to withstand the high temperature conditions, ie, $>1000°C$, inherent in activation processes. The thermal activation process is illustrated in Figure 1 for the production of activated carbon from bituminous coal (23,24).

Bituminous coal is pulverized and passed to a briquette press. Binders may be added at this stage before compression of the coal into briquettes. The briquetted coal is then crushed and passed through a screen, from which the on-size material passes to an oxidizing kiln. Here, the coking properties of the coal particles are destroyed by oxidation at moderate temperatures in air. The oxidized coal is then devolatilized in a second rotary kiln at higher temperatures under steam. To comply with environmental pollution regulations, the kiln off-gases containing dust and volatile matter pass through an incinerator before discharge to the atmosphere.

The devolatilized coal particles are transported to a direct-fired multihearth furnace where they are activated by holding the temperature of the furnace at about $1000°C$. Product quality is maintained by controlling coal feed rate and bed temperature. As before, dust particles in the furnace off-gas are combusted in an afterburner before discharge of the gas to the atmosphere. Finally, the granular product is screened to provide the desired particle size. A typical yield of activated carbon is about 30–35% by weight based on the raw coal.

The process for the thermal activation of other carbonaceous materials is modified according to the precursor. For example, the production of activated carbon from coconut shell does not require the stages involving briquetting, oxidation, and devolatilization. To obtain a high activity product, however, it is important that the coconut shell is charred slowly prior to activation of the

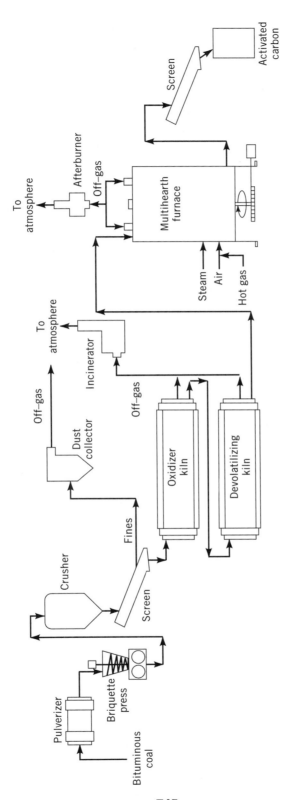

Fig. 1. Thermal activation of bituminous coal.

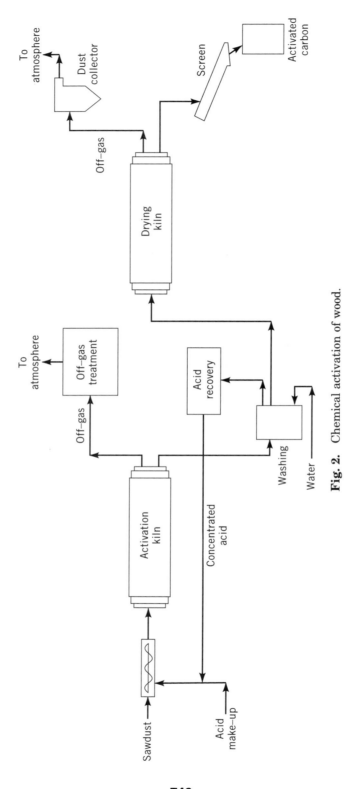

Fig. 2. Chemical activation of wood.

char. In some processes, the precursor or product is acid-washed to obtain a final product with a low ash content (23,25).

3.2. Chemical Activation Processes.   In contrast to the thermal activation of coal, chemical activation is generally carried out commercially in a single kiln. The precursor, usually wood, is impregnated with a chemical activation agent, typically phosphoric acid, and the blend is heated to a temperature of 450–700°C (26). Chemical activation agents reduce the formation of tar and other by-products, thereby increasing carbon yield. The chemical activation process is illustrated in Figure 2, for the production of granular activated carbon from wood (23,27).

Sawdust is impregnated with concentrated phosphoric acid and fed to a rotary kiln, where it is dried, carbonized, and activated at a moderate temperature. To comply with environmental pollution regulations, the kiln off-gases are treated before discharge to the atmosphere. The char is washed with water to remove the acid from the carbon, and the carbon is separated from the slurry. The filtrate is then passed to an acid recovery unit. Some manufacturing plants do not recycle all the acid but use a part of it to manufacture fertilizer in an allied plant. If necessary, the pH of the activated carbon is adjusted, and the product is dried. The dry product is screened and classified into the size range required for specific granular carbon applications. Carbon yields as high as 50% by weight of the wood precursor have been reported (26).

Other Manufacturing Processes.   Different chemical activation processes have been used to produce carbons with enhanced adsorption characteristics. Activated carbons of exceptionally high surface area (>3000 m^2/g) have been produced by the chemical activation of carbonaceous materials with potassium hydroxide (28,29). Activated carbons are also produced commercially in the form of cloths (30), fibers (31), and foams (32) generally by chemical activation of the precursor with a Lewis acid such as aluminum chloride, ferric chloride, or zinc chloride.

3.3. Forms of Activated Carbon Products.   To meet the engineering requirements of specific applications, activated carbons are produced and classified as granular, powdered, or shaped products. Granular activated carbons are produced directly from granular precursors, such as sawdust and crushed and sized coconut char or coal. The granular product is screened and sized for specific applications. Powdered activated carbons are obtained by grinding granular products. Shaped activated carbon products are generally produced as cylindrical pellets by extrusion of the precursor with a suitable binder before activation of the precursor.

4. Shipment and Storage

Activated carbon products are shipped in bags, drums, and boxes in weights ranging from about 10 to 35 kg. Containers can be lined or covered with plastic and should be stored in a protected area both to prevent weather damage and to minimize contact with organic vapors that could reduce the adsorption performance of the product. Bulk quantities of activated carbon products are shipped in metal bins and bulk bags, typically 1–2 m^3 in volume, and in railcars and tank

Table 2. **U.S. Producers of Activated Carbon and Their Capacities**[a]

Producer	Capacity, $\times 10^6$ kg ($\times 10^6$ lb)	
Barnebey & Sutcliffe, Columbus, Ohio	13.6	(30)
Calgon Carbon, Catlettsburg, Ky.	63.5	(140)
Calgon Carbon, Pearlington, Miss.	18.1	(40)
Norit Americas, Pryor, Okla.	15.8	(35)
Norit Americas, Marshall, Texas	45.3	(100)
Royal Oak, Romeo, Fla.	9.1	(20)
Westvaco, Covington, Va.	22.7	(50)
Westvaco, Wickliffe, Ky.	22.7	(50)
Total	*210.8*	(465)

[a] From Ref. 33.

trucks. Bulk carbon shipments are generally transferred by pneumatic conveyors and stored in tanks. However, in applications such as water treatment where water adsorption does not impact product performance, bulk carbon may be transferred and stored as a slurry in water.

5. Economic Aspects

U.S. producers of activated carbon are listed in Table 2.

Demand in 1999 was 168×10^6 kg (370×10^6 lb); in 2000 it was 173×10^6 kg (381×10^6 lb). The forecast for 2004 is 206×10^6 kg (454×10^6 lb). Demand equals production plus imports less exports. Imports in 2000 totaled 49×10^6 kg (109×10^6 lb) (33).

Growth is expected at the rate of 4.5% through 2004. Demand is driven by environmental considerations. In many applications, activated carbon has best available technology status. However, some alternative systems may be more cost effective (33).

Price history for the years 1995–2000 was a high of \$0.50/kg (\$1.10/lb) gran. and a low of \$0.20/kg (\$0.45/lb) (33).

6. Specifications

Activated carbon producers furnish product bulletins that list specifications, usually expressed as a maximum or minimum value, and typical properties for each grade produced. Standards helpful in setting purchasing specifications for granular and powdered activated carbon products have been published (34,35).

7. Analytical Test Procedures and Standards

Source references for frequently used test procedures for determining properties of activated carbon are shown in Table 3. A primary source is the *Annual Book of*

Table 3. **Source References for Activated Carbon Test Procedures and Standards**

Title of procedure or standard	Source
Standard Definitions of Terms Relating to Activated Carbon	ASTM D2652
Apparent Density of Activated Carbon	ASTM D2854
Particle Size Distribution of Granular Activated Carbon	ASTM D2862
Total Ash Content of Activated Carbon	ASTM D2866
Moisture in Activated Carbon	ASTM D2867
Ignition Temperature of Granular Activated Carbon	ASTM D3466
Carbon Tetrachloride Activity of Activated Carbon	ASTM D3467
Ball-Pan Hardness of Activated Carbon	ASTM D3802
Radioiodine Testing of Nuclear-Grade Gas-Phase Adsorbents	ASTM D3803
pH of Activated Carbon	ASTM D3838
Determination of Adsorptive Capacity of Carbon by Isotherm Technique	ASTM D3860
Determining Operating Performance of Granular Activated Carbon	ASTM D3922
Impregnated Activated Carbon Used to Remove Gaseous Radio-Iodines from Gas Streams	ASTM D4069
Determination of Iodine Number of Activated Carbon	ASTM D4607
Military Specification, Charcoal, Activated, Impregnated	Ref. 36
Military Specification, Charcoal, Activated, Unimpregnated	Ref. 36
AWWA Standard for Granular Activated Carbon	Ref. 34
AWWA Standard for Powdered Activated Carbon	Ref. 35
BET Surface Area by Nitrogen Adsorption	Refs. 6,8,9,37
Pore Volume by Nitrogen Adsorption or Mercury Penetration	Refs. 10,38–40
Particle Density	Ref. 41

American Society for Testing and Materials (ASTM) Standards (42). Other useful sources of standards and test procedures include manufacturers of activated carbon products, the American Water Works Association (AWWA) (34,35), and the Department of Defense (36).

8. Health and Safety Factors

Activated carbon generally presents no particular health hazard as defined by NIOSH (43). However, it is a nuisance and mild irritant with respect to inhalation, skin contact, eye exposure, and ingestion. On the other hand, special consideration must be given to the handling of spent carbon that may contain a concentration of toxic compounds.

Activated carbon products used for decolorizing food products in liquid form must meet the requirements of the *Food Chemical Codex* as prepared by the Food & Nutrition Board of the National Research Council (44).

According to the National Board of Fire Underwriters, activated carbons normally used for water treatment pose no dust explosion hazard and are not subject to spontaneous combustion when confined to bags, drums, or storage bins (45). However, activated carbon burns when sufficient heat is applied; the ignition point varies between about 300 and 600°C (46).

Dust-tight electrical systems should be used in areas where activated carbon is present, particularly powdered products (47). When partially wet activated carbon comes into contact with unprotected metal, galvanic currents can be set up; these result in metal corrosion (48).

Manufacturer material safety data sheets (MSDS) indicate that the oxygen concentration in bulk storage bins or other enclosed vessels can be reduced by wet activated carbon to a level that will not support life. Therefore, self-contained air packs should be used by personnel entering enclosed vessels where activated carbon is present (49).

9. Environmental Concerns

Activated carbon is a recyclable material that can be regenerated. Thus the economics, especially the market growth, of activated carbon, particularly granular and shaped products, is affected by regeneration and industry regeneration capacity. The decision to regenerate an activated carbon product is dependent on the cost, size of the carbon system, type of adsorbate, and the environmental issues involved. Large carbon systems, such as those used in potable and wastewater treatment, generally require a high temperature treatment, which is typically carried out in rotary or multihearth furnaces. During regeneration, carbon losses of 1 to 15% typically occur from the treatment and movement of the carbon (50). However, material loss is compensated for by the addition of new carbon to the adsorber system. In general, regeneration of spent carbon is considerably less expensive than the purchase of new activated carbon. For example, fluidized-bed furnace regeneration of activated carbon used in a 94,600 m^3 per day water treatment system cost only 35% of new material (51). For this system, regeneration using either infrared or multihearth furnaces was estimated to be more expensive but still significantly less so than the cost of new carbon.

Because powdered activated carbon is generally used in relatively small quantities, the spent carbon has often been disposed of in landfills. However, landfill disposal is becoming more restrictive environmentally and more costly. Thus large consumers of powdered carbon find that regeneration is an attractive alternative. Examples of regeneration systems for powdered activated carbon include the Zimpro/Passavant wet air oxidation process (52), the multihearth furnace as used in the DuPont PACT process (53,54), and the Shirco infrared furnace (55,56).

Other types of regenerators designed for specific adsorption systems may use solvents and chemicals to remove susceptible adsorbates (57), steam or heated inert gas to recover volatile organic solvents (58), and biological systems in which organics adsorbed on the activated carbon during water treatment are continuously degraded (59).

10. Liquid-Phase Applications

Liquid phase applications account for 82% of total activated carbon. They include potable water, 31%; industrial and municipal wastewater, 22%; sweetener decolorization, 11%; groundwater, 9%; household uses, 6%; food and beverages, 5%; mining, 4%; pharmaceuticals, 3%; miscellaneous, including chemical processing, 9% (33). Activated carbons for use in liquid-phase applications differ from gas-phase carbons primarily in pore size distribution. Liquid-phase carbons have significantly more pore volume in the macropore range, which permits liquids to diffuse more rapidly into the mesopores and micropores (60). The larger pores also promote greater adsorption of large molecules, either impurities or products, in many liquid-phase applications. Specific-grade choice is based on the isotherm (61,62) and, in some cases, bench or pilot scale evaluations of candidate carbons.

Liquid-phase activated carbon can be applied either as a powder, granular, or shaped form. The average size of powdered carbon particles is 15–25 μm (61). Granular or shaped carbon particle size is usually 0.3–3.0 mm. A significant factor in choosing between powdered and nonpowdered carbon is the degree of purification required in the adsorption application. Granular and shaped carbons are usually used in continuous flow through deep beds to remove essentially all contaminants from the liquid being treated. Granular and shaped carbon systems are preferred when a large carbon buffer is needed to withstand significant variations in adsorption conditions, such as in cases where large contaminant spikes may occur. A wider range of impurity removal can be attained by batch application of powdered carbon, and the powdered carbon dose per batch can be controlled to achieve the degree of purification desired (60) (see ADSORPTION, LIQUID SEPARATION).

Batch-stirred vessels are most often used in treating material with powdered activated carbon (63). The type of carbon, contact time, and amount of carbon vary with the desired degree of purification. The efficiency of activated carbon may be improved by applying continuous, countercurrent carbon–liquid flow with multiple stages (Fig. 3). Carbon is separated from the liquid at each stage by settling or filtration. Filter aids such as diatomaceous earth are sometimes used to improve filtration.

Granular and shaped carbons are used generally in continuous systems where the liquid to be treated is passed through a fixed bed (63,64). New binder technology produces shaped carbon bodies having key properties beyond the best level that has been accomplished with other binders (65). Compounds are

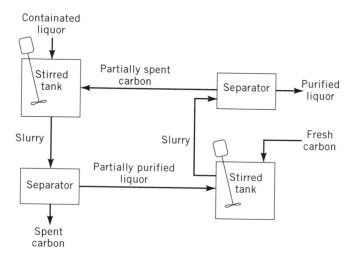

Fig. 3. Multistage countercurrent application of powdered activated carbon.

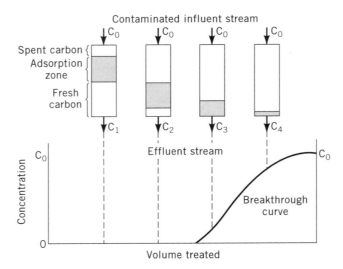

Fig. 4. Adsorption zone and breakthrough curve for fixed bed of granular or shaped activated carbon.

adsorbed by the carbon bed in the adsorption zone (Fig. 4). As carbon in the bed becomes saturated with adsorbates, the adsorption zone moves in the direction of flow, and breakthrough occurs when the leading edge of the adsorption zone reaches the end of the column. Normally at least two columns in series are on line at any given time. When the first column becomes saturated, it is removed from service, and a column containing fresh carbon is added at the discharge end of the series. An alternative approach is the moving bed column (64). In this design the adsorption zone is contained within a single column by passing liquid upward while continuously or intermittently withdrawing spent carbon at the bottom and adding fresh carbon at the top.

10.1. Potable Water Treatment.   Treatment of drinking water accounts for about 31% of the total activated carbon used in liquid-phase applications (33). Rivers, lakes, and groundwater from wells, the most common drinking water sources, are often contaminated with bacteria, viruses, natural vegetation decay products, halogenated materials, and volatile organic compounds. Normal water disinfection and filtration treatment steps remove or destroy the bulk of these materials (66). However, treatment by activated carbon is an important additional step in many plants to remove toxic and other organic materials (67–69) for safety and palatability. An efficient and economical use of activated carbon for treating chemical-contaminated drinking water has been described (70).

10.2. Industrial and Municipal Wastewater Treatment.   Wastewater treatment consumes about 22% of the total U.S. liquid-phase activated carbon (33), and governmental regulations are expected to increase demand over the next several years. Wastewater may contain suspended solids, hazardous micro-organisms, and toxic organic and inorganic contaminants that must be removed or destroyed before discharge to the environment. In tertiary treatment systems, powdered, granular, or shaped carbon can be used to remove residual toxic and other organic compounds after the primary filtration and secondary biological treatment (71). Powdered carbon is also used in the PACT process by direct addition of the carbon to the secondary biological treatment step (53) (see WATER, INDUSTRIAL WATER TREATMENT; WATER, MUNICIPAL WATER TREATMENT).

10.3. Sweetener Decolorization. About 11% of the liquid-phase activated carbon is used for purification of sugar (qv) and corn syrup (33). White sucrose sugar is made from raw juice squeezed from sugar cane or sugar beets. The clarified liquor is decolorized using activated carbon, or ion-exchange resins (72). High fructose corn sweeteners (HFCS) are produced by hydrolysis of corn starch and are then treated with activated carbon to remove undesirable taste and odor compounds and to improve storage life. The demand for HFCS rose sharply in the 1980s primarily because of the switch by soft drink producers away from sucrose (73).

10.4. Groundwater Remediation.   Concern over contaminated groundwater sources increased in the 1980s, and in 1984 an Office of Groundwater Protection was created by the EPA (33). Groundwater remediation accounts for 9% of total liquid-phase usage (33). There are two ways to apply carbon in groundwater cleanup. One is the conventional method of applying powdered, granular, or shaped carbon to adsorb contaminants directly from the water. The other method utilizes air stripping to transfer the volatile compounds from water to air. The compounds are then recovered by passing the contaminated air through a bed of carbon (74,75).

10.5. Food, Beverage, and Cooking Oil. Approximately 5% of the liquid-phase activated carbon is used in food, beverage, and cooking oil production (33). Before being incorporated into edible products, vegetable oils and animal fats are refined to remove particulates, inorganics, and organi cont aminants. Activated carbon is one of several agents used in food purification processes. In the production of alcoholic beverages, activated carbon removes haze-causing compounds from beer, taste and odor from vodka, and fusel oil from whiskey (72). The feed water for soft drink production is often treated with carbon to capture undesirable taste and odor compounds and to remove free chlorine remaining from disinfection treatment. Caffeine is removed from coffee beans by extraction with organic solvents, water, or supercritical carbon dioxide prior to roasting. Activated carbon is used to remove the caffeine from the recovered solvents (73).

10.6. Pharmaceuticals. Pharmaceuticals account for 3% of the liquid-phase activated carbon consumption (33). Many antibiotics, vitamins, and steroids are isolated from fermentation broths by adsorption onto carbon followed by solvent extraction and distillation (72). Other uses in pharmaceutical production include process water purification and removal of impurities from intravenous solutions prior to packaging (73).

10.7. Mining. The mining industry accounts for only 4% of liquid-phase activated carbon use, but this figure may grow as low-grade ores become more common (33). Gold, for example, is recovered on activated carbon as a cyanide complex in the carbon-in-pulp extraction process (72). Activated carbon serves as a catalyst in the detoxification of cyanides contained in wastewater from cyanide stripping operations (64). Problems caused by excess flotation agent concentrations in flotation baths are commonly cured by adding powdered activated carbon (72).

10.8. Miscellaneous Uses. Activated carbon removes impurities to achieve high quality. For example, organic contaminants are removed from solution in the production of alum, soda ash, and potassium hydroxide (72). Other applications include the manufacture of dyestuffs, glycols, amines, organic acids, urea, hydrochloric acid, and phosphoric acid (73). Miscellaneous uses including chemical processing account for 9% of usage (33). Several relatively low volume activated carbon uses comprise the remaining 9% of liquid-phase carbon consumption (33). Small carbon filters are used in households for purification of tap water. Oils, dyes, and other organics are adsorbed on activated carbon in dry cleaning recovery and recycling systems. Electroplating solutions are treated with carbon to remove organics that can produce imperfections when the thin metal layer is deposited on the substrate (72). Medical applications include removal of toxins from the blood of patients with artificial kidneys (73) and oral ingestion into the stomach to recover poisons or toxic materials (72,76). Activated carbon also is used as a support for metal catalysts in low volume production of high value specialty products such as pharmaceuticals, fragrance chemicals, and pesticides (77).

11. Gas-Phase Applications

Gas-phase applications of activated carbon include separation, gas storage, and catalysis. Although only 20% of activated carbon production is used for gas-phase applications, these products are generally more expensive than liquid-phase carbons and account for about 40% of the total dollar value of shipments. Most of the activated carbon used in gas-phase applications is granular or shaped. Gas phase applications account for 18% of total activated carbon. They include air purification, 42%; automotive emission control, 21%; solvent vapor recovery, 14%; cigarette filters medium, 2%; miscellaneous, 21% (33). Separation processes comprise the main gas-phase applications of activated carbon. These usually exploit the differences in the adsorptive behavior of gases and vapors on activated carbon on the basis of molecular weight and size. For example, organic molecules with a molecular weight greater than about 40 are readily removed from air by activated carbon (see ADSORPTION, GAS SEPARATION).

11.1. Solvent Recovery. Most of the activated carbon used in gas-phase applications is employed to prevent the release of volatile organic compounds into the atmosphere. Much of this use has been in response to environmental regulations, but recovery and recycling of solvents from a range of industrial processes such as printing, coating, and extrusion of fibers also provides substantial economic benefits.

The structure of activated carbons used for solvent recovery has been predominantly microporous. Micropores provide the strong adsorption forces needed to capture small vapor molecules such as acetone at low concentrations in process air (78). In recent years, however, more mesoporous carbons, specifically made for solvent recovery, have become available and are giving good service, especially for the adsorption of heavier vapors such as cumene- and cyclohexanone that are difficult to remove from micropores during regeneration (78). Regeneration of the carbon is performed on a cyclic basis by purging it with steam or heated nitrogen.

11.2. Gasoline Emission Control. A principal application of activated carbon is in the capture of gasoline vapors that escape from vents in automotive fuel systems (79). Under EPA regulations, all U.S. motor vehicles produced since the early 1970s have been equipped with evaporative emission control systems. Most other auto producing countries now have similar controls. Fuel vapors vented when the fuel tank or carburetor are heated are captured in a canister containing 0.5 to 2 L of activated carbon. Regeneration of the carbon is then accomplished by using intake manifold vacuum to draw air through the canister. The air carries desorbed vapor into the engine where it is burned during normal operation. Activated carbon systems have also been proposed for capturing vapors emitted during vehicle refueling, and activated carbon is used at many gasoline terminals to capture vapor displaced when tank trucks are filled (80). Typically, the adsorption vessels contain around 15 m^3 of activated carbon and are regenerated by application of a vacuum. The vapor that is pumped off is recovered in an absorber by contact with liquid gasoline. Similar equipment is used in the transfer of fuel from barges (81). The type of carbon pore structure required for these applications is substantially different from that used in solvent recovery. Because the regeneration conditions are very mild, only the weaker adsorption forces can be overcome, and therefore the most effective pores are in the mesopore size range (82). A large adsorption capacity in these pores is possible because vapor concentrations are high, typically 10–60%.

11.3. Adsorption of Radionuclides. Other applications that depend on physical adsorption include the control of krypton and xenon radionuclides from nuclear power plants (83). The gases are not captured entirely, but their passage is delayed long enough to allow radioactive decay of the short-lived species. Highly microporous coconut-based activated carbon is used for this service.

11.4. Control by Chemical Reaction. Pick-up of gases to prevent emissions can also depend on the chemical properties of activated carbon or of impregnants. Emergency protection against radioiodine emissions from nuclear power reactors is provided by isotope exchange over activated carbon impregnated with potassium iodide (84). Oxidation reactions catalyzed by the carbon surface are the basis for several emission control strategies. Sulfur dioxide can be removed from industrial off-gases and power plant flue gas because it is oxidized to sulfur trioxide, which reacts with water to form nonvolatile sulfuric acid (85,86). Hydrogen sulfide can be removed from such sources as Claus plant tail gas because it is converted to sulfur in the presence of oxygen (87). Nitric oxide can be removed from flue gas because it is oxidized to nitrogen dioxide. Ammonia is added and reacts catalytically on the carbon surface with the nitrogen dioxide to form nitrogen (88).

11.5. Protection Against Atmospheric Contaminants. Activated carbon is widely used to filter breathing air to protect against a variety of toxic or noxious vapors, including war gases, industrial chemicals, solvents, and odorous compounds. Activated carbons for this purpose are highly microporous and thus maximize the adsorption forces that hold adsorbate molecules on the surface. Although activated carbon can give protection against most organic gases, it is especially effective against high molecular weight vapors, including chemical warfare agents such as mustard gas or the nerve agents that are toxic at parts per million concentrations. The activated carbon is employed in individual canisters or pads, as in gas masks, or in large filters in forced air ventilation systems. In airconditioning systems, adsorption on activated carbon can be used to control the buildup of odors or toxic gases like radon in recirculated air (89).

Inorganic vapors are usually not strongly adsorbed on activated carbon by physical forces, but protection against many toxic agents is achieved by using activated carbon impregnated with specific reactants or decomposition catalysts. For example, a combination of chromium and copper impregnants is used against hydrogen cyanide, cyanogen, and cyanogen chloride, whereas silver assists in the removal of arsine. All of these are potential chemical warfare agents; the Whetlerite carbon, which was developed in the early 1940s and is still used in military protective filters, contains these impregnants (90). Recent work has shown that chromium, which loses effectiveness with age and is itself toxic, can be replaced with a combination of molybdenum and triethylenediamine (91). Oxides of iron and zinc on activated carbon have been used in cigarette filters to absorb hydrogen cyanide and hydrogen sulfide (92). Mercury vapor in air can be removed by activated carbon impregnated with sulfur (93). Activated carbon impregnated with sodium or potassium hydroxide has long been used to control odors of hydrogen sulfide and organic mercaptans in sewage treatment plants (94). Alkali-impregnated carbon is also effective against sulfur dioxide, hydrogen sulfide, and chlorine at low concentrations. Such impregnated carbon is used extensively to protect sensitive electronic equipment against corrosion by these gases in industrial environments (95). An activated carbon air filter for use in room air cleaners comprising a carbon-coated corrugated paper with a long life and high efficiency has been described (96).

11.6. Process Stream Separations. Differences in adsorptivity between gases provides a means for separating components in industrial process gas streams. Activated carbon in fixed beds has been used to separate aromatic compounds from lighter vapors in petroleum refining process streams (97) and to recover gasoline components from natural and manufactured gas (98,99).

Molecular sieve activated carbons are specially made with restricted openings leading to micropores. These adsorbents are finding increasing use in separations utilizing pressure swing adsorption, in which adsorption is enhanced by operation at high pressure and desorption occurs upon depressurization (100). Larger molecules are restricted from entrance into the pores of these carbons and, therefore, are not retained as strongly as smaller molecules. The target product can be either the adsorbed or unadsorbed gases. Examples include separation of oxygen from air and recovery of methane from inorganic gases in biogas production. Hydrogen can be removed from gases produced in the catalytic cracking of gasoline, and carbon monoxide can be separated from fuel gases. Use of pressure swing techniques for gas separation is an area of growing interest in engineering research.

The Hypersorption process developed in the late 1940s used a bed of activated carbon moving countercurrent to gas flow to separate light hydrocarbons from each other and from hydrogen in refinery operations. The application is of interest because of its scale, treating up to 20,000 m³/h of gas, but the plants were shut down within a few years, probably because of problems related to attrition of the rapidly circulating activated carbon (101). It should be noted, however, that in recent years moving-bed and fluid-bed adsorption equipment using activated carbon has been successfully employed for solvent recovery (102).

11.7. Gas Storage. Adsorption forces acting on gas molecules held in micropores significantly densify the adsorbed material. As a result, activated carbon has long been considered a medium for lowering the pressure required to store weakly adsorbed compressed gases (103). Recent work with modern high capacity carbons has been directed toward fueling passenger cars with natural gas, but storage volume targets have not yet been attained (104). Natural gas storage on activated carbon is now used commercially in portable welding cylinders (105). These can be refilled easily at about 2000 kPa and hold as much gas as a conventional cylinder pressurized to 6000 kPa (59 atm).

11.8. Catalysis. Catalytic properties of the activated carbon surface are useful in both inorganic and organic synthesis. For example, the fumigant sulfuryl fluoride is made by reaction of sulfur dioxide with hydrogen fluoride and fluorine over activated carbon (106). Activated carbon also catalyzes the addition of halogens across a carbon–carbon double bond in the production of a variety of organic halides (77) and is used in the production of phosgene from carbon monoxide and chlorine (107,108).

BIBLIOGRAPHY

"Active Carbon" under "Carbon" in *ECT* 1st ed., Vol. 2, pp. 881–899, by J. W. Hassler, Nuchar Active Carbon Division, West Virginia Pulp and Paper Co., and J. W. Goetz, Carbide and Carbon Chemicals Corp.; "Activated Carbon" under "Carbon" in *ECT* 2nd ed., Vol. 4, pp. 149–158, by E. G. Doying, Union Carbide Corp., Carbon Products Division; "Activated Carbon" under "Carbon (Carbon and Artificial Graphite)" in *ECT* 3rd ed., Vol. 4, pp. 561–570, by R. W. Soffel, Union Carbide Corp.; in *ECT* 4th ed., Vol. 4, pp. 1015–1037, by Frederick S. Baker, Charles E. Miller, Albert J. Repik, E. Donald Tolles, Westvaco Corporation Charleston Research Center; "Activated Carbon" in *ECT* (online), posting date: December 4, 2000, by Frederick S. Baker, Charles E. Miller, Albert J. Repik, E. Donald Tolles, Westvaco Corporation Charleston Research Center.

CITED PUBLICATIONS

1. H. Jüntgen, *Carbon* **15**, 273–283 (1977).
2. Brit. Pat. 14,224 (1900), R. von Ostrejko; Fr. Pat. 304,867 (1900); Ger. Pat. 136,792 (1901); U.S. Pat. 739,104 (1903).
3. J. W. Hassler, *Forest Products J.* **8**, 25A–27A (1958).
4. J. W. Hassler, *Activated Carbon*, Chemical Publishing Co., Inc., New York, 1963, 1–14. A comprehensive account of the development and use of activated carbon products to about 1960.
5. R. V. Carrubba, J. E. Urbanic, N. J. Wagner, and R. H. Zanitsch, *AIChE Symp. Ser.* **80**, 76–83 (1984).
6. B. McEnaney and T. J. Mays, in H. Marsh, ed., *Introduction to Carbon Science*, Butterworths, London, 1989, 153–196. A good introduction to carbon science in general.

7. H. Marsh and J. Butler, in K. K. Unger, J. Rouquerol, K. S. W. Sing, and H. Kral, eds., *Characterization of Porous Solids, Proceedings of the IUPAC Symposium (COPS I)*, Bad Soden a.Ts., FRG, Apr. 26–29, 1987, Elsevier, Amsterdam, The Netherlands, 1988, 139–149.

8. K. S. W. Sing and co-workers, *Pure Appl. Chem.* **57**, 603–619 (1985).

9. S. Brunauer, P. H. Emmett, and E. Teller, *J. Am. Chem. Soc.* **60**, 309–319 (1938).

10. S. J. Gregg and K. S. W. Sing, *Adsorption, Surface Area, and Porosity*, 2nd ed., Academic Press, London, 1982, 303 pp. An indispensable text on the interpretation and significance of adsorption data.

11. H. Marsh, D. Crawford, T. M. O'Grady, and A. Wennerberg, *Carbon* **20**, 419–426 (1982).

12. *Jpn. Chem. Week* **30**, 5 (Mar. 16, 1989).

13. M. M. Dubinin, *J. Colloid Interface Sci.* **46**, 351–356 (1974).

14. K. S. W. Sing, *Carbon* **27**, 5–11 (1989).

15. J. Zawadzki, in P. A. Thrower, ed., *Chemistry and Physics of Carbon*, Vol. 21, Marcel Dekker, Inc., New York, 1989, 147–380. *Chemistry and Physics of Carbon*, published in 23 volumes through 1991, is a primary source of excellent review articles on carbon, many relevant to activated carbon.

16. D. Atkinson, A. I. McLeod, K. S. W. Sing, and A. Capon, *Carbon* **20**, 339–343 (1982).

17. J. M. Solar, C. A. Leon y Leon, K. Osseo-Asare, and L. R. Radovic, *Carbon* **28**, 369–375 (1990).

18. K.-D. Henning, W. Bongartz, and J. Degel, *19th Biennial Conference on Carbon*, Penn State University, Pa., June 25–30, 1990, extended abstracts, pp. 94, 95.

19. Product data bulletins from activated carbon manufacturers, Calgon Carbon Corp., 1990, American Norit Co., 1990, and Westvaco Corp., 1988.

20. T. Wigmans, *Carbon* **27**, 13–22 (1989).

21. F. Derbyshire and M. Thwaites, *Proceedings of the 4th Australian Coal Science Conference*, Brisbane, Australia, Dec. 3–5, 1990, pp. 372–379.

22. U.S. Pat. 3,976,597 (Aug. 24, 1976), A. J. Repik, C. E. Miller, and H. R. Johnson (to Westvaco Corp.).

23. W. Gerhartz, Y. S. Yamamoto, and F. Thomas Campbell, eds., *Ullmann's Encyclopedia of Industrial Chemistry*, 5th ed., Vol. A5, VCH Publishers, New York, 1986, pp. 124–140. Good descriptions of activation processes.

24. Product literature on Pittsburgh activated carbon, Pittsburgh Coke & Chemical Co. (now Calgon Carbon Corp.), Pittsburgh, Pa., ca 1960.

25. U.S. Pat. 4,014,817 (Mar. 29, 1977), B. C. Johnson, R. K. Sinha, and J. E. Urbanic (to Calgon Corp.).

26. A. Cameron and J. D. MacDowall, in J. M. Haynes and P. Rossi-Doria, eds., *Principles and Applications of Pore Structural Characterization, Proceedings of the RILEM/CNR International Symposium*, Milan, Italy, Apr. 26–29, 1983, J. W. Arrowsmith, Ltd., Bristol, UK, 1985, 251–275.

27. R. C. Bansal, J.-B. Donnet, and F. Stoeckli, *Active Carbon*, Marcel Dekker, Inc., New York, 1988, p. 8. A modern treatise on activated carbon based on a comprehensive review of the literature.

28. U.S. Pat. 4,082,694 (Apr. 4, 1978), A. N. Wennerberg and T. M. O'Grady (to Standard Oil Co.).

29. T. Kasuh, D. A. Scott, and M. Mori, *Proceedings of an International Conference on Carbon*, The University of Newcastle upon Tyne, UK, Sept. 18–23, 1988, pp. 146–148.

30. Product literature on activated carbon cloth, Charcoal Cloth Ltd., UK, 1985, and on C-tex products, Siebe Gorman & Co., Ltd., UK, 1985.

31. Product literature on KYNOL activated carbon fibers and cloths, GUN EI Chemical Industry Co., Ltd., Japan, 1987; Product literature on AD'ALL activated carbon fibers, Unitika, Ltd., Japan, 1989.

32. Product literature on KURASHEET activated carbon foam sheets, Kuraray Chemical Co., Ltd., Japan, 1987.

33. "Activated Carbon, Chemical Profile,"*Chemical Market Reporter*,April 9, 2001.

34. *AWWA Standard for Granular Activated Carbon*, ANSI/AWWA B604, American Water Works Association, Denver, Colo., 1991, 32 pp.

35. *AWWA Standard for Powdered Activated Carbon*, ANSI/AWWA B600, American Water Works Association, Denver, Colo., 1990, 32 pp.

36. *Department of Defense Military Specifications*, MIL-C-0013724D(EA), Sept. 22, 1983; MIL-C-0013724D(EA) Amendment 1, Mar. 5, 1986; and MIL-C-17605C(SH), Mar. 22, 1989.

37. S. J. Gregg and K. S. W. Sing, *Adsorption, Surface Area, and Porosity*, 1st ed., Academic Press Inc. (London) Ltd., London, 1967, 308–355.

38. H. M. Rootare, *Advanced Experimental Techniques in Powder Metallurgy*, Plenum Press, New York, 1970, 225–252. A comprehensive review of the use of mercury penetration to measure porosity.

39. G. Horvath and K. Kawazoe, *J. Chem. Eng. Jpn.* **16**, 470–475 (1983).

40. M. M. Dubinin and H. F. Stoeckli, *J. Colloid Interface Sci.* **56**, 34–42 (1980).

41. C. Orr, Jr., *Powder Technol.* **3**, 117–123 (1970).

42. *Annual Book of ASTM Standards*, 15.01, Section 15, American Society for Testing and Materials, Philadelphia, Pa., 1989.

43. *1985–1986 Registry of Toxic Effects of Chemical Substances*, Vol. 2, National Institute for Occupational Safety and Health, U.S. Department of Health and Human Services, Washington, D.C., 1987, p. 1475.

44. National Research Council, Assembly of Life Sciences, Division of Biological Sciences, Food and Nutrition Board, and Committee on Codex Specifications, *Food Chemicals Codex*, 3rd ed., National Academy Press, Washington, D.C., 1981, pp. 70, 71.

45. American Society of Civil Engineers, American Water Works Association, and Conference of State Sanitary Engineers, *Water Treatment Plant Design*, American Water Works Association, Inc., New York, 1969, p. 297.

46. J. W. Hassler, *Purification with Activated Carbon*, 3rd ed., Chemical Publishing Co., Inc., New York, 1974, p. 353. Contains much of the information given in reference 4 but with more emphasis on the commercial uses of activated carbon.

47. Ref. 46, pp. 84, 85.

48. U.S. Environmental Protection Agency, *Process Design Manual for Carbon Adsorption*, Swindell-Dressler Co., Pittsburgh, Pa., 1971, 3–68.

49. Material safety data sheets on activated carbon products, available from the manufacturers, 1991.

50. W. G. P. Schuliger, *Waterworld News* 4(1), 15–17 (1988).

51. R. M. Clark and B. W. Lykins, Jr., *Granular Activated Carbon—Design, Operation, and Cost*, Lewis Publishers, Inc., Chelsea, Mich., 1989, 295–338.

52. P. N. Cheremisinoff and F. Ellerbusch, *Carbon Adsorption Handbook*, Ann Arbor Science Publishers, Inc., Ann Arbor, Mich., 1978, 539–626. An excellent reference book on activated carbon, ranging from theoretical to applied aspects.

53. Ref. 52, pp. 389–447.

54. Product literature on PACT systems, Zimpro/Passavant, Inc., Rothschild, Wis., 1990.

55. Ref. 51, p. 51.

56. W. E. Koffskey and B. W. Lykins, Jr., *J. Am. Water Works Assoc.* **82**(1), 48–56 (1990).

57. A. Yehaskel, *Activated Carbon—Manufacture and Regeneration*, Noyes Data Corporation, Park Ridge, N.J., 1978, 202–217. A dated, but still useful summary of key patent literature.

58. P. N. Cheremisinoff, *Pollut. Eng.* **17**(3), 29–38 (1985).

59. R. G. Rice and C. M. Robson, *Biological Activated Carbon—Enhanced Aerobic Biological Activity in GAC Systems*, Ann Arbor Science Publishers, Ann Arbor, Mich., 1982, 611 pp.

60. R. A. Hutchins, *Chem. Eng.* **87**(2), 101–110 (1980). A particularly useful paper on liquid-phase adsorption.

61. M. Suzuki, *Adsorption Engineering*, Kodansha Ltd., Tokyo and Elsevier Science Publishers B.V., Amsterdam, The Netherlands, 1990, pp. 11, 35–62.

62. T. F. Speth and R. J. Miltner, *J. Am. Water Works Assoc.* **82**(2), 72–75 (1990).

63. F. L. Slejko, ed., *Adsorption Technology*, Marcel Dekker, Inc., New York, 1985, 23–32. A good account of the theory, design, and application of adsorption systems.

64. Ref. 52, pp. 8–19.

65. U.S. Pat. Appl. 20030022787(Jan. 30, 2003),P. D. A. McCral and co-workers.

66. American Water Works Association, *Water Quality and Treatment*, 3rd ed., McGraw-Hill Book Co., New York, 1971, 1–216.

67. W. J. Weber, Jr. and B. M. Van Vliet, in I. H. Suffet and M. J. McGuire, eds., *Activated Carbon Adsorption of Organics from the Aqueous Phase*, Vol. 1, Ann Arbor Science Publishers, Inc., Ann Arbor, Mich., 1980, pp. 15–41. A comprehensive, two volume treatise with many key references.

68. J. L. Oxenford and B. W. Lykins, Jr., *J. Am. Water Works Assoc.* **83**(1), 58–64 (1991).

69. I. N. Najm and co-workers, *J. Am. Water Works Assoc.* **83**(1), 65–76 (1991).

70. U.S. Pat. Appl. 20020030020(March 14, 2002).J. Moorehead andJ. T. Lodeco.

71. G. Culp, G. Wesner, R. Williams, and M. V. Hughes, *Wastewater Reuse and Recycling Technology*, Noyes Data Corp., Park Ridge, N.J., 1980, pp. 343–432. A useful review of wastewater treatment with activated carbon.

72. Ref. 46, pp. 87–125, 274–292.

73. *The Economics of Activated Carbon*, 3rd ed., Roskill Information Services Ltd., London, 1990, pp. 92–135.

74. L. W. Canter and R. C. Knox, *Ground Water Pollution Control*, Lewis Publishers, Inc., Chelsea, Mich., 1985, pp. 89–125.

75. Environmental Science and Engineering, Inc., *Removal of Volatile Organic Chemicals from Potable Water—Technologies and Costs*, Noyes Data Corp., Park Ridge, N.J., 1986, pp. 23–40.

76. M. Smisek and S. Cerny, *Active Carbon—Manufacture, Properties, and Applications*, Elsevier Publishing Co., New York, 1970, pp. 290–294.

77. A. J. Bird, in A. B. Stiles, ed., *Catalyst Supports and Supported Catalysts*, Butterworths, Stoneham, Mass., 1987, pp. 107–137.

78. P. J. Luft and P. C. Speers, Paper 52c, *AIChE Summer National Meeting*, Aug. 19–22, 1990.

79. P. J. Clarke and co-workers, *SAE Trans.* **76**, 824–837 (1967).

80. Product literature on hydrocarbon vapor recovery systems, John Zink Co., Tulsa, Okla., 1990.

81. J. Hill, *Chem. Eng.* **97**, 133–143 (1990).

82. H. R. Johnson and R. S. Williams, *S.A.E. Technical Paper No. 902119*, International Fuels and Lubricants Exposition, Tulsa, Okla., Oct. 23, 1990.

83. D. W. Moeller and D. W. Underhill, *Nucl. Saf.* **22**, 599–611 (1981).

84. M. L. Hyder, *Comm. Eur. Communities [Rep.] EUR 1986, EUR 10580, Gaseous Effluent Treat. Nucl. Install.*, 451–462 (1986).

85. F. J. Ball,S. L. Torrence, and A. J. Repik, *APCA J.* **22**, 20–26 (1972).

86. P. Ellwood, *Chem. Eng.* **76**, 62–64 (1969).
87. J. Klein and K.-D. Henning, *Fuel* **63**, 1064–1067 (1984).
88. E. Richter, *Catal. Today* **7**, 93–112 (1990).
89. M. A. Brisk and A. Turk, *Proc. APCA Ann. Meet., 77th* **2**, 84–93 (1984).
90. U.S. Pat. 2,920,050 (Jan. 5, 1960), R. J. Grabenstetter and F. E. Blacet (to U.S. Dept. of Army).
91. U.S. Pat. 4,801,311 (Jan. 31, 1989), E. D. Tolles (to Westvaco Corp.).
92. U.S. Pat. 3,460,543 (Aug. 12, 1969), C. H. Kieth, V. Norman, and W. W. Bates, Jr. (to Ligget & Meyers Corp.).
93. R. K. Sinah and P. L. Walker, *Carbon* **10**, 754–756 (1972).
94. W. D. Lovett and R. L. Poltorak, *Water and Sewage Works* **121**, 74–75 (1974).
95. G. N. Brown,M. A. Lunn,C. E. Miller, and C. D. Shelor, *Tappi J.* **66**, 33–36 (1983).
96. U.S. Pat. Appl. 20010052224(Dec. 20, 2001),S. M. R. Gelderlond andJ. Marra(to U.S. Philips Corporation).
97. S. Dunlop and R. Banks, *Hydrocarbon Process.* **56**, 147–152 (1977).
98. G. F. Russell, *Petrol. Refiner* **40**, 103–106 (1961).
99. T. Scott, *Gas. J.* **303**, 300–307 (1960).
100. E. Richter, *Erdol Kohle, Erdgas, Petrochem.* **40**, 432–438 (1987).
101. C. Berg, *Chem. Eng. Prog.* **47**, 585–590 (1951).
102. *Gastak Solvent Recovery System*, product literature, Kureha Chemical Industry Co., Ltd., New York, 1990.
103. H. Briggs and W. Cooper, *Proc. Roy. Soc. Edinburgh* **41**, 119–127 (1920–1921).
104. J. Braslaw, J. Nasea, and A. Golovoy, *Alternative Energy Sources: Proceedings of the Miami Int. Conf. on Alternative Energy Sources*, 4th ed., Ann Arbor Science Publishers, Ann Arbor, Mich., pp. 261–270, 1980.
105. U.S. Pat. 4,817,684 (Apr. 4, 1989), J. W. Turko and K. S. Czerwinski (to Michigan Consolidated Gas Co.).
106. U.S. Pat. 4,102,987 (July 25, 1978), D. M. Cook and D. C. Gustafson (to The Dow Chemical Company).
107. H. Jüntgen, *Fuel* **65**, 1436–1446 (1986).
108. H. Jüntgen, *Erdol Kohle, Erdgas, Petrochem.* **39**(12), 546–551 (1986).

FREDERICK S. BAKER
CHARLES E. MILLER
ALBERT J. REPIK
E. DONALD TOLLES
Westvaco Corporation Charleston Research Center

CARBON BLACK

1. Introduction

Carbon black is a generic term for an important family of products used principally for the reinforcement of rubber, as a black pigment, and for its electrically conductive properties. It is a fluffy powder of extreme fineness and high surface area, composed essentially of elemental carbon. Carbon black is one of the most stable chemical products. In a general sense, it is the most widely used nano-material with its aggregate dimension ranging from tens to a few hundred nanometers (nm), and imparts special properties to composites of which it is a part. Plants for the manufacture of carbon black are strategically located worldwide in order to supply the rubber tire industry, which consumes 70% of carbon black production. About 20% is used for other rubber products and 10% is used for special nonrubber applications (1). World capacity in 2001 was estimated at >8 million metric tons (1). The U.S. capacity was ~2 million metric tons. Over 42 grades, listed in ASTM 1765-01 (2), are used by the rubber industry. Many additional grades are marketed in the nonrubber markets.

Carbon blacks differ from other forms of bulk carbon such as diamond, graphite, cokes, and charcoal in that they are composed of aggregates having complex configurations, quasigraphitic in structure, and are of colloidal dimensions. They differ from other bulk carbons in being formed from the vapor phase by homogeneous nucleation through the thermal decomposition and the partial combustion of hydrocarbons. Carbon black is the product of a technology incorporating state-of-the-art engineering and process controls. Its purity differentiates it from soots that are impure by-products from the combustion of coal and oils and from the use of diesel fuels. Carbon blacks are essentially free of the inorganic contaminants and extractable organic residues characteristic of most forms of soot.

A number of processes have been used to produce carbon black including the oil-furnace, impingement (channel), lampblack, the thermal decomposition of natural gas, and decomposition of acetylene (3). These processes produce different grades of carbon black and are referred to by the process by which they are made, eg, oil-furnace black, lampblack, thermal black, acetylene black, and channel black. A small amount of by-product carbon from the manufacture of synthesis gas from liquid hydrocarbons has found applications in electrically conductive compositions. The different grades from the various processes have certain unique characteristics, but it is now possible to produce reasonable approximations of most of these grades by the oil-furnace process. Since >95% of the total output of carbon black is produced by the oil-furnace process, this article emphasizes this process (1).

2. History of Manufacture

Carbon blacks' use as a pigment dates back to prehistoric times. Wall paintings from Paleolithic caves are the earliest known use. The Egyptians used carbon black to pigment paints and lacquers. In China, ~3000 BC, carbon black for pigment use was made by burning vegetable oils in small lamps and collecting the carbon on a ceramic lid.

Prior to 1870 the dominant carbon black manufacture was by the lampblack process where oil from animal or vegetable sources was burned in a shallow pan with a restricted air supply. Starting in 1870, natural gas began to be used as the feedstock for carbon black manufacture. The resulting blacks were much darker and better covering than lampblacks. Over a couple of decades, the channel process was developed in which small gas flames burning in restricted air supply impinged on iron channels. The black adhered to the cool channel surface and was recovered by scraping it from the channel. Carbon yields were poor—a few percent. In part this was from the inefficiency of methane as a feedstock, but it also reflects the very poor capture efficiency of the early channel black process. Reportedly, the smoke plumes from channel black plants could be seen for 50 miles. The last channel black plant in the United States was closed in 1976. Two plants remain in the former Soviet Union, and a related but much evolved process is still operated in Germany.

A critical event in the development of the carbon black industry was the discovery of the benefits of carbon black as a reinforcing agent for rubber in 1904 (4). As the automobile became ubiquitous during the decade of the 1920s, the application in pneumatic tires grew rapidly and soon by-passed other applications, causing rapid growth in consumption. During the 1920s, two other processes were introduced, both using natural gas as feedstock, but having better yields and lower emissions than the channel process. One was the thermal black process in which a brick checker-work alternately absorbs heat from a natural gas air flame, and then gives up heat to crack natural gas to carbon and hydrogen. The other process was the gas furnace process that is no longer practiced.

The oil-furnace process was first introduced by Phillips Petroleum at its plant in Borger, Texas, in 1943. This process rapidly replaced all others for the production of carbon black for use in rubber. In this process fuel is burned with air in a primary combustion flame that contains excess air. A heavy, highly aromatic oil is then atomized in the hot gases leaving the primary combustion flame. A portion of the oil is burned by the excess oxygen providing the heat to maintain temperature and pyrolyze the remainder of the oil. In a modern version of the oil furnace process, carbon yields range from 65% downward depending on the surface area of the product. Product recovery is essentially 100% as a result of high efficiency bag filters. The overwhelming majority of carbon black reactors today are based on the oil furnace process.

The wide adoption of radial tires during the decades of the 1970s and 1980s caused a major contraction in demand for blacks for tire use as the expected life of an automobile tire moved from 20,000 miles with bias ply tires to over 40,000 miles with radial tires. This brought about considerable consolidation in the carbon black industry, particularly in North America and Europe.

3. Properties and Characterization

The structure of carbon black is schematically shown in Figure 1. The primary dispersable unit of carbon black is referred as an "aggregate" that is a discrete, rigid colloidal entity. It is the functional unit in well-dispersed systems. The aggregate is composed of spheres that are fused together for most carbon blacks. These spheres are generally termed as primary "particles" or "nodules". These nodules are composed of many tiny graphite-like stacks. Within the nodule the stacks are oriented so that their c axis is normal to the sphere surface, at least near the nodule surface.

The carbon blacks are characterized by their chemical compositions, microstructure, morphologies, and the physical chemistry of the surface. Morphology is a set of properties related to the average magnitude and frequency distribution

Aggregate **Primary particle (nodule)**

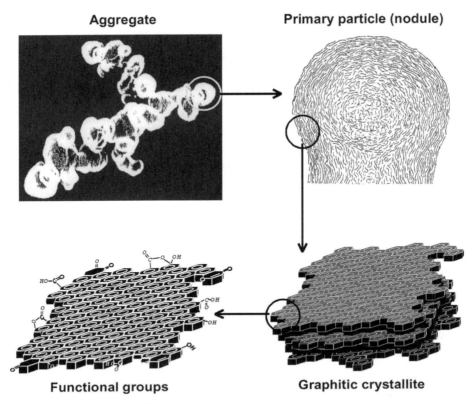

Functional groups **Graphitic crystallite**

Fig. 1. Structure of carbon black.

of the nodule diameter, aggregate diameter, and the way nodules are connected in the aggregates.

3.1. Chemical Composition. Oil-furnace blacks used by the rubber industry contain >97% elemental carbon. Thermal and acetylene black consist of >99% carbon. The ultimate analysis of rubber-grade blacks is shown in Table 1. The elements other than carbon in furnace black are hydrogen, oxygen, sulfur, and nitrogen. In addition there are mineral oxides, salts and traces of adsorbed hydrocarbons. The hydrogen and sulfur are distributed on the surface and the interior of the aggregates. The oxygen content is located on the surface of the aggregates as C_xO_y complexes.

Since carbon blacks are produced from hydrocarbon materials, the dangling bonds at the edges of the basal planes of graphitic layers are saturated mostly by hydrogen. The graphitic layers are large polycyclic aromatic ring systems.

Oxygen-containing complexes are by far the most important surface groups. The oxygen content of carbon blacks varies from 0.2 to 1.5% in mass for furnace blacks to 3 to 4% for channel blacks. Some speciality blacks used for pigment purposes contain larger quantities of oxygen than normal furnace blacks. These blacks are made by oxidation in a separate process step using nitric acid, ozone, air, or other oxidizing agents. They may contain from 2 to 12% oxygen. The oxygen-containing groups influence the physicochemical properties, such

Table 1. **Chemical Composition of Carbon Blacks**

Type	Carbon, %	Hydrogen, %	Oxygen, %	Sulfur, %	Nitrogen, %	Ash, %	Volatile, %
furnace rubber-grade	97.3–99.3	0.20–0.80	0.20–1.50	0.20–1.20	0.05–0.30	0.10–1.00	0.60–1.50
medium thermal	99.4	0.30–0.50	0.00–0.12	0.00–0.25	NA[a]	0.20–0.38	
acetylene black	99.8	0.05–0.10	0.10–0.15	0.02–0.05	NA[a]	0.00	<0.40

[a] Not available = NA.

as chemical reactivities, wettability, catalytic, electrical properties, and adsorbability. Oxidation improves dispersion and flow characteristics in pigment vehicle systems such as lithographic inks, paints, and enamels. In rubber-grade blacks surface oxidation reduces pH and changes the kinetics of vulcanization, making the rubber compounds slower curing.

A convenient method for assessing the extent of surface oxidation is the measurement of volatile content. This standard method measures the weight loss of the evolved gases on heating up from 120 to 950°C in an inert atmosphere. The composition of these gases consists of three principal components: hydrogen, carbon monoxide, and carbon dioxide. The volatile content of normal furnace blacks is <1.5%, and the volatile content of oxidized special grades is 2–22%.

The origin of the volatile gases is the functional groups attached to carbon black, especially those on the surface. Surface oxides bound to the edges of the of carbon layers are phenols, hydroquinones, quinones, neutral groups with one oxygen, carboxylic acids, lactones, and neutral groups containing two oxygens (5,6). Figure 2 shows an idealized graphite surface layer plane with the various functional groups located at the periphery of the plane. Carbon blacks with few oxygen groups show basic surface properties and anion exchange behavior (7,8).

In addition to combined hydrogen and oxygen, carbon blacks may contain as much as 1.2% combined sulfur resulting from the sulfur content of the aromatic feedstock that contains thiophenes, mercaptans, and sulfides. The majority of the sulfur is not potentially reactive as it is inaccessibly bound in the interior of carbon black particle and does not contribute to sulfur cross-linking during the vulcanization of rubber compounds.

The nitrogen in carbon blacks is the residue of nitrogen heterocycles in the feedstocks. Thus carbon blacks derived from coal tars have far more nitrogen than petroleum derived blacks.

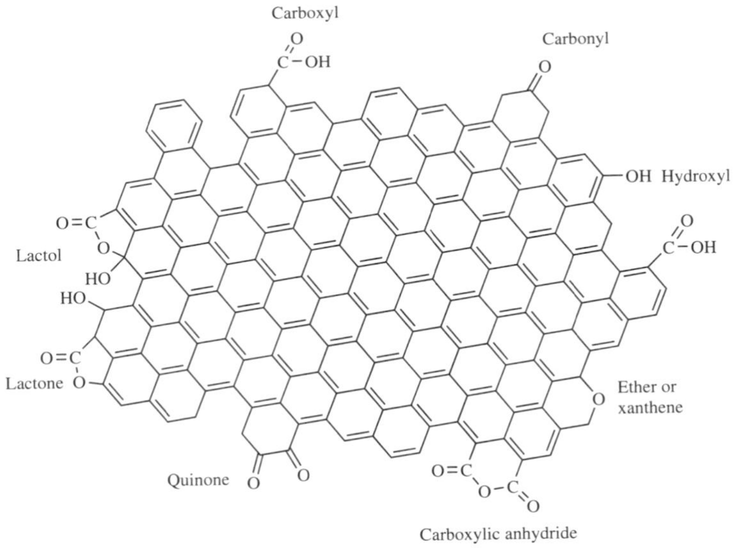

Fig. 2.   Aromatic layer plane with functional groups.

The ash content of furnace blacks is normally a few tenths of a percent but in some products may be as high as one percent. The chief source of ash is the water used to quench the hot black from the reactors during manufacture and for wet pelletizing the black.

3.2. Microstructure: Molecular and Crystallite Structure. The arrangement of carbon atoms in carbon black has been well established by X-ray diffraction methods (9,10). The diffraction patterns show diffuse rings at the same positions as diffraction rings from pure graphite. The suggested relation to graphite is further emphasized as carbon black is heated to 3000°C. The diffuse reflections sharpen, but the pattern never approaches that of true graphite. Carbon black has a degenerated graphitic crystalline structure as defined above. Whereas graphite has three-dimensional order, as seen in the model structures of Figure 3, carbon black has two-dimensional order. The X-ray data indicate that carbon black consists of well-developed graphite platelets stacked roughly parallel to one another but random in orientation with respect to adjacent layers. As shown in Figure 3, the carbon atoms in the graphitic structure of carbon black form large sheets of condensed aromatic ring systems with an interatomic spacing of 0.142 nm within the sheet identical to that found in graphite. However, the interplanar distances are quite different. While graphite interplanar distance is 0.335 nm, which results in a relative density of 2.26, the interplanar distance of carbon black is larger, in the range of 0.350−0.365 nm, as a consequence of the random planar orientations or so-called turbostratic arrangement. The relative density of commercial carbon blacks are 1.76−1.90 depending on the grade. About one-half of the decrement in density is attributed to stacking height, L_c in the crystallites. X-ray diffraction data provide estimates of crystallite size. For a typical carbon black, the average crystallite diameter, L_a, is ∼1.7 nm and average L_c is 1.5 nm, which corresponds to an average of four layer planes per crystallite containing ∼375 carbon atoms.

It was originally suggested that these discrete crystallites were in random orientation within the particle. This view was later abandoned when electron microscopy of graphitized and oxidized carbon blacks indicated more of a

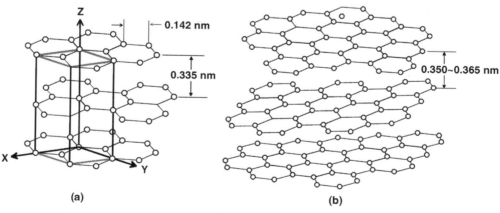

Fig. 3.   Atomic structural models of (**a**) graphite, and (**b**) carbon black.

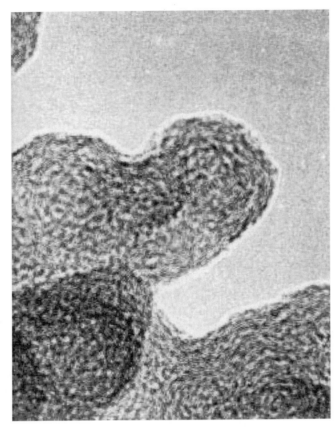

Fig. 4. High resolution (3,000,000 ×) electron micrograph of N220-grade carbon black. Courtesy of W. M. Hess.

concentric layer plane arrangement that can be described by a paracrystalline model. This structure has been confirmed by the use of high resolution phase-contrast electron microscopy that made possible the direct imaging of graphitic layer planes in carbon black (11). Figure 4 shows a phase-contrast electron micrograph of carbon black at high resolution that displays the marked concentric arrangement of the layer planes at the surface and around what appear to be growth centers.

More recently, the microstructure of the carbon black surface has been investigated by means of scanning tunneling microscopy (STM) (12,13). Figure 5 shows the STM images obtained in the current mode for graphite, graphitized carbon black and normal carbon black N234. Compared to graphite, the structure of carbon blacks graphitized for 24 h at a temperature of 2700°C in an inert atmosphere still remains in a certain inperfect state, shown by different tunneling current partterns in the organized domains. The surface structure of carbon black can be classified in two types: organized domains and unorganised domains. The organized domains occupy the majority of the carbon black surface, and its size generally decrease with decreasing particle size.

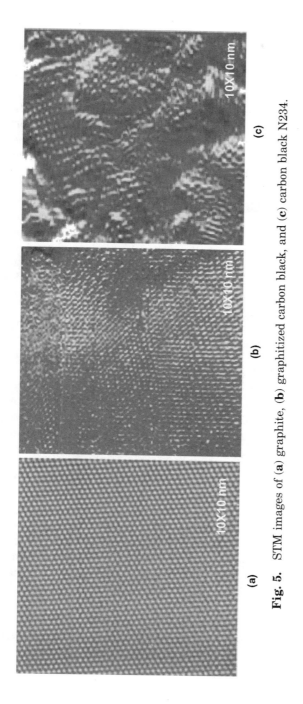

Fig. 5. STM images of (**a**) graphite, (**b**) graphitized carbon black, and (**c**) carbon black N234.

3.3. Morphology.   Morphologically, carbon blacks differ in primary "particle" or nodule size, surface area, aggregate size, aggregate shape, and in the distribution of each of these.

Primary "Particle" (Nodule) Size.   Although the smallest discrete entity of carbon black is the aggregate, the "particle" size, and its distribution is one of the most important morphological parameters with regard to its end-use applications, even though the particles do not exist as discrete entities except for thermal black. The "particle" size is of critical importance to the specific surface of the carbon blacks and has been taken as the principal parameter for grade classification of rubber blacks in ASTM. In almost all types of carbon black, the primary particles within a single aggregate are similar. However, the types of blacks can differ in the uniformity of the primary particles of different aggregates. While many types do show a quite narrow range of primary particles, others are clearly quite broad mixtures of aggregates of differing primary particle size or specific surface area. The electron microscope is the universally accepted instrument for measuring particle size, aggregate size, and aggregate morphology. Typical electron micrographs of rubber-grade carbon blacks are shown in Figure 6.

"Particle" size measurement is based on the visual electron microscope counts for several thousand particles on electromicrographs of known magnification (ASTM D3849). Automated image analyzers provide measurements of a variety of particle parameters.

Surface area.   The surface area is one of the most important features influencing the performance of carbon blacks. It is an extensity factor that determines the interfacial area between carbon black and the medium in which a given volume of black is dispersed.

The surface area can be calculated from particle size measured with transmission electron microscope (TEM) (ASTM D3849). Generally, for rubber grade, the surface areas determined by TEM are in reasonable agreement with surface areas determined by nitrogen adsorption measurements. However, for those carbon blacks that have highly developed micropores, such as special pigment blacks and blacks used for electrical conductivity, the surface areas calculated from their particle diameters are smaller than those calculated from gas absorption as the internal surface area in the micropore is excluded.

While the TEM images contain very detailed statistical information, such images are expensive and time consuming to obtain. Direct measurement of specific surface area is much faster and cheaper. Such measurements are most easily made by gas- and liquid-phase adsorption techniques that depend on the amount of adsorbate required to form a surface monolayer. If the area occupied by a single-adsorbate molecule is known, a simple calculation will yield the surface area.

A low temperature nitrogen adsorption method, based on the original method of Brunauer, and co-workers (BET) (14), has been adopted by ASTM as standard method D 6556. It is not sensitive to changes in the surface chemistry of carbon black such as those resulted from surface oxidation and presence of trace amount of tarry material. With a molecular diameter of <0.5 nm, nitrogen is small enough to enter the micropore space so that the surface measured by BET is the total area, inclusive of micropore. For some applications such as, eg, rubber reinforcement, the internal surface area in the micropore with <2-nm

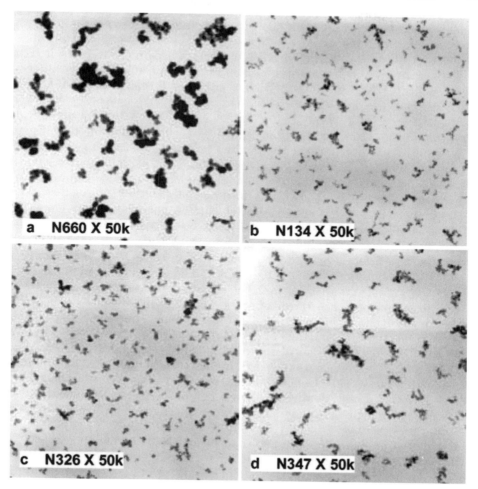

Fig. 6. Electron micrographs of rubber-grade carbon blacks.

diameter is inaccesible to large rubber molecules, thus it plays no part or has a negative effect on rubber reinforcement. The specific surface area that is accessible to rubber is defined as "external" surface area. This is conveniently measured by a multilayer nitrogen adsorption, also defined in ASTM 6556 and known as the statistical thickness surface area (STSA) (15).

Liquid-phase adsorption methods are also widely used. The adsorption of iodine from potassium iodide solution is the standard ASTM method D1510. The surface area is expressed as the iodine number whose units are milligrams of iodine adsorbed per gram of carbon blacks. The test conditions such as adsorbate concentration and the amount of carbon black sample used are specified in such a way that the values of iodine numbers turn out to be about the same as the values for surface areas in square meters per gram that are measured by nitrogen adsorption for nonporous and nonoxidized furnace carbon blacks. The iodine number is raised by porosity and decreased by surface oxygen or adsorbed organics. Still it is the most easily measured surface area estimate and is used extensively, especially for process control.

Another standard industry method for surface area measurement is based on the adsorption of cetyltrimethylammonium bromide (CTAB) from aqueous solution, which is ASTM method D3765 and has largely been replaced by STSA in the last decade.

Aggregate Morphology (Structure). The aggregate morphology is another important characteristic that influences performance. The term "structure" is widely used in the carbon black and rubber industries to describe the aggregate morphology. It was originally introduced in 1944 (16) to describe the ensemble of aggregates that is a stochastic distribution of the number and arrangement of the nodules that make up the aggregates.

Structure comparisons of grades with different surface areas cannot be made. It is now known that the properties associated with structure are associated principally with the bulkiness of individual aggregates. Aggregates of the same mass, surface area, and number of nodules have high structure in the open bulky and filamentous arrangement and a low structure in a more clustered compact arrangement. Therefore, the structure is now used to describe the relative void volume characteristics of grades of black of the same surface area. Structure is determined by aggregate size and shape, and their distribution. They are geometrical factors that affect aggregate packing and the volume of voids in the bulk material. Therefore, in composite systems, structure is also a principal feature that determine the performance of carbon black as a reinforcing agent and as a pigment (17). In liquid media, structure affects rheological properties such a viscosity and yield point. In rubber, viscosity, extrusion die swell, modulus, abrasion resistance, dynamic properties, and electrical conductivity are affected by structure.

The direct method for structure measurement of carbon black is TEM (ASTM D3849). This method is unique in furnishing information about the aggregate size, shape and the distribution of these. Typical electron micrographs of rubber-grade carbon blacks are shown in Figure 6. There is an enormous range in aggregate size. The size of the aggregates is generally related to the size of the particles. The shapes of the aggregates have infinite variety from tight grape-like clusters to open dendritic or branched arrangements to fibrous configurations.

A useful method for determining relative aggregate sizes and distributions is by centrifugal sedimentation. For a sphere, the diameter can be derived from the sedimentation rates in the gravitational field according to Stokes equation. For the nonspherical particles such as carbon black aggregates, an equivalent Stokes diameter, D_{st} can be obtained as the diameter of a sphere of carbon black having the same settling behavior. A convenient instrument for these measurements is the Joyce Loebl disk centrifuge photosedimentometer (DCP) (18). Large aggregates sediment at a faster rate than smaller ones. The sedimentation rate is also influenced by the bulkiness of the aggregates. At constant volume or mass, a bulky aggregate sediments more slowly than a compact aggregate because of frictional drag. The DCP curve is characteristic of the black structure but the measured diameters need to be viewed with suspicion.

Table 2 lists average D_{st} values and the weight mean diameter, D_{wm}, for the aggregates calculated from their estimated volumes measured by TEM. There is

Table 2. **Carbon Black Morphology**

ASTM designation	Particle size[a], D_{wm},[b] nm	Aggregate size[a], D_{wm},[b] nm	D_{st}[c], nm	Surface area[a], m^2/g
N110	27	93	76–111	143
N220	32	103	95–117	117
N234	31	109	74–97	120
N326	41	108	98	94
N330	46	146	116–145	80
N339	39	122	96–125	96
N351	50	159	127	75
N375	36	106	91	105
N550	93	240	220–242	41
N660	109	252	227–283	34
N774	124	265	261	30
N990	403	593	436	9

[a] Measured by TEM.
[b] D_{wm} = weight mean diameter = $\Sigma nd^4/\Sigma nd^3$.
[c] Stokes diameter by centrifugal sedimentation from various sources.

reasonable agreement between the two diameters. Aggregate size distributions from centrifugal sedimentation analysis are very useful for assessing the differences in this characteristic within a given grade or at constant surface area.

By far the most intensively investigated measure of structure is the maximum packing fraction. In fact, the amount of void at the maximum packing has almost become synonymous with carbon black "structure". At least two approaches are widely used. The first determines the amount of liquid that is needed to just fill all the spaces between aggregates when the aggregates are pulled together by the surface tension forces of that liquid. This measurement is done by means of an absorptometer with dibutyl phthalate (DBP) as the liquid (ASTM D 2414). This is based on the change in torque during mixing of carbon black and the liquid as there is a sharp increase in viscosity of the mixture, when it changes from free flowing powder to a semiplastic continuous paste. The viscosity, hence the torque, will drop as the liquid is continually added, due to the lubrication effect. The volume of DBP needed for a unit mass of carbon blacks to reach a predetermined level of torque is termed the DBP number. In order to eliminate the effects of pelletizing conditions, the DBP absorption test has been modified to use a sample that has been pre-compressed at a pressure of 165 MPa (24,000 psi) and then broken up four successive times (24M4) (ASTM D 3493). This procedure causes some aggregate breakdown and is claimed to more closely approximate the actual breakdown that occurs during rubber mixing. The DBP numbers measured for compressed samples are also termed crush DBP (CDBP) number.

The maximum random packing for mono-sized spheres is a volume fraction of 68 percent, which is equivalent to a DBP number of 25 mL/100 g of carbon black.

The types of blacks with the lowest DBP number, and therefore highest packing, have values ~30–35 mL/100 g. The types with the highest structure, and therefore lowest packing, have DBP number of ~140 mL/100 g. Grades with significantly higher DBP number do not have solid primary particles and the fluid is absorbed both within the primary particles and between the aggregates.

The second way of estimating the maximum packing fraction is mechanical compression. Most commonly this is practiced in uniaxial compression. The test is termed the void volume measurement (ASTM D6086). With both liquid and mechanical compaction, the measured value can be influenced to some degree by either the speed of the compaction, ie, the time the aggregates have to rearrange, and the final pressure to which they are subjected.

Tinting strength, adopted by ASTM as D3265, is another industry method used for the classification of carbon blacks. In this test, a small amount of carbon black is mixed with zinc oxide and an oil vehicle to produce a black or gray paste. The reflectance of this paste is measured and compared to the reflectance of a paste made with a reference black. The ratio of the reference black paste reflectance to the sample black multiplied by 100 is the tint strength. It provides a rough estimate of the reinforcing potential of carbon black in rubber. Tint strength is closely related to carbon black morphology. The carbon blacks with smaller particle size, ie, larger surface area and smaller size of the aggregates, ie, lower structure show a greater ability to cover the typically larger size zinc oxide particles, giving higher tinting strength. The tinting strength is also related to the aggregate size distribution. The narrower the aggregate size distribution, the higher is the tinting strength.

3.4. Surface Activity. Surface activity is also an important factor in performance. This factor can, in a chemical sense, be related to different chemical groups on the carbon black surface. In a physical sense, variations in surface energy determine the adsorptive capacity of the carbon blacks and their energy of adsorption. However, compared with the morphologies, a satisfactory description of surface properties of carbon black is still lacking because only a limited number of tools have been available to assess the carbon black surface in terms of ensemble properties.

With regard to rubber reinforcement, the surface activity of carbon black has traditionally been measured by bound rubber. Bound rubber, sometimes termed "carbon gel", is defined as the rubber portion in an uncured compound that cannot be extracted by a good solvent of the polymer due to the adsorption of the rubber molecules onto the filler surface. This phenomenon has been studied extensively and is recognized as a typical feature of carbon black surface activity. For given polymer systems and for the carbon black with comparable surface area, the higher the bound rubber content, the higher is the polymer-filler interaction, and hence the higher is the surface activity of carbon black. Generally speaking, bound rubber is a parameter that is simple to measure, but the factors that influence the test results are highly complicated. It has been recognized that carbon black–polymer interaction leading to the formation of bound rubber involves physical adsorption, chemisorption, and mechanical interaction.

It is generally found that the surface energy of carbon blacks have a much greater effect on the mechanical properties of filled elastomers than the chemical composition, particularly when general-purpose hydrocarbon rubbers are concerned. The surface energy, γ, is defined as the work, necessary to create a unit new surface of liquid or solid. This energy is caused by different types of cohesive forces, such as dispersive, dipole–dipole, induced dipole–dipole, and hydrogen-bond forces. In the case of all these cohesive forces being involved in independent ways, the surface energy can be expressed as the sum of several components, each corresponding to a type of molecular interaction (dispersive, polar, hydrogen bond, etc). Since the effect of the dispersive force is universal, the dispersive component of the surface free energy, γ^d, is particularly important. If a solid substance can have only dispersion interaction with its environment, its surface energy, γ_s, is identical with its dispersive component, γ_s^d. For most substances, the surface energy of a solid is the sum of γ_s^d and γ_s^{sp} that is the sum of the other components of surface energy and is termed "specific component" or "polar" component. The higher the γ_s^d of the carbon black, the stronger is the interaction between carbon black and non- or less-polar polymers such as hydrocarbon rubbers. The higher polar component of the surface energy leads to higher interaction with polar polymer or polar groups in the polymer chains.

Several methods to measure the solid surface energy can be used for carbon black. However, inverse gas chromatography (IGC) has recently been shown to be one of the most sensitive and convenient methods for measuring carbon black surface energy (19). In IGC, the filler to be characterized is used as the stationary phase and the solute injected is called a probe. When the probe is operated at infinite dilution, the adsorption energy of the probe on the carbon black surface, and hence the surface energy of the black can be calculated from the net retention volume (19). If, however, the surface is energetically heterogeneous, the values of parameters obtained from IGC measurement are mean values over the whole surface of the fillers, but they are "energy-weighted", ie, the high energy sites play a very important role in the determination of adsorption parameters measured (20). When the probe is operated at finite concentration, the adsorption isotherms of the probes on carbon black surface can be generated from the pressure dependence of the retention volume and, hence, the distribution of free energy of adsorption of the probe chemicals can be derived from the isotherm (20).

3.5. Other Methods for Carbon Black Characterization.   There are many other test methods used to characterize carbon blacks for quality control and specification purposes. Table 3 lists some of these methods that, with a few exceptions, have been adopted by ASTM.

4. Classification

Carbon blacks have been classified by their production process, by their production feedstocks such as acetylene blacks, by their application field, such as rubber blacks, color blacks, electric conductive blacks, and by properties of end use products such as high abrasion furnace black (HAF) and fast extrusion furnace black (FEF). From their applications, the carbon black is classified into two groups: one used for rubber products, and another for non-rubber applications. They are referred as special blacks. Generally, special carbon blacks cover a wider range of morphology and surface chemistry than rubber blacks.

Table 3. **Special Analytical Test Methods for Carbon Black**

Test method	Standard	Comment
iodine adsorption, mg/g	ASTM D1510	amount of iodine adsorbed from aqueous solution as a measure for the specific surface area; not applicable for oxidized or highly porous carbon blacks
N_2 surface area, and external surface area, m^2/g	ASTM D6556	determination of the total surface area (NSA) by B.E.T. theory of multilayer gas adsorption behavior using multipoint determinations and the external surface area based on the statistical thickness surface area (STSA) method
CTAB surface are, m^2/g	ASTM D3765	amount of cetytrimethylammonium bromide adsorbed from aqueous solution as measure of specific nonporous (outer) surface area
aggregate dimension and aggregate size distribution	ASTM D3849	determination of aggregate dimensions (unit length, width, etc) by electron microscope image analysis
aggregate size distribution		diameters of equivalent solid spheres that sediment at same rate as aggregates during centrifuging
DBP absorption, mL/100 g	ASTM D2414	determination of the void volume with dibutyl phthalate in a special kneader as measure of structure
24 M4-DBP absorption, mL/100 g	ASTM D3493	determination of DBP absorption after four repeated compressions at 165 MPa (24,000 psi)
compressed volume index	ASTM D6086	determination of compressed volume of carbon black under a specified compression force.
jetness		light absorption of a carbon black paste in linseed oil; determination by visual comparison against standard blacks or by measuring the absolute light emission
tint strength, %	ASTM D3265	ability of a carbon black to darken a white pigment in an oil paste. The tinting strength is the reflectance of the tested carbon black paste with respect to the reflectance of the reference carbon black paste.
volatiles, %		weight loss when calcined at 950°C for 7 min
heating loss (moisture), %	ASTM D1509	weight loss on drying at 125°C for 1 h
pH	ASTM D1512	pH of an aqueous slurry of carbon black; pH is mainly influenced by surface oxides
extractables, %	ASTM D1618	amount of material that can be extracted by a boiling solvent, usually toluene
extractables, %	ASTM D4527	determination of the total material extracted from carbon black by toluene under specified conditions. The procedure is also applicable to other solvents

Table 3   (*Continued*)

Test method	Standard	Comment
ash content, %	ASTM D1506	amount of noncombustible material after burning the carbon black at 675°C
sulfur content, %	ASTM D1619	
sieve residue, %	ASTM D1514	amount of coarse impurities that cannot be purged through a testing sieve by water
pour density, g/L	ASTM D1513	measure for the densification of carbon black
tamped density, g/L		similar to bulk density; however, void volume is reduced by temping
pellet crush strength	ASTM D5230	automated individual pellet crush strength
pellet crush strength	ASTM D3313	individual pellet crush strength
pellet mass strength	ASTM D1937	pellet mass strength
pellet size distribution	ASTM D1511	determination by means of sieve shaker
fines content, %	ASTM D1508	only for pelletized blacks; percentage passing through a sieve of 125 μm (mesh) width

The rubber industry is by far the major consumer of carbon black. For rubber grades, a classification system issued by ASTM is based essentially on particle size, structure and their effect on the cure rate of filled rubber compounds which is related to the degree of surface oxidation (ASTM D1765). It is composed of a letter followed by three numbers. The N series are for the normal-curing furnace and thermal blacks and S for "slow-curing" blacks with higher degree of oxidation, such as channel black and oxidized furnace blacks that are acidic in nature. The first number of three digit suffix identifies particle size and is inversely related to the surface area. The range of particle sizes from 0 to 500 nm has been grouped into 10 categories, covering surface areas from 0 to 150 m^2/g. The remaining two digits are assigned arbitrarily by the carbon black manufacturers. A selected list of typical properties, taken from ASTM D1765 of rubber-grade carbon blacks, is shown in Table 4. In addition to the assigned ASTM N-numbers, the list includes structure, surface areas, and tint data. The structure–surface area relationships of these grades, called the carbon black spectrum, is illustrated in Figure 7, which shows a diagram of DBP numbers for compressed samples versus the nitrogen surface areas.

5. Carbon Black Formation

The formation of particulate carbon involves either pyrolysis or incomplete combustion of hydrocarbon materials. Enormous literature has been published to describe the mechanism of carbon black formation, from a series of lectures by Michael Faraday at the Royal Institution in London in the 1860s (21), to a recent

Table 4. **Typical Properties of Rubber-Grade Carbon Blacks**

ASTM classification	Iodine No., mg/g	DBP No. mL/100 g	CDBP No[a], mL/100 g	NSA, m^2/g	STSA m^2/g	Tint strength,%
N110	145	113	97	127	115	123
N115	160	113	97	137	124	123
N120	122	114	99	126	113	129
N121	121	132	111	122	114	119
N125	117	104	89	122	121	125
N134	142	127	103	143	137	131
N135	151	135	117	141		119
S212		85	82	120	107	115
N220	121	114	98	119	106	116
N231	121	92	86	111	107	120
N234	120	125	102	119	112	123
N293	145	100	88	122	111	120
N299	108	124	104	104	97	113
N315		79	77	89	86	117
N326	82	72	68	78	76	111
N330	82	102	88	78	75	104
N335	92	110	94	85	85	110
N339	90	120	99	91	88	111
N343	92	130	104	96	92	112
N347	90	124	99	85	83	105
N351	68	120	95	71	70	100
N356	92	154	112	91	87	106
N358	84	150	108	80	78	98
N375	90	114	96	93	91	114
N539	43	111	81	39	38	
N550	43	121	85	40	39	67
N582	100	180	114	80		
N630	36	78	62	32	32	
N642	36	64	62	39		
N650	36	122	84	36	35	
N660	36	90	74	35	34	
N683	35	133	85	36	34	
N754	24	58	57	25	24	
N762	27	65	59	29	28	
N765	31	115	81	34	32	
N772	30	65	59	32	30	
N774	29	72	63	30	29	
N787	30	80	70	32	32	
N907		34		9	9	
N908		34		9	9	
N990		43	37	8	8	
N991		35	37	8	8	

[a] For compressed samples.

intensive review (22). Since Faraday's time, many theories have been proposed to account for carbon formation, but controversy still exists regarding the mechanism.

Mechanisms of carbon black formation must account for the experimental observations of the unique morphology and microstructure of carbon black. These include the presence of nodules, or particles, multiple growth centers

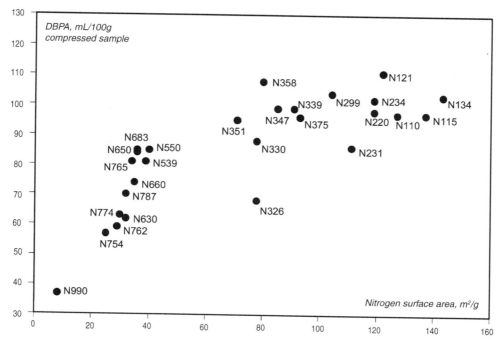

Fig. 7. Rubber grades carbon black spectrum.

within some nodules, the fusion of nodules into large aggregates, and the para-crystalline or concentric layer plane structure of the aggregates. It is generally accepted that the mechanism of formation involves a series of stages as follows:

- Formation of gaseous carbon black precursors at high temperature: This involves dehydrogenation of primary hydrocarbon molecular species to atomic carbon or primary free radical and ions that condense to semisolid carbon precursors (or polynucleararomatic sheet) and/or formation of large hydrocarbon molecules by polymerization, which then is dehydrogenated to particle precursors.

- Nucleation: Because of increasing mass of the carbon particle precursors through collision, the larger fragments are no longer stable and condense out of the vapor phase to form nuclei or growth centers.

- Particle growth and aggregation: In the system, three processes go on simultaneously; condensation of more carbon precursors on the existing nuclei, coalescence of small particles into larger ones, and formation of new nuclei. Coalescence and growth seem to predominate. The products of this stage are "proto-nodules".

- Surface growth: Surface growth includes the processes in which the small species attach to or deposit on the surfaces of existing particles or aggregates, forming the nodules, and aggregates with their characteristic onion microstructure. The surface growth represents ~90% of total carbon yield. It is responsible for the stability of the aggregates because of the continuous carbon network formation. Aggregates are formed and cemented in this stage.

- Agglomeration: Once no more carbon is forming and aggregation ceases, aggregates collide and adhere from van der Waals forces but there is no material to cement them together, hence they form temporary structures.
- Aggregate gassification: After its formation and growth, the carbon black surface undergoes reaction with the gas phase, resulting in an etched surface. Species such as CO_2, H_2O, and of course any residual oxygen attack the carbon surface. The oxidation is determined by gas phase conditions, such as temperature, oxidant concentration and flow rates.

Practically, the carbon black morphology and surface chemistry can be well controlled by changing the reaction parameters. For furnace carbon blacks, the reaction temperature is the key variable that governs the surface area. The higher the temperature, the higher is the pyrolysis rate and the more nuclei are formed, resulting in an earlier stop of the growth of the particles and aggregates due to the limitation of starting materials at given feedstock. Therefore, with higher reaction temperature, that can be achieved by adjusting air rate, fuel rate and feedstock rate, the surface area of carbon black can be increased. Addition of alkali metal salts into the reactor can modify the aggregation process, influencing carbon black structure. At the reactor temperature, the salts of alkali metals, such as potassium, are ionized. The positive ions adsorb on the forming carbon black nodules and provide some electrostatic barrier to internodule collisions, resulting in lower structure (23).

The timescale of carbon black formation varies substantially across the range of particle sizes found in commercial furnace blacks. For blacks with surface areas $\sim$120 m^2/g, the carbon black formation process from oil atomization to quench takes <10 ms. For blacks with surface areas $\sim$30 m^2/g, formation times are a few tenths of seconds.

6. Manufacture

6.1. Oil-Furnace Process. The oil-furnace process accounts for >95% of all carbon black produced in the world. It was developed in 1943 and rapidly displaced prior gas based technologies because of its higher yields and the broader range of blacks that could be produced. It also provides highly effective capture of particulates and has greatly improved the environment around carbon black plants. As indicated in the mechanism discussion, it is based on the partial combustion of residual aromatic oils. Because residual oils are ubiquitous and are easily transported, the process can be practiced with little geographic limitation. This has allowed construction of carbon black plants all over the world. Plants are typically located in areas of tire and rubber goods manufacture. Because carbon black is of relatively low density, it is far less expensive to transport feedstock than to transport the black.

Over the 50 years since its invention, the oil-furnace process has undergone several cycles of improvement. These improvements have resulted in improved yields, larger process trains, better energy economy, and improved product performance. A simplified flow diagram of a modern furnace black production line is shown in Figure 8 (24). This figure is intended to be a generic diagram and contains elements from several operators' processes. The principal pieces of equipment are the air blower, process air and oil preheaters, reactors, quench tower, bag filter, pelletizer, and rotary dryer. The basic process consists of atomizing the preheated oil in a combustion gas stream formed by burning fuel in preheated air. The atomization is carried out in a region of intense turbulent mixing. Some of the atomized feedstock is combusted with excess oxidant in the combustion gas. Temperatures in the region of carbon black formation range from 1400 to >1800°C. The details of reactor construction vary from manufacturer to manufacturer and are confidential to each manufacturer. Leaving the formation zone, the carbon black containing gases are quenched by spraying water into the stream. The partially cooled smoke is then passed through a heat exchanger where incoming air is preheated. Additional quench water is used to cool the smoke to a temperature consistent with the life of the bag material used in the bag filter. The bag filter separates the unagglomerated carbon black from the by-product tail gas which contains nitrogen, hydrogen, carbon monoxide, carbon dioxide, and water vapor. It is mainly nitrogen and water vapor. The tail gas is frequently used to fuel the dryers in the plant, to provide other process heat, and sometimes is burned to manufacture steam and electric power either for internal plant use or for sale.

The fluffy black from the bag filter is mixed with water, typically in a pin mixer, to form wet granules. These are dried in a rotary dryer, and the dried product is conveyed to bulk storage tanks. For special purposes, dry pelletization in rotary drums is also practiced. Most carbon black is shipped by rail or in bulk trucks. Various semibulk containers are also used including IBCs and large semibulk bags. Some special purpose blacks are packed in paper or plastic bags.

While the reactor and its associated air-moving and heat-exchange equipment are where the properties of the black are determined, they tend to be dwarfed by the bag collectors, the dryers, and particularly the storage tanks.

Feedstocks. Feedstocks for the oil-furnace process are heavy fuel oils. Preferred oils have high aromaticity, are free of suspended solids, and have a minimum of asphaltenes. Suitable oils are catalytic cracker residue (once residual catalyst has been removed), ethylene cracker residues, and distilled heavy coal tar fractions. Other specifications of importance are freedom from solid materials, moderate to low sulfur, and low alkali metals. The ability to handle such oils in tanks, pumps, transfer lines, and spray nozzles is also a primary requirement.

The pricing of carbon black feedstocks depends on their alternate market as residual fuel oil, especially that of high sulfur No. 6 fuel oil. The actual price is determined by the supply–demand relationships for these two markets. Feedstock cost contributes ~60% of the total manufacturing cost. The market price of carbon black is strongly dependent on the feedstock cost as shown in Figure 9.

Reactor. The heart of a furnace black plant is the furnace or reactor where carbon black formation takes place under high temperature, partial combustion conditions. The reactors are designed and constructed to be as trouble-free as possible over long periods of operation under extremely aggressive conditions. They are monitored constantly for signs of deterioration in order to ensure

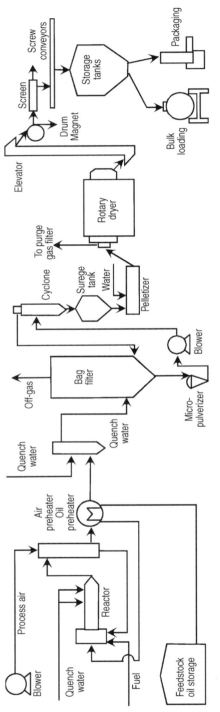

Fig. 8. Flow diagram of oil furnace black process.

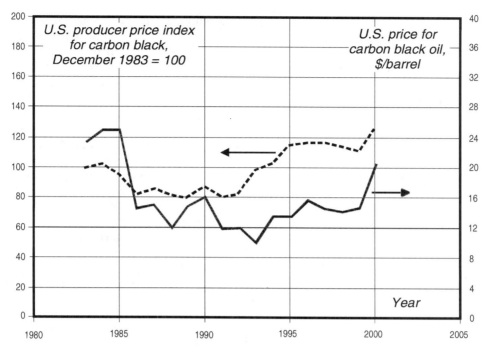

Fig. 9. Carbon black price and raw material cost in the United States (1983–2001) (1).

constant product quality. The wide variety of furnace black grades for rubber and pigment applications requires different reactor designs and sizes to cover the complete range, though closely related grades can be made in the same reactor by adjusting input variables. Reactors for higher surface area and reinforcing grades operate under high gas velocities, temperatures, and turbulence to ensure rapid mixing of reactant gases and feedstock. Lower surface area and less reinforcing grades are produced in larger reactors at lower temperatures, lower velocities, and longer residence time. Table 5 lists carbon formation temperatures, and residence times for the various grades of rubber blacks.

A key development in the carbon black reactor technology was the development of the zoned axial flow reactor for reinforcing blacks in the early 1960s (23). The reactor consists of three zones. The first zone is a combustion zone in which

Table 5. **Reactor Conditions for Various Grades of Carbon Blacks**

Black	Surface area, m^2/g	Temperature, °C	Residence time, s	Maximum velocity, m/s
N100 series	145	1800	0.008	
N200 series	120		0.010	180–400
N300 series	80	1550	0.031	
N500 series	42		1	30–80
N700 series	25	1400	1.5	0.5–1.5
N990 thermal	8	1200–1350	10	10

fuel and air are completely burned to produce combustion gases with excess oxygen. This gas flow is accelerated to high velocity in a throat zone with intense turbulent mixing. The feedstock is injected either into this throat zone or just ahead thereof. The reacting gases issue from the throat into a second cylindrical zone as a turbulent diffusion jet. Depending on the desired black, the jet may be allowed to expand freely, or may be confined by bricking. Downstream of the reaction zone is a water quench zone. The throughput of a single reactor train varies from manufacturer to manufacturer and with grade of black. The largest reactors in operation have capacities of over 30,000 metric tons/year. Many producers operate smaller reactors in parallel. Reactors are typically designed to make a series of related blacks. Air and gas may be introduced to the primary combustion zone either axially, tangentially, or radially. The feedstock can be introduced into the primary fire either axially or radially in the high velocity section of the mixing zone. The high velocity section may be venturi-shaped or consist of a narrow diameter choke. Plants may have from one to several operating trains.

Carbon black reactors are made of carbon steel shells lined with several courses of refractory. The most severe services are in the combustor and in the throat zone. Different manufacturers take different approaches to these elements, some using exotic materials or selected water cooled metal surfaces, others using conventional materials and limiting temperatures to what their materials can stand. Most manufacturers achieve refractory life of one to several years. For the rubber grade carbon blacks, at least three different reactor designs must be used to make this range of furnace blacks. Figures 10 and 11 show the designs of commercial reactors based on the patent literature.

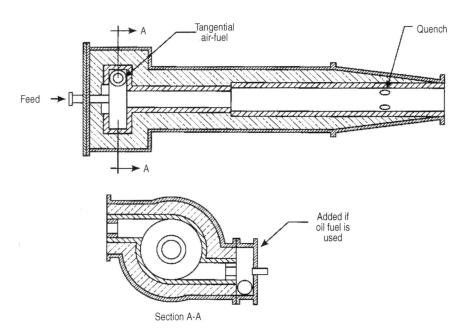

Fig. 10.   Reactor for N300–N200 carbon blacks (25).

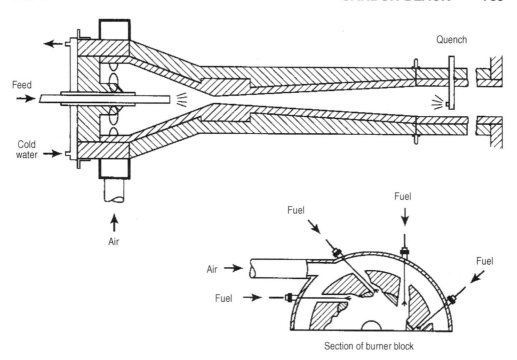

Fig. 11.   Reactor for tread blacks (26).

The quality and yield of carbon black depend on the quality and carbon content of the feedstock, the reactor design, and the input variables. Surface area in particular is controlled by adjusting the temperature in the reaction zone. Structure is adjusted by introducing potassium into the combustion gas. This may be done in any of a variety of ways.

The energy utilization in the production of one kilogram of oil-furnace carbon black is in the range of $9-16 \times 10^7$ J, and the yields are 300–660 kg/m^3 depending on the grade. The energy inputs to the reactor are the heat of combustion of the preheated feedstock, heat of combustion of natural gas, and the thermal energy of the preheated air. The energy output consists of the heat of combustion of the carbon black product, the heat of combustion and the sensible heat of the tail gas, the heat loss from the water quench, heat loss by radiation to atmosphere, and the heat transferred to preheat the primary combustion air.

6.2. The Thermal Black Process.   Thermal black is a large particle size, low structure carbon black made by the thermal decomposition of natural gas, coke oven gas, or liquid hydrocarbons in the absence of air or flames. Its economic production requires inexpensive natural gas. Today it is among the most expensive of the blacks regularly used in rubber goods. It is used in rubber and plastics applications for its unique properties of low hardness, high extensibility, low compression set, low hysteresis, and excellent processability. Its main uses are in O-rings and seals, hose, tire innerliners, V-belts, other mechanical goods, and in cross-linked polyethylene for electrical cables.

The thermal black process dates from 1922. The process is cyclic using two refractory-lined cylindrical furnaces or generators ~4 m in diameter and 10 m high. During operation, one generator is being heated with a near stoichiometric ratio of air and off-gas from the make generation whereas the other generator, heated to an average temperature of 1300°C, is fed with natural gas. The cycle between black production and heating is 5 min alternating between generators, resulting in a reasonably continuous flow of product and off-gases to downstream equipment. The effluent gas from the make cycle, which is ~90% hydrogen, carries the black to a quench tower where water sprays lower the temperature before entering the bag filter. The effluent gas is cooled and dehumidified in a water scrubber for use as fuel in the heating cycle. The collected black from the filters is conveyed to a magnetic separator, screened, and hammermilled. It is then bagged or pelletized. The pelletized form is bagged or sent to bulk loading facilities.

There are thermal black plants in Canada, the United States, the United Kingdom (1) and Russia. Two common grades are manufactured. These are Medium Thermal Black, N990, and Fine Thermal Black N880.

6.3. Acetylene Black Process. The high carbon content of acetylene (92%) and its property of decomposing exothermically to carbon and hydrogen make it an attractive raw material for conversion to carbon black. Acetylene black is made by a continuous decomposition process at atmospheric pressure and 800–1000°C in water-cooled metal retorts lined with refractory. The process consists of feeding acetylene into the hot reactors. The exothermic reaction is self-sustaining and requires water cooling to maintain a constant reaction temperature. The carbon black-laden hydrogen stream is then cooled followed by separation of the carbon from the hydrogen tail gas. The tail gas is either flared or used as fuel. After separation from the gas stream, acetylene black is very fluffy with a bulk density of only 19 kg/m^3. It is difficult to compact and resists pelletization. Commercial grades are compressed to various bulk densities up to 200 kg/m^3.

Acetylene black is very pure with a carbon content of 99.7%. It has a surface area of ~65 m^2/g, an average particle diameter of 40 nm, and a very high but rather weak structure with a DBP number of 250 mL/100 g. It is the most crystalline or graphitic of the commercial blacks. These unique features result in high electrical and thermal conductivity, low moisture adsorption, and high liquid absorption.

A significant use of acetylene black is in dry cell batteries where it contributes low electrical resistance and high capacity. In rubber it gives electrically conductive properties to heater pads, tapes, antistatic belt drives, conveyor belts, and shoe soles. It is also useful in electrically conductive plastics such as electrical magnetic interference (EMI) shielding enclosures. Its contribution to thermal conductivity has been useful in rubber curing bags for tire manufacture.

6.4. Lampblack Process. The lampblack process has the distinction of being the oldest and most primitive carbon black process still being practiced. The ancient Egyptians and Chinese employed techniques similar to modern methods collecting the lampblack by deposition on cool surfaces. Basically, the process consists of burning various liquid or molten raw materials in large, open, shallow pans 0.5–2 m in diameter and 16 cm deep under brick-lined flue enclosures with a restricted air supply. The smoke from the burning pans passes through low velocity settling chambers from which the carbon black is cleared by motor-driven ploughs. In more modern installations the black is separated by cyclones and filters. By varying the size of the burner pans and the amount of combustion air, the particle size and surface area can be controlled within narrow limits. Lampblacks have similar properties to the low surface area oil-furnace blacks. A typical lampblack has an average particle diameter of 65 nm, a surface area of 22 m^2/g, and a DBP number of 130 mL/100 g. Production is small, mostly in western and eastern Europe. Its main use is in paints, as a tinting pigment where blue tone is desired. In the rubber industry lampblack finds some special applications.

6.5. Impingment (Channel, Roller) Black Process. From World War I to World War II the channel black process made most of the carbon black used worldwide for rubber and pigment applications. The last channel black plant in the United States was closed in 1976. Operations still exist and are even being expanded in Europe. The demise of channel black was caused by environmental problems, cost, smoke pollution, and the rapid development of oil-furnace process grades that were equal or superior to channel black products particularly for use in synthetic rubber tires.

The name channel black came from the steel channel irons used to collect carbon black deposited by small natural gas flames impinging on their surface iron channels. Today tar fractions are used as raw material in addition to natural gas. In modern installations channels have been replaced by water cooled rollers. The black is scraped off the rollers, and the off-gases from the steel box enclosed rollers are passed through bag filters where additional black is collected. The purified exhaust gases are vented to the atmosphere. The oils used in this process must be vaporized and conveyed to the large number of small burners by means of a combustible carrier gas. Yield of rubber-grade black is 60% and 10–30% for high quality color grades.

The characteristics of roller process impingement blacks are basically similar to those of channel blacks. They have an acidic pH, a volatile content of ~5%, surface area of ~100 m^2/g, and an average particle diameter of 10–30 nm. The smaller particle size grades are used as color (pigment) blacks, and the 30 nm grade is used in rubber.

6.6. Recycle Blacks. The pyrolysis of carbon black containing rubber goods has been promoted as a solution to the accumulation of waste tires. In the processes in question, tires are pyrolyzed in the absence of oxygen, usually in indirect fired rotary kiln type units. The rubber and extender oils are cracked to hydrocarbons that are collected and sold as fuels or petrochemical feedstocks. The gaseous pyrolysis products are burned as fuel for the process. Steel tire cord is removed magnetically and the remainder of the residue is milled into a "pyrolysis black". This contains the carbon black, silica, and other metal oxides from the rubber and some newly created char. Typically these materials have 8–10% ash, and contain a lot of coarse residue. Most are difficult to pelletize. They have on average, the reinforcing properties of a N300 black but because they are a mixture of N600 and 700 blacks with N100 and N200 blacks they are not particularly suitable for either reinforcing or semireinforcing applications. To date they find application in relatively nondemanding uses such as playground and floor mats.

6.7. Surface Modification of Carbon Blacks.

For most of its long history, the carbon black industry had concentrated on the morphlogy as the key factor controlling product performance and grade differentiation. Recently, the importance of the composition of the interface between the carbon blacks and the medium in the composite in which the carbon black is used has been recognized.

The early stages of surface modification can be traced back to 1940s and 1950s. The approaches include physical adsorption of some chemicals on carbon black surface, heat treatment, and frequently oxidation. During 1980s and 1990s, some work on plasma treatment was reported. For chemical and polymer grafting modifications, a great deal of academic work was done in 1950s and 1960s in France, United States, and Japan, using surface oxygen groups as functional groups. However, because of rapid development applications of carbon black in different areas and the challenge from other reinforcing particles in its traditional applications, the surface modification technology for carbon black has been developing very rapidly over the last decade. These include surfactant treated surfaces, chemically modified surfaces, and deposition of other phases during or after black formation. Today there is active commercial development and new product introduction in all areas.

Attachments of the Aromatic Ring Nucleus to Carbon Black. Two approaches characterize this area. A number of patents have been issued to Cabot Corporation (27,28), which describe that the decomposition of a diazonium compound derived from a substituted aromatic or aliphatic amine results in the attachment of a substituted aromatic ring or chain onto the surface of the carbon black. This results in a stable attachment that is not sensitive to moisture. Examples show attachment of amines, anionic and cationic moieties, polysulfide moieties that can be attached into an elastomer network, and alkyl, polyethoxyl, and vinyl groups. Practically, the surface chemistry and physical chemistry can be tailored according to the applications of carbon blacks. Some applications are claimed in aqueous media for dispersion (29), in oil based coatings and inks for dispersion (30), and in rubbers for reduction of hysteresis and wear resistance improvement (31). The initially attached groups can also function as sites for further chemical substitution. Another approach has been developed by Xerox Corporation in which oligomers of the polymer are prepared using stable free radical polymerization and these are attached to the carbon black surface by reaction of the stable radical (32).

Attachments to the Aromatic Ring Structure Through Oxidized Groups. The acidic surface groups that result from surface oxidation of carbon black are natural synthons for the attachment of functionality. Generally, chemistry is done through either phenolic or carboxylic acid groups on the surface. Some of these groups are present in most blacks, but their density can be increased by treating with various oxidants such as ozone, nitric acid, or hypochlorite (33–35). Compared to the previous class, these C–O attachments are somewhat more labile, being particularly susceptible to hydrolysis. The concept of using phenolic groups as points of attachment for conventional silane treating agents has been described in several patents with the particular aim of attaching polysulfide moieties that can be vulcanized into elastomer networks for hysteresis reduction (36). Recently, patents have been issued on using the acidic sites on carbon black surfaces as points of reaction of amines. In the particular case in point, the attachment was used to improve compound stability and dispersion in conductive plastics applications (37).

Metal Oxide Treatments.   The carbon black industry has worked on ways to respond to the challenge of silica in tire treads for low rolling resistance (replacing all or some of the carbon black). Cabot has filed on and widely published a class of dual phase fillers in which silica or other metal oxides and carbon are coformed in a carbon black like reactor (38). In the particular product they describe, the carbon black and silica are intimately intermixed on a scale that is about the same size as the carbon black crystallite. In more recent variants, materials where the silica location is more on the exterior of the particle are described (39). In these materials, the silica is the minor constituent. The main characteristics of these carbon blacks are their lower filler–filler interactions. Filler–polymer interactions are also increased, but by incorporating coupling agents these interactions can be adjusted as required. These materials are used as fillers for low rolling resistance, higher wet skid resistance and improved wear resistance in tire treads when used with conventional sulfide–silane coupling agents (40,41), or as fillers for silicone rubber when used with alkyl silane and vinyl silane agents (42). The patent literature suggests that other applications have been considered as well (43). Patents have also been issued on coated carbon black by depositing silica on the black surface in an aqueous solution of sodium silicate by adjusting pH with acid (44,45).

7. Economic Aspects

7.1. Manufacturers and Production.   Starting with the oil crisis of 1973, consumption of carbon black in the United States decreased to 1.2 million tons in 1986 (1). A number of events had contributed to decreased consumption by the rubber and tire industries including tire radialization, increased tire mileage, downsizing of tires, and increased imports of foreign cars. The negative influence of these events have pretty much run their course, and during the last two decades there has been a modest growth in carbon black production. Production for the period 1971–2000 is shown in Table 6.

Table 6. **U.S. Production of Carbon Black (1971–2000)**[a]

Year	Millions of metric tons
1971	1.380
1976	1.415
1981	1.285
1986	1.200
1991	1.230
1994	1.475
1997	1.660
2000	1.665

[a] Ref. 1.

Table 7. **U.S. Carbon Black Manufacturers and Capacities, 2000**[a]

	Furnace black		Total carbon black	
	Millions of metric tons	% of capacity	Millions of metric tons	% of capacity
Cabot Corporation	0.451	24	0.451	24
Columbian Chemicals Co.	0.361	19	0.361	19
Continental Carbon Co.	0.254	14	0.254	13
Degussa-ECI	0.463	25	0.484	25
Sid Richardson Carbon Co.	0.340	18	0.340	18
other	0.000		0.014	1
Total	*1.869*	*100*	*1.903*	*100*

[a] Ref. 1.

The shrinkage in demand has resulted in a restructuring of the carbon black-industry. Several of the principal multinational oil companies have left the business including Ashland, Cities Service Co., Phillips, and Conoco. Some plants have changed ownership. Decreased margins, rising feedstock and environmental compliance costs have led to further restructuring in the late 1990s and early 2000 time period. ECI and Degussa consolidated their operations in 2002. Today's U.S. industry consists of five principal producers. Rated capacities of the five U.S. manufacturers is shown in Table 7. Cabot Corp., ECI-Degussa and Columbian Chemicals are the leading producers, followed by Continental Carbon Co. and Sid Richardson.

World carbon black rated capacities are shown in Table 8. North America has the largest capacity, and Africa and the Middle East have only a small production. The growth areas are predicted to be the Asian and Eastern European markets.

Table 8. **World Carbon Black Capacities of July 1, 2001 (1)**

	Millions of metric tons	% of total
North America[a]	2.4	28
South America	0.5	6
Western Europe	1.5	17
Japan	0.8	9
other Asia	2.0	24
other[b]	1.3	16
Total	*8.5*	*100*

[a] Including Mexico.
[b] Including eastern Europe, Africa, and the Middle East.

8. Health and Safety Factors

8.1. Health. There is a long history of health studies, many of them sponsored by the carbon black manufacturing industry and a number of authoritative publications on carbon black SH&E aspects. In particular, Patty's Handbook chapter on Industrial Hygiene and Toxicology of carbon black is a recent authoritative reference (46). Mortality and worker health have been extensively studied, often in studies sponsored by the carbon black industry. These studies have included mortality (47,48) and health (49–50) in workers in the carbon black industry against unexposed controls in the same industry as well as those in nonindustry populations. The North American mortality studies (47) have logged ~55,000 worker exposure years since the first studies were first undertaken in 1939. They show no elevation in death rate among carbon black workers and no elevation in cancers of the respiratory organs because of occupational exposure to carbon black. In the United Kingdom a study of workers in five carbon black plants (48) shows that among this population there is an elevation of the death rate from cancers of the respiratory organs, which is significant versus the U.K. population, but the study indicated clearly that there was no correlation of the incidence of disease with exposure to carbon black. In other words, some other factors rather than carbon black exposure appeared to be the source of the excess cancers. Studies of worker health among both U.S. and European workers have been completed in the last decade and publications of the results of these studies continue to appear. These show no evidence of clinically significant health effects due to occupational exposure to carbon black. While there are some differences in the exposure metric between the North American Studies and the European studies, they show very similar effects. Over the course of time covered by these various studies, there has been a marked improvement in the work place air quality.

In 1995, IARC (International Agency for Research on Cancer) revised its evaluation of carbon black from category 3, "not classifiable as to its carcinogenicity to humans" to category 2B, "possibly carcinogenic to humans" based on lung tumor formation in two long term inhalation studies in rats (51–53). Comparable studies in mice and golden hamsters have failed to find any elevation of tumor incidence of lung or any other tissue (54). Other insoluble respirable materials have shown similar positive responses in rats, but not in other rodent species.

There are two major areas of carcinogenic concern with carbon black. Of these, the one that has historically attracted the most attention is the few tens to hundreds of pasts per million (ppms) of polynuclear aromatic hydrocarbons (PAH) that are adsorbed on the surface of most blacks. An extract made by exhaustive extraction of these materials with aromatic solvent has been shown to cause skin tumors in rodents. The PAH on carbon black is very tightly adsorbed and is not liberated by biological fluids. Hence, this material is believed to have little or no bioavailability (55). This is supported in the companion studies to those referenced by IARC (52,56). Attempts were made to get a dose response to PAH by using PAH free black and doping it with high levels of PAH. There was no statistically significant difference in the tumor incidence or type between the two groups.

The second carcinogenic mechanism seems to be common to all inert respirable insoluble particles. Positive carcinogenic responses have been shown with TiO_2, talc, carbon black, and diesel soot in the female rat (57). There is evidence that a common mechanism is at work in all these cases involving a process of damage to the lung epithelium, inflammatory response, saturation of the body's defense mechanisms, proliferation of new epithelial cells, recruitment of activated polymorphonuclear (PMN) cells, secretion of radical forming species by the PMN cells, and damage to the DNA in the dividing epithelial cells. Studies by Driscoll and co-workers are building a strong case for this mechanism and are ongoing (58). These data indicate that the distinction between the rat and other rodents is the nature of the inflammatory response that appears quite intense in the rat compared to other species. If indeed this proves to be the explanation, a strong case can be made for a threshold and establishment of a no observed adverse effect level (NOAEL). It is also clear that in the rat, an overload mechanism is at work in which above a certain critical loading, the mechanisms which normally clear dust particles from the lung lose effectiveness. Again this suggests a threshold and a no observed adverse effect level.

Carbon black inhalation is currently regulated by Occupational Safety and Health Administration (OSHA) in the United States at 3.5 mg/m^3 (total dust), by the Health and Safety Executive (HSE) in the United Kingdom at 3.5 mg/m^3 (inhalable dust), by the MAK Commission in Germany at 6.0 mg/m^3 (inhalable dust) for an 8 h time-weighted average (46). Reviews of these occupational exposure levels are currently underway by the MAK Commission, the HSE, and the American Conference of Governmental Industrial Hygienists (ACGIH). It is unlikely that any of these reviews will be completed before 2005.

8.2. Safety.   Carbon blacks will burn in air if ignited and once ignited, are difficult to extinguish. In bulk storage, local hot spots can exist for very long periods. Great care needs to be exercised where a smoldering fire is suspected as there can be accumulations of carbon monoxide in enclosed spaces.

Carbon black dust clouds in air are not considered flammable. Carbon blacks have a high ignition energy requirement, and entrained dust clouds do not propagate flame, nor exhibit substantial overpressures. The reason is that they presumably have essentially no combustible volatile matter. Carbon black in air can be incinerated but only with difficulty, requiring long burnout times.

9. Environmental Concerns

The carbon black industry takes extreme efforts to confine the product during all stages of manufacturing and transport. Highly efficient bag filters are used to collect the product. After collection the fluffy carbon black is densified and pelletized to minimize dusting during shipment and use by customers. The process gas leaving the bag filter contains primarily water, nitrogen, carbon monoxide, carbon dioxide, and hydrogen. There are also traces of hydrogen sulfide, carbon disulfide, carbonyl sulfide, and various nitrogen-containing species. A portion of these gases is burned for internal plant fuel, and the residual gas is generally burned in a flare or incinerator. Where local conditions warrant, the remaining gas may be used to generate power or steam, either for the plant itself, or for merchant sale.

Like all other operators of combustion equipment, carbon black plants are subject to the usual pressures for reduced sulfur and nitrogen oxide emissions. It appears that the use of lower sulfur feedstock is the most economic way of reducing sulfur emissions. Redesign of combustion equipment for nitrogen oxide reduction is showing some promise. The primary NO_x issues arise from the combustion of tail gas since the carbon black production process is exceedingly fuel-rich.

10. Uses

The U.S. consumption of carbon black in 2000 by various market sectors is shown in Table 9. About 89% of total consumption is in the rubber industry and 70% for tires. About 10% is consumed for other automotive products and 9% for rubber products unrelated to the automotive industry. The automotive industry accounts for 80% of consumption and 11% of the blacks is for nonrubber uses. Its main applications are related to pigmentation, ultraviolet (uv) absorption, and electrical conductivity of other products such as plastics, coating, and inks. These carbon blacks are also termed special blacks (1).

10.1. Rubber-Grade Carbon Blacks. Carbon black is a major component in the manufacture of rubber products, with a consumption second only to rubber itself. It is by far the most active rubber reinforcing agent owing to its unique ability to enhance the physical properties of rubbers. Table 10 lists the principal rubber grades by their N-number classification, general rubber properties, and typical uses.

The consumption of the various carbon black grades can be divided into tread grades for tire reinforcement and nontread grades for nontread tire use and other rubber applications. Table 11 shows the distribution of production of types for these uses. A typical passenger tire has several compounds and uses five to seven different carbon black grades.

Table 9. **U.S. End Use Consumption of Carbon Black in 2000 (1)**

	Millions of metric tons	Percent of total
automotive rubber uses tire and tire products	1.17	70
belts, hoses and other automotive products	0.17	10
industrial rubber products	0.16	9
non-rubber uses	0.18	11
Total	*1.67*	*100*

Table 10. **Application of Principal Rubber-Grade Carbon Blacks**

Designation	General rubber properties	Typical uses
N110, N121	very high abrasion resistance	special tier treads, airplane, off-the-road, racing
N220, N299, N234	very high abrasion resistance, good processing	passenger, off-the-road, special service tire treads
N339, N347, N375, N330	high abrasion resistance, easy processing, good abrasion resistance	standard tire treads, rail pads, solid wheels, mats, tire belt, sidewall, carcass, retread compounds
N326	low modulus, good tear strength, good fatigue, good flex cracking resistance	tire belt, carcass, sidewall compounds, bushings, weather strips, hoses
N550	high modulus, high hardness, low die swell, smooth extrusion	tier innerliners, carcass, sidewall, innertubes, hose, extruded goods, v-belts
N650	high modulus, high hardness, low die swell, smooth extrusion	tire innerliners, carcass, belt, sidewall compounds, seals, friction compounds, sheeting
N660	medium modulus, good flex fatigue resistance, low heat buildup	carcass, sidewall, bead compounds, innerliners, seals, cable jackets, hose, soling, EPDM compounds
N762	high elongation and resilience, low compression set	mechanical goods, footwear, innertubes, innerliners, mats

The behavior of different grades in rubber is dominated mainly by surface area, structure (DBPA), and surface activity. All these parameters play a role in rubber reinforcement through different mechanisms, such as interfacial interaction between rubber and carbon black, occlusion of the polymer in the internal voids of the aggregate, and the agglomeration of carbon black aggregates in the polymer matrix.

One of the consequences of the incorporation of carbon blacks into a polymer is the creation of an interface between a rigid solid phase and a soft elastomer phase. For rubber-grade carbon blacks, whose surfaces exhibit very little

Table 11. **Carbon Black Production by Grade in United States for 2000 (K metric tons) (1)**

N330 high abrasion	0.623
N550 fast extruding	0.138
N762 semireinforcing	0.129
N660 general purpose	0.356
N110 super abrasion	0.061
N220 intermediate super abrasion	0.170
N990 thermal	0.014
Total	*1.493*

porosity, the total area of the interface depends on both filler loading and the specific surface area of the filler. Due to the interaction between rubber and filler two phenomena are well documented: the formation of bound rubber and a rubber shell on the carbon black surface. Both are related to the restriction of the segmental movement of polymer molecules.

The effect of filler structure on the rubber properties of filled rubber has been explained by the occlusion of rubber by filler aggregates (59). When structured carbon blacks are dispersed in rubber, the polymer portion filling the internal void of the carbon black aggregates, or the polymer portion located within the irregular contours of the aggregates, is unable to participate fully in the macro-deformation. The partial immobilization in the form of occluded rubber causes this portion of rubber to behave like the filler rather than like the polymer matrix. As a result of this phenomenon, the effective volume of the filler, with regard to the stress–strain behavior and viscoelastic properties of the filled rubber, is increased considerably.

The filler aggregates in the polymer matrix have a tendency to associate to agglomerates, especially at high loadings, leading to chain-like filler structures or clusters. These are generally termed secondary structure or, in some cases, filler network, even though the latter is not comparable to the continuous polymer network structure. The formation of filler network is dependent on the intensity of interaggregate attractive potential, the distance between aggregates and polymer–filler interaction (60).

From the point of view of carbon black morphology, high surface area produces high reinforcement as reflected in high tensile and tear strengths, high resistance to abrasive wear, higher hysteresis, and poorer dynamic performance, while high structure leads to higher viscosity, lower die swell, and high modules. The hysteresis is a measure of the loss of mechanical energy as heat during cyclical deformation of a rubber body. Listed in Table 12 are the effects of surface area and structure that can be a guideline for choice of carbon blacks according to the processability and property requirements of rubber products.

A present day challenge to carbon black technologists is to optimize the balance between tire wear, tire hysteresis that determines the rolling resistance, and wet skid resistance. It is now recognized that besides the morphology, while the wear resistance is closely related to the polymer–filler interaction and dispersion of the carbon blacks, the hysteresis is mainly determined by filler

Table 12. **Effect of Carbon Black Morphologies on the Properties of Filled Compounds**

Surface area increase	Rubber properties	Structure increase
higher	abrasion resistance	depend on severity
higher	hardness	higher
higher	tensile strength	lower
not main factor	modulus	higher
lower	elongation	lower
lower	rebound	not main factor
higher	viscosity	higher
lower	dispersibility	higher
not main factor	dimensional stability	higher

networking or agglomeration. For tread compounds of tire, depression of filler networking results in lower hysteresis at the temperatures from 50 to 80°C, leading to low rolling resistance, and in higher hysteresis at lower temperature that is in favor of wet skid resistance. Some progress on this problem has been made by using new furnace designs and other process variables that broaden the aggregate size distributions and lower the tint strength while maintaining surface area and structure (61,62). A substantial improvement in global tire performance can be achieved through surface modification of carbon blacks, such as carbon–silica dual phase fillers (40,41,63).

10.2. Special-Grade Carbon Blacks. Besides reinforcement for rubber, the principal functions that carbon black imparts to a compound material are color, uv damage resistance, electrical conductivity, nondegradation of polymer physical properties, and ease of dispersion. The carbon blacks used for these purposes are classified as special-grade blacks. Smaller volume applications exploit other principal attributes, such as chemical inertness, thermal stability, and an open porous structure. The secondary attributes include chemical and physical purity, low affinity for water adsorption, and ease of transportation and handling.

In 2000, 11% of the U.S. consumption of carbon black was special blacks. About 51% of special blacks are used in plastics, 32% in printing inks, 5% in paint, 3% in paper, and 9% in miscellaneous applications (1).

Dispersion.   The ability to disperse the special grades is an important consideration in almost all applications. The customer's milling costs can be comparable to the purchase price of the carbon black. In other cases, the inability to achieve an excellent dispersion impairs the ability to realize the full performance of the blacks or it creates other undesirable characteristics. For example, a black plastic part will not appear as dark nor have a smooth surface if the black cannot be fully dispersed.

The term "dispersion" is used in several ways in actual applications. It may refer to the amount of work that required to achieve a specified level imperfections or it may refer to the level of imperfections per se. When a black is referred to as "easily dispersed" it can mean that relatively low shear mixers can achieve a compound that is relatively free of imperfections. Or it may mean that with a given mixing protocol the compound achieves an imperfection count below a specified level.

The "level of imperfections" is most often assessed in one of two ways. Most often it refers to the size distribution of surface imperfections on an extruded or injection molded part. It may also refer to the rate of pressure buildup on an extrusion screen pack.

Undispersed material is one of three types. The first is non-carbon contaminants that come either in the feedstock or the result of corrosion of the manufacturing train. The second are carbon contaminants that occur when the feedstock droplets fail to evaporate. If the residual material in the droplet fails to evaporate, but is pyrolyzed in place, the contaminant is referred to as a coke ball. If the droplet reaches the wall before it is completely pyrolyzed, the resulting build up is termed wall coke and it can flake off and contaminate the carbon black product. Finally, the aggregates of carbon black can remain agglomerated or undispersed. Too often the concepts of contaminant free and the difficulty of separating agglomerates into aggregates are not clearly distinguished. Detailed microscopy is required to determine the nature of defects and therefore resolve if the problem is contamination or true difficulty in dispersion.

In most cases the properties of special carbon black aggregates that engender performance are antagonistic to achieving a good dispersion. High pigmenting strength for both light and uv implies high surface area. High surface area increases the difficulty of dispersion. Achieving an electrically conductive network at low concentration implies a black with high structure. High structure, per se, is an asset in dispersion. However, most carbon black formation technologies will not make high structure blacks unless the surface area is also high and the trade-off means that the best conducting grades are difficult to disperse.

Pigmentation.   Carbon black is an excellent pigment. It is considerably more effective in absorbing light than any other material on either a weight or a cost basis. It is also environmentally stable.

Black pigments are used both alone and with other pigments. In the first the measure of performance is termed mass tone and is simply the ability to prevent the transmission or reflection of light. The second is called tint tone and it is the ability to soften or darken other colors.

How black is black? Two aspects can be important. The first is the total amount of white light that is either transmitted or reflected by a black material. The second is variation in these values with the frequency or color of the light. In almost all cases, relatively higher red light adsorption is valued as it gives the material a blue "undertone". Various measures are used to equate blue tone with overall absorption. These color metrics are based on subjective judgments and are most highly developed for automotive paints.

Surface area and structure, or aggregate size, play a role in determining color strength. Pigmenting carbon blacks are used at quite low concentrations. Single particle Mie scattering theory is reasonably applicable. In applying the theory, an aggregate of carbon black is viewed as a composite of carbon and the resin or polymer. The refractive index of the carbon is taken as 1.84(1-0.46i) (64). And the index of refraction of the composite particle is the volume average of the carbon black and polymer values. The difficulty in using single particle scattering theory is in evaluating the size and carbon content of the "scattering particle". Various methods are used to estimate aggregate size and to divide the CDBP value into "intra-" and "inter-" aggregate parts. The "intra-aggregate" part is considered to form part of the composite scattering particle.

Because aggregate and primary particle size cannot be controlled independently in carbon black reactors, and because the ability to achieve complete dispersion depends on morphology, the coloring properties of carbon blacks are often treated empirically.

The theory works best when carbon black is used with other pigments, ie, in tinting applications. In these applications, both theory and experiments indicate that smaller primary particles and aggregates absorb more light and have a bluer tone.

Table 13 shows the entire range of properties of blacks sold for color applications. Blacks with large primary particle sizes are used in news inks and other

Table 13. **Types and Applications of Special Pigment Grades of Carbon Blacks**

Type	Surface area, m^2/g	DBP number, mL/100 g	Volatile, %	Uses
		Normal grades		
high color	230–560	50–120	2	high jetness for alkyl and acrylic enamels, lacquers, and plastics
medium color	220–220	70–120	1–1.5	medium jetness and good dispersion for paints and plastics; uv and weathering protection for plastics
regular color	80–140	60–114	1–1.5	for general pigment applications in inks, paints, plastics, and paper; gives uv protection in plastics, high tint, jetness, gloss, and dispersibility in inks and paints
	46	60	1.0	good tinting strength, blue tone, low viscosity; used in gravure and carbon paper inks, paints, and plastics
	45–85	73–100	1.0	main use is in inks; standard and offset news inks
low color	25–42	64–120	1.0	excellent tinting black–blue tone; used for inks-gravure, one-time carbon paper inks; also for paints, sealants, plastics, and cements
thermal blacks	7–15	30–35		tinting-blue tone; plastics and utility paints
lamp blacks	20–95	100–160	0.4–0.9	paints for tinting-blue tone
		Surface oxidized grades		
high color	400–600	105–121	8.0–9.5	used for maximum jetness in lacquers, coatings, plastics, fibers, record disks
medium color, long flow	138	55–60	5	used in lithographic, letterpress, carbon paper, and typewriter ribbon inks; high jetness, excellent flow, low viscosity, high tinting strength, gloss, and good dispersability
medium color, long flow	96	70	2.5	used for gloss printing and carbon paper inks; excellent jetness, dispersibility; tinting strength, and gloss in paints
low color	30–40	48–93	3.5	used for tinting where flooding is a problem; easy dispersion

nondemanding applications. The extremely fine blacks are used in high performance enamels and automotive paints.

An emerging application is as the colorant in jet printer inks. The requirement of extremely high quality dispersion meant that the original black inks were dyes. The image permanence of carbon blacks lead to the displacement of dyes when techniques were developed that assured the stability of excellent, water-based dispersion.

The blacks used for uv protection of polymers have very small primary particle sizes. The industry standard for uv protection is primary particle sizes of <20 nm.

Conductivity. Carbon black is added to polymer or resin compounds to achieve electrical conductivity. If their concentrations are high enough, carbon black aggregates will form interconnected paths through the compound material. These networks can have resistivities in the range of $1-10$ Ω-cm.

Concentrations that are too low will not form percolation networks. The Jansen equation (65) predicts that the critical concentration occurs when volume of polymer equals four times the volume of the CDBP value. The compounds are continuous when the amount of polymer is the CDBP value or greater and contain continuous carbon networks when the volume of polymer is between the crush DBP value and four times that amount.

In general it is desirable to achieve percolation at low loading to reduce the negative effects of the presence of the carbon black on the physical properties of the compound. Therefore blacks with high CDBP values are used in conducting application. Figure 12 shows the percolation threshold of a number of carbon blacks with widely varying CDBP values.

One of the most demanding applications for conducting compounds is as a layer around the center conductor of a high voltage cable. The purpose of the

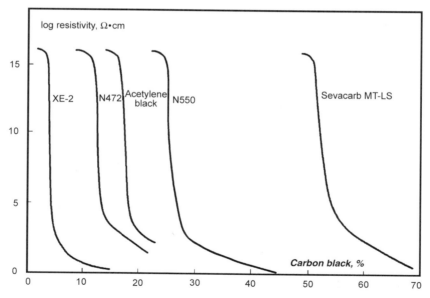

Fig. 12. Electrical resistivity-concentration curves for various carbon blacks (66).

compound is to smooth any surface imperfection that would otherwise result in high electric field gradients and the breakdown of the insulating layers. These compounds have the highest loading of carbon blacks to achieve the lowest possible resistance. Two very important secondary requirements are that the black is quite hydrophobic, free of inorganic salts, and has extremely low contaminant levels.

Other applications of carbon black filled polymers do not require as high a level of conductivity. The purpose is the dissipation of static electricity. These types of compounds are used in carrying containers for microelectronic components and in containers and pipes for combustible liquids such as gasoline. The newest application is adding enough conductivity to automobile panels to enable coatings of electrostatically charged paint particles.

Other Uses. Carbon blacks are used to make highly porous structures for catalyst supports. Most supports are metal oxides rather than carbon, but in certain circumstances a chemically inert support is required. The open structure of the carbon black aggregates allows control of the pore size distribution within the consolidated carbon body.

Fuel cell electrodes are a particularly interesting catalyst application. The active layers of the proton conducting membrane fuel cells must have the ability to transport both gases and hydrogen ions to the catalyst as well as conducting water away. Mixtures of carbon blacks with special polymers create the desired hydrophilic and hydrophobic pore structures.

The so-called boundary layer capacitors are also electrochemical devices that rely carbon blacks to achieve the appropriate pore structures.

Special carbon blacks are also used in batteries, with the largest use in old technology nonrechargeable cells.

BIBLIOGRAPHY

"Carbon Black" under "Carbon" in *ECT* 1st ed., Vol. 3, pp. 34–65, and Suppl. 1, pp. 130–144, by W. R. Smith, Godfrey L. Cabot, Inc.; "Acetylene Black" in *ECT* 1st ed., Vol. 3, pp. 66–69, by B. P. Buckley, Shawinigan Chemicals Ltd.; "Carbon Black" under "Carbon" in *ECT* 2nd ed., Vol. 4, pp. 243–282, by W. R. Smith, Cabot Corp., D. C. Bean, "Acetylene Black", Shawinigan Chemicals Ltd.; "Carbon Black" under "Carbon" in *ECT* 3rd ed., Vol. 4, pp. 631–666, by E. M. Dannenberg, Cabot Corp. in *ECT* 4th ed., Vol. 4, pp. 1037–1074, by Eli M. Dannenberg, Consultant and Lyn Paguin and Harry Gwinnell, Cabot Corporation; "Carbon Black" in *ECT* (online), posting date: December 4, 2000, by Eli M. Dannenberg, Consultant and Lyn Paguin, Harry Gwinnell, Cabot Corporation.

CITED PUBLICATIONS

1. J. F. Aucher, T. Kälin, and Y. Sakuma, "CEH Marketing Research Report, Carbon Black," Chemical Economic Handbook—SRI International, January 2002.
2. *2001 Annual Book of ASTM Standard.* Vol. 9, American Society for Testing and Materials, Philadelphia, Pa., 2001.

3. G. Kühner and M. Voll, in J.-B. Donnet, R. C. Bansal, and M.-J. Wang, eds., *Carbon Black, Science and Technology*, Chapt. 1, Marcel Dekker, Inc., New York, 1993.

4. E. M. Dannenberg, *Rubber World Mag. Spec. Pub.—Rubber Div. 75th Anniv.* (1907–1984).

5. J. Janzen, *Rubber Chem. Technol.* **56**, 669 (1982).

6. J.-B. Donnet and A. Voet, *Carbon Black, Physics, Chemistry, and Elastomer Reinforcement*, Chapt. 4, Marcel Dekker, Inc., New York, 1976.

7. H. P. Boehm, *Farbe und Lack* **79**, 419 (1973).

8. C. A. Leon, J. M. Solar, V. Calemma, and L. R. Radovic, *Carbon* **30**, 797 (1992).

9. C. R. Houska and B. E. Warren, *J. Appl. Physics* **25**, 1503 (1954).

10. B. E. Warren, *Phys. Rev.* **59**, 693 (1941).

11. L. L. Ban and W. M. Hess, *Abstr. 9th Bienn. Conf. Carbon* 162 (1969).

12. J.-B. Donnet and E. Custodero, *Carbon* **30**, 813 (1992).

13. M.-J. Wang, S. Wolff, and B. Freund, *Rubber Chem. Technol.* **67**, 27 (1994).

14. S. Brunauer, P. H. E. Emmett, and J. Teller, *J. Am. Chem. Soc.* **60**, 310 (1938).

15. B. C. Lippens and J. H. de Boer, *J. Catalysys* **4**, 319 (1965); R. W. Magee, *Rubber Chem. Technol.* **68**, 590 (1995).

16. C. W. Sweitzer and W. C. Goodrich, *Rubber Age* **55**, 169 (1944).

17. I. Medalia, in E. K. Sichel, ed., *Carbon Black—Polymer Composites*, Chapt. 1, Marcel Dekker, Inc., New York, 1982.

18. E. Redman, F. A. Heckman, and J. E. Connelly, Paper No. 14, Rubber Division, ACS Meeting, Chicago, Ill., 1977.

19. M.-J. Wang, S. Wolff and J.-B. Donnet, *Rubber Chem. Technol.* **64**, 714 (1991).

20. M.-J. Wang, and S. Wolff, *Rubber Chem. Technol.* **65**, 890 (1992).

21. M. Faraday, in W. Crooker, ed., *The Chemical History of a Candle*, Viking Press, New York, 1960.

22. R. C. Bansal and J.-B. Donnet, in J.-B. Donnet, R. C. Bansal, and M.-J. Wang, eds., *Carbon Black, Science and Technology*, Chapt. 2, Marcel Dekker, Inc., New York, 1993.

23. U.S. Pat. 3,010,794 (1961), G. F. Frianf and B. Thorley (to Cabot Corp.).

24. D. Rivin, *Rubber Chem. Technol.* **56**, 709 (1982).

25. U.S. Pat. 2,564,700 (1951), J. C. Krejci (to Phillips Petroleum Co.).

26. U.S. Pat. 3,490,869 (1970), G. L. Heller (to Columbia Carbon Co.).

27. U.S. Pat. 5,554,739 (1996), J. A. Belmont (to Cabot Corp.).

28. U.S. Pat. 5,851,280 (1998), J. A. Belmont, R. M. Amici, and C. P. Galloway (to Cabot Corp.).

29. U.S. Pat. 5,672,198 (1997), J. A. Belmont (to Cabot Corp.).

30. U.S. Pat. 5,713,988 (1998), J. A. Belmont and C. E. Adams (to Cabot Corp.).

31. U.S. Pat. 6,494,946 (2002), J. A. Belmont, R. M. Amici, and C. P. Galloway (to Cabot Corp.).

32. U.S. Pat. 5,545,504 (1996), B. Keoshkerian, M. K. Georges, and S. V. Drappel (to Xerox).

33. R. C. Bansal and J.-B. Donnet, in J.-B. Donnet, R. C. Bansal, and M.-J. Wang, eds., *Carbon Black, Science and Technology*, Chapt. 4, Marcel Dekker, Inc., New York, 1993.

34. U.S. Pat. 4,366,138 (1982), E. Eisenmenger, R. Engel, G. Kuehner, R. Reck, H. Schaefer, M. Voll (to Degussa AG.).

35. U.S. Pat. 2,439,442 (1948), F. H. Amon, and F. S. Thornhill (Cabot Corp.).

36. U.S. Pat. 5,159,009 (1992), S. Wolff and U. Görl (to Degussa AG.).

37. U.S. Pat. 5,708,055 (1998), G. A. Joyce and E. L. Little (to Columbian Chemical Co.).

38. U.S. Pat. 5,830,930 and U.S. Pat. 5,877,238 (1998 and 1999) K. Mahmud, M.-J. Wang, and R. A. Francis (to Cabot Corp.); U.S. Pat. 5,904,762 and U.S. Pat. 6,211,279 (1999 and 2001), K. Mahmud and M.-J. Wang (to Cabot Corp.).

39. U.S. Pat. 6,364,944 (2002), K. Mahmud, M.-J. Wang and Y. Kutsovsky (to Cabot Corp.).

40. M.-J. Wang, K. Mahmud, L. J. Murphy, and W. J. Patterson, *Kauts Gummi Kunsts.* **51**, 348 (1998).

41. M.-J. Wang, Y. Kutsovsky, P. Zhang, L. J. Murphy, S. Laube, and K. Mahmud, *Rubber Chem. Technol.* **75**, 247 (2002).

42. U.S. Pat. 6,020,402 (2000), J. N. Anand, J. E. Mills, and S. R. Reznek (to Cabot Corp.).

43. U.S. Pat. 5,919, 855 (1999), T. Reed and K. Mahmud (to Cabot Corp.).

44. U.S. Pat. 5,679,728 (1997), T. Kawazura, H. Kaido, K. Ikai, F. Yatsuyanagi, and M. Kawazoe (to Yokohama Rubber Co.).

45. U.S. Pat. 5,916,934 and U.S. Pat. 6,197,274 (1999 and 2001), K. Mahmud and M.-J. Wang, S. R. Reznek, and J. A. Belmont (to Cabot Corp.).

46. R. J. McCunney, H. J. Muranko, P. A. Valberg, in E. Bingham, B. Cohrssen, and C. H. Powell, eds., *Patty's Toxicology*, 5th ed., Vol. 8, John Wiley and Sons, Inc., New York, 2001.

47. J. Robertson and K. J. Inman, *J. Occup. Environ. Med.* **38**, 569 (1996).

48. K. Gardiner, W. N. Trethowan, J. M. Harrington, I. A. Calvert, and D. C. Glass, *Ann. Occup. Hyg.* **36**, 447 (1992)

49. P. Harber, H. Muranko, S. Solis, A. Torossian, and B. Merz, *J. Occup. Environ. Med.* **45**, 144 (2003).

50. K. Gardiner, M. Van Tongeren, J. M. Harrington, *Occup. Environ. Med.* **58**, 496 (2001).

51. IARC Monograph 65, pp. 149–262, IARC, Lyon, France, 1996.

52. U. Heinrich, L. Peters, O. Creutzenberg, C. Dasenbrock, and H. G. Hoymann, in U. Mohr, D. L. Dugworth, J. L. Mauderly, and G. Oberdorster, eds., *Toxic and Carcinogenic Effects of Solid Particles in the Respiratory Tract*, ILSI Press, Washington, D.C., 1994, pp. 433–441.

53. J. L. Mauderly, M. B. Snipes, E. B. Barr, S. A. Belinsky, J. A. Bond, A. L. Brooks, I.-Y. Chang, Y. S. Cheng, N. A. Gillett, W. C. Griffith, R. F. Henderson, C. E. Mitchell, K. J. Nikula, and D. G. Thomassen, HEI Research Report Number 68, Health Effects Institute, Cambridge Mass., 1994.

54. K. J. Nikula, M. B. Snipes, E. B. Barr, W. C. Griffith, R. F. Henderson, and J. L. Mauderly, in U. Mohr, D. L. Dungworth, L. L. Mauderly, and G. Oberdorster, eds., *Toxic and Carcinogenic Effects of Solid Particles in the Respiratory Tract*, ILSI Press, Washington D.C., 1994, pp. 566–568.

55. D. Rivin and J. H. Atkins, *Carbon* **25**, 135 (1987).

56. U. Heinrich, R. Fuhst, S. Rittinghausen, O. Creutzenberg, B. Bellman, W. Koch, and K. Levsen, *Inhal. Toxicol.* **7**, 533 (1995).

57. G. Oberdorster, *Regulatory Toxicol. Pharmacol.* **27**, 123 (1998).

58. K. E. Driscoll, J. M. Carter, B. W. Howard, D. G. Hassenbein, W. Papelko, R. B. Baggs, and G. Oberdorster, *Toxicol. Appl. Pharmacol.* **136**, 372 (1996).

59. A. I. Medalia, *J. Colloid Interf. Sci.* **32**, 115 (1970).

60. M.-J. Wang, *Rubber Chem. Technol.* **71**, 520 (1998).

61. U.S. Pat. 4,071,496 (1978), G. Kraus and H. R. Hunt (to Phillips Petroleum Co.); U.S. Pats. 4,241,022 (1980) (to Phillips Petroleum Co.); 4,267,160 (1982) (to Phillips Petroleum Co.).

62. U.S. Pat. 4,478,973 (1984), S. Misono and H. Suzuki (to Tokai Carbon Co.).

63. Eur. Pat. 0 877 047 A1 (1998), G. Labouze and S. Cohen (to Michelin); Eur. Pat. 0 890 607 A1 (1999), G. Labouze and S. Cohen (to Michelin); Eur. Pat. 0 997 490 A1

(2000), Y. Mizuno and M. Uchida (to Sumitomo Rubber Industries); Eur. Pat. 1 254 786 A1 (2002) C. E. Drvol, S. L. Taylor, T. D. Linster, M. B. Rodger, and F. Ozel (to Goodyear).

64. A. I. Medalia and L. W. Richards, *J. Colloid and Interf. Sci.* **40**, 233 (1972).
65. J. Janzen, *J. Appl. Phys.* **46**, 966 (1975).
66. N. Probst, in J.-B. Donnet, R. C. Bansal, and M.-J. Wang, eds., *Carbon Black, Science and Technology*, Chapt. 8, Marcel Dekker, Inc., New York, 1993.

Meng-Jiao Wang
Charles A. Gray
Steve A. Reznek
Khaled Mahmud
Yakov Kutsovsky
Cabot Corporation

CARBON DIOXIDE

1. Introduction

Carbon dioxide [124-38-9], CO_2, is a colorless gas with a faintly pungent odor and acid taste first recognized in the sixteenth century as a distinct gas through its presence as a by-product of both charcoal combustion and fermentation. Today carbon dioxide is a by-product of many commercial processes: synthetic ammonia production, hydrogen production, substitute natural gas production, fermentation, limestone calcination, certain chemical syntheses involving carbon monoxide (qv), and reaction of sulfuric acid with dolomite. Generally present as one of a mixture of gases, carbon dioxide is separated, recovered, and prepared for commercial use as a solid (dry ice), liquid, or gas.

Carbon dioxide is also found in the products of combustion of all carbonaceous fuels, in naturally occurring gases, as a product of animal metabolism, and in small quantities, about 0.03 vol %, in the atmosphere. Its many applications include beverage carbonation, chemical manufacture, firefighting, food freezing, foundry-mold preparation, greenhouses, mining operations, oil well secondary recovery, rubber tumbling, therapeutical work, welding, and extraction processes. Although it is present in the atmosphere and the metabolic processes of animals and plants, carbon dioxide cannot be recovered economically from these sources.

2. Physical Properties

Some values of physical properties of CO_2 appear in Table 1. An excellent pressure–enthalpy diagram (a large Mollier diagram) over 260 to 773 K and 70–20,000 kPa (10–2,900 psi) is available (1). The thermodynamic properties of saturated carbon dioxide vapor and liquid from 178 to the critical point, 304 K, have been tabulated (2). Also given are data for superheated carbon

Table 1. **Properties of Carbon Dioxide**

Property	Value
sublimation point at 101.3 kPa[a], °C	−78.5
triple point at 518 kPa[b] °C	−56.5
critical temperature, °C	31.1
critical pressure, kPa[b]	7383
critical density, g/L	467
latent heat of vaporization, J/g[c]	
at the triple point	353.4
at 0°C	231.3
gas density at 273 K and 101.3 kPa[a], g/L	1.976
liquid density	
at 273 K, g/L	928
at 298 K and 101.3 kPa[a] CO_2, vol/vol	0.712
viscosity at 298 K and 101.3 kPa[a], mPa·s(=cP)	0.015
heat of formation at 298 K, kJ/mol[d]	393.7

[a] 101.3 kPa = 1 atm.
[b] To convert kPa to psia, multiply by 0.145.
[c] To convert J/g to Btu/lb, multiply by 0.4302.
[d] To convert kJ/mol to Btu/mol, multiply by 0.9487.

dioxide vapor from 228 to 923 K at pressures from 7 to 7,000 kPa (1–1,000 psi). A graphical presentation of heat of formation, free energy of formation, heat of vaporization, surface tension, vapor pressure, liquid and vapor heat capacities, densities, viscosities, and thermal conductivities has been provided (3). Compressibility factors of carbon dioxide from 268 to 473 K and 1,400–69,000 kPa (203–10,000 psi) are available (4).

Available data on the thermodynamic and transport properties of carbon dioxide have been reviewed and tables compiled giving specific volume, enthalpy, and entropy values for carbon dioxide at temperatures from 255 K to 1088 K and at pressures from atmospheric to 27,600 kPa (4,000 psia). Diagrams of compressibility factor, specific heat at constant pressure, specific heat at constant volume, specific heat ratio, velocity of sound in carbon dioxide, viscosity, and thermal conductivity have also been prepared (5).

Equations for viscosity at different temperatures, pressures, and thermal conductivity have also been provided (5). The vapor pressure function for carbon dioxide in terms of reduced temperatures and pressure is as follows:

$$\log P_R = 4.2397 - \frac{4.4229}{T_R} - 5.3795 \log T_R + 0.1832 \frac{P_R}{T_R^2}$$

where P_R equals reduced pressure, which equals P/P_c (P_c, critical pressure 7.38 MPa or 72.85 atm), and T_R equals reduced temperature, which equals T/T_c (T_c, critical temperature, 304.2 K) (6). This equation gives accurate vapor pressure values from the triple point to the critical point. A table of reduced density values for carbon dioxide covering the range of reduced pressures from 0.3 to 500 kPa (0.044–72.5 psi) and reduced temperatures from 0.712 to 20 K is also supplied (6). Enthalpy values for carbon dioxide in the critical region, and with

Table 2. **Vapor Pressure of Solid Carbon Dioxide**^{a}

Temperature,°C	Pressure, Pa	Temperature,°C	Pressure, Pa
−188°C	1.333×10^{-4}	−148°C	1.333×10^1
−182°C	1.333×10^{-3}	−136°C	1.333×10^2
−175°C	1.333×10^{-2}	−120°C	1.333×10^3
−167°C	1.333×10^{-1}	−100°C	1.333×10^4
−159°C	1.333	−65°C	1.333×10^5

^{a}To convert Pa to mm Hg, divide by 1.333×10^2.

temperatures from 423 to 923 K and pressures from 0 to 20 MPa (0–200 atm), have been computed (7,8).

Diagrams of isobaric heat capacity (C_p) and thermal conductivity for carbon dioxide covering pressures from 0 to 13,800 kPa (0–2,000 psi) and 311 to 1088 K have been prepared. Viscosities at pressures of 100–10,000 kPa (1–100 atm) and temperatures from 311 to 1088 K have been plotted (9).

Vapor pressure data for solid carbon dioxide are given in Table 2 (10). The sublimation temperature of solid carbon dioxide, 194.5 K at 101 kPa (1 atm), was selected as one of the secondary fixed points for the International Temperature Scale of 1948.

The solubility of carbon dioxide in water is given in Figure 1 (11). Over the temperature range 273–393 K, the solubilities at pressures below 20 MPa (200 atm) decrease with increasing temperature. From 30 to 70 MPa (300–700 atm) a solubility minimum is observed between 343 and 353 K, with solubilities increasing as temperature increases to 393 K. Information on the solubility of carbon dioxide in pure water and synthetic seawater over the range 268 to 298 K and 101–4,500 kPa pressure (1–44 atm) is available (12,13).

The following tables of properties of carbon dioxide are available: enthalpy, entropy, and heat capacity at 0 and 5 MPa (0 and 50 atm, respectively) from 273 to 1273 K; pressure–volume product (PV), enthalpy, and isobaric heat capacity (C_p) from 373 to 1273 K at pressures from 5 to 140 MPa (50–1,400 atm) (14).

A more recent compilation includes tables giving temperature and PV as a function of entropies from 0.573 to 0.973 (zero entropy at 0°C, 101 kPa (1 atm) and pressures from 5 to 140 MPa (50–1400 atm) (15). Joule-Thomson coefficients, heat capacity differences (C_p–C_V), and isochoric heat capacities (C_v) are given for temperatures from 373 to 1273 K at pressures from 5 to 140 MPa.

3. Chemical Properties

Carbon dioxide, the final oxidation product of carbon, is not very reactive at ordinary temperatures. However, in water solution it forms carbonic acid [463-79-6], H_2CO_3, which forms salts and esters through the typical reactions of a weak acid. The first ionization constant is 3.5×10^{-7} at 291 K; the second is 4.4×10^{-11} at 298 K. The pH of saturated carbon dioxide solutions varies from 3.7 at 101 kPa (1 atm) to 3.2 at 2,370 kPa (23.4 atm). A solid hydrate [27592-78-5], $CO_2 \cdot 8H_2O$, separates from aqueous solutions of carbon dioxide that are chilled at elevated pressures.

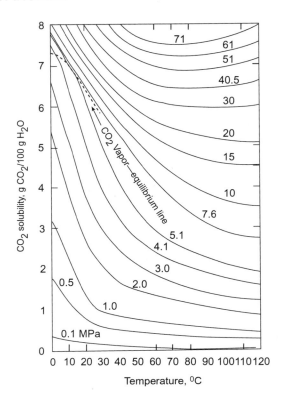

Fig. 1. Solubility of carbon dioxide in water at various pressures in MPa. To convert MPa to atm, multiply by 10.

Although carbon dioxide is very stable at ordinary temperatures, when it is heated above 1700°C the reaction forming CO proceeds to the right to an appreciable extent (15.8%) at 2500 K. This reaction also proceeds to the right to a limited extent in the presence of ultraviolet light and electrical discharges.

$$2\,CO_2 \rightleftharpoons 2\,CO + O_2$$

Carbon dioxide may be reduced by several means. The most common of these is the reaction with hydrogen.

$$CO_2 + H_2 \longrightarrow CO + H_2O$$

This is the reverse of the water–gas shift reaction in the production of hydrogen and ammonia (qv). Carbon dioxide may also be reduced catalytically with various hydrocarbons and with carbon itself at elevated temperatures. The latter reaction occurs in almost all cases of combustion of carbonaceous fuels and is generally employed as a method of producing carbon monoxide.

$$CO_2 + C \longrightarrow 2\,CO$$

Carbon dioxide reacts with ammonia as the first stage of urea manufacture to form ammonium carbamate [1111-78-0].

$$CO_2 + 2\,NH_3 \longrightarrow NH_2COONH_4$$

The ammonium carbamate then loses a molecule of water to produce urea [57-13-6], $CO(NH_2)_2$. Commercially, this is probably the most important reaction of carbon dioxide and it is used worldwide in the production of urea (qv) for synthetic fertilizers and plastics (see AMINO RESINS; CARBAMIC ACID).

3.1. Radioactive Carbon. In addition to the common stable carbon isotope of mass 12, traces of a radioactive carbon isotope of mass 14 with a half-life estimated at 5568 years are present in the atmosphere and in carbon compounds derived from atmospheric carbon dioxide. Formation of radioactive ^{14}C is thought to be caused by cosmic irradiation of atmospheric nitrogen. The concentration of carbon-14 [14762-75-5] in atmospheric carbon dioxide is approximately constant throughout the world. The ratio of $^{14}CO_2$ [51-90-1] to $^{12}CO_2$ in the atmosphere, although constant for several hundred years, has decreased in the last sixty years by about 3% because of the influx of carbon dioxide in the atmosphere from the burning of fossil fuels, ie, coal, petroleum, and natural gas. Procedures have been developed for estimating the age of objects containing carbon or carbon compounds by determining the amount of radioactive ^{14}C present in the material as compared with that present in carbon-containing substances of current botanical origin (16). Ages of materials up to 45,000 years have been estimated with the radiocarbon dating technique.

Carbon dioxide containing known amounts of ^{14}C has been used as a tracer in studying botanical and biological problems involving carbon and carbon compounds and in organic chemistry to determine the course of various chemical reactions and rearrangements. It has also been used in testing gaseous diffusion theory with mixtures of CO_2 and $^{14}CO_2$ at elevated pressures (17).

3.2. Environmental Chemistry. Carbon dioxide plays a vital role in the earth's environment. It is a constituent in the atmosphere and, as such, is a necessary ingredient in the life cycle of animals and plants.

In animal metabolism, oxygen from the atmosphere reacts with sugars in the body to produce energy according to the overall formula:

$$C_6H_{12}O_6 + 6\,O_2 \longrightarrow 6\,CO_2 + 6\,H_2O + energy$$

The by-product CO_2 is released to the atmosphere. In plant metabolism, carbon dioxide from the air is taken into the leaves of the plant. Using energy from light, carbon dioxide reacts with water in the presence of enzymes to produce sugar. This reaction, photosynthesis, is the reverse of the above reaction.

The balance between animal and plant life cycles as affected by the solubility of carbon dioxide in the earth's water results in the carbon dioxide content in the atmosphere of about 0.03 vol %. However, carbon dioxide content of the atmosphere seems to be increasing as increased amounts of fossil fuels are burned. There is some evidence that the rate of release of carbon dioxide to the atmosphere may be greater than the earth's ability to assimilate it. Measurements from the U.S. Water Bureau show an increase of 1.36% in the CO_2 content of the atmosphere in a five-year period and predictions indicate that at present the content may have increased by 25% (see AIR POLLUTION).

The effects of such an increase, if it occurs, are not known. It could result in a warmer temperature at the earth's surface by allowing the short heat waves from the sun to pass through the atmosphere while blocking larger waves that reflect back from the earth. If the earth's average temperature were to increase by several degrees, portions of the polar ice caps could melt causing an increase in the level of the oceans, or air circulation patterns could change, altering rain patterns to make deserts of farmland or vice versa.

On the other hand, it has been demonstrated that the addition of CO_2 to greenhouses increases the growth rate of plants so that an increase in the partial pressure of CO_2 in the air could stimulate plant growth making possible shorter growing seasons and increased consumption of carbon dioxide from the air. CO_2 is also used in water-treatment applications. Because it is significantly safer than mineral acids, it can be used to reduce the alkalinity of treated water.

4. Manufacture

Sources of carbon dioxide for commercial carbon dioxide recovery plants are (*1*) synthetic ammonia and hydrogen plants in which methane or other hydrocarbons are converted to carbon dioxide and hydrogen ($CH_4 + 2\,H_2O \longrightarrow CO_2 + 4\,H_2$); (*2*) flue gases resulting from the combustion of carbonaceous fuels; (*3*) fermentation in which a sugar such as dextrose is converted to ethyl alcohol and carbon dioxide ($C_6H_{12}O_6 \longrightarrow 2\,C_2H_5OH + 2\,CO_2$); (*4*) lime-kiln operation in which carbonates are thermally decomposed ($CaCO_3 \longrightarrow CaO + CO_2$); (*5*) sodium phosphate manufacture ($3\,Na_2CO_3 + 2\,H_3PO_4 \longrightarrow 2\,Na_3PO_4 + 3\,CO_2 + 3\,H_2O$); and (*6*) natural carbon dioxide gas wells.

4.1. Ammonia and Hydrogen Plants. More carbon dioxide is generated and recovered from ammonia and hydrogen plants than from any other source. Both plants produce hydrogen and carbon dioxide from the reaction between hydrocarbons and steam. In the case of hydrogen plants, the hydrogen is recovered as a pure gas. For ammonia plants the hydrogen is produced in the presence of air, controlled to give the volume ratio between hydrogen and nitrogen required to synthesize ammonia. In order to produce either product it is necessary to remove the carbon dioxide (18). The annual synthetic ammonia production capacity in the United States was about 15.9×10^6 t in 2001 (see AMMONIA). For each ton of ammonia produced, more than a ton of carbon dioxide is generated. Hence the available carbon dioxide supply from this source is several times as large as the total commercial production of carbon dioxide. A substantial amount of the carbon dioxide recovered from ammonia plants is used for urea production.

4.2. Flue Gases. In a typical plant for producing gaseous carbon dioxide from coke, coal, fuel oil, or gas the fuel is burned under a standard water-tube boiler for the production of 1400–1800 kPa (200–260 psi) steam. At 613 K flue gases containing 10–18% carbon dioxide leave the boiler and pass through two packed towers where they are cooled and cleaned by water. The gases are then passed through a booster blower into the base of the absorption tower. The

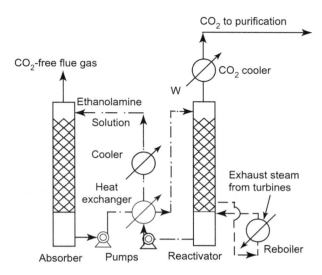

Fig. 2. Recovery of carbon dioxide using a Girbotol recovery unit.

recovery system shown in Figure 2 is the Girbotol amine process (19) although an alkaline carbonate system may also be used (Fig. 2). In the tower, carbon dioxide is absorbed selectively by a solution of ethanol amines passing countercurrent to the gas stream (see ALKANOLAMINES). The carbon dioxide-free flue gases pass out of the top of the tower into the atmosphere, the carbon dioxide-bearing solution passes out from the bottom of the tower, through pumps and heat exchangers, into the top of a reactivation tower. Here heat strips the carbon dioxide from the amine solution and the reactivated solution returns through the heat-exchanger equipment to the absorption tower. Carbon dioxide and steam pass through the top of the reactivation tower into a gas cooler in which the steam condenses and returns to the tower as reflux. The carbon dioxide at this point is available as a gas at a pressure of about 200 kPa (2 atm). If liquid or solid carbon dioxide is desired it may be further purified for odor removal before compression.

The steam balance in the plant shown in Figure 2 enables all pumps and blowers to be turbine-driven by high pressure steam from the boiler. The low pressure exhaust system is used in the reboiler of the recovery system and the condensate returns to the boiler. Although there is generally some excess power capacity in the high pressure steam for driving other equipment, eg, compressors in the carbon dioxide liquefaction plant, all the steam produced by the boiler is condensed in the recovery system. This provides a well-balanced plant in which few external utilities are required and combustion conditions can be controlled to maintain efficient operation.

4.3. Fermentation Industry. Large quantities of carbon dioxide are present in gases given off in the fermentation of organic substances such as molasses, corn, wheat, and potatoes in the production of beer, distilled beverages, and industrial alcohol. These gases may contain impurities such as aldehydes, acids, higher alcohols, glycerol, furfural, glycols, and hydrogen sulfide. Two processes are in general use for removing these contaminants and preparing the carbon dioxide for use. In one process, this is accomplished by the use of activated-carbon adsorbers; the other process is a chemical purification process (see BEER; BEVERAGE SPIRITS, DISTILLED; ETHANOL; FERMENTATION). A very small percentage of commercial CO_2 is produced by this process.

The Backus process (20,21) uses active carbon. Carbon dioxide gases from the top of the fermentors are collected in a low pressure gas holder to even out the flow. Roots-Connersville-type blowers force the gases through Feld scrubbers where they are washed with water to remove the bulk of entrained material, alcohols, aldehydes, etc. The washed gases pass through active-carbon purifiers, which adsorb the balance of impurities, and then to the compressors. The adsorption process gives off heat, which is removed by water coils embedded in the carbon. Periodically, the carbon beds must be reactivated to remove accumulated impurities. This is accomplished by passing live steam through the carbon bed and water coils. After steaming, the carbon beds are dried by passing air through them; they are then ready for reuse. In general, two sets of active-carbon purifiers are used, one on-stream and the other being reactivated (see ADSORPTION, GAS SEPARATION).

The Reich process (22,23) uses chemical processes to remove impurities. Carbon dioxide from the fermentors is bubbled through a wash box, or catchall, which removes entrained liquids and mash. The gases pass through three packed scrubbing towers. In the first, dilute alcohol solution is passed countercurrent to the gases for de-alcoholizing the gases. The other two towers use water for further removal of alcohol. The alcohol is returned to the alcohol plant for distillation or use in the fermentors. The washed gas passes through a gas holder and blower, which boosts it through the balance of the purification plant. The first stage in this section is a potassium dichromate washer for the oxidation of organic impurities and the removal of hydrogen sulfide. Next, the gas is passed countercurrent to concentrated sulfuric acid for dehydration and dichromate removal. Entrained sulfuric acid is removed by a dry solid ash tower and the residual oxidized material is removed by countercurrent scrubbing with a light oil. The gas is then ready for compression. In some installations the first stage of compression is inserted before the dichromate washer and the purification operations are carried out at 600–800 kPa (87–116 psi).

4.4. Lime-Kiln Operation. Gases containing up to 40% carbon dioxide from the lime kiln pass through a cyclone separator, which removes the bulk of entrained dust. The gas is then blown through the two scrubbers, which remove the finer dust, cooled, and passes into an absorption tower. Here carbon dioxide may be recovered by the sodium carbonate or Girbotol process.

4.5. Sodium Phosphate Manufacturing. Some pure carbon dioxide gas is available as a by-product in plants manufacturing sodium phosphate from sodium carbonate [497-19-8] and phosphoric acid [7664-38-2]. Two carbon dioxide plants were installed prior to 1962 to utilize this by-product gas.

4.6. Natural Gas Wells. Natural gas, containing high percentages of carbon dioxide, has been found in a number of locations including New Mexico, Colorado, Utah, and Washington. Several small plants have been in operation for a number of years producing commercial solid and liquid carbon dioxide from these sources (see GAS, NATURAL). Several large CO_2 plants are in operation using well gas as a source of CO_2.

There are a number of methods of recovering carbon dioxide from industrial or natural gases. The potassium carbonate and ethanolamine processes are most common. In all these processes the carbon dioxide-bearing gases are passed countercurrent to a solution that removes the carbon dioxide by absorption and retains it until it is desorbed in separate equipment. All of these processes are in commercial use and the most suitable choice for a given application depends on individual conditions. Water could be used as the absorbing medium, but this is uncommon because of the relatively low solubility of carbon dioxide in water at normally encountered pressures. The higher solubility in the alkali carbonate and ethanolamine solutions is the result of a chemical combination of the carbon dioxide with the absorbing medium.

Sodium Carbonate Process.   This process of recovering pure carbon dioxide from gas containing other diluents, such as nitrogen and carbon monoxide, is based on the reversibility of the following reaction:

$$Na_2CO_3 + H_2O + CO_2 \rightleftharpoons 2\,NaHCO_3$$

This reaction proceeds to the right at low temperatures and takes place in the absorber where the carbon dioxide-bearing gases are passed countercurrent to the carbonate solution. The amount of carbon dioxide absorbed in the solution varies with temperature, pressure, partial pressure of carbon dioxide in the gas, and solution strength. Operating data on this reaction have been obtained by numerous investigators (24). The reaction proceeds to the left when heat is applied. The reaction takes place in a lye boiler. A heat exchanger preheats the strong lye approaching the boiler and cools the weak lye returning to the absorber. A lye cooler further cools weak lye to permit the reaction to proceed further to the right in the absorber. The carbon dioxide gas and water vapor released from the solution in the boiler pass through a steam condenser where the water condenses and returns to the system. The cool carbon dioxide proceeds to the gas holder and compressors. The absorber is generally a carbon–steel tower filled with coke, Raschig rings, or steel turnings. The weak solution is distributed evenly over the top of the bed and contacts the gas on the way down. Some plants operate with the tower full of sodium carbonate solution and allow the gas to bubble up through the liquid. Although this may afford a better gas-to-liquid contact, an appreciable amount of power is required to force the gas through the tower.

The lye boiler is usually steam heated but may be direct-fired. Separation efficiency may be increased by adding a tower section with bubble-cap trays. To permit the bicarbonate content of the solution to build up, many plants are designed to recirculate the lye over the absorber tower with only 20–25% of the solution flowing over this tower passing through the boiler. Several absorbers may also be used in series to increase absorption efficiencies.

The sodium carbonate process is used in a number of dry-ice plants in the United States, although its operating efficiency is generally not as high as that of processes using other solutions. These plants obtain the carbon dioxide from flue gases as well as lime-kiln gases.

Potassium Carbonate Process.   The potassium carbonate process is similar to the sodium carbonate process. However, as potassium bicarbonate

[298-14-6] is more soluble than the corresponding sodium salt, this process permits a more efficient absorption than the other. The equipment layout is the same and the operation technique is similar.

There are several variations of the potassium carbonate process. The hot potassium carbonate process does not involve cooling of the solution flowing from the boiler to the absorber (25). Absorption takes place at essentially the same temperature as solution regeneration. In its simplest form this process uses an absorption column, a regeneration column, a heated boiler or reboiler, and a circulating pump. This arrangement minimizes energy requirements and capital costs but the higher absorbing temperature does not allow the carbon dioxide removal to be as complete as with lower temperatures. One modification that improves removal is the use of a split stream flow to the absorber. Part of the solution is cooled and used at the top of the absorber. The balance of the solution, uncooled, is added part way down the absorber. The combined solution from the absorber then flows to the regenerator. In another modification two-stage absorption and regeneration is used. The carbonate solution from the absorber flows to the regeneration column. Part of the solution is withdrawn from the column at some intermediate point and pumped, uncooled, to an intermediate point in the absorber. The remaining solution undergoes a more complete regeneration and is cooled and pumped to the top of the absorber.

Three commercial processes that use these various hot carbonate flow arrangements are the promoted Benfield process, the Catacarb process, and the Giammarco-Vetrocoke process (26–29). Each uses an additive described as a promoter, activator, or catalyst, which increases the rates of absorption and desorption, improves removal efficiency, and reduces the energy requirement. The processes also use corrosion inhibitors, which allow use of carbon–steel equipment. The Benfield and Catacarb processes do not specify additives. Vetrocoke uses boric acid, glycine, or arsenic trioxide, which is the most effective.

These processes have been used in many plants to remove carbon dioxide from ammonia synthesis gas or natural gas. They are most effective if the gas stream being treated is at elevated pressures (1700 kPa (17 atm) or higher). This increases the carbon dioxide partial pressure so that the hot potassium carbonate solution absorbs a substantial amount of carbon dioxide. The stripping tower or regenerator operates at or near atmospheric pressure.

Ethanolamines may also be added to carbonate solutions to improve their performance (30). Vapor pressure-equilibrium data for the K_2CO_3–$KHCO_3$–CO_2–H_2 system are given (31).

Girbotol Amine Process.   This process developed by the Girdler Corporation is similar in operation to the alkali carbonate processes. However, it uses aqueous solutions of an ethanolamine, ie, either mono-, di-, or triethanolamine. The operation of the Girbotol process depends on the reversible nature of the reaction of CO_2 with monoethanolamine [141-43-5] to form monoethanolamine carbonate [21829-52-7].

$$2\ HOCH_2CH_2NH_2 + H_2O + CO_2 \rightleftharpoons (HOCH_2CH_2NH_3)_2CO_3^{2-}$$

The reaction proceeds in general to the right at low temperatures (300–338 K) and absorbs the carbon dioxide from the gas in the absorber as shown in

Figure 2. The amine solution, rich in carbon dioxide, passes from the bottom of the tower through a heat exchanger, where it is preheated by hot, lean solution returning from the reactivator. On passing into the reactivator the solution passes countercurrent to a stream of carbon dioxide and steam, which strips the carbon dioxide out of the solution. By the time the solution reaches the bottom of the tower, where heat is supplied by a steam-heated or direct-fired reboiler, it has been reactivated. This hot solution (373–423 K) passes out of the tower, through the heat exchanger and cooler, and returns to the absorber tower.

The amine process or one of the various carbonate processes is used in the majority of CO_2-removal applications. At low pressures the amine process has clear advantages in removal efficiency and installed cost. At higher pressures the increased carbon dioxide partial pressures favor a higher CO_2 content in the absorbing solution, whether carbonate or amine, which permits a lower steam usage per unit of carbon dioxide stripped from the absorbent. Until recently, amine systems have not been able to take full advantage of the benefits of higher absorber pressure because of corrosion problems. The CO_2-rich amine solution is inherently more corrosive than potassium carbonate solutions so that the stronger the amine solution and the greater the carbon dioxide content the worse the corrosion potential. To hold corrosion to a minimum, amine solution designs have been limited to a maximum solution strength of 20 wt% with a maximum CO_2 content of about 0.0374 m^3/L (5 ft^3/gal) of solution. Even then, some plants experience severe corrosion problems requiring replacement of some carbon-steel equipment with stainless steel. Carbonate solutions pick up more CO_2 per unit of solution than the amines, resulting in a lower circulation rate and a correspondingly lower heat regeneration requirement.

Overall comparison between amine and carbonate at elevated pressures shows that the amine usually removes carbon dioxide to a lower concentration at a lower capital cost but requires more maintenance and heat. The impact of the higher heat requirement depends on the individual situation. In many applications, heat used for regeneration is from low temperature process gas, suitable only for boiler feed water heating or low pressure steam generation, and it may not be useful in the overall plant heat balance.

Union Carbide has developed Amine Guard, which essentially eliminates corrosion in amine systems (32–35). It permits the use of substantially higher amine concentrations and greater carbon dioxide pick-up rates without corrosive attack. This results in an energy requirement comparable to that of the carbonate process and allows the use of smaller equipment for a specific CO_2-removal application thereby reducing the capital cost.

Solubility of carbon dioxide in ethanolamines is affected by temperature, amine solution strength, and carbon dioxide partial pressure. Information on the performance of amines is available in the literature and from amine manufacturers. Values for the solubility of carbon dioxide and hydrogen sulfide mixtures in monoethanolamine and for the solubility of carbon dioxide in diethanolamine are given (36,37). Solubility of carbon dioxide in monoethanolamine is provided (38). The effects of catalysts have been studied to improve the activity of amines and provide absorption data for carbon dioxide in both mono- and diethanolamine solutions with and without sodium arsenite as a catalyst (39). Absorption kinetics over a range of contact times for carbon dioxide in monoethanolamine have also been investigated (40).

Sulfinol Process.   The Sulfinol process was developed during the 1960s to remove carbon dioxide and other acidic gas from gas streams at high partial pressures. It uses a circulating solution with a flow pattern similar to those in the amine and carbonate processes. Regeneration occurs at low pressure, and heat is exchanged between the regenerated solution and the solution from the absorber. The Sulfinol solution is a mixture of sulfolane [126-33-0] (tetrahydrothiophene 1,1-dioxide), an alkanolamine, and water.

The process is capable of achieving higher solubilities of CO_2 in the solution without the corrosion problems encountered with amine systems before the advent of Amine Guard. The Sulfinol process is used in over 50 plants worldwide; nevertheless, it is used less often than the amine or carbonate processes.

Rectisol Process.   This is one of several processes that use a solvent for removing carbon dioxide from gas streams. In the Rectisol process, the solvent is usually methanol and the operating temperature of the absorber is about 273 K. The process uses an absorption column and stripping column with intermediate pumps and heat exchangers. It is necessary to cool the process gas to the absorbing temperature. Split flow of the solvent stream to the absorber is used, with partially stripped solvent from the midsection of the stripper being used at the top of the absorber. Stripping is accomplished by pressure release, heat, and contact with an inert stripping gas, which is blown through the column.

The processes using physical absorption require a solvent circulation proportional to the quantity of process gas, inversely proportional to the pressure, and nearly independent of the carbon dioxide concentration. Therefore, high pressures could favor the use of these processes. The Recitsol process requires a refrigeration system and more equipment than the other processes. This process is primarily used in coal gasification for simultaneous removal of H_2S, COS, and CO_2.

Purisol Process.   This is a solvent process which uses N-methyl-2-pyrrolidinone [872-50-4] as the solvent and benefits from high pressure, 7000 kPa (69 atm) or higher. All commercial installations are at pressures above 4000 kPa (39 atm).

Fluor Process.   This is a solvent process that uses propylene carbonate [108-32-7]. Propylene carbonate has a high solubility for CO_2, a low solubility for other light gases, is chemically stable, and noncorrosive to carbon steel. Physical solvent plants represent a higher capital cost than the carbonate or amine process plants but may result in reducing operating costs where high pressures are involved. Seven plants using the Fluor process for removal of acidic gases were in existence by 1970; five natural gas plants, one hydrogen, and one ammonia plant.

Pressure Swing Adsorption.   A number of processes based on Pressure Swing Adsorption (PSA) technology have been used in the production of carbon dioxide. In one version of the PSA process, CO_2 is separated from CH_4 using a multibed adsorption process (41). In this process both CH_4 and CO_2 are produced. The process requires the use of five adsorber vessels. Processes of this type can be used for producing CO_2 from natural gas wells, landfill gas, or from oil wells undergoing CO_2 flooding for enhanced oil recovery (see ADSORPTION, GAS SEPARATION).

4.7. Methods of Purification. Although carbon dioxide produced and recovered by the methods outlined above has a high purity, it may contain traces of hydrogen sulfide and sulfur dioxide, which cause a slight odor or taste. The fermentation gas recovery processes include a purification stage, but carbon dioxide recovered by other methods must be further purified before it is acceptable for beverage, dry ice, or other uses. The most commonly used purification methods are treatments with potassium permanganate, potassium dichromate, or active carbon.

Potassium Permanganate. Probably the most widely used process for removing traces of hydrogen sulfide from carbon dioxide is to scrub the gas with an aqueous solution saturated with potassium permanganate [7722-64-7]. Sodium carbonate is added to the solution as buffer. The reaction is as follows:

$$3\,H_2S + 2\,KMnO_4 + 2\,CO_2 \longrightarrow 3\,S + 2\,MnO_2 + 2\,KHCO_3 + 2\,H_2O$$

The precipitated manganese dioxide and sulfur are discarded. The solution is used until it becomes spent or so low in potassium permanganate that it is no longer effective and is discarded and replaced. It is customary to place two scrubbers in series, with the liquid flow countercurrent to the gas flow, to more efficiently use the permanganate solution. When the solution in the first scrubber is spent, with respect to the gas, the positions of the scrubbers are reversed and the spent scrubber is recharged with fresh solution.

Two types of scrubbers are used. The simpler consists of a vessel half or two-thirds full of solution. The gas feeds into the bottom of the vessel and bubbles up through the solution. The other scrubber is a small packed tower through which the gas stream is passed countercurrent to a recirculating shower of potassium permanganate and soda ash solution. The latter requires a circulating pump and a solution mix chamber, but has the advantage of reducing the pressure drop through the equipment to a minimum. The solution is used until spent and then discarded. Two scrubbers of this type may also be used to improve efficiency.

Potassium Dichromate. This method is similar in application to the potassium permanganate method.

$$K_2Cr_2O_7 + 3\,H_2S + H_2O + 2\,CO_2 \longrightarrow 3\,S + 2\,Cr(OH)_3 + 2\,KHCO_3$$

The precipitated chromic hydroxide and sulfur are discarded. This process is used to purify carbon dioxide from fermentation in the Reich process and as a final cleanup after the alkali carbonate or ethanolamine recovery processes (22,23).

Active Carbon. The process of adsorbing impurities from carbon dioxide on active carbon or charcoal has been described in connection with the Backus process of purifying carbon dioxide from fermentation processes. Space velocity and reactivation cycle vary with each application. The use of active carbon need not be limited to the fermentation industries but, where hydrogen sulfide is the only impurity to be removed, the latter two processes are usually employed (see ACTIVATED CARBON).

4.8. Methods of Liquefaction and Solidification. Carbon dioxide may be liquefied at any temperature between its triple point (216.6 K) and its critical point (304 K) by compressing it to the corresponding liquefaction pressure, and removing the heat of condensation. There are two liquefaction processes. In the first, the carbon dioxide is liquefied near the critical temperature; water is used for cooling. This process requires compression of the carbon dioxide gas to pressures of about 7600 kPa (75 atm). The gas from the final compression stage is cooled to about 305 K and then filtered to remove water and entrained lubricating oil. The filtered carbon dioxide gas is then liquefied in a water-cooled condenser.

The second liquefaction process is carried out at temperatures from 261 to 296 K, with liquefaction pressures of about 1600–2400 kPa (16–24 atm). The compressed gas is precooled to 277 to 300 K, water and entrained oil are separated, and the gas is then dehydrated in an activated alumina, bauxite, or silica gel drier, and flows to a refrigerant-cooled condenser (see DRYING AGENTS). The liquid is then distilled in a stripper column to remove noncombustible impurities. Liquid carbon dioxide is stored and transported at ambient temperature in cylinders containing up to 22.7 kg. Larger quantities are stored in refrigerated insulated tanks maintained at 255 K and 2070 kPa (20 atm), and transported in insulated tank trucks and tank rail cars.

Solidification. Liquid carbon dioxide from a cylinder may be converted to "snow" by allowing the liquid to expand to atmospheric pressure. This simple process is used only where very small amounts of solid carbon dioxide are required because less than one-half of the liquid is recovered as solid.

Solid carbon dioxide is produced in blocks by hydraulic presses. Standard presses produce blocks $25 \times 25 \times 25$ cm, $50 \times 25 \times 25$ cm, or $50 \times 50 \times 25$ cm. A 25-cm cube of dry ice weighs 23 kg, allowing for about 10% sublimation loss during storage and shipment (some 27-kg blocks are also produced). Dry ice is about 1.7 times as dense as water ice, whereas its net refrigerating effect on a weight basis is twice that of water ice. Automation and improved operating cycles have increased dry-ice press capacities so that one $50 \times 50 \times 30$ cm press can produce more than thirty metric tons of dry-ice blocks per day (42).

Liquid carbon dioxide from a supply tank at 700 kPa (7 atm) and 227 K is fed to the press chamber through an automatic feed valve. The pressure in the press is maintained slightly above the triple point (480–550 kPa or 70–80 psi). The quantity fed to the press may be controlled by a timer, or a device that measures the level of liquid in the press chamber. The pressure is reduced and the evolved CO_2 vapor is returned to a recycle system. When the pressure falls below the triple point (518 kPa or 75 psi), the liquid CO_2 solidifies to form carbon dioxide snow. Heat-exchange units are used to cool the liquid CO_2 with cold vapors from the press. About 50% of the liquid fed to the press remains as snow when the pressure has dropped close to atmospheric. The hydraulic rams then press the snow to a solid block of dry ice. The block moves along a conveyor and is cut by band saws into four blocks, which are subsequently carried through automatic weighing and packaging machines.

Carbon dioxide is ordinarily dehydrated during the liquefaction cycle to prevent freeze-ups in the condenser and flow valves in the liquid lines. In some cases brittle or crumbly blocks of dry ice have been formed. This difficulty has been overcome either by varying the residual moisture content of the liquid carbon dioxide, or by injecting minute quantities of colorless mineral oil or diethylene glycol into the liquid carbon dioxide entering the press. If the dry ice is to be used for edible purposes, the additive must meet FDA specifications.

Although liquid carbon dioxide may be stored without loss in tanks and cylinders, dry ice undergoes continuous loss in storage because of sublimation. This loss can be minimized by keeping the dry ice in insulated boxes or bins. Special insulated rail cars and trucks are used for hauling dry-ice blocks. Most plants produce the material at the time it is sold to avoid storage losses and rehandling costs.

5. Economic Aspects

In 1998, 35×10^6t of gaseous carbon dioxide was consumed. Liquid carbon dioxide consumption was 6.6×10^6t. Gaseous carbon dioxide is used for enhanced oil recovery and, thus, it exceeded liquid use.

In the United States, 27.9×10^6t of gaseous carbon dioxide was used in the oil fields and 5.8×10^6t was used in urea production. Western Europe consumed 3.5×10^6t. Growth in the U.S. and Western Europe is expected at the rate of 3–4% through 2003. In Japan, a 2% growth rate is expected (43).

Much more carbon dioxide is generated daily than is recovered (44). The decision whether or not to recover by-product carbon dioxide often depends on the distance and cost of transportation between the carbon dioxide producer and consumer. For example, it has become profitable to recover more and more carbon dioxide from CO_2-rich natural gas wells in Texas as the use of carbon dioxide in secondary oil recovery has increased. The production levels for enhanced oil recovery are generally not reported because of the captive nature of the application.

6. Health and Safety Factors

Although carbon dioxide is a constituent of exhaled air, high concentrations are hazardous. Up to 0.5 vol% carbon dioxide in air is not considered harmful, but carbon dioxide concentrates in low spots because it is one and one-half times as heavy as air. Five vol% carbon dioxide in air causes a threefold increase in breathing rate and prolonged exposure to concentrations higher than 5% may cause unconsciousness and death. Ventilation sufficient to prevent accumulation of dangerous percentages of carbon dioxide must be provided where carbon dioxide gas has been released or dry ice has been used for cooling.

The ACGIH TLV (TWA) is 5000 ppm (9000 mg/m^3) with a STEL of 30,000 ppm (54,000 mg/m^3). However, the documentation of the threshold limit values and biological exposure indices stated that "medically fit" persons in special circumstances may tolerate daily exposure to 1.5% (15,000 ppm). The definition of "medically fit" excludes all persons over the age of 65 and persons with current endocrine disorders. The National Institute for Occupational Safety and Health has recommended a TWA of 5000 ppm as a limit, with short-term excursions up to 3% (30,000 ppm) for 10 min (45).

7. Uses

A large portion of the carbon dioxide recovered is used at or near the location where it is generated as an ingredient in a further processing step. In this case, the gaseous form is most often used. Low temperature liquid and solid carbon dioxide are used for refrigeration. Where the producer and the consumer are distant, carbon dioxide may be liquified to reduce transportation cost and revaporized at the point of consumption.

About 51% of the carbon dioxide consumed in the United States is used in the food industry. It is generally purchased in liquid form but may be used in any form. It is generally used for food freezing or chilling. Numerous patents on applications and equipment for these applications have been received.

Approximately 18% of carbon dioxide output is used for beverage carbonation. Both soft drinks and beer production consume the largest quantity of CO_2 for carbonation (see CARBONATED BEVERAGES).

About 10% of the carbon dioxide produced is for chemical manufacturing. Sold as a liquid, it is used as a raw material, for inerting and pressurizing, and for cooling. Other applications include metal working (4%) and oil and gas recovery (6%).

7.1. Dry Ice. Refrigeration of foodstuffs, especially ice cream, meat products, and frozen foods, is the principal use for solid carbon dioxide. Dry ice is especially useful for chilling ice cream products because it can be easily sawed into thin slabs and leaves no liquid residue upon evaporation. Crushed dry ice may be mixed directly with other products without contaminating them and is widely used in the processing of substances that must be kept cold. Dry ice is mixed with molded substances that must be kept cold. For example, dry ice is mixed with molded rubber articles in a tumbling drum to chill them sufficiently so that the thin flash or rind becomes brittle and breaks off. It is also used to chill golf-ball centers before winding. A device and process for generating carbon dioxide snow has been reported (46).

Dry ice is used to chill aluminum rivets. These harden rapidly at room temperature, but remain soft if kept cold with dry ice. It has found numerous uses in laboratories, hospitals, and airplanes as a convenient and readily available low temperature coolant.

7.2. Liquid Carbon Dioxide. The rapid increase in the use of liquid carbon dioxide is the result of new applications as well as improved facilities for transporting, storing, and handling liquid carbon dioxide. Carbon dioxide manufacturers have developed refrigerated bulk-liquid storage systems that they install and maintain for large consumers. These systems are available in sizes from 2 to 50 tons and have an insulated storage tank, maintained at 2080 kPa (20.5 atm) and 255 K by a Freon refrigeration unit, with a refrigeration coil in the upper part of the storage tank. An external vaporizer is provided when gaseous carbon dioxide is needed. Liquid-level and pressure relief valves, and a safety rupture disk, are provided. The entire assembly is enclosed in a sheet-steel housing mounted on a steel base, and requires nominal power. A 12-ton unit is supplied with a 1500-W (two-horsepower) refrigeration unit. Vaporizers can be heated by electricity or steam. One kg of steam vaporizes approximately 6 kg of CO_2. The storage tanks are refilled by the CO_2 supplier, either by tank truck or rail car delivery.

Ready availability and easy application of bulk liquid carbon dioxide have caused it to replace dry ice in many cases. Liquid CO_2 can be stored without loss and is easily measured or weighed. Liquid carbon dioxide is also used, along with dry ice, for direct injection into chemical reaction systems to control temperature.

Liquid carbon dioxide provides the most readily available method of rapid refrigeration and is used for rapid chilling of loaded trucks and rail cars before shipment. A two to four minute injection of liquid carbon dioxide into a loaded ice cream truck causes the temperature to drop as much as 70°C, flushing the warm air out of the truck and leaving a layer of carbon dioxide snow in the truck, which sublimes slowly to provide additional refrigeration. This greatly reduces the load on the truck's mechanical refrigeration system and eliminates the time lag in cooling the truck contents to a safe storage temperature. Test chambers for environmental studies have been effectively cooled at temperatures to −80°C with liquid carbon dioxide, either by direct injection, or by using the liquid to chill circulating refrigerant liquids. The speed with which chilling may be obtained and the low equipment cost required have been the main factors in the selection of liquid carbon dioxide for this service.

Liquid carbon dioxide has been used for many years in the Long-Airdox blasting system for mining coal. A steel cartridge containing liquid carbon dioxide is placed in a hole drilled in the coal seam. A heating mixture in the cartridge is ignited electrically. This vaporizes the carbon dioxide, causing the pressure to increase enough to burst a steel rupture disk and release the carbon dioxide, which shatters the coal. The cartridge is then recovered and reused.

Liquid carbon dioxide is used as a source of power in certain applications. The vapor pressure of liquid carbon dioxide (7290 kPa or 72 atm at 294 K) may be used for operating remote signaling devices, spray painting, and gas-operated firearms. Carbon dioxide in small cylinders is also used for inflating life rafts and jackets.

Fire-extinguishing equipment, ranging from hand-type extinguishers to permanent installations in warehouses, chemical plants, ships, and airplanes, uses liquid carbon dioxide. In addition to its snuffing action liquid carbon dioxide exerts a pronounced cooling effect helpful in fire extinguishing. It may be used on all types of fires and leaves no residue, but care must be exercised to safeguard against suffocation of personnel.

Carbon dioxide is sometimes added to irrigation water, in the same manner as fertilizer ammonia, in hard water regions. Carbon dioxide is also used with other gases in treating respiratory problems and in anesthesia.

In addition to chemical synthesis and enhanced oil recovery, gaseous carbon dioxide is used in the carbonated beverage industry. Carbon dioxide gas under pressure is introduced into rubber and plastic mixes, and on pressure release a foamed product is produced. Carbon dioxide and inert gas mixtures rich in carbon dioxide are used to purge and fill industrial equipment to prevent the formation of explosive gas mixtures.

The addition of small amounts of carbon dioxide to the atmosphere in greenhouses greatly improves the growth rate of vegetables and flowers.

Carbon dioxide is widely used in the hardening of sand cores and molds in foundries. Sand is mixed with a sodium silicate binder to form the core or mold after which it is contacted with gaseous carbon dioxide. Carbon dioxide reacts with the sodium silicate to produce sodium carbonate and bicarbonate, plus silicic acid, resulting in hardening of the core or mold without baking.

The use of carbon dioxide gas for shielded arc welding with semiautomatic microwire welding equipment has led to welding speeds up to 10 times those obtainable within conventional equipment. No cleaning or wire brushing of the welds is required (45) (see WELDING).

Carbon dioxide gas is used to immobilize animals prior to slaughtering them (46). In addition to providing a humane slaughtering technique, this results in better quality meat. The CO_2 increases the animal's blood pressure, thereby increasing blood recovery. The increased accuracy obtainable in the killing operation reduces meat losses because of cut shoulders.

As a weak acid (in aqueous solution) carbon dioxide neutralizes excess caustic in textile manufacturing operations. It does not injure fabrics and is easy to use. Carbon dioxide is also used for neutralizing alkaline wastewaters, treating skins in tanning operations, and carbonating treated water to prevent scaling.

Carbon dioxide is used as a chemical reagent in the manufacture of sodium salicylate, basic lead carbonate or white lead, and sodium, potassium, and ammonium carbonates and bicarbonates.

A device or sachet for maintaining or modifying the atmosphere in a package suitable for meat, beef, lamb, pork, etc, the device including at least two containing layers, a moisture activated chemical substance contained by the containing layers, and at least one absorbent layer has been reported. Moisture within the package is soaked up by the absorbing layer. The moisture is then transferred from the absorbing layer to permeate through a micro-porous layer as water vapor and activate the chemicals. The gas produced by the chemicals then passses through the sachet and into the package's atmosphere. This modifies the atomosphere to ensure that there is an adequate quantity of CO_2 for achieving the desired shelf life of the packaged food (49).

BIBLIOGRAPHY

"Carbon Dioxide" in *ECT* 1st ed., Vol. 3, pp. 125–142, by R. M. Reed and N. C. Updegraff, The Girdler Corporation; in *ECT* 2nd ed., Vol. 4, pp. 353–369, by R. M. Reed and E. A. Comley, The Girdler Corporation; in *ECT* 3rd ed., Vol. 4, pp. 725–742, by W. R. Ballou, C&I Girdler Inc.; in *ECT* 4th ed., Vol. 5, pp. 35–53, by Ronald Pierantozzi, Air Products and Chemicals, Inc.; "Carbon Dioxide" in *ECT* (online), posting date: December 4, 2000, by Ronald Pierantozzi, Air Products and Chemicals, Inc.

CITED PUBLICATIONS

1. L. N. Canjar and co-workers, *Hydrocarbon Process.* **45**(1), 139 (1966).
2. *Matheson Gas Data Book*, The Matheson Co., Inc., East Rutherford, N.J., 1961, 81–82.
3. C. L. Yaws, K. Yu. Lai, and C. Kuo, *Chem. Eng.* **81**, 115 (1974).
4. B. J. Kendall and B. H. Sage, *Petroleum (London)* **14**, 184 (1951).
5. L. H. Chen, *Thermodynamic and Transport Properties of Gases, Liquids and Solids*, McGraw-Hill Book Co., Inc., New York, 1959, 358–369.
6. J. T. Kennedy and G. Thodos, *J. Chem. Eng. Data* **5**(3), 293 (1960).
7. L. B. Koppel and J. M. Smith, *J. Chem. Eng. Data* **5**(3), 437 (1960).
8. P. E. Liley, *J. Chem. Eng. Data* **4**(3), 238 (1959).
9. I. Granet and P. Kass, *Petrol. Refiner* **31**(11), 137 (1952).
10. R. E. Honig and H. O. Hook, *RCA Rev.* **21**, 360 (1960).
11. W. S. Dodds, L. F. Stutzman, and B. J. Sollami, *Chem. Eng. Data Serv.* **1**(1), 92 (1956).
12. P. B. Stewart and P. Munjal, *J. Chem. and Eng. Data* **15**, 67 (1970).
13. P. Munjal and P. B. Stewart, *J. Chem. and Eng. Data* **16**(92), 1970 (1971).
14. D. Price, *Ind. Eng. Chem.* **47**, 1649 (1955).
15. D. Price, *Chem. Eng. Data Serv.* **1**(1), 83 (1956).
16. W. F. Libby, E. C. Anderson, and J. R. Arnold, *Science* **190**, 227 (1949).
17. H. A. O'Hern, Jr. and J. J. Martin, *Ind. Eng. Chem.* **47**, 2081 (1955).
18. U.S. Pat. 18,958 (Sept. 26, 1933), R. R. Bottoms (to The Girdler Corp.) (reissue of 183,901).
19. A. V. Slack and G. R. James, eds., *Ammonia Part II*, Vol. 2, Marcel Dekker, New York, 1974.
20. U.S. Pat. 1,493,183 (May 6, 1924), A. A. Backus (to U.S. Industrial Alcohol Co.).
21. U.S. Pat. 1,510,373 (Sept. 30, 1924), A. A. Backus (to U.S. Industrial Alcohol Co.).
22. U.S. Pat. 1,519,932 (Dec. 16, 1924), G. T. Reich.
23. U.S. Pat. 2,225,131 (Dec. 17, 1940), G. T. Reich.
24. J. H. Perry, ed., *Chemical Engineers Handbook*, 3rd ed., McGraw-Hill Book Co., Inc., New York, 1950, p. 702.
25. U.S. Pat. 2,886,405 (May 12, 1959), H. E. Benson and J. H. Field (to U.S. Government).
26. Ital. Pat. 587,522 (Jan. 16, 1959), G. Giammarco (to S. A. Vetrocoke).
27. U.S. Pat. 2,840,450 (June 24, 1958), G. Giammarco (to S. A. Vetrocoke).
28. Ital. Pat. 518,145 (Mar. 4, 1955), G. Giammarco (to S. A. Vetrocoke).
29. U.S. Pat. 3,086,838 (Apr. 23, 1963), G. Giammarco (to S. A. Vetrocoke).
30. U.S. Pat. 3,144,301 (Aug. 11, 1964), B. J. Mayland (to The Girdler Corp.).
31. J. P. Bocard and B. J. Mayland, *Pet. Refiner.* **41**(4), 128 (1962).
32. K. F. Butwell, E. N. Hawkes, and B. F. Mago, *Chem. Eng. Prog.* **69**(2), 57 (1973),
33. *Nitrogen* (96), 33 (1975).
34. *Oil Gas J.* **73**(11), 107 (1975).
35. *Nitrogen* (102), 40 (1976).
36. J. I. Lee, F. D. Otto, and A. E. Mather, *J. Chem. Eng. Data* **20**(2), 161 (1975).
37. J. I. Lee, F. D. Otto, and A. E. Mather, *J. Chem. Eng. Data* **17**(4), 465 (1972).
38. J. D. Lawson and A. W. Garst, *J. Chem. Eng. Data* **21**(1), 20 (1976).
39. P. V. Danckwerts and K. M. McNeil, *Trans. Inst. Chem. Engr.* **45**, T32 (1967).
40. E. Sada, H. Kumazawa, and M. A. Butt, *AIChE J.* **2**(1), 196 (1976).
41. U.S. Pat. 4,077,779 (1978), S. Sircar and J. W. Zoudlo.
42. *Chem. Eng.* **62**, 120 (1955).

43. *Chemical Economics Handbook*, SRI International, Menlo Park, Calif., June 2000.

44. S. Terra, *EPRI J.*, 22 (July/Aug., 1978).

45. C. D. Leikauf in E. Bingham, B. Cohrssen, and C. H. Powell, Eds., *Patty's Toxicology*, 5th ed., Vol. 5, John Wiley & Sons, Inc., New York, 2001.

46. U.S. Pat. 6,543,251 (April 8, 2003), T. H. Gasteyer, III and co-workers (to Praxair Technology, Inc).

47. U.S. Pat. 5,042,262 (1991), K. L. Burgers, R. F. Gyger, and G. D. Lany (to Liquid Carbonic Corp.); U.S. Pat. 4,955,206 (1990), G. D. Lang and B. Zyer; U.S. Pat. 4,428,535 (1984), J. M. Venetucci.

48. *Animal Immobilization with CO₂*, Bull. No. 11, General Dynamics Corporation, Chicago, Ill., 1962.

49. U.S. Pat. 6,592,919 (July 15, 2003), A. E. Matthews, T. E. Snowball, and R. Darnett (to Sealed Air (NZ) Limited).

RONALD PIERANTOZZI
Air Products and Chemicals, Inc.

CARBON DISULFIDE

1. Introduction

Carbon disulfide [75-15-0] (carbon bisulfide, dithiocarbonic anhydride), CS_2, is a toxic, dense liquid of high volatility and flammability. It is an important industrial chemical and its properties are well established. Low concentrations of carbon disulfide naturally discharge into the atmosphere from certain soils, and carbon disulfide has been detected in mustard oil, volcanic gases, and crude petroleum. Carbon disulfide is an unintentional by-product of many combustion and high temperature industrial processes where sulfur compounds are present.

Carbon disulfide was first prepared nearly two hundred years ago by heating sulfur with charcoal. That general approach was the only commercial route to carbon disulfide until processes for reaction of sulfur and methane or other hydrocarbons appeared in the 1950s. Significant commercial production of carbon disulfide began around 1880, primarily for agricultural and solvent applications. Both the physical and chemical properties of carbon disulfide are utilized in industry. Commercial uses grew rapidly from about 1929 to 1970, when the principal applications included manufacturing viscose rayon fibers, cellophane, carbon tetrachloride, flotation aids, rubber vulcanization accelerators, fungicides, and pesticides. Production of carbon disulfide in the United States has declined in recent years. Other chemical fibers and films, as well as environmental and toxicity considerations related to carbon tetrachloride, have had significant impact on the demand for carbon disulfide.

2. Physical Properties

Pure carbon disulfide is a clear, colorless liquid with a delicate etherlike odor. A faint yellow color slowly develops upon exposure to sunlight. Low-grade commercial carbon disulfide may display some color and may have a strong, foul odor because of sulfurous impurities. Carbon disulfide is slightly miscible with water, but it is a good solvent for many organic compounds. Thermodynamic constants (1), vapor pressure (1,2), spectral transmission (3,4), and other properties (1,2,5–7) of carbon disulfide have been determined. Principal properties are listed in Table 1.

Table 1. **Properties of Carbon Disulfide**

Property	Values			References
	General			
melting point, K	161.11			5
latent heat of fusion, kJ/kg[a]	57.7			5
boiling point at 101.3 kPa[b], °C	46.25			2
flash point at 101.3 kPa[b], °C	-30			8
ignition temperature in air, °C				2
10-s lag time	120			
0.5-s lag time	156			
critical temperature, °C	273			2
critical pressure, kPa[b]	7700			2
critical density, kg/m^3	378			2
solubility H2O in CS^2				9
at 10°C, ppm	86			
at 25°C, ppm	142			
dielectric constant	2.641			10
Liquid at temperature, °C	0°C	20°C	46.25°C	
density, kg/m^3	1293	1263	1224	2
specific heat, J/kg·K[a]	984	1005	1030	2
latent heat of vaporization, kJ/kg[a]	377	368	355	1
surface tension, mN/M (=dyn/cm)	35.3	32.3	28.5	2
thermal conductivity, W/m·KW/(m·K)		0.161		2
viscosity, mPa·s(= cP)	0.429	0.367	0.305	2
refractive index, n^D	1.6436	1.6276		11
solubility in water, g/kg soln	2.42	2.10	0.48	2
vapor pressure, kPa[b]	16.97	39.66	101.33	2
Gas at temperature, °C[c]	46.25	200	400°C	
density, kg/m^3	2.97	1.96	1.37	1
specific heat, J/kg·KJ/(kg·K)[a,d]	611	679	730	6
viscosity, mPa·s(= cP)	0.0111	0.0164	0.0234	7
thermal conductivity, W/m·KW/(m·K)	0.0073			7
Thermochemical data at 298 K[a]				
heat capacity, C^{0}p, J/mol·KJ/(mol·K)[a]		45.48		12
entropy, S^0, J/mol·KJ/(mol·K)[a]		237.8		12
heat of formation, H$^0{}_f$, kJ/mol[a]		117.1		12
free energy of formation, G$^0{}_f$, kJ/mol[a]		66.9		12

[a] To convert J to cal, divide by 4.184.
[b] To convert kPa to atm, divide by 101.3
[c] At absolute pressure, 101.3 kPa.
[d] $C_p/C_v = 1.21$ at 100°C (2).

Carbon disulfide is completely miscible with many hydrocarbons, alcohols, and chlorinated hydrocarbons (9,13). Phosphorus (14) and sulfur are very soluble in carbon disulfide. Sulfur reaches a maximum solubility of 63% S at the 60°C atmospheric boiling point of the solution (15). Solubility data for carbon disulfide in liquid sulfur at a CS_2 partial pressure of 101 kPa (1 atm) and a phase diagram for the sulfur–carbon disulfide system have been published (16). Vapor–liquid equilibrium and freezing point data are available for several binary mixtures containing carbon disulfide (9).

Under extremely high pressures of about 5.5 GPa (5.4×10^4 atm) and temperatures up to 175°C, a black, solid form of carbon disulfide has been observed (17).

3. Chemical Properties

The low flash point temperature of -30°C at atmospheric pressure and wide flammability range of carbon disulfide deserve special attention (18). The flash point is lowered if the pressure is decreased or the oxygen content enriched. The flammability limits or explosive ranges depend on conditions of temperature, pressure, and geometry of the enclosure. Flammability limits of 1.06–50.0 vol % carbon disulfide in air are reported for upward propagation and 1.91–35.0 vol % for downward propagation in a 75-mm diameter glass tube (19). The upper flammability limit can be significantly decreased by dilution with carbon dioxide (20). Maximum explosive force occurs at a 4–8% concentration of carbon disulfide in air, at which a maximum absolute pressure increase of 730 kPa (7.2 atm) has been measured (21).

Hot surfaces and electric sparks are potential ignition sources for carbon disulfide. The ignition temperature depends on specific conditions, and values from 90 to 120°C in air have been reported (2,22). Data on carbon disulfide oxidation and combustion have been summarized (18). Oxidation products are generally sulfur dioxide [7446-09-5] and carbon dioxide [124-38-9]:

$$CS_2 + 3\,O_2 \longrightarrow 2\,SO_2 + CO_2 \tag{1}$$

Carbonyl sulfide and carbon monoxide [630-08-0] can also form under certain conditions.

Thermodynamic calculations for reactions forming carbon disulfide from the elements are complicated by the existence of several known molecular species of sulfur vapor (23,24). Thermochemical data have been reported (12). Although carbon disulfide is thermodynamically unstable at room temperature, the equilibrium constant of formation increases with temperature and reaches a maximum corresponding to 91% conversion to carbon disulfide at about 700°C. Carbon disulfide decomposes extremely slowly at room temperature in the absence of oxidizing agents.

Carbon disulfide chemistry is thoroughly described in several publications, which include many references (15,25–28). Several important reactions are mentioned here.

Carbon disulfide is essentially unreactive with water at room temperature, but above about 150°C in the vapor phase some reaction occurs forming carbonyl sulfide (carbon oxysulfide) [463-58-1] and hydrogen sulfide [7783-06-4]. Carbonyl sulfide is an intermediate in the hydrolysis reaction:

$$CS_2 + H_2O \longrightarrow COS + H_2S \tag{2a}$$

$$COS + H_2O \longrightarrow CO_2 + H_2S \tag{2b}$$

At temperatures of 300–600°C in the presence of an activated alumina catalyst, carbon dioxide and hydrogen sulfide are formed in almost quantitative yields (29):

$$CS_2 + 2 H_2O \longrightarrow CO_2 + 2 H_2S \tag{3}$$

This is a desirable side reaction in the first catalytic reactor of the Claus sulfur recovery process.

Carbon disulfide slowly reacts with alkali hydroxides to form trithiocarbonates and alkali carbonates:

$$3 CS_2 + 6 KOH \longrightarrow 2 K_2CS_3 + K_2CO_3 + 3 H_2O \tag{4}$$

Trithiocarbonates can also be prepared from aqueous alkali sulfides:

$$CS_2 + K_2S \longrightarrow K_2CS_3 \tag{5}$$

Industrially important dithiocarbonates (xanthates) result from reaction with various alcoholic alkalies:

$$CS_2 + NaOH + C_2H_5OH \longrightarrow C_2H_5OC(S)SNa + H_2O \tag{6}$$

Of great commercial significance is preparation of sodium cellulose xanthate [9032-37-5] solution (viscose) by reaction with alkali cellulose:

$$CS_2 + R_{cellulose}ONa \longrightarrow R_{cellulose}OC(S)SNa \tag{7}$$

Cellulose is subsequently regenerated from the viscose solution in sulfuric acid and carbon disulfide is liberated. These are the basic steps in manufacturing viscose rayon. The production of regenerated cellulose is estimated to account for more than 75% of the total carbon disulfide consumption worldwide (see FIBERS, REGENERATED CELLULOSICS).

Carbon disulfide and chlorine react in the presence of iron catalysts to give carbon tetrachloride [56-23-5] and sulfur monochloride [10025-67-9]:

$$2 CS_2 + 6 Cl_2 \longrightarrow 2 CCl_4 + 2 S_2Cl_2 \tag{8}$$

This is followed by a second reaction where sulfur monochloride becomes the chlorinating agent:

$$CS_2 + 2 S_2Cl_2 \longrightarrow CCL_4 + 6 S \tag{9}$$

Reactions 8 and 9 have been used in the large-scale production of carbon tet-rachloride since the early 1900s. As a result of decreased demand for carbon tetrachloride, this process is no longer used in the United States (see CARBON TETRACHLORIDE).

If bromine is used in equation 8, carbon tetrabromide [558-13-4] is formed. With a minor amount of iodine present, and in the absence of iron catalyst, carbon disulfide and chlorine react to form trichloromethanesulfenyl chloride (perchloromethyl mercaptan [594-42-3]), CCl_3SCl, which can be reduced with stannous chloride or tin, and hydrochloric acid to form thiophosgene (thiocarbo-nyl chloride [463-71-8], $CSCl_2$, an intermediate in the synthesis of many organic compounds (see SULFUR COMPOUNDS).

Carbon disulfide reacts with concentrated ammonia to give ammonium thiocyanate [1762-95-4] and ammonium trithiocarbonate [13453-08-2] in a reac-tion promoted by alumina catalysts:

$$2\,CS_2 + 4\,NH_3 \longrightarrow NH_4SCN + (NH_4)_2CS_3 \qquad (10)$$

At approximately 160°C, some of the ammonium thiocyanate is converted to thiourea [62-56-6], H_2NCSNH_2, in low yield. With alcoholic ammonia, ammo-nium dithiocarbamate [513-74-6] forms:

$$CS_2 + 2\,NH_3 \longrightarrow NH_2C(S)SNH_4 \qquad (11)$$

Carbon disulfide reacts with primary and secondary amines to yield substituted ammonium salts of N-substituted dithiocarbamic acids, $RNHC(S)SNH_3R$ and $R_2NC(S)SNH_2R_2$:

$$CS_2 + 2\,RNH_2 \longrightarrow RNHC(S)SNH_3R \qquad (12)$$

Industrially important alkali salts result if alkali hydroxide is present, such as the reaction with dimethylamine [124-40-3] forming sodium dimethyl dithiocar-bamate [128-04-1]:

$$CS_2 + (CH_3)_2NH + NaOH \longrightarrow (CH_3)_2NC(S)SNa + H_2O \qquad (13)$$

Analogous reactions form sodium methyldithiocarbamate [137-42-8] from methylamine, and disodium ethylenebis(dithiocarbamate) [142-59-6] from ethyle-nediamine. Iron, manganese, and zinc salts can be prepared from the sodium salts; heavy metals form characteristically colored compounds with dithio-carbamates.

Ethylenediamine reacts with carbon disulfide in alcoholic solution to give ethylenethiourea (2-imidazolidinethione [96-45-7]).

In boiling excess aniline, thiocarbanilide (1,3-diphenyl-2-thiourea [102-08-9]) is formed:

$$CS_2 + 2\ C_6H_5NH_2 \longrightarrow SC(NHC_6H_5)_2 + H_2S \tag{14}$$

If sulfur is present, another industrially important compound results, 2-mercapto benzothiazole (2-benzothiazolethiol [149-30-4]).

Carbonyl sulfide is a coproduct of many carbon disulfide oxidation reactions. Some examples are

$$CS_2 + MgO \longrightarrow COS + MgS \tag{15}$$

$$CS_2 + ClSO_3H \longrightarrow COS + SO_2 + S + HCl \tag{16}$$

$$CS_2 + (NH_2)_2CO \longrightarrow COS + NH_4SCN \tag{17}$$

With alkyl mercuric hydroxide:

$$CS_2 + RHgOH \longrightarrow COS + RHgSH \tag{18}$$

Carbon disulfide reacts with Grignard reagents to prepare the corresponding dithiocarboxylic acids:

$$CS_2 + RMgBr \longrightarrow RC(S)SMgBr \longrightarrow RC(S)SH \tag{19}$$

For example, dithiobenzoic acid [121-68-6] results from the reaction of carbon disulfide, phenyl bromide, ether, and magnesium.

Sodium azidodithiocarbonate [38093-88-8] is prepared by the reaction of aqueous sodium azide [26628-22-8] at 40–50°C:

$$CS_2 + NaN_3 \longrightarrow NaSCSN_3 \tag{20}$$

Sodium azidodithiocarbonate decomposes with evolution of nitrogen gas on addition of iodine, thus providing a useful qualitative test for the presence of residual carbon disulfide in aqueous solutions (25).

Carbon disulfide reacts with alkanols or dialkyl ethers at 250–500°C over activated alumina catalyst to give dialkyl sulfides. For example, methanol yields dimethyl sulfide [75-18-3].

Hydrogen reduces carbon disulfide at high temperatures by the following reactions:

$$CS_2 + 2\ H_2 \rightleftharpoons 2H_2S + C \tag{21}$$

$$CS_2 + 4\ H_2 \rightleftharpoons 2H_2S + CH_4 \tag{22}$$

Hydrogenation at lower temperature and in the presence of catalysts yields organic sulfur compounds. With a reduced nickel catalyst at 180°C, methanedithiol

[6725-64-0] is formed:

$$CS_2 + 2\,H_2 \longrightarrow CH_2(SH)_2 \qquad (23)$$

With a cobalt catalyst at 250°C, methanethiol (methyl mercaptan [74-93-1]) results:

$$CS_2 + 3\,H_2 \longrightarrow CH_3SH + H_2S \qquad (24)$$

Dimethyl sulfide [75-18-3], thioethers, thioformaldehyde [865-36-1], and thiophene [110-02-1] are among other possible carbon disulfide hydrogenation products.

4. Manufacture

The earliest method for manufacturing carbon disulfide involved synthesis from the elements by reaction of sulfur and carbon as hardwood charcoal in externally heated retorts. Safety concerns, short lives of the retorts, and low production capacities led to the development of an electric furnace process, also based on reaction of sulfur and charcoal. The commercial use of hydrocarbons as the source of carbon was developed in the 1950s, and it was still the predominate process worldwide in 1991. That route, using methane and sulfur as the feedstock, provides high capacity in an economical, continuous unit. Retort and electric furnace processes are still used in locations where methane is unavailable or where small plants are economically viable, for example in certain parts of Africa, China, India, Russia, Eastern Europe, South America, and the Middle East. Other technologies for synthesis of carbon disulfide have been advocated, but none has reached commercial significance.

4.1. Charcoal–Sulfur Process. Sulfur vapor reacts with charcoal at temperatures of 750–900°C to form carbon disulfide:

$$C + S_2 \longrightarrow CS_2 \qquad (25)$$

Sulfur vapor is an equilibrium mixture of several molecular species, including S_8, S_6, and S_2. The equilibrium shifts toward S_2 at higher temperatures and lower pressures. The overall reaction is endothermic and theoretically consumes 1950 kJ/kg (466 kcal/kg) of carbon disulfide when the reactants are at 25°C and the products are at 750°C. Most of the heat input goes into dissociation of sulfur vapor to the reactive species, S_2. Equation 25 is slightly exothermic when the reactants are at a constant temperature of 750°C.

Charcoal–sulfur processes need low ash hardwood charcoal, prepared at 400–500°C under controlled conditions. At the carbon disulfide plant site, the charcoal is calcined before use to expel water and residual hydrogen and oxygen compounds. This precalcination step minimizes the undesirable formation of hydrogen sulfide and carbonyl sulfide. Although wood charcoal is preferred, other sources of carbon can be used including coal (30,31), lignite chars (32,33), and coke (34). Sulfur specifications are also important; low ash content is necessary to minimize fouling of the process equipment.

Various reactor designs have been proposed, including fluidized beds (35–37), a whirlpool-type system (38), and a moving refractory bed to superheat the sulfur (39).

Retort Process.   Retorts for producing carbon disulfide are typically oval or cylindrical vessels approximately 1 m in diameter by 3 m high, constructed from chrome alloy steel or cast iron (40,41). Normally one to four retorts are installed in a single furnace, fired with coal, gas, or oil. Alternatively, external electric heaters can be used. The precalcined charcoal is intermittently charged to the top of the retort through a special valve arrangement. Sulfur is added continuously near the bottom of the retort. The sulfur may be first vaporized and superheated to about 700°C in a pipe-coil heat exchanger located in the furnace. Carbon disulfide forms as the sulfur vapor rises through the hot charcoal at 850–900°C. Carbon disulfide, excess sulfur, and other vapors exit from the top of the retort through a duct. Nonreactive ash consolidates with charcoal dust and sifts down to the bottom of the retort from where the residue is periodically removed. Depending on the quality of the raw materials, deposits on the inside walls of the retort must be scraped off approximately monthly. Retorts must be replaced every 1–2 years due primarily to corrosive attack from sulfur vapor. Production capacities are typically up to about 5 tons of carbon disulfide per day per retort with external sulfur vaporization, or 1–3 tons per day with liquid sulfur feed.

The vapor leaving the retort consists of carbon disulfide along with smaller amounts of free sulfur, hydrogen sulfide, and carbonyl sulfide. Sulfur is condensed and recycled. Carbon disulfide is next condensed and then distilled to yield a pure product. In some adaptations, gases leaving the primary condensation are treated by mineral oil absorption to remove residual carbon disulfide selectively, which is later recovered by stripping. The distilled carbon disulfide is treated with lime or dilute sodium hydroxide to neutralize any remaining hydrogen sulfide or other acidic impurities, and the product is finally washed with water. Depending on local circumstances, the tail gas, containing principally hydrogen sulfide, is either flared, incinerated, treated with caustic soda solution to make coproduct sodium hydrogen sulfide [16721-80-5], NaSH, or sent to a Claus sulfur recovery unit. Raw material and energy usages per kg of carbon disulfide product are approximately 0.92–0.95 kg sulfur, 0.22–0.25 kg charcoal, and 8.4–10.0 MJ (2000–2400 kcal) fuel.

Electric Furnace Process.   In this process charcoal and sulfur continuously react in a resistance-type electric furnace (40,42–44). One such furnace described in the literature is a cylindrical, refractory-lined vessel roughly 5 m in diameter by 10 m high. Lump charcoal is fed at the top through a gas-lock valve. Electric current supplied to two or four electrodes in the base of the furnace generates heat in passing through the bed of charcoal between opposing electrodes. The electrodes may either be radially or axially placed in the cylindrical furnace. Liquid sulfur enters the furnace at multiple locations in the sidewall near the base, where it quickly vaporizes and heats to 800–1000°C. Carbon disulfide forms in the lower section of the furnace. As the vapors rise, heat is transferred to incoming charcoal in the upper section. Product purification is by methods similar to those used in the retort process, except that entrained dust is normally removed from the vapor prior to the sulfur condensation step. Because heat is formed internally instead of being conducted through a thick wall, large-capacity reactors are feasible. Vessels can last many years, depending largely on the integrity of the lining. Sulfur and charcoal usages are similar to those of the retort process. Electric power consumption is approximately 4.0–4.8 MJ (1.1 – 1.3 k/W · h) per kg of carbon disulfide, about half the external energy required for the retort process. Electric furnaces were first employed for carbon disulfide around 1900 but did not gain wide acceptance until the 1940s.

4.2. Hydrocarbon–Sulfur Process. The principal commercial hydrocarbon is methane from natural gas, although ethane, and olefins such as propylene (45, 46), have also been used.

Methane [74-82-8] reacts with sulfur essentially without side reactions:

$$CH_4 + 2\,S_2 \longrightarrow CS_2 + 2\,H_2S \tag{26}$$

At 400–700°C, equilibrium exceeds 99.9% (24). About 5–10% excess sulfur is usually maintained in the reaction mixture to promote high methane conversion and to minimize by-product yield. Carbon disulfide is also formed by the following reaction that is 80% complete at equilibrium at 700°C (47):

$$CH_4 + S_2 \longrightarrow CS_2 + 2\,H_2 \tag{27}$$

Reaction 27 is usually negligible in practice because of the less favorable equilibrium and the usual presence of excess sulfur, which favors reaction 26.

Other hydrocarbons can be used. Stoichiometrically, ethane [74-84-0] is preferable to methane since its lower hydrogen/carbon ratio results in a smaller yield of coproduct hydrogen sulfide:

$$2\,C_2H_6 + 7\,S_2 \longrightarrow 4\,CS_2 + 6\,H_2S \tag{28}$$

Propane [74-98-6] and heavier paraffins tend to form undesired products, although reaction conditions can be controlled to minimize coke formation (48).

Propylene [115-07-1] can be used in a properly designed system (45,46):

$$2\,C_3H_6 + 9\,S_2 \longrightarrow 6\,CS_2 + 6\,H_2S \tag{29}$$

Extensive research has been conducted on catalysts that promote the methane–sulfur reaction to carbon disulfide. Data are published for silica gel (49), alumina-based materials (50–59), magnesia (60,61), charcoal (62), various metal compounds (63,64), and metal salts, oxides, or sulfides (65–71). For a silica gel catalyst the rate constant for temperatures of 500–700°C and various space velocities is (72)

$$\log k_c = 10.9 - 131.4/[2.303(RT)] \tag{30}$$

The overall activation energy for equation 30 is 131.4 kJ (31.4 kcal) per mole. Without catalysts, high yields are claimed under certain conditions for using methane (73,74) or olefin (75–77) feedstocks.

For equation 26, starting with methane and solid sulfur at 25°C, and ending with gaseous products at 600°C, the reaction is endothermic and requires 2.95 MJ/kg (705 kcal/kg) of CS_2. The reaction of methane and sulfur vapor in the diatomic form is actually exothermic (23,78). Superheating of the sulfur is claimed to be preferable (79), and series operation of reactors offers a means of reducing process temperatures at which the sulfur dissociates (80).

A disadvantage of the hydrocarbon–sulfur process is the formation of one mole of hydrogen sulfide by-product for every two atoms of hydrogen in the hydrocarbon. Technology for efficient recovery of sulfur values in hydrogen sulfide became commercially available at about the same time that the methane–sulfur process was developed. With an efficient Claus sulfur recovery unit, the hydrocarbon–sulfur process is economically attractive.

In a modern carbon disulfide plant, all operations are continuous and under automatic control. On-stream times in excess of 90% are obtainable. The process is in three steps: melting and purification of sulfur; production and purification of carbon disulfide; and recovery of sulfur from by-product hydrogen sulfide (50).

High purity sulfur with low ash and organic content is desirable. If the raw sulfur is in solid form, it is first melted and filtered or otherwise treated for purification. Liquid sulfur must be handled between the melting point (132°C) and about 150°C to avoid its peculiar high viscosity range (see SULFUR).

The hydrocarbon gas feedstock and liquid sulfur are separately preheated in an externally fired tubular heater. When the gas reaches 480–650°C, it joins the vaporized sulfur. A special venturi nozzle can be used for mixing the two streams (81). The mixed stream flows through a radiantly-heated pipe coil, where some reaction takes place, before entering an adiabatic catalytic reactor. In the adiabatic reactor, the reaction goes to over 90% completion at a temperature of 580–635°C and a pressure of approximately 250–500 kPa (2.5–5.0 atm). Heater tubes are constructed from high alloy stainless steel and reportedly must be replaced every 2–3 years (79, 82–84). Furnaces are generally fired with natural gas or refinery gas, and heat transfer to the tube coil occurs primarily by radiation with no direct contact of the flames on the tubes. Design of the furnace is critical to achieve uniform heat around the tubes to avoid rapid corrosion at "hot spots."

The reaction products are cooled to about 132°C in a waste-heat boiler before going to a sulfur separator where recirculating liquid sulfur scrubs the incoming gases (85). The collected sulfur is recycled to the reactor heater. The sulfur-free process stream is cooled to about 38°C prior to entering a counter-current column where carbon disulfide is absorbed in a circulating oil stream and hydrogen sulfide goes overhead. Carbon disulfide is recovered from the oil in a separate stripping tower, and the lean oil is cooled and recycled. The carbon disulfide stream from the top of the stripping tower contains hydrogen sulfide and oil, which are removed in a series of two fractional distillation towers to produce the final carbon disulfide product. The liquid CS_2 product is sometimes washed with water or dilute alkali to neutralize traces of hydrogen sulfide, but in a well-controlled plant this precautionary step can often be omitted.

The gas leaving the absorber contains mainly hydrogen sulfide with small amounts of methane, carbon disulfide, and inerts (nitrogen). That stream is sent to a Claus sulfur recovery unit (see SULFUR REMOVAL AND RECOVERY). In the Claus unit, part of the hydrogen sulfide is burned in air to form sulfur dioxide and water. The sulfur dioxide and the remaining hydrogen sulfide are then catalytically converted to elemental sulfur and water. Sulfur is condensed in one or more waste-heat boilers and recycled as liquid. The Claus unit must be specially designed to handle high concentration (90+%) of hydrogen sulfide. Gas leaving the Claus unit may be incinerated or subjected to tail gas treatment, depending on the desired efficiency or local environmental conditions.

Water that has been in contact with carbon disulfide can be steam-stripped to 5 ppm or less carbon disulfide before discharge (86). Safety release vents throughout the process are routed to a flare. Gas displaced from vessels and other equipment containing carbon disulfide can be incinerated, or the carbon disulfide can be collected by oil absorption or active carbon adsorption. Mechanical safety features in sections of the plant handling liquid carbon disulfide can include centrifugal pumps having double mechanical seals with oil seal fluid, or long-shaft vertical pumps. Special canned pumps have also been used. Oversized electrical equipment can be used to minimize hot surfaces that could ignite carbon disulfide vapor (86).

Raw material usages per ton of carbon disulfide are approximately 310 m^3 of methane, or equivalent volume of other hydrocarbon gas, and 0.86–0.92 ton of sulfur (87,88), which includes typical Claus sulfur recovery efficiency. Fuel usage, as natural gas, is about 180 m^3/ton carbon disulfide excluding the fuel gas assist for the incinerator or flare. The process is a net generator of steam; the amount depends on process design considerations.

Several variations of the basic process are possible (89–91). One proposed simplification is elimination of the oil absorption and stripping steps by operating the reactor at a higher pressure of 1000–2000 kPa (10–20 atm) and with a reaction zone space velocity of about 1000 h^{-1} (92). Gas leaving the reactor sulfur separator goes directly to the distillation columns where hydrogen sulfide and other light-boiling compounds are taken overhead in the first tower, and carbon disulfide bottoms are then distilled in the second column. Alternatively, the reactor can be operated at lower pressure with the product gas compressed prior to entering the recovery section (93).

4.3. Other Processes. Sulfur vapor reacts with other hydrocarbon gases, such as acetylene [74-86-2] (94) or ethylene [74-85-1] (95), to form carbon disulfide. Higher hydrocarbons can produce mercaptan, sulfide, and thiophene intermediates along with carbon disulfide, and the quantity of intermediates increases if insufficient sulfur is added (96). Light gas oil was reported to be successful on a semiworks scale (97). In the reaction with hydrocarbons or carbon, pyrites can be the sulfur source. With methane and iron pyrite the reaction products are carbon disulfide, hydrogen sulfide, and iron or iron sulfide. Pyrite can be reduced with carbon monoxide to produce carbon disulfide.

The reaction of hydrogen sulfide and methane has a calculated equilibrium of 67% at 1100°C and 86.5% at 1288°C (47,98):

$$CH_4 + 2\,H_2S \rightleftharpoons CS_2 + 4\,H_2. \tag{31}$$

Side reactions reduce the yield (99). Proposed processes for obtaining carbon disulfide from hydrogen sulfide and methane include a high temperature plasma (100) and low temperature operation with a catalyst and oxygen (101).

Hydrogen sulfide and carbon react at 900°C to give a 70% yield of carbon disulfide (102,103). A process for reaction of coke and hydrogen sulfide or sulfur in an electric-resistance-heated fluidized bed has been demonstrated on a laboratory scale (104). Hydrogen sulfide also forms carbon disulfide in reactions with carbon monoxide at 600–1125°C (105) or carbon dioxide at 350–450°C in the presence of catalysts (106).

Sulfur dioxide [7446-09-5] and methane react to form carbon disulfide in a yield of 84% at 850°C in the presence of certain catalysts (107). Sulfur dioxide and anthracite at 900–1000°C produce very high yields (108).

Carbonyl sulfide can be either a starting or intermediate material (108–110), or it can be used as a fluidizing gas in a carbon fluid-bed process (111). Making carbon disulfide from boiler flue gas by catalytically reducing SO_2 with CO to COS, and then converting COS to CS_2 over an alumina catalyst has been proposed (112).

5. Handling, Shipment, and Storage

Transportation of carbon disulfide is controlled by federal regulations (113). Acceptable shipping containers include drums, tank trucks, special portable tanks, and rail tank cars. Barges have been used in the past. The United States Department of Transportation classifies carbon disulfide as a flammable liquid and a poison. For ship transport, carbon disulfide must be marked as a marine pollutant (114). All air transport, cargo, or passenger, is forbidden (115).

Carbon disulfide is normally stored and handled in mild steel equipment. Tanks and pipes are usually made from steel. Valves are typically cast-steel bodies with chrome steel trim. Lead is sometimes used, particularly for pressure relief disks. Copper and copper alloys are attacked by carbon disulfide and must be avoided. Carbon disulfide liquid and vapor become very corrosive to iron and steel at temperatures above about 250°C. High chromium stainless steels, glass, and ceramics may be suitable at elevated temperatures.

Contact of carbon disulfide with air should be avoided because the combination of high volatility, wide flammability range, and low ignition temperature results in a readily combustible mixture (116). Carbon disulfide must be stored in inert-blanketed, closed tanks. Normally carbon disulfide is transferred from vessels or tank cars through a downpipe by displacement with an inert material such as water or nitrogen (8,117). Direct pumping requires special equipment and precautions, hence that method of transfer is less common. Carbon disulfide is normally padded with water in storage tanks and bulk containers. Nitrogen blanketing has been used in cold climates. A combination of water with nitrogen above it can also be used, but in any case the total space not occupied by liquid carbon disulfide must be filled with inert material. The tanks themselves should be underwater or surrounded by dikes capable of holding the total tank contents. All equipment containing carbon disulfide should be located well away from potential sources of ignition, which include open flames, frictional heat, sparks, exposed electric light bulbs, and bare steam pipes. Good ground or floor-level ventilation should be provided because carbon disulfide vapor is heavier than air and can accumulate in low areas. Each piece of stationary equipment should be individually grounded by easily-visible conductors. Automatic monitoring of grounding continuity, with electrical detection and alarms, is recommended.

Temporary grounding connections should be provided for movable equipment, such as drums or tank cars. Drums should be kept in a shaded but ventilated area with provision for cooling by water spray or other means if temperature exceeds ca 30°C. All of the equipment and operations preferably should be outdoors.

Small carbon disulfide fires can be smothered with carbon dioxide. Large fires can be controlled with certain types of foams or by a fog or spray of water with attention to proper impoundment of the contaminated water runoff. Without containment, the runoff may transport toxic fumes and fire or explosion hazards to other areas, such as sewers. Caution should be used in fighting a carbon disulfide fire because the flame is nearly invisible and a product of combustion is toxic sulfur dioxide. Special fire-fighting procedures have been published (118,119). Guidelines for emergency response to leaks, spills, and fires are available from the Department of Transportation (120). Cleanup of spills or disposal of wastes containing carbon disulfide must be managed in accordance with government regulations (22,121,122).

6. Economic Aspects

Depending on energy and raw material costs, the minimum economic carbon disulfide plant size is generally in the range of about 2000–5000 tons per year for an electric furnace process and 15,000–20,000 tons per year for a hydrocarbon-based process. A typical charcoal–sulfur facility produces approximately 5000 tons per year. Hydrocarbon–sulfur plants tend be on the scale of 50,000–200,000 tons per year. The production capacities of known U.S. hydrocarbon–sulfur based plants are listed in Table 2 (123).

Demand for carbon disulfide was 71×10^3 t in 2001. Projected demand for 2005 is 71×10^3 t, ie, 0% growth is expected. Demand equals production plus imports (in 2001, 0.9×10^3 t) less exports (in 2001, 6.8×10^3 t). For the period 1996–2001, prices range from \$465/t to \$485/t (123).

Production of carbon disulfide expanded rapidly after World War II to supply the growing needs of the viscose rayon industry, which consumes about 0.31 ton CS_2 per ton rayon. The high plant capacities obtainable with the methane–sulfur route resulted in consolidation of the carbon disulfide industries in the United States and Western Europe, where a few producers now account for the bulk of the capacity. Some rayon manufacturers produce their own carbon disulfide. Rayon enjoys an extensive international market that can affect local

Table 2. **U.S. Manufacturers of Carbon Disulfide**[a]

Producer	Location	Estimated annual capacity, 10^3 t
Akzo Noble Chemicals	Axis, Ala.	113
PPG Industries	Natrium, W. Va.	27
Auto Fina Chemicals	Houston, Texas	18
Total		*158*

[a] Ref. 123.

CS_2 manufacturers. Competition from noncellulosic synthetic fibers has caused a drop in rayon production in the United States since the mid-1960s. One rayon plant in the United States closed in 1989 as a result of environmental concerns. This pattern of modern viscose rayon plants replacing aging facilities that cannot be economically upgraded is apt to be repeated in other parts of the world. In a development that could have far-ranging implications, a viscose rayon producer is constructing a solvent spun cellulosic fiber plant using an amine oxide solvent rather than carbon disulfide (124,125).

Demand for carbon disulfide has apparently bottomed out at approx. 72.5×10^3 t over the last few years. Rayon is CS's largest market, but represents only 4% of the synthetic fiber market. This usage seems to stable.

Carbon disulfide is used in the manufacture of rubber vulcanization accelerators. Production of these accelerators required 12.7×10^3 t in 2001. Demand in agricultural chemicals is increasing slightly because of the use of Metam sodium as a replacement for methyl bromide.

Use of carbon disulfide for manufacture of cellophane [9005-81-6] had dropped dramatically because of competition from plastic films. The decline has stabilized. Cellophane remains strong in speciality applications like hard candy and cigar wrappings.

Carbon disulfide for manufacture of carbon tetrachloride increased in the 1950s and 1960s to supply the key raw material for chlorofluorocarbon refrigerants and aerosol propellants. Because of ecological and health concerns, carbon tetrachloride consumption began to decline in the mid-1970s. That use for carbon disulfide will suffer under a United Nations proposal to phase out carbon tetrachloride and chlorofluorocarbons to protect the earth's ozone layer (126). During 1991 the only remaining carbon tetrachloride plant in the United States that employed the carbon disulfide route was permanently shut down.

7. Specifications and Quality Control

Modern plants generally produce carbon disulfide of about 99.99% purity. High product quality is ensured by closely controlled continuous fractional distillation. Reagent and U.S. Federal specifications, and typical commercial-grade quality are listed in Table 3.

8. Health, and Safety Factors

Care must be exercised in handling carbon disulfide because of both health concerns and the danger of fire or explosions. Occupational exposure potentially may involve as many as 20,000 workers in the United States (129). Ingestion is rare, but a 10 mL dose can prove fatal (130). Contact usually occurs by inhalation of vapor. However, vapor and liquid can be absorbed through intact skin and poisoning may occur by the dermal route. Repeated contact of liquid carbon disulfide with the skin can cause inflammation and cracking because carbon disulfide removes protective waxes and oils. Extended skin contact results in blistering and possibly second- and third-degree burns. Precautions should be taken to

Table 3. **Carbon Disulfide Specifications**

Property	Method	Technical industry, typical	Technical, U.S. Federal[a]	Reagent, ACS[b]
specific gravity	pycnometer	1.270–1.272[c]	1.262–1.267[d]	
residue	dry at 60°C	0.002% max	10 mg/100 mL	0.002% max
color	APHA, Pt–Co 500 std	<20[e]	special test	10 max
boiling range, °C	ACS distillation	45.5–47.5	45.5–47.5	1°C incl. 46.3±0.1°C
foreign sulfide	lead acetate test	negative		
foreign sulfide	special tests		no discoloration of copper	
H$_2$S and SO$_2$	iodine color test			passes
water, %	Karl Fischer		no turbidity	0.05% max

[a] Ref. 127.
[b] Ref. 128. USP specifies ACS reagent-grade.
[c] 15/4°C.
[d] 20/20°C.
[e] Light transmission vs water = 98% minimum is sometimes used as a specification.

avoid breathing of vapors or mists that may contain carbon disulfide. Contact with skin or eyes should also be avoided, and adequate safety gear should be worn, including goggles, impervious gloves, and appropriate clothing. Contact lenses should be removed before going into any area where exposure to carbon disulfide might occur. A chemical cartridge respirator with an organic vapor cartridge can offer protection up to 50 ppm carbon disulfide in air. Above that level or in entering areas of unknown vapor concentrations, a self-contained breathing apparatus should be worn, with careful attention given to possible explosion hazards (131).

The odor threshold of carbon disulfide is about 1 ppm in air but varies widely depending on individual sensitivity and purity of the carbon disulfide. However, using the sense of smell to detect excessive concentrations of carbon disulfide is unreliable because of the frequent co-presence of hydrogen sulfide that dulls the olfactory sense.

Immediate effects of overexposure to carbon disulfide vapors range from headache, dizziness, nausea, and vomiting to life-threatening convulsions, unconsciousness, and respiratory paralysis. For an exposure time of 30 min, 1150 ppm carbon disulfide in air results in serious symptoms, 3210 ppm is dangerous to life, and 4815 ppm is fatal (132). Prolonged and repeated exposure to carbon disulfide vapor can affect both the central and peripheral nervous systems. Manifestations of long-term overexposure may include headache, vertigo, irritability, nervousness, depression, mental derangement, memory loss, muscular weakness, fatigue, insomnia, eating disorders, gastrointestinal disturbances, impaired vision, diminished reflexes, numbness, and difficulty walking. Certain workers exposed to a time-weighted average of 11 ppm experienced headaches and dizziness, and those with an average of 186 ppm had additional complaints of nervousness, fatigue, sleep problems, and weight loss (133). Repeated exposure to relatively high concentrations of carbon disulfide has long been known to cause serious neurological and psychological impairments. In recent years, previously unrecognized and more subtle toxic effects of repeated lower level exposures became evident. This led OSHA in 1989 to reduce permissible concentration limits to 4 ppm (12 mg/m^3) maximum time-weighted average for 8-h exposure and 12 ppm (36 mg/m^3) maximum for 15 min short-term exposure (130). Compliance with both limits is preferably achieved by engineering and work practice controls, although respirators are acceptable in certain specific operations. OSHA states the new limits should substantially reduce the risk of both cardiovascular disease and adverse reproductive effects associated with carbon disulfide. Analysis of urine specimens for carbon disulfide metabolites by an iodine–azide test (134) and other methods (135) can indicate overexposure.

Health hazards linked to carbon disulfide are extensively covered (128). Also available are epidemiological studies (136–138), general reviews containing many references (139–142), and a Material Safety Data Sheet (143).

A method for the remediation of soil containing carbon disulfide via oxidation has been reported (144).

9. Uses

United States applications of carbon disulfide are rayon production (44%), agriculture and other chemicals (35%), rubber chemicals (18%), cellophane and other regenerated cellulosics (3%) (123).

Carbon disulfide is used to make intermediates in the manufacture of rubber vulcanization accelerators, including MTB (2-mercaptobenzothiazole [149-30-4]), thiocarbanilide, and dimethyl dithiocarbamate salts (see RUBBER CHEMICALS). Thiocarbanilide is also used in dyes. Thiophosgene (thiocarbonyl chloride), made from carbon disulfide, is a useful intermediate in the synthesis of many organic sulfur compounds. Xanthates (dithiocarbonates) produced from carbon disulfide are widely employed as flotation chemicals for metal sulfide ores in the mining industry. The solvent properties of carbon disulfide find a wide range of industrial uses, including various dehydration, extraction, reaction, and separation applications. A useful laboratory chemical, carbon disulfide is a reactant in the synthesis of many compounds and a solvent in Friedel-Crafts reactions. Its solvent properties are useful in spectroscopy and in solubilizing phosphorus and sulfur.

Pharmaceutical intermediates, such as thiocarbanilide and thiocyanates, are prepared from carbon disulfide. Methionine [59-51-8], an essential amino acid, is manufactured from carbon disulfide intermediates. At one time carbon disulfide was commonly used in combination with chlorinated hydrocarbons as a grain fumigant, but that application was suspended in 1985 by the United States Environmental Protection Agency. Carbon disulfide is a starting material for several fungicides, soil fumigants, and insecticides or their intermediates, such as trichloromethanesulfenyl chloride, disodium ethylenebis(dithiocarbamate), sodium methyldithiocarbamate, ammonium dithiocarbamate, and thiocyanate salts. The thiocyanates also have many nonagricultural uses.

BIBLIOGRAPHY

"Carbon Disulfide" in *ECT* 1st ed., Vol. 3, pp. 142–148, by H. O. Folkins, The Pure Oil Company; in *ECT* 2nd ed., Vol. 4, pp. 370–385, by H. O. Folkins, The Pure Oil Company; *ECT* 3rd ed., Vol. 4, pp. 742–757, by R. W. Timmerman, FMC Corporation; in *ECT* 4th ed., Vol. 5, pp. 53–76, by David E. Smith and Robert W. Timmerman, FMC Corp.; "Carbon Disulfide" in *ECT* (online), posting date: December 4, 2000, by David E. Smith and Robert W. Timmerman, FMC Corporation.

CITED PUBLICATIONS

1. L. J. O'Brien and W. J. Alford, *Ind. Eng. Chem.* **43**, 506 (1951).
2. E. W. Washburn, ed., *International Critical Tables*, Vol. 3, McGraw-Hill Book Company, New York, 1928, pp. 23, 213, 231, 248; Vol. 4, p. 447; Vol. 5, pp. 114, 215, 227; Vol. 7, p. 213.
3. B. J. Zwolinski and co-workers, *Catalog of Selected Ultraviolet Spectral Data, Serial No. 100*, Manufacturing Chemists' Association Research Project, Texas A & M University, College Station, Tex., 1965.
4. B. J. Zwolinski and co-workers, *Catalog of Selected Infrared Spectral Data, Serial No. 321*, Thermodynamics Research Center Data Project, Texas A & M University, College Station, Tex., 1967.
5. O. L. I. Brown and G. G. Manov, *J. Am. Chem. Soc.* **59**, 500 (1937).
6. K. A. Kobe and E. G. Long, *Petrol. Refin.* **29**(1), 126 (1950).
7. R. W. Gallant, *Hydrocarbon Process.* **49**(4), 132 (1970).
8. *Carbon Disulfide. Chemical Safety Data Sheet SD-12*, Manufacturing Chemists' Association, Washington, D.C., 1967.
9. A. Seidell, *Solubilities of Organic Compounds*, Vol. 1, D. Van Nostrand, New York, 1941, pp. 238, 584; Vol. 2, 10–11.
10. S. Budavari, ed., *The Merck Index*, 11th ed., Merck & Co., Rahway, N.J., 1989, p. 1821.
11. *Beilstein Handbuch der Organischen Chemie*, Vol. 3, J. Springer, Berlin, Germany, 1921, p. 198.
12. D. R. Stull, E. F. Westrum, and G. C. Sinke, *The Chemical Thermodynamic Properties of Organic Compounds*, John Wiley & Sons, Inc., New York, 1969, p. 220.
13. A. Seidell and W. F. Linke, *Solubilities of Inorganic and Organic Compounds*, D. Van Nostrand, New York, 1952, p. 101.
14. *Supplement to Mellor's* Comprehensive Treatise on Inorganic and Theoretical Chemistry, Vol. VIII, Suppl. III, *Phosphorus*, Longman Group Ltd., London, 1971, 172–173.
15. *CS2*, brochure, Stauffer Chemical Co., Westport, Conn., 1975.
16. F. J. Touro and T. W. Wiewiorski, *J. Phys. Chem.* **70**, 3534 (1968).
17. E. G. Butcher, J. A. Weston, and H. Gebbie, *J. Chem. Phys.* **41**, 2554 (1964).
18. J. H. Meidl, *Flammable Hazardous Materials*, Glencoe Press, Beverly Hills, Calif., 1970, pp. 25–40, 173–178.
19. A. G. White, *J. Chem. Soc.* **121**, 1244 (1922).
20. G. Peters and W. Ganter, *Angew. Chem.* **51**, 29 (1938).
21. C. Bondroit, *Rev. Universelle Mines Metall Mec.* **15**, 197 (1939).
22. *Emergency Action Guides, Carbon Bisulfide*, Association of American Railroads, Bureau of Explosives, Washington, D.C., pp. C2.1(1–6), 1988.
23. K. K. Kelley, *Contributions to the Data on Theoretical Metallurgy, U.S. Bureau of Mines Bulletin 406*, 1937, Chapt. VII.

24. D. R. Stull, *Ind. Eng. Chem.* **41**, 1968 (1949).

25. A. D. Dunn and W. D. Rudorf, *Carbon Disulphide in Organic Chemistry*, Ellis Horwood Ltd., Chichester, UK; Halsted Press, div. of John Wiley & Sons, New York, 1989.

26. M. Yokoyama and T. Imamoto, *Synthesis* (10), 797–824 (Oct. 1984).

27. G. Gattow and W. Behrendt, *Topics in Sulfur Chemistry, Vol. 2: Carbon Sulfides and Their Inorganic and Complex Chemistry*, Georg Thieme, Stuttgart, Germany, 1977.

28. F. Thoemel, *Chem.-Ztg.* **11**(10), 285–296 (1987).

29. R. F. Bacon and E. S. Boe, *Ind. Eng. Chem.* **37**, 469 (1945).

30. R. M. Levit and co-workers, *Khim. Technol. Serougleroda* **1970**(2), 63 (1972).

31. F. Molyneux, *Ind. Eng. Chem.* **54**(7), 50 (1962).

32. E. A. Sondreal, A. M. Cooley, and R. C. Ellman, *U.S. Bur. Mines Rep. Invest*, 6891 (1967).

33. Russ. Pat. 139,310 (Sept. 4, 1960), I. Z. Sorokin and co-workers.

34. Jpn. Pat. 7,228,319 (July 27, 1972), Y. Harada and M. Miyamoto (to Asahi Chemical Industry Co. Ltd.).

35. U.S. Pat. 2,700,592 (Jan. 25, 1955), T. D. Heath (to Dorr Co.).

36. U.S. Pat. 2,443,854 (June 22, 1948), R. P. Ferguson (to Standard Oil Development Co.).

37. Brit. Pat. 833,562 (Apr. 27, 1960), H. S. Johnson and J. Reid (to Shawinigan Chemicals Ltd.).

38. H. Sperling, *Chem. Tech. Berlin* **8**, 405 (1956).

39. Brit. Pat. 642,557 (Sept. 6, 1950), (to The Dow Chemical Company).

40. J. F. Thorpe and M. A. Whiteley, *Thorpe's* Dictionary of Applied Chemistry, 4th ed., Vol. II, Longmans, Green and Co., London, 1938, 328–344.

41. C. M. Thacker, *Hydrocarbon Process.* **49**(4), 124–128; **49**(5) 137–139 (1970).

42. E. R. Taylor, *Trans. Am. Electrochem. Soc.* **1**, 115–117 (1902).

43. *Ibid.* **2**, 185–188 (1902).

44. *Chem. Eng.*, 174–177 (Jan. 1951).

45. *Sulphur* **96**, 45 (1971).

46. U.S. Pat. 2,369,377 (Feb. 13, 1945), C. M. Thacker (to The Pure Oil Co.).

47. H. O. Folkins, E. Miller, and H. Hennig, *Ind. Eng. Chem.* **42**, 2202 (1950).

48. U.S. Pat. 3,927,185 (Dec. 16, 1975), M. Meadow and S. Berkowitz (to FMC Corp.).

49. R. C. Forney and J. M. Smith, *Ind. Chem. Chem.* **43**, 1841 (1951).

50. U.S. Pat. 2,568,121 (Sept. 18, 1951), H. O. Folkins, C. A. Porter, E. Miller, and H. Hennig (to The Pure Oil Co.).

51. C. M. Thacker and E. Miller, *Ind. Eng. Chem.* **36**, 182 (1944).

52. Y. P. Tret'yakov, *Zh. Chim. Abstr.*, 6L47 (1962).

53. U.S. Pat. 2,330,934 (Oct. 5, 1943), C. M. Thacker (to The Pure Oil Co.).

54. U.S. Pat. 2,428,272 (Oct. 7, 1947), C. M. Thacker (to The Pure Oil Co.).

55. U.S. Pat. 2,492,719 (Dec. 27, 1949), C. M. Thacker (to The Pure Oil Co.).

56. U.S. Pat. 2,666,690 (Jan. 19, 1954), H. O. Folkins, E. Miller, and H. Hennig (to FMC Corp.).

57. U.S. Pat. 2,709,639 (May 31, 1955), H. O. Folkins, E. Miller, and H. Hennig (to FMC Corp.).

58. U.S. Pat. 2,565,215 (Aug. 21, 1951), H. O. Folkins and E. Miller (to The Pure Oil Co.).

59. R. A. Fisher and J. M. Smith, *Ind. Eng. Chem.* **42**, 704 (1950).

60. U.S. Pat. 2,616,793 (Nov. 4, 1952), H. O. Folkins and E. Miller (to FMC Corp.).

61. U.S. Pat. 2,712,982 (July 12, 1955), K. W. Guebert (to The Dow Chemical Company).

62. Belg. Pat. 630,584 (Aug. 1, 1963), M. Preda.

63. U. Sborgi and E. Giovanni, *Chim. Ind.* (Milan) **31**, 391; Ital. Pat. 457,263 (May 12, 1950).

64. F. Giovanni, *Ann Chim. Appl.* **39**, 671 (1949).
65. U.S. Pat. 2,536,680 (Aug. 21, 1951), H. O. Folkins and E. Miller (to The Pure Oil Co.).
66. U.S. Pat. 2,668,752 (Feb. 9, 1954), H. O. Folkins and E. Miller (to FMC Corp.).
67. U.S. Pat. 2,411,236 (Nov. 19, 1946), C. M. Thacker (to The Pure Oil Co.).
68. U.S. Pats. 2,712,984, 2,712,985 (July 12, 1955), K. W. Guebert (to The Dow Chemical Company).
69. U.S. Pat. 2,187,393 (Jan. 16, 1940), M. DeSimo (to Shell Development Co.).
70. W. J. Thomas and B. John, *Trans. Inst. Chem. Eng.* **45**(3), T119 (1967); *Chem. Eng. (London)*, 207 (1967).
71. W. J. Thomas and S. C. Naik, *Trans. Inst. Chem. Eng.* **48**(4–6), 129 (1970).
72. G. W. Nabor and J. M. Smith, *Ind. Eng. Chem.* **45**, 1272 (1953).
73. U.S. Pat. 2,882,130 (Apr. 14, 1959), D. J. Porter (to FMC Corp.).
74. U.S. Pat. 3,087,788 (Apr. 30, 1963), D. J. Porter (to FMC Corp.).
75. U.S. Pat. 3,436,181 (Apr. 1, 1969), J. Berthoux, J. P. Quillet, and G. Schneider (to Société Progil).
76. Fr. Pat. 1,493,586 (Sept. 1, 1967), J. Berthoux, J. P. Quillet, and G. Schneider (to Société Progil).
77. U.S. Pat. 3,699,215 (June 13, 1972), P. Gerin, L. Louat, and J. P. Quillet (to Société Progil).
78. J. R. West, *Ind. Eng. Chem* **42**, 713 (1950).
79. U.S. Pat. 2,857,250 (Oct. 21, 1958), R. W. Timmerman, A. G. Draeger, and J. W. Getz (to FMC Corp.).
80. U.S. Pat. 2,882,131 (Apr. 14, 1959), J. W. Getz and R. W. Timmerman (to FMC Corp.).
81. U.S. Pat. 3,876,753 (Apr. 8, 1975), J. L. Manganaro, M. Meadow, and S. Berkowtiz (to FMC Corp.).
82. H. W. Haines, Jr., *Ind. Eng. Chem.* **55**, 44–46 (1963).
83. M. D. S. Lay, M. W. Sauerhoff, and D. R. Saunder, in W. Gerhartz, ed., *Ullmann's Encyclopedia of Industrial Chemistry*, 5th ed., VCH Verlagsgesellschaft, Germany, 1986, pp. 185–195.
84. U.S. Pat. 2,661,267 (Dec. 1, 1953), H. O. Folkins, E. Miller, and H. Hennig (to FMC Corp.).
85. P. W. Sherwood, *World Petrol.* **31**, 62 (1960).
86. J. Kuhn and W. Stehney, *Sulphur* **151**, 30–32 (1980).
87. *Sources and Production Economics of Chemical Products*, 2nd ed., McGraw-Hill, New York, 1979, 142–144.
88. W. L. Faith, D. B. Keyes, and R. L. Clark, *Industrial Chemicals*, 3rd ed., John Wiley & Sons, Inc., New York, 1965, 223–228.
89. *Sulphur* **76**, 28–32 (May/June 1967).
90. P. W. Sherwood, *World Petrol.* **31**, 62–68 (1960).
91. U.S. Pat. 2,708,154 (May 10, 1955), H. O. Folkins and E. Miller (to FMC Corp.).
92. U.S. Pat. 3,079,233 (Feb. 26, 1963), C. J. Wenzke (to FMC Corp.).
93. U.S. Pat. 3,250,595 (May 10, 1966), D. R. Olsen (to FMC Corp.).
94. Brit. Pat. 265,994 (Feb. 15, 1926), J. Komlos.
95. U.S. Pat. 1,907,274 (May 2, 1933), T. S. Wheeler and W. Francis (to Imperial Chemical Industries Ltd.).
96. H. E. Rasmussen, R. C. Hansford, and A. N. Sachanen, *Ind. Eng. Chem.* **38**, 376 (1946).
97. C. M. Thacker, *Hydrocarbon Process.* **49**(5), 137–139 (1970).
98. H. I. Waterman and C. Van Vlodrop, *J. Soc. Chem. Ind. (London)* **58**, 109 (1939).
99. U.S. Pat. 2,468,904 (May 3, 1949), C. R. Wagner (to Phillips Petroleum Co.).
100. *Sulphur* **74**, 39 (1968).

101. Russ. Pat. 458,204 (Sept. 7, 1982), S. V. Gerei and co-workers.
102. U.S. Pat. 1,193,210 (Aug. 1, 1916), A. Walter.
103. Brit. Pat. 314,060 (June 23, 1928), H. Oehne (to Chemische Fabrik Kalk Ges.).
104. U.S. Pat. 3,009,781 (Nov. 21, 1961), H. S. Johnson and A. H. Anderson (to Shawinigan Chemicals).
105. U.S. Pat. 2,767,059 (Oct. 16, 1956), W. E. Adcock and W. C. Lake (to Stanolind Oil and Gas Co.).
106. Fr. Pat. 1,274,034 (Oct. 20, 1961), (to Hamburger Gaswerke GmbH and Salzgitter Industriebau GmbH).
107. N. P. Galenko and co-workers, *Gaz. Prom.* **5**(12), 46 (1960).
108. C. W. Siller, *Ind. Eng. Chem.* **40**, 1227 (1948).
109. A. Stock, W. Siecke, and E. Pohlard, *Ber. Dtsch. Chem. Ges. B* **57**, 719 (1924).
110. Ger. Pat. 938,124 (Jan. 26, 1955), H. Welz (to Farbenfabriken Bayer).
111. U.S. Pat. 4,695,443 (Sept. 2, 1987), A. M. Leon (to Stauffer Chemical Co.).
112. U.S. Pat. 4,122,156 (Oct. 24, 1978), J. R. Kittrell and C. W. Quinlan (to New England Power Co.; Northeast Utilities Service Co.).
113. *Code of Federal Regulations*, Title 49, U.S. Department of Transportation, Washington, D.C., 1990; *Fed. Reg.* **55**, 246 (Dec. 21, 1990).
114. *International Maritime Dangerous Goods Code*, Vol. II, International Maritime Organization, London, 1989.
115. *IATA Dangerous Goods Regulations*, 31st ed., International Air Transport Association, Montreal, Canada, 1990.
116. *Carbon Disulfide: Properties, Handling and Storage*, FMC Corp., Princeton, N.J., 1982.
117. *Unloading Flammable Liquids from Tank Cars, Manual Sheet TC-4*, Manufacturing Chemists' Association, Washington, D.C., 1969.
118. *Fire Protection Guide On Hazardous Materials*, 9th ed., National Fire Protection Association, Boston, Mass., 1986, pp. 325M-24, 491-M47.
119. *Hazardous Chemical Data*, Vol. II, U.S. Coast Guard, Department of Transportation, Government Printing Office, Washington, D.C., 1984.
120. *Emergency Response Guidebook*, Department of Transportation, Government Printing Office, Washington, D.C., 1987.
121. *Engineering Handbook For Hazardous Waste Incineration*, United States Environmental Protection Agency 68-03-3025, Washington, D.C., 1981, 3–8.
122. *Treatment And Disposal Methods for Waste Chemicals*, United Nations, Geneva, Switzerland, Dec. 1985.
123. "Carbon Disulfide," Chemical Profile, *Chemical Market Reporter*, August 5, 2002.
124. *Chem. Mark. Rep.*, 7 (May 28, 1990).
125. *Eur. Chem. News*, 33 (May 28, 1990).
126. *Eur. Chem. News*, 6 (July 2, 1990).
127. *Carbon Disulfide, Technical-Grade, United States Federal Specification, Document No. O-C-131*, Government Printing Office, Washington, D.C., Jan. 29, 1952 and Apr. 14, 1954.
128. *Reagent Chemicals, American Chemical Society Specifications*, 7th ed., American Chemical Society, Washington, D.C., 1986, p. 201.
129. *Criteria for a Recommended Standard, Occupational Exposure to Carbon Disulfide*, DHEW(NIOSH) Publication No. 77-156, U.S. Government Printing Office, Washington, D.C., 1977.
130. J. M. Arena, *Poisoning*, Charles C Thomas, 3rd ed., Springfield, Ill., 1976, p. 84.
131. *Industrial Exposure And Control Technologies For OSHA Regulated Hazardous Substances* Vol. I, U.S. National Technical Information Service, Springfield, Va., 1989, 366–369.

132. E. Bingham, in E. Bingham, B. Cohrssen, and C. H. Powell eds., *Patty's* Toxicology, 5th ed., Vol. 3 John Wiley & Sons, Inc., New York, 2001, p. 505.

133. F. W. Mackison, R. S. Sticoff, and L. J. Partirdge, Jr., eds., *NIOSH / OSHA Occupational Health Guidelines for Chemical Hazards*, DHHS(NIOSH) Publication No. 81–123, U.S. Government Printing Office, Washington, D.C., 1981.

134. D. Djuric, N. Surducki, and I. Berkes, *Br. J. Ind., Med.* **22**, 321 (1965).

135. M. Ogata and T. Taguchi, *Indust. Health* **27**(1), 31–35 (1989).

136. M. Nurminen and S. Hernberg, *Brit. J. Indust. Med.* **42**(1), 32–35 (1985).

137. P. M. Sweetnam, S. W. C. Taylor, and P. C. Elwood, *Brit. J. Indust Med.* **44**(4), 220–227 (1987).

138. B. MacMahon and R. R. Monson, *J. Occ. Med.* **30**(9), 698–705 (1988).

139. A. J. Finkel, *Hamilton and Hardy's* Industrial Toxicology, John Wright, PSG Inc., Boston, Mass., 1983, 262–266.

140. R. Lilis, in W. N. Rom, ed., *Environmental and Occupational Medicine*, Little, Brown and Co., Boston, Mass., 1983, pp. 318, 342–343, 360, 627–630.

141. R. E. Gosselin, R. P. Smith, and H. C. Hodge, *Clinical Toxicology of Commercial Products*, 5th ed., Williams and Wilkins, Baltimore, Md., 1984, pp. III 90–94.

142. J. Wojtczak-Jaroszowa and S. Kubow, *Med. Hypoth.* **30**(2), 141–150 (1989).

143. P. J. Igoe, *Material Safety Data Sheet No. 350, Carbon Disulfide*, Rev. C, Genium Publishing Corp., Schenectady, N.Y., Apr. 1989.

144. U.S. Pat. 6,283,675 (Sept. 4, 2001), R. Dulsey and co-workers (to Crompton Corporation).

David E. Smith
Robert W. Timmerman
FMC Corporation